2nd International Joint Conference on Advanced Engineering and Technology (IJCAET 2017) and International Symposium on Advanced Mechanical and Power Engineering (ISAMPE 2017)

MATEC Web of Conferences Volume 159 (2018)

Bali, Indonesia
24 – 26 August 2017

Editors:

B.H. Setiadji
A.L. Han
A. Widodo
J.D. Setiawan
O. Kurdi
J.U.D. Hatmoko

ISBN: 978-1-5108-6104-6

Printed from e-media with permission by:

Curran Associates, Inc.
57 Morehouse Lane
Red Hook, NY 12571

Some format issues inherent in the e-media version may also appear in this print version.

Printed by Curran Associates, Inc. (2018)

For additional information, please contact EDP Sciences – Web of Conferences
at the address below.

EDP Sciences – Web of Conferences
17, Avenue du Hoggar
Parc d'Activité de Courtabœuf
BP 112
F-91944 Les Ulis Cedex A
France

Phone: +33 (0) 1 69 18 75 75
Fax: +33 (0) 1 69 28 84 91

contact-edps@webofconferences.org

Additional copies of this publication are available from:

Curran Associates, Inc.
57 Morehouse Lane
Red Hook, NY 12571 USA
Phone: 845-758-0400
Fax: 845-758-2633
Email: curran@proceedings.com
Web: www.proceedings.com

TABLE OF CONTENTS

edp sciences

MATEC Web of Conferences

Statement of Peer review

In submitting conference proceedings to MATEC Web of Conferences, I certify to the Publisher that I adhere to the **Policy on Publishing Integrity** of the journal in order to safeguard good scientific practice in publishing.

1. All articles have been subjected to peer review administered by the proceedings editors.
2. Reviews have been conducted by expert referees, who have been requested to provide unbiased and constructive comments aimed, whenever possible, at improving the work.
3. Proceedings editors have taken all reasonable steps to ensure the quality of the materials they publish and their decision to accept or reject a paper for publication has been based only on the merits of the work and the relevance to the journal.

Title, date and place of the conference

The 2nd International Joint Conference on Advanced Engineering and Technology (IJCAET) and International Symposium on Advanced Mechanical and Power Engineering (ISAMPE)
August 24-26, 2017
Bali, Indonesia

Proceedings editor(s):

Bagus Hario Setiadji, Ph.D
Dr. Han Ay Lie
Dr. Eng Achmad Widodo
Joga Dharma Setiawan, Ph.D
Ojo Kurdi, Ph.D
Jati Utomo Dwi Hatmoko, Ph.D

Date and editor's signature

February 02, 2018

Housing Typology of Bali Aga Architecture in Sukawana Village: Developments and Challenge

Ni Made Yudantini[1*], *Kadek Agus* Surya Darma[2]

[1,2]Architecture Department, Engineering Faculty, Udayana University, Kampus Bukit-Jimbaran, Bali

Abstract. The Bali Aga is an Indigenous people of Bali scattered in remotes areas such as mountains, hills, along the Batur Lake, and highlands. The Bali Aga people still practices the original tradition and differ from other Balinese people who mostly live in mainland area which differentiated by the reign of Majapahit Kingdom. Bali Aga refused Majapahit's influences, in contrast other Balinese followed the Majapahit rules that known as Bali Apanaga. Bali Aga still conserves original tradition but cannot avoid the increasing of development in technology and communication. Sukawana Village is one of Bali Aga village still conserve the Indigenous culture and tradition that located in highland of Kintamani Regency. The research aims to seek how the people adhere the tradition, norm, and values; and to explore the housing typology in Sukawana Village that faces on influence of development and threaten by globalization. The research primarily involves a qualitative method with includes an extensive literature review regarding the Bali Aga village and its distribution in Bangli Regency. The field research collected data through on-site evaluation and observation in Sukawana Villages in 2017, interviews to village leader and randomly questions to the householders. The research finding is the changing occurred in Sukawana Village such as using of new material and changes of indoor housing pattern.

1 Introduction

There are two types of Balinese people based on the history of Majapahit Kingdom's influences. First type is called Bali Aga, who's the Indigenous Balinese people that called as "the Mountain Balinese" [1] which apply the original tradition without having influence from Majapahit era. Second type is Bali Apanaga [2] who's implement tradition with having influence from Majapahit Kingdom and mostly located in mainland area of Bali. Although both embrace same Hindu Religion, Bali Aga have their own uniqueness of spatial of architecture including village pattern and layout of the housing. They still conserve the tradition however they cannot avoid the increasing of development in technology and communication that affect the way of thinking. This influences impact on the changes of housing lay-out and appearances.

* Corresponding author: nmyudantini@unud.ac.id

1.1 Research Aim and Scope

The research aims to explore the housing typology of Bali Aga village, specifically in Sukawana Village that faces on influence of development and threaten by globalization. There many changes have been occurred in Sukawana Village such as the use of new material on the building and changes of lay-out housing pattern. The investigation of housing typology aims to understand traditional housing architecture in the village as an effort for conservation of the heritage. The scope of research focuses on case study of the typology of housing pattern in Sukawana Village, Kintamani Sub-district, Bangli Regency. The research gathers information of traditional housing pattern and its changes from past to present through comparison from previous studies.

1.2 Research Methodology

The research primarily involves a qualitative method with includes an extensive literature review regarding the Bali Aga history, distribution of Bali Aga villages, and its development of traditional architecture including tangible and intangible aspects that influenced by the origin tradition of Bali Aga. The field research collected data through on-site evaluation and observation in Sukawana Villages that focus on village pattern and housing typology and its changes. The interviews to the village leader aimed to obtain fundamental information regarding village history, tradition, and socio-culture condition of the village, tangible and intangible aspects that contribute to the architecture. The research also questioned to the people in the village randomly regarding their housing function traditionally and its changes of building appearances. The questions also aimed to ascertain the consciousness of villagers in conserving and maintaining their traditional norms and values into the housing. The method of on-site evaluation and observation aim to seek how the people adhere tradition, norms, and values, and it also investigate to discover housing pattern and its changes. The observation has been done in 2017. The research found that the changes clearly appear on some house's layout and appearances.

2 Results and Discussions

The research provides information regarding the existence of Bali Aga villages in Bali Province which scattered across Bali Island. There have been studied the distribution of Bali Aga villages mostly in the highlands area. Yudantini (2015) studied that there are sixty two of Bali Aga villages still exist in Bali [3]. These are based on Bali Province inventory in 1988 to 1989 and the fieldwork of Carole Muller in 1980s. Many scholars have been studied about Bali Aga villages such as Krause in 1912 described the beautiful of Bali Island and depicted the people socio-culture life in Bangli Regency; Covarrubias (1974) described Balinese life, caste, organization, traditional villages, ceremonies as well as art and crave; Mead and Bateson (1936-1939) observed Bayung Gede Village which research focus on anthropology. There are many other scholars who contribute in the research of Bali Aga villages in Bali.

Bangli Regency is located in the middle of Bali Island. The only Bangli Regency does not have a beach among nine regencies in Bali Province. Topographically it ranges from 100 – 2,152 metres above sea level, and has an average rainfall around 400 mm per year. The Bangli Regency is bordered by the Buleleng Regency in the north, Karangasem Regency in the east, Klungkung Regency in the south, and Gianyar and Badung regencies in the west. Bangli Regency encompasses approximately 52,081 hectares, or 9.25% of the total area of

Bali Province (563,666 ha). Bangli Regency is divided into four sub-districts and 72 villages. Based on Population Census 2010, the population in Bangli Regency is 215,353 people with 50,840 households and a density of 413 per km² [4]. The people work mainly in agriculture (58.85%), industry (14.77%), trades (10.26%), services (7.13%), and in mining, utility, transportation, finance and construction (8.99%). There are around 25 Bali Aga villages scattered in Bangli Rengency that located along Batur Lake, in the foot of Batur Mountain, and some villages located in plain area [5, 6] (see Figure 1).

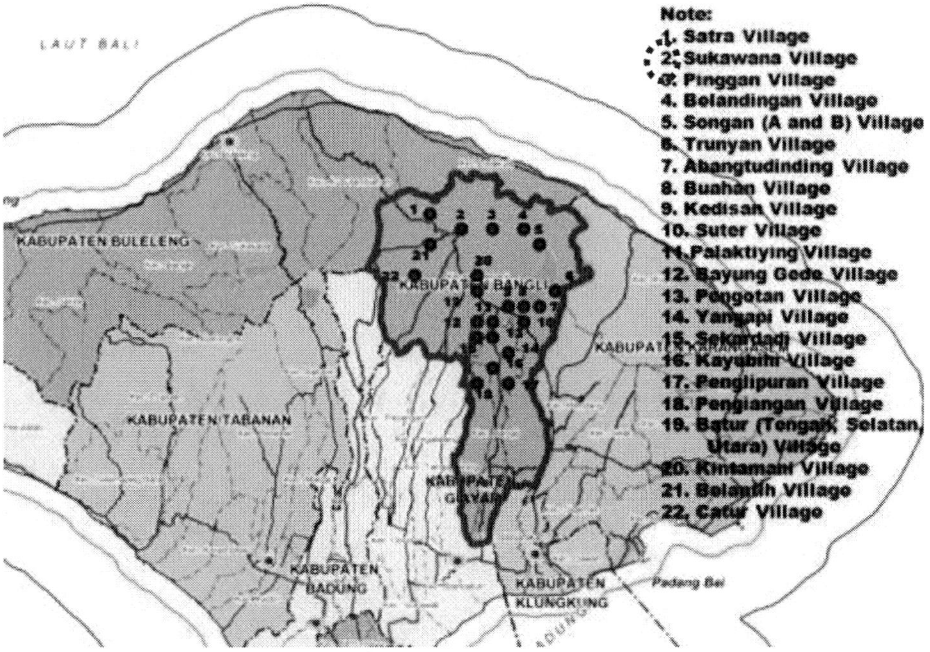

Fig. 1. Mapping of Bali Aga Villages Distribution in Bangli Regency
Source: Yudantini, 2015

Sukawana Village located in the hills and has rather steep topography. The village has an area of 3.361 hectare, inhabited by more than 900 householders and mostly as farmers and livestock. Traditionally, Sukawana Village governance is called *ulu apad* system as a main characteristic of government system of Bali Aga. The village pattern adopts a linear pattern; it is tied by village temples (Bale Agung Temple, Puseh Temple and Dalem Temple), settlement and its infrastructure and cemetery.

2.1 Housing Typology of Sukawana Village

It has been observed that the housing pattern in Sukawana Village adopts a linear pattern with one unit housing yard is aligned in rows comprises of six to eight households (Figure 2). Traditionally, one unit house yard consists of traditional building called *sakaroras* and a family temple or *merajan* as household shrine. The *sakaroras* (twelve pillars) comprises of holy room, bed room, traditional kitchen, and family temple. The traditional housing of *sakaroras* has a rectangular form with 12 or 8 or 6 pillars (*saka*) supporting a steep gable roof constructed from five layers of bamboo shingles (*sirap*) with a wall constructed from woven bamboo (*bedeg*) with one door oriented to the centre of yard (*natah*) (Reuter, 2002).

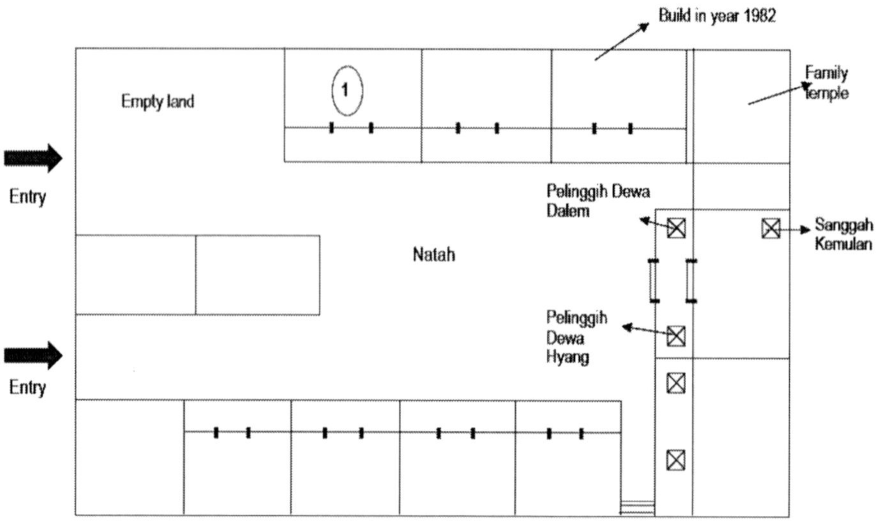

Fig. 2. One unit housing in Sukawana Village
Source: Yudantini, 2015

The house floor is covered by soil except the entry room, called *geladag/ampik/ampen* that is coated by a wooden slab. Rooms in the *sakaroras* are divided by several functions. There are *trojogan, lubangan gede, lubangan beten,* traditional kitchen/*paon, slatan kaja,* and *slatan kelod*. The *trojogan*, derived from the word *ojog* meaning toward, functions as a children's bed room and as a guest room. The *lubangan gede* is a male bed room, and the family also dines in this room. The *lubangan beten* is part of the traditional kitchen or *paon* and also functions as a food store. The traditional kitchen or *paon* is a place for the daily cooking food for the family and includes a hearth fire and a water jar for cooking and drinking. *Slatan kaja* (*Pajegan*) is a room for giving birth and for the storage of heirlooms. In this room is also an offering platform (*pelangkiran*) to celebrate the ancestors (*Bhatara Guru*, *Bhatara Surya*, and *Sang Hyang Komara*). Finally the *slatan kelod* is a room for keeping jewelry or other worthy goods (Figure 3 and 4).

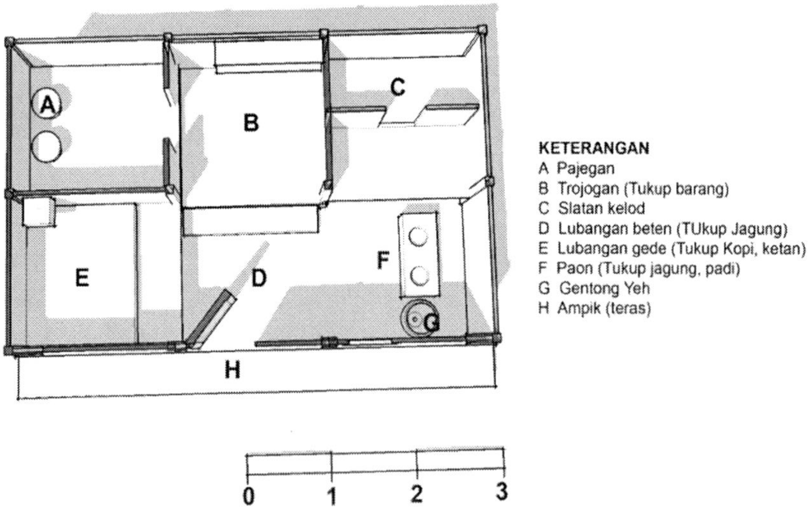

Fig. 3. Lay out Plan of traditional housing in Sukawana Village
Source: Reuter, 2002

Fig. 4. Traditional housing appearances
Source: Observation, 2017

2.2 Developments and Challenges

The increase of development and technology give impact on increasing villager's economic. It cannot be avoided that the people in the village want to have better appearances on their housing. The observation found some change happening in Sukawana Village. Some villager have been changed the housing layout into simple layout and the use of new material such as concrete, ceramic instead of local material (Figure 5). Other extended their room function to the *ampik* or terrace. This becomes challenge that the people in Sukawana Village want to change traditional or local material to modern or new material to have new and modern appearances. Although the people adopt new style (modern style) in the lay out of the house, the people still conserve and implement a holy room in their houses as a place to worship to the ancestors (Figure 6). In general, the unit housing pattern remain still extant with linear pattern.

Fig. 5. The use of new material in the house, and simple lay-out with no *ampik*/terrace
Source: Observation, 2017

The change cannot be avoided in Sukawana's housing appearances. However, the people still obligate not to change the housing pattern that apply the linear pattern and remain the same dimension of space. This strategy of the people in maintain their traditional value sustain the environment to provide the open space for public activities. This could be adopted this strategy in modern housing pattern through implement the traditional value to conserve and maintain the local genius, and also as local identity of Bali.

Fig. 6. Changes in lay-out and holy room remain
Source: Observation, 2017

3. Conclusion

Traditional housing is unique as a place to live in. Pattern, form, dimension, and material as well as customs become characteristic of Bali Aga culture. Sukawana as one of village of Bali Aga remain conserve their traditions although they face globalisation. As the increase of the people economic and improvement the way of living, the Bali Aga people also cannot avoid the changing. This is becoming the challenges of traditional architecture especially in Indigenous village. It is urgently need awareness of the people to maintain their tradition, norm and value that as a local characteristic and identity that worthy for Balinese traditional architecture.

References

1. Reuter, Thomas A. *Custodians of the Sacred Mountains; Culture and Society in the Highlands of Bali*, University of Hawai'i Press, Honolulu (2002).
2. Dwijendra, Ngakan Ketut Acwin. *Arsitektur & Kebudayaan Bali Kuno; Berdasarkan kajian Desa-Desa Tradisional di Bali*, Udayana University Press collaboration with CV. Bali Media Adhikarsa, Denpasar (2009).
3. Yudantini, NM. *Bali Aga Cultural Landscape Challenges: Conserving the Balinese Traditional Landscape for Future Balinese Indigenous Villages (Bali Aga) and Communities*, a Doctoral Thesis, Deakin University, Australia (2015).
4. Badan Pusat Statistik Kabupaten Bangli. Kabupaten Bangli Dalam Angka 2016 (2016).
5. Pemerintah Daerah Propinsi Bali. *Peraturan Daerah No.16 Tahun 2009 Tentang Rencana Tata Ruang Wilayah Provinsi Bali 2009-2029* (2009).
6. Muller, Carole. *Bali Aga Villages, field work in the 1980s*. Wals Bay Press (2011).

An investigation of spatial arrangement, form, and structural system of traditional houses in Pedawa indigenous village - Bali

Tri Anggraini Prajnawrdhi[1,*]
[1]School of Architecture, University of Udayana, Jimbaran Campus, Bali-Indonesia

Abstract. The Balinese traditional architecture has become valuable inheritance from the ancestor which still exists and is preserved until now, it is a symbol of local culture and reflects its characteristics. However, overpopulation has become one of the major challenges for Bali to cope with, and the impact of this phenomena significantly transform the traditional houses both its use and its form. The uniqueness of the architecture of Pedawa village has not yet been investigated in previous research. This paper aims to explore the detail of spatial arrangement, form, and structural system of the traditional houses, and its conservational effort. The case study method is employed for this research. Both primary and secondary data from interview, site observation, and literature data are analysed with qualitative analysis. The results show that the locals make every effort to sustain their values. The unique spatial arrangement, form, and structural system of the houses are strongly influenced by local customs; lifestyles; beliefs, and the environment. Hence, challenges are due to lack of understanding about conservation by the locals of their values and lack of natural resources; lifestyles; beliefs sand the environment.

1 Introduction

The Balinese traditional architecture is one identity of Bali which has been preserved. There are many types of traditional architecture in Bali according to its area or region. The traditional houses from one region to the other have different shapes and forms, such as their sizes; functions; ornaments, and materials. Traditional indigenous architectures are also different from the common traditional architecture widely known as Balinese traditional architecture. However, they adopt similar concepts from the past. The important thing is the strong relationship between humans, the nature, and the gods known as the universal concepts of Tri Hita Karana. This concept maintains the harmonious relationships amongst them. It is the foundation of most indigenous village settlements, traditional houses, and villages across Bali.

The traditional house is a symbol of local culture and reflects its characteristics. A traditional house could reflect the characteristic of the local culture which includes norms, values, behavioural patterns, artefacts, and activities [2]. The expression of the local residents' sociocultural system is reflected on their architecture; therefore the architecture is based on their local norms, values, and behaviour [11, 15]. According to Samadhi [10]

* Corresponding author: anggieprajnawrdhi@unud.ac.id

there are seven Balinese Hindu conceptions of space and their core periphery continuum order in the contemporary Balinese spatial formation: 1) psycho-cosmic concept; 2) Tri Hita Karana; 3) Tri Angga; 4) Kaja-Kelod; 5) Kangin-Kauh; 6) Sanga Mandala; 7) local knowledge systems (Tri Pramana and Tri Masa). These conceptions reflect their degree of importance within the spatial formation.

Pedawa village is one of the old indigenous villages in Bali which has an ancient heritage from megalithic era. The uniqueness of the architecture in Pedawa village has not yet been investigated in previous research about this village. This paper aims to explore the detail of spatial arrangements, forms, and structural systems of the traditional houses. At the same time, it shows that the characteristics of Pedawa house are different compared to the well-known Balinese traditional house typology. Further, this paper also reveals the effort of the locals to sustain their values and the challenges they cope during the conservational process.

2 Pedawa village overview

This village is located in Banjar District, Buleleng Regency, it is about 30 km to the west of Singaraja. Located on the highland about 450-800 above sea level, this village is considered to have colder temperature than the villages in the mainland. The name of Pedawa is derived from Pada Wang which means everyone is the same, Pedawa also means Panjak Dewa (God's helper). This village is one of the 62 indigenous villages in Bali. It has existed before Hindu and Buddha religions came to Bali during the Majapahit Kingdom [5].

This village has two ancient heritages in the form of sarcophagus in Jinjit and Ingsakan area as proof that this village has existed in the megalithic era. It has its own uniqueness based on its nature, architecture, and tradition which has not been explored in depth yet. In addition, the uniqueness of this village is that there are no castes in the community like other traditional villages in Bali generally. Furthermore, the characteristics of public temples in Pedawa are different from those in other traditional villages in terms of land division concept. The well-known concept for public temple land division in Bali is Tri Mandala (three divisions of the area of public/profane, semi public and private/ sacred area). Though, the temples in Pedawa only use the concept of Rwa Bhineda (two divisions of sacred and profane) and Eka Bhuana (no divisions of sacred and profane).

There are 62 indigenous villages identified based on the research of Carole Muller in 1980 and the inventory of the Government of Bali, Department of Public Works with the project entitled The Conservation of Villages and Environment Planning in 1988/1989 [5], but only few of them were explored in architectural point of view. Some research about Pedawa has been done by some scholars. The research was in the field of agriculture conducted by Udayana University (www.pps.unud.ac.id); research on the management of funding in adat village and dinas by Tera Padmani, *et.al* in 2015; research on health by Budiarta and Suka Aryana in 2012; research on education of primary school children by Komang Arini *et.al* in 2013; and some unpublished field surveys [8].

3 Method

The case study method is applied for this research. This method offers a descriptive analysis to analyse the relationship between the research object and the related determinant factors such as history and community and individual behaviour across the sites. This method helps the researcher to find a comprehensive way to explore the object of research deeply [13].Therefore, this method is very suitable for research in social disciplines. The strength of this method is that all data could be gathered from different data collection methods or a combination of several data collection methods both qualitatively and quantitatively. This research applies some interview with 12 selected respondents and observes 24 indigenous houses as the primary data. The secondary data were gathered from the report on the physical data of the village, village map, local regulation, custom regulation named awig-awig, and some relevant literature. The use of multiple source of data enhances the quality of analysis and optimizes the result [6]. Using two different methods of data collection would improve the objectivity of data before the analysis process [9].

4 Results and discussion

Traditional house in Pedawa village has two different types. The first one is called Bandung Rangki which has 18 wooden columns and the second one is called Sri Dandan which has 16 wooden columns. The former is the authentic Pedawa traditional house, while the latter is the influence from neighbouring village of Cempaga. Therefore, the majority of traditional houses in Pedawa are Bandung Rangki. Both Bandung Rangki and Sri Dandan have similar functions within the house which reflect, looked at from its spatial arrangement.

4.1 Spatial arrangement

Spatial arrangement of the traditional house both Bandung Rangki and Sri Dandan is based on the daily activities of the owner. Sulistyawati mentions that a traditional house in Bali has two functions which includes a place for living and an activity such as the area for sleeping, eating and resting; and the second function is for psychological needs such as doing ritual activities and local traditions [12, 15]. Further, Balinese traditional houses are classified into four attributes which include: sociology; symbolic; morphology; and function [7]. It is clear that Balinese traditional houses as a place for different kinds of activities for both physical and psychological activities of the owner.

The traditional houses in Pedawa, both Bandung Rangki and Sri Dandan, have a similar spatial arrangement based on the activities of the owner. There are six categories in daily and special activities of the villages.

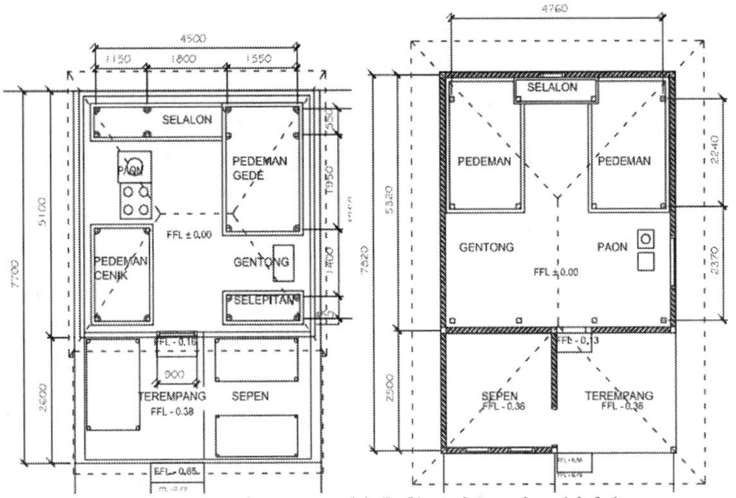

Fig. 1. Bandung Rangki (left), Sri Dandan (right)

The spatial arrangements (see figure 1) are divided into six categories: {1} Area for resting and sleeping consist of two different bed sizes called Pedeman Gede (for parents and has bigger size) and Pedeman Cenik (for children with smaller size). These two beds are made of wood and these beds columns are the structure of the house. Other function for the Pedeman Gede is to place offering during ritual ceremonies; {2} The activity of cooking is accommodated with a kitchen called Paon. The kitchen has three different types of burner made of Tanah Polpolan (soil/clay), the three different burners are: (a) Paon Tuak (the biggest one, for making palm sugar), (b) Paon Jakan (medium one, for cooking rice and meat) and (c) Irun (the smallest one, for heating food); {3} The social activities in this house are performed in the inner part of the house as the empty space called Tengan Umah (for hosting relatives only) and the front part of the house serving as veranda called Terempang (for hosting guest); {4}To accommodate ritual activities, this house has Pelangkiran placed on the top of Pedeman Gede, where the owner puts the offerings for their ancestor on a daily basis. Other ritual activities are also done outside the house with a special shrine made of bamboo called Sanggah Kemulan Nganten (see Figure 1 right); {5} To store the equipment and other household stuff and also groceries, these houses have areas called Sepen located next to the veranda. Further, there is also a place called Gentong to store kitchen equipment, cooked food, and water. The unprocessed rice and food from the farms are stored outside the house in a special building as a rice barn called Jineng. It is placed just in front of each house and commonly right in front of the front door of the house; {6} During the funeral ceremony, the corpse will be put on the Pedeman Gede and some offering along with the corpse. The corpse is bathed and laid on the Pedeman Gede before bringing it to the graveyard for cremation ceremony. The point to be made here is that traditional houses in Pedawa are able to express its characteristics. As mentioned by Broadbent [2] that the characteristics of the traditional house express its local culture including norms, values, behavioural pattern, artefacts, and activities. Nowadays some changes of the spatial arrangement occur due to changes of lifestyle and growing number of family members. Even though it changes, the culture and especially ritual activity are very well maintained.

4.2 Form and structure

The form and structure of these two houses adopted are very simple. The form of the house is using simple basic shape of form to be harmonized with the nature. The shape of roof adopted from a triangle which means the shape of mountain as a place of gods. Further, the triangle is steeper than a typical traditional Balinese house. It is due to the climate in highland which has more rains compared to the mainland. The shape of the floorplan is also using basic shape of rectangle. This shape provides an easier way for the people to arrange the spatial arrangement of the house based on their daily and special activities. Simple shape of these houses reflect simple ways of thinking and doing activities.

Further, the material for these two houses and the rice barns in the past were from local materials easily found within the surrounding areas. The local woods, bamboo and grass roof (called Alang-Alang) were commonly used for Bandung Rangki and Sri Dandan houses. The wooden columns that support the houses are integrated with the bed both for Pedeman Gede and Pedeman Cenik. All structural materials are made from local wood and bamboo. The roof and wall structures are made of bamboo. There are simple structure adopted for these houses. The wood and bamboo joining all the constructions were using simple cuts and wooden pegs called *lait* and also using bamboo ropes. Therefore the renovation to some part of the materials would not damage other parts or the whole structure. The structure of these houses are shown in detail in Figure 2. These types of construction system are simple and long lasting. Even though there were changes in the materials used, these simple construction principles are still applied to modern materials, and are well sustained for the future generation.

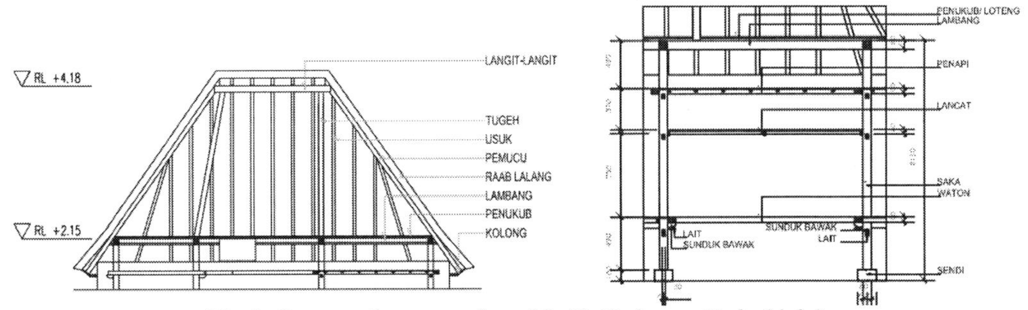

Fig.2. Structural system of roof (left), Pedeman Gede (right)

These two houses use human scale based on the concept of harmonious relationship between the human and the nature. This scale is grounded on the spatial design which has to consider the relationship between microcosm (human) and macrocosm (universe) which is the core conception of all the structures [10]. Therefore all the traditional indigenous houses in this village are based on this concept. By using human scales, the buildings nicely fit in with the nature without any attempt to dominate the nature.

4.3 House transformation, sustainability, and its challenges

The lesson learned from Pedawa traditional houses is that significant changes of the traditional houses today happen. People tend to have modern houses rather than old traditional indigenous houses. Further, natural resources are also one main reason such as the lack of good quality of bamboo and wood for house structures. Some houses are totally demolished and new modern houses are built, but some make transformation of the old houses. The changes of the houses such as (1) Replacing materials of the roof and wall structure for more modern, stronger and more affordable houses; (2) Replacing the floor material of local clay (Tanah Polpolan) to tile and bricks. The changes have brought some significant impacts to the thermal comfort in the house as new materials could not help to protect the people from cold weather. Further, the funeral activates for bathing the corpse could not be done due to the change of floor materials. However, two important functions that are sustained by the community and do not face any changes in the two traditional houses are the Pelangkiran and Sanggah Kemulan Nganten. These two ritual functions are the most important unchangeable functions. Rituals are the most important cultural attributes both in private or public uses and have created a specific spatial structure and it became the centre of orientation and identification for the community [4, 11]. The sustainability of this indigenous houses are strongly based on belief and religion of the community. Even though the community struggles to sustain their houses as a physical part, but they maintain their belief and religion.

As mentioned in the National Regulation No 11 of 2011, traditional houses are part of traditional villages that need to be protected. Therefore, the need to sustain traditional houses is very important as part of traditional villages themselves. It is important to widespread the knowledge to all society about the protection of traditional houses as cultural resources. Some efforts have been made to sustain the traditional houses in Pedawa by some group in the community. The efforts are: they work together to fix some traditional damaged houses; making family shrines; and making one original house for informational purposes. These efforts have brought awareness for the preservation of the cultural values in the community. To face the challenges for the future, it is suggested that the need to sustain the cultural resources is not only on the physical aspects but also more to the aspect of value and reinterpretation of the value and spirit into new built form based on the current context and situation. As mentioned by Olgyay in Altomonte [1], to promote sustainable design in a built environment, architecture must therefore assume a further dimension, conscientiously responding to the context where it is built and to the environment as a whole. Cuthbert [3] points out that architectural forms in Bali are not fundamentally architectural typologies, but are based on social typologies such as caste, class, and capital. However, since Pedawa does not recognize any caste and that everyone is socio-religiously the same, then there are no typologies for the houses, and it would be easier to sustain the existing cultural values.

5 Conclusion

The results show that the space arrangement of the traditional houses is closely linked to the local tradition and the lifestyle of the population in the past, while the transformation of the spatial arrangement is also the result of the change of life style and demographic in this village. However, the transformation of the spatial arrangement of the two houses still preserve the space for rituals as a proof that the villagers still have the same belief as their ancestors. The spatial arrangement, and the choice of form, material and structural system of Bandung Rangki and Sri Dandan refer to the local climate; the ability of local craftsmanship; affordability; local belief; local custom; and local natural and materials in the past. The building detail and technique have some similarity to the general Balinese traditional architecture typology. Thus, the spatial arrangement, form, material, and structural system of the traditional houses in Pedawa are used to accommodate the local's daily activities, protecting them from the local climate which serves as a form of the local identity. Lastly, the main important factor of the sustainability of this house is based on the belief and religion of the local community.

References

1. S. Altomonte, Environmental Education for Sustainable Architecture, R.E.S. 1.2 (2009)
2. G. Broadbent, R. Bunt, C. Jencks, Signs, Symbols and Architecture, John Wiley & Sons. Chichester (1980)
3. A. Cuthbert, Paradise Lost, Sanity Gained, J.K.B. **1**. 2 (2012)
4. R. Knowles, Rhythm and Ritual, Maintaining the Identity of a Place, J.T.D.S, **94**, (1996)
5. C, Muller, Bali Aga Villages; field work in the 1980s, Walsh Bay Press (2011)
6. M. McCutcheon, J.R. Meredith, 1993, Conducting case study research in operation management, J .O. M **11**, 239-256 (1993)
7. A. P. Parimin, Fundamental Study on Spatial Formation of Island Village: Environmental Hierarchy of Sacred Profane Concept in Bali (PhD Thesis) Universitas Osaka, Japan (1986)
8. T.A Prajnawrdhi, Tantangan konservasi pada Rumah Bandung Rangki dan Sri Dandan di Desa Bali Aga Pedawa, Buleleng-Bali, Semnas Heritage, Cirebon (2016)
9. D. Silverman, Doing Qualitative Research, **4th**.SAGE London, (2013)
10. T. N. Samadhi, Conservations of Principles or Objects? J. D.T .A **28**, 2 (2000)
11. C. N. Schulz, Intentions in Architecture. The M.LT Press, Cambridge Massachusetts. Genius Loci, Electa/Rizolly, New York (1977)
12 D. M.McCutcheon, J.R Meredith, Conducting case study research in operation management, J. O. **11**, 239-256 (1993)
13. Sulistyawati, Preservasi Lingkungan Perumahan Pedesaan dan Rumah Tradisional Bali di Desa Bantas, Kabupaten Tabanan', Udayana, P3M, Universitas Denpasar (1985)
14. R.K. Yin, Case Study Research: Design and Methods, 5th, SAGE, Los Angeles (2009)
15. Y. Ahmad, 'The Scope and Definitions of Heritage: From Tangible to Intangible', International Journal of Heritage Studies, vol. 13, no. 3, pp. 292-300 (2006)

Sustainable Streetscape Design Guideline based on Universal Design Principles

Bangun IR Harsritanto*[1]

[1] Departement of Architecture, Engineering Faculty, Universitas Diponegoro, Semarang, Indonesia

Abstract. People access the streets to fulfill their daily demands regardless of their gender, age, and abilities. The streetscape is an urban public space which built to facilitate the basic needs of people as social beings. Universal design accommodates the wide range of human diversity in physical appearance, perception, cognitive abilities, body sizes, and shapes. By designing for this diversity, the streetscapes are becoming more user friendly, functional and sustainable. The purpose of this study is to suggest streetscape design guideline based on universal design principles for giving proper directions of sustainable streets development. Literatures study and cases study were used as research methods. The findings indicated that the new guideline based on universal design can be suggested to streetscapes design directions on many countries.

1 Introduction

Streets roled an important part of open public space and become most influenced symbols of the public area. Streets have functional, social, and leisure needs of people have been associated with economic growth, people physical health, and community senses. Scholars suggest idea of the street as social space instead of only a channel for the movement [1]. People access the street, regardless their sexes, age and abilities for many reasons. The streetscape is an urban fabric that built to facilitate people's basic need as social being. Since people use most of their time on the streets, the role of proper designed streets is important in urban development.

Universal design builds facility designs which provide the full range of potential users, including people with disabilities, various mobility, and other special needs. Universal design suggested integration in streetscape design from origin to destination for the widest possible range of prospect users. Universal design regarded all possible obstacles that may exist in buildings, sidewalks, paths, roads, transportation terminals, and vehicles. The universal design principles may bring the streetscapes become more functional, user friendly, and sustainable.

The studies of universal design application can be done in two ways, the first is done by existing product appraisal and the second by adding more parameters to be next appraisal items. This study referred to the second type of universal design study which added more parameters to be next appraisal items [2].

* Corresponding author: bangunirh@arsitektur.undip.ac.id

The purpose of this study is to suggest streetscape design guideline based on universal design principles for giving proper directions of sustainable streets development.

2. Methods

Literatures study and cases study were used as research methods. This study carries several literatures of streetscapes design guidelines which already referenced by many cities (Barton, Islington, Landcome, Torbay, Scottsdale, Abu Dhabi, Guide dog, etc.), literature of universal design based studies that adopted in many built environment facilities (Center of Universal design, Shehu, Null, Kadir, Goldsmith, et al), and some of related papers to build the evaluation sheet. Furthermore, the constructed evaluation sheet will be assessed to urban design aspect at case studies in Indonesia.

3. Analysis

Descriptive analyses were conducted for all study variables. Comparative study holds a simple research by comparing at least two literatures to bring a result. This study uses multi disciplinary in a new guideline.

4. Discussion and Results

4.1 Universal Design

Universal design is a design philosophy which removed distinctions among various abilities by adhering to four major principles. The principles identified by Ron Mace [3] as follows:

1. Accessible
2. Safety
3. Supportive
4. adaptable

The center for universal design at North Carolina expanded the four principles list into seven principles of universal design. Furthermore. The latest detailed seven principles of universal design become one of the grand theories in this study. They described as:

1. Equitable Use
2. Flexibility in Use
3. Size and Space for Approach and Use
4. Simple and Intuitive Use
5. Low Physical Effort
6. Perceptible Information
7. Tolerance for Error

4.2 Urban Streetscape

Urban streetscape is a natural and built fabric of street. Urban Streetscape designs affects the quality and visual effect of the street. It includes street surfaces, buildings, building fixtures and the fittings which facilitate their function to landscape schemes. Urban streetscape created a good environment for people to gather or travel. Social and economic interactions are communal goals of a good street design [4] which also sustainable design. The streets have There are five principal functions of the streets [5], such as : access, movement, place, drainage, utilities, parking; and lighting.

This study summarized the 7 existing street guidelines, they are :

1. Inclusive design: street for life
2. Islington street book
3. Inclusive streets: design principles for blind and partially sighted people
4. Abu dhabi urban street design manual
5. Landcome street design manual
6. Scottsdale city specialized street design guidelines
7. Torbay specialized street guideline

There were fundamental issues of this new guideline study (see table 1), such as :

1. the previous studies were aimed for tourism facilities [7,11,12] that brought artificial disneyfication of street. This disneyfication of tourism may limits the urban society's natural expression [6].
2. the previous studies were aimed to localize [9,10] that unsustainable for street fabric from and to the site.
3. the previous guidelines were not apply universal design [4,8] ,so that people with disabilities will be treated as a different kind of person. The facilities were designed specifically only for them mentioned on the literature [13]. The designer should learn about the various needs of people and reduce dependency to other people helps [14].

Table 1. Street guidelines comparisons

Literature	Purpose	Result	Cons	Pros
Inclusive Urban Design [6]	Creating environments that maximize people's quality of life-based on dementia people preference research	Dementia and elderly friendly street	Concern on dementia-elderly	Dementia and elderly friendly, strong memory/history aspect
Islington Streetbook A borough-wide streetscape manual [7]	Built an attractive, safe and accessible streetscape by setting guidelines and specifications	Tourism streetbook	Concern on local tourism	Promote tourism, minimal material use
Design principles for blind and partially sighted people [8]	Accommodate the differing needs and expectations of all those who use the street space especially for blind and partially sight people	Blind and partially blind sighted street	Concern on blind-dog walker	Blind people dan dog walker friendly
Abu Dhabi Urban Street Design Manual [9]	To design streets that create a safe environment for all users, recognizing varying levels of driver education and cultural differences	estidama's principle on Abudhabi	Desert climate response	Concern on residential area

Literature	Purpose	Result	Cons	Pros
Street Design Guideline for Landcome [10]	Built environment providing routes for vehicles and public transport and accommodate utility services and drainage system	Australia streetbook guideline	Concern on residential area	Legibility force, design community-based
City of Scottsdale Streetdesign guidelines [11]	Establishing a strong identity and distinctive visual character for scottsdale road as a "signature" corridor	Tourism streetbook	Concern on local tourism	Signage concern, transit oriented
Torbay Streetdesign guidelines [12]	Create attractive,pedestrian environment , pleasant spaces that encourage greater pedestrian activity - with improved safety and health benefits to the community, and environmental benefits	Tourism streetbook	Concern on local tourism	Local conservation item, sustainability
this dissertation [2]	To build streetscape that can be used by all user regardless sexes, ability and ages based on principle of universal design	Universal design based urban streetbook	Concern on urban streetscape	Resumed, add urban identity concept, universal design based

4.3 The Result

This study developed and studied the answer for 3 urgent problems which stated above as research questions and originality. The urgent problems itself derivated from 7 previous streetscape research evaluation with principles of universal design and comparison between each research detailed characteristics.

Later on, the study of the sustainable streetscape design guideline will be constructed as:

1. urban design oriented specialized street design guidelines. This point indicated that every parameter and element were suited to urban design scope
2. universal design principles based study. The four and seven principles of universal design were integrated to all points of this study
3. addition of urban identity factors. Every part of the streetscapes shall has a theme and identity which sustained and linked to others part of the urban area. This identity may helps urban people memorized their cities.

The study already constructed 89 items of sustainable street developments on 8 design sections, such as : parking lots, bus stops/subway gates, linkage designs, street furnitures, street utilities, street materials, signages, and pedestrian spaces [2]. The design sections are universal (every streets have those design sections), however the quality and identity of each streets may be vary (depend on the surrounding environment). The more accessible, safety, adapatable, and supporting degree resulted higher quality and better sustainable environment, since it will be used continuesly by many people.

5. Conclussion

This study has constructed and developed new guidelines from 7 streetscape guideline books which based on universal design principles. The study case of parking lot, bus stop/subway gate, linkage design, street furniture, street utilities, street materials and pedestrian areas can be applied directly to streetscapes in many countries. The scale is the level of guidelines application, started from the the partial until the most sustainable streetscape design related to the universal design which may resulted the more user friendly regardless their sexes, abilities, ages, etc.

References

1. V. Meht , *Lively Streets: Exploring The Relationship Between Built Environment And Social Behavior* (2006)
2. B. Indrakusumo, *Design Characteristics of Specialized Streets for Indonesia through Cases of South Korea* (2016)
3. F. Aslaksen, S. Bergh , O.R. Bringa, E.K. Heggem; *Universal design: Planning and Design for All* (1997)
4. E. Burton, M. Lynee, *Inclusive Urban Design: Street for Life* (2006)
5. Department of Transport, *Manual for Street* (2007)
6. B.A. Putra, *Cultural Representation of Vernacular Housing in Jambi* (2013)
7. Islington Council, *Islington Streetbook A borough-wide streetscape manual* (2005)
8. Guide Dogs on Behalf of Action for Blind People, *Inclusive Streets:Design Principles for Blind and Partially Sighted People* (2010)
9. Abu Dhabi Urban Planning Council , *Abu Dhabi Utility Corridor Design Manual ver* 1 (2017)
10. Landcome, *Street Project guideline* (2008)
11. Otak Team, *Scottsdale Road Streetscape Design Guidelines* (2008)
12. Torbay council ,*Torbay Streetscape Design Guidelines (*2004)
13. V. Shehu, *Universal Design Strategies On Developing Country, Case Study: Tirana* (2011)
14. M. Guimares, *An Assessment Of Understanding Universal Design Through Online Visual Resources and Role-Playing Simulation Exercises* (2005)

On the Sound Absorption Improvement of Water Hyacinth and Coconut Husk based Fiber Reinforced Polymer Panel

Erni Setyowati[1], *Iwan* Yahya[2,*], *Edi* Supriyo[3], *Intan* C. Romadhona[2], *Aris* Minardi[2]

[1]Architecture Dept., Faculty of Engineering, Diponegoro University, Jl. Prof. Soedarto, SH, Tembalang, Semarang
[2]The Iwany Acoustics Research Group (iARG) Physic Dept, Sebelas Maret University, Jl. Ir. Sutami 36A Kentingan, Solo
[3]Chemical Engineering Dept., Vocational Faculty, Diponegoro University, Jl. Prof. Soedarto, SH, Tembalang, Semarang

Abstract. This paper emphasizes on the sound absorption improvement of the water hyacinth and coconut husk based fiber reinforced polymer (FRP) panel. The dry water hyacinth and coconut husk were milled and sieved before it mixed with polyester and methyl ethyl ketone peroxide as the catalyst. The composition is 200 ml, 25 grams, and 20 ml for the polyester, fiber, and catalyst respectively. The mixture casting is by using a square tray to form a flat FRP panel through open air drying without any compaction process. Three different approaches i.e. multiple quarter wavelength resonators, air cavity, and front fibrous layer inclusion are implemented to the FRP to improve its sound absorption performance. The entire measurements conducted by using transfer function based impedance tube technique refer to ASTM E-1050-98 standard. B&K 4206. It found that the proposed approaches are increase sound absorption coefficient of the bulk FRP significantly. The best performance occurred on the coconut husk based FRP when the 8 of quarter wavelength resonators combined with the fibrous layer. Sound absorption increased accordingly to above of 0.7 starting from 1.5 kHz. On the other side, the air cavity shows similar influence on the entire test sample due to Helmholtz resonance mechanism.

1 Introduction

Indonesia is the world most coconut producing country. This plantation grown and cultivated in all islands over the country and used for many different purposes and applications. The entire parts of the plant bring valuable benefits for peoples ranging from foods, construction materials, drugs, home appliances, alternative energy resource, plantation media, and for packaging. Apart from its benefits, coconut husk is one of the plant valuable part was not yet optimally utilized.

* Corresponding author: iyahya@mipa.uns.ac.id

Just like coconut husk, water hyacinths also grow in many lakes along the islands of Indonesia but not yet optimally utilized for creating valuable products. Currently, it used as raw materials for supply to handicraft and furniture industries. Even more, this plant causes serious environmental problems in many regions of the country.

According to its fibrous and porous structure, both coconut husk and water hyacinth are possible to use as sound absorbing materials. As reported in the literature, these two natural materials show a potential acoustical performance. Setyowati et al. have conducted experimental analysis for measuring the sound transmission loss and sound absorption coefficient of the materials. It found that the water hyacinths performs well at the range of frequency 2.5 kHz to 3.25 kHz with 57-59 dB of sound transmission loss, while the absorption coefficient of coconut husk based sample reaches the peak value of 0.9 at 2.5 kHz[1]. The other research conducted by Fouladi et. al to compare the absorption coefficient of coconut fiber from the traditional market and industrial product made of coconut fiber with binder through analysis after Delaney-Bazley and Biot-Allard model. The absorption coefficient was successfully improved. The sound absorption coefficient shifted to a lower frequency band from 1.36 kHz to 578 Hz as the thickness of the samples increased from 20 mm to 45 mm[2].

As for prediction purposes, Ramis et al. developed equations for prediction and simulation models of acoustics systems[3]. The other improvement approach such as air gap, fibrous material inclusion, and surface perforation conducted by Lee et al. and Zulkifli et al. [4], [5]. Previous work of Lee and Chen analyzed the improvement strategy of the multi layered sound absorbing material in regards to perforated plates, air spaces and porous materials[6]. The use of natural fibrous materials as the potential sound absorber also recommended by Ersoy and Kucuk, Mohanty et al. and Mueller & Krobjilowski [7–9].

Even though the natural fibrous and porous materials have potential advantages, but most of them are not naturally compact and low in mechanical strength to use directly in noise control applications. According to this reason, researchers used several types of binder and additive substance to increase the mechanical strength and performance. This technique is running successfully but also resulting another constraint concerning the decrease of its acoustical performance. As the natural fibers mixed with any chemical or natural binder and then compacted through heating and or pressing technique, the materials becoming more stiff, harder, less porous and very reflected. This paper emphasized on the acoustical performance improvement strategy of the water hyacinth and coconut husk based fiber reinforced polymer (FRP) panel in regards to the use of quarter wavelength resonator, air cavity, and fibrous surface layer.

2 The Object of Study

The raw water hyacinth and coconut husk materials are collected from Rawa Pening lake and traditional market in Semarang respectively. After cleaning process, it dried in open air before milling and sieving steps. The dry water hyacinth and coconut husk fiber flour separately mixed with polyester as the binder and methyl ethyl ketone peroxide as the catalyst. Both have the same composition of 200 ml polyester, 25 grams fiber, and 20 ml catalyst. And as for the test sample preparation, the mixture casting was conducted by using a square shaped tray before it dried in open air for hardening purpose without compaction or pressing process. The resulting flat FRP were cut into some circularly shaped test sample sized 30 mm in diameter. Table (1) shows the detailed geometrical dimension and the treatment of the entire test samples.

Table 1. The geometrical dimension and treatment of test samples

Code	Material	Diameter	Thickness
A	Water hyacinth	30 mm	15 mm
B			30 mm
C	Coconut husk		15 mm
D			30 mm
Treatment code			
H	Number of quarter wavelength resonators (4, 8, 9)		
F	Foam front layer		
Cav	Cavity (10 mm)		

3 Methods

The transfer function based two microphones impedance tube technique after ASTM E-1050 conducted for measuring the sound absorption coefficient[10]. Each test samples treated with three different way as listed in Table (1). The small tube set up of the B&K 4206 impedance tube used in the experiment with frequency span up to 6.4 kHz.

For investigation of resonator number influence on the sound absorption performance, the circular shaped FRP was drilled to form quarter wavelength resonator structure inside the specimens. In this case, there are three different number of the resonator as listed in Table (1). The first eight resonators have the same diameter of 3 mm while the 9th is 8 mm. For investigating of the back cavity and porous front layer impact, the experiment conducted with and without 10 mm cavity depth and 10 mm front layer. The front porous layer made of commercial Dacron layer. The test samples are illustrated in figure (1).

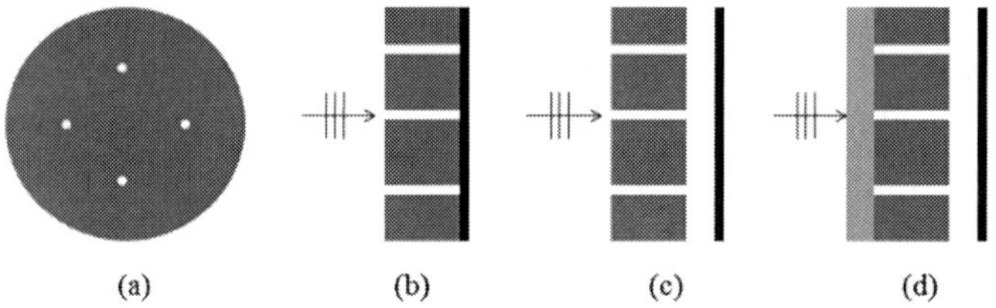

Fig. 1. The test sample top view (a) and configuration variations: (b). quarter wavelength resonator, (c). cavity backed FRP, and (d). cavity backe FRP with porous front layer.

The entire experiment is fully computer controlled where B&K 4206 impedance tube connected to the four channels B&K Pulse Analyzer LAN-XI type. The dedicated B&K Material Testing Software used to control entire experiment steps. Random noise is generated from the loudspeaker of impedance tube and propagate along the tube before it hit the test sample surface. The two ¼ inch 4187 B&K microphones capture both incidents and reflected wave before it decomposed by using transfer function analysis. Since the impedance tube assumed to be has an infinite impedance on the far end, energy decay is must be totally associated with the absorption performance of the test sample.

The frequency responce function H_{12} and reflection coefficient R can be expressed as follows,

$$H_{12} = \frac{P_2}{P_1} = \frac{e^{jkh} + e^{-jkh}}{e^{jk(h+s)} + e^{-jk(h+s)}} \qquad (1)$$

and

$$R = \frac{H_{12} - e^{jks}}{e^{jks} - H_{12}} e^{j2k(h+s)} \quad (2)$$

where P_1 and P_2 are sound pressure level captured by the two microphones in position one and two respectively. k is wave number while h and s are the distance from the first microphone to sample and the distance between microphones respectively. The coefficient absorption α can be calculated by:

$$\alpha = 1 - [R]^2 \quad (3)$$

4 Results and Discussion

The influence of quarter wavelength resonator inclusion on the sound absorption performance illustrated in figure (2). Sound absorption performance of the entire four specimens before treatment is shown in figure (2a) while the treatment results depicted in figure (2b) and (2c). There is no significant improvement occurred according to thickness increment on the untreated water hyacinth based FRP, while the different phenomenon exists on the untreated coconut husk FRP. The untreated or bare water hyacinth FRP has the maximum sound absorption coefficient of 0.2 for the entire frequency range, while the coconut husk based FRP has the better performance with a maximum of 0.4 for the 15 mm thickness sample and 0.85 for the 30 mm sample. It occurred on frequency 1.7 kHz and 2.8 kHz respectively.

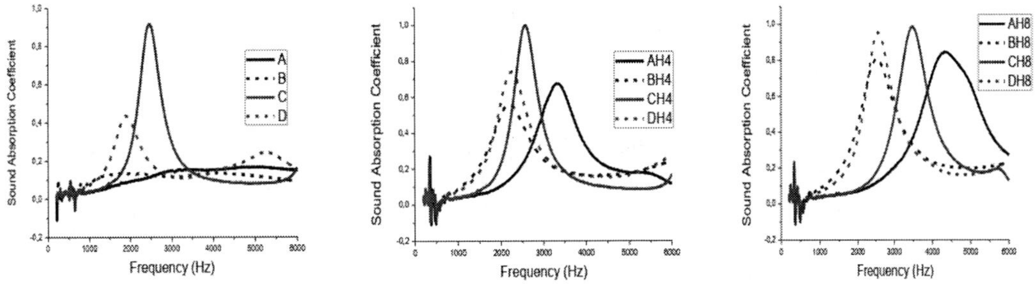

(a) (b) (c)

Fig. 2. Influence of resonator number on the sound absorption

It is clear from the results in figure (2) that water hyacinth based FRP is more reflected compared to the coconut husk based FRP. It happened because of the difference in the fiber size of the both materials. Water hyacinth has a smaller dimension and finer fiber compared to the one of coconut husk. As it mixed and blended with polymer and catalyst, the entire inter-fiber spaces are fully loaded with the mixture of polymer and catalyst. The resulted in FRP becoming more compact and dense. It explains why the water hyacinth based FRP sound absorption performance is lower than FRP made of coconut husk fiber.

The influence of resonance mechanism illustrated very clear in figure (2b) and (2c). It all agreed with the theoretical base that the longer resonator is resulting the lower resonant frequency. It causes the sound absorption performance shifted to the lower frequency band. As the number of resonators increased, the sound absorption tends to shifted to little bit higher frequency band, and the sound absorption coefficient value also increased. This phenomenon occurred as the viscous damping also occurred on the entire four test samples according to the resonators small diameter. The sound absorption performance in the higher frequency range gained accordingly. Typical of the air cavity and front porous layer impact on the sound absorption performance of the both water hyacinth (A) and coconut husk (C) based FRP illustrated in figure (3).

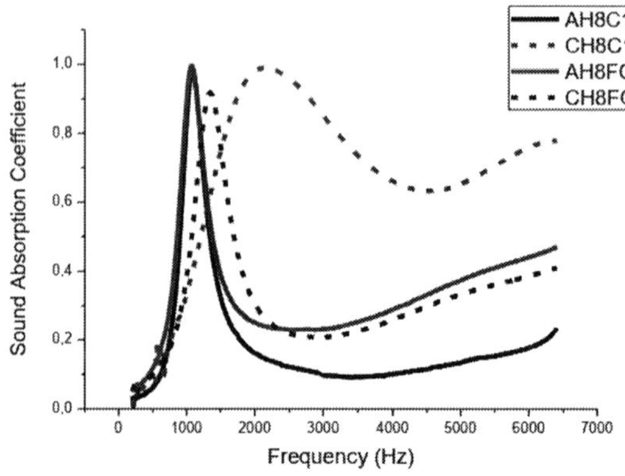

Fig. 3. Back cavity and front porous layer effect on the sound absorption

Helmholtz resonance mechanism occurs on both FRP as it coupled with an air cavity. The sound absorption coefficient reaches its best value of 0.98 on 1 kHz for the water hyacinth and 0.95 for the coconut husk on the frequency 1.2 kHz. Since the geometrical dimension of the air cavity is similar for the entire test samples, the sound absorption shifted to the almost similar frequency as well. The water hyacinth based FRP perform on the lower frequency due to its internal structure which is denser compared to FRP made of coconut husk fiber. According to its soft structure, the coconut husk based FRP brings significant advantage as the front porous layer attached to the cavity backed CH8C1. As seen on the graph, the test sample CH8FC1 has a very promising broadband sound absorption profile. In this case, viscous damping associated with the porous and fibrous structure on the front layer occurs simultaneously with the resonance damping mechanism. When the back cavity is attached to the FRP that has an array of quarter wavelength resonators on it, the coupled structure response changes accordingly. It is becoming a new structure that works as a multiple neck Helmholtz resonator covered by porous and fibrous front layer. In the same time, the entire necks of the new coupled structure also have its frequency response as an array of half wavelength resonators. It explains why the sound absorption coefficient has an almost full spectrum profile and it shifted to the little bit higher frequency band compared to the sample test CH8C1. The entire finding in this work proofs the advantages of natural fiber reinforced polymer panel made of natural resources. Natural fiber resources such as water hyacinth and coconut husk have a very potential economic value. It can be use as the raw materials of the high-performance, sustainable sound absorbing panel for various and wide range of noise control applications.

5 Conclusion

The cavity backed natural fiber reinforced polymer panel (NFRP) composite made of water hyacinth, and coconut husk has the valuable possibility for replacing the synthetic fiber composite mainly for building noise control applications. Broadband and high-performance sound absorber are very possible to make in the form of coupled structure where Helmholtz and half wavelength resonance occurred simultaneously with the viscous damping. Coconut husk based FRP is most powerful compared to the one of water hyacinth. The discrepancies on the ability to absorb sound waves energy mostly associated with geometrical size and mechanical properties fiber structure of the materials. Water hyacinth has the smaller geometrical dimension, and as it mixed with the polymer and catalyst, it is resulting in a denser and reflecting FRP panel.

This research funded by the Diponegoro University Research Grant Number: 276-47/UN7.5.1/PG/2017 and partially supported by The Iwany Acoustics Research Group (iARG) of Sebelas Maret University especially on the laboratory test process. The funding institution does not have any intervention on the entire works. The authors would like to thanks to all colleagues and students and collaborating partners for their valuable supports.

References

1. E. Setyowati, E. E. Pandelaki, and E. Supriyo, JST (to be published).
2. M. H. Fouladi, M. Jailani, and M. Nor, Appl. Acoust., **72**, 35–42 (2011).
3. J. Ramis, R. Rey, J. Alba, L. Godinho, and J. Carbajo, Matriales de Construccions. **March**, (2014).
4. J. Lee, and G. W. Swenson, NCE.. **38**, 109-117 (1992).
5. R. Zulkifli, M. J. M. Nor, M. F. M. Tahir, A. R. Ismail, and M. z. Nuawi, J. Appl. Sci., **20**, 3709–3714 (2008).
6. F. C. Lee and W. Chen, J. Sound Vib., **248**, 621–634 (2001).
7. S. Ersoy and H. Kucuk, Appl. Acoust., **70**, 215–220 (2009).
8. A. K. Mohanty, A. Wibowo, M. Misra, and L. T. Drzal, Composites, **35**, 363–370 (2006).
9. D. H. Mueller and A. Krobjilowski, J. Ind. Text., **33**, 111–130 (2003).
10. *ASTM E 1050-98, Standard test method for Impedance and absorption of acoustical materials using tube two microphones and digital frequency analysis system*. (1998).

Dynamic Response of RC Cantilever Beam by Equivalent Single Degree of Freedom Method on Elastic Analysis – A Review on Transformation Factors and Dynamic Magnification Factors

Sri Tudjono[1,*], and *Patria* Kusumaningrum[2]

[1]Diponegoro University, Civil Engineering Department, Semarang, Indonesia
[2]ITB, Civil Engineering Department, Bandung, Indonesia

Abstract. The response of multi-degree-of-freedom (MDOF) structure can be correlated to the response of an equivalent single-degree-of-freedom (SDOF) system, implying that the response is controlled by a single, unchanged mode shape. This equivalent SDOF method is eminent as an approximate method of dynamic analysis. In this study, equivalent SDOF method analysis is carried out on RC cantilever beam subjected to dynamic blast loading to review the transformation factors (TFs) provided by TM5-1300 code.

1 Equivalent SDOF Approach

This simple approach of equivalent Single Degree of Freedom (SDOF) method to analyze structural elements subjected to blast loads has first been performed by Mays and Smith [1]. A lumped mass of an equivalent SDOF system is proposed to predict the displacement of simplified structures subjected to blast by applying transformation factors of load (K_L), stiffness (K_S), and mass (K_M) to the SDOF analysis. This method follows the TM5–1300 US Army Code [2]. Based on this code, amplitude and time of peak response can be obtained from the idealized charts based on natural period (T) of respective structure.

SDOF method on one way element was studied by Crawford et al. [3] for single column. Big differences between the displacement curves obtained from the numerical analysis and the SDOF analysis were reported. Study on two-way elements was carried out by Morison [4] on walls and slabs. TFs proposed by TM5-1300 for SDOF analyses were reviewed using the yield line theory of plates. Errors were found in some of TFs of two way elements some reaching as much as 50%. In this study, TFs provided by TM5-1300 code were reviewed on RC cantilever beam.

1.1 Elastic Analysis of Cantilever RC Beam

* Corresponding author: tudjono@gmail.com

Multi degree of freedom (MDOF) system can be represented using SDOF system by introducing equivalent dynamic systems. In equivalent dynamic systems, K_L, K_S, and K_M are introduced into the system. Triangular blast pressure (Fig. 1a) applied with respect to time is assumed to be uniformly distributed along the height of beam (Fig. 1b). It is then converted into a lateral load (Fig. 1c) applied at the beam's tip after being multiplied by K_L.

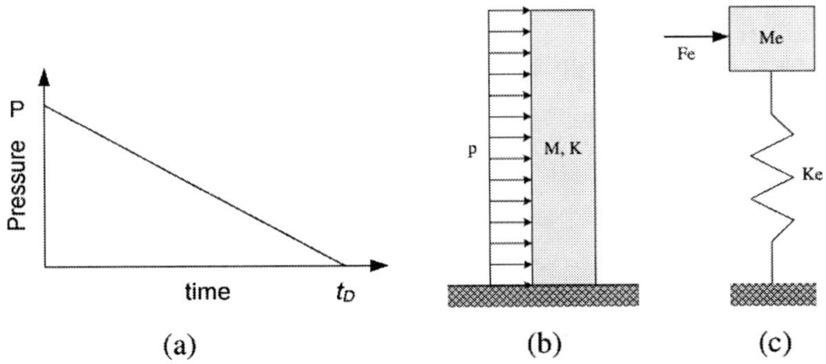

Fig. 1. (a) Triangular Blast Pressure (b) MDOF System (c) Equivalent SDOF System

The governing equation of the SDOF system may be given in terms of mass m, damping c, stiffness k, force F and displacement u as follows.

$$m\ddot{u} + c\dot{u} + ku = F(t) \tag{1}$$

while the governing equation of equivalent SDOF system is given as

$$m_e\ddot{u} + c\dot{u} + k_e u = F_e(t) \tag{2}$$

The equivalent terms of mass m_e, stiffness k_e and load F_e are as follows

$$\begin{aligned} m_e &= K_M . m \\ k_e &= K_S . k \\ F_e(t) &= K_L . F(t) \end{aligned} \tag{3}$$

K_M, K_S and K_L refer to the mass, the stiffness and the load transformation factors respectively. The transformation factors, theoretically, are values used to transform the MDOF problems into SDOF system by modeling the deflection as a shape function of $\phi(x)$. The deflection functions of a cantilever RC beam are determined from the boundary conditions. The deflection function assumed for the first mode is represented as

$$\phi(x) = 1 - \cos\left(\frac{\pi x}{2l}\right) \tag{4}$$

where x is the distance from fixed support whereas l is the beam 's length.

The equivalent mass, stiffness and force are as given in Eq. 5. This equation is equal to Eq. 3.

$$m_e = \int_l m(x)\phi(x)^2\, dx$$
$$k_e = \int_l EI(x)\phi''(x)^2\, dx \quad (5)$$
$$F_e = \int_l q(x)\phi(x)\, dx$$

Thus, K_L, K_S, and K_M are theoretically obtained as :

$$K_M = \frac{\int_l \phi(x)^2\, dx}{l} = \frac{\int_l \left(1-\cos\left(\frac{\pi x}{2l}\right)\right)^2 dx}{l} = \frac{3}{2} - \frac{4}{\pi} \approx 0.227 \quad (6)$$

$$K_S = \frac{\int_l \phi''(x)^2\, dx}{l^3} = \frac{\int_l \left(\left(\frac{\pi x}{2l}\right)^2 \cos\left(\frac{\pi x}{2l}\right)\right)^2 dx}{l^3} = \frac{\pi^4}{32} \approx 3.044 \quad (7)$$

$$K_L = \frac{\int_l \phi(x)\, dx}{l} = \frac{\int_l \left(1-\cos\left(\frac{\pi x}{2l}\right)\right) dx}{l} = 1 - \frac{2}{\pi} \approx 0.3634 \quad (8)$$

The displacement function is derived using the direct integration method of equivalent SDOF system [5]. It is obtained as a function of the displacement and the velocity at time step *i* and the applied forces at time step *i* and *i+1*.

$$u_{i+1} = u_i\left[e^{-\xi\omega_e\Delta t}\left(\frac{\xi\omega_e}{\omega_{De}} sin\omega_{De}\Delta t + cos\omega_{De}\Delta t\right)\right] + \dot{u}_i\left[\frac{e^{-\xi\omega_e\Delta t}}{\omega_{De}} sin\omega_{De}\Delta t\right] +$$
$$F_{ei}\left[\frac{1}{\omega_e k_e\Delta t}\left\{2\xi - (\omega_e\Delta t + 2\xi)e^{-\xi\omega_e\Delta t}cos\omega_{De}\Delta t\right\} + \left(\frac{1-\xi(\omega_e\Delta t+2\xi)}{\omega_{De}k_e\Delta t}\right)e^{-\xi\omega_e\Delta t}sin\omega_{De}\Delta t\right] + \quad (9)$$
$$+F_{e(i+1)}\left[\frac{1}{\omega_e k_e\Delta t}\left\{(\omega_e\Delta t - 2\xi) + 2\xi e^{-\xi\omega_e\Delta t}cos\omega_D\Delta t\right\} + \left(\frac{2\xi^2-1}{\omega_{De}k_e\Delta t}\right)e^{-\xi\omega_e\Delta t}sin\omega_D\Delta t\right]$$

Similarly, by using the same procedures as the displacement equation, the velocity function is derived as follows [5]:

$$\dot{u}_{i+1} = -u_i e^{-\xi\omega_e\Delta t}\left[\frac{\xi^2\omega_e{}^2 + \omega_{De}{}^2}{\omega_{De}}\right] sin\omega_{De}\Delta t + \dot{u}_i e^{-\xi\omega_e\Delta t}\left[cos\omega_{De}\Delta t - \frac{\xi\omega_e}{\omega_{De}} sin\omega_{De}\Delta t\right] +$$
$$F_{ei}\left[-\frac{1}{k_e\Delta t} + e^{-\xi\omega_e\Delta t}\left\{\frac{1}{k_e\Delta t}cos\omega_{De}\Delta t - \left(\frac{\omega_e{}^2}{\omega_{De}k_e} + \frac{\xi\omega_e}{\omega_{De}k_e\Delta t}\right)sin\omega_{De}\Delta t\right\}\right] - \quad (10)$$
$$F_{e(i+1)}\frac{1}{k_e\Delta t}\left[cos\omega_{De}\Delta t + \frac{\xi\omega_e}{\omega_{De}} sin\omega_{De}\Delta t - 1\right]$$

2 Displacement Analysis of the Cantilever RC Beam Subjected to Blast Load

In this study, 35 cantilever RC beams of different sizes are investigated (Table 1) and subjected to uniformly distributed pressure varies from 0.01 MPa to 0.05 MPa, provided that the beam is still in elastic condition. Loading duration t_D is varied from 5 to 15 *ms* to obtain higher variation of t_D/T ratios. K_L of 0.4 based on TM5-1300 is first applied to the model [2].

Table 1. Model Descriptions of Cantilever Beams

B	*H*	*L*	*T*	*B*	*H*	*L*	*T*
(mm)	(mm)	(m)	(ms)	(mm)	(mm)	(m)	(ms)
400	400	2	19.2	300	400	2	25.5
250	650	2	30.9	400	300	2	19.2
650	250	2	12.3	300	300	2	25.5
250	500	2	30.9	400	1000	2	19.3
500	250	2	15.5	1000	400	2	8.7
250	400	2	30.9	400	800	2	19.3
400	250	2	19.1	800	400	2	10.3
250	350	2	30.8	400	600	2	19.3
350	250	2	19.1	600	400	2	13.2
250	300	2	30.8	400	500	2	19.2
300	250	2	25.4	500	400	2	15.6
250	250	2	30.8	500	500	2	15.6
300	800	2	25.6	600	600	2	13.3
800	300	2	10.3	600	500	2	13.2
300	600	2	25.6	500	600	2	15.6
600	300	2	13.2	250	600	2	30.9
300	500	2	25.6	600	250	2	13.1
500	300	2	13.2				

It is shown in Fig. 2 that for the case of a cantilever RC beam subjected to blast load over a range of applied pressures, the peak displacement which occurs at the tip of the beam plotted against t_D/T follows a power function. The results also show that under elastic condition, for structures with the same t_D/T ratio, a 5 *ms* longer of loading duration t_D, results in higher peak displacement at the cantilever tip of up to 5 times.

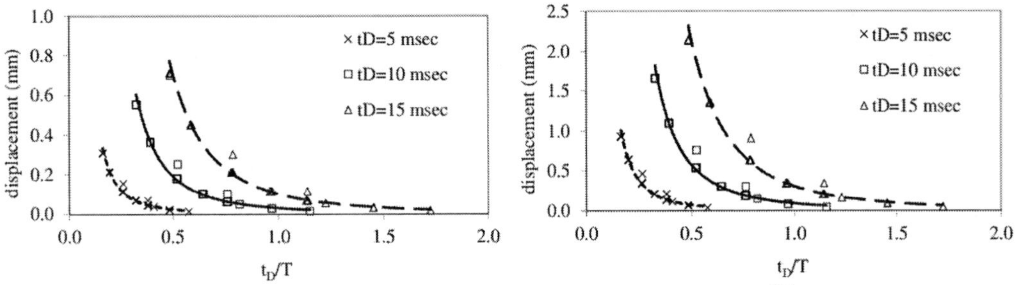

(a) (b)

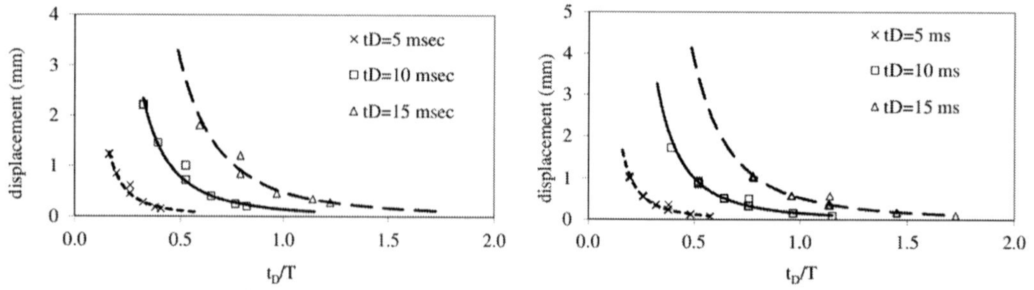

(c) (d)

Fig. 2. Peak displacements of beams subjected to blast pressures of (a) 0.01 MPa, (b) 0.03 MPa, (c) 0.04 MPa and (d) 0.05 MPa

2.1 Dynamic Magnification Factor (DMF)

DMF is the ratio of the displacement under dynamic loading $F(F_0,t)$ to the displacement under the static loading F_0. The DMF value of a beam differs in time following trigonometric function. DMF equation is derived [5] and given, based on its intervals of time and loading duration t_D as follows :

$$DMF = \frac{u}{u_{static}} = 1 - \cos\omega_e t + \frac{1}{\omega_e t_D}\sin\omega_e t - \frac{t}{t_D}, \text{ for } 0 \le t \le t_D \tag{11}$$

$$DMF = \frac{u}{u_{static}} = \frac{1}{\omega_e t_D}\{\sin\omega_e t - \sin\omega_e (t - t_D)\} - \cos\omega_e t, \text{ for } t \ge t_D \tag{12}$$

As shown in Fig. 3, the maximum value of DMF depends upon the ratio of t_D / T. The higher the ratio of t_D /T, the DMF obtained is greater. Once the ratio of t_D/T reaches a value larger than 1, the slope of the curve reduces until the DMF reaches the maximum value of 2. The DMF analysis can therefore be used as to validate the results obtained from the theoretical equivalent SDOF analysis. Fig. 3 also shows a comparison between the DMF value obtained from the structural dynamic theory and from SDOF analysis. It is observed that the DMF of SDOF results are in good agreement with the theoretical values.

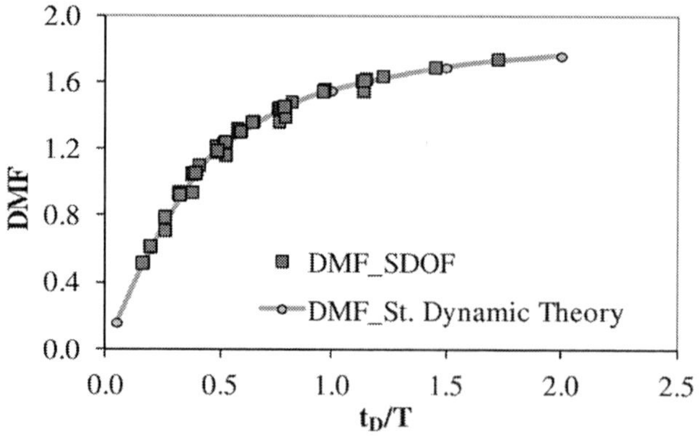

Fig. 3. Maximum DMF

2.2 Transformation Factor (TF)

On the basis of load transformation factor (K_L), TM5-1300 proposed a value of 0.40. It slightly overestimates the K_L value derived by the structural dynamics theory which provides a unique KL value equal to 0.363 (Fig. 4). With regards to the mass transformation factors (K_Ms), TM5-1300 proposed a K_M value of 0.26, again, higher than K_M derived using structural dynamics theory which is given as 0.227.

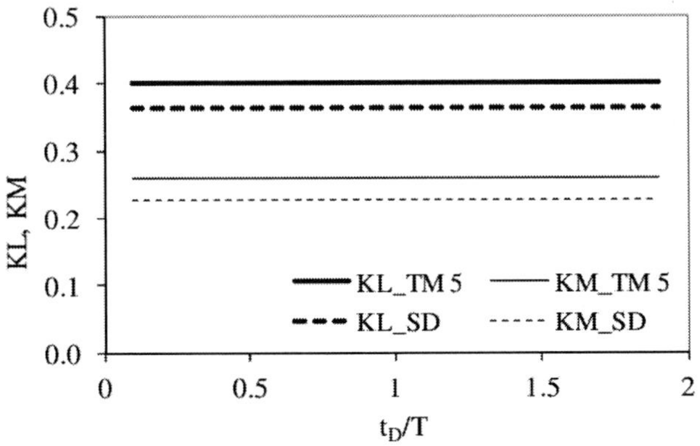

Fig. 4. Transformation Factors

3 Conclusion

With respect to DMF, the DMFs obtained from the SDOF analysis are found to be in good agreement with those obtained from the structural dynamics theory. On load and mass transformation factors, TM5-1300 provides an upper bound value, meanwhile the structural dynamics theory gives a lower bound value. The TM5-1300 may use different approach with respect to deflection function and may include additional safety factor of 10 to 15% for the purpose of blast resistant design.

References

1. C. G. Mays, P. D. Smith, *Blast effects on buildings* (Thomas Telford Publication, London, 1995)
2. TM5-1300, *Structures to resist the effects of the accidental explosions* (US Department of Army, New Jersey, 1990)
3. J. E. Crawford, J. L. Malvar, J. W. Wesevich, J. Valancius, D. R. Aaron, *Retrofit of RC structures to resist blast effects.* ACI Struct J, **94**, 371-377 (1997)
4. C. M. Morison, *Dynamic response of walls and slabs by single-degree-of-freedom analysis-a critical review and revision,* Int J Impact Eng, **32**, 1214-1247 (2006)
5. P. Kusumaningrum, *Numerical Modeling of RC and ECC Encased RC Columns Subjected to Close-in Explosion* (PhD Thesis, NUS, Singapore, 2010)

Development of Back-calculation Model for Aggregate Gradation Determination Using Fractal Theory

Bagus Hario Setiadji, Supriyono and *Djoko* Purwanto

Department of Civil Engineering, Diponegoro University, Jl. Prof. H. Soedarto, SH., Semarang, Indonesia 50275

Abstract. Several studies have shown that fractal theory can be used to analyze the morphology of aggregate materials in designing the gradation. However, the question arises whether a fractal dimension can actually represent a single aggregate gradation. This study, which is a part of a grand research to determine aggregate gradation based on known asphalt mixture specifications, is performed to clarify the aforementioned question. To do so, two steps of methodology were proposed in this study, that is, step 1 is to determine the fractal characteristics using 3 aggregate gradations (i.e. gradations near upper and lower bounds, and middle gradation); and step 2 is to back-calculate aggregate gradation based on fractal characteristics obtained using 2 scenarios, one- and multi-fractal dimension scenarios. The results of this study indicate that the multi-fractal dimension scenario provides a better prediction of aggregate gradation due to the ability of this scenario to better represent the shape of the original aggregate gradation. However, careful consideration must be observed when using more than two fractal dimensions in predicting aggregate gradation as it will increase the difficulty in developing the fractal characteristic equations.

1 Introduction

In developing a design mix formula of asphalt mixture, there is a quality control at each stage of the procedures, starting from preparing materials, testing the materials, determining aggregate gradation and so on. However, it is often encountered that although all parameters at each stage are controlled according to their requirements, the end result partially cannot meet the expectation. One cause of this non-conformance is the aggregate gradation used.

Generally, the selection of aggregate gradation for an asphalt mixture is done by trial and error, and the control of the control of the gradation is only conformity to the specification. This means that it could be more than one possible variation of aggregate gradation that can be selected to make the asphalt mixture, and different variations of the aggregate gradation will give different asphalt mixture properties.

To overcome this problem, this research is proposed to develop a back-calculation model using fractal theory, to obtain aggregate gradation that can produce asphalt mixture with optimal performance. Fractal theory is a manner to quantitatively describe the geometric complexity and the ability to fill the space of a particular object. Fractal geometry has become a keyword to better understand the concept of the dimension of the space of an object in a complex and irregular form. The study and application of the fractal theory of aggregate geometry have been conducted by researchers (1) [1, 2], however, the question that arises then is whether a fractal value can actually represent a unique aggregate gradation. This study is actually expected as an attempt to predict aggregate gradation. The back-calculation model developed in this study is an important part of a major research in developing a model for predicting aggregate gradation (which unknown parameter) based on the specification of the asphalt mixture properties (which is considered as known or given parameters).

2 Fractal Dimension

Aggregate gradation is a pattern with irregular and complex form or arrangement. The arrangement of aggregate gradations is often difficult to analyze conventionally, so the qualitative analysis is often a logical choice for analyzing aggregate gradation morphology [3]. At present, the fractal theory is one of the most effective quantitative approaches in explaining the complexity of aggregate gradation arrangements.

According to Mandelbrot [4], in general, fractal forms are self-similar, meaning that every small part of a fractal can be viewed as small-scale replication of the overall form. Mandelbrot [5] introduced a concept of the fractal to characterize the shape and roughness of an object and an irregular and rock-like surface. The level of roughness characteristics of the surface of the object is then called the fractal dimension.

Fractal dimension is used to describe the morphology of an irregular shape by characterizing the structure in one, two, and three-dimensional spaces using the fractal dimensions D1, D2, and D3 [6]. The fractal dimension is determined by using a geometric rank relationship between each geometric dimension (mass or volume, area and perimeter for three-, two- and one-dimensional spaces, respectively), with the long-term characteristic of an aggregate.

For an aggregate gradation, if the aggregate distribution is analyzed with fractal dimension, Equation 1 will be obtained.

$$P(r) = \frac{r^{3-D} - r_{min}^{3-D}}{r_{max}^{3-D} - r_{min}^{3-D}} \tag{1}$$

in which:

$P(r)$ = percentage passing of sieve size

r_{max}, r_{min} = maximum and minimum sizes of aggregate

Different gradation curves can be obtained in accordance with the given value D, and vice versa. In aforementioned equation, the r_{min} can be ignored because of its small value, therefore, Equation 1 can be re-written as Equation 2.

$$P(r) = \left(\frac{r}{r_{max}}\right)^{3-D} \tag{2}$$

Currently, the main method used to calculate the fractal dimension of aggregate gradation is by calculating the slope according to the dual logarithmic axis ln $P(r)$ and ln (r/r_{max}) using the linear regression equation. At this stage, the fractal dimension derived from the slope is used as the basic parameter, as seen in Equation 3.

$$\ln P(r) = (3 - D) \ln (r / r_{max}) + a \quad (3)$$

in which, a is constant.

3 Research Methodology

There are four main steps conducted in this study. They are:

- Determination of aggregate properties and gradation.
  In this study, the kind of the mixture used was asphaltic concrete – wearing course (AC-WC), which composed of dense-graded aggregate. To ensure that the mixture was made of high-quality aggregate material and composed of suitable gradation, aggregate properties measurement and sieve analysis were conducted and then the properties and gradation were evaluated to ensure that they could conform to the specifications of Directorate General of Highway [7].
  Once the suitable aggregate material was selected, three aggregate gradations (named as Gradation A, Gradation B, and Gradation C) were selected. These gradations should be located within the upper and lower bounds of aggregate specification [7]. The criteria of the three gradations selected were as follows:
  a. Gradation A: gradation with all percentage passing is 0.4% lower than that of the upper bound.
  b. Gradation B: middle gradation (i.e. gradation with all percentage passing is average between percentage of passing of upper and lower bounds).
  c. Gradation C: gradation with all percentage passing is 0.4% higher than that of the lower bound.
  The 0.4% increase or decrease of percentage passing of Gradation A and C is only intended to differentiate that both gradations is not upper and lower bounds.
- Determination of Fractal Dimension Based on Aggregate Gradation.
  In this step, three fractal dimensions were determined from three different aggregate gradations (A, B and C) using Equation 3. In this research, two scenarios were proposed in determining fractal dimension, that is, (i) one fractal dimension which represented all sieve sizes; (ii) multi-fractal dimensions in which each fractal dimension reflected one group of sieve size with similar trend/slope.
- Backcalculation of Aggregate Gradation Based on Fractal Dimension.
  Two scenarios in the previous step have their own benefit and drawback, in terms of accurateness and easiness to develop. Using these scenarios, aggregate gradation was back-calculated.
- Evaluation of back-calculated aggregate gradation towards specification.
  Once the aggregate gradation was obtained, an evaluation was conducted to examine whether the back-calculated gradation could conform to the specification [7].

4 Results and Analysis

Table 1 shows the results of aggregate properties measurement and their corresponding requirements as stated in Indonesia General Specification [7]. As shown in the table, all properties of selected aggregate material could fulfill all the requirements.

In Table 2 and Fig. 1, three aggregate gradations were selected. It can be seen that all gradations were within the upper and lower bounds of the specification. The percentage passing of all three gradations in Table 2 was used as a reference to measure the accurateness of back-calculated aggregate gradations developed in this study.

Table 1. Properties of Aggregate

Aggregate properties	Test Method	Requirements*)	Results
Properties of coarse aggregate			
Affinity to asphalt (%)	ASTM D1664-80	> 95	99
Abrasion (%)	ASTM C131-06	< 40	24
Abrasion (%)	ASTM C131-06	< 40	24
Water Absorption (%)	ASTM C127 – 07	< 3	1.341
Specific gravity	ASTM C127 – 07	> 2.5	2.73
Properties of fine aggregate			
Water Absorption (%)	ASTM C127 – 07	< 3	2.229
Specific gravity	ASTM C127 – 07	> 2.5	2.679
Sand Equivalent (%)	ASTM D2419-14	> 50	93.1

Remarks: *) requirement as stated in General Specification version by DHG [7]

Table 2. Several Gradation Used in This Study

Gradation Types	Passing rate at different sieve size (mm), in %									
	19	12.50	9.50	4.75	2.36	1.18	0.60	0.30	0.15	0.075
Upper bound	100	100	90	63	39.1	25.6	19.1	15.5	13	10
Gradation A	100	99.95	89.95	62.95	38.91	25.5	19	15.4	12.9	9.9
Gradation B	100	95	81	53	33.55	22.3	16.05	12.25	9.51	7.01
Gradation C	100	90.4	72.04	43.04	28.42	19.08	13.16	9.23	6.16	4.22
Lower bound	100	90	72	43	28	19	13	9	6	4

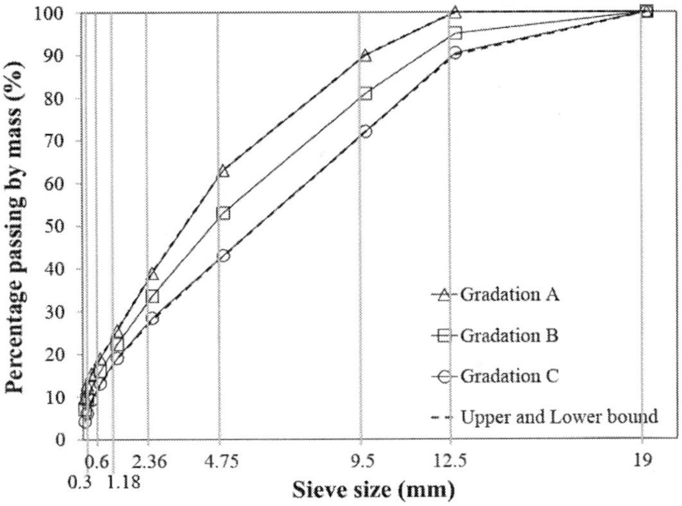

Fig 1. Gradation Curve of Different Gradation Used

Based on these three gradations, fractal dimensions were determined using two scenarios. Figs. 2 and 3 show the fractal characteristic that developed based on one fractal dimension and multi-fractal dimension scenarios, respectively. In this study, two fractal dimensions were used to represent a multi-fractal dimension scenario due to its easiness in developing linear equation from the fractal dimension in future works.

In two fractal dimension scenario, both lines have different fractal dimension D. The fractal dimensions for line 1 and line 2 are called as D1 and D2, respectively. The use of line 1 will separate data of two maximum sieve sizes with the highest deviation against estimated data (see Fig. 2) from the rest. Line 1 only consists of two data, therefore, this line always produces a coefficient of determination equals to unity (see Fig. 3).

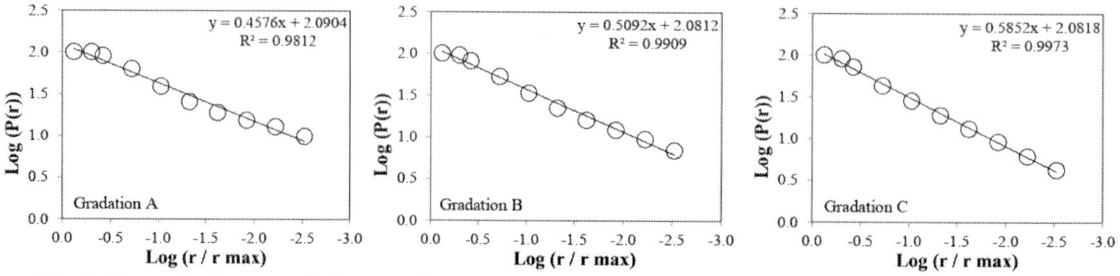

Fig 2. Fractal Characteristic Based on One Fractal Dimension

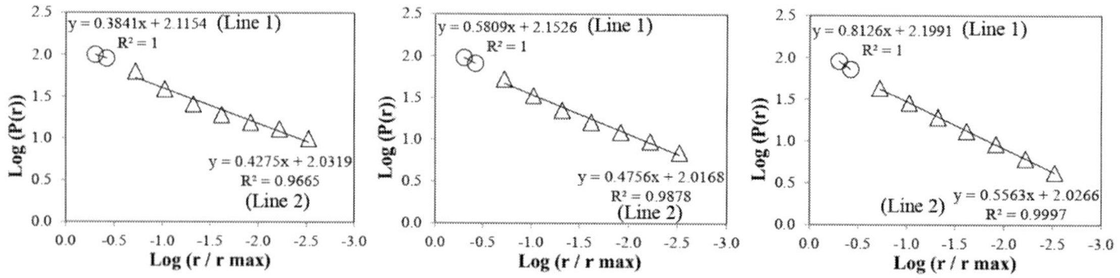

Fig 3. Fractal Characteristic Using Multi-Fractal Dimension

To examine the suitability of the scenario towards the accurateness of aggregate gradation, the linear equation in each scenario was used to predict the percentage passing of each sieve size in the gradation. The results of this back-calculation procedure are depicted in Figs. 4 and 5.

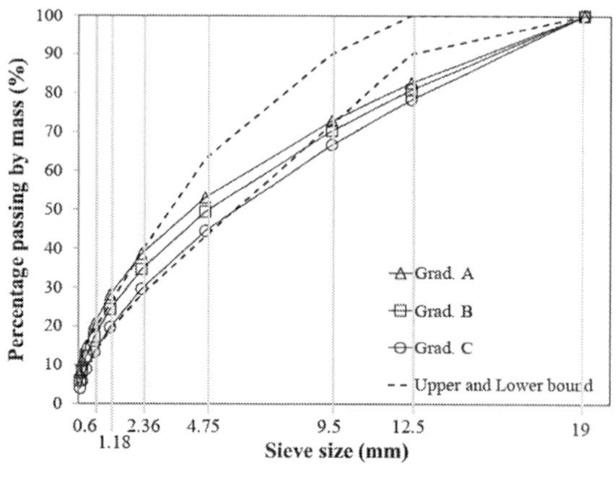

Fig 4. Gradation Curve Based on First Scenario (One Fractal Dimension)

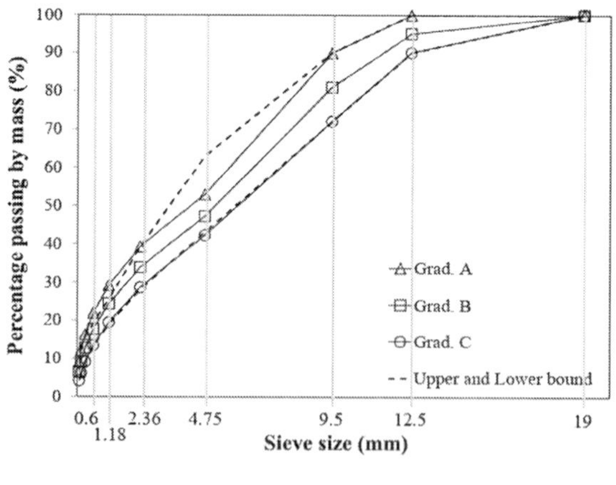

Fig 5. Gradation Curve Based on Second Scenario (Multi-Fractal Dimension)

As shown in Fig. 4, the back-calculated aggregate gradations from one fractal dimension underestimate the original gradation, especially on several sieve size (9.5 mm to 19 mm). Different results are shown in Fig. 5 where the use of multi-fractal dimensions could correct the drawback of the use of one fractal dimension, especially on large sieve sizes. This could be contributed to the ability of multi-fractal dimension scenario to better represent the shape of the original aggregate gradation. However, careful consideration must be observed when using more than two fractal dimensions in predicting aggregate gradation as it will increase the difficulty in developing the fractal characteristic equations.

Once the most suitable scenario in determining aggregate gradation is selected, now it is necessary to determine the procedure in developing linear equations in the scenario. The following procedure might be proposed:

- Fractal dimension is obtained based on the specification of asphalt mixture, especially voids in the mix (VIM). It considers as a fractal dimension for line 2 and called as D2 (see Fig. 3).
- Do sieve analysis of aggregate material.
- Determine percentage passing of nominal maximum size (i.e. one size below the largest size), percentage passing of one- and two-size below nominal maximum sieve size.
- Derive the fractal characteristic equation of line 2 (as seen in Fig. 3) using D2 and percentage passing of two-size below nominal maximum sieve size.
- Derive the fractal characteristic equation of line 1 (as seen in Fig. 3) using percentage passing of nominal maximum and one-size below nominal maximum sieve sizes.
- Predict aggregate gradation using equations of lines 1 and 2.

5 Conclusions

This paper presented a method to predict aggregate gradation using fractal dimension. Two scenarios were proposed in this paper to predict aggregate, i.e. one and multi-fractal dimensions. The results showed that multi-fractal dimension scenario (in this case, two fractal dimension) could produce a better prediction of gradation than that of one fractal dimension scenario due to the ability of this scenario to represent the shape of original aggregate gradation.

References

1. A.Y. Yin, X.H. Yang, Z. Yang, *2D and 3D Fracture Modeling of Asphalt Mixture with Randomly Distributed Aggregates and Embedded Cohesive Cracks,* IUTAM Symposium on Multiscale Problems in Stochastic Mechanics, Procedia IUTA, Vol. **6**, 114-122 (2013)
2. D. Weizhi, G. Xuedong, G. Liang, *Study on the Fractal Gradation Optimization of ATB-25*, the 2nd Int. Conf. Electron. Mech. Eng. Inf. Technology Study (2012)
3. P. Wang, Y. Bi, J. Pei, J. Zhang, B. Xue, J. Cai, X. Zhang, Aggregate Gradation Design Method Based on Fractal and Non-Cooperative Game Theory, the 10th Int. Conf. Road Airf. Pave. Tech. (2017)
4. B.B. Mandelbrot, *The Fractal Geometry of Nature* (W.H. Freeman, San Francisco, 1982)
5. B.B. Mandelbrot, *Fractals - Form, Chance, and Dimension.* San Francisco (W.H. Freeman, San Francisco, 1977)
6. S. Tang, Y. Ma, C. Shiu, *Modelling the Mechanical Strength of Fractal Aggregates*, Colloids and Surfaces A: Physicochemical and Engineering Aspects, 7-16 (2001)
7. Directorate General of Highway, *General Specification of Hot-mixed Asphalt Section 6.3, Revision 2,* Ministry of Public Works (2010) (in Bahasa Indonesia)

Modelling and analysis the effect of unbalanced moment directions on behaviour of edge column-slab connections

I Ketut Sudarsana[1,*], *I Gede Gegiranang* Wiryadi [2], and *Gede Adi* Susila [1]

[1]Universitas Udayana, Civil Engineering Department, Kampus Bukit Jimbaran-Bali, Indonesia
[2]Graduate Student, Universitas Udayana, Civil Engineering Department, Kampus Bukit Jimbaran, Indonesia

Abstract. This research investigates the effect of unbalanced moment directions on the behaviour of edge column slab connections using a finite element analysis. The analyses were done on subassembly edge column slab connections that were designed according to Indonesian Concrete Standard (SNI 2847:2013). Three unbalanced moment directions were considered namely perpendicular, parallel and inclined 45° to the slab free edge. The concrete damage plasticity (CDP) and truss elements in Abaqus were utilized to model and analyse the behaviour of concrete and reinforcement bars, respectively. The modelling techniques were first validated using an experimental result available in the literature. There are five parameters in the CDP model need to be validated to get convergent results with the experimental data. Using the CDP validated parameters, then seven specimen models were analysed under combined shear force and an unbalanced moment in three directions. The ratio of M/V was kept constant of 0.3. The results show that the punching failure capacity of connections having an unbalanced moment inclined 45° is smaller than that of an unbalanced moment perpendicular to the slab free edge, but higher than that of an unbalanced moment parallel to the slab free edge. The patterns of concrete strain are consistent with the moment directions. All tension rebars passing through column sections yield at the connection failures.

1 Introduction

Punching shear failures around a column is a complex phenomenon. It becomes more complex when unbalanced moment has to be transferred from the slab to the column that cannot be avoided at edge column slab connections. The unbalanced moment can be due to span discontinuity at the slab free edge, or lateral loads caused by earthquake and wind loads. Numerous experimental researches and numerical analysis have been done to investigate the punching failure of the edge column slab connections, however, experimental research needs advanced equipment and quite expensive to investigate a

* Corresponding author: ksudarsana@unud.ac.id

parameter, therefore, finite element analysis become an alternative solution as done by [1] using finite element based program, Abaqus. It is found that using concrete damage plasticity (CDP) method in Abaqus to model nonlinear behaviour of concrete give the analysis results similar to that of experimental results. A good correlation in deformation, capacity and crack pattern were also found by [2].

Most building codes in the world recommend that in accounting the effect of unbalanced moments due to earthquake load can be taken in parallel direction to building axes, however, it is still worth to investigate the effect of unbalanced moment directions since there are not many data available in the literature. This paper describes the results of finite element analysis on punching behaviour of edge column slab connections due to the effect of shear force and unbalanced moment applied in three different directions. The research investigates the shear capacity, slab deflections, concrete strain as well as tensile stress of the slab reinforcement.

2 Analysis Method

A nonlinear analysis using a finite element based software, Abaqus, was conducted on seven specimen models as shown in Table 4 to investigate the effect of the unbalanced moment directions on punching shear phenomena in the edge column slab connections. Nonlinear behaviour of concrete in Abaqus can be modelled using Concrete Damage Plasticity (CDP) Model, which needs to define five parameters namely dilatation angle (ψ), eccentricity (ε_c), the ratio of the tensile to the compressive meridian (K_c), the ratio between biaxial compressive strength and the uniaxial compressive strength (σ_{bo}/σ_{co}) and viscosity (μ). Detail theory of CDP Model can be found in [3,4,5]. The CDP Model has been used previously by [1,2,6,7,11], it was shown that the fifth parameters of CDP model have to be adjusted according the concrete strength.

2.1 Verification model

Numerical analysis based on structural dimensions, materials, elements, and other parameters in the CDP model of Abaqus was first verified using an experimental specimen (slab XXX) tested by [8]. Detail process on verification model can be found in [9]. The verification was done iteratively by analysing 22 variation models to seek a correct value of all CDP parameters. It is found that the value of dilatation angle (ψ) of 40°, eccentricity (ε_c) of 0.1, K_c of 2/3, the ratio (σ_{bo}/σ_{co}) of 1.16, and the viscosity (μ) of 0.00005.

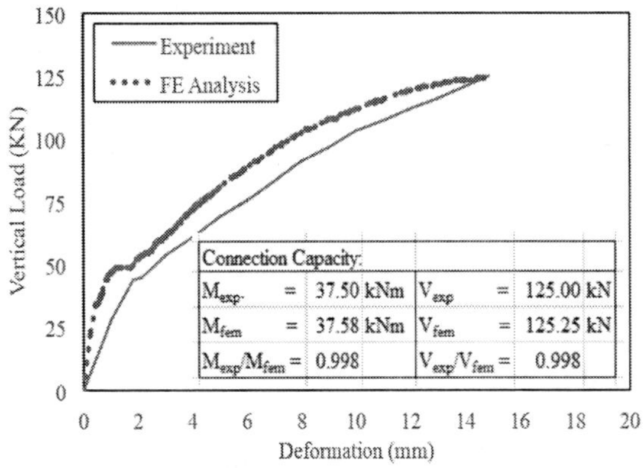

Fig. 1. Experimental and FE analysis deformation of slab XXX.

The mode of failure was based on compression and tension; the tension behaviour of concrete was based on tension strain. The mesh size was 60 mm. The analysis shows that the connection capacities were predicted well with the ratio of experimental to FE analysis (M_{exp}/M_{fem} and V_{exp}/V_{fem}) of 0.998 as shown in Figure 1. Therefore, the techniques were then used to carry out further analysis in this study.

2.2 Specimen models

The specimen models were subassembly edge column plate connections obtained from a properly designed three storey building having column to column span of 3 meter in both direction following the requirement of [10]. Figure 2 shows the specimen dimensions, reinforcement, and support positions. The specimens were pin supported along three slab edges on the opposite directions of vertical force. All loads were applied through the column stubs. The unbalanced moments were obtained from application of two equal horizontal forces as shown in Figure 2(c). The ratio between a shear force and an unbalanced moment (M/V) was kept constant of 0.3 with three variations on moment directions namely 0° (Model E00), 45° (Model E45) and 90° (Model E90) that are perpendicular, incline 45° and parallel to slab free edge, respectively. The vertical load (V) and an unbalanced moment (M) were applied simultaneously to keep the aspect ratio of *M/V* constant up to failure identified.

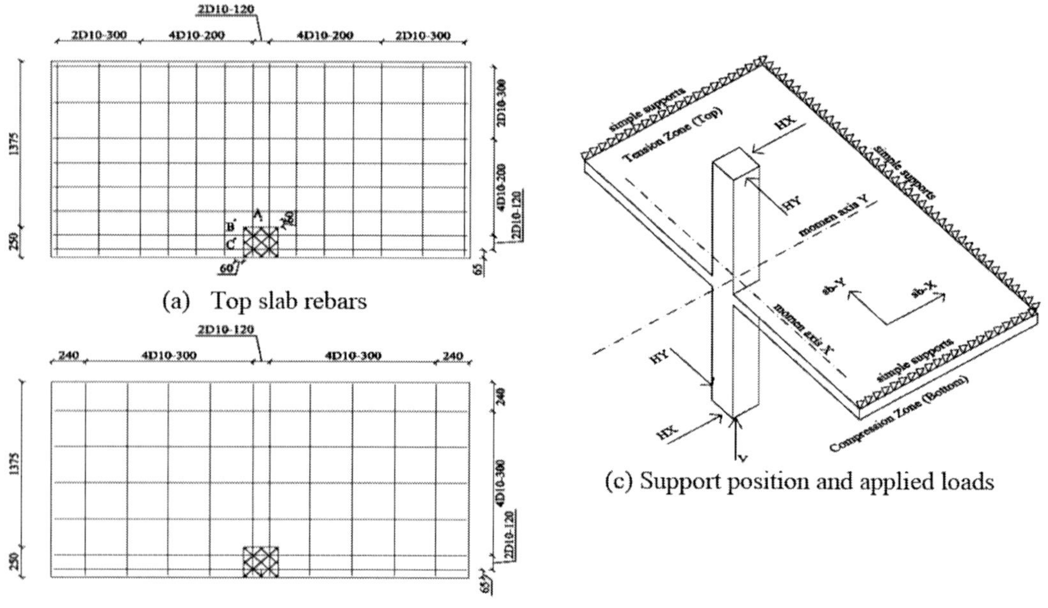

(a) Top slab rebars

(b) Bottom slab rebars

(c) Support position and applied loads

Fig. 2. Dimension and slab reinforcement of the specimen models.and support position and applied loads

2.3 Material properties and FE Modelling

Properties of concrete and reinforcement bars used in the analysis are shown in Table 1. In the absence of test data for stress-strain diagram of concrete, the Hognestad Model (1951) was used. The reinforcement was assumed to be elastoplastic material.

Table 1. Properties of materials.

Concrete properties		Rebars properties	
Compressive strength (f'_c), MPa	: 30	Yield stress (f_y), MPa	: 400
Tension strength (f_{ct}), MPa	: 1.81	Ultimate stress (f_u), MPa	: 600
Elastic modulus (E_c), MPa	: 25742	Elastic Modulus (E_s), MPa	: 200000
Strain at peak compression stress (ε_{co})	: 0.0023	Yield strain (ε_y)	: 0.002
Strain at peak tension stress (et)	: 0.00007	Ultimate strain (ε_u), MPa	: 0.10

Concrete was modelled using a three-dimension stress element (3D stress) with eight nodes that is defined in Abaqus as C3D8R (8-Node linear brick, reduced integration with hourglass control). These elements were used to reduce the integration for nonlinear analysis. The reinforcement was modelled as a 2-node linear 3-D truss element. According to [11], modelling of reinforcement using one-dimension elements gives equivalent results to model it using solid elements. Concrete and rebars were assumed having a perfect bond that was simulated using embedded region technique in Abaqus.

3 Results and Discussion

3.1 Connections capacity and slab deformation

Deformations of slab in three locations (A, B, and C) identified in Figure 2(a) are plotted against the vertical loads (V) up to connection failure as shown in Figure 3 together with the model's capacities. It shown that the connection capacity is affected by the unbalanced moment direction. The smallest capacity is obtained when the moment direction parallel to the slab free edge.

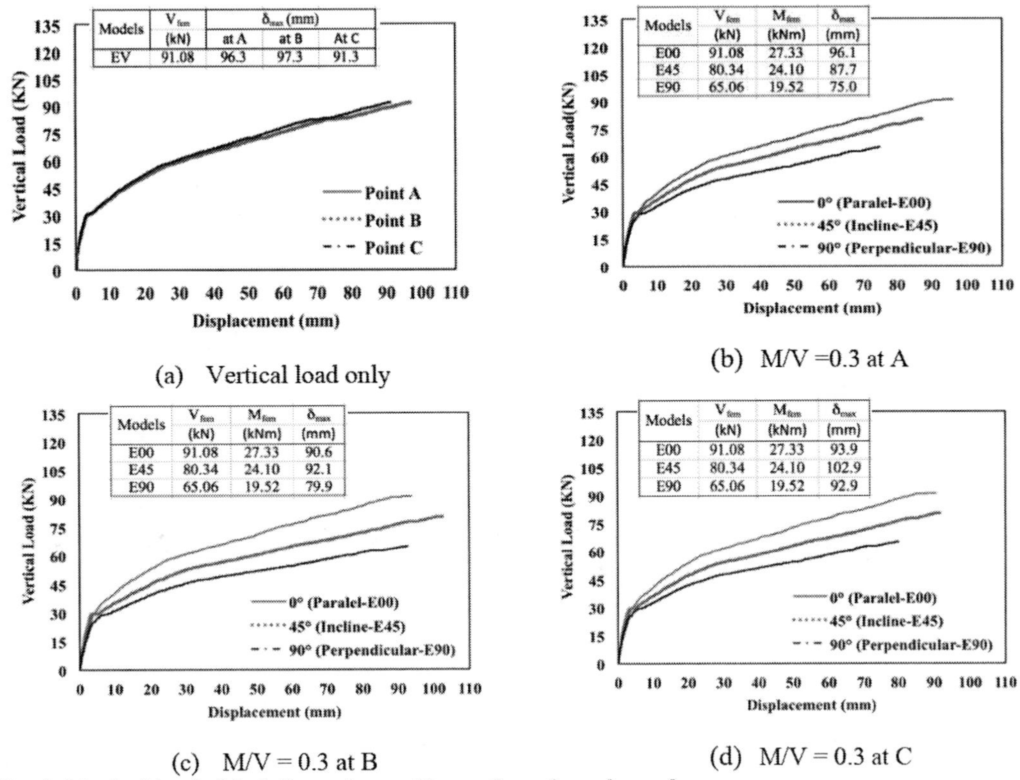

(a) Vertical load only

(b) M/V =0.3 at A

(c) M/V = 0.3 at B

(d) M/V = 0.3 at C

Fig. 3. Vertical load-slab deformation at 60 mm from the column faces.

3.2 Concrete strain in slab

Punching failure of the connections can also be identified base on the strain of the concrete on slab. The slab principle strains are shown in Figure 4. It is shown that the pattern of concrete principal strain follows the moment directions. The highest concrete strain is occurred at the column faces.

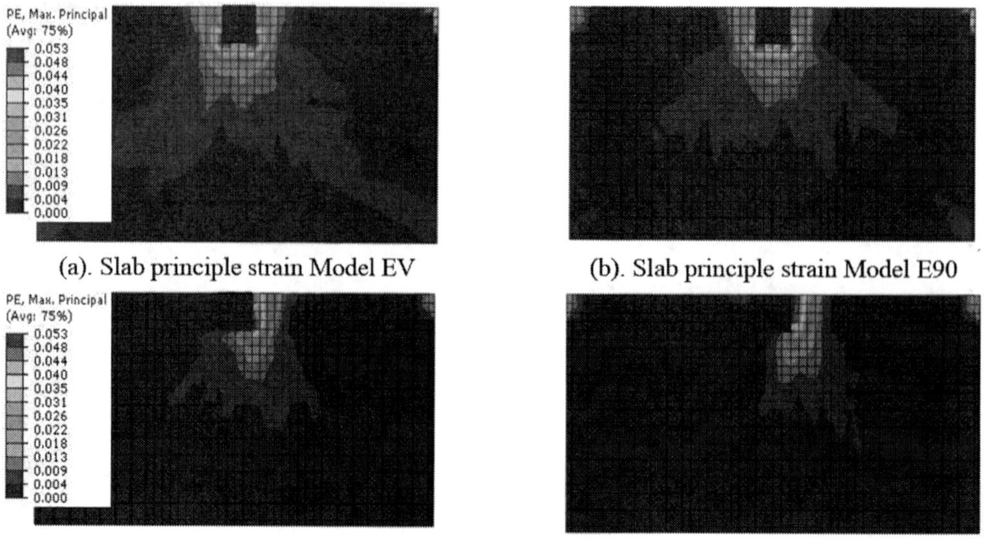

(a). Slab principle strain Model EV
(b). Slab principle strain Model E90
(c). Slab principle strain Model E45
(d). Slab principle strain Model E00

Fig. 4. The concrete principle strain in slab for Model EV, E90, E45 and E00.

3.3 Rebar stress of slab

The stress on slab reinforcement was measures at 60 mm from column faces for both perpendicular and parallel rebars to slab free edge. Figure 5 and 6 provide the tensile stress of the rebars.

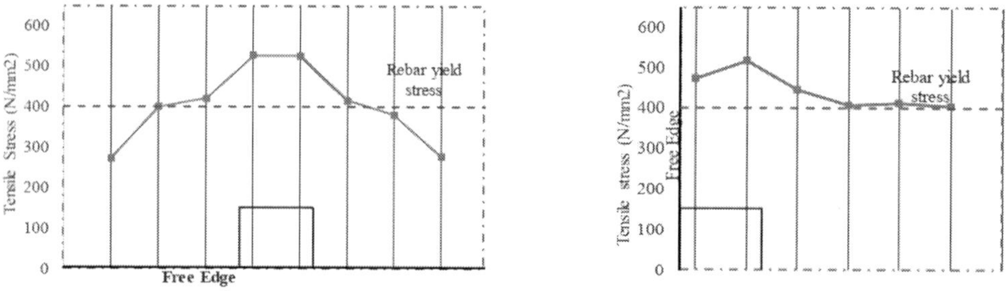

(a) Rebar perpendicular to the slab free edge
(b) Rebar paralel to the slab free edge

Fig. 5. Stress on tension rebars of the slab on Model EV.

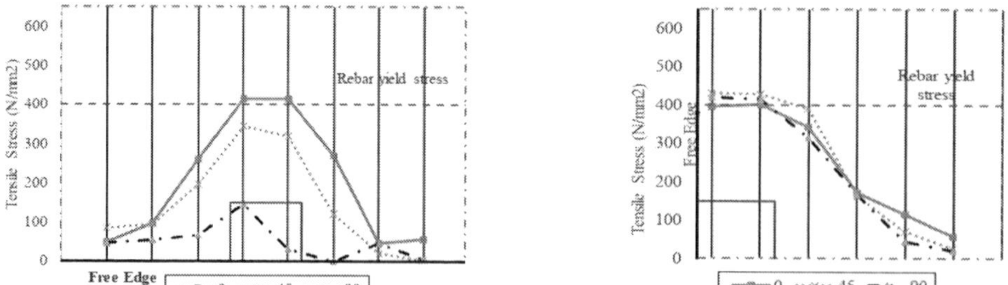

(a) Rebar perpendicular to the slab free edge
(b) Rebar paralel to the slab free edge

Fig. 6. Stress on tension rebars of the slab in Model E00, E45 and E90.

All tension rebars perpendicular to the slab free edge within the column strip reach yield stress for all loading conditions except for those with unbalanced moment parallel to slab free edge. For tension rebars parallel to the slab free edge within the column strips, almost all of them yield at the connection failure except the specimen with pure bending parallel to the slab free edge. This confirms the experimental results obtained by [12] that all rebars passing through the column yield at the punching failure occurred. The tension forces result in from the vertical load and unbalanced moment are resisted dominant by the rebars parallel to the unbalanced moment directions. However, for the connections with the direction of unbalanced moment incline 45°, both parallel and perpendicular rebars to moment direction resist almost equal forces. This indicates that rebars in column strip region contribute punching shear resistance of flat plate as already proposed by [13]. However, this effect has not been adopted in [10] to calculate the punching shear resistance.

4 Conclusions

The results of numerical analysis using CDP model in Abaqus shows that the punching capacity of the connections decreases from the moment direction of 0°, 45° and 90° to the slab free edge. The slab concrete strain and reinforcement tensile stress tend to follow the moment directions. Most of the rebars passing through the edge columns yield at the punching failure. The punching failure capacity of connections having 45° unbalanced moment is smaller than the connections with unbalanced moment perpendicular to free edge, but higher than the connections with unbalanced moment parallel to the free edge. The patterns of concrete strain are consistent with the moment directions. All tension rebars passing through the column sections yields at the connection failures.

5 References

1. A.S. Genikomsou, M.A. Polak, Eng. Str. **98**, 38–48 (2015)
2. A.Tambusay, P. Suprobo, Faimun, A. A. Amiruddin, Appl. Mech. and Mat. **851**,714-719 (2016)
3. ABAQUS, *Analysis user's manual 6.14*, (Dassault Systems Simulia Corp.,Providence, RI, USA; 2014)
4. J. Lubliner, J. Oliver, S. Oller, E. Oñate, Int. J. Solids Str. **25**(3):299–326 (1988).
5. J. Lee, G.L. Fenves, J. Eng. Mech., **124**(8), 892–900 (1998).
6. S.V. Chaudhari, M.A. Chakrabarti, Int. J. Comp. Appl. **44**(7), 14–18 (2012).
7. Y. Sümer, M. Aktaş, M., J. Str. Mech. **1**(3),149–155 (2015).
8. E.F. El-Salakawy, M.A. Polak, M.H. Soliman, Can. J. Civ. Eng. **25**, 526–538 (1998).
9. I K. Sudarsana, I G. A. Susila, I G. G. Wiryadi, J. Spektran. **5** (2) 102-111 (2017)
10. SNI 2847:2013. *Persyaratan beton struktural untuk bangunan gedung,* (BSNI, Jakarta, 2013)
11. T.S.Viswanathan, G.G. Mohan, A.S. Santhi, Int. J. Civ. and Str. Eng. **2**,3, (2012).
12. I K.Sudarsana, N.J Gardner, *2nd ACF Conference*. (Nusa Dua, Bali-Indonesia, 2006).
13. N.J. Gardner, CSCE J. **23** (2) 502-210 (1996).

Effect of Cement Variation on Properties of CLC Concrete Masonry Brick

Erwin Sutandar[1,*], *Asep* Supriyadi[1], and *Cek Putera* Andalan[1]

[1] Universitas Tanjungpura, Civil Engineering Department, Prof. Dr. Hadari Nawawi Street, Pontianak, West Kalimantan, Indonesia

Abstract. One of the methods used to reduce the weight of a construction is by reducing the weight of the walls of the building. In such a case, a wall made of red brick has a volume weight of 1,500–2,000 kg/m^3, and concrete masonry bricks made of CLC have a volume weight of 400–1,800 kg/m^3. So, in comparison, concrete masonry bricks have a volume weight that is ≤ 50% of that of red brick. In the manufacturing of concrete masonry bricks, one variant is CLC (Cellular Lightweight Concrete), produced using a mixture of cement, sand, chemical admixture and water, with the filler material in the form of air generated as microscale soap bubbles (microbubbles), also known as foam agent. In the manufacturing of concrete masonry bricks, the cement as a binder material clearly affects the physical and mechanical properties of the bricks produced. This research is conducted to investigate the effect of the amount or composition of the cement used on the physical and mechanical properties of concrete masonry bricks. The composition is varied among 200, 250, 300, 350 and 400 kg/m^3 of cement usage.

1 Introduction

For the purpose of lightweight construction in soft soil or peatland areas, innovation is required to reduce the self-weight of the construction. One way to reduce this is by reducing the weight of the walls, which are generally made of red brick having a volume weight of 1,500–2,000 kg/m^3 [5], cement brick with a volume weight of 950–1,000 kg/m^3 [6][7] or CLC masonry concrete brick with a lighter volume weight of 400–900 kg/m^3[1]. This CLC masonry concrete brick also has the effect of reducing the reinforcement of a structure, as a result of the lower section dimensions of the structural elements, and it reduces the amount of concrete, avoiding the usage of sand in the plaster, making the building more economical [6][7].

CLC masonry concrete brick can be used as an alternative to red brick, to reduce environmental pollution and global warming. CLC masonry concrete brick is environmentally friendly. The energy consumed in the production of this type of masonry lightweight concrete brick is less than for the production of red brick, and it also neither produces pollution nor creates toxic products or have any impact on the environment [12]. This is because CLC masonry lightweight concrete brick is made of a mixture of cement, sand, water, and chemical admixture, to which is then added a foam that is stable in an

* Corresponding author: erwin_sutandar@yahoo.com

ordinary concrete mixer. The addition of foam to the CLC mixture creates millions of voids or small cells in the material, which is why it is given the name of CLC masonry concrete brick.

Cement, as one of the basic ingredients of CLC masonry concrete bricks, is a binder that aggregates granules into a solid form. In using cement as a binding agent, the physical and mechanical properties of the CLC masonry concrete brick are determined by the composition of the cement to be used, and determining the optimum composition in terms of cement consumption when manufacturing the lightweight concrete brick will ensure the optimal results. Therefore, research is needed to compare the physical and mechanical properties of the CLC masonry concrete brick that is produced.

2 Material and Dimensions of CLC Masonry Concrete Brick

2.1. Cement

Commercially available cement is used, namely Holcim cement type III **(i.e. ASTM C 150-95a & SNI 15-2049-1994 & BS 12:1989)** with the strength of the initial high Portland Composite Cement (PPC) and a **specific gravity** of 2,950 kg/m^3 [11]. The physical and chemical properties of the PCC cement can be seen in Table 1.

Table 1. Physical and chemical properties of cement [3][11]

A. Compression Strength	
3 day	min 13 MPa
7 day	min 20 MPa
28 day	min 28 MPa
Fineness	280 m^2/kg
B. Time Settings	
Initial	45 mins
Final	375 mins
C. **Chemical properties**	
Silicon Dioxide (SiO_2)	23.04%
Aluminium Oxide (Al_2O_3)	7.40%
Ferric Oxide (Fe_2O_3)	3.36%
Calcium Oxide (CaO)	57.38%
Magnesium Oxide (MgO)	1.91%
Sulphur Trioxide (SO_3)	2.0%
Loss on Ignition (LOI)	3.94%
Free Lime	0.56%
Insoluble Residue	10.96%

2.2 Fine Aggregate

Fine aggregate as one of the components in cement must undergo a series of tests on the properties and characteristics of the material. In this study, the type of fine aggregate used is yellow sand (SNI-03-2834–2002) and the qualification limits determined for the fine aggregate are classified into four groups based on the degree of smoothness of the sand material [13]. This research includes sand zone III with a fines modulus of 3.030 [3].

2.3 Water

The water used is taken from the local water company with a pH 6–7 (**SNI 03-2874-2002** and meeting the requirements of **PBBI 1971 NI-2**).

2.4 Foaming Agent

Foamed concrete is produced by either a pre-foaming method or a mixed foaming method. The pre-foaming method involves the separate production of a base mix cement slurry (cement paste or mortar) and a stable preformed aqueous addition (foam agent with water), and this foam is then thoroughly blended into the base mix. In mixed foaming, the surface active agent is mixed with the base mixture ingredients and, during the process of mixing, the foam is produced, resulting in a cellular structure in the concrete as shown in Fig. 1. The preformed foam can be either wet or dry. Wet foam is produced by spraying a solution of foaming agent over a fine mesh, has a 2–5 mm bubble size, and is relatively less stable. Dry foam is produced by forcing the foaming agent solution through a series of high-density restrictions and forcing compressed air simultaneously into the mixing chamber. Dry foam is extremely stable and has a size smaller than 1 mm. Table 2 shows the properties of foamed concrete [1].

Table 2. Typical properties of foamed concrete [1]

Dry density (kg/m^3)	Compression strength (MPa)	Modulus of elasticity (E-value) (GPa)	Thermal conductivity (3% moisture) (W/mK)	Drying shrinkage (%)
400	0.5-1.0	0.8-1.0	0.10	0.30-0.35
600	1.0-1.5	1.0-1.5	0.11	0.22-0.25
800	1.5-2.0	2.0-2.5	0.17-0.23	0.20-0.22
1000	2.5-3.0	2.5-3.0	0.23-0.30	0.15-0.18
1200	4.5-5.5	3.5-4.0	0.38-0.42	0.09-0.11
1400	6.0-8.0	5.0-6.0	0.50-0.55	0.07-0.09
1600	7.5-10	10.0-12.0	0.62-0.66	0.06-0.07

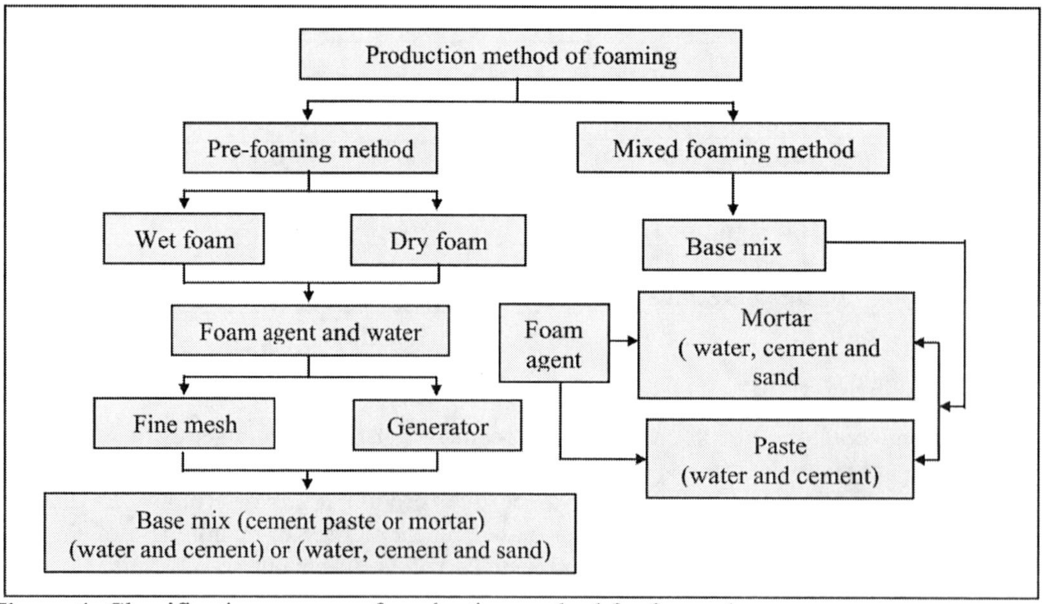

Figure 1. Classification process of production method for foamed concrete

The container holding the foaming agent must be kept airtight and under temperatures not exceeding 25°C. Once diluted in 80 parts of potable water, the emulsion must be used as soon as possible. The weight of the foam solution used is a minimum of 80 g/l, the containment solution used is as close as possible to 10 litres in volume, to check the weight (density) of the foam. Under no circumstances must the foaming agent should not contact with any oil or fat, chemical or other material that might harm its function (oil has an influence on the surface tension of water) [6] [10]. The foaming agent used is a brand under the trademark ADT.

2.5 Chemical Admixture

The chemical admixture is usually used in small amounts in the concrete mix. Its use is intended to improve certain properties of the mixture. The materials used are a high-range water-reducing admixture, wherein the material can reduce the water demand of the cement by up to a maximum of 15%. The product Sikament LN from Sika Company is a type of chemical additive used as a water reducer and to speed up the hardening of mortar that requires an immediate settling time or an accelerator. Sikament LN is an admixture classified as ASTM C494-92 type F [2]. With Sikament LN, the composition uses a dose rate of 0.40% of the total cement requirement of 300 kg, so the quantity needed for this experiment is as much as 1.2 kg.

2.6 Dimensions of the test objects

The nominal dimensions of the CLC brick of the United States standard are as follows: length: 600 mm, height: 200 mm, width: 75 mm. And the dimensions for modulus of elasticity testing diameter 150 mm and height 300 mm [5][6].

3 Experimental

This stage of the study involves the design of the mixture that will be used in the creation of samples of CLC masonry lightweight concrete brick. This variation plan is created as a guide to the composition of the mixture that will be investigated in this research and is presented in Table 3.

Table 3. Composition of the basic materials of samples [3][8]

No	Ingredients	Number /m^3					Units
		V1	V2	V3	V4	V5	
1	Cement	200	250	300	350	400	kg
2	Sand	500					kg
3	Foam agent and water	64.8					litres
4	Water	150					kg
5	Sikament LN	1.2					kg

4 Results and Discussion

From the research that has been done, the findings obtained based on the test results, i.e. according to ASTM and ACI standards, can be seen in Table 4.

Table 4. Comparison of the results for different amounts of cement in the making of CLC masonry concrete brick

No	Parameters	Cement 200 kg/m³	Cement 250 kg/m³	Cement 300 kg/m³	Cement 350 kg/m³	Cement 400 kg/m³
1	Success rate of CLC brick making (%)	33	100	100	100	100
2	Slump flow (mm)	280	280	275	275	280
3	Drying shrinkage (%)	31.13	0	0	0	0
4	Porosity (%)	0	23.93	27.64	42.30	25.38
5	Sound resistance (db)	0	49.38	34.94	31.10	50.57
6	Thermal conductivity (W/mK)	0	0.55	0.57	0.47	0.55
7	Permeability (cm/second)	0	2.77 x 10^{-5}	1.62 x 10^{-5}	2.60 x 10^{-5}	2.96 x 10^{-5}
8	Modulus of elasticity (MPa)	0	51.00	45.74	98.64	75.23
9.	Dry volume weight at the age of 28 days (kg/m³)	0	851	895	749	842
10.	Compressive strength (MPa)	0	0.82	1.28	0.56	0.68
11.	Absorption (%)	0	16.02	29.11	37.85	28.69

5 Conclusion

The different compositions studied included the use of 200 kg/m^3, 250 kg/m^3, 300 kg/m^3, 350 kg/m^3 and 400 kg/m^3 of cement. From the research conducted, it was found that the cement composition affects the physical and mechanical properties of concrete masonry bricks, including the success rate of CLC brick making, volume weight, compressive strength and others.

The research results show that the CLC masonry concrete brick variants that qualified as brick CLC included only the cement proportions of 250 kg/m^3, 300 kg/m^3, 350 kg/m^3 and 400 kg/m^3, while the cement sample with 200 kg/m^3 cannot be recommended for use as a successful mixture for producing CLC masonry bricks because of the drying shrinkage seen at the age of just 1 day.

The use of different cement compositions of 200 kg/m^3, 250 kg/m^3, 300 kg/m^3, 350 kg/m^3 and 400 kg/m^3 has an influence on the physical and mechanical properties of the CLC masonry concrete bricks produced, as can be seen in Table 4. The results of the research show that the average volume weight is 749–895 kg/m^3. The lightest variant is the sample using 350 kg/m^3 of cement. The average compressive strength after 28 days is 0.56–1.28 MPa, where the highest compressive strength is found in the sample using 300 kg/m^3 of cement. The average modulus of elasticity after 28 days is in the range 45.74–98.64 MPa, and the highest value of the modulus of elasticity was when using 350 kg/m^3 of cement.

If we compare the results from Table 4 with previous studies shown in Table 2, then we can see that the results obtained for the average volume weight (749 to 895 kg/m^3), compressive strength (0.56 to 1.28 MPa), thermal conductivity (0.47 to 0.57 W/mK) and drying shrinkage (0%). The compressive strength and volume weight are similar, while the results for the thermal conductivity, modulus of elasticity, and drying shrinkage are different.

Therefore, increase in the use of cement in the manufacturing of CLC masonry concrete brick is found not to produce an increase in compressive strength and also resulted in a more severe volume weight of the CLC masonry concrete brick. From this review of the mechanical and physical properties obtained using varying amounts of cement, the best (optimum) result is achieved with the use of 350 kg of cement.

We give thanks to the Faculty of Engineering Tanjungpura University Pontianak West Kalimantan for support and funding, and to all parties involved in this research.

References

[1] Hamad, A. J. (2014). Material, production, properties and application of aerated lightweight concrete: Review. *International Journal of Materials Science and Engineering, 2* (2), 152-157.

[2] ACI. (1991). *ACI Manual of Concrete Practice Part 211.2-91, Standard Practice for Selecting Proportions for Structural Lightweight Concrete.* Farmington Hills, MI, USA: ACI.

[3] Supriyadi, A., et al. (2016). *Comparison of the masonry concrete brick CLC (Cellular Lightweight Concrete) physical and mechanical properties with variations of sand gradation as the base material.* Dipa Research, Department of Civil Engineering, Faculty of Engineering, University of Tanjungpura, Pontianak.

[4] Kumawat, G., Maru, S., & Kumar Pandey, K. (2016). Cost comparison of RCC structure using heading CLC blocks with burnt clay bricks. *International Journal of Advanced Research, 4* (7), ISSN 2320-5407.

[5] Kumawat, G., Maru, S. (2016). Analysis and comparison of RCC structure using heading CLC block with burnt clay bricks. *International Journal of Engineering Research and General Science, 4* (3), ISSN 2091-2730.

[6] Krishna Bhavani, K. (2012). Cellular light-weight concrete blocks as a replacement of burnt clay bricks. *International Journal of Engineering and Advanced Technology (IJEAT), 2* (2), ISSN: 2249-8958.

[7] Kayyali, A., & Haque, M. N. (1997). A New Generation of Structural Lightweight Concrete. In V. M. Alhotra (Ed.), *Proceedings of Third CANMET/ACI International Conference on Advances in Concrete Technology,* ACI SP 171. Auckland, New Zealand.

[8] Ardyansyah, M. (2014). *Study of the manufacture of lightweight brick CLC (Cellular Lightweight Concrete) with high levels of coal fly ash as a partial substitution of cement.* MT thesis. Sebelas Maret University, Jawa Tengah.

[9] Murdock, U., & Brook, K. M. (1999*). Concrete materials and practice* (4th ed.). Jakarta: Erlangga.

[10] Prakash, C. E. (2013). Properties of aerated concrete blocks. *International Journal of Scientific and Engineering Research, 4* (1), ISSN 2229-5518.

[11] SNI 7064:2014 (2014). *Portland Cement Composites.* National Standardization Agency.

[12] Nandi, S., Chatterjee, A., Samanta, P., & Hansda, T. (2016). Cellular concrete and its facets of application in Civil Engineering. *International Journal of Engineering Research, 5* (Special 1), 37-43, ISSN:2319-6890.

[13] Mulyono, T. (2004). *Concrete technology.* Yogyakarta: Andi.

 https://doi.org/10.1051/matecconf/201815901009

Comparison of Genetic Algorithm and Harmony Search Method for 2D Geometry Optimization.

Mohammad Ghozi[1,*] and *Anik* Budiati[2]

[1,2]Bhayangkara Surabaya University, Engineering Department, Surabaya, Indonesia

Abstract. There are many applications of Genetic Algorithm (GA) and Harmony Search (HS) Method for solving problems in civil engineering design. The question is, still, which method is better for geometry optimization of a steel structure. The purpose of this paper is to compare GA and HS performance for geometric optimization of a steel structure. This problem is solved by optimizing a steel structure using GA and HS and then comparing the structure's weight as well as the time required for the calculation. In this study, GA produced a structural weight of 2308.00 kg to 2387.00 kg and HS scored 2193.12 kg to 2239.48 kg. The average computational time required by GA is 607 seconds and HS needed 278 seconds. It concludes that HS is faster and better than GA for geometry optimization of a steel structure.

1 Introduction

The Genetic Algorithm (GA) method has been the designer's tool of many steel structures [1,2]. GA was developed and combined with chromosome repairing strategy on solving size optimization of steel structures [7]. On the other hand, Harmony Search (HS) came with a robust performance for determining the size optimization of steel structures [4-6], [8-9]. Both strategies are powerful for solving size optimization problems with minimization objective function due to several engineering constraints. With advantages of both methods, it is important to compare the performance between GA and HS especially for solving geometry optimization of a steel structure.

2 Methods

Methods used in this paper will be divided into three-part as 1) Research Method, 2) Steel Structure Model, and 3) Objective function. In the research method, there will be described the method used to solve the problem in this paper. The structure model description also will be explained and then the objective function of the optimization problem is outlined in the next subchapter.

* Corresponding author: mghozi@ubhara.ac.id

2.1 Research Method

The aim of this research is to compare the best method between Genetic Algorithm (GA) and Harmony Search (HS) for 2D steel structure geometric optimization. The scopes of this research are the best weight and fastest computing time between GA and HS for optimizing one geometric structure model. This research is initiated by literature study, followed by deploying GA and HS to optimize the geometric structure model. Each method runs 10 times optimization process for accommodating natural behaviour of both methods. The Finite element method is organized to calculate stress and joint deflection [3]. Analysing data of objective function plot, weight and computing time are then proceed. The research method's flowchart is then shown in Fig. 1

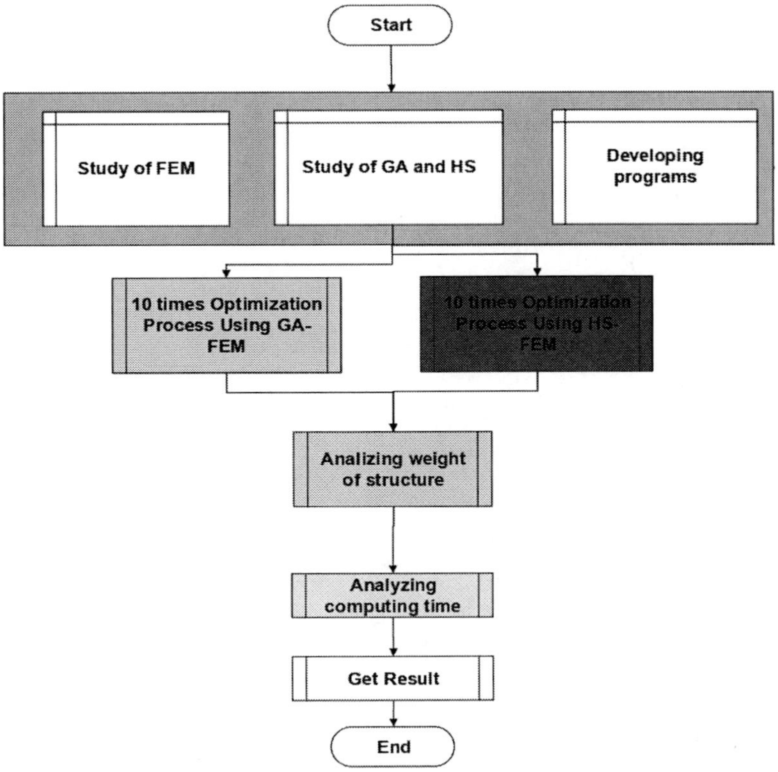

Fig. 1. Flowchart of Research Method

The GA follows flowchart as described in [7] and the HS uses the program as described in [8,9]. The FEM deployed in this research is developed according to the theory as described in [3]. Flowchart of GA and HS can be seen in Fig. 2. The parameters used by GA method in this paper are Generation = 100, Population = 100, Mutation Probability = 0.05, Crossover Probability = 0.85, Roulette Wheel Mode as Selection and Elitism = 10%. HS method uses Pitch Adjusting Rate = 0.5, Harmony Memory Size = 30, Harmony Memory Considering Rate = 0.95, Iteration = 10000. The optimization process uses PC with Intel Core i5 2400 3.1 GHz processor, 8 GB DDR3 memory and AMD R7 200 series 2GB DDR3 as VGA Card.

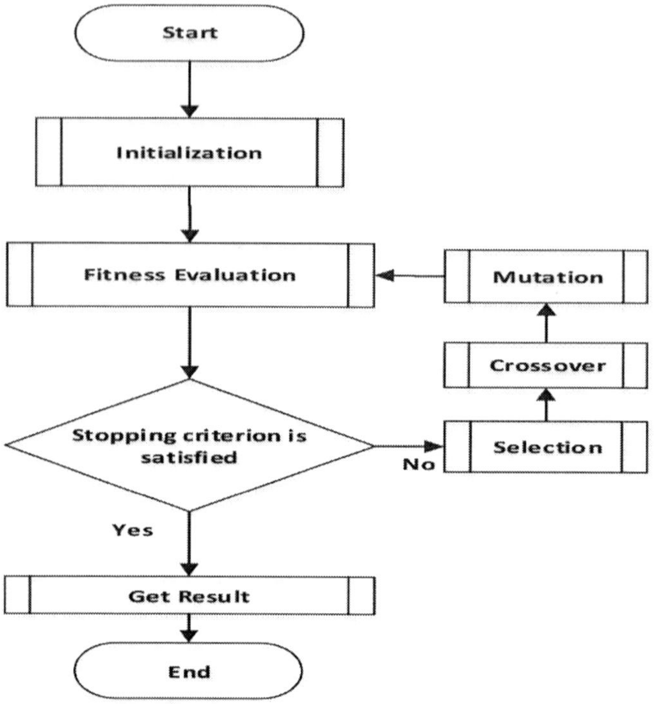

(a)

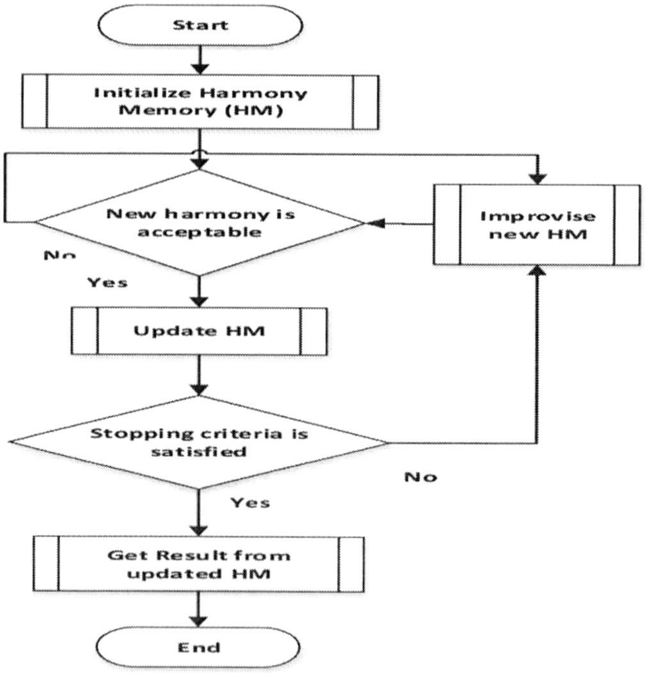

(b)

Fig. 2. Flowchart of: (a) GA [7], (b) HS [8]

2.2 Steel Structure Model

The structure model used in this paper is a 2D structure with 10 elements and 6 joints [2]. Vertically downward loads of 100 kips (445.374 kN) are applied on joint 2 and 4 (see Fig.3). The structure model data are Modulus of Elasticity, E = $1x10^4$ ksi (6.89 x 10^4 MPa), ρ = 0.10 lb/in (2,770 kg/m^3). The allowable maximum stress is 25 ksi (172.25 MPa) and the allowable joint deflection is 2 inch (50.8 mm). The available cross section areas are 10.45, 11.61, 12.83, 13.74, 15.35, 16.90, 16.96, 18.58, 18.90, 19.93, 20.19, 21.80, 22.38, 22.90, 23.41, 24.77, 24.96, 25.03, 26.96, 27.22, 28.96, 29.61, 30.96, 32.06, 33.03, 37.03, 46.58, 51.41, 74.19, 87.09, 89.67, 91.61, 99.99, 103.22, 109.03, 121.29, 128.141, 141.93, 147.74, 170.96, 193.54 and 216.12 (in cm unit). The allowable moved joints are joint 1 and 3.

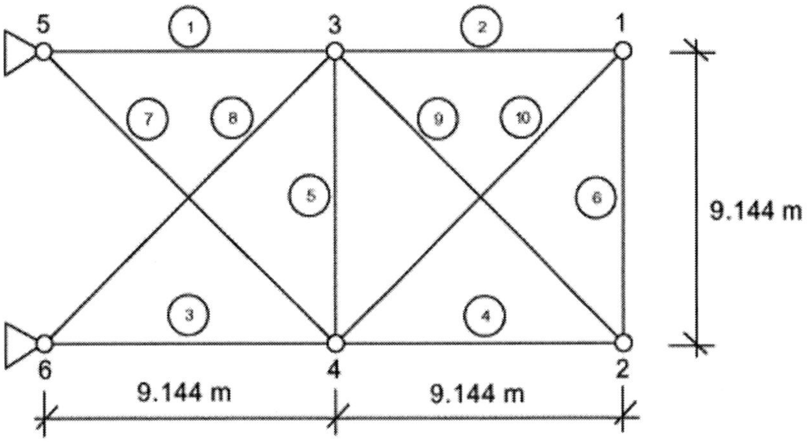

Fig. 3. Structure Model [2].

2.3 Objective Function

The objective function of optimization process in this research is:

$$Minimize\ W = \sum \rho\ l_e A_e \quad (1)$$

Subjected to :

$$Stress \quad \sigma_e \leq \sigma_{max} \quad (2)$$

$$Deflection \quad \Delta_j \leq \Delta_{max} \quad (3)$$

Where :

W = weight of structure (kg), ρ = unit weight of steel, l_e = length of element, A_e = area of cross section. σ_e =stress of $element_e$, σ_{max} = allowable stress, Δ_j = deflection of $joint_j$, Δ_{max} = allowable deflection of $joint_j$.

Penalty function strategy is used in this research. The penalty functions are :

$$Sc_e = 1\ if\ \sigma_e \leq \sigma_{max}\ or\ Sc = 0\ if\ \sigma_e \leq \sigma_{max} \quad (4)$$

$$Dc_j = 1\ if\ \Delta_j \leq \Delta_{max}\ or\ Sc = 0\ if\ \Delta_j \leq \Delta_{max} \quad (5)$$

And the fitness equation is :

$$Fitness_c = W + 10\ (\sum Sc_e) \quad + 100\ (\sum Dc_j) \tag{6}$$

Where : $Fitness_c$ = Fitness value of $chromosome_c$

3 Result and Discussion

The 10 times optimization process using GA and HS have already conducted this research. HS produced structure's weight 2193.12 kg and GA got 2308.00 kg as best weights (see Table 1). This means HS can produce 4.9% lighter than GA's best weight. For obtaining 2500 kg as optimum weight [2], HS requires 890 until 1781 iterations. GA demands 1900 to 4500 evaluations for finding 2500 kg as optimum weight (see Fig. 4). This concludes that HS is 53% to 60% faster than GA.

Table 1. Comparison of weight and computational time between GA and HS.

	GA Method		HS Method	
Section	**Weight (kg)**	**Time (second)**	**Weight (kg)**	**Time (second)**
Best	2308.00	602	2193.12	268
Average	2361.50	607	2213.22	278
Maximum	2387.00	615	2239.48	289

The GA and HS trajectory's objective functions are then displayed in Fig. 4.

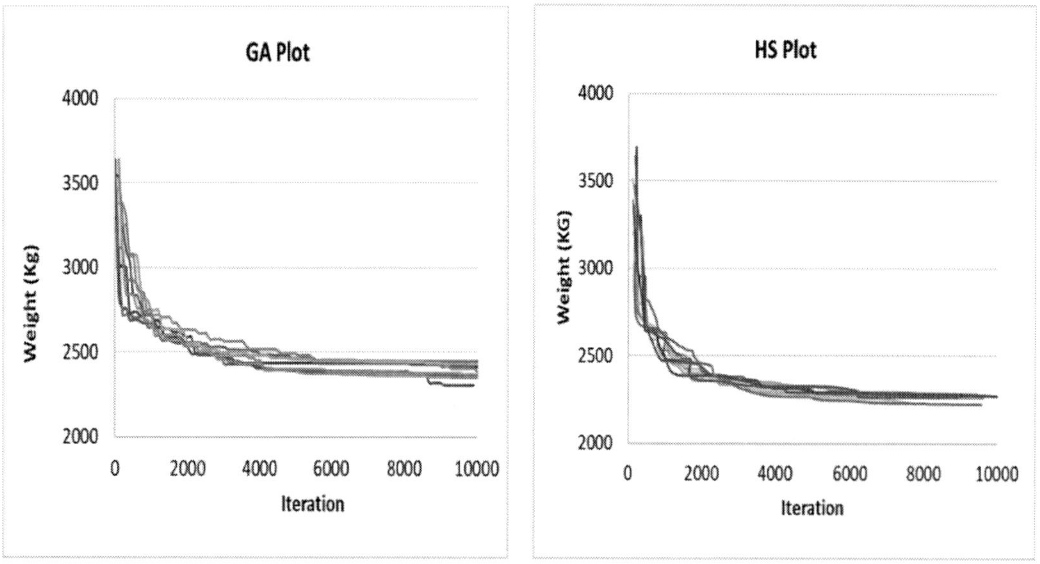

(a) (b)

Fig. 4. Objective function plot: (a) GA's Plot; (b) HS's Plot.

4 Conclusion

Comparison of GA and HS performance for geometry optimization of a steel structure has been implemented in this project. The comparison is overcome by optimizing one steel structure model using GA and HS and then comparing the weight of the structure. The time required for the computation is also be compared. GA produced structure's weight 2308.00 kg to 2387.00 kg and HS scored 2193.12 kg to 2239.48 kg. The GA's average computation time is 607 seconds and HS's time is 278 seconds. It is concluded that HS is 53% faster than GA and HS also yields a lightweight steel structure of 4.9% than GA for geometric optimization of a steel structure.

This research was financially supported by Competitive Research Grant Committee of Bhayangkara Surabaya University, 2016.

References

1. D.E. Goldberg, *Evolutionary Algorithm İn Search, Optimization and Machine Learning*, (Addition Wesley Publishing Co. Inc, USA, 1989)
2. S. Rajeev, C.S. Krishnamoorthy, Jour. Stru. Eng., 118 (1992).
3. R. Cook, *Finite Element Modeling for Stress Analysis*, (John Wiley & Sons, USA, 1995)
4. Z.W. Geem, H.K. Joong, G.V. Loganathan, Sim. **76** (2001)
5. A.J. Khalifa, *Design Optimization Of Steel Framed Structures To AISC- LRFD Using Harmony Search Algorithm*, (Gaza Islamic University, Gaza, 2011)
6. S.A. Patil, D. A. Patel, Int. Jour. Eng. Sci. Inn. Tech., **2** (2013)
7. A. Budiati, M. Ghozi, Acad. Res. Int., **4**, 493 (2013)
8. M. Ghozi, Acad. Res. Int., **5**, 52 (2014)
9. M. Ghozi, P. Suprobo, Int. Jour. App. Eng. Res., **10**, 33206 (2015)

Interval Confinement in Compression Zone to Evaluated Beams Performance Subjected Monotonic Loads

Yulita Arni Priastiwi[1,*], *Iswandi* Imran[2], Nuroji[1], and *Rudi* Yuniarto Adi[1]

[1]Civil Engineering Department, Faculty of Engineering, Diponegoro University, Semarang, Indonesia
[2]Structural Engineering Research Group, Civil Engineering Department, Bandung Institute of Technology, Bandung, Indonesia

Abstract. Confinement is one way that can be used to improve the performance of reinforced concrete structures, mainly related to ductility. The parameter of the distance between the confinement becomes an important thing that must be studied its effect on ductility produced by a structural element. This study aims to study the effect of different distance between the confinement in compression zone in the beam at the plastic hinge area to the displacement and the behavior of the beam when it was given monotonic loading. The specimen model which is a simplified form of the plastic hinge area up front column will be fitted with a confinement in the compression zone which is attached to the shear reinforcement with different distances of 0, 70, 125 mm. Also made a beam with a crossties confinement spaced 125 mm for comparison. The presence of a centralized load in the middle of the span is intended to obtain the largest moment and shear areas in the plastic hinge. The test results showed that the installation of 125 mm intervals for confinement in the compression zone resulted in a higher ductility of 11-18% against the beam without confinement than the 70 mm interval which only increased by 2.78%. The hoops confinement produces higher ductility than crossties confinement for the same confinement interval. The increased confinement interval from 70 to 125 mm not yet significantly affect the moment capacity of the beam and the collapse was still dominant in the bending collapse although the distance between the stirrups and the confinement was slightly widened.

1 Introduction

Confinement is one way that can be used to improve the ductility of a reinforced concrete structure element [1]. Concrete with good and adequate confinement will be ductile and be high deformability especially in the post-elastic range [2]. Some research results indicate that the strength and ductility of reinforced concrete structural elements especially in the

* Corresponding author: yulita_tiwi@ymail.com

column are closely related to the lateral strain given by the stirrups with the spaced reinforcement as one of the determinants.

The distance of the stirrup will increase the unconfined volume of concrete and the possibility of spalling [4]. Minimum transverse reinforcement requirements are performed to avoid loss of power in case of spalling [5], and the presence of more tightly spaced restraint may increase the energy absorption rate of the column due to lateral loads gave [6].

This study aims to study the behavior of the plastic hinge beam area given confinement in the compression zone and mounted attached to stirrup for a backstop of shear. Different intervals between confinements are the parameters to be studied for their effect on the ductility of the plastic hinge region of the beam when given monotonic loading.

2 Methods

Four half-full scale specimens of the beam with size 150 x 300 x 4000 mm were made. The beam was given magnification in the middle of the span was called stub to obtain a simplified form of the plastic hinge area in the beam on the face of the column. The designs on the three specimens apart from the longitudinal reinforcement and stirrup for a backstop of shear also added confinement in compression zone of the beam. Confinement shaped hoops as well as crossties are entirely in compression zone of the beam and mounted attached to the stirrup. Confinement spacing is based on the distance needs of the shear stirrups for specimens without confinement (s = 70 mm) and widened for other specimens. The design of the specimen beam is shown in Fig. 1 and Fig. 2, while for the calculation of the spacing of the stirrups according to the prevailing standards in Indonesia [7], where the shear at the plastic hinge location is calculated based on the strongest possible moment value of the largest endpoint at the site and the distance between the stirrups according to the requirements specified for the plastic hinge area in order to avoid shear failure of the beam.

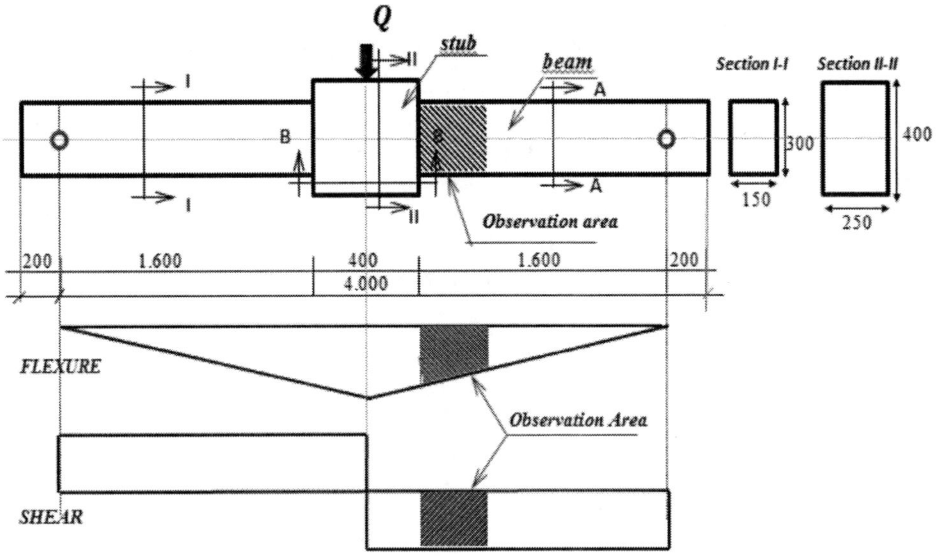

Fig. 1. Area Observation of Specimen

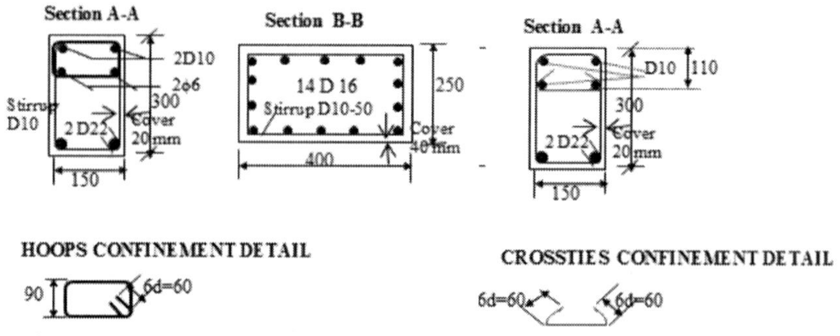

Fig. 2. The Section of Specimen

The calculation of shear reinforcement distance for beam without confinement using Eq. (1)

$$s = \frac{A_v . f_y . d}{V_s} \tag{1}$$

where s = the spacing of stirrup, Av = the area of leg stirrup, f_y = the yield of reinforcement, d = the effective height of the beam, and V_s = nominal shear force provided by shear reinforcement.

2.1 Detail of Specimens

All specimens in this study were designed with 25 MPa concrete quality with a maximum aggregate of 10 mm with the intention that the aggregate granules can enter the sidelines of reinforcement so that the concrete is not porous. The specimens use the deform reinforcement for either the main longitudinal reinforcement, stirrup and the confinement reinforcement, whereas the additional longitudinal reinforcement used to ϕ 6 mm plain reinforcement.

The tensile reinforcement was used 2D22 (f_y = 476 MPa), and the compressive reinforcement used 2D10 (f_y = 444 MPa). For the stirrup and confinement was used the reinforcement D10 (f_y = 444 MPa). At the magnification of the beam called the stub, it is installed 14D16 (f_y = 402 MPa), while the additional reinforcement is used 2ϕ6 has f_y = 430 MPa. The distance of the shear stirrup (s) based on the calculation for the beam without confinement in the plastic hinge area was 70 mm, which is then made the minimum distance for the placement of the stirrups as well as the distance between the confinement (s_c) of the specimen (BN and BC-70), while other specimens are obtained by how to widen the minimum distance. The specimen details are shown in Fig. 3 and Table 1.

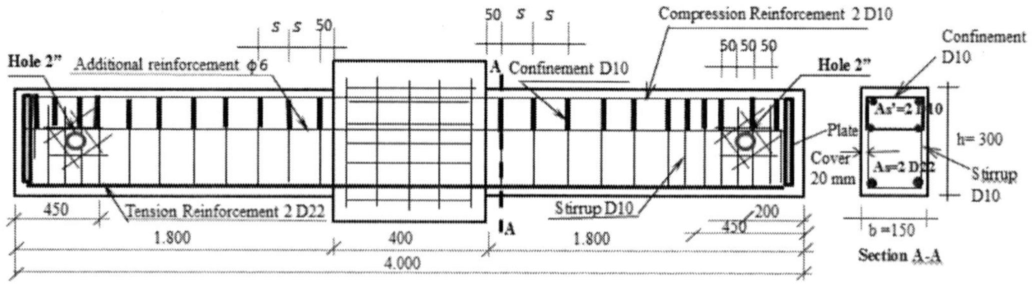

Fig. 3. Reinforcement in Specimen

Table 1. Detail of Specimens

Type	ρ (%)	ρ' (%)	*s* (mm)	s_c (mm)
BN	1.72	0.40	70	0
BC-70	1.72	0.40	70	70
BC-125	1.72	0.40	125	125
BK-125	1.72	0.40	125	125

2.2 The Experimental Set up

The specimens are placed on the loading frame with support by joints and rolls. The support was placed at the point as high beam of 200 m from each edge of the beam-free end. A 500 kN hydraulic jack and load cell of similar capacity is used to provide central loading at the center of the beam span. The test was carried out with monotonic loading given in two loading stages: load control system and continued with displacement control system. The load control system was provided at 0.2-ton intervals until loading is approximately 75% of the maximum theoretical calculation load and continued with the displacement control system until the ultimate condition of the beam is reached.

A Linear Variables Displacement Transducers (LVDT) was installed in the middle of the span to measure the deflection due to a given load. The resulting loading and displacement data will be recorded in the data logger to be processed and analyzed. In addition, LVDT was also installed in the horizontal direction on the top and bottom of the beam to measure the rotation that occurs. In the longitudinal reinforcement or concrete was also installed strain gauge to measure the strain that occurs due to loading. All data will be recorded in the data logger and will be analyzed until the moment-deflection relationship graph was obtained, while the strain information will be used to measure the magnitude of the curvature-moment in the beam specimens. In the curvature equation, the compressive strain on the concrete (ε_c) and the tensile strain on the reinforcing steel (ε_s) will be used and the curvature calculated according to Eq. 2 with *d* is the distance between the two strains. The test set up refers to the research of Yulita et al [8].

$$\varphi = \frac{\varepsilon_c + \varepsilon_s}{d} \tag{2}$$

3 Result and Discussion

3.1 Moment-Displacement Respons

The test results in the load capacity (kN) which is then changed in the moment capacity (kN-m) in the cross-section illustrated with the displacement in Fig. 4 and also in Table 2.

Table 2. Moment and ductility of specimens

Specimen	M_u (kN)	M_y (kN)	δ_y (mm)	D_{ry} (%)	K_e $=M_y/\delta_y$	δ_u (mm)	μ $=\delta_u/\delta_y$	enhancement
BN-70	72.07	62.30	21.61	1.35	2.88	64.90	3.00	-
BC-70	71.28	64.42	22.58	1.41	2.85	69.70	3.09	2.78
BC-125	69.70	64.68	22.94	1.43	2.82	81.46	3.55	18.24
BK-125	72.60	65.47	22.92	1.43	2.86	76.64	3.34	11.34

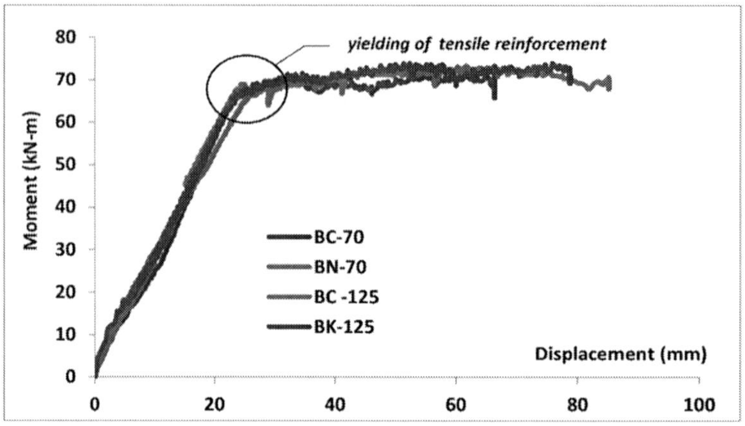

Fig. 4. Moment-displacement relationship on all four specimens

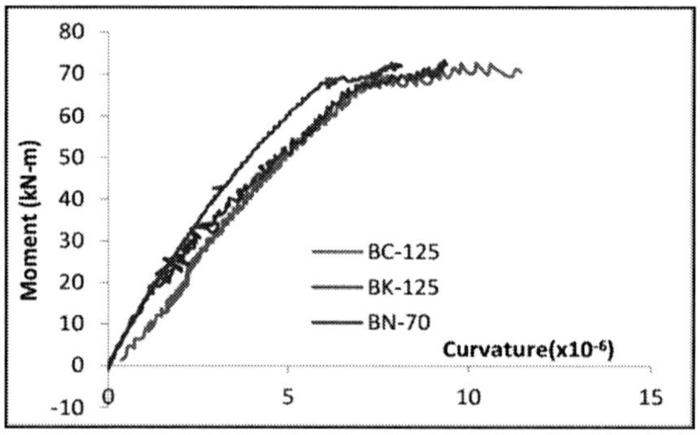

Fig. 5. Moment-curvature relationship

The test results showed that on the four specimens (BC-70, BN-70, BC-125, and BK-125) there was a dominant bending collapse pattern with the direction of the vertical crack upwards to the compression area. The heaviest damage occurred in accordance with the planned area of column face. No significant damage was found in the support or columns.

From Fig. 4 depicting the moment-displacement relationship of the four specimens in the middle of the beam span, it was seen that the four specimens have elastic stiffness curves which are the ratio of moment and displacement when the near yielding condition is approximately 2.8; while the value of Drift Ratio *(DR)* at yield (which is the value of deviation per length) indicating that the presence of confinement in the compression zone of beam can increase the drift ratio of the beam and the addition of confinement spacing from 70 mm to 125 mm actually increases the drift ratio. The distance 125 mm is more effective than the confinement distance of 70 mm. Having increased moment and displacement when yielding for the confinement of the beam, and the distance made wider than 70 mm to 125 mm does not significantly reduce the resulting moment.Although not yet the final result for the increase of ductility but it has been seen that spacing between confinement and wider stirrings is more effective for increasing the ductility of the beam as shown in Table 1 where the 125 mm spacing results in higher ductility compared to 70 mm distance which increases 11-18% against the beam without confinement, whereas the interval 70 mm only increased by 2.78%.

The test results show, although not yet reached the boundary condition seen that the change of distance from the confinement does not significantly affect the increase in the moment capacity that occurs. The increase in moments from yielding to near ultimate conditions only increased by about 1%, which occurs either on hoops and crossties or on the beam without confinement. The test also shows the same moment level, the curvature of the beam with confinement in the compression zone is larger than the beam without confinement for either confinement hoops or crossties as shown in the fig. 5.

4 Conclusion

Based on the test results it can be concluded that the addition of the confinement from 70 mm to 125 mm on confinement in compression of beam indicates that the distance between confinement 125 mm more effectively increase the ductility and drift ratio of the beam compared to the distance of 70 mm and the beam without confinement. The difference in the distance between confinement from 70 mm to 125 mm is less significant affecting the change in moment capacity on the beam and for the same moment level, the presence of confinement in compression zone increases the curvature of the beam than the beam without confinement in both hoops confinement and crossties confinement.

Thank you to the Faculty of Engineering Diponegoro University through the Department of Civil Engineering for the Grant Assistance Basic Research which has been given to support this research.

References

1. O. Bayrak and S. A. Sheikh, Confinement Reinforcement Design Considerations for Ductile HSC Columns, *ASCE Journal of Structural Engineering*, **V.124**, No.9, p. 999-1010 (1998)
2. O. Bayrak and S.A. Sheikh, Plastic Hinge Analysis, *ASCE Journal of Structural Engineering*, **V.127**, No.9, p. 1092-1100 (2001)
3. S. R. Razvi and M. Saatcioglu, Strength and Deformability of Confined High-Strength Concrete Columns, *ACI Struct. Journal*, **V. 91**, No. 6, p.678 – 686 (1994)
4. H. Tanaka and R. Park, Prediction of The Ultimate Longitudinal Compressive Concrete Strain at Hoop Fracture Using Energy Considerations, Library Niwa, ISSN: 0110-0718, p.290-305 (1987)
5. ACI Committee 318, *Building Code Requirements for Structural Concrete (ACI 318-11) and Commentary*, American Concrete Institute, Farmington Hills, Michigan, USA,p.503 (2011)
6. M. Saatcioglu and D. Baingo, Circular High Strength Concrete Columns Under Simulated Seismic Loading, ASCE *Journal of Structural Engineering,* **V.125**, No.3, p. 272-280 (1999)
7. SNI 2847-2013, Requirements for Structural Concrete Building in Indonesian. BSN (2013).
8. Y.A. Priastiwi, I. Imran, and Nuroji, The effect of different shapes of confinement in compression zone on beam's ductility subjected to monotonic loading, Procedia Eng. **V125**, p. 918–924 (2015).

Development of seismic risk microzonation map for Semarang due to Semarang fault earthquake scenarios with maximum magnitude 6.9 Mw

Windu Partono[1,*], *Masyhur* Irsyam[2] , and *Sri* Prabandiyani Retno Wardani[1]

[1]Civil Engineering Department, Diponegoro University, 50275 Semarang, Indonesia
[2]Civil Engineering Department, Bandung Institute of Technology, 40132 Bandung, Indonesia

Abstract. Research on seismic microzonation of Semarang is still ongoing. The first seismic microzonation research for this area was performed on 2015. Seismic microzonation map was developed by implementing deterministic 1-D site response analysis at 190 boring locations. Lasem fault was considered to be the main earthquake source which taken into account for seismic microzonation research of Semarang. The second research for developing seismic microzonation map was performed on 2016. Following the research conducted by Team for Updating Seismic Hazard Maps of Indonesia 2016, Rawapening Fault, Ungaran Fault, Weleri Fault, Demak Fault and Semarang Fault are five new shallow crustal fault earthquake sources identified in this research. This paper presents the development of seismic risk microzonation map caused by Semarang fault earthquake scenarios with maximum 6.9 Mw. Deterministic 1-D site response analysis was performed at 288 boring locations for developing seismic risk microzonation map. This research was performed by implementing response spectral matching and site propagation analysis. Due to inadequate data caused by Semarang fault earthquake, response spectral matching was implemented in this research to obtain modified acceleration time histories. All acceleration time histories were developed from five different earthquakes with magnitude 6.05 – 6.9 Mw and maximum epicentre distance 15 Km.

1 Introduction

Research on seismic microzonation of Semarang is still ongoing. The first research of seismic microzonation for this area was performed on 2015 by [1] and following the work conducted by Team for Revision of Seismic Hazard Maps of Indonesia 2010 (TRSHMI 2010). Seismic microzonation map was developed by implementing deterministic 1-D site response analysis at 190 boring locations. Lasem fault, an active and closest shallow crustal fault, was considered to be the earthquake source scenario conducted for site response analysis. Following the research conducted by Team for Updating of Seismic Hazard Maps of Indonesia (TUSHMI) 2016, five new shallow crustal fault earthquake sources were

* Corresponding author: windu_bapake_dila@yahoo.com

identified (Rawapening Fault, Ungaran Fault, Weleri Fault, Demak Fault and Semarang Fault).

This paper presents the development of seismic risk microzonation map based on the surface spectra acceleration values (PGA, Spectra 0.2s and Spectral 1s) calculated using 1-D site response analysis. All surface spectra acceleration were developed due to Semarang fault (reverse fault mechanism) earthquake scenarios with maximum magnitude 6.9 Mw and maximum epicentre distance 15 Km. Surface PGA, spectral 0.2 s and spectral 1s were calculated at 288 boring locations. Fig. 1(a) shows the position of Semarang and Lasem fault traces and the position of all 288 boring locations against Semarang fault trace.

This research was performed by conducting five important steps: 1) geological, geophysical and soil investigations and analysis for developing soil profile and bedrock elevation 2) collecting acceleration time histories data from worldwide historical earthquake records due to reverse fault mechanism earthquakes with magnitude 6 - 7 Mw and maximum distance 15 km, 3) developing modified acceleration time histories by conducting response spectral matching analysis, 4) conducting shear wave propagation analysis using modified time histories and calculated using 1-D site response analysis for obtaining surface acceleration time histories and surface response spectra and 5) developing seismic risk microzonation maps.

2 Experiment details

2.1 Geological and geotechnical investigations

Depth of engineering bedrock and soil properties profiles from top until bedrock elevation are two important informations for performing site response analysis. Prediction of bedrock elevation for the study area was performed using single station feedback Microtremor [1]. The prediction of bedrock elevation was implemented by conducting three component ambient vibrations investigation at 218 locations [1 and 2]. Top soil properties investigations were conducted at 288 boring locations with minimal 30 meters depth. Fig. 1(b) shows the distribution of bedrock elevation for the whole area of the city.

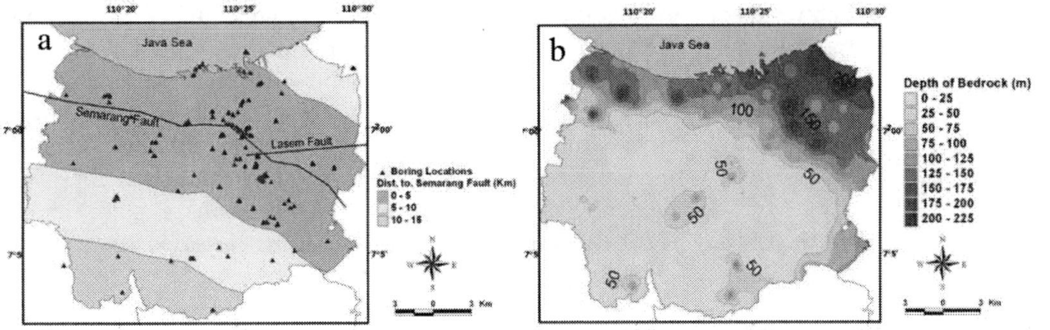

Fig. 1. Semarang fault and Lasem fault trace and boring locations (a) and contour map of bedrock elevation (b).

2.2 Time histories data collection and response spectral matching analysis

Due to inadequate and limited records of Semarang fault earthquake, historical earthquake records with magnitude from 6 to 7 Mw and maximum distance 15 km were collected for reverse mechanism shallow crustal fault. Five selected earthquakes (Iwate Japan 2008, Chuetsu-oki Japan 2007, Chi-Chi Taiwan 1999, Northridge 01 California 1994, Northridge

02 California 1994) were used in this research for site response analysis and developing seismic risk map of the study area. All acceleration time histories data were collected from PEER NGA-West 2 Databases. Table 1 shows all acceleration time histories from 5 different earthquake events used in this study.

Table 1. Earthquake sources data used in this study.

Event	Station	Magnitude (Mw)	Epicentre Distance (km)
Northridge-02 (1994)	Arleta - Nordhoff Fire Sta	6.05	1.48
	Newhall - Fire Sta	6.05	7.36
	LA - Century City CC North	6.05	18.34
Chi-Chi, Taiwan-03 (1999)	TCU084	6.2	3.68
	TCU089	6.2	5.93
	TCU076	6.2	13.04
Northridge-01 (1994)	Arleta - Nordhoff Fire Sta	6.69	3.3
	Beverly Hills - 14145 Mulhol	6.69	9.44
	LA - Brentwood VA Hospital	6.69	12.92
Chuetsu - oki, Japan (2007)	Nagaoka	6.8	3.97
	Kashiwazaki City Takayanagicho	6.8	10.38
	Yan Sakuramachi City Watershed	6.8	12.98
Iwate, Japan (2008)	IWTH24	6.9	3.1
	IWT011	6.9	8.41
	Kurihara City	6.9	12.83

All acceleration time histories used in this study need to be matched with the target spectrum of Semarang fault earthquake scenario. The target response spectrum were developed using deterministic analysis and conducting three different attenuation functions proposed by [3, 4 and 5]. The target spectrum used for spectral matching analysis was predicted at bedrock elevation and following the same method conducted by [6]. All target spectrums were calculated for reverse mechanism shallow crustal fault earthquake sources.

Response spectral matching analysis was conducted following the same method proposed by [6] for producing modified (matched) acceleration time histories. The matched acceleration time histories were predicted using two components acceleration time histories (East-West/EW and North-South/NS directions). Initial time histories must have similar shape with matched time histories (time histories calculated from response spectral matching). Fig. 2(a) shows an example result of spectral matching calculated for Iwate Japan earthquake 6.9 Mw with epicentre distance 3.1 Km (EW direction). Fig.2(b) shows initial and matched time histories for EW direction acceleration time histories due to Iwate earthquake with magnitude 6.9 Mw and epicentre distance 3.1 Km.

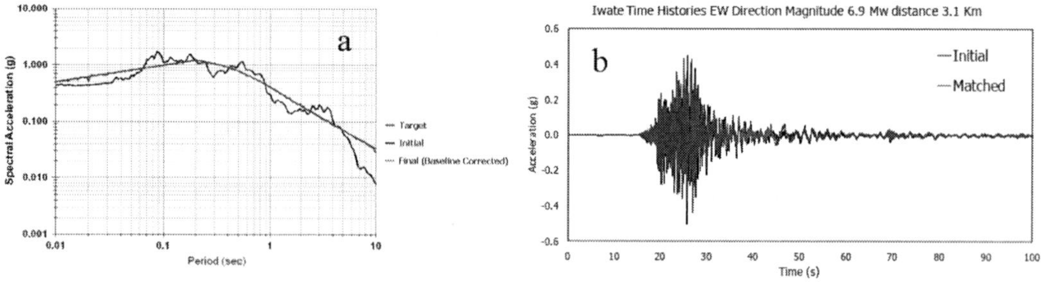

Fig. 2. Response spectral matching analysis (a) and two time histories (initial and matched) (b) for Iwate earthquake 6.9 Mw and distance 3.1 Km East/West (EW) directrion.

2.3 Site response analysis

Site response analysis was performed at 288 boring positions for obtaining surface acceleration time histories and surface response spectra acceleration. Site response analysis due to a specific earthquake was implemented using a pair of (NS and EW) matched time histories calculated from spectral matching analysis. All 288 soil investigation locations were distributed into three different distance groups (0-5 Km, 5-10 Km and 10-15 Km). Site response analysis for this study was implemented using soil profile model as can be seen on Fig. 3(a). VS1, VS2 and VSn on this figure represents shear wave velocity for soil layer 1, layer 2 and layer n respectively.

Site response analysis was conducted by taking the assumption that all boundaries of soil layer are horizontal and the response of all soil layers for all boring locations are predominantly caused by shear wave propagating vertically from bedrock elevation to surface elevation. In this study the general site response analysis was carry out using equivalent linear approach and conducting the model developed by [7 and 8] and utilizing free software NERA [9].

Surface response spectra was calculated using a method proposed by Goleshorki, R. 2013 [10] using two components surface response spectra (EW and NS) directions. Fig. 3(b) shows an example of two surface response spectra (NS and EW) and total spectra calculated from both NS and EW spectra at Lok33BH1 due to Iwate earthquake scenario with magnitude 6.9 Mw and epicentre distance 3.1 Km. The seismic maps (PGA, Spectral 0.2s and Spectral 1s) were then developed by plotting all 288 surface response spectra calculated from site response analysis. For spectral zoning evaluation all 288 response spectra data were then transferred and distributed into three different zones or categories which represents similar values of corresponding earthquake effect [11].

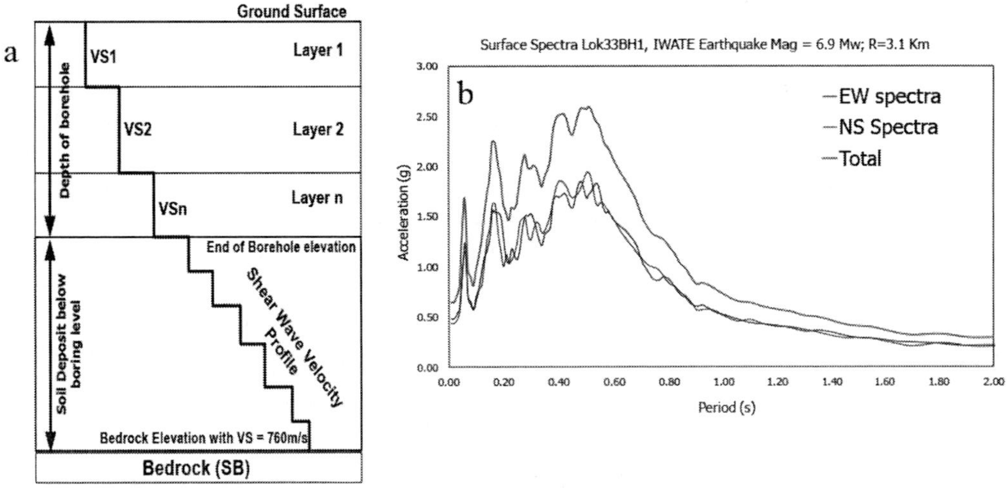

Fig. 3. Soil profile model for site response analysis (a) and example surface response spectra results calculated at Lok33BH1(b).

3 Results and discussions

Based on all surface PGA values calculated at 288 boring locations the distribution of PGA due to earthquake scenario with magnitude 6.9 Mw can be seen on Fig. 4(a). As shown in this figure the highest PGA values are distributed within the closest area to fault trace with values in between 0.7g to 0.9g. The lowest PGA values are distributed in the southern part of the study area in between 0.2g to 0.4g. Fig. 4(b) shows the distribution of surface PGA due to 6.2 Mw earthquake scenarios. Based on those two examples surface PGA map it can be seen that the highest surface PGA values due to Semarang fault earthquake scenarios are distributed at the closest area to fault trace with the value in between 0.7g to 0.9g.

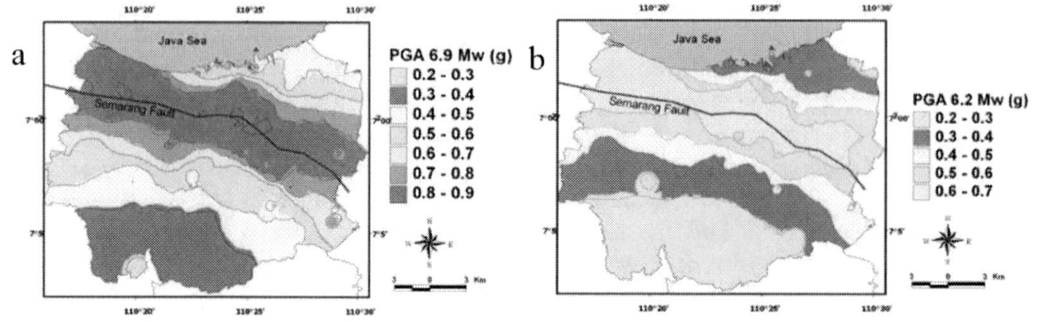

Fig. 4. Surface PGA map due to 6.9 Mw (a) and 6.2 Mw (b) Semarang fault earthquake scenarios.

For evaluating and comparing the distribution of all 288 surface PGA due to Semarang fault earthquake with Peak Surface Acceleration (PSA) values calculated using [12] all surface PGA and PSA values are then distributed in terms of epicentre distance to fault trace. Fig. 5(a) shows the distribution of surface PGA due to 6.2 Mw and 6.9 Mw earthquake scenarios and PSA values (calculated using [12]) in terms of epicentre distance. The surface PGA values due to 6.2 Mw and 6.9 Mw earthquakes are greater than PSA values for all locations with maximum distance 5 Km to fault trace. However the PGA values are smaller than PSA values for all locations located greater than 5 Km to fault trace.

Following the same criteria proposed by [11] risk map of the study area was developed using average value of surface PGA caused by five earthquake scenarios with magnitude 6.05 Mw, 6.2 Mw, 6.69 Mw, 6.8 Mw and 6.9 Mw. Based on the average PGA calculated at 288 locations, the minimum surface PGA is 0.27g and maximum surface PGA is 0.85g. The criteria for developing low, medium and high risk level area were carry out using both minimum and maximum PGA values. The lowest risk level area was identified within the area with PGA values less than 0.46g. The highest risk level area was identified within the area with PGA values greater than 0.74g and the medium risk level area was identified within the area with PGA values in between 0.46g until 0.74g. Fig. 5(b) shows earthquake risk map of Semarang due to Semarang fault earthquake scenarios with magnitude in between 6.05 Mw to 6.9 Mw. Based on risk map as can be seen in Fig. 5(b) the lowest risk level area was identified at the southern part and small area in the north-eastern part of the city. The highest risk level area is identified at the closest distance area to fault trace.

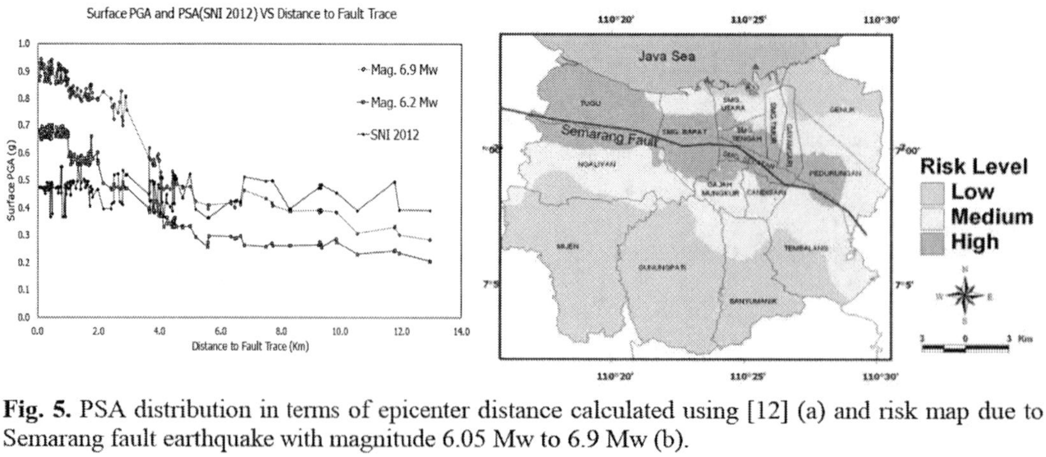

Fig. 5. PSA distribution in terms of epicenter distance calculated using [12] (a) and risk map due to Semarang fault earthquake with magnitude 6.05 Mw to 6.9 Mw (b).

4 Conclusions

Risk microzonation map for Semarang due to Semarang fault earthquake scenarios with maximum magnitude 6.9 Mw has already developed. The risk map was developed using combination of response spectral matching and site response analysis. Response spectral matching analysis was performed due inadequate data related with Semarang fault earthquake events. Site response analysis was performed at 288 boring locations using a pair of (EW and NS) acceleration time histories calculated from spectral matching analysis.

Based on risk map developed from this study it seems that the lowest risk level due Semarang fault earthquake was identified at the southern part and north-eastern part of the city. The highest risk level due to Semarang fault earthquake was identified within the closest area to Semarang fault trace with maximum 5 Km distance from fault trace.

References

1. W. Partono, S.P.R. Wardani, M. Irsyam and S. Maarif, *Development of seismic microzonation maps of Semarang*, Indonesia, Jurnal Teknologi, **77:11**, 99-107 (2015)
2. W. Partono, S.P.R. Wardani, M. Irsyam and S. Maarif, *Seismic Microzonation of Semarang, Indonesia Based on Site Response Analysis using 30 M Soil Deposit Model*, Jurnal Teknologi, **78:8-5**, 31-38 (2016)
3. D.M. Boore and G.M. Atkinson, Earthq. Spec., Vol. **24**, No. 1, 99-138, Earthq. Eng. Res. Inst. (2008)
4. K.W. Campbell and Y. Bozorgnia, Earthq. Spec., Vol. **24**, No. 1, 139-171, Earthq. Eng. Res. Inst. (2008)
5. B.S.J. Chiou and R.R. Youngs, *NGA Model for Average Horizontal Component of Peak Ground Motion and Response Spectra,* (PEER 2008/09, Pacific Engineering Research Centre, College of Engineering, University of California, Berkeley, 2008)
6. N.A. Abrahamson, *Non Stationary Spectral Matching Program RSPMATCH>PG&E*, (Internal Report, 1998)
7. W.D. Iwan, *On a Class of Models for the Yielding Behaviour of Continuous and Composite Systems*, (Journal of Applied Mechanics, ASME, **34**, New York, 1967), 612-617
8. Z. Mroz, *On the Description of Anisotropic Work Hardening*, (Journal of Mechanics and Physics of Solids, **15**, Elsevier, Rhode Island USA, 1967), 163-175
9. J.P. Bardet and T. Tobita, *NERA A Computer Program for Nonlinear Earthquake Site Response Analysis of Layered Soil Deposits*, (Department of Civil Engineering University of Southern California, 2001)
10. I.W. Sengara, M.A. Yulman and A. Mulia, *Seismic Time-History Ground-Motions for a specific Site in Jakarta*, Jurnal Teknologi, **77:21**, 127-136 (2015)
11. TC4-ISSMGE, 1999, *Manual for Zonation on Seismic Geotechnical Hazard*, Revised edition, Technical Committee For Earthquake Geotechnical Engineering (TC4) of International Society of Soil Mechanics and Geotechnical Engineering (ISSMGE) 209
12. Standar Nasional Indonesia, *Tata Cara Perencanaan Struktur Bangunan Gedung dan Non Gedung* (SNI-03-1726-2012) (2012)

Project Cost Estimation Based on Standard Price of Goods and Services (SHBJ)

Toriq Arif Ghuzdewan*, *Brigitta* Petra Kartika Narindri

* Civil and Environmental Engineering Department, Gadjah Mada University, Indonesia

Abstract. Cost estimation is determining the probable cost of a project, which is critical in project development. For government projects in Indonesian construction industry, it is common to use a standard price of goods and services (called SHBJ) publised by the government as a cost reference. This research examines whether the SHBJ provide sufficient and reliable data. A case study from a contractor's bidding document (RAB) and the SHBJ from the city of Yogyakarta are employed. The result shows that cost estimation based on the SHBJ and the contractor's RAB gives 12% difference. Using SHBJ data from years 2010 – 2016 a regression equation is obtained as Yi = 62,156 X + 2,003,962 which can be used to estimate project cost in the future. This research also suggests some improvments for the SHBJ.

1 Introduction

Cost estimation is determining the probable cost of a project, which is critical in project development. Cost estimation is needed by all parties of project such as owner who to obtain reasonable price; engineer to develop enginner's estimate and contractor to bid a project.

There are several methods of cost estimation such as expert judgment, analogous estimate, parametric estimate, detailed estimate, etc. [1]. Expert judgment is determining cost estimates based on expert's experience from the past. Analogous estimate calculates based on similar projects. Parametric estimate applies parameter such as area per square meter. While detailed estimate calculate based on detail components of the project. Each method has its own certain level of accuracy, feasibility estimate 25-30%, appropriation 15-25% and detailed estimate 5 - 15% [2]. Cost accuracy is not easy to determine, but it can be detected from large cost overrun or underrun [3].

In construction projects parametric and detail estimates are the most commonly used methods. Here is an example of parametric estimate for a building with area of 1.000 m2 and cost per square meter Rp. 5.000.000; so the estimated cost will be 1,000 m2 x Rp. 5,000,000 = Rp. 5.000.000.000. Whereas in detail estimate the calculation is based on the volumes of all components of the project multiplied by the unit prices. So cost will be equal to volume times the unit price. The unit prices are determined from a standard unit price and depend on the cost of resources (materials, wages, equipment, etc.).

* Corresponding author: toriq@ugm.ac.id

There several sources of data as cost references such as catalogs, surveys, historical data, etc. Reliable data is very important since inaccurate data can cause loss [4]. Standar price of goods and services (SHBJ or Standar Harga Barang dan Jasa) published by the government annually is widely used. The prices in the SHBJ are the highest price limit allowed in a government project. For construction sector the SHBJ provide prices of materials and wages and the unit price per square meter for building projects [5].

This research aims to examine data of the SHBJ, i.e. the price of materials and wages and the unit price per square meter, whether the SHBJ has accomodated all data required for developing cost estimate and whether they provide fair prices. This research also utilises SHBJ from 2010 to 2016 to obtain regression equation for cost prediction. Lastly this research suggests improvment of the SHBJ.

2 Methodology

This research uses a case study from a contractor's bidding document (called RAB) which is a detailed estimate. The RAB consists of several tables, i.e. Recapitulation, Bill of Quantity, Unit Price Analysis and Price of Material and Wages. Calculations in the RAB basically following steps as shown in Figure 1 and can be explained as follows:

- Detemine the price of material dan wages
- Calculate the unit price, usually using SNI standard
- Calculate the bill of quantity
- Cost recapitulation

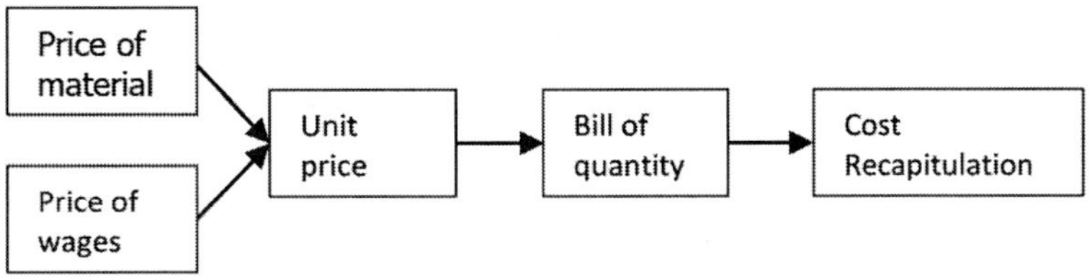

Fig. 1. Steps in cost calculation

To calculate cost estimate based on the SHBJ, data from the SHBJ are inputted into the table of Price of Material and Wages. If data are not available in the SHBJ then data from the contractor's bidding is applied. The contractor's bidding is in 2015, so to obtain prices for different years, annual indexes of construction prices are used. Table 1 shows construction indexes published by the Central Bureau of Statistics. Price of material in year i is calculated as follow:

$$\textit{Price (year i)} = \textit{(indexyear-i / index 2015) x Contractor's price} \quad (1)$$

For example, a material iron strip is not found in the SHBJ 2010 and the contractor's price is Rp. 12.000. Then the price in 2010 is calculated as (100 / 130,27) x Rp. 12,000 = Rp. 9.211.

Table 1. Construction Index

Year	2010	2011	2012	2013	2014	2015	2016
Index	100	103,78	108,64	111,89	122,70	130,27	132,43

3 RESULT AND DISCUSSION

3.1 Project in the Case Study

The case study project is a two-story office building with an area of 1,258 m2 located in Gunung Kidul distric, Yogyakarta. The building is a reinforced concrete structure with lightweight steel roof truss. The components of the building consist of 14 groups of work, some of which are preparation, earthwork, masonry, concrete, roof works, etc. The RAB is Rp. 2,743,289,872 before tax and it is in 2015.

3.2. Project Costs Based on the SHBJ

Based on the price of materials and wages from the SHBJ 2015 it is found that the cost of the project is Rp. 3,076,787,720. Compared to the contractor's bid Rp. 2,743,289,872 there is a difference of 12%. This difference is reasonable as it is in range of 15 – 25 % detailed estimate accuracy [2].

The SHBJ provides cost of area per square meter for simple building and complex building. For the year 2015 cost per square meter for the simple building is Rp. 3,522,142 and for complex building is Rp. 4,931,000. The building in the case study can be classified as complex building as it is more than 500 m2 and two-story, so the price is 1,258 m2 x Rp. 4,931,000.00 = Rp. 6,203,198,000. Compared to the contractor's bidding (Rp. 2,743,289,872) it is 126% more expensive. This difference is very large so it needs to be checked further. Another research also found that cost per square meter of the SHBJ is over estimate up to 90% and suggested improvement of the standard [6]. If applying the classification of simple building, the building price would be 1.258 m2 x Rp. 3.522.142 = Rp. 4,430,855,714 or 62% deviation and it is also consider quite large different.

Table 2 shows the SHBJ's cost per square meter in 2010 – 2016. By multiplying these costs with the area of the buildings, costs of the building for these years can be obtained. Figure 2 compares these costs with the project costs based on the SHBJ's price of materials and wages.

Table 2. SHBJ's Cost per square meter for complex building in Gunung Kidul Distric

Year	Cost per m2 (Rp)
2010	4,322,000.00
2011	4,322,000.00
2012	4,322,000.00
2013	4,582,000.00
2014	4,696,500.00
2015	4,931,000.00
2016	5,367,000.00

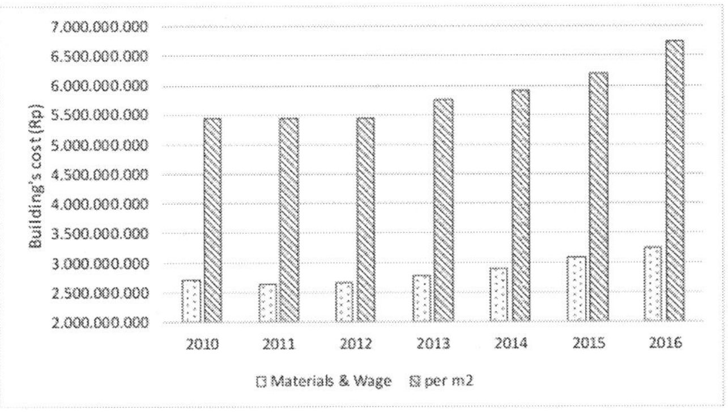

Fig. 2. Costs based on the SHBJ's price of materials and wages v.s. Price of area per square meter

3.3 Annual Project Price and Regression Equation

Figure 3 shows project prices based on the SHBJ's material and wages prices in 2010 to 2016. Price in 2011 declined 2,48% caused by the lower price of some materials such as crushed stone, tile, wood, wall paint, door lock and water tap. Then since 2012 the price increased 0.83%, 4.44%, 4.23%, 6.14% and 6.00% yearly.

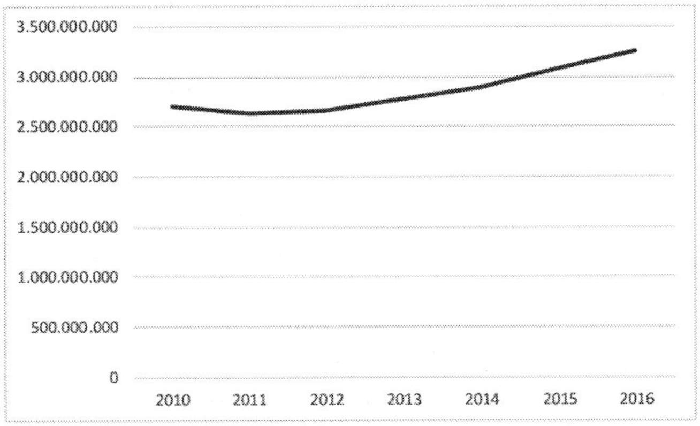

Fig. 3. Annual Project Prices based on the SHBJ

Based on project costs in 2010-2016 a linear regression can be drawn and an equation is obtained as Yi = 62,156 X + 2,003,962. Yi is cost per square meter in year-i, while X is year-i – 2009. This equation can be used to estimate project cost in the future. For example to estimate building cost in 2020 the cost per square meter can be calculated as Yi = 62,156 (2020 – 2009) + 2,003,962 = 62,156 (11) + 2,003,962 = 2,687,678. If the the building's area is 1,258 m2 then the price can be estimated as Rp. 2,687,678/ m2 x 1,258 m2 = Rp 3,381,098,924.

3.4 Suggestion for Improving SHBJ

From 126 material in the contractor's RAB, 35 items are not available in the SHBJ such as steel plate, plasterboard edge, dolken wood, door hinge, hooks, roster, natural stone, bathtub, floor drain, conblock, weathershield paint, and ridge. This indicate that the SHBJ are not complete. New or innovative materials such as readymix, precast, and other advanced materials have not been accomodated yet in the SHBJ.

Some of material prices in the SHBJ have significant difference from the contractor's prices such as wood, coating and alcasite which have difference up to 24 - 36% and other materials even have difference greater more than 200% such as ridge, tap, floor and nail. The prices of these materials need to be rechecked.

Some materials have different names on the SHBJ compare to that in the contractor's document such as for sandpaper, door lock, gravel, common rafter, nail, etc. Although these differences can be understood, but standardization of names would be beneficial for clarity. Similarly some materials have different units such as meter vs piece, m2 vs sheets, m3 vs piece, etc. Standardization of units is also important for ease and to avoid errors in cost calculation.

4 Conclusions

The SHBJ provides two references for cost estimasting, i.e. the price of materials and wages and the standar price per square meter to estimate building projects. The price of materials and wages in the SHBJ provides quite accurate result as can be seen from the case study that it results in difference of about 12% against the contractor's RAB. Whereas the standar price per square meter gives quite a large difference of up to 62% - 126%. The standar price per square meter therefore need to examined further to provide reasonable price. Based on the analysis of annual prices from the SHBJ in 2010-2016, a linear regression equation is obtained, i.e. Yi = 62,156 X + 2.003.962 which can be used to estimate cost per square meter in the future. The equation can be applied for projects in Gunung Kidul District of Yogyakarta, but for other places should be adjusted with location factor. The study also suggest some improvement for the SHBJ such as to compelete resources which are not available yet, more accurate price, and standardization of resource's names and units.

References

1. G.W.R. Duncan, *A Guide To The Project Management Body Of Knowledge*, PMI USA, (1996).
2. G.J. Ritz, *Total Construction Project Management*, International Ed edition, McGraw-Hill Inc., US., (1994)
3. I. Soeharto, *Manajemen Proyek*, Erlangga, (1997)
4. G.A. Garret, *Cost Estimating Methods and Sources of Risk*, (2008)
5. SHBJ DIY, *Standar Harga Barang dan Jasa, Peraturan Gubernur DIY*, (2010 – 2016)
6. M. Abduh and B. Kurniawan, KoNTekS 2, (2008).

Right Turn Coeficient Evaluation Analysis on Intersection capasity formula at Indonesia Highway Capasity Manual 1997 Using Vehicle Tracking Simulation

Basuki Kami Hari [1,*], and *Purwanto* Djoko[1]

[1] Researcher at Civil Engineering Department, Diponegoro University, Semarang, Indonesia

Abstract. The aims of this study are to evaluate the the right-turn coefficients at intersection capasity formula by Indonesia Highway Capacity Manual (MKJI 1997). Therefore, it is necessary for analysis that examines right-turn coefficient values in intersection capacity function. The simulation with many various geometric intersections defined by many scenarios using vehicle tracking application program, which increase the space maneuver vehicle at the intersection includes: 1) widen approach lane; 2) add a right-turn lane; and 3) backward median in each approach. Intersection performance Analysis will produce degree of saturation that is affected by right-turn coefficient according to the simulated vehicle maneuver using new formula. The study can be concluded that the intersection changes the right-turn factor adjustment which was originally set with the value based on linear function being a log function. According to the new function, Prt Less 2,5% right turn vehicle make significant decrease of intersection capacity, that why, it makes many conflict based on opposite traffic from other approach. In the other side, if Prt more then 2,5%, it makes increase intersection capacity because intersection space used to right turn vehicles.
Keywords: right-turn coefficient, vehicle maneuver, IHCM 1997

1 Introduction

Semarang as capital city of Central Java has pretty high vehicle movement activity, and to provide a good transportation climate, it needs a good intersection to serve vehicle movement which pass through it. Intersection in Semarang in many cases cannot provide a good traffic flow movement yet, especially when it comes to freight vehicle movement. It is indicated by delay, which is probably caused by bad traffic control or inadequate intersection/right-turn geometric [2, 4]. Therefore, it needs further analysis about intersection and right-turn performances and freight vehicle maneuver by taking case study at Semarang Northern Ring Road. Right-turn which represent main intersections for freight vehicle movement route.

* Corresponding author: basuki.kh@gmail.com

2 Methodology

Analysis starts by doing traffic counting survey and geometric data collecting on each case study sites. Intersection's performance is analyzed using SIG-I – SIG-V forms [3] from Manual Kapasitas Jalan Indonesia 1997 to generates intersection's performance parameters such as capacity, delay and degree of saturation for each intersection's approach. Right-turn performance is analyzed using queue modelling method to generates right-turn intensity value, average queue waiting time and average number of vehicle in the queue [4, 5]. After doing performance analysis on existing intersections and right-turn, vehicle maneuver simulation can be performed using vehicle tracking simulation from Autodesk which will generates vehicle maneuver route and its ellapsed time to do that maneuver [1]. Based on those analysis and simulation, several optimization steps are taken to repair or improve intersection and right-turn performances especially due to heavy vehicle maneuver by changing its existing geometrics to be efficiently able to serve vehicle maneuver on it. Optimization result then evaluated to see the performance difference as well as the conclusions drawn based on the analysis and optimization.

3 Data Analysis

3.1 Performance Analysis at Northern Ring Road Intersection

Traffic survey data is processed in order to obtain traffic flow data for each hour interval and the biggest traffic flow data value will be used as intersection's traffic peak hour [2, 3]. In Northern Ring Road Intersection, traffic flow peak hour is at 4.15-5.15 pm. Northern Ring Road Intersection performance analysis generates intersection's capacity (C), degree of saturation (DS), vehicles queue and delay (D) as shown in Table 1.

Table 1. Intersection Performance Analysis Parameters

Approach	Traffic Flow	Effective Width	Green Ratio	Saturated Flow	Capacity	Degree of Saturation	Queue Length	Delay
	Q	We	GR	S	C	DS	QL	D
East	319,9	7	0,144	2778,750	400,781	0,798	45,714	59,856
South	1881,5	11,5	0,327	6096,150	1992,972	0,944	135,565	50,806
West	291,8	7	0,144	3087,500	445,313	0,655	40,000	49,739
North	1349,6	10,5	0,365	5566,050	2033,749	0,664	89,524	31,839

Based on that result, Northern Ring Road Intersection cannot serve the existing traffic flow yet, which is showed by degree of saturation for south approach = 0,94 (DS>0,85).

3.2 Heavy vehicle maneuver simulation on existing geometric condition

Figure 1 below has shown vehicle maneuver simulation result on each Northern Ring Road Intersection's approach. Based on vehicle tracking simulation, Northern Ring Road Intersection's existing geometric can serve heavy vehicle maneuver on it, but with several maneuver simulation test it could be optimized to get a more efficient maneuver route by changing intersections geometry aiming for reduced heavy vehicle maneuver ellapsed time to enchance intersections performance.

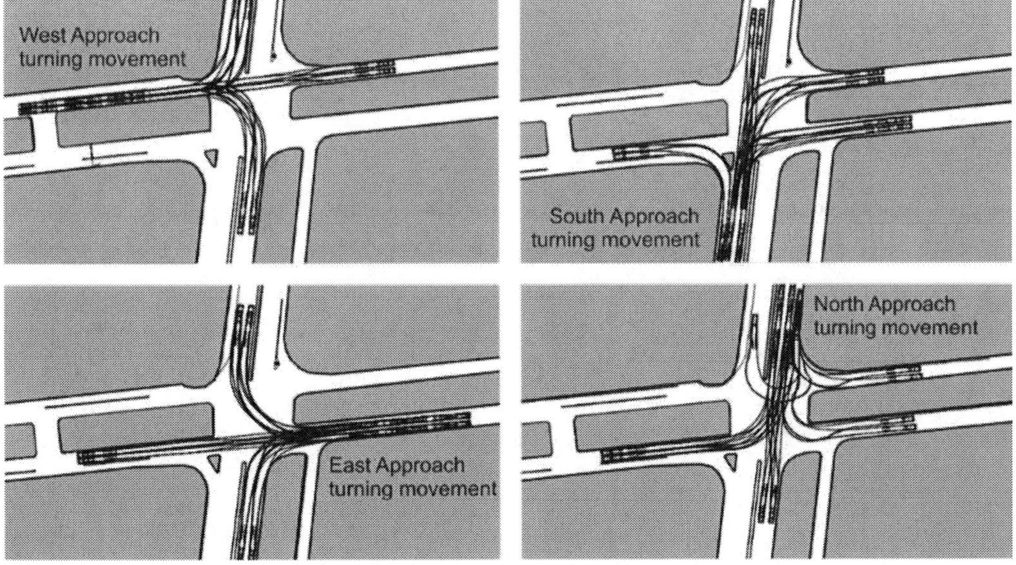

Fig. 1. Vehicle Maneuver Simulation on Existing Northern Ring Road Intersection Geometric in each approach

3.3 Intersection and Right-turn Optimization

Intersection optimization to serve the existing traffic flow is done by rearranging its signal time and make a turn around restriction. Table 2 shows intersection performance analysis result after signal time optimization.

Table 2. Intersection Performance Analysis Parameters with Signal Time Optimization

Approach	Traffic Flow	Effective Width	Green Ratio	Saturated Flow	Capacity	Degree of Saturation	Queue Length	Delay
	Q	We	GR	S	C	DS	QL	D
East	694,7	7	0,288	2778,750	801,5625	0,866	38,486	50,88
South	622,2	11,5	0,231	6096,150	1406,804	0,442	135,565	38,00
West	448,7	7	0,173	3087,500	534,375	0,840	31,257	59,25
North	964,8	10,5	0,346	5566,050	1926,71	0,501	82,076	30,33

After signal time optimization, to optimize it even more due to heavy vehicle maneuver geometric optimization can be done as follows: 1) Bridge widening from 41,3 meters to 51,3 meters; 2) North and east median tip adjustment. Figure 2 shows geometric optimization for Northern Ring Road Intersection [2].

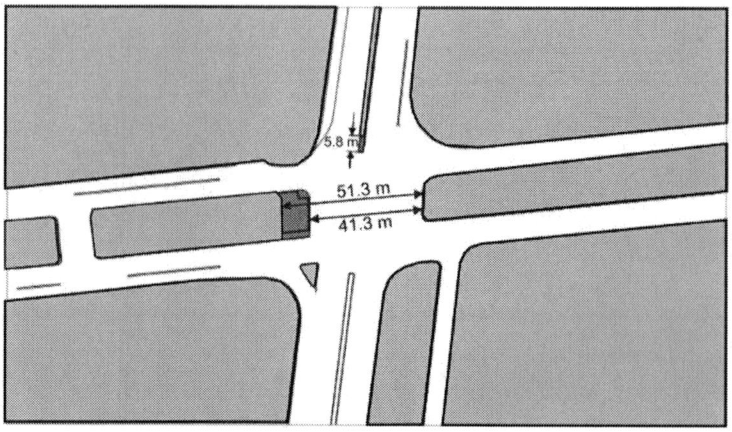

Fig. 2. Geometric Optimization Plan for Northern Ring Road Intersection

With reduced turn right maneuver ellapsed time, delay on each approach is reduced too and affecting in increased turn right factor (turn right vehicle got a reduced ellapsed time to do the maneuver due to geometric optimization and it makes intersection can afford more number of turning right vehicle. It is in line with the concept of right turn magnification factor (Frt) according to *Manual Kapasitas Jalan Indonesia* (MKJI 1997) [3] that turning. Vehicle maneuver elapsed time difference caused by geometric optimization shown in Table 3.

Table 3. Vehicle Maneuver Ellapsed Time Comparison Before and After Geometric Optimization for Northern Ring Road Intersection

Heavy Vehicle Maneuver	Ellapsed Time (s)	
	Existing	Optimized
West-RT	27,7	25,4
North-RT	28,1	26,9
East-RT	25,6	24,0
South-RT	31,5	30,6

4 Discussion

4.1 Intersection Performance Comparison

Right vehicle has a tendency to cut out the center line of the road before turning right so the number of turning right vehicles could be increased. By increased right turn factor, saturated flow is also increased and it affecting in increased intersection approach's capacity. This additional capacity ends up reducing intersection's degree of saturation. Northern Ring Road Intersection performance analysis result after geometric optimization is shown in Table 4. As comparison for that value, Table 5 below shows the performance parameters comparison between only signal time optimized intersection and after geometric optimization is taken.

Table 4. Performance Analysis Result of Geometric Optimized Northern Ring Road Intersection

Approach	Traffic Flow	Effective Width	Green Ratio	Saturated Flow	Capacity	Degree of Saturation	Queue Length	Delay
	Q	We	GR	S	C	DS	QL	D
East	694,7	7	0,288	2778,750	801,5625	0,866	38,486	50,88
South	622,2	11,5	0,231	6096,150	1406,804	0,442	135,565	38,00
West	448,7	7	0,173	3087,500	534,375	0,840	31,257	59,25
North	964,8	10,5	0,346	5566,050	1926,71	0,501	82,076	30,33

Table 5. Performance Parameters Comparison Between Before and After Geometric Optimization on Northern Ring Road Intersection

Approach	Degree of Saturation		Queue Length		Delay	
	Existing	Optimized	Existing	Optimized	Existing	Optimized
East	0,798	0,866	45,714	38,486	59,856	50,88
South	0,944	0,442	135,565	135,565	50,806	38,00
West	0,655	0,840	40,000	31,257	49,739	59,25
North	0,664	0,501	89,524	82,076	31,839	30,33

Based on that optimizations, it signifies that intersection performance is increased by rearranging traffic signal time, restricting turn around movement and repairing intersection's geometric in order to optimizing heavy vehicle maneuver on intersection.

4.2 Frt Function

According to MKJI 1997 [3] there is graph between right turning factor (Frt) with right turn ratio (Prt). The graph is in the form of a linear line upwards. The value is obtained from the quotient between the number of vehicle volumes that turn right and the total volume of vehicles in each approach. From the calculation in the graph of Prt and Frt relationship as Figure 3 as follows:

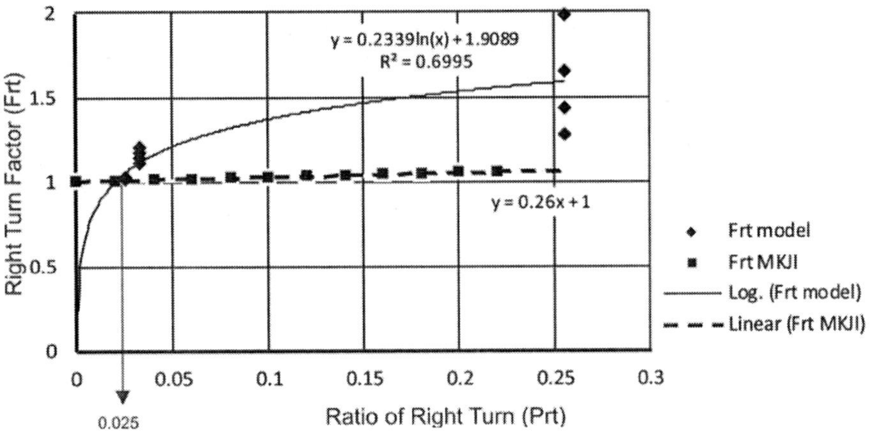

Fig. 3. The Relation Between Frt and Prt

The dotted line shows the graph of the relationship between Prt and Frt according to MKJI 1997 in the form of linear [3]. While on the black line shows the graph of the relationship between Prt and Frt according to survey results in the field in the form of logarithm. It can be seen from the graph that the right turn ratio is 2.5%, then Frt is worth. It means that if there is one right turn vehicle, right turn ratio is worth 2,5%. On the graph can also be concluded that the value of Frt is not worth 0. If there are a few vehicles that turns right, then it can decrease the performance of intersection capacity. Otherwise, if there are many vehicles that turn right, it can improve the performance of intersection capacity or the intersection has accommodated the right turn well. So, it can be concluded that the northern Ring Road intersection changes the right turning factor adjustment which was originally set with the value of 1 being a function y = 0.2339ln (x) + 1.9089 with R2 = 0.6995.

5 Conclusion

Existing intersection results on Northern Ring Road Intersection performance analysis shows value of DS (degree of saturation) for east approach is 0,80, south approach 0,94, west approach 0,66 and north approach is 0,66. DS of south approach exceeding the maximum standard value of intersection's DS which is 0,85. Vehicle maneuver simulation on existing intersection and using vehicle tracking simulation gives results as follows, Northern Ring Road Intersection geometrics can still be optimized to serve vehicle maneuver more efficiently in order to improve intersection's performance. So, it can be concluded that the northern Ring Road intersection changes the right-turn factor adjustment which was originally set with the value based on linear function Frt=0,26Prt+1 being a log function Frt= 0.2339ln (Prt) + 1.9089 with R2 = 0.6995. According to the new function, Prt less then 0,025 the meaning is that 2,5% vehicle in right turn depend from all traffic approach. Less 2,5% right turn vehicle make significant decrease of intersection capacity, that why, it makes many conflict based on opposite traffic from other approach. In the other side, if Prt more then 2,5%, it makes increase intersection capacity because intersection space used to right turn vehicles. The formula need verification and validation step to according on study cases in variate intersection type. Basically, it is on going research to get the best function.

Reference

1. Autodesk, *Vehicle Tracking Manual Drive,* http://help.autodesk.com/view/VEHTRK/2017/ (2017)
2. Direktorat Jenderal Bina Marga, *Standar Perencanaan Geometrik Untuk Jalan Perkotaan,* Jakarta (2004)
3. Direktorat Jenderal Bina Marga, *Manual Kapasitas Jalan Indonesia. Kementerian Pekerjaan Umum,* Jakarta (1997)
4. May, A. D., *Traffic Flow Fundamentals*, New Jersey: Prentice Hall (1990)
5. Ortúzar, J. D., & Willumsen, G. L., *Modelling Transport*, West Sussex: John Wiley & Sons (2011)

The Influence of Freight Generation Production Characteristics of the Internal – Regional Zone Commodities on Sustainable Freight Transportation Highway Network System

Juang Akbardin[1, 3*], *Danang* Parikesit[2], *Bambang* Riyanto[3], *Agus Taufik* Mulyono[2]

[1]Department of Civil Engineering, Indonesia University of Education, Bandung – Indonesia
[2]Department of Civil Engineering, Gadjah Mada University, Yogyakarta – Indonesia
[3]Department of Civil Engineering, Diponegoro University, Semarang – Indonesia

Abstract. The increasing of commodity freights production in the internal – regional zone is a national development strategy to create inter-zone consolidation to accelerate the productivity of an internal – regional zone staple commodity. Sustainable freight transportation development requires a strategic direction based on the commodity freights production dominant characteristics in an internal-regional zone. This research has been done to estimate the distribution model and route assignment model of staple alternative commodity. Estimated generation production of commodity freights movement in production zones is influenced by zone demographic variables, infrastructure, and means of transportation. Gravity method is used to estimate the distribution model and route assignment model by calibrating it with Newton-Raphson method. The application and development of the gravity method in the modelling analysis will determine the estimation effect of sustainable freights transportation needed. The computational simulation result is an approach of mathematical model illustration for complex and systematic estimation. Therefore, the modelling results can provide the direction of sustainable highway construction on the highway network system, will determine the effectiveness and efficiency of traffic movement, and distribute the movement.
Keywords: Trip Production, Highway Network System, Freight Transportation

* Corresponding author: akbardien@upi.edu

1 Introduction

The need for staple consumption is increasing in accordance with the population growth in a region. Meanwhile, the production of staple commodities declines in general due to the decrease of the staple area production, rice-cultivated paddy fields. Central Java region has a diversity of land that produces alternative staples. The tubers commodity in the Central Java province has sufficient amount of production. Production of tubers and their yields can be used as an alternative staple which has been depending on rice production so far. Production of tubers and their products requires the movement magnitude estimation of the commodity production. The production movement of the tubers commodity and their products evolves in accordance with the demands of the products and the economic laws of supply and demand. The rise of movements resulting from the economic activity of tubers commodity production and its manufactured goods led to an increase in vehicular traffic on the production lines of tuber commodity and its manufactured goods production zones. Production zones generally have highway network systems that are not on the main routes of provincial and state distribution systems. The increase of tuber commodity production, used as alternative staple goods, urges the existing local highway network system to be able to serve the traffic development due to the movement increase, caused by the alternative staple commodity production escalation. So, to support the smooth goods distribution system requires the highway service performance that is able to encourage the increased production of strategic commodities and basic necessities. [2-5]

2. Gravity Model

Generating movement is the number of moves generated by a zone of origin (Oi) and the number of moves that are attracted to each destination zone (Dd) included in the study area.

2.1. Modeling of Freight Transportation

Gravity model that gives the formula of freight transportation total volume based on the commodity type transported from one place to the other place is as follows (equation 1)

$$T_{idk} = \frac{S_{ik} \cdot D_{dk}\, f_{id}}{\sum D_{dk} \cdot f_{id}} \quad \ldots(1)$$

Where:
T_{idk} = the number of k commodity produced in the i region and sent to the d region
S_{ik} = the total amount of shipments of k commodity from the i region
D_{dk} = the total amount of demand of k commodity in d region
f_{id} = the friction/ barriers factor (= 1 / $d_{id}\lambda$)
λ = parameter

This model uses the gravity concept introduced by Newton developed from the gravity law analogy.[1]

$$F_{id} = G \cdot \frac{m_i \cdot m_d}{d_{id}^2} \quad .(2)$$

with G as the gravitational constant of the movement attraction, as well as distance, time, or cost as an accessibility measurement. Thus, for transportation purposes, the GR model is formulated as: (Equation 3)

$$T_{id} = k \cdot \frac{O_i \cdot O_d}{d_{id}^2} \quad ..(3)$$

In which k is a constant
So, in mathematical equation, GR model can be formulated as below:

$$T_{id} = O_i \cdot D_d \cdot A_i \cdot B_d \cdot f(C_{id}) \quad ..(4)$$

2.2. Modelling of Transportation based on Traffic Flow

$\hat{V}$ total flows on certain roads is the sum of the movement between the zones within the study area that use these roads, shown in equation (5).

$$\hat{V}_l = \sum_{i=1}^{N} \sum_{d=1}^{N} T_{id} \cdot p_{id}^l \quad ...(5)$$

Commodity moved between the zones within the study area is represented by a model of transportation demand with Gravity Model Opportunity (GO). In which, T_{id} total movement with the i starting point zone and the d destination zone for all commodity movements purposes are indicated by equation (6)

$$T_{id} = \sum_{k=1}^{K} T_{id}^k \quad ...(6)$$

where the commodity that move from i zone to d zone is shown on equation (7)

$$T_{id}^k = b_k \cdot O_i^k \cdot D_d^k \cdot A_i^k \cdot B_d^k \cdot f_{id}^k \quad ...(7)$$

By substituting equation (7) to equation (5), the basic equation for the transportation demand estimation model with traffic flow data is shown on equation (8) [1]

$$V_l = \sum_{k=1}^{K} \sum_{i=1}^{N} \sum_{d=1}^{N} \left(b_k \cdot O_i^k \cdot D_d^k \cdot A_i^k \cdot B_d^k \cdot f_{id}^k \, p_{id}^l \cdot \right) \quad ...(8)$$

3. METHODOLOGY

The modelling approach of sustainable freight transportation which is very complex and extensive is simplified by defining the dominant determined variables.

3.1. Variable of Generating Movement

Dependent variables
Y1 = Oi = Generating Movement of Vegetable Commodity

Independent variables
X1 = population,
X2 = GRDP (Gross Regional Domestic Product),
X3 = Freights based on tubers commodity and their manufactured products IO,
X4 = the length of state highways in the Central Java districts or cities,
X5 = the length of provincial highways in the Central Java districts or cities,
X6 = the length of local highways in the Central Java districts or cities,
X7 = highways in good condition in the Central Java districts or cities,
X8 = highways in moderate condition in the Central Java districts or cities,
X9 = damaged highways in the Central Java districts or cities,
X10 = severely damaged highways in the Central Java districts or cities,
X11 = the number of freights vehicles in the Central Java districts or cities,
X12 = the number of freights vehicles owned by individuals,
X13 = the number of freights vehicles owned by company
b_0 = a constant
[1-5]

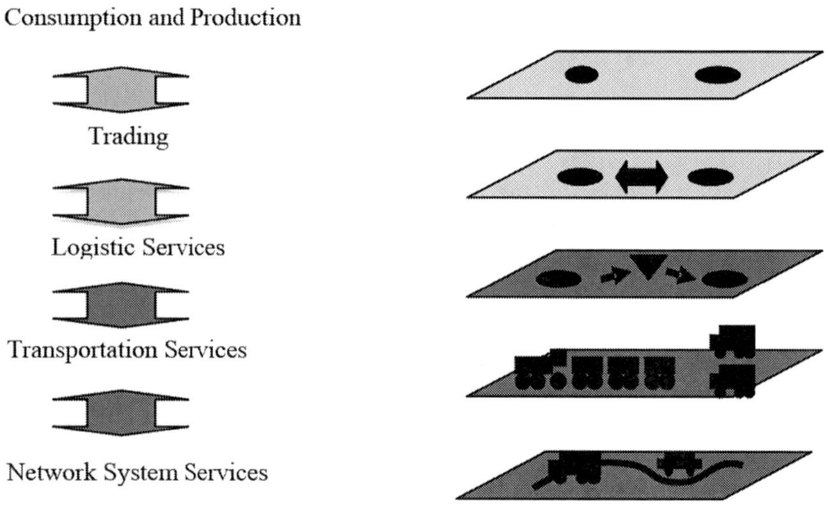

Fig 1. Logical Structure of Freights Transportation

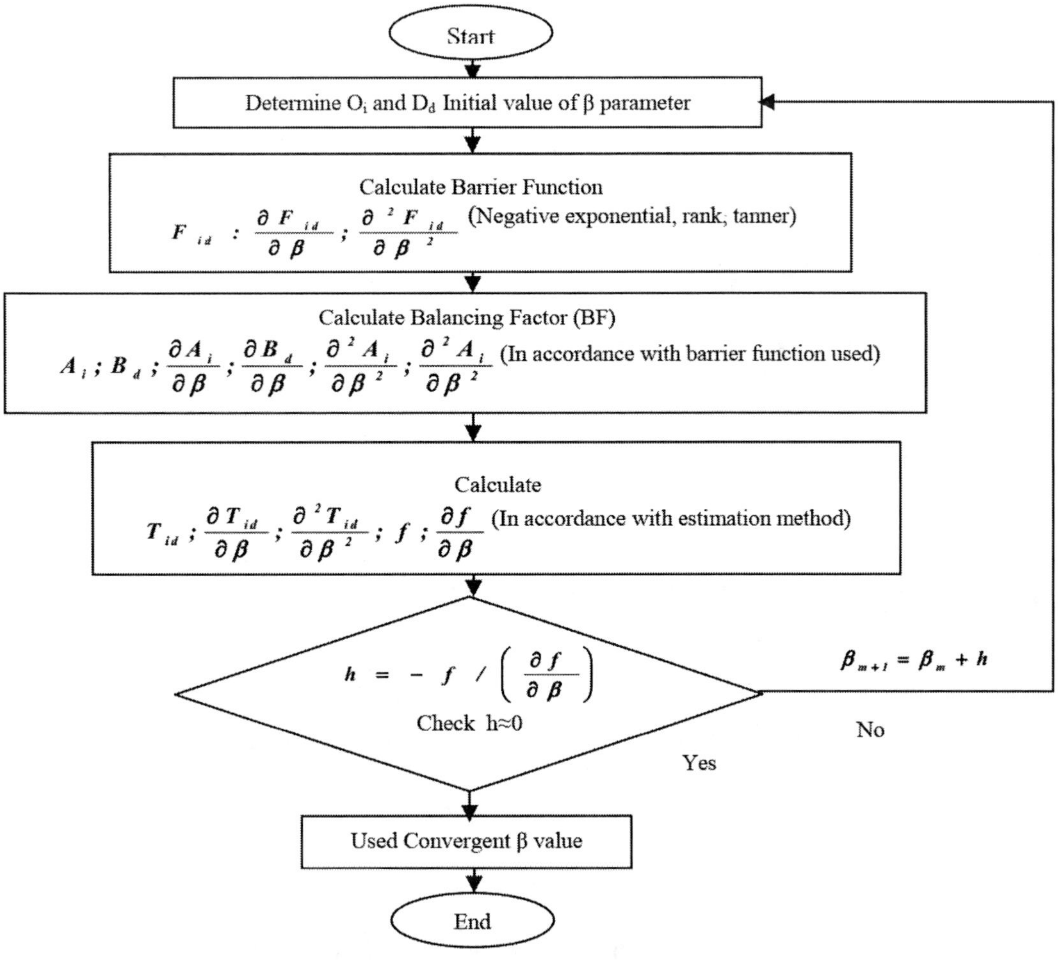

Fig 2. Newton-Rhapson Calibration Method Flow Chart

4. RESULTS AND DISCUSSIONS

Generation modelling movements of tubers commodity and their poducts as alternative staples using multiple regression method is shown in equation (9)

$$\text{Ln Yi} = 1{,}94 + 0{,}990 \text{ Ln X1} + 0{,}194 \text{ Ln X2} + 0{,}0123 \text{ Ln X3} + 0{,}0247 \text{ Ln X4} + 0{,}0106 \text{ Ln X5} - 0{,}113 \text{ Ln X6} + 0{,}0196 \text{ Ln X7} - 0{,}0249 \text{ Ln X8} - 0{,}0036 \text{ Ln X9} - 0{,}0451 \text{ Ln X10} - 0{,}108 \text{ Ln X13} \quad ...(9)$$

The mathematical model equation obtained has been tested with the model significance level where the determination test $R^2 = 0.87$. Furthermore, the model obtained is simulated based on the data of alternative staple commodity producers' zones with matlab computation method 12 as shown in figure 3. The movement distribution of the alternative staple commodity products using gravity method with each zone model simulation is indicated by Disire Line distribution of movement. [1,4,5]

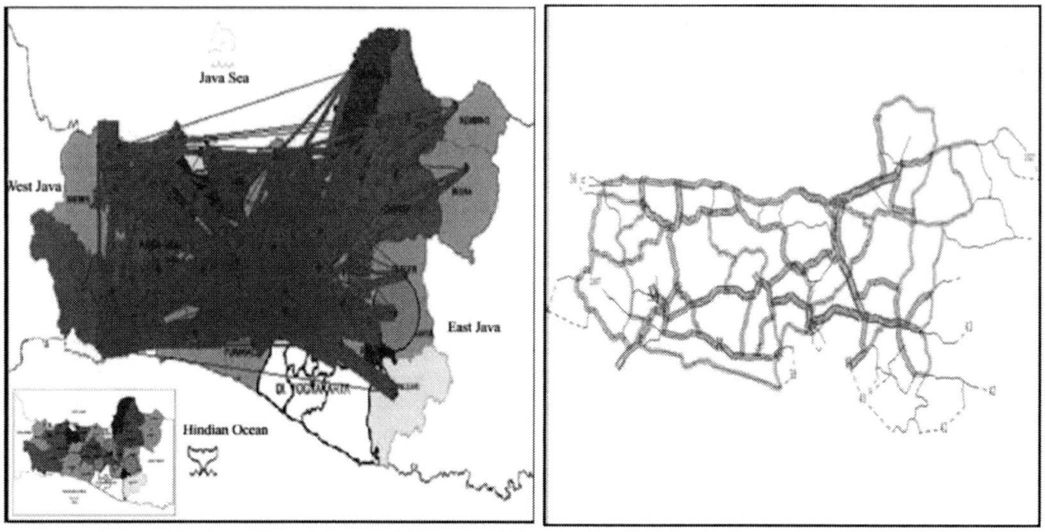

Fig 3. The Simulation of Movement Distribution and Traffic Loading due to alternative staple commodity production according to zone and track path

The data analysis and modeling simulation result shows that the generating movements in alternative staple commodity production zones has adequate significant influence on the highway network system built, particularly in the alternative paths of the main traffics movement distribution. [1-5].

Conclusion

The generating movement effects of increased alternative staple commodity production in the regional-internal zone on the highways network system in Central Java shows the saturation degree in the connecting highway network system and the alternative path is increased. It is influenced by the commodity producers zone characteristics that are outside the main distribution highways, and thus it is necessary to expand the highways capacity to support the distribution system on these highways.

References

1. T. Z. Ofyar. Perencanaan dan Permodelan Transportasi, Second Edition, ITB Bandung(2000)
2. Jin, Y., Williams, I., Shahkarami, M.,. Integrated regional economic and freight logistics modelling: results from a model for the Trans-Pennine Corridor, UK, European Transport Conference (2005)
3. Davydenko, IY, Tavasszy, LA, Meijeren, JJ & Ivanova, O. Estimation of freight flow and trip generation according to transport purpose in urban and regional logistics networksMETRANS Urban Freight Conference Proceedings (pp. 1-17). Long Beach: METRANS. (2011)
4. V. Holguín, J., M. Jaller, L. Destro, X. Ban, C. Lawson, and H. S. Levinson, Freight Generation, Freight Trip Generation, and Perils of Using Constant Trip Rates, In *Transportation Research Record: Journal of the Transportation Research Board, No. 2224*, Transportation Research Board of the National Academies, Washington, D. C., pp. 68-81, (2011)
5. Davydenko, Igor Y.; Tavasszy, Lóránt A.; Smeets, Peter S.G.M. Commodity Freight and Trip Generation by Logistics Distribution Centers Based on Sectorial employment Data,91st Annual TRB Meeting, Washington DC (2012).

Developing a Prototype of Early Warning System of Delay Risks for Public Projects (EWaSDRiP)

Jati Utomo Dwi Hatmoko [1,*], *Hery* Priyono[1], *Mochamad* Agung Wibowo[1], and *Riqi* Radian Khasani[1]

[1]Civil Engineering Department, Diponegoro University, Semarang - Indonesia

Abstract. Delays may occur during pre-construction or construction phases, or both. The aim of this research is to develop a prototype of an early warning system of delay risks for public projects in pre-construction phase (EWaSDRiP). These delays are attributed to owner, consultant, project characteristics, and external factors. Data was collected through literature review, observations on project reports, interview, and questionnaire surveys with 30 respondents from government client teams, contractors and consultants. The respondents were asked to validate the indicators, and estimate their weight and the degree of accuracy. A 1 to 10 Likert scale reflecting worst to excellent is used to score the state of each of the delay indicators within three zones, i.e. red (scale 1-3), yellow (4-6), green (7-10). The delays in red zone indicates that they pose dangerous threats to the project and must be monitored closely, delays within yellow zone indicate moderate threats to the project, while delays in green zone means that they are unlikely to delay the project. EWaSDRiP gives a visual indication of which zone most delays are distributed. This prototype is potentially beneficial for the owner or government client team to prepare strategic actions for the relevant pre-construction delays accordingly.

1 Introduction

Delays have long been recognised as one of main problems in public construction projects. Their impacts, financially and socially, are significant to project stakeholders as well as to society. Quite often delays prevent the owner or users to take benefits of the project as scheduled.

Delays may occur during pre-construction or construction phases, or both. While many literature are mainly focused on delays during construction phase (e.g. [1-4]), delays occurring in pre-construction phase may also impact quite significantly on the project completion time [5, 6]. Delays due to problems in public project procurement, for example, may prevent the project to start on schedule, resulting in late project completion. It is also very common that the late start of a project may put the winning contractors in difficult situation due to unrealistic remaining project durations. On the other side, it is not

* Corresponding author: jati.hatmoko@ft.undip.ac.id

uncommon that project owners (government clients) have no comprehensive understanding at early stage on whether or not the projects of which they are responsible will be completed meeting the target date. Therefore, an instrument to assess and predict prior to project execution is required [7-9]. Besides predicting the project final status based on conditions in pre-construction phase, this tool will also be beneficial to give confidence to the government client team about the project. The aim of this research is to develop a prototype of an early warning system of delay risks for public projects in pre-construction phase (EWaSDRiP).

2 Delay risks in Public projects

A study of delay risks in public project found the top three delay factors, i.e. delays related to information delivery and exchange between related parties, lack of competencies of owner, and lack of capabilities of supervision consultant [10]. Koushki et.al [11] highlighted three main causes of delays by project owners, i.e. changes to orders, financial constraints and lack of experience of owners in the construction business. Chabota et.al [12] found that delayed payments were the main causes of the project delays, followed by financial processes, financial difficulties, economic problems, contract modifications, material procurement, personnel issues, changes in drawings, equipment inadequacy, lack of supervision, construction errors, poor coordination on site, changes in specifications and labor disputes and strikes. This result is similar with the finding of Aibinu and Odeyenka [13] who found the contractor's financial factor as one of the main causes of delays among 44 delay factors identified from construction manager perspective. While Messah et.al [14] found main factors causing delays during construction phase of building projects as agreed by contractors, owners, and supervisor consultants, i.e. lack of manpower, unavailability or late delivery of materials, unavailability of construction equipment, and non-contract payments.

Table 1. Selection of project delay risks in pre-construction phase

Delay Risks	Reference
Owner	
Imbalanced ratio of project supervisor and number of project supervised	[13, 15, 16, 32]
Late project procurement	[16, 17, 18, 31, 32]
Late budget of project agreed	[15, 16, 19, 31, 32]
Technical specification incomplete	[11, 14, 16, 29, 32]
Incompetence of head of project	[2, 16, 20, 32]
Consultant	
Incompetence of consultant	[10, 15, 28, 29, 32]
Late drawing approval	[11, 15, 16, 29, 31]
Design errors	[16, 21, 22, 29, 31]
Incomplete design drawing	[14, 15, 16, 29, 31]
Poor quality of supervision	[14, 15, 28, 29, 32]
Project Characteristic	
Project scale	[15, 16, 28, 30, 32]
Complex project	[15, 16, 19, 32]
Unrealistic project duration	[15, 16, 19, 29, 31]
Contract is signed late	[14, 15, 28, 31, 32]
Remote project location	[15, 28, 29, 30, 32]
Site not ready	[15, 23, 28, 29, 31]
External factors	
Unavailability of certain materials in the market	[15, 16, 19, 29, 31]
Inclement weather condition	[15, 16, 28, 29, 31]
Uncertainty due to new regulation related to project	[14, 15, 16, 29, 31]

From various studies of delay risks mentioned earlier, it can be concluded that delays risks can be originated from owner, contractor, consultant, project characteristics, and external factors. Table 1 shows the results of identified delay risks in pre-construction phase, which later are used as indicators of EWaSDRiP.

3 Early warning system and use of EWaSDRiP

In general, an early warning system is designed to alert the occurrence of undesirable events to minimize the incidents, such as natural disasters, political chaos, wars, accidents, and delays [24, 25]. Early warning systems can be applied in risk assessment, safety risk, mitigation delay, project performance and problem solving on construction projects (e.g. [26, 27]).

EWaSDRiP is intended to be used during pre-construction phase by particularly project owner (Fig 1). The use of this instrument prior to the construction phase will allow the project owner to predict the final status of the project earlier based on the condition and information available during pre-construction phase.

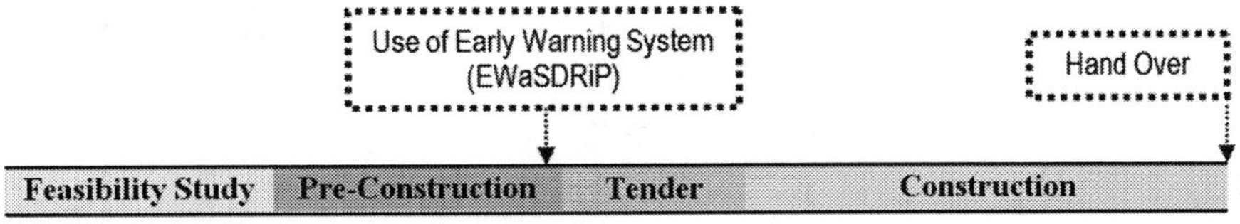

Fig. 1. Use of Early Warning System (EWaSDRiP) in a project timeline

4 Research method

Data of project delays was collected through literature review, observations on project reports, interview, and questionnaire surveys with 30 respondents, which consists of government client teams (20%), contractors (33%), and consultants (47%). Most of respondents (>54%) have more than ten years of experience. Collections of delays identified were classified based on when they might occur, i.e. pre-construction and construction phases, and only delays in pre-construction phase were then used as indicators for the instrument. These delays were later attributed to owner, consultant, project characteristics, and external factors. The respondents were asked to validate the indicators obtained from literature review, and estimate the weight of each indicator based on the impact to the project completion, and the state the degree of accuracy of each indicator. This degree of accuracy indicates the extent of accurateness of each indicator in predicting delay risks which can occur during construction phase.

5 Results

Table 2 shows the instrument of EWaSDRiP. A 1 to 10 Likert scale reflecting worst to excellent is used to score the state of each of the delay indicators within three zones, i.e. red (scale 1-3), yellow (4-6) and green (7-10) (see Table 1). Any potential delays with scores fall into red zone simply mean that they pose dangerous threats to the project and must be watched closely. Potential delays within yellow zone indicate moderate threats to the project, while the state of potential delays in green zone means that they are unlikely to delay the project. When the score of each delay is plotted on Table 1, it gives a visual indication of which zone most delays are distributed. To calculate the total impact of each delay to the project (T), the score (S) is multiplied by the weight (W) and accuracy (A). The cumulative total (CT) of the total impact (T) basically represents the prediction of the state of project completion, which can be classified into three categories, i.e. high possibility to be delayed (0<CT<4), moderate possibility to be delayed (4≤CT<7), low possibility to be delayed (7≤CT<10).

Table 2. The prototype of early warning system of delay risks for public projects (EWaSDRiP)

Indicator	Weight (W)	Accuracy (A)	Score (S)										Total (T) (T=W*A*S)
			Red Zone			Yellow Zone			Green Zone				
			worst →									excellent	
			1	2	3	4	5	6	7	8	9	10	
Owner													
Ratio of project supervisor and number of project supervised	0.066	0.92			●								0.182
Construction Project procurement on time	0.056	1			●								0.168
Budget of project agreed on time	0.052	1			●								0.156
Completeness of technical specification	0.052	1				●							0.208
Competence of head of project (PPK)	0.054	1								●			0.432
Consultant													
Designer Consultant													
Track record of consultant	0.032	0.92			●								0.088
Approval of Drawing	0.029	0.79				●							0.092
Design error	0.064	1			●								0.192
Completeness of design drawing	0.055	0.94				●							0.207
Supervision Consultant													
Quality of supervision	0.034	0.72			●								0.073
Project Characteristics													
Project scale / value of project	0.047	0.92					●						0.216
Level of complexity	0.071	1					●						0.355
Project duration	0.066	1			●								0.198
Contract is signed on time	0.049	0.83			●								0.122
Project location	0.057	0.92				●							0.210
Readiness of site	0.041	0.79				●							0.130
External factors													
Availability of certain materials in the market	0.069	1								●			0.552
Weather condition	0.068	1			●								0.204
New regulation related to project	0.029	0.7								●			0.162
	Cumulative Total (CT)												3.947

6 Simulating EWaSDRiP

To give a clearer picture of how EWaSDRiP works, a simulation has been done by assuming a hypothetical 4-storey building project to be observed (see Table 1). The project was assumed to be late in the procurement process, due to long process of budget agreement with the legislative. The contract was finally signed in late August, practically leaving less than four months for the contractor to finish the project. The design was regular, with materials locally available. The head of project was a civil servant with civil engineering background and 20 years of experience, however due to recent job rotation, he was lacking of staffs to supervise this project. The project was located at rather remote area. Since the project started in September, lots of rain was expected during the construction. The design drawing had many errors and was often incomplete. The supervision consultant had bad reputation for its quality supervision.

Based on the description, the project is then assessed using EWaSDRiP, resulting a cumulative total (CT) equals to 3.947, which falls into the red zone ($0<CT<4$). This means that this project has high possibility to be delayed, and must be monitored extensively. The distribution of black dots in the instrument representing scores of indicators, and visually helps the owner to recognise indicators which must be monitored closely to avoid delays. This simulation proves that EWaSDRiP does work and is useful for project owner to find delay risks early, and to predict the final status of the project completion, so that immediate actions can be taken.

7 Conclusions

This research has proposed a prototype of an early warning system of delay risks for public construction projects. This prototype is the first of its kind which is expected to give a significant contribution in tackling acute problems of delays in public projects in Indonesia. This instrument is beneficial for the owner or government client team to prepare strategic actions for the relevant pre-construction delays accordingly. Further research will be carried out by implementing it to public projects as case studies and to evaluate the applicability as well as the accuracy of the instrument.

References

1. J. U. D. Hatmoko and S. Scott, *Constr. Manag. Econ.* **28**,1 (2010)
2. J. U. D. Hatmoko and R. R. Khasania, *Civ. Eng. Dimens.* **18**,2 (2016)
3. J. K. Larsen, G. Q. Shen, S. M. Lindhard, and T. D. Brunoe, *J. Manag. Eng.* **32** (2015)
4. O. Bagaya and J. Song, *J. Manag. Eng.* **32** (2016)
5. T. Ophiyandri, D. Amaratunga, C. Pathirage, and K. Keraminiyage, *Int. J. Disaster Resil. Built Environ.* **4** (2013)
6. H. Samarghandi, S. M. M. Tabatabaei, P. Taabayan, A. M. Hashemi, and K. Willoughby, *J. Constr. Dev. Ctries.* **21**,1 (2016)
7. W. He, Y. Duan, L. Deng, and W. Zhou, *J. Perform. Constr. Facil.* **30**,1 (2016)
8. L. Y. Ding *et al.*, *Autom. Constr.* **36** (2013)
9. Z. Yuan, Y. Wang, and C. Sun, *J. Intell. Fuzzy Syst.* **32**, 1 (2017)
10. L. T. Van, N. M. Sang, and N. T. Viet, *ARPN J. Sci. Technol.* **5**, 2 (2015)

11. P. A. Koushki, K. Al-Rashid, and N. Kartam, *Constr. Manag. Econ.* **23**, 3 (2005).
12. C. Kaliba, M. Muya, and K. Mumba, "Cost escalation and schedule delays in road construction projects in Zambia," *Int. J. Proj. Manag.* **27**, 5 (2009).
13. A. A. Aibinu and H. A. Odeyinka, **132**, 7 (2006)
14. Y. A. Messah, T. Widodo, and M. L. Adoe, *J. Tek. Sipil* **2**, 2 (2013)
15. R. F. Aziz and A. A. Abdel-Hakam, *Alexandria Eng. J.* **55**, 2 (2016)
16. J. Mccord, M. Mccord, P. T. Davis, M. Haran, and W. J. Rodgers, *J. Financ. Manag. Prop. Constr. Iss Int. J. Manag. Proj. Bus.* **20**,1 (2015)
17. A. M. Odeh and H. T. Battaineh, *Int. J. Proj. Manag.* **20**, 1 (2001)
18. Y. Frimpong, J. Oluwoye, and L. Crawford, *Int. J. Proj. Manag.* **21**, 5 (2003)
19. A. Adam, P.-E. B. Josephson, and G. Lindahl, *Eng. Constr. Archit. Manag.* **24**, 3 (2017)
20. B. Mpofu, E. G. Ochieng, C. Moobela, and A. Pretorius, *Eng. Constr. Archit. Manag.* **24**, 2 (2017)
21. M. Sambasivan and Y. W. Soon, *Int. J. Proj. Manag.* **25**, 5 (2007)
22. S. A. H. Tumi, A. Omran, and A. H. K. Pakir, *The International Conference on Economics and Administration*, 265–272 (2009)
23. J. U. D. Hatmoko and R. R. Khasani, *Applied Mechanics and Materials*, **845** (2016)
24. a Mert, M. Erd, and Y. Fahjan, *Bull. Earthq. Eng.* **1**, 272 (2004)
25. Z. Yuan, Y. Wang, and C. Sun, *J. Intell. Fuzzy Syst.* **32**, 1 (2017)
26. W. He, Y. Duan, L. Deng, and W. Zhou, *J. Perform. Constr. Facil.* **30**, 1 (2016)
27. L. Y. Ding *et al.*, *Autom. Constr.* **36** (2013)
28. H. Samarghandi, S. M. M. Tabatabaei, P. Taabayan, A. M. Hashemi, and K. Willoughby, *J. Constr. Dev. Ctries.* **21** (2016)
29. H. Doloi, A. Sawhney, K. C. Iyer, and S. Rentala, *Int. J. Proj. Manag.* **30**, 4 (2012)
30. A. Kazaz, S. Ulubeyli, and N. A. Tuncbilekli, *J. Civ. Eng. Manag.* **18**, 3 (2012)
31. M. Sambasivan, T. J. Deepak, A. N. Salim, and V. Ponniah, *Eng. Constr. Archit. Manag.* **24**, 2 (2017)
32. A. Shebob, N. Dawood, R. K. Shah, and Q. Xu, *Iss Int. J. Manag. Proj. Bus.* **19**,1 (2012)

Development of institutional funding model of deep discount bond and land lease on a toll road project

Randika Dwirahman[1], *Yusuf* Latief[1], and *Ayomi* Dita Rarasati[1*]

[1]Civil Engineering Department, Faculty of Engineering, Universitas Indonesia, Depok, Indonesia

Abstract. Toll road is one of the most important parts of connectivity infrastructure. However, the current length of the existing toll roads is still not sufficient compared to the increasing number of vehicles in Indonesia. Nevertheless, the needs for infrastructure funding including for toll roads are not comparable to the availability of state budget. Innovations in funding schemes were required to meet such needs such as Deep Discount Bond (DDB) and Land Lease (LL). To ensure these schemes are running appropriately, an institutional funding model is needed to organize the stakeholders involved. Therefore, five institutional funding models were developed based on the conditions of the DDB issuer (project or corporate finance), LL executor [State Assets Management Institutions (LMAN) or Special Purpose Company (SPC)], and the number of SPC in the models. In addition, ten institutional success factors were identified. Data was collected using in-depth interviews consisting of three parts and analyzed using the Relative Importance Index to rank the institutional success factors found and using the Multi Criteria Analysis to choose the most effective institutional funding model. The institutional success factors that were found to have great values were those related to Government. At the same time, the most effective funding model is a corporate finance scheme, where LL is executed by LMAN, and a single SPC is used.

1 Introduction

The World Economic Forum in its 2016 Global Competitiveness Report revealed that the existence of efficient and growing infrastructure within a country can ensure economic activities in various sectors that can develop that country. One part of the infrastructure with a large portion of development plan is toll roads. Nevertheless, many countries are experiencing a serious gap between their needs and the availability of funding for that purpose [1]. There are alternative schemes that can be applied, i.e. DDB and LL. Bonds is one of the most popular infrastructure debt finance instruments in developed economies such as the UK and the USA [2]. Other cases showed that the USA leased lands to gain income to fund their transportation infrastructure [3]. DDB and LL are two financing schemes that have not been implemented in Indonesia. Therefore, each stakeholder involved has to be educated on their role to ensure that the schemes succeed, meaning that

* Corresponding author: ayomi@eng.ui.ac.id

guidelines are required [4]. The institutional model must be made based on the success factors to ensure DDB and LL schemes are running well. Therefore, this research aims to identify the role of the stakeholders for DDB and LL joint funding schemes and the institutional success factors, and to determine the most effective institutional model by involving private parties.

2 Literature review

2.1 Institutional

Institutional is a condition required to govern the relationship pattern between the stakeholders involved [5]. In other words, it is a human relationship pattern set with regulations and a combination of law, politics, and management [6]. Some said that it is an important factor for innovation [7]. Based on those definitions, it can be concluded that the institutional model will represent stakeholder relationships based on law, politics, and management in order to be able to innovate. Therefore, DDB and LL schemes as innovative funding should be based on institutional models as a guide for stakeholders to be involved in order to ensure the success of the schemes. Based on the literature review, ten institutional success factors were identified as the basis to develop the institutional funding models, i.e.: Effectiveness of Government (X1) [8]; Ease to Start a Business (X2) [8]; Protection of Investors (X3) [8]; Stakeholder Dialogue (X4) [9]; Sustainability of Financial Condition (X5) [10]; Complexity of Project Financing (X6) [11]; Adequate Financial Package (X7) [12]; Reliable Consortium (X8) [12]; Local Financing (X9) [13]; Effective Management (X10) [14].

2.2 Deep Discount Bond

Bond is fixed income securities issued in relation to debt agreements [15]. Based on this definition, bonds are able to provide fixed income periodically via bond coupon or bond interest rate. The four essential components of a bond are coupon, maturity date, bond liquidity, and bond price [16]. DDB was first issued in America in early 1980s. DDB has only one cash in its cash flow, i.e. at the time of maturity. DDB is said to be typical with the Zero Coupon Bond, defined as cheap bond price listed by the issuer of the discount [17]. The advantage for the bond holder is the difference between the purchased price when the bonds are issued and the price of bond par value on its maturity date.

2.3 Land Lease

The fundamental definition of land lease is leasing land for business affairs to obtain an income. Land lease income is obtained from businesses such as properties or utilities [18]. Land lease deployment would involve a complicated process because there is a lag time between the leased land and the development of infrastructure [19]. An incomplete infrastructure will inhibit the market from rent the land. This condition will encourage land owners to hold their land until the infrastructure is complete so that the land value increases in line with their earnings [20]. In this research, the land leased is beside the toll roads. The land is acquired not only for the use of toll road itself, but also to be leased after the toll road is completed.

2.4 Alternatives of institutional funding model

Five alternatives of institutional funding model of DDB and LL for the toll road project were developed, from which the most effective institutional model will be selected. Afterwards, it will be compared with the existing condition of toll roads in order to identify the gap. The main stakeholders in this funding model are sponsor company; SPC, a subsidiary of sponsor company; and LMAN, a government institution on land stewardship. Each alternative institutional model is developed based on three conditions: issuer of DDB (SPC or sponsor company); executor of LL (SPC or LMAN); and the number of SPC (one or two). If the issuer of DDB is SPC, then the model will become project finance. However, if the issuer is a sponsor company itself, then it becomes corporate finance. Specifically for the third alternative, it is a model that consists of two SPC. The first SPC has a duty to fund projects by issuing DDB while the second SPC has a duty to execute LL and build the toll roads. The conditions of each model are listed in Table 1.

Table 1. Institutional Model Alternatives

Condition	**Alternatives**				
	1st	**2nd**	**3rd**	**4th**	**5th**
Number of SPC					
One SPC (*) or Two SPC (**)	*	*	**	*	*
DDB Issuer					
Project Finance (*) or Corporate Finance (**)	*	*	*	**	**
LL Executor					
SPC (*) or LMAN (**)	**	*	*	**	*

3 Research Method

Each variable will represent each of the research objectives. This research uses three stage in-depth interviews to answer these three research objectives. The terms of reference will be a guide during each stage of the in-depth interview and contain questions to experts according to the context of each research objective. The first in-depth interview is aimed at learning the roles stakeholders involved in DDB and LL schemes and analyzing them using the qualitative analysis. The second in-depth interview is aimed at identifying the institutional success factors and ranking them using the Relative Importance Index (RII) analysis. The results of this RII rank will be used as the basis for the Multi Criteria Analysis (MCA) on the results of the third in-depth interview to select the most effective alternative institutional model.

The experts invited for the in-depth interview are ten professionals with experience in infrastructure, especially toll roads. These experts are required to have a minimum of 10 years of experience and master degrees, and they come from diverse backgrounds including academia, government, toll road companies, infrastructure company guarantor, and financing company for infrastructure. Each institutional success factor will be assessed by experts in the second in-depth interview based on their importance. Ranks are calculated by using the following formula.

$$RII = \frac{5.n5 + 4.n4 + 3.n3 + 2.n2 + 1.n1}{5N} \quad (1)$$

The symbols n5, n4, n3, n2, and n1 represent the following: number of respondents answering very important, important, moderately important, slightly important, and not important, respectively, while 'N' notation is the total number of respondents.

The institutional success factors based on the RII will be converted into weights at the MCA. Higher RII ranks mean greater weights on the MCA. Each alternative institutional model will then be assessed based on these institutional success factors. In turn, each institutional success factor will be valued in its pages with institutional scale 1-10 on any alternative models. The average results of the expert's answers will be multiplied by the weighting of MCA which is then aggregated into a final value on each model. The institutional model with the highest content of success factors based on the MCA or MCA's highest-value model will be the model selected.

4 Results

4.1 Rank of institutional success factors

Based on the results of RII calculation, the factor that gets the first rank is 'Effective Management' with a score of 0.9143. The second rank is occupied by 'Effectiveness of Government' with a score of 0.8857. The third rank is occupied by 'Protection of Investors' with a score of 0.8571. The three lowest factors are by 'Adequate Financial Package', 'Complexity of Project Financing', and 'Local Financing' with a RII score of 0.6857; 0.6000; and 0.5714 respectively.

Table 2. Rank and Weights of Institutional Success Factors

Factors	RII Value	Rank	Weight	Factors	RII Value	Rank	Weight
X1	0.8857	2	16%	**X6**	0.6000	9	4%
X2	0.7143	7	7%	**X7**	0.6857	8	5%
X3	0.8571	3	15%	**X8**	0.7714	5	11%
X4	0.8000	4	13%	**X9**	0.5714	10	2%
X5	0.7429	6	9%	**X10**	0.9143	1	18%

4.2 Scoring of institutional funding model of DDB and LL

The scoring of the institutional funding model using MCA based on their weighting results from the RII calculation is shown in Table 2. From these results, the fourth alternative is a model with the highest score of 7.0479. This model uses the corporate finance scheme, a single SPC and LMAN as LL executor. The institutional model with the second highest score is the fifth alternative with a score of 6.8993. We can see that models with corporate finance scheme get higher score than models with project finance. It means that project finance is still not suitable to be implemented in Indonesia. The third highest score is obtained by the second alternative with a score of 6.6571. These results indicate that LMAN is better than SPC in executing LL scheme. The institutional model with the lowest score is the third alternative with a score of 6.4536. This indicates that the use of two SPC is not effective to be implemented.

5 Discussions

5.1 Stakeholder analysis for DDB and LL schemes

Issuing DDB with corporate finance scheme is needed to get the guarantee of bonds from the government. The guarantee is an advantage as it will increase DDB rating as rated by the Credit Rating Agency before it is marketed in Indonesia Stock Exchange through securities companies. The research results show that LMAN is a government agency stakeholder best to execute LL scheme. Not only tasked to manage the land assets to be leased in LL scheme, LMAN also has another role in accordance with the objectives of this institution establishment. On toll road development, LMAN can reimburse bailouts issued by SPC to accelerate the process of acquiring land. According to Indonesian regulations, land acquisition is the government's responsibility (National Land Agency) using State Budget which requires a lengthy procedural bureaucracy before those funds can be used. Therefore, the SPC can fund the land acquisition before they are reimbursed by LMAN as a state instrument. Another LMAN role in supporting LL execution is that this institution can deposit funds without being influenced by State Budget periodization. Funds in LMAN should not be returned to the state at the end of December as per State Budget deadline. This role can be very helpful for stakeholders related to the development of toll roads, considering that the construction of toll roads is a project that exceeds one year in its implementation.

5.2 Institutional success factors

Three institutional success factors with the highest value, namely 'Effective Management', 'Effectiveness of Government', and 'Protection of Investors', are factors that focus on the government. The results are in line with the concept that infrastructure development in Indonesia is the responsibility of the government. Therefore, the government is the most important stakeholder in infrastructure development especially toll roads.

5.3 Most effective institutional funding model

Based on MCA results, the institutional funding model selected as the most effective model is the 4th alternative using corporate finance for issuing DDB, LMAN as the LL executor, and consisting of a single SPC. Corporate finance means that the project is funded by a sponsor company. The use of corporate finance schemes does not eliminate the presence of SPC on the project, which becomes important to maintain the project from a technical point of view. On the other side, project finance scheme is still difficult to apply in Indonesia, making it less commonly used because of the mismatch of regulations as one of the reasons. Based on the Financial Service Authority regulations, DDB issuance needs an enterprise financial reporting terms during the last three years. However, SPC are on average new companies and therefore would not be able to meet this requirement.

Only one of the five alternatives consists of two SPC. Based on MCA results, this alternative gets the lowest score. Higher score of corporate finance than project finance makes the use of two SPC no longer effective. In addition, higher score of LMAN than SPC as LL executor reduces SPC's responsibilities. Based on these two reasons, the presence of one SPC is enough to manage the project though it has a large scale. It is worth noting that despite the positive sides of having two SPC in terms of management and coordination, their ineffectiveness is greater than their benefits.

According to Article 11 (1) of Indonesian Law No. 2 of 2012, land acquisition for public interest must be held by the government and will be owned by government itself. On the other hand, paragraph (2) stated that State-owned Enterprises are able to own the land acquired by the government. Based on this law, SPC is not allowed to acquire land for public interest. Basic regulations for executing LL are the Regulations issued by the Head of National Land Authority No. 9 of 1999 and Minister of Finance Regulation No. 68 of 2014. Based on the first regulation, state owned lands are able to give their ownership and few of its authority to government banks, religious affairs agencies, or social agencies selected by the government. In addition, according to the Minister of Finance regulations, state-owned property, which one of them is lands, can be used by the government to gain revenue by leasing them to other parties. Therefore, by taking this research's context, the lands beside toll roads can be used to gain revenue from leaser's mortgage payments. Because SPC cannot meet the required regulations, LMAN gets higher score in executing LL. In reverse, LMAN has a good position because of its status as part of the government and enables it to own lands. In addition, LMAN is an agency under the authority of the Directorate General of State Assets Management in Ministry of Finance. State-owned properties are also managed by the same directorate general. Therefore, this will give an advantage to LMAN in executing LL.

6 Conclusions

Every stakeholder has its own roles in DDB and LL scheme. The roles become the basic for developing the institutional funding model to ensure the DDB and LL schemes are running as expected. The institutional success factors were identified as parameters to score the five alternative models. Factors with the highest score are those related to the government, which shows that the government is the key stakeholder to develop infrastructure especially toll roads. The most effective institutional funding model is a model using corporate finance for issuing DDB, LMAN as the LL executor, and consisting of a single SPC.

The authors would like to express gratitude to the University of Indonesia for granting its support through PITTA scheme No. 742/UN2.R3.1/HKP.05.00/2017 for the completion of this research.

References

1. A.A. Maw, F. Nakamura, T. Okamura, *Eastern Asia Society for Transportation Studies*, **6**, 63 (2007)
2. P.T.I. Lam, Y.H. Chiang, S.H. Chan, *J. Manage. Eng.,* **27**, 4 (2011)
3. G.T. Johnson, L.A. Hoel, *J. Transp. Eng.*, **113**, 1 (1987)
4. S. Syawal, *Dismissal of Jakarta Monorail Project.* (UIN Syarif Hidayatullah, Jakarta, 2011)
5. H.P. Tserng, J.S. Russell, C.W. Hsu, C. Lin, *J. Constr. Eng. Manage.*, **138**, 2 (2012)
6. N.S. Grigg, *J. Manage. Eng.*, **21**, 4 (2005)
7. M.P. Hekkert, P. Kukk, E. H. M. Moors, *Res. Policy*, **45**, 8 (2016)
8. J.S.L. Lam, P.M. Panayides, F. Parola, *Transp. Res.*, **71**, 110-127 (2015)
9. L.M. Ferri, M. Pedrini, V. Pilato, *J. Clean Prod.*, **136**, 226-236 (2016)
10. R. Agrawal, A. Gupta, M. C. Gupta, *Manage. Res. Review*, **36**, 11 (2013)
11. B. Ashuri, K. Moostan, *J. Manage. Eng.*, **31**, 6 (2015)
12. X. Zhang, *J. Constr. Eng. Manage.*, **131**, 10 (2005)
13. X. Zhang, *J. Constr. Eng. Manage.*, **131**, 1 (2005)
14. P.R. Davis, J. Liu, P.E.D. Love, M. Regan, J. Smith, *J. Manage. Eng.*, **31**, 5 (2015)
15. M. Regan, P.E.D. Love, J. Smith, *J. Infrastruct. Syst.*, **19**, 3 (2013)
16. A. Berndt, S. Yeltekin, *J. Monetary. Econ.*, **73**, 119-136 (2015)
17. F.K. Reilly, K.C. Brown, *Investment Analysis & Portfolio Management* (South-Western Cengage Learning, Mason, 2012)
18. A.D. Rarasati, L. Sihombing, A. Wibowo, Y. Latief, *Integrated Solutions for Infrastructure Development* (ISEC Society, Sarawak, 2016)
19. E.C. Hui, *Prop. Manage.*, **22**, 2 (2004)
20. R.Y.C. Tse, *Urban Studies,* **35**, 8 (1998)

The Comparison of Bond Strength between Geopolymer Concrete and OPC Concrete for Plain Reinforcing Bars

Nuroji[1*] , *Daniel* Herdian Primadyas[2] , *Ilham* Nurhuda[1] , and *Muslikh*[3]

[1]Dept. Of Civil Engineering, Diponegoro University, Indonesia
[2]Master Program in Civil Engineering, Diponegoro University, Indonesia
[3]Dept. Of Civil Engineering, Gajah Mada University, Indonesia

Abstract. This paper describes the research on bond behavior of plain reinforcing bars in geopolymer and normal concrete. The geopolymer concrete in this research was made of class F fly ash taken from Tanjung Jati Electric Steam Power Plant (PLTU) with Sodium Hydroxide (NaOH) and Sodium Silicate (Na_2SiO_3) as alkaline activator, added in the mixture. The effect of bar size was studied by varying the bar diameter in range 10 mm to 19 mm. Each bar was casted in the center of concrete blocks made of geopolymer as well as normal concrete. Pull-out tests were carried out to the specimens that have reached 28 days of age. The test results show that the bond behavior of geopolymer concrete differs substantially from normal concrete, where geopolymer concrete has a higher bond strength when compared to normal concrete with identical concrete strengths.

1 Introduction

Current data shows that Portland cement concrete making material produces about seven percent of carbon dioxide (CO_2) [1]. The concentration of CO_2 in the atmosphere can dissolve in rain water to form carbonic acid (H_2CO_3) and falls to earth as acid rain. Hence, to reduce the production of CO_2 and to make environment friendly concrete, it is necessary to look for alternate materials to replace part, or all of Portland cement in concrete. In the past decade, research is focused on the study on geopolymer concrete to develop a more environmentally friendly concrete material. Geopolymer concrete is a type of alumina-silicate product and shows good bonding properties. In a fly ash-based geopolymer binder, fly ash reacts with the alkaline solution to create an alumina-silicate binder; no cement is used [2]. Numerous research showed that geopolymer can act as a binding agent in concrete [3,4]. Fly ash is one of the materials which is used to produce geopolymer concrete [5]. In the making process of geopolymer concrete, water and binding chemicals such as sodium hydroxide (NaOH) and sodium silicate (Na_2SiO_3) are required.

As a construction material, the use of concrete is often composed with steel reinforcement, where the reinforcing bar in concrete is to resist tensile stress that cannot be carried by concrete. However, to establish the compatibility action to enable force transfer from the concrete to reinforcing bars or vice versa, a good bond between the steel bars and

*Corresponding author : nuroji.undip@gmail.com

the concrete becomes mandatory. The bond between reinforcing bars and concrete has been acknowledged as a key to the proper performance of reinforced concrete structures [6].

Previous studies accessed the pull-out test to investigate the behavior of bond between reinforcing bar and geopolymer concrete with a variation in concrete compressive strength and for a range of deformed bar diameters [2]. Alternately, the study of Sarker [7] investigating the bond strength of geopolymer versus conventional concrete reported that the performance of pull-out tests on reinforced geopolymer concrete yielded in a conclusion that a higher bond strength for geopolymer concrete exist when compared to OPC concrete.

The research conducted by Kim and Park (2015) studied the bond using the pull-out tests to investigate the behavior of embedded reinforcements in geopolymer concrete for various compressive strengths and a variation in deformed reinforcement diameter [2]. This paper will elucidate the research result on the bond strength of plain reinforcing bars embedded in geopolymer concrete, compared and OPC.

2 EXPERIMENTAL STUDIES

2.1 Materials

This research employed fly ash from PLTU Tanjung Jati Jepara as a binder. The type of fly ash depends on the amount of silicon dioxide (SiO_2), aluminium oxide (Al_2O_3), and ferrite oxide (Fe_2O_3). The chemical composition of the fly ash used in this study is as shown in Table 1. It shows that the amount of SiO_2, Al_2O_3, and FeO in the fly ash make up to more than 83% of the total weight. Therefore, according to ASTM C618 [8], the fly ash can be classified as type F.

Sodium Silicate (Na_2SiO_3) and Sodium Hydroxide (NaOH) is used as alkaline activator. The ratio between Na_2SiO_3 and NaOH is 2:1.

Table 1. *Fly Ash* Composition (%)

Na2O	MgO	Al_2O_3	SiO_2	SO_3	K_2O	CaO	TiO_2	FeO	CuO
1.59	2.86	25.96	46.52	1.13	2.77	5.89	1.36	11.81	1.12

All the aggregates used as concrete-mix in this research were conditioned as saturated surface dry (SSD) before mixing to control the water content of the mix. The composition of the mixture of Geopolymer concrete can be seen in Table 2. The nominal reinforcing bars diameter were10 mm, 12 mm, 16 mm, and 19 mm, and three specimens were prepared for every diameter category. The steel bar specimens were tested in the laboratory to determine their effective diameter, yield strength and ultimate strength as shown in Table 3.

Table 2. Mix Proportion

Material	Mass (kg/m^3)
Coarse Aggregate	1008
Fine Aggregate	672
Fly Ash (Type F)	468

Material	Mass (kg/m^3)
NaOH (8M)	84
Na_2SiO_3	168

Table 3. Properties of Reinforcement Bar

Diameter	Effective Diameter (kg/m^3)	Yield Strength (MPa)	Ultimate Strength (MPa)
10	9.6-	305.09	468.27
12	11.80	311.44	481.09
16	15.23	334.77	505.72
19	18.75	341.03	538.879

2.2 Preparation of Specimens

In this experiment, pull-out tests were conducted to determine the bond strength behavior. Each bar was embedded in a 150 x 150 x 150 mm concrete cube with a bonded length of 7D, D being the nominal diameter of the embedded steel bar. To measure the compressive strength of concrete, concrete cylinders 100 x 200 mm in size, were casted. To achieve the desired development length of 7D, the contact surface between the concrete and reinforcing bar at the top and bottom ends were separated by using PVC pipe to create a frictionless area between the steel bar and the surrounding concrete (fig. 2). The geopolymer concrete specimens were kept moist for 7 days by using wet gunny-sack covers, and were further stored in room condition until 28 days. All specimens were tested at the age of 28 days.

2.3 Test Setup

The test was conducted in the Material and Construction Laboratory, Diponegoro University in Semarang. The incremental load was subjected to the specimen by a tensile force, while the base plate was restrained against movement as shown in figure 2.

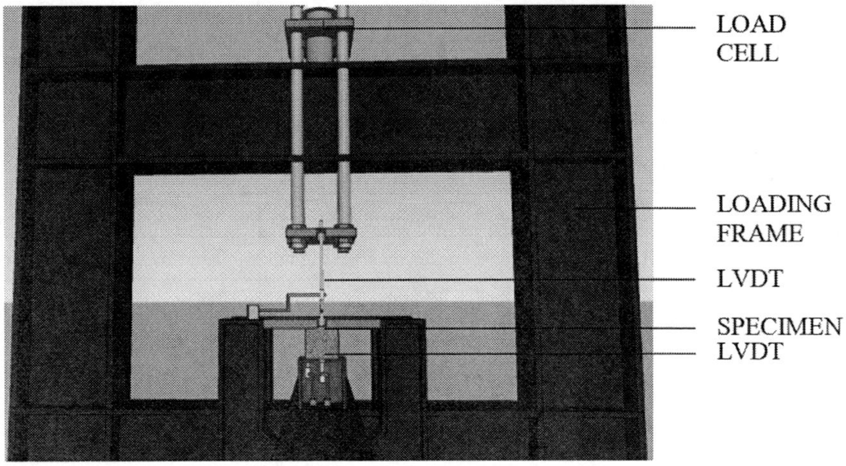

Fig. 1. Experimental Set Up

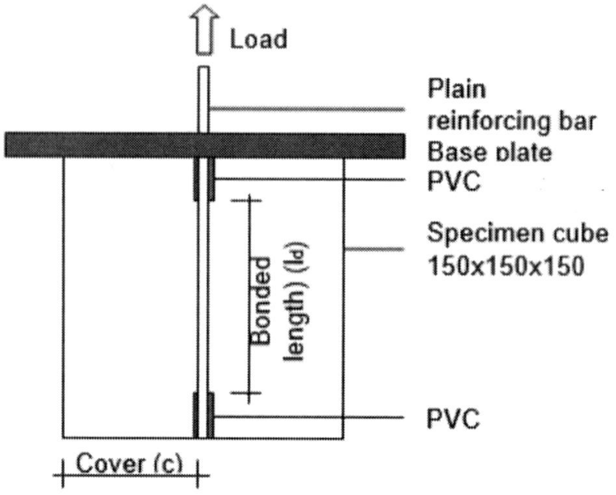

Fig. 2. Pull Out Test Scheme

3 RESULT AND DISCUSSION

The bond strength was derived from the maximum load recorded during the test, the average bond strength can then be calculated by dividing the maximum load by the area of contact surface between the reinforcement and the concrete as shown in equation 1.

$$\tau = \frac{F}{\pi \, d_b \, L_d} \quad (1)$$

Where F, d_b and L_b are force, diameter and bonded length of the bar, respectively. The test results and the calculation of average bond strength are given in **Table 4.**

Table 4. Bond Test Result

Specimen	f_c' (MPa)	Bar diameter (mm)	c/d_b	$L_d = 7$ d_b (mm)	Bond Area (mm^2)	Avg. Max. Load (kN)	Avg. Bond Strength (MPa)
GEO-10	32.45	9.60	7.31	67.20	2026.70	18.66	9.21
GEO-12	32.45	11.80	5.86	82.60	3062.05	27.11	8.86
GEO-16	32.45	15.23	4.42	106.61	5100.91	61.63	12.12
GEO-19	32.45	18.75	3.50	131.25	7731.26	77.30	10.00
OPC-10	30.56	9.60	7.31	67.20	2026.70	9.02	4.45
OPC-12	30.56	11.80	5.86	82.60	3062.05	11.64	3.80
OPC-16	30.56	15.23	4.42	106.61	5100.91	30.78	6.03
OPC-19	30.56	18.75	3.50	131.25	7731.26	35.86	4.64

It can be clearly seen in Table 4 that even though the compressive strength of geopolymer concrete is slightly higher than the OPC concrete, but the average bond strength of geopolymer concrete is twice higher compared to the bond strength of OPC concrete.

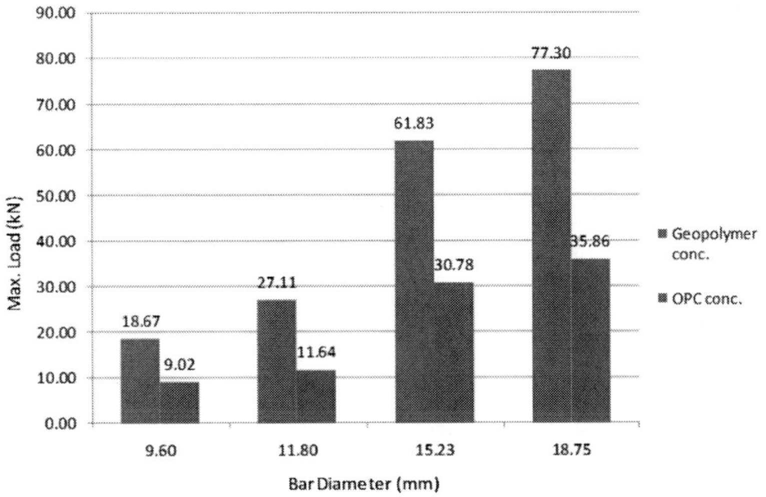

Fig. 3. Maximum load vs bar diameter of geopolymer and OPC concrete

Figure 3 shows the relationship between the maximum load and bar diameter for geopolymer concrete as well as OPC concrete. The relationship shows that the larger the bar diameter, the higher the maximum load to pull of the bars were. This phenomenon was observed for both the geopolymer concrete and OPC concrete. This behavior could be explained due to the fact that the large bar diameter has a larger contact area, when compared to the bars with a smaller diameter, the force transfer through the interface from the steel bar to concrete increases, with the increase of bar diameter. However, the relationship between the maximum load and bar diameter as can be seen in figure 3 approaches a linear pattern, although the interface area is not linear to the bar diameter, where in this research the interface area is determined to be πd_b (7 d_b). It can be concluded that the maximum load is not only affected by bar diameter, but also influenced by other parameters. The area of interface between steel bar and concrete and passive confinement of concrete play an important role in bond mechanism.

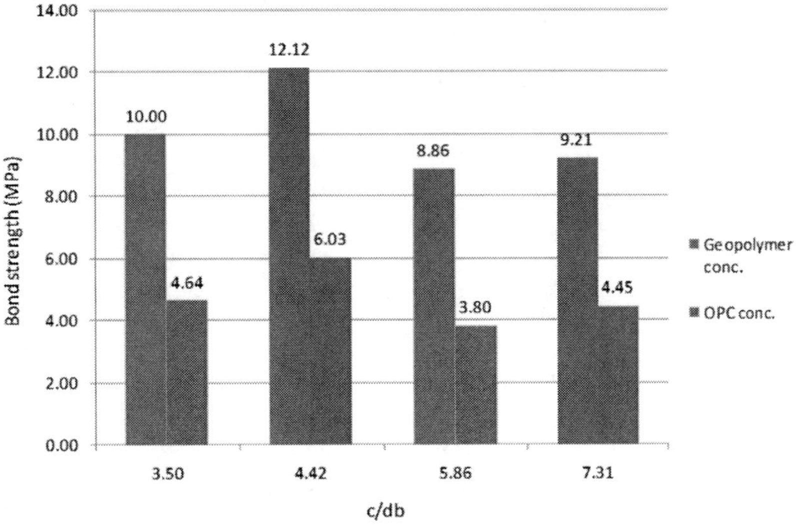

Fig. 4. Bond strength vs cover bar diameter ratio for geopolymer and OPC concrete

Figure 4 shows the relationship between the bond strength and concrete cover-to-bar-diameter ratio (c/d_b) for geopolymer concrete and OPC concrete. Where c being the length of the concrete cover, measured from the most out bar diameter to the boundaries of the concrete specimen, and d_b being the bar diameter. The bond strength of the geopolymer specimen with a c/d_b ratio of 7.31 is slightly higher than that for the specimens with the c/d_b ratio of 5.86. Further, the specimen with the c/d_b ratio of 4.42 has higher bond strength than that of the specimens with a ratio c/d_b equal 3.50. The phenomenon is almost the same with the OPC concrete specimen. The observation to the specimens show that the failure mechanism of all specimens is characterized by shear failure, no concrete splitting or radial cracks occur for both geopolymer concrete and OPC concrete.

4. CONCLUSIONS

The bond strength between plain steel bar for both geopolymer and OPC concrete was investigated in this research. The following conclusions can be drawn form the result:

1. The bond strength of geopolymer concrete is two times higher than the bond strength for OPC concrete, even though the both compressive strengths are relatively being almost identical.
2. The bond strength increase as a function of an increases in concrete cover-to-diameter ratio (c/d_b) from 3.50 to 4.42. But the higher the ratio, the lower the strength. The bond area and thickness of concrete cover are the major contributing factors to this behavior.
3. The increase in bonded length (l_d) from 67.2 mm to 106.61 mm enhances the bond strength. However, the increases in bonded length (l_d) from 106.61 mm to 131.25 mm yields in a lower bond strength, caused by the absence of passive confining provided by the concrete.
4. Failure mechanism of all specimens are characterized by shear failure, for both the geopolymer and OPC concrete.

References

1. K. Sathish, S. T. Blessen and C. Alex, "An Experimental Study of the Properties of Glass Fiber Reinforced Geopolymer Concrete," *International Journal of Engineering Research and Applications*. Vol. 2, no. 6, pp. 722-726 (2012).
2. Jee-Sang Kim and Jong Ho Park, "An Experimental Bond Properties of Reinforcements Embedded in Geopolymer Concrete." World Academy of Science, Engineering and Technology International Journal of Structural and Construction Engineering Vol:9, No:2, (2015)
3. D. Hardjito, "Properties of Geopolymer Concrete with Fly Ash as it's Source Material," *The 21th Biennial Conference of Concrete Institute,* (2005).
4. D. Hardjito and B.V. Rangan, "Development and Properties of Low Calcium Fly Ash Based Geopolymer Concrete," Curtin University of Technology, Perth, Australia, (2005).
5. J. Davidovits, *Geopolimer: Chemistry and Application*. (2008).
6. ACI Committee 408, *Bond and Development of Straight Reinforcing Bars in Tension (ACI 408R-03)*, American Concrete Institute, (2003).
7. P.K. Sarker,*Bond Strength of Reinforcing Steel Embedded in Fly Ash based Geopolymer Concrete*, Materials and Structures, Vol. 44, Issue 5, pp 1021–1030. (2011).
8. ASTM,*Standard Specification for Coal Fly Ash and Raw or Calcined Natural Pozzolan for Use in Concrete,* ASTM C618, West Conshohocken, PA, (2012).

Behavior of Geopolymer Concrete Confined by Circular Hoops

Muslikh[1,*], *N. K.* Anggraini[2], *D.* Hardjito[3], and *Antonius*[4]

[1]Dept. of Civil and Environmental Engineering, Gadjah Mada University, Yogyakarta, Indonesia
[2]Master Student of Civil Engineering, Diponegoro University, Semarang, Indonesia
[3]Professor of Civil Engineering, Petra Christian University, Surabaya, Indonesia
[4]Professor of Civil Engineering, Sultan Agung Islamic University, Semarang, Indonesia

Abstract. This paper discusses the behavior of geopolymer concrete subjected to passive confinement under compression loads. The confinement is induced by the use of lateral hoops, assembled from un-deformed reinforcing bars. To compare the effect of confinement, identical specimens were produced using conventional concrete with the similar concrete compressive strength. The cylinder specimens were 100 mm in diameter and 200 mm in height, and the hoops were placed on the outer most fibers of the cylinders, perpendicular to the line of loading, with no concrete cover. The parameters analyzed in this study were the steel bar to concrete volumetric ratio, the hoop spacings and the steel yield stresses. The experimental results show that unconfined geopolymer concrete were very brittle compared to the unconfined Portland cement concrete. The strength enhancement (*K* value) of the confined geopolymer concrete was higher than *K* value of Portland cement concrete. Confined geopolymer concrete also has better deformability compared to the confined Portland cement concrete. The average confinement effectiveness of geopolymer concrete also has a higher value than that commonly used in the Indonesian Concrete Standard (SNI), that is 4.1. The results were further assessed to the most recent experimental test results conducted in this area.

Keywords: geopolymer, confinement, hoops, stress-strain

1 Introduction

Inelastic deformability of reinforced concrete columns is essential for overall strength and stability of structures during a strong earthquake. The inelastic deformation of the structural column can be significantly achieved by confining the concrete core [1], or by increasing the matrix bonding of concrete materials such as fiber concrete [2,3]. On the other hand, the current development of concrete technology also leads and prioritized to the eco-friendly concrete. Geopolymer concrete is known as an environmentally friendly and energy efficient material. Ekaputri [4] reveals that geopolymer concrete is the best technology in overcoming the problem of national coal ash waste crisis in Indonesia. Compared to other types of concrete, geopolymer concrete produces the least amount of CO_2 emissions, advanced materials with many advantages (fireproof, rustproof and high strength). The process of making geopolymer concrete does not require much energy, as does the cement making process which requires at least temperatures up to 800°C. By heating ± 60°C for one full day can be produced high strength geopolymer concrete. Therefore, geopolymer concrete can reduce greenhouse gas emissions.

*Corresponding author: muslikh2007@gmail.com

Currently, research on geopolymer concrete is also evolving. Mechanical geopolymer concrete has excellent ability to withstand sulfate attacks, has relatively low creep and shrink properties, making it suitable for structural applications [5-8].

The stress-strain model of geopolymer concrete has also begun to be proposed, among them by Nguyen et al. [9] and Diaz-Loya [10]. However, Diaz-Loya revealed that geopolymer concrete is more brittle when compared with normal concrete.

Compared with steel fibrous concrete, the use of geopolymer concrete within the structure is still very limited. As a result, geopolymer concrete design equations are still not widely developed. Several geopolymer concrete research on structures are still performed and developed, such as column structure, and precast structures [11-13]. The triaxial behavior of geopolymer concrete was also evaluated by Giasuddin et al. [14], where based on the test results, indications that the geopolymer concrete brittle behavior changes to ductile when active confined are applied. Another way to increase the ductility of geopolymer concrete is by installing CFRP confinements [15]. The results of the investigation Ganesan et al. [16] concluded that an increase in concrete strength of confined geopolymer concrete is known to be not significantly different from that of confined normal concrete.

In this study, an evaluation of the behavior of strength and ductility of geopolymer concrete confined by a round section hoops has been carried out. Experimental tests were conducted with the primary objective of evaluating the collapsed behavior of confined geopolymer concrete by reviewing the design parameters of reinforcement bars such as the volumetric ratio (ρ_h) and spacing (s).

2 Experimental Program

2.1 Materials

Concrete materials are designed into two types, i.e. geopolymer concrete as the main material (code: GC) and normal concrete (code: NC) as comparison. The confined concrete specimen is controlled by the unconfined concrete, whereas for the geopolymer concrete, the UGC code is given. The normal concrete is coded as UPC. The coarse aggregates have a maximum diameter size of 19 mm and are designated as local sand from the area in Central Java. The composition of the concrete mixture is shown in Table 1. The plain confining steel reinforcement has a diameter (d) of 5.5 and 6 mm, with a yield stresses (f_y) of 514 and 466 MPa respectively.

Table 1. Composition of Concrete Mix Design

Material	Geopolymer Concrete (kg/m^3)	Normal Concrete (kg/m^3)
Coarse Aggregates	1209.6	1209.6
Fine Aggregates	806.4	806.4
Fly Ash	561.6	-
Cement		445.6
Natrium Hydroxide Fluid (8 Molar)	100.8	-
Natrium Silicate Fluid	201.6	-

2.2 Specimens and instrumentations

The specimen is a concrete cylinder 100 mm in diameter and 200 mm in height. In the test area two LVDT (linear vertical displacement transducers) mounted in the vertical direction are used to measure the deformation of the specimen in the direction of the load. The FLA-6-11 type strain gages are installed on the confinement reinforcement. The cables from both LVDT and strain gages are then connected to the data logger.

2.3 Data Execution and Data Analysis of the Experimental Results

The concentrically loading is conducted on the specimens, with a constant increment, and monotonic mode. The compression testing machine has an effective capacity of 1800 kN and was deformation controlled. The analysis of stress and strain on confined concrete were based on the calculations proposed by Antonius [17]. The increased strength of confined concrete (K) is defined as the ratio between confined concrete stresses at the maximum response (f'_{cc}) to the unconfined concrete stress (f'_{co}).

3 Results and Discussion

The value of stresses, strains and increased strengths of confined concrete ($K = f'_{cc} / f'_{co}$) of all specimens are presented in Table 2.

3.1 Failure modes and geopolymer effects

Fig 1 shows the failure modes of both confined normal concrete and confined geopolymer concrete. In general, the collapse behavior of the two types of concrete does not show a significant difference. The confinement caused a non-brittle failure mode, the direction of cracks running vertical (columnar) and the crack propagation is observed to be distributed evenly around the surface of the specimen.

Table 2. Values of experimental result

Specimen	fa_{ct} (MPa)	Hoops			f'_{co} (MPa)	f'_{cc} (MPa)	ε'_{cc}	ε_{c85}	$K = \frac{f'cc}{f'co}$	
		d-s (mm)	ρ_h	f_y (MPa)					K	$K_{average}$
UGC		-	-	-		36.28	0.0026	-	-	-
GC1						62.76	0.0059	0.0234	1.73	
GC2		5.5-60	0.017	514		62.76	0.0059	0.0227	1.73	1.73
GC3						63.33	0.0032	0.0173	1.74	
GC4						66.50	0.0032	0.0163	1.83	
GC5		6-60	0.020	466		62.21	0.0026	0.0183	1.71	1.73
GC6	32				36.3	59.58	0.0048	0.0197	1.64	
GC7						48.20	0.0030	0.0078	1.33	
GC8		5.5-100	0.010	514		46.66	0.0040	0.0069	1.29	1.32
GC9						48.39	0.0044	0.0089	1.33	
GC10						46.57	0.0048	0.0079	1.28	
GC11		6-100	0.012	466		48.55	0.0036	0.0070	1.34	1.33
GC12						49.26	0.0043	0.0077	1.36	
UPC		-	-	-		21.69	0.0025	-	-	-
NC1	20				21.7	35.94	0.0077	0.0159	1.66	
NC2		5.5-60	0.017	514		37.87	0.0076	0.0136	1.57	1.63
NC3						35.86	0.0074	0.0173	1.65	

Confined geopolymer concrete Confined normal concrete

Fig 1. Failure modes

Following the visual observations to the failure behavior, a comparison of stress-strain behavior between the unconfined geopolymer concrete and unconfined normal concrete is conducted. The method chosen in this study is by normalizing the compressive strength of each type of concrete to the standard cylindrical concrete compressive strength (f'_c). Fig 2 shows the data assessment. It can be seen that the curves prior to reaching the peak, coincides. The initial stages even show an exact behavior between the confined and unconfined specimens. For the post-peak behavior, it is seen that the unconfined geopolymer concrete is more brittle, it can be concluded that the existence of reinforcement becomes very important in geopolymer concrete.

The data in Table 2 underlines the above-mentioned statements, since the unconfined concrete peak strain comparison between the geopolymer concrete and normal concrete are almost identical, with values of 0.0026 and 0.0025 respectively.

The stress-strain behavior evaluation between the confined geopolymer and normal concrete is shown in Fig 3. From this figure, it is demonstrated that the confined concrete peak stress (f'_{cc}) of each specimen was normalized to the unconfined concrete peak stress (f'_{co}). The normalization also automatically reveals the increase in strength of confined concrete (K). Comparisons were made between specimens GC1 versus NC1, GC2 versus NC2 and GC3 vs NC3. Based on the data in Table 2, the values of K and deformability of the confined geopolymer concrete are higher than the values obtained from the confined normal concrete.

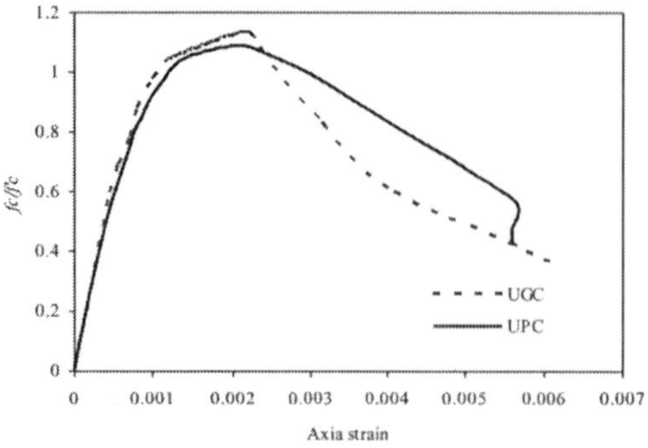

Fig 2. Comparison of unconfined concrete

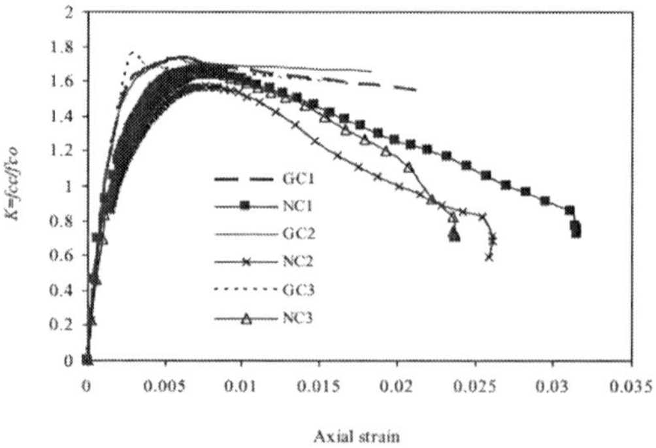

Fig 3. Comparison of confined concrete

3.2 Effect of parameter change on the confined concrete design

Basically, changing the confinement reinforcement spacing will have implications on the shifting of the confinement reinforcement volumetric ratio values. As with confined normal concrete, closer spacing of confinement reinforcement, higher volumetric ratios and confinement reinforcement yield stress, will yield in the increase of the *K* value (Table 2), while the behavior of the post-peak for confined concrete will exhibit a more pronounced slope gradient change and a more ductile specimen behavior.

Fig 4, 5, 6 and 7 illustrate the effect of confinement reinforcement parameters. The confined geopolymer concrete behavior demonstrates the utility of highly effective confinement and possesses analogous behavior properties to the confined normal concrete. This behavior is very advantageous if the confinement utilization of geopolymer concrete is applied to earthquake resistant concrete column structures.

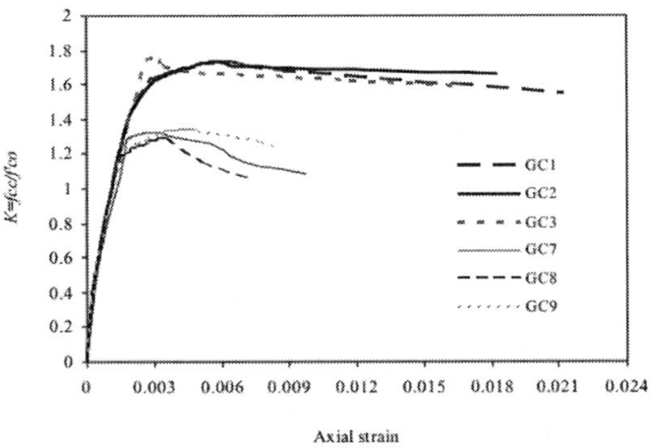

Fig 4. Effect of tie spacing

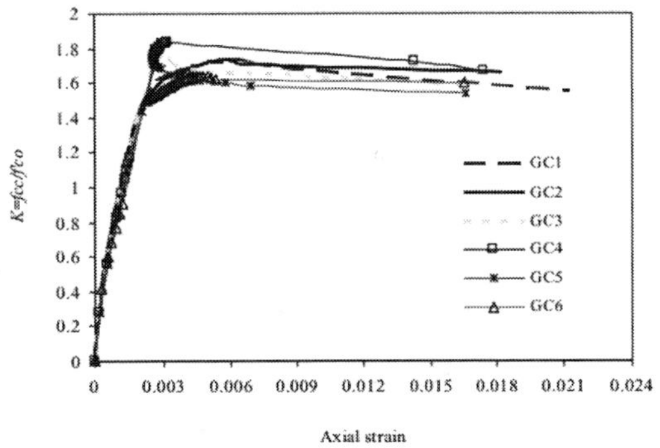

Fig 5. Effect of tie spacing

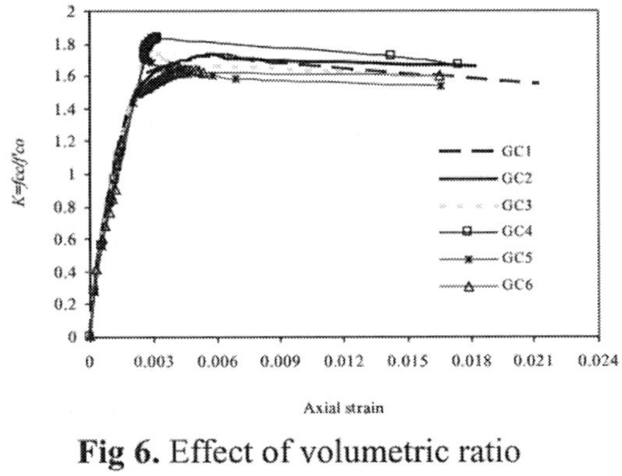

Fig 6. Effect of volumetric ratio

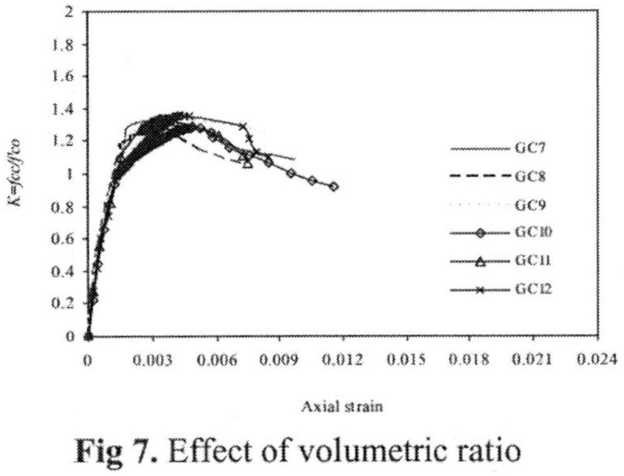

Fig 7. Effect of volumetric ratio

3.3 Effectiveness of confinement

The effectiveness of the confinement adopted by the current Indonesian National Standard (SNI 2847-2013) for structural concrete requirements [18], has a magnitude of 4.1. The value is basically based on the data of normal concrete test results [1]. Furthermore, the experimental results data in this study (Table 2) were processed to obtain the confinement effectivity value (k) of each specimen, using equation (1) below [1]:

$$k = \frac{f'_{cc} - f'_{co}}{f_2} \qquad (1)$$

In the above equation, the peak confined concrete stress (f'_{cc}) and unconfined concrete stress (f'_{co}) are taken from Table 2, and the lateral stress f_2 is calculated based on the equation below [16]:

$$f_2 = \frac{2.A_s.f_y}{s.d_c} \tag{2}$$

Fig 8 shows the calculated results of the confinement effectiveness for each specimen. The data are also compared to the confinement effectiveness value adopted in SNI (k = 4.1). Based on the figures, except for the GC8 and GC10 specimens, the *k* values of all specimens are higher than the confinement effectiveness based on SNI. These results show that the circular cross-sectional hoop is very effective in increasing the strength of confined concrete. Furthermore, the confinement effectiveness in confined geopolymer concrete is higher than the confinement effectiveness for normal concrete.

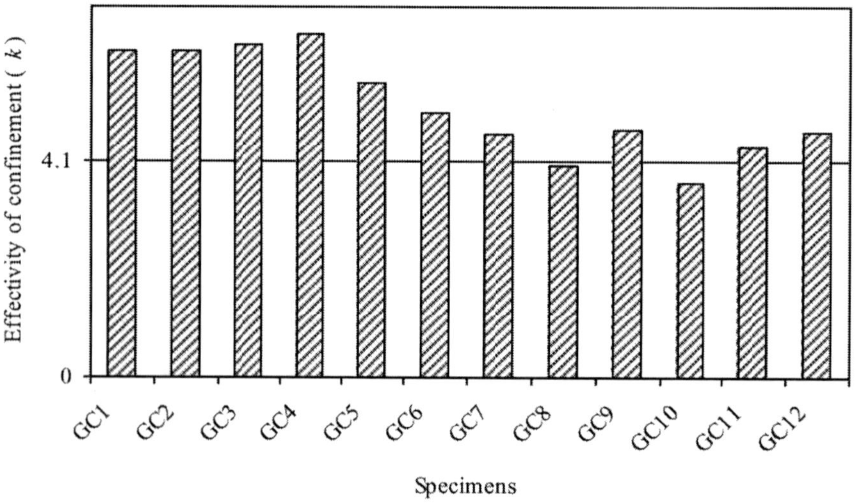

Fig 8. Effectivity of confinement

4 Conclusions

Geopolymer concrete has excellent mechanical properties and is environmentally advantageous because the material does not consume cement. The CO_2 that is produced as residue during the cement-making process can be minimized maximally. These conditions make geopolymer concrete as one of the environmentally friendly construction materials. However, geopolymer concrete is brittle compared to normal concrete using cement, so the installation of confinement reinforcement on geopolymer concrete becomes very important in order to increase ductility, especially for earthquake resistant structures. The value of the geopolymer concrete peak strain remains within the range of the values customary assumed in normal concrete. The effect of confinement reinforcement reduction, higher volumetric ratios and higher reinforcement yield stresses on confined geopolymer concrete, will result an increase in strength of the confined geopolymer concrete, and improves the material ductility. It is also concluded that this behavior is similar to the effect of changing the reinforcement parameters in the confined normal concrete. Further, it can be concluded that the confinement effectiveness for geopolymer concrete has a higher degree of effectiveness as compared to the confinement of normal concrete. This research has opened up new insights and shows a good prospect in the direction of structural element's research developments for geopolymer concrete, in the future.

Acknowledgment. The implementation of this research is a collaboration of researchers from Gadjah Mada University, Diponegoro University, Petra Christian University, Sepuluh Nopember Institute of Technology, Sultan Agung Islamic University. The support received for this research is gratefully acknowledged.

References

1. Antonius, Asian Journal of Civil Engineering, **15**, 245 (2014).
2. P. Paultre, R. Eid, Y. Langlois, Y. Lévesque, Journal Of Structural Engineering, **136**, 1225 (2010).
3. Antonius, A. Widhianto, D. Darmayadi, Gata D. Asfari, Asian Journal of Civil Engineering, **15**, 655 (2014).
4. J.J. Ekaputri, S. Junaedi, Wijaya, Procedia Engineering, **171**, 572 (2017).
5. A. Noushini, F. Aslani, A. Castel, R.I. Gilbert, B. Uy, S. Foster, Cement and Concrete Composites, **73**, 136 (2016).
6. D. Hardjito, S.E. Wallah, D.M.J. Sumajouw, B.V. Rangan, ACI Materials Journal, **101**, 467 (2004).
7. D. Hardjito, S.E. Wallah, D.M.J. Sumajouw, B.V. Rangan, Australian Journal of Structural Engineering, **6**, 1 (2005).
8. L.K. Turner, F.G. Collins, Construction and Building Materials, **43**, 125 (2013).
9. K.T. Nguyen, N. Ahn, T.A. Le, K. Lee, Construction and Building Materials, **106**, 65 (2016).
10. E.I. Diaz-Loya, E.N. Allouche, S. Vaidya, ACI Materials Journal, **108**, 300 (2011).
11. D.M.J. Sumajouw, D. Hardjito, S.E. Wallah, B.V. Rangan, Journal of Material Science, **42**, 3124 (2007).
12. P.K. Sarker, Materials and Structures, **42**, 715 (2009).
13. N.A. Lloyd, B.V. Rangan, *2nd International Conf. on Sustainable Construction Materials and Technologies*, Ancona, Italy (2010).
14. H.M. Giasuddin, J.G. Sanjayan, P.G. Ranjith, International Journal of Civil, Environmental, Structural and Architectural Engineering, **7**, 521 (2013).
15. W. Lokuge, W. Karunasena, Journal of Composite Materials, DOI:10.1177/0021998315597553, 1 (2015).
16. N. Ganesan, R. Abraham, S. Deepa Raj, D. Sasi, Construction and Building Materials, **73**, 326 (2014).
17. Antonius, International Journal of Technology, **6**, 217 (2015).
18. Indonesian National Standard, *Requirements of Structural Concrete for Building*, SNI-2847-2013 (in Indonesian).

MATEC Web of Conferences **159**, 01019 (2018) https://doi.org/10.1051/matecconf/201815901019
IJCAET & ISAMPE 2017

Development of risk-based standardized WBS (Work Breakdown Structure) for quality planning of high rise building architectural works

Mirradewi Rianty[1,*], *Yusuf* Latief[1], and *Leni Sagita* Riantini[1]

[1]Project Management Program, Departement of Civil Engineering, Universitas Indonesia, Depok 16424, Indonesia

Abstract. Work Breakdown Structure (WBS) is a breakdown of project work into smaller components so it can be better managed. Although each project is unique, most buildings can be standardized in their activities to enable the provision of a stronger forecast for project management. Quality performance is also important to be controlled and risk consideration approach is now required for overall quality management process in the updated ISO 9001. Therefore, the development of risk-based WBS for quality planning is proposed. The objective of the study was to develop risk-based WBS for high-quality building architectural works. The research consisted of several stages with qualitative risk analysis method. The result indicates that standardized WBS consists of 5 primary levels and 2 complementary levels, with 14 dominant risk variables on quality performance, and recommended risk responses as the development.

1 Introduction

Rework intensity on architectural works is found to be the most often in construction industry [1], there are many cases on high-rise apartment building in Jakarta, Indonesia indicate bad quality of building leaks and cracks [2]. Bad quality performance relates with the poor reliability of the planning process, where the contractors are unable to manage the increasing complexity of the projects [3]. In the latest developments of international standard of quality management ISO 9001, the approaches of risk considerations are required for the whole process to improve results through proactive precautions [4].

The development of Work Breakdown Structure (WBS) is the first step in the planning process after defining project requirements. The definition of activity is the most critical in planning process, that a project manager should spend more effort to identify the project activity using the WBS [5]. WBS is a breakdown of project works into smaller components so it can be better managed and measured against the ultimate completion [6]. Although each project is unique, most building construction works can be standardized to enable the provision of basic activities to have robust estimates for project management [7]. Therefore,

* Corresponding author: mirradewi.rianty@gmail.com

the development of risk-based standardized WBS of high rise building architectural works for quality plan is proposed. The objectives of this research are to identify standardized WBS for high-rise building construction (case study of apartment building's architectural works), to identify risks from the standardized WBS that may cause quality incompliances, and to develop risk-based standardized WBS for architectural works

The scope of the study is limited to the following: (1) It is directed to practitioners of the construction project, (2) The research conducted on high-rise construction project in the last 5 years at Jabodetabek area Indonesia, and (3) The performance of the project referred in this study is the achievement of the quality performance of the construction work

2 Risk based-standardized WBS for quality planning

The approach followed by the project team to develop WBS revolves around the application of previous WBS with some changes, progressive breakdown of the work required for the project, and the development of WBS-based deliverables with focus on the basic function of final product [8]. Instead of developing WBS for each project, it is sometimes appropriate to develop a general WBS for typical project, then the necessary segment can be modified. It is suitable for organizations which perform similar projects [9]. Construction activities are highly predictable and there are consistent rules that governs the selection of activities, so it is feasible to develop standard activity [10].

A quality plan must determine all procedure and measurement (tests, inspections, etc.) to assess progress towards goal requirements [11]. WBS can be used as a planning approach to improve project performance, the practice was reported to reduce rework on quality performance—mostly associated with reduced change of scope on work, and increased control on site for industrial construction projects [3]. The WBS standardization framework is also recommended to support integrated planning and supervision [12].

WBS-based planning with risk considerations was also developed for software projects, which were found to increase modification flexibility so as to reduce risk of change in planning [13]. The WBS-based risk identification on real estate projects also indicates that each stage of the task is more intuitive, the risks are easier to identified and controlled [14].

3 Research methodology

The methodology in this research was based on the following: (1) To identify the standardized WBS, the data are collected from analysis of archives, benchmarks, and interview of experts with Delphi method; (2) The risk identification derived from the standardized WBS category and the variables are obtained from literature analysis, then questionnaire are distributed to respondents from construction background. From returned questionnaires, a qualitative risk analysis is conducted using PMBOK risk matrix to seek dominant risk variables; (3) To develop risk-based standardized WBS, the highest ranked risk variables are analysed for their preventive and corrective actions through pattern recognition, and also through RBSxWBS matrix, the mapping of risk variables (Risk Breakdown Structure) which has impact on WBS item. RBSxWBS matrix is using dominant risk variables for the RBS input, and Pareto method as an approach to select certain dominant architectural work for the WBS input. Result of the risk responses is used to develop the standardized WBS. There are 6 experts interviewed using questionnaire.

4 Findings

4.1 Standardized WBS

The result of standardized WBS consists of 5 primary levels and 2 complementary levels.

The primary levels are:

- WBS Level 1 Project Name: The highest level and representation of the entire project.
- WBS Level 2 Work Section: The decomposition of the construction elements of the project which as a whole is the entire high-rise building project.
- WBS Level 3 Physical Area: Experts suggest the work section of high-rise building is divided by physical area or location vertically to indicate storeys and also horizontally to indicate tower zone.
- WBS Level 4 Sub-Work Section: It is the further decomposition of architectural work, it is: Wall, Floor, Ceiling, Door & Window, Door & Window Hardware, Façade, Roof, Sanitair, and Others.
- WBS Level 5 Work Package: It is the lowest level of WBS components, it represents the degree to which the performance of each work package can be assigned to an individual or organization.

The complementary levels are:

- WBS Level 6 Activity: It is added to the WBS template to help the project manager identify the work to be done on the lower level of work package. Project managers can add as much of the activity as they need to understand the details of the work to be done in order to properly deliver the project.
- WBS Level 7 Resources: This level divide the resources needed to perform the activity, it is the materials, equipment, and labor.

The activity level is highly dependent on the alternative of design or method. The alternative must be set first in order to breakdown the activity up to the resources, thus it is not a hierarchy within the WBS structure, but as a determinant to detail the activity.

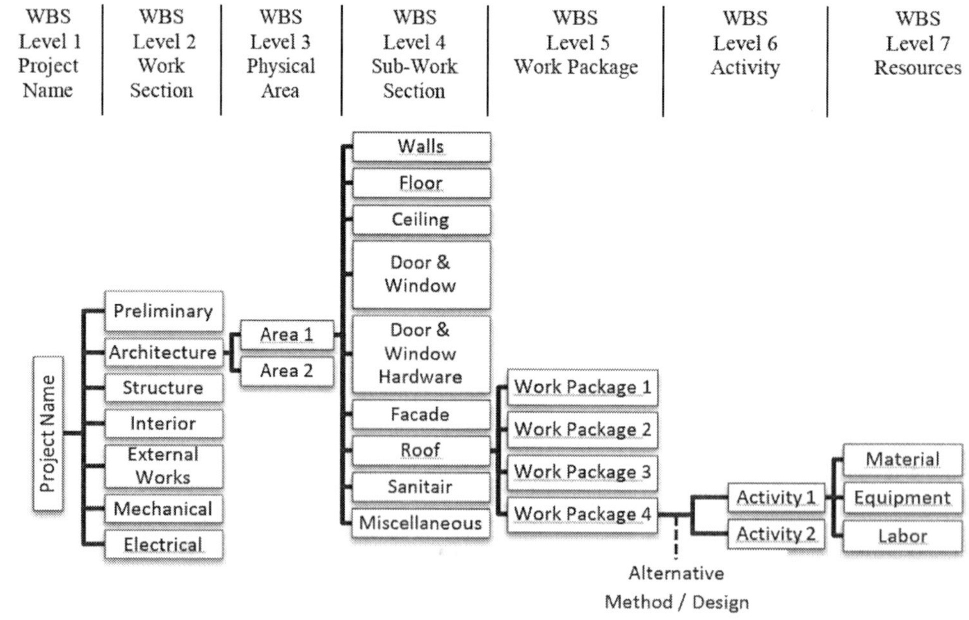

Fig. 1. Diagram tree of architectural works standardized WBS

4.2 Dominant risks on quality performance

To identify the risk variable, the standardized WBS framework was used as risk category towards quality performance from the last primary level of 'Level 5: Work Package', then 'Alternative Method/Design', 'Level 6: Activity', and 'Level 7: Resources'. There are 36 risk variables that are narrowed to 14 dominant risk variables after qualitative risk analysis and the highest risk score obtained from the category of WBS Level 7 Labor Resources.

Table 1. Dominant risk variables on quality performance

RISK VARIABLE		SCORE	RANK
RISK CATEGORY: WBS LEVEL 5 WORK PACKAGE			
X3	Subcontractor qualification is not in accordance with planning	0.100	12
X6	Subcontractor productivity is not in accordance with planning	0.106	9
RISK CATEGORY: ALTERNATIVE METHOD/DESIGN			
X8	Applied method does not match site condition	0.091	16
X9	Applied method is not in accordance with planning	0.085	19
RISK CATEGORY: WBS LEVEL 6 ACTIVITY			
X10	Installation does not conform the specification or construction drawing	0.107	8
X11	Installation is not in accordance with procedure	0.088	18
RISK CATEGORY: WBS LEVEL 7 MATERIAL RESOURCES			
X14	Material usage exceeds planning	0.097	14
X15	Quality of applied material is not in accordance with planning	0.104	10
RISK CATEGORY: WBS LEVEL 7 EQUIPMENT RESOURCES			
X20	Planned equipment productivity lower than actual needs	0.080	24
X25	Equipment specification is not in accordance with planning	0.069	29
RISK CATEGORY: WBS LEVEL 7 LABOR RESOURCES			
X26	Lower labor productivity	0.155	2
X27	Unsuitable manpower planning (wrong specialization)	0.173	1
X31	Labor is not ready to implement new methods	0.121	3
X33	Labor qualification is not according to plan	0.120	4

4.3 Risk-based standardized WBS

The analysis resulting risk responses that can be distinguished into 5 different categories:

(1) *Addition to related WBS*: Items added to related WBS element which is exposed to risk. It can be included into the WBS element by some consideration, whether the risk response is a specific item for a particular job so it is always be done and largely affecting project costs. This decision may vary depending on the project conditions.

(2) *Addition to another WBS*: Items added to another sub-work package (Level 4) different than the related Architecture WBS which is exposed to risk, or additions to Preliminary, Structure, or M/E WBS (Level 2). Infrastructure facilities as operational support and managerial activity considered as overhead item can be included in Preliminary.

(3) *Addition to managerial item*: Items required to execute a project or a risk response related to project managerial from the initial stage to project closure.

(4) *Addition to activity requirement*: Items added to requirements, which can be incorporated into Work Instructions, specification, or contracts. It is also related to organizational policy, whether the risk response does not largely affecting cost so it can be included in Work Instructions, or when the work is transferred to another party then the price-forming should consider the risk response.

(5) *Affecting WBS coefficient*: Risk response can also affect the coefficients related with the structure of resources. The coefficient on the material relates to waste and material composition, the coefficient on the equipment relates to equipment capacity, and the coefficient on labor relates to labor productivity. In this case, if the risk response affects

the WBS coefficients, it is important for the project executor to concern regarding the costing strategy since it is tied to the structure of unit price, so that the risk response can be considered for bidding whether it is included in unit price structure or not.

The pattern recognition analysis resulting 23 Preventive Actions and 18 Corrective Actions as risk responses. Table 2 and Table 3 shows the sample of category mapping.

Table 2. Risk response category mapping for preventive action

NO	PREVENTIVE ACTION	CATEGORY					RECOMMENDATION
		1	2	3	4	5	
P1	Incorporate the requirements of the specialization and numbers of labor in the contract		•	•			Contract / WBS Preliminary: Project Document & Administration

Table 3. Risk response category mapping for corrective action

NO	CORRECTIVE ACTION	CATEGORY					RECOMMENDATION
		1	2	3	4	5	
C1	Perform rework on the failed part or entirely			•			Managerial

There are 14 dominant risk variables for the RBS input, and there are Wall works, Door & window works, and Floor works, as the dominant sub work section resulting from Pareto analysis for the WBS input of RBSxWBS matrix. The results are 14 risk responses, and Table 4 shows the sample from the Sub Work Section: Wall

Table 4. Risk response category mapping for architectural works (RBSxWBS Matrix)

NO	RBS	WBS	RISK RESPONSE	CATEGORY					RECOMMEN-DATION
				1	2	3	4	5	
Sub work section: Wall									
1	X6, X26	Work Package-Brick wall	Activity: Wall Crashing & Repairing; Material res.: Concrete filler, Wiremesh	•					Activity on Brick wall work package

Here is the sample of standardized WBS which is enhanced by the risk responses obtained for *Sub work section : Wall* of interior brick wall.

Table 5. Risk-based standardized WBS for interior brick wall

* : Risk response (development of the standardized WBS)

WBS L. 4	WBS L. 5	Alt. Method/ Design	WBS L. 6 Activity	WBS L. 7 Resources	*Requirement*
Wall	Brick Wall	Interior Brick wall	Brick wall installation		* Practice beam & column /9m²; Anchor/5 row installing
			Bonding plastering		
			Mortar plastering		
			Cement plastering		
			*Primer coating		
			Interior painting		
			*Wall crashing & repairing;	*Material: Concrete filler, Wiremesh	

Conclusion

The first research objective indicates the standardized WBS consists of 5 primary level and 2 complementary level. The second research objective generates 14 dominant risks and the highest risk score obtained from the category of WBS Level 7 Labor Resources. The third research objective shows that not every risk responses can be adopted directly into related WBS structure, there are 5 different categories and the application becomes project executor decision which is influenced by corporate policy or project conditions. The risk-based standardized WBS can be utilized for the basis of quality planning to help project executor identifies project works to the smallest items and setting quality objective.

The authors would like to thank the financial support provided by Universitas Indonesia through the PITTA 2017 funding scheme under Grant no. 862/UN2.R3.1/HKP.05.00/2017 managed by the Directorate for Research and Public Services (DRPM) Universitas Indonesia.

References

1. W. S. Andi, Y. Hendarlim, Faktor-Faktor Penyebab Rework pada Pekerjaan Konstruksi. Civil Engineering Dimension/CED, **7.1**, 22-29 (2005)
2. F. Fakih, Banyak Rusunawa Bocor dan Retak, Ahok akan Hati-Hati Berikan Tender, Retrieved 01 November 2016 from https://www.merdeka.com/jakarta/banyak-rusunawa-bocor-dan-retak-ahok-akan-hati-hati-berikan-tender.html (2016)
3. S. Ponticelli, W. O'Brien, F. Leite, 5th International/11th Construction Specialty Conference, 1-10 (2015)
4. ISO 9001 and Risk, http://isotc.iso.org/, Retrieved 01 October 2016 from http://isotc.iso.org/livelink/livelink/fetch/2000/2122/-8835176/-8835848/8835872 /8835883/ISO9001andRisk.docx (2016)
5. O. Zwikael, Critical Planning Processes in Construction Projects, Cons. Innov., **9**, 372-387 (2009)
6. T. Leemann, Managing the Chaos of Change, The J. of Bus. Strategy, **23.5**, 11-15 (2002)
7. K. Blyth, J. Lewis, A. Kaka, Developing Framework for Standardized Work Programme for Building Project, Cons. Innov., **4**, 193-210 (2004)
8. J.V. Tonder, M. Bekker, Analysis of a Methodology to Obtain a Work Breakdown Structure Built Up From Interdependent Key Project Deliverable Packages. African Rhythm Project Management Conference. Johannesburg South Africa, 1-13 (2002)
9. B. Atkin, Stereotypes and themes in building designs: insights for model builders, Cons. Mngmt. and Eco., **11**, 119-30 (1993)
10. C. Gray, J. Little, The classification of work packages to determine the relationship between design and construction, Occasional paper No. 18, Department of Construction Management and Economics University of Reading (1985)
11. J.M. Nicholas, Project Management for Business and Engineering Principles and Practice (Elsevier, Oxford, 2004)
12. Y.M. Ibrahim, A. Kaka, G. Aouad, M. Kagioglou, Framework for a generic work breakdown structure for building projects, Const. Innovation, **9.4**, 388-405 (2009)
13. S.T. Lai, A WBS-Based Plan Changeability Measurement Model for Reducing Software Project Change Risk, Lecture Notes on Software Engineering, **2.1**, 94-99 (2014)
14. L. Su, WBS-based Risk Identification for the Whole Process of Real Estate Project and Countermeasures, National Conference on Information Technology and Computer Science (2012)

Components of Indonesian Traditional Market: Design Tools for Revitalisation

Ferry Hermawan[1,*], *Himawan* Indarto[1], and *Ismiyati*[1]

[1]Department of Civil Engineering, Diponegoro University, Professor Sudarto, SH Tembalang, 50275, Indonesia

Abstract. The lesson learnt of revitalization of market buildings is still a dilemma for traders as the main actors of traditional Indonesian market life. The root of the revitalization issue has not been properly mapped objectively, due to the complex role of the stakeholders involved in it. This research uses case study method with a qualitative approach by semi-structured interviews and observation of typology of Indonesian traditional market activities in five metropolitan cities of Java. This study uses four samples of traditional markets, one in Jakarta and three revitalized markets in Semarang. The mapping of the component is composing by field observation and Focus Group Discussion (FGD) with a civil engineer and two architects. The results of this study obtained justification that the revitalization of traditional Indonesian market if mapped components are factors that affect the success or failure of a revitalization process when compared to some case studies of market revitalization in Jakarta and Semarang. This research suggests to the local decision makers to avoid the paradigm of budget-based development but, also to adapt socio-engineering aspects by considering the social interaction of traditional market traders in the Indonesian context.

1 Introduction

Traditional Markets as part of public buildings have unique characteristics in the management and interaction of stakeholder work in the place. The social interaction of market stakeholders becomes the inherent soul of market operations even in the process of its revitalization [1,2]. The CIB Research Roadmap becomes an indication of how the practice of the public building construction industry becomes a dilemma that the trend of sustainable building issues is no longer popular since 2012 [3]. In the context of the construction industry in Indonesia, especially in traditional market buildings as part of the public building sector, it shows indications of unhealthy economic competition with the modern market since 2007 [4]. According to the concept of sustainability, the condition of traditional Indonesian market which is non-competitive and begin to be abandoned by the community is a form of unsustainable practice on the pillars of social-economy. Furthermore, the market situation that is not properly set up is also a form of inequality for the traditional market building design that is not adaptive to the changing of socioeconomic

* Corresponding author : hermawan.ferry@gmail.com or ferry.hermawan@live.undip.ac.id

phenomenon. Case studies of traditional market revitalization in Jakarta was lesson learned for future market building arrangement. An adaptive market building design for a comfortable and flood-free environment situation is a fundamental need for a public building. A building, including traditional markets, in addition to meet the technical requirements of the building code it has to have added value as well as the concept of sustainable building [3]. Therefore, research on how sustainable market design become a necessity to address the challenges of changes in buyer behaviour and trade competition in the micro-economic sector.

1.1 Engineering design of revitalisation traditional market in urban area: a case of Jakarta

The design of buildings in Indonesia experiences a phase of change adapted from several concepts of building structures from several countries. However, environmental conditions such as earthquakes and other types of local construction materials are accommodated in the Building Code and the National Standard (SNI) of materials and planning procedures. The location of buildings in Indonesia located in the Ring of Fire is the main factor considered in the design. Although traditional market engineering designs cannot be separated from building standard rules for earthquake resistant, we cannot rule out how the context of a traditional market is in the community. Thus, there are several things should be considered in the design of market building revitalization. Unfortunately, the building practitioners and stakeholders involved are not sensitive enough to take into account the social impacts of traditional market revitalization processes. In general, the revitalization of traditional markets in Indonesia is still focused on spending project budgets and replacing old buildings.

The civil engineering and architectural aspects are two fundamental aspects of the traditional market revitalization process in this research. Hermawan [1] has described the process of revitalizing traditional markets that have been successful in Jakarta. The KGM market building is designed with the principles of energy-efficient, modern and public accessible tropical buildings. The boundary condition of the KGM market design. Historically, the location of the KGM market is in flood-prone and densely populated areas. The architectural concept used is mezzanine and semi-basement to camouflage the number of floors so it does not look too high. Structurally, the KGM market building applies a pre-cast element to accelerate the construction process with guaranteed accuracy.

Learning from case studies of KGM market as one of the public buildings in Jakarta lies in how the strategic approach are performed by contractors. The socialization process has a significant role for the construction process. The key to successful socialization is how the contractor communicates the design of the market building.

Although market traders are not a party with knowledge on building structures, but as end-users they play a role in the approval process of project owner (client) design because they involve in the determination of selling price of the kiosks. Interaction between stakeholders, Local Authority, Contractor and End-user in the indicator of KGM market revitalization success. The balance of interests of the three parties is a 'sustainable' measure in the context of market in Jakarta. Furthermore, the technical design concept (engineering design) of market revitalization design in Jakarta is more profitable because it uses the contract system of Design and Build (D & B). Thus, the gap between planning and construction implementation is small because there is shift in the interaction process of design implementation during socializing with the market traders.

1.2 Experience of traditional market revitalisation in Semarang

The experience of revitalizing traditional markets in Semarang City becomes the comparison of urban market case study. Semarang has a city typology similar to Jakarta. Traditional market location becomes the determinant for its operational in post revitalization. Based on market building standards in Indonesia, several design components are set to include the dimensions of the market building structure: the capacity of the merchant to be accommodated, the dimensions of access to and from the market and its complementary infrastructure [5].

Successful revitalization is described not only during the initial process but also post-construction. This can also be measured through the hustle of a market. Thus, Rizal and Alhamidi [6] tested the correlation of eight factors developed from several empirical studies such as Market Distance to Main Street, Visibility from Main Street, Market Parking Area, Area Density, Number of Market Traders, Total Access, and Number of Floors of market buildings. However, according to Rizal and Alhamidi [6] statistically proves that there are 4 significant factors correlated to the market operations in Semarang city, namely the density of the region, the number of traders, the amount of access to / exit from the market building and the number of public transport routes that serve toward the market.

The experience of market revitalization in Semarang city shows that from three revitalized markets, there are some design components that have not been met in Detail Engineering Drawing (DED) when compared to the applicable SNI Community Market [5]. Regarding the standard revitalization of the traditional market arrangement according to SNI Community Market [7], the application of the zoning system and the provision of facilities are good and proper. Nevertheless, there are several things required to be updated and added, because the previous condition has not met with the standards.

1.3 Beyond to design tools for revitalisation: a challenge in Indonesian public building

Building Standard or Code is an ideal form that can be accommodated in the design stages. However, some revitalization of traditional market cases remains a dilemma for traders as the main actors of traditional Indonesian market life [1,2]. The root of the revitalization issue has not been properly mapped objectively, due to the complex role of the stakeholders involved in it. As the developed environment is social science, the engineering matter still exists among revitalization. Not only the physical matter but also the life aspect to be resolved along the life cycle project. The needs of design tools in the design aspect due to revitalization of some projects in Java Island [5,6]

The limitation of Building Code exists in the social aspect of revitalization. The general perspective of the design tools of public projects is still on the scope of engineering such as the reliability of structure and material [8]. The social aspect of revitalization is multi-disciplinary and crossing sector which is identified from the same platform among stakeholders involved. The design platform means for every party from top management to the front liner at local authorities. Since the local autonomy on Post-Reformation Order, Local Government has different role in public building sector. Thus, the Design Tools should be generated explicitly on the design code.

1.4 Research method

The case study method for the traditional market context in Java was used to map the design components of the market building revitalization with a qualitative approach by semi-structured interviews and observation of typology of Indonesian traditional market activities in five metropolitan cities of Java. This research used four samples of traditional markets, one in Jakarta and three revitalized markets in Semarang. The mapping of the component was composed by field observation and Focus Group Discussion (FGD) with experienced engineers (more than 15 years), a civil engineer (Civil 1-HS) and two architects (Architect 1- ABS and Architect 2-WS). The element of revitalization had been figured out through full transcript of FGD.

2 Results and discussion

2.1 Indonesian traditional market from concept to practice

The concept of traditional Indonesian market has developed since Ancient Mataram [9]. Artefacts supporting the argument are in the reliefs of Borobudur temple. The results of FGDs with civil engineers and architects illustrated some perspectives on traditional markets, in terms of concept, characteristics and current developments.

According to the concept of the city in Java that has been developed up to date, it was strongly influenced by the Hindu concept. As described by Architects (1-ABS) that: "As we look back from the ancient city arrangement in Java, in Javanese dimension, market is one of 4 elements as written in mandala, based on the Hinduism concept of "Gatra Tunggal"; the first one is the Palace or the "Keraton", the second is the "Alun-Alun" (city square), the third is market, and the fourth is temple (for Hindu) or mosque (for Islamic Mataram) ".

Further, from the architectural perspective, Indonesian market emerges from social interaction which makes items or merchandise, such as fabrics, groceries and other daily needs much cheaper. The market building is also humble, down to earth, and constructed from local materials. In the past, most of Indonesian traditional market grew as an organic market, it could be under the trees and it appeared in the intersection. The people built the small stalls between there area, and it created a vernacular building pattern (Non-Engineered Building).

However, there are two styles of the traditional market in Java developed currently. First, the colonial building style model. There are some traditional markets planned by Dutch architects, such as Pasar Gede, as it adopts local architecture. Pasar Gede has many openings and open air. As there may be many buyers (especially women) with heavy groceries, there are several ramps so that the grocery can be dragged. Secondly, the modern model of building concept like Mall. The current modern markets try to imitate the concept of mall buildings, but without Air Conditioning (AC) on market to avoid high maintenance for renters, who seek for affordable rent.

Another perspective of civil engineer identifies the characteristics of traditional Indonesian market in two types, wet market and iconic market. He stated that "...... traditional markets have some degrees; wet markets which really serve local people to shop daily needs, such as Karang Sari market in Semarang, and iconic markets such as Bringharjo in Yogyakarta, Peterongan in Semarang, and Pasar Gede in Solo". According to the civil engineer (Civil 1-HS), for wet markets focus on the function to maintain low cost and direct interaction between vendors and buyers. While the iconic market is more likely to function for tourism. Therefore, the difference in function will affect how the market building should be designed because it affects the attractiveness of the buyers.

The challenge of revitalizing a traditional market is to reorganize the components of social interaction within. The components are more difficult than the physical design of market buildings, because social interactions cannot be measured directly by the building engineer (Architect 2-WS). Knowledge of traditional market interaction can only be obtained from research experience and conversation with market traders and stakeholders involved in market management. The revitalization design experience considering socio-engineering aspects also occurs in the KGM market in Jakarta [1,10]. In fact, technical aspects will be successful after social aspects pass through the process of socialization and design consultation if market traders agreed with the contractor and local authority.

2.2 Component of Indonesian traditional market

Through the interpretive approach of the FGD results and the DED document from the KGM market case study in Jakarta and three sample revitalised traditional market in Semarang [1,5], traditional market components can be described as in Fig 1.

Revitalization is an activity that has two main dimensions, namely social aspects and physical aspects. First, the social aspect is a form of social interaction of traditional market communities that describe *trust*, *social capital* and *loyalty* [11,12]. These three elements of social aspect are traditional market sociology phenomena determining the direction of market development before and after revitalization. Trust is developed in a relatively long time because sociologically, traditional market traders have a strong relationship with each other. In fact, under the economic crisis alone, the rotation of Indonesia's traditional market economy is not affected [4]. Based on the KGM market revitalization experience in Jakarta, the determinant of the successful implementation of the revitalization design and the socialization phase depends on how good communication with the representatives of the market traders. The social aspects of the revitalization design component consist of the pattern of market activity, the purpose of market development and the value added after the revitalization. Second, the physical aspect is a reflection of the engineering aspect of the building, which is influenced by climatic conditions, parking arrangement, merchant relocation management and the role of local leader in making decisions related to market revitalization process. The KGM market revitalization experience in Jakarta is a form of interaction between the role of local government and experienced contractors. The key decision of the revitalization process is determined by how the government is behaving towards the environmental change process due to the revitalization of buildings and the social interaction of traders during pre- and post-construction.

3 Conclusions

The design tools of revitalization should be considered by social and physical aspects. This article justifies that social aspect should be considered as a design tool in the future. The lesson learned from KGM project in Jakarta and revitalized markets in Semarang proves that the concept of local market, the structure of local leaders, the structural or building construction aspects, the purpose of revitalization accommodate people needs of the market and added value in the post-revitalization. Revitalisation suggests to the local decision makers to avoid the paradigm of budget-based development and to revisit the manual design of Indonesian Tradition market to be more adaptive with character of traditional market, technical assisting to Government (National or Local) to craft the Design Guide for Revitalization which is more adaptive to local problems, local needs and local content.

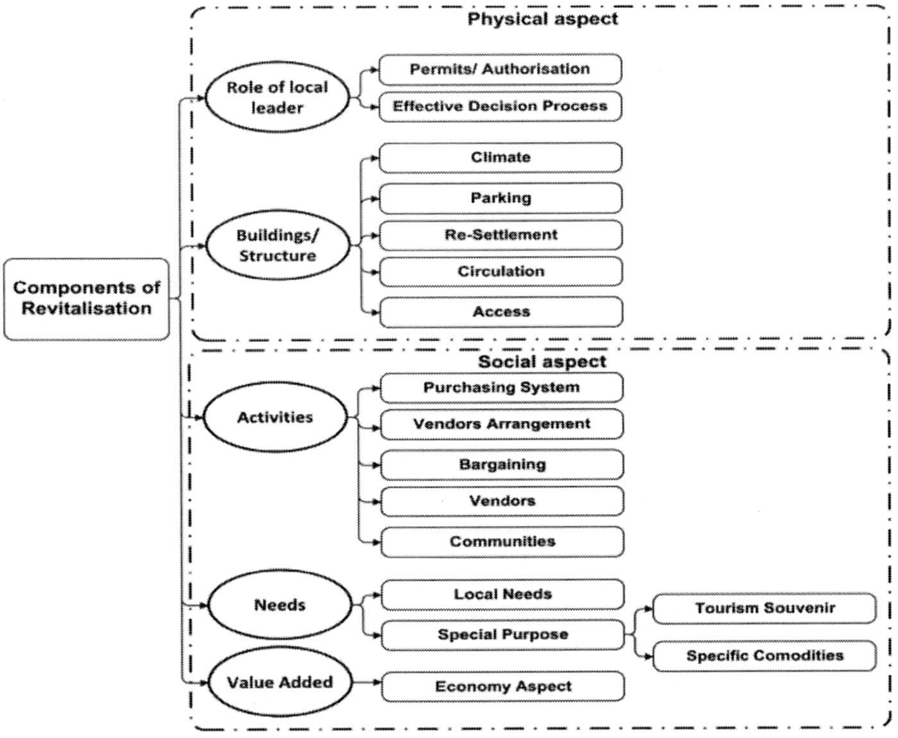

Fig. 1. Component of Indonesian Traditional Market Revitalisation

References

1. F. Hermawan, Ph.D. Thesis, Coventry University, UK (2015)
2. F. Hermawan, F. Kistiani, Ismiyati, Research Report of RPP, Diponegoro University, Year 1 (2016)
3. CIB, *CIB Research Roadmap- Sustainable Construction*, International Council for Research and Innovation in Building and Construction (2016)
4. Semeru, Newsletter, **22**/ June (2007)
5. G. Anggraini, D. Amalia, Final-Project, Department of Civil Engineering, Diponegoro University, Indonesia (2016)
6. R.M. Rizal, N.R. Alhamidi, Final-Project, Department of Civil Engineering, Diponegoro University, Indonesia (2017)
7. BSN, *Standar Nasional Indonesia Pasar Rakyat*, Jakarta (2015)
8. F. Hermawan, H. Indarto, R. Soetanto, Procedia Engineering, **171**, 323-332 (2017)
9. T.S. Nastiti, *Pasar di Jawa pada Masa Mataram Kuna Abad VIII – XI Masehi*, Bandung (2003)
10. F. Hermawan, R. Soetanto, Proceeding ICONIC 2014, Nuremberg, Germany (2014)
11. S. Suryadarma, A. Poesoro, Akhmadi, S. Budiyati, M. Rosfadhila, A. Suryahadi, Food Policy, **35** (79-86) (2010)
12. J. Smith, D. Maye, B. Ilbery, Royal Geographical Society, **46**, 122-128 (2014)

River Conservation Using SECEEL Approach: A Case Study on East Flood Way River, Semarang, Central Java, Indonesia

Ignatius Sriyana[1,*]

[1] Lecturer in Civil Engineering Department, Faculty of Engineering, Diponegoro University, Tembalang Campus, Semarang, Indonesia 50275

Abstract. Current practices on river conservation are solely depending on social engineering, cultural, economic and environmental approaches without taking law enforcement into account. This issue becomes critical because river conservation is frequently experiencing conflict without assertive law enforcement. One of the cases is the presence of abundant structures on river bank despite of rules against such practices. This violation is rooted on law enforcement aspect. Therefore, integrated approach incorporating law enforcement is important to be studied. The purpose of this study is to test the effect of this integrated approach named Social Engineering, Culture, Economic, Environment comfort and Law enforcement (SECEEL) against river conservation on river bank. The analysis method is quantitative approach using Partial Least Squares Structural Equation Modeling (PLS – SEM) program involving 150 respondents as a sample of approximately 400 habitants in East Flood Way river, Semarang. The result shows that higher social engineering value and economic expediency together with lower rule of law are creating higher environment comfort level along the river bank. Meanwhile, more positive cultural values will decrease the level of environment comfort on the area. Interestingly, the higher the comfort level of the environment along the river will decrease the value of conservation.

1 Introduction

As stipulated in Indonesian Government Regulation (PP) No. 38 Year 2011, Chapter 2 Article 5 Verse 4 and 5, the river functions as a path for running water and as a venue for the life of river ecosystems. Furthermore, riverbanks act as a buffer space for river and land ecosystems that prevent disruption between each other [1]. To promote optimal river utilization, it is necessary to manage the river in a proper way, one of which is to carry out river conservation which includes river protection and the prevention of river water pollution in a sustainable manner involving the community and the government [2]. If the river management is carried out appropriately then the results would be more effective and sustainable [3].

* Author: sriyana808@gmail.com

Some problems on the river management are causing violations in the riverbanks, the diminishing dimension of the embankment due to human activities and waste disposal to the river resulted in pollution of river water and disruption of water flow. These existing problems are caused by low public awareness and lack of attention regarding river conservation's implementation. This violation can adversely affect the continuity of the river in the future, which in turn will be detrimental to human's life.

Current practices on river conservation are only utilizing social engineering, cultural, economic [4] and environmental comfort approaches without taking into account law enforcement variable. This issue becomes a critical point because river conservation implementations are frequently experiencing conflict without assertive law enforcement [5]. On the field, warning signs for prohibiting community to build structure on river bank area have already installed, however, violations are still abundant. Integrated approach incorporating social engineering, cultural, economic, environment comfort and law enforcement is important to be studied.

The purpose of this research is to test SECEEL (Social Engineering, Culture, Economic, Environment comfort and Law enforcement) approach against conservation value on river banks. Several parameters are tested, including the effect of economic benefits, social values, cultural values and law enforcement to the environment comforts.

2 Materials and Research Method

2.1 Research Location

The research location is in East Flood Way (*Banjir Kanal Timur*) river in Semarang of approximately 2 km in length and the width of 174 meters (including riverbank). The northern boundary coordinates are at x = 438560, y = 9230059, eastern boundary at x = 438462, y = 9228616, southern boundary at x = 437955, y = 9226296 and western boundary at x = 438239, y = 9228304.

The data used are basic map in the form of administrative map, satellite image, river map along with river area (based on Government Plan on Spatial and Territory Utilization of Semarang City for 2017) and field survey on 2017.

Fig. 1. Map of East Flood Way.

2.2 Research Method

The analysis method used in this research was quantitative methods using software PLS - SEM.3.0 [6]. The stages of this study included preparation stage (field orientation, building questionnaires, etc.), data collection phase of both primary and secondary data, data processing and continued data analysis.

Researcher directly observed and interviewed the residents of the river using questionnaires to obtain primary data. Number of respondents involved was 150 respondents compared to the population of approximately 400 respondents located along the East Flood Way (*Banjir Kanal Timur*) river. Meanwhile, the secondary data was obtained from the Ministry of Public Works and Public Housing, Directorate General of Water Resources, Major River Basin Pemali - Juana Region in Semarang.

3 Analyses and Discussion

SECEEL approach, which tests the effect of economic benefit, social values, cultural values of society and law enforcement to the comfort of the environment on river banks and the influence of comfort to the river conservation value, is using multivariate statistical analysis including convergence validity test, discriminant validity test, construct reliability test, path and hypothesis testing analysis (see Fig. 2 below).

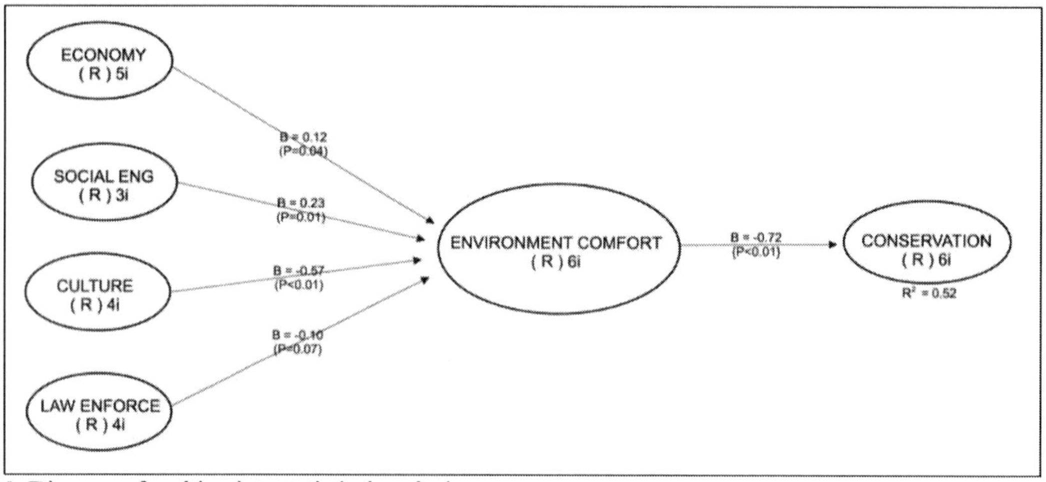

Fig. 2. Diagram of multivariate statistical analysis.

Based on Fig. 2, it can be interpreted that the model and coefficient values contained in the diagram are fit (APC = 0.348, P <0.001, ARS = 0.740, P <0.001 and AVIF = 6.258, Good if <5), it means the model is appropriate for the analytical tool in the study. Variations of predictor variables (Economic, Social, Cultural and Legal) may account for 96% of variation in criterion variables (Environment Comfort). Meanwhile, variations in the predictor variables of environment comfort account for 52% of variation in the conservative criterion variables. The following analysis results are obtained from the software.

The first hypothesis stating that the greater the economic benefits that can be enjoyed, the higher the level of environment comfort in the East Flood Way river is proven (p = 0.04 <0.05). This is apparent that in the river banks, street traders have built kiosks which are used to run several businesses like car workshop, salon, shops, radio service, vegetable traders, bird merchants, etc. This has violated Indonesian Government Regulation (PP) no.

38 year 2011, article 22, verse 2, which state within the river border, there are embankments for the interest of flood control. Protection of embankment bodies is done by prohibiting the planting of plants other than grass, erecting buildings and reducing dimension of the embankment.

The second hypothesis states that the higher the social value, the higher the level of environment comfort in the riverbanks of the East Flood Way river is proven ($p = 0.01 \leq 0.05$). The residents at the research sites appear to have strong ties, mutual cooperation spirit, hand in hand, helpful mind and have formed the board of neighborhood, where the people from outside follows the citizens who occupy riverbanks. It tends to cause social unrest in the time of executing the river conservation program.

The third hypothesis states that the more positive the cultural values of the environment, the lower the comfort level in the East Flood Way river is proven ($p = 0.01 < 0.05$). This is evidenced that most residents of riverbanks have built a negative culture of dumping domestic waste in the river for their comfort. This act will pollute the river and clearly violating Government Regulation (PP) no. 38 year 2011, article 27 and paragraph 1.

The fourth hypothesis states that the lower the law enforcement, the higher the level of environment comforts in the riverbanks of the East Flood Way river is proven ($p = 0.07 < 0.1$, significant with $\alpha = 0.1$) [7]. It shows the weakness of law enforcers in implementing sanctions for perpetrators who have committed violations to the river management rule.

The fifth hypothesis stating that the higher level of environment comfort in river banks will further decrease conservation value is proven ($p = 0.01 < 0.05$). The impact will be in the form of potential conflict on implementing river conservation, either river protection or river pollution prevention activities, against the environment therefore require greater costs to overcome them.

4 Conclusions and Recommendations

4.1 Conclusions

Based on the analysis and discussion above, it can be concluded as follows:

1. SECEEL approach showed that the more economic benefits that can be enjoyed and the higher the social value resulted in the higher the level of environment comfort, the more positive the cultural value of the environment will be lower the comfort level, the lower the law enforcement will be resulting in high level of environment comfort, and the higher the level of environmental comfort on the river banks will further diminish the value of conservation.
2. Persistently weak law enforcement in implementing sanctions for perpetrators who have committed violations against river management rule is proven to be one of detrimental factor on river conservation.
3. Lack of attention on means of supporting the implementation of river management activities by the Government.

4.2 Recommendations

Below are several recommendations in order to realize the sustainability of river functions:

1. Law enforcement must act assertively, as early as possible, in gradual manner sequencing first, second and final warning letter in case of violation of the law (e.g. erecting building on river banks, dumping in rivers, etc.).
2. Routine river management needs to be carried out, for example, protection and prevention of river pollution activities, by involving the community.

3. There needs to be cooperation and mutual coordination between the government (Central, Provincial and District/City) in implementing river management in accordance with their respective authorities.

The author wishes to express sincere gratitude to Water Resources Laboratory, Civil Department, Faculty of Engineering, Diponegoro University for allowing the author to use the laboratory facilities. Same gratitude is addressed to the Ministry of Public Works and Public Housing, Directorate General of Water Resources, Major River Basin Organization Pemali - Juana Region in Semarang for providing data needed in this study.

References

1. Indonesian Government Regulation (PP) No. 38 Year 2011 about "River"
2. J. S. Antunes do Carmo, Natural Responses to Changes in Morphodynamic Processes Caused by Human Action in Watercourses: A Contribution to Support Management, *International Journal of Disaster Risk Reduction,* **24,** 109–118 (2017)
3. M. A. Evans, J. Hall, E. Penning-Rowsell, P. Sayers, C. Thorne and A. Watkinson, Future Flood Risk Management in the UK, *Proceedings of the Institution of Civil Engineers Water Management* **159**, 53–61 (2006)
4. C. Spörri, M. Borsuk, I. Peters, P. Reichert, The Economic Impacts of River Rehabilitation: A Regional Input-Output Analysis, *Ecological Economics* **62,** Issue 2, 341-351 (2007)
5. P. J. Cosgrove, L. C. Hastie, Conservation of Threatened Freshwater Pearl Mussel Populations: River Management, Mussel Translocation and Conflict Resolution, *Biological Conservation* **99**, Issue 2, 183-190 (2001)
6. I. Ghozali, *Model Persamaan Struktural: Konsep dan Aplikasi dengan Program AMOS19.0* (2002)
7. C. J. Barrow, River Basin Development Planning and Management: A Critical Review, *World Development* **26**, No. 1, 171-186 (1998)

Determining Factors for Implementing Green Supply Chain Management in the Construction Industry: A Literature Review

Mochammad Agung Wibowo[1], *Naniek Utami* Handayani[2*], *Anita* Mustikasari[2], *Asri* Nurdiana[1], and *M. Nur* Soleh[1]

[1]Department of Civil Engineering, Faculty of Engineering, Diponegoro University, Semarang, Indonesia 50275
[2]Department of Industrial Engineering, Faculty of Engineering, Diponegoro University, Semarang, Indonesia 50275

Abstract. The construction industry plays a significant role in improving the quality of the built environment, but its activities also impact the wider environment. Supply chain management (SCM) aims to increase the efficiency of the construction process through improvements in the quality, cost, and time frame of projects. Green SCM has as its purpose reducing environmental impact by opting for eco-friendly materials starting from the design phase in the choice and procurement of materials to be used in construction projects. This study aims to adopt GSCM concepts of the manufacturing industry for application in the construction industry. A literature review was conducted to learn from previous research about the implementation of GSCM in manufacturing to serve as a basis for adopting of this concept for the purposes of building criteria for the implementation of GSCM in the construction industry.

1 Introduction

The contributions of the construction industry to infrastructure development in Indonesia have encouraged equitable economic growth. However, the construction industry has the reputation of contributing bad impact on the environment [1]. Effective management systems are essential to ensuring that projects are completed on time and within budget while helping project managers monitor progress [2, 3]. At the same time, stakeholders need efficiency in the activity flow of construction projects compatible with a lean focus and the reduction of energy consumption and waste [4].

The supply chain in the construction industry is different from the manufacturing industry [5]. However, the unavoidable problem of fragmentation should be solved so that inefficiency in construction can be reduced or eliminated. Supply chain management is implemented to enhance efficiency and accuracy as reflected in quality, cost, and delivery time. GSCM prioritizes the minimization of waste and energy use and the procurement of environmentally friendly materials. The implementation of GSCM in construction projects

* Corresponding author: naniekh@ft.undip.ac.id

is meant to bring the concept of eco-friendliness into every process, from upstream to downstream and through all of the PLC phases. The modeling of eco-friendly concepts at all phases in the project life cycle with the aim of creating green supply chains (GSCs) in construction projects requires further study. Therefore, this study will conduct research on the implementation of GSCs in construction, based on project life cycle. The concept of GSCs in manufacturing can be adopted in construction projects. The implementation of GSCs in a construction project should apply indicators to each PLC phase which need to be assessed from the beginning until the end of the construction process. The purpose of this research is to adopt GSCM concepts of the manufacturing industry for application in the construction industry.

2 Green Supply Chain Management

2.1 The Concept of Green Supply Chain Management

Green supply chain management (GSCM) is defined as the integration of sustainable practices into upstream and downstream supply chain management6. GSCM consists of green product design, green materials management, green manufacturing processes, green distribution and marketing, and reverse logistics (RL) [6].

Green Product Design opens up opportunities to diminish any environmental impacts inherent in the establishment of new products and production processes. Green product design consists of two primary conceptions: environmentally conscious design (ECD) and life-cycle assessment (LCA). LCA is defined by its focus on analysing the environmental burdens of processes and products over their entire life cycle [7][8]. ECD refers to concepts and practices related to designing products while keeping environmental considerations in mind.

Green materials management is a way to substitute less hazardous activities or materials that replace more harmful ones. It is classified into two processes: green materials selection and green materials sourcing. The criteria for materials that can be processed in green materials management include that materials used in a product must be easy to separate, be adaptable, or can be used to streamline existing processes [8-9].

Green manufacturing processes consist of three main phases: resource utilization decrement, waste decrement, and emission decrement. Green manufacturing process aims to minimise the total amount of waste at the manufacturing phase indirectly by decreasing energy and resource utilization [10]. Emission decrement has two essential characteristics: control and prevention. Control means emissions and waste are trapped, stored, treated, and disposed of using pollution control equipment. Prevention means decreasing, changing, or preventing emissions and waste together through better materials substitution, recycling, reuse, or process innovation [6].

Green Distribution and Marketing refers to the promotion or advertising of products, changes in production processes, or packaging changes that are weighed in terms of environmental criteria [11].

Reverse logistics, another significant component of GSCM, has the objective of collecting, distributing, and managing products until they are delivered to customers. Reverse logistics encompass the recycling, reusing, and remanufacturing of materials. Benefits of recycling include extraction of basic materials, diminished energy usage, and reduced air and water pollution [6]. Examples of reuse include the utilization of reusable packaging or shipping materials [12].

Green SCM Frameworks in the Construction Industry

Based on research of GSCM in manufacturing, this paper proposes to elaborate the implementation of GSCM in construction projects by adopting the concept of GSCM used in manufacturing to the construction industry but with various adjustments and modifications. This application of GSCM to construction is expected to reduce waste, energy usage, and negative impacts of construction projects on the environment.

2.2.1 The GSCM Concept Applied to the Construction Industry

GSCM is defined as the integration of sustainable practices into upstream and downstream supply chain management to bring long-term benefits through the application of responsible environmental and social behavior in all supply chain members [6-13].

The framework of green supply chain process implementation in the construction industry will now be explained in greater detail.

The Green design has as a benefit to diminish environmental impact resulted from the establishment of construction design and processes. Green design consists of two main elements: LCA to strengthen construction development so the overall negative environmental impact of construction is reduced, and the concept of ECD to design construction activities while keeping environmental considerations firmly in mind.

The Green materials management refers to ways of substituting more benign activities or materials for potentially hazardous ones. The criteria used in selecting materials that can be processed in green materials management include that materials used in construction should be easy to separate, adaptable, or useful in streamlining existing processes.

The Green construction aims to maximise the conservation of resources and reducing construction activities that incur negative impacts for the environmental while achieving the goal of savings in the four resource areas (energy, land, water, and materials) in addition to ensuring environmental protection.

The Green Marketing Management consists of green customer education, green customer social responsibility, and green image promotion. The aim of green marketing management is to promote or advertise project construction using environmental criteria.

The Reverse logistics are defined as a set of activities to manage materials that potentially increase waste streams via strategies consisting of reuse, recycling, and remanufacture. Recycling is described as a procedure to reuse materials that may otherwise be considered waste in a form other than the primary use. Reuse means a material is used again for the same purpose or may find a new product life in a distinct function.

2.2.2 The Dimensions of the GSCM Concept in the Construction Industry

The proposed elements of each dimension of GSCM concepts in the construction industry have been broken down into several components. There are thirteen dimensions in Green Design concept which are Eco-Design [14-16], Innovation Capability [17], Environment Collaboration with Customer [18-22], Product Safety [23], Occupational Health and Safety [23], Pollution Prevention [23], Resource Conservation and Waste Management [23], Appropriate Site Development [24], Energy Efficiency and Conservation [23-25], Water Conservation [24-25], Material Resource and Cycle [24], Indoor Health and Comfort [24], and Building and Environment Management [24].

In Green Materials Management concept, there are three dimensions: Procurement / Purchasing [26-28], Government Regulation and Legislation [29, 30-31], and Environment Certification [18-19, 22, 32-33]. While in Green Construction Processes, the dimensions are Production Cost [14, 34], Waste Management [35-13, 10], Pollution Control / Environment Treatment [15, 36-37], Quality Control [38], Employment Practices [37, 39], Resource Management [40-41], and Health and Safety [15].

In Green Marketing Management concept, Green Management / Marketing Policy [17], Green Corporate Social Responsibility [42-43], Green Customer Education [25], and Green Image Promotion [25] are recognized as its dimensions. Whereas, the concept of Reverse Logistics has three concepts: Recycling [6, 12, 18, 32], Reuse [6, 12, 18, 32, 42], and Remanufacture [14, 21, 26, 39, 48].

3 Research Methodology

A literature review is conducted to manage the diverse knowledge pertaining to a specific research topic [44]. The researchers conducted a literature-review-based study to address the research objectives [45], whereby a literature review is a decisive tool that supports the researcher in the following ways: by mapping, consolidating, and evaluating the existing academic structure of the identified field by developing the scope of further opportunities through identifying key research gaps in the existing body of research.

4 Discussion

The GSCM concepts are applied to manufacturing industries for the purpose of environmental sustainability. This concept is being applied to the construction industry to reduce the impact of construction on the environment. In the early stages of research, the researchers conducted a literature review of the GSCM concept in the manufacturing industry and tried to adapt the findings for use in the construction industry. The dimensions of GSCM in the construction industry are divided into five concepts: Green Design, Green Materials Management, Green Construction Processes, Green Marketing Management, and Reverse Logistics. Each of these concepts represents a measurement dimension of green supply chains.

These elements will be reviewed based on PLC in the construction industry. At the initiation phase, the owner should consider green initiation to a project. Green initiation includes green building, green construction, and all concepts that minimize waste and energy. At the design phase, planners have to think about green product design and green materials management. The construction phase is the most complex stage of the project life cycle, and it is also one of the phases that determines the success of GSCM implementation. At the construction phase, all stakeholders involved have a responsibility for the implementation of GSCM concepts. The GSCM indicators at this phase are green materials management and green construction. At the operation and maintenance phase, the owner should take green operation and maintenance concepts into consideration.

Each GSCM concept is composed of multiple dimensions. The dimensions of the green initiation concept are eco-design and environmentally conscious design. The dimensions of green product design are design, innovation capability, product safety, environmental control, and building and environmental management. The dimensions of green materials management are materials planning, procurement/purchasing, materials storage, materials handling, green materials selection, green materials sourcing, and government regulations and legislation. The dimensions of green construction are green transportation, on-site management and planning, site operation, residual, environment society, and reverse logistics. The dimensions of green operation and maintenance are green building and green management policy.

GSCM in the manufacturing industry can be adopted in the construction industry, it includes the concepts of green product design, green materials management, green manufacturing processes, green distribution and marketing, and reverse logistics [6]. From the literature study, the GSCM concepts in the construction industry are green initiation, green product design, green materials management, green construction, and green operation and maintenance. The GSCM concepts are similar in the manufacturing and construction industries in that environmentally friendly concepts should be applied from the beginning to the end phases and/or from upstream to downstream.

Based on the studies conducted, it is known that in order to reduce waste, energy consumption, and negative impact on the environment in the construction industry, GSCM concepts should be integrated from the initiation phase to the operation and maintenance phase.

5 Conclusion

Based on literature review, this paper has proposed a framework for the implementation of GSCM in construction industries. The framework includes the concepts and the dimensions of the GSCM model that was adapted from manufacturing industry. This study is rooted in a background of managed SCM in construction that has increased efficiency and productivity in construction projects. These benefits can be obtained in the construction industry by reducing waste, energy used, and negative impacts on the environment. The concepts of GSCM in construction include green initiation, green design, green materials management, green construction processes, and green operation and maintenance. These concepts can be integrated throughout the PLC from the initiation phase to the operation and maintenance phase.

References

1. J.S. Guzmán, M. Marrero, M.V.M. Delgado, A.R. De Arellano, Waste Manage **29**, 9 (2009)
2. R.H. Mohring, F. Stork, Math Meth Oper Res **52**, 501 – 515 (2000)
3. P. Schonsleben, M.K. Vodicka, K. Bunse, F.O. Ernst, CIRP Annals - Manuf Tech **59,** 477–480 (2010)
4. P. Ball. J, Manuf. Tech. Manage. **26,** 3 (2015)
5. Y. Rezgui, J. Miles, J. Constr. Eng. Manag **136**, 5 (2010)
6. M. Ghobakhloo, S.H. Tang, N. Zulkifli, M. Ariffin, M, Int. J. Innov Manage. Tech. **4**, 1 (2013)
7. M. Buyle, J. Braet, A. Audenaert, Renew. Sust. Energ. Rev. **26**, 379-388 (2013)
8. A. Gungor, S.M. Gupta, Comput. Ind. Eng. **36**, 4 (1999)
9. A.A. Hervani, M.M. Helms, J. Sarkis, Benchmarking: Int. J. **12**, 4 (2005)
10. S.K. Srivastava, Int. J. Manag. Rev. **9**, 1 (2007)
11. M.J. Cox, Earth Env **3** (2008)
12. J. Sarkis, J. Clean Prod. **11**, 4 (2003)
13. P. Dadhich, A. Genovese, N. Kumar, A. Acquaye, Int. J. Prod. Econ. **164** (2015)
14. Q. Shi, J. Zuo, R. Huang, J. Huang, S. Pullen, Habitat Int. **40**, 1-8 (2013)
15. K. Govindan, S. Rajendran, J. Sarkis, P. Murugesan, J. Clean Prod. **98**, 66-83 (2015)
16. H.C. Zhang, T.C. Kuo, H. Lu, S.H. Huang, J. Manuf. Syst. **16**, 5 (1997)
17. C.-C. Chen, H.-S. Shih, H.-J. Shyur, K.-S. Wu, Comput. Math. Appl. **64**, 8 (2012)
18. D. Holt, A. Ghobadian, J. Manuf. Tech. Manag. **20**, 7 (2009)

19. A.H. Hu, C.-W. Hsu, IEEE Int. C. (2006)
20. R.D. Klassen, S. Vachon, Prod. Oper. Manag. **12**, 3 (2003)
21. A. Paulraj, Bus. Strateg. Environ. **18**, 7 (2009)
22. S. Vachon, Int. J. Prod. Res. **45**, 18-19 (2007)
23. Y.-F. Wang, S.-P. Chen, Y.-C. Lee, C.-T. S. Tsai, Int. J. Hosp. Manag. **34**, 263-273 (2013)
24. GBCI, GREENSHIP (2010)
25. G. Noci, Eur. J. Purch. Supply Manag. **3**, 2 (1997)
26. P. Evangelista, M. Huge-Brodin, K. Isaksson, E. Sweeney, Implications for small business (2012)
27. S.U. Hoejmose, J. Grosvold, A. Millington, J. Purch. Supply Manag. **20**, 4 (2014)
28. W. Yang, Y. Zhang, Y, J. Bus. Manage. Econ. **3**, 5 (2012)
29. B.M. Beamon, Logist. Inf. Manage. **12**, 4 (1999)
30. A. Hafezalkotob, Transport. Res. Part E Logist. Transport. Rev. **97**, 228-250 (2017)
31. H. Min, W.P. Galle, Int. J. Oper. Prod. Man. **21**, 9 (2001)
32. P. Rao, D. Holt, Int. J. Oper. Prod. Man. **25**, 9 (2005)
33. Q. Zhu, J. Sarkis, J. Clean Prod. **14**, 5 (2006)
34. T.E. Glavinich, P. DE, BOOK TOOLS (2008)
35. Q. Zhu, J. Sarkis, K. Lai, Transport. Res. Part E **44**, 1–18 (2008)
36. A. Awasthi, S.S. Chauhan, S.K. Goyal, Int. J. Prod. Econ. **126**, 2 (2010)
37. C. Bai, J. Sarkis, Int. J. Prod. Econ. **124**, 1 (2010)
38. C.-W. Hsu, A.H. Hu, J. Clean Prod. **17**, 2 (2009)
39. K. Muduli, K. Govindan, A. Barve, D. Kannan, Y. Geng, Resour. Conserv Recy. **76**, 50-60 (2013)
40. C.-W.R. Lin, H-Y.S. Chen, Comput. Ind. **55**, 2 (2004)
41. C. Wu, D. Bames, J. Clean. Prod. **112**, 2114-2132 (2016)
42. S.-H. Chun, H.J. Hwang, Y.-H. Byun, Procedia-Soc. Behv. Sci. **186** (2015)
43. J. Moon, Sustain. Dev. **15**, 5 (2007)
44. D. Tranfield, D. Denyer, P. Smart, British J. of Manag, **14**(3), 207-222. (2003).
45. A. Fink, Thousand Oaks, Sage (1998).

Sustainable Community Based Water Supply at Salatiga by Use of Rapfish Method

Suharyanto[1,*], *Angelica Deasy* K[2], and *Sudarno*[3]

[1]Department of Civil Engineering, Eng. Faculty, Diponegoro University, Semarang, Indonesia
[2]Public Works Department, Salatiga Regency, Indonesia
[3]Department of Environmental Engineering, Eng. Faculty, Diponegoro University, Semarang, Indonesia

Abstract. Sustainable water supply has been the United Nation's goal as stipulated in Sustainable Development Goals (SDG's). In 2030, all people must have acess to safe and affordable drinking water. In Salatiga Indonesia, the water supply company (PDAM) can now only serves 65.64% of the city's population. There are about 11.14% of the population served by community based water supply (CBWS). In line with the SDG's, The Government of Indonesia's program to have 100% coverage of water supply. The limitation on PDAM coverage has led the increase number of CBWS. This tendency has raised the question on the sustainability of CBWS. This paper presents the sustainability evaluations of some CBWS in Salatiga using Rapfish method. In the method, sustainability is evaluated using multi-dimensional scalling based on aspects of ecological, technological, social, institutional, and economical. The strategy to improve sustainability is identified using SWOT analysis. The study performed at 4 (four) CBWS at Kecamatan Sidorejo, Salatiga. It represents the CBWS using deep water sources and spring water sources. It shows that all CBWS analysed are barely sustainable on the dimension of ecological, technological, and social. The dimension on institutional and economical are in the status of no sustainability. Therefore, the CBWS need some improvement and reinforcement especially on institutional and economical.

1 Introduction

Sustainable water supply has been the United Nation's goal as stipulated in Sustainable Development Goals (SDG's). In 2030, all people must have acess to safe and affordable drinking water which are sustainable [1]. It is therefore needs to Support and strengthen the participation of local communities in improving water and sanitation services. In Salatiga Indonesia, the water supply company (PDAM) can now only serves about 65.64% of the city's population [2]. There are about 11.14% of the population which are served by community based water supply (CBWS) [2]. In line with the SDG's, The Government of Indonesia's program to have 100% coverage of water supply and the limitation on PDAM coverage services has led the increase number of the community based water supply. The role of CBWS in Indonesia have long been recognised and now becoming more significant. This tendency has raised the question on the sustainability of CBWS.

Currently, there are two framework in the analysis of the "health" status of water supply company (PDAM) which are published by BPPSPAM (Board of Development for Drinking Water Supply) and The Minsitry of Internal Affairs (Depdagri). However, those method are applicable for more formal companies such as PDAM. The evaluation framework for CBWS has not yet been defined.

This paper presents the sustainability evaluations of some CBWS in Salatiga City by use of Rapfish method [3]. The objectives of this research is to identify the sustainability status of some CBWS and formulate strategy for CBWS to achieve its sustainability by use of SWOT analysis [4].

* Corresponding author: suharyanto20@yahoo.co.id

2. Literatures Review

2.1 Community Based Water Supply

Community based water supply (CBWS) has long been recognised in fulfiling water needs especially in remote areas where government water supply (PDAM) are not present. Currently, the Government of Indonesia has recognised the inseperable role of CBWS to meet people's water requirements in combined with the coverage of PDAM. CBWS is a water supply system which are constructed by the government but its operation management including its agreed tariffs performed by a community. The objectives of CBWS is to boost the coverage of water supply services and to stimulate community to participate in the services. In order to guarantee its successes, the participation of community in a CBWS must be started since its plannig, design, and its operation phase. A CBWS will cover the operation and maintenance costing. The government, however, still required to facilitate the preparation of management team for a CBWS, trained and prepared the management team.

A sustainable CBWS is a CBWS which can sustain its services continuously for long term. The management team of a CBWS should be able to manage its operation, collect tarrifs from users, and able to overcome the operation problems. A sustainable CBWS must satisfy the following criterias [5]:

1. In operation and fullfill its services.
2. Provide the services as planned in good quality, in sufficient quantity, accessible, reliable, continuous, and beneficials to users.
3. Can operate and function well for long term in accordance to the life-cycle of the designed system.
4. The management should attract and involving the community/users through formal institusion (community's institution such as Paguyuban), able to closely coordinate with gorvernment and other institutions (private) as needed.
5. The cost for operation, maintenance, repair, replacement, and administrative services are affordable and can be met with the user's fees or other sustainable financial supports.
6. Can be operated and maintained in the communitys' level of expertiese and resources, and accesed to outside support (technical, training, and monitoring).
7. Does not give negative environmental impacts.

2.2. Rapfish Method

The Rapfish method uses 5 dimension i.e., ecology, technology, social, institutional, and economy [6]. Each dimension is represented by usually 6-12 relevant attributes [7]. Application of The Rapfish method for each CBWS starts from the valuation (score) of each attribute. Multi dimensional scaling (MDS) analysis is then performed to access the status of sustainability in each dimension and overall sustainabillty status based on all dimension. The following are steps in the analysis using Rapfish method [8].

1. Attribute Scoring
   Each attribute in each dimension indicating sustainability of water supply is given score 1 to 5.
2. MDS Scaling and analysis.
   The status of sustainability of each attribute is represented in the scale of 0 to 100 (from bad to very good). The result of MDS analysis is the state of a CBWS score of sustainability in each dimension and in overall. The category of sustainability is Not sustainable (score < 25), Less Sustainable (score between 25 to 50), sustainable (score between 50 to 75), and very sustainable (score well over 75) [9].
3. Attribute Leverage
   The attribute leverage is to assess which attribute have more significant roles in supporting sustainability in each dimension. It uses Jacknife like technique, where each time each attribute is dropped in the analysis. The percentage contribution of attribute can be assessed by the resulting root mean square (RMS) change in ordination when the attribute is dropped. The higher the RMS change when an attribute is dropped, shows the influence of this attribute is bigger. This is ussually called pareto calculation [10].
4. Monte Carlo Simulation
   This analysis is to evaluate the influence of any deviation in the score valuation, discrepancies, and any subjectivity in the MDS proceses, including on the selection of attribute). The deviation between MDS versus Monter Carlo simulation should show comparable results, otherwise it indicates the existence of stresses in the MDS analysis [11].

2.3. Criterias for Sustainability

In 1999, the internal affair minister of Indonesia has published criterias for evaluating performances of PDAM as part of performance's audit for government companies [12]. The criterias used are financial, operational, and administrative aspects. Each aspect consists of several relevant attributes. Applied for each PDAM , based on the status or level on each attributes and some weighting on the scoring, the status or the condition of PDAM can be classified into Very Good, Good, Sufficient, Barely Sufficient, Not Good.

As the implementation of government's regulation No. 16/2005 about drinking water supply, the government has established BPPSAM (government board for development of drinking water supply). Currently BPPSPAM has published criterias to evaluate the "health status" of PDAM. It uses 4 (four) aspects, i.e., Financial, Services to Customers, Operational, and Human Resources [13]. Each aspect is further evaluated by several relevant attributes. Based on the state of each attribute in every aspect, the PDAM can be classified into three category of being "healthy", "less healthy", and "not healthy".

The above two frameworks of evaluation, however, applicable only to formal company such as PDAM where its management system has been prepared for accountable and professional services. For evaluation of a CBWS, it needs more open evaluation framework. In this paper, it uses a Rapfish method (Rapid Appraisal for fishery) [14]. The method is initially used and applied in fishery industries. However, the opennes of the method can be applied in any natural resources field with some justification on the evaluated aspects and attributes.

3 Research Method

The research method uses primary and secondary data collected from each CBWS studied. Primary data consist of questionair survey, interview, and field observation. Secondary data consist of data on legal aspects of the CBWS, its organisation (structures, rules, SOP, job description, etc.), activities of organisation, administrative and financial activites, consumer's role, and its reporting.

The analysis is performed using Rapfish method to access its sustainability and SWOT analysis to evaluate which strategy to improve each CBWS sustainability.

In this paper, some attributes used to access in each dimension is formulated by considering attributes on Permendagri and BPPSPAM. The following are attributes in each dimension which are used. In each attribute, it further supported by several parameters.

Table 1. Attributes in each dimenstion used

No	Dimension/ Attribute	No.	Dimension/Atttribute	No.	Dimension/Atttribute
I	**Ecological**	**III**	**Social**	**V**	**Economical**
1.	Water Quality	1.	Community's Behaviour	1.	Profit
2.	Water Consumption	2.	Social Participation	2.	Revenue Effectivity
3.	Available Water discharge	3.	Maintenance of infrastructures	3.	Development Allocation
4.	Environment Protection on water source	4.	Ability to maintain infrastructures	4.	Community's Wellfare
5.	Maintenance water source	5.	Ability to pay	5.	Cost for OM
6.	Impact to Environment	6.	Willingness to pay		
7.	Number of rainy days	7.	Consumers' Earning		
8.	Continuity	**IV**	**Institutional**		
II	**Technological**	1.	Legality of management		
1.	Complexity	2.	Cooperation		
2.	Spareparts	3.	SOP		
3.	Operation and Maintenance	4.	Job Description and Planning		
4.	Non Revenue Water	5.	Working Schedule		
5.	Customer's increase	6.	Internal and External Reporting		
6.	Coverage	7.	Services to customer		
7.	Customer coverage	8.	Satisfaction of customers		

4 Analysis and Discussion

4.1. Description of CBWS studied

The research is performed to 4 (four) CBWS in two Kelurahan at Salatiga City, each represents a deep well and a spring water sources such as shown in Table 2. It is expected to demonstrate whether the source of water has some influences on the sustainability of a CBWS.

Table 2. List of Community Based Water Supply (CBWS) at Salatiga Studied

No	CBWS	Kelurahan	Source of Water	Users (house)	SOP	Reporting	Management team	Daily Operational	Returned Quest.
1.	Tirta Barokah	Sidorejo Lor	Deep Well	168	No	Internal, it has 2,660 US$ in their account	Head, Secretary, Treasury, Technician, Field staffs.	Head, Treasury, Field staffs.	60
2.	Tirta Sejahtera	Sidorejo Lor	Spring	120	Available, but not fully implemented	Internal, it has 350 US$ in their acount	Head, Secretary, Treasury, Field staffs.	Head, Treasury, Field staffs.	49
3.	Tirta Kamandanu	Kauman Kidul	Deep Well	160	No	No Internal reporting	Head, Secretary, Treasury, Technician,	Head do all	20
4.	Tirto Agung	Kauman Kidul	Spring	40	No	No Internal reporting	Head, Secretary, Treasury, and Technician	Head handles all	39

4.2. Results of MDS in Rapfish Method

The results of MDS analysis for each dimension is shown in Figure 1 to 5. In Figure 1, based on the dimension Ecology, all CBWS are in good sustainability. It indicates that the management put efforts in maintaining and protecting water sources and its water abstraction are below yield capacity.

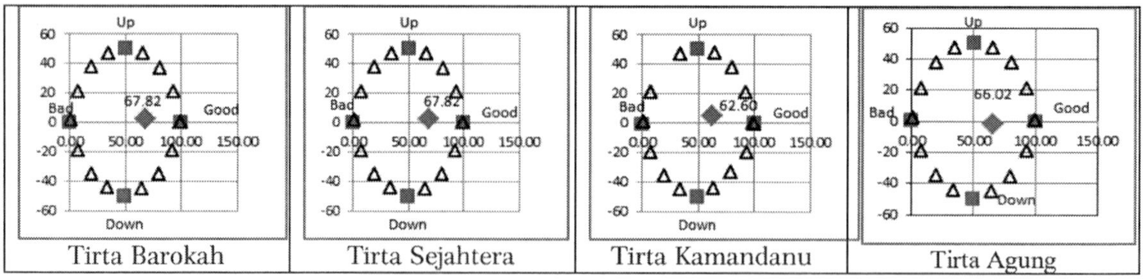

Fig 1. Sustainability Status Based on Ecological Dimension

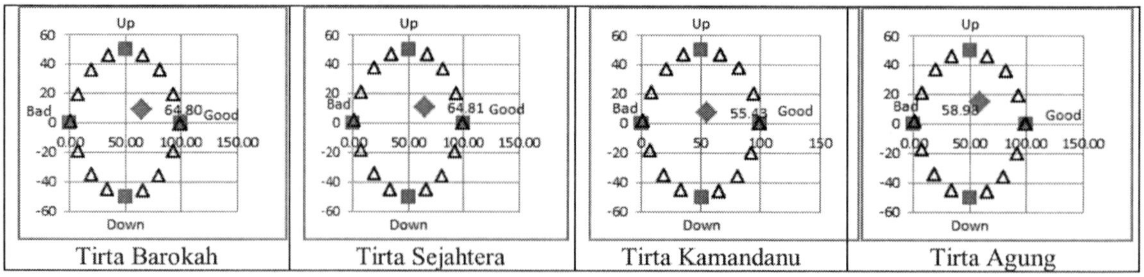

Fig 2. Sustainability Status Based on Technological Dimension

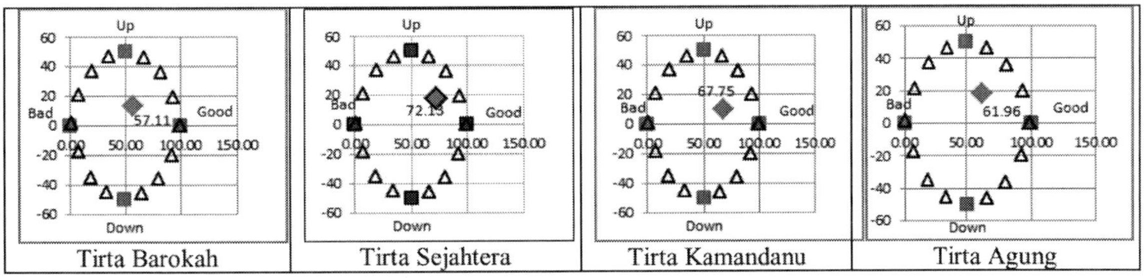

Fig 3. Sustainability Statuses based on Social Dimension.

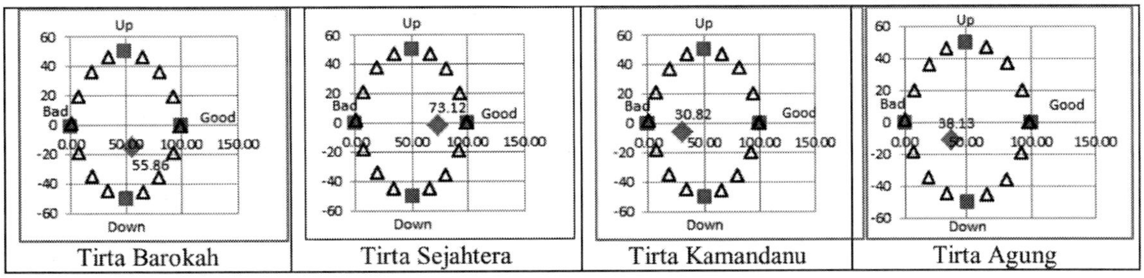

Fig 4. Sustainability Status Based on Institutional Dimension

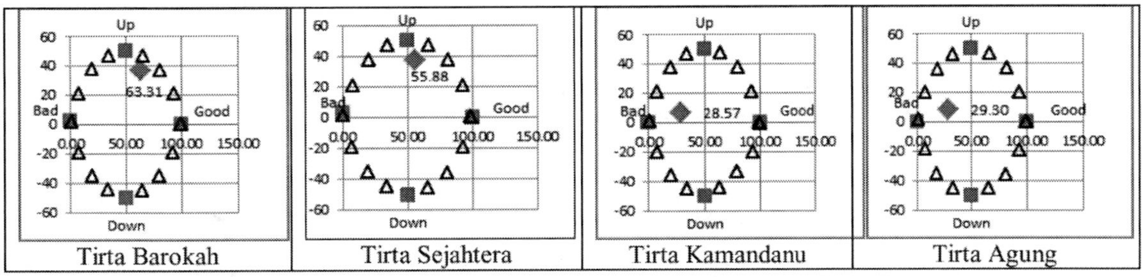

Fig 5. Sustainability Status Based on Economic Dimension.

From the point of view Technological Dimension (Figure 2), Tirta Barokah and Tirta Sejahtera poses good (sustainable status). CBWS Tirta Kamandanu and Tirta Agung has reached sustainable status, which are barely over 50.

From Social Dimension (Figure 3), all the CBWS are in good sustainable status. It shows that the community support the existence of all CBWS. They are active in maintaining its facilities, and eager participating for the maintenance, as well as tariffs' payment. This situation is a very good modal for developing and increasing the sustainability of any CBWS.

Figure 4 shows the Institutional Dimension. It shows that Tirta Barokah and Tirta Sejahtera are in sustainable status (over 50 score). Meanwhile, Tirta Kamandanu and Tirta Agung are very low (not sustainable status), ecah obtain 30.82 and 38.64. This result is also reflected that the management team does not functioning properly. There is only one person who is active currently.

Furthemore, the dimension of Economic such as shown in Figure 5, shows that Tirta Barokah and Tirta Sejahtera are in sustainable status. However, Tirta Kamandanu and Tirta Agung has low sustainability (Not Sustainable status), since it has low score of 28.57 and 29.30 respectively.

Overall, Tirta Barokah and Tirta Sejahtera are in sustainable status with MDS score are all over 50 which allow them to develop further. The Tirta Kamandanu and Tirta Agung are still having problem internally (institutional and economic), which cause them to be in the status of not sustainable

4.3. Sensitivity Using Monte Carlo Simulation

The Monte Carlo simulation result show minor discrepancies (less than 1.5 % of discrepancies) compared with the result of MDS analysis. It shows that the attributes and its scoring are consistent the results does not influenced by any random error.

5 Conclusion and Recommendation

1. It shows that the MDS method within the Rapfish method can be comprehensively applied in accessing sustainability area wide including in water supply.
2. The type of source water (deep well or springs) does not have significant (determining factor) guaranteeing sustainable CBWS. Rather, it more dependent to its management practices.
3. CBWS Tirta Barokah (using deep well) and Tirta Sejahtera (using springs water) are two CBWS which are currently can be said to be on the right track of sustainability as indicated by the score of each dimension reaches over 50. They are also having some distingusing features as well which are the modal for further development. The strategy recommended is to focus on maintaining the management practices and more improvements on its operational aspects in order to gain more strengths.
4. CBWS Tirta Kamandanu (using deep well) and Tirta Agung (using springs) show that these CBWS has not run well as indicated that the score in dimension of institution and economics are far below 50. It needs urgent improvement on the dimension of institution (organisation, job description, and its management, etc.) and on the dimension of economic. The recommended strategy is changing the current management practices overcoming the weaknesses, seek assistance and facilitation from outside institutions (government institution).

References

1. UNDP, Sustainable Development Goals : 17 Goals to Transform our World, http://www.un.org/sustainabledevelopment/water-and-sanitation/ (accessed 20 May 2017).
2. Badan Pusat Statistik. Kota Salatiga Dalam Angka (2016).
3. Tony J. Pitcher. RAPFISH : Rapid Appraisal Technique for Fisheries, and its Application to the Code of Conduct for Responsible Fisheries. Food and Agriculture Organization of the United nations, Rome (1999).
4. Sayed M. N. U., Zifu Li, Heinz-Peter mang. A strengths, weaknesses, opportunities, and threats analysis on integrating safe water supply and sustainable sanitation systems, Journal of Water, Sanitation and Hygiene for Development **4**(3):437-448· September (2014).
5. Brikké, François dan Bredero, Maarten. Linking Technology Choice with Operation`and Maintenance in The Context of Community Water Supply and Sanitation. Health Organization and IRC Water and Sanitation Centre (2003).
6. Castro, Vivian. Sustainable Community Management of Urban Water and Sanitation Schemes.Water and Sanitation Program – Africa, World Bank. Nairobi, Kenya (2009).
7. Fauzi A and S Anna. Pemodelan Sumberdaya Perikanan dan Kelautan Untuk Analisis Kebijakan. Penerbit Gramedia Pustaka Utama. Jakarta (2005).
8. Castro, Vivian. Sustainable Community Management of Urban Water and Sanitation Schemes.Water and Sanitation Program – Africa, World Bank. Nairobi, Kenya (2009).
9. Hasim., A. Sapei., S. Budiharsono and Y. Wardiatno. Analisis Status Keberlanjutan Untuk Pengembangan Pengelolaan Pada Danau Limboto Provinsi Gorontalo. Prosiding Seminar Nasional Limnologi VI. Program Studi Pengelolaan Sumberdaya Alam dan Lingkungan. Institut Pertanian Bogor. Bogor (2012).
10. Kusbimanto, I.W. Sitorus, S.R.P. Machfud. Poerwo. I.F.P, and Yani M. Analisis Keberlanjutan Pengembangan Prasarana Transportasi Perkotaan di Metropolitan Minasata Provinsi Sulawesi Selatan . Jurnal Jalan-Jembatan. Penerbit Puslitbang Jalan dan jembatan Badan Litbang, Kementerian Pekerjaan Umum. **30**(1). (2013).
11. Kavanagh, P. Rapid Appraisal of Fisheries (Rapfish) Project. Rapfish Software Description (For Microsoft Exel). University of British Columbia. Fishries Centre. Vancouver (2001).
12. Kepmendagri No. 47 Tahun 1999 tentang Pedoman Penilaian Kinerja PDAM.
13. Kemen PU BPPSPAM. 2010. Petunjuk Teknis Penilaian Kinerja PDAM.

Indonesian Highway Capacity Guidance (IHCG) correction factor for signalized intersection

Amelia K. Indriastuti and *Eko* Yulipriyono
Civil Engineering Department, Diponegoro University, Prof. Sudharto Avenue, Semarang, Indonesia

Abstract. Indonesian Highway Capacity Guidance or IHCG is a manual developed in 2014 to accommodate traffic characteristics which are no longer appropriate to the preceding manual (Indonesian Highway Capacity Manual or IHCM, established in 1997). From a preliminary study, there are still some differences between theoretical analysis (using IHCG) with the actual conditions. This study aims to obtain IHCG correction factor based on a comparison between predicted performance parameter of a signalized intersection (using IHCG) and measured data, in order to optimize IHCG development. Intersection performance parameters analyzed are the number of queue vehicles (N_{Q1} and N_{Q2}), and queue length (QL). The model of the correction factor was determined with regression analysis, using DS value as the independent variable, and N_{Q1}, N_{Q2}, or QL differences value as the dependent variables. Implementation of the proposed correction factor of queue length on nearby intersections resulted that the corrected guidance could produce more accurate queue length against measured one by 12% deviation.

1 Introduction

Indonesian Highway Capacity Manual or IHCM [1] is a manual used to analyze the capacity and performance of roads in Indonesia since 1997. Along with the changes in traffic characteristics, especially due to the rapid growth of motorcycle population, this reference is no longer appropriate to the actual conditions, as stated in some references [2] and in IHCG [3]. To adapt IHCM to current conditions, Indonesian Highway Capacity Guidance or IHCG [3] is published in 2014.

A preliminary survey was carried out on Gajah Intersection [4], a four-leg signalized intersection in East Semarang. This survey was conducted to analyze how traffic conditions at intersections are viewed from the point of view of IHCM and IHCG. It showed that there were differences between theoretical analysis (predicted using IHCM and IHCG) with the real conditions. However, using IHCG, the results were closer to the actual conditions compared to the results of IHCM, but some significant differences were still found. This indicates that IHCG is quite better than IHCM, but some reference value still requires adjustment to the field conditions. Therefore, this research was conducted to obtain IHCG correction factor based on a comparison between IHCG results with measured data, in order to optimize IHCG development.

2 Method

The case study was at Gajah Intersection in Semarang City. The observation was conducted at peak time with normal settings (without manual interference, neither Area Traffic Control System/ATCS nor police assistance). IHCG parameters analyzed were the number of queue vehicles (N_{Q1} and N_{Q2}), and queue length (QL), as follows.

$$N_{Q1} = 0.25 \, x \, C \, x \left\{ (DS-1)^2 + \sqrt{(DS-1)^2 + \frac{8}{c}(DS-0{,}5)} \right\}, \text{ for DS} > 0.5$$

$$\text{And } N_{Q1} = 0 \text{ for DS} \leq 0.5 \quad (1)$$

$$N_{Q2} = c \, x \, \frac{1-GR}{1-GRxDS} x \frac{Q}{3600} \quad (2)$$

$$QL = (N_{Q1} + N_{Q2}) \, x \, \frac{20}{W} \quad (3)$$

N_{Q1} is the number of passenger car unit (pcu) that remain from the previous green phase; N_{Q2} is the number of queuing pcu that arrive during the red phase; QL is queue length (in m); C is capacity of signalized intersection; DS is degree of saturation or the ratio of flow (Q) and capacity (C); GR is the green ratio (green time per total cycle time); and W is the entry width.

N_{Q1} and N_{Q2}, and QL from IHCG analysis then compared with the N_{Q1}, N_{Q2}, and QL from field observation, using Paired Sample T-Test method for normally-distributed data or Wilcoxon Test for nonnormally-distributed data [5]. For parameters with significant differences, the correction factor was modeled using regression method.

In IHCG, Degree of Saturation (DS) is a parameter of signalized intersection that mostly used to conclude the field condition. DS also has influences on other parameters, including N_{Q1}, N_{Q2}, and QL. Therefore, the model of the correction factor was determined using regression analysis, using DS value as the independent variable, and N_{Q1}, N_{Q2}, or QL value as the dependent variables.

The proposed model will be implemented to see whether the model is suitable for other intersections. Among three parameters, this paper was focused on the implementation of the proposed queue length correction factor.

3 Data

3.1 Geometric

Gajah Intersection lies in Eastern Semarang, in a commercial area with a low class of side friction. Geometric condition and dimension are as shown in Table 1.

Table. 1. Geometric condition of Gajah Intersection

Approach		Width (m)			Median (m)	Parking distance
		Approach	Entry	LTOR		
East	Brigjen. Sudiarto Rd. (to Purwodadi)	12.5	10.0	2.5	0.5	n.a
South	Lamper Tengah St.	4.0	4.0	2.0	n.a	n.a
West	Brigjen. Sudiarto Rd. (to downtown)	10.5	6.5	4.0	0.5	n.a
North	Gajah Raya St.	7.0	6.0	1.0	n.a	n.a

Note : LTOR = Left Turn on Red; n.a = not available

3.2 Phase setting and cycle time

Gajah Intersection is arranged with four phases of movement, as follows:

a. First phase : all movement from east approach and straight movement from west approach
b. Second phase : right-turning movement from east approach and all movement from west approach
c. Third phase : all movement from north approach
d. Fourth phase : all movement from south approach

Thus, left-turning movements from North, East, and West are LTOR (Left Turn on Red). The pattern in a single cycle time is green – yellow – red – yellow, in 150 seconds, as shown in Table 2.

Table 2. Phase setting and cycle time of Gajah Intersection (in seconds)

Phase	Green time	Amber time	All red time	Red time	Cycle time
1	60	2	2	82	150
2	20	2	2	122	150
3	25	2	2	117	150
4	25	2	2	117	150

3.3 Flow rate

Traffic data used in the study were the flow-rate data (Q), the number of vehicles retained by the previous green phase (N_{Q1}) and the number of vehicles approaching and stopped on the red phase (N_{Q2}). Traffic counting was carried on using the video camera on certain location so that all data needed could be measured.

4 Result

4.1 Intersection performance comparison between analytical (IHCG) and measured data

Intersection performance parameters reviewed were the number of queue vehicles (N_{Q1} and N_{Q2}), and queue length (QL). Analytical data were predicted using IHCG analysis. Meanwhile, measured data were data from the field which are managed so that they have the same unit as IHCG parameters. The comparison between those data is shown in Table 3.

It was found that between the results of predicted (IHCG) and measured data, the variation of N_{Q1} differed up to 300 pcu, the variation of N_{Q2} was up to 9 pcu, and the variation of PA was up to 1900 meter.

The statistical difference test or comparative analysis was performed with the t-Two Paired Sample t-Test for normally distributed data and with Wilcoxon test for non-distributed data. From the difference test on the N_{Q1} parameter, 25% of data were significantly diverse. On parameter N_{Q2}, 41% were significantly different. On parameter QL, 50% were significantly dissimilar.

It can be seen that there were still some differences between IHCG calculation result and field observation for all parameters of N_{Q1}, N_{Q2}, and QL. These differences can be minimized by using correction factor (FC), which includes the correction factor of queue vehicle number (FC_{NQ1} and FC_{NQ2}), and the correction factor of queue length (FC_{QL}).

Table 3. Intersection performance comparison between analytical (IHCG) and empirical data [4]

No. of set data	Flow	DS	N_{Q1}		N_{Q2}		QL	
			Predicted	Measured	Predicted	Measured	Predicted	Measured
	pcu/h		pcu	pcu	pcu	pcu	m	m
1	428	0.68	0.6	0.0	16.8	16.1	80.8	67.5
2	492	0.79	1.4	0.0	19.7	19.7	98.2	47.0
3	342	0.55	0.1	0.0	13.1	13.1	61.5	25.0
4	541	0.86	2.4	0.0	21.9	14.5	113.8	41.0
5	421	0.67	0.5	0.0	16.4	9.3	78.9	33.5
6	729	0.46	0.0	0.0	22.3	24.3	89.3	92.5
7	853	0.54	0.1	0.0	27.1	29.9	108.8	80.0
8	885	0.55	0.1	0.0	28.4	30.3	114.2	75.0
9	118	0.52	0.0	0.0	4.6	6.9	43.1	25.0
10	60	0.27	0.0	0.0	2.3	2.8	21.1	12.5
11	68	0.30	0.0	0.0	2.6	3.1	24.0	17.5
12	71	0.31	0.0	0.0	2.7	4.4	25.1	17.5
13	1094	0.64	0.4	0.0	36.8	34.3	138.7	90.0
14	1331	0.78	1.2	0.0	48.3	29.8	184.9	102.5
15	1281	0.75	1.0	0.0	45.7	27.4	174.4	110.0
16	1166	0.68	0.6	0.0	40.1	22.6	151.8	65.0
17	110	0.48	0.0	0.0	4.2	5.4	39.6	67.5
18	156	0.69	0.6	0.0	6.2	6.7	63.3	47.0
19	125	0.55	0.1	0.0	4.9	5.9	46.6	25.0
20	129	0.56	0.2	0.0	5.0	4.9	48.3	41.0
21	464	0.74	0.9	0.0	18.4	16.4	89.8	45.0
22	606	0.96	7.7	1.8	25.1	17.0	152.8	68.0
23	1429	0.89	3.6	2.7	55.6	51.6	237.0	147.5
24	183	0.80	1.4	0.0	7.4	3.1	82.2	12.5
25	1174	0.69	0.6	0.0	40.5	20.1	153.3	70.0
26	171	0.75	1.0	0.0	6.9	4.0	73.4	45.0

Note: *pcu* (passenger car unit) is a metric used in Transportation Engineering. A *pcu* is a measure of the impact that a mode of transport has on traffic variables (such as headway, speed, density) compared to a single standard passenger car.

4.2 Model of IHCG correction factors

The IHCG correction factors, which includes the queue vehicle number correction factor (FC_{NQ1} and FC_{NQ2}) and the queue length correction factor (FC_{QL}), were developed by comparing the predicted data (N_{Q1}, N_{Q2}, and QL) to the measured data. Thus, the differences between the predicted data and the measured data on Gajah Intersection were analyzed using regression approach. The preliminary result of the correction factor modeling is shown in Table 4.

The application of this correction factor was as follows: corrected IHCG parameter value = IHCG parameter value x (1 - FC). Thus, corrected $N_{Q1} = N_{Q1}$ x $(1 - FC_{NQ1})$. Using this equation, the corrected N_{Q1} value from IHCG prediction on Gajah Intersection was closer to the measured N_{Q1} up to 2% in average. Corrected $N_{Q2} = N_{Q2}$ x $(1 - FC_{NQ2})$. The corrected

N_{Q2} value from IHCG prediction was less differ to the measured N_{Q2} up to 30% on average. Finally, corrected QL = QL x (1 – FC_{QL}), that brought the average difference between analytical and measured queue length only about 20%.

Table 4. Preliminary model of correction factor [4]

Variables	Model		Limitation
N_{Q1} – DS	FC_{NQ1} =	0	DS < 0.85
	FC_{NQ1} =	$-0.003\ DS^4 + 0.069\ DS^3 - 0.515\ DS^2 + 1.629\ DS - 0.928$	$0.85 \leq DS < 1.00$
	FC_{NQ1} =	$0.024\ DS^3 - 0.213\ DS^2 + 0.630\ DS + 0.331$	$DS \geq 1.00$
N_{Q2} – DS	FC_{NQ2} =	$2784\ DS^4 - 4576.3\ DS^3 + 2759.1\ DS^2 - 721.5\ DS + 68.712$	DS < 0.55
	FC_{NQ2} =	$-407.03\ DS^3 + 797.92\ DS^2 - 515.28\ DS + 109.84$	$0.55 \leq DS < 0.80$
	FC_{NQ2} =	$897.76\ DS^3 - 2333.2\ DS^2 + 2013\ DS - 576.25$	$0.80 \leq DS < 1.00$
	FC_{NQ2} =	$0.1026\ DS^3 - 0.7564\ DS^2 + 1.8792\ DS - 1.2186$	$DS \geq 1.00$
QL – DS	FC_{QL} =	$578.53\ DS^5 - 1417.5\ DS^4 + 1321.2\ DS^3 - 578.59\ DS^2 + 117.61\ DS - 8.4937$	DS < 0.85
	FC_{QL} =	$1.6347\ DS^4 - 11.05\ DS^3 + 26.99\ DS^2 - 27.677\ DS + 10.541$	$DS \geq 0.85$

4.3 Implementation on other intersections

The implementation on other intersections nearby (Bangkong, Milo, and Sompok Intersection) was focused on the queue length correction factor. The result was presented in Figure 1. This figure compares the measured and predicted queue lengths using IHCM, IHCG, and IHCG corrected methods.

From Figure 1, it can be seen that the implementation of IHCG correction factor on these intersection gives predicted queue length that was more similar to the measured queue length, compare to IHCM and IHCG. In Bangkong Intersection, the average deviation of the predicted queue length was -0.29 (using IHCG corrected), 0.28 (using IHCG) and 0.70 (using IHCM). In Milo Intersection, the average deviations were 0.23, 1.40, and 0.99 for predicted queue length using IHCG corrected, IHCG and IHCM, respectively. In Sompok Intersection, the average deviation of predicted queue length was -0.29, 0.31, and 2.04 for IHCG corrected, IHCG and IHCM successively. The lower value of average deviation means the closer value of predicted data to measured data. The minus sign (-) means that the prediction is underestimated, vice versa. On average, the proposed queue length correction factor of IHCG could produce more accurate queue length against measured one by 12% deviation.

The result of these models are quite good for Gajah Intersection and the three nearby intersections (Bangkong, Milo, and Sompok Intersection), but further research by involving other types of intersections is still needed to know whether this model can be applied on other typical intersection and to find out the limitation of intersection category in order to obtain a more general model for signalized intersections.

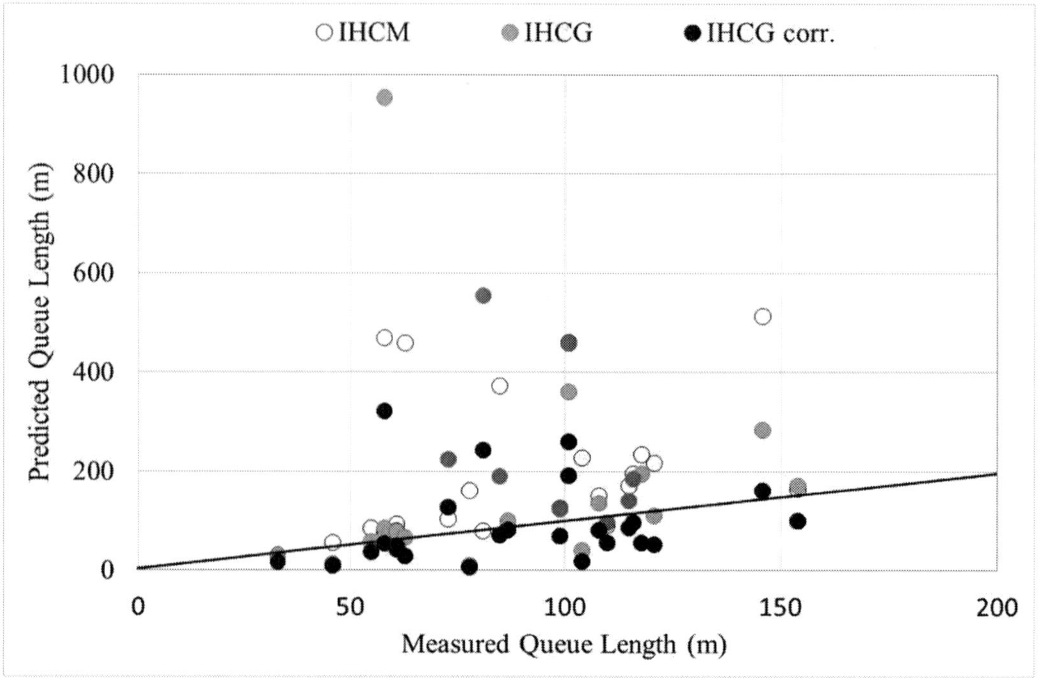

Figure 1. Comparison of predicted and measured queue lengths at several intersections

5 Conclusion

This paper presented a proposed correction to existing Indonesian Highway Capacity Guidance (IHCG) in order to improve the accurateness of the guidance in predicting queue length in intersections. It resulted that the corrected guidance could produce more accurate queue length against measured one by 12% deviation.

References

1. Department of Public Work, *Indonesian Highway Capacity Manual* (Department of Public Work, 1997) in Bahasa Indonesia
2. H. Iskandar, *Basic Capacity for Freeway* (Jurnal Jalan – Jembatan, Volume 29 No.1, April, 2012) in Bahasa Indonesia
3. Department of Public Work, *Indonesian Highway Capacity Guidance* (Department of Public Work, 2014) in Bahasa Indonesia
4. S. Budi, G. Sihite, E. Yulipriyono, and A.K. Indriastuti, *Comparison of Signalized Intersection Performance based on Indonesian Highway Capacity Guidance 2014 and Direct Observation* (Jurnal Karya Teknik Sipil, Volume 6 No 2, 2017) in Bahasa Indonesia
5. R.E. Walpole, R.H. Myers, S.L. Myers, and K. Ye, *Probability and Statistics for Engineers and Scientists* (Prentice Hall, 2011)

Critical Success Factors in Developing Collaborative Design-Build Project Team to Improve Project Performance

Afrizal Nursin[1*], *Yusuf* Latief[1], and *Ibrahim*[1]

[1]Civil Engineering Department, Faculty of Engineering, University of Indonesia, Depok, Indonesia

Abstract. To complete a construction project, it takes the cooperation and collaboration of stakeholder involved in construction projects to obtain optimum project performance. Problems that frequently occur in construction projects is the lack of cooperation, lack of trust, ineffective communication, lack of visualization of the planned buildings and different relationships between stakeholders that project performance is not optimal. To overcome these problems need to develop a model of collaboration between stakeholders in a construction project. The purpose of this research is to identify the critical success factors of the collaboration of design and build construction project team so that can understand the influence of collaboration model to collaboration performance with the geographic, professional, stakeholder, and ethnic cultural boundary spanners in design and build construction project in Indonesia. This research uses factor analysis and variable relation with Structural Equation Modeling (SEM) method. The results found that geographical and ethical boundaries are influencing collaborative performance significantly.

1 Introduction

Collaboration and partnerships between owners and others have formed a full field of inquiry over the last decade. Hu studied the performance improvement of collaborations in construction and said that active collaborative development among all stakeholders and project life cycles is critical to improving the performance of construction projects [1]. Collaboration is the approach between stakeholders to secure the best decisions and results about design and construction choices. Collaborative of design and planning will maximize positive iterations and reduce negative iterations [2].

The construction industry has faced problems with lack of cooperation and trust [3], ineffective communication [4], [5], lack of visualization of planned buildings [6] and conflicting relationships among stakeholders [7]. Previous research on construction collaborations has shown new forms of collaborative approaches, such as partnerships, creating common spaces in construction problem solving [8], communication [9], trust [3], commitment and transparency [10].

* Corresponding author: afrizalnursin@gmail.com

The study of collaborations found three boundary spanners that influence the success of project team collaboration ie geographic, professional, and stakeholder [11].

In this research, the fourth critical success factor of collaborative design and build construction projects is the cultural boundary spanner. Discussion of cultural boundaries is essential for research where many of the problems of collaboration and cooperation of construction project teams are hampered due to cultural differences, whether corporate culture [12] or ethnic culture [13]. Erwin in his research explains the factors that can integrate ethnic cultures such as historical togetherness, external threats, leader agreement, homogeneity in socio-cultural relations, and the interdependence of stakeholders [14].

The purpose of this research is to identify the critical success factors of collaboration of design and build construction project team so that can understand the influence of collaboration model to collaboration performance with the geographic, professional, stakeholder, and ethnic cultural boundary spanners in design and build construction project in Indonesia

2 Literature review

2.1 Collaboration

Collaboration is an activity that involves many people with various interests to achieve mutually satisfying outcomes. Collaboration is known by many names, some popular names include "problem solving", "consensus", "desire-based negotiation", "win-win solution", "mutual benefit", and "principled negotiation" [15]. The purpose of the collaboration is to manage the dispute so that the result is more constructive than destructive. The results spawned effective communication, problem-solving, and improved relationships [16], [17]. Benefits and advantages of collaboration include: improving profits by sharing expertise among business units or companies; Practical cost reduction; Enhance decision-making through sharing deep insights; Innovation through sharing ideas; Improving the ability to pursue targets [18]. Owners control full collaboration through sharing information at the beginning of the project process most likely to achieve desired results: fast, efficient, effective, and related building costs. Owners must be the part that drives these changes, by leading the creation of collaboration, cross-functional teams composed of design, construction, and professional management facilities [19]. The process of collaboration consists of preparation, seeking consensus agreements, and after the final agreement of deliberation [20]. The collaborative strategy consists of bringing the parties together in collaboration, providing understanding, determining the desired level of success, explaining project collaboration, identifying possible barriers, identifying non-collaborative components, and collaborative commitment of members [20]. The leadership style of the collaboration consists of the leadership of organizational structure, important information leadership, leadership in decision-making authority, leadership by accountability and industry, and leadership in the best possible way [21]. The process of collaboration, strategy, leadership style, and collaboration models are the independent variables in this study **(X)**. The performance of collaboration is the dependent variable in this study **(Y)**.

2.2 Collaboration Types / Models

The collaboration types made the model of collaboration that consists of (1) the type of collaboration base on the process, then named as process collaboration [22], (2) the type of collaboration base on the task-dependent, named as task dependent collaboration model [23], (3) the type of collaboration that full of integration, then named full integration model [24],

(4) consortium model [25], information democracy and multi-organizational project model [26]. The performance of collaboration is the dependent variable in this study **(Y)**.

2.3 Collaboration Performances

Collaboration performance in this research gets the well know of the project team, get trust of the project team, good communication, encourage to contribute to solving problems, create a solid and well-connected project team, high flexibility space in the dynamics of project implementation, innovative methods of high innovation in the team drive [27], as well as a strong and passionate sharing of knowledge sharing contributes experience based on expertise, and there is an intercultural customization, the same feeling in the nation and language [14].

3 Methodologies

This research collected data questionnaires from 127 respondents who dabbled in building construction and infrastructure project used an object of this research. The instrument in this study consist of two variables, namely the dependent variable (collaboration performances) and independent variables (collaboration model and boundary spanners).

The instrument of data processing in this research consists of a questionnaire that will be used to analyze the influence of indicators and this relationship with SEM (Structural Equation Modeling) method. In this research will be conducted some analysis, consisting of statistical analysis of homogeneity test using SPSS, relationship analysis with SEM PLS, and analysis of inner model evaluation.

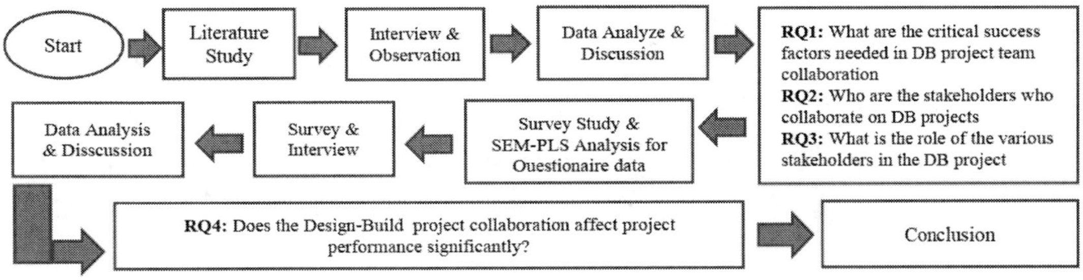

Fig. 1. Research Methodologies

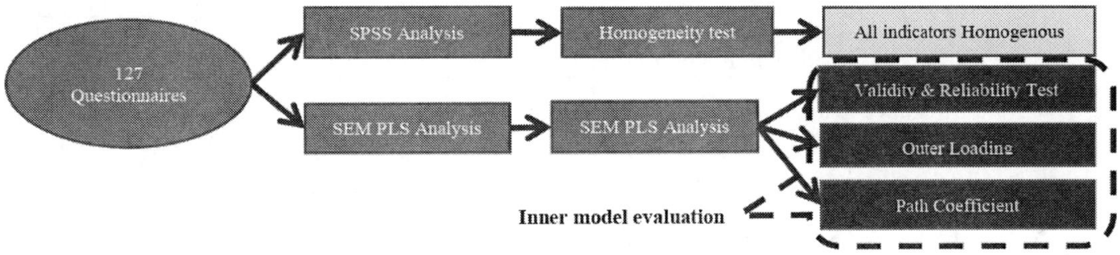

Fig. 2. Data Analysis

4 Results

The analysis was conducted in two stages, namely the study of collaboration variables model relationship with collaboration performance and continued with the report of the boundary indicator relationship model with the collaboration model. The results of the study that has been discussed presented in the form of the tabulation of the significance of Path coefficient and Outer loading of two models of relationship that is the model of performance and boundary spanners on the model, as follows:

Table 1. Influences ranking of collaboration model indicators to collaboration performance

Rank of Influences	Collaboration Model Indicators	T Statistics (|O/STDEV|)
1	M3.1.2 <- Organizational Structure	140.404
2	M3.4.2 <- Basis of accountability and industry	110.594
3	M3.1.1 <- Organizational Structure	72.634
4	M1.3.3 <- After the agreement	57.960
5	M2.1.2 <- Bringing the parties to collaborate	57.203

Table 2. Influences ranking of boundary spanners indicator to collaboration model

Rank of Influences	Boundary Spanner Indicators	T Statistics (|O/STDEV|)
1	M4.7.1 <- Geographic Boundary Spanner	138.479
2	M4.10.4.8 <- Ethnic Culture Boundary spanner	59.853
3	M4.7.2 <- Geographic Boundary Spanner	53.324
4	M4.1.1 <- Collaboration process/project Model	40.957
5	M4.10.4.9 <- Ethnic Culture Boundary spanner	38.944

Table 3. Relationship Ranking of collaboration model variables to collaboration performance

Rank of Relationship	Variable Relationship	Original Sample (O)	T Statistics (|O/STDEV|)
1	After the agreement -> Cooperation in the project	0.551	7.421
2	The best way to work -> Process/project Collaboration model	0.660	7.028
3	After the agreement -> Bringing the parties to collaborate	0.481	6.932
4	After agreement -> Collaborative commitment of members	0.738	6.920
5	Find agreement -> Identify possible obstacles	0.501	6.627
6	Process/project collaboration model -> Collaboration performances	0.403	3.897

Tabel 4. Relationship Ranking of boundary spanners to collaboration model

Rank of Relationship	Variable Relationship	Original Sample (O)	T Statistics (|O/STDEV|)
1	Professional Boundary Spanners -> Process/project Collaboration Model	0.275	1.812
2	Geographic Boundary Spanners -> Process/project Collaboration Model	0.241	1.751
3	Stakeholder Boundary Spanners -> Process/project Collaboration Model	0.160	0.794
4	Ethnic Culture Boundary spanners -> Process/project Collaboration Model	0.039	0.269

5 Discussions

5.1 Influences ranking of collaboration model indicators to collaboration performance

Inter-organizational network (Organizational Structure), Performance Indicators Common goals (Basis of accountability and industry), Scattered across organizations (Organizational Structure), Conduct agreements (After agreement), and Explanation of cooperation methods (Carrying the parties collaborating) should get attention so that collaboration performance is going well because these are the most influential indicators with significant T statistical value.

5.2 Influences ranking of boundary spanners indicators to collaboration model

Provide shared workspace (Geographic boundary spanners), Equal feeling in the nation (Ethnic cultural boundary spanners), Provide shared social space (Geographic boundary), more structured of Project Indicators (Process collaboration model), and Equal feeling in the language (Ethnic cultural boundary) these are the most influential indicators.

5.3 Relationship Ranking of collaboration model variables to collaboration performance variables

Process/project collaboration model affects Collaboration performances with coefficient = 0.403 and significant T statistic = 3.897 > 1.96. This shows the significant effect of the relationship. Boundary number 1.96 shows the level of significance of the relationship between variables where if the value of statistic T is smaller than 1.96 then the relationship between these variables is not significant, on the contrary, if the value of T statistic is bigger than 1.96 then the relationship between these variables is Significant [28].

5.4 Relationship Ranking of boundary spanners variables to collaboration model variables

Relationship the professional boundary spanners indicators affect the process/project collaboration model on the design-build project is rejected even though it is related to the coefficient = 0.275 but not significant T statistic = 1.812 < 1.96 this indicates that the professional boundary spanners have not significantly influenced the collaboration model.

Relationship of geographical boundary spanners indicators affect the process/project collaboration model on the design-build project rejected even though there is a relationship with the coefficient = 0.241 but not significant T statistic = 1.751 < 1.96.

Relationship of stakeholder boundary spanner indicators affects the process/project collaboration model on design-build project rejected even though there is a correlation with coefficient = 0.160 but not significant T statistic = 0.794 < 1.96.

Relationship ethnic cultural boundary spanners indicator affect the process/project collaboration model on the design-build project is rejected even though there is a relationship with the coefficient = 0.039 but not significant T statistic = 0.269 < 1.96.

6 Conclusions

a. Critical success factors of project collaboration performance are as follows: Provide shared workspace (Geographic boundary spanners), equal feeling in the nation (Ethnic cultural boundary spanners), provide social space together (Geographic boundary spanners) Structured project implementation (Process/project collaboration model) and equal feelings in language (Ethnic cultural boundary spanners). Giving more attention to these factors on design-build projects is expected to improve collaboration performance.

b. The result of testing of process/project collaboration model has a significant influence on collaboration performance with path coefficient value 3.897 > 1.96, then tested the impact and relationship between process/project collaboration model on boundary spanners indicators; geographic, stakeholders, professional and ethnic cultures. The results of testing of collaboration models with boundary spanners resulted in significant influence, especially on geographic and ethnic boundary spanners.

The authors would like to thank the financial support provided by Universitas Indonesia through the PITTA 2017 funding scheme under Grant number 861/UN2.R3.1/HKP.05.00/2017 Managed by the Directorate for Research and Public Services (DRPM) Universitas Indonesia.

References

1. Hu. W, *International Conference on Information Management, Innovation Management and Industrial Engineering,* (2008)
2. Howel. A. M, *J. Lean. Constr.*, **2**, 47 (2005)
3. G. Winch, *J. Manage. Constr.,* (2010)
4. A. Kadefors, *J. Project Manage.*, **22**, 3, 175-82 (2004)
5. T. K. Gustavsson, O. S, *J. IT. Const.*, **17**, 520-534 (2012)
6. Ö. Wikforss, A. L, *Itcon.*, **12**, 337-345 (2007)
7. J. Runberger, *Doctoral Dissertation, KTH School of Architecture* (2012)
8. M. Bresnen, N. M, *Const Manage. Econ.*, **18**, 2, 229-37 (2000)
9. Eriksson. P. E, *Const Manage. Econ.*, **28** (9), 905-917 (2010)
10. M Forman, S. L, *Proceedings of the 6th Nordic Conference on Construction Economics and Organisation: Shaping the Construction/Society Nexus,* **2** (2011)
11. J. Nyström, *J. Constr. Manage. Econ.*, **23**, 5, 473-481 (2005)
12. An-Shuen Nir, J.-F. D.-C. *J. Business. Manage.,* **6** (19), 5927-5938 (2012)
13. Lise Lotte Homgreen, I. A, *J. Cross-Cultural Manage.*, 339-356 (2013)
14. Erwin, M, 48-50 (2012)
15. Burgess, G. *Encyclopedia of Conflict Resolution* (1997)
16. L. Coser, *The Functions of Social Conflict* (1956)
17. M. Deutsch, *The Resolution of Conflict: Constructive and Destructive Processes* (1973)
18. Hansen M.T., N. N, *MIT Sloan Management Review* **46** (1), 22-30 (2004)
19. Roundtable, C. o. *The Construction Users Roundtable*. WP-1202 (2004)
20. G. Giesen, *Creating Collaboration: A Process That Works* (2007)
21. T. Xu, D. B, *J. Project Manage.,* **21**, 45-53 (2005)
22. Richman, H. S. *The Power of Collaboration* (2001)
23. Shawn Callahan, M. S, www.fullcire: http://www.anecdote.com (2006)
24. T. L. Foundation, *The Johnson Center for Philanthropy and Nonprofit Leadership at Valley State University* (2008)
25. Verganti, G. P, *Harvard Business Review*, 01-16 (2008)
26. J . Conner, *Human Resource Planning*, **26** (3) (2003)
27. T. K. Gustavsson, *7th Nordic Conference on Constr. Econ. Organisation* (2013)
28. Hengky Latan, I. G (2012)

Optimization of irrigation water use to increase the benefit of agricultural products

Edy Anto Soentoro[1] *, *Erlangga* Perwira[2], *Yadi* Suryadi[1], *and* Winskayati[3]

[1] Water Resources Engineering Research Group, FCEE, Institut Teknologi Bandung, Indonesia.
[2] Master Program in Water Resources Management, FCEE, Institut Teknologi Bandung, Indonesia.
[3] Citarum River Basin Authority, Ministry of Public Works and Housing, Indonesia.

Abstract. Shortage of irrigation water supply in dry season prevents many farmers from growing their crops, and the annual benefit from agricultural products will decrease as much as the area of irrigation fields which have lack of water. The objective of this study is to determine the maximum benefit from agricultural products based on water availability, by determining the appropriate cropping pattern and maximum planting areas through linear programming. The case-study location is at Leuwi Kuya Irrigation Region. Planting schedule is selected based on minimum water shortage from simulation of 6 alternative planting schedules. Then, the best pattern of cropping (planting method and the total area) is determined using linear programming. Optimization is carried out in 3 scenarios with various planting methods (conventional and SRI), minimum irrigation water demand (class-area system), and schedule for beginning of the 3-growing seasons annually. Result of this study is the optimal area of the irrigated region that can be planted based on the water availability. The maximum benefit is 89 billion rupiahs, using SRI planting method and distribution of three groups of irrigation fields in water supply schedule.

Key words: Optimization, linear programming, irrigation, planting area.

1 Introduction

Shortage of irrigation water supply in dry season has caused some areas of paddy field to have lack of water and eventually causing many farmers to lose their crops. The annual benefit from agricultural product of farmer decreases as much as the area of irrigation fields which do not get enough water. Decades ago when the Irrigation Region ("*Daerah Irigasi*") had been built by the Government, the river flow discharge through a weir was enough to supply irrigation water for wet and dry seasons. Due to land-use change in the upstream of the watershed over decades, river flows supplying irrigation water became unstable, causing flood in wet season and small discharge in dry season. Thus, there are not enough irrigation water to supply irrigation area in the region that has been built. So, the government money for building infrastructures in the Irrigation Region is "useless" during dry season, and some of the farmers cannot plant their crops. One effort to overcome the problem above is by finding the optimal cropping pattern and planting area based on water availability.

* Corresponding author: edyas209@gmail.com

The objective of this study is to optimize irrigation water use to increase the benefit of agricultural products using best cropping pattern and linear programing. Result of the study will determine the area of irrigation region which can get enough irrigation water, the crop type, and how much benefits that can be obtained from agricultural products for each growing season, based on water availability.

The study location is Leuwi Kuya Irrigation Region (*Daerah Irigasi Leuwi Kuya*), West Java Province covering 2,535 ha of paddy fields. Infrastructure of Leuwi Kuya Irrigation Region is in good condition, there is no sedimentation problem in canals that obstructs the water distribution process. Data used in this study are regulations related to pattern and planting system, rainfall and flow discharges, irrigation area, agricultural production, and commodity prices data.

The existing PS (planting schedule) in Leuwi Kuya Irrigation Region has water shortage in dry season, many farmers fail to grow their crops, as well as reducing the benefit of agricultural products. During one-year period, there are 3 times of growing season (GS) for two or three crop types (paddy-paddy-"*palawija*") in the Irrigation Region. The planting schedule (PS) is the beginning of the first growing season of a one-year period, and it is determined based on the balance between water requirement and water availability. Simulation procedure of water balance in every half month (e.g., October-1 or the first half of October) is applied to select the best time of planting schedule (PS).

2 Methodology

There are three stages of analysis in this study to optimize the irrigation water use to increase the benefit of agricultural products.

1. Determine the best planting schedule (PS) of 6 alternatives PSs based on the minimum water requirement through simulation procedure.
2. Apply Linear Programing [1, 3] on the best PS, and maximize the area of irrigation field that gets enough water from 3 scenarios of Pattern and Cropping Order (PCO), and,
3. Compare each scenario in terms of cropping intensity and the resulting benefits.

Detailed of stage analysis are as follows:

2.1 Stage 1: Determine the best Planting Schedule (PS)

At this stage, the best planting schedule (PS) is chosen from 6 alternatives of PSs, beginning from October-1 (first half of October) to December-2 (second half of December). The selection is through simulations that are based on the minimum water shortage of each alternative. The planting methods that applied in the study are conventional and SRI methods. Optimization by Linear Programing is carried out based on 3 scenarios as follows,

- Scenario 1: Conventional method for growing the crop is applied on the best planting schedule (PS) from simulation of 6 PS alternatives.
- Scenario 2: SRI method [4] for growing the crop is applied on the best PS.
- Scenario 3: SRI method is applied on one PS that is not the best, to know the effect of method change on pattern determination and planting system.

There are three growing season (GS) in one-year planting schedule. In the first and second GS (GS-1 and GS-2), the entire irrigated field is planted with paddy, while in the third GS (GS-3) it is planted with *"palawija"* such as corn or soybean. Period of GS-1, GS-2, and GS-3 are 3.5, 3.5, and 3 months respectively. In order to reduce total required water supply at intake, a best practice of group system ("*system golongan*") is used. The irrigation region is divided into 3 groups of irrigation fields in which each planting schedule is different a half month to each other. The division is based on the scheme of the canal and box structure to regulate water flows.

2.2 Stage 2: Maximize the planting area based on water availability.

At this stage, optimization is carried out by linear programing to determine maximum area that can get enough water based on water availability for each half-month during a one-year period. The calculation is starting from the beginning of planting schedule, and is done for every half-month period. A sample of the linear programming optimization model is to determine the optimum area that cat get enough water ($AR_{j,1}$, $AR_{j,2}$, and $AR_{j,3}$) for period 3 (j=3, November-2) with group of irrigation field: k=1, 2, and 3, is as below:

$$Max\ Z = Rp.B * (AR\ j,1 + AR\ j,2 + AR\ j,3)$$

Subject to:

$$VREQ\ 1 * AR\ j,1 + VREQ\ 2 * AR\ j,2 + VREQ\ 3 * AR\ j,3 \leq VAVL\ j$$
$$AR\ j,1 + AR\ j,2 + AR\ j,3 \leq AR\ total$$
$$AR\ j,1 = AR\ j\text{-}1,1$$
$$AR\ j,2 = AR\ j\text{-}1,2$$
$$AR\ j,1 \geq 0;\ \ AR\ j,2 \geq 0;\ \ AR\ j,3 \geq 0$$

$Rp.\ B$: net benefit for paddy crop (Rp. million/ ha)
$AR\ j,k$: total area (in (Ha) of group k irrigation field, period j which can get enough water
$VREQ\ j$: volume of water requirement at intake in the jth half month (m^3/s) for group area j
$VAVL\ j$: volume of water available at intake in the jth half month (m^3/s)

2.3 Stage 3: Compare the optimization results.

This third stage is to continue what has been obtained from the optimization process in the second stage, by choosing scenarios that provide the highest value of irrigation benefits in terms of benefits which are generated.

3 Result of study and discussion

Simulation method is used to determine the optimal planting schedule (PS) in Leuwi Kuya Irrigation Region from 6 alternatives of proposed planting schedules, while the existing planting schedule (PS) is the PS alternative-1. The optimal PS is evaluated based on the water required, and availability of irrigation water. After simulation by shifting per half-month from the beginning of planting schedule, the most optimal planting schedule is PS alternative-2, which has the smallest required discharge or volume of water during the year.

3.1 Scenario 1

In scenario 1, the conventional planting method is applied on the best alternative of planting schedule (PS), it is the PS -2 (starting at the 2nd half month of October), with the smallest water shortage. Results obtained from optimization is shown in table and picture and below:

Table 1. Actual planting areas, Planting coverage and the Benefits of Scenario -1

Growing Season (GS)	Group-1 irrigation field	Group-2 irrigation field	Group-3 irrigation field	Group-4 irrigation field	Potential planting area	Actual planting area	Planting coverage	Benefit
	(ha)	(ha)	(ha)	(ha)	(ha)	(ha)	(%)	Billion Rp.
GS -1	293	1149	362	731	2535	2535	100	27.046
GS -2	293	1149	362	731	2535	2535	100	27.046
GS -3	293	16	0	0	2535	309	12	4.536
Total in 1 year					7604	5378	212	58.628

Source: Optimization result

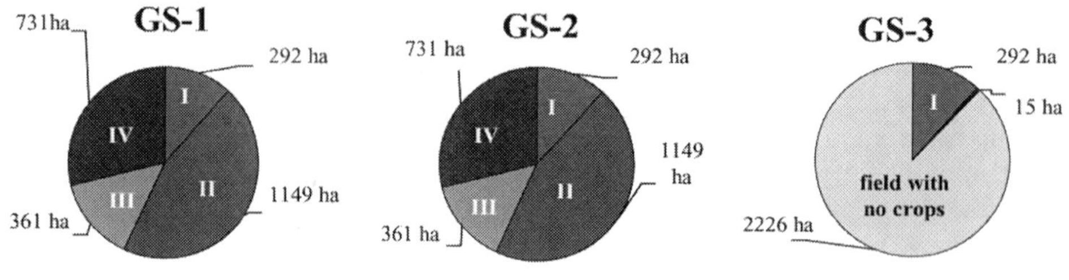

Fig. 1. Planting Area for each Group on Scenario-1 based on Water Availability

3.2 Scenario 2

In scenario 2, it is attempted to increase the income of farmers through rice planting by System of Rice Intensification (SRI) method, which is more profitable than conventional methods. Based on the results of a study that has been conducted by the Research and Development Agency of the Ministry of Public Works [5], SRI method can reduce water demand drastically depending on the location where the SRI method is applied. Nevertheless, based on the calculation results, there is still a period of water shortage in GS-1, so optimization must be done. The SRI method of paddy crop is applied in GS-1 and GS-2, while in GS-3 it is soybean crop. The optimized cultivation pattern is PS alternative-2, which in scenario 1 has been optimized but with different cultivation methods. From the optimization process, the obtained results are as in table and picture below:

Table 2. Actual planting areas, Planting coverage and the Benefits of Scenario -2

Growing Season (GS)	Group-1 irrigation field	Group-2 irrigation field	Potential planting area	Actual planting area	Planting coverage	Benefit
	(ha)	(ha)	(ha)	(ha)	(%)	Billion Rp.
GS -1	431	2104	2535	2535	100	40.987
GS -2	431	2104	2535	2535	100	40.987
GS -3	309	0	2535	309	12	4.536
Total in 1 year			7604	5378	212	86.510

Source: Optimization result

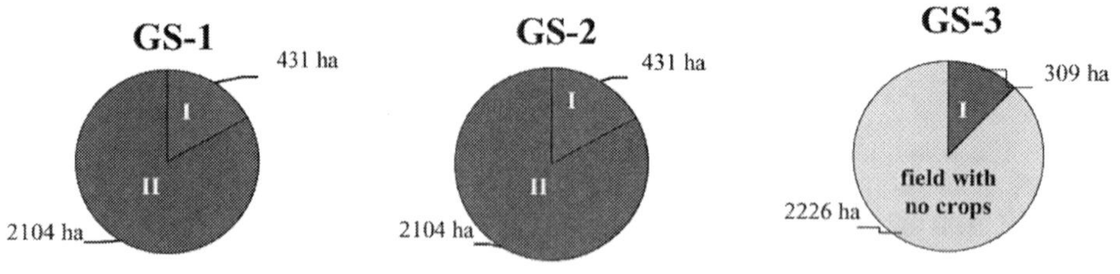

Fig. 2. Planting Area for each Group on Scenario-2 based on Water Availability.

It can be concluded from the above results that comparing with conventional methods, SRI method with the same planting schedule (PS-2: October-2) can give benefits by almost two times, although there is no increase in the planting coverage. In addition, implementation of SRI method can reduce the number of irrigated water group, from 4 group to become 2 groups only, thus making water division easier.

3.3 Scenario 3

The scenario 3 is almost the same with scenario 2, in effort to increase farmer's profit through planting paddy with System of Rice Intensification (SRI) method. The difference from scenario 2 are the planting schedule and the number of group irrigated field. Scenario -3 used the PS-1 instead of the PS-2, and the irrigated region is divided into 3 groups of irrigated fields. The selected PS-1 (the existing condition) in scenario 3 is to furtherly increase the benefits that can be obtained by farmers through the water availability for GS-3 of PS-1 which is more than that of PS-2 on Scenario-2. The choice of 3 groups of irrigated fields is to minimize the total water requirement from the intake gate. Here is the recapitulation of optimization results in scenario 3.

Table 3. Actual planting areas, Planting coverage and the Benefits of Scenario -3

Growing Season (GS)	Group-1 irrigation field	Group-2 irrigation field	Group-3 irrigation field	Potential planting area	Actual planting area	Planting coverage	Benefit
	(ha)	(ha)	(ha)	(ha)	(ha)	(%)	Billion Rp.
GS -1	144	286	2104	2535	2535	100	40.987
GS -2	144	286	2104	2535	2535	100	40.987
GS -3	144	286	82	2535	512	20	7.527
Total in 1 year				7604	5582	220	89.501

Source: Optimization result

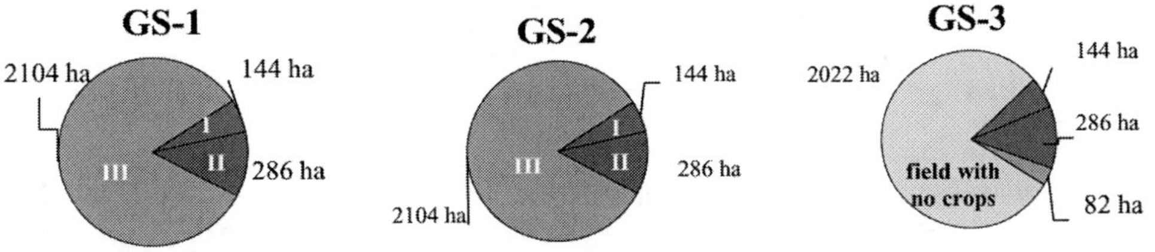

Fig. 3. Planting Area for each Group on Scenario-3 based on Water Availability

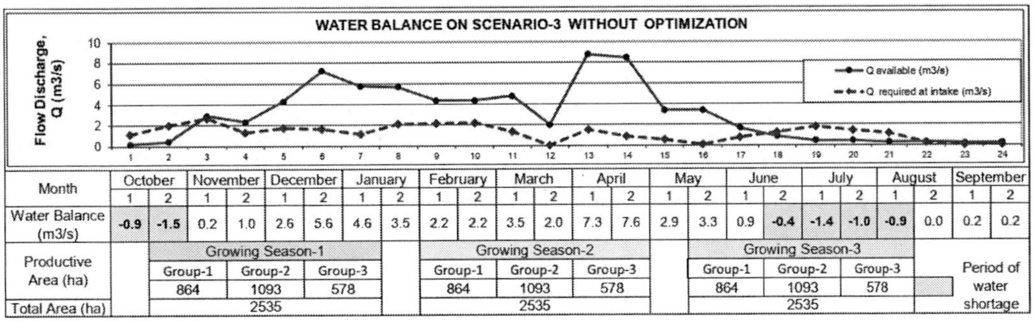

Source: Optimization result

Fig. 4. Water Balance on Scenario 3 without Optimization

Results of this study showed that there is a difference in the number of group division and the maximum planting coverage in this 3rd scenario with the situation in scenario 2 above. With the beginning of planting schedule which is back to October 1 as the existing conditions, and SRI method on growing season (GS) -1 and 2, the planting index for paddy crop remains 200% for GS-1 and GS-2. Planted with soybean crops in GS-3, there was an increase of actual planting area (productive area) from 308.65 ha to 512,72 ha, due to the availability of water. The planting coverage increased 8% and the benefit increased up to 3 billion rupiah per year.

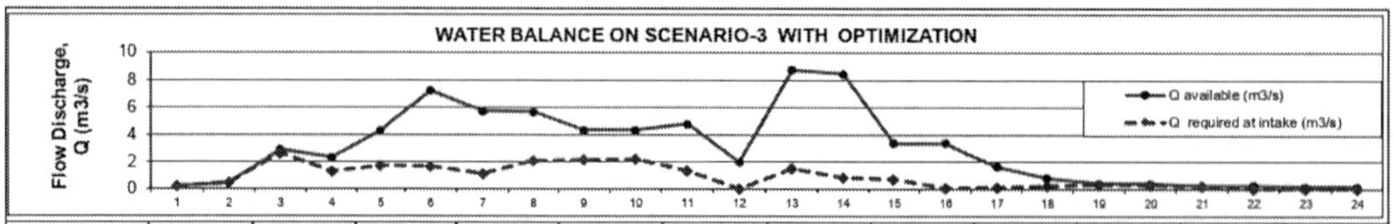

Month	October		November		December		January		February		March		April		May		June		July		August		September	
	1	2	1	2	1	2	1	2	1	2	1	2	1	2	1	2	1	2	1	2	1	2	1	2
Water Balance (m3/s)	0.0	0.0	0.2	1.0	2.6	5.6	4.6	3.5	2.2	2.2	3.5	2.0	7.3	7.6	2.6	3.3	1.5	0.6	0.1	0.1	0.0	0.3	0.2	0.2
Productive Area (ha)		Growing Season-1							Growing Season-2							Growing Season-3								
		Group-1		Group-2		Group-3			Group-1		Group-2		Group-3			Group-1		Group-2		Group-3				
		144		286		2104			144		286		2104			144		286		82				
Total Area (ha)		2535							2535							512								

Source: Optimization result

Fig. 5. Water Balance on Scenario 3 with Optimization

3.4 SRI method as an alternative solution

This study showed clearly that applying the SRI method in irrigation practice resulted more benefits for farmers than that obtained from conventional method. In the wide scale development of SRI method, however, remains some obstacles that must be solved, such as the availability of organic fertilizers and compost. If SRI method will be applied into alternative farming activities on a large scale, then the supply of manure, compost and other organic fertilizer making materials should be done in an integrated manner.

4 Conclusion

The following conclusions can be drawn:

1. The optimization method above can increase the benefit of agricultural products by maximizing area of irrigation field that can be fulfil its water demand, based on water availability and scenario of planting pattern.
2. The best planting method is SRI (System of Rice Intensification), with three groups of water division and three growing season (GS) that begin at early October.
3. The difference in the number of profits between scenario-1 (conventional planting methods) is only 58 billion rupiahs compared to scenario-2 and scenario-3 (SRI planting method) which is 86 billion and 89 billion rupiahs due to the different planting methods used. Scenario-3 is higher in value compared to scenario 2 because the water supply problem in growing season GS-3.

References

1. Bazaraa, M.S., J.J. Jarvis and H.D. Sherali, Linear Programming and Network flows, John Willey & Sons Co., (1990)
2. Kuswara, Basic Ideas and Practice of System of Rice Intensification (SRI) Method - Ecological Farming, FIELD Foundation Indonesia, (2003)
3. Mays, Larry W., Yeoh Koung Tung, Hydro-systems Engineering and Management, McGraw-Hill Book Company, (1992)
4. Purwasasmita, M., and A. Sutaryat, Paddy SRI organic Indonesia, Penebar Swadaya publisher, (2014)
5. Team of Irrigation Agency, Water-saving in paddy cultivation with System of Rice Intensification (SRI) method, *Research Series on Irrigation*, the Research and Development Agency of the Ministry of Public Works, (2006)

Green House Gases (GHG's) Emission Reduction Measures and Verification Challenge at Transport Sector

Haryono Huboyo[1*], *Hari* Wibowo[2], *and Winda* Retna Sari[3]

[1] Department of Environmental Engineering, Diponegoro University Jl.Prof.Soedarto, SH Tembalang Semarang Indonesia
[2] Directorate of IGRK and MPV, Ministry of Environment and Forestry, Indonesia
[3] United Nation Development Program (UNDP), Indonesia

Abstract. The Ministry of Transportation Republic of Indonesia adopted several co-benefits activities related to climate change mitigation action through 2010-2014. In order to claim these activities results as climate change mitigation action, Ministry of Environment and Forestry need to verify them. The verification covers data accuracy, consistency, transparency and completeness of data quality and information. The author also verify the other parameters such as the calculation method, managerial system, monitoring system and funding support. In this program, several mitigation actions were implemented such as Area Traffic Control System (ATCS), traffic impact control, parking management, Bus Rapid Transit (BRT) system, smart driving, non-motorized transport, shifting from private vehicles to rail transport, rejuvenating of navigation facilities, aircraft fleet rejuvenation, renewal of the operation and maintenance system of airplane and improvement of flight navigation system during take-off and landing. The absolute difference results between claim and verification for all climate change mitigation actions based on Government Regulation (PP) No. 61/2011 were 13.44% - 43.5%. While for mitigation action beyond PP 61/2011, the absolute difference were 0% - 67.31%. There are no mitigation action satisfying all verification criteria. Based on these verification activities, The Ministry of Transportation should refine the method of mitigation action.

1 Introduction

National mitigation action to abate GHGs emission facilitated through Presidential Act No. 61/2011[1] will be revised shortly. National target for GHGs emission reduction as stated in the National Action Plan for Reducing GHGs Emission Reduction (RAN-GRK) to attain "Green Indonesia" already has been set. The target is 30% GHGs emission reduction by 2030. In order to get effective mitigation action, the implementation of the previous program of mitigation actions should be clarified. Verifications system is a tool to know the effectiveness of the executed programs.

The scope and measurement-reporting-verification (MRV) supervision on mitigation action for GHGs reduction needs support and commitment. This issue is still on negotiation stage worldwide [2]. Both developed and non-developed countries have their own views on how to verify their achievement. The developing countries are encouraged to commit on implementing definite mitigation action on GHGs emission reduction by using best available resources in their development [3]. MRV system, which is developed domestically for each country before implementing future recognized global MRV, will raise transparency and build trust inter-countries, funders. This will ultimately ascertain international treaty performance [4]. Verification process will be considered effectively if the process is carried out at proper scale as well as measurement scope based on its initial design [5]. Conformity is the significant part during the verification process. The verification will evaluate several significant mitigation actions which may be highlighted for future development [6]. Ministry of Transportation Republic of Indonesia adopted several co-benefits activities related to climate change mitigation action through 2010-2014

This study will evaluate the achievement of national mitigation actions for reducing GHGs emission from Ministry of Transportation through verification process. Transport sector is the main contributors of anthropogenic GHGs emission in Indonesia. Thus reducing this emission will have positive impact at national scale.

2 Methodology

2.1 Scope of verification

In verifying each mitigation actions the author did two general steps i.e: a desk review of official report and interview with represented officer of climate change mitigation for each Directorate of Ministry of Transportation. The author reviewed on data accuracy, consistency, transparency also the conformity of the mitigation action to its original planned (PP No.61/2011) as featured in Fig. 1

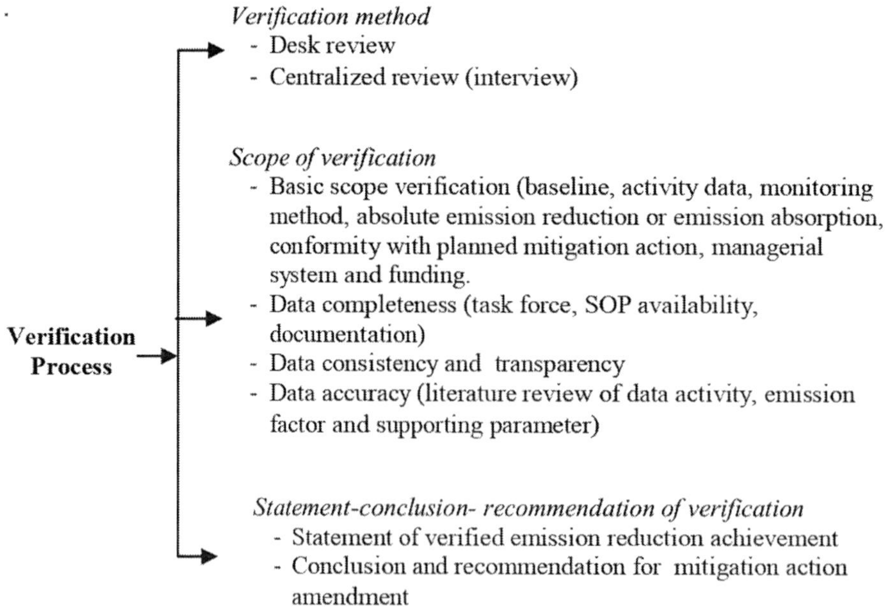

Fig. 1. Scope of verification process.

2.2 Reference Method

The referenced methods used for each Directorate of Ministry of Transportation are as follows:.

Table 1. Methods used for calculating emission reduction

Directorate	Referenced methods
Directorate General of Land Transportation (DGLT)	Technical Guidance on Measurement, Evaluation and Reporting (PEP RAD 2013 Bappenas)
Directorate General of Railway Transportation (DGRT)	Calculation method for National Action Plan for Greenhouse Gas Emission Reduction. Directorate General of Railway Transportation
Directorate General of Sea Transportation (DGST)	USEPA – ICF, 2009 Current Methodologies in Preparing Mobile Sources-Related Source Inventories
Directorate General of Civil Aviation Transportation (DGCAT)	Doc 9988 ICAO: Guidance on the Development of States' Action Plans on CO2 Emissions Reduction, 2nd edition 17 March 2014

Note: USEPA: United States Environmental Protection Agency
ICAO: International Civil Aviation Organization

The method used by DGLT is national methodology established by National Agency of Development Planning (Bappenas). While DGRT used their own method which had not been verified externally. Other Directorates adopted international reference method

3 Results and Discussion

3.1 List of Mitigation Actions

Based on desk review and clarification the author list mitigation actions referred to PP 61/2011 and beyond PP 61/2011. Mitigation actions under PP 61/2011 are as follows: Intelligent Transportation System (ITS) implementation, Traffic Impact Control/TIC, Parking management, Congestion Charging dan Road Pricing, BRT reformation, Public fleet rejuvenation, Gas Fuel Converter Kit, Smart driving or eco-driving, Non-Motorized Transport, City railway networking in Bandung, Railway double track development, Electric train procurement, Modification of Diesel train to Diesel-electric train, Mass Rapid Transportation in Jakarta, Airport train to Soekarno Hatta airport, Monorail development in Jakarta, Road preservation.
Mitigation actions beyond PP.61 Tahun 2011 are as follows: Railway double track at North-Java lane, Airplane fleet rejuvenation, Improvement of System and operation maintenance of airplane, Direct routes implementation of Air Navigation (RNAV) 5, Required Navigation Performance (RNP) 10, Navigation of Continous Climb and Descent Operation program i.e Standard Arrival - Standard Instrument Departure (STAR-SID-RNAV1), RNP Approach Procedure (RNP 0.3, RNP 0.1), Rejuvenation of Aids to Marine Navigation.

3.2 Review on data completeness, consistency, transparency and accuracy

Reviewing data quality and information of mitigation action achievement was based on three aspects i.e team structure, standard operational procedure and documentation work.

The author interviewed represented officers in each Directorate who knows or responsible for the mitigation action of emission reduction. While the author did interview the author collected supported data which were kept at their directorate. Periodical restructuring officer as usually occur in state institutions might prevent us to get the right person to be interviewed.

Annual forum in the Centre for Sustainable Transport Management (PPTB) in Ministry of Transportation for constructing mitigation action report is not categorized as evaluation forum of mitigation action due to it is not specific for each Directorate. Generally, there were persons in charge (PIC) of mitigation action to coordinate mitigation action. However these PICs were not supported by a solid team with definite job description therein except in DGCAT. This circumstance was caused by common view that the mitigation action of climate change is not real program (it is just co-benefit of other program). The PICs in DGLT kept the data consistency by using guidance prepared by Bappenas, yet due to data acquisition (particularly activity data) from local officers were not well managed then in the course of calculation, several calculation adjustment which beyond of the referenced method, took place very often.

Monitoring of the program has been already done by the Ministry of Transportation although several parameters were missing such as registered number of urban vehicles since 2013. The biggest challenge is how to maintain continuous monitoring as in the guidance book (PEP Bappenas) numerous data should be collected. If not there should have simplicity of calculation method, without sacrifice the method accountability, to gather complete data from the local officers. The PICs in DGRT had an effort to keep data consistency by using primary data measured at the branch office of state railway enterprise (PT.KAI) as well as secondary data collected by representative institution such as central statistical bureau (BPS). DGRT has specific person to verify the method and implement the method however due to restructuring personnel in DGRT recently, it is difficult to find the supported data of calculation.

PICs of DGCAT maintained data consistency by adopting ICAO standard for their referenced method. Nevertheless, collecting activity data from local airlines was a difficult task. The simple data such as airplane fleet rejuvenation were not used by DGCAT therefore method modification was used instead. This would deny the achievement claim as they adopted different method beyond what they stated originally.

PICs of DGST already kept the data consistency both on conformity of method as well as data collection method. DGST only propose GHGs emission reduction from rejuvenation of Aids to Sea Navigation thus the activity data were well collected from Navigation District offices. Overall, the related report and data sources for implementing the mitigation actions were not officially managed therefore several data calculation seems to use assumption for their calculation or using target achievement as the data sources. The use of data assumption in the calculation was driven by:

a. There is no study yet for the parameters which being used.

b. There is no effective transferred information of activity data from local officers to Ministry of Transportation.

Information on funding sources of each mitigation action is important part for verification. Amount of fund for several mitigation actions were not detected. The report of budget accountability performance of state institution (LAKIP) was the main source of the funding source for each mitigation action. Moreover, several mitigation actions had co-external funding sources such as from private company which may not recorded in LAKIP.

3.3 Verification of absolute emission reduction

Based on desk review and clarification the author verified mitigation actions referred to PP 61/2011 and beyond PP 61/2011 as listed in Table 5. Two big challenges in their achievement to be claimed as verified mitigation actions. First is lack of recorded activity data from the level which the mitigation actions take place. Usually local officers have such above task. Secondly is the verified methodology they used in their calculation. As long as the methodology the used are based on traceable and reliable literatures, then it will have verified methodology.

Table 2. Evaluation of mitigation action documentation

MITIGATION ACTION	2010		2011		2012		2013		2014	
	C	V	C	V	C	V	C	V	C	V
Land Transportation										
ITS implementation	0.0006	0.0014	0.0046	0.0035	0.0453	0.0343	0.0954	0.0724	0.117	0.0932
Traffic Impact Control/TIC							0.00007	[illegible]	0.00008	n.a
Bus Rapid Transit (BRT) reformation	0.0208	0.0102	0.0226	0.0111	0.0226	0.0111	0.0239	0.0117	0.0529	0.0261
Smart driving or eco-driving	3.7E-04	6.2E-05	3.8E-04	n.a	0.0005	8.7E-05	0.0006	9.8E-05	0.0008	1.2E-04
Non-Motorized Transport	-				-		0.0015	n.ap	0.0017	n.ap
Railway Transportation										
Electric train procurement					0.0148	0.078	0.0144	0.076	0.0085	0.045
TOTAL EMISSION REDUCTION (PP 61/2011)	0.02177	0.01166	0.02757	0.0146	0.0832	0.12348	0.13587	0.16259	0.18098	0.16442
Railway Transportation										
Railway double track at North-Java lane									0.0043	0.01
Aviation Transportation										
Airplane fleet rejuvenation	0.51144	n.a	0.58272	n.a	0.16606	n.a	0.20714	n.a	0.25148	n.a
Improvement of System and operation maintenance of airplane.	-		0.0549	n.a	0.0605	n.a	0.0462	n.a	0.0554	n.a
Direct routes implementation, RNAV5, RNP10	-		0.11958	n.a	0.13393		0.15			
	-		0.05979	n.a	0.06697		0.075			
Navigation of Continous Climb and Descent Operation (STAR-SID-RNAV1)	-		0.47833		0.53573		0.60002			
RNP Approach Procedure (RNP 0.3, RNP 0.1)	-		0.0287		0.03214		0.036			
Rejuvenation of Aids to Marine Navigation	-					-			0.1179	0.0076
TOTAL EMISSION REDUCTION (Non PP 61/2011)	0.51144		1.32402		0.99533		1.11436		0.42908	0.0176
TOTAL EMISSION REDUCTION	0.5332	0.0117	1.3516	0.0146	1.0785	0.12349	1.2502	0.1626	0.61006	0.18202

Notes:

- : Reliable method, complete activity data-supported parameters
- : Reliable method, incomplete activity data-supported parameters
- : Unreliable method, complete activity data-supported parameters
- : Unreliable method, incomplete activity data-supported parameters

n.a : not available, n.ap: not applicable
C: Claimed, V: Verified

Among all mitigation actions, only mitigation action from DGST which calculate CO_2 equivalent as recommended by Ministry of Environment and Forestry. Other directorates only calculated CO_2 reduction.

4 Conclusion and Recommendation

There are no mitigation action satisfying all verification criteria. Partial fulfillment of verification criteria was achieved for several mitigation actions as they lack of calculation reference. On average the absolute difference between claim to verified data in 2010 - 2014 is about 88%. Based on these verification activities, The Ministry of Transportation should refine the method of mitigation action.

Untuk promote verifiable mitigation action, the Ministry of Transportation should conduct Quality Assurance (QA) and Quality Control (QC) on their mitigation actions. Documented SOP should be imposed to prevent uncontrolled scope of mitigation action. This SOP should be robust and flexible to anticipate the dynamics of mitigation action implementation in the field.

The completeness of supporting data such as baseline emission, managerial system, funding support documentation should be disclosed comprehensively so as to get reliable mitigation actions. To have an efficient mitigation action, the Ministry of Transportation should prioritize mitigation actions which may have huge amount of emission reduction for their measurement and monitoring program. This is get optimum amount of claimed emission reductions

References

1. Presidential Act. National Action Plan for GHGs Emission Reduction. (2011).
2. J.Morlot, J. Ellis J; S.Moarif. Measurable, Reportable and Verifiable Mitigation Actions and Support, A summary of OECD/IEA analyses for COP 15. Available at www.oecd.org/env/cc/mrv (2015)
3. Gu, F. Teng, Y. Wang. Sector Mitigation Policies and Methods in China: Measurable, Reportable, and Veriable Mechanisms. Adv. in Climate Change Research **2**: 115-123 (2011)
4. P. Hogan, A. Falconer, V. Micale, A.Vasa, Y. Yu, X. Zhao. Tracking Emissions and Mitigation Actions: Current Practice in China, Germany, Italy,and the United States. Report of Climate Policy Initiative (2012)
5. L. Larmuseau. MRV: -NAMA's -Future and challenges. 11th Annual IEA-IETA-EPRI Workshop on Greenhouse Gas Emission Trading. Paris October (2011).
6. N Singh, J Finnegan, K Levin. MRV 101: Understanding Measurement, Reporting, and Verification of Climate Change Mitigation.

   https://www.wri.org/sites/default/files/MRV_101_0.pdf. Accessed by December 2017.(2017).

Life-Cycle Assessment of Crude Palm Oil Produced at Mill J, PT XYZ, Sumatera Island using Eco-indicator 99

Pertiwi Andarani[1,a], *Winardi* Dwi Nugraha[1], *Desinta* Sawitri[1], and *Wiwik* Budiawan[2]

[1]Dept. of Environmental Engineering, Faculty of Engineering, Diponegoro University, Indonesia
[2]Dept. of Industrial Engineering, Faculty of Engineering, Diponegoro University, Indonesia

Abstract. The Crude Palm Oil industry has now become the largest agricultural industry in Indonesia. Nevertheless, the growth of CPO industry could also bring negative impacts on the environment if the company does not control their emissions and discharges properly. Life-cycle Assessment (LCA) is one of the tools that can assess the environmental impacts due to CPO production activities. This study aims to assess the potential environmental impacts arising from the CPO production system at Mill J, PT XYZ, Sumatera Island by using Eco-indicator 99. Based on this study, in 2015, the process in plantation and mill contributed to climate change category was 0.013 DALY or after normalized 202 Pt. Meanwhile, the land use category has 395 PDF*m^2yr or 30.8 Pt. Meanwhile, all of the other categories were less than 30.8 Pt, hence, the highest impact of this CPO production system is climate change at the activities in industrial estate (fertilizers usage) and industry (emitted from waste water of palm oil mill).

1 Introduction

Some countries in Southeast Asia such as Indonesia, Malaysia, and Thailand become the largest producers of CPO in the world. The CPO and its derivatives production in Indonesia were 32.5 million tons in 2015 [1]. Meanwhile, palm oil plantation area reached 9 million hectares [1]. Various products can be derived from CPO such as food, cosmetics, biofuels, and other commodities. Palm oil grows in tropical climates and becomes the competitor of tropical rain forest. However, further expansion of palm oil plantation should consider the environmental impacts arise from anthropogenic activities.

The environmental impacts of palm oil production system are from not only the plantation, but also the palm oil mill. Palm oil mill discharges waste water and solid waste. If those wastes are not properly managed, potential environmental pollution could not be prevented. In order to identify and assess the environmental impacts of a product, a method widely used is life-cycle assessment (LCA). CPO production system from cradle-to-gate consists of nursery, palm oil plantation or estate, palm oil mills, water treatment plant, waste

[a] Corresponding author: andarani@ft.undip.ac.id

water treatment plant, boiler, etc. The impacts are varied based on the production system used, such as carcinogenic, respiratory organics, respiratory inorganic, climate change, radiation, ozone layer, ecotoxicity, acidification/eutrophication, land-use, minerals and fossil fuel uses.

Some studies related to LCA of CPO have been conducted worldwide. Depend on the scope, objective, location, and the production system, the results of studies could be varied. The largest impact of CPO production in Indonesia is climate change [2,3]. CPO-based biodiesel also has 66% larger environmental impact than Jatropha-based CPO according to assessment by using Simapro Software [4].

PT XYZ has several palm oil mills in Sumatera and Kalimantan Island. One of the mills located in Sumatera Island is Mill J. The Mill J is well known for its development on composting process. The solid waste has been treated, but further improvement needs to be planned. Cleaner production is an option to improve the production system to minimize the environmental impacts. Therefore, as the first step, this study aims to assess the environmental impacts from the CPO production system at Mill J. LCA can be used to achieve this objective. The Simapro Software has been known for its reliability to calculate environmental impacts based on the product's life-cycle.

2 Methods

2.1 Crude Palm Oil Production System at Mill J

The activities in Mill J were done in oil palm estate (plantation) and palm oil mill. The area of estate is approximately 203 Ha. There are two kinds of planting-year, i.e. in 1993 and 1995. Replanting will be conducted after the existing plants reach 25 years old. The operation in this estate includes harvesting, herbicides and fertilizer application, land application (come from palm oil mill effluent), and pest control. In the palm oil mill, fresh fruit bunches (FFB) are processed to become CPO. The other supporting system of this production includes water treating plant (WTP), waste water pond, boiler, and composting. The by-product of this CPO production system is palm kernel that will be transported to palm kernel oil mill.

2.2 Scope

In this LCA study, cradle-to-gate principle has been used to assess the environmental impacts of the CPO production system at Mill J. The system boundary of the selected production system can be seen in Fig.1. In this system, there are two stages, i.e. background stage and foreground stage. The background stage is the process indirectly related to the system. The input both energy and materials will affect the environmental impacts of the product. Meanwhile, the foreground stage is the process directly related to the production system. This includes all of the process starting from the oil palm plantation (the output is FFB) until the product (CPO) transported to CPO storage. This also includes waste, such as empty fruit bunches (EFB), shell, fiber, and palm oil mill effluent (POME). The by-product of this system (Kernel) is out of this study scope.

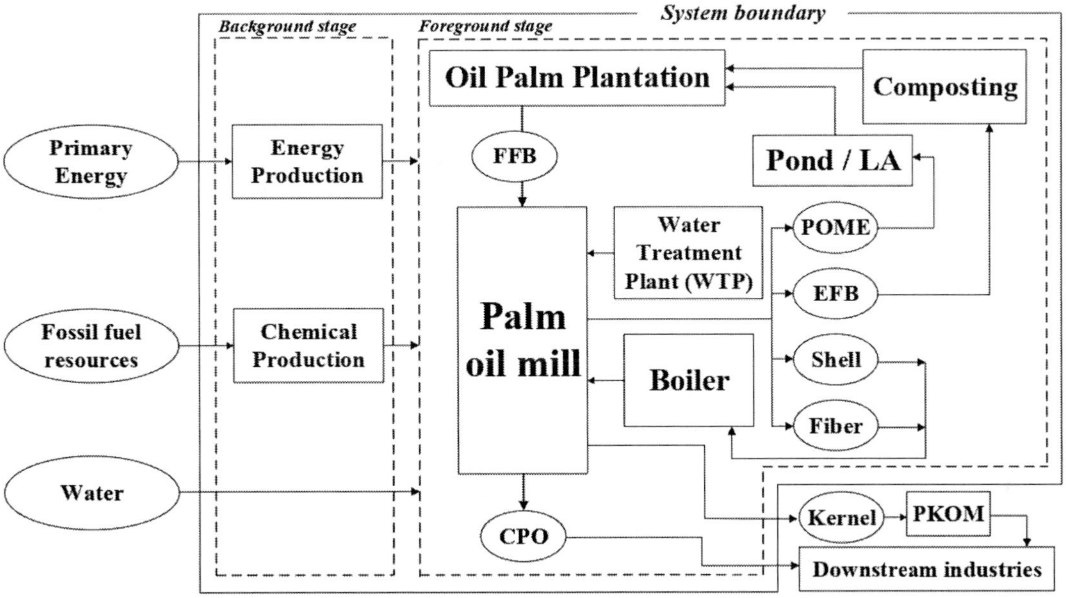

Fig. 1. System boundary of CPO production system at Mill J. CPO: Crude Palm Oil; POME: Palm Oil Mill Effluent; EFB:Empty Fruit Bunch; PKOM: Palm Kernel Oil Mill; LA: Land Application

2.3 Simapro v.7.1

The Simapro v.7.1 has been used to assess the environmental impacts of this CPO production system. Eco-Indicator 99 was selected as the life-cycle impact method because of ISO 14040 recommendation. All of the relevant process and affecting the environment was inputted to the scope of this study. The impacts of transportation of all chemical agents such as fertilizer and herbicides were not included in this study, as well as the human aspect.

3. RESULTS AND DISCUSSION

3.1 Life-Cycle Inventory (LCI)

Inventory Data is collected based on direct observation in the field, as well as data from estate and mill. In addition to the main product (CPO), other outputs from the production process are POME, EFB, fiber, and shell. The generated POME will be forwarded to the waste water pond, and flow to the plantation area as a liquid fertilizer or land application. Fiber and shell are the fuel source of the boiler in the plant to produce steam. EFB will be processed into compost that will substitute the use of chemical fertilizers. The average FFB produced per hectare is about 30 tons FFB / ha / year. Data of the chemical agents as well as the fuels used in the system has been collected and depicted in Table 1. The production capacity of the palm oil mill is 30 tons/hour. In the process, it takes a variety of inputs ranging from materials, energy, transport, and so forth. In Table 1, there are some data from the Mill J needed in this study. All of these data were inputted into SimaPro software.

Table 1. Plantation and Mill Data for input.

Parameter	**2015**
Fertilizer - N - P205 - K_2O - MgO - B_2O_3	 0.828 ton 44.112 tons 99.69 tons 3.839 tons 6.025 tons
Compost	1,552,180 kg
LA	120,236 m^3
Herbicide - Glyphosate - Metsulfuron Methyl - Paraquat Dichloride - Triclopyr Butoxy	 348.63 kg 8.308 kg 46.368 kg 17.433 kg
Diesel fuel consumption for FFB transport	14,844 liters
Total area	203.3 Ha
FFB yield	26,067.84 tons
Distance to mill	20 km
Truck load per ritation	4.92 tons
Truck ritation per year	1,003
Transport emission	9.44 $kgCO_2$ per ton FFB
Emission 1. Air - Nitrogen - K_2O - CO_2 from herbicide - CO_2 from diesel use 2. Soil - N_2O	 2,740.68 kg 56,823 kg 4,616.05 kg 70,331 kg 953,283 kg

Parameter	**2015**
FFB processed	102,537 tons
Electricity from genset	366.56 MWh
Electricity from steam turbin	2,077.96 MWh
Diesel fuel for mill	154,452 liters
Shell	5,306 tons
Fibre	11,535 tons
Empty Fruit Bunch (EFB)	21,533 tons
Water usage	148.123,95 m^3
Chemical agent - $CaCO_3$ - Soda Ash - Alum - B 120 + - B 171 + - NaOH - H_2SO_4	 162.800 tons 19.464 tons 23.20 tons 0.306 tons 0.191 tons 0.23618 tons 0.00591 tons
Truck ritation for CPO	840
Produksi CPO	5,020.60 tons
Distance to CPO Storage	250 km
CPO truck ritation per year	840
CPO truck loading per ritation	23.99 ton
POME	57,554 ton
Steam Input	146,333,773 MJ
BOD Raw Effluent	1,408 ppm
COD influent	5,513 ppm
GHG from operational - Methane - CO_2 fossil diesel from production energy - CO_2 from chemical agent usage	 575,540 kg 20,526,541.4 kg 36166.8 kg

Source : PT XYZ, 2016

3.2 Life-Cycle Impact Assessment

3.2.1 Characterization

Characterization is an assessment of the magnitude of the substance that contributes to the impact category. In each category the impacts have different units because they are related to different damages. For example, Disability Adjusted Life Years (DALY) is for human health; Potential Affected Fraction (PAF*m^2yr) and Potentially Disappeared Fraction (PDF*m^2yr) are for ecosystem quality, and MJ Surplus is for resources.

The activities in the oil palm estate in 2015 had the largest impact on the climate change (0.00229 DALY) and the smallest impact on the ozone layer (9,33E-10 DALY). Moreover the land use impact also had the largest value among the ecosystem qualities (89.9 PDF*m^2yr). Similarly, in the palm oil mill, the largest impact was on the climate change (0.0103 DALY). However, the ecotoxicity in the mill (621 PAF*m^2yr) was larger than the land use (395 PDF*m^2yr) and acidification/eutrophication (49,6 PDF*m^2yr).

3.2.2 Normalization

Normalization is intended to create a uniform unit for all impact categories, so that all impact categories use a same and comparable unit. In 2015, the plantation that had the most significant impact category was climate change (0.149). The second largest impact category was land use (0.0175) while the third largest impact category is inorganics respiratory (0.00404). Furthermore, in the mill, the results of normalization for climate change, land use, and carcinogens were 0.672, 0.077, and 0.072, respectively.

The result of weighting from the plantation process in 2015 shows that the largest impact category was climate change. The climate change impact was due to land conversion from forests to oil palm plantations. Furthermore, the fertilizers and herbicides usage indirectly emitted N_2O. It also delivered emissions to the ground. The transportation sector also brought the impact on the air, such as for plantation activities. The emission that had the largest impacts magnitude of climate change on plantation activities was N_2O. The fertilizers application caused the emissions; one of them was urea fertilizer. The second impact that contributed to the industrial process was land use. In the process of plantation activities, there was land use change from forest to a large palm oil plantation of 200 hectares. In 2015, the process in plantation and mill contributed to climate change category was 0.013 DALY or after normalized 202 Pt. Meanwhile, the land use category had 395 PDF*m^2yr or 30.8 Pt.

3.2.3 Single Score

Single Score aims to classify impact categories that have been weighted based on the activity or process that contributes to the impact. It can be inferred that the process of *Fresh Fruit Bunch* or activities in the plantation has the largest impacts, such as landuse, acidification/eutrophication, climate change, respiratory inorganics, carcinogens, and fossil fuel depletion. The cause of the high impact of climate change was the emission value of N_2O emitted indirectly from the fertilizers usage using chemical fertilizers. N_2O emissions might also come from palm oil mill, i.e. the use of inputs derived from chemicals during the production process. In 2015, N_2O emission was as much as 215 Pt. Apart from the N_2O emissions, the impacts of climate change can also be affected by POME. POME actually has the potential to produce biogas such as methane gas (CH_4) but it also generates CO_2 emissions. In 2015, CH_4 emissions was 2.42 Pt. CO_2 emissions also arose from the use of chemicals and from transportation sector in the mill. It is expected that the biogas can potentially be utilized as a renewable energy source as well as to reduce the rising environmental impacts.

3.3 Comparison to other studies

Compared to the study conducted by Hansen [2], who examined the life-cycle assessment (LCA) on the production of crude palm oil in Malaysia, there are some differences in the magnitude of environmental impacts. The impacts that were largest than 1.0 were on the fossil fuel depletion (19.5 Pt at the activities at the plantation), respiration inorganics (10.9 Pt at the plantation and 13.9 Pt at the Mill), climate change (2.7 Pt at the plantation), and acidification/eutrophication (1.8 Pt at the plantation). The system boundary applied by Hansen's research began with plantation (nursery and harvesting), industry, and disposal processes. Their research used SimaPro v 5 LCA software and impact method of Eco-indicator 99 [2]. Functional unit (FU) had been adjusted to the production of 1000 kg CPO (output of 5000 kg FFB) in Malaysia. The differences could be due to the land use change, yield of FFB, and the fertilizer production. It can be seen that climate change impacts in Mill J are much larger than those in Malaysia according Hansen's study. In other study in

Malaysia, Wicke et al. [6] assumed that FFB yield at their case plantation study was 25 ton FFB per hectare per year. Both of these studies [2 and 6] have relatively large impact on respiratory inorganics due to chemicals utilization.

Meanwhile, Sampattagul et al. [7] studied the life-cycle assessment of palm oil biodiesel in Thailand that had a system boundary starting from the plantation process, biodiesel production, and utilization of biodiesel. SimaPro software was used in their research but the method used to calculate the impact was the Environmental Design of Industrial Products (EDIP) [7]. The comparison between Sampattagul et.al's study [7] and the current research is discussed more deeply in regarding the impact of plantation activities, because in Sampattagul et al.'s study did not calculate the impact that generated in the processing of palm oil. The impact of the activities at the plantation in Thailand showed 1.03 x 10^{-4} Pt. The greatest impact was the release of toxic substances into the air, water, and soil; radioactive waste; and global warming. Radioactive waste was generated from chemical production, such as pesticides, glyphosate, boron, and kieserite. The environmental impact are almost identical to those of Mill J, from chemical fertilization and herbicide processes.

4. CONCLUSIONS

In 2015, the process in plantation and mill contributed to climate change category was 0.013 DALY or 202 Pt. Meanwhile, the land use category had an impact of 395 PDF*m^2yr or 30.8 Pt. The cause of the high impact of climate change was the emission value of N_2O derived from the existing plantation process (to produce fresh fruit bunches). The impact was generated due to chemical fertilizers application. The optimization of chemical fertilizers usage should be done as well as increasing the yield of fresh fruit bunch to improve the efficiency. Furthermore, one of the environmental impacts that can be treated based on this study is CH_4 emission from palm oil mill effluent (POME). It can be processed to biogas, which is useful for energy. Otherwise, it can contribute to the greenhouse gas emission, which increase the impact of climate change. Therefore, a feasibility study of CH_4 emission treatment should be conducted. Another case study in Mill P, the biogas generated from POME has been successfully produced and, hence, the surplus energy can be obtained [8].

References

1. GAPKI. Reflection of Palm Oil Industry and Its Prospect in 2016 [Internet]. Pers Conference GAPKI. 2016 [cited 2016 Apr 20]. Available from: http://www.gapki.or.id/Page/PressReleaseDetail?guid=39f6f3f2-0419-42d4-8d1b-9524871d3cf2
2. S. Hansen, Int. J. Life Cycle Assess. **12**, 50–59 (2005)
3. A. Wijono, *Environmental Assessment and Palm Oil Potency Mapping as a Renewable Energy Source in Indonesia* (BPPT, Tangerang, 2013)
4. N. Nazir, D. Setyaningsih, *7th Biomass Asia Workshop, Jakarta* (2010)
5. H. Baayen, *Eco-indicator 99 Manual for Designers* (Ministry of Housing, Spatial Planning and the Environment, The Netherlands, 2000)
6. B. Wicke, V. Dornburg, M. Junginger. Biomass Bioenergy **32**, 1322–1337 (2008).
7. S. Sampattagul, P. Nutongkaew, T. Kiatsiriroat Int. J. Renew. Energy **6** (1), 1–14 (2011).
8. P. Andarani, W.D. Nugraha, Wieddya. *AIP Conference Proceedings* **1823, 020064** (2017)

Recovery Practice of Unsorted Solid Waste: from Landfill towards Economic Benefits in Semarang, Indonesia

Mochamad Arief Budihardjo[1,*], *Indah* Fajarini Sri Wahyuningrum[2]

[1]Department of Environmental Engineering, Universitas Diponegoro, Jl. Prof. H. Soedarto, S.H., Tembalang, Semarang, Indonesia

[2]Department of Accounting, Universitas Negeri Semarang, Jalan Sekaran, Gunung Pati, Semarang, Indonesia

Abstract. The field study in Jatibarang landfill in Semarang was conducted by quantifying the solid waste that had been collected and transferred to the landfill to obtain the amount of the solid waste generated by the residents of Semarang. The gross weight of the waste was calculated by randomly selecting and weighing the shipments on the scale before the load was dumped by the vehicle. The solid waste received by Jatibarang landfill is about 651.5 tons per day, with a density of 245 kg/m^3. The composition of domestic waste in Jatibarang landfill is 77.7% organic waste, 14.19% plastics, 5.31% mixed paper, and 1.36% others. The inorganic wastes in Jatibarang landfill still retain economic value of IDR 141 million per day when fully recovered. The current recovery practices are performed by scavengers who live in the surrounding area of Jatibarang landfill.

1 Introduction

The city of Semarang is the capital of the Central Java Province, one of the most developed and densely populated provinces in Indonesia. As a big city in a developing country, the local authorities of Semarang face environmental problems mainly related to the solid waste generated by its residence. Similar to cities in other developing countries, the solid waste management in Semarang consists of collection, transport, and final processing, which is basically landfilling [1, 2].

Most of the solid waste generated in Semarang municipality ends up at Jatibarang landfill. The solid waste transported to landfill generally consists of a mixture of organic and inorganic matter, as the practice of solid waste separation at the source for recycling purposes has not been well implemented. The organic substances will undergo the decomposition process, but the inorganic substances which mostly consist of plastic, paper, and fabric will retain their original forms for long periods of time. The accumulation of inorganic substances over time can cause the landfill to reach its capacity limit too soon [3].

* Corresponding author: m.budihardjo@ft.undip.ac.id

Jatibarang landfill performs some optimization efforts, such as recycling the old zone of the landfill to extend its service time. However, as the percentage of the solid waste being generated is increasing every day, another approach should be considered to be implemented at Jatibarang landfill. One visible way to reduce the amount of the solid waste to be landfilled is by conducting a recovery practice of inorganic waste in the landfill. By reducing the volume of the waste dumped at the landfill, the life span of the landfill can be extended. Furthermore, the inorganic waste still has potential economic worth if handled and treated properly [4, 5]. Therefore, this paper aims to investigate the amount and characteristics of the solid waste disposed of at Jatibarang landfill and analyze the economic benefits of the recoverable material.

2 Methods

The research was conducted at the Jatibarang landfill area by quantifying the solid waste that has been collected and transferred to the landfill to obtain the amount of the solid waste generated by the residents of Semarang City. The gross weight of the waste was calculated through randomly selecting and weighing the shipments on the scale before the load was dumped by the vehicle [6]. The tare weight was the weight of truck container.

The number, volume, and source of the waste of each vehicle were recorded. Waste characterisation was also conducted during solid waste sampling. Solid waste is divided into organic and inorganic waste plastic (HDPE, PE, and PP), paper, fabrics and miscellaneous) [7, 8]. Interviews also were done to depict participation level of the residents living adjacent the landfill area and existing scavengers who involved in the recovery process, potential of plastic recycler, and to find out the market value of the recovered material.

3 Results and Discussion

3.1. Amount of solid waste

The amount of the solid waste generated by Semarang city residents which is handled by the city's local authorities fluctuates between days. In order to get a precise amount, the sampling was conducted for 8 consecutive days which served to cover the solid waste generation during both work days and the weekend. As shown in Figure 1, the amount of the solid waste transferred to the Jatibarang landfill varied about 600-750 tonnes per day. The lowest amount was found to be for Sunday which was 609 tonnes. This result gave the average daily amount of 651.5 tonnes per day of the solid waste received by the Jatibarang landfill. The density of the solid waste was found to be 245 kg/m^3.

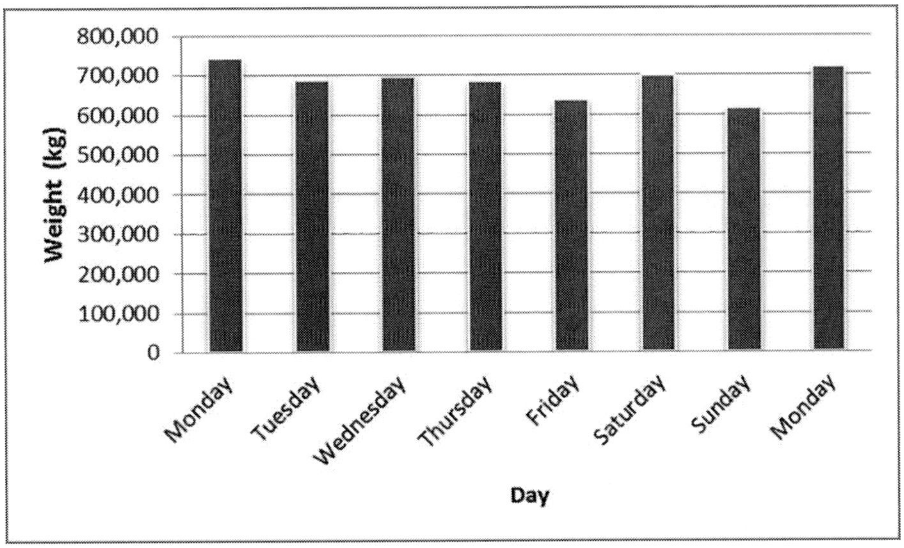

Fig 1. Solid waste received by Jatibarang landfill

3.2. Composition of solid waste

The composition of domestic waste in the Jatibarang landfill is shown in Figure 2. The prevailing composition of organic waste (in wet weight percentages) includes food waste, kitchen waste, and garden waste, etc. which was biodegradable, with a total proportion of 77.7%. The proportion of the recoverable waste, including some types of plastics, ran up to 14.19% and mix paper which was 5.31%. The category 'others', including batteries and other indistinguishable waste, was only 1.28%. The plastic category was then broken down into more specific classifications which were HDPE, PE, PP and PPs. This plastic categorisation was based on the initial survey to the local plastic recycler to increase the economic value of the recovered plastic since each type of plastic has a different price and the lowest price is for the mixing/unsorted plastic.

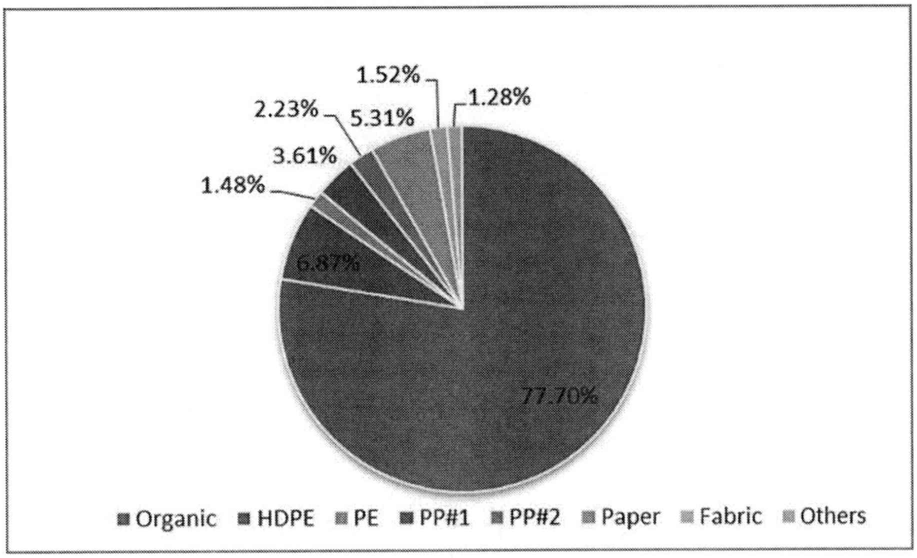

Fig 2. The composition of solid waste

3.3. Economic value of the recoverable materials

The solid waste that is being transferred to the Jatibarang landfill still retains some economic value since it contains some recyclable materials. The economic value of the waste varies according to the composition of the waste and the required treatment of the recoverable material prior to selling. Based on the survey of the local recycler industries, the individual price for each category is shown in Table 1. Economic value that shown in Table 1 was calculated from multiplication of the average daily amount solid waste per day (kg), the percentage of proportion each types/product, and the price of each product. The result shows that the solid waste dumped into Jatibarang Landfill potentially can be recovered with value about IDR 141 million per day.

Table 1. Price of the recovered selected inorganic waste

No	Product	Price (IDR/Kg)	Economic value (IDR)
1	HDPE	1,200	53,709,660.00
2	PE	3,000	28,926,600.00
3	PP#1	1,300	30,574,895.00
4	PP#2	1,200	17,434,140.00
5	Paper (mix)	300	10,378,395.00
		Total	141,023,690,00

3.4. Local Community Involvement

There are hundreds of people (scavengers), most of them local resident who work to recover sellable materials from incoming solid waste in Jatibarang landfill. They sort the waste which still has value to sell to the local recycle collector resided in surrounding Jatibarang landfill. Their activities actually help to reduce waste to be piled up in the landfill zone. However, those people work less thorough, only waste that is still highly valuable and visible (on the surface) that they recover. The rest of the solid waste that is actually still being able to be reprocessed and has economic value becomes residue. This residue will rest in landfill zone and take space of the landfill.

However, the present of scavengers in landfill also create come issues. Since they are unmanaged, they tend to rush into the garbage truck when the garbage truck starts to dump the load. They disturb the truck activity and delay the disposal process. Once the garbage truck finish unloading, they still try to get more material and delay the activity of the bulldozer to push and spread the waste into dumping area.

4 CONCLUSION

The average daily amount of 651.5 tonnes per day of the solid waste received by Jatibarang landfill with density of 245 kg/m^3. The composition of domestic waste in the Jatibarang landfill is 77.7% organic waste, 14.19% some types of plastics, 5.31% mix paper, and 1.36% others (including batteries and other indistinguishable waste). The solid waste dumped into Jatibarang Landfill potentially can be recovered with value about IDR 141 million per day. Some of the waste has already been recovered by scavengers who mainly reside adjacent Jatibarang landfill. However, their activities are not managed properly by

Jatibarang landfill authorities and might delay the unloading and spreading process of the solid waste.

References

1. E. Damanhuri, I.M. Wahyu, R. Ramang, T. Padmi, J. Mater. Cycles Waste Manage. **11**, 3 (2009)
2. L.A. Guerrero, G. Maas, W. Hogland, Waste Manage. **33**, 1 (2013)
3. S. Suthar, P. Singh, Sustain. Cities Soc. **14** (2015)
4. J. Singh, I. Ordonez, J. Clean. Prod. **134** (2016)
5. L. Rigamonti, M. Grosso, J. Møller, V.M. Sanchez, S. Magnani, T.H. Christensen, Resou. Conser. Recy. **85** (2014)
6. M.A. Mir, P.T. Ghazvinei, N.M.N. Sulaiman, N.E.A. Basri, S. Saheri, N.Z. Mahmood, A. Jahan, R.A. Begum, N. Aghamohammadi, J. Environ. Manage. **166** (2016)
7. M.E. Edjabou, M.B. Jensen, R. Götze, K. Pivnenko, C. Petersen, C. Scheutz, T.F. Astrup, Waste Manage. **36** (2015)
8. C. Areeprasert, J. Asingsamanunt, S. Srisawat, J. Kaharn, B. Inseemeesak, P. Phasee, C. Khaobang, W. Siwakosit, C. Chiemchaisri, Enrgy Proced. **107** (2017)

Methane enrichment on biogas generated from anaerobic digester using coconut shell-based activated carbon

Ridwan Hakim[1,*], *Nyoman* Suwartha[1] , and *Cindy* Priadi[1]

[1]Environmental Engineering Program, Department of Civil Engineering, Universitas Indonesia, Indonesia

Abstract. This study evaluates methane enrichment on biogas generated from Anaerobic Digester (AD) through CO_2 adsorption process so that biogas can be used as fuel for vehicle engines, power plants, and natural gas substitutes. The experiment was observed by passing biogas synthesis (45% CH_4 + 55% CO_2) and biogas from cattle manure ±59.7% CH_4, ± 37.1% CO_2 and ±3.2% other gases) in spontaneously pressurized adsorption column. In addition, observation of CO_2 adsorption capacity at various pressure and fixed temperature (27°C) was performed using pure CO_2 (±98%). Methane in biogas has been successfully purified to 92% at 0.5 L/min flowrate and 79.6 seconds retention time. The adsorbent will be saturated after gas flowing for 60 and 80 minutes for synthesis biogas and biogas from AD on the amount of adsorbent of 266 grams. A change of surface area of activated carbon (AC) after thermal regeneration at 160°C for 2 hours was 7.51% and regeneration efficiency was 67%. The adsorption process followed Freundlich isothermal. This process can be feasible alternative technology to meet the need for biogas with high levels of methane in small-scale AD.

1 Introduction

Energy content of biogas is directly proportional to the levels of methane in biogas. Biogas typical composition is 50-80% methane, 15-50% CO_2, N_2 0-5%, 0-1% O_2 and 0-1% other gases [1]. A process of reducing CO_2 levels (upgrading) is required in biogas to obtain methane with a high enough purity. The CO_2 content in biogas can significantly reduce the calorific value of biogas because CO_2 is an inert compound. Biogas upgrading process is highly dependent on the utilization of biogas that will be done. Some applications of biogas utilization require biogas to have a high enough methane content that can meet the burning energy required. For example, application of biogas in Europe for injection in domestic gas pipelines and motor vehicle fuels or generators requires a minimal concentration of methane in biogas is 93% [2] and 95% [3], [4]. In most developing countries, biogas production facilities are still on a small scale, less than 10 m^3 [5]. Small-scale biogas purification technology is required to encourage the use of biogas as a renewable energy source. Therefore, the development of a simple and efficient biogas purification technology becomes

* Corresponding author: ridwan.hakim51@ui.ac.id

a thing to be developed nowadays. The adsorption technology can be applied to small-scale digester because it is relatively cheaper in operation and can be operated easily [6]. AC has been widely adopted for gas separation processes including biogas upgrading [7].
This study analyzed CO_2 removal efficiency, CO_2 adsorption isotherms, regeneration efficiency, and the characteristics of the solid granular AC made from coconut shell as an adsorbent of methane enrichment process using synthetic biogas and cattle manure biogas. The efficiency of the removal and characteristics of the AC was revealed at before and after the regeneration of AC. Thus, this research got design criteria of methane enrichment technology on biogas by adsorption method using coconut shell based active carbon adsorbent.

2 Experimental setup

2.1 Gas separation kinetics

The adsorption column is 25.4 mm in diameter or 1 inch maximum height of 150 cm and the sampling port is every 25cm. With variation of gas flow rate from 0.5 to 1.5 L / min will get variation of resistance time 15.19–91.16 second. Adsorbate in this research is biogas synthesis in laboratory scale and biogas result of Anaerobic Digester in pilot scale. The synthesis biogas used is a gas mixture with a composition of 45% CH_4 and 55% CO_2.

The gas retention time variation is 15.19, 30.39, 45.58, 60.77, 75.97 and 91.16 for 0.5 L/min flow rate. At a flow rate of 1 L / m the retention time variation is 7.60, 15.19, 22.79, 30.39, 37.98 and 45.58 seconds. At a flow rate of 1.5 L / the retention time variation is 5.06, 10.13, 15.19, 20.26, 25.32 and 30.39 seconds. Methane concentration data at difference retention time matched with first, second, and third order kinetics plots [8].

2.2 Adsorption isotherm

The isothermal adsorption test was performed at 0-2 bar pressure range and temperature 27°C. The pure carbon dioxide gas from pressurized gas cylinders flows into the Dozing Cylinder which will regulate and measure the amount of gas to be injected into the Sampling Cylinder. Cylindrical sampling is an alloy metal cylinder with a diameter of 0.25 inches and a length of 20cm containing 3.48 grams adsorbent.

2.3 Regeneration efficiency

Regeneration of the activated carbon is performed when adsorbent is saturated. The regeneration efficiency was obtained by comparing the breakthrough time of activated carbon before (Tb_o) and after regeneration (Tb_r). Breakthrough time was obtained by flowing gas up to 100 minutes with a flowrate of 0.5L / min and 79 seconds retention time. Thermal regeneration is done by heating at a temperature of 160°C for 2 hours [9] with an ambient atmosphere.

$$\text{Regeneration Efficiency (RE)} = (Tb_o / Tb_r) \times 100\% \quad (1)$$

2.4 Pilot scale experimental

Biogas used is obtained from 7 m^3 Anaerobic Digester with cattle manure substrate at PT SWEN Inovasi Transfer, Bogor. The biogas composition is ±59.7% CH_4, ±37.1% CO_2 and

±3.2% other gas. Breakthrough time in the pilot scale test was obtained by flowing gas up to 100 minutes with a flowrate of 0.5 L/min and 79 seconds retention time.

2.5 Surface characteristics of activated carbon

Adsorbent is granular coconut shell based activated carbon on size 12–20 mesh. Changes in the characteristics and surface area due to regeneration were carried out with SEM JOEL JSM-6390A and Surface Area Analyzer Quantachrome Nova e1200.

3 Results and Discussion

3.1 Gas Separation Kinetics

Data of methane concentration at different retention time have been obtained. Figure 1 is a comparison of the retention times (Tr) versus methane content in the first, second, and third order kinetics plots.

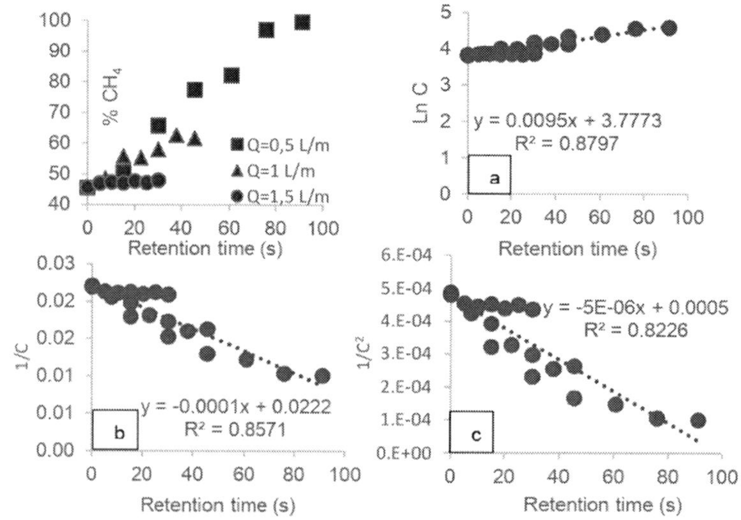

Fig 1. The first, second, and third kinetical adaption plots are (a), (b) and (c), respectively

The determination coefficient value (R^2) for the first, second and third order kinetics plots is 0.8797, 0.8571 and 0.8226 respectively. Based on these values, it can be concluded that the gas separation process that runs with first-order kinetics. Equation 2 is the kinetic equation of adsorption rate applicable for increasing methane content in biogas using coconut shell activated carbon adsorbent with C and Co is purified and raw methane concentration on biogas.

$$\text{Ln } C = 0{,}0095\ Tr + \text{Ln } Co \tag{2}$$

3.2 Adsorption Isotherm

Adsorption capacity at various pressures matched with the isothermal plots Langmuir, Freundlich and BET by regression of the straight-line equation of the three types of

$$\text{Ln } a = 0.7558 \text{ Ln } (P/Po) - 1.3681 \tag{3}$$

adsorption isothermal [10]. Figure 2 shows that the isothermal plots of Langmuir, BET and Freundlich have determination coefficient of 0.9605, 0.8771, and 0.9957 respectively. CO_2 adsorption with coconut shell AC follows Freundlich isothermal with the following equation: where q and P/Po is amount of adsorbate per unit adsorbent (mmol/gr) and relative pressure to atmospheric pressure. Similar results were obtained by rashidi et al. [11] using Norit® AC.

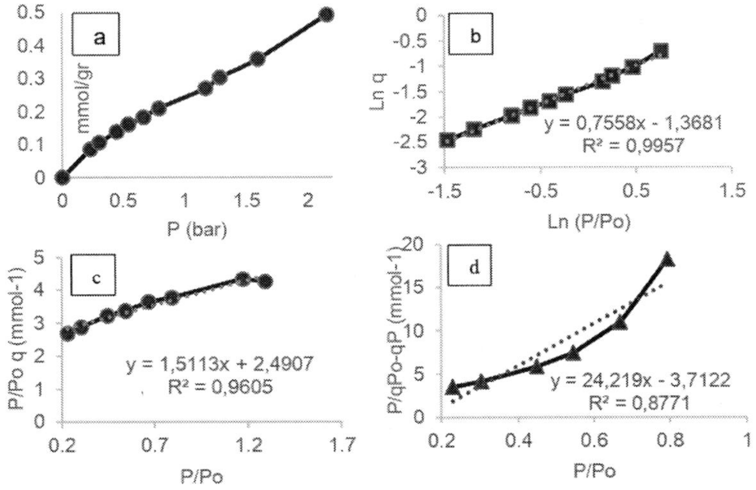

Fig 2. (a) Adsorption of CO_2 at 27°C; (b) Isotherm Freundlich;(c) Langmuir;(d) BET

3.3 Regeneration Efficiency

Figure 3 shows that the breakthrough time of fresh and regenerated activated carbon is 60 minutes and 40 minutes, respectively. Regeneration efficiency of activated carbon is 67% based on Equation 1. The low value of regeneration efficiency indicates that there has been an incomplete regeneration process.

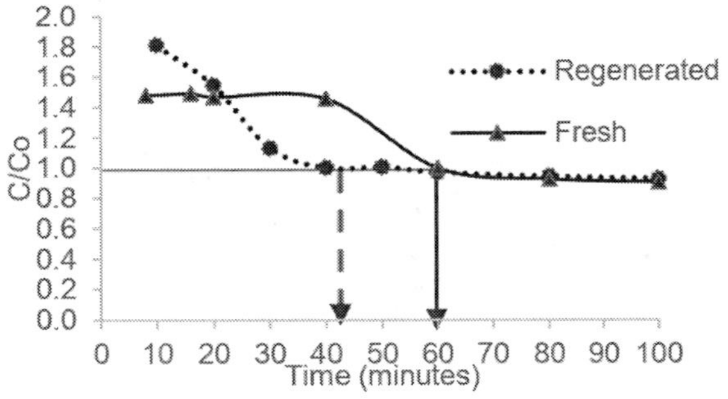

Fig 3. Comparison of fresh and regenerated activated carbon breakthrough time

3.4 Pilot Scale Experimental

This pilot scale analysis use of the same conditions as laboratory scale aims to enable pilot-scale results to be compared with laboratory-scale. The pilot scale breakthrough time in Figure 4 shows that the breakthrough time of CO_2 adsorption process with biogas from AD and biogas synthesis shows slightly different result. In pilot scale, breakthrough time occurred at 80 minutes, whereas on laboratory scale breakthrough time occurred in 60 minutes, this is due to the raw gas methane concentrations in the AD biogas much higher than in the synthesis biogas. The concentration of methane on the AD biogas is ± 60%, while the concentration of methane in the synthesis biogas is 45%, there is a difference of about

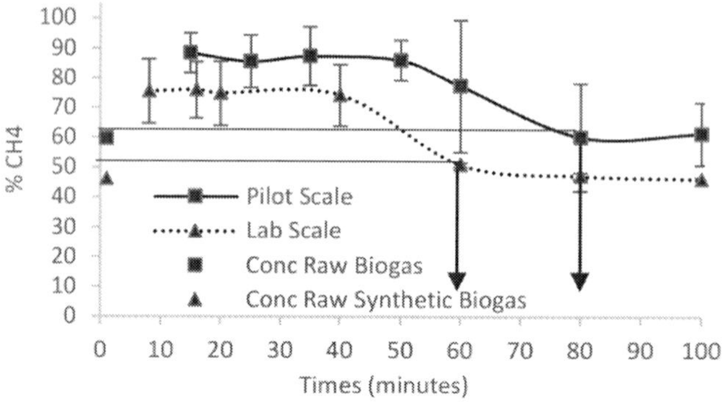

Fig 4. Breakthrough time curve of pilot and lab scale adsorption

15% methane levels at the initial state of adsorption. Higher levels of methane in the initial state make the amount of CO_2 that must be adsorbed by activated carbon to reach pure methane becomes reduced, so that with the same amount of activated carbon produces a longer breakthrough time.

Based on Figure 4, the concentration of methane in column outlets can be maintained at ±89% in pilot-scale and ±74% in laboratory scale until breakthrough time reached. There is an increase in methane content (C-Co) in pilot-scale of ±28.8% and ±29.2% on laboratory scale. Methane content increasing that occurs at laboratory and pilot scale adsorption do not show any significant difference. This further reinforces that a similar process occurs in adsorption with biogas synthesis and AD biogas, so the design criteria obtained on laboratory scale testing can be applied well on a pilot scale.

3.5 Surface Characteristics of Activated Carbon

The results of SEM characterization that can be seen in Figure 5 show the difference of average pore diameter and degree of surface roughness between fresh and regenerated activated carbon. In the regenerated activated carbon (Figure 5b and 5d) have a rougher

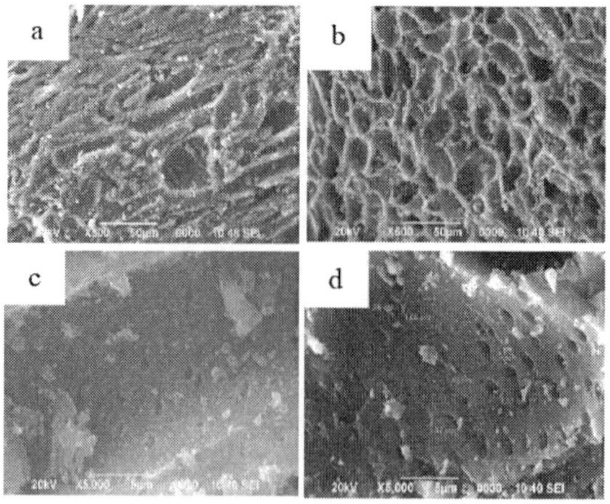

Fig 5. Results of SEM Characterization of Active Carbon Surfaces. (a) and (c) are activated carbon before regeneration 500× and 5,000×, respectively. (b) and (d) are activated carbon after regeneration (500× and 5,000×)

surface and higher macro pore number than fresh activated carbon (Figures 5a and 5c). In addition, the regenerated also has a larger mean macro-pore diameter (±1.3 μm) compared with the fresh one (±0.7 μm). This shows the change of surface characteristics due to thermal regeneration.

BET surface area of fresh activated and regenerated carbon is 916.78 m^2/gram and 847.96 m^2/gram. The changes in mesopores and micropores are not significant, this is indicated by a decrease in surface area of only 68.8 m^2/ gram or 7.51%. The decrease in surface area will directly affect the regeneration efficiency, because will also decreases the space for adsorbate molecule on the surface.

The fact that regeneration efficiency is only 67% indicates that the main cause of low regeneration efficiency is not due to changes in the surface structure (7.51%). The main factor of low regeneration efficiency is probably caused by uncompleted regeneration process with uncontrolled heating atmosphere.

4 Conclusion

Based on the research can be concluded that the adsorption column with coconut shell-based activated carbon adsorbent can increase the methane level by 28.8% for 80 minutes on the AD biogas with 79 seconds retention time. The adsorption process of CO_2 follows Freundlich adsorption isothermal. Thermal regeneration with heating for 2 hours at 160°C in an uncontrolled atmosphere provides poor regeneration results.

This work was financially supported by Ministry of Research Technology and Higher Education Republic through grant International Indexed Publications for Student Final Project (PITTA) No : 824/UN2.R3.1/HKP.05.00/2017 and Faculty of Engineering Universitas Indonesia.

References

[1] A. Petersson, "Biogas cleaning," The Handbook of Biogas (2013)
[2] W. Urban, "Biomethane injection into natural gas networks," The Biogas Handbook, (2013)
[3] P. Kaparaju dan J. Rintala, "Generation of heat and power from biogas for stationary applications: boilers, gas engines and turbines, combined heat and power (CHP) plants and fuel cells," The biogas Handbook (2013)
[4] M. S. Son, "Biomethane for transport applications," The Biogas Handbook (2013)
[5] K. Surendra, D. Takara, A. G. Hashimoto dan S. K. Khanal, "Biogas as a sustainable energy source for developing countries: Opportunities and challenges," enewable and Sustainable Energy Reviews (2014)
[6] Rashed. M, Karim M, M.Rahman, Torii A. M. "Methane enrichment of biogas by carbon dioxide fixation with calcium hydroxide and activated carbon," Journal of the Taiwan Institute of Chemical Engineers (2016)
[7] Q. Zhao, E. Leonhardt, C. MacConnell, C. Frear dan S. Chen, "Purification technologies for biogas generated by anaerobic digestion," CSANR Research (2010)
[8] Jean-Pierre Simonin "On the comparison of pseudo-first order and pseudo-second order rate laws in the modeling of adsorption kinetics," Chemical Engineering Journal, 2016.
[9] S. Shahkarami, R. Azargohar, A. K. Dalai dan J. Soltan, "Breakthrough CO_2 adsorption in bio-based activated carbons," Journal of environmental science (2015)
[10] Dong-June Seo, Zhichao G, Hirotaka F, Takao F, Akiyoshi S "Breakthrough CO_2 adsorption in bio-based activated carbons," Renewable Energy (2015)
[11] Rashidia N.A., Yusup S., Borhan A ." Isotherm and Thermodynamic Analysis of Carbon Dioxide on Activated Carbon," Procedia Engineering (2016)

Analysis of Groundwater Quality Surrounding Municipal Solid Waste Landfill: Banyuurip Landfill, Magelang, Indonesia

Mochamad Arief Budihardjo[1,*], *Purwono*[1] , and *Annisa* Selfi Nugraheni[1]

[1]Department of Environmental Engineering, Universitas Diponegoro, Jl.Prof.H.Soedarto, S.H.Tembalang, Semarang, Indonesia

Abstract. Banyuurip landfill generates leachate containing suspended and dissolved matter, including some heavy metals, a result of waste degradation and decomposition of organic materials. Leachate can infiltrate the soil then contaminate groundwater. Once the leachate infiltration reaches the soil and groundwater, it will contaminate the surrounding environment. In this study, the contamination of leachate into ground water was investigated by measuring the concentration of Manganese (Mn), Iron (Fe) and Free Ammonia (NH_3-N) in leachate and dug wells surrounding the landfill. The concentration of Mn, Fe and NH_3-N in leachate was found to be 5.42 mg/L, 6.16 mg/L, and 1652 mg/L, respectively. The concentration in the surrounding dug wells located within 500 m from the landfill varied from 0.00 – 0.86 mg/L for free ammonia, 0.00 – 0.31 mg/L for total Fe, and 0.00 – 0.26 mg/L for total Mn. The concentration of Mn, Fe and NH_3-N decrease with increasing distance between the landfill and wells.

1 Introduction

Solid waste landfill servicing the Magelang municipality is located at Banyuurip, Magelang. Total domestic solid waste received by the Banyuurip landfill is about 4.5 tonne/day, which consists of organic matter, paper, plastic, metal, rubber, wood, glass and other materials. The composition is dominated by approximately 70% organic matter and 10% of plastic.

Banyuurip landfill generates leachate containing suspended and dissolved matter as a result of waste degradation and decomposition of organic materials, consisting of both organic and inorganic materials. The organic matter includes hydrocarbons and humic acid, and the inorganic matter include calcium, magnesium, natrium, kalium, ammonium, iron, manganese, chloride, sulphate, cadmium, chromium, copper, lead, nickel and zinc [1].

The primary substances found in leachate are Mn, Fe and NH_3-N. Manganese and Fe could lead to turbidity, colour and bad water taste, while NH_3-N can cause odour and different tasting water [2]. Leachate can infilter through soil causing contamination to

* Corresponding author: m.budihardjo@ft.undip.ac.id

groundwater. Once the leachate infiltration reaches a well, it will also pollute the well water [3].

Landfill leachate quality is affected by the composition of the waste and meteorological conditions, such as rainfall, landfill operation, chemical and biological reactions during the decomposition processes [4]. The leachate may seep into the soil and reach the groundwater through the landfill liner which consists of compacted clay [5]. The groundwater contamination can be evaluated by performing water quality testing on a sample of wells surrounding the landfill [6].

In this study, infiltration of three leachate parameters, Mn, Fe and NH_3-N, into the surrounding dug wells located in area within 500 m from the landfill are investigated. The concentration of these parameters were analyzed and results compared with the distance between landfill and the dug wells in which the samples were taken.

2 Methods

This study was undertaken at Banyuurip Landfill, Magelang, Central Java, Indonesia (Figure 1). The samples were collected from leachate collection outlets and dug wells owned by local residents located no more than 500 m from the landfill. The data collected from the site includes coordinates of well, diameter of the wells, water table, and the well distance from the landfill.

A water sample collected from the dug wells was analyzed in the laboratory to determine the concentration of total Mn, total Fe and NH_3-N. Geology and hydrogeology conditions, as well as porosity and permeability of the soil at both locations were also determined. Laboratory data was analyzed focusing on Mn, Fe and NH_3-N.

Fig. 1. Location of Banyuurip Landfill, Magelang.

3 Results and Discussion

3.1 Landfill Leachate Quality

Banyuurip landfill leachate appearance had a brown to black color, contained a lot of total suspended solid matter and had a bad odor, as a result of high concentration of NH_3-N. Table 1 shows Banyuurip landfill leachate characteristics on total Mn, total Fe and NH_3-N parameters.

Table 1. Leachate Quality

No	Parameter	Concentration (mg/L)
1	Total Mn	5.42
2	Total Fe	6.16
3	NH_3-N	1652

Concentrations of all three parameters in the landfill leachate were high, with NH_3-N being most abundant with up to 1652 mg/L. NH_3-N in the leachate can be traced back to solid waste characteristics, 70% of solid waste being organic material. The main composition of the organic matter is Carbon (C), Hydrogen (H), Oxygen (O) and Nitrogen (N) [1], and most of the nitrogenous compounds in the leachate are in the form of free ammonia. The leachate was highly concentrated with free ammonia.

Total Fe in the Banyuurip landfill leachate was 6.16 mg/L, which was higher than the drinking water quality threshold of 0.5 mg/L. Iron in the landfill leachate comes primarily from metal waste which comprises of about 1-2% of total waste sent to the Banyuurip landfill. Total Mn in the leachate was 5.42 mg/L. Similar to Fe, Mn also comes from metal waste dumped at the landfill.

3.2 Groundwater Quality

Total Mn concentration in the surrounding wells was between 0.00 – 0.26 mg/L. Total Mn was found in almost all wells. The relationship between the distance and total Mn concentration is shown in Figure 2.

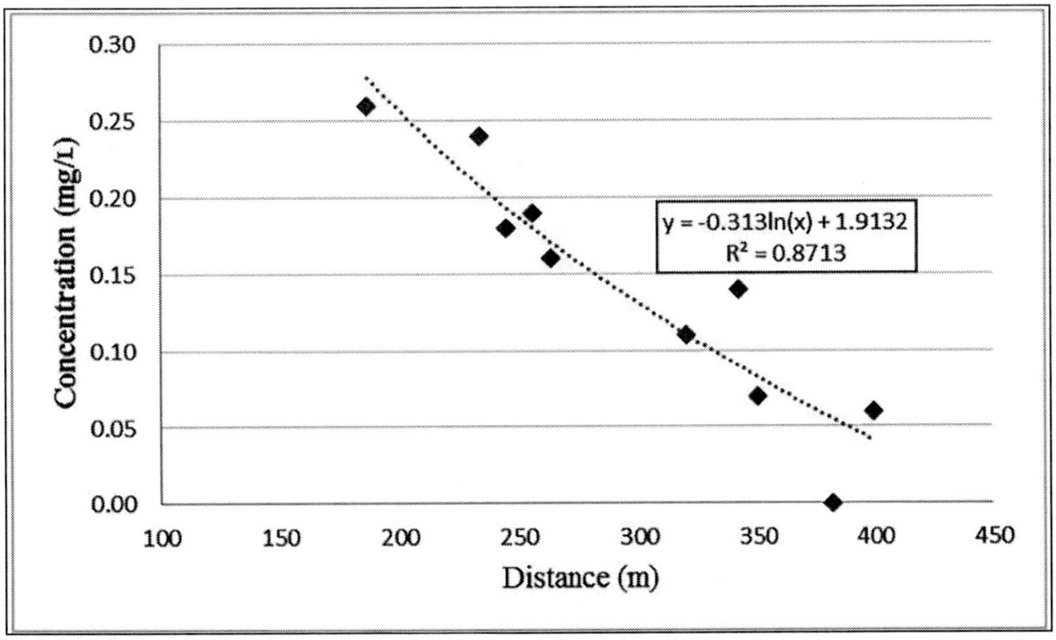

Fig. 2. Relationship between landfill-well distance and Mn concentration

The $R^2 = 0.8713$ suggests that there is a very strong and significant correlation between landfill-to-well distance and total Mn concentration. When the distance between the landfill and dug well increases, the total Mn concentration decreases. The Mn in the soil and physio–chemical–biological conditions could alter Mn concentration at the sub-surface. Manganese concentration in the groundwater above 0.10 mg/L can cause color, odor or change of water taste. Moreover, very high concentrations of Mn is toxic to the neural system and may trigger infant mortality [7].

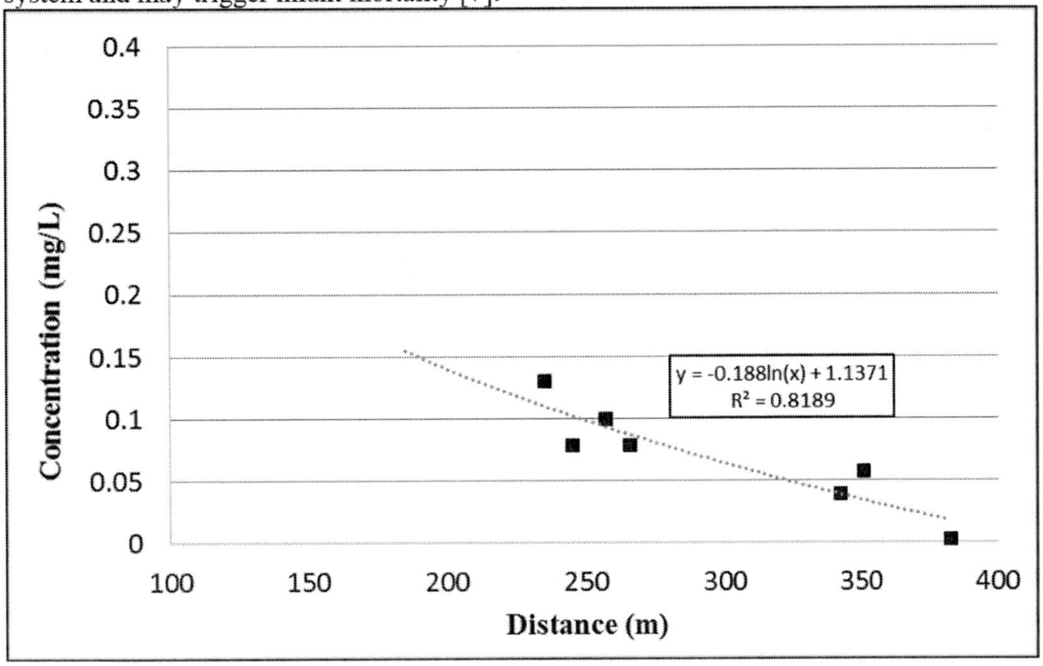

Fig. 3. Relationship between landfill-well distance and Fe concentration

Iron is classified as a secondary pollutant because its effects are more related to aesthetics rather than human health concerns [8]. When the concentration of Fe is high, it will generate a bad taste in the water, yellow color, stimulate growth of Fe bacteria and turbidity. Total Fe concentration in the dug wells are between 0.00 – 0.31 mg/L. In dug wells located more than 350 m from the landfill, the concentration of Fe is very low and almost undetectable in the groundwater samples. The relationship between distance and total Fe concentration is illustrated in Figure 3.

As seen in Figure 3, the R^2 = 0.8189 suggests that there is a strong and significant relationship between landfill-to-well distance and total Fe concentration. The concentration of Fe in the dug wells decreases as distance increases between the landfill and dug well. A correlation coefficient of 0.8189 indicates that about 76.73% of the total Fe in the groundwater can be explained by the distance variable. The remaining 28.11 % is affected by other factors.

The NH_3-Nconcentration ranging from 0.00 to 0.86 mg/L, is lower than the drinking water standard of 1.5 mg/L. The relationship between the distance and NH_3-N concentration is plotted in Figure 4. The R^2=0.8844 suggests that the concentration of NH_3-N in the groundwater is mostly affected by the distance variable. Concentration of NH_3-N above the standard, results in water odor and unpleasant taste. Very high ammonia levels can cause damage to cells and body membrane irritation [9].

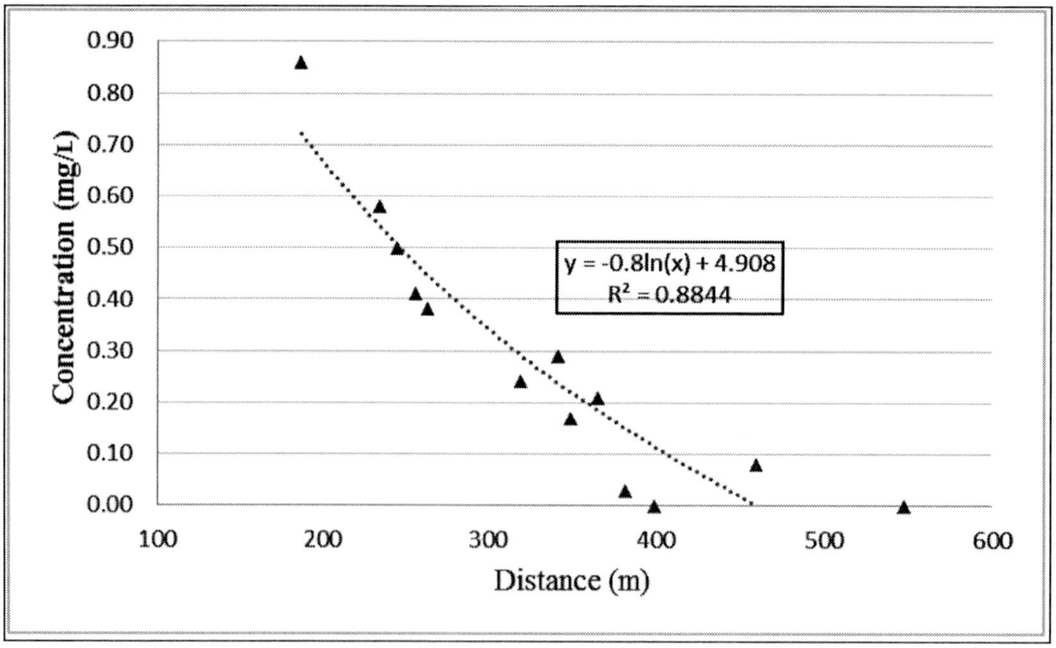

Fig 4. Relationship between landfill-well distance and NH_3-Nconcentration

4 CONCLUSION

This study was performed to investigate potential migration of Mn, Fe and NH_3-N from landfill leachate into the groundwater. The concentration of these parameters in the ground water were determined by taking water samples from dug wells located within 500 m from the landfill. The concentration of Mn, Fe and NH_3-N in the leachate was found to be 5.42 mg/L, 6.16 mg/L and 1652 mg/L. The concentration of these three parameters in the dug wells varied depending on the distance between the dug wells and landfill. The Mn concentration ranged from 0 to 0.26 mg/L, the Fe and NH_3-N concentration are 0 to 0.31

mg/L and 0 to 0.86 mg/L, respectively. The concentration of all these three parameters in the ground water decreases with increasing distance from the landfill.

References

1. R.B. Brennan, M.G. Healy, L. Morrison, S. Hynes, D. Norton, E. Clifford, Waste Manage. **55** (2016).
2. M. Loizidou, E. Kapetanios, Sci. Total Environ. **128**, 1 (1993)
3. M.M.A. El-Salam, G.I. Abu-Zuid, J. Adv. Res. **6**, 4 (2015)
4. R. Slezak, L. Krzystek, S. Ledakowicz, Waste Manage. **43** (2015)
5. M.A. Budihardjo, Adv. Mater. Sci. Eng. (2016).
6. S. Singh, N.J. Raju, W. Gossel, P. Wycisk, Arabian J. Geosci. **9**, 2 (2016).
7. L.A. Schaider, , R.A. Rudel, , J.M. Ackerman, , S.C. Dunagan, J.G. Brody, Sci. Total Environ. **468** (2014)
8. S.K. Sharma, J. Kappelhof, M. Groenendijk, J.C. Schippers, J. Water Supply Res. T. **50**, 4 (2001)
9. I.M Suffet, D. Khiari, A. Bruchet, Water Sci. Technol. **40**, 6 (1999)

Predicting Lead (Pb) Contaminant Migration through Landfill Liner

Purwono[1] and *Mochamad* Arief Budihardjo[1,*]

[1]Department of Environmental Engineering, Universitas Diponegoro, Jl.Prof.H.Soedarto, S.H. Tembalang, Semarang, Indonesia

Abstract. Temanggung Landfill is equipped with a liner system that consists of gravel, geotextile, geomembrane, and compacted clay. However, there is still potential for contaminant mobilization through liners to the surrounding environment, especially the soil beneath the landfill. The purpose of this research is to simulate the movement of lead (Pb) in landfill, vertically at every depth. Contaminant migration from landfill across time can be predicted using POLLUTEv7 software, a one-dimensional model that can describe contaminant mobilization affected by dispersion, advection, diffusion, and retardation phenomena. Over a period of 15 years simulated by the model, the concentration of Pb increased continuously. The initial concentration of Pb above the liner was 0.56 mg/l. This level dropped to 0.35 mg/l after passing the liner in the first year, then increased to 0.4106 mg/l in the second year, and kept increasing up to 0.5071 mg/l in the 15th year.

1 Introduction

The characteristics of municipal solid waste leachate are influenced by the type of solid waste buried in landfill. Most of the solid waste in developing countries going to landfill is in the form of mixed and unsorted waste. Some hazardous waste are also dumped in landfills [1]. During the decomposition process of solid waste, leachate will be produced and flow to the bottom of the landfill [2]. Typically, a landfill is equipped with a semi-permeable liner that functions to prevent leachate escaping to surrounding soil [3]. However, a small amount of leachate still passes through the liner and migrate into the soil due to seepage or liner leakage [4].

Temanggung Landfill started operations in 2012, and was equipped with a liner system that consisted of gravel, geotextile, geomembrane, and compacted soil. That system functioned to stop leachate escaping from the landfill area. Temanggung Landfill mainly relied on the geomembrane layer as a leachate barrier. However, since the geomembrane does not endure pressure and accumulation of leachate containing heavy metals over a prolonged period, there is some possibility of it puncturing or cracking.

Leachate is generally toxic because it contains heavy metals and various other organic pollutants that can contaminate soil and ground water. Because of electronic waste, a

* Corresponding author: m.budihardjo@ft.undip.ac.id

predominant heavy metal in landfill leachate is lead (Pb) [5]. Thus, this study focuses on predicting Pb migration from landfill over a period of 15 years, using average Pb concentrations in some solid waste landfills. The purpose of this study is to predict lead (Pb) concentrations in the soil beneath the landfill using Pollutev7 software, a one and a half dimensional model to the advection-dispersion equation [6]. Some studies have previously used POLLUTEv7 software to effectively predict contaminant movement from landfill. This software has also been used in landfill research in other countries, such as China [7] and Canada [8].

2 Methods

Preliminary tests were conducted to measure Pb concentrations in leachate. The leachate sample was collected from the leachate treatment plant inlet. Pb concentrations in the inlet were assumed to represent the leachate concentration at the bottom of the landfill. Water samples were also taken from two monitoring wells surrounding the landfill. Other information such as soil characteristics of the Temanggung Landfill (permeability, porosity, dry density, and grain size), the types of soil layer, their location and layer thickness, were also collected.

All information was entered in the POLLUTEv7 software (Fig. 1). To get a reliable result, all parameters must be complete, including the thickness of each layer. The POLLUTEv7 software is user-friendly and developed from a one and a half dimensional model. The software also allows for some default values, in anticipation of incomplete data. However, the results are more reliable if all the parameters required by the software are measured from the field.

The information was then used as model inputs and processed using the POLLUTEv7 software (Figure 1). After running the model, the software provided a list of tables and graphs showing the calculated results, which can be the amount of contaminant concentration to time, the amount of contaminant concentration to depth, and the depth to time. There were some assumptions and limitations in the study. For example, it assumed that the ground water movement is in steady state condition, the soil types in the landfill are homogenous, and concentration of contaminants is constant.

Fig. 1. A screenshot of the POLLUTEv7 Software

3 Results and Discussion

3.1 Properties of the Liner

Temanggung Landfill has been equipped with a leachate barrier that consists of various materials. A gravel layer with permeability of $1x10^{-1}$ m/s acts to drain leachate and protection for leachate collection pipe. The geotextile is laid underneath the gravel layer, and is followed by a geomembrane layer that has a very low permeability of $5x10^{-10}$ m/s. The geotextile functions to protect the geomembrane. Below the geomembrane, there is compacted soil layer and native soil 1 and 2. While the permeability value for compacted soil is quite low at $1.109x10^{-7}$ m/s, small amounts of leachate can still pass this layer. The properties of all layers are presented in Table 1.

Table 1. Initial Data

No	Landfill layer	Permeability (m/s)	Coef. Hydrodinamic Dispersion (D) (m^2/s)
1.	Gravel 3/5	1.00 e-01	51.43497152
2.	Geotextile	6.50 e-02	2.630784222
3.	Geomembrane	5.0 e-10	1.3491 4e-08
4.	Compacted Soil	1.109 e-07	4.67379 e-06
5.	Native Soil 1	3.2387 e- 06	0.000109061
6.	Native Soil 2	6.3834 e- 06	0.000122617

3.2 Findings

The initial Pb concentration above the liner was 0.56 mg/L. Once past the liner system, the concentration of Pb decreased to 0.395 mg/L, and continued to decrease to 0.358 mg/L at a depth of 7.1 m. However, due to accumulation, contaminant levels of Pb were predicted to increase year to year at each depth level, from year 1 to year 15. In the 1st year, the Pb concentration at the depth of 7.1 m was 0.358 mg/L. In the 2nd year, Pb concentration was predicted to increase to 0.4106 mg/L, and to keep increasing up to 0.5071 mg/L in the 15th year. The increase in concentration is caused by accumulation of Pb and the process of diffusion and dispersion in soils [9]. Annual increments of concentration varied; however, they became very small after the 8th year. Changes in Pb concentrations of Pb are depicted in Figure 2.

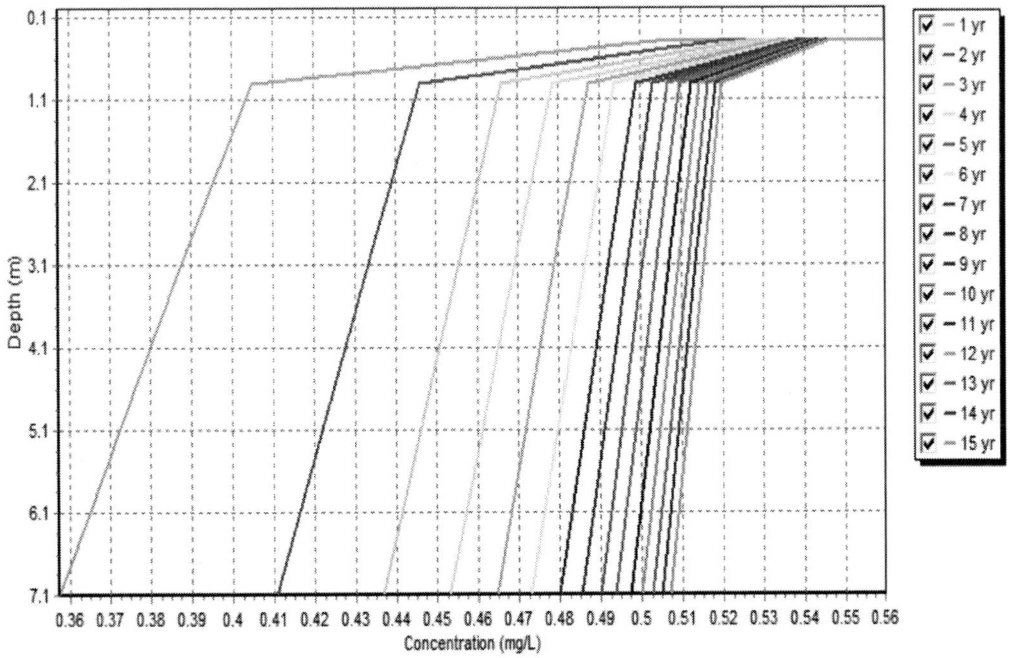

Fig 2. Predicted Pb Concentrations at Different Depths over a Period of 15 Years

Assuming that the Pb concentration in leachate remains constant, Pb concentration in the soil at 7.1 m depth will not reach 0.51 mg/L within 15 years. However, if there are any changes in Pb concentration in leachate, the current model will no longer hold and another simulation will need to be run.

Involving a model to predict the Pb concentration in the soil underneath a solid waste landfill over the year would become a real challenge for landfill's authorities. In order to avoid any misleading result, a careful consideration should be taken into account when performing the simulation. The Pb concentration in leachate should be monitored in regular basis, such as in every month or two months followed by simulating the model base on the concentration obtained during monitoring activities. Any potential change of other parameters that possibly altering the result should also be considered in the developed model. The result of the simulation then should be validated accordingly by performing some field samplings and re-adjusting the model once any escalating on the deviation of the result occurred.

4 Conclusion

Leachate containing Pb in the Temanggung Landfill can be simulated using a one and a half dimensional model using POLLUTEv7. The simulation is very useful for policymakers and the Temanggung Landfill authority who are interested in predicting Pb contamination. The concentration of Pb at the bottom of the landfill was found to be 0.56 mg/L. The Pb concentration was then predicted to decrease to 0.415 mg/L after passing the main leachate barrier, which is a geomembrane layer. The concentration of Pb in every soil layer was predicted to increase from year to year, because of Pb accumulation at every depth. However, a routine monitoring of Pb concentration in the leachate collection system should be conducted followed by model simulation to avoid any invalid results.

References

1. Q. Sui, W. Zhao, X. Cao, S. Lu, Z. Qiu, X. Gu, G. Yu, J. Hazard. Mater. **323** (2017)
2. M.M.A. El-Salam, G.I. Abu-Zuid, J. Adv. Res. **6**, 4 (2015)
3. M.A. Budihardjo, Adv. Mater. Sci. Eng. (2016).
4. V. Tsarpali, S. Dailianis, Aquat. Toxicol. **124** (2012)
5. C. Amor, E. De Torres-Socías, J.A. Peres, M.I. Maldonado, I. Oller, S. Malato, M.S. Lucas, J. Hazard. Mater. **286** (2015)
6. T.L.T. Zhan, C. Guan, H.J. Xie, Y.M. Chen, Sci. Total Environ. **470** (2014)
7. H. Xie, Z. Lou, Y. Chen, A. Jin, T.L. Zhan, X. Tang, Geotext. Geomembranes. **36** (2013)
8. A. Hoor, R.K. Rowe, Waste Manage. **32**, 5 (2012)
9. G. Zeng, J. Wan, , D. Huang, L. Hu, C. Huang, M. Cheng, W. Xue, X. Gong, R. Wang, D. Jiang, J. Hazard. Mater. **339** (2017)

Leaves Composting Process and the Influence of Rumen Content and Bran Addition

Mochamad Arief Budihardjo[1,*], *Endro* Sutrisno[1], *Purwono*[1] and *Maulia* Fatimah[1]

[1]Department of Environmental Engineering, Universitas Diponegoro, Jl.Prof.H.Soedarto, S.H. Tembalang, Semarang, Indonesia

Abstract. Leaves on the Diponegoro University campus were sent to the university's solid waste facility for composting. A modification of the normal composting process was implemented to speed up the process. In this study, the content of cattle rumens and rice bran were added to accelerate the composting process. Three samples with different composition were prepared (sample code: L, L-R, and L-B). Sample L contained leaves only, sample L-R comprised of leaves with rumen content added and sample L-B consisted of leaves with the addition of rice bran. The results indicated that a maximum temperature of 60.2°C was reached in the L-B sample, while the temperature of compost and L-R was about 30°C. The lowest pH was 6.76 found on the L sample, with sample L-B and L-R having pH of 6.7 and 6.9, respectively. The quality of the mature compost of all variation was about the same, but the compost in sample L-B produced the highest C/N ratio of the three composts, about 20% higher.

1 Introduction

The processing of solid waste contributes significantly to environmental pollution. In some developing countries, the concepts of solid waste management practise, such as reuse and recycling, has not been implemented effectively [1]. To date, the handling of solid waste comprises primarily of the collection of waste and its transfer to landfill where it is dumped. Without any reduction, reusing or recycling practices in place, the amount of solid waste to be sent to landfill has increased over time [2, 3].

The University is one of the biggest contributors to municipal waste due to its various activities, such as administration, teaching and cleaning. Solid waste that is produced by Diponegoro University contains a significant amount of leaves, about 40% of the total waste by weight. The leaves are sent to the waste facilities in the Diponegoro University area to be composted. However, the composting process of leaves takes time and the land availability is limited [4]. Therefore, a modification of the composting process with the purpose to accelerate the process, by adding the rumen contents of cattle and rice bran, would have great advantages. The ruminal content of cattle was chosen as it is rich with microbes [5], while rice bran is commonly used in Asian countries as a bulking agent in the composting process [6].

* Corresponding author: m.budihardjo@ft.undip.ac.id

Therefore, this study investigates the effect of the addition of these two components on the process of composting leaves. The composting experiment was conducted over 30 days. Temperature and pH of the composting piles were measured regularly, in order to understand the influence of rice bran and the content of cattle rumen on the composting process.

2 Methods

In order to conduct the experiment, three compost samples with different composition were prepared (sample codes: L, L-R, and L-B). Sample L contained leaf waste only, sample L-R comprised of leaves with rumen content added and sample L-B consisted of leaves with the addition of rice bran. Cow ruminal content was collected from the local slaughter house and rice bran was supplied by the local rice mill.

The conventional composting process of leaves (L) is compared with samples L-R and L-B. Temperature and pH of all compost piles were measured daily. The composting was done using the MacDonald method which uses boxes and an aerobic system. Watering was done regularly to keep the moisture content of compost piles constant. After 30 days the compost products were analyzed for parameters such as physical appearance, organic content, total nitrogen (N), C/N ratio, total phosphorus (P), total potassium (K) and water content.

3 Results and Discussion

3.1 Temperature

The temperature was measured on a daily basis during the composting process. Figure 1 shows the temperature fluctuation during the composting period. From all three samples there was only one sample that experienced the ideal thermophilic phase, sample L-B. The thermophilic phase occurred between the 2nd and 5th day. The maximum temperature of samples L and L-R during the composting period was about 30°C.

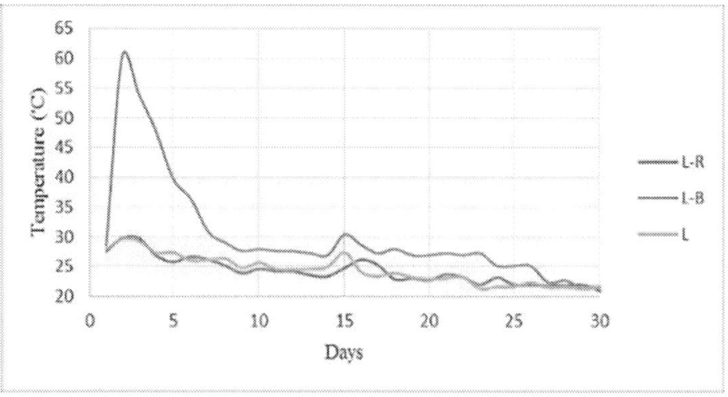

Fig 1. Temperature changes during composting

The presence of the rice bran caused L-B variation reach thermophilic conditions. Rice bran functioned as a nutrient for composting bacteria which influenced the activities of the bacteria in the processing of organic materials during composting, thus accelerating the materials' decomposition. Whereas the samples L-R and L did not reach the thermophilic temperature, with temperatures ranging from about 23°C to 30°C during the composting process.

3.2 PH

A pH measurement of the compost piles was performed on a daily basis. The results of the pH monitoring for each sample is shown in Figure 2. At the beginning of the decomposition process, the pH of each compost sample dropped during the thermophilic phase, indicating high microbe activity in decomposing the organic material. The pH gradually increased after the maximum temperature was reached and gradually became neutral. The changes in pH could be due to the microbial activity, i.e. various bacteria and fungus require different acid environments to decompose organic matter. This pH decrease was consistent with the increase of compost temperature in the thermophilic phase.

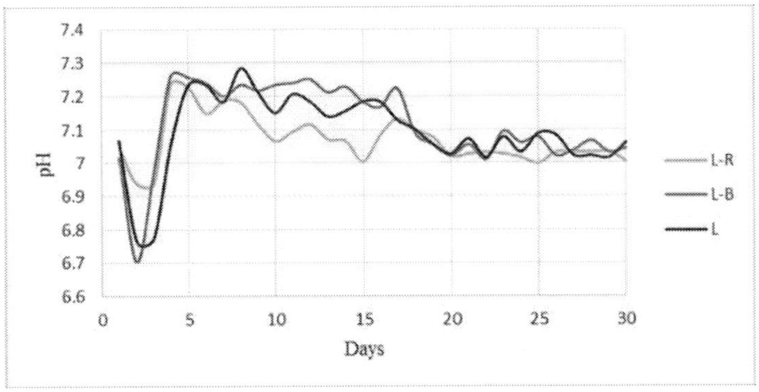

Fig. 2. Average pH in compost pile

After the thermophilic phase, the pH gradually increases to produce weak basic conditions. The activity of the thermophilic bacteria creates ammonia and nitrogen gases, causing the rise pH. The highest pH recorded was between days 4 and 8 for all compost samples. In the next phase, a nitrification process occurs where the nitrogen is found in the form of nitrate. This will gradually decrease the pH to a constant pH, indicating composting maturity. The pH for all the compost samples were within the optimum range for composting (pH5 – 8) [7].

3.3 Mature Compost

The physical appearance of the mature compost sample appeared to be similar. The compost did not have an unpleasant odour but smelt like soil. The colour of compost was brownish black, similar to soil. Table 1 shows the quality of the compost product for each sample. Overall, the final compost for samples were similar in quality. The C-organic content ranged from 3.42% to 6.9% with total N 0.18% to 0.44%. Sample L-B had the highest C/N ratio of 19, L and L-R had C/N ratios of 15.68 and 14.6, respectively.

Sample L-B had the fastest decomposition process when compared to samples L-R and L. Once the ideal thermophilic temperature was reached in sample L-B, the decomposition process occurred more rapidly due to the nutritional and bulking properties of the rice bran, thus accelerating the whole composting process. Visual observation (colour and physical appearances), indicated the sample L-B reached maturity at about 24 days, whereas the sample L-R reached maturity at 27 days and L had not reached maturity by day 30

Table 1. The quality of compost after 30 days

PARAMETER	VARIATION		
	L	L-B	L-R
C-Organic	6.90	5.41	3.42
N-Total	0.44	0.37	0.18
C/N Ratio	15.68	14.60	19.00
P-Total	0.65	0.63	0.54
K-Total	0.20	0.16	0.22
Water content (%)	15.96	15.31	14.95
pH	7.56	7.57	7.32
Temperature	25	25	25

4 CONCLUSION

Three different variations in the composting of leaves; composting of leaves only, leaves with rice bran and leaves with rumen content, were investigated. During the composting process, temperature and pH were monitored. The addition of rice bran optimised the composting process. The quality of the mature compost with regards to organic content, total N, C/N ratio, total P and total K were also examined. Physically, all the mature compost samples had similar appearance with brownish black colour and smelt like soil. The compost with rice bran added reached maturity in 24 days, while the compost with leaves only and addition of rumen contents required more time to reach maturity.

References

1. T.T.P. Nguyen, D. Zhu, N.P. Le, Habitat Int., **48** (2015)
2. M.K. Younes, Z.M. Nopiah, N.A. Basri, H.Basri, M.F. Abushammala, M.Y. Younes, Waste Manage. **55** (2016)
3. S. Suthar, P. Singh, Sustain. Cities Soc. **14** (2015)
4. A. Makan, Waste Manage. **42** (2015)
5. L.L.C. Guidoni, R.V. Marques, R.B. Moncks, F.T. Botelho, M.F. da Paz, L.B. Corrêa, É.K. Corrêa, J. Environ. Manage. **207** (2018)
6. K. Shrestha, , E.M. Adetutu, P. Shrestha, K.B. Walsh, K.M. Harrower, A.S. Ball, D.J. Midmore, Bioresource Technol. **102**, 17 (2011)
7. K. Kato, N. Miura, Bioresource Technol. **99**, 9 (2008)

Analog Study of Fluid Flow on Deformation Band at Petani Formation, Riau, Indonesia

Tiggi Choanji[1,*], *Novia* Rita[2] , *Yuniarti* Yuskar[1], *Aulia* Pradana[3]

[1] Department of Geological Engineering, Universitas Islam Riau, Pekanbaru, Riau, Indonesia
[2] Department of Petroleum Engineering, Universitas Islam Riau, Pekanbaru, Riau, Indonesia
[3] IAGI Riau, Indonesia

Abstract. Deformation band at Petani Formation included in the type of disaggregation band, where deformation on this formation resulted from tectonics stress at Plio – Pleistocene affecting sandstone layer that being deformed and resulted strain localization feature, filling with iron oxide sand. The measurement for this research was using grain size analysis, distribution analysis of shear trends, and measurement of permeability for each deformation band. The result shows that grains size dominantly fine sand – medium sand (0.149 mm – 0.297 mm), followed by the distribution of deformation trends showing NW – SE direction, and also good in permeability (3.4 mD– 188.4mD). It can be concluded that the structural features of deformation band can also create good permeability for fluid flow.

1 Introduction

Deformation bands are the most common strain localization features found in porous and sedimentary sandstones, including quartz precipitates, gravitational and tectonic landslides that affect sandstones in hydrocarbon and aquifer reservoirs. This occurs in different types of tabular deformation zones where the granules undergo a shift, rotation and/or fracture reorganization during dilation, shear, and/ or compaction processes. The band deformation with the most common shear component accommodates a sliding offset up to millimeter or cm. Deformation bands can occur in the form of single structures or cluster zones, and are the major zone deformation element in porous rocks. Factors such as porosity, mineralogy, grain size and grain shape, lithification, stress conditions and loading control the type of band deformation that is formed. Of the various types, deformation bands of phyllosilicate and deformation of cataclastic bands show the greatest decline in permeability, and thus have the greatest potential to influence fluid flow. Sorting bands, where non-cathartic, granular flow is the dominant mechanism, show little influence on fluid flow Aided by chemical compaction or cementation [1].

* Corresponding author: tiggich@eng.uir.ac.id

Character deformation band in Kampar Region has a special character that needs to be mapped and analyzed in detail, this is quite different from the studies commonly found in Tertiary rocks.

2 Geology Regional

2.1 Geology of Research Area

The research area covered with Petani Formation that composed of mudstone, containing carbon, lignite, silt and sandstones. Based on the geological map sheet of Pekanbaru at scale 1: 250.000, at Tapung Hulu and Batubelah Kampar Region, there are three types of formation developed around research area which are Petani Formation, Telisa Formation, and Minas Formation [2]. Structures geology developed in the research area include elongated anticline folds with west-southeast direction. Folds are exposed the lithology formation. This is influenced by the effects of Indian-Australia plates subducted Eurasian Plate, resulting in an inversion tectonic process during the sedimentation of Petani Formation. Kusau Makmur area, Tapung Hulu, Kampar regency, Riau and Batu Belah, Kampar regency, Riau has a considerable distance (+ 35 km) but has a similarly deformed band characteristics. This indicates that the formation of the band's deformation are formed quite widely and evenly and allegedly due to tectonic processes[3,4].

We can assume the conditions on subsurface of the reservoir under certain conditions can be reflected from the existence of rock outcrops exposed to the surface[5,6]. Therefore, it is important to understand the condition of reservoir quality changes in the condition of the deformed band layer, whether the band deformation becomes a barrier or becoming capable of flowing fluid between two layers.

Research area geologically included at Petani Formation that deposited in late Miocene – Pliocene. From two sites of location in Tapung Hulu District and Batu Belah District, deformation band is formed, and if drawn between the two positions is closely related to the tectonic pattern formed at the Plio-Pleistocene age, with the northwest-southeast direction.

2.2 Deformation Band

Deformation Bands are a common type of local strain in porous rocks that are deformed and formed in the form of single structures, clusters, and in fault zones due to fractures[7][8]. Faults are generally considered to be fractures or surfaces across which there is appreciable relative displacement, or to be zones of such fractures or surfaces. Further, small faults with displacements less than a few meters generally are treated as minor, or as secondary structures relative to major faults nearby[7]. In sandstone, deformation band/band deformation is limited to porous granular media, particularly porous sand and sandstone[9]. There are several important characteristics that distinguish deformation bands from ordinary fractures (such as slip surfaces or extension fractures). First, they are thicker and exhibit smaller offsets than classical slip surfaces of comparable length. Also, whereas cohesion is lost or reduced across ordinary fractures, most deformation bands maintain or even increase cohesion. Furthermore, deformation bands often exhibit a reduction in porosity and permeability, whereas both slip surfaces and tension fractures are typically associated with a permeability increase. Strain hardening behaviour, commonly associated with deformation band formation, also contrasts to the strain softening associated with classical fractures. Kinematically, deformation bands can be classified as dilation bands, shear bands, compaction bands or hybrids of these types [10].

Although a kinematics-based classification is logical, it is also useful to classify deformation bands in terms of the dominant deformation mechanism operating during their formation. Deformation mechanisms depend on internal and external conditions such as mineralogy, grain size, shape, sorting, cementation, porosity and stress state. Different mechanisms produce bands with different petrophysical properties. Thus, such a classification is particularly useful where permeability and fluid flow are an issue. The dominant deformation mechanisms are: (1) granular flow (grain boundary sliding and grain rotation)[11]; (2) cataclasis (grain fracturing and grinding or abrasion) [10 -12]; (3) phyllo-silicate smearing[15]; (4) dissolution and cementation[16].

3 Methodology

The methodology used for measuring the deformation band and its connectivity was conducted on several steps. Firstly, taking the eight core samples at field station and conduct measurement on field site such as length, width, thickness, trends, description of the samples by using field geological tools and cluster the data to define the difference between single deformation band and group of deformation band (show in graphic displacement vs thickness in logarithmic scale [17] and also doing measurement in grain distribution of each samples. Then, we plot all measurement trends (105 sets data) on stereographical projection for giving two-dimensional or projection image of the surface of a sphere as a place of geometry orientation of plane and line. This stereographic projection used for solving problems related to geometry of the magnitude of direction and angle in the geological analysis of geological structures because this projection can describe the geometry of the position or the orientation of the plane and line in the field [18]. As we can see the trends of deformation band as geological structure that define stress and relative tectonic direction from measurement of stereographic projection, next we will measure the flow in each size at laboratorium to define the permeability barrier by Permeability measurements using a tool called Gas Permeameter.

Measurement using this tool there are several procedures that must be done, which the core samples should be in a clean and dry condition to get the best results. The sample is cleaned with solvent and dried in an oven before the gas permeability measurement, Core samples will be made in cylindrical core plug that must be perpendicularly sized + / - 0.005 inches to get the best results. After that the process continued to prepare the instrument and ensure there are no obstructions in the chore holder or end stems. Then make sure the tubing fittings are in comfortable condition and leak free. Furthermore, when loading the friable core, a 200-mesh disk-shaped screen is placed at the top and bottom of the sample to prevent sand migration from intake holders of flow meters. After that set the all panels to bypass position, and then perform the core measurements (length, height, and area), next is inserting the cores into the system core test system, then set the pressure of the compressor with the nitrogen and note the pressure, and the flow rate. Then the value of permeability will be obtained by using the darcy equation. The value of this permeability will be plot on graphic distribution to see the flow value of the deformation band.
The equations as follow :

$$K = \frac{\mu Q L}{A (P_1 - P_2)} \tag{1}$$

where K is shown value of Permeability (darcy) , Q shows the flow rate in cc/sec, μ is viscosity, A show value of area in cm2, L is length (cm) , and P is Pressure, atm. So by using this equation, effect from the deformation band between two layer can be defined clearly.

4 Result

The deformation bands in the research area (**Fig. 1**) on the left showed two different characteristics in deformation band, where on Fig 1a shows the characteristic of this deformation band consist of loose fine sand to medium sand (**Fig. 2**), yellowish brown color, good sorting, dominated by quartz mineral and iron oxide, good porosity. deformation band was perpendicular to the layer and also cutting each other, and from stereographic projection shows that stress on the area dominated by NE-SW direction (**Fig 3**).

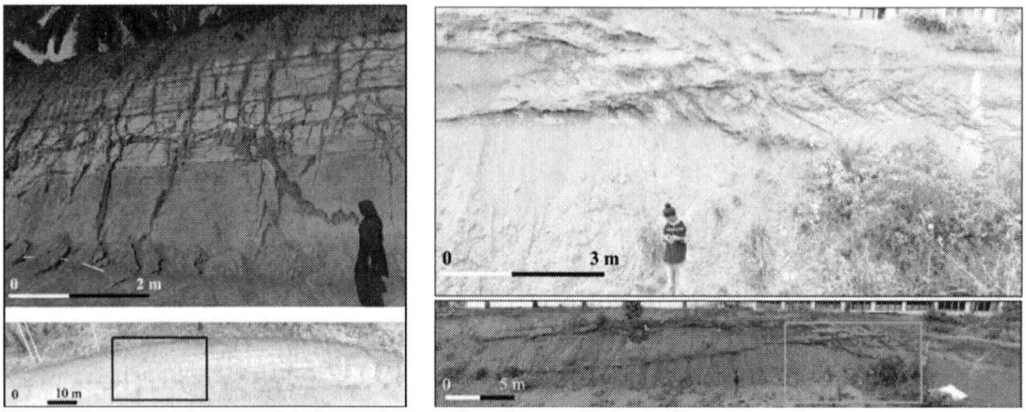

Fig. 1. Structural feature of deformation band at Tapung Hulu (left) and Batu Belah, Kampar (right).

On **Fig 1** on the right, shows the characteristic of this deformation band consists of loose fine sand, reddish brown colour, good sorting, also dominated by iron oxide. deformation band was more gently dip to the layer showing different thickness band that wider between the layer, and from stereographic projection shows that stress on the area dominated more W-E direction (**Fig. 3**).

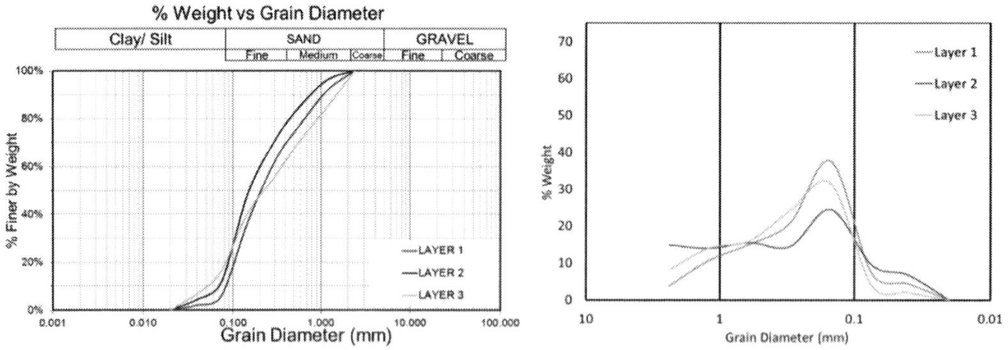

Fig 2. Distribution % weight vs grain diameter analysis and cumulative percentage for define dominant size of grain on different lithology sample.

From reconstruction of historical geology, the deformation was affected by the tectonic process with several phases. Based on the results of the 105 plots of data shows the condition of movement dominated by tectonic movement that trend southwest-northeast resulting deformation band that perpendicular to the stress (**Fig. 3**). Based on[1], pattern that occurs in this area in line with the disaggregation band. This band formed due to the final cataclastic process which then lifted to the surface and produces a tension fracture which then the fracture is filled by sediment material from deposition after tectonic process. Field data show that deformation bands occur as isolated structures, linked systems, complex zones of multiple, interconnected deformation bands.

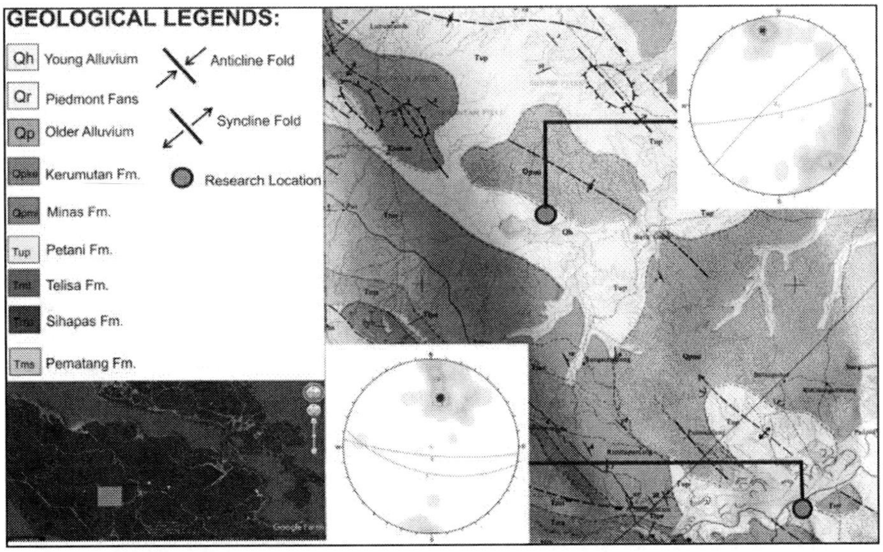

Fig 3. Stereographic projection of deformation band overlay with geological map.

Permeability result showing that deformation bands having permeability ranges from 5.4 mD – 188.4mD (Tables 1). Permeability value result show from northern area site are higher than on the southern site. Tapung Hulu, permeability value ranges from 34 mD–188.4 mD. On Batu Belah area show permeability value ranges from 5.4 mD – 13.1 mD (Table 1).

Based on this different ranges permeability value, probably there can be some factors that can be concluded, such as different level of stress event on this two area, because if there is any difference in the direction of the stress, its also possible to influence process development of deformation band, and also the mineralogical differences, the grain size, displacement and thickness of deformation band also need to be considered when looking at different permeability values between these two locations.

Table 1. Result data of permeability on samples.

Samples	Location	Q high (cc/sec)	Qlow (cc/sec)	High DP (atm)	Low DP (atm)	K (mD)
TP-1	Tapung Hulu	2.53	1.74	0.041	0.025	73.3
TP-2	Tapung Hulu	2.37	1.74	0.082	0.072	94.5
TP-3	Tapung Hulu	2.28	1.75	0.225	0.211	54.2
TP-2	Tapung Hulu	3.17	1.75	0.245	0.234	188.4
TP-2	Tapung Hulu	2.025	1.74	0.327	0.315	34
C2	Batu belah	1.0	0.92	0.361	0.352	13.1
C3	Batu belah	1.3	1.252	0.598	0.585	5.4
C3	Batu belah	1.4	1.35	0.51	0.502	9.3

$\mu = 0.018$ (Viscosity of Nitrogen Gas)

5 Conclusion

So based on this research there are several conclusion can be defined from this research, which is :

1. The deformation band characteristic in research area shows that the pattern is more considered to be disaggregation band after comparing on two location that resulted from stress release after tectonic occurs.
2. Permeability result showed that deformation band at tapung hulu (northern site) having higher permeability than Batu Belah (southern site), this considered to be direction and level of stress and that happens in this two area that resulting different permeability value on this two location.

This research was conducted based on grant from Lembaga Penelitian Universitas Islam Riau. We are trully thank you to this collaboration research between Dept. of Geological Engineering and Dept. of Petroleum Engineering at Universitas Islam Riau.

References

1. Aydin, A. Small faults formed as deformation bands in sandstone. Pure and Applied Geophysics PAGEOPH, **116**, 4–5, 913–930, (1978)
2. Aydin, A. Small faults formed as deformation bands in sandstone. Pure and Applied Geophysics PAGEOPH, **116**, 4–5, 913–930. (1978)
3. Aydin, A., Borja, R. I., & Eichhubl, P. Geological and mathematical framework for failure modes in granular rock. Journal of Structural Geology, **28,** 1, 83–98, (2006)
4. Aydin, A., & Johnson, A. M. Analysis of faulting in porous sandstones. Journal of Structural Geology, **5**, 1, 19–31. (1983)
5. Choanji, T. Indikasi Struktur Patahan Berdasarkan Data Citra Satelit dan Digital Elevation Model (DEM) di Sungai Siak, Daerah Tualang dan Sekitarnya Sebagai Pertimbangan Pengembangan Pembangunan Wilayah. Jurnal Saintis, **16**, 2, 22–31, (2016).
6. Choanji, T. Slope Analysis Based On SRTM Digital Elevation Model Data: Study Case On Rokan IV Koto Area And Surrounding. Journal of Dynamics, **1**, 2, (2017)
7. Choanji, T., & Indrajati, R. *Analysis of Structural Geology based on Sattelite Image and Geological Mapping on Binuang Area, South Kalimantan*. Geosea XIV Congress And 45th Iagi Annual Convention 2016 (GIC 2016) **935,** (2016)
8. Clarke, M.C.G; Kartawa, W.; Djunuddin, A.; Suganda, E.; Bagdja, M. Peta Geologi Lembar Pekanbaru. Bandung. (1982)
9. Davis, G. H. *Structural geology of the Colorado Plateau region of southern Utah, with special emphasis on deformation bands*. Geological Society of America, (1999)
10. Eichhubl, P., Hooker, J. N., & Laubach, S. E. Pure and shear-enhanced compaction bands in Aztec Sandstone. Journal of Structural Geology, **32** ,12,, 1873–1886, (2010)
11. Fossen, H., & Bale, A. Deformation bands and their influence on fluid flow. AAPG Bulletin, **91,** 12, 1685–1700. (2007)
12. Fossen, H., Schultz, R. A., Shipton, Z. K., & Mair, K. Deformation bands in sandstone: a review. Journal of the Geological Society, **164**, 1–15. (2007)
13. Gibson, R. G. Fault-zone seals in siliciclastic strata of the Columbus Basin, offshore Trinidad. AAPG Bulletin, **78**, 9, 1372–1385, (1994)
14. Knipe, R. J., Fisher, Q. J., Jones, G., Clennell, M. R., Farmer, A. B., Harrison, A., White, E. A. Fault seal analysis: successful methodologies, application and future directions. Norwegian Petroleum Society Special Publications, **7**, 15–38. (1997)
15. Putra, D. B. E., & Choanji, T. Preliminary Analysis of Slope Stability in Kuok and Surrounding Areas. Journal of Geoscience, Engineering, Environment, and Technology, **1**, 1, 41–44. (2016)
16. Ragan, D. M. *Structural geology: an introduction to geometrical techniques (4th ed). Cambridge University* Press. (2009)
17. Twiss, R. J., & Moores, E. M. *Structural Geology*. New York, United States: W.H.Freeman & Co Ltd. (2nd Revise), (2007)

Palaeoecological analysis of Sentolo Formation based on foraminifera fossil assemblage from stratigraphic cross-sectional measurements in Pengasih, Kulon Progo, Yogyakarta

Anis Kurniasih[1,*], *Muhammad* Idham Fauzan[1], *Tri* Winarno[1], and *Reddy* Setyawan[1]

[1]Departement of Geological Engineering Universitas Diponegoro, Semarang, Indonesia

Abstract. Sentolo formation that is included in the Kulon Progo stratigraphic series has a wide distribution. Although there have been a number of studies on it, more detailed research is needed, as there are still many unexplored outcrop locations. Moreover, there are still a lot of questions about the stratigraphic relationship between Sentolo Formation and Oligocene-Miocene volcanic series of Kulon Progo. Based on this background, this study is focused on the existence of foraminifera fossil in Sentolo Formation as an indicator of palaeoecological conditions at the time Sentolo Formation was deposited. This research is very important to get a detail understanding about the indicators that are influential at the time of deposition process of Sentolo Formation in research area. Based on palaeoecological analysis, the palaeoecology at the time Sentolo Formation deposited was normal marine continental slope waters. Meanwhile, the community structure of foraminifera indicates a stable water conditions in a marine environment that was not disrupted by volcanism, so it becomes a good factor for the growth of foraminifera. Whereas the presence of volcaniclastic material mixed with carbonate which is occurred in the lower part of Sentolo Formation were interpreted as reworked of older volcanic rock.

1 Research Background

Foraminifera is a single-celled organism that inhabits marine waters. Although it is a microorganism, the dead foraminifera can be preserved in sediments and rocks for hundreds or even millions of years. In geology, foraminifera fossils are widely used to interpret the relative age and depositional environments of rocks. Foraminifera occur since the Paleozoic Period until recently and has undergone evolution so that many types of foraminifera became the index of certain geological age. The foraminifera fossil in rocks can also be used as an indicator of ecological conditions when the fossil-containing rocks are formed in an environment or referred as paleoecology.

The Sentolo formation that is included in the Kulon Progo stratigraphic series is fairly wide spread. Researchers such as [1-3] have studied stratigraphic aspects of the Sentolo Formation. Although there have been number of studies on the Sentolo Formation, more detailed research is needed, as there are still many unexplored outcrop locations. The study can be focused on the existence of foraminifera fossil in Sentolo Formation rocks as a guideline of paleoecological conditions at the time the Sentolo Formation was deposited. This research is very important to identify the condition of aquatic ecology at the time of foraminifera lived and the relation between volcanism process that happened in Oligocene - Miocene with deposition of Sentolo Formation in research area.

The ecological condition of aquatic environment measured directly to water samples, which include carbonate, salinity, and temperature. However, this measurement method is not possible in the paleo-aquatic environment, because the existing waters must have undergone a change of condition. Therefore, paleoecological analysis is applicable to sediments or rocks deposited in an environment and one of its methods is by analyzing the abundance of foraminifera fossil contained in the rocks or sediments.

2 Methods

1. Foraminifera analysis

The planktonic foraminifera classification is based on [4] and for benthic foraminifera based on [5, 6]. Once the foraminifera are identified, the presentation of each taxon in every sample is displayed in foraminifera distribution chart.

2. Planktonic and benthic foraminifera ratio P/B Ratio)

The ratio of individual presence of planktonic and benthic foraminifera is used as an indicator of the water depth.

3. Community structure

The community structure of foraminifera is expressed in diversity index, similarity, and dominance index. Diversity is the number of species or taxon contained within a community. It is difficult to determine diversity. Therefore, the method of calculating diversity in this study is conducted using the Shannon - Wiener index [7] or H, with the following criteria:

$H < 1$ = low diversity

$1 < H < 3$ = medium diversity

$H > 3$ = high diversity

The uniformity index is the pattern of distribution of each taxon whether it is evenly or not. If the uniformity index is relatively high, the existence of each type of foraminifera in a waters is in a uniform condition. The uniformity index is expressed in Evennes Index (e), with the criterion of $e < 0.3$ showing the uniformity is low; $e = 0.3 - 0.6$ is moderate; and $e > 0.6$ is high [7].

The dominance index is used to obtain information about the domination of a taxon in a particular community. It is ranges from zero to one. If the value is close to 1 then the foraminifera community is dominated by a particular species and if it is close to 0 then no dominant species [7].

4. MRT triangular diagram

Triangular diagrams are based on the three sub-ordo of foraminifera abundance, which are distinguished by their test compositions; agglutinated, porcelain, and hyaline. The triangular diagram may be an indicator of carbonate content, dissolved oxygen, and salinity.

5. Research Location

The research location covers the area of Pengasih, Kulon Progo Regency. Overall, there are seven detailed stratigraphic measurement locations scattered in the southwest and northeast part of the study area.

3 Results and Discussion

3.1 Lithostratigraphy and Biostratigraphy of Sentolo Formation

In general, the stratigraphic cross-sectional measurement is more focused on the Sentolo Formation, especially at the lowest part where there is direct contact with the older formation of the Old Andesite Formation (OAF) and the middle part of the formation. Stratigraphic measurements were carried out at 7 locations spread in two areas, in the southwest and north-east. There are 5 sections in the southwest; i.e. Serang1, Serang2, Serang3, Clereng, and Sidomulyo. While in the northeast there are 2 sections, Kiskendo and Bubrah, both sections have contact with OAF.

Overall, the lithostratigraphic character of Sentolo Formation in research area from the bottom most layer is as follows:

1. The lower part of Sentolo Formation

The lower part of Sentolo Formation can be observed in the Kiskendo and Bubrah sections. Based on its megascopic appearance, the lower stratigraphy of the Sentolo Formation in the study area is composed of sandstones, as well as calcarenite and calcilutite unconformly laid above volcanic breccia of OAF member.

From the petrographic observation of calcarenite sample JGK1, taken from Kiskendo section, microscopically is composed by abundant skeletal grain comprising planktonic foraminifera. In addition, there is also non-skeletal grain of feldspar, which is estimated to be the result of volcanic rock reworking deposited along with foraminifera.

2. The middle part of Sentolo Formation

A different character of Sentolo Formations sedimentary rock layers found on Serang 1, Serang 2, Serang 3, Clereng, and Sidomulyo section. Based on the strike and dip of the rock layers, these sections are younger than Bubrah and Kiskendo sections. In general, the rock layers in the sections are composed of calcarenite and calcilutite, which are intercalated and contain ichnofossils. The microscopic characteristic seen in SS-1 rock samples from Serang1 section are the abundance of the foraminifera microfossil contents. The mineral feldspar as non-skeletal grain as found in the sample on the older track is not present in this rock sample.

Biostratigraphic analysis conducted in this study included determination of the relative age of rocks using [8] foraminifera biozonation. The determination of the planktonic microfossil relative age of each rock samples can be observed in Table 1.

3.2 Paleoecological Analysis Based on Foraminifera Fossil Assemblage

3.2.1 Paleobathymetry

Paleobatimetric analysis was performed using planktonic and benthic foraminifera ratio (P/B Ratio), which then classified using [9] classification to obtain the bathymetry accordingly. The marine environment determination was conducted based on [10]. The paleobathymetry analysis of each samples can be seen in Table 2.

Table 1. Relative age determination based on planktonic foraminifera analysis.

Section	Sample	Relative age	Index fossil
Kiskendo	JGK 1	Early–Middle Miocene (N7-N8)	*Globigerinoides diminitus*
Bubrah	KB 1	Early–Middle Miocene (N7-N8)	*Globigerinoides diminitus*
	KB 2	Early–Middle Miocene (N7-N8)	*Globigerinoides diminitus*
	KB 3	Early–Middle Miocene (N7-N8)	*Globigerinoides diminitus*
Serang1	SS 1	Middle Miocene (N12)	*Globorotalia fohsi*
Serang2	SS 2.1	Middle – Late Miocene (N7-N18)	(-)
	SS 2.2	Middle Miocene - Pliocene (N13-N19)	(-)
Serang3	SS 3	Middle – Late Miocene (N13-N18)	(-)
Sidomulyo	TS 1	Middle Miocene (N12)	*Globorotalia fohsi*
Clereng	JJC 1	Middle – Late Miocene (N13-N15)	(-)
	JJC 2	Middle – Late Miocene (N9-N17)	(-)

Table 2. Paleobathymetry analysis using P/B Ratio.

Section	Sample	Plankton percentage (%)	Water depth (m) according to [9]		Bathymetry according to [10].
Kiskendo	JGK 1	63,63	680-825 m		Upper Slope
Bubrah	KB 1	55,50	550-700 m		Upper Slope
	KB 2	74,00	700-1100 m		Upper Slope
	KB 3	75,12	700-1100 m		Upper Slope
Serang 1	SS 1	73,56	680-825 m		Upper Slope
Serang 2	SS 2.1	78,20	700-1100 m		Upper Slope
	SS 2.2	67,87	680-825 m		Upper Slope
Serang 3	SS 3	66,14	680-825 m		Upper Slope
Sidomulyo	TS 1	63,04	680-825 m		Outer Shelf
Clereng	JJC 1	30,18	100-600 m		Upper Slope
	JJC 2	60,39	680-825 m		Upper Slope

3.2.2 Community *structure of foraminifera*

The distribution of benthic foraminifera fossil population on 11 samples of rock representing all stratigraphy cross-sectional of Sentolo Formation in research are calculated using PAST program as seen in Table 3 to obtain its community structure.

Based on the calculation of benthic foraminifera fossil community mentioned above, it is known that the number of taxa that is determined in all samples ranges from 4 to 14 taxa. The population of foraminifera individuals is very abundant so as to describe stable water conditions that enable the foraminifera community to live properly.

In addition, the results of the calculation of the dominance index on all samples close to zero, indicating that there were no taxon dominates among other taxa. The diversity index value of all samples entirely falls within the range of 1 to 3, which means a moderate degree of foraminifera diversity. The last parameter is the index of similarity or evenness that varies in all samples. Almost on all samples, the evenness is moderate to high. While the evenness which less than 0.3 indicates the low uniformity, as in sample JJC1. Samples with high evenness are generally samples with small individuals, though none of the taxon dominates. From these analyses, it can be conclude that the condition of waters at the time of foraminifera benthic lived was stable.

Table 3. The result of community structure analysis of benthic foraminifera using PAST software.

	SS3	SS2.1	SS2.2	SS1	KB1	KB2	KB3	JGK1	JJC1	JJC2	TS1
Taxa	11	4	7	10	10	10	8	9	14	7	4
Individuals	108	46	80	55	85	65	54	68	229	101	85
Dominance (D)	0.20	0.30	0.19	0.19	0.17	0.23	0.17	0.17	0.43	0.31	0.38
Shannon (H)	1.90	1.28	1.76	1.89	1.98	1.79	1.85	1.89	1.37	1.42	1.15
Evenness(e)	0.61	0.90	0.83	0.66	0.72	0.60	0.80	0.74	0.28	0.59	0.79

To strengthen the results of community structural analysis, benthic foraminifera fossils found in each sample were grouped according to their test wall composition and analyzed using triangular MRT diagrams according to [11]. In the Fig. 1 it is seen that the classification of the waters condition where foraminifera lived or when the deposition of Sentolo Formation was normal marine continental slope waters.

Based on the results of the analyses can be interpreted that the condition of the waters at the time of fossil foraminifera lived is in stable and optimum conditions. If it is associated with volcanic activity during the Oligocene - Miocene in the research area, then it is likely that at the time of the foraminifera lived the activity of volcanism has stopped.

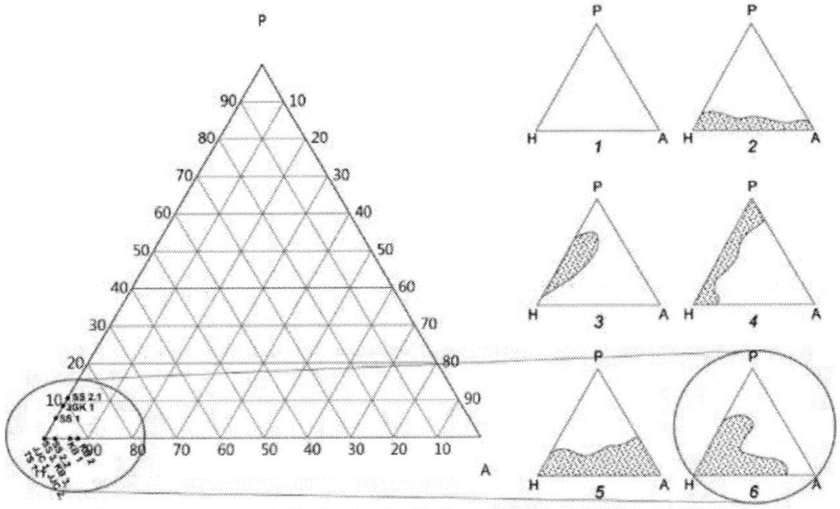

1 - fresh waters; 2 - hyposaline lagoons; 3 - normal marine lagoons and carbonate platforms; 4 - hypersaline lagoons; 5 - shelf seas; 6 - normal marine continental slope waters; types of tests: P - porcelaneous; H - hyaline; A - agglutinated (modification after [12])

Fig. 1. Triangular diagram plotting of benthic foraminiferal assemblage

4 Conclusion

Based on paleobathymetry and paleoecology of all samples, it can be concluded that Sentolo Formation was deposited at normal marine continental slope waters. Meanwhile, the community structure of foraminifera analysis indicates a stable water conditions with optimum level of carbonate content, oxygen supply, salinity, and temperature. According to the analysis, the palaeoecological condition at the time Sentolo Formation was deposited in the research area was a marine environment that was not disrupted by volcanic activity, so it becomes a good factor for the growth of foraminifera. Whereas the presence of volcaniclastic material mixed with carbonate which is occured in the lower part of Sentolo Formation (KB1, SS2.1 and SS2.2), were interpreted as reworking processes of older volcanic rocks.

The authors thank to Ilham Hani Pratama, Hanif Indra Wicaksana, Bagus Rachmad, Dewi Mindasari, Yan Bachtiar, Siti Rofikoh, and Kuncahyo Tantri as field and discussion partners. The thanks also go to parties who have been willing to provide corrections, criticisms, suggestions and input, so that the research and writing of this paper can be resolved.

References

1. R. W. Van Bemmelen, *The Geology of Indonesia, Volume IA, General Geology of Indonesia and Adjencent Archipelago, 2nd Edition* (Government Printing Office, The Hague, 1949)
2. D. Kadar, *Neogene Planktonic Foraminiferal Biostratigraphy of the South Central Java Area Indonesia* (Geological Research and Development Centre Directorate General of Geology and Mineral Resources Ministry of Mines and Energy, Bandung, 1986)
3. W. Rahardjo, Sukandarrumidi, H. M. G. Rosidi, *Peta Geologi Lembar Yogyakarta, Jawa skala 1:100.000* (Directorate General of Geology and Mineral Resources Ministry of Mines and Energy, Bandung, 1977)
4. J.A. Postuma, *Manual of Planktonic Foraminifera* (Elsevier Publishing Company, Amsterdam, 1971)
5. E. Boltovskoy, Mar. Geol., **26(1)**:139-175 (1978)
6. A.R. Loeblich, and H. Tappan, *Foraminiferal Genera and Their Classification* (Springer, New York, 1988)
7. E.P. Odum, *Fundamentals* of *ecology Third Edition,* (W.B. Saunders Company, Philadelphia, 1971)
8. W. H. Blow, *Proc. of International Conference on Planktonic Microfossils* ***1***, 199-422 (1969)
9. T. F. Grimsdale and F. P. C. M. Van Morkhoven, *Proc. 4th. World Petroleum Congress* 473–491 (1955)
10. H. L. Tipsword, F. M. Setzer, F. L. Smith Jr., G. C. Assoc. Geol. Soc. Transaction, **16**, 119 - 130 (1966)
11. J.W. Murray, *Ecology and paleoecology of benthic foraminifera* (Longman Harlow, Essex, 1991)
12. B. Valchev, *50 years University of Mining and Geology "St. Ivan Rilski Annual*, 189 – 194 (2003)

Application of Aquifer Vulnerability Index (AVI) method to assess groundwater vulnerability to contamination in Semarang urban area

Thomas Triadi Putranto[1*], *Narulita* Santi[1], *Dian Agus* Widiarso[1], and *Dimas* Pamungkas[1]

[1]Geological Engineering Department, Diponegoro University, 52075 Semarang, Indonesia

Abstract. An approach to assess groundwater to contamination is developing groundwater vulnerability map. Groundwater vulnerability is formulated as an intrinsic relative, the dimensionless property of a groundwater system that depends on its sensitivity to unfavourable human and natural impacts. Thus, to assess groundwater vulnerability to contamination in Semarang urban area is the objective of this research by using Aquifer Vulnerability Index (AVI). This method considered of two parameters related to the unsaturated zone, i.e. thickness (d) of each sedimentary layer in the unsaturated zone and Estimated hydraulic conductivity (K) of these sedimentary layers. There are five levels of groundwater vulnerability by using AVI method, i.e. extremely low, low, moderate/medium, high, and extremely high. The thinner the layer covers the aquifer, and the higher the value of the hydraulic conductivity of sediment layers will be increasingly vulnerable to pollution.

1 Introduction

Semarang, as an urban city and a capital city of the Central Java Province in Indonesia, is facing some environmental problems (e.g. declining groundwater level and its quality, land subsidence, seawater intrusion, and flooding) as an impact of groundwater overexploitation [1]. Without population and pollution management control, the amount of per capita safe water available for daily need is gradually reducing with time [2]. One technique to assess groundwater contamination is developing groundwater vulnerability map [3-7]. Thus, to assess groundwater vulnerability to contamination in Semarang urban area is the objective of this research by using Aquifer Vulnerability Index (AVI). Groundwater vulnerability can be formulated as an intrinsic relative, the dimensionless property of a groundwater system that depends on its sensitivity to unfavourable human and/or natural impacts. It means that groundwater vulnerability is combining a function of hydrogeological factors (e.g., aquifer characteristics and its properties in unsaturated zone) and human activities.

* Corresponding author: putranto@ft.undip.ac.id

2 Geologic and hydrogeologic setting

The study area is located in Semarang urban city. Geologic setting of Semarang consists of from the oldest to the youngest which are sedimentary deposits (Kalibeng Formation, Kerek Formation, and Damar Formation), volcanic product (Kaligetas Formation, Kaligesik Formation, Jongkong Formation, and Gajahmungkur Formation), and alluvium deposit [8]. The sedimentary deposits spread in the centre of Semarang. Kerek formation (middle Miocene age) consists of alternating claystone, marl and limestone. Claystone partly interlayered with siltstone or sandstone Kalibeng Formation (late Miocene-Pliocene) contains massive marl in the upper part, locally carbonaceous marls intercalating with tuffaceous sandstone and limestone. Damar Formation is mainly non-marine deposit consist of tuffaceous sandstone, breccia, conglomerate which is located in the centre of Semarang. The volcanic products are mainly located in the south of Semarang. They consist of breccia, lava, tuff and laharic breccia. Moreover, the latest deposits (Holocene, Lower Pleistocene) are in a marine environment form the coastal plain of Semarang. The bulk of the basin sediments contain alluvium (Qa). Alluvium is the result of surficial deposits from the coastal plain, river, and deltaic system deposits as shown in Figure 1.

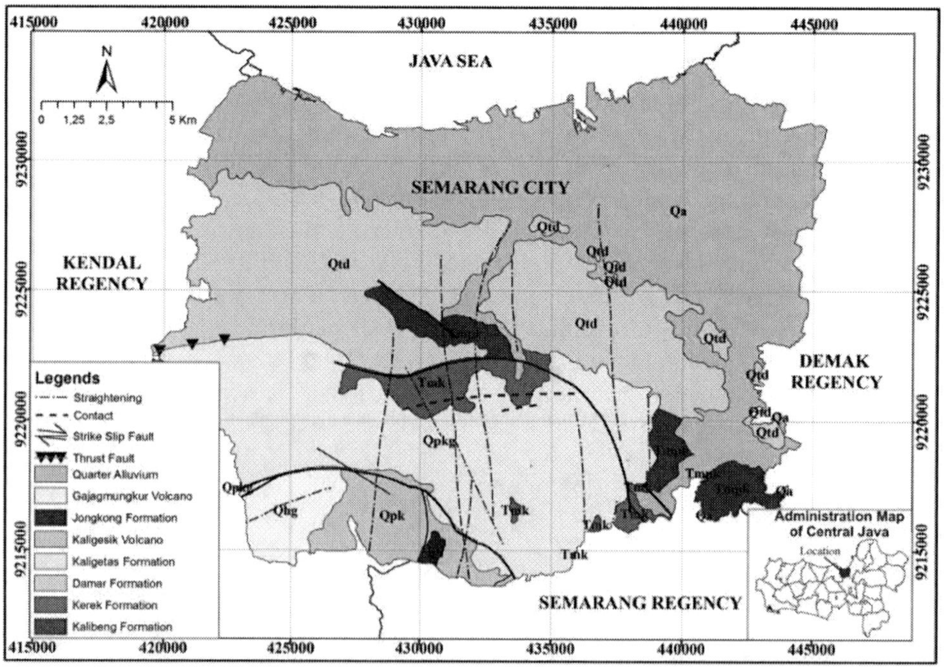

Fig. 1. Regional geological map of Semarang [3].

According to the regional hydrogeological map of Semarang [9], in the northern part, the aquifer is mainly flowing in the intergranular system. This system has the extensive aquifer with highly to the productive aquifer. The main lithology in this aquifer system is alluvial deposits. Meanwhile, in the south of Semarang, aquifer mainly flows both through fissures, and interstices which extensive spreading and moderately productive, but locally also available productive aquifer.

3 Methods

To assess groundwater vulnerability to contamination, Aquifer Vulnerability Index (AVI) method [10] was applied to protect groundwater. The AVI rating system model performed for the evaluation of intrinsic vulnerability. This method considered of two parameters related to the unsaturated zone, i.e.

i. Thickness (d) of each sedimentary layer in the unsaturated zone
ii. Estimated hydraulic conductivity (K) of these sedimentary layers

The thickness of each sedimentary later was measured by levelling water level in some dug wells (around 308 wells) using water level indicator. Moreover, Geo-resistivity survey was applied to construct sedimentary layers. The resistivity of the rocks was measured by injected currents and the resulting differences at the surface [11]. There were around 46 locations to measure geo-resistivity in Semarang urban area. Since K values may not be visible for each sedimentary layer, a table of estimated values was required in the study area as shown in table 1 [10 &12]. The hydraulic resistance (c) that results from the product of the two parameters using the equation below gives the aquifer vulnerability index:

$$c = \Sigma d/K \tag{1}$$

Table 1. Hydraulic conductivity from various Sediment [10 & 12].

Sediment type	Hydraulic conductivity
Gravel	1,000 m/d
Sand	10 m/d
Silty sand	1 m/d
Silt	10^{-1} m/d
Fractured till, clay or shale (0-5 m from ground surface)	10^{-3} m/d
Fractured till, clay or shale (5-10 m from ground surface)	10^{-4} m/d
Fractured till, clay or shale (10 m from ground surface, but weather based on colour, brown or yellow)	10^{-4} m/d
Sand – silt – clay (massive or mixed)	10^{-5} m/d
Massive till or mixed sand-silt-clay	10^{-6} m/d

Using equation 1 above, the groundwater vulnerability method using AVI can be generated from each geo-resistivity points. However, the calculated c or log c is related to Table 2 to define the vulnerability level.

Table 2. Relationship of Aquifer Vulnerability Index (AVI) to Hydraulic Resistance [10].

Hydraulic resistance (c)	Log (c)	Vulnerability
0 – 10	<1	Extremely high
10 – 100	1-2	High
100 – 1,000	2-3	Moderate/Medium
1,000 – 10,000	3-4	Low
>10,000	>4	Extremely low

4 Results and Discussion

Based on the hydrogeological mapping as shown in Figure 2, the minimum and maximum of groundwater depth, as unsaturated zones, are 0.03 m and 34 m respectively. Following the topography elevation, the minimum depth is found in the north to the east of Semarang, i.e. Genuk Sub-District while the deepest is located in the south-west and south-east of Semarang, i.e. Ngalian Sub-District and Tembalang Sub-District respectively. Figure 3 shows the class of groundwater depth. Five classes are less than 2 m, 2-5 m, 5-10 m, 10-20 m, and more than 20 m depth. The average groundwater depth is located in the centred of Semarang area that is 5-10 m.

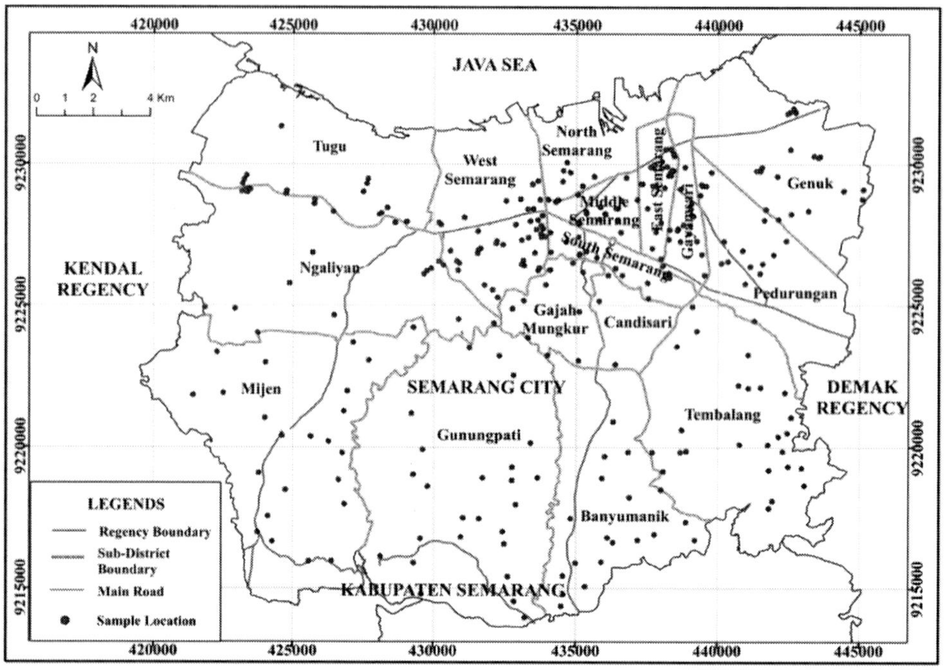

Fig. 2. Map of groundwater level measurement.

Based on the resistivity measurement, the subsurface consists of unconsolidated material/alluvium such as clay to sand, sedimentary rocks (sandstone, limestone, and claystone) and volcanic products (breccia, tuff breccia, tuff, tuffaceous sandstone, and lava). Alluvium is mainly located in the north and the east of Semarang whereas the sedimentary and volcanic rocks, are present in the centre and the south of Semarang. Thus, the hydraulic conductivity is the range 10^{-6} to 10^{3} m/d. The highest hydraulic conductivity indicates excellent permeability, i.e. gravel to sand. Claystone is the lowest of the coefficient of permeability in the study area.

There are five levels of groundwater vulnerability by using AVI method (Figure 4), i.e. extremely low, low, moderate/medium, high and extremely high. Extremely low to low vulnerable zones describe that the thickness of unsaturated zones is in the range 10-20 m and above 20 m thick while the conductivity is very low (10^{-6} to 10^{-3} m/d). These areas are in Banyumanik, Gungungpati, and Ngalian Sub-District. This zone is located mainly in the south and west of Semarang.

The moderate/medium vulnerable has the unsaturated thickness of 5-10 m depth and the hydraulic conductivity 10^{-3} m/d. The area is located in the east of Semarang (locally Semarang Utara Sub-District and Genuk Sub-District).

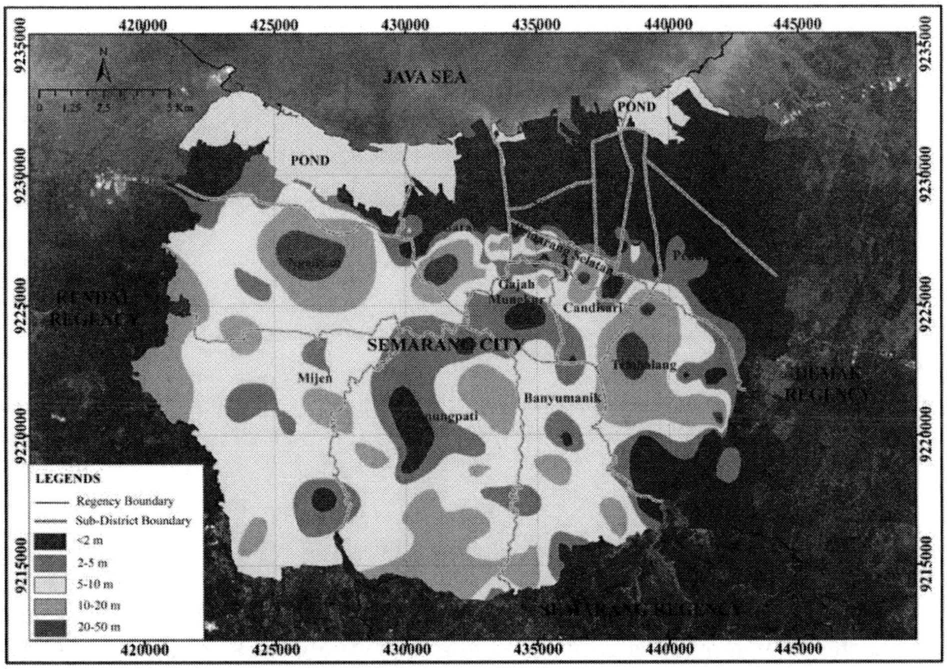

Fig. 3. Groundwater depth of Semarang urban area.

Moreover, the high to extremely high vulnerability zones are located in the west (Mijen Sub-District), the centre (Semarang Barat, Gajah Mungkur), in the north (Semarang Utara), in the east (Genuk) and the south-east (Tembalang) of Semarang urban area. The unsaturated zones are in the range 2-5 m and less than 2 m. The hydraulic conductivity is the range 10-10^3 m/d.

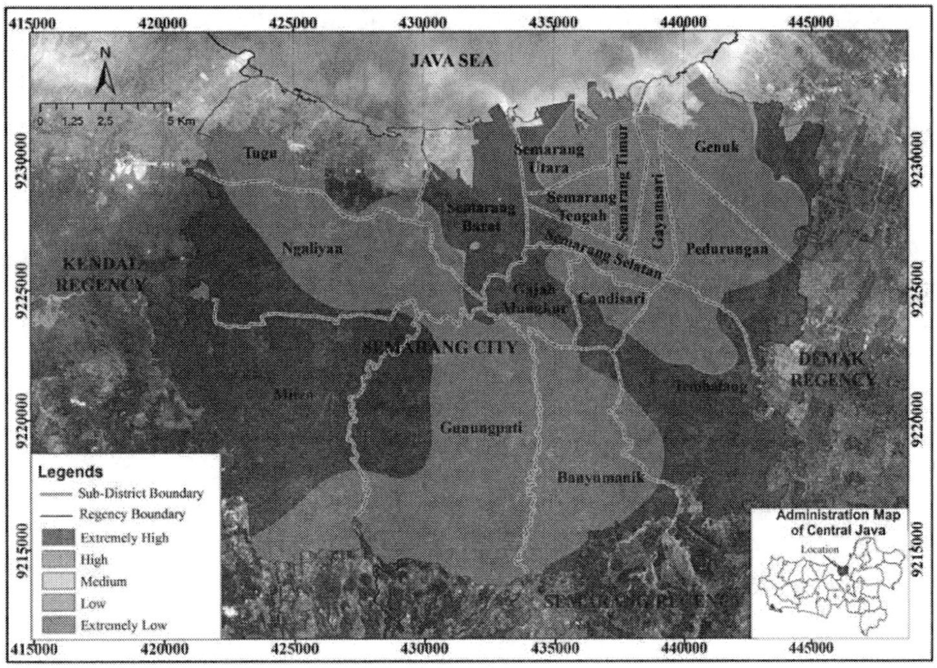

Fig. 4. Groundwater vulnerability map using AVI method.

5 CONCLUSIONS

Based on the AVI method, the thinner the layer covers the aquifer, and the higher the value of the hydraulic conductivity of sediment layers will be increasingly vulnerable to pollution. There are five levels of groundwater vulnerability to contamination in Semarang urban area, i.e. extremely low, low, moderate, high, and extremely high.

References

1. Putranto, T.T., Hidajat W.K., Susanto, N., Developing groundwater conservation zone of unconfined aquifer in Semarang, Indonesia, IOP Conf. Series: Earth and Environmental Science 55 (2017)
2. Maria, J.A., Liliana F., Alcide, P., Luis, N., Laura, G., Lúcia, G., Bernhard, M., Fernando, R., José M.M, Helder, I.C., Environmental groundwater vulnerability assessment in urban water mines (Porto, NW Portugal), J. Water, **8**(11), 499 (2016)
3. Robins, N.S., MacDonald, A.M., and Allen, D.J. The vulnerability paradox for hard fractured Lower Palaeozoic and Precambrian rocks. The Groundwater Vulnerability Assessment and Mapping International Conference, Ustron′, Poland (2004)
4. Foster, S., Hirata, R., and Andreo, B. The aquifer pollution vulnerability concept: aid or impediment in promoting groundwater protection? *Hydrogeology Journal*, **21**, pp.1389–1392. (2013)
5. Foster, S.S.D. Fundamental concepts in aquifer vulnerability, pollution risk and protection strategy. In: van Duijvenbooden W & van Waegeningh HG (eds.) Vulnerability of soils and groundwater to pollution. TNO Committee on Hydrological Research, The Hague, *Proceedings and Information*, **38**: 69–86 (1987)
6. Foster, S.S.D. Groundwater recharge and pollution vulnerability of British aquifers: a critical review. In: Robins NS (ed.) Groundwater Pollution, Aquifer Recharge and Vulnerability. *Geological Society, London, Special Publications*, **130**: 7–22 (1998)
7. Foster, S.S.D. Aquifer pollution vulnerability concept and tools – use, benefits and constraints. *The Groundwater Vulnerability Assessment and Mapping International Conference*, Ustron′, Poland (2004)
8. Thanden RE, Sumadirdja H, Richards PW, Sutisna K and Amin TC. Geological map of The Magelang and Semarang sheet, Jawa, Geological research and development centre, Bandung (1996)
9. H D Said and Sukrisno. Hydrogeological map of Indonesia scale 1:250.000 Sheet VII Semarang (Jawa), Directorate of Environmental Geology, Bandung (1998)
10. Stempvoort, D.V., Ewert, L., Wassenaar, L. Aquifer Vulnerability Index: A GIS compatible method for groundwater vulnerability mapping. Canadian Water Resources Journal, Canada (2013)
11. Kirsch, R. Groundwater geophysics a tool for hydrogeology 2nd ed. Springer-Verlag Berlin Heidelberg (2009)
12. Freeze R. A. and J. A. Cherry. Groundwater. Prentice-Hall Inc., Englewood, Cliffs, NJ. (1979)

The Influence of Fault and Stress Contributed On Overpressure Mechanism For Neogen Formation (Mundu, Wonocolo, Ngrayong) East Java Basin, Indonesia

Fahrudin[1,*], *Hadi* Nugroho[1], *Tri* Winarno[1], *Zul* Hayuddin[1], *Joko* Suprayetno[1], *Rully* Firmansyah[2], *Muhajir*[2], and *Ade* Kadarusman[3]

[1]University of Diponegoro, Semarang, Indonesia
[2]Join Operating Body Pertamina PetroChina East Java (JOB-PPEJ), Indonesia
[3]Independent Consultant, Indonesia

Abstract. Rembang Madura Kangean Sakala (RMKS) fault zone and W-E structures of Kendeng Zone are the main structural elements in back arc basin, East Java, Indonesia. There are many perspectives about the overpressure zone at the basin. This research focus on the overpressure zone that related with formation and tectonic evolution. The study infer that high overpressure is explained by regional and local mechanisms. The results of analysis of log derived porosities at Karang Mudi-1 (KM-1) well indicate that high overpressure and overpressure was Neogen Formation (Mundu, Wonocolo, Ngrayong). Pore pressure prediction at KM-1 exerted Eaton's method. There are two overpressure zones at depth 2.680 – 4.900 fTVD (B Zone) and 5.600 – 8.400 fTVD (D Zone). High overpressure D Zone are regionally caused by flexural structure (anticline and fracture) from transpression setting (West – East RMKS). High overpressure in KM-1 is transpression setting early which the overpressure follow the pop-up boundary. Area of high overpressure always related with fault thrust. Static mechanism is possible in the present day interplate stress field. Present the maximum horizontal stress was NEE – SWW. Local mechanisms occurs illitization at D zone from depth 6.800 – 8.400 fTVD. D Zone is unloading mechanism (illitization and lateral transfer open fracture).

1. Introduction

Area of the active margin has specific tectonic and sediment characterization that generate a unique overpressure zone. This zone is also influenced by the history of tectonic evolution. The East Java Basin is an active sedimentary basin that has several high, normal, low overpressure zones [1]. The tectonic and active sedimentary related with overpressure. In case of the Watukosek fault [2] and the basement configuration appears to control the presence of overpressure [3]. Fault is driven by pressure and a source of fluid at structure

* Corresponding author: fahru_gabru@yahoo.com

area on the compressonal tectonic belts, while initiated by fracture propagation [4]. The fractures or faults caused the lateral transfer of fluid (Yardly and Swarbick, 2000 in [5]). The Oligo-Miocene to Recent basin was filled with shallow marine deposits, some of which are known to be "overpressured" [6]. As a result of the compressional inversion, these strata are gently folded with normal and reverse faults cutting the inversion anticline crests [7]. Overpressured lower Miocene clays probably equivalent to the Tuban, Tawun and Kujung Formations [8]. Overpressure mechanisms include loading, unloading and tectonic control mechanisms.

Unloading refers to any process that tends to reduce the effective pressure. The process in unloading mechanism is the increase of fluid volume in the pore (fluid expansion) or by the loading transfer from the matrix to the fluid in the pores of clay mineral diagenesis such as smectite to illit or kaolin to illit and gas generation (oil cracking). Fluid transfer in the sandstone is also included in the process that can occur in the unloading mechanism, where the sandstone is isolated in the shale lithology unit. Overpressure zone is also associated with mud volcano. The shallow overpressured zone beneath the transition zone could partially reflect the presence of open fractures associated with a strike slip fault and relate with basal detachment fault [9].

This study conducted in order to better understand and other perspectives about the overpressure zone in East Java Basin associated with the formations and the tectonic evolution. Tectonic evolution at active margin and continental collision maybe give contributions this case. How influence fault and insitu stress controlled mechanism of overpressure.

2. Methods

This paper describes regional stress, regional geology, distribution of overpressure, and the seismic around KM-1 well. That seismics include inline 1500, cross-line 5686, and arbitrary line is presented the structure model around this well. Then, we interpret data of the logs, drilling problems, and leak of test (LOT). The logs data are gamma ray, density, sonic and resistivity logs. Log data is used to predict the value of sub-surface pressures with the help of Drillworks Predict software. These log data are also used to interpret composite lithology properties such as the interpretation of sediment compaction associated with Normal Compaction Trend (NCT) and sonic log trends. The lithology property is then associated with the pressure graph that is one of the ways in identifying overpressure characteristics. The pressure graph data has shown the existence of overpressure and normal pressure zone to depth. This research also supported by XRD, petrography, RO analysis and SEM-EDS data.

3. Results

The East Java Basin onshore is an extensional inversion basin with different basement characterizations. The basement was the result of collision from transition zone of Sundaland with Gondwana microcontinent on Late Createceous. There are two characterizations of the basement structures, basement with NEE-SWW direction and E-W direction. It is reflected in the reverse fault of the subsurface (derived from seismic data) on the geological map (**Fig. 1a**). The western part of East Java Basin has a structure with NEE-SWW direction and the eastern part has an E-W structure. Geological cross sections show that the basement control causing the highland of Kendeng Zone and the lowland/highland of Rembang Zone (**Fig. 1b**). The higher part of Rembang Zone is the overpressure area and the lower one is at normal pressure. Subsidence/sagging and the early of tectonic compression is dominated by the deposition of shale (**Fig. 1c**). The pre-existing fault structure of fracture basement and rifting structure (normal fault) on the Plio-Pleistocene tectonic reactivated and uplifted.

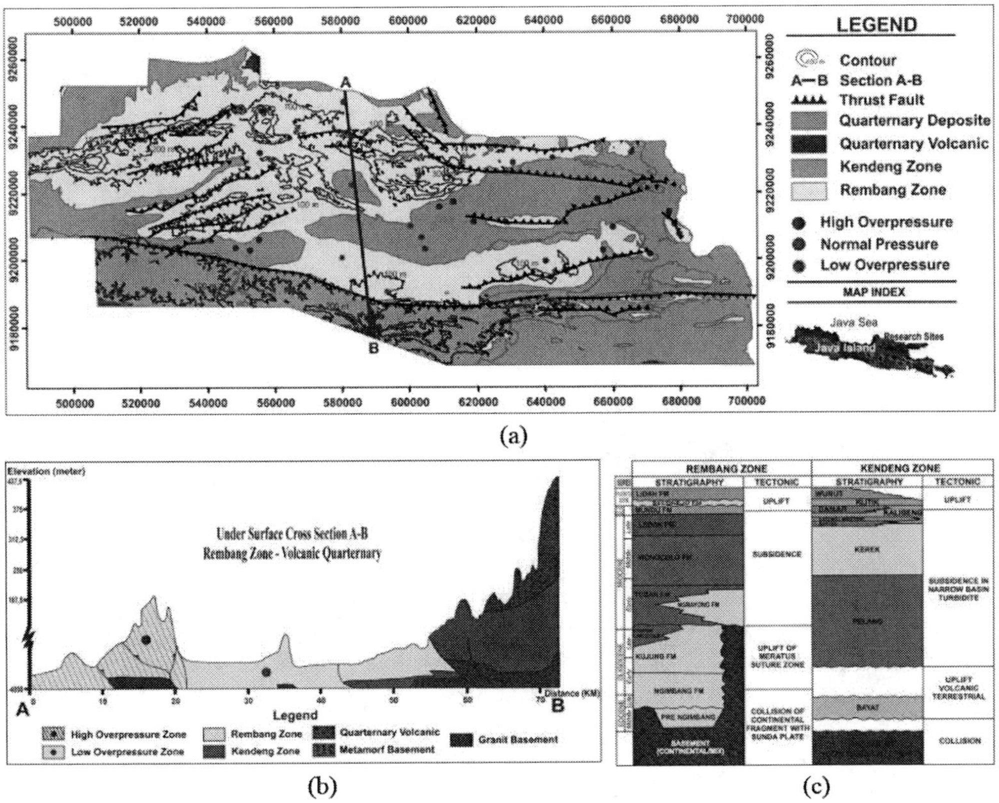

Fig. 1. (a). Regional Geology **(b).** Geological Cross Section **(c).** Stratigraphy of Rembang Zone.

Mechanisms for the high fluid overpressure are probably coused by basement control or structural patterns. There are two large structures found in the basin, the RMKS structure in the form of sinistral strike slip fault and Kendeng fault in the form of thrust fault. The overpressure zones are clustered into two zones, the western zone with the area of NW-SE direction and the eastern zone with the area of E-W direction. Both zones are truncated by RMKS fault. The current stress rating with a value of 8 – 7 mm/year was recorded in the Kendeng Zone, while in the Rembang Zone is about < 7 mm/year. The stress direction of the NE-SW is due to subduction in the southern of Java. The Kendeng fault is an active fault as well as the eastern RMKS part is also active. This stress contributes to lateral stress that can control the possibility of pressure in the eastern part of the basin (**Fig. 2**). In EF cross section shows the thickness of the Ngrayong Formation that is different, above the body of the limestone grows relatively thin when compared with the wings part of Tuban Formation (**Fig. 3a**). In line AB, showing the thickness of the Ngrayong Formation is different, indicating there is flexural due to fault, folding and subsidence. This indicates instability in the Ngrayong zone, this is correlated with pore pressure data and it is predicted that, this zone is a high overpressure area. The line seismic AB indicates there is compression to form a thrust fault and fold. The line seismic CD also indacates there is the beginning of transpressional compression fault. Around the faults in the AB seismic line there are X Well and KM-1 Well (**Fig. 3a**).

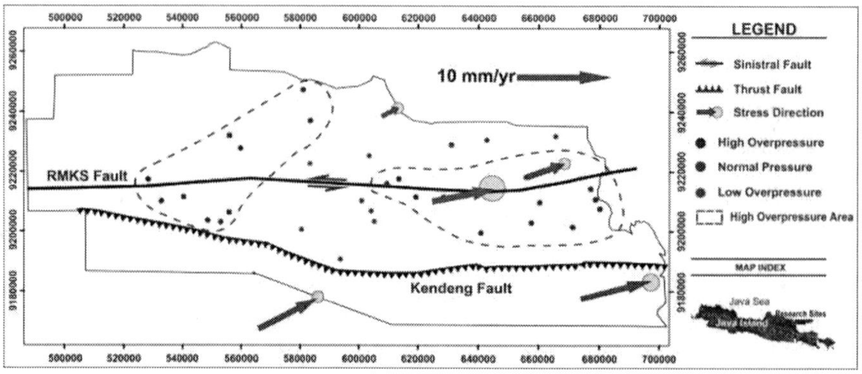

Fig. 2. Regional Stress and Major Structure.

Normal pore pressure analysis is performed on two logs i.e resistivity log and sonic log. In Eaton's method, the vertical effective stress, σ'_v, is estimated from the measured sonic transit time, Δt, using the equation:

$$\sigma'_v / \sigma'_n = (\Delta t_n/\Delta t)^3 \quad (1)$$

where σ'_n and Δt_n are the vertical effective stress and sonic transit time on the normal compaction curve for hydrostatic pore pressure at the depth of measurement. There are two overpressure zones, (Figure 3b), as follows: B Zone, this zone is an overpressure found in the middle to the lower part of Mundu Formation (2,680 - 4,900 ftTVD). D Zone, this is an overpressure zone which form build up or called with high overpressure. This overpressure is found in the upper Wonocolo Formation to the lower Ngrayong Formation (5,600-8,400 ftTVD). Gamma-ray logs show the dominant lithology of claystone (shale), but at the depth of 5700 ftTVD, 7500-7800 ftTVD and 9200-9400 ftTVD show sandstone lithology. Effective pressure also shows varying trends in depth as it is controlled by intensive sedimentary compaction. Chemical compaction takes place predominantly in this zone due to rising temperatures. Authigenic mineral precipitation is quite abundant with distinctive characteristics. Based on XRD analysis, mineral composition on KM 1 Well are illite and smectite with the high percentage found on Wonocolo and Ngrayong Formation. This condition followed by the decreasing of the smectite. The petrographic thin sections show that the precentage of clay minerals are >30% on Wonocolo and Ngrayong Formation. SEM analysis at 2280 ftTVD with the composition of sheet structure of illite, and the illite at 6600 ftTVD show the different structure, which more dense. The result of RO analysis is the early mature stage (0.58%), where the organic matter has shown a change to the oil phase. This stage occurs in the lower Ngrayong Formation at about 9000 ftTVD depth.

4. Discussions and Conclusions

The overpressure prediction analysis on the KM-1 Well indicates the presence of overpressure zones from sonic log trend analysis with the Normal Compaction Trend (NCT) which are B zone and D zone. The red circles represent overpressure zones in KM-1 Well, the green circles are the low overpressure zones, and the blue circles are the normal pressure zones (**Fig. 1a**). The area nearby RMKS fault has higher overpressure value than the further area (normal pressure and low overpressure zones). The location of KM-1 Well shown to explain that the area is surrounded by EW reverse fault. Regionally (**Fig. 2**), the eastern area tend to have overpressure zone and its spreading horizontally, while the overpressure zone in the western area tend to spread with NE – SW directions. The difference between these two overpressure zones are truncated by the major structures of RMKS fault.

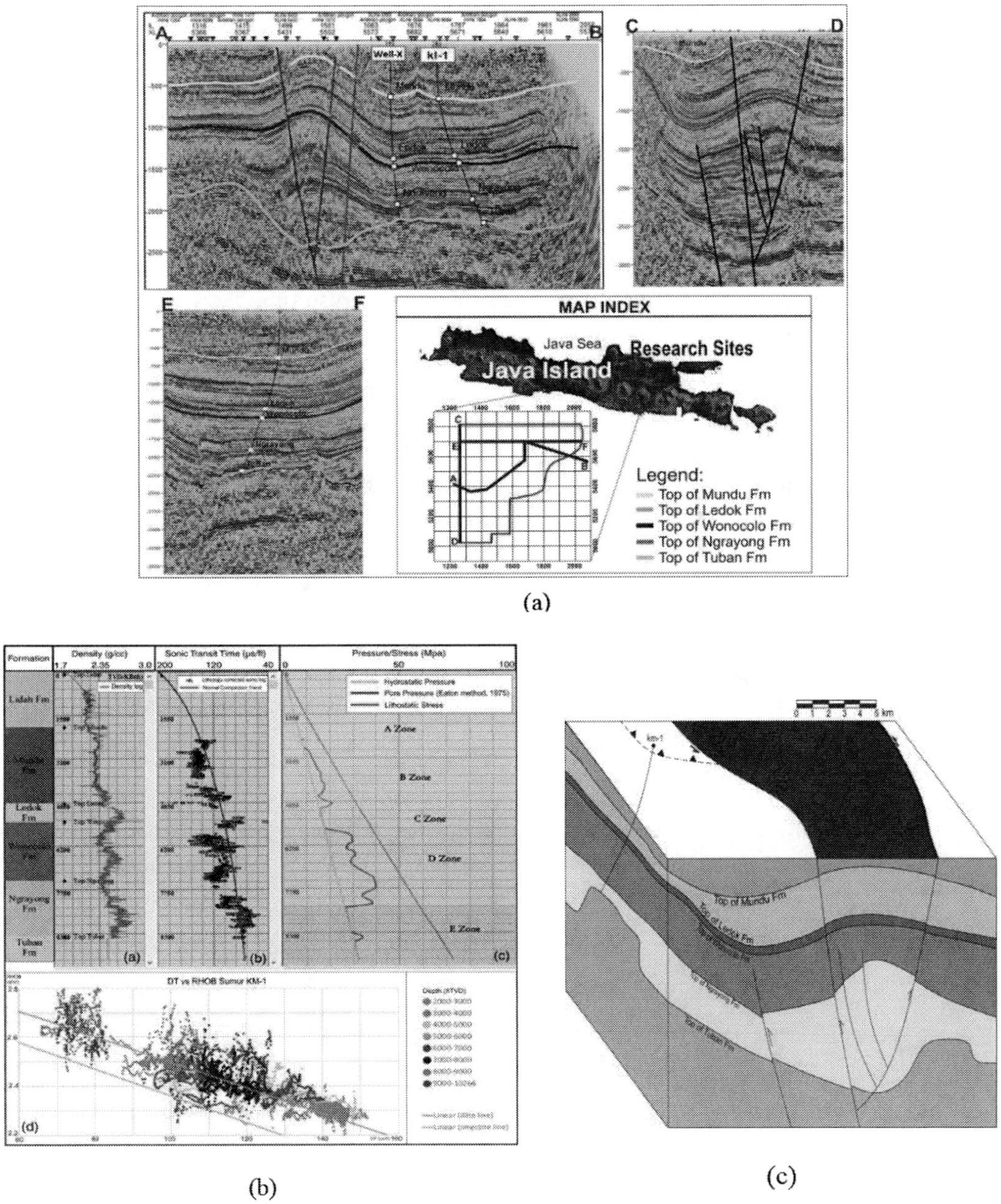

Fig. 3. (a). Picking Horizon Fault of Seismic Line AB, CD, and EF of the KM-1 Well **(b).** Overpressure zone contains B zone and D zone **(c).** 3D Modelling of Geological Structure and Overpressure Zone on the KM-1 Well.

The different overpressure direction is controlled by tectonic evolution of the North East Java Basin. The characteristics of geological structures of East Java Basin show the difference of two main patterns i.e E-W pattern on eastern part and NEE-SWW pattern in western part. So this shows the similarity between the spread of overpressure zone in the research area. In the western part it is a spreading zone that tends to have NEE-SWW directions that represent regional geological structure characteristics with a meratus pattern. The influence of geological structures in the form of faults have a very dominant contribution to the presence of ovepressure in this study area. The overpressure zone modeling which shows the vertical view of the sides and at the top 3 dimensions from the well of KM-1 (**Fig. 3c**). This is obtained from fault analysis by picking on seismic line AB, CD and EF. It can be seen that the zone around the KM-1 Well has an overpressure zone that is controlled by a reverse fault. The vertical presence of the geological structure shows the flower structure through the Tuban Formation - Wonocolo Formation. The three formations are controlled by a reverse fault that cause block one move up against to the other block and produce different thicknesses due to the influence of geological structures. This is of course correlated with the pore pressure prediction analysis found in zone B and zone D indicating the presence of ovepressures in Ngrayong and Wonocolo Formations. That fault is component of RMKS faults which have transpression setting that is still active now. The effect of the structure produces an unstable zone resulting in an overpressure zone at KM-1 well. The contribution of this faults to the mechanism of overpressure is interpreted that the RMKS faults are pathways which result in the existence of lateral fluid transfer between the rock layers at the KM-1 Well.

Based on the analysis, it can be concluded that overpressure zone in this area influenced by unloading mechanism which is controlled by transpression setting of RMKS fault and present stress at this fault. The presence of geological structures in the form of RMKS fault and Kendeng faults as a result of tectonic inversion occur in Early Miocene getting up until now reached 6-7 cm/year that produce an overpressure zone.

5. References

[1] A. Koulali, S. McClusky, S. Susilo, Y. Leonard, P. Cummints, P. Tregoning, I. Meilano, J. Efendi, A.B. Wijarnarto. The kinematics of crustal deformation in Java from GPS observations : Implications for fault slip partitioning. Earth and Planetary Science letters. **458,** 72 (2017).

[2] A. Mazzini, A. Nermoen, M. Krotkiewski, Y. Podladchikov, S. Planke, H. Svensen. Strike slip faulting as a trigger mechanism for overpressure release through piercement structures. Implications for the Lusi mud volcano, Indonesia. Marine and Petroleum Geology J. **26**, 1751 (2009).

[3] Agus M.R, Fahmi H, Lambok M.H, Neil R.G, Wuryadi S, Maikel A, Fatrial B, Kharisma E, Rully F, Risdi M.Z, Meyman Y.G, Sihman M, Purwo H.S, Agus H.P. Importance of understanding geology in overpressure prediction : The Example of The East Java Basin. Procc. IPA Thirty Seventh Annual Convection & Exhibition. **IPA13-G-152** (2013).

[4] R.J. Davies. Birth of a mud volcano : East Java, 29 May 2006. Durham DH1 3LE. (2006).

[5] Bowers, G. L, *Detecting Overpressure*. Applied mechanics Technologies. Houston, Texas, U.S (2002).

[6] Osborne, M.J, and Swarbrick, R.E. Mechanisms for generating overpressure in sedimentary basins : A reevaluation. AAPG Bulletin, **81**, 1023-1041 (1997).

[7] Mattews, S.J and Bransden P.J.E. Late Cretaceous and Cenozoic tectonostratigraphic development of the East Java Sea Basin, Indonesia. Marine and Petroleum Geology J, **12**, 499-510 (1995).

[8] JOB-PPEJ and LPPM Undip, Final Report of Study Geomechanic, unpublished (2016)

[9] T. Tsuji, R. Kamei, R. Gerhard Pratt. Pore pressure distribution of a mega-splay fault system in the Nankai Trough subduction zone : Insight into up-dip extent of the seismogenic zone. Earth and Planetary Science letters. **396,** 176 (2014).

Characteristics of Lateritic Nickel Mineralization In Mid Part of Madang and Serakaman Areas, Sebuku Island, South Kalimantan

Yoga Aribowo[1,*], *Yudi* Syahputra[1], and Dian Agus Widiarso[1]

[1]Department of Geological Engineering Diponegoro University, Semarang. 50275. Indonesia

Abstract. The research was conducted in Madang and Serakaman Tengah area, Sebuku Island Subdistrict, Kotabaru Regency, South Kalimantan Province which is one of the nickel potential areas in Indonesia. The aim of this research is to know the characteristic and distribution of laterite nickel mineralization. The rocks present in the study area are serpentinized dunite, serpentinized harzburgite, gabbro, silicified gabbro, tuff, and basalt. Methods used in this research were surface geological mapping, rock observation and sampling from outcrop and drill core representing each laterite horizon from limonite horizon to bedrock. Laboratory analysis consist of X-Ray Fluorescene (XRF) analysis is used to determine the abundance of certain chemical elements and compound which characterized the mineralization stage zonation in the laterite profile. The laterite deposite in the study area can be divided based on physical and chemical properties into four zones; red limonite, yellow limonite, saprolite, and bedrock. Saprolite is dominated by a group of hydrocylicic minerals (serpentine) so it can be predicted that the laterite types are developing laterite oxide and laterite silicate types.

1 INTRODUCTION.

Indonesia is one of the largest nickel producer countries in the world (Geomagz, 2013). In 2016, nickel reserves in Indonesia are estimated to reach 221 million tons derived from laterite nickel and nickel sulphide (Databoks, 2016). Most of the nickel produced in Indonesia is nickel laterite.

One of the areas that have the potential for lateritic nickel deposits is Sebuku Island, Kotabaru Regency, South Kalimantan. The nickel mining exploration activity in this area is owned by PT. Sebuku Iron Lateritic Ores (PT.SILO). This research was conducted on Madang and Serakaman Tengah area which is under exploration to know the existence of lateritic nickel in the area. The aim of this research is to know the characteristic of laterite nickel mineralization.

* Corresponding author: yogaribowo@gmail.com

2 RESEARCH METHODS

The methodology used is by surface mapping and description of drilling well samples representing each laterite horizon ranging from limonite horizon to bedrock. X-Ray Fluorescene (XRF) analysis is used to determine of elements content found in the laterite profile.

3 RESULT AND DISCUSSION

3.1 Lithology

The lithology of the study area can be divided into 6 units, namely serpentinized dunite unit, serpentinized harzburgite unit, gabbro unit, silicified gabbro unit, tuff unit, and basalt unit.

3.1.1 Serpentinized dunite unit

Rocks generally have fractures filled with serpentine minerals, in some locations found in massive. Rocks have dominated by a weak magnetic level. Physically, serpentinized dunite have minerals composition of olivine, piroxene, and serpentine with dominated of olivine > 90%. The rocks have average chemical composition of MgO 37.08%, SiO_2 40.13%, Cr_2O_3 0.39%, Ni 0.41%, and Co 0.01%. Has a dispersal area 30% of the research area.

3.1.2 Serpentinized harzburgite unit

Rocks generally have fractures filled with serpentine minerals. These rocks have minerals composition of olivine, piroxene with dominated orthopiroksen, brucite and serpentine. The rocks have an average chemical composition of MgO 34.94%, SiO_2 40.28%, Cr_2O_3 0,33%, Ni 0.4%, and Co 0.00%. These rocks have weak magnetism. Has a dispersal area 25% of the research area.

3.1.3 Gabbro unit

This rock has a composition of dominated of plagioclas, piroxene, and a little bit of olivine. The rocks have an average chemical composition of MgO 14.91%, SiO_2 49.34%, Cr_2O_3 0.39%, Ni 0.11%, and Co 0.01%. Rocks have magnitude ranges from weak to moderate. Magnetism is interpreted because of the presence of magnetite minerals that are present on some spots of rock. Has a dispersal area 5% of the research area.

3.1.4 Silicified gabbro unit

Rock has a texture of holocrystalline, fine-grained, porfiritic, and subhedral. This rock has a composition of mineral dominant silica minerals such as quartz caused by silicified. Other minerals are plagioclase, pyroxene, slightly olivine. The rocks have an average chemical composition of MgO 5.03%, SiO_2 59.28%, Cr_2O_3 0.00%, Ni 0.00%, and Co 0.00%. These rocks have a weak magnetism. Has a dispersal area 10% of the research area.

3.1.5 Basalt Unit

These rocks are generally found in have weathered level III to IV. Has a texture holocrystalline, fine-grain, having afanitic granularity, with anhedral crystal form. These rocks have a composition in the form of the dominated plagioclase, pyroxene, and olivine. The rocks have an average chemical composition MgO 31.77%, 45.05% SiO_2, 0.15% Cr_2O_3, 0.10% Ni, and Co 0.00%. In the basalt weatherized was also found gibbsite and hematite. Has a dispersal area 5% of the research area.

3.1.6Tuff Unit

The tuff unit was found in the outcrop of the Serakaman Tengah river and the hilly area of Mid Serakaman. Locally discovered silicified tuff. Tuff in the research area has weathered so that minerals are found in the form of mineral goetite and there are vein filled with magnetite minerals. Has a dispersal area of approximately 25%

3.2 Geomorphology

Geomorphology of the study area can be divided into 4 units based on Van Zuidam (1983), ie the denudational terrain landform unit (0°-3° slope), denudational hills landform unit (8°-16° slope), structural hills landform unit (3°- 16° slope), and intrusion hill landform unit (8°-21° slope).

3.3 Geological structure

The results analysis of the joint measurements of the Madang and the Middle Serakaman areas indicate the relative mainline of the southwest-northeast direction.

3.4 Characteristics of nickel laterite in research area

In general, laterite profiles can be distinguished into limonite zone, saprolite zone, and bedrock zone (Ahmad, 2008).

3.4.1 Red limonite zone

Based on its physical, red limonite has a dark red color and is sometimes brownish red, has a thickness of 1-3.5 meters, loose, soft hardness level, has a high weathering rate, medium-strong magnetism, clay-gravel size. The mineral composition of the red limonite zone is composed of hematite and goetite with hematite mineral dominated. The hematite has a red color that controls the color of the laterite in the red limonite zone (Figure 2a). Red limonite zone has Fe content of about 41.3 - 48.7%, Mg about 0.2-0.6%, Si of 0.9-2.5%, Al by 4.0-8.4%, and Ni 0.3-0.7%.

3.4.2 Yellow limonite zone

Yellow limonite zone has a yellow to brownish color, has a thickness of 2-8.5 meters, loose, soft hardness, has a high degree of weathering, moderate magnetism, fine grain size (clay-silt). Mineral composition in the yellow limonite zone consists of goetite and manganese with dominated of goetite, locally also found clay, quartz, serpentine, and carbonate minerals based on drilling logging data in Serakaman Tengah area. The dominance of goetite minerals affects the color of the laterite in this zone. Goetit has a yellow color so that the soil produced in this zone is also yellow (Figure 2b). Yellow limonite zone has Fe content of about 38.7-50%, Mg about 0.2-0.7%, Si of 1.0-4.7%, Al of 2.5-8.1%, and Ni 0.7-0.9%.

3.4.3 Saprolite zone

Saprolite zone in the study area has a green to yellowish green color, has a thickness of 0.2-7 meters, sand-boulder grain size. Rock texture of origin in this zone can still be observed. This zone has low to moderate magnetism, the weathering rate in this zone varies from low to moderate. The mineral composition in this zone consists of serpentine, goetite, clay minerals, olivine, manganese, and locally found quartz in the Middle Serakaman region. Serpentine is the dominant composition of this zone. This is influenced by existing rock origin in the research area. Rock origin in the research area is ultramafic rocks that have serpentinized. The green color in this zone is affected by the presence of serpentine minerals (Figure 2c). The saprolite zone has a Fe content of 10-21%, Mg of 1.5-14.7%, Si of 2.9-17.8%, Al of 1-7.6%, and Ni of 1.0-2 , 2%.

3.4.4Bedrock zone

Bedrock zone at the study area is ultramafic rocks consisting of serpentinized dunite and serpentinized harzburgite. The mineral composition in this zone consists of serpentine, olivine, and pyroxene (Figure 2d). The bedrock zone has a Fe content of about 3.9-8.3%, Mg of 19.6-29.6%, Si 16.9-20.7%, Al of 0-0.7%, and Ni of 0 , 2-0.4%.

4 CONCLUSION

Characteristics of lithology in the study area consisted serpentinized dunite, serpentinized harzburgite, gabbro, silicified gabbro, tuff, and basalt.

Geomorphologyl characteristics consist of denudational terrain landform unit (0°-3° slope), denudational hills landform unit (8°-16° slope), structural hills landform unit (3°- 16° slope), and intrusion hill landform unit (8°- 21° slope).

Characteristics of geological structure in the study area consisted joints of which the main direction is southwest-northeast.

The laterite characteristics can be divided into 4 zones ie, red limonite zone, yellow limonite zone, saprolite zone, and bedrock zone.

The lateritic chemical characteristics are divided into, the red limonite zone has Fe content of about 41.3 - 48.7%, Mg about 0.2-0.6%, Si of 0.9-2.5%, Al by 4.0-8.4%, and Ni 0.3-0.7%, yellow limonite zone has a Fe content of about 38.7-50%, Mg about 0.2-0.7%, Si of 1.0-4.7%, Al of 2.5-8, 1%, and Ni 0.7-0.9%. In the saprolite zone has a Fe content of about 10-21%, Mg of 1.5-14.7%, Si of 2.9-17.8%, Al of 1-7.6%, and Ni of 1.0- 2.2%, and on Bedrock has Fe content of about 3.9-8.3%, Mg of 19.6-29.6%, Si 16.9-20.7%, Al of 0-0.7 %, And Ni by 0,2-0,4%.

Ni element at the initial depth has a relatively small content, getting down, the Ni on saprolite zone becomes larger, then in the bedrock zone Ni content declines again. At the initial depth Fe has a considerable content, but getting down, it becomes smaller.

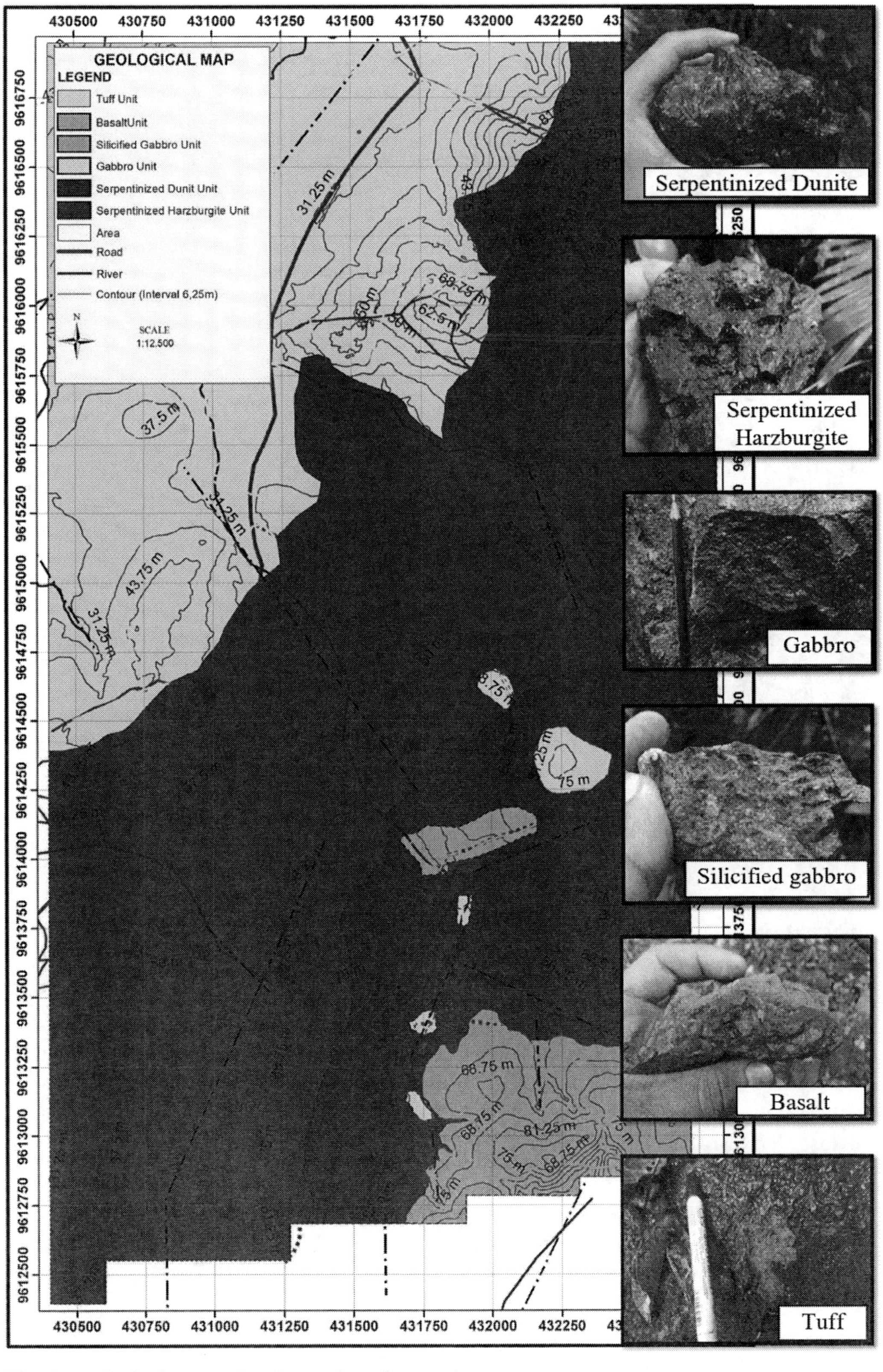

Fig. 1. Geological map and rock samples of research areas.

https://doi.org/10.1051/matecconf/201815901038

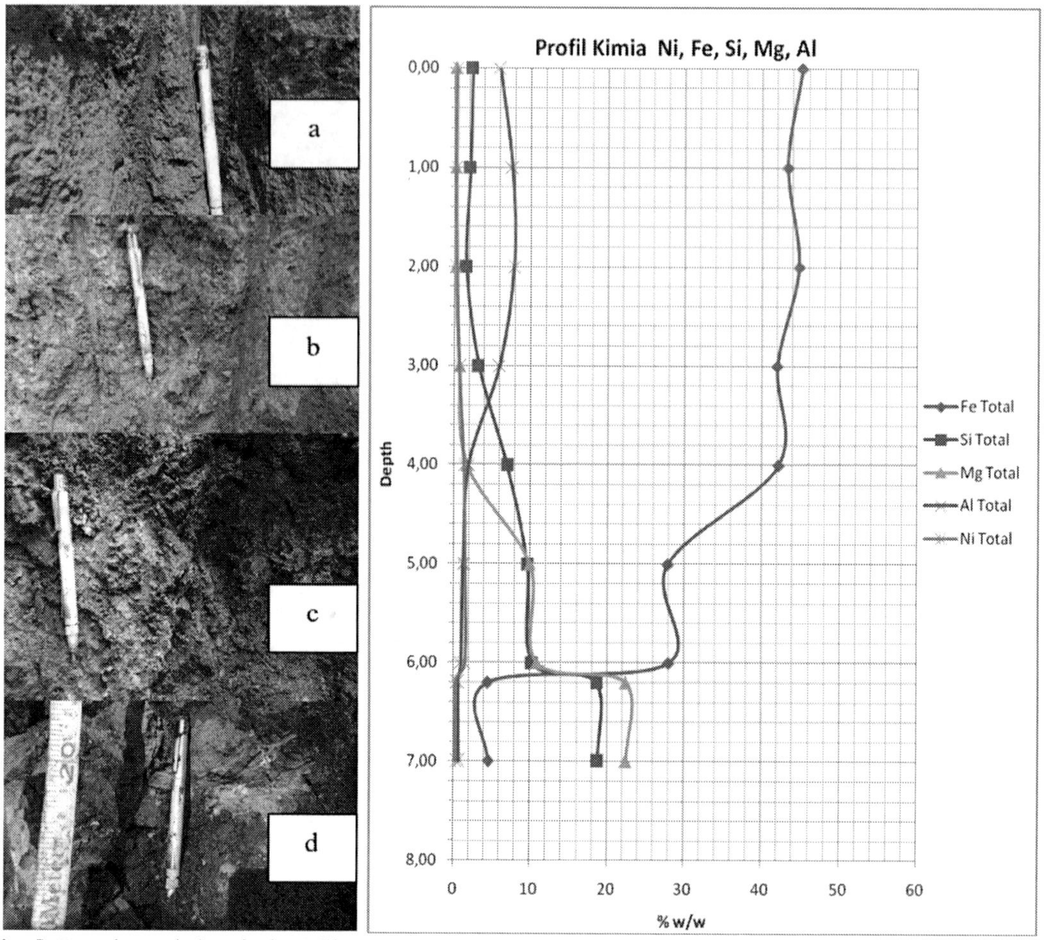

Fig. 2. Laterite and chemical profile.

REFERENCE

1. Ahmad, W.. *Nickel Laterites: Fundamental of chemistry, mineralogy, weathering processes, formation, and exploitation*., P48, VALE Inco- unpublished (2008)
2. British Geological Survey, *Nickel: Definition, Mineralogy, and Deposits*. Natural Environment Research Council, p 16-34 (2008).
3. Elias, M., *Nickel Laterite Deposits-Geological Overview, Resources and Exploration*, Australia: CSA Australia Pty Ltd. (2005)
4. Gill, R., *Igneous Rocks and Processes: A Practical Guide*, UK: Wiley-Blackwell John Willey & Sons Ltd., (2010).
5. Le Maitre, R.W.. *Igneous Rocks: A Classification and Glossary of Term*. 2nd Edition. Cambridge University Press (2002)

Analysis of Piping at Kedung Uling Earthfill Dam, Wonogiri Regency, Central Java Province, Indonesia

Najib[1*], *Agus* Setyawan[2], *Indriyani* Ardyarini[1], *Dwiyanto* Joko Suprapto[1]

[1]Geological Engineering Department, Faculty of Engineering, Diponegoro University, Semarang 50275, Indonesia
[2]Physics Department, Mathematics and Science Faculty, Diponegoro University

Abstract. Kedung Uling earthfill dam is located between Eromoko and Wuryantoro district, Wonogiri Regency, Central Java, Indonesia. Water leaked from the dam was found and could decrease volume of dam. The objectives of this study are to identify piping and to recommend for solving piping problem. In order to determine the piping direction and litological pattern, investigation of 10 boreholes with various depth has been carried out. This investigation aim to determine characteristics of soil and rock by analyzing 45 cm of thickness of undisturbed samples and to perform permeability test. Permeability test have been conducted by falling head method with 3 different depths. Leaking at the dam was analyzed by Slide 6.0 software. The results indicated that litology of dam have been identified as claystone, sand and tuff. The piping passed under the eastern abutment of the dam is the accumulation of groundwater in the aquifers formed by 8-10 meters thick layer of sand. The total debit of piping is 17,774 cm^3/sec. The Factor of safety of piping is under the minimum of standard namely 3,6. Grouting is one of alternative method to minimize the piping. Grouting was simulated to reduce permeability value and increase soil strength. By grouting, the total debit of piping was reduced to 3.314 cm^3/sec.

1 Introduction

Kedung Uling dam was built in the colonial netherland era which is located between Eromoko and Wuryantoro district border, Wonogiri regency, Central Java Indonesia. This dam was built in 1917, which is lain between Jatirejo and Tempuran river. By the function as irigation, this dam has 479.000 m^3 storage capacity with the 596 hectare extensive paddies field. Another functions of this dam is for local tourism and fisheries fresh water. Kedung Uling dam has decreasing quality from its structure resulting sliding and settlement in the body of dam in 2013-2014. Besides, the problem which is found in the dam is leaking due to water flow come out to the inside of dam. The water flow erode fine materials resulted in potential unstability of dam. If this erosion happens, there is seepage line, which is called piping, between downstream and upstream dam.

*Corresponding Author: najib@ft.undip.ac.id

Piping or leaking in the dam can be detected using some methods such as multi tracer tests [1,2] and bore hole data[1,3], geological mapping, geophysical survey and laboratory analysis[4]. Fractures can be detected due to water infiltration by using geophysical approach [5]. Bedrock such as uncompacted claystone can generate settlement of dam [6]. Uromeihy and Barzegari [7] conducted leaking dam evaluation based on the cracks and simulated groundwater flow and carried out insitu tests to estimate permeability value. They suggested to use curtain grouting to overcome leaking of dam. The objective of this research is to understand piping factor of safety and recommendation for overcoming the piping.

2 Method

2.1. Ground observation

The objective of this observation was identified soil and rock at surrounding Kedung Uling dam. In addition, this observation also observed geological condition such as cracks and sliding or seepage.

2.2. Core drilling

Drilling has been arranged in 10 points with 10 meter, 12 meter and 15 meter in depth. Drilling has been done with single tube barrel core for getting undisturbed samples used for laboratory tests. That samples were taken 45 cm at the variation depth from 1,5-1,95 meter, 5,5-5,95 meter and at the end of drilling.

2.3. Permeability test

The objective of this test was observed the soil and rock permeability coeficient. This test has been performed in the field with the falling head method and has been tested with the depth interval at 4,5-5 meter, 8,5-10 meter and 13,5-15 meter.

2.4. Bore log and corelation

This data has been collected from sample of 10 core drillings and recorded in 10 bore log with the depth 10 meter, 12 meter and 15 meter. Every 1 meter depth sample has been taken from boring to identified soil characteristics and has been tested in soil mechanics laboratory. Log discription consists of soil, rock, color, grain size, characteristics of soil and rock. Bore log is corelated for each point based on litology in every depth. The result has been used for subsurface dam modeling.

2.5. Piping flow model

From *K* value, dam geometry and dam materials can be simulated of the model and analyzed by using slide 6.0 software for understanding value of safety factor of dam piping. After factor of safety is acquired, the analysis of dam from piping can be simulated. The equation for calculating of piping safety factor is shown below [8]:

$$FS_{piping} = \frac{Icr}{In} \tag{1}$$

where FS_{piping} minimum is 4 (safe), Ln is gradien hidroulic of discharge, Lcr is gradien hidroulic from filling materials. Lcr can be observed by using equation 2[8]:

$$Lcr : \frac{SG - 1}{1 + e} \quad (2)$$

where *SG* is specific gravity (based on the depth of bore hole) and e is void ratio.

Lcr gets from laboratory test which is soil volume weight that float is devided with water volume weight. Ln is different of height h1-h2 divided by L. h1 is highest hidroulic head and h2 is hidroulic head that the high less than h1. Meanwhile, L is the length of distance of piping or distance between h1 and h2.

Piping curve is direction of water flow when the seepage happens in the dam. This curve use for giving information where the water flow of seepage, so that it can be known where the piping happened resulted unstable of dam and it can be happened sliding in the body of dam. The piping curve can be simulated by using slide 6.0. From this software, we can input value of permeability from every materials at the dam modeling, and afterward can be calculated where the flow of piping.

If piping happens, grouting can install in dam and the depth of grouting can be calculated by using equation 3[8]:

$$H = h + k.h \quad (3)$$

where H is depth of grouting (in meter), h is height of dam (in meter) , k is constanta 0.8-1.2. h is measured from the top to the upper part of bed rock of foundation.

3 Results and discussion

3.1. Geological condition

Based on the surface and drilling observation, dominant litology of the research area is grey to black of clay with characteristics of soil is loose to stiff. Intercalation sand and yellowish white limestone pebble is found in some limited research area. The tickness of soil depends on the countour and weathering stage. Based on the surface observation, the area of dam is categorized as flat with the slope close to 0°-2°. This site are categoried flat morphology. There were found cracks top of the dam (fig 1.a) and sliding (fig 1.b) on the embankment of dam.

Fig 1. (a) Cracks at the top of dam and (b) sliding at the embankment of dam.

3.2. Core drilling

There are ten core drillings conducted in research area. Those drillings are devided in two locations. Driling was carried out at the sliding locations with six core drillings and was conducted 25 meter from sliding location with four core drillings. Those locations can be seen in fig 2. The depth of core drillings have been done 15 meter, 12 meter and 10 meter in depth. Those drilling results have been used for laboratory analysis and for modeling of piping stability analysis. The modeling based on the log corelation of bore hole 1 through bore hole 10.

3.3. Piping curve analysis

Piping curve is the direction of flow water coming out when the dam encounter seepage. This curve used for giving information where the leak happen. This curve was simulated by using slide 6.0. In this program, permeability value from each materials was set based on the laboratory analysis. Afterward, it was calculated to find where the flow of seepage takes place (fig 3). Seepage flow direction head to east of Kedung Uling dam and reach to the dam body (fig 4). This means the dam has vulnerable to piping because water flows to the area of dam body. Piping curve which pointing up posibly caused by soil critical hydrolic gradient. If the condition of piping pointing up, this means piping pressure same with density of soil. Therefore, effective pressure is zero. If the soil hydrolic gradient exceed critical hydrolic gradient, then posibly grains soil which is empty will pushed by seepage flow resulted in piping. Seepage curve headed to east of dam has total hidroulic gradient value is 0,190 (fig 4). By increasing higher value of total hydroulic gradient, the dam will not safe.

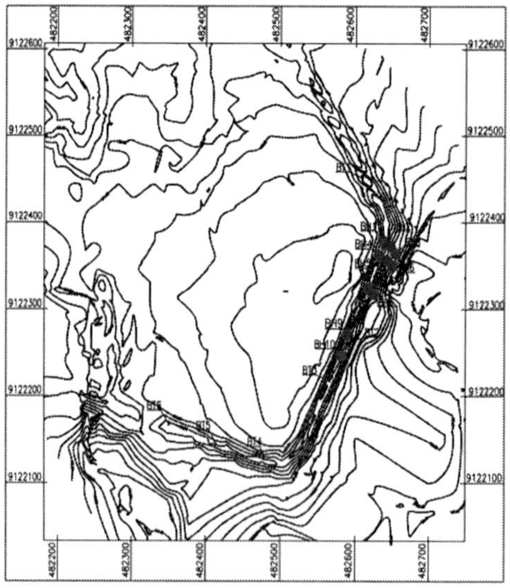

Fig 2. Core drilling location (dots are core drilling location).

3.4. Analysis of piping discharge

Piping discharge has been calculated from falling head test in the field and afterward has been modeled by slide 6.0(fig 4). Fig. 4 shows piping discharge value with direction of water outflow. Based on analytical results using slide 6.0, discharge of piping is 1,5357 m^3/day or 17,774 cm^3/sec per meter. Value of piping discharge permission in a dam is under 0,1m^3/sec or around less than 1% Q of river. Piping discharge in Kedung Uling dam is categorized as vulnerable and induces potential for sliding. This result coinside with field investigation that slidings are found at the embankment of dam. This piping discharge is highly influence stability of body of dam. Piping discharge at this model is categorized as large value. By encountering piping, the water in dam will loose largely resulting in piping lines at the ground. As the result, the dam could not resist the loading of water and could trigger sliding at the embankment of dam.

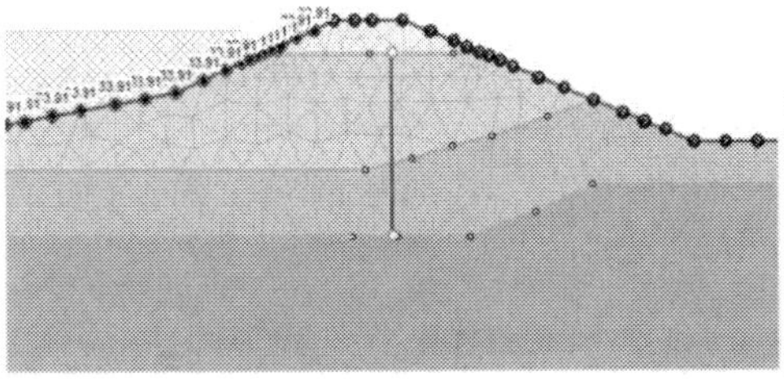

Material Name	Color	Model	KS (m/s)	K2/K1	K1 Angle	Soil Type
Filling impervious clay material		Simple	2.21e-007	1	0	General
Filling old impervious clay material		Simple	2.21e-006	1	0	General
Filling old impervious grey clay material		Simple	2.21e-005	1	0	General
Bedrock		Simple	3.96e-007	1	0	General

Fig 3. Permeability coeficient value for each materials

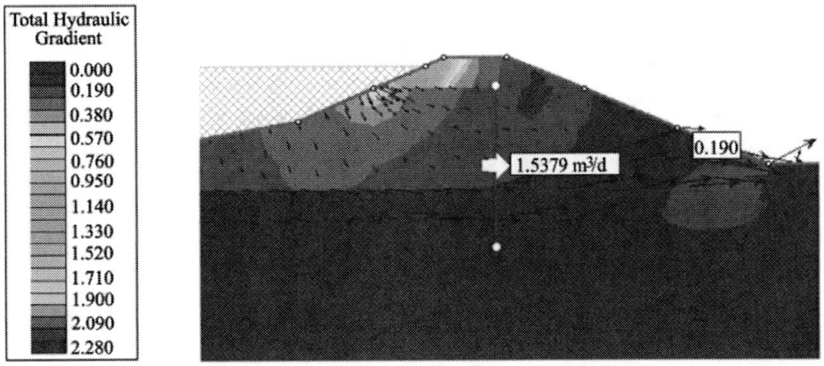

Fig 4. Direction of seepage modeling and seepage discharge value. Total hidroulic gradient value is 0,190.

3.5. Analysis of piping factor of safety

One of the main problems in dam is piping at the body of dam. This could happen if the dam overcomes different of water level and if the flow come into the soil around the dam. This flow affects stability of building of dam due to erosion of fine materials resulted in piping. Piping analysis indicates piping found based on factor of safety of piping. Based on equation 1, factor of safety of piping (FS piping) is 3.6 (unsastisfy). From the calculation of factor of safety of piping, horizontal line is highly influence in piping, if L is deflected, piping will smaller and will reduce discharge of piping. There is indication of piping in Kedung Uling dam body that makes the erosion at the subsurface based on the calculation resulted in sliding at this dam.

3.6. Analysis of piping factor of safety

From the analysis of piping showed that factor of safety of piping is not satisfying the standard of factor of safety. This means Kedung Uling dam is not safe. Mitigation can be managed by grouting. Before grouting is simulated, discharge of piping at the foundation at the bedrock is 17,774 cm^3/sec. This means the seepage is swift resulted in friction among materials bringing by flowing water. Direction of flow curve (fig. 4) also show that the flow reach the body of dam resulting piping at the inside body of dam. After grouting, piping discharge has change quite significant and the piping flow is not reach the body of dam (fig 5). Discharge piping changed to 3,134 cm^3/sec. This result shows that grouting can reduce permeability value and heightening strength of soil. Thus, grouting can minize discharge of piping. Grouting can be set up in the body of dam which is determined by using the equation of grouting depth [8].

Grouting depth is calculated by using equation 3. Based on the calculation, depth of grouting is between 23,4 meter to 28,6 meter. Grouting depth is set up from top of dam to the end of dam foundation. The reason is slurry sement can fill pores or cracks causing piping. Discharge and total hydrolic gradient value change after grouting at the body of dam. After grouting, spesific gravity and porosity will change smaller due to soil pore is filling by grouting materials. Calculation of factor of safety of piping after grouting use equation 1. The result shows piping factor of safety after grouting is 7,5. This results show that factor of safety of dam is higher than that of standard (more than 4) and the dam is safe from piping.

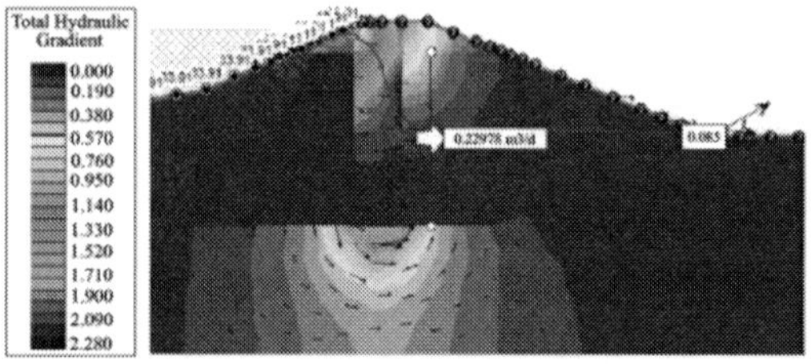

Material Name	Color	Unit Weight kN/m³	Strength Type	Cohesion kN/m²	Phi
Filling impervious clay material		18.3	Mohr-Coulumb	15.7	18.3
Filling old impervious clay material		22.2	Mohr-Coulumb	15.2	11.92
Filling old impervious grey clay material		20.17	Mohr-Coulumb	14.42	14.97
Bedrock		20	Mohr-Coulumb	7.1	30.59
Grouting1		18.3	Mohr-Coulumb	20	18.3
Grouting 2		22.3	Mohr-Coulumb	20	11.92
Grouting 3		20.17	Mohr-Coulumb	20	14.42
Grouting 4		20	Mohr-Coulumb	20	30.59

Fig 5. The model of dam after grouting.

4 Conclusions

This paper clarified the piping at the Kedung Uling earthfill dam and the method for overcoming piping problem. The discharge of piping in the bedrock of foundation before grouting is 17,774 cm^3/second and after grouting is 3,134 cm^3/second. Factor of safety of piping at the dam is not satisfy 3,6 (less than 4). To strengthen factor of safety of piping at the body of Kedung Uling is recommended using grouting. After grouting factor of safety change to 7,5.

We thankfully appreciate for any supports from PT. Selimut Bumi Adhi Cipta and Kemen PUPR Dirjen SDA BBWS Bengawan Solo, which helped us in data and discussion. Research has supported by another APBN DPA SUKPA LPPM Diponegoro University, April 2017, Fiscal year 2017.

References

1. H.Dong, J.Chen, X. Li, Eng.Geol.**212** (2016)
2. D. Battaglia, F. Birindelli, M. Rinaldi, E. Vettraino, A. Bezzi Eng.Geol.**205** (2016)
3. M.H.Ghobadi,G.R. Khanlari, H. Djalaly, Eng.Geol. **82**,7(2005)
4. G. Berhane, M. Amare, T. Gebreyohannes,K.Walraevens, J.Afr.Earth Sci.**129** (2017)
5. A. Loperte, F. Soldovieri, A. Palombo, F. Santini, V. Lapenna, Eng.Geol.**211** (2016)
6. H.Cetin, M. Laman, A. Ertunc, Eng.Geol. **56**, 17(2000)
7. A.Uromeihy, G. Barzegari, Eng.Geol. **91**, 9(2007)
8. H.Hardiyatmo, *Mekanika Tanah 2 (*5th Edition,Gadjah Mada University, 2010)

The analysis of volcanic activity influences at the lower and middle part of Sentolo Formation, Kulon Progo using petrographic method

Tri Winarno[1,*], *Jenian* Marin[1], *Ilham Hani* Pratama[1], and *Anis* Kurniasih[1]

[1]Geological Engineering, Diponegoro University, Indonesia

Abstract. The research area is a part of South Serayu Mountain zone, specifically Bregada Kulon Progo. The Bregada Kulon Progo is a product of volcanic activity in the past, which can be seen by the volcanic rock called Old Andesite Formation. The end of the volcanic activity is characterized by the deposition of carbonate rocks which were part of Jonggrangan Formation and Sentolo Formation. The lower part of Sentolo Formation is composed by the mix of carbonate and volcanic materials. The research is carried out by two methods. The first is field method, which is implemented by geological mapping and rock sampling. The second is laboratory method, which is implemented by petrographic method. The petrographic method is implemented by making thin section of the rocks, then observing the thin section with polarization microscope to determine the composition of the rocks. The rock samples are mainly from the lower and middle part of Sentolo Formation, which probably still got the influence of volcanic activity. The benefit of the research is knowing the influence of volcanic activity in the carbonate rocks of Sentolo Formation. The presence of volcanic materials in the lower part of Sentolo Formation, can be predicted by two processes. First, the volcanic activity supplied those materials directly, and second, from the rework of the volcanic materials then mixed with carbonat materials of Sentolo Formation.

1 Introduction

1.1 Background

Bregada Kulon Progo, the location of this research, is a part of the physiography of the Southern Mountains [1,5], based on other sources Bregada Kulon Progo belongs to the South Serayu Mountains [6]. Bregada Kulon Progo is the product of the volcanism activity during the Oligocene to Early Miocene. The volcanism activity produced the formation of volcanic deposition called Old Andesite Fomation (OAF).

The end of volcanism activity is marked by the growth of reefs namely Jonggrangan Formation. During the Early Miocene to Middle Miocene the Jonggrangan Formation was

[*] Corresponding author: triwingeo@gmail.com

formed. The growth of reefs at Bregada Kulon Progo was followed by the deposition of clastic carbonate rocks called Sentolo Formation. Sentolo Formations from bottom to top are composed by tuffaceous marl, limestone, and marly sandstone [6]. Sentolo Formation was deposited unconformously above the Old Andesite Formation and then covered by the Young Merapi Deposits and the Alluvial Sediment unconformously [8].

The lower part of the Sentolo Formation is composed by a mixture of carbonate materials and volcanic materials. The presence of volcanic material at the bottom of Sentolo Formation leads to the presumption of volcanic activity which supply the volcanic material into Sentolo Formation or the existence of rework material from the old Andesite Formation which was deposited into the Sentolo Formation, resulting in a mixing on the lower part of Sentolo Formation. Based on the description, a study was conducted to study the lower and middle stratigraphy of Sentolo Formation, based on differences or variations in the composition of the formation.

1.2 Location

The research area is located at Sidomulyo Village, Pengasih District (X = 407000-411000 TM and Y = 9134000-9137000), and Giripurwo Village, Girimulyo District (X = 407000-411000 TM dan Y = 9138000-9141000), Kulon Progo Regency, Yogyakarta Province, Indonesia (Fig. 1).

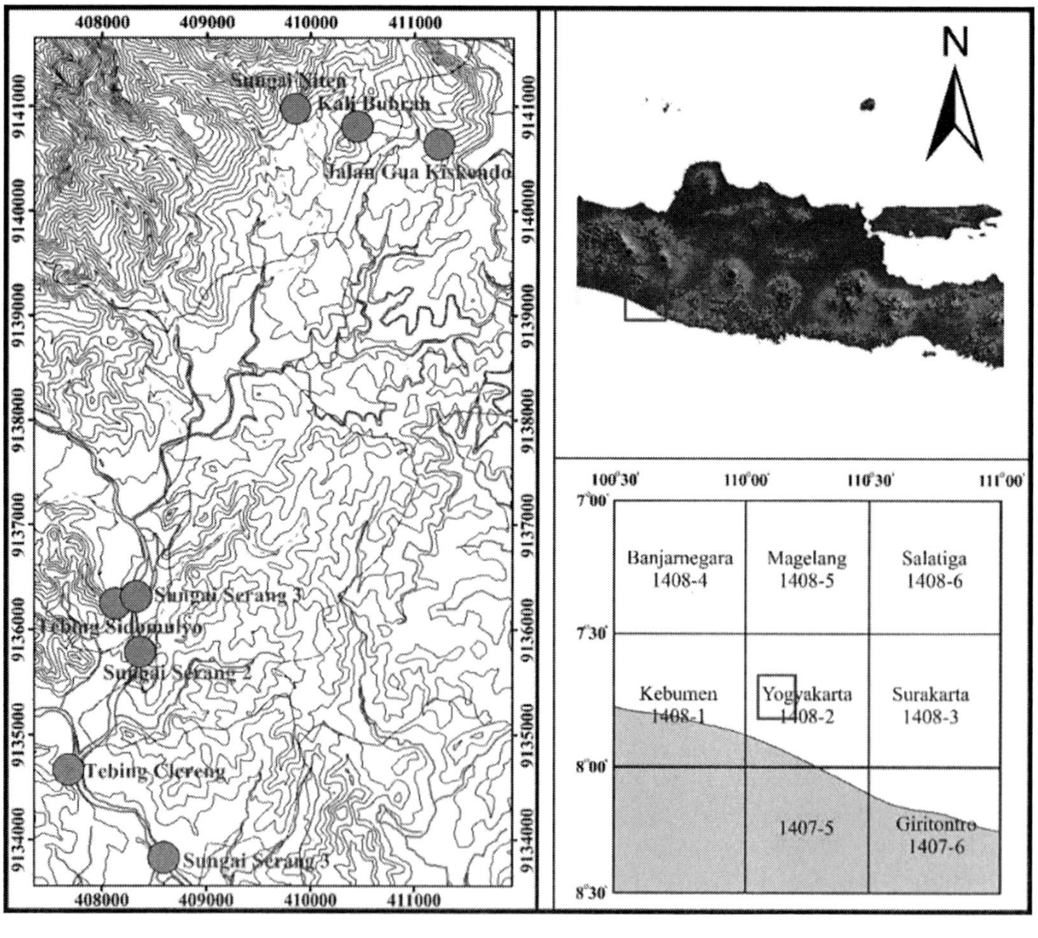

Fig. 1. Research area location

2 Literature Review

The research area is part of regional geology of Kulon Progo, particularly Sentolo Formation. The Sentolo Formation is deposited above The Old Andesite Formation. It is interfingering with Jonggrangan Formation which also deposited above The Old Andesite Formation. The Sentolo Formation consists of limestone and marly sandstone. The bottom of the formation consists of conglomerate piled by tuffaceous marl with tuff insertion. Towards the upper part of the formation, the rocks gradually transform into layered limestone. The thickness of this formation is about 950 m [8].

3 Methods

This research was using petrographic method to analyze the rocks in thin section. The rock samples from the research area were polished to make the thin sections. The thin sections were analyzed using polarization microscope. This analysis aims to determine the character of the rock and the microscopic composition of the rock composition of each sample.

4 Discussion

Observation of changes in the composition of volcanic materials in the study area, particularly in the lower Sentolo Formation to the middle Sentolo Formation is held by sampling the rocks on each of the stratigraphic measurement paths assumed to be the lower or middle part of the Sentolo Formation. Rock samples from each measured stratigraphic path were observed with petrographic analysis.

At the bottom of the Sentolo Formation there is a volcanic material input inside the rock Volcanic material supply derived from erosion on the body of volcano (post-volcanism) is seen on the Niten River and Kali Bubrah tracks. On both tracks were found volcanic rocks debris deposit. The products of the erosion were sandstone deposits composed of volcanic material and the presence of igneous rock fragments floating on the sandstone body (Fig. 2)

Fig. 2. The outcrops of debris sandstones on the Niten River showing the supply of volcanic material after the process of volcanism (post-volcanism).

The volcanic materials were interpreted not directly derived from the volcanic eruption but from the erosion of pre-exixting volcanic materials of Old Andesite Formation. It was proved by the presence of foraminifera fossils at the rocks. Organisms such as foraminera could not live in the environment where there was volcanic activity.

Based on the results of petrographic observations can be seen that there was a huge supply of volvanic material at the bottom of Sentolo Formation with a small organism diversity (Fig. 3).

It can be interpreted that the bottom of Sentolo Formation is not directly deposited by carbonate material, because at the time of the deposition of Sentolo Formation, the sea conditions have not been in a clear and stable environment,so the organisms were unable to live well.

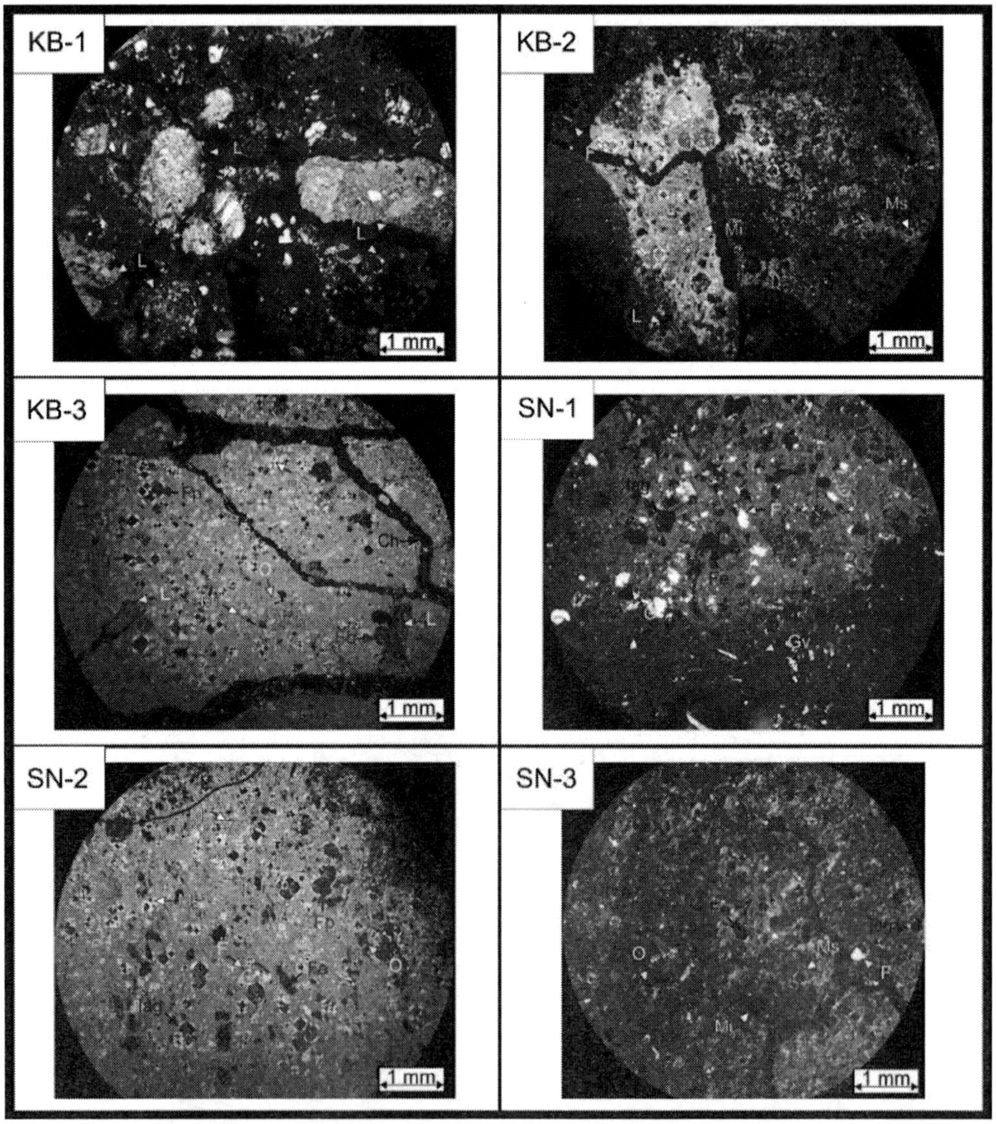

Fig. 3. The thin section of rocks from Lower Sentolo Formation (Fp= Planctonic Foraminifera; Fb= Benthic Foraminifera; O= Opaque Mineral; L=Lithic; F=Feldspar; Gv=Volcanic Glass; Mi=Micrite; Ms=Microspar; Ch=Channel Porosity; Fe=Fenestral Porosity; Iag=Intragranular Porosity).

Towards the middle of the Sentolo Formation the supply of volcanic materials has begun to diminish until nothing remains, although some thin sections show the existence of volcanic glass material with a very wide array of organisms (Fig. 4).

Overall measured stratigraphic measurement path from the bottom of the Sentolo Formation toward the middle of Sentolo Formation show the decrease of volcanic material abundances. This change in composition is due to the occurrence of equilibrium, either sea level rise or reduction of tectonic processes and volcanic activity, so the environment condition was suitable for the development of organisms.

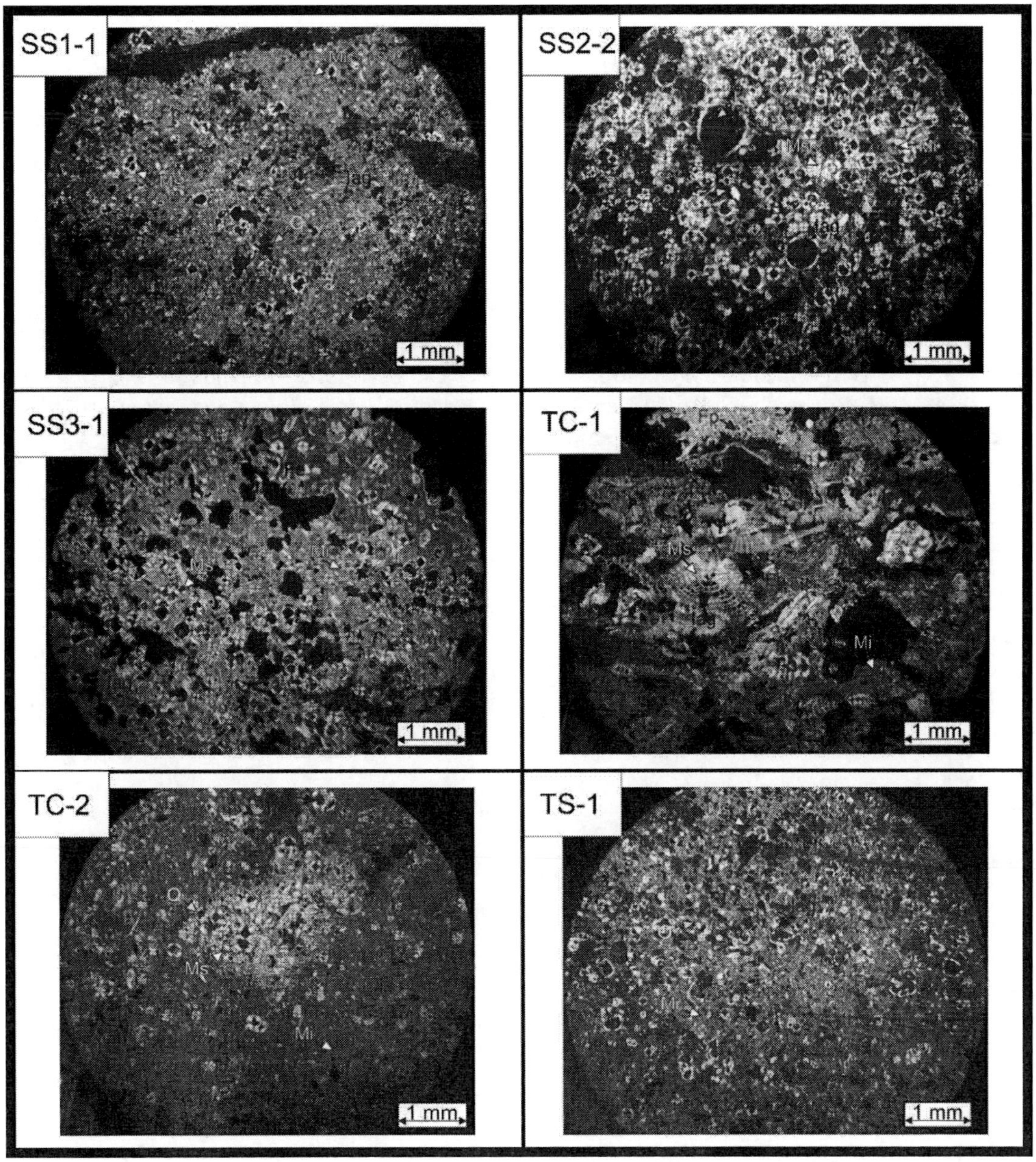

Fig. 4. The thin section of rocks from Middle Sentolo Formation (Fp= Planctonic Foraminifera; Fb= Benthic Foraminifera; O= Opaque Mineral; L=Lithic; F=Feldspar; Gv=Volcanic Glass; Mi=Micrite; Ms=Microspar; Ch=Channel Porosity; Fe=Fenestral Porosity; Iag=Intragranular Porosity).

5 Conclusions

1. The volcanic material in Sentolo Formation is abundant in the bottom part of the formation and decreasing toward the middle part of the formation.
2. The volcanic materials in the Sentolo Formation derived from the erosion of volcanic rocks of Old Andesite Formation and not directly from volcanic activity.

References

1. A. Harjanto, Jurnal Ilmiah MTG **4**, 2 (2011)
2. H.G. Hartono, S. Pambudi, *Prosiding Seminar Nasional ReTII ke-10,* 134-139 (2015)
3. H. Pringgoprawiro, Contribution Dept. Of Geology Institute Technology Bandung **64**, 5-21 (1988)
4. M.F. Al Hafizh, Akmaluddin, *Prosiding Seminar Nasional Kebumian Ke-9,* 76 - 80 (2016)
5. R.H. Smyth, R. Hall, G.J. Nichols, *The Geological Society of America Special Paper 436*, 199-222 (2008)
6. R.W. van Bemmelen, *The Geology of Indonesia Vol. 1A* (The Hague: Government Printing Office, 1949)
7. S. Maryanto, Jurnal Geologi dan Sumber Daya Mineral **16**, 129-139 (2015)
8. W. Rahardjo, Sukandarrumidi, H.M. Rosidi, *Peta Geologi Lembar Yogyakarta, Jawa, Skala 1:100.000* (Puslitbang Geologi, Bandung, 1995)

Analysis of Landslide Disaster Impact Identification Using Unmanned Aerial Vehicle (UAV) and Geographic Information System (GIS) (Case Study: Ngesrep Sub District, Semarang City)

Yudo Prasetyo[1], *Nurhadi* Bashit[2], and *Reyhan* Azeriansyah[3]

[1,2,3] Remote Sensing and Photogrammetry Lab, Geodesy Engineering Dept, Faculty of Engineering, Jl. Prof. Soedarto SH, Tembalang, Semarang, 50277, Central Java, Indonesia,
yudo.prasetyo@ft.undip.ac.id

Abstract. Landslide is a natural disaster that commonly happens in Indonesia, especially in Region of Semarang that geologically has hilly topography. In Semarang city, there are 22 Sub Districts classified as landslide potential areas, which one of them is Sub District Ngesrep, based on to BPBD Semarang. The disaster of landslides can cause human injuries and loss in infrastructure, life, and assets. Disaster management requires identifying for the impact of landslide disaster at a location. One of the methods to identifying the impact of landslide disaster uses UAV technology. UAV technology can be used to collect, map, extract information of landslide and build Digital Model in surface or elevation based on overlapping imageries. Elevation data from UAV are combined with data of rainfall, land cover and geological which will produce the map of the potential landslide disaster. The map of the potential landslide disaster is combined with the result of land cover digitation to determine the impact of landslide disaster.

Keyword: Digital Model, Land Cover, Landslide, UAV

1. Introduction

Indonesia region is located in the tropical climate area with two seasons that is summer and winter with characteristics of weather changes, temperature and wind direction is quite extreme. Climatic conditions in Indonesia and topographic conditions of the surface and rocks that are relatively diverse both physically and chemically will result in fertile soil conditions. The condition can cause some bad consequences for humans such as the occurrence of hydro meteorological disasters such as floods, landslides, forest fires, and drought.

• Corresponding author: yudo.prasetyo@ft.undip.ac.id

Landslide is a natural disaster that commonly happens in Indonesia, especially in Region of Semarang that geologically has hilly topography. In Semarang city, there are 22 sub-districts classified as landslide potential areas that one of them is Sub District Ngesrep based on to BPBD Semarang. In 2016, in the Bukitsari region of Sub District Ngesrep, cases of the landslide had been happening on October 4, 2016. This disaster not only causes human injuries but loss in infrastructure, life, and assets. Topography with the moderately steep condition and relatively high-intensity of rainfall become main factors for a landslide in Sub District Ngesrep.

Disaster management needs to identify impacts for the landslide to make decision process and prevent this from happening again. One method to identify landslide impacts is using unmanned aerial vehicle technology (UAV). UAV is one solution to identify the impact of landslide disaster. UAV produce imagery with low altitude flight, the effectiveness of time in gaining images, low-cost budget than using satellite imagery, and less cloud disruption. This paper presents the use UAV to collect data of landslide area.

The data processed into Digital Surface Model (DSM) and Digital Elevation Model (DEM) that provides elevation information based on images overlapping. DSM and DEM data are processed to obtain surface slope data. The result is combined with data of rainfall, land cover and geological which will produce the map of landslide potential. The map of landslide potential is combined with the result of land cover digitation to determine the impact of landslide disaster.

2. Materials and Methods

2.1 Materials

In this paper using primary and secondary data as follows: an UAV Images, a Photograph of the location of the landslide impact, a Landsat Image, The Recording of GPS data and questionnaire, a Physical Building Map, a Land Cover Map, a Slope map, a Land Type Map, a Potential Water Ground Map, and The Rainfall data.

2.2. Methods

2.2.1 Unmanned Aerial Vehicle (UAV) Technology

The UAV (Unmanned Aerial Vehicle) has become a popular platform to facilitate this method by using sensors to capture aerial photograph data. The pilot and automatic remote control can take the photos manually. UAV have advantages such as low altitude flight, effectiveness of time in gaining images, low-cost budget and build a map disaster areas that have high risk. Disadvantages of UAVs are in terms of small format cameras, image coverage and low resolution [3].

Aerial photographs taken by UAV are processed such as orthorectification, mosaicing, 3D modeling of DSM and DEM so that the conversion of DEM data into slope data [1]. Orthorectification is a process of geometric correction of satellite imagery or aerial photographs to correct geometric errors of an image originating from topographic, sensor geometry and other errors. The result of orthorectification is an orthoimages. The orthoimages have been geometrically corrected from removing the effects of image distortion induced by the sensor, viewing perspective and relief for the purpose of creating a planimetrically correct image.

The aerial photograph mosaic is a combination of two or more overlapping aerial photographs that making a combination image with displaying images that have a large area coverage [5]. Combining overlap or sidelap photo parts does mosaicing an image. Mosaicing results are done making 3D DSM and DEM models until DEM data is converted to slope data.

2.2.2 AHP (Analytical Hierarchy Process)

The Analytical Hierarchy Process is a general theory of measurement to make a decision support model developed by [2] where this decision support model will describe problem multi factor or multi criteria problems into a hierarchy. The AHP has a special concern with multi criteria structure, its measurement and on dependence within and between the groups of elements of its structure. It has found its widest applications in multi criteria decision-making, planning, and resource allocation and in conflict resolution. AHP makes a complex problem can be solved into several groups to form a hierarchy so that the problem will seem more structured and systematic [4]. AHP must first determine the purpose of the case problem; specify criteria and alternatives, then according to the following steps:

1. Making a matrix of criteria comparison with the value that has been entered.
2. Making the priority vector weights.
3. Calculate lambda.
4. Calculate the Consistency Index (CI).
5. Finding Ratio Consistency (CR), Level of consistency if CR value < 0.1.

3. Results and Discussion

3.1. Orthophoto Mosaic

The orthophoto mosaic is generated from aerial images that are made up of 1387 images with a covered area of 3.75 km^2 so that it can be seen in fig 1. The orthophoto mosaic itself shows the wide of the area damaged by landslide disaster. The average flight height of this aerial photo acquisition is 211 m. Based on the comparison of the average flight height with DJI Phantom 3 camera resolution of 3.61 mm; the spatial resolution obtained is 7.87 cm/pixel.

Fig. 1. The Orthophoto

3.2. Analysis of Landslide Impacts

The UAV images can be utilized for classification of areas affected by landslide disaster. The result of the classification there is 2 areas affected by a landslide that can be seen in fig 2. Using UAV methods so that the acquisition data of objects can be easily obtained can use mapping of disaster potential areas. In fig 2 (a) can be seen an image from UAV shows a building affected by a landslide disaster. In the house, almost half of the buildings affected by landslide disaster and endanger near of building. In area 1 is a dense residential location so that it may endanger the safety of people living in the area. In the Sub District Ngesrep made possible the main priority in mitigation landslide because the location is one of the densely populated areas in the location potential to landslide disaster. In fig 2 (b) can be seen the locations affected by the landslide disaster are areas not including settlements. Area 2 is a location affected by landslides so that the area is not expected to be a residential area.

(a) Area 1 (b) Area 2

Fig. 2. Area Affected Landslide

In fig 2 can be seen there are 2 locations affected by the landslide that is the first area around to 1,026.339 m² and the second area that affected by landslide disaster around to 521 m². The authorities can utilize an UAV technology to mapping the risk areas affected by landslides. The UAV can be used to collect landslide disaster data from another location. All types of landslides are covered by the database, which stores not only main attributes, but also various complementary data, including data on landslide causes, impacts, and mitigation. The database can be utilized for decision making in determining the mitigation of landslide disaster. The authorities may create a regulation to regulate settlements in locations prone to landslide disaster in order to avoid human injuries and loss material.

3.3. ANALYSIS OF LANDSLIDE HAZARD Based on BNPB (Badan Nasional Penanggulangan Bencana)

The map of landslide hazard based on BNPB can be seen on fig 3. Landslide hazard based on BNPB is a simplification of levels contained in SNI. Level of landslide hazard is divided into 3 levels so as to allow zone division differences that including within a certain category. In fig 3 illustrates estimated of landslides hazard according to the low level has a wider area. It happens because the estimated level of landslides hazard according to the low level and the lowest level joins into one zone so as to show the wider area.

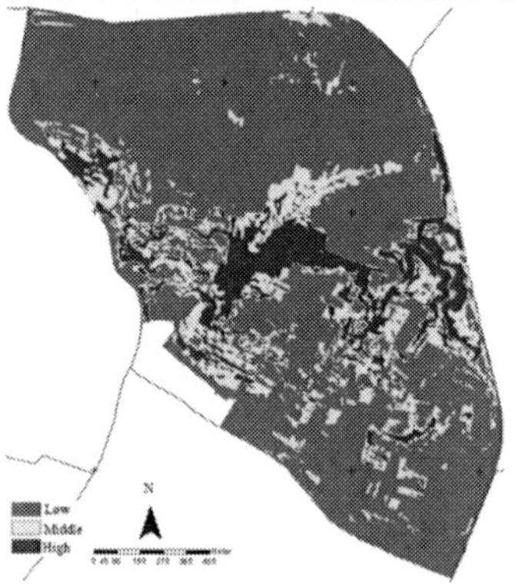

Fig. 3. The Map of Landslide Hazard Based on BNPB

The map of landslide hazard base on BNPB is divided into three levels that are low, medium and high. Estimated of landslides hazard according to the low level around 182.64 ha (71.85%), according to the moderate level around 41.471 ha (16.31%) and according to the high level around 30.094 ha (11.84%) of the total area of 254.205 ha.

3.4. Landslide Risk Landscape Analysis

The map of risk evaluation landslide can be seen in fig 4 with classification into 3 levels that are low, medium and high. In Sub District Ngesrep has several locations that have high risk. The risk of a landslide disaster is present in steep slopes. There is more than 10 percent of the area located at high-risk locations for landslides. Location in steep slopes should be monitored to avoid the impact of landslide disaster. Authorities and decision makers can use a map of risk evaluation landslide so they are considered in development plans and/or that appropriate risk mitigation measures are implemented.

The landslide risk for an object or an area must be calculated with reference to a given time frame for which the expected frequency or probability of occurrence of an event of intensity higher than a minimum established value is evaluated [6].

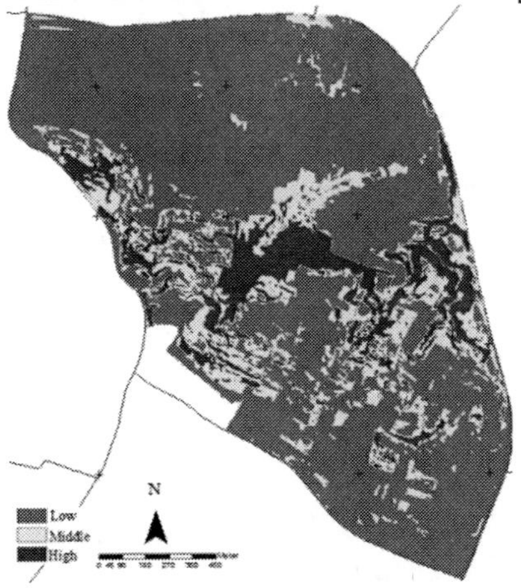

Fig. 4. The Map of Risk Evaluation Landslide

3.5. Analysis Of Landslide Hazard Impacts On Landslide Risk Landscape

The classification of areas affected by the landslide disaster is merged combination with the zoning risk of landslide disaster such as fig 5. Merging can be identified about the accuracy of the landslide risk zonation result on the impact of the affected area.

(a) Area 1 (b) Area 2

Fig. 5. Combined a Map of Landslide Hazards and Landslide Affected Areas

Fig 5 shows the area 1 where there is zoning risk of landslide disaster consisting of three classes with according to the low level around 141.3 m^2 (13.77%), according to the moderate level around 326.812 m^2 (31.18%) and according to the high level around 522.5 m^2 (50.91 %) of the total affected area of 1026.3 m^2. The area 2 where there is zoning risk of landslide disaster consisting of one class with according to the high level around 520.82 m^2 (100%) of the total affected area of 520.82 m^2.

4. Conclusion

UAV technology can be used to mapping the risk areas affected by landslides. The orthophoto mosaic generated from aerial images that are made up of 1387 images with a covered area of 3.75 km^2 so that it can be seen in fig 1. The orthophoto mosaic itself shows the wide of the area damaged by landslide disaster. Based on the comparison of the average flight height with DJI Phantom 3 camera resolution of 3.61 mm, the spatial resolution obtained is 7.87 cm/pixel.

Mapping of disaster potential areas can be used by using UAV methods so that the acquisition data of objects can be easily obtained. UAV images shows a building affected by a landslide disaster. In the Sub District Ngesrep made possible the main priority in mitigation landslide because the location is one of the densely populated areas in the location potential to landslide disaster.

The authorities can utilize UAV technology to mapping the risk areas affected by landslides. UAV can be used to collect landslide disaster data from another location. All types of landslides are covered by the database, which stores not only main attributes, but also various complementary data, including data on landslide causes, impacts, and mitigation. The database can be utilized for decision making in determining the mitigation of landslide disaster. The authorities may create a regulation to regulate settlements in locations prone to landslide disaster in order to avoid human injuries and loss material.

In this research, AHP method has been used to determine landslide hazard impacts on landslide risk. AHP can be used for a special concern with multi criteria structure, its measurement and on dependence within and between the groups of elements of its structure. So that AHP results can be utilized for decision-making. AHP results in area 1 generate 3 classes zoning risk of landslide disaster consisting with according to the low level around 141.3 m^2 (13.77%), according to the moderate level around 326.812 m^2 (31.18%) and according to the high level around 522.5 m^2 (50.91 %) of the total affected area of 1026.3 m^2. While in area 2 generate 1 class zoning risk of landslide disaster consisting with according to the high level around 520.82 m^2 (100%) of the total affected area of 520.82 m^2.

References

1. Cooper, M., Robson S. J ASPRS E **25**, **201-210**, California (2006)
2. Prasetyo Y, Wikantika K, Sumantyo J.T.S. (2012). J. AGSE. E **2**, **221-229** (2012)
3. Pulliveli, *Analysis of Close Range Photogrammetry* (New Portland Publishing, New York, 2005)
4. Soeta'at, *Fotogrammetri Analitik*. (Universitas Gadjah Mada. Yogyakarta, 1994)
5. Wolf P.R., *Elemen Fotogrametri* (Gadjah Mada University Press, Yogyakarta, 1983)
6. Zhou, X., Chang, N.-B., Li, S. J. Sensors, E **9**, **1876-1912** (2009)

The Spatial Analysis to Choose a Location to be Designated as a Landfill Site

Zaflis Zaim[1*], *Febby* Asteriani[2]

[1,2]Urban & Regional Planning Department, *Universitas Islam Riau*, Indonesia
[*]Doctoral Student, Faculty of Engineering, *Universitas Diponegoro* Semarang Indonesia

Abstract. Pekanbaru city has a landfill site (TPA). The local government estimated that it could only accommodate disposal for the next 2 years and a half. The aims of this research are: to identify a location suitable to be designated as a landfill site and considering to the land use change. The methods it employed were the overlay technique in Geographic Information System and Technique for Order Preference by Similarity to Ideal Solution (TOPSIS) as non-spatial analysis. Research findings found 4 feasible locations using GIS, i.e: Palas Sub-district and Rumbai Bukit Sub-district in Rumbai; Tebing Okura Sub-district in Rumbai Pesisir; Sail Sub-district in Tenayan Raya; and Labuh Baru Sub-district in Payung Sekaki District. From the TOPSIS, a number of locations with a specific rank were found such as: Rumbai, Rumbai Pesisir, Tenayan Raya & Tampan District. This research have one recommendation namely that Rumbai is the most suitable district to be designated as a landfill site.

1 Introduction

The TPA has been established since 1987 situated in Muara Fajar. It produces 1.816 m^3 of waste per day from 12 districts and has over capacity. There are 18.062 residents who working in agriculture sector [3], and product i.e: paddy, maizena flour, cassava, crops, chili and fruits. According to Borges, Fragoso, et al: land use is the key for human activity that drives socio-economic development in rural regions [4]. Therefore, the agriculture area in the urban fringe need protection to guarantee a sustainable agriculture land management. That is to determine the best position to establish lasting monuments and other valuable place [5]. This study fills this gap and was conducted with the objective to determine a location suitable to be designated as a landfill site. It will make an alternative method in landfill site selection & geomorphic study in planning [1].

2 Method & Analysis

The spatial analysis overlaid maps to describe some phenomena using Arcview. This research has define 9 criteria which were devided into 2 categories (see table 1 and 2).

* Corresponding author: zaflis@eng.uir.ac.id

Table 1: The Regional Feasibility Category

Code	Parameter	Weight	Criteria	Score
K1	Slope	5	>15% 5 - 15% 0 - 5%	1 2 3
K2	Geological Condition	4	Not situated within the active fault zone Situated within the active fault zone	3 1
K3	Distance to the Water Body	3	>100 m <100 m	3 1
K4	Distance to the Residential Area	5	>1500 m <1500 m	3 1
K5	Distance to the Agricultural Area	3	>150 m from the agricultural area <150 m from the agricultural area	3 1
K6	Distance to the Protected Area	2	Situated outside the protected area Situated within the protected area	3 1
K7	Distance to the Outskirts of the City	5	>1000 m <1000 m	3 1

Table 2: The Filter Feasibility Category

Code	Parameter	Weight	Criteria	Score
K8	Rainfall intensity	2	>1000 mm/year 500-1000 mm/year <500 mm/year	1 2 3
K9	Flood-Prone Area	2	There are no flood hazards. The probability of flood is more than 25 years. The probability of flood is less than 25 years.	3 2 1

Source: Results of the Analysis & SNI (*Indonesia National Standard*) number: 03-3241-1994.

3 Research Findings

3.1 Spatial Analysis of the Selection of the Landfill Location

3.1.1. The Regional Feasibility Criteria

3.1.1.1. Slope

The ideal slope for a landfill ranges from 0 to 20%. There are two colors, the pink zone refers to the area which is not suitable to be designated as a landfill which cover the following districts: some areas in Rumbai, Rumbai Pesisir & Tenayan Raya.

3.1.1.2. Protected Area

In the map, there are two zones, namely the green zone, which is located within the protected areas, and the yellow zone, which is located outside the protected area.

3.1.2. The Overlay of the Regional Feasibility and Filter Feasibility Map

Based on the overlay result, Rumbai District especially in Palas and Rumbai Bukit Sub-district, is considered a zone suitable to be designated as a landfill site. Palas & Rumbai Bukit Sub-district have a slope ranging from 2 to 15% which falls into the slightly sloping category. The potential location 3 (Tebing Okura) has a slope ranging from 2 to 8%.

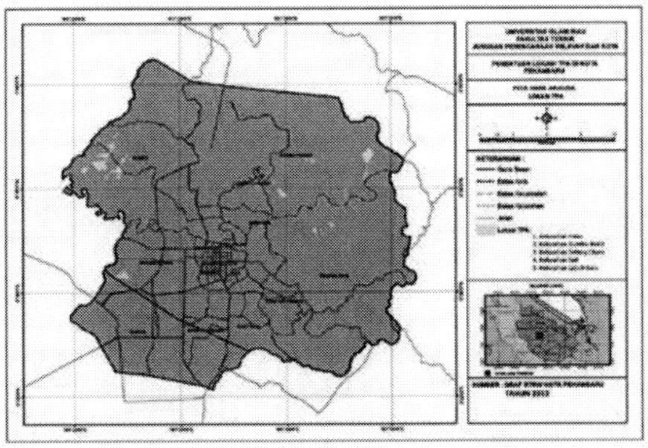

Fig 1. The Map Generated by the Analysis Illustrating as Potential Locations.

Table 3. Scores for the Regional Feasibility and Filter Feasibility Criteria

No	Score Classification	Sub-district				
		Palas	R.Bukit	T.Okura	Sail	L.Baru
Regional Feasibility						
1	Slope	15	15	15	15	15
2	Geological Condition	12	12	12	12	4
3	Distance to the Water Body	9	9	9	9	9
4	Distance to the Residential	15	15	15	15	5
5	Agricultural Area	9	9	9	9	9
6	Protected Area	4	4	4	4	4
7	Distance to the Outskirts of the city	15	15	15	15	15
Filter Feasibility						
8	Rainfall Intensity	6	6	6	6	6
9	Flood-Prone Area	6	6	6	6	6
Total		91	91	91	91	73

Total = Score * Criterion Weight

3.2.The Non-Spatial Analysis for Landfill Site Selection

To support the results of spatial analysis, TOPSIS method that classifies districts on the ranking basis was performed as alternative landfill locations, also to strengthen the basis for decision making as well. There are weights and criteria set to determine the selected alternative as alternative landfill locations. Each category has a criterion weight between 1 and 5 namely the categories very bad (1), bad (2), fairly good (3), good (4), excellent (5).

3.2.1. Developing the Decision Matrix

The matrix column states that the attributes are comprised of the existing criteria, whereas the matrix row states that the alternative is the assessment data on the zones suitable to be used as a landfill to be compared.

$$D = \begin{bmatrix} x_{11} & x_{12} & \dots & x_{1n} \\ x_{21} & x_{22} & \dots & x_{2n} \\ x_{m1} & x_{m2} & \dots & x_{mn} \end{bmatrix} \tag{1}$$

D = Decision matrix n = total criteria m = Total number of alternatives.
Symbols $x_{11} \dots x_{34}$ = indicate the performance of the alternatives.

3.2.2. Calculating the total score for x (the alternative score for each criterion)

$$\text{Total } X_1 = \sqrt{A_1{}^2 + \ldots A_{12}{}^2}$$

$$= \sqrt{3^2+3^2+3^2+3^2+3^2+2^2+2^2+3^2+3^2+3^2+3^2+2^2} = 9.644$$

A1=Bukit Raya; A2=Lima Puluh; A3=Marpoyan Damai; A4=Payung Sekaki; A5=Pekanbaru Kota; A6=Rumbai Pesisir; A7=Rumbai; A8=Sail; A9=Senapelan; A10=Sukajadi; A11=Tampan; A12 =Tenayan Raya.

3.2.3. Developing the normalized decision matrix

Once the decision matrix has been developed, the next step was to develop the normalized decision matrix R, in which the elements are determined using:

$$r_{ij} = \frac{x_{ij}}{\sqrt{\sum_{i=1}^{m} x_{ij}^2}} \quad (2)$$

where i = 1, 2, 3,.... m; and j = 1, 2, 3,..., n

An example of the normalized matrix calculation (matrix R) for K_1 in Bukit Raya:

$$R_{1,1} = \left(\frac{A_{1.1}}{\sum X_1}\right) = \left(\frac{3}{9.644}\right) = 0.311$$

3.2.4. Developing the weighted normalized decision matrix

With the weight $w_j = (w_1, w_2, \ldots w_n)$, where w_j is the weight of the jth criterion and $\sum_{j=1}^{n} w_j = 1$, thus the weight normalisation of the matrix V is:

$$V_{ij} = W_{ij} * r_{ij} \quad (3)$$

$V_{1,1}$ = Criterion Weight * $R_{1,1}$

= 5 * 0.311 = 1.555 (this calculation for Criterion-1 in Bukit Raya District)

3.2.5. Determining the positive and negative ideal solution matrices

The positive ideal solution is denoted by the symbol A+, while the negative ideal solution is denoted by the symbol A-.

$$A^+ = \{(\max v_{ij} \mid j \in J), (\min v_{ij} \mid j \in J), \ i = 1, 2, 3, \ldots, m\}$$

$$= v_1^-, v_2^-, v_3^-, \ldots v_n^-$$

$$A^- = \{(\min v_{ij} \mid j \in J), (\max v_{ij} \mid j \in J), \ i = 1, 2, 3, \ldots, m\}$$

J = {j = 1, 2,.., n and J is the benefit criterion. J'= {j = 1, 2.., n} & J' is the cost criterion.

3.2.6. Calculating Separation

D^+ is the alternative distance (in the Euclidean perspective) from the positive ideal solutions

$$D_I^+ = \sqrt{\sum_{i=1}^{n} (vj^+ - vij)^2}, \text{ with } j = 1,2,3....,n$$

D^- is the alternative distance from the negative ideal solution. It is described as:

$$D_i^- = \sqrt{\sum_{i=1}^{n} (vj^- - vij)^2}\text{ , with j = 1, 2, 3,...., n}$$

v_{ij} = element of the weighted normalized decision matrix V
v_1^+= element of the positive ideal solution
v_1^- = element of the negative ideal solution
An example of the distance calculation for positive (D^+) and negative (D^-).

$$D^+ = \sqrt{\sum_{i=1}^{n} (vj^+ - vij)^2}$$

$$= \sqrt{(A_1^+ - V_{1.1})^2 + (A_2^+ - V_{1.2})^2 + ...A_9^+ - V_{1.9})^2}$$

$D^-$the alternative distance from the negative ideal solution, which is defined:

$$D^- = \sqrt{\sum_{i=1}^{n} (vj^- - vij)^2}$$

$$= \sqrt{(A_1^- - V_{1.1})^2 + (A_2^- - V_{1.2})^2 + ...A_9^- - V_{1.9})^2}$$

3.2.7. Calculating the relative proximity to the ideal positive solution

$$V_1^+ = \left(\frac{Di^-}{Di^+ + Di^-}\right) \text{ , } 0 \leq D_i^+ \leq 1 \qquad (4)$$

V_i^+ is the relative proximity from the Ith alternative to the positive ideal solution.

$$V_1 = \frac{2.153}{2.153+1.570} = 0{,}578$$

Table 4. Ranking of the Alternatives

Code	Score	Alternative
V7	0.780	Rumbai District
V6	0.691	Rumbai Pesisir District
V12	0.657	Tenayan Raya District
V11	0.578	Tampan District
V1	0.578	Bukit Raya District
V4	0.482	Payung Sekaki District
V2	0.475	Lima Puluh District
V3	0.469	Marpoyan Damai District
V8	0.412	Senapelan District
V9	0.391	Sail District
V5	0.378	Pekanbaru Kota District
V10	0.365	Sukajadi District

3.3. Recapitulation of the Potential Locations

Based on the recapitulation results of the spatial analysis (GIS) and TOPSIS, there are 3 districts with the same score, namely Rumbai, Rumbai Pesisir and Tenayan Raya.

Table 5. The Recapitulation of the Potential Locations for Landfill Site according to Districts

Rank	Spatial Analysis (SIG)	Score	Non-Spatial Analysis (TOPSIS)	Score
1	Rumbai District	91	Rumbai District	0.780
2	Rumbai Pesisir District	91	Rumbai Pesisir District	0.691
3	Tenayan Raya District	91	Tenayan Raya District	0.657
4	Payung Sekaki District	73	Tampan District	0.578

3.4. The Indicator of Land Use Change for Landfill-Related Decision Making

The considerations can be supported through physical observation of the urban land especially in terms of the trend of land use changes. The process of land use changes can be explained based on the map. After enactment of the regional division policy in 1987, there are 3 types of land use shifting patterns identified, namely concentrated in CBD; the linear and the leap frog pattern [2]. From the map below, it is shown that land use changes are not so massive and fast in the northern part of the City. Land use changes are still focused on the southern part of the city or in the south of Siak River. This can be caused by the distribution of primary functions of the City which are widely spread in South of the city thus affecting the acceleration of the physical residential area. The northern areas, namely Rumbai Bukit & Palas Sub-districts can be considered as the location for new landfill site.

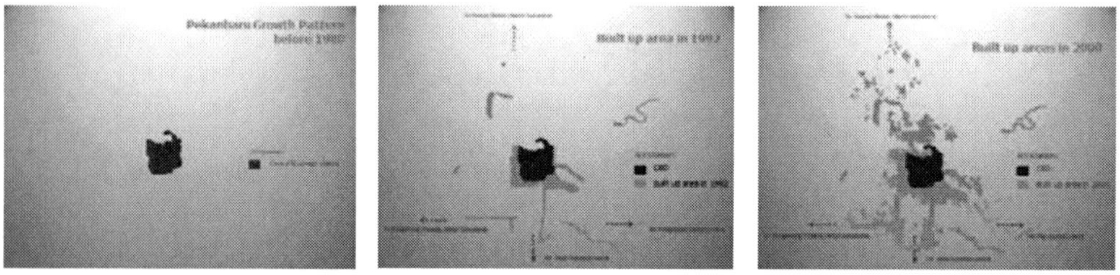

(1980) (1992) (2000)

Fig 2. The Process of Land Use Changes in Pekanbaru

4. Conclusions

Nowadays, GIS technology applied to find the best solution such as land use issues. GIS-based mapping methods can suggest a more specific location that GIS software & using digital maps can collect data, select and process spatial data through an information analysis. TOPSIS only ranks the values of criteria for each alternative, where all the alternatives are comprised of districts. As one of advanced engineering, GIS are recommended to solve a problems in determination of the ideal location of landfill site. We suggest to make cooperation between local government to provide a common landfill. This effort can reduce the problems of landfill site in fast growing cities.

References

1. Adeli, Z & Ali Mohammad K., Appl of geomorphology in urban planning: Case study in landfill site selection, Procedia Social & Behavioral Sciences **19**, p.662–667 (2011).
2. Malik, Hapriadi., *Physical Growth of Pekanbaru City after Expansion*; Thesis at Graduate School Gadjah Mada University Yogyakarta, unpublish, p.96 (2002).
3. Badan Pusat Statistik., *Pekanbaru in Figures*, BPS Pekanbaru (2012).
4. Borges, P.J., Fragoso, R., Garcia-Gonzalo, J., Borges, J.G., et al., Assessing impacts of Common Agricultural Policy changes on regional land use patterns with a decision support system; An Appl in Southern Portugal, Forest Policy & Economics **12** (2010).
5. Motamed. A, Moghimi., Application of Geomorphology in Urban Planning, (1998).

Development of two components acceleration time histories for Semarang, Indonesia, due to Semarang fault earthquake scenarios using 30 meters soil deposit model

Windu Partono

Civil Engineering Department, Diponegoro University, 50275 Semarang, Indonesia

Abstract. Development of surface acceleration time histories is important for dynamic analysis of structure design and evaluation. Acceleration time histories usually developed from seismograph records due to specific earthquake event. Following the research conducted by Team for Revision of Seismic Hazard Maps of Indonesia 2010 and 2016, Lasem fault and Semarang fault are two closest and dangerous shallow crustal fault earthquake sources which must be taken into account for seismic mitigation of Semarang. This paper presents the development two components surface acceleration time histories for Semarang caused by Semarang fault earthquake scenarios, with magnitude from 6 Mw to 7 Mw and maximum epicentre distance 15 Km. This research was performed by conducting deterministic hazard analysis, response spectral matching and site response analysis to obtain a pair of modified acceleration time histories. Site response analysis was performed by conducting 30 meters soil deposit model by taking the assumption that the position of bedrock elevation is 30 meters below the surface layer. Modified acceleration time histories were developed from a pair time histories (North-South/NS and East-West/EW direction) collected from worldwide historical earthquakes. Modified time histories were developed due to limited time histories data caused by Semarang fault earthquake source.

1 Introduction

Developement of surface acceleration time histories is important for dynamic structure analysis. Acceleration time histories usually developed from seismograph records due to specific earthquake event. Following the research conducted by Team for Revision of Seismic Hazard Maps of Indonesia (TRSHMI) 2010 and 2016, Lasem Fault and Semarang Fault are two closest and dangerous earthquake sources which must be taken into account for seismic hazard and seismic mitigation of Semarang. Lasem fault is a typical strike slip mechanism earthquake source however Semarang fault is a typical reverse mechanism seismic source. To evaluate the hazards of this city, all structure must be designed and evaluated using the ground motion developed from both earthquake sources. This paper presents the development of two components surface acceleration time histories for Semarang caused by Semarang fault earthquake source scenarios, with magnitude from 6

Mw to 7 Mw and maximum epicentre distance 15 Km. This research was performed by conducting deterministic hazard analysis, response spectral matching and site response analysis at 288 boring locations to obtain two components modified acceleration time histories at surface level.

Site response analysis was performed by conducting 30 meters soil deposit model by taking the assumption that the position of bedrock elevation is 30 meters below the surface layer and the shear wave velocity at bedrock elevation is 760 m/s. The soil deposit model was implemented following the same research conducted by [1]. This 30 meters model was implemented due to inadequate soil dynamic characteristics data below 30 meter soil elevation until bedrock elevation. Research on predicting bedrock elevation has already performed by [2] using single station feedback seismometer.

Modified acceleration time histories were developed from two components (a pair) time histories (North-South/NS and East-West/EW) directions and collected from worldwide historical earthquakes databases. All acceleration time histories used in this research were collected from worldwide historical earthquakes with similar mechanism to Semarang fault earthquake mechanism. Modified time histories were developed due to inadequate time histories data caused by Semarang Fault earthquake source.

This research was performed by implementing four major steps: 1) conducting geological, geophysical and soil investigation and analysis for developing soil profile, shear wave velocity profile and bedrock elevation 2) collecting a pair of acceleration time histories data (NS and EW) from worldwide historical earthquake records due to reverse mechanism shallow crustal fault source earthquakes with magnitude 6 - 7 Mw and maximum epicentre distance 15 km, 3) developing a pair of modified acceleration time histories by conducting response spectral matching analysis, 4) conducting shear wave propagation analysis using modified acceleration time histories and conducting 30 meters soil deposit model for obtaining surface acceleration time histories and surface spectra acceleration.

2 Experiment details

2.1 Geological and geotechnical investigations

Following the work conducted by TRSHMI 2010 for updating national seismic hazard maps, Lasem fault and Semarang Fault are two closest and dangerous seismic sources for Semarang. Those two seismic sources were identified to be the main shallow crustal fault source earthquakes that can significantly influence the hazard of the city. Fig. 1(a) shows the position of Semarang and Lasem fault trace.

Soil dynamic parameters of all soil layers above the bedrock elevation and bedrock elevation are two important parameters used for site response analysis. Prediction of bedrock elevation for Semarang was performed using single station feedback Microtremor [2 and 3] by conducting three component ambient vibrations investigation at 218 locations. Fig. 1(b) shows the contour map of bedrock elevation developed from 218 locations. Fig. 2(a) shows microtremor test equipment for bedrock elevation prediction measurements. Based on contour map of bedrock elevation, shear wave velocity test for rock samples were performed by [1] using Ultrasonic Pulse test. The rock samples were collected from soil boring investigations with maximum 40 meters depth. Fig. 1(b) shows five soil boring investigations for verifying bedrock elevation and collecting rock samples for shear wave velocity measurement test. Fig.2(b) shows Ultrasonic Pulse test equipment for predicting shear wave velocity of rock sample.

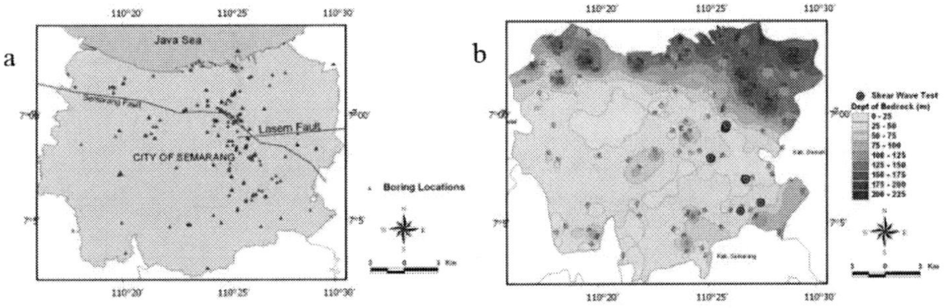

Fig. 1. Semarang fault and Lasem fault trace and boring locations (a) and contour map of bedrock elevation (b).

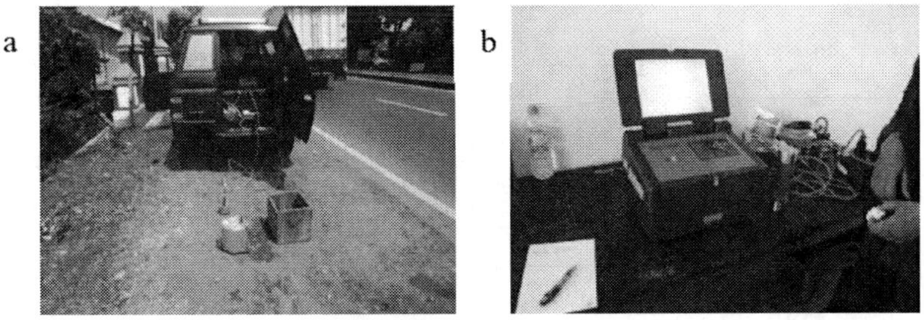

Fig. 2. Microtremor test (a) and Ultrasonic Pulse test (b).

2.2 Acceleration time histories data collection

Due to limited earthquake records of Semarang fault earthquake, historical earthquake records with magnitude from 6 to 7 Mw and maximum distance 15 km were collected from worlwide databases. All acceleration time histories were collected for reverse mechanism shallow crustal fault earthquake events. Five selected earthquakes (Iwate Japan 2008 magnitude 6.9 Mw, Chuetsu-oki Japan 2007 magnitude 6.8 Mw, Northridge-01 California 1994 magnitude 6.69 Mw, Chi-Chi Taiwan 1999 magnitude 6.2 Mw and Northridge-02 California 1994 magnitude 6.05 Mw) were used in this research for site response analysis. All acceleration time histories data (pair of NS and EW directions) were collected from Pasific Earthquake Engineering Research (PEER) NGA-West 2 Databases. Table 1 shows all acceleration time histories from 5 different earthquake events used in this study.

2.3 Response spectral matching

All acceleration time histories used in this study need to be matched with the target spectrum of Semarang fault earthquake scenarios. All target response spectrums were developed using deterministic hazard analysis by conducting three different attenuation functions proposed by [4, 5 and 6]. The target spectrums used for spectral matching analysis were predicted at bedrock elevation. All target spectrums were calculated for reverse mechanism shallow crustal fault earthquake sources.

Response spectral matching analysis was conducted following the same method proposed by [7] and producing modified (matched) acceleration time histories. The matched acceleration time histories were predicted using all acceleration time histories (fifteen pairs of N/S and E/W time histories) produced by 5 (five) earthquake events. Initial time histories must have similar shape with matched time histories. Fig. 3(a) shows an example result of spectral matching calculated for Chi-Chi Taiwan (EW direction) with magnitude 6.2 Mw and epicentre distance 5.93 Km. Fig.3(b) shows initial and matched time histories for Chi-Chi Taiwan (EW direction) with magnitude 6.2 Mw and epicentre distance 5.93 Km.

Table 1. Earthquake sources data used in this study.

Event	Station	M (Mw)	R (km)
Northridge-02 (1994)	Arleta - Nordhoff Fire Sta	6.05	1.48
	Newhall - Fire Sta	6.05	7.36
	LA - Century City CC North	6.05	18.34
Chi-Chi, Taiwan-03 (1999)	TCU084	6.2	3.68
	TCU089	6.2	5.93
	TCU076	6.2	13.04
Northridge-01 (1994)	Arleta - Nordhoff Fire Sta	6.69	3.3
	Beverly Hills - 14145 Mulhol	6.69	9.44
	LA - Brentwood VA Hospital	6.69	12.92
Chuetsu - oki, Japan (2007)	Nagaoka	6.8	3.97
	Kashiwazaki City Takayanagicho	6.8	10.38
	Yan Sakuramachi City Watershed	6.8	12.98
Iwate, Japan (2008)	IWTH24	6.9	3.1
	IWT011	6.9	8.41
	Kurihara City	6.9	12.83

M = seismic magnitude R = epicentre distance

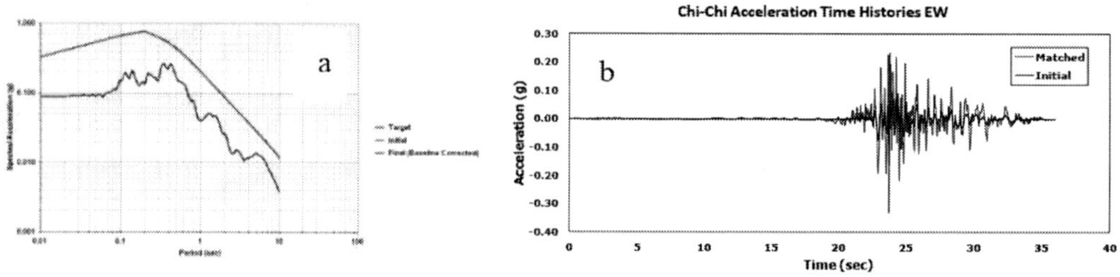

Fig. 3. Spectral matching result (a) and acceleration time histories (initial and matched) EW direction (b) calculated for Chi-Chi Taiwan earthquake 6.2 Mw and epicentre distance 5.93 Km.

2.4 Site response analysis

Site response analysis was performed at 288 positions for obtaining surface time histories and surface response spectra. Site response analysis due to a specific earthquake was performed using NS and EW matched time histories calculated from spectral matching analysis. Site response analysis was implemented using 30 meters soil deposit model, an alternatif model used to reduce the problems related with soil characteristics data for deep bedrock elevation. Site response analysis was implemented by conducting equation (1) where 'ρ', 'G' and 'η' are soil density, shear modulus and soil viscosity repectively ([8], [9] and [10]).

$$\rho\frac{\delta^2 u}{\delta t^2} = G\frac{\delta^2 u}{\delta z^2} + \eta\frac{\delta^3 u}{\delta z^2 \delta t} \quad (1)$$

As it can be seen on Fig. 1(b) the bedrock elevation on the center and northern part of the study area were predicted more than 50 meters depth. For building construction and foundation design, soil boring investigations on the northern part of the study area usually performs until 60 meters depth. It will be spent alot of time and money to performs boring investigation until more than 200 meters depth. To reduce the problems related with soil properties for deep bedrock elevation, site response analysis in this study was performed using 30 meters soil profile model. Fig. 4 shows 30 meters soil profile model used for site response analysis.

To obtain surface spectra acceleration which can be used for structural dynamic analysis, two spectra acceleration (SA) EW direction and NS direction should be modified to produce SA_{total}. SA_{total} can be calculated using equation (2) (Goleshorki, R. 2013 [11]) where SA_{EW} and SA_{NS} are two spectra acceleration (EW and NS directions) calculated from two acceleration time histories EW and NS directions,

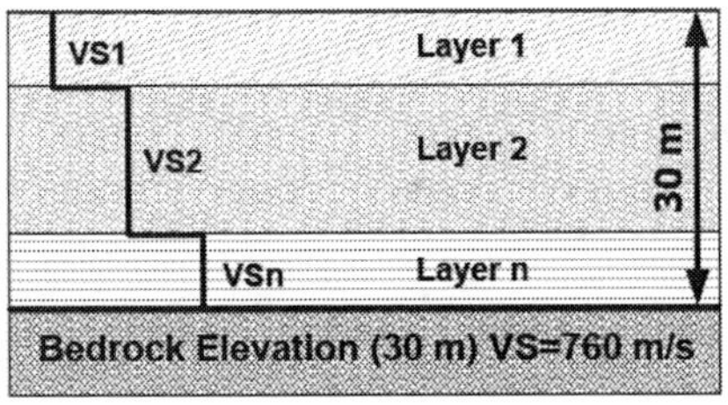

Fig. 4. Soil profile model for site response analysis.

$$SA_{total} = \sqrt{SA_{EW}^{\ 2} + SA_{NE}^{\ 2}} \tag{2}$$

3 Results and discussions

Based on response spectral matching and site response analysis conducting for fifteen acceleration time histories, fifteen pair (NS and EW directions) surface acceleration time histories were developed for this study area. All fifteen pair surface acceleration time histories can be used for dynamic analysis of structures. Fig. 5 shows two examples surface acceleration time histories (EW and NS directions) developed from Northridge-01 earthquake event with magnitude 6.69 Mw and epicentre distance 3.3 Km. Fig. 6 shows two examples surface acceleration time histories (EW and NS directions) developed from Chi-Chi Taiwan earthquake with magnitude 6.2 Mw and epicentre distance 3.86 Km.

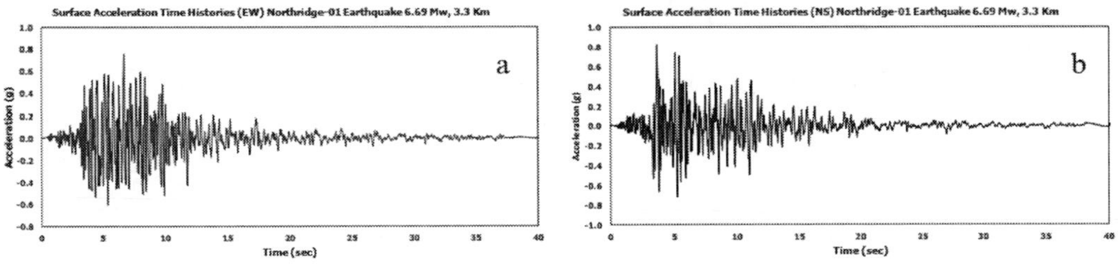

Fig. 5. Surface acceleration time histories EW direction (a) and NS direction (b) for Northridge-01 earthquake with magnitude 6.69 Mw and epicentre distance 3.3 Km.

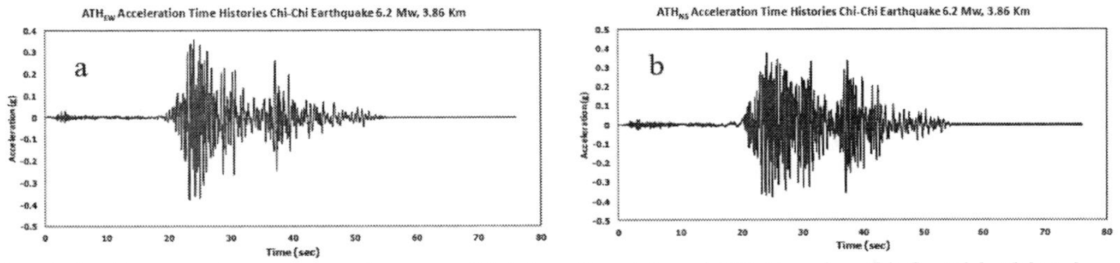

Fig. 6. Surface acceleration time histories EW direction (a) and NS direction (b) for Chi-Chi Taiwan earthquake with magnitude 6.2 Mw and epicentre distance 3.86 Km.

4 Conclusions

Two components surface acceleration time histories for Semarang due to Semarang fault earthquake with magnitude from 6.05 Mw to 6.9 Mw and epicentre distance maximum 15 Km has already developed. All acceleration time histories was developed using 30 meters soil deposit model. The model was implemented to reduce the problems related with soil investigation and soil dynamic parameters at the northern part of the study area. Fifteen pair surface acceleration time histories were developed following two major analysis response spectral matching and site response analysis.

Due to limited information and data related with original or initial acceleration time histories caused by Semarang fault earthquake all acceleration time histories were collected from PEER NGA-West 2 Databases. Five selected earthquakes (Iwate Japan 2008 magnitude 6.9 Mw, Chuetsu-oki Japan 2007 magnitude 6.8 Mw, Northridge 01 California 1994 magnitude 6.69 Mw, Chi-Chi Taiwan 1999 magnitude 6.2 Mw and Northridge 02 California 1994 magnitude 6.05 Mw) can be used for structural dynamic analysis and design for the study area.

References

1. W. Partono, *Seismic Microzonation Map Design for Semarang City by Developing Seismic Hazard Program and Adjusting Building Fragility Condition*, Doctor Disertation Civil Engineering Department Diponegoro University, (2015)
2. S.P.R. Wardani, M. Irsyam, W. Partono and S. Maarif, *Bedrock elevation measurement using ambient vibrationsand ultra-sonic pulse test*, Geotechnics for Castratophic Flooding Events – iai (ed), Taylor and Francis Group, London, (2015)
3. W. Partono, S.P.R. Wardani, M. Irsyam and S. Maarif, "Development of seismic microzonation maps of Semarang, Indonesia, Jurnal Teknologi", **77:11**, 99-107 (2015)
4. D.M. Boore and G.M. Atkinson, Earthq. Spec., Vol. **24**, No. 1, 99-138, Earthq. Eng. Res. Inst. (2008)
5. K.W. Campbell and Y. Bozorgnia, Earthq. Spec., Vol. **24**, No. 1, 139-171, Earthq. Eng. Res. Inst. (2008)
6. B.S.J. Chiou and R.R. Youngs, *NGA Model for Average Horizontal Component of Peak Ground Motion and Response Spectra,* (PEER 2008/09, Pacific Engineering Research Centre, College of Engineering, University of California, Berkeley, 2008)
7. N.A. Abrahamson, *Non Stationary Spectral Matching Program RSPMATCH>PG&E*, (Internal Report, 1998)
8. W.D. Iwan, *On a Class of Models for the Yielding Behaviour of Continuous and Composite Systems*, (Journal of Applied Mechanics, ASME, **34**, New York, 1967), 612-617
9. Z. Mroz, *On the Description of Anisotropic Work Hardening*, (Journal of Mechanics and Physics of Solids, **15**, Elsevier, Rhode Island USA, 1967), 163-175
10. J.P. Bardet and T. Tobita, *NERA A Computer Program for Nonlinear Earthquake Site Response Analysis of Layered Soil Deposits*, (Department of Civil Engineering University of Southern California, 2001)
11. I.W. Sengara, M.A. Yulman and A. Mulia, "Seismic Time-History Ground-Motions for a specific Site in Jakarta, Jurnal Teknologi", **77:21**, 127-136 (2015)

MATEC Web of Conferences **159**, 01044 (2018) https://doi.org/10.1051/matecconf/201815901044
IJCAET & ISAMPE 2017

Spatial Studies and juridical utilization of vacant land and abandoned land control in efforts of regional authority in semarang city

Sawitri Subiyanto, *Fauzi* Janu Amarrohman

Geodesy Departemen Faculty of Engineering, Diponegoro University. Jl. Prof. Soedarto, SH, Tembalang, Semarang, ZIP code 50275, Indonesia

Abstract. In the constellation of regional autonomy, regional land management problems that substantially contains the local authorities in the land sector is still not getting the understanding and implementation of a satisfactory response. These issues are then more prominent at the time of issue of PP 11/2010 on Control and Utilization of Abandoned Land. In accordance with the description above, this study aims to conduct a study Spatial And Juridical Vacant Land Utilization and Abandoned Land Control in Semarang City. Spatial information about vacant land and abandoned land information is required. In identifying vacant land used high-resolution imagery that QuickBird Satellite Imagery in 2010 and Pleides for 2016. Juridical studies carried out on the abandoned land to find out the status of the land and how many years of neglect by using secondary data from BPN and equipped with field surveys. For data processing was performed using methods of remote sensing and Geographic Information System. The results of this study can be used to determine how widespread availability of vacant land and abandoned land including amendments as an indicator of the direction of regional growth. Statistical testing conducted to determine the correlations between vacant and abandoned land Semarang.

1 Background

The authority of the utilization affairs tasks and problem resolution vacant land andabandoned land can not be implemented fully by the majority of the district / city. Land abandonment in urban areas is the result of multi-dimensional drivers of which some are relevant at the local scale [1], while others at regional or national scales [2]. If ever there was the district / city that has tried to implement this authority is still limited to set up a team or committee that is dealing with this issue, so it is still a decree establishes the establishment, not operationalized. Though the authority organizing this affair clearly been mentioned in Annex I of Government Regulation No. 38 of 2007 that include:

a. Iventory and identification of vacant land for the use of seasonal food crops.
b. Determination of parcels of land as vacant land that can be used to season food crops together with other parties under the agreement;
c. The determination of the parties that need the land for crops with emphasis on the local community;
d. Facilitation of cooperation agreements between holders of land rights to those who will use the land in front of / known by the village chief / headman and local district with an agreement for the two planting seasons;

e. Handling issues that arise in the utilization of vacant land if one party does not meet the obligations in the agreement.

These issues are then more prominent at the time of issue of PP 11/2010 on Control and Utilization of Abandoned Land. In this legislation no mention of the terminology of vacant land, but the land that indicated neglected and abandoned land. In this case that meant the abandoned land is land that has been granted the right by the state in the form of HM, HGU, HGB, HP and HPL or basic control of land, which is not cultivated, not used, or not used in accordance with the circumstances or the nature and purpose of basic rights or mastery.

In the context of economic, land abandonment has resulted in missed opportunities to bring in the economic potential of the land. As an illustration of the opportunity lost in the lands abandoned due to land disputes and conflicts are as follows:

"The BPN RI data show that the number ofagrarian disputes in Indonesia until 2007, there were 4,581 cases; as many as 858 cases of agrarian conflicts; and agrarian cases pending in the courts as much as 2,052 cases. In terms of area, productive land object of the dispute that can not be exploited and used in an optimal area: 607 886 ha. The economic value of the land being disputed by: 6.07886 billion m2 x Rp. 15.000, - (SVTO lowest land) = Rp. 91,182,900,000,000, - While forecasts opportunity lost from the land that is not utilized due to the status of the dispute reached 146.804 trillion rupiah (exceeds state budget)[3].

The argument above shows that there is a crucial issue of vacant and abandoned land that can not be resolved. Spatial and juridical information on vacant land and abandoned land and land prices as a factor information reinforcement is needed in order to empower and control the implementation of the regional authority efforts in the area of land and food resilience Semarang.

2 Problem Formulation

The formulation of the problem in this research there are:

1. What are the contributing factors to the empty and abandoned land as well as enforcement efforts and utilization of abandoned land?
2. What is the estimated opportunity lost of vacant land and abandoned land that is not utilized it in 2016 at Semarang?

3 Data / Material research

Data that is used in data processing, among others:

1. Spatial Data
   a). Quickbird image was corrected in 2010
   b). Pleides image was corrected in 2016
   b). Administrative map of Semarang
2. Non-Spatial Data
   Judicial land data 2016 were obtained from Badan Pertanahan Nasional.

4 Research Areas Description

This research study area is in the city of Semarang, Central Java, Republic of Indonesia. Semarang city is the capital of Central Java Province is the only city in Central Java, which can be classified as a metropolitan city. As the provincial capital, Semarang be the parameters of progress other cities in Central Java Province. Semarang city development progress can not be separated from the support areas in the vicinity, such as the City

Ungaran Demak regency, Salatiga and Kendal. Land use in the city of Semarang from year to year changes that lead from agriculture to non-agriculture, this is a natural phenomenon of the development of the city. The proportion of land use in the city of Semarang's largest is the type of land use for settlements (33.12%), this suggests that the land still has a dominant function as a domestic service. Based on the land use map shows that the distribution of land use settlements were spur-lane main roads located mainly in the city center. The large proportion of residential land area indicate the magnitude of the demands of public service and this proves that the city of Semarang totally urbanized.

GeographicallySemarang City located between 6°50' -7°10 'LS and 109°35'- 110°50' BT with an area of 373.70 km2 with boundaries as follows:

1. North Boundary : Java Sea
2. South Boundary : Semarang District
3. East Limit : Demak
4. West Limit : Kendal.

Semarang city composed of 16 districts and 177 villages with a total area of 373.7 km2 with a population of 1,351,246 inhabitants. Districts that have the most extensive area that is subdistrict Mijen (62.15 km2) while the districts with the smallest area is subdistrict Candisari (5.56 km2). Semarang altitude varies, between 0.75 meters to 348.00 meters above the shoreline. The study area shown in figure 1.

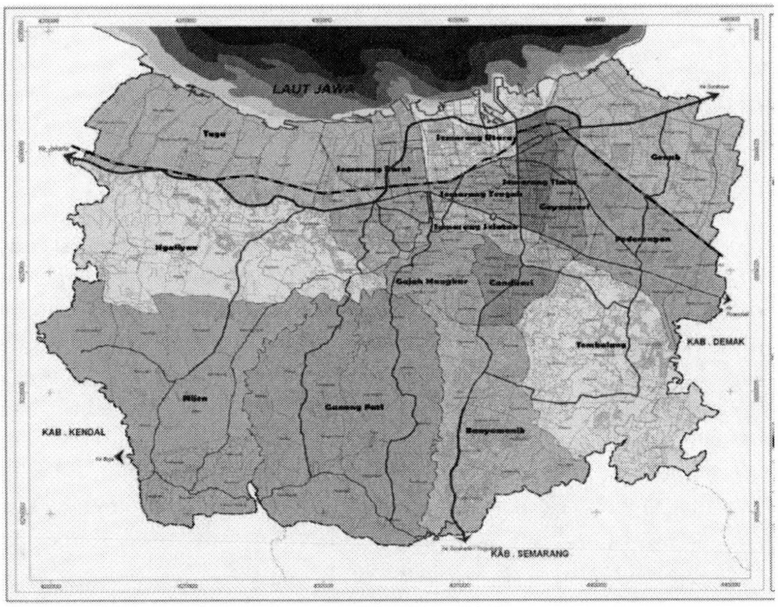

Fig. 1. Administration Map dan Semarang City Land Use

5 Implementation Research

Spatial data used are of high-resolution satellite imagery was corrected in 2016, administrative maps to identify the availability of vacant land and abandoned land that has the potential to be built into a variety of land use. Classification of vacant land and abandoned land in question includes vacant land and abandoned land that are not used are already reserved. Classification method wasteland and abandoned land uses high-resolution image interpretation methods and methods to assess changes in the availability of vacant land and abandoned land 2016. Phase of the research shown in the flowchart in figure 2 below.

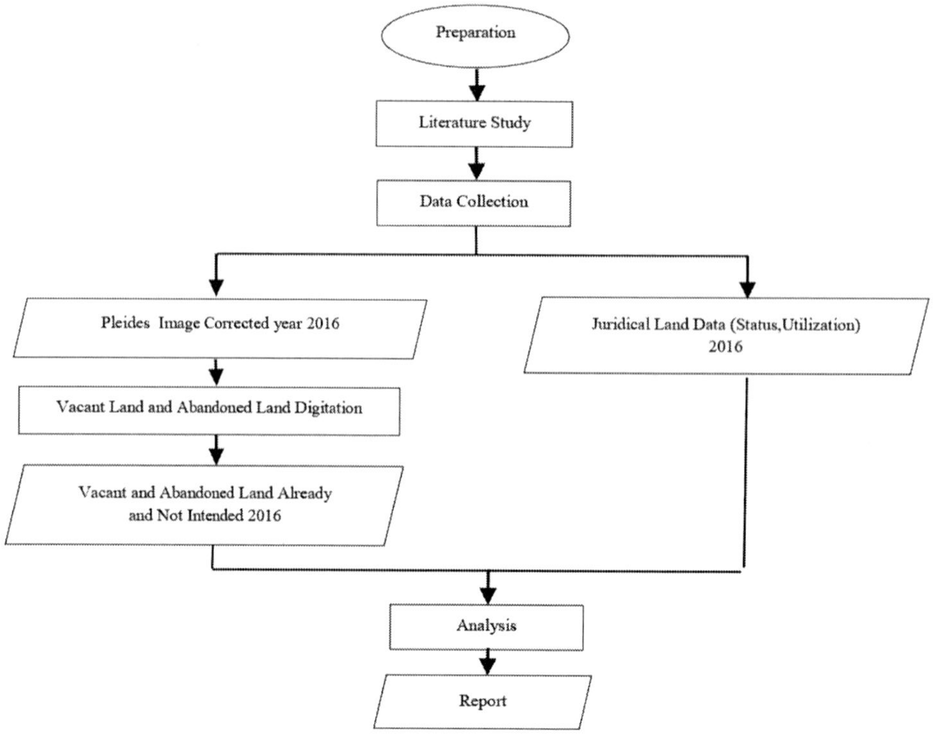

Fig. 2. Research Flowchart

5.1 Early Stage Data Processing

This research data processing carried out stages of digitization on the satellite image and data processing the availability of vacant land and abandoned land of Semarang. The method used to identify vacant land and abandoned land is the visual interpretation and digitization process Imagery in Pleides image which has been corrected geometric with Quickbird Image later in the interpretation of visually by its characteristic appearance of the surface of the land use and digitization to get the spread of vacant land and abandoned land. The use of scenes encompassing the beginning of the vegetation growing season enhance the accuracy of abandoned land classification [4]. Visual interpretation aims to distinguish one theme to another theme based on the appearance depicted in the image. The parameters used in the visual interpretation is the hue or color, the roof shape, size, texture, associations. The digitation of vacant land shown in figure 3.

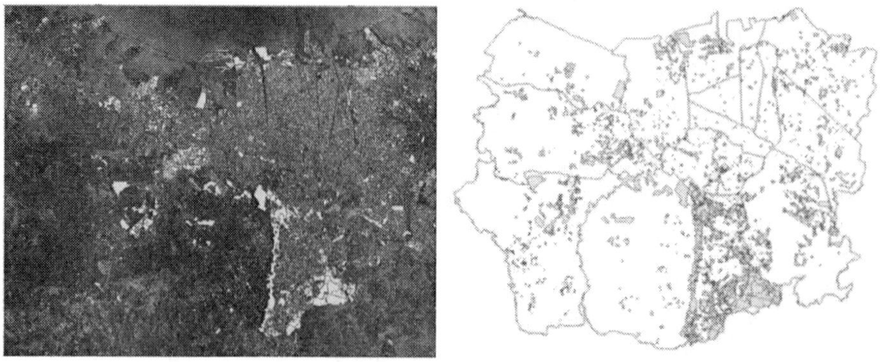

Fig. 3. Vacant land digitation result

The size of vacant land shown in Table 1 below.

Table 1. Vacant land in each district

No	District	Size (Ha)	No	District	Size (Ha)
1	Tugu	3100,272	9	Ngaliyan	4493,430
2	Tembalang	4145,276	10	Mijen	5381,080
3	Semarang Utara	1427,463	11	Gunungpati	6148,800
4	Semarang Timur	561,732	12	Genuk	2733,250
5	Semarang Tengah	535,296	13	Gayamsari	643,487
6	Semarang Selatan	614,525	14	Gajah Mungkur	941,386
7	Semarang Barat	2416,450	15	Candisari	661,336
8	Pedurungan	2198,652	16	Banyumanik	3149,286

The graphic of vacant land in Semarang shown in figure 4 below.

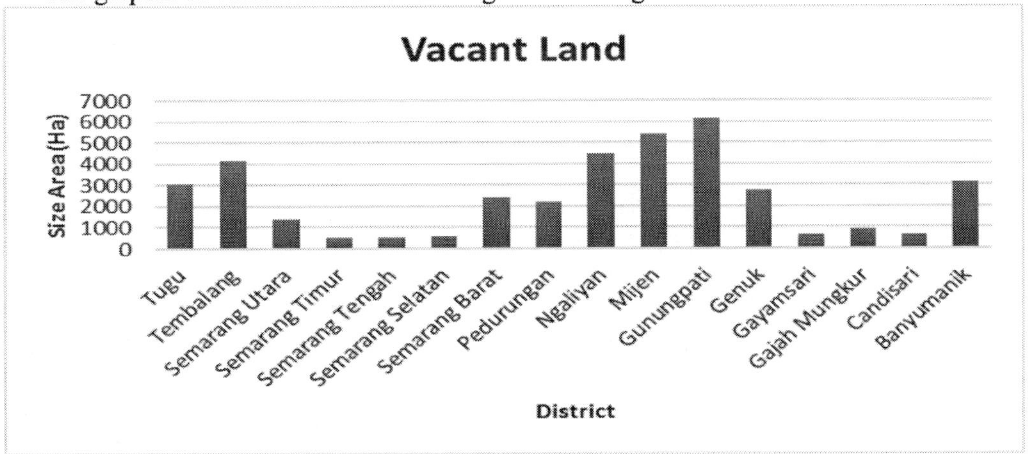

Fig. 4. Graphic of vacant land in Semarang

6 Basic Consideration of Issuance of Government Regulation No. 11 Year 2010

Government Regulation No. 11 of 2010 issued by the main considerations, first, that the condition of land abandonment is getting a greater gap between the social, economic, and welfare and the quality of the environment; Second, the regulatory instrument in the form of legislation that already exist, namely Regulation No. 36 of 1998 on Control and Utilization of Abandoned Land along with its implementing regulation can no longer be used as a reference completion of demolition and utilization of abandoned land. But politically there is actually a greater purpose and spacious and located with the issuance of Regulation No. 11 Year 2010 is to support the successful implementation of the National Agrarian Reform Program. Where the abandoned land has been set as one of the objects of Agrarian Reform in the State land beside former West Rights and / or Swapradja and other state land, State land, the former HGU and HGB that have expired period of validity of the right and was not renewed; as well as former State Land Forest Area.

6.1 Abandoned Land Definition

In the government regulated is not clearly regulated formulation of what is the definition of abandoned land. Both in the General Provisions Article 1, which usually describes the terms used in the subsequent chapters. As a comparative understanding of abandoned land can be found in Article 1 point 5 of Government Regulation No. 36 of 1998 which has been revised, the land abandoned by the holders of land rights, the holder of management rights, or those who have obtained a basic land tenure [permission / decision / certificate from the competent authority is the basis of tenure] but has not yet obtained the right to land in accordance with the legislation in force. Deeper explanation of the meaning of the abandoned land, the explanation can be found in Article 1 of Regulation No. 11 of 2010, that is, if the land is: a. not sought; b; not used or c. not used according to the circumstances or the nature and purpose of their rights. d. basic mastery of the existing soil when the soil:a. not sought; b; not used or c. not used according to the circumstances or the nature and purpose of their rights. d. basic mastery of the existing soil when the soil: a. not requested right; not sought; or c is not used; d. or not used in accordance with the requirements or conditions stipulated in the location permit, the decree granting the right, the decree release of forest area, and / or in the permit / decision / other letters from the authorities.

6.2 Criteria Object of Abandoned Land

PP No. 11 of 2010 that is the new object abandoned land determined as specified in Article 2. Then the criteria defined in Article 3 of land that does not include the control object abandoned land are land HM or HGB on behalf of individuals who inadvertently not used according to the circumstances or the nature and purpose of their rights; and land held by the government, either directly or indirectly and had the status of whether or not the status of State/Regional inadvertently not used according to the circumstances or the nature and purpose of their rights. Land rights of individuals who form HM or HGB is inadvertently not used according to the circumstances or the nature and purpose of his rights because the rights holder does not have the ability to undertake economic terms, use, or utilize. While specific land controlled by the government *(land assets of the Government*) are not intentionally used in accordance with the circumstances or the nature and purpose of their rights, due to the limited budget of the State / region to promote, use, or utilize.

Thus, it can be deduced, if land is HM or HGB individual or land controlled by the government (*Land Asset government*), the holder of rights has the ability to economically and/or Agency (Center/Regions) have a sufficient budget to promote, use, or use the land according to the situation or the nature and purpose of its rights, it will increase the status of previously abandoned land is not an object, it will be the object of curbing the abandoned land.

References

1. Pointereau, P., Coulon, F., Girard, P., Lambotte, M., Stuczynski, T., Sanchez Ortega, V., Del Rio, A. *Analysis of Farmland Abandonment and the Extent and Location of Agricultural Areas That Are Actually Abandoned or Are in Risk to Be Abandoned*. In: Anguiano, E., Bamps, C., Terres, J. (Eds.), Institute for Environment and Sustainability, Joint Research Centre, European Commision (2008)
2. Terres, J.-M., Scacchiafichi, L.N., Wania, A., Ambar, M., Anguiano, E., Buckwell, A., Coppola, A., Gocht, A., Källström, H.N., Pointereau, P. *Farmland abandonment in Europe: identification of drivers and indicators, and development of a composite indicator of risk*. Land Use Policy **49**, 20–34 (2015)
3. Badan Pertanahan Nasional, *Penelitian Penetapan Harga Dasar Tanah di Perkotaan*, Diktat Puslitbang BPN, Jakarta (2009)
4. Prishchepov, A.V., Radeloff, V.C., Dubinin, M., Alcantara, C., *The effect of Landsat ETM/ETM+ image acquisition dates on the detection of agricultural land abandonment in Eastern Europe*. Remote Sens. Environ **126**, 195–209 (2012b)

Development of collaboration in sustainable agribusiness cluster

Tomy Perdana[1], *Mahra* Arari H.,[1] *Fernianda* Rahayu H.[2], *Tetep* Ginanjar[2], *Ajeng* Sesy N.P.[2]

[1] Department Agro Socio-Economic, Faculty of Agriculture, Universitas Padjadjaran, Sumedang, Indonesia
[2] Agricultural Logistics and Supply Chain System (AGRILOGICS) Research Group, Faculty of Agriculture, Universitas Padjadjaran, Sumedang, Indonesia.

Abstract. There is tendency towards sustainable agricultural production to fulfill food security. It ensures that agricultural production is carried out in a sustainable way. Indonesia as developing country has rapid growing population. It will impact to the food security and food demand. That condition is challenging to develop sustainable agricultural production. Meanwhile, the sustainable agricultural production needs interaction and collaboration throughout the agricultural supply chain that involve multi-stakeholders. This study was conducted in Gemah Ripah 2 farmer group in Kuningan District which has collaboration activities with multi-stakeholders (i.e. Universitas Padjadjaran, Bank Indonesia region Cirebon, local government in Kuningan District, and Universitas Kuningan) in the development of sustainable agribusiness cluster. The sustainable agribusiness is a combination between animals (cattle, sheep, and fish), and plants (rice, shallot, chili, and secondary plants). Each subject on sustainable agribusiness (livestock and plants) is linkage and complement for each other. This study was aimed to analyze the potential of collaboration process in sustainable agribusiness cluster. Drama theory approach can illustrate the episodes that showed the dilemmas in the collaboration of sustainable agribusiness cluster, whereas a collaboration framework is used to initiatives for resolving dilemmas.

1 INTRODUCTION

Agricultural is critical for both economic growth and poverty alleviation and dominated by smallholders. Commonly, smallholders are often faced with several hamper, i.e., lack of basic information, inadequate market information, high operating cost, lack of business unit, limited value added to farm production in the rural areas, etc. Development sustainable in agribusiness is about a balancing farming system, social and environment issues, and financing access. Sustainable agribusiness is one of the developing techniques that is utilization of natural resources in agricultural system to fullfill people needs and also to maintain quality of ecological environment as well as it needs integration between livestock and agricultural that has linkage with all stakeholders. The relationship between stakeholders are pull factor in the agricultural and sustainable economic growth on rural areas (Pasandaran et al., 2006). In addition, Sustainable agribusiness is a phenomenon in developing agricultural supply chain, regard with it, the development of sustainable agribusiness in developing countries is exacerbated by poverty, leading to tremendous problems of employment, infrastructures and transportation, food supply chain, and environment protection (De Bon, et al., 2010).

[1]Corresponding author: tomy.perdana@unpad.ac.id/ mahra.arari@gmail.com
[2] tetepginanjar@gmail.com/ferniandarahayu@gmail.com/ ajengsesynurpratiwi@gmail.com

The application of sustainable agribusiness can be done by developing agribusiness cluster. Business clusters are complex and informally integrated systems of entrepreneurship, business skills, knowledge, research and development, innovation, technology, commercialization, production, quality management, marketing, international trade, productivity, investment, convergence, competitiveness, profitability, and growth. The sustainable agribusiness cluster is the integration between ecological agricultural and sustainable competitive advantage, emphasize the aspects of geographical and networks of stakeholders in agribusiness supply chain (Porter, 2000; Toma, 2009; Geldes et al., 2015) and deemed able to overcome the weakness of the smallholders in rural areas to respond the competition and globalization (Matopoulos et al., 2005). Meanwhile, the ecological agribusiness emphasis on the sustainable use of inters resources (e.g. ecological, economics, and social) rather than the use of external resources (Van't Hooft, 2013).

The agribusiness cluster is connecting farmers, local entrepreneurs (involved in supplier and channel value chains), financial services and business development services in the target region (Toma, 2009). In Addition, agricultural supply chain in developing countries are commonly involve multi-stakeholders i.e., farmers, supplier or trader, market, and other stakeholders (government, university, etc.). The collaboration is one of the social interaction that has several activities to achieve a goal with understanding and responsiblity of each part. Collaboration of sustainable agribusiness cluster development needs synergy, integrity, empathy, and enthusiasm from all stakeholders to gain a better idea in services innovation, collaboration action, and product image in consumers as well as each party having the same position.

Development collaboration in sustainable agribusiness cluster has several dilemmas that hamper their collaboration and should be eliminated the collaboration goals. To reduce the dilemmas, each party cannot depend only on rationality but also have to consider the influence of emotions, hidden agenda, and irrationality (Howard, 1999). Collaboration for agribusiness actors have two emotion sides, there are; positive emotion (empathy, integrity, enthusiasm, and synergy) and negative emotion (blaming, regret, anger, un-desire, and hostility) and those are strategies to know further and influence the preference from other supply chain actors. In regard to development of the understanding about emotions, in agribusiness collaboration needs to develop efforts to support the succession of the collaboration. According to O'malley (1998), sustainable value cannot be creating when each actor tries to achieve their own profit.

Based on that statement, Gemah Ripah 2 farmer group has already developed collaboration with Universitas Padjadjaran and other stakeholders. This study was aimed to analyze the potential of collaboration process in sustainable agribusiness cluster development between the main stakeholders. This study contribute to the current knowledge of sustainable agribusiness cluster by focusing on a specific sequence of episodes in which stakeholders interact to resolve different of interest and conflicts.

2 METHODOLOGY

The research was conducted in Gemah Ripah 2 farmer group, Japara village, Japara sub-district, Kuningan District trough qualitative descriptive survey. About 20 respondents was interviewed to grasp a deep understanding regarding the development of sustainable agricultural system in Gemah Ripah 2 farmer group, which has been collaboration with Universitas Padjadjaran, and local government. Key informant was interviewed to validate the data. The case was chosen due to the development of sustainable agribusiness process that is impacted several dilemmas between Gemah Ripah 2 farmer group, university, and local government. The data was analyzed and modeled by using confrontation Manager with drama theory approach.

Drama theory is an analysis about the importance of irrationality, preference change, emotion, and political aspects of choice in interdependent decision making situations. Drama theory uses dilemmas to analyze, predict, and understand the pressure on characters to change their positions, preferences, and common reference frame. Drama theory is a method to deal with paradoxes of rationality or dilemmas of interaction decision making include metagame and hypergame analysis (Bennett et al., 2001; Howard, 1971). Drama theory is used to analyze how a framework could be changed to the other framework in a series of episodes (Howard et al., 1992). Framework changing takes place due to the presence of dilemmas. Dilemmas happened if there are hampering to achieve an objective. The objective is reflected in the form of position (open future scenario) that tries to convenience other stakeholders to accept the position, with promise or threat if necessary.

3 RESULT AND DISSCUSION

3.1 Sustainable Agribusiness Cluster in Kuningan District

Sustainable agribusiness cluster in Kuningan District-West Java has been developed since 2015. The development of the cluster is the result of collaboration between Gemah Ripah 2 farmer group and Universitas Padjadjaran supported by Bank Indonesia Cirebon region, local government of Kuningan district, and Universitas Kuningan. The main actors of this collaboration are Universitas Padjadjaran as academicians and researcher, Gemah Ripah 2 farmer group as an implementation object, and local government (including Bank Indonesia region Cirebon and local government (*Pemda*) in Kuningan District). Collaboration of sustainable agribusiness cluster implemented the integration farming system of agriculture (rice), fisheries (tilapia and nilem fish), and livestock (cattle and padjadjaran sheep).

The collaboration is aimed to increase competitive advantage and value added, hereinafter, decreasing poverty and increasing welfare is a goal of this collaboration. The concept of this sustainable agribusiness cluster is to serve high quality products for consumers so that farmers and market can get highest profit. On the other hand, the development of collaboration was set to achieve sustainability between multi-stakeholders and multi-commodities. The actors are collaborating by sharing the roles and resources to support the enterprise developed. Each actor has a support action such as field assistance, demonstrating plots, and visit study to improve the farming capacity of the farmer's community.

3.2 Dilemma Analysis

The concept is developing multi-commodities which performed several benefits, i.e., improving social benefit for farmers, improving economic and environmental advantages, improving knowledge both for farmers, university and local government in the implementation of sustainable agribusiness. Meanwhile, implementation of sustainable agribusiness cluster is inflicting positive and negative impact for stakeholders. The collaboration of sustainable agribusiness cluster faced several dilemmas for all stakeholders.

Every party in sustainable agribusiness cluster tends to achieve their own objectives. Farmers refused to implement sustainable agribusiness and prefer to implement traditional farming that did not give more benefits. Meanwhile, Universitas Padjadjaran as an academician needs support from several stakeholders in the implementation of sustainable agribusiness cluster. Universitas Padjadjaran should develop sustainable agribusiness cluster based on research and combined with environmental and social condition. Local government needs to ensure that this program is eligible with government program.

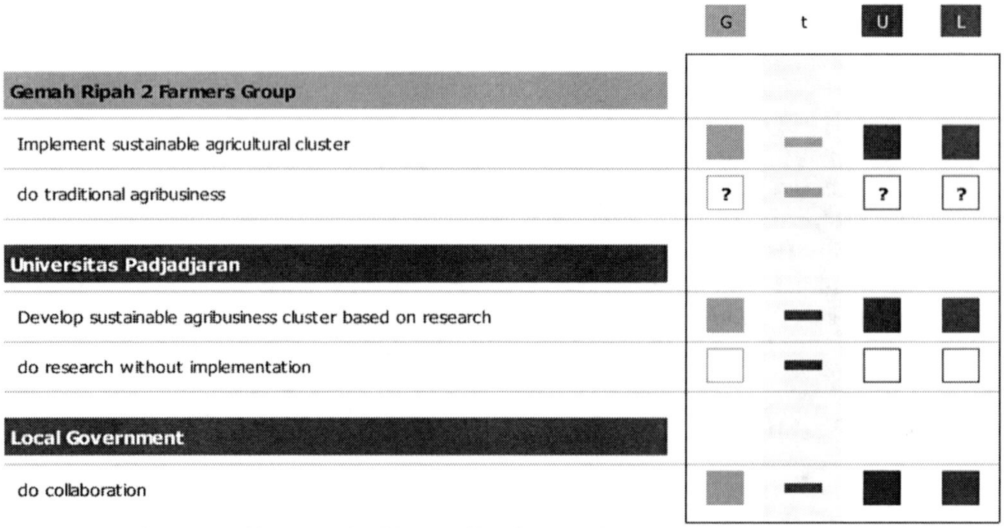

Figure 1. Option Board in Sustainable Agribusiness Cluster

Additional Information;
G – Gemah Ripah 2 Farmer Group **U** – Universitas Padjadjaran **L** – Local Government
: Adopted : Left open **t** – Threatened Future **:** Rejected **?** : Doubted

In regard to the collaboration, each party tries to issues their threatened future; Gemah Ripah 2 farmer group may or may not implement sustainable agribusiness cluster, because Gemah Ripah 2 farmer group doubted to Universitas Padjajdaran and local government. Universitas Padjadjaran will increase agribusiness system based on research and in the practice of sustainable agribusiness; Universitas Padjadjaran does assistance and facilitate to help farmers in the implementation of sustainable agribusiness. Furthermore, Local Government might do collaboration, it is because local government doubted to Universitas Padjadjaran in the implementation of sustainable agribusiness cluster continuously.

3.2.1 Gemah Ripah 2 Farmer Group

Gemah Ripah 2 farmer group faced cooperation dilemmas with respect to Universitas Padjadjaran and local government. Gemah Ripah 2 farmer group doubted that Universitas Padjadjaran and local government would assist and support farmers in the practice of sustainable agribusiness. Gemah Ripah 2 farmer group must gain trust from Universitas and local government that Gemah Ripah 2 farmer group is capable to implement sustainable agribusiness. Project course of action for Gemah Ripah 2 farmer group are all stakeholders should agree to Gemah Ripah 2 farmer group position, this group would incur in carrying out these commitments and inevitably carry them out. Gemah Ripah 2 farmer group should apply positive emotion towards Universitas Padjadjaran and local government. By convincing Universitas Padjadjaran and local government that Gemah Ripah 2 farmer group position is better for Universitas Padjadjaran that might not only do research without implementation but also collaboration with farmer group. Gemah Ripah 2 farmer group might show the advantages to Universitas Padjadjaran of sticking to Gemah Ripah 2 farmer group position is greater or more credible than universitas padjadjaran and local government supposes, to make it serious, Gemah Ripah 2 farmer group should apply positive emotion toward Universitas Padjadjaran and local government.

3.2.2 Universitas Padjadjaran

Universitas Padjadjaran has a dilemma in the implementation of sustainable agribusiness cluster in Kuningan District. Universitas Padjadjaran has trust dilemma with respect to Gemah Ripah 2 farmer group. Universitas Padjadjaran doubts Gemah Ripah 2 farmer's group would actually implement its proposals. Universitas Padjadjaran must make Gemah Ripah 2 farmer group trustworthy. Universitas Padjadjaran analyzes Gemah Ripah 2 farmer group concerns to see why, after agreeing to Universitas Padjadjaran position, Gemah Ripah 2 farmer group might do traditional agribusiness again and left sustainable agribusiness concept. Universitas Padjadjaran should show that Universitas Padjadjaran position is greater or more incredible that Gemah Ripah 2 farmer group supposes. Universitas Padjadjaran also needs to show the advantage to Gemah Ripah 2 farmer group of sticking to Universitas Padjadjaran's position are great or more credible, than Gemah Ripah 2 farmer group supposes. Project emotion of Universitas Padjadjaran should show mistrustful or neutral toward Gemah Ripah 2 farmer group.

3.2.3 Local Government

Local government has a dilemma towards Gemah Ripah 2 farmer group in the implementation of sustainable agribusiness. Local government has trust dilemma with respect to Gemah Ripah 2 farmer group. Local government doubted that Gemah Ripah 2 farmer group would actually implement its proposal. Local government must make Gemah Ripah 2 farmer group trustworthy. Local government analyzes Gemah Ripah 2 farmer group concerns to see why, after agreeing to local government position that Gemah Ripah 2 farmer group might do traditional agribusiness and refuse to implement sustainable agribusiness. Local government should show greater position or more credible than Gemah Ripah 2 farmer group supposes. Local government also needs to show the advantages to gemah Ripah 2 farmer group of sticking to local government's position are greater than Gemah Ripah 2 farmer group supposes. Local government should show mistrustful or neutral emotion toward Gemah Ripah 2 farmer group. This emotion is to ensure that Gemah Ripah 2 farmer group and local government will be capable to develop sustainable agribusiness by doing collaboration with Universitas Padjadjaran.

4 CONCLUSION

This paper describes about sustainable agribusiness cluster, and this collaboration between Gemah Ripah 2 farmer group, Universitas Padjadjaran, and local government. In regard to this paper, each party has dilemmas about implementation of sustainable agribusiness; the dilemmas are about trustworthy and cooperation dilemmas in the implementation of sustainable agribusiness. The multi-stakeholders who were involved in the collaboration of sustainable agribusiness cluster should develop better relationship, maintain communication between each actor, and the actors realize that this collaboration has long term relationship to gain more profit through collaboration in the development of sustainable agribusiness cluster. However, some parties are still doubt and lack of trust to the other party to implement the collaboration and show the position that might interfere the collaboration.

5 SUGGESTION

Development of sustainable agribusiness by involving multi-stakeholders should develop several strategies, i.e., development intensive communication between Gemah Ripah 2 farmer group, Universitas Padjadjaran, and local government. According to Duncan et al. (1998), communication in the development of collaboration is the foundation and critical element in the collaboration. Intensive communication should develop better relationship and improve trustworthy among parties. Discussion about the process of development should reduce the problem and build trustworthy. According to Handayani et al. (2015), develop supply chain relationship among actors is one of the important part to improve food supply and gain value co-creation.

Warm thanks to Ministry of Research, Technology, and Higher Education of the Republic Indonesia and Universitas Padjadjaran which founded the research through Iptek *Bagi Wilayah* (IbW) *Perguruan Tinggi* (PT) Local government-CSR with the tittle IbW "Development of agribusiness cluster with integration between vegetables and sheep padjadjaran" and Central bank of Indonesia Cirebon region, local government Kuningan district, and Universitas Kuningan as partner in the implementation of the IbW research in Kuningan District from 2016-2018.

6 REFERENCES

[1] Pasandaran E, Djajanegara A, Kariyasa K, and Kasryno F. Integrasi Tanaman Ternak di Indonesia, Badan penelitian dan Pengembangan Pertanian. Jakarta. (2006)

[2] De Bon, H., Parrot, L. and Moustier, P. Sustainable urban agriculture in developing countries. A review. *Agronomy for sustainable development*, **30** (1), pp.21-32. (2010)

[3] Porter, M. E. Location, competition, and economic development: Local clusters in a global economy. *Economic development quarterly*, **14** (1), 15-34. (2000).

[4] Toma E. Agribusiness Cluster – Between Theory and Practices, *Scientific Papers Management, Economic Engineering in Agriculture and Rural Development*. **Vol 9 No 3.** 58. (2009).

[5] Geldes, C., Felzensztein, C., Turkina, E. and Durand, A. How does proximity affect interfirm marketing cooperation? A study of an agribusiness cluster. *Journal of Business Research*, **68** (2), pp.263-272. (2015)

[6] Matopoulos, A., Vlachopoulou, M., Manthou, V. and Manos, B. A conceptual framework for supply chain collaboration: empirical evidence from the agri-food industry. *Supply Chain Management: an international journal*, **12** (3), pp.177-186.(2007)

[7] Van't Hooft, K. Livestock, the World, and the Dutch-Part II. *Institute for Agricultural Market Studies (IKAR) Russia Dairy Industry Profile http://www. ikar. ru/eng (retrieved)*. (2013)

[8] Howard, N. Confrontation Analysis: How to Win Operations rather than War.*CCRP, Department of defense, Washington, DC.* (1999).

[9] O' Malley, P. Value Creation and Business Success.*The Systems Thinker*, **9** (2). (1998).

[10] Bennet, P.G. Bryant, J. and Howard, N. 'Drama theory and confrontation analysis', in Rosenhead, J.V. and Mingers, J. (Eds.): *Rational Analysis for a Problematic World Revisited*, pp.225-248, Wiley, New York. (2001)

[11] Howard, N. Paradoxes of Rationality: Games, Metagames, and Political Behavior, MIT Press, Cambridge, MA. (1971)

[12] Howard, N., Bennett, P.G., Bryant, J.W., and Bradley, M. Manifesto for a theory of drama and irrational choice.*Journal of Operational Research Society*, **44**, 99-103.(1992).

[13] Duncan, Tom, and Sandra E. Moriarty. "A communication-based marketing model for managing relationships." *The Journal of marketing* : 1-13. (1998)

[14] Handayati, Yuanita, Togar M. Simatupang, and Tomy Perdana. "Value Co-creation in Agri-chains Network: An Agent-Based Simulation." *Procedia Manufacturing* **4**: 419-428. (2015)

Coordination and Collaboration Functions of Disaster Management Centers for Humanitarian Logistics: A Case Study at Merapi Eruption

Naniek Utami Handayani[1*], *Anita* Mustikasari[1]

[1] Department of Industrial Engineering, Faculty of Engineering, Diponegoro University, Semarang, Indonesia 50275

Abstract. Although the humanitarian logistics coordination model in Indonesia is fairly effective, the role of each player remains unclear. For example, some of the activities that are sufficient to be done by one agency are in fact done by other institutions as well. This has resulted in inefficiency and ineffectiveness regarding on the handling of disaster relief goods. Therefore, it is essential to determine the role of each stakeholder involved in the management of natural disasters in order to improve the model. This study was conducted to provide the proposed model of humanitarian logistics coordination for developing the framework of coordination between stakeholders in the management of natural disasters, with the involvement of logistics players through the RASIC Chart method.

1 Introduction

Disaster is a term that refers to a physical disorder that affects the system as a whole and threatens its priorities and goals [1,2]. The importance of disaster management should encourage the successful implementation of an emergency response effort that starts with strategic planning [2,3]. Disaster response is generally composed of four phases [4], namely mitigation, preparation, response, and recovery.

Indonesia is one of the countries located in the disaster zone. The requirement of an integrated natural disaster management system is satisfied by the disaster logistics management system. Logistics management of disasters requires support via information about the needs of humanitarian goods, a system that is able to quickly provide the information needs of victims, the distribution of the sacrificial site, and listing the logistics needs of disaster victims at the scene of the disaster as well at any evacuation point. In order to improve the performance of natural disaster relief operations, logistics and inventory management activities are necessary. The role of logistics systems is to provide the necessary relief items quickly and efficiently. Though the humanitarian logistics coordination model in Indonesia is fairly effective, the role of each player remains unclear. Coordination among relief agencies is an important aspect to be noted. Lack of coordination and cooperation among humanitarian agencies is one of the obstacles in humanitarian logistics [5]. Coordination and cooperation among organizations prevents

* Corresponding author: naniekh@gmail.com

huge losses in terms of resources and materials [6]. Coordination problems also occur in the handling of natural disasters in Indonesia. Currently, the management of natural disaster emergency responses involves the Disaster Mitigation Agency (BPBD), the Department of Social Services, Indonesian Army, Department of Public Works, Department of Health Services, Indonesian Red Cross, Department of Communication and Informatics, Department of Finance, Department of Education, Head District, SAR, NGOs, Indonesian Agency for Meteorological, Climatological and Geophysics, and volunteers. The model of collaboration and coordination between the stakeholders still has some drawbacks, such as emergency responses that do not involve those who are competent in the field of logistics.

This study was conducted to provide the proposed model of humanitarian logistics coordination for developing the framework of coordination between stakeholders in the management of natural disasters, with the involvement of logistics players. This framework is expected to be a reference in establishing the involvement of stakeholders and the roles of each party to improve the effectiveness and efficiency of the management of natural disasters. In this paper, deductive methods are used to map stakeholders which are tied to disaster management and clarified through the RASIC Chart.

2 Research Method

A RASIC Chart will be used in this research. It has functions for recognizing the rules and competences of each humanitarian logistics stakeholder in Indonesia in a process that ensures that ownership, approval, support, consultation, and communication responsibilities are assigned [6]. The advantages of a RASIC chart is ensuring an obvious relationship between activities and resources, emphasizing distinct competences for every resource, and providing suggestions for communication planning [7]. *Responsible* represents the stakeholders who are responsible for accomplishing the activity, while *approves* represents the stakeholders who have authority to take an action. *Support* means the stakeholders who offer resources. Stakeholders who have to know the outcome of the implementation of the activities are referred to as *informed*. Finally, *consulted* refers to stakeholders who adopt the implementation suggestions indirectly [8]. A RASIC chart will ensure stakeholders that are required to know do indeed know and stakeholders who are not required to know will not [9]. Data are compiled by interviewing and surveys about coordination among the Department of Social Services, Indonesian Army, Department of Public Works, Department of Health Services, Indonesian Red Cross, Department of Communication and Informatics, Department of Finance, Department of Education, Head District, SAR, NGOs, Indonesian Agency for Meteorology, Climatology and Geophysics, and volunteers through the BPBD Magelang branch.

3 Results and Analysis

3.1 Job Analysis of Each Stakeholder in Humanitarian Logistics

In order to improve the humanitarian logistics system in Merapi Eruption, it is better to begin by defining all the stakeholders of the system and also assessing their consciousness level regarding a natural disaster threat. For instance, the BPBD is defined as a government stakeholder, who has coordinator, practitioner and commander functions in disaster management and emergency. The coordination function of the BPBD involves it coordinating with all stakeholders in case there is no disaster. The BPBD uses its practitioner function when a minor disaster occurs, for instance a landslide in which one or two houses are damaged. When there is a major disaster, the BPBD has the function as commander to coordinate and collaborate with all stakeholders. The scope of the BPBD includes mitigation, preparedness, response, and recovery. The Department of Social Services is obliged to organize, provide technical guidance, and supervise the enforcement of the mitigation, preparedness, response, and recovery activities for natural disaster victims. The Indonesian Army/Police has the functions of helping the BPBD throughout the technical process with SAR in evacuate victims. The Department of Public Works has the responsibility for constructing infrastructure for access during and post disasters. The Department of Health Services focuses on handling the health factors of disaster risk management by providing medicines and medical personnel in disasters. Meanwhile, the Indonesia Red Cross has the responsibility of helping SAR and the Department of Health Services give first aid to disaster victims such as blood transfusion. The Department of Communication and Informatics has the responsibility of providing information during disaster while the Department of Finance has authority over issues of budget costs in all processes of the disasters phase. The Department of Education supports the education of people in disaster risk management. The Indonesian Agency for Meteorology, Climatology and Geophysics has the responsibility for providing information related to disaster risk including affected potential areas, while Head of District work in managing the logistics center and people during the disaster phase. NGOs and volunteers, being Non-Government institutions, supports all the stakeholders related to disaster risk management. All the stakeholders of the Humanitarian Logistics system should in particular have a similar level of awareness and involvement concerning the system and its capabilities. Learning and training play a vital role in order to create a disaster resilient community with an effective emergency management system, by helping to increase the community awareness. The cluster concept is another key factor that forms part of improving the operational effectiveness of an emergency management system. Clusters can be extremely helpful in establishing the lacking coordination between different actors of the emergency system in Merapi Eruption.

A RASIC chart is used for representing the role of each stakeholder concerning humanitarian logistics in Indonesia according to the four phases on disaster risk management: mitigation, preparedness, response, and recovery. The mitigation phase is an essential activity for local governments to identify and coordinate efforts of stakeholders to reduce and eliminate risks associated with disasters. The preparedness phase includes evolving urgency procedures and stakeholder competence to ensure an effective and efficient response that tackles the impacts of a disaster. The primary objective of an effective response to a disaster is rescuing people and reducing damage so that aid is provided during or immediately following a disaster in order to care for the people affected. The recovery phase consists of rehabilitation in a short period and reconstruction in a long period, with the purpose of repairing important support systems and restoring life back to normal. The measurement indicators of disaster risk management phases are shown in Table 1 and were adopted from [10].

In the mitigation phases, BPBD has a function as coordinator of all stakeholders. The Department of Finance has the functions of analyzing and giving approval to all measurement indicators, which includes the budget cost. In addition, developing a master plan for disaster risk management and mitigation is supported by all stakeholders excluding Head District. Head District only gets result information from BPDB. While, developing a master plan for disaster risk management become measurement indicator that supported by almost all stakeholders except Dept. of Communication and Informatics as approves and District Head as informed.

Table 1. RASIC Chart on The Mitigation Phase

R-Responsible *A-Approves* *S-Supports* *I-Informed* *C-Consulted* *NA-Not Available*	**Organization**	Disaster Mitigation Agency (BPPD)	Dept. of Social Services	Indonesian Army/Police	Dept. of Public Works	Dept. of Health Services	Indonesian Red Cross	Dept. of Communication and Informatics	Dept. of Finance	Dept. of Education	District Head	SAR	NGOS	Volunteers	Indonesian Agency for Meteorology, Climatology and Geophysics
Measurement Indicators		**Mitigation Phase**													
Training and education on disaster risk management		R	S	I	NA	NA	NA	I	A	S	R	R	C	S	R
Analyzing risks to measure the potential areas for disasters		R	I	NA	I	NA	NA	NA	A	NA	R	NA	C	NA	R
Zoning and land use controls		R	NA	NA	R	NA	NA	NA	NA	NA	R	NA	NA	NA	R
Insuring roads and bridges to reduce the financial impacts of disasters		NA	NA	NA	S	NA	NA	NA	A	NA	NA	NA	NA	NA	NA
Developing a master plan for disaster risk management		R	S	S	S	S	S	S	A	S	I	S	S	S	S
Developing disaster risk management information system		R	S	NA	NA	NA	S	R	NA	NA	NA	NA	S	S	NA
Developing engineering design standards for resilient roads and bridges		R	S	S	R	NA	NA	NA	A	NA	I	NA	NA	NA	NA
Providing timely and effective information related to disaster risk		R	S	NA	NA	NA	S	R	NA	NA	S	NA	NA	NA	R
Constructing retarding basins, barriers, culverts, levees, and drainage		R	NA	S	R	NA	NA	NA	A	NA	I	NA	NA	NA	NA

The preparedness phases as shown on Table 2, the BPBD has a coordinator function in almost all of the measurement indicators excluding the developing strategies for public education which the Department of Education is responsible for. As regards all measurement indicators for the preparedness phases, the BPBD and the Department of Social Services work on them for developing disaster emergency management systems, budgeting and maintaining disaster emergency supplies, and conducting risk management exercises to train personnel and test capabilities.

In the response phase, several measurement indicators are analyzed using RASIC Chart. Activating the disaster emergency operations plans and operations centers and evacuating threatened populations and vehicles be the responsibly of Indonesian Army, Indonesian Red Cross and SAR. Operating shelters and provision of mass care are supported by Indonesian Army, Dept. of Health Services and Indonesian Red Cross and be the responsibly of BPPD, Dept. of Social Services, District Head, NGOs and Volunteers. Estimating economic damage is only the responsibly of BPBD. While Dept. of Finance only approves regarding on budget cost. BPBD and Dept. of Social Services also be responsible on establishing procedures to prevent and suppress secondary hazards, documenting lessons learned and best practices, and implementing effective logistics management. Dept. of Education, NGOs, and Volunteers merely support it. Implementing effective coordination with other stakeholders and providing information on disaster-affected areas to public be the responsibly of BPBD and Dept. of Communications and Information supported by District Head.

In the recovery phase, almost all of measurement indicators be the responsibly of Dept. of Public Work. Those measurement indicators are considering sustainability in post-disaster reconstruction, shortening reconstruction time by applying quick mobilization, selecting reconstruction contractors earlier, constructing temporary roads and bridges, implementing execution plan for post-disaster reconstruction, applying lean construction in post-disaster reconstruction, realigning roads and relocating bridges to lower disaster hazard locations, and acquiring stakeholders' approval on reconstruction projects. Cleaning disaster debris and documenting lessons learned and best practices be the responsibly of BPBD supported by Volunteers and Dept. of Education. While, the function of Dept. of Finance in this phase is remain the same as the previous phase, that is concerning in approval of budget cost.

Table 2. RASIC Chart on The Preparedness Phase

R-Responsible *A-Approves* *S-Supports* *I-Informed* *C-Consulted* *NA-Not Available*	**Organization**	Disaster Mitigation Agency (BPPD)	Dept. of Social Services	Indonesian Army/Police	Dept. of Public Works	Dept. of Health Services	Indonesian Red Cross	Dept. of Communication and Informatics	Dept. of Finance	Dept. of Education	District Head	SAR	NGOS	Volunteers	Indonesian Agency for Meteorology, Climatology and Geophysics
Measurement Indicators	**Preparedness Phase**														
Recruiting personnel for disaster emergency services		R	NA	R	NA	NA	R	NA	NA	NA	NA	S	S	R	NA
Developing disaster emergency management systems		R	R	NA	NA	NA	NA	NA	NA	NA	NA	NA	NA	NA	NA
Developing strategies for public education		NA	R	NA	NA	NA	R	R	NA	R	NA	NA	S	S	NA
Budgeting for disaster emergency equipment		R	S	NA	NA	NA	NA	NA	A	NA	NA	NA	NA	NA	NA
Maintaining disaster emergency supplies		R	R	NA	NA	NA	NA	NA	NA	NA	NA	NA	NA	NA	NA
Locating places for emergency operation centers		R	S	S	S	S	S	S	A	S	R	S	S	S	S
Developing prediction and warning communications		S	NA	NA	NA	NA	NA	S	NA	NA	S	NA	NA	NA	R
Conducting risk management exercises to train personnel and test capabilities		R	R	R	NA	NA	R	NA	NA	NA	NA	R	S	R	NA
Using technology to identify and assess disasters, and damaged facilities		S	NA	NA	NA	NA	NA	NA	NA	NA	NA	NA	NA	NA	R
Developing coordination procedures with stakeholders		R	NA	S	NA	NA	S	R	NA	NA	S	S	S	S	NA

3.2 Merapi Eruption Coordination Model for Humanitarian Logistics

According to an interview with BPBD and the Magelang and literature review [11], the success factors of humanitarian logistics management can divided into three variables, i.e., strategical, tactical, and operational. Each variable has several indicators. Strategical i.e., sustainability, cooperation, networking, standardization aids logistics, safety, independence and fairness, caring. Tactical, i.e., coordination, collaboration, involvement of stakeholders, assessment and planning, management of humanitarian aids logistics, inventory management. Operational, i.e., quick response, flexibility, efficiency of management of humanitarian aids logistics, availability of disaster relief goods, and availability of volunteers.

On the operational level - the coordination and collaboration on management of natural disasters - Merapi Eruption has worked well, but there is still a lack of disaster relief management by any stakeholders involved. This needs to be supported with supervision in the location, so that relief goods can be channeled more effectively and efficiently. Coordination on disaster relief management needs to be supported by the existing information systems being up to date regarding the living layout of the recipients of the relief goods, the barracks, the number of refugee needs, availability of goods, and how to distribute the goods. The manual coordination which is currently done has led to slow responses in handling victims so the coordination process should be supported by an integrated information system.

4 Conclusions

Concerning humanitarian logistic systems in Merapi Eruption involves certain institutions. The BPBD acts as a coordinator for the functions of the mitigation and preparedness phase, as practitioner in response phase, and as a commander in the recovery phase. The Department of Social Services has the responsibility of helping the BPBD in the mitigation and preparedness phase. Meanwhile the Department of Public Works contributes in the recovery phase as practitioners in all activities related to restoration and construction post disaster. Regarding budgeting costs, funding can be secured by any stakeholder under the supervision of the Department of Finance. Besides that, the significant stakeholder who works on the mitigation phase is the Indonesian Agency for Meteorology, Climatology and Geophysics. It is also concluded that there are several stakeholders who do not contribute much directly in disaster risk management such as the Department of Education. A RASIC chart is a method for analyzing the role of stakeholders involved in disaster relief and for indicating the relationships among them in terms of humanitarian logistics in Indonesia. In this research, we demonstrated the primary model for this using the case of Merapi Eruption and analyzed the roles and relationships of each stakeholder, considering their major responsibility. For next research, it is suggested that work is done on integrating an information system between each stakeholder to establish a better humanitarian logistic system in Indonesia.

References

1. L. Van Wassenhove, J. Oper. Res. Soc. **57**, 5 (2006)
2. A. Cozzolino, SSBM (2012)
3. R. Tomasini, L. Van Wassenhove, and L. Van Wassenhove, SPRINGER (2009)
4. G. O'Brien, P. O'Keefe, Z. Gadema, J. Swords, Int. J. **19**, 498-508 (2010)
5. A. Thomas, L.R. Kopczak, FHI (2005)
6. R. Kaynak, A.T. Tuğer, Procd. Soc. Behv. **109**, 432-437 (2014)
7. R.D. Reid, Qual. Prog. **38**, 11 (2005)
8. T. Octavia, I. G. A Widyadana, H. C. Palit, Springer Singap. (2016)
9. R. D. Reid, Qual. Prog. **36**, 4 (2003)
10. M. Mojtahedi, B. L. Oo. Int. J. Disas. Risk Re. **21**, 35-43 (2017)
11. H. Abidi, S. Leeuw, M. Klump, Int. J. Bus. Manag. Invent. (2013)

E-Translator Kawi to Bahasa

Oka Sudana [1,*], *Darma* Putra[2], *M.*Sudarma[3], *Rukmi* S.Hartati[3], *Ryan* Pradnya Prastika[2] *and* *Ayu* Wirdiani[2]

[1]Study Program of Doctoral Engineering Science, Faculty of Engineering, Udayana University
[2]Information Technology Departement, Faculty of Engineering, Udayana University
[3]Electrical and Computer EngineeringDepartement, Faculty of Engineering, Udayana University
Jalan Raya Bukit Jimbaran, Kabupaten Badung - Bali, 80361

Abstract. Kawi Language is a kind of language which has developed in Java during the Hindu-Buddhist era in the ancient Indonesian archipelago. As the time passed, Kawi began to come to extinction. To overcome this, preservation of Kawi language can be done by enhancing the understanding of culture through the use of technology. Technology that can be used for cultural preservation is one of NLP based applications such as Kawi Language to Bahasa Translator Application which is developed in this research. The translator application is an application that translates from one language to another. The result of translation of the translator application is a language that is easy to understand by the people or commonly used language. This translator application is developed to processes Kawi Language text into bahasa based on search method which each word inserted. The stemming method used is Bobby Nazief & Mirna Adriani Algorithm using Kawi Language rule. The translated Kawi Language text can be a word, sentence or paragraph with the result of the Text in Bahasa as well as words, sentences or paragraphs.

1 Introduction

Kawi language is a kind of language that once has developed on the island of Java in the era of the Hindu-Buddhist Kingdom in the ancient Indonesian Archipelago and used in the writing of literary works. Kawi Language is now approaching extinction because of its users are getting less.

Preservation of Kawi Language can be done by making translator which is the application of Natural Language Processing. Natural Language Processing (NLP) is a branch of Artificial Intelligence (AI) that focuses on natural language processing. Translator application is widely available, but there is no application that can translate Kawi Language into Bahasa (Indonesian). The Translator Application of Kawi Language to Bahasa is considered necessary due to the absence of an electronic-based dictionary or translation that still uses a conventional dictionary so that translation becomes longer and more difficult to learn by people who do not really understand Kawi Language.

* Corresponding author: agungokas@unud.ac.id

Translator apps are widely available and can be easily found with the help of the Internet. Translator application can be built on desktop, web, and mobile. There are several translators that have been created using various methods such as the Translator Application of Balinese Language into Bahasa, based on Android with searching process using binary search [1]. Electronic dictionary application of Bahasa and Javanese sentences using Markov Model [2], English to Bangla Language using rule-based methods [3], an English to Balinese text translator using a rule-based approach [4], an Android-based Bahasa to Javanese language translator [5], an Android-based Translator Application of Javanese Language to Bahasa [6], An Indian to English Translator using a transfer-based method [7], an English to Marathi translator using the rule-based method [8].

The translator application is an application that can be combined with the cultural aspects, especially language. Technology which is combined with the culture will have a very positive impact especially in terms of learning, especially in language learning. Language is an important part of the organization of human life in the aspects of manners and the value of humanity itself [9]. Humans can grow, learn, and can develop their culture, preserve their culture or even develop their culture through language. These reasons indicate the necessity of Translator Application the Kawi Language to Bahasa. This application can process input in the form of Kawi Language word to produce output in the form of translation in Bahasa. Translator apps created on a web basis. Translator application will help interpret and understand Kawi Language easier.

2 Methods

2.1 Natural Language Processing

The principle of natural language is a form of representation of a message which wishes to be communicated between humans. The main form of representation is in the form of sound or speech but often expressed in writing also. The essence of natural language processing is the sentence deciphering or often called parsers. This parser works to read sentences, word by word and determine what types of words may follow the respective word. There are several applications that can be created using NLP such as, machine translation, speech recognition, speech synthesis, natural language interfaces to databases, information retrieval, information extraction, question answering, text summarization, dialogue systems [10]. Machine translation refers to the translation of the text from one language to another. The Translator of Kawi to Bahasa is a type of machine translation in the NLP. The Translator Application of Kawi to Bahasa can be categorized as machine translation because the software serves to replace humans in translating the source language text into target. Human intervention is not necessary when the translation process is being done because all the processes have been pre-programmed with Kawi Language based rules.

2.2 Tokenization

Tokenization is the process of dividing text in the form of a sentence, paragraph or document, into a token or a specific section. This process is done by searching spaces after each word. All generated tokens are stored in a separate file for further processing. Tokenization is the task of cutting the order of characters and units of a document which is defined into sections and discarding certain characters such as punctuation. The tokenization process places the word limit. The endpoint of a word and the beginning of the next word are called boundary words. This process is also referred to as word segmentation [11].

2.3 Stemming

Stemming is a process or way of finding the base word of a word. Stemming itself serves to eliminate morphological variations attached to a word by removing affixes to the word so that later obtained a correct word according to the morphological structure of language [12]. Stemming itself is part of the Information Retrieval process. Information Retrieval has two processes, indexing and searching. The Indexing process consists of four sub processes: Word Token (converts document into a set of terms by deleting all characters in punctuation found on the document and converting the term set into a lowercase), Stop Word Removal (The process of deleting words that often appear in documents Such as: and, or, not etc.), Stemming is the process of converting a full word into a base word and Word Weighting is the process of weighting each term in a document [13].

The stemming process in text of Bahasa is different from stemming in English text. The process required in English text stemming is only the process of removing suffixes, whereas in text of Bahasa it eliminates suffixes, prefixes, and confixes. Stemming algorithms for several languages have been developed, such as Porter Algorithm for English text, Algorithm Porter for text in Bahasa, Bobby Nazief and Mirna Adriani Algorithms for text in Bahasa [14]. The Bobby Nazief and Mirna Adriani algorithm refers to the morphological rules in Bahasa, which classify the affixes, ie allowable affixes or unauthorized affixes [15]. These groupings include prefixes, back affixes (suffixes), middle affix (infix) and the combination of prefix and suffix (confix) [16].

The stemming process of Kawi language is a process or way of finding the basic word of a Kawi word. Stemming Kawi word serves to eliminate morphological variations inherent in a Kawi word by removing affixes to the Kawi words according to the rules of Kawi language so that later on can be a correct word according to the structure of morphology of the true Kawi language. The stemming of the Kawi word is based on the stemming algorithm of Bobby Nazief and Mirna Adriani using the prefix, infix, and Kawi language suffixes. The method of Bobby Nazief and Mirna Adriani stemming algorithm is used in this application because this algorithm always checks to the database every word which is succeed omitted prefix, infix or suffix. Checking into the database of each cutting process makes the results of Bobby Nazief and Mirna Adriani algorithms close to perfect.
The stemming process of the Kawi word will take the word which has been tokenized before which then eliminates the prefix, infix, and suffix contained in Kawi language rules to produce the base word. The process of stemming the word Kawi which takes the word that has been tokenized initially.

2.4 The Word Searching Process

The word searching process is a process to search the word in Bahasa, at the database based on the base Kawi word that has been obtained by the results of the stemming process. Word search process using looping to find the until the following word is found out. The process of searching a word which originally took the base Kawi language that has been stemmed for matching the result in Bahasa.

2.5 The Word Searching Process

The word combining process is a process to combine words that have been translated into sentences. The process of combining words begins with taking the Bahasa word that has been searched in the database which then combined using the rules of Bahasa.

3 Experiment and Result

The testing phase will be conducted by taking samples of Kakawin Ramayana *Sargah* 1 consisting of 64 stanzas. The Kakawin will be inserted into the text area "Kawi Language Text" in the application as in Fig. 1.

Fig. 1 Kawi Text

The text which has already inputted will be executed for tokenizing and stemming process. The stemming process applies Bobby Nazief and Mirna Adriani Algorithms that are enhanced by Kawi language rules. Bobby These algorithms combined with Kawi language rules produce stemming products with a level of 82.27% accuracy.

Fig. 2 Tokenizing and Stemming Result

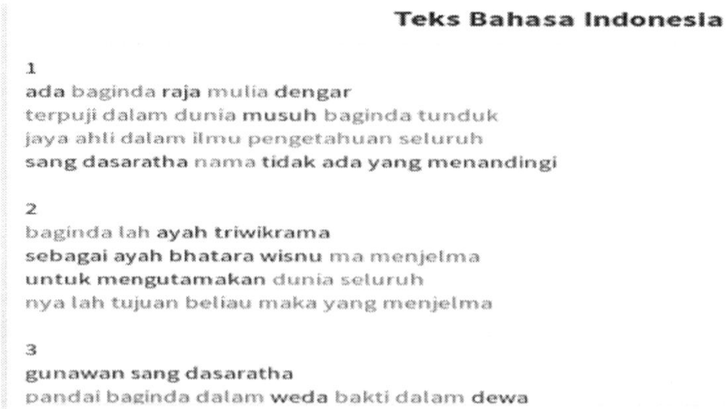

Fig. 3 Text Result

Fig. 2 is the result of the tokenizing and stemming process. The first sentence will be divided to produce a word or phrase. The result of a word that has been altered will be stemmed to find the base word of a word. Stemming eliminates morphological variations embedded in a word by removing affixes such as prefixes, infixes, and suffixes to the word, to obtain a correct word according to the correct Indonesian morphology structure.

After going through some process then The text result in Bahasa will appear as shown in Fig. 3. The translation in Bahasa results have more than one meaning such as the blue word. The blue colour is the link used to select the word. The selection feature is used to select another word which is the meaning of the source text. The selection of the other words can be used if there are some meanings of the word Kawi which are more than one. Selection of other words also serves to adjust the word on the sentence so as to produce a good sentence in Bahasa.

The structure of sentences which have been resulted from the translation process is still not so perfect, for it needs some correction of sentences according to grammar or rules of Bahasa. However, this application does not use rule based methods such as English to Bangla Translator Application, English to Balinese text translator, English to Marathi Translator who keeps the basic rules that contain all knowledge of the problems occurred [3, 4, 8]. The Kawi Language Translator Application to Bahasa does not use rule based method because it only corrects spelling on the result of a bad sentence. Sentences generally form a series of words arranged in accordance with applicable rules. The word is the smallest element of the syntactic structure of a language, while the sentence is its greatest element [16]. The E-Translator Kawi to Bahasa can translate words, sentences and even paragraphs with a maximum word count of 400 words. It was used to translate the Kawi language, which is mostly used in Kakawin. The structure of the Kakawin's sentence is different from the Balinese, Bahasa or other language sentence structure. This is because Kakawin is a story which blended poetry that can be sung, for that the translator has an improvement process of sentence structure after successfully translated to Bahasa. Different from Android based translator application from Balinese into Bahasa using Binary Search Method, the Android based translator application from Balinese into Bahasa using Binary Search Method is a translator that translates Balinese into Bahasa. The translated source language is the Balinese which has the same sentence structure as Bahasa so it easily translates and produces a good sentence even though it translates for each word [1].

4 Conclusion

The E Translator Application of Kawi to Bahasa is made with web-based. The process begins with tokenization, stemming, word translation to database and merging the words translation into a sentence. The Stemming process uses an algorithm developed by Bobby Nazief and Mima Adriani which is enhanced by using the rules in Kawi. Bobby Nazief and Mirna Adriani's algorithm which is enhanced by Kawi language rules is resulting a level of 82.27% accuracy in stemming phase. The E Translator Application of Kawi to Bahasa which has been made is able to translate words, sentences and even paragraphs with a maximum word count of 400 words. Accuration of word translation (dictionary) is close to 100%. However, the result of the sentence translation, there are still much to be adjusted, because the actual meaning of the sentence or verse in Kawi Language often has the allegory or metaphor meaning.

References

1. Sudana, A.K.O., I.K.A. Purnawan, and N.M.R.M. Dewi, Android Based Translator of Balinese into Indonesian using Binary Search Method. International Journal of Software Engineering and Its Applications, **8(6): p. 165-182** (2014)
2. Afifah, N., T.B. Santoso, and M. Yuliana, Pembuatan Kamus Elektronik Kalimat Bahasa Indonesia dan Bahasa Jawa untuk Aplikasi Mobile Menggunakan Interpolation Search. EEPIS Final Project, (2010)
3. Francisca, J., M.M. Mia, and S.M. Rahman, Adapting rule based machine translation from english to bangla. Indian Journal of Computer Science and Engineering (IJCSE), **2(3): p. 334-342** (2011)
4. Deni Pratama, I.P. and A. Muliantara, Perancangan dan Implementasi Sistem Penerjemah Teks Bahasa Inggris ke Bahasa Bali dengan Menggunakan Pendekatan Berbasis Aturan (Rule Based). Jurnal Ilmu Komputer, **5(1)** (2012)
5. Rohman, F. and P.W. Buana, Rancang Bangun Penerjemah Bahasa Indonesia ke Bahasa Jawa Berbasis Android. Merpati. **3(1)**
6. Budi Setiawan, A., P. Wira Buana, and I. Sukarsa, Aplikasi Translator Bahasa Jawa Ke Bahasa Indonesia Berbasis Android. Merpati, **2(3)** (2014)
7. Gehlot, A., et al., Hindi to English Transfer Based Machine Translation System. arXiv preprint arXiv:1507.02012, (2015)
8. Godase, A. and S. Govilkar, A Novel Approach for Rule Based Translation of English to Marathi. Advanced Computational Intelligence: An International Journal, **2(4)** (2015)
9. Chaer, A., Morfologi bahasa Indonesia: pendekatan proses, Rineka Cipta (2008)
10. Preeti and B.K. Sidhu, Natural Language Processing. International Journal Computer Technology and Applications, **4(5): p. 8** (2014)
11. Munot, N. and S.S. Govilkar, Rule Based Approach for Abstractive Text Summarization. Journal of Research, Science and Technologies, **1(6)** (2016)
12. Rajput, B.S. and A. NilayKhare, A survey of Stemming Algorithms for Information Retrieval, IOSR Journal of Computer Engineering, **1(17): p.76-80** (2015)
13. Jayanthi, R. and C. Jeevitha, An approach for effective text pre-processing using improved Porters stemming algorithm. Int J Innov Sci Eng Technol, **2(7): p.797-807** (2015)
14. Pramudita, H.R., Penerapan Algoritma Stemming Nazief & Adriani dan Similarity pada Penerimaan Judul Thesis. Data Manajemen dan Teknologi Informasi (DASI), . **15(4): p. 15** (2014)
15. Asian, J., H.E. Williams, and S.M. Tahaghoghi. Stemming indonesian. in Proceedings of the Twenty-eighth Australasian conference on Computer Science-Volume 38. Australian Computer Society, Inc (2005)
16. Adriani, M., et al., Stemming Indonesian: A confix-stripping approach. ACM Transactions on Asian Language Information Processing (TALIP), **6(4): p. 1-33** (2007)
17. Sulastra, J.S., Perancangan Penganalisis Struktur Kalimat Bahasa Indonesia dengan menggunakan Constraint-Based Formalism. Lontar Komputer: Jurnal Ilmiah Teknologi Informasi, **5(2)** (2014)

Housing Satisfaction and Willingness to Move to Low-cost Rental Apartments of Slum Dwellers in Semarang Urban Area

Asnawi Manaf[1,*], *Hadi* Wahyono[1] , *Ita* Puspita Sari[1] and *Dias* Aprilia[1]

[1]Urban and Regional Planning Department, Diponegoro University, Semarang, Indonesia

Abstract. This article examines one of the approaches commonly taken by the government to reduce the number of slums in urban areas, namely the provision of subsidised/affordable rental apartments. This article focuses on the assessment of to what extent the approach is relevant to the perceived problem (needs) of the target group (low income family) and what are the socio-economic characteristics of those living in slums and what are their preferences for moving to affordable rental apartments or remains to stay in the current slums. Since the problems are complex and dynamic, the case study research methods has been utilised and the low-income families living in slums in surrounding areas of Rusunawa Kaligawe (affordable rental apartment) in Semarang have been used as the unit of analysis.

1 Introduction

In recent years, various efforts and approaches have been implemented in helping the poor to fulfil their basic needs, particularly in shelter sectors and its supporting basic infrastructure such as safe water, electricity and sanitation. This article describes a study that examines one of the approaches commonly taken by the government recently to reduce the number of slums in urban areas, namely the provision of subsidised/affordable rental apartments. Certainly, the policies that are being intensively implemented these days need to be reviewed. This article it is designed primarily for policymakers, demonstrating that this housing issue is important to look at as a complex and 'multifaceted' issue. With a proper understanding of the problems at hand, the policies to be taken will be more relevant and effective. This is closely related to the topics that are often discussed lately about sustainable development, especially in the field of Housing and Urban Development.

This article focuses on the assessment of to what extent the approach is relevant to the perceived problem (needs) of the target group (low income family) and identifying their socio-economic characteristics and how this affects their preferences for living in slums (substandard settlements) or moving to adequate affordable rental apartments (Rusunawa).

2 Literature Review

* Corresponding author: asnawi.manaf@pwk.undip.ac.id

To limit the scope of the study, this paper will substantively describe several terms associated with the slums areas, housing satisfaction and housing mobility based on several studies of existing literature. [1] defines slums as areas in a city that are usually inhabited by the poor, who do not have security of land tenure, and as areas that have the characteristics of informal settlements with very low quality. In line with this, [2] states that the slum is associated with poor people, so that understanding the various programmes to minimise slums, such as slum upgrading, subsidised/affordable rental apartment programmes, etc., requires an understanding of poverty.

In contrast to the slums (sub-standard areas), [3] describes how adequate housing can provide protection from all threats and provide a safe and comfortable place to live. It can also be culturally acceptable, secure tenure, have availability of basic facilities for health and other basic needs, be affordable, and have easy access to workplaces and public facilities. The cost and time of travel to work and service facilities are not too burdensome on the family budget. The house is considered affordable if the amount of the costs incurred to the house and transport is less than 45% [4]. In general, the difficulty faced by poor families in achieving adequate housing is affordability due to lack of income or inability to pay. In addition, the increasingly high price of land, especially in urban areas will, make it increasingly difficult for poor families to afford the available house in the market place.

To deal with this issue, especially in handling slums in urban areas, there are some concepts that can be applied, which include: Resettlement, Slum Upgrading and Urban Renewal/Urban Redevelopment [5,6]. In Indonesia the government has introduced construction of cheap and affordable rental apartments for low-income people, especially those living in urban areas, where it is increasingly difficult to obtain cheap and affordable land.

Although the schemes that have been introduced have been able to help poor and low-income families, there are still segments of families who could not access the program. There are various issues and challenges of housing development in Indonesia both from the supply side (provider) and from the demand side or the target group of the programme [7,8]. One thing to be studied here is related to the socio-economic characteristics of target groups of the programme that influence how they perceive the problems, expectations and priorities for residence and satisfaction with their existing living conditions.

Residential satisfaction is a complex construct [9]. When analysing the housing problem and housing satisfaction, it is important to understand the term 'housing norm'. 'Housing norms' is a fundamental concept in the theory of housing adjustment in a family and the social rules of individuals and households to live in the house structure is closely linked to the characteristics of groups and communities [10]. The concept of the 'housing norm' explains the motivation of households to adjust or to adapt their housing condition to fulfil the perceived housing deficits normatively or socially [11].

If the condition of the house is judged unable to meet its needs normatively (in terms of culture or family), then it is called the 'Normative Housing Deficit' [12]. If the deficit is very prominent (sharp), then there is a need to respond through three possible behaviours, namely: residential adaptation, family adaptation, and residential mobility [11]. Adaptation appears when the desired living environment could not be selected, so communities adapt to the residential environment and settlements that were initially considered less comfortable [13]. Residential mobility occurs if the resident satisfaction levels are currently low [14], so households will move into new housing that fits their needs.

Associated with the response to the housing condition, it is important to note that a family has different considerations and priorities in determining the way that they respond to meeting the housing needs. [15] explains that the use of residential facilities is closely related to length of stay, reasons for residence, and ownership status. The longer a person lives in a residential environment, then the greater the need for space. Likewise, the lifestyle of each individual will affect their tendency in the pattern of residence utilisation. This individual tendency can be discerned from how they spend time with family, mobility and consumption type, as well as desire to interact with the social environment.

3 Research Objective and Method

As mentioned above, this article focuses on the assessment of to what extent the approach is relevant to the perceived problem (satisfaction) of the target group (low-income family) and what are their socio-economic characteristics and how this affects their preferences for moving to adequate affordable rental apartments (Rumah Susun Sewa---Rusunawa) or to remain living in slums (substandard settlements). The slum areas located in around a radius 3 km of Rusunawa Kaligawe (affordable rental apartments) in Semarang City, Central Java Province have been selected as the research location. 50 low-income families living in these areas have been selected as the respondents through the random sampling method.

To analyse the level of satisfaction of slum dwellers with their dwellings, the Residential Satisfaction Index (RSI) method has been used. The questionnaire consists of several variables that have been used, and these variables are grouped into five sections of satisfaction, namely satisfaction with physical condition of housing (section 1); satisfaction with location of housing (section 2); satisfaction with ownership (section 3); satisfaction with physical environment (section 4); and satisfaction with social environment (section 5). The level of housing satisfaction has been measured by using a five-point Likert scale, with "1" for very dissatisfied, "2" for dissatisfied, "3" for slightly satisfied, "4" for satisfied, and "5" for very satisfied.

The total score that has been obtained from 50 respondents is then categorised again into the new five-point Likert scale. The new range of interval (which in this case is 40) of the five-point Likert scale is determined by calculating the maximum and minimum scores as follows:

- Maximum score : The maximum score that can be obtained in this case (5) will be multiplied by the number of respondents
- Minimum score : The minimum score that can be obtained in this case (1) will be multiplied by the number of respondents.

$$The\ new\ range\ of\ Interval\ = \frac{Maximum\ score - Minimum\ Score}{Likert\ Scale} \quad (1)$$

$$The\ New\ Range\ of\ Inteval = \frac{(5x50)-(1x50)}{5} \quad (2)$$

$$= \frac{250-50}{5} = 40 \quad (3)$$

Subsequently, RSI will reclassify through the minimum score (50) and maximum score (250) with the interval as follows (very satisfied: 211-250); (satisfied: 171-210); (slightly satisfied: 131-170); (dissatisfied: 91-130) and (very dissatisfied: 50-90).

4 Finding and Discussion

4.1 Socio-Economic and Demographic Characteristics of the Slum Dweller

The following socio-economic and demographic characteristics of the respondent considerations while doing the data analysis:

- 28% of households have widow or widower status, and there are even family heads with unmarried status (2%). This is in line with the age of the head of the family, where many of them are over 50 years old (34%). With the marital status and the age of the household, we can say that they are almost no longer productive.

- The majority (68%) of respondents are native inhabitants, while 32% are identified as migrants. Most of them (68%) rely on their income by working in the informal sector (non-fixed income). They work as entrepreneurs, street vendors, and cheap labourers. This can be explained by their relatively low level of education. The majority of respondents completed only elementary (34%) and junior high school (34%); there are even respondents who never completed education at all (12%).

In general, the level of family income per month is not much different between the respondents; almost all families have income between Rp. 1,000,000-Rp. 2,000,000 per month. This income level is far below the Semarang City Minimum Regional Wage on 2017 (Rp. 2,125,000). With this level of income, the World Bank grouped them into the category of poor families with income in the 1-2 deciles of 10 deciles of income of the Indonesian people classification. Income of decile 1 (less than IDR 1,200,000 per family per month) and decile 2 (less than IDR 1,800,000 per family per month) are the two lowest income groups. Respondents with the income of deciles 1 and 2 will almost certainly not be able to have adequate housing without any support from the government.

4.2 Measurement of Residential Satisfaction

In addition to measuring the priority needs scale, this study also measures the degree of satisfaction of residents with their living quarters and neighbourhoods as follows:

- Satisfaction Level with Physical Condition of Shelter

The variable that gets the highest satisfaction is the number and functions of the rooms in the dwelling, while the size of dwelling is the variable that has the smallest score, as illustrated in Table 1. This indicates that for dwellers, the size of the dwelling is not a very urgent need to be fulfilled. This low level of satisfaction with the size of these dwellings is understandable because of the very limited area of land they have.

- Satisfaction Level with Location of Housing

The location variable that has the highest level of satisfaction is the proximity to health facilities; this can be seen in Table 1. Another thing that can be seen is the satisfaction with the aspect of closeness to the facilities of worship and trade. The high level of satisfaction of access to religious facilities is explained by the fact that the facilities of worship in general become one means of, or place for, social interaction.

- Satisfaction with Physical Environment

The highest level of respondent satisfaction is the satisfaction of the neighbourhood safety regarding crime (Table 1). The physical environment variables that are considered not to be satisfied are 'safe residential environments from floods, traffic jams, etc'.

- Satisfaction with Social Environment

Most respondents were satisfied with all of the variables for social environment (Table 1). Ranking the social environment features by total score of satisfaction level shows that respondents are most satisfied with their relationship with their neighbours, while the same score is achieved by existence of relatives/families. Overall, the satisfaction with the social environment indicates that the respondents are generally satisfied with the social environment within the housing area.

4.3 The Average Value of Satisfaction Level of Aspects of Existing Dwelling

The aspect of 'social environment' has the highest level of satisfaction (197); followed by the physical condition of housing (187); the location of housing (186); the physical environment (152); and lastly, house ownership (143). This can be understood because with all the limitations they have, their daily lives cannot be separated from interactions with neighbours and relatives. Many activities (PKK, Arisan, Karang Taruna, etc.) carried out will also hopefully be able to strengthen the relationship between dwellers. This will certainly have implications for their responds whether to stay or prefer to move to a better dwelling, in this case is willingness to move to the low-cost rental apartment.

Table 1. Residential Satisfaction

Variables	Score					Total Score	Level of Satisfaction
	1	**2**	**3**	**4**	**5**		
Physical Conditions of Housing							
Adequate size of shelter	2	19	17	9	3	184	Satisfied
Adequate number & function of rooms	3	17	15	14	1	189	Satisfied
Good quality of building materials	1	24	13	11	1	188	Satisfied
Location of Housing							
Ease of access to service centre	7	7	13	26	4	184	Satisfied
Ease to the workplace	15	8	15	22	5	189	Satisfied
Ease of access to public transport	22	11	12	22	4	188	Satisfied
Ease of access to health facilities	5	2	13	30	5	193	Satisfied
Ease of access to place of worship	0	2	12	28	8	192	Satisfied
Proximity to trading facilities	0	3	12	26	9	191	Satisfied
House/Land Ownership							
Secure residential/land ownership	3	19	11	16	1	143	Slightly Satisfied
Physical Environment							
Safe residential environments from floods, traffic jams, etc.	8	25	10	6	1	117	Dissatisfied
Adequate of employment opportunities	2	8	17	21	2	163	Slightly Satisfied
Environmental safety of the crime	1	5	15	26	3	175	Satisfied
Quiet residential neighbourhood	1	6	19	21	3	169	Slightly Satisfied
Clean residential environment	3	19	19	7	2	136	Slightly Satisfied
Social Environment							
Neighbourhood relations	0	1	5	33	11	204	Satisfied
The existence of relatives and family	1	2	8	31	8	193	Satisfied
The existence of social activities and regular meetings in the neighbourhood	0	2	13	25	10	193	Satisfied

The data indicates that the majority (46%) of respondents had already thought of moving, but there is no definite plan of when they will do so. The big answers in this category indicate that they understand there is a deficit in their residence but they can still accept the poor living conditions in which they live today. Even as many as 36% of respondents did not think of moving at all. Only 8% are seriously planning to move and starting to save money to move to other places they want.

From the answers of the respondents we can also see that only 28% are interested in living in affordable rental apartments, while the rest (72%) are not interested. From their point of view, it is comfortable living in their existing dwelling because it is close to relatives Those answers are closely related to the high level of satisfaction with the social environment of the dwelling, the fact that it is uncomfortable to climb stairs in the apartment, the distance to the workplace becoming greater, a preference to live in landed houses and do not have to pay additional costs if they live in low-cost rental apartment.

5 Conclusion

Based on the findings of this study, this article argues that the issue of housing, particularly for the poor, is a complex and 'multifaceted' problem. On the one hand, the majority of residents understand and feel that their living conditions are still inadequate, but others feel that they still want to stay in their inadequate residential and living environment. In general, slum dwellers are not interested in living in Rusunawa Kaligawe, even though the existence of Rusunawa Kaligawe could be a more viable alternative dwelling.

The slum dwellers have different socio-economic characteristics, and this is related to the housing norms (value and expectation). These have affected the satisfaction, the priority and the approach they choose in fulfilling their housing needs. It has nothing to do with the government's ability to provide cheap or affordable housing, as there are many other factors (variables) that need to be considered. The aspect of 'social environment' has the highest level of satisfaction. It means that the social environmental condition is part of the main priorities considered in their approach to fulfilling their housing needs. With all the limitations they have, they cannot separate their daily lives from interactions with neighbours and relatives.

The satisfaction levels of the substandard settlement and intention to move were not in line. Therefore, the study recommends the reassessment of those features of residential satisfaction for others types of housing and other types of socio-economic characteristics of the target group. Last but least, this study concludes that putting the responsibility for the housing problem for the people only on the supply side will not always match the characteristics of prospective target groups. Therefore, adequate attention is needed to the characteristics of the needs of the target group.

References

1. A. M. Khalifa, Habitat Int. *35*(1), 40-49. (2011)
2. J. Minnery et. al., *Habitat Int.,* **39**, 162-169, (2013)
3. S. Deka, Operational Performance of State Housing Agencies in Assam a Case Study of Assam State Housing Board Since 1987, (2003)
4. T. Litman *Victoria Transport Policy Institute.,* (2011)
5. V. Mukhija, Cities, **18(4)**, 213-222, (2001)
6. C. Pugh, Cities, **11(3)**, 159-180, (1994)
7. A. Manaf. *The International Journal of Environmental, Cultural, Economic and Social Sustainability,* **7(6),** 337-350, (2011)
8. A. Manaf. Suharmono, H. Yuzal, M. Fisher. *Housing, Care and Support,* **19(3/4),** 81-92, (2016)
9. M.A. Mohit, A.M.M.A.K. Raja, Plan. Malaysia J. **12(3),** 47-66, (2014)
10. L. Shi, *Doctoral dissertation at University of Stellenbosch.* (2005)
11. E. W. Morris, S.R. Crull, M. Winter, J. Marriage Fam., 309-320, (1976)
12. E. W. Morris, M. Jacubczak, *Housing and society*, **15(1)**, 41-55, (1988)
13. R. A. Oppoku, A.G.A. Muhmin. Habitat Int., **34(2),** 219-227, (2010)
14. S. R. Crull, M. E. Bode, E.W. Morris, *Housing and Society*, **18(3)**, 53-64, (2015)
15. Knox, Integrated City Strategy and Implementation Plan, *Knox City Council*, (2015)

The potential biodiesel production from *Cerbera odollam* oil (Bintaro) in Aceh

Khairil[1,*], *Aulia* Rizki[1], Iskandar[1], Jalaluddin[1], *A.S.* Silitonga[1,2,3], *H.H.*Masjuki[3], *T.M.I.*Mahlia[1,4]

[1]Syiah Kuala University, Faculty of Engineering, Department of Mechanical Engineering, 23111,Darussalam-Banda Aceh, Indonesia
[2]Politeknik Negeri Medan, Department of Mechanical Engineering, 20155, Medan, Sumatera Utara, Indonesia
[3]University of Malaya, Faculty of Engineering Department of Mechanical Engineering, 50603 Kuala Lumpur, Malaysia
[4]Universiti Tenaga Nasional, Faculty of Engineering Department of Mechanical Engineering, 43000, Kajang, Selangor, Malaysia.

Abstract. Biodiesel production from non-edible vegetable oils is an effective way to conquer the linked problems with edible oils such as food versus fuel and other environmental impacts. *Cerbera odollam* oil is one of these possible non-edible feed stocks for future biodiesel production. This study evaluated the potential biodiesel production from *cerbera odollam*. The seed was collected and extracted from Aceh, Indonesia. Moreover, biodiesel has been produced using degummed (H_3PO_4) and two step acid catalyst (HCl) and alkaline catalyst (KOH). The results of properties of the *cerbera odollam* methyl esters show that such as viscosity was about 847.9 mm^2/s, density was 3.1578 kg/m^3, flash point was 214.0 °C, acid value was 0.4 mg KOH/g, oxidation stability was 6.35 h, FAME content was 97.77 % w/w and heating value was 40.49 MJ/kg. After analysing these properties, it has been found that there is a huge chance to produce biodiesel from this seed which complies with the limits of ASTM 6751 and EN 14214 specifications and therefore it can boost the future production of biodiesel from non-edible sources.
Keywords: *Cerberaodollam*, esterification, transesterification, biodiesel.

1. Introduction

The diminishing supply of fossil fuels reserves and the growing environmental concerns have made renewable energy an exceptionally attractive alternative energy source for the future [1,2]. Biodiesel is a promising alternative fuel for diesel fuel. It is defined as the mono-alkyl esters of long chain fatty acids derived from vegetable oils or animal fats and alcohol with or without a catalyst. It is renewable, biodegradable non-toxic, portable, readily available and eco-friendly fuel [3]. The Indonesian government has developed alternative fuel to tackle the energy crisis to solve the economic revival of society. Therefore, to the best of the author's knowledge, limited study on the possibility of *cerbera odollam* oil as a potential non-edible oil feed stock for biodiesel production has been reported in the literature.

1.1 Objective of this study

The purpose of this paper is to produce biodiesel from *cerbera odollam* oil using pre-treatment degummed oil, homogeneous acid HCl catalyst and alkaline KOH catalyst

*) Corresponding author : khairil@msn.com

followed by an investigation of some physical and chemical properties of the produced methyl ester (COME). The success of this study will promote more investigation of this feed stock for biodiesel production and eventually could yield promising and massive new raw material for biodiesel production on a large scale.

1.2 Description of *Cerbera Odollam*

Cerbera odollam belongs to mangrove family. It is native to the mangrove swamps of Asia, Australia, Madagascar, and the western islands of the Pacific Ocean. It is a medium to large tree with large, glossy, heart-shaped leaves that are conspicuously veined and long-stemmed. The flowers are large and greenish; the sexes are separate. The fruit is oval and about the size of a large husked coconut, brown and rough-surfaced as shown in Fig.1. *Cerbera odollam* trees are abundant due to it is used as a coastal crop, greening and decorating the city park and requires no special maintenance [4]. *Cerbera odollam* seeds contain oil yield is about 43- 64% [5].

(a) (b) (c)

Fig. 1 *Cerbera odollam* tree (a), fruit (b)and seeds (c)

2. Material and Methods

2.1 Materials and Réagents

The *Cerberaodollam* seeds was supplied from Banda Aceh, Indonesia through personal communication in order to obtain Crude *Cerbera Odollam* Oil (CCOO). Reagent chemical such as methanol, H_3PO_4, HCl, KOH were used. Additionally, the qualitative filter paper (Filtres Fioroni, France) of 150 mm size was used to purify the crude oil from solidified matters.

2.2 Methodology

2.2.1 Degumming process

In this process, 0.7 % (vol) of phosphoric acid (H_3PO_4 20% concentration) was added to crude c*erbera odollam* oil at a temperature of 70°C for 30 minutes in a glass beaker. This mixture was separated in separation funnelby density separation in which the phosphate compounds drop down at the bottom. The refined oil produced from this process was reported to be light yellow.

2.2.2 Acid esterification

*) Corresponding author : khairil@msn.com

In this process, 100 gr of Crude *Cerbera Odollam* Oil (CCCO) was entered into a two-necked boiling flask equipped with a magnetic stirrer, reflux condenser and thermometer at room temperature. The sample of CCCO was mixed with 1:8 molar ratios of methanol and the mixture was stirring at a speed of 300 rpm. Then, hydrochloric acid (HCl) was added to the sample drop by drop using a connected pipe to the flask. The amount of HCl catalyst added was 1% (v/v). During the reaction, the mixture was stirring at constantly speed of 300 rpm using a magnetic stirrer for 3 hours at 65°C. The esterified oil was then removed from the reactor and entered into a separation funnel and left for 24 hours to remove the upper layer which includes extra and methanol from the lower layer of the first esterified CCCO.

2.2.3 Alkaline transesterification

In this process, *cerbera odollam* esterifed oil (COEO) was mixed with 29.76 gr of methanol while stirring at speed of 300 rpm. Then 2 gr of KOH of alkaline catalyst using in this process was diluted which added to the oil and maintained at 60°C for 2 hours. The oil was then removed from the reactor and entered into a separation funnel left for 24 hours to remove the upper layer of *cerbera odollam* methyl ester (COME) oil from extra methanol and glycerol formed at the lower layer.

2.2.4 Characterization of crude Cerbera odollam oil and methyl ester

The crude *cerbera odollam* was found to be dark brown yellow. The physicochemical properties were tested according to the ASTM D6751 and EN 14214 standards. Several tests such as kinematic viscosity, density, flash point, acid value, oxidation stability, FAME content and heating value. Moreover, the fattyacid methyl ester (FAME) content in the biodiesel was measured by gas chromatography based on EN 14103 method which determined using the following equation [6]:

$$FAME = [(\sum A - A_{EI})/A_{EI}] \text{ x } [(C_{EI} \text{ x } V_{EI})/ \text{m}] \text{ x } 100\ \% \qquad (1)$$

Where *FAME* represents the fattyacid methyl ester content (%), $\sum A$ represents the total gas chromatography peak area of methyl ester, A_{EI} represents the gas chromatography area of internal standard, methyl heptadecanoate, C_{EI} represents the concentration of the standard solution (mg/ml), V_{EI} represents the volume of the standard solution (ml) and m represents the amount of the biodiesel sample (mg). The methyl ester yield was determined using the following equation [7] :

$$Methyl\ ester\ yield\ (\%) = [(FAME \text{ x } W_{ME})/ W_{CO}] \text{ x } 100\ \% \qquad (2)$$

Where *FAME* represents the fattyacidmethyl ester content (%), W_{ME} represents the weight of the *cerbera odollam* methyl ester (g) and W_{CO} represents the weight of the crude *cerbera odollam*oil (g). The equipment used to analyse physiochemical properties of crude *cerbera odollam* oil is listed in Table 1.

*) Corresponding author : khairil@msn.com

Table 1. List of equipment used for biodiesel properties

Property	Equipment	Standard method	Accuracy
Kinematic viscosity	Stabinger viscometer™ SVM 3000	ASTM D7042	± 0.01 mm^2/s
Density	DM40 LiquiPhysics™ density meter	ASTM D127	± 0.1 kg/m^3
Acid value	Automation titration rondo 20	ASTM D664	±0.001mg KOH/g
Flash point	NPM 440 Pensky-martens flash point	ASTM D93	± 0.1°C
Heating value	6100EF Semi auto bomb calorimeter	ASTM D240	± 0.001 MJ/kg
FAME content	QP2010 Ultra, Shimizu	EN 14103	± 0.008% or 0.0008 min
Oxidation stability	873 Rancimat	EN 14112	± 0.01 h

3. Results and discussion

3.1 Physico-chemical properties of *Cerberaodollam* methyl ester

The physical and chemical properties of the *cerbera odollam* methyl ester (COME) produced were analyzed using ASTM specification: kinematic viscosity at 40°C (ASTM D445), density at 15°C (ASTM D1298), flash point (ASTM D93), acid value (ASTMD664), heating value (ASTM D240) etc. The physical and chemical properties of COME are listed in Table 2. The viscosity of the CPME is 4.56 mm^2 /s, thus meeting the requirements of both the ASTM D6751 and EN-14214 biodiesel standards, which prescribe viscosity ranges that should be between 1.9–6.0 and 3.5–5.0 mm^2 /s, respectively. The density of methyl ester usually varies between 860 and 900 kg/m^3. The density of the COME is 878.10 kg/m^3. The acid value of COME was observed to be 0.47 %. These results are in agreement with the value specified in ASTM D664.The heating value of *cerbera odollam* methyl ester was found 39.46 MJ/kg respectively.

3.2 Fatty acid methyl ester of COME

The fatty acid methyl ester of the COME is shown in is shown in Fig.2 and a picture for the produced biodiesel is shown in Fig. 3. It can be seen that the COME biodiesel is mostly composed of oleic acid (19.33%), linoleic acid (20.15%), linolenic acid (9.82) and palmitic acid (17.10%). The total unsaturated fatty acid content of the COME is found to be 54.08% which contributes to the improve deoxidation stability [8]. The composition of fatty acids in methyl ester may affect oxidation stability. This can be explained that the content of unsaturated fatty acids are sensitive to oxidative degradation, as well as unsaturated fatty compounds have higher flash points than saturated fatty compounds [9,10].

*) Corresponding author : khairil@msn.com

Table 2.Physicochemical properties of *cerbera odollam* methyl ester

Property	COME	Limit ASTM D6751	Limit EN 14214
Kinematic viscosity at 40°C (mm^2/s)	847.9	880	860-900
Density at 15°C (kg/m^3)	3.1578	1.9–6.0	3.5–5.0
Acid value (mg KOH/g)	0.4	Max. 0.5	Max. 0.5
Heating value (MJ/kg)	40.490	35	–
Flash point (°C)	214.0	Min. 130	Min. 101
Oxidation stability (h)	6.35	3	6
FAME content (%)	97.77	–	96.5

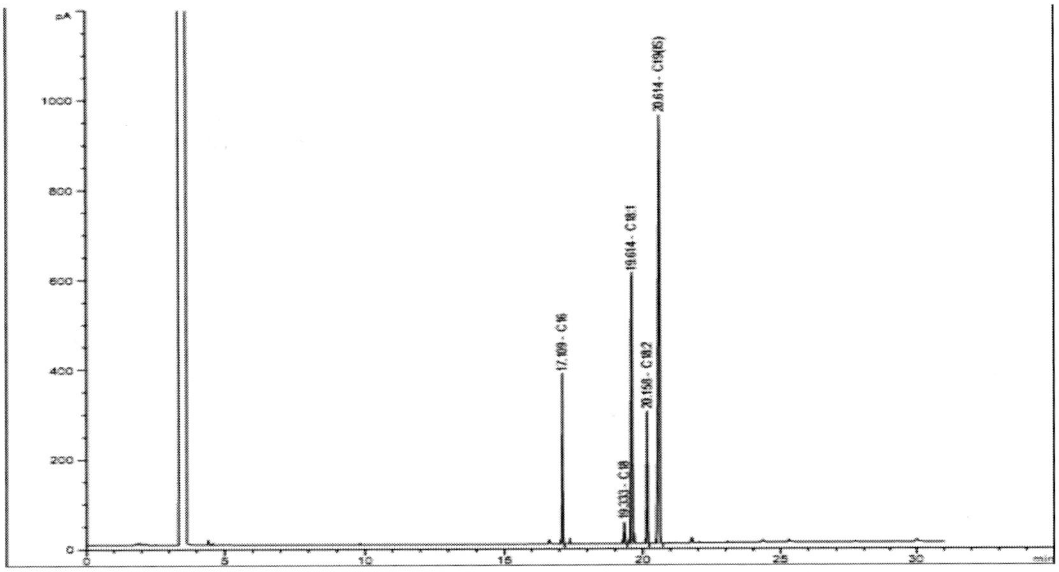

Fig. 2 Gas cromathography of c*erbera odollam* methyl ester

Fig. 3 *Cerbera odollam* biodiesel

*) Corresponding author : khairil@msn.com

4. Conclusion

Cerbera odollam oil is one of non-edible feedstocks which can replace fossil diesel. The evaluation of biodiesel production using two step acid esterification and alkaline transesterification was performed in order to obtain high methyl ester yield. In this study, the FAME of *cerbera odollam* biodiesel attained 99.77 % w/w and the measured properties were within the limits EN14214 and ASTM D 6751 standards. The results show that all of the reaction variables which clearly influenced the acid esterification and alkaline transesterification in a positive manner. From all of these encouraging out comes, and given further work testing on diesel engine performance, crude *cerbera odollam* oil could be recommended as a supplementary non-edible oil feedstock for biodiesel production.

The authors graciously acknowledge for the financial support provided from Syiah Kuala University Banda Aceh, Indonesia through the Penelitian Professor No.: 04/UN11.2/PP/PNBP/SP3/2017. The authors also graciously acknowledge for a part of the financial support provided by Politeknik Negeri Medan, Medan, Indonesia, under the Research and Community Service Unit (UPPM-2017) and also Universiti Tenaga Nasional (Grant No.: 20160101FRGS). Besides this, the authors would also like to thank Dr. Surya Dharma and Dr. Abdi Hanra Sebayang for their valuable help and support.

References:

1. J. Yan, Y. Yan, Encyclopedia of Sustainable Technologies, Elsevier, Oxford,**1**, (2017)
2. N. Damanik, H.C. Ong, W.T. Chong, A.S. Silitonga, Energ Sources Part. A, **1** (2017)
3. R.L. Patel, C.D. Sankhavara, Renew. Sust. Energ Rev,**71** (2017).
4. U. Anisa Rahmi, 2011, Kajian proses produksi biodiesel dari minyak bintaro dengan transerterifikasi, Tugas Akhir, IPB Bogor.
5. Imahara H, Minami E, Hattori M, Murakami H, Matsui N. and Saka S, Current situation and properties of oil/fat resources for biodiesel production, The 2nd International Conference on SEE (2006) 1-5.
6 S. Dharma, H.H. Masjuki, H.C. Ong, A.H. Sebayang, A.S. Silitonga, F. Kusumo, T.M.I. Mahlia, Energ Convers. Manage, **115** (2016)
7. F. Kusumo, A. Silitonga, H.C. Ong, H. Masjuki, T. Mahlia, A comparative study of ultrasound and infrared transesterification of Sterculia foetida oil for biodiesel production, Energy Sources, Part A: Recovery, Utilization, and Environmental Effects, (2017) 1-8.
8. G. Martínez, N. Sánchez, J.M. Encinar, J.F. González, Fuel properties of biodiesel from vegetable oils and oil mixtures. Influence of methyl esters distribution, Biomass and Bioenergy, 63 (2014) 22-32.
9. G. Knothe, Fuel Proces, Technol, **86** (2005).
10. S. Schober, M. Mittelbach, Europ. J. Lipid Science Technol, **106** (2004)

*) Corresponding author : khairil@msn.com

Fatigue Testing and Evaluation of Fatigue Strength under Multiaxial Stress State; Why do we need fatigue testing?

Takamoto Itoh*, *Fumio* Ogawa and *Takahiro* Morishita

Ritsumeikan University, Department of Mechanical Engineering, College of Science and Engineering, 1-1-1 Nojihigashi, Kusatsu-shi, Shiga, 525-8577, Japan

Abstract. Types of multiaxial fatigue tests and their experimental results are presented in this paper. There are typical three types in multiaxial fatigue tests: the combining push-pull and reversed torsion loading test using hollow cylinder specimen, the biaxial tension-compression test using cruciform specimen and the inner pressure applied the push-pull loading test using the hollow cylinder specimen. In the combining a push-pull loading and a reversed torsion loading test, failure life under non-proportional loading in which principal directions of stress and strain were changed in a cycle was shortened compared to proportional loading in which those are fixed. Fatigue lives were well-correlated using a non-proportional strain range considering the effect of strain path and material dependence. In the biaxial tension-compression test, the failure life decreased with increase of the principal strain ratio. In the inner pressure applied the push-pull loading test, cyclic deformation behaviour due to complex loading paths of multiaxial fatigue tests with the inner pressure associated with push-pull and rev. torsion acted to reduce the failure lives. Experimental investigation of multiaxial failure life and elucidation of their governing mechanism is essential and it can broaden the applicability of structural components.

1 Introduction

Investigation of fatigue properties is essential for design of structural components. In practical application, structures are subjected to complex multiaxial load. Therefore, the understanding of multiaxial fatigue properties of materials is important. Indeed, failure lives are overestimated when the effect of multiaxiality is neglected. Multiaxial fatigue testing usually has been carried out using a hollow cylinder specimen by applying push-pull loading and a reversed torsion loading and the applicability of multiaxial stress and strain parameters has been discussed [1-5]. However, a principal strain ratio (ϕ) and a principal stress ratio (λ) ranges performable by the testing method are $-1\leq\phi\leq-\nu$ and $-1\leq\lambda\leq0$, where ν is the Poisson's ratio. Structural components sometimes undergo fatigue damage at principal strain/stress ratios in excess of the above range under service loading. In

* Corresponding author: itohtaka@fc.ritsumei.ac.jp

multiaxial fatigue, loading is classified into two types: proportional and non-proportional loadings. The former is the loading in which principal directions of stress and strain are always fixed in a cycle and the latter has phase shift of axial and torsional stresses/strains. Figure 1 shows three types of most representative testing methods classified by loading and a shape of specimens as well as the principal stain and stress ratio ranges which can be performed in each type of test. Type A is the push-pull and rev. torsion test using hollow cylinder specimen, Type B is the biaxial tension-compression test using cruciform specimen and Type C is the inner pressure applied push-pull and reversed torsion test using the hollow cylinder specimen.

Principal stain and stress ratio ranges which can be performed in each type of test are also summarized in Figure 1. Type A has been widely used, but the principal strain/stress ratio ranges performable by this testing are $-1\le\phi\le-\nu$, $-1\le\lambda\le0$. Types B and C can perform the multiaxial fatigue test under full ranged principal strain/stress ratio ranges of $-1\le\phi$, $\lambda\le+1$. However, Types B has no change in principal directions of stress and strain since the directions always are fixed into the direction of applied loading. Only Type C can perform the multiaxial fatigue test in the full ranged principal strain/stress ratio ranges with non-proportional loading. In this study, the results of multiaxial fatigue tests based on Types A-C are presented and the effect of multiaxiality and nonproportionality on failure lives are discussed.

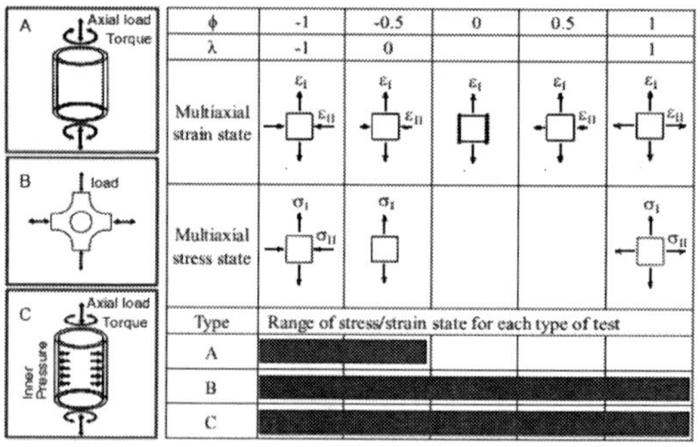

Fig.1. Type of tests with specimens and multiaxial stress state.

(a) Testing machine (b) Shape and dimensions of test specimen (mm)

Fig.2. Testing machine and test specimen of push-pull and reversed torsion.

2 Push-pull and reversed torsion tests using cylinder specimen

In this section, results of multiaxial fatigue tests using push-pull and reversed torsion tests of hollow cylinder specimens of SUS316NG (rolled stainless steel for nuclear plants) [6] are presented. Figure 2 (a) shows schematic of an employed electrical servo controlled hydraulic fatigue testing machine for push-pull and reversed torsion loading. Outer diameter of hollow cylinder specimen in gauge part was 12 mm, while inner diameter was 9 mm as shown in Figure 2 (b). Strain controlled fatigue tests were performed at room temperature. Figure 3 shows the strain paths and strain waveforms: strain paths were a push-pull loading and a circle loading. In the push-pull loading test, principal directions of stress and strain are fixed. Therefore, that is proportional loading tests. On the other hand, the circle loading test is non-proportional loading test because axial stress and shear stress have phase difference of 90°. In the circle loading test, axial and shear strain ranges are the same value based on von Mises; $\varepsilon=\gamma/\sqrt{3}$. Strain rate was set to 0.1%/s based on von Mises' equivalent plastic strain. The criterion for specimen failure was the decrease of stress amplitude to 3/4 of its maximum value or a breakage of specimen. In Figure 4 (a), a correlation of failure life using an equivalent total strain based on von Mises is shown. Thick solid line in the figure was drawn based on data of the push-pull loading test. Thin solid lines on both sides of the line indicates a factor of 2 band. Failure life N_f of the circle loading test is 1/5 of that of the push-pull loading test. It is also reported that the reduction in N_f due to non-proportional loading has strong correlation with increase in stress due to non-proportional loading and it depends on materials tested [6].

(a) Push-pull loading (b) Circle loading

Fig.3. Strain loadings: strain path and strain waveform.

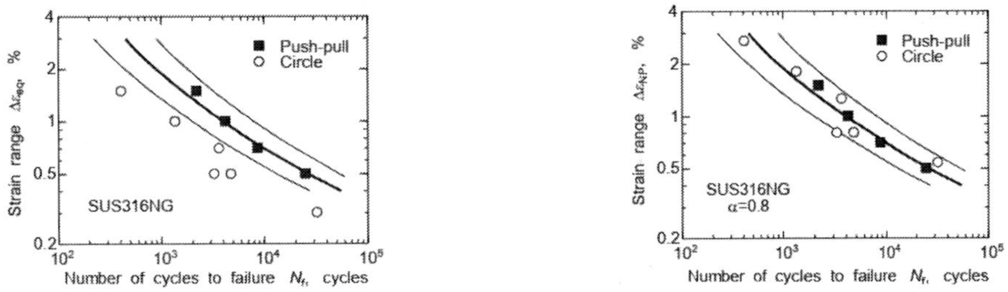

(a) Equivalent total strain amplitude based on von Mises (b) Non-proportional strain range.

Fig.4. Correlation of N_f with strain parameter.

The reason why failure life of non-proportional loading test is correlated in conservative regime by $\Delta\varepsilon_{eq}$ is that $\Delta\varepsilon_{eq}$ does not take into account the effect of strain path dependence of failure life under non-proportional loading. Therefore, it is important to consider the effect of strain path. Itoh et al. proposed the following strain based parameter for evaluation of non-proportional multiaxial low cycle failure life [6],

$$\Delta\varepsilon_{NP}=(1+\alpha f_{NP})\,\Delta\varepsilon_{I} \tag{1}$$

where $\Delta\varepsilon_{NP}$ is non-proportional strain range that considers material dependence and strain paths. α is material constant for evaluating the amount of additional hardening by non-proportional loading. f_{NP} is non-proportional factor that expresses severity of nonproportional loading and $\Delta\varepsilon_I$ is the maximum principal strain range. f_{NP} is defined as,

$$f_{NP}=\frac{\pi}{2\varepsilon_{I\max}\cdot L_{path}}\int_C \varepsilon_I(t)\cdot\left|\boldsymbol{e}_1\times\boldsymbol{e}_R\right|ds \tag{2}$$

where $\varepsilon_I(t)$ is maximum principal value at time t defined by the following equation.

$$\varepsilon_I(t)=\mathrm{Max}\left[\left|\varepsilon_1(t)\right|,\left|\varepsilon_3(t)\right|\right] \tag{3}$$

$\varepsilon_1(t)$, $\varepsilon_3(t)$ are the maximum principal vector and minimum principal vector of strain. The maximum value of $\varepsilon_I(t)$ during a cycle is defined as the maximum principal value $\varepsilon_{I\max}$. $\boldsymbol{e}_1$ is a unit vector directing to $\varepsilon_{I\max}$, $\boldsymbol{e}_R$ is a unit vector directing to $\varepsilon_I(t)$, ds is the infinitesimal trajectory of the loading path. L_{path} is the whole loading path length during a cycle.

Figure 4 (b) shows a correlation result of failure life using the non-proportional strain range. N_f of the circle loading tests are correlated within factor of 2 bands and failure lives are well-correlated using non-proportional strain range.

3 Biaxial tension-compression test using cruciform specimen

In this section, the results of biaxial tension-compression test using cruciform specimen [7] is mentioned. Strain controlled biaxial tension-compression test was performed for type 304 stainless steel cruciform specimen at 923 K. Figure 5 shows a schematic showing of testing apparatus and a control flow. The apparatus has four actuators and four servo controllers to generate widely ranged biaxial stress states. Two extensometers were attached to the specimen through the windows of the electric resistant furnace. Amplitude of the principal strain in *y*-direction (ε_y) was fixed and that in *x*-direction (ε_x) was changed to achieve the principal strain ratios (ϕ) ranged from −1 to 1, where the principal strain ratio is defined as $\varepsilon_x/\varepsilon_y$. A fully reversed triangle waveform with strain rate $\dot{\varepsilon}_y$=0.1%/s was adopted, where $\dot{\varepsilon}_y$ is the principal strain rate in *y*-direction. Specimen failure was defined as the cycle at which the tensile stress amplitude normal to the crack decreases to 95% from saturation. The ϕ=−1 test corresponds to a reversed torsion test using cylindrical specimen, while the ϕ=−0.5 test corresponds to a uniaxial tension/compression test. The ϕ=1 test corresponds to an equi-biaxial tension-compression test.

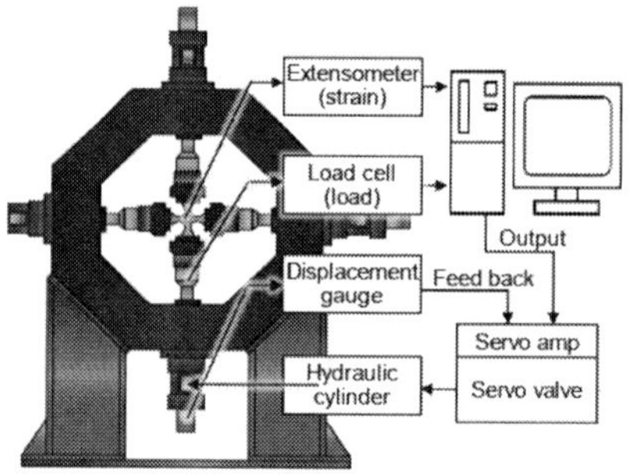

Fig.5. Biaxial tension compression testing machine and control flow of the test apparatus

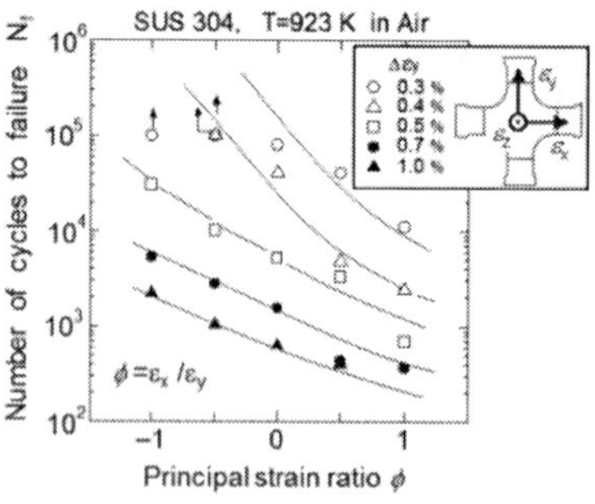

Fig.6. Relationship between failure life and principal strain ratio.

Figure 6 shows the failure life against principal strain ratio. It is clear the failure life decreases with increase of the principal strain ratio and decreasing rate gets larger as strain range decreases. Multiaxial low cycle fatigue strain parameters have been proposed for the correlation of multiaxial fatigue data. To take examples of them, maximum principal strain range, Mises' equivalent strain range, Γ-plane parameter (BMK strain [8]), Γ*-plane parameter (LE strain [9]), the equivalent strain based on crack opening displacement (COD strain [10]) exist. Particularly, the COD strain range correlates the fatigue data within a factor of two band and the scatter of data is smaller compared to that of LE strain (data correlations are omitted here).

4 The inner pressure applied push-pull and reversed torsion test using the hollow specimen

In this section, the results of type C test [11] are presented. Apparatus equipped with three actuators was used (Figure 7). The actuator in the upper side applies inner pressure and the two actuators in the lower side apply the push-pull and the reversed torsion loading to the hollow cylinder specimen. The maximum pressure that can be applied to the inner surface of the specimen is 200 MPa. The maximum axial load for push-pull is ±50 kN, while the maximum torque for reversed torsion is ±250 N·m. This testing apparatus can perform tests under wide ranged multiaxial stress ratios. Hollow cylinder specimen of which dimensions are 12 mm inner diameter, 14 mm outer diameter and 8.5 mm gauge part was used for testing.

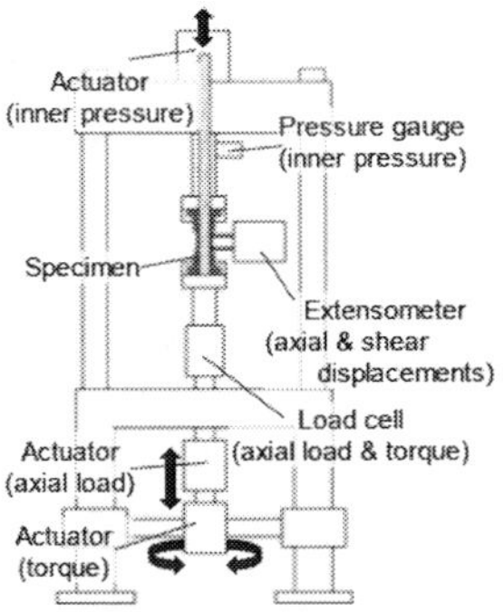

Fig.7. Multiaxial fatigue testing machine for push-pull and reversed torsion tests with inner pressure.

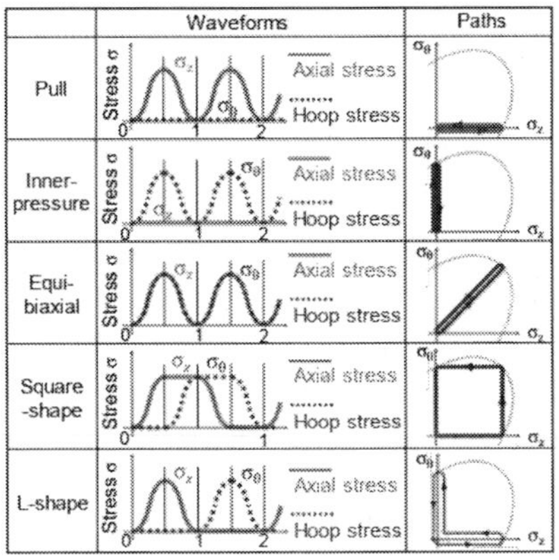

Fig.8. Loading waveforms and paths.

Figure 8 shows 5 types of loading waveforms and loading test paths. The Pull is a cyclic loading where only axial stress σ_z is applied. The Inner-pressure is a cyclic inner pressure test in which only hoop stress σ_θ is applied. These two loadings are classified as the uniaxial loading. The Equi-biaxial, the Square-shape and the L-shape are multiaxial loading tests in which combined cyclic pull loading and inner pressure are applied. In the Equi-biaxial test, axial stress σ_z and hoop stress σ_θ are applied simultaneously. In the Square-shape test, axial stress σ_z and hoop stress σ_θ are applied as trapezoidal waveforms with 90 degree phase difference so that the loading path is shown by a square shape in a σ_z-σ_θ coordinates. In the L-shape test, axial stress σ_z and hoop stress σ_θ are applied alternately so that the shape of loading path becomes the shape of the letter "L". The failure life N_f was determined as the cycle at which the maximum inner pressure was reduced by the oil leak due to an initiation of through crack or rupture of the specimen. In addition, inner pressure of 1 MPa was applied during the Pull test to unify definition of the N_f.

Figure 9 (a) shows a relationship between maximum Mises' equivalent stress and N_f at inner surface, while Figure 9 (b) shows a relationship between mean hoop strain $\varepsilon_{\theta\ mean}$ and failure life ratio N/N_f under the Pull and the Square-shape tests. N_f of Square-shape test is the shortest, and that of the L-shape test is the second shortest. N_f of the Inner pressure, the Pull test and the Equi-Biaxial test are almost the same. In Figure 9 (b), $\varepsilon_{\theta\ mean}$ under the Pull test decreased due to Poisson's effect, while $\varepsilon_{\theta\ mean}$ under the Square-shape test increased until near the N_f. Although omitted here, the observation of the specimens after fatigue tests revealed that the diameters on the gauge part after the Square-shape test and the L-shape test were larger than those after other loading test. On the other hand, the lengths of all cracks were almost the same. Apparently, N_f was decreased because of the increase of the amount of plastic strain in the Square-shape test and the L-shape test. Cyclic deformation behavior due to complex strain paths of multiaxial fatigue tests with the inner pressure associated with push-pull and rev. torsion acts to reduce N_f.

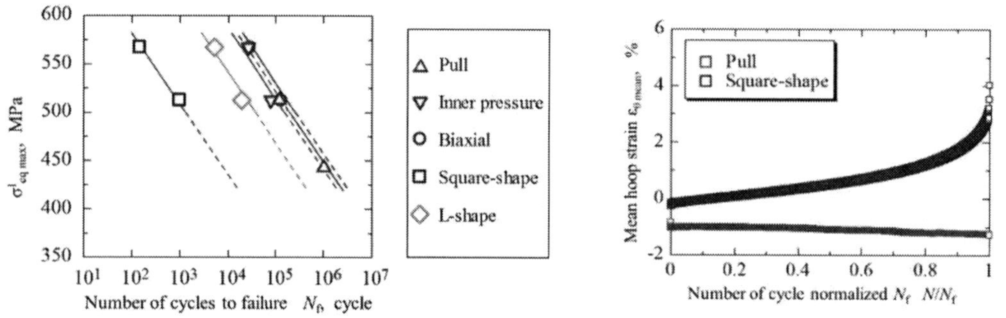

(a) Maximum Mises stress at inner surface *vs* N_f (b) Hoop mean strain $\varepsilon_{\theta mean}$ and N/N_f.

Fig.9. Correlation of failure life and property of cyclic defamation.

5 Conclusion

In this paper, types of multiaxial fatigue tests and multiaxial fatigue properties of structural materials were presented. Strain paths, principal strain ratio, non-proportionality of loading strongly govern the multiaxial failure life. Therefore, the development of multiaxial fatigue test methods and elucidation of mechanisms governing multiaxial fatigue failure is critically important to enable the accurate life estimation of actual structural components.

References

1. S. H. Doong, D.F. Socie, I.M. Robertson, J. Eng. Mater-T ASME **112** (1990)
2. C. H. Wang, M. W. Brown, Fatigue Fract. Eng. M. **16**, (1993)
3. D. F. Socie and G. B. Marquis, *Multiaxial fatigue*, SAE International, (1999)
4. L. Pejkowski, D. Skibicki, J. Sempruch, J. Mech. Eng. **60**, (2014)
5. F. Berto, A. Campagnolo, P. Lazzarin, Fatigue Fract. Eng. M. **38**, 5 (2015)
6. T. Itoh, K. Murashima, T. Hirai, J. Soc. Mater. Sci, Jpn. **56**, 157 (2007)
7. T. Itoh, M. Sakane, M. Ohnami, J. Eng. Mater. Technol., Trans. ASME **116**, 90 (1994)
8. F.A. Kandil, M.W. Brown, K.J. Miller, Mech. Behav. Nucl. Appl. Stainless Steel at Elevated Temp. (1982)
9. R.D. Lohr, E.G. Elison, Fatigue Eng. Mater. Struct. **3**, 1 (1980)
10. N. Hamada, M. Sakane, M. Ohnami, 3rd Int. Conf. on Biaxial/Multiaxial Fatigue **2** (1989)
11. T. Morishita, Y. Takada, T. Itoh, Fract. Struct. Integr. **41**, 71 (2017)

Experimental performance studies of small wall room air conditioner for R410A retrofitted with R290

Berkah Fajar TK.[1*] *and Geraldus* Nicolas Dipto I.[2,*]

[1]Department of Mechanical Engineering, Diponegoro University, Prof.Sudharto, SH. Street, Tembalang-Semarang 50275, Indonesia
[2]Department of Mechanical Engineering, Diponegoro University, Prof.Sudharto, SH. Street, Tembalang-Semarang 50275,. Indonesia

Abstract. An original R410A wall room air conditioner was retrofitted with R290 for performance experiments. The R410A wall room air conditioner has a COP and a refrigerating capacity 6.0597and 1929.3 W respectively. In this study, the compressor R410A wall room air conditioner compressor was replaced with an R22 which is compatible. Mass R290 charged into the wall room air conditioner was 33.88%, 38.12% and 42.35%. Refrigerating capacity obtained was 1837 W, 1848.2 W and 2008.4W, while the COP was 5.2014, 5.0294, and 5.3275. The results obtained the R410A wall room air conditioner performance better than if the air conditioning machine retrofitted with R290.

1 Introduction

Chlorofluorocarbons (CFCs) and hydrochlorocarbons (HCFCs) can affect both stratospheric ozone and climate change, whereas hydrocarbons (HFCs) can affect climate change [1]. Therefore, it is important to raise awareness about preserving the ozone layer. In 1981 UNEP (*United Nations Environment Programme*) started negotiating over international measures to prevent ozone layer depletion through Vienna Convention. This convention creates efforts to preserve the well-being of human kind and the environment from the effects of ozone layer depletion, promotes cooperation on international research, observes ozone layer condition, and exchanges information regarding the matter. Then, in September 1987, Vienna convention is followed with the ratification of Montreal Protocol which contains regulation monitoring production, consumption, and trading of ODS (Ozone Depleting Substance) [2].

Mao-Gang He and associates study theoretically and experimentally for propane and its mixture as a replacement for R134a which was used in a large capacity chest freezer. Theoretically, R290 is 87.7% and 54.2% higher than R134a in terms of the mass and volumetric refrigerating capacities, while R290/R600a mixture (90/10 wt%) is 48.3% and 2.4% higher than R134a in terms of the volumetric refrigerating capacity and COP. The experimental test shows power consumption of R290 is 26.7% lower than R134a and the obtained optimal mass ratio from R290/R600a mixture is 93.75/6.25 wt% with 27.5%

*Corresponding author: fajarberkah10@gmail.com

lower in terms of power consumption than R134a [3]. In Marco Bortolini and associates research about the performance of retrofit process R404a with R410A and R407f in a refrigerating system which the expected result shows that R410A and R407f could substitute R404a. It is observed that moderate temperatures (-5; -10 C) R407f and R410A can achieve higher COP and refrigerating capacity than R404a which R407f has higher refrigerating capacity. However the COP is lower than R410A [4]. According to M. Mohanraj and associates research about R290/R600a mixture with mass ratio of 45.2:54.8 compare to R134a in domestic refrigerator with single evaporator and variation of ambient temperature namely 24, 28, 32, 38, and 43°C, is obtained which propane mixture has 11.1%, 11.6% and 13.2% lower power consumption and 3.25-3.6% higher COP based on pull down and on time ratio [5]. C. S. Choudhari and S. N. Sapali are also conducting research on analytical possibility to substitute R22 with R290 with the application of standard vapor compression cycle of evaporation temperature ranging from 25°C - 10°C and condensation temperature of 45°C which refer the analytical calculation. REFPROP 9.0 is utilized to obtain COP of R290 which is lower than R22, however the COP is expected to be higher by developing certain system design that compatible with R290 characteristics [6]. Referring to previous references, current research aims to study phenomena that occur by comparing the performances of R410A-charged split-type air conditioner with R290 where both R290 mass variations are analyzed to determine its optimal charging mass in order to achieve the same performance as R410A. This retrofit process is expected to be considered as one of the efforts to substitute R410A into R290.

2 Experimental Setting and Procedure

2.1 Experimental Setting

The experimental apparatus consists of one unit of split-type AC with 5000 BTU rated cooling capacity, data acquisition device, and measurement instruments, namely pressure gauges, temperature sensors (lm35), and a digital scale. The experiment starts with 340 grams of R410A (in accordance with the AC's name plate). After all the data from R410A have been recorded by data acquisition, various mass of R290 were charged into the system which were 21,18%, 25,41%, 29,65%, 33,88%, 38,12%, and 42,35% based on total mass of R410A on the name plate. Changing the default compressor into R22 compressors is considered in this experiment because of previous research by Sutandi shows that R22 compressor is compatible to R290 [7]. Each variation is measured for 90 minutes with 2 minutes interval for each measured data. Measurement starts along with the system starts and stops after the system reaches a steady state. Last variation involves changing the current compressor with default compressor (R410A) then charge the system with R290 based on condenser pressure of 270 PSI.

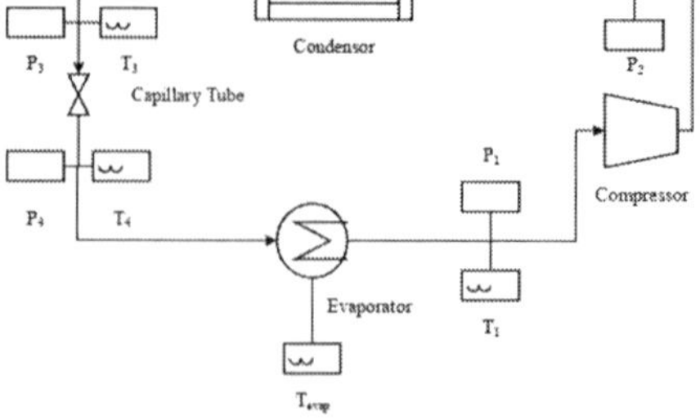

Figure 1. Schematic diagram of the experiment

From figure 1, the test apparatus and its measuring instruments are shown to illustrate the experiment. The following details describe each measuring instruments and their position:

1. One temperature sensor, LM35, is located where low temperature of air flows from evaporator (illustrated as T_{evap}) and the other are located after the refrigerant absorbed heat and pass through the evaporator. One pressure gauge is also located after the refrigerant pass through evaporator (illustrated as T_1 and P_1).
2. One temperature sensor, LM35, along with one pressure gauge is located at the outlet side where the refrigerant is compressed by the compressor (illustrated as T_2 and $P_{2)}$.
3. One temperature sensor, LM35, is located at the outlet of air flow from the condenser and the other one is located little distance away to measure the ambient temperature.
4. Two temperature sensors, LM35, and two pressures gauges are located at the inlet and outlet side of the capillary tube (illustrated as T3, P3, T4, and P4).

2.2 Experimental Procedure

After adjusting and setting the measurement devices, assembly process is conducted and leak test is also conducted on the testing apparatus afterward. If a leak is not detected, the system will be charged with R410A which its mass is adjusted according to its name plate. Re-checking the AC after conducting the test is mandatory so the system can perform normally. Then R290 with all its variations are charged into the AC gradually and are followed by replacing R410A compressor with R22/R290 compressor. The charged mass of R290 is measured with a digital scale by means of placing the refrigerant tank on top of the digital scale. Various variations, namely 21,18%, 25,41%, 29,65%, 33,88%, 38,12%, and 42,35% is obtained by adjusting and converting some variations that have been obtained and analyzed by Sutandi's research [6]. Before operating the system, always do a leak test first, then the experiment is ready. After collecting all data from various variations and default system, all those data are analyzed using Refprop to determine the enthalpy of each component.

3 Equation

A refrigerating effect is a difference of enthalpy on the outlet and inlet side of evaporator. The refrigerating effect can be calculated as following:

$$\frac{\dot{Q}}{\dot{m}} = h_1 - h_4 \tag{1}$$

While the work of the compressor can be calculated as follows:

$$\frac{\dot{W}_C}{\dot{m}} = h_2 - h_1 \tag{2}$$

Heat transfer rate per unit mass in condensor is total heat that is rejected by the condensor onto the surrounding per unit mass of refrigerant flowing.

$$\frac{\dot{Q}_{out}}{\dot{m}} = h_2 - h_3 \tag{3}$$

Efficiency of an air conditioner can be expressed by Coefficient of Performance (COP or β). It can be stated as the ratio between refrigerating capacity and work of the compressor.

$$\beta = \frac{\dot{Q}_{in}/\dot{m}}{\dot{W}_C/\dot{m}} = \frac{h_1 - h_4}{h_2 - h_1} \tag{4}$$

4 Results and Analysis

4.1 Refrigerating Capacity

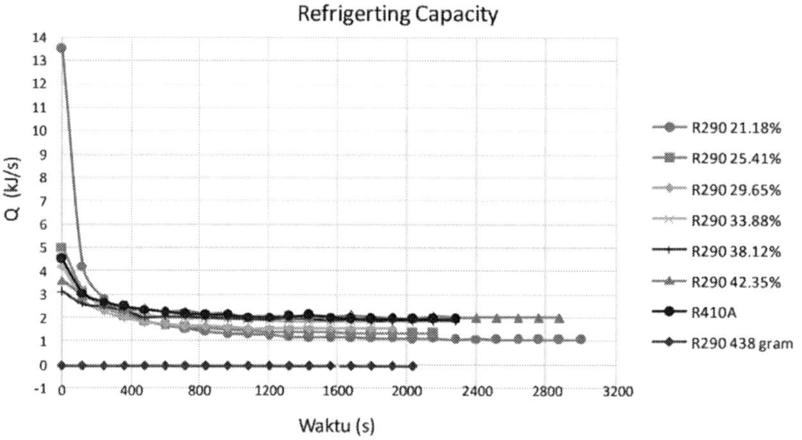

Figure 2. Graphic of refrigerating capacity of the evaporator in split-type air conditioner

From figure 2 above, it can be observed that the highest value of refrigerating capacity in steady state is resulted from 42.35% mass of R290 which has the value of 2.0084 kJ/s, while the lowest value of refrigerating capacity is resulted from 438 grams of R290 mass with an amount of -0.0831 kJ/s. It can be seen from the graphic above that the refrigerating capacity slowly declined over time. This phenomenon happens because the enthalpy from the outlet side of evaporator keeps declining, however the enthalpy on the inlet side keeps inclining over time. R410A has lower refrigerating effect, about 178.64 kJ/kg, than all variants of R290 mass which have values from 311.32 to 333 kJ/kg (except for R290 438 grams), however R410's value of refrigerating capacity has an amount of 1.929 kJ/s, slightly lower than R290 mass of 42.35%. This phenomenon is caused by the value of R410A mass flow rate which is considered to be the highest mass flow rate among all R290 mass variants with a value of 0.0108 kg/s. Therefore in terms of refrigerating capacity mass variations of R290 that can equal R410's refrigerating capacity are R290 33.88%, 38.12%, and 42.35%.

4.2 Compressor Power Consumption

From the figure 3, the highest value of compressor power consumption is achieved by R290 42.35% mass variation with an amount of 0.377 kJ/s, while the lowest value is achieved by R290 21.18% mass variation with an amount of 0.312 kJ/s.

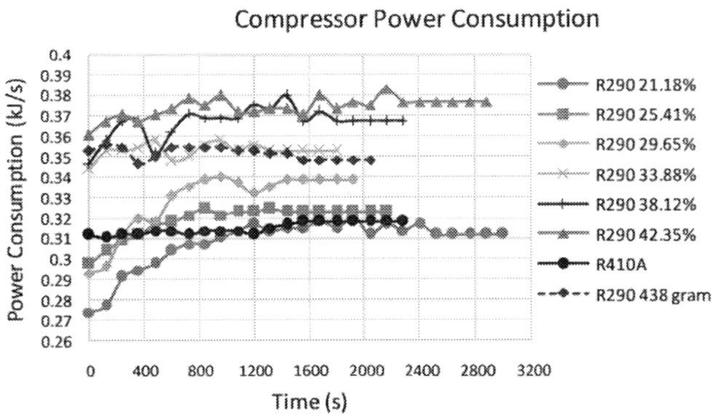

Figure 3. Graphic of compressor power consumption in split-type air conditioner

Each variation has similar trend which is gradually inclined then fluctuated over time (except for 438 grams of R290), due to the increased of enthalpy difference at the discharge and suction side of the compressor. R410A has relatively lower compressor work per unit of mass which has the value of 29.48 kJ/kg than the compressor work per unit mass of all R290 mass variants which have values from 59.51 to 96.48 kJ/kg. Even though the mass flow rate of R410A is considered to be the largest of all variants, R410A has lower compressor work per unit mass, hence the power consumption of R410A is also relatively low. Therefore in terms of compressor power consumption, the variants of R290 mass that can equal or even lower than the compressor power consumption of R410 are R290 21.18% and 25.41% which have the value of 0.312 kJ/s and 0.3231 kJ/s, respectively.

4.3 Coefficient of Perfomance

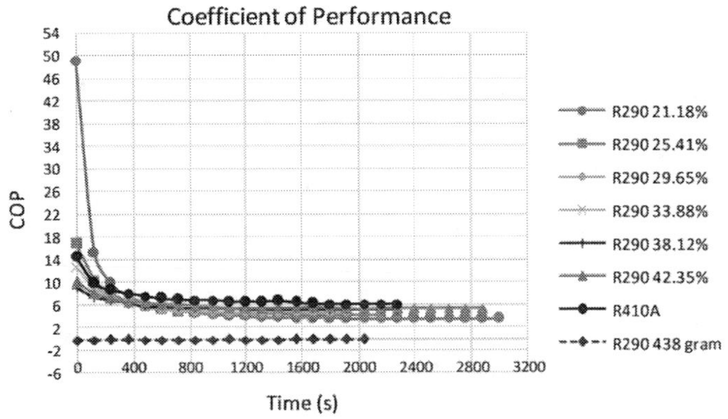

Figure 4. Graphic of coefficient of performance in split-type air conditioner *Split*

From the figure 4, it can be observed that the highest value of coefficient of performance in steady state is resulted from R410A as default system which has the value of 6,0597, while the lowest value of coefficient of performance is resulted from 438 grams of R290 mass which has the value of 0,23839. Each variation has a similar trend which is gradually inclined over time until a steady state is achieved. This phenomenon occurs due to the value of refrigerating effect per unit mass which gradually decrease over time, however the compressor work per unit mass gradually increases as well. Hence in terms of coefficient of performance, the variants of R290 mass that can equal or even lower than the R410A performance are R290 33,88%, R290 38,12% and R290 42,35% with a value of 5,2014, 5,0294, and 5,3275, respectively.

5 Conclusions

1. The coefficient of performance (COP) obtained from R410A test has the value of 6.097 in a steady state while the refrigerating capacity and compressor power consumption have the value of 1.9293 kJ/s and 0.3183 kJ/s, respectively.
2. After retrofitting the air conditioner from R410A into R290 either with changing the R410A compressor with the compatible one or changing it back to R410 compressor, the result analysis shows that in steady state, R290 33,88%, 38,12%, and 42,35% have values of COP that close to the COP of R410A. Those values are 5.2014, 5.02940, and 5.3275, respectively. In terms of refrigerating capacity, the variations of R290 mass that can equal the performance of R410A are R290 33.88% with 1.8373 kJ/s, 38.12% with 1.8482 kJ/s, and 42.35% with 2.0084 kJ/s. From the compressor power consumption, it is observed that R290 21.18% with 0.3120 kJ/s and 25.41% with 0.3231 kJ/s are close to the result of R410A test. Therefore, experimenting on the air conditioner and conducting retrofit of R410A into R290 is proven to give effect on the air conditioner system in terms of COP, refrigerating capacity, compressor power consumption, and heating capacity.
3. After conducting various test of every variation of R290 mass, it can be concluded that the charged mass of 33.88% R290 shows that it can be considered to be used in retrofitting air conditioner from R410A into R290 even though its compressor power consumption is higher than R410A. Due to lower charged mass of R290, 33.88% is much safer than one with 38.12% or 42.35%.

References

[1] ASHRAE.. *ASHRAE Handbook Fundamentals (SI Edition)*(2005)
[2] Ozone.unep.org/en/treaties-and decisions
[3] M. Gang He, et al. *Application of Natural Refrigerant Propane and Propane/Isobutane in Large Capacity Chest Freezer*. Applied Thermal Engineering 70 (2014) 732 - 736 (2014)
[4] M. Bortolini, et al. *Retrofitting of R404a Commercial Refrigeration Systems Using R410a and R407f Refrigerant*. International Journal of Refrigeration 55 (2015) 142 - 153 (2015)
[5] M. Mohanraj, et al. *Experimental Investigation of R290/R600a Mixture as an Alternative to R134a in a Domestic Refrigerator*. International Journal of Thermal Sciences 48 (2009) 1036 – 1042 (2009)
[6] C. S. Choudhar, S. N. Sapali. 2017. *Performance Investigation of Natural Refrigerant R290 as a Substitute to R22 in Refrigeration Systems*. Energy Procedia 109 (2017) 346 – 352.
[7] T. Sutandi. *Experimental Study of Vapor Compression Refrigerator Performance in Variation of Propane-type Refrigerant Mass as a Substitute to R22*. Universits Diponegoro (2014)

Simulation and analysis of the aeroelastic-galloping-based piezoelectric energy harvester utilizing FEM and CFD

Ismoyo Haryanto[1,*], *Achmad* Widodo[1] , *Toni* Prahasto[1], *Djoeli* Satrijo[1], *Iswan* Pradiptya[2], and *Hassen* Ouakad[2]

[1]Mechanical Engineering Department, Faculty of Engineering, University of Diponegoro, Tembalang, Semarang, Indonesia

[2]Mechanical Engineering Department, College of Engineering Sciences and Applied Engineering King Fahd University of Petroleum & Minerals, Dhahran, 31261, Kingdom of Saudi Arabia

Abstract. Due to a large oscillation amplitude, galloping can be an admissible scenario to actuate the piezoelectric-based energy harvester. In the case of harvesting energy from galloping vibrations, a prismatic bluff body is attached on the free end of a piezoelectric cantilever beam and the oscillation occurs in a plane normal to the incoming flow. The electrical power then can be extracted from the piezoelectric sheet bonded in the cantilever structure due to the dynamic strain. This study is proposed to develop a theoretical model of a galloping-based piezoelectric energy harvester. A FEM procedure is utilized to determine dynamic characteristics of the structure. Whereas the aerodynamic lift and drag coefficients of the tip bluff body are determined using CDF. The results show that the present method gives precise results of the power generated by harvester. It was found that *D*-section yields the greatest galloping behavior and hence the maximum power.

1 Introduction

The lifespan limitation of small batteries is one of the main challenges in the standalone systems/self-power devices application such as a wireless sensor network [1]. Structural vibration is one of the most used mechanical domain which enables to be converted into electrical field by utilizing several well-known energy harvester. Piezoelectric-based vibration energy harvester has been used as favorable transduction mechanism by many researchers due to its simplicity, efficient and wide range operational frequencies. There are many studies have been published aiming the investigation of piezoelectric energy harvester as an electronic devices power supply, either as redundant charging system or primary power supply [2,3,4,5]. There are increasing research interest of aeroelastic-based energy harvesters including vortex-induced-based and galloping-based energy harvesters. Galloping is one of the classical aeroelastic instabilities phenomena which can be

* Corresponding author: ismoyo_h@undip.ac.id

characterized as a low-frequency and large-amplitude oscillation. As an aeroelastic phenomenon, galloping is generated by a coupling between the forces of aerodynamic, elastic and inertia. Because causing a large oscillation amplitude, galloping can be an admissible scenario to actuate the piezoelectric-based energy harvester. In the case of harvesting energy from galloping vibrations, a prismatic bluff body is attached on the free end of a piezoelectric cantilever beam and the oscillation occurs in a plane normal to the incoming flow. The electrical power then can be extracted from the piezoelectric sheet which bonded in the cantilever structure due to the dynamic strain.

2 Analytical modeling of energy harvester

Figure 1 shows a prismatic bluff body tip mass having a *D*-, square, or triangular -section is mounted at the tip of a flexible cantilever beam. When the bluff body is facing an incoming wind flow, then the occurring of wind-induced vibration causes the box to gallop in the transverse direction, leading to self-excited bending vibrations in the beam. Energy is harvested due to the motion of the piezoelectric sheet bonded in the cantilever beam, whose terminals are connected across a load resistance.

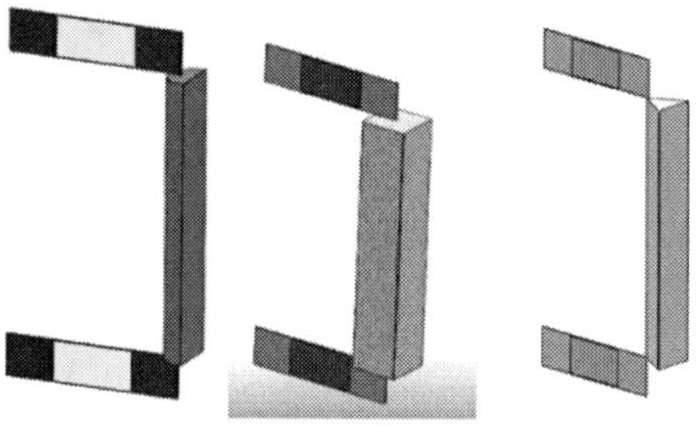

(a) (b) (c)

Fig. 1. A schematic illustration of energy harvester with tip body having section of (a) *D*, (b) square and (c) triangular.

2.1 Aerodynamic model

Figure 2 shows a section which is exposed to an air flow with a mean speed U and undergoing both transverse and angular displacements denoted by w and θ, respectively. The generated lift (L) and drag (D) forces are then formulated as

$$L = ½\, \rho(U^2 + (\dot{w}(L_b))^2)b_{tip}C_L \quad (1)$$

$$D = ½\, \rho(U^2 + (\dot{w}(L_b))^2)b_{tip}C_D \quad (2)$$

where ρ is the air density , b_{tip} is the bluff body section width, $\dot{w}$ is the transversal velocity, and C_L and C_D are the lift and drag coefficient, respectively. The aerodynamic analysis is conducted for determining C_L and C_D with the effective angle of attack of α. On the tip body the effective angle of attack is given by

$$\alpha = \tan^{-1}\left(\dot{w}(L_b)/U\right) + w'(L_b) \quad (3)$$

Projection of the lift and drag components along the vertical direction gives the excitation force F_{tip} in the downward as

$$F_{tip} = F_z L_{tip} = \left(L\cos\alpha + D\sin\alpha\right)L_{tip} = ½\,\rho U_{rel}^2 L_{tip} C_y \tag{4}$$

where $U_{rel} = \sqrt{U^2 + \left(\dot{w}(L_b)\right)^2}$, $C_y = C_L \cos\alpha + C_D \sin\alpha$ and L_{tip} is length of bluff body.

Assuming small values of the velocity ratio $\dot{w}/U$, the force coefficient can be expanded in powers of the angle of attack as follows

$$C_y(\alpha) = \sum_{j=0}^{n} a_j \alpha^j = \sum_{j=0}^{n} a_j \left(\frac{\dot{w}}{U}\right)^j \tag{5}$$

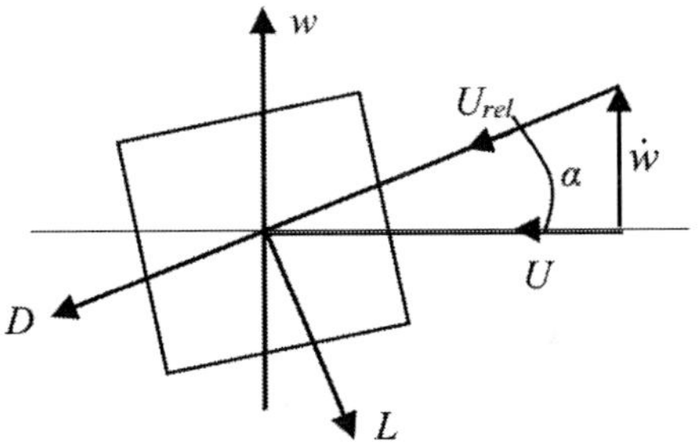

Fig. 2. Aerodynamic loads acting on the section of bluff body.

2.2 Structural Analysis

Consider a cantilever beam bonded with piezoelectric sheets. Both of the beam and the piezoelectric sheets are considered as Euler-Bernoulli Beam elements. So the elemental stiffness matrix $[k]$ and mass matrix $[m]$ of beam element are

$$[k] = \frac{EI}{l_p^3}\begin{bmatrix} 12 & 6l_p & -12 & 6l_p \\ 6l_p & 4l_p^2 & -6l_p & 2l_p^2 \\ -12 & -6l_p & 12 & -6l_p \\ 6l_p & 2l_p^2 & -6l_p & 4l_p^2 \end{bmatrix} \;;\; [m] = \frac{\rho A l_p}{420}\begin{bmatrix} 156 & 22l_p & 54 & -13l_p \\ 22l_p & 4l_p^2 & 13l_p & -3l_p^2 \\ 54 & 13l_p & 156 & -22l_p \\ -13l_p & -3l_p^2 & -22l_p & 4l_p^2 \end{bmatrix} \tag{6}$$

In which $EI = E_b I_b + 2E_p I_p$ is the flexural rigidity and $\rho A = b_b(\rho_b t_b + 2\rho_p t_p)$ is the mass per unit length of a bonded Beam-Piezoelectric Element. In the present study suffix '*b*' is used for regular beam element, suffix '*p*' is used for piezoelectric element. If w indicates the vertical displacements of the beam, $[\overline{M}]$ and $[\overline{K}]$ are global mass and stiffness matrices of the structure, respectively, then the equation of motion for free vibration of entire structure is

$$[\overline{M}]\{\ddot{w}\} + [\overline{K}]\{w\} = 0 \tag{7}$$

Solving Equation (7) gives natural frequencies, ω_n, and mode shapes, ϕ, of the system.

2.3 Equation of motion

The coupled matrix differential equations for a piezoelectric energy harvester, namely electromechanic equation, can be formulated with respect to the generalized coordinate u and the voltage generated by the piezoelectric sheets V as

$$M\ddot{u} + Ku + \Theta V = F_{tip}\phi(L_b) \tag{8}$$

$$\Theta^T u + \frac{1}{C_P} V + R_L \dot{V} = 0 \quad (9)$$

where M and K are matrices of generalized mass and generalized stiffness of the system, respectively. Whereas Θ is the electromechanical coupling coefficient defined as $\Theta = -E_p d_{31} b_p (t_p + t_b)/2$, where d_{31} is strain coefficient of the piezoelectric. In Equation (9) $C_P = e_{33} A_p / t_p$ is the capacitance of the piezoelectric sheets (at constant strain), where e_{33} is dielectric constant of the piezoelectric. By defining a state vector containing the generalized displacement, generalized velocity, and charge, $X = [X_1 \quad X_2 \quad X_3]^T = [u \quad \dot{u} \quad q]^T$ the equations of motion (Equations (8) and (9)) can be rewritten as

$$\begin{aligned} \dot{X}_1 &= X_2 \\ \dot{X}_2 &= -\left(M^{-1}K\right)X_1 - \left(M^{-1}\Theta\right)X_3 - (M^{-1}\phi(L_P))F_{tip} \\ \dot{X}_3 &= -\left(\frac{1}{R_L}\Theta^T\right)X_1 - \left(\frac{1}{R_L C_P}\right)X_3 \end{aligned} \quad (10)$$

Equation (10) can be solved using standard numerical method to obtain time response of displacement, velocity and electrical charge of harvester.

3 Results

Table 1 shows the geometrical and material properties of cantilever beam and piezoelectric which are applied in this study. A tip mass of 0.08 kg was used for generating galloping. The CFD analysis was performed to obtain the variation of lift and drag coefficients (C_L and C_D) at angles of attack ranging from -45° to 45° and at Reynolds number of 10800.

Table 1. Geometrical and material properties harvester.

Parameter	Cantilever Beam	Piezoelectric
Material	Aluminium	Piezoelectric PZT-5H
Length (m)	$l_b = 0.161$	$l_p = 0.0724$
Width (m)	$b_b = 0.038$	$b_p = 0.0362$
Thickness (m)	$t_b = 0.00635$	$t_p = 0.267 \times 10^{-3}$
Young's modulus (N/m^2)	$E_b = 6.9 \times 10^{10}$	$E_p = 6.2 \times 10^{10}$
Density (kg/m^3)	$\rho_b = 2700$	$\rho_p = 7800$
Damping Constants	$\alpha = 0.001$, $\beta = 0.0001$	-
Stress Constant (Vm/N)	-	$g_{31} = 8.5 \times 10^{-3}$
Strain Constant (C/N)	-	$d_{31} = -320 \times 10^{-12}$
Dielectric Constant (nF/m)	-	$e_{33} = 33.65$

The natural frequency of the present energy harvester was found to be 21.90 Hz. Figure 3 shows a comparison of measured and predicted output voltage when air speed 4.47 m/s and load resistance 37 kΩ for a triangular cross section. It can be seen that the trend of the predicted voltage closely approximates that of the measured voltage and the steady-state behavior correlates very well with predictions. Furthermore, this method is applied to predict the characteristic of energy harvester with bluff body of *D*- and rectangular section.

Figures 4a and 5a show samples of the velocity field distribution for Reynolds number of 10800 obtained using CFD to determine C_L and C_D for *D*- and rectangular section, respectively. Having been C_L and C_D determined the aerodynamic coefficient C_y can be calculated. The approximation of C_y then can be conducted utilizing Equation (4). Table 2 shows the constant coefficients for approximate C_y for *D*- and square section. Meanwhile, the comparison of calculated and approximated of C_y are shown by Figure

4b and 5b. Figure 6 and 7 show the real part of first eigenvalue and the time history of the output voltage plots for the *D*- and square section for a load resistance of 37 kΩ and an air speed of 4.47 m/s. It can be noted that the galloping onset for *D*- and square section are 1.85 m/s and 3.05 m/s, respectively.

Table 2. Constant coefficient of polynomial for C_y approximation

Section	a_0	a_1	a_2	a_3	a_4	a_5	a_6	a_7	a_8	a_9
D-section	0	0.57	-0.05	0.82	0.41	-24.0	-1.00	83.3	0.76	-73.5
Square	0	1.70	0	-21.0	0	48.0	0	-35.0	-	-

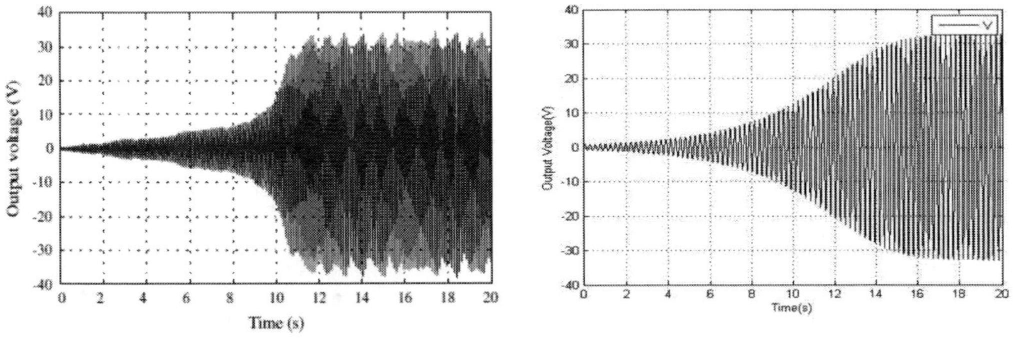

(a) (b)

Fig. 3. (a) Measured output voltage [5], (b) predicted output voltage for a triangular section when U = 4.47 m/s and R = 37 kΩ.

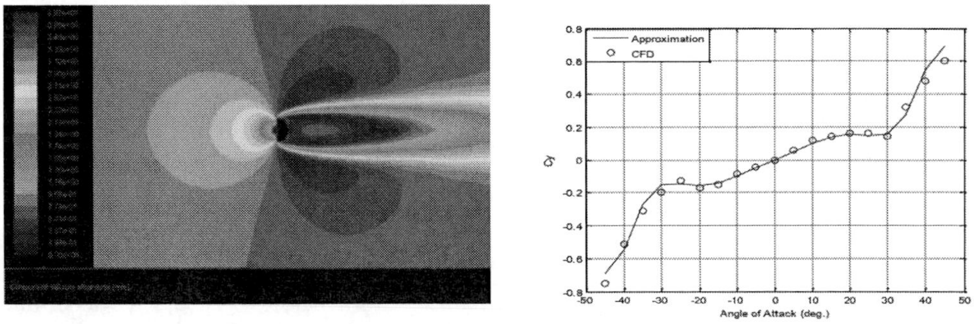

(a) (b)

Fig. 4. (a) Velocity field distribution, (b) steady force coefficient C_y for *D*- section.

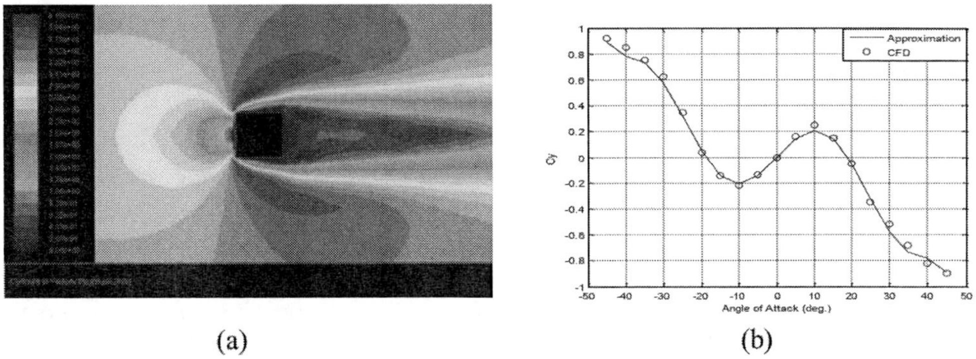

(a) (b)

Fig. 5. (a) Velocity field distribution, (b) steady force coefficient C_y for square section.

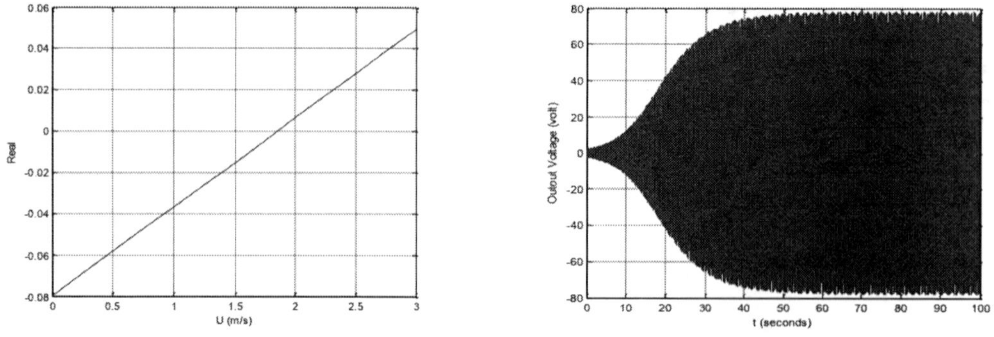

(a) (b)

Fig. 6. (a) Real part of the first eigenvalue, (b) time histories of the harvested voltage when $R = 37$ kΩ and $U = 4.47$ m/s for *D*-section.

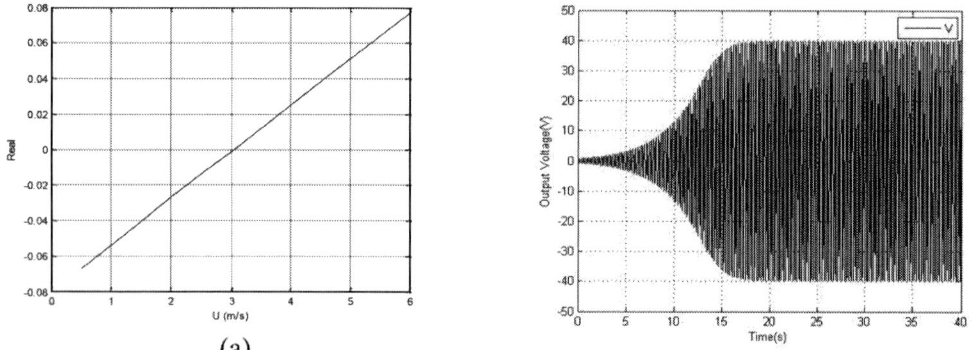

(a)

(b)

Fig. 7. (a) Real part of the first eigenvalue, (b) time histories of the harvested voltage when $R = 37$ kΩ and $U = 4.47$ m/s for square section.

4 Conclusions

A piezoelectric energy harvester, based on the concept of aeroelastic galloping was developed in this work. A finite element model was formulated to evaluate the dynamic characteristics of the system. Meanwhile, the lift and drag coefficients of the galloping section are evaluated using CFD. Of the geometries studied, the *D*- section was found to yield the lowest galloping onset and the greatest galloping behavior and hence the maximum power. Meanwhile, the shortest transient time response was found by the square section of the bluff body. The proposed design is considered to offer a viable solution for a standalone power source for remote wireless sensors.

The authors acknowledge the support of Messrs. Abid Rahman, Rendi Saputro and Yusran Anwar in conducting the Finite Element models and some computations.

References

1. A. Abdelkefi, "Aeroelastic energy harvesting: A review," *International Journal of Engineering Science,* **100**, 112 (2016)
2. S. Anton, R. and H. Sodano, A., "A review of power harvesting using piezoelectric materials (2003–2006)," *Smart Materials and Structures,* **16**, R1 (2007)

3. A. Erturk, W. G. R. Vieira, C. De Marqui, and D. J. Inman, "On the energy harvesting potential of piezoaeroelastic systems," *Applied Physics Letters,* **96**, p. 184103, 2010.
4. S. Roundy and P. K. Wright, "A piezoelectric vibration based generator for wireless electronics," *Smart Materials and Structures,* **13**, 1131 (2004)
5. J. Sirohi and R. Mahadik, "Piezoelectric Wind Energy Harvester for Low-Power Sensors," *Journal of Intelligent Material Systems and Structures,* **16**, 2011 (2011).
6. A. Abdelkefi, A. H. Nayfeh, and M. R. Hajj, "Effects of nonlinear piezoelectric coupling on energy harvesters under direct excitation," *Nonlinear Dynamics,* **67**, 1221 (2012)

An Integration of PSO-based Feature Selection and Random Forest for Anomaly Detection in IoT Network

Bayu Adhi Tama[1], *Kyung-Hyune* Rhee[1*]

[1]Information Security and Internet Applications Lab., IT Convergence and Application Engineering, Pukyong National University
Daeyon Campus, 45, Yongso-ro, Nam-Gu. Busan, South Korea 48513

Abstract. The most challenging research topic in the field of intrusion detection system (IDS) is anomaly detection. It is able to repeal any peculiar activities in the network by contrasting them with normal patterns. This paper proposes an efficient random forest (RF) model with particle swarm optimization (PSO)-based feature selection for IDS. The performance model is evaluated on a well-known benchmarking dataset, i.e. NSL-KDD in terms of accuracy, precision, recall, and false alarm rate (FAR) metrics. Furthermore, we evaluate the significance differences between the proposed model and other classifiers, i.e. rotation forest (RoF) and deep neural network (DNN) using statistical significance test. Based on the statistical tests, the proposed model significantly outperforms other classifiers involved in the experiment.

1 Introduction

The present escalation of Internet of Things (IoT) devices and services has changed our daily life dramatically. Many applications are built based on IoT technologies, i.e. smart cities, smart health care, smart home and vehicular networks [1]. Apart from these benefits, attackers may take this such opportunity to launch malevolent code or program to the network. According to [2], security is a key barrier of the implementation of IoT network and services. This is because IoT works with different standard and protocol forming a heterogeneous network. Moreover, IoT devices prevalently produce a huge amount of data so it might become a big threat as malicious users can intercept the data while it is transmitted.

As the development of IoT devices increase, insecure information processing might immediately affects to the whole IoT network. The jeopardy of information disclosure in public space will increase caused by the broadly development of IoT. As presented in Figure 1, security architecture in IoT is divided into three layers, i.e. perception layer, transportation layer, and application layer [3] [4]. Transportation layer comprises network access security which is responsible for attack detection and prevention. An intrusion detection system (IDS) is one security solution which can be deployed in the transportation

*Corresponding author: khrhee@pknu.ac.kr

layer. IDS copes with security threats, i.e. DoS/DDoS attack, wireless LAN attack, middle attack which affect the transportation security of IoT.

An IDS possesses an ability to detect and repeal any malicious activities in the network before they cause a strenuous failure. Moreover, it is prevalently comprised of two approaches, i.e. signature detection and anomaly detection. Among the two techniques, anomaly detection has gained much interest since it is able to detect new types of attacks by characterizing the pattern that deviates significantly from normal network profiles. Nevertheless, this approach still suffers from higher false alarm rate (FAR) [5] [6]. Since the main issue of anomaly detection is the solving of two-class classification problem, the constructed classifiers should have a good performance, i.e. higher predictive accuracy while preserving lower FAR.

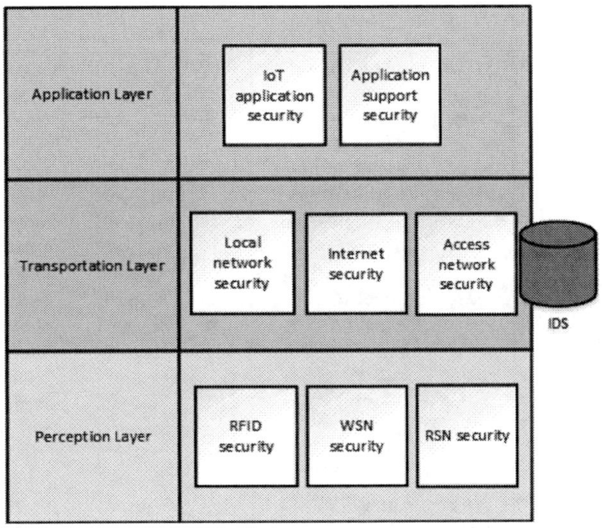

Fig. 1. IoT security framework.

In this paper we incorporate a feature selection technique using particle swarm optimization (PSO) and a classification algorithm so called random forest (RF) [7] [8] for anomaly detection in IoT network. The contributions of the paper are as follows.

- The performance differences of RF are compared with other similar ensemble learners, i.e. rotation forest (RoF) [9] and deep neural network (DNN) [10] using two statistical tests. This attempt is yet to explore in the IDS research so far (see Table 1 for details).
- The grid search is applied to the RF in order to obtain the best learning parameters. This allows RF to outperform other similar ensembles.
- Based on the experimental results, the proposed model performs better than other models found in the literature.

2 Related Work

In this section we present the related work of anomaly detection using ensemble learning. A plethora techniques have been proposed, yet we exclude other techniques, i.e. single classifier. Table 1 summarizes the existing literature and the current work by categorizing them into several features, i.e. ensemble scheme, base classifiers, dataset, performance metrics, and statistical test.

3 Method

3.1 Overview of Random Forest

Random forest (RF) is an ensemble classifier for classification and regression. It is another variant of bagging ensemble proposed by Breiman [7]. In some cases, it performs better than boosting and faster than bagging and boosting [7]. The original version of RF could be taught as a version of bagging where the base classifier is a random tree (RT) [19]. However, it is considered as an ensemble learning where the base classifier is decision tree.

Table 1. Prior researches on IDS using ensemble learning

Study	Ensemble Scheme	Base Classifiers	Dataset	Performance Metrics	Statistical Test
[11]	Weighted ensemble	Classification and regression trees (CART), Bayesian networks	KDD Cup 1999	Accuracy	No
[12]	Majority voting	Neural network, support vector machine, and multivariate regression splines	KDD Cup 1999	Accuracy	No
[13]	Weighted ensemble	Decision tree, support vector machine	KDD Cup 1999	Accuracy	No
[14]	Boosting	Decisions stumps	KDD Cup 1999	Precision, false alarm rate	No
[15]	Product rule	NA	Private	Area under ROC curve (AUC)	No
[16]	Min, Max, and product rule	k-means, v-SVC	KDD Cup 1999	Precision, false alarm rate	No
[17]	Voting	Neural network, decision tree	KDD Cup 1999	TP rate, FP rate, Precision, Recall, and F1 measure Accuracy	No
[18]	Bagging	Multilayer perceptron, radial basis function	Private	accuracy, precision,	No
This Study	Random forest	Decision tree	NSL-KDD	recall, and false alarm rate	Yes

Furthermore, RF is a classifier comprising a pool of tree-structured classifiers, each tree grown with respect to a random vector, which are independent and identically distributed. Each tree in the ensemble gives a vote for the most popular class of input vector [19]. The diversity of RF could be obtained by sampling from feature set, from the data set, or just varying randomly some parameters of the decision tree [19]. There are two parameters in RF: the number of variables to be selected in each node which is commonly kept constant in all nodes, and the number of trees, that build the forest. We adopted an efficient implementation of distributed RF in R with H2O package [20].

3.2 Experimental Setup

3.2.1 Validation Technique

In most cases, validation techniques (resampling) are used to evaluate the performance of classification algorithms. A resampling strategy so-called *k*-fold cross validation is employed. The dataset *D* is partitioned into k subsamples of equal size. In the *n*-th of the *k* iterations, the *n* subsample is used for testing, whilst the union of the remaining subsamples are used for training. We take into account $k = 10$ or *10*-fold cross validation.

3.2.2 Statistical Significance Test

Performance differences among classifiers are assessed using two statistical test, i.e. Quade and Quade post-hoc test [21]. It is quite powerful than Friedman test in case of number classifiers to be compared is less than five ($k < 5$). The null hypothesis (*H0*) is that there are no performance differences among classifiers, whilst alternative hypothesis (*HA*) means that at least one classifier's performance differs from at least one other classifier's performances.

4 Result and Discussion

First of all, the result of feature selection is presented in this section. Different number of particles (n) are taken into account in order to obtain the best configuration of features from the NSL-KDD. Originally, the dataset comprises 41 features; 25,192 instances (20% of training samples); and two classes, i.e. normal and malicious. The selected-features at each experiment is then assessed based on classification analysis using REPTree algorithm. A total 10 experiments are conducted in this study. The results of feature selection experiment are presented in Table 2. It is obvious that PSO with $n = 2$ is the best combination for feature selection. A total of 37 features are successfully obtained with 99.67% accuracy.

Table 2. The result of PSO-based feature selection.

Number of particles (*n*)	Selected features	Accuracy (%)	Number of particles (*n*)	Selected features	Accuracy (%)
2	37	99.67	100	6	98.09
5	12	99.3	200	7	99.06
10	19	99.56	500	7	99.07
20	5	99.02	1000	8	99.07
50	6	98.98	2000	8	99.07

Afterward, we show and discuss the average of performance results of all classifiers involved in our experiment, i.e. rotation forest (RoF) and deep neural network (DNN), and the proposed model (RF). A classifier, namely decision tree (J48) is employed as base classifier of RoF. Parameter settings of DNN are as follows. The $rate = 0.01$, $l1 = 1e\text{-}05$, $l2 = 1e\text{-}05$, $max\_w2 = 10$, $rate\_annealing = 2e\text{-}06$, and number of hidden layer is 3 with 500 nodes at each layer. Figure 2 contrasts the average performance of all classifiers in terms of accuracy, precision, and recall, while Figure 3 compares the performance in terms of FAR. It is revealed that the proposed model outperforms RoF and DNN in terms of all performance metrics. It does not only enhance the detection accuracy, but also it reduces FAR significantly compared to other classifiers.

In addition, the benchmark of classifier's performance using Quade and Quade post-hoc test are shown in Table 3 and Table 4, respectively. Regarding the result of Quade test, it might be concluded that the performance differences among classifiers are highly significant ($p < 0.01$), thus we can reject the H0. Consecutively, by referring Table 4, it is worth-mentioning that the proposed method's performance is very significantly different ($p < 0.01$) compared to RoF in terms of all evaluation metrics.

Table 3. The results of significant test using Quade.

Metrics	F	*p*-value
Accuracy	14.542	0.0001744
Precision	18.966	3.703e-05
Recall	17.798	5.436e-05

Table 4. Pairwise comparisons using Quade post-hoc.

	Accuracy		Precision		Recall	
	RoF	DNN	RoF	DNN	RoF	DNN
DNN	0.0300	-	0.0012	-	0.0142	-
PROPOSED	0.0073	4.1e-05	9.4e-06	0.0387	1.2e-05	0.0045

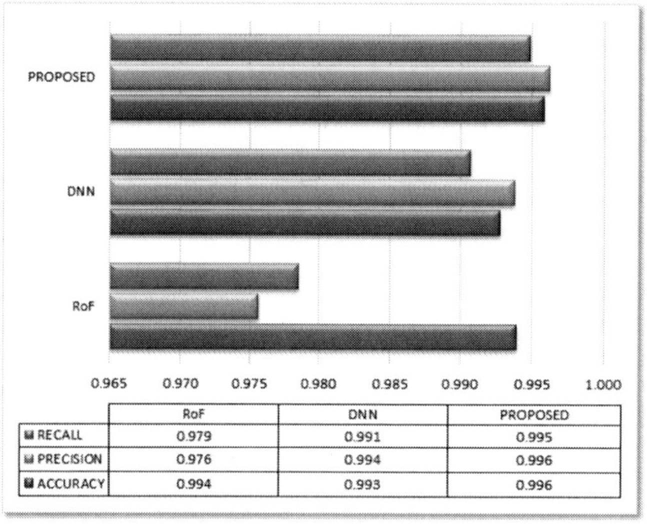

Fig. 2. Performance average of all classifiers in terms of accuracy, precision, recall metric.

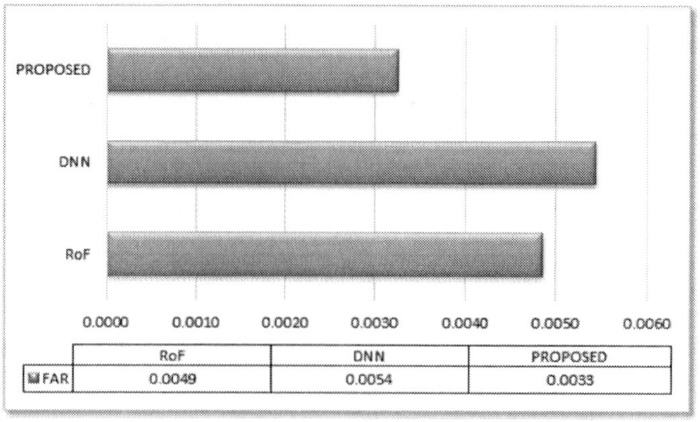

Fig. 3. Performance average of all classifiers in terms of FAR metric

5 Conclusion

Anomaly detection has been actively researched in IDS. We proposed an effective anomaly detection by incorporating PSO-based feature selection and random forest model. The performance differences among classifiers were validated using two statistical tests. Based on the experimental results using NSL-KDD dataset, it could be revealed that the proposed model outperforms others two classifiers, i.e. rotation forest (RoF) and deep neural network (DNN) in terms of all performance metrics.

This work was supported by the National Research Foundation of Korea (NRF) grant funded by the Korea government (MSIT) (No. NRF-2014R1A2A1A11052981), and partially supported by the MSIT (Ministry of Science and ICT), Korea, under the ITRC (Information Technology Research Center) support program (IITP-2017-2015-0-00403) supervised by the IITP (Institute for Information & Communications Technology Promotion).

References

1. L. Atzori, A. Iera, G. Morabito, Computer networks **54**, 2787 (2010)
2. S. Sicari, A. Rizzardi, L.A. Grieco, A. Coen-Porisini, Computer Networks **76**, 146 (2015)
3. L. Liu, S. Lai, *in International Conference onWireless Communications, Networking and Mobile Computing* (2006), pp. 1–4
4. Q. Jing, A.V. Vasilakos, J. Wan, J. Lu, D. Qiu, Wireless Networks **20**, 2481 (2014)
5. B.A. Tama, K.H. Rhee, in *Advanced Multimedia and Ubiquitous Engineering* (Springer, 2017), pp. 452–458
6. B.A. Tama, K.H. Rhee, Neural Computing and Applications pp. 1–11 (2017)
7. L. Breiman, Machine learning **45**, 5 (2001)
8. B.A. Tama, Journal of Information Processing Systems **11**, 165 (2015)
9. J.J. Rodriguez, L.I. Kuncheva, C.J. Alonso, IEEE transactions on pattern analysis and machine intelligence **28**, 1619 (2006)
10. Y. LeCun, Y. Bengio, G. Hinton, Nature **521**, 436 (2015)
11. S. Chebrolu, A. Abraham, J.P. Thomas, Computers & security **24**, 295 (2005)
12. S. Mukkamala, A.H. Sung, A. Abraham, Journal of network and computer applications **28**, 167 (2005)
13. S. Peddabachigari, A. Abraham, C. Grosan, J. Thomas, Journal of network and computer applications **30**, 114 (2007)
14. W. Hu, W. Hu, S. Maybank, IEEE Transactions on Systems, Man, and Cybernetics, Part B (Cybernetics) **38**, 577 (2008)
15. J.B. Cabrera, C. Gutiérrez, R.K. Mehra, Information Fusion **9**, 96 (2008)
16. G. Giacinto, R. Perdisci, M. Del Rio, F. Roli, Information Fusion **9**, 69 (2008)
17. S.S.S. Sindhu, S. Geetha, A. Kannan, Expert Systems with applications **39**, 129 (2012)
18. M. Govindarajan, R. Chandrasekaran, Computer networks **55**, 1662 (2011)
19. L.I. Kuncheva, *Combining pattern classifiers: methods and algorithms* (John Wiley & Sons, 2004)
20. A. Candel, V. Parmar, E. LeDell, A. Arora, *Deep learning with h2o* (2015)
21. D.J. Sheskin, *Handbook of parametric and nonparametric statistical procedures* (CRC Press, 2003)

Analysis of calcium carbonate polymorphs deposited in water piping system and the effect of tartaric acid additive

W. A. Putranto [1,*], *Y.* Umardhani [1], Sulistyo [1], Yurianto [1], and *A. P.* Bayuseno [1]

[1] Department of Mechanical Engineering, Faculty of Engineering, Diponegoro University, Tembalang Campus, Semarang 50275, Indonesia

Abstract. Calcium carbonate ($CaCO_3$) formed in a water piping system was investigated in the presence of chemical additives tartaric acid (0.00 and 10.00 ppm) and various temperatures ((27 and 50°C). The flow rate inside pipe (35 ml/min) were selected. Solutions of $CaCl_2$ and Na_2CO_3 were prepared in water with equimolar to Ca^{2+} concentration of 3000 ppm. The induction time of scale nucleation varied from 24 min to 44 min. An increasing temperature of the solution resulted in more $CaCO_3$ scale, mass, while the higher tartaric acid made the reduced mass of scales by 90%. SEM/EDS analysis verified $CaCO_3$ with a plate like morphology. Also the XRPD Rietveld method provided the confirmation of a major phase of calcite and vaterite minerals followed by the minor of aragonite in the absence of tartaric acid and at room temperature. However, the presence of 10 ppm tartaric acid yielded the increasing result of calcite, while aragonite precipitation was hampered under the influence of tartaric acid. The presence of tartaric acid at increasing temperature of 50°C could delay formation of calcite, whereas aragonite could be formed significantly. Thus the study showed the capacity of the tartaric acid in influencing $CaCO_3$ crystallization.

1 Introduction

The mineral deposits of $CaCO_3$ are commonly found in all types of water resources such as surface water, ground water, brackish water, seawater and wastewater. Generally, the $CaCO_3$ crystallization can be found in many piping systems such as cooling water [4]. The scaling of pipes can reduce heat transfer performance and lead to high cost of maintenance [5]. The crystallization phenomenon follows three processes: supersaturation, nucleation and crystal growth. The saturated solution is the pre-condition and the driving force of the crystallization process which affects the rate of nucleation [6]. $CaCO_3$ scale may be formed in three polymorph crystals: namely calcite, vaterite and aragonite, of which calcite is the most stable form at any temperatures [1, 7]. Vaterite and aragonite, at given appropriate conditions, may eventually transform into calcite. [8]. Studies on the $CaCO_3$ scaling in pipes are always associated with crystallization, which can be affected by many factors such as fluid flow rates, temperature and presence of chemical additives [9].

This paper discusses characteristics of calcium carbonate deposits formed in a water piping system due to the presence of additive of tartaric acid and under influence of various temperatures and the constant flow rate. SEM (Scanning Electron Microscope) / EDX (Energy Dispersive X-ray Spectrometry) was employed for examining the crystal deposits

* Corresponding author : wap_mesin@yahoo.co.id

formed in the solution, while XRPD (X-ray powder diffraction) Rietveld was used to validate the crystalline form of calcium carbonate.

2 Materials and Methods

2.1 $CaCO_3$ crystal formation in solution

Scaling experiments were performed using an experimental laboratory unit, which was equipped with a small diameter pipe section in which the laminar solution is flown by a water pump [2]. Equimolar solutions of $CaCl_2$ and Na_2CO_3 were prepared for the crystal forming solution corresponding to Ca^{2+} concentration of 3000 ppm. The $CaCl_2$ solution was made by dissolving $CaCl_2$ crystals in distilled water at room temperature and Na_2CO_3 solution was prepared in the same way. The tartaric acid ($C_4H_6O_6$) crystal as an additive (10 ppm) was added to the Na_2CO_3 solution. The reaction of the scaling process can be expressed as (Eq.1):

$$CaCl_2(aq) + Na_2CO_3(aq) \rightarrow CaCO_3(s) + 2NaCl(aq) \quad (1)$$

In the experimental run, five litters each of $CaCl_2$ and Na_2CO_3 solutions were initially added into a vessel (1 and 2) respectively. The solutions from vessel (1 and 2) were then transported into a vessel (3 and 4) used the pump system (5 and 6). The solutions of $CaCl_2$ and Na_2CO_3 on vessel (3 and 4) were simultaneously entered into the coupon house (7) which contained five coupons made of copper. The solutions of $CaCl_2$ and Na_2CO_3 were expected to react in the coupons (which were designed specifically for the growth of the $CaCO_3$ scale). The solution out of the coupon house was continuously monitored by a conductivity meter for three hours, i.e. until the end of the experimental run. At the end of the experiment, valve for connecting the coupon house and the vessel (3 and 4) was closed for one hour. After one hour the coupons were removed from the coupon house and dried in the oven at 80 ^oC for one hour. The coupons were weighed after cooled (difference weight of the coupon after and before testing was the mass of the scale).

The nucleation and crystal growth of $CaCO_3$ was observed by measuring the change in the solution conductivity. The first nucleation form was signed by a sudden decrease in the value of conductivity of the solution (induction time), indicating that a large number of ions forming the crystals begin to leave the solution to form crystals.

2.2 Characterization of $CaCO_3$ scale

Scale precipitated on the coupons of samples was then examined by SEM/EDX for morphology and elemental analysis.The SEM test was employed to determine the morphology of crystals, likely hexagonal (vaterite), orthorhombic (aragonite) and trigonal (calcite). The EDX test was used to determine the chemical composition percentages of the scale formed.

Further, the XRPD Rietveld method was selected for phase composition examination. Scan parameters chosen for measurements were: 10-90°, 2θ, 0.020 steps and 15 s/steps. Peak position and peak intensity were identified using the feature finder or on the screen in the software. The data obtained were compared with the ratio of PC based search data base program, Philips X'Pert Software (Philips Electronics NV) is used to detect the possibility of crystalline phase in the sample. In this approach, the peak position is automated using the diffraction minima in the second derivate of diffraction traces. The peak positions and heights were checked against XRPD standard for $CaCO_3$. The XRPD Rietveld program in the software validated the presence of crystal formed qualitatively and quantitatively.

3 Results and Discussion

3.1 Induction time and mass of scale

In the present research, stable concentrations could be fixed at 3000 ppm of Ca^{2+} in the solution by means of constant stirring in all experiments. Also, experiments were conducted with the laminar flow rate (Reynolds number, NRE ± 180), of which the flow rate 35 ml/min, two variations of temperature (27 and 50°C) and also two variations of tartaric acid concentrations (0.00 and 10.00 ppm) were selected. The induction time of $CaCO_3$ scale formation is given in Figure 1.

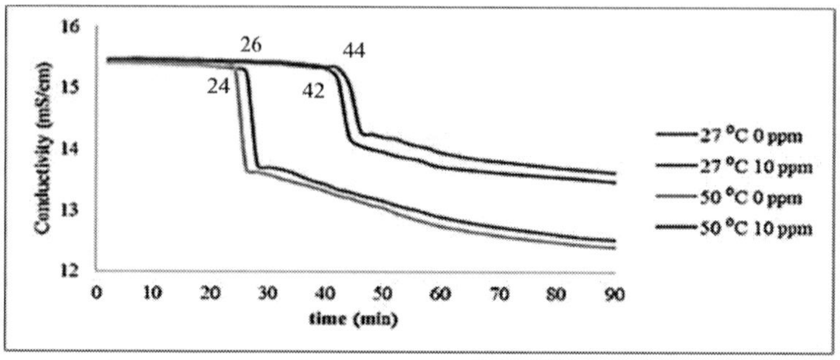

Fig. 1. Plot of conductivity versus time of scaling at flow rate 35 ml/min.

It is demonstrated that the reaction between $CaCl_2$ and Na_2CO_3 solutions occurred as can be seen from the decline in conductivity of the solution coming out of the coupon house. The conductivity of the solution at first stabilized and then decreased suddenly. The time between mixing the solution to decrease in the conductivity could be fluctuated in accordance with the parameters investigated. This time interval is commonly called as the induction time. Obviously, the presence of additive (10 ppm) made the increasing induction time, thereby delaying the crystal growth of calcium carbonate.

Table 1. Mass of the scales produced during the experiments

No	Mass of scales (wt. gr)	Additive at 27 ^{O}C		Additive at 50 ^{O}C	
		0 ppm	10 ppm	0 ppm	10 ppm
1	Calcium Carbonate	0.1608	0.0166	0.1771	0.0153

The $CaCl_2$ and Na_2CO_3 solutions were met in the coupons and reacted to form $CaCO_3$ scale. This can be seen that the mass of scale production is presented in Table 1. The scale mass results could be varied depending on the variation of testing parameters. The mass product reduced in the presence of additive, while increasing temperature led to increasing the mass scales.

3.2 Effects of temperature and additive in the crystal forming solution

The temperature and additive (absence and presence of tartaric acid) affect significantly the induction time and mass scale formed. The higher the temperature caused to the decreasing induction time, simply increasing the mass scale would be produced. In contrast, the presence of additive could be delayed the crystal formation and reduced the mass of scales (Figure 1, Table 1). Correspondingly, the parameters (27°C, 0 ppm) versus (50°C, 0 ppm) provided the decreasing of induction time from 26 to 24 minutes, while the mass on the scale increasing from 0.1608 to 0.1771 gr (9.2 %). This occurs because the higher temperature leads to the more energy in the molecules or ions and produces a faster rate of reaction.

Instead the higher concentrations of tartaric acid were added, the mass scale would be the less. In this way, the parameters (27°C, 0 ppm) versus (50°C, 10 ppm) lead to the increasing of induction time from 26 to 42 minutes, but mass scale decreased from 0.1608 to 0.0153 gr (90.5 %). This is because more core crystals is prevented from reaching a critical size, thereby reducing the number ions forming crystals. Here the mass crystal reduced consequently. Moreover, temperature affect the crystallization of $CaCO_3$ scale. At temperatures < 30 ^{O}C, the pure amorphous $CaCO_3$ (ACC) may transform into calcite via vaterite, while at the temperature of ≥ 40°C, the pure ACC may turn into aragonite through vaterite [9].

3.3 Characterization of the $CaCO_3$ scale

In the present study, $CaCO_3$ crystal morphology can be examined using SEM (Figure 2). It shows a clear difference in the formation of $CaCO_3$ crystals. At (27°C, 0.00 ppm) on left, crystal morphology was observed on the spot, namely vaterite (ball irregular) and calcite (rhombohedral). However, at (50°C, 10 ppm) on right, aragonite crystals are shown (flower shape) including calcite. The high temperature and the high ppm tartaric acid may alter the crystal morphology of vaterite into aragonite. Consequently, tartaric acid is highly effective for use in changing the crystal morphology of the $CaCO_3$ crystals.

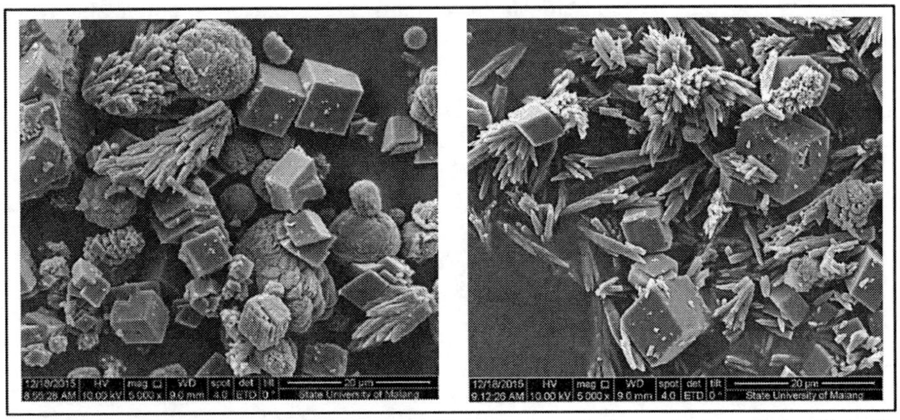

Fig. 2. Morphology of $CaCO_3$ crystals produced at 27°C in the absence of tartaric acid (left), and at 50°C in the presence of tartaric acid 10 ppm (right).

The chemical composition of the $CaCO_3$ crystal can be examined using the EDX data. It shows that the major elements of Ca, C and O belongs to the $CaCO_3$ crystal in the presence 10.00 ppm tartaric acid. The weight percent (wt. %) of chemical elements can be calculated as follows: Ca = 47.49 wt. %, C = 10.19 wt. %, O = 40.33 wt. % . The difference between the weight percent of the actual theory as follows Ca = 7.49 wt.%, C = 1.81 wt.% and O = 7.67 wt.%. From EDX analysis results, it can be concluded that the crystals formed is $CaCO_3$ scale, because each weight percent of Ca, C and O corresponds to a ratio of 1 : 1 : 3.

Phase composition of $CaCO_3$ can be examined using XRPD data (Figure 3). XRD analysis can determine the crystal phase formed by comparing the data experiment with the database $CaCO_3$ phase composition can be examined using XRPD data (Figure 3) and also by comparing the data experiment with the database (JCPDS). The XRPD test results are in the form of stripes (peak). The different plot (below) provides the direct experimental evidence of phase crystals formed in the presence of tartaric acid 10 ppm and the different temperature (27 and 50°C). The presence of the tartaric acid at a temperature of 27°C (left) provided vaterite and calcite with the strongest peak at 27.03° 2θ (PDF # 96-901-5899) and 29.38° 2θ (PDF # 96-210-0993). The presence of tartaric acid (10 ppm) at a temperature of 50°C (right) made the crystal phase of vaterite transformed into aragonite (PDF # 96-500-0086) and calcite. Mixing calcium chloride and sodium carbonate salt solutions at a temperature ≥40°C yield to deposit the first phase of ACC results in the coupons within a few minutes turned into aragonite. The rapid changes similar to the findings of subsequently aragonite transformed into calcite (the most thermodynamically stable polymorph) within a few hours [10]. Overall XRD analysis of the crystals of the scale was obtained in accordance with the powder diffraction data for calcium carbonate.

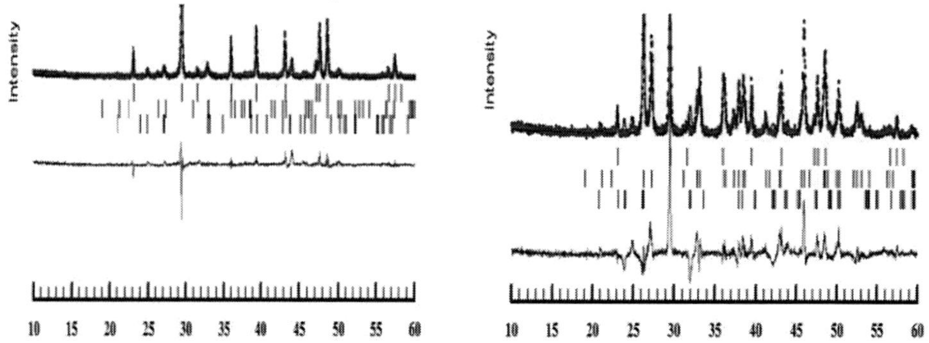

Fig. 3. XRPD Rietveld plots of crystals obtained at 27°C in the absence of tartaric acid (left), and at 50°C in presence of tartaric acid 10 ppm (right). Peaks are given color: blue (calcite), red (aragonite) and green (vaterite).

Table 2. Phase composition of carbonate scale obtained at the laminar flow rate of 35 ml/min

No	Phase composition (wt. %)	Additive at 27 °C		Additive at 50 °C	
		0 ppm	10 ppm	0 ppm	10 ppm
1	Calcite	80.19	81.78	30.42	23.79
2	Aragonite	7.30	-	69.58	15.81
3	Vaterite	12.52	18.22	-	60.41

Table 2 presents that calcite is the only major mineral formed at a low temperature of 27°C in the absence and presence of additive. Apparently calcite is more stable in the low temperature, while the temperature of 50°C makes the reduction of the amount of calcite formed in the solution accordingly. More metastable phases of such aragonite and vaterite may be easily formed in this temperature, whether the absence and presence of additive are in the solution.

4 Conclusion and Remarks

The experiment of calcium carbonate scale formation ($CaCO_3$) has been conducted. $CaCO_3$ scale formation in pipes was affected by the solution temperature and tartaric acid. The temperature negatively affects the induction time, but the tartaric acid positively affects the induction time. The higher the temperature caused the shorter the induction time. Shorter induction time for higher temperature might be caused by more scale forming component being able to react to form the scale. On the other hand, the higher the tartaric acid concentration caused longer induction time. This may be due to the disruption activity of the tartaric acid molecules on the formation of $CaCO_3$ crystals, hence the scale formation is inhibited. Tartaric acid in ppm amount and temperature was found to change the crystal phase of $CaCO_3$ (vaterite to aragonite). $CaCO_3$ scale formation by the addition of the tartaric acid at the temperature of 27°C produce vaterite phase (ball irregular) and slightly calcite phase (rhombohedral). With the addition of the tartaric acid at the temperature of 50°C to produce aragonite phase (flower shape) and a little amount of calcite. Of all the results showed tartaric acid can be used as an effective antiscalant to the scale of calcium carbonate in the various water piping system.

Authors grateful to the Laboratory of Physical Metallurgy Diponegoro University that has given the place during the experiment. The research grant provided by the Diponegoro University is also much appreciated. The authors would like to appreciate the comments useful and the kind suggestions from the anonymous reviewers

References

1. J. Garcia-Carmona, J. Gomez-Morales, J. Fraile-Sainz, R. Rodriguez-Clemente, Morphological characteristics and aggregation of calcite crystals obtained by bubbling CO_2 through a $Ca(OH)_2$ suspension in the presence of additives. Powder Technology, **130**, 307-315 (2003)
2. S. Muryanto, A. P. Bayuseno, W. Sediono, W. Mangestiyono, Sutrisno. Development of a versatile laboratory project for scale formation and control, Education for Chemical Engineers, **7**, e78-e84. (2012).
3. D. J. Tobler, J. D. Rodriguez Blanco, K. Dideriksen, K. K. Sand, N. Bovet, L. G. Benning, L. S. Stipp. The effect of aspartic acid and glycine on amorphous calcium carbonate (ACC) structure, stability and crystallization. Procedia Earth and Planetary Science, **10**, 143 – 148. (2014).
4. Tzotzi,C., Pahiadaki,T., Yiantsios,S.G., Karabelas, A.J.,Andritsos,N. A study of CaCO3 scale formation and inhibitionin RO and NF membrane processes. Journal of Membrane Science, **296**, 171-184. (2007).
5. Chao, Y., Horner, O., Hui, F., Lédion, J., Perrot, H. Direct detection of calcium carbonate scaling via a pre-calcified sensitive area of a quartz crystal microbalance. Desalination, **352**, 103–108. (2014).
6. Sousa, M.F., Bertran,C.A. New methodology based on staticlight scattering measurements for evaluation of inhibitors for in bulk crystallization. Journal of Colloid and Interface Science, **420**, 57-64. (2014).
7. Kladi, A., Klepetsanis, P.G., Østvold, T., Kontoyiannis, C. G., Koutsoukos, P.G. Crystal growth of calcium carbonate in seawater. The effect of temperature and of the presence of inhibitors, in: Z. Amjad (Ed.), Advances in Crystal Growth Inhibition Technologies, Kluwer Academic Publishers, New York etc., 85–106. (2002).
8. MacAdam, J., Parsons, S. A. Calcium carbonate scale control, effect of material and inhibitors. Water Sci. Technol, **49**, 153–159. (2004).
9. Tobler, D. J., Rodriguez Blanco, J. D., Dideriksen, K., Sand, K. K., Bovet, N., Benning, L. G., Stipp, L. S. The effect of aspartic acid and glycine on amorphous calcium carbonate (ACC) structure, stability and crystallization. Procedia Earth and Planetary Science, **10**, 143 – 148. (2014).
10. Altay, E., Shahwan, T., Tanoglu, M. Morphosynthesis of CaCO3 ad different reaction temperatures and the effect of PDDA, CTAB, and EDTA on the particle morphology and polimorph stability. Powder Technology, **178**, 194-202. (2007).

Rear Wheel Drive Recumbent Bicycle for Urban Transportation in a Tropical Emerging Country

Bambang Suhardi[1], *Ilham* Priadythama[2,*], and *Pringgo Widyo* Laksono[3]

[1]Laboratory of Work System Design and Ergonomics, Industrial Engineering Department, Universitas Sebelas Maret, Ir. Sutami Street No. 36A, Jebres Surakarta, Indonesia.
[2]Laboratory of Product Planning and Design, Industrial Engineering Department, Universitas Sebelas Maret, Ir. Sutami Street No. 36A, Jebres Surakarta, Indonesia.
[3]Center of Technological Study and Development and Industrial Collaboration, Universitas Sebelas Maret, Ir. Sutami Street No. 36A, Jebres Surakarta, Indonesia.

Abstract. Bicycle for urban transportation in emerging countries still relies on "upright" bicycle which is unchanged since the 20th century. In a tropical emerging country such as Indonesia, a bicycle must be ergonomically designed because it may be used in a long run. On the other hand, bicycle rider may face various traffic conditions. Additionally, weather makes trip become heavy and takes a lot of energy. This research aims to develop and test a new bicycle concept for an urban transportation in the tropical emerging country. A multi-criteria decision method was used to select a right bicycle concept. Then the concept was detailed into a bicycle design. We used biomechanics model and simulated finite element method to verify its frame strength. After the design verified, we built a full bicycle prototype and compare it with a commercially available urban bicycle with equal specifications. The comparison tests include physiological tests and a subjective test for several terrain and traffic conditions. The new concept is an ergonomic bicycle based on rear wheel drive recumbent bike. Based on the overall test result, it can be concluded that the new bicycle concept is promising and still appropriate for the next development.

1 Introduction

A Bicycle is a clean transportation mode which can significantly reduce air pollution in an urban area [1]. Study of bicycle development for an alternative transportation and sport has been being conducted for many years and has resulted more aerodynamic and ergonomic bicycle [2]. The application of bicycle sharing system has also done for several cities such as in America and Australia [3]. However, bicycle for urban transportation in emerging countries still relies on "upright" bicycle which is unchanged since the 20th century. In a tropical emerging country such as Indonesia, a bicycle is used to deliver worker from small-town around the city to the center of the city. For that purpose, a bicycle must be

* Corresponding author: priadythama@gmail.com

ergonomically designed because it may be used for a long run and high-frequency, otherwise some issues may occur for bicycle overuse. For a long run riding, bicycle rider may face Carpal Tunnel Syndrome risk [4]. Spinal and pelvis disorder may also be affected by high-frequency riding [5]. Moreover, upright posture for high-frequency bicycle pedalling may cause urogenital disorder and even infertility [6].

From outskirt to city, bicycle rider may face various traffic conditions from long quiet to very dense with "stop and go" road. Additionally, the distance and temperature make the trip become heavy and takes a lot of energy and progressively reduce physical endurance [7]. When it combined with low infrastructure quality, safety risk may increase [8].

This research aims to develop and test a new bicycle concept for an urban transportation in the tropical emerging country. Many types of bicycle existed with their advantages and weaknesses [9], then the right type must be chosen for a design reference. A multi-criteria selection approach is needed since there are many criteria related to bicycle usage [10]. The reference must be developed into a prototype and evaluated to see its performance.

2 Methods

Our research consists of four major stages. The first stage is *Concept Selection*. In this stage, Criteria were collected from 22 scientific resources, then reduced and confirmed by 8 experts consist of enthusiast user, bicycle technician, academician, and bicycle research team. The criteria were confirmed using there point scale valuation and cut off point method [11], then remaining criteria were weighted using the valuation. Bicycle type concepts from 4 sources [9,12,13,14] were collected as reference alternatives. The reference may be selected from non-urban bicycle type. Then the alternatives were scored to find the top three concepts. After that, using the possibility of further development consideration [15], and by improving weak criteria, the best reference concept was chosen. Finally, the urban bicycle concept and its riding posture were visualized using Catia V5.

The next stage is *Detail Development*. Visual design cannot be directly transformed into a prototype. It must be detailed first with specifications. We choose a collaborative approach which involved custom bicycle manufacturer to specify the proposed urban bicycle components. After that, there is a *Design Verification* stage. In building a bicycle prototype, frame design is very crucial. In this stage, we didn't want to have a complex frame. A simple but safe construction is our priority. A biomechanics model of bicycle rider was used to determine input force parameter for finite element simulation. We compared circular and rectangular cross-section. We also compared several cross-section sizes of the frame to find the best (strong but lightweight) frame. The simulation was conducted in Autodesk Inventor 2016.

The last stage is *Prototype Evaluation*. In this stage, we built a rideable urban bicycle concept prototype. In order to evaluate its performance, we conducted a comparison test between the concept and commercially available urban bicycle (a modern roadster type). We carefully selected the brand and model type to provide the closest specification to the concept. Then we designed a test with 4 type of terrains, a light traffic flat road (LTFR), a light traffic, up and down-road (LTUR), a heavy traffic flat road (HTFR), and a damaged road (DR). We selected 20 cyclists from male students (aged 20±2) with 173±3cm tall height. For the test, we compute Cardiovascular Load (%CVL) of the subjects separately for each road type. %CVL can be a measure of workload of light and heavy activity [16]. %CVL can be computed by following formula:

$$\%CVL = \frac{100\%(\text{Working Pulse Rate} - \text{Initial Pulse Rate})}{(\text{Maximum Pulse Rate} - \text{Initial Pulse Rate})} \quad (1)$$

3 Result and Analysis

3.1 The Best Concepts Reference

There are top three urban bicycle concept reference, Modern Roadster, 27" Mountain Bike, and Short Wheelbase Recumbent Bike. More detail information can be read in our previous research [17]. Bicycle for the tropical emerging country is characterized by almost all of the criteria, for example, heavy traffic, the bicycle must be easy to handle or good in maneuver, for long distance riding in hot temperature, the bicycle must be efficient in pedalling and comfort in riding. After applying further development consideration which may improve the weak criteria score of the three reference, the total final scores become 2.7823 for Modern Roadster, 2.6973 for 27"MTB, and 2.7857 for Short Wheelbase Recumbent Bike. Therefore, Short Wheelbase Recumbent Bike is the chosen reference for new urban bicycle concept.

3.2 Concept Visualization

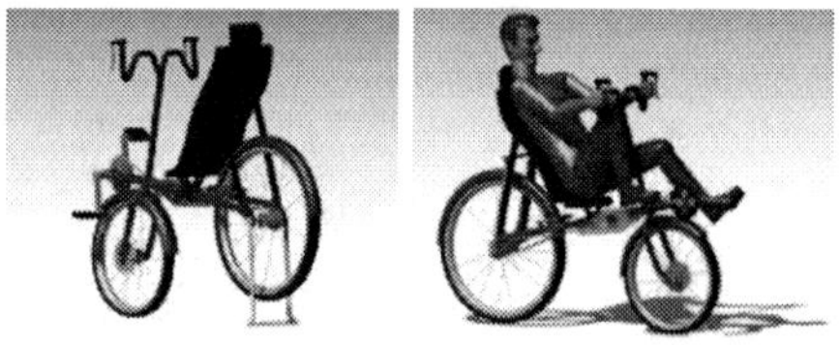

Fig. 1. Visualization of Urban Bicycle Concept. Left: rear-side view. Right: side view with rider

This bicycle concept (Fig.1.) have some differences from a typical commercially available recumbent bicycle. First, the riding posture is set more upright by 50⁰ from a horizontal line with 15⁰ for its frame. Second, the handlebar is using W shape. W shape will give more space for the knee while pedalling.

3.3 Bicycle Specification

Table 1. Specification result after CPD.

Component	Previous Specification	Specification After CPD	Specification After CPD (Lightweight)
Shifter	7 speed	3 Speed	3 Speed
Front Brake	Mechanical Discs Brake	Mechanical Discs Brake	Mechanical Discs Brake
Rear Brake	V-Brake	V-Brake	V-Brake
Brake Lever	Reverse Type	Road Type	Reverse Type
Pedal	Al Alloy, Sealed Bearing	Al Alloy, Sealed Bearing	Al Alloy, Sealed Bearing
Crankset	Al Alloy, 48T, single speed	Al Alloy, 48T, single speed	Al Alloy, 42T, single speed
Bottom Brackets	Square Shaft, Sealed Bearing	Square Shaft, Sealed Bearing	Square Shaft, Sealed Bearing
Chain	7 speed	Single speed	10 Speed, Hollow Link
Front Rim	Al Alloy, 20", 32H	Al Alloy, 20", 36H	Al Alloy, 20", 32H
Rear Rim	Al Alloy, 700c, 32H	Al Alloy, 700c, 36H	Al Alloy, 700c, 36H
Front Hub	32H, 6 bolts Disc Brake	36H, 6 bolts Disc Brake	32H, 6 bolts Disc Brake
Rear Hub	cassette 7 speed, 32H	Internal Gear 3 speed	Internal Gear 3 speed
Front Tire	20" x 1.5	20" x 1.5	20" x 1.25
Rear Tire	700 x 28c	700 x 28c	700 x 25c
Cassette	7 speed	none	none
Derailleur	7 speed	none	none
Head Set	1 3/8 Oversized	1 3/8 Oversized	1 3/8 Oversized
Handle Bar	Custom	Custom, W shaped, Al Alloy	Bull Horn, Al Alloy

Component	Previous Specification	Specification After CPD	Specification After CPD (Lightweight)
Seat	Custom	Custom, plywood	Custom, plastic-Al Composite
Chain Tensioner	Custom	Custom, Nylon	RD Pulley, CNC Al Alloy
Main Frame	Custom	Custom, JIS G 3472 Round	Custom, JIS G 3472 Rectangular
Fork	Custom	Custom, JIS G 3472 Round	Al Alloy 20" Folding Bike Fork
Parking Stand	Custom	Custom, steel, upright	Custom, Steel, skewed

There are several changes in the specification of bicycle concept from the previous (as set in visual design) to after collaborative process [18] (Table 1.). The main change is on drive train which is the 7-speed groupset is no longer suitable and replaced by 3-speed Internal Gear. Internal Gear makes drivetrain become simpler and reduce chain length.

3.4 Finite Element Model

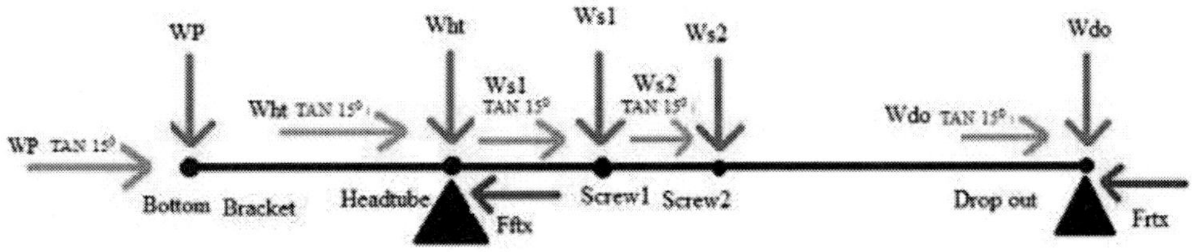

Where,
WP : Force exerted by foot to pedal
Wht : Force exerted by hand to head tube
Ws1 : Force exerted by body to seat at screw 1
Ws2 : Force exerted by body to seat at screw 2
Wdo : Force exerted by body to rear drop out
Frtx : Reaction force at horizontal direction

Fig. 2. Biomechanics model for a rider which rides the bicycle

Fig. 3. Finite Element Simulation Example Result

Table 2. Summary of Finite Element Result.

Cross Section (mm)	Wall Thickness (mm)	Safety Factor (ul)	Displacement (mm)	Mass (kg)
Ø44.5 (round)	2.80	>15	0.019	4.97
Ø44.5 (round)	3.05	14	0.034	5.16
Ø44.5 (round)	4.80	>15	0.020	6.30
30x60 (rectangular)	1.30	>15	0.057	4.00
25x60 (rectangular)	1.40	>15	0.138	3.30

On Fig. 2. Some assumption has been made for distributing body weight into several constraints [19,20]. We distribute body segment weight in two closest constraints and the magnitude is proportional to its distance from the constraints. The finite element simulation result can be seen in Fig. 3. And their summary is in Table 2. From the table, it can be seen that 30x60mm rectangular steel tube is the lightest bicycle material which still has good strength and stiffness. This frame weight is still 1kg more weight than average steel roadster frame on the market which is around 3kg.

3.5 Physiological Test Result

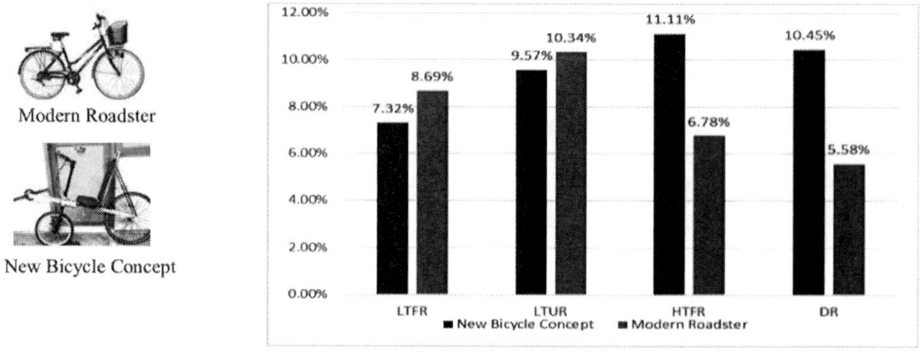

Fig. 4. Physiological (Cardiovascular Load) Test Result

Fig. 4. Shows the result of %CVL for 20 subjects on 4 different terrain types, which is processed using %CVL formula (1). From the graphic, it can be seen that the new bicycle concept can outperform the roadster on light traffic and up and down-road. These results prove that short wheelbase recumbent bicycle has more pedalling efficiency than roadster due to the rear wheel size and riding posture. Unfortunately, for heavy traffic and damage road, it becomes worse due to bicycle balance. Heavy traffic and damage road made the riders stop the bike several times and these cases gave difficulties for riding the recumbent. We found that subject still need many efforts when starting the recumbent even though we have trained them well. Our investigation led to rider's habits of using upright type bicycle which could not be washed out completely during the test.

4 Conclusion

This paper presents a comprehensive development of new urban bicycle concept, from requirement definition to prototype building. The new concept can outperform modern roadster type bicycle in light traffic road and up and downhill terrain. For stop and go usage, such as on heavy traffic road, the new concept still needs improvement in bicycle balance. Based on the overall test result, it can be concluded that the new bicycle concept is promising and still appropriate for the next development.

Acknowledgement: This paper and presentation is funded by Hibah Penelitian Unggulan Perguruan Tinggi from (PUPT) Ristek Dikti 2016 and Hibah Peningkatan Kapasitas Laboratorium Penelitian (PKLP) PNBP Universitas Sebelas Maret 2017.

References

1. C. Johansson, B. Lövenheim, P. Schantz, L. Wahlgren, P. Almström, A. Markstedt, M. Strömgren, B. Forsberg, J.N. Sommar, Impacts on air pollution and health by changing commuting from car to bicycle. Science of the Total Environment, **584-585**, 55-63 (2017)
2. H.K. Epema, S. van den Brand, W. Gregoor, J.D.G. Kooijman, H.P. Pereboom, D.C. Wielemaker, C.-J. van der Zweep Bicycle Design: A different approach to improving on the world human powered speed records. Procedia Engineering, **34**, 314-318 (2012)

3. M. Ahillen, D. Mateo-Babiano, J. Corcoran, Dynamics of bike sharing in Washington, DC and Brisbane, Australia: Implications for policy and planning. International Journal of Sustainable Transportation, **10**, 441-454 (2016)
4. D.C. Rehak, Cyclist's Hands Overcoming Overuse Injuries. Hughston Helath Alert, **15**, 3 (2003)
5. J.M. Muyor, P. Minarro, A. Lopez, F. Alacid, Spinal Posture of Thoracic and Lumbar Spine Pelvic Tilt in Highly Trained Cyclist. Journal of Sport Science and Medicine, **10**, 355-361 (2011)
6. I. Leibovitch, and Y. Mor, The Vicious Cycling: Bicycling Related Urogenital Disorder. European Urology, **47**, 277-287 (2005)
7. H. Otani, M. Kaya, A. Tamaki, P. Watson, R.J. Maughan, Effects of solar radiation on endurance exercise capacity in a hot environment. European Journal of Applied Physiology, 116, 769–779 (2016)
8. J. DiGioia, K.E. Watkins, Y Xu, M. Rodgers, R. Guensler, Safety impacts of bicycle infrastructure: A critical review. Journal of Safety Research, **61**, 105-119 (2017)
9. E. Kocabiyik. *Engineering Concept in Industrial Product Design with A Case Study of Bicycle Design*. Izmir Institute of Technology (2004)
10. T. Buchert, S. Neugebauer, S. Schenker, K. Lindow, R. Stark, Multi-criteria decision making as a tool for sustainable product development - Benefits and obstacles. Procedia CIRP, 26, 70-75 (2015)
11. M.C.Y. Tam, V.M.R. Tummala, An application of the AHP in vendor selection of a telecommunications system. Omega, **29**, 171-182 (2001)
12. M. McGoldrick, *Choosing The Right Bike for You*. http://gobiking.ca/resourceshow-to/choose-the-right-bicycle-for-you/ (2009)
13. A.L. Scwab, J.D.G. Koijman, On The Design of Recumbent Bicycle with A Perspective on Handling Qualities. IDETC/CIE (2012)
14. Fairuz, *Bike to Work Polygon Heist 1.0 Hybrid Bicycle with Affordable Price* (in Indonesian Language) https://fairuzelsaid.wordpress.com/2011/03/27/byke-to-work-polygon-heist-1-0-sepeda-hibrid-harga-terjangkau/ (2013)
15. K.T. Ulrich and S.D. Eppinger. *Product Design and Development, 5th Edition*. McGraw-Hill (2012)
16. L. Herdiman, S. Susmartini, N. Adiputra, K. Tirtayasa, Application of Appropriate Technology in Ankle Joint to Below Knee Prosthetic Reduce Working Pulse on Transtibial Amputee. *7th WACBE World Congress on Bioengineering*, 154-157 (2015)
17. B. Suhardi, I. Priadythama, M. Arief, Urban Bicycle Conceptual Development Considering Ergonomic Criteria (in Indonesia language), **12**, Spektrum Industri, 231-240 (2014)
18. B. Suhardi, I. Priadythama, C. W. Fitriyani, Design Specification Determination with Collaborative Approach for Indonesian Short Wheelbase Recumbent Bicycle, Applied Mechanics and Materials, **815**, 298-303 (2015)
19. I. Priadythama, B. Suhardi, V. G. Rengganis, Standard Tube Selection for Recumbent Bicycle Frame Using Simulated Finite Element Analysis. *Joint International Conference of Electric Vehicular Technology and Industrial, Mechanical, Electrical and Chemical Engineering (ICEVT & IMECE)*, 5E1-1 (2015)
20. I. Priadythama, B. Suhardi, I. Adiasa, Further study on a short wheel base recumbent bike frame using simulated finite element analysis, *International Conference of Industrial, Mechanical, Electrical, and Chemical Engineering (ICIMECE)*, (2016)

SYNTHESIS OF BARIUM SULFATE IN THE VARIATION OF TEMPERATURE AND ADDITIVE CONCENTRATION

Waleed Ali[1], *Wahyu* Putranto[1], *Stefanus* Muryanto[1], and *Athanasius* Bayuseno[1]

[1]Department of Mechanical Engineering, Faculty of Engineering, Diponegoro University, Tembalang Campus, Semarang 50275, Indonesia

Abstract. Barium sulfate (BaSO4) with fine sizes is important for many applications. Barium Sulfate (BaSO4) is suitable for applications as diverse as whiteness, inertness and high density. Many different approaches have been reported for the preparation of nanoparticles BaSO4 including the addition of different additives. Aims of this study to synthesize barium sulfate (BaSO4) from natrium sulfate and barium chloride with the modification of the mixing speed, temperature and additive concentration variations. This study conducted in Diponegoro University Laboratorium. The sample were characterized by a number of different methodologies, including XRD and SEM-EDS. The higher the temperature of the mass of the crust that formed more and more. The results showed a temperature of 40°C was obtained crust mass greater than at 30°C. The addition of the additive citric acid to 20 ppm does not have enough influence to lower the formation of crust mass. From the SEM image, it is clear that the particle size is less than 10 μm.

1 Introduction

Synthesis of inorganic powder with ultrafine size, surface properties controlled and controlled morphology attract increasing interest for use are important in many fields. Barium Sulfate (BaSO4) is suitable for applications as diverse as whiteness, inertness and high density [1]. Barium sulfate is a kind of important inorganic chemical products as packaging and additives in painting, coating, plastics and pharmaceuticals fiber [2-4]. Barium sulfate crystals can be formed from a chemical reaction as follows:

$$BaCl_2\,(l) + Na_2SO_4\,(l) \rightarrow BaSO_4\,(s) + 2NaCl\,(l) \quad (1)$$

The reaction liquid / liquid is the primary method of preparation of nano-BaSO4. This method has several divisions such as direct rainfall, the separation of micro-emulsion [5] membrane [6], the reactor microchannels [7]. Preparation $BaSO_4$ particles have been studied to assess the effect of mixing, the model precipitation, the stirrer speed and position feedback on particle size distribution, crystal growth and morphology [8,9].

Many different approaches have been reported for the preparation of nanoparticles BaSO4 including the addition of different additives [10,11] induction with monolayer and microemulsion [12]. Addition of additives and induction with LB monolayer can cause significant changes in morphology, but the size is generally within the micron scale than the scale of nanometers. The size and morphology of the corresponding obtained in W / O micro emulsion [13] or reverse micelles approach can be well controlled by adjusting the molar ratio of water to surfactant. However, the results of the product is rather low due to poor solubility of the salt in a conventional micro emulsion. There are also some reports about the preparation of organo-modified $BaSO_4$, but $BaSO_4$ obtained in the micron scale. Although the preparation of $BaSO_4$ organo capped was previously described, [14] are not suitable conditions for the industry. Preparation of barium sulfate nano particles using tetradecanoic acid, citric acid, heksadekanoat acid and stearic acid as a modifier was studied [2].

When an inorganic filler with surface functional groups reached a polymer matrix, can be expected to improve the stiffness and impact properties simultaneously [1]. However, the surface of the inorganic particles (eg, BaSO4) is hydrophilic, so it is not easy to disperse these particles in a lipophilic matrix. For this reason, it is important to modify the surface of BaSO4. One of the few ways to achieve good dispersivitas is to modify the surface hydrophilic particles so that the lipophilic nature. For the purpose of these various coupling agents, who are able to introduce specific functional groups to the surface of the particles, have been used titanate coupling agent, silane coupling agents and organophosphonic acid [15]. Aims of this study to synthesize barium sulfate (BaSO4) from natrium sulfate and barium chloride with the modification of the mixing speed, temperature and additive concentration variations.

2 Material and Method

2.1 Chemicals and instruments

The starting material used in this study are Barium chloride, sodium sulphate and citric acid. They are used without further purification. Barium sulfate is characterized by using X-ray Diffraction (XRD) patterns recorded on a diffractometer XRD BaSO4 and SEM-EDS.

2.2 Sample preparation

Direct precipitation (DP) Procedure: 4.447 g barium chloride (BaCl2) was added to 1000 ml H2O and 2.585 g of sodium sulfate (NaSO4) was added to 1000 ml H2O and added citric acid as an additive in various concentrations (5 and 20 ppm), Drogpwise solution was added into the flask while stirring at room temperature with mechanical stirring strong dispersant at 200 rpm. Steady drop rate is 20 drops min-1. The precipitate is separated from the mother liquor by centrifuging at 3000 rpm for 20 minutes. And then the supernatant solution was discarded and the solid redispersed in deionized water. This process is repeated three times to rinse particles. After the last centrifugation, the precipitated particles are dried in a microwave oven with a temperature varying 30 and 40°C.

3 Result and Discussion

3.1 X-ray Diffraction

XRD test to prove that the scale of the research result was really the barium sulfate (BaSO4) scale Figure 1 shows the XRD patterns of barium sulfate. This shows that nanoscale barium sulfate is mainly composed of barite. All peaks can be indexed as orthorhombic structure typical BaSO4, with crystalline cell constants a = 7.144 Å, b = 8.865 Å, c = 5.445 Å, which is basically in accordance with the reported values (JCPDS No. 80-0512) [2]. Crystal size of the sample estimated from XRD peak line widths.

Identification of the powder as a phase-burnt and calcined done using XRD. Average crystal size of the powder was measured by X-ray line-extension technique using Scherrer formula BaSO4 XRD patterns of nano powders prepared at different calcination treatment is shown in Figure 1.

In this study we have been exposed to nanoparticles of barium sulfate (BaSO4) of sodium sulphate and barium chloride with variations in temperature and concentration of additives. The more the concentration of additive affects the particle size increases. Results were different from the research conducted [16] showed that the polymaleic acid with a concentration of 4 ppm is able to reduce the formation of CaSO4 crust on process and growth.

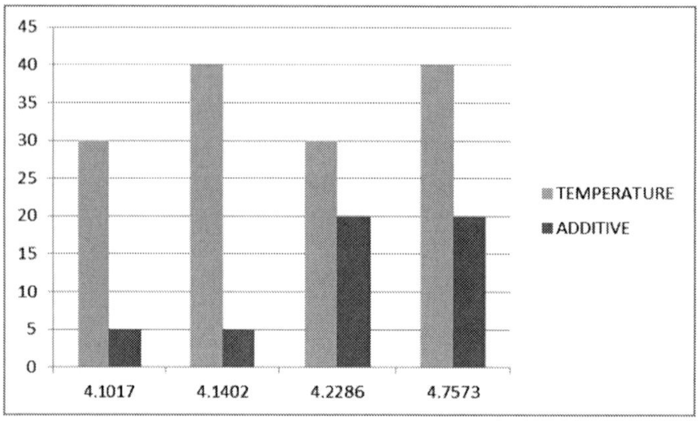

Fig. 1. The XRD pattern of particle in different crystallography orientation at different temperature and concentration additive

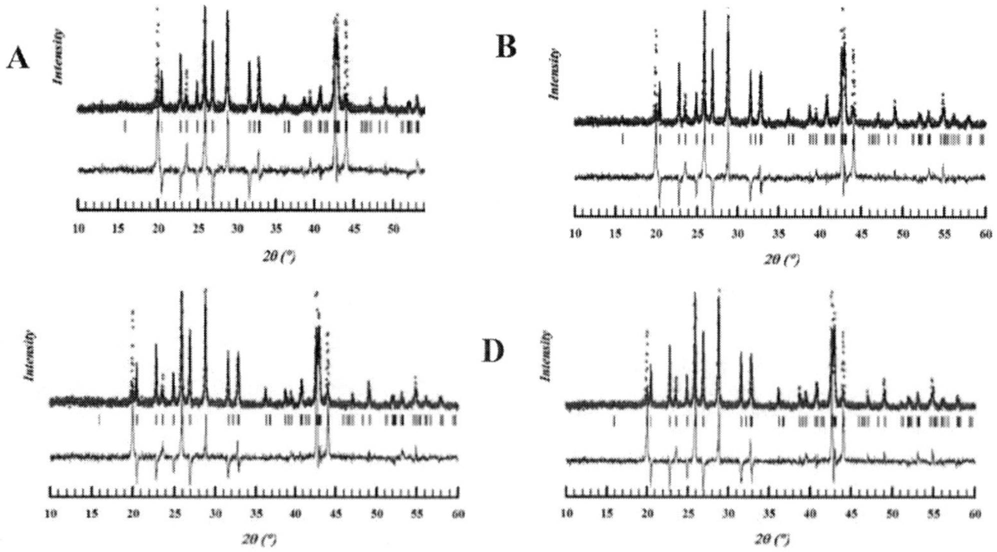

Fig. 2. XRD patterns of nanocrystalline $BaSO_4$ at different temperature and concentration additive (A : 30°C and 5 ppm ; B : 30°C and 20 ppm ; C : 40°C and 5 ppm ; D : 30°C and 20 ppm)

The addition of additives was able to suppress the formation of vaterite so the crust that dominate the form of calcite [13]. The use of additives for substances that are very complex, very important in adjusting the crystal habit and purity, when low concentrations it will affect the kinetic nucleation and crystal growth. It is assumed that the additive serves to inhibit crystal growth by slowing the rate of crystal growth, increasing the heterogeneous nucleation, control and stabilize precipitation polymorph. This affects a number of additives in salt deposition related to adsorption on the surface. One way to prevent a crust that is by keeping the anion-cation forming the crust remains in the solution. Scale inhibitor is a chemical that serves to keep the crust-forming anion-cation remains in the solution, so expect no precipitation occurs [17].

Figures 2. (a, b, c, and d) represent the formation of BaSO4 crust at different temperatures of 30 and 40°C with the addition of the same additive at the same concentration (5 ppm and 20 ppm). The crust precipitates of BaSO4 have ions which is solid and has a tendency to form crust precipitates as well as calcium sulfate (CaCO3), gypsum or calcium sulphate (CaSO4.2H2O). The formed crust is included in the type of hard scale that has the brightest primary properties, and if there is impurities (oil or iron oxide) it will become rather dark and almost insoluble in acid. BaSO4 crust precipitates are generally formed in oil fields whose processes are affected by changes in pressure, temperature and dissolved salts. Results were different from the research conducted showed that the polymaleic acid with a concentration of 4 ppm is able to reduce the formation of CaSO4 crust on process and growth [16].

3.2 SEM-EDS

The SEM (Scanning electron microscopy) test was conducted to examine the crystal morphology while the microanalyser test (EDS) was aimed to find out the crystal composition and. SEM was used to analyze the morphology and size of the synthesized BaSO4 particles [18]. Water content greatly affects the shape of BaSO4 particles. When the water content is maximal, cubic particles can be obtained.

Figure 3.(a, b, c, and d) are the result of crystallization test BaSO4 particles with adding additive. The photos show that the crystal are approximately same, in pate like form. Gypsum phase of the barium sulfate scale has a monoclinic crystal form. Monoclin is a characteristic feature of the BaSO4 crystal form. Referring to the crystal system, the crystal system has different rib lengths ($a \neq b \neq c$), and angles $\alpha = \gamma = 90$ ° and $\beta \neq 90$ °. The shape of a flat-shaped crystal BaSO4 resembles a board shape, in which the width is thick not too far away.

From the results of the SEM test BaSO4 particles indicates that the addition of a larger additive can suppress the formation of calcite / monoclinic phase which is a type of hardscale phase. When these crystals form and settle in the pipes it will produce a crust that is difficult to clean from a piping system. While the other two types of crystals, namely aragonite and vaterite, is a type of softscale that is easier to clean when attached to the wall in the pipe [19].

Fig. 3. The SEM image of the Barium suplhate nanoparticles at different temperature and concentration additive (A : 30°C and 5 ppm ; B : 30°C and 20 ppm ; C : 40°C and 5 ppm ; D : 30°C and 20 ppm)

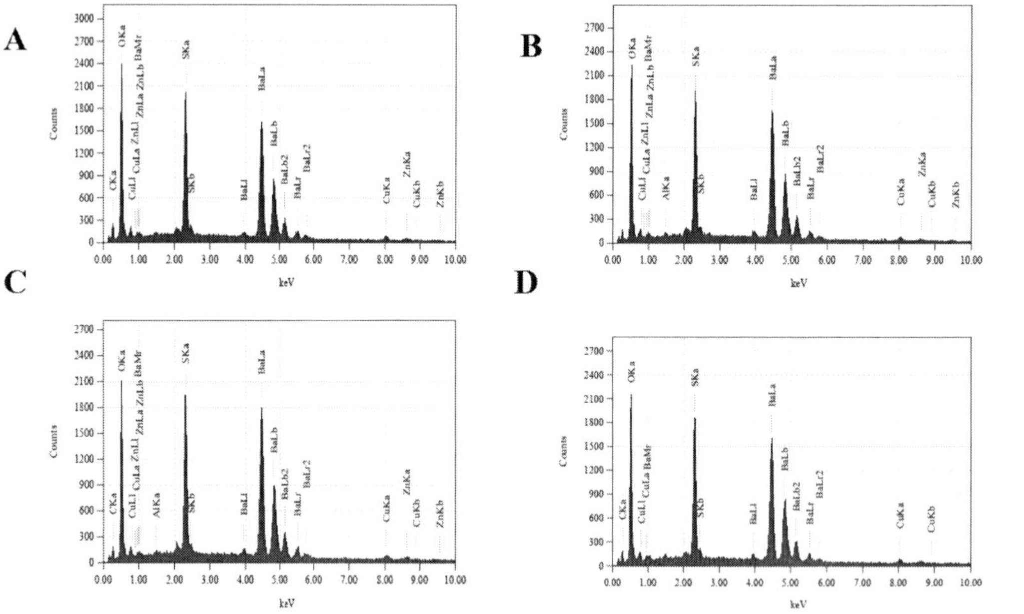

Fig. 4. The EDS spectrum of the Barium suplhate nanoparticles at different temperature and concentration additive (A : 30°C and 5 ppm ; B : 30°C and 20 ppm ; C : 40°C and 5 ppm ; D : 30°C and 20 ppm)

Figure 4 shows the energy dispersive spectrum of the $BaSO_4$ particles with additive citric acids which clearly show the presence of C, O, S, Cu, Zn, and Ba elements. The results of the study did not show any citric acid particle in EDX pattern. This is because the addition of citric acid is too little. Meanwhile, there are Cu and Zn in the EDX pattern. Maybe Cu and Zn were released from the container used for crystallization. Cu and Zn derive from the product of corrosion.

4 Conclusion and Remarks

The higher the mass of the crust is formed more and more. The result shows a temperature of 40° C obtained by crust mass greater than 30° C. In addition of 5 ppm additive, crust mass at temperature 30° C is 4.1017 gr while at temperature 40° C is 4.1402 gr. In addition of 20 ppm additive, the crust at 30° C is 4.2286 gr and at 40° C is 4.7573 gr. Based on these results it is evident that the addition of citric acid as an additive to 20 ppm does not have enough effect to decrease the mass formation of the crust. From the SEM image, it is clear that the particle size is less than 10 μm. BaSO4 particle dispersive spectrum with added citric acid, which clearly indicates the presence of C, O, S, Cu, Zn, and Ba elements. The absence of citric acid seen in the EDX pattern is due to the use of too little citric acid. While the presence of Cu and Zn should not exist, it is possible to contaminate Cu and Zn on the containers used in crystalline formation.

Authors grateful to the Laboratory of Physical Metallurgy Diponegoro University that has given the place during the experiment. The research grant provided by the Diponegoro University is also much appreciated. The authors would like to appreciate the comments useful and the kind suggestions from the anonymous reviewers

References

1. Bala, H., W. Fu, Y. Guo, J. Zhao and Y. Jiang *et al.*, Eng. Aspects, **274**: 71-76 (2006).
2. Shen, Y., C. Li, X. Zhu, A. Xie and L. Qiu *et al.* J. Chem. Sci., **119**: 319-324 (2007).
3. Wu, G., H. Zhou and S. Zhu, Mater. Lett., **61**: 168-170 (2007).
4. Kucher, M., T. Beierlein and M. Kind, AIChE J., **54**: 1178-1188 (2008).
5. Adityawarman, D., A. Voigt, P. Veit and K. Sundmacher, Chem. Eng. Sci., **60**: 3373-3381 (2005).
6. Wu, G., H. Zhou and S. Zhu, Mater. Lett., **61**: 168-170 (2007).
7. Wang, Q.A., J.X. Wang, M. Li, L. Shao and J.F. Chen *et al.*, Chem. Eng. J., **149**: 473-478 (2009).
8. Judat, B. and M. Kind, J. Colloid Interf. Sci., **269**: 341-353 (2004).
9. Chen, T., Neville, A., Yuan, M., J. Crustal Growth, **275**: e1341-e1347 (2004).
10. Jones, F., A. Oliviera, G.M. Parkinson, A.L. Rohl and A. Stanley *et al.*, J. Crys. Growth, **262**: 572-580 (2004)
11. Wang, F., Xu, G., Zhang, Z., Xin, X., Eng. Aspects, **259**: 151-154 (2005).
12. Nagaraja, B.M., H. Abimanyu, K.D. Jung and K.S. Yoo, J. Colloid Interf. Sci., **316**: 645-651 (2007).
13. Chen, G., G. Luo, J. Xu and J. Wang, Powder Technol., **153**: 90-94 (2005).
14. Sui, X., Y. Chu, S. Xing and C. Liu, Mater. Lett., **58**: 1255-1259 (2004).
15. Sheng, Y., B. Zhou, J. Zhao, N. Tao and K. Yu *et al.*, J. Colloid Interf. Sci., **272**: 326-329 (2004).
16. Martinod, A., Euvard, M., Foissy, A., Neville, A., Desalination, **220**, 345-352 (2007).
17. Reddy, M.M., Hoch, A.R. J. Colloid and Interface Sci. **235**(2): 365-370 (2001).
18. L. Qi, J. Ma, H. Cheng, Z. Zhao, Eng, Aspect **10**; 117-126 (1996).
19. Salah, N., S.S. Habib, Z.H. Khan, S. Al-Hamedi and S.P. Lochab, J. Luminescence, **129**: 192-196 (2009).

Influence of hull entrance angle "Perintis 750 DWT", toward ship resistance: the case study for design development "Perintis 750 DWT."

Eko Sasmito Hadi[1,1], *Parlindungan* Manik[1], and *Muhammad* Iqbal[1]

[1]Department of Naval Architecture, Diponegoro University, Semarang 50275, Indonesia

Abstract. Geographically Indonesia which most of its territory consists of the sea, causing the necessity of strengthening maritime facilities and infrastructure to conduct activities of distribution of goods and sea transportation facilities, especially ships as a vehicle of sea transportation to maintain inter-island connectivity. Indonesian government builds national connectivity (sea tolls) by the development of pioneer ship design "Kapal Perintis.". In this study, the vessel Perintis 750 DWT is still lacking in the EEDI (Energy Efficiency Design Index) aspect as required by IMO, so it is necessary to modify the hull part of the vessel to reduce the resistance. The modified hull is carried out by varying the angle of entrance bow of the ship ± 12° in step of 3°. Modifications were made using Delftship software, Computational Fluid Dynamic, called TDyn and verified by towing tank test at Indonesian Hydrodynamics Laboratory (IHL) Surabaya. Every change of the angle of entrance bow in 3 degrees, the ship's resistance will change more and less 3.5%. The smaller angle of entrance bow of the vessel the smaller ship's resistance.

1 Introduction

The geographical condition of Indonesia which mostly consists of sea, causing the necessity of strengthening maritime facilities and infrastructures to carry out activities of distribution of goods and sea transportation facilities, especially ships as a marine vehicle to maintain inter-island connectivity.

One of the developments of national connectivity regarding maritime is the development of pioneer ships [1]. To support the acceleration of domestic economic growth and improve inter-island connectivity in remote, deepest and outside areas, and to ensure the availability of primary commodities and the growth of trade and industry centers, the government implements the construction of pioneer vessels. One of them is the procurement of 5 units of ship type 750 DWT. The 750 DWT pioneer ship has a length of 58.5 meters, a width of 12 meters, a speed of 12 knots, and a passenger capacity of 265 people [2].

This type of pioneer ships still lacks the EEDI (Energy Efficiency Design Index) aspect as required by IMO [3], so it is necessary to modify the hull part of the vessel to reduce the resistance and energy of prime movers of the ship, which impact technically and

1 Corresponding author: ekosasmito@ft.undip.ac.id

economically operational vessel. Optimization or modification of the hull form is one way to make ship's resistance lower [4], which will affect the use of more energy efficient.

Some concepts about the method of reduction of ship resistance by changing the entire hull form of the ship are by using the optimization method. The use of an integrated optimization module with CFD software is helpful in optimizing the overall shape of the hull with the primary function of minimizing the value of ship resistance [5]. The ship hull optimization method with the focus on the bow and stern parts or a combination of bow and stern has tried on the KSUEZMAX ship. The parametric function of modification for hull forms with the variation of SAC shape and section shape, while for stern hull also involves the variety of SAC shape and section shape [6].

The method of optimizing the ship's hull shape on a bulk carrier vessel 66,000 DWT with parametric modification function of SAC and section shape of DLWL, capable of reducing wave resistance and added resistance, both on the calm and wave sea [7].

In this research, the modified hull is carried out by varying the angle of entrance bow of the vessel ± 12 degrees in a step of 3 degrees. Modifications were made using Delftship software, Computational Fluid Dynamic software, called TDyn and verified by towing tank test at Indonesian Hydrodynamics Laboratory (IHL) Surabaya.

2 Methodology

This study aims to examine the effect of changes in bow entrance angle of the Pioneer 750 DWT vessel to the overall value of the ship resistance component and to determine the entrance angle corresponding to the hull shape of the 750 DWT Pioneer vessel with the least resistance value. The calculation of ship resistance is carried out by CFD method with Tdyn software. For the initial model, CFD calculations verified by the experimental test for ship resistance in towing tanks at the Indonesian Hydrodynamics Laboratory (IHL) Surabaya. The comparison between ship and ship model is 1: 18. The ship's dimensions and ship models are shown in Table 1. The initial model line plan along with the ship's bow entrance angle shown in Figure 1. The model and 3D model of the vessel shown in Figure 2.

The bow entrance angle of the ship's initial model is 22.72 degrees. This angle is then varied every 3 degrees up to ± 12 degrees so that the ship model changes up to 9 models including the initial model. The speed of the tested vessel is at 12 - 17 knots or in the range Fr 0.26 - 0.36. The resistance calculation uses Tdyn software which simulates the ship at certain speed travels in calm water and predicts its resistance. The calculation of ship resistance using Tdyn has been carried out by Samuel et al. [8] and Iqbal et al. [9]. Their CFD results are relatively good with the empirical equation proposed by Jamaluddin [10].

Table 1. Main dimension of ship and ship model of pioneer 750 DWT vessel.

Dimension	Full Scale	Model Scale
Length of Water Line (LWL)	58.96 m	3.276 m
Breadth (B)	11.60 m	0.645 m
Depth (H)	4.50 m	0.250 m
Draft (T)	2.90 m	0.161 m
Volume Displacement	1107.30 m^3	61.527 m^3

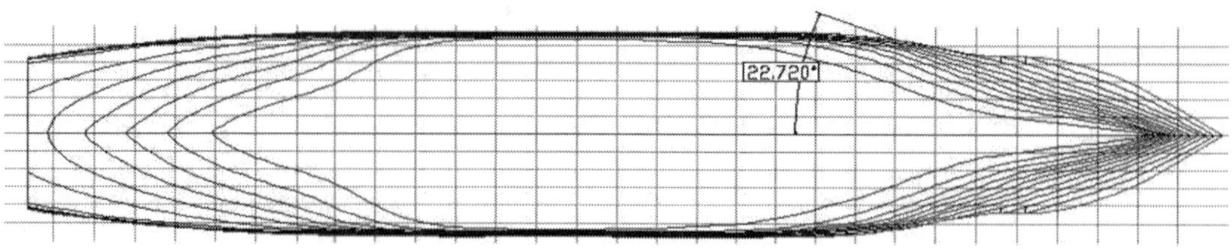

Fig. 1. Lines plan and bow entrance angle.

Fig. 2. Testing of ship models in LHI towing tanks.

3 Results and discussion

The results of CFD calculations on the initial model validated with the experimental test results in towing tank. Figure 3 shows that CFD calculations have relatively accurate results with an error less than 1% compared to the preliminary results. Furthermore, the CFD set up and meshing size in each boundary will be used for the varied models. Mesh sizes in each boundary are ship boundary 0.009, free surface boundary 0.05, and other boundaries 0.1.

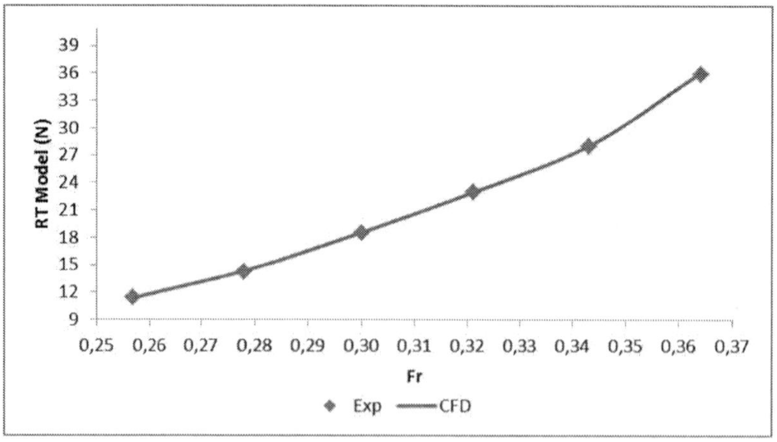

Fig. 3. The Comparison is of CFD and experimental test of ship model resistance.

Comparison of ship resistance of all models shown in Fig. 4. The bow entrance angle of the initial model, which has an angle of 22.72 degrees was varied every 3 degrees up to ±

12 degrees. The results show that the smaller of bow entrance angle, the more modest ship's resistance. Every change of the angle of entrance bow in 3 degrees, the ship's resistance will change more and less 3.5%. This will means that the magnitude or the bow entrance angle of the vessel has a linear effect on the importance of the resistance. However, the minimum of bow entrance angle should be assessed for this ship because it will affect the size of the tank and compartment inside the vessel.

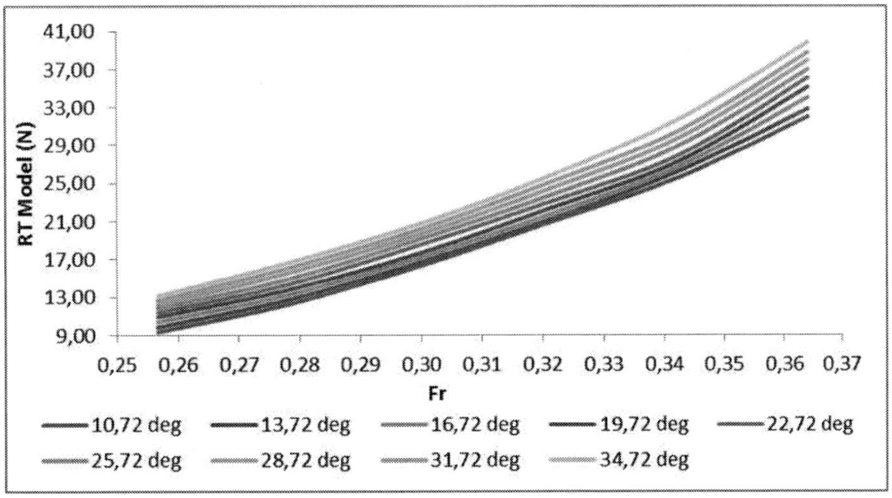

Fig. 4. The comparison is of ship model's resistance.

Table 2. The component is of ship model's resistance.

Component of Ship Resistance	10.72 deg	22.72 deg	34.72 deg
Wave Resistance (RW)	7.56 N	9.70 N	11.67 N
Viscous Resistance (RV)	17.81 N	18.37 N	18.98 N
Total Resistance (RT)	25.38 N	28.07 N	30.65 N

As shown in Table 2, the components of total resistance which are consist of viscous and wave resistance are presented to determine which elements are most affected by the bow entrance angle. The results of the ship resistance component introduced are the initial model and the model that was having the smallest and largest entrance angle at the service speed of 16 knots (Fr 0.343). Each component of viscous and wave resistance tends to increase as the bow entrance angle increases. Viscous resistance tends to increase by a slight margin (±3%), while the wave resistance increases by a considerable margin (±21%). This comparison proves that changes in the bow entrance angle of the vessel are more influential to the wave resistance.

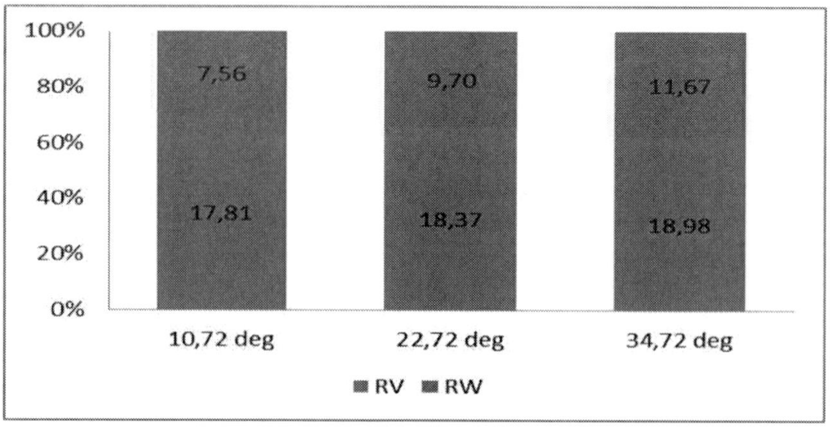

Fig. 5. Comparison of viscous and wave resistance at service speed (Fr 0.343)

Figure 5 presents the magnitude of the viscous and wave resistance ratios for each total resistance at Fr 0.343. For each model, viscous resistance is more dominant than the wave resistance. For this reason, because the ship is displacement and slow ship type. However, at the same speed, the proportion of wave resistance to its total resistance varies for different entrance angle. For the model with the smallest entrance angle, which is 10.72 degrees, the portion of wave resistance is 29.8% of the total ship resistance, followed by 34.57% for the initial model with the angle of 22.72 degrees and 38.06% for the most significant angle model, 34.72 degrees. As the wave resistance is low at small bow entrance angle, Figure 6 shows that at the same speed (Fr 0.343) the lowest bow entrance angle has the low wave elevation.

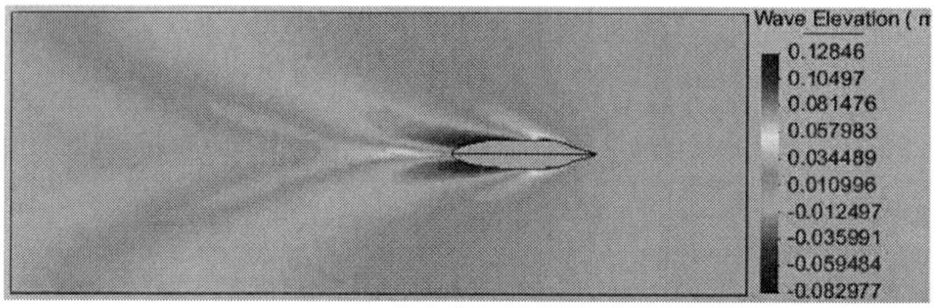

(a)

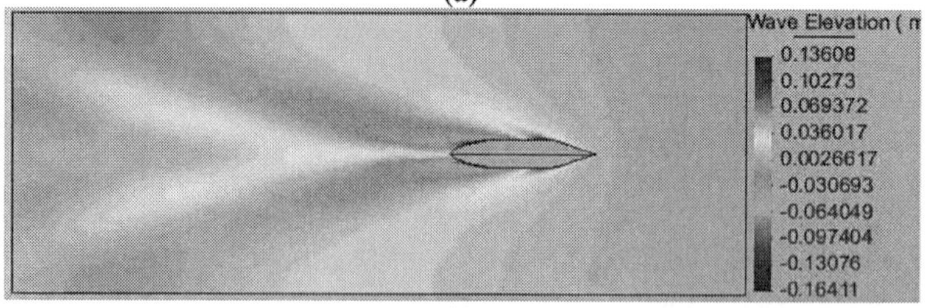

(b)

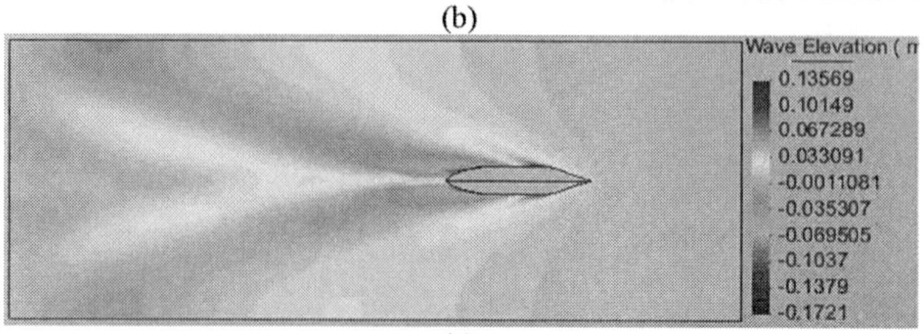

(c)

Fig. 6. Comparison of wave elevation at service speed (Fr 0.343). (a) Bow entrance angle 10.72° (b) bow entrance angle 22.72° (c) bow entrance angle 34.72°

4 Conclusion

At the service speed (Fr 0.343), the resistance of initial ship model with an entrance angle 22.72 degrees is 28,067 N. While ship model with the smallest angle, 10.72 degrees, the ship's resistance is 25.7 N or decreased up to 8.43%. For ship model with the most significant edge, which is 34.72 degrees, the ship's resistance is increased up to 13.57% or 31.878 N. Every change of the angle of entrance bow in 3 degrees, the ship's resistance will change more and less 3.5%. Reduced angle of entry will have an impact on the reduction of ship resistance.

References

1. L. Adam, Politica, Kebijakan Konektivitas Maritim Di Indonesia (Maritime Connectivity Policy in Indonesia, **6 (1)**, 20-41, (2015)
2. Directorate General of Sea Transportations. "Dukung Program Tol Laut, 50 Kapal Perintis Dan 3 Kapal Perambuan Mulai Dibangun." http://www.dephub.go.id/berita/baca/dukung-program-tol-laut,-50-kapal-perintis-dan-3-kapal-perambuan-mulai-dibangun/?cat=QmVyaXRhfHNlY3Rpb24tNjU= (2015)
3. Z. Bazari, M2 Ship Energy Efficiency Regulations and Related Guidelines IMO Train the Trainer (TTT) Course on Energy Efficient Ship Operation Module 2 – Ship Energy Efficiency Regulations and Related Guidelines MODULE 2 Ship Energy Efficiency Regulations and Related Guidelines Module Aims and Objectives, http://www.imo.org/en/OurWork/Environment/PollutionPrevention/AirPollution/Documents/Air pollution/M2 EE regulations and guidelines final.pdf (2016)
4. ABS. "Ship Energy Efficiency Measures; Status and Guidance." American Bureau of Shipping. http://www.eagle.org/eagleExternalPortalWEB/ShowProperty/BEA Repository/References/Capability Brochures/ShipEnergyEfficiency. (2013)
5. F. Huang, C. Yang, Hull form optimization of a cargo ship for reduced drag, Journal of Hydrodynamics, **28 (2)**, 173-183, (2016)
6. J-H. Park, J-E. Choi, H-H. Chun, Hull-form optimization of KSUEZMAX to enhance resistance performance, International Journal of Naval Architecture and Ocean Engineering, **7 (1)**, 100-114, (2015)
7. J-W. Yu, C-M. Lee, I. Lee, J-E. Choi, Bow hull-form optimization in waves of a 66,000 DWT bulk carrier, International Journal of Naval Architecture and Ocean Engineering, 9 **(7)**, 499-508, (2017)
8. Samuel, M. Iqbal, I.K.A.P. Utama, An investigation into the resistance components of converting a traditional monohull fishing vessel into catamaran form, Int. J. of Tech, **6 (3)**, 432–441, (2015)
9. M. Iqbal, Samuel, Traditional catamaran hull form configurations that reduce total resistance, Int. J. of Tech, **8 (1)**, 85-93, (2017)
10. A. Jamaluddin, I.K.A.P. Utama, B. Widodo, A. Molland. Experimental and numerical study of the resistance component interactions of catamarans. J. of Eng. for the Maritime Environment, **227**, 51–60, (2012)

Structural analysis for in-service gas pipeline lowering using numerical method

Mochamad Safarudin[1*], *Joga Dharma* Setiawan[2]
[1]Mechanical Engineering, Sekolah Tinggi Teknologi Mandala, Indonesia
[2]Mechanical Engineering Department, Faculty of Engineering, Diponegoro University, Indonesia

Abstract. Construction of new highways, buildings, airport runways and other facilities is often planned at locations where aboveground pipelines are present. Relocating such lines can be extremely expensive in terms of shutdown time and new pipeline materials. Lowering this existing pipeline can have big cost benefits. The line can be lowered while remaining in service with no lost production and the cost of such lowering an existing pipeline section is relatively cheap. In this paper, the calculation method with both analytically and numerically are discussed and explained in a 28 in pipeline lowering process while keep the pipeline is safe and still in-service.

1. Introduction

Lowering a segment of pipeline is an easy job but it can have a huge cost. Pipeline can be lowered while in-service without loss of production and the cost is not too expensive [1, 2].

New building constructions, highways, railways, airport runways etc are often planned on the location where above ground pipeline exists. Relocating such a pipeline can incur high cost with downtime and new pipeline materials (valves, fittings, flanges etc) accounted [3,4]. The alternative is to lower the pipeline with sufficient safety protection to handle the problem.

In this paper, an analytical analysis is performed to assess how to lower the pipeline to the desired depth in a gas pipeline segment suffering from lack of buoyancy control in a large swampy area. The pipeline segment also was not buried to the sufficient depth during the construction. This analysis is a real case experienced by an in-service gas pipeline. A numerical stress analysis is also carried out using non-linear Finite Element Analysis software Abaqus [5] to this 28 in pipeline by modeling several load cases to evaluate pipe structure response against pipeline profile change. A numerical analysis is also performed to lower the pipeline to the desired depth underground.

2. Analytical analysis

An optimum profile of the pipeline must be calculated first to ensure a smooth transition in the pipeline lowering process and to keep the stress in a pipe still in the allowable limit. An assessment should also be performed to determine the construction method in physically lowering the pipeline to follow this optimum profile.

Engineering and design of the new profile must be very precise to avoid sudden drop of pipe resulting in pipe rupture or buckling.

*Corresponding author: mochsafarudin@gmail.com

The bending radius of lowered pipeline as shown in Fig. 1 is

$$R = \frac{x^2+y^2}{2y} \tag{1}$$

The minimum bending radius for pipeline profile is

$$R = \frac{EC'}{S} \tag{2}$$

The depth and horizontal distance of pipeline to avoid undesired stress are

$$y = \frac{EC\prime}{S} - \sqrt{\left(\frac{EC\prime}{S}\right)^2 - x^2} \tag{3}$$

$$x = \sqrt{2y\frac{EC\prime}{S} - y^2} \tag{4}$$

$$R = \frac{A\sin\left[90 - \cos^{-1}\frac{B^2+C^2-A^2}{2BC}\right]}{\sin\left[2\cos^{-1}\frac{B^2+C^2-A^2}{2BC}\right]} \tag{5}$$

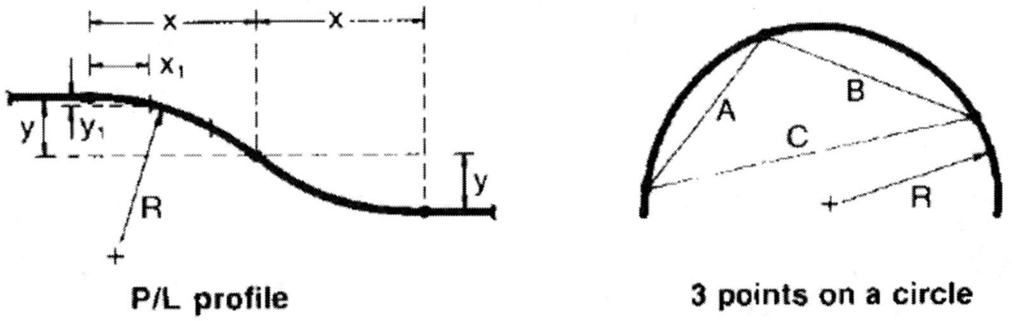

Fig. 1 Optimum profile of pipeline after lowering

where: R = bending radius m; x = horizontal distance, m; y = deflection or dropped from horizontal condition, m; S = bending stress, Mpa; C' = one-half diameter of pipe, m; E = elastic modulus of pipe material, MPa; A, B and C = triangle dimension, m

From the eqs. 1 to 5, it can be seen that the bending radius R does not depend on wall thickness and cross section of the pipe.

3. Finite element method

Analysis was conducted using Finite Element Method to evaluate the response of the structure in linear static conditions. Finite Element model was created as a 1 dimensional beam based on geometry from the field. For models 4 to 7, soil models as rigid surface were added. The summary of model input for FEA is tabulated in Table 1.

https://doi.org/10.1051/matecconf/201815901058

Table 1 Model input for linear static analysis using FEA

Model	Length (m)	Mid segment (m)	Internal pressure (psig)	Upward disp (m)	Downward disp (m)	Gravity	Concrete weight
1	200	25	1000	0.8	2.3	No	No
2	200	50	1000	0.8	2.3	No	No
3	200	100	1000	0.8	2.3	No	No
4	200	-	1000	-	0.9	Yes	No
5	200	-	1000	-	2	Yes	No
6	200	-	1000	-	0.9	Yes	Yes
7	100	-	1000	-	0.9	Yes	Yes

Some models are illustrated in Fig.2 and Fig.3.

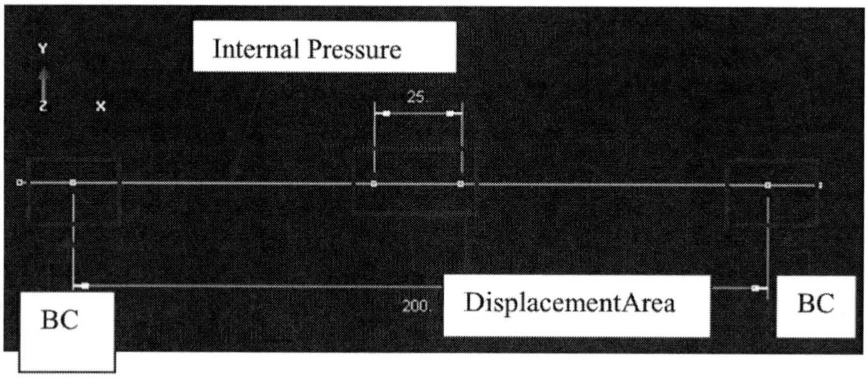

Fig. 2 Boundary conditions and loading for model 1

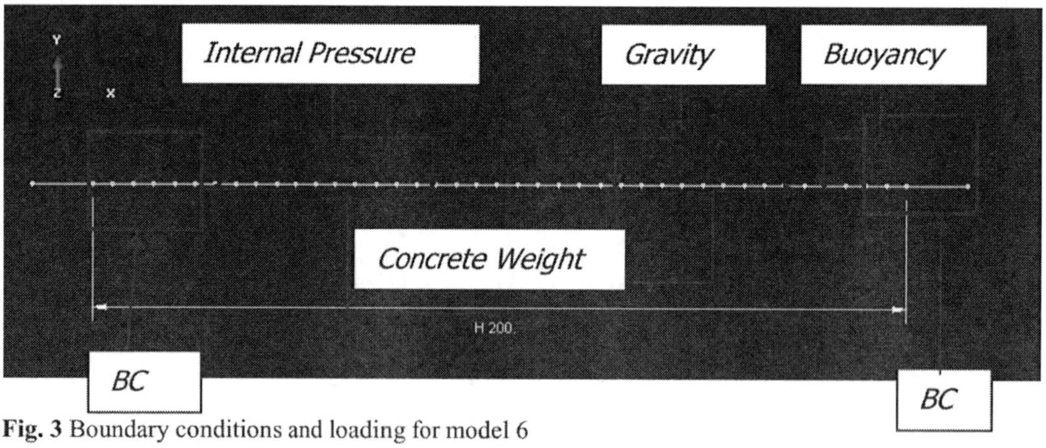

Fig. 3 Boundary conditions and loading for model 6

Boundary conditions for all models are fixed in the x, y, z direction and pipeline longitudinal axis rotation (x axis). Actual internal pressure loading is applied and in-situ conditions are assumed as mid-segment displacement as represented in model 1 to model 7.

4. General assumptions

Some general assumptions are made for the analysis. Analysis is carried out in the linear range and static condition. Pipe material has a *Specified Minimum Yield Strength* of 65,000 psi (448 MPa) and outer diameter of 28 inch (0.711 m). Soil surface is assumed to be rigid. New pipeline depth is 0.9 m at the pipe axis for maximum pipe stress of 0.5 SMYS (case 1) and 2 m for maximum pipe stress of 0.75 SMYS (case 2). Pipeline condition before lowering is buried to the one-half diameter of pipe (356 mm), fluid soil density of 1442 kg/m^3. Concrete weight is 1776 kg. Buoyancy is due to marsh (soil and water).

5. Result and analysis

5.1 Results for analytical method

Table 2 shows the summary of analytical result for pipeline lowering calculation. New pipeline profile for case 1 is illustrated in Fig.4.

Table 2 Summary of analytical result

Case	E, GPa	C_1 (m)	SMYS (MPa)	SR $_{all}$	σ_{all} (Pa)	Bend Radius (m)	Drop from horizontal (m)	X (m)	X_{total} (m)
1	200	0.36	448	0.5	224	317	0.45	17	34
2	200	0.36	448	0.75	224	211	1.00	21	41

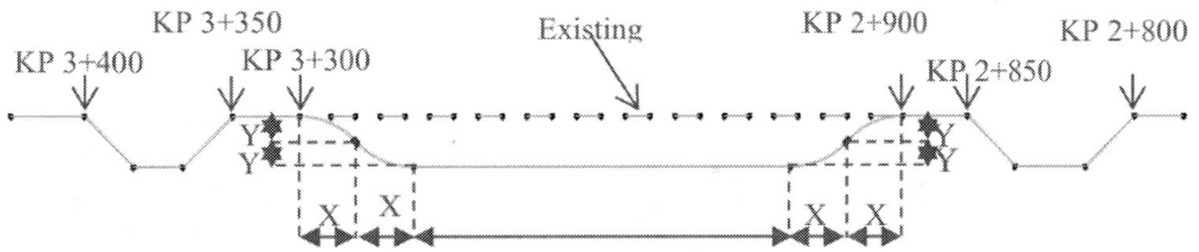

Fig. 4 Pipeline profile

For case 1, the maximum is found to be 0.91 at 2934 to KP+3267.

5.2 Linear static analysis using Finite Element Analysis

Linear static simulations are conducted for the corresponding models using Non-Liner Finite Element Analysis software Abaqus. The outputs required from the simulation are the maximum stresses for all loading conditions. Summary of linear static analysis results showing maximum stresses for every loading condition in all models are shown in Table 3.

Example of displacement and stress distribution occurs in pipeline model 1 after final loading is shown in Fig.5 while the displacement and stress distribution occurs in pipeline model 4 is illustrated in Fig.6.

Table 3 Linear static analysis results

Loading I		
Linear static analysis	**Maximum Stress**	**Stress Ratio**
Model 1, 2,3, 4, 5, 6 and 7	156 MPa	0.35
Loading II		
Linear static analysis	**Maximum Stress**	**Stress Ratio**
Model 1	162 MPa	0.36
Model 2	166 MPa	0.37
Model 3	187 MPa	0.42
Model 4	224 MPa	0.50
Model 5	336 MPa	0.75
Model 6	163 MPa	0.36
Model 7	170 MPa	0.38
Loading III		
Linear static analysis	**Maximum Stress**	**Stress Ratio**
Model 1	218 MPa	0.49
Model 2	255 MPa	0.57
Model 3	420 MPa	0.94

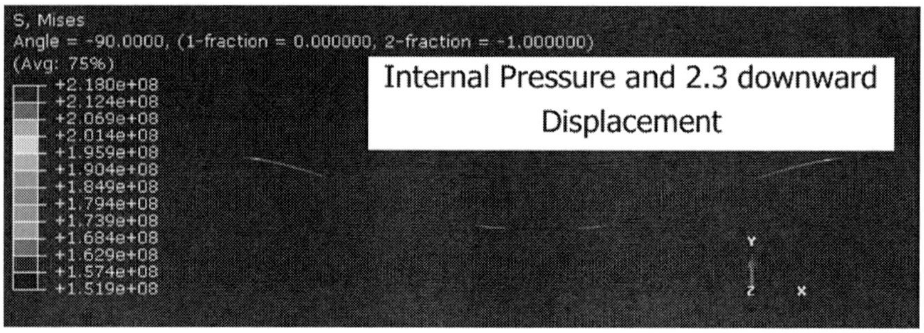

Fig. 5 Stress distribution for model 1

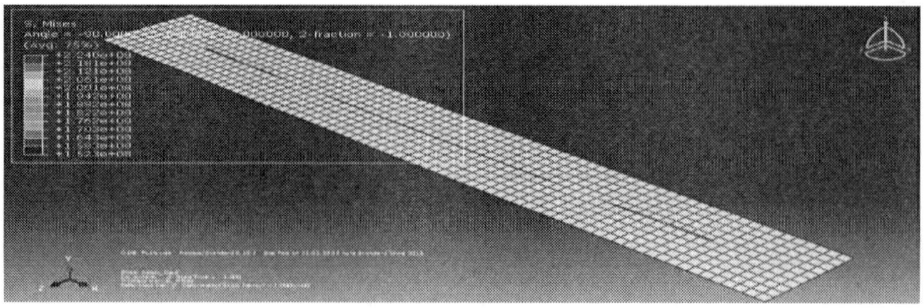

Fig. 6 Displacement and stress distribution in model 4

6. Comparison of results

The horizontal distance of pipeline that can be lowered from analytical method can be compared to the horizontal distance obtained from linear static analysis using FEM for case 1 as tabulated in Table 4.

Table 4 Comparison of horizontal distance using analytical and FEM for case 1 (0.5 SMYS)

Analysis	Horizontal Distance, m	
	Stress Ratio=0.5	**Stress Ratio=0.75**
Analytical	34	41
Numerical (MEH)	36.5	50

7. Conclusions and discussions

From the analysis results shown for every case, it can be concluded that:

1. Maximum stress occurs on model 3 when applying 2.3 m downward displacement with the magnitude of 420 MPa and stress ratio of 0.94.
2. FE analysis results for model 4 and model 5 show that with stress ratio limit of 0.5 and 0.75 the pipeline lowering can be executed to 0.9m and 2m respectively.
3. FE analysis for model 6 shows that by applying combination of internal pressure, gravity, buoyancy and concrete weight loads the pipeline can be lowered to 0.9m with new profile and maximum horizontal distance of 60 m with stress ratio of 0.36.
4. FE analysis result for model 7 shows that by applying combination of internal pressure, gravity, buoyancy and concrete weight loads the pipeline can only be lowered to 0.6 m depth with the stress ratio of 0.38.
5. Based on analysis results and engineering judgment considering appropriate code/standard, the 28" pipeline can be lowered to 2 m depth at pipe axis by experiencing maximum stress of 0.75 SMYS.
6. Structural analysis with FE method results are slightly difference compared to analytical results due to non-linear geometry/material considered in FE method.
7. Structural analysis with FE method can be conducted to determine maximum distance and depth of pipeline lowering to avoid overstress that can cause pipe rupture or permanent buckling.
8. Construction method executed in the field should consider the result from this analytical/numerical method.

References

1. Da-Ming Duan, Thomas Jurca and Charles Zhou, "A Stress Check Procedure for Pipe Lowering-In Process during Pipeline Construction, DOI:10.1115/IPC2014-33134, 10th International Pipeline Conference, 2014, Paper No. IPC2014-33134, pp. V001T03A06
2. Fan Zhang, Ming Liu, Yong-Yi Wang and William A. Bruce, "Stress Analysis of Lifting and Lowering-In Process", DOI:10.1115/IPC2014-33237, 10th International Pipeline Conference, 2014, Paper No. IPC2014-33237, pp.V001T03A012
3. McAllister, E.W, Pipeline Rules of Thumb Handbook, Elsevier, (2009)
4. ASME, Gas Transmission and Distribution Piping Systems, ASME Code for Pressure Piping, B31.8, (2012)
5. Abaqus Analysis User Manual, Dassault Systems

Resistance and Stability Analysis for Catamaran Fishing Vessel with Solar Cell in Calm Water

Teguh Putranto[1,*], *Wasis Dwi* Aryawan[1] , *Hesty Anita* Kurniawati[1], *Dony* Setyawan[1] , and *Sri Rejeki Wahyu* Pribadi[1]

[1]Department of Naval Architecture, Faculty of Marine Engineering, Institute of Technology Sepuluh Nopember

Abstract. The use of solar cell as the moving force for the ship's tour seems to be an important theme in action to create a sea transportation that is environmentally friendly as contribution of the Energy Efficiency Design Index (EEDI) for the decrease of pollutant levels. The electrical energy produced by solar cell is not as much energy from fuels so that the solar energy powered ship that can only be followed with a limited range and speed. This type of multi hull design like catamaran that has two symmetrical hull which is possible to have an more expansive main deck, small boat boundaries and good stability. The ship deck shape is going to contribute the extension of solar panels that can be used as a generator of power needed at the time of calculation of powering the ship. The analysis of the stability of the ship carried out numerically bu using references from the IMO regulations which require minimum value stability arm at certain angles. Load cases are definitely varied in 2 (two) conditions and compared in full and light load. The goal of this research is catamaran fishing vessel that have battery, solar panel, electric motor and all of supporting solar energy moving with speed 5 knots.

1 Introduction

By the topic of efficient energy uses, the world of shipbuilding is currently vying to conduct more studies on the ship design which is believed to have a small resistance compared with monohull [1]. One of the conceptual design is a type of catamaran which is often used by passenger ship for example as a tourism vessel in the area Sarangan, Bedugul Lake and other water attractions with calm water conditions. As shown in Figure 1, the environmental conditions in the unspoiled tourist spot that needs to be kept preservation of natural way in order to reduce or eliminate air pollution occurred. International Maritime Organization (IMO) has published some of regulations on the limitation of air pollutants produced by a ship called the Energy Efficiency Design Index (EEDI).

All ship types which is either single or double hull have a good stability if only metacenter point is above the center of gravity [2]. However, if the righting arm of stability is smaller or even negative, then the ship will not have a reserve moment so that the ship will be sink.

* Corresponding author: theories@na.its.ac.id

Therefore, IMO regulations already require the magnitude of arms that have to be had a ship so that safety and comfortable of passenger can be secured [3].

In calm wates, the determination within demihull on the small vessel appeared to have a significant effect on the total resistance of ship. The volume of the submerged hull on the catamaran and single hull is assumed in the same weight. One of the factors that affect the ship weight is the material used where Fiber Reinforced Plastics (FRP) have lightweight material and is suitable for vessel moving in the low speed [4]. The use of a catamaran hull for tourist ships can increase speed by 30% of the speed of the single hull ships. It certainly would have an effect on driving energy use [5].

Fig. 1. Sarangan Lake.

Energy obtained from fuel had started to be decreased by considering the pollution aspect. Natural resources cannot be upgraded so that it is necessary to look for the environmentally friendly renewable energy [6]. A small passenger vessel using solar panels as an energy for propulsion can be seen in Figure 2.

Although the ship dynamics is not considered later on, the ship design have to be analyzed the fluid structure interaction. It means that the shear force and bending moment caused by still water condition can influence the dimension of stiffener and thickness of ship structure [7].

Fig. 2. Catamaran vessel with solar cell energy.

The popular types of ship hull, for example usually monohull and catamaran, have been proven to provide some of advantages in the sea transportation. For the ship that needs more space or area in main deck as payload but the displacement of ship is small precisely, the design of catamaran vessel can be used to be applied for the ship design. The study of symmetrical and asymmetrical of catamaran vessel had been described that the magnitude of resistance does not significantly affect in the low speed ship. For the high speed ship, the effect of fluid interference along hulls does relatively affect the magnitude of total resistance. If it is reviewed in aspect of the ship draft, the symmetrical hull draft is smaller than the asymmetrical hull [8].

2 Theory

This research is generally carried out by using numerical method for the stability and resistance analysis. This vessel has solar panel, battery, electric motor, and propulsion system which support for the sustainable energy. The source of energy comes from the solar energy that is converted to be electric energy to rotate the propeller.

2.1 Catamaran Vessel Model

Catamaran vessel have 2 (two) hull types which are symmetrically or asymmetrically. In this research, the ship model uses catamaran symmetrically type because from the previous research, this model give some advantages in aspect of stability and resistance.

2.2 Stability Calculation Analysis

Intact stability calculation is carried out to make sure the evidence that the ship is safe in seaworthiness condition [9]. The calculation of righting arm which is used as parameter of stability does produce a metacenter point. In order to obtain this result, the Krylov I provides the equation to calculate it. The length of righting arm (GZ) could be used to check whether the ship stability is accepted or rejected which it is definitely regulated in the Intact Stability Code, IMO Regulation A. 749 (18).

2.3 Resistance Calculation

Because ship moves in low speed, the van oortmerssen calculation for ship resistance can be used. The ship speeds are varied in 0, 1, 2, 3, 4, and 5 knots. The Van Oortmerssen formula is usually used by condition the ship resistance in low speed. Beside it, the total resistance of ship could be calculated by using Holtrop and Mennen equation. This method can be divided in viscous and wave making resistance consideration. The Holtrop and Mennen equation is generally used to calculate the ship resistance in any ship type and speed.

Two methods are used to verify the total resistance of ship obtained from those methods. The magnitude of resistance could be used to estimate the magnitude of main engine power of ship which the efficiency of shaft propeller is neglected because of the simply of propulsion system . So that the Effective Horse Power (EHP) obtained from multiplication between resistance and ship speed is same as the Break Horse Power (BHP).

3 Result and Discussion

The catamaran vessel worked in this research have main dimension as follow:

Length of Water Line (LWL)	= 4.35 m
Breadth (B)	= 2.55 m
Draft (T) at full load	= 0.45 m
Freeboard (Fb)	= 0.35 m
Depth (H)	= 0.85 m
Capacity of Passenger	= 8 persons

The location of solar panel is put above the catamaran fishing vessel because it considers the absorb of solar energy. Figure 3 shows the catamaran fishing vessel for this research.

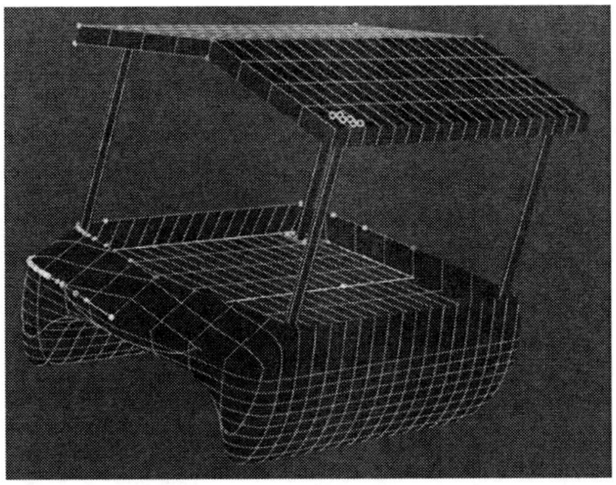

Fig. 3. Surface view of catamaran vessel using solar panel.

3.1 Intact Stability Analysis

Intact stability can be approached by the length of metacenter to gravity (MG) because this length is always affected the reserve moment of ship when the ship is in heel condition. Figure 4 shows graph of statics stability of this catamaran fishing vessel. At the initial position with heel 0 degree, the GZ length is zero. At the moment where heel is about 40 degrees, the GZ length is in the maximum value of 0.35 m and the empty condition is provided such result.

Figure 5 shows the graph of GZ curve for full load condition. The GZ maximum is occurred in the heel angle about 33 degrees. After 33 degrees, the GZ length decreases significantly to zero value. This is exactly critical point where the ship stability changes to be an unstable condition and it is very dangerous.

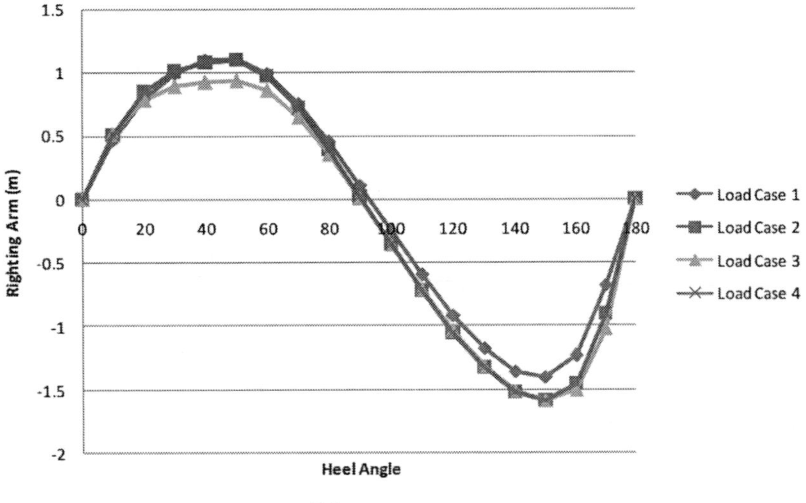

Fig. 4. The GZ curve at no passenger condition.

This condition can cause the ship capsize. Because this catamaran fishing vessel does not sail in the wave water condition, the ship capsize caused by wave is impossible occurred. The possibility of ship sinking is caused by the position or distribution of passenger focused on the one side of ship. Table 1 shows that the ship stability analysis of two load cases. The load cases consists of no passenger (LC1) and full passengers (LC2). From the result, each of parameter has accepted criteria for each load cases. If there is a parameter that is not fulfills the IMO regulation, the criteria would be definitelyrejected because it means that the catamaran fishing vessel would be fail.

Table 1. The ship stability analysis based on the load cases compared by IMO criteria.

Parameter	LC1	LC 2	IMO	Criteria
$e_{0.30}$o	0.21	0.63	≥ 0.055	Accepted
$e_{0.40}$o	0.28	0.83	≥ 0.09	Accepted
$e_{30,40}$o	0.07	0.20	≥ 0.03	Accepted
h_{30}o	0.40	1.20	≥ 0.2	Accepted
ϕmax	30^0	33^0	$\geq 25^0$	Accepted
GM_0	1.26	4.20	≥ 0.15	Accepted

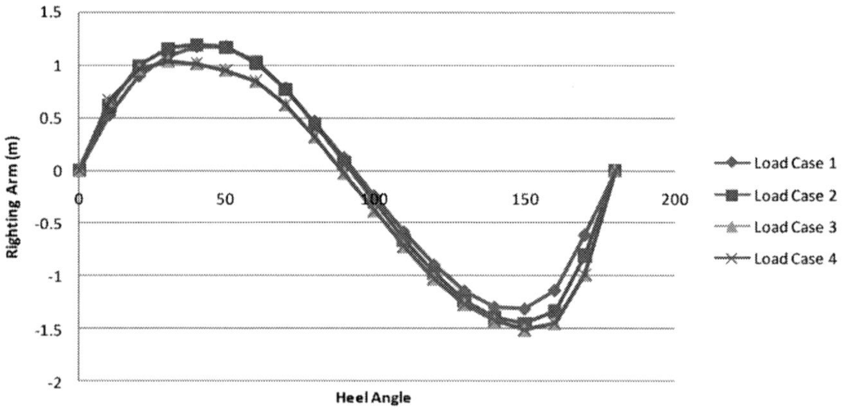

Fig. 5. The GZ curve at full load condition.

3.2 Resistance Analysis

There are some methods for calculating the catamaran fishing vessel resistance. One of method for calculating the resistance is Van Oortmerssen Method. This method is suitable for small displacement ships especially for catamaran fishing vessel which have involved the main resistance components such as viscous resistance and wave making resistance. This method is like as Holtrop Method. The result of two methods is compared to show that the pattern of graph is relatively same and ensure that the methods is recommended to be used. The greater resistance is occurred in the faster of ship speed. The maximum of total resistance of catamaran fishing vessel is about 1.8 kN in ship speed 5 knots.

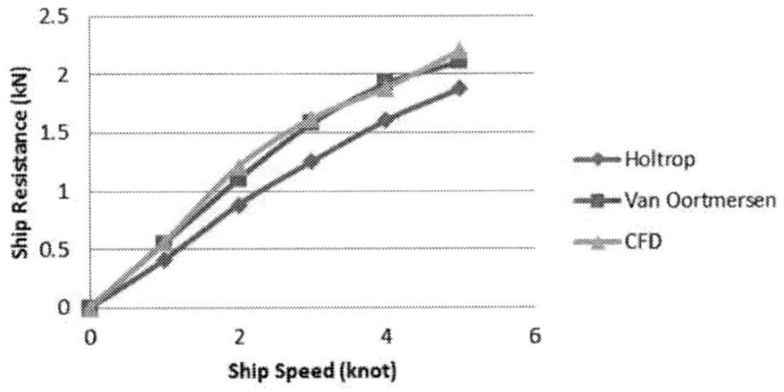

Fig. 6. Total resistance in various of speed at catamaran vessel.

Need main engine power depends on the total resistance of ship. For the total resistance is 1.5 kN, the main engine power needed for ship speed 5 knots is 3.75 kW. The specification of main engine is by using the electric motor because the power is produced from solar energy. The magnitude of electric motor power is 2000 Watt so that is needed two electric motors precisely.

4 Conclusion

The conclusions of this research can be concluded as follows:

1. The positions of metacenter that indicate the stability of ship is still located above gravity point either the full load condition or the empty condition. The maximum GZ at full load condition is about 1.2 m and one at empty load condition is about 0.4 m;
2. The vessel can be operated because electric motor power is greater than resistance. From the result and discussion, the resistance is about 2.043 kN in maximum speed, but the electric motor is about 2000 Watt;

Acknowledgements

Researcher would like to express the appreciation for LPPM ITS that have given the fund and support through Penelitian Pemula Grant in 2016-2017.

References

1. A. Jasionowski, D. Vassalos, OE, **3** (2009)
2. R. Bhattacharyya, *Dynamics of marine vehicles*, **1** (1978)
3. D. B. Danisman, JOE, **91** (2014)
4. J. Date, S. R. Turnock, *Ship Science Report*, **2** (1999)
5. S. H. Lee, Y. G. Lee, S. H. Kim, JOE, **34** (2007)
6. R. Pujiwat, *Analisa Teknis dan Ekonomis Perencanaan Water Bus Dengan Tenaga Surya di Banjir Kanal Barat Jakarta*, **1** (2008)
7. T. Putranto, K. Suastika, J. Gunanta, IPTEK , **2** (2017)
8. T. Putranto, A. Sulisetyono, KAPAL, **12** (2015)
9. T. Putranto, A. Sulisetyono, IJAER, **12** (2017)

Synthesis Sensitive Layer of Ethylene Gas Sensor Based Tin Oxide Nanoparticles Using Water as Solvent In Precipitation Method

Erica Caesariaty Harni Prima Nabena[1], *Brian* Yuliarto[1,2*] , Nugraha[1,2] , and *Muhammad* Iqbal[1]

[1]Advance Functional Materials Laboratory, Departement of Engineering Physics, Fakultas Teknologi Inndustri, Institut Teknologi Bandung, Bandung 40132, Indonesia.
[2]Research Center for Nanoscience and Nanotechnology (NRCN), Institut Teknologi Bandung, Bandung 40132, Indonesia

Abstract. Ethylene gas is a gas naturally released by fruits. The maturity level of these fruits could be predicted from the amount of ethylene around them. To maintain the freshness of these fruits, the concentrations of surrounding ethylene need to be monitored. Therefore, suitable gas sensor ethylene were still in progress to get good respond. In this paper systhesis of pure SnO2 nanoparticle from $SnCl_2.2H_2O$ and water solution in precipitation method was held. The tin oxide powder was obtained in good distribution with nanoparticle size. This nanoparticle powder was formed into thick film, using ethylene glycol as solvant. To observe this sensor characteristics, several test was held in various conditions. Gas testing used pure ethylene gas show better respond at higher than room temperature but recovery time still unsatisfying.

1 Introduction

Since 1934, fruit's porter known ethylene was emitted by tropical fruit while shipping them from Jamaica to Europe. In plants, ethylene was emitted by flowers, seeds, especially fruits. In fruits, ethylene was indicator of initiation ripening. There are two type of fruits, climacteric and non-climacteric. Most of fruits harvested in Indonesia are climacteric type. In the climatric fruit, the concentration of ethylene gas will raise up until it reaches the peak. So, they continue ripen as ethylene existence. When fruits had uncontroll ethylene concentration environment, it will spoiled. Less than 1 ppm by volume of ethylene can trigger the process of climacteric fruit ripening. This fact made important to monitoring ethylene gas during storage and distribution. If these fruit products are left too long, they would get fermented and rotten. Fruits can emit ethylene in range ppb scale to around 140 ppm [1, 2].
Ethylene is an invisible, colorless, and odorless gas. As in container or the other transport, many other organic substance may exist, then potential sensing methods need good

* Corresponding author: brian@tf.itb.ac.id

sensitivity as well as good selectivity. Then, we must understand the characteristics of ethylene gas. Ethylene molecular weight is 28.05 g mol^{-1}. As its specific weight (1.178 kg m-3 at 15°C) is similar to that of air (1.225 kg m-3 at 15°C), ethylene freely diffuse in storage [2].

Several technologies have been proposed in order to monitoring concentration of ethylene. Hence, there is a need for simple, compact, inexpensive, portable, and reliable ethylene sensors that could be used parallel in controlling ethylene concentration systems. The technologies that have been commonly used include gas chromatography [3], optical sensors [4], colorimetric sensors [5], amperometric sensors [6], photoluminescence sensors [7], gravimetric sensors [8], and chemoresistive and capacitive sensors [9, 10]. From technologies has been mentioned, chemoresistive can meet criteria desired

One of material frequently used as sensitive layer in gas sensor is tin oxide (SnO_2), because of its properties including fast–response speed, high chemical stability, and prominent selectivity. SnO_2 is a n-type semiconductor metal oxide with band gap 3,6 eV at 300 K, making it favorable material for sensing device [11]. As n-type materials major carriers in SnO_2 is electrons. In the air condition, when layers of nanoparticles absorbs oxygen, a depletion layer of species O- was formed. As more oxygen was absorbed, the depletion region increase and the conduction region decrease in thickness. [9] When reductive gas like ethylene was presented, the active sensitive layer with oxygen adsorbates will react with the gas and permit electrons inject back into the active material, causing an increase in electronic conductivity or decrease in resistivity. The depleted electrons were replenished back to the SnO_2 surface, thereby increasing the conduction region and reducing depletion region. While this process occur, the sensing performance was determined by the diffusion of gas throughout the pores of the sensing layer. To improve the sensor performance, some research did focusing on controlling SnO_2 nanostructure, develop method of producing the sensitive layer improve the porosity of the material, which obtain better performance in gas sensing behaviors [12]. One of modification was fabricating SnO_2 in nanoscale so that interaction area between etilen and sensitive layer become larger. This condition effected much detected gas adsorbed into SnO_2 and produce higher sensitivity. [13]

Some research about ethylene gas sensor based metal oxide already done several time, as we could see below :

1) Passive capacitive gas sensor based SnO_2 (10-15 nm) detected 0-100 ppm of ethylene in room temperature with sensitivity 39% [9].
2) Thick film gas sensor from WO_3/SnO_2 nanoparticle was made to detect 2-8 ppm ethylene at optimum operating temperature of 300°C with best respond at 37,5%[10].

From informations above, SnO2 was successfully detect gas ethylene in lowconcentration. The development capacitive gas sensor was rare, then we develop pure SnO_2 nanoparticles with chemo resistive gas respond to detect ethylene. The methode of synthesis was precipitation, which promising decrease the particle size to decrease optimum operating temperature.

2 Detail Experiment

Fig.1 shows a flowchart of the method used for synthesis of SnO_2 sensitive layer and the characterization. This section was divide to some parts. Powder preparation, powder characterization, thick film fabrication, and characterization gas sensor.

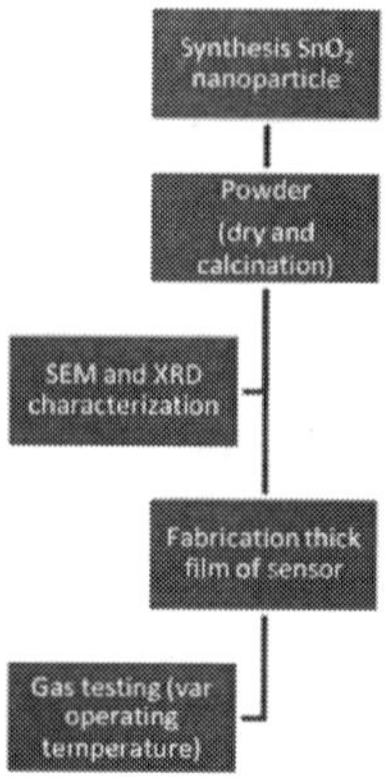

Fig. 1 Experimental Flowchart

2.1 Powder Preparation

SnO_2 nanoparticles obtain were prepared by precipitation method using 0,4 gram $SnCl_2.2H_2O$ in 40 ml water as solvent and amount of NaOH was dropped as precipitant agent to stirred solution until pH condition wanted. This solvant colored like watery milk, not so white. Proof that from this solution will get nanoparticle powder. As we know from data solubility of metal hydroxide compound, that $Sn(OH)_2$ will dissolve around pH 3.5 to 11. This research using condition above pH 11. This condition permit the solution into sol condition, which the result is not like ordinary precipitation. Then, the obtain precipitant was centrifuged and washed several time with aqua bides until neutral condition obtained. The samples was dried in furnace a night to remove water, then calcinated for 2 hours. Powder obtain will be characterized to observe the properties.

2.2 Powder Characterization

SnO_2 nanoparticles was investigated by scanning electron microscopy (SEM, Hitachi 9100 0005) to analyze morphology and structure which operating in 5 kV.
The crystal structures of tin dioxide powder was analyzed using X-ray diffraction (XRD) using a Advanced Bruker with Cu-Kα radiation (1.54439 Å) and LynxEye detector. The pattern of sample was collected from 20° to 80° as 2θ with step size of 0.019°.

2.3 Thick Films Fabrication

The obtained powder was added by ethylene glycol to form white paste. Before this paste was smeared on alumina. Firstly, alumina was coated by silver paste to get electrodelayer. Using doctor blade method white paste was smeared on alomina. To remove the impurity drying samples at 200°C for 2 hours. Then we get thick film senstive layer of gas sensor.

2.4 Characterization Gas Sensor

Gas sensing performance of SnO_2 based gas sensor was evaluated with pure ethylene gas in test chamber. The electrical resistances in air and in present of ethylene were measured at

100 ppm from room temperature to 200 °C. The exposure time was for 10 minutes. The response towards ethylene were calculated and expressed in percentage as follows :

$$\%S = \frac{|R_{gas} - R_{air}|}{R_{gas}} \quad (1)$$

Where, R_{air} and R_{gas} are the sensor resistance in air and in the presence of gas target, respectively at the same operating temperature.

3 Result and Discussion

3.1 Morphology and Structure Characterization

In **fig 2** was shown the XRD pattern of samples having tetragonal rutile structure (Joint Committee on Powder Diffraction Standards Card no 41-1445). Peaks that showed indicate materials was a tetragonal rutile crystal structure. Where the diffraction peaks are (110), (101), and (211). [14] Based on peak (211), using Scherre's equation obtain crystallite size 12,61 nm. The result of crystallite size was better than using 1-buthanol as solvent with same calcination temperature [15].

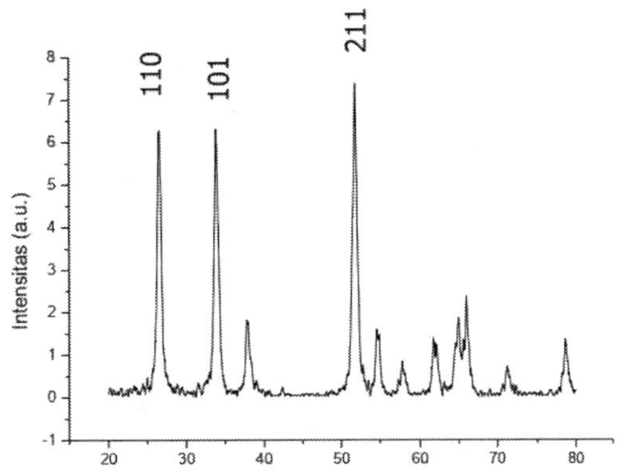

Fig. 2 XRD result of SnO_2

From **fig 3** SnO_2 samples has good homogenity and distribution, with average particle size 30 nm. This nanoparticle material size provide many site for reaction between gas target and oxygen ion. Although, agglomeration happen in many place as one of SnO_2 problem. From this result of characterization we decide to make SnO_2 thick film as sensitive layer and test it in chamber.

Fig. 3 SEM result of SnO_2

3.2 Sensor Gas Testing

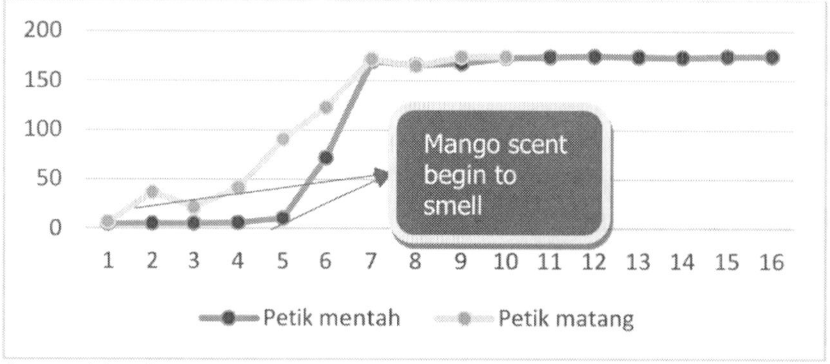

Fig. 4 Ethylene emition from a mango.

The mechanism of detection ethylene at the surface sensitive layer SnO_2 thick film can observe by chemical surface interaction. Firstly, initial physisorption occured, then ethylene gas bonded with O^{-}_{ads} to formed C_2H_4O and released e^-. The change in resistance would occur simultaneously with exposure to ethylene gas in the gas sensor. We used two operation temperature as comparison, in low temperature and in high temperature. From fig. 1 we know that mango started ripe stage while ethylene concentration around 20 ppm indicated by smell mango flavor. And the point above 150 ppm of ethylene mango reach full ripe stage, usually at this stage we called rotten. So we tested the sensor on 40-100 ppm based on concentration of ethylene while mango best at flavour.

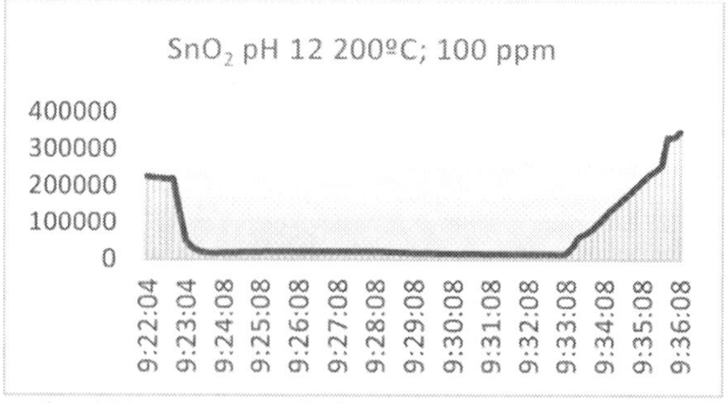

Fig. 5 Gas sensing respond to 100 ppm pure ethylene at 200°C

Firstly, we tested SnO_2 gas sensor in 100 ppm pure ethylene at moderate temperature (50°C) and get S = 19.8%. Then, we would like to get better sensitivity. Fig 5 show the respond of SnO_2 gas sensor when we increase the temperature became 200°C obtain S = 92.68%. If we compared to ref 10, material we made much better, because get 92.68% at 200°C. But, as higher operating temperature we used the recovery time show unstatisfying result. At moderate temperature, need 1.5 minutes to back in air condition. But at higher temperature need 3 minutes to back, but the resistance not stable as we see in figure 5, the resistance still increase. It was happened because at high temperature sensor the chemisorption dominate the interaction as said before [16]. Then we asume, there was ethylene trap in material site.

4 Conclusion

We had been successfully synthesized the SnO_2 nanoparticles with average size 30 nm by precipitation method using $SnCl_2.2H_2O$ and NaOH as precipitate agent in water solution. The formation of crystalline tin dioxide nanoparticles had been confirmed by XRD patterns and FE-SEM. Through gas testing we were informed that increase the temperature comparable with sensitivity. But took longer to recovery back in air condition. Portable gas sensor to monitoring rippening fruit needs sensor's material that high sensitivity and fast respond also recovery in low temperature.

References

1. E. Brigit, J. M. Schnorr and T. M. Swager, Angewandte Chemie International Edition. **51.23** (2012).
2. S. Janssen, K. Schmitt, M. Blanke, M. L. Bauersfeld, J. Wollenstein and W. Lang. Phil. Trans. R. Soc. A**, 372: 20130311** (2014).
3. N.A. Zaidi, M.W. Tahir, M.J. Vellekoop, and W. Lang. MDPI Sensors **17, 2283** (2017).
4. Q.Tan, X. Pei, S. Zhu, D. Sun, J. Liu, C. Xue, T. Liang, W. Zhang, and J. Xiong. MDPI Sensors **13(4)** (2013).
5. P. Cabanillas-Galán, L. Farmer, T. Hagan, M. Nieuwenhuyzen, S. L. James and M. C. Lagunas. Inorg. Chem. **47** (2008).
6. L. R. Jordan, P. C. Hauser and G. A. Dawson. Anal. Chem. **69** (1997).
7. A. M. Marti, Y. J. Nijem, K. J. Chabal and Balkus Jr. Microporous Mesoporous Mater. **174**. (2013).
8. B. Graewe, A. Rang, C. Schalley, J. Haubrich and J. Bargon. Sens. Actuators, B: Chem. **119**. (2006).
9. M. Agarwal, M. Balachandran, S. Shrestha and Varahramyan. Journal of Nanomaterials. **Article 145406**. (2012).
10. Y. Pimtong-Ngam, S. Jiemsirilers and S. Supothina. Sensors and Actuators A, **139.** (2007).
11. G. Korotcenkov. Sensors and Actuators B. **107** (2005).
12. X. Y. Feng, J. Jiang, H. Ding, R. M. Ding, D. Luo and X. T. Huang. Sens. Actuators B. **183.** (2013).
13. B. Yuliarto, L. Nulhakim, M.F. Ramadhani, M. Iqbal, A. Nuruddin, IEEE Sensors Journal. **15 (7)**. (2015).
14. B. Yuliarto, G. Gumilar, D. Zulhendri, Nugraha and N. Septiani. Acta Physica Polonica A. **131**. (2017).
15. K. Nejati, Cryst. Res. Technol. **47**. (2012).
16. A. Helwig, G. Muller, G. Sberveglieri and M. Eickhoff. Journal of Sensors. **Article ID 620720** (2009).

Projection Matrix Design for Co-Sparse Analysis Model Based Compressive Sensing

Endra Oey[1,*], *Dadang* Gunawan[1] , and *Dodi* Sudiana[1]

[1]Wireless and Signal Processing (WASP), Department of Electrical Engineering, University of Indonesia, Depok City, West Java 16424, Indonesia

Abstract. Co-sparse analysis model based-compressive sensing (CAMB-CS) has gained attention in recent years as alternative to conventional sparse synthesis model based (SSMB)-CS. The equivalent operator as counterpart of the equivalent dictionary in the SSMB-CS is introduced in the CAMB-CS as the product of projection matrix and transpose of the analysis dictionary. This paper proposes an algorithm for designing suitable projection matrix for CAMB-CS by minimizing the mutual coherence of the equivalent operator based on equiangular tight frames design. The simulation results show that the CAMB-CS with the proposed projection matrix outperforms the SSMB-CS in terms of the signal quality reconstruction.

1 Introduction

Compressive sensing (CS) as a new paradigm in signal acquisition has gained popularity over the last decade after it was introduced in [1-2]. CS acquires the signal directly in already compressed form by projecting it into a well-designed projection matrix. CS framework has been applied in many applications such as imaging applications, internet of thing, data security, and more [3-4]. A conventional CS systems works based on the sparse synthesis model of signal where a signal can be synthesized from a few atoms of a synthesis dictionary [5]. The alternative model is co-sparse model where sparse analysis coefficients can be obtained by multiplying the signal and an analysis dictionary (operator) [6]. Co-sparse analysis model based (CAMB)-CS has attracted attention in recent years because it outperforms the synthesis model as shown in [7-8].

Three main problems of CS are how to build a dictionary, design a proper projection matrix and reconstruct the signal from CS. The famous KSVD algorithm and its extensions have been commonly used to build a synthesis dictionary [9-10] also the improvements by exploiting additional structure of sparse coefficients can be found in [11-12]. The analysis version of KSVD [13] and sparsifying transforms learning algorithms have been used to build an operator [14-15]. The Convex and Relaxation, Greedy, and Bayesian algorithms are used for signal reconstruction in synthesis based CS [16] as well as the counterpart algorithms for analysis based CS [6, 17].

* Corresponding author: oey.endra51@ui.ac.id

While how to design optimal projection matrix for sparse synthesis model based (SSMB)-CS has been widely proposed such as in [18-19] but for CAMB-CS has not received attention. This paper addresses how to design a projection matrix for CAMB-CS, use it to perform CS on a natural image and compare the image reconstruction performance to SSMB-CS.

2 SSMB-CS And CAMB-CS

In the SSMB-CS, the signal $x \in \Re^{N\times 1}$ is synthesized from a sparse linear combinations of the dictionary columns $\Psi \triangleq [\psi_1\, \psi_2 \ldots \psi_L\,] \in \Re^{N\times L}$:

$$x = \sum_{i=1}^{L} \theta_i \psi_i \triangleq \Psi\theta \tag{1}$$

if $L > N$ Ψ is called *overcomplete synthesis dictionary*, $\theta \in \Re^{N\times 1}$ is sparse coefficients with $\|\theta\|_0 = K$, $K < N$ and $\|\theta\|_0$ is number of non-zero elements in θ. The CS is performed by multiplying the signal x and the projection matrix $\Phi \in \Re^{M\times N}$ then $y = \Phi x$, where $y \in \Re^{M\times 1}$ is a compressive measurement vector and $M < N$. The reconstructed signal $\hat{x}$ can be obtained from y by solving the following constrained problem :

$$\min \left\|\hat{\theta}\right\|_0 \; s.t. \; y = \Phi\Psi\hat{\theta} \text{ and } \hat{x} = \Psi\hat{\theta} \tag{2}$$

The problem in (2) is NP-hard and has combinatorial complexity but can be approximately solved using the Convex and Relaxation, Greedy, or Bayesian algorithms [16].

In CAMB-CS, the analysis coefficients $\alpha \in \Re^{R\times 1}$ are obtained by multiplying the operator $\Omega \in \Re^{R\times N}$ and the signal $x \in \Re^{N\times 1}$ then $\alpha = \Omega x$. If $R > N$, Ω is called *overcomplete operator*. The *co-support* $\Lambda \subseteq \{1 \ldots R\}$ of x is the index rows of Ω that are orthogonal to it, so $\Omega_\Lambda x = 0$ where $|\Lambda| = c$ is *co-sparsity* of the signal x and $\|\alpha\|_0 = R - c$ is number of non-zero elements in α. The reconstructed signal $\hat{x}$ can be obtained from CAMB-CS by solving the following constrained problem :

$$\min \|\Omega\hat{x}\|_0 \; s.t. \; y = \Phi\hat{x} \tag{3}$$

The problem in (3) can be solved by using the SSMB-CS counterpart algorithms for CAMB-CS [17].

3 Projection Matrix Design

In the SSMB-CS, the equivalent dictionary $D \in \Re^{M\times L}$ is defined as $D = \Phi\Psi$ where $G_s = D^T D$ is the Gram matrix of D, g_{s-ij} is the (i,j)th elements of G_s and $Z_s \triangleq diag\left(g_{s-11}^{-1/2} \ldots g_{s-ll}^{-1/2} \ldots g_{s-LL}^{-1/2}\right)$. The normalized equivalent dictionary is $\overline{D} \triangleq DZ_s$ and the normalized Gram matrix is $\overline{G}_s = \overline{D}^T\overline{D}$ such that $\overline{g}_{s-ll} = 1$, $\forall l$. The mutual coherence of D is defined as $\mu(D) = \max_{i \neq j} |\bar{g}_{s-ij}|$ where $\mu(D)$ is bounded with $\mu_B \le \mu(D) \le 1$ and $\mu_B \triangleq \sqrt{\frac{L\text{-}M}{M(L-1)}}$ is Welch bound [20]. The t*-averaged mutual coherence* $\mu_t(D)$ was introduced in [18] as $\mu_t(D) := \frac{\sum_{i \neq j, 1 \le i, j \le L} (|\bar{g}_{s-ij}| \ge t) |\bar{g}_{s-ij}|}{\sum_{i \neq j, 1 \le i, j \le L} (|\bar{g}_{s-ij}| \ge t)}$ where $0 \le t < 1$ and indicator function $(|\bar{g}_{s-ij}| \ge t) = 1$ if the condition is true and otherwise is zero. The common projection matrix design for SSMB-CS is based on how to make $\mu_t(D)$ as small as possible. It is done by making G as close as possible to a target Gram matrix G_t that has desired properties such as ETF Gram matrix [20]. The projection matrix design is performed by solving:

$$\min_{\Phi, G_t} \left\| G_t - \Psi^T \Phi^T \Phi \Psi \right\|_F^2 \tag{4}$$

where $\| \ \|_F$ denotes the *Frobenius* norm and it can be solved based on shrinkage method [18] or alternating projection [19]. The equivalent operator $O \in \Re^{M \times R}$ is introduced for CAMB-CS as $O = \Phi \Omega^T$ where $G_c = O^T O$ is the Gram matrix of O, g_{c-ij} is the (i, j)th elements of G_c and $Z_c \triangleq diag\left(g_{c-11}^{-1/2} \ \dots \ g_{c-ll}^{-1/2} \dots \ g_{c-LL}^{-1/2}\right)$. The normalized equivalent operator is $\bar{O} \triangleq O Z_c$ and the normalized Gram matrix is $\bar{G}_c = \bar{O}^T \bar{O}$ such that $\bar{g}_{c-ll} = 1, \ \forall l$. The projection matrix design for CAMB-CS is performed by solving:

$$\min_{\Phi, G_t} \left\| G_t - \Omega \Phi^T \Phi \Omega^T \right\|_F^2 \tag{5}$$

This paper adapted algorithm in [19] to solve (5) by using the following algorithm which is denoted as OGS algorithm.

OGS Algorithm

Initialization:

Ω - Operator;
$\Phi^{(0)}$ - Initial Projection Matrix;
K - Number of Iterations;
Set $k = 1$.

Begin: For $1 \le k \le K$, calculate (6) until (12):

$$O^{(k-1)} = \Phi^{(k-1)} \Omega^T \tag{6}$$

$$\bar{O}^{(k-1)} = O^{(k-1)} Z_c \tag{7}$$

$$\bar{G}_c^{(k-1)} = \left(\bar{O}^{(k-1)}\right)^T \left(\bar{O}^{(k-1)}\right) \tag{8}$$

$$\overline{G}_{c-e}^{(k)}(i,j)=\begin{cases}1, & i=j\\ \overline{G}_c^{(k-1)}(i,j), & \left|\overline{G}_c(i,j)\right|\le\xi\\ sign\left(\overline{G}_c^{(k-1)}(i,j)\right)\xi & otherwise\end{cases} \tag{9}$$

eigenvalue decomposition of $\overline{G}_{c-e}^{(k)}$:

$$\overline{G}_{c-e}^{(k)}=V_e\begin{bmatrix}\Lambda_e & \mathbf{0}\\ \mathbf{0} & \mathbf{0}\end{bmatrix}V_e^T \tag{10}$$

where V_e is orthonormal matrix and $\Lambda_e=diag[\lambda_1,\ldots\lambda_r,\ldots,\lambda_R]$ with $\lambda_r\ge\lambda_{r+1},\forall r$

$$O_{etf}^{(k)}=\frac{U}{\sqrt{M}}\left[\sqrt{\left(\sum_{r=1}^{M}\lambda_r\right)}I_M \quad \mathbf{0}\right]V_e^T \tag{11}$$

$$\Phi^{(k)}=O_{etf}^{(k)}\left(\Omega^T\right)^{+} \tag{12}$$

where $\left(\Omega^T\right)^{+}$ is pseudo-inverse of $\left(\Omega^T\right)$.

Go to ***Begin*** with $k\rightarrow k+1$ to continue the iterative procedure.

End: End the algorithm, output $\Phi=\Phi^{(k-1)}$.

4 Results and Discussion

This paper used 1000 training-images in LabelMe training data set [21-22] where 20 non-overlapping 8×8 patches are taken randomly from each image and each patch is re-arranged as a vector of 64×1. This training patches $P\in\Re^{64\times20000}$ were used to build synthesis dictionary $\Psi\in\Re^{64\times96}$ by using KSVD algorithm [9] and operator $\Omega\in\Re^{96\times64}$ by using the algorithm in [15]. Algorithm in [19] which is denoted with BLH and OGS algorithm were used for SSMB-CS and CAMB-CS projection matrix design respectively. Both algorithms used the same Gaussian random matrix $\Phi^{(0)}\in\Re^{M\times64}$ as initial projection matrix, number of iterations $K=1000$ and $\xi=\mu_B$. The test image I^{test} was divided into 8×8 non-overlapping patches where each patch convert to vector $x_j\in\Re^{64\times1}$, $j=1,2,\ldots J$ where J is number of patches in the tes image. CS was performed on those patches to obtain $y_j\in\Re^{M\times64}$. The OMP [23] and its counterpart Greedy algorithm GAP [6] were used for SSMB-CS and CAMB-CS respectively to obtain each reconstructed patch $\hat{x}_j\in\Re^{64\times1}$ and the whole reconstructed patches are arranged to get the reconstructed image $\hat{I}$. The Peak Signal-to-Noise Ratio (PSNR) was used to measure image reconstruction accuracy. It is defined as $\text{PSNR (dB)}=10\log_{10}\left(\frac{W\times H\times1^2}{\sum_{w=1}^{W}\sum_{h=1}^{H}\left(\hat{I}_{w,h}-I_{w,h}^{test}\right)^2}\right)$ where W and H are width and height of image.

Table 1 shows the reconstruction (in PSNR (dB)) comparison of SSMB-CS and CAMB-CS for several standard test images with Compression Ratio (CR) = Ratio Measurement Number (RMN) = $\frac{M}{N}$ = 31.25 %. Figure 1 shows the PSNR versus CR from 18.75 % until 81.25 % for Brick test image.

Table 1. Reconstruction comparison of SSMB-CS and CAMB-CS for CR = 31.25 %.

Projection Matrix	**Test Images (Size)**			
	Moon (1986 x 1986)	**Brick (768 x 512)**	**Barbara (720 x 576)**	**Peppers (512 x 512)**
	PSNR (dB)			
Random SSMB-CS	27.48	20.86	22.52	25.14
Random CAMB-CS	31.46	23.61	23.93	28.45
BLH SSMB-CS	27.22	19.72	21.71	24.03
OGS CAMB-CS	**33.98**	**25.95**	**25.89**	**31.25**

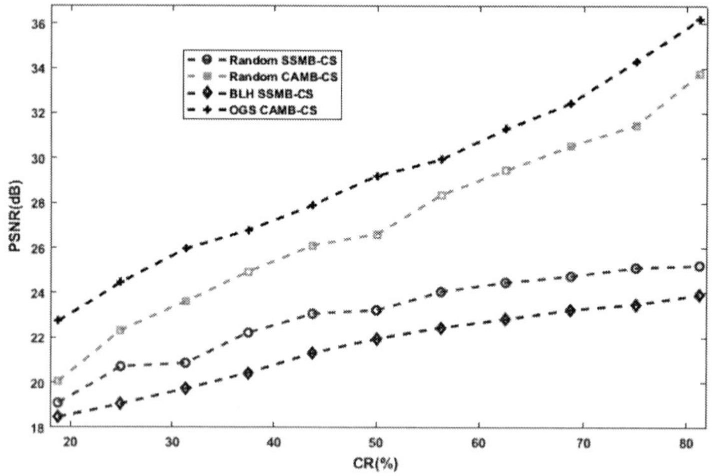

Fig. 1. PSNR (dB) versus CR (%) of Brick test image for SSMB-CS and CAMB-CS.

From Table 1 and Figure 1, it is clear that the proposed algorithm in this paper (OGS CAMB-CS) outperforms the random projection matrix and the previous algorithm (BLH SSMB-CS). It is noted that the reconstruction time for CAMB-CS is comparable to SSMB-CS. Reconstruction time of Barbara test image, as an example, for CR = 31.25 % are 4.22 s, 6.19 s, 4.45 s and 6.67 s for Random SSMB-CS, Random CAMB-CS, BLH SSMB-CS and OGS CAMB-CS respectively.

5 Conclusion

In this paper, the projection matrix design algorithm for CAMB-CS was proposed to improve image reconstruction accuracy. The results show that CAMB-CS outperforms the SSMB-CS in terms of PSNR of the image reconstruction. Further improvement can be attempted in future work by designing projection matrix and operator learning simultaneously.

This work was supported by the PITTA-Grant 2017 University of Indonesia.

References

1. D.L. Donoho, Compressed Sensing, IEEE Trans. Inf. Theory **52(4)**, 1289-1306 (2006).
2. E.J. Candès, J.R, T. Tao, Robust Uncertainty Principles: Exact Signal Reconstruction From Highly Incomplete Frequency Information, IEEE Trans. Inf. Theory **52(2)**, 489-509 (2006).
3. J.A.M Lorenzo, R. Obermeier, Sensing Matrix Design via Mutual Coherence Minimization for Electromagnetic Compressive Imaging Applications, IEEE Trans. Comp. Img **3(2),** 217-229 (2017).
4. S.Qaisar, R.M. Bilal, W. Iqbal, M. Naureen, S. Lee, Compressive Sensing: From Theory to Applications, a Survey, ComNet J **15(5)**, 443 – 456 (2013).
5. A. M. Bruckstein, D. L. Donoho, M. Elad, From Sparse Solutions of Systems of Equations to Sparse Modeling of Signals and Images, SIAM Rev **51(1)**, 34–81 (2009).
6. S. Nam, M. Davies, M. Elad, R.Gribonval, The Cosparse Analysis Model and Algorithms, Appl. Comput. Harmon. Anal **34(1)**, 30–56 (2013).
7. O. Endra, D. Gunawan, Comparison of Synthesis-Based and Analysis-Based Compressive Sensing, In Proceedings of IEEE International Conference QiR, 167-170 (2015).
8. R. Saiprasad, Y. Bresler, Data-Driven Learning of a Union of Sparsifying Transforms Model for Blind Compressed Sensing, IEEE Trans. Comp. Img. **2(3)**, 294-309 (2016).
9. M. Aharon, M. Elad, A. Bruckstein, The KSVD: An Algorithm for Designing Overcomplete Dictionaries for Sparse Representation, IEEE Trans. Sig. Proc, **54(11)**, 4311-4322 (2006).
10. I. Kviatkovsky, M. Gabel, E. Rivlin, I. Shimshoni, On the Equivalence of the LC-KSVD and the D-KSVD Algorithms, IEEE Trans. Pattern Analysis and Machine Intelligence, **39(2)**, 411-416 (2017).
11. M.F. Duarte, V. Cevher, R.G. Baraniuk, Model-based compressive sensing for signal ensembles, In Proceedings of 47th Annual IEEE Allerton Conference on Communication, Control, and Computing, 244-250 (2009).
12. C. Hegde, P. Indyk, L. Schmidt, Approximation-tolerant model-based compressive sensing, In Proceedings of 25th Annual ACM-SIAM Symposium on Discrete Algorithms, 1544-1561 (2014).
13. R. Rubinstein, T. Peleg, M. Elad, Analysis K-SVD: A Dictionary-Learning Algorithm for the Analysis Sparse Model. IEEE Trans. Sig. Proc, **61(3)**, 661-677 (2013).
14. S. Ravishankar, Y. Bresler, Learning Sparsifying Transforms, IEEE Trans. Sig. Proc **61(5)**, 1072-1086 (2013).
15. B. Hou, Z. Zhu, G. Li, A. Yu, An Efficient Algorithm for Overcomplete Sparsifying Transform Learning with Signal Denoising, Math. Problems. in Eng. (2016).
16. Y. Arjoune, N. Kaabouch, H.E. Ghazi, A. Tamtaoui, Compressive Sensing: Performance Comparison of Sparse Recovery Algorithms, In IEEE 7th Annual Computing and Communication Workshop and Conference (CCWC), 1-7 (2017).
17. R.Giryes, A Greedy Algorithm for the Analysis Transform Domain, Neurocomputing **173,** 278-289 (2016).
18. M. Elad, Optimized Projections for Compressed Sensing, IEEE Trans. Sig. Proc, **55(12)**, 5695–5702 (2007).
19. H. Bai, S. Li, X. He, Sensing Matrix Optimization Based on Equiangular Tight Frames with Consideration of Sparse Representation Error, IEEE Trans. Multimedia, **18(10)**, 2040-2053 (2016).
20. T. Strohmer, R.W. Heath, Grassmannian Frames with Applications to Coding and Communication, Appl. Comp. Harmon. Analysis, **14(3)** (2003).
21. B. C. Russell, A. Torralba, K. P. Murphy, W. T. Freeman, LabelMe: A Database and Web-Based Tool for Image Annotation, Comput. Vis. J. **77(1)**, 157–173 (2008).
22. R. Uetz, S. Behnke, Large-Scale Object Recognition with CUDA-Accelerated Hierarchical Neural Networks, In Proceedings of IEEE ICIS, 536-541 (2009).
23. J. A. Tropp, A.C. Gilbert, Signal Recovery from Random Measurements via Orthogonal Matching Pursuit, IEEE Trans. Inf. Theory **53(12)**, 4655–4666 (2007).

Machine health prognosis based on multi-regime condition monitoring signals

Achmad Widodo[1,*], *Toni* Prahasto[1] , *Djoeli* Satrijo[1], *Budi* Setiyana[1], *Sugiyanto*[1], *Ismoyo* Haryanto[1] and *Didik* Djoko Susilo[2]

[1]Department of Mechanical Engineering, Universitas Diponegoro, Semarang Indonesia
[2]Department of Mechanical Engineering, Universitas Sebelas Maret, Surakarta, Indonesia

Abstract. This paper presents a proposal for machine health prognosis based on multi regime condition monitoring (CM) signals. The basis idea is performing deeply analysis of CM signals that possibly includes steady state signals – that is normal condition, and developing transient signal that represents some faults exist in the machine. Trigonometric features is extracted from such signals and some energy vectors was used to calculate the health index of machine. Prognosis is then performed on the machine which has the lowest health index, that means the worst condition of the machine. RUL prediction is addressed to estimate the remaining life of the machine up to breakdown. The proposed method gives relatively promising results of RUL prediction that possibility give some times for maintenance actions before catastrophic failure occurs.

1 Introduction

Machine condition monitoring (MCM) cannot be bargained of its implementation in industry due to the widely benefits taken from safety aspect to business profits. The consequence of loss of profits due to unplanned downtime of industrial critical assets have forced the engineers to change their traditional maintenance strategy to be condition-based and more intelligent. Through the intelligent maintenance, the industries hope to optimize the operational life span – that is, the remaining useful life (RUL) – as well as quality, security and environmental standards. It also optimizes the availability of the assets, thus avoiding high downtime costs [1].

In recent years, a significant amount of research work has been undertaken to develop models that can be used to estimate RUL in industrial machineries to support the intelligent maintenance system. Selecting the appropriate techniques of prognosis for a particular application is critical to obtain the ultimate success of prognosis implementation. For example, the articles reported research work of prognosis system is listed in Refs. [2-11] and many more have already been reported in the literatures. Even so the such works obtained high appreciation in the state of the art in a fields, it do little to help operators in typical industry selecting the appropriate method for their specific needs. The reported

* Corresponding author: awid@undip.ac.id

works tend to focus on the technical (or theoritical) merits and demerits of each approach rather than the practical implementation issue.

Therefore, this paper contributes the strategy of prognosis system to estimate the RUL in such of simple way and applicable. The proposed method consists of recognition of regime signal acquired by condition monitoring device, data preprocessing, feature extraction, health assessment and prognosis strategy.

2 Method and Materials

The method employed in this paper includes signal processing and feature extraction of vibration signals. Some of signal processing techniques thru fast-fourier transform (FFT), discrete wavelet transform (DWT) and emperical mode decomposition (EMD) were used for data preparation which is valuable for health assessment process as well as feature extraction.

2.1 Signal Processing and Feature Extraction

FFT is a common tool for spectrum analysis to obtain dominant frequency components present in time domain signals and to assign them of various physical phenomena. The signal transformation is formulated in the simple form as follows:

$$X(k)=\sum_{n=0}^{N-1}x(n)e^{-\frac{i2\pi kn}{N}} \tag{1}$$

Basically, it tranforms the signals from time domain into frequency domain which is capable for digitization signals analysis.

DWT transforms the signal from time domain into a different form, that is, a series of wavelet approximation and details in time-scale domain as follows

$$dwt(j,k)=\frac{1}{\sqrt{2^j}}\int x(t)\psi^*\left(\frac{t-k2^j}{2^j}\right)dt \tag{2}$$

where $\psi^*(t)$ denotes conjugate wavelet function, j is number of level.

EMD method introduced by Huang et al [12] is a relatively new approach to signal analysis. It was developed from assumptions of: (i) the signal is an integration of intrinsic components of oscillations, (ii) every linear and non-linear mode has at least one maximum and one minimum, (iii) the signal is characterized by a time series between successive extrema. Each signal is decomposed into a finite number of intrinsic mode function (IMF) through a shifting process and then examined using some conditions as described in [12].

Feature extraction was conducted on decomposed signals by DWT at defined level, i.e., level 4, using classical statistical and trigonometric method. The aim of feature extraction is to obtain the salient features that clearly reflect failure progression in meaningful way. Trigonometric feature is calculated using standard deviation of inverse hyperbolic sinus and inverse tangent as follows

$$F_{t1}=\sigma\left(\log\left[x_j+(x_j^2+1)^{1/2}\right]\right) \tag{3}$$

$$F_{t2}=\sigma\left(\frac{i}{2}\log\left(\frac{i+x_j}{i-x_j}\right)\right) \tag{4}$$

The mentioned data processing method were employed to signal reconstruction which is assumed there is no information lost from the signal. Furthermore, an energy index model was calculated from the reconstructed signal for generating health assessment through correlation analysis (CA). CA is modeled as cosines distance of two vectors that related to the energy of the reconstructed signals.

$$CA = \frac{\boldsymbol{\varepsilon}_1 \cdot \boldsymbol{\varepsilon}_n}{|\boldsymbol{\varepsilon}_1||\boldsymbol{\varepsilon}_n|} \tag{5}$$

where $\boldsymbol{\varepsilon}_1$ and $\boldsymbol{\varepsilon}_n$ are vectors of energy index model calculated from signal regime normal (steady state) and faulty state, respectively. The values of CA ranges between zero and one, higher CA means a higher correlation between evaluated signals.

Furthermore, the prediction of RUL is approached by a non-linear state space models as a finite-dimensional system as follows [12]

$$\begin{aligned} x(t+1) &= f(t, x(t), u(t), w(t); \theta) \\ y(t) &= h(t, x(t), u(t), v(t); \theta) \end{aligned} \tag{6}$$

where $y(t)$, $u(t)$ are output predictors and input, $w(t)$, $v(t)$ and θ are sequences of independent random varaibles and a vectors of unknown parameters, respectively. The predictor is then constructed from [13]

$$\begin{aligned} \hat{y}(t|\theta) &= g(t, Z^{t-1}; \theta) \\ Z^t &= (y^t, u^t) = (y(1), u(1), ..., y(t), u(t)) \end{aligned} \tag{7}$$

The method of machine health prognosis based on multi-regime condition monitoring signals is summarized in Fig. 1. This method includes system capability of selecting regime based on a certain health status and a specific operating conditions. Some procedures involved in this method such as regime clustering, feature extraction, information reconstruction, health index determination, health and risk status, diagnosis and prognosis of RUL.

Moreover, the procedure of information reconstruction from the MCM signals is depicted in Fig 2. The previous method were employed to obtain the information from the regimed signals for machine health prognosis.

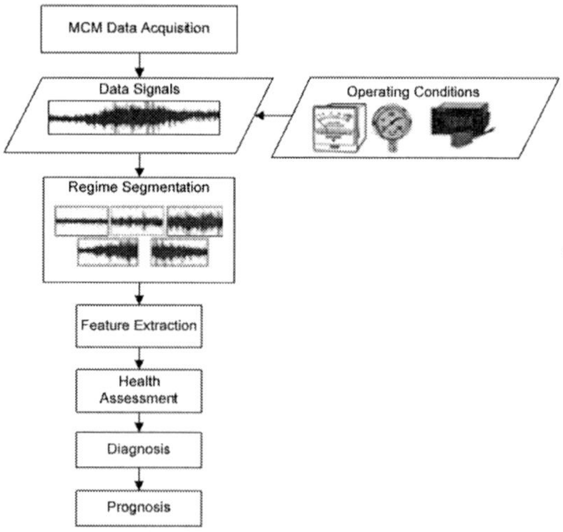

Fig. 1 Multi-regime machine health prognosis.

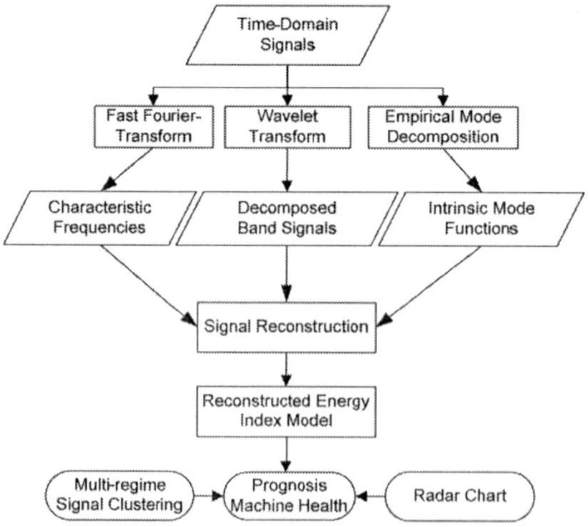

Fig. 2 Information reconstruction of signals.

Condition monitoring signal collected from data acquisiton devices is segmented into several regimes which is defined by user. Feature extraction is performed using trigonometric features as formulated in equations (3) and (4). Machine health state index is calculated based on energy of the regimed signals after signal reconstruction as shown in Fig. 2.

The acquired time-domain signal is processed using FFT, wavelet transform and EMD to obtain characteristic frequencies, decomposed signals and IMF, respectively. Reconstruction information is conducted based on processed signal which is free from background noise. Energy index model is then generated for machine health state index according to equation (5). This index is presented in radar chart for machine condition monitoring purpose. Finally, prognosis machine is performed using data from machine health state and regimed-signal.

2.2 Data

The data used in this research is originally developed by PRONOSTIA, an experimental platform from FEMTO-ST Institute, France. This test-bed is dedicated to test and validate fault detection, diagnosis and prognosis of rolling element bearing [14,15]. This platform allows to perform accelerated degradations (run-to-failure) of bearing at constant and/or variable bearing conditions, while gathering on-line condition monitoring data. The tested bearings could have various type of faults such as inner race, outer race, cage and rolling element.

The conditions of the measurement data are summarized in Table 1. Vibration signals were acquired from data acquisition card using sampling frequency 25.6 kHz with recording samples of 2560 samples in time duration 10 second.

Table 1. Datasets of vibration measurement using PRONOSTIA.

Datasets	Condition#1 (1800rpm, 4000N)	Condition#2 (1650rpm, 4200N)	Condition#3 (1500rpm, 5000N)
Learning set	Bearing1_1 Bearing1_2	Bearing2_1, Bearing2_2	Bearing3_1 Bearing3_2

3 Results and Discussion

Vibration signal presented in Fig. 3 is assumed coming from MCM device that has been installed in the system. At early stage, the signal represents normal condition recorded from rolling element bearing housing, but grows suddenly become transient till the end of the life. The recorded signal is then devided into several regimes from normal to transient regime that shows some degradations have experienced in the machine. For example, seven regimes of signal have been simulated and preprocessed using previous methods including FFT, DWT and EMD. Each regime gives three energy models extracted from FFT, DWT and EMD, so therefore there are 21 energy model vectors from which health assessment can be derived.

Applying equation (5), the calculation of CA represents health index of the machine presented in radar chart. Higher CA means machine in normal condition, otherwise the machine in abnormal (degradation) condition that needs maintenance actions. Fig.4 shows the lowest health index comes from bearing data C5, this means the abnormality has been detected. Further analysis should be attempted for such signals to estimate how much time left till the end life of this bearing. This process actually known as prognosis of RUL.

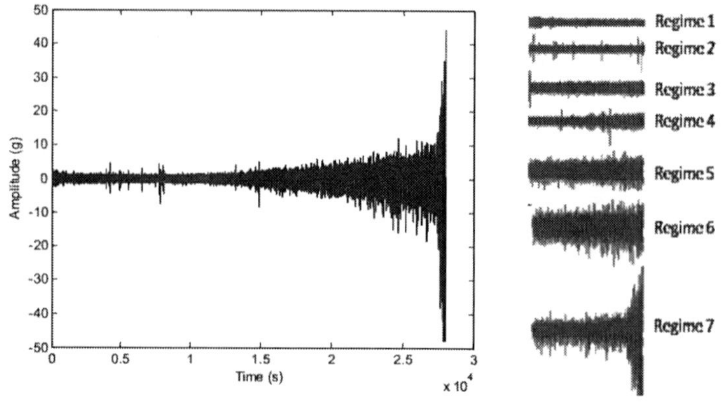

Fig. 3 Vibraton signal acquired from bearing1_1: (a) run-to-failure signal; (b) regimed signal.

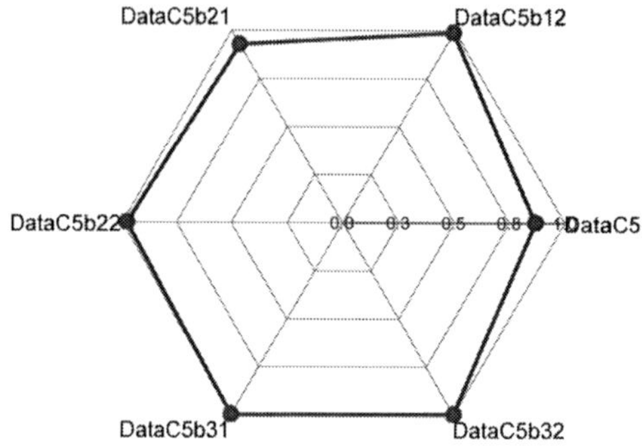

Fig. 4 Radar chart of health index.

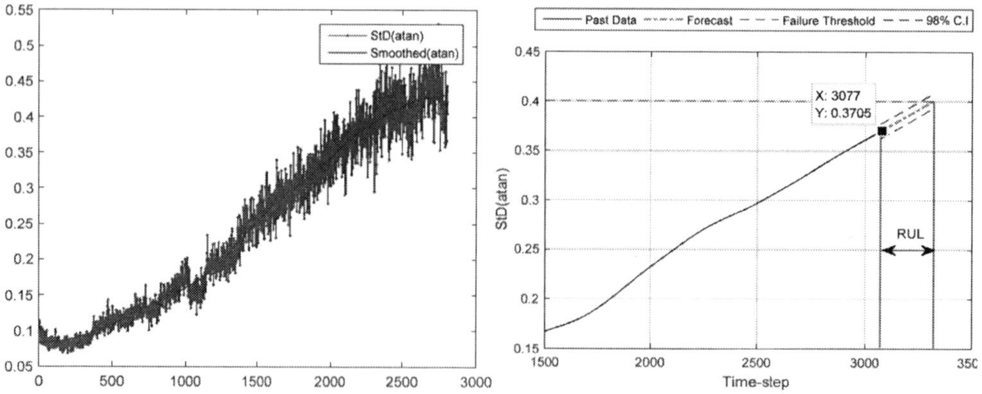

Fig. 5 (a) Feature of $\sigma(atan)$; (b RUL prediction $\sigma(atan)$.

Trigonometric features are then extracted from the abnormal signals using equations (3) and (4) and yield standard deviation of inverse hyperbolic sinus and inverse tangent. Fig. 5a shows, for example, feature of standard deviation of inverse tangent. Prognosis of machine is conducted based on these features by means of time series prediction using non-linear state-space models in equations (6) and (7).

Prognosis model was trained by using 1000 past time data points for building a dynamic forecast model. The threshold of failure is assumed and set up 0.6 and 0.4 for $\sigma(asinh)$ and $\sigma(atan)$ features, respectively. The predictors learned the past time data dan predict the future data points based on determined model. Fig. 5b shows the result of RUL prediction of $\sigma(atan)$ features using 250 step-ahead horizon presented by the estimated time of predictors when hits the determined threshold. As results, feature of $\sigma(asinh)$ and $\sigma(atan)$ gives 246 and 243 time-steps of RUL prediction, respectively. In real application, the predicted RUL allows the actions of maintenance before the system reachs catastrophic failures.

4 Conclusion

This paper deals with prediction of RUL based on multi-regime condition monitoring signals. The acquired signals was partitioned to separate the steady-state or normal condition and the transient mode that represents of faults exist in the machine. Feature extraction has been performed on regimed signal to generate health index through CA calculation. The lowest index shows the worst condition of machine, therefore, prognosis is employed to predict RUL of machine. The proposed method gives relatively promising results of RUL prediction that possibility give some times for maintenance actions before catastrophic failure occurs.

Authors gratefully acknowledge financial support from the Grant for the Strategic Research of Faculty of Engineering, Diponegoro University 2017.

References

1. V.T. Tran, Bo-Suk Yang, Int. J. FMS **2**(1), 61-71 (2009)
2. P.A. Scarf, Eur. J. Op. R **99** (**3**), 493–506 (1997)
3. S.J. Engel, B.J. Gilmartin, K. Bongort, A. Hess, *IEEE Aero. Conf. Proc.***6**, 457–469 (2000)
4. G. Vachtsevanos, F. Lewis, M. Roemer, A. Hess, B. Wu, *Intel. Fault Diag. and Prog. for Eng. Syst.* (John Wiley and Sons Inc., Hoboken, New Jersey, 2006)
5. J. Luo, M. Namburu, K. Pattipati, L. Qiao, M. Kawamoto, S. Chigusa, *IEEE Inc.*, Piscataway, NJ 330–340 (2003)
6. A. Hess, G. Calvello, P. Frith, S.J. Engel, D. Hoitsma, *IEEE Aerospace Conference*, 1–19 (2006)
7. H. Qiu, J. Lee, J. Ling, G. Yu, J. Adv. Eng. Inf **17,** 127-140 (2003)
8. T. Wang, J. Lee, *Proc. of 62th Meeting of the MFPT Society: Failure Prev. for Syst. Avail.*, 87-98 (2008)
9. J. Lee, J. Ni, D. Djurdjanovic, H. Qiu, H. Liao, Comp. Ind. **57**, 476–489 (2006)
10. A. Widodo, B.S. Yang, ESWA **38**, 2592-2599 (2011)
11. A. Widodo, M.C. Shim, W. Caesarendra, B. S Yang, ESWA **38**, 11763-11769 (2011)
12. N.E. Huang, Z. Shen, S.R Long, M.C. Wu, H.H. Shih, Q. Zheng, N.-C. Yen, C.C. Tung, H.H. Liu, *Proc. of the Royal Soc. of London. Series A: Math. Phy. & Eng. Sci.* **454**, 903-995 (1998)
13. P. Nextoux, R. Gouriveau, K. Medjaher, E. Ramasso, B. Morello, N. Zerhouni, C. Varnier, *IEEE Int. Conf. on Progn. and Health Manag.*, Denver, USA (2012)
14. L. Ljung, *Syst. Ident.: Theory for the User* (Prentice Hall, Englewood Cliffs, NJ, 1987)
15. NASA. Prognostics data repository. http://ti.arc.nasa.gov/tech/dash/pcoe/prognostic-data-repository/

Induction Motor Centrifugal Blower Health Diagnostic Based on Color Segmentation of Thermal Image and Vibration Signal Feature

Panut Widodo[1,*], *Gunawan* Dwi Haryadi[1], and *Achmad* Widodo[1]

[1]Departement of Mechanical Engineering Faculty of Engineering, UNDIP, 50275 Semarang, Indonesia

Abstract. The rotating machinery requires condition monitoring which its measurement without being intrusive operation, especially on the equipment needed to continue running. One such machinery is a centrifugal blower induction motor. Infrared thermography and vibration are important and effective technologies to diagnose of health condition it without destructive or disturb of operations. The diagnostics of induction motor are based on the analysis results data onto vibration and processing thermal image. This paper focused on thermography image processing based on color segmentation which it will produce ROI (region of interest) images. The ROI image is extracted based on HSV color and shape feature. Feature extraction is intended to determine value of mean, standard deviation, kurtosis, skewness and entropy HSV and shape features (area, perimeter, metric, and eccentricity). The highest RMS (root mean square) vibration data is used as reference to classify data into normal and abnormal. Parameters that can be used to classify normal and abnormal conditions based on data analysis are standard deviation Hue, kurtosis HS, skewness HSV, entropy HSV and metric.

1 Introduction

Centrifugal blower is one of the rotating machines used in power plant system. It serves to blow air from outside which is used as cooling for exhaust frame and Gas Turbine shell. This centrifugal blower is driven an induction motor. These parts can cause a vibration and abnormal temperature in the system, and it can make the machine failure. To prevent such condition, it needs to have a condition based monitoring (CBM). CBM is intended to prevent unplanned breakdowns, maximize the plant availability and reduce the danger associated. Condition monitoring of induction motor can be done online and off line, invasive and non-invasive. Invasive and non-invasive methods of CBM are used to measure motor vibration, temperature, speed, torque variations and Motor current stator analysis (MCSA) for detection of motor abnormalities [1].

One of the methods being developed is condition monitoring based on temperature measurement using infrared thermography (IRT)[2]. IRT is non-destructive, non-invasive

*Corresponding author: pntwidodo@gmail.com

method, an online or offline and non-contact type condition monitoring technique [3]. Health diagnosis of induction motors can be done by evaluating the image pattern generated because of the transformation of input energy into waste heats. IRT can use as a valuable tool to fault diagnosis in induction motors which relying on infrared data combined with the heat transfer theory. IRT has been widely used in various fields like industry, meteorology, environment, medicine, architecture and other technical fields[2].

Image texture evaluation method is one of the methods of image analysis. The primary problem of texture evaluation is the feature extraction that it can describe the particular feature of the image numerically. It can utilize statistical methods of thermal image analysis. This method can be used for condition monitoring and health diagnosis of induction motor [4]. Image segmentation is a crucial part in image processing before implementing the next step such as feature extraction and classification [5][6]. It has many different kinds of methods are histogram threshold, clustering, edge based, region growing, fuzzy, and neural-network methods [7]. Clustering techniques identified analogous clusters of pixels in the feature space like RGB, HSV, and LUV color space [6].

In this paper proposed condition monitoring and health diagnostics of induction motor of the centrifugal blower based on data analysis of thermal images and vibration. Thermal images data processed by color segmentation method that continued feature extraction process by color HSV and shape feature which will produce feature of normal and abnormal condition. Induction motor health diagnostics is based on statistical analysis result of boundary condition of ROI image features.

2 Basic Theory

2.1 Color segmentations

Segmentation is done to split an image into specific sections for further analysis, resulting in a higher level illustration of the image pixels like objects or foreground and background. Segmentation based on colour in certain cases, is the simplest and most obvious way of distinguishing between objects and background [6]. One of the image segmentation methods is clustering. Clustering techniques identified homogeneous clusters of pixels in the feature space such as RGB and HSV color space [7]. RGB color space is identical with original image colors, so to get HSV needed conversion RGB to HSV color space[5][6].

The RGB colour space consists of three components colors (Red, Green, blue), where all components have 8-bit depth. Every component is interpreted as unsigned integers in range (0; 255). It is based on the portion of the electromagnetic spectrum visible to humans (approximate range of wave lengths in 400–700 nm)[5][6].

The HSV color model defines color in terminology of three components : hue (H), saturation (S), and value (V). It is an approximately uniform perceptual color space. Hue is the dominant wavelength of the colour, used to distinguish colors and determine redness, etc. Saturation is the 'purity' of colour (in the sense of the amount of light white mixed with it). Value is the brightness of the colour (also known as luminance) [5][6].

2.2. Feature extraction of image

Feature extraction is the process to regain the most important data from the raw data to find parameters that defines the shape of a character exactly and uniquely [8]. It can be classified into 3 groups are color feature extraction, shape and texture features [9]. Extraction feature of HSV and RGB is example color feature extraction. The calculation of the perimeter and area are meaningful for binary images. The binary image containing one or more object, where **p** (x, y) = 1 if a pixel is part of the object or foreground and **p** (x,y) = 0 for background pixels [6]. From area and perimeter can be calculated form factor or called 'Metric'[5]. Eccentricity is the value of comparison between distance between the ellipse minor and major. Eccentricity and metric value have range 0 to 1. Calculation of eccentricity (e) and metric (M) can be used Equation 1. Where 'a' is mayor axis, 'b' is the minor axis, 'A' is area and 'C' is the perimeter.

$$e = \sqrt{1-\frac{b^2}{a^2}}, M = \frac{4\pi A}{C^2} \quad (1)$$

3 Methodology

3.1 Data acquisition process

Collecting data in this study is conducted by measuring the induction motor centrifugal blower contained in the power generation unit. This is done when the induction motor operates. Specifications of induction motor of the centrifugal blower are 100 HP, 1180 A, 460 V AC 3 phase/60 Hz, and diameter motor 411 mm.Thermography images data taken by the infrared thermography (IRT) camera Flir E-40, and vibration using vibscanner.

The infrared camera manufactured by Flir E-40 can measure the temperature on a range scale of -20 ^{0}C to 650 ^{0}C. The thermal lens produces visual image of 640 x 480 pixels and captures images of 320 x 240 pixels. The distance between induction motor and thermal imager were range of 2.0 metres, with ambient temperature was 35-41 ^{0}C. The emissivity value was set at 0.95 as recommended material of cast iron [2].

Vibration on this object is carried out with 3 positions of horizontal position, vertical and radial at both ends. The result of vibration measurement is Root-Mean Square (RMS in mm/s) data record, while flir E-40 produces thermal image 320 x 240 pixels. This thermal image will be crop to get the observed region of interest (ROI).

3.2 Data classifications, HSV color and shape feature extraction

Classification data based on vibration data according to ISO 10816-3 standard. After cropping, the thermal image implemented RGB color segmentation by K-mean clustering to produce of Region of interest (ROI) image. The pixel features are clustered to group them into 4 regions of similar color. From number 4 cluster images selected one image of ROI representing temperature increases that is redness image region. This ROI calculated of HSV characteristic (mean, standart deviation, kurtosis, skewness and entropy) and shape feature (area, perimeter and eccentricity). Image statistics are investigated and studied based on first order statistics that capture certain statistical regularities of ROI thermal image [10], following Equation 2-5, where $\sum\sum$ p(x,y) represents the summation of all pixel values of the component HSV and MxN is size of the image.

$$Mean\,(\mu) = \frac{1}{MxN}\sum_{x=1}^{M}\sum_{y=1}^{N} p(x,y),\ Std.\text{dev.}(\sigma) = \sqrt{\frac{1}{MxN}\sum_{x=1}^{M}\sum_{y=1}^{N}(p(x,y)-\mu)^2} \quad (2)$$

$$Entropy\,(E_n) = \frac{1}{MxN}\sum_{x=1}^{M}\sum_{y=1}^{N} p(x,y)\log_2(p(x,y)) \quad (3)$$

$$Skewness\,(s) = -\frac{1}{MxN}\sum_{x=1}^{M}\sum_{y=1}^{N}\left|\frac{p(x,y)-\mu}{\sigma}\right|^3 \quad (4)$$

$$Kurtosis(k) = -\frac{1}{MxN}\sum_{x=1}^{M}\sum_{y=1}^{N}\left|\frac{p(x,y)-\mu}{\sigma}\right|^{4} - 3 \tag{5}$$

Next step is plotting the features of statistical HSV and shape feature. Afterwards selected features that can distinguish between normal and abnormal conditions. Final step is determining the boundary condition between normal and abnormal using statistical logic method. Boundary values obtained by calculation value of mean, min, max, standard deviation, delta (D), final minimum and final maximum of selected feature data.Value of delta (D) is average difference between minimum and maximum value in each class.

$$final\max = \max + D,\ \text{final}\ \min = min - D \tag{6}$$

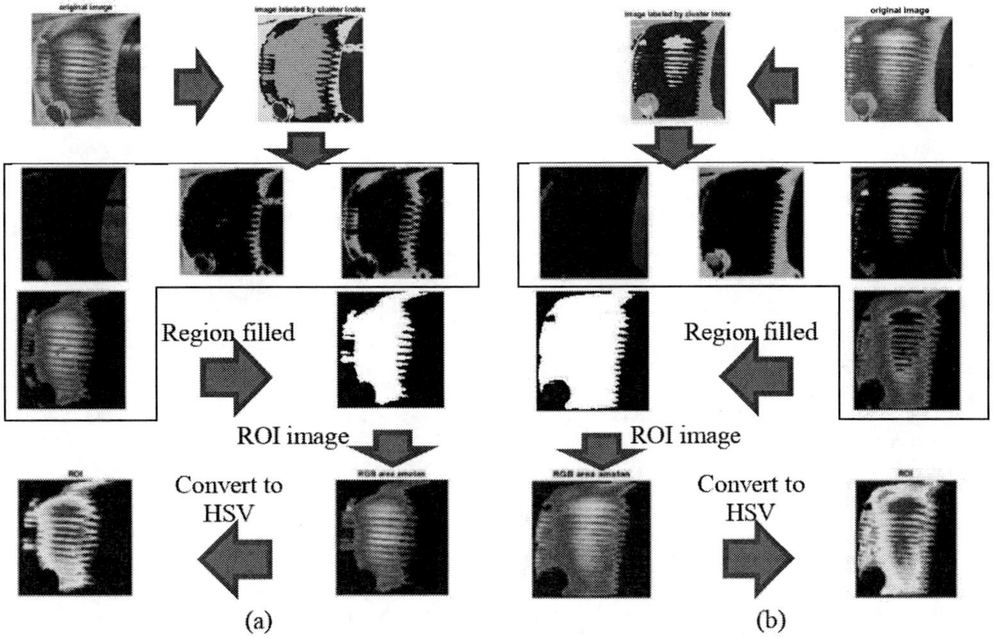

Fig. 1. Color segmentation process, decide to ROI: (a) Normal condition, (b) Abnormal condition

4 Result and Discussion

Induction motor of the centrifugal blower used as subject of this observation according to ISO 10816-3 can be categorized into a group 1 Rigid. Classification based on motor diameter 411 mm (motor diameter more than 315 mm). Normal condition if the root mean square (RMS) values has up 2.3 mm/s. If then RMS >2.3 mm/s being categorized as abnormal condition. From number 50 data can be classified into number 40 data of normal and 10 data of abnormal condition. Feature of the highest vibration shown in Figure 2a.

The ROI image determination process and convert to HSV image shown in Figure 1. ROI, which is an RGB image converted into HSV image, extracted to calculate the statistical of each component of HSV. ROI will be transformed into binary image by threshold method, then extracted to calculate area, perimeter and eccentricity. Referring to determination and the resulting feature, not all features can be used as reference to differentiate induction motor conditions. Figure 2b shows random patterns for mean HSV of hot area of ROI which are not precise patterns because the conditions of motor induction cannot be distinguished clearly. Feature mean of HSV has similarity between normal and abnormal condition. As well as standard deviation of S and V, kurtosis V, area, perimeter and eccentricity features. Feature that can differentiate its condition attend standard deviation of Hue, kurtosis HS, skewness HSV, entropy HSV and metric.

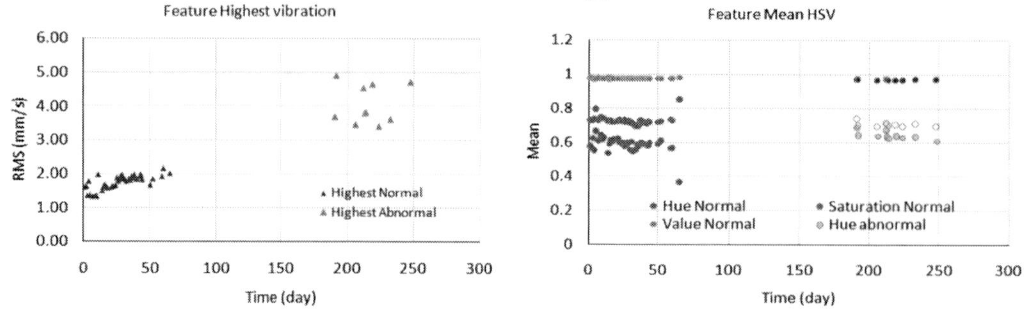

(a) (b)

Fig. 2. Feature; (a) Result of highest vibration, (b) Mean of each component HSV

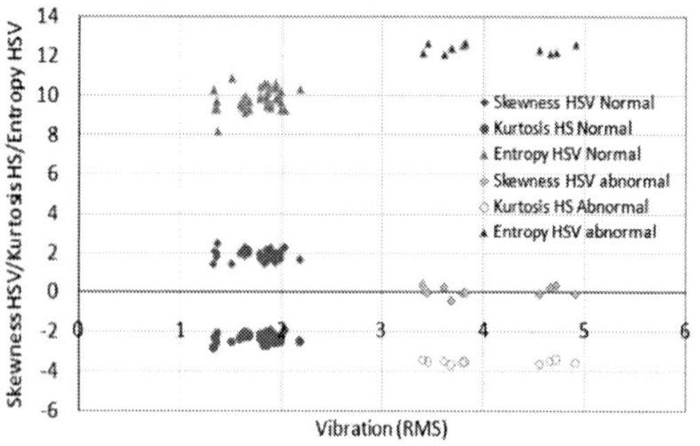

a. Feature vibration – kurtosis HS/skewness HSV/entropy HSV (KSE HSV)

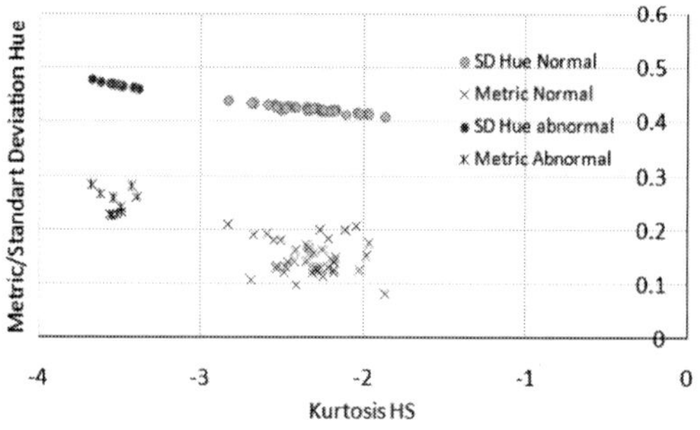

b. Feature kurtosis HS – metric or SD Hue

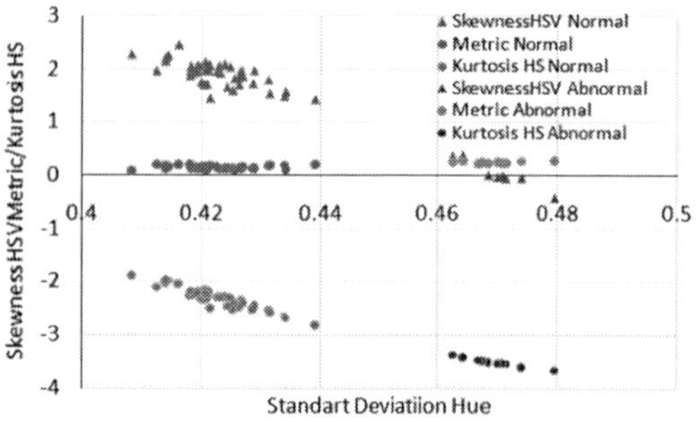

c.Feature SD Hue – skew HSV/metric/kurt HS

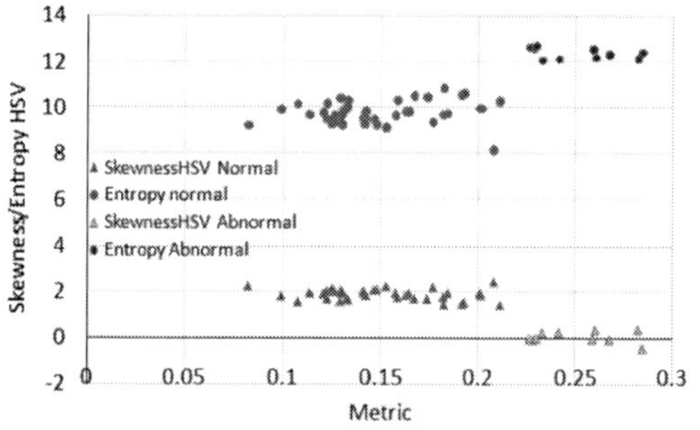

d.Feature metric – skewness or entropy HSV

Fig. 3. Feature differentiate between normal and abnormal condition

HSV parameter values obtained by summing the component values of H, S and V, like skewness HSV is skewness value of component H, S and V. While for kurtosis HS obtained from sum of the kurtosis H and kurtosis S value. In order to determine the good features to be used as reference differentiate of normal and abnormal conditions, it is necessary to simulate the plot of each feature parameter. Not only juxtaposed with time only. A good feature pattern is shown in Figure 3.a-d feature pattern RMS-KSE HSV, RMS-SD H, RMS-Metric, kurtosis HS-Metic etc. Along with the addition of vibration value then the value of entropy, standard deviation hue, metric will rise while for kurtosis and skewness will decrease towards negative value.

Finally, the result of determination of boundary conditions value (final maximum/F.max and minimum/F.min) according Equation 6, can be shows Table 1. Not overlapping value of boundary condition between normal and abnormal condition.

Table 1. Determination result of boundary condition value of each parameters.

Condition of induction motor	RMS		SD Hue		Skewness HSV		Kurtosis HS		Entropy HSV		Metric	
	F.Min	F.Max	F.Min	F.Max	F.Min	F.Max	F.Min	F.Max	F.Min	F.Max	F.Min	F.Max
Normal	1.1125	2.3885	0.4009	0.4443	1.1447	2.6523	-3.0072	-1.6317	7.9164	11.6549	0.0539	0.2122
Abnormal	2.7370	5.3910	0.4562	0.4829	-0.5850	0.7187	-3.7544	-3.2926	11.8459	12.9362	0.2262	0.3027

5 Conclusion

Based on this study, HSV color segmentation on the thermal image can be used to diagnose normal and abnormal conditions of induction motor. Parameters that can distinguish between normal and abnormal conditions are standard deviation Hue, skewness HSV, kurtosis HS, entropy HSV and metric. Induction motor of centrifugal blower is categorized as abnormal condition if RMS > 2.3885 mm/s, standard deviation Hue ≥ 0.4562, skewness HSV ≤ 0.7187, Kurtosis HS < -3.0072, Entropy HSV ≥ 11.8459, and metric ≥ 0.2262.

The author gratefully acknowledges the support of measuring instruments and research objects from PT. Indonesia Power Semarang on this research

References

[1] D. G. Ece and M. Başaran, "Condition monitoring of speed controlled induction motors using wavelet packets and discriminant analysis,"*Expert System Application"*, **vol. 38**, pp. 8079–8086 (2011)

[2] S. Bagavathiappan, B. Lahiri, T. Saravanan, J. Philip, and T. Jayakumar, "Infrared thermography for condition monitoring , A review," *Infrared Phys. Technol.*, **vol. 60**, pp. 35–55 (2013)

[3] M. S. Jadin and S. Taib, "Recent progress in diagnosing the reliability of electrical equipment by using infrared thermography," *Infrared Phys. Technol.*, **vol. 55**, pp. 236–245 (2012)

[4] M. Fidali, "An idea of continuous thermographic monitoring of machinery, " in *Proceedings of the 2008 International Conference on Quantitative Infrared Thermography*, pp. 61–18 (2008)

[5] C. Solomon and T. Breckon, *Fundamentals of Digital Image Processing: A Practical Approach with Examples in Matlab*, **vol. 3**. Chichester, UK: John Wiley & Sons, Ltd, (2010)

[6] R. Gonzalez and R. Woods, *Digital image processing,* (2002)

[7] H. Wang and D. Suter, "Color Image Segmentation Using Global Information and Local Homogeneity," *Proceeding VIIth Digital Image Computing: Techniques and Applications,* pp. 10–12 (2003)

[8] G. Kumar and P. K. Bhatia, "A detailed review of feature extraction in image processing systems," in *Fourth International Conference on Advanced Computing and Communication Technologies*. IEEE Computer Society, pp. 5–12 (2014)

[9] S. A. Medjahed, "A Comparative Study of Feature Extraction Methods in Images Classification," *I.J. Image, Graph. Signal Process.* **7**, 3 (2015)

[10] S. Selvarajah and S. R. Kodituwakku, "Analysis and Comparison of Texture Features for Content Based Image Retrieval," *International Jurnal Latest Trends Computing 2,* 1 (2011)

Design studies of inner and outer embedded Permanent Magnet for hybrid electric vehicles

Aravind CV[1,*], *Ramani* Kannan[2], *Heng Jing* Lei[1], *Joga Dharma* Setiawan[3]

[1]School of Engineering, Taylor's University, Malaysia 47500
[2]Department of Electrical and Electronics Engineering, University Teknologi PETRONAS, Malaysia
[3]Diponegro University, Indonesia

Abstract. Hybrid vehicles require high torque for propel, hence permanent Magnet machines are highly suiting for the improvement in the torque density. The paper focus on designing built-in interior permanent magnet (IPM) synchronous machine for hybrid electric drive. With the permanent magnet switched from rotor to stator and the characteristics over a wide range of speed operation is studied. The results obtained though performance analysis shows that at 130 rpm high torque with power peaking at around 900 rpm. Both the inner and outer machine are studied using numerical study tool for performance analysis for the application mentioned above. The inner magnet rotor design has provide a better magnetic flux flow due to the larger flux linkage between the permanent magnet and stator pole. Both type of machines are evaluated for torque where the machine with inner magnet provide a higher torque density of 4.94% as compared to the outer magnet machines.

1 Introduction

Unlike internal combustion engine (ICE) vehicles, the mechanical losses are converted into heat and dissipate to surrounding which does not practice the concept of energy efficiency [1-2]. Hybrid vehicles operate based on permanent magnet machine where magnetic flux linkage generated from the overlapping of magnetic field between permanent magnet and electromagnet [3]. The brushless permanent magnet DC machines have advantages over other machines such as simpler to maintain, more durable, and compact, less likely to suffer reduction in torque performance. The issues occurred where the fixed magnetic field in the stator could not increase the torque density due to difficulty in controlling the magnetic flux. [4-5]. A new type of design that includes two rotors one inside the machine and the other outer rotor enable applications in future wherein the operating mode is feasible in either simultaneous or independent operations using control techniques. A comparative design analysis using torque density value as evaluation parameter is presented to suiting to a hybrid electric vehicle.

* Corresponding author: aravindcv@ieee.org

2 Research Methodology

2.1 Permanent Magnet placed in the Rotor

Figure 1 (a) - Figure 1(c) shows the hybrid drive brushless single rotor interior permanent magnet machine with permanent magnet placed in rotor which is the conventional model used in the electric vehicle including that of the first generation of Toyota hybrid configurations. It encompasses magnet inside the rotor, bringing disadvantages of mechanical stress and loading effect on the vehicle mass. However, for variable speed motion inner rotor type is recommended which makes it a viable case for the hybrid vehicle with the shape of the magnet optimized in the third generation of vehicle types [6-7].

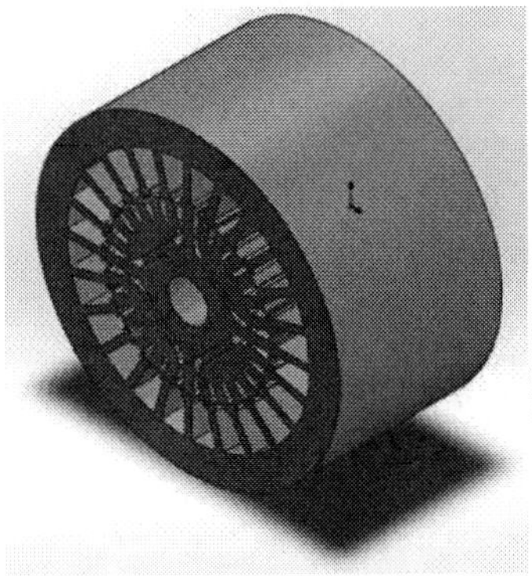

(a) 3D view.

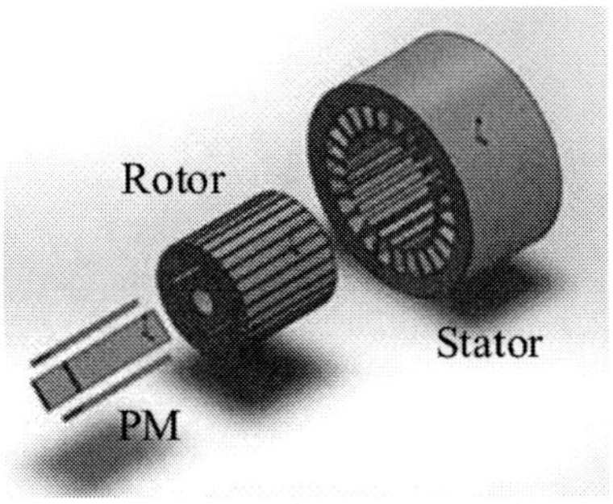

(b) Exploded view.

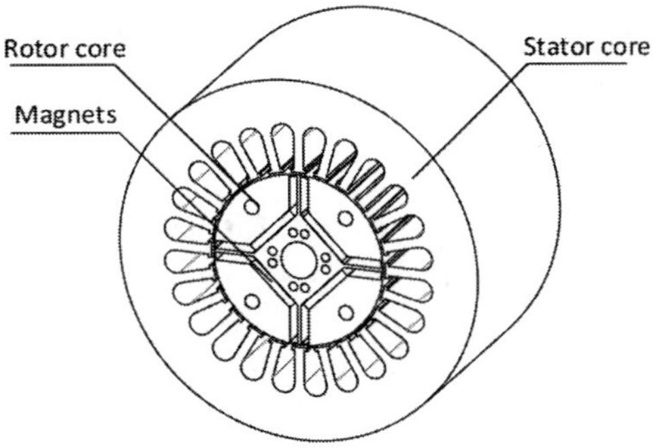

(c) Front view

Fig. 1. Inner permanent magnet motor design.

2.2 Permanent Magnet placed in the Stator

The hybrid drive brushless single rotor interior permanent magnet machine with permanent magnet placed in stator is as shown in Figure 2(a) – Figure 2(c) with the rotor light-weight and hence the motion faster. However, the magnetic and the electric loadings in the stator makes it need to be temperature tolerant, which is one of the challenging in this design. But the rotor being less inertia makes it to become faster than that of the motor with inner magnet.

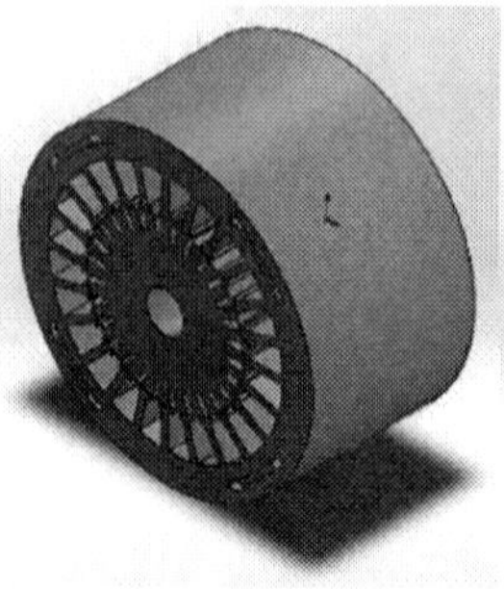

(a) 3D view.

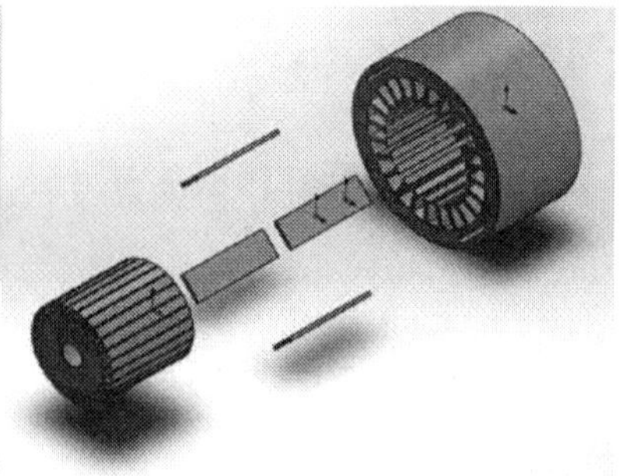

(b) Components.

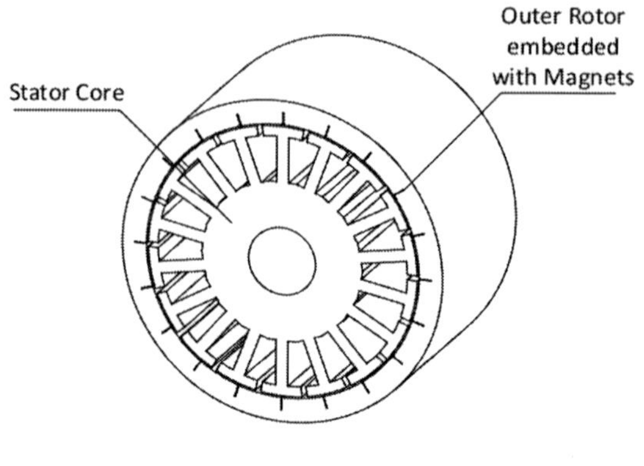

(c) Front view.

Fig. 2. Outer permanent magnet motor design

3 Design Analysis

3.1 Numerical Design

Finite element analysis simulation used to configure the machine to derive the basic characteristic of the machine that helps to identify the characteristics of the machine before fabrication process. The construction of the machine model done in the beginning by using numerical software tool. In this investigations the models generated are limited to two-dimensional and with our previous works [8-11] the results are closer in accuracy with that of the three dimensional model. Figure 3 shows the model machine modeled in JMAG® software for transient magnetic field analysis.

3.2 Evaluation Parameters

In order to evaluate the machine using numerical tools, various performance parameters are utilized for the analysis as in this section.

3.2.1 Motor Constant (K_m)

Motor constant (K_m) refer to the potential of a machine to convert mechanical power to electrical power and is expressed in Eq. (1), with I being the injected current, R being the resistance of the conductor.

$$K_m = \frac{K_t}{\sqrt{I^2 R}} \quad (1)$$

3.2.2 Torque Constant (K_t)

Torque constant (K_t) is known as the torque generated (T_d) over the input current (I) is as shown in Eq. (2)

$$K_t = \frac{T_d}{I} \tag{2}$$

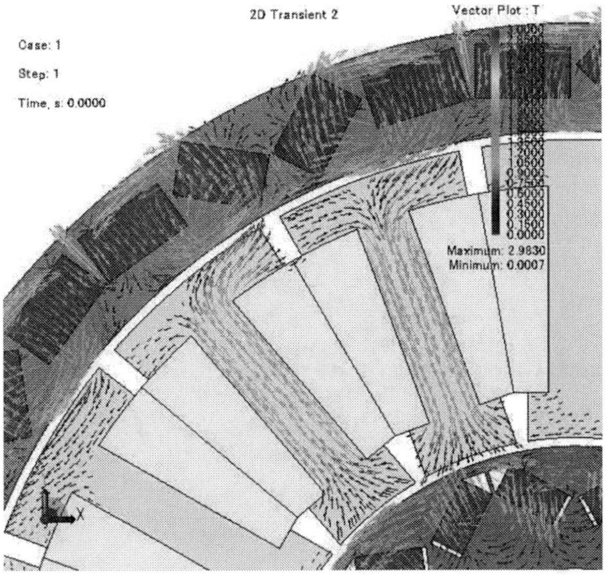

Fig. 3. Flux flow of DRSSPM in FEA tool.

3.2.3 Torque per Unit Volume (T_m)

Torque per unit volume (T_m) is similar to torque density where it is used as a measure for the torque producing capability of a particular machine. The torque per unit volume is mainly used during the design evaluation stage where the torque density of the mechanical design components can be determined as in Eq. (3), with L_{stk} the stack length, D is the outer diameter of the machine.

$$T_m = \frac{T_d}{(\pi D^2/4)\,L_{stk}} \tag{3}$$

3.2.4 Motor Constant Square Density (G)

Motor constant square density (G), is used to evaluate the overall machine design performance considering the torque, volume, and power of a machine [6-9] as in Eq. (4).

$$G = \frac{{K_m}^2}{(\pi D^2/4)\,L_{stk}} \tag{4}$$

4 Results and Discussions

The starting torque of the machine is the minimal torque required to start the motor, which is 5 Nm as shown in Figure 4, with the torque of the machine seems to drop linearly at the torque above 130 rpm. The peak value gives the maximum operating torque the machine could develop at the nominal speed. As the speed is increased from the nominal value (the same as power developed is reaching peak) the torque drops (from start-up of the vehicle to reaching constant speed) According to Figure 5, the starting torque of the machine is the minimal torque required to start the motor is approximately 5 Nm.

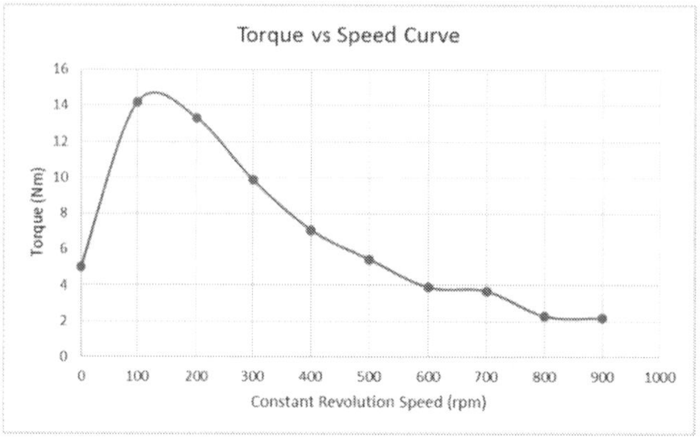

Fig. 4. Torque vs Speed characteristic (outer rotor).

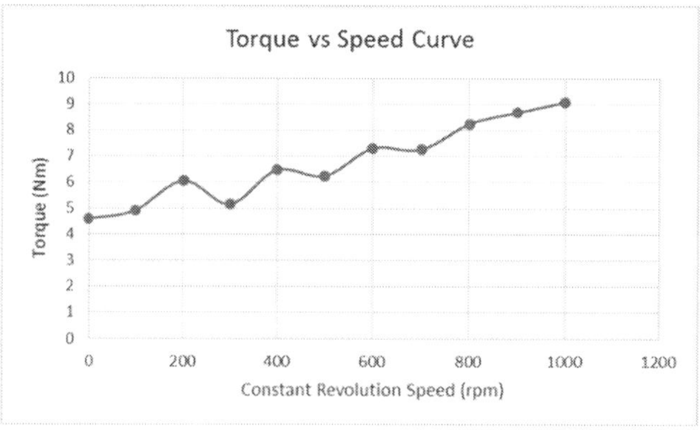

Fig. 5. Torque vs Speed characteristic (inner rotor)

Table 1. Comparative Evaluations

Parameter	Outer Magnet	Inner Magnet
Current [Amperes]	15	15
Volume [m^3]	5.4 x 10^{-5}	5.4 x 10^{-5}
Average Torque [Nm]	14.189	4.9145
Motor Torque [Nm/mm^3]	409259.26	429074.074
Torque Constant [Nm/A]	1.4735	1.5449
Motor Constant Square Density	2851.85	2992.59
Improvement in MCSD	-	4.94%

The torque performance of built-in magnet in rotor machine provide a higher torque performance compared to the modified machine. Inner magnet rotor design provide a better magnetic flux flow and flux linkage between the permanent magnet and stator pole. Both factors have resulted in increasing the torque producing capability of the machine. The inner rotor machine design provides a higher torque to volume ratio performance in comparison with that of the outer rotor magnet machine. But the average torque in case of the outer rotor machine is about three times that of the inner motor which makes it suiting for the torque requirement such as the electric and hybrid electric where a high torque density makes the vehicle to run for longer period with the regenerative capability.

5 Conclusions

Both inner magnet and outer magnet orientation in interior permanent magnet machine are studied. The machine with inner magnet provide a higher torque density of 4.94% as compared to the outer magnet machine. Inner magnet rotor design provide a better magnetic flux flow and flux linkage between the permanent magnet and stator pole than that of its counter-part. But the average torque in case of the outer rotor machine is about three times that of the inner motor which makes it suiting for the torque requirement of electric vehicle. This is a promising factor where using magnet in both the stator and the rotor and optimizing it to saturation level might bring new directions of machine design in electric vehicles.

References

1. X. Zhu, M. Cheng. A Novel Stator Hybrid Excited Doubly Salient Permanent Magnet Brushless Machine for Electric Vehicles, J of Elect. Eng. Tech., **1**, 2, 185–191 (2006)
2. C. C. Chan, K. T. Chau, J. Z. Jiang. Vehicles Drives for Electric Vehicles, IEEE Trans. On Indl. Electronics, **43**, 2 (1996)
3. Krishnan, Ramu, Switched reluctance motor drives: modeling, simulation, analysis, design, and applications, CRC press, (2001)
4. T. D. Kumar, A. Nagarajan, Design of Switched Reluctance Motor for Elevator Application, J. Asian Sci. Res., **3**, 258-267, (2013)
5. Erwan Sulaiman, Takashi Kosaka, Design Improvement and Performance Analysis of 12S-10P Permanent Magnet Flux-Switching Machine with Field Excitation Coil J. E. Sys. **8**, 4, 425-432, (2012)
6. Aravind, CV, Lam Pin Wen, Design Evaluations of the New Double Rotor Interior Permanent Magnet Machine, J. Engg. Sci. Tech., Eureca2014, **Special Issue**, 73-84, (2015)
7. Yee Wan Wong, Lee Wai Chong, Rajparthiban Kumar Rajkumar, Rajprasad Kumar Rajkumar, A New State-of-Charge Estimation Method for Valve Regulated Lead Acid Batteries, J. Engg. Sci. Tech. **12(3)**, 584-595, (2017)
8. Vaithilingam, C.A, Norhisam M, Zare M.R., Aris.I., Marhaban M.H, Computation of Electromagnetic Torque in a Double Rotor Switched Reluctance Motor Using Energy Methods, Energies, **5**, 4008-4026, (2012)
9. Grace I, Aravind CV, Rozita T, Samuel Bright, Optimized Toolbox for the design of Rotary Reluctance Motor,' Proceedings of IEEE Student Conference on Research and Development Conference, Malaysia, doi: 10.1109/STUDENT.2011.6089318, 1-6, (2011)
10. Aravind CV, Norhisam M, Aris.I, Marhaban M.H, M. Nirei, Electromagnetic Design and FEM Analysis of a Novel Dual Air-gap Reluctance Machine, Prog. in Electromagnetic Res., **140**, 523-544, (2013)
11. Aravind CV, M. Norhisam, R.N.Firdaus, I. Aris, M. H. Marhaban, Computation of Magnetic Characteristics of the Double Rotor Switched Reluctance Motor using Flux Tube Analysis, J. of Appl. Electro-magnetics and Mech., **21**, 3, (2013).

Porosity Control and Mechanical Properties of Porous Ceramic Material from Plered Region

Sulistyo S.[1,*], *Rizky*[2]

[1,2]Departemen Teknik Mesin, Fakultas Teknik, Universitas Diponegoro
Jl. Prof. Sudharto, SH., Tembalang-Semarang 50275, Telp. +62247460059

Abstract. Plered region is located in around Purwakarta area, west java, Indonesia. Plered is famous as ceramic product especially ceramic tiles, decorative ceramics, tableware and roof tiles. This paper investigated the manufacturing of porous ceramic from Plered and evaluated the mechanical properties especially bending strength. The investigation include the controlling of porosity using pore former during manufacturing, investigation of chemical composition and evaluated the fracture strength using bending test. Evaluation of porosity ceramic product use an archimedes method and scanning electron microscopy and the bending strength of porous ceramic using three point bending test. The porous ceramic was tailored by using compaction process and sintering process at temperature of 1250 C, the pore former (rice starch) was added on the material ceramic to control of porosity. The bending strength use standard ASTM C1161-13 which sample has cross section of 6x8 mm^2 and 90 mm long. The results show that the porosity increases with the number of pore formers in a mixture of clay, the porosity of sinter ceramic was 7 vol%,12 vol%, 17 vol% at using pore former 5 wt%, 15 wt% and 25 wt % respectively. The bending strength of sinter ceramic was declined with increasing porosity. The bending strength of sintered ceramic was 58 MPa, 49.5 MPa and 34.7 MPa at porosity of 7 vol%, 9 vol% and 14 vol% respectively.

1 Introduction

Plered region is located in around Purwakarta area, west java, Indonesia. Plered is famous as ceramic product especially ceramic tiles, decorative ceramics, tableware and roof tiles [1]. Currently, the function of ceramic plered can be used for engineriing material such as ceramic filter, thermal barier materials and catalytic materials [2-4]. Fabrication of porous ceramic can be done by using compaction process and continued by sintering process [3]. The compaction can be use to increase green compact strength and certain shape while the sintering process can increase the strength porous body due to forming bond of interparticle ceramic occurs.

*Corresponding author: listyo2007@gmail.com

The porous ceramic can be controlled by adding pore former likes a rice starch. The rice starch will burn during sintering process to form void as pore in the ceramic body. The starch must be controlled to obtain the certain porosity of ceramic body. The During operation of porous ceramic sometimes the body should support the certain load. The load must be capable to sustain in the ceramic body. The strength of porous ceramic trend to decrease with the increaasing of porosity [5]. The porous ceramic strength should be appropriate with the environment condition. For example, the range of ceramic porous as an electrode of solid oxide fuel cell (SOFC) is limited between 20 – 60 vol % [6-7]. This condition is to serve the electrode can be function well as load support, catalytic substrate and a good gas transport [7].

2 Materials and Methods

Plered bowlder is dried in the room temperature. The bowlder was crushed by using mortar into small particles. The small particles (powders) were dried in oven at temperatur 110°C for 2 hours. The dried powders were sieved by using mesh size of 100. The pore former used local rice starch of 100 mesh size. Manufacturing of bending strength specimen, Plered powders was added rice starch as pore former. The content of rice starchs in the Plered Powders are 5 %, 10 %, 25 % by weight respectively. The water was used as binder. The compaction use load of 15 kN. The sintering temperature of specimen was controlled at 1250°C using holding time of 1 hour. The final size of specimens were smoothed by using sand paper to accordance standard ceramic bending test ASTM 1161-13. The size of bending test specimen is 6x8 mm^2 and 90 mm long using Wance hydraulic universal testing machine (China).The composition of Plered powder was tested by ICP-AES, optical emission spectroscopy (Japan), the specimens porosity was done by using Archimedes metods and the bending strength was tested by universal machine test

3 Result and Discussion

Chemical composition of Plered powder was presented in table 1. The weight procentage of chemical content of Si (silicone), Al (aluminum), Fe (iron), K (Kalium) , Mg (magnesium), U (uranium) and others are 49.50 %, 44.75 %, 3.25 %, 1.00 %, 0.53 %, 0.46 % and balance respectively. The highest content is component of Si and the lowest is U.

Table1. Chemical composition of Plered powder by weight precentage

	Si %	Al %	Fe %	K %	Mg %	U %	others
Plered powders	49.50	44.75	3.25	1.00	0.53	0.46	balance

Table1 show the contents of Si, Al, Fe, K, Mg, the Plered powder can be classified as kaolin clay [8,9]. These composition of Plered powder can affect the temperature heating during sintering process especially the component of Al, Si and Fe content [1,4,9]. The higher of Al can increase the temperature sintering to get the strength of Plered substrate [9]. The temperature sintering was done at temperature of 1250° C with soaking time of 1 hr. This temperature is appropriate to increase the strength of the green compact without a damage of the specimen. The Figure 1 shown the specimen of the bending test. It seen that the specimen has been succesfully to be form into specimen bending test standard without crack and warpage. The size of the specimen is 6x 8 mm^2 and 90 mm long. The sintering temperature of 1250° C with soaking time of 1 hr are capable to increase the strength of the green compact. The diffusion process of the inter ceramic powder occured during sintering process [4]. The temperature sintering of 1250 C can be used as reference of the plered ceramic process.

The bending test was done by universal testing machine. The size of the specimen must be followed the standard ASTM 1161-13.

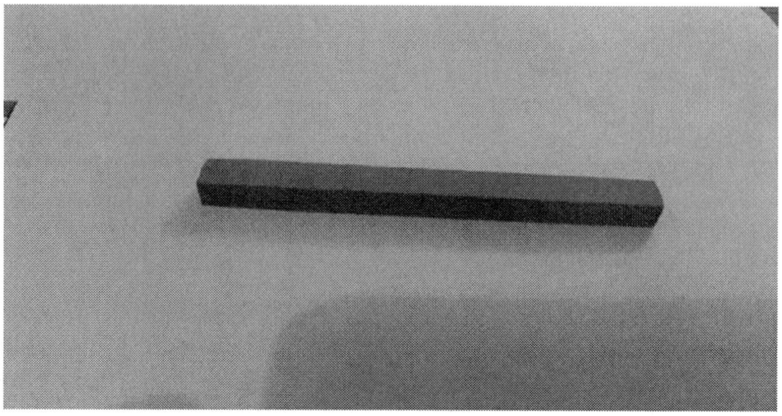

Fig. 1. The specimen of bending test

Figure 2 shows the relation between porosity and bending strength. The porosity of specimen can be controlled by adding of pore former (rice starch). The pore former will escape from the specimen during sintering process to form the void. The void in the specimen was called a pore [3,5]. The adding pore former in the substrate can increase the value of porosity specimen [2,7]. The pore former increase in the substrate, during sintering process, the pore former will be vanished to form more porosity.

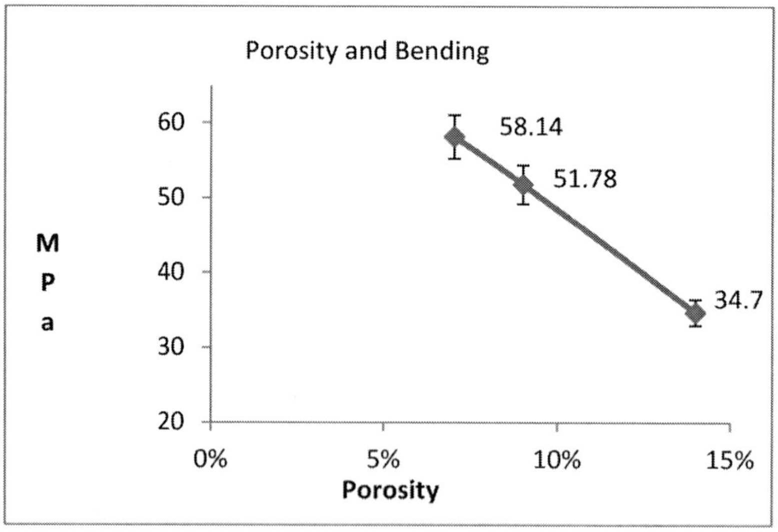

Fig. 2. Relation between porosity and bending strength

The value of specimen porosity will increase.. The specimen of bending strength of the plered clay was controlled by the porosity during testing.

The figure 2 show the relation between porosity and the bending strength. At porosity of 7 vol %, the bending strength is 58.14 MPa, while at porosity of 14 vol%, the value of bending strength is 34.7 MPa. The increase of porosity in the body will decline the bending strength [5]. The increase of the porosity can be easier to form big void due to an external load. The external load will be easier to merge small void into big void. This big void is as source of the crack initiating that will decline the strength. The porosity increase, the strength of the specimen decline [5]. This information can be used as the guidance of the filter product manufacturing process in the future.

Figure 3 show the microstructure of porous ceramic. The observation use scanning electronic microscopy (SEM). From the picture can be shown the particle size and the pore size.

Fig. 3. Microstructure of ceramic plered using SEM

The distribution of of the pore is relatively homogen. The size pore is relatively small around 0.28 μm-0.54 μm. Generally the pore morphology that was formed in the specimen is randomly. The temperature sintering of 1250°C can consolidate the clay particles. It can be seen in the Fig.3. The strength of porous material can be controlled by adding pore former as shown the figure 2. The bonding of interpartcles is be enough to give the strength of specimens [2,7].

4 Conclusion

The conclusion can be obtained after the discussion. The rice starch local can be used as pore former to make the pore material, the increase rice starch can increase the porosity.The bending strength of the pore material is affected by the sum of porosity in the materials. The incresing of pore material will decline the bending strength.The temperature of 1250° can be use as sintering temperature to get consolidation of the Plered clay.

Reference

[1] A. K. Pakpahan, Procedia Economics and Finance, vol.4, International Conference on Small and Medium Enterprises Development with a Theme (ICSMED 2012) , pp. 44-53 (2012)
[2] O. Yu, Materials Science is a copyright of Springer, vol. 51, no. Ukraina, pp. 847-853 (2016)
[3] Chao wang, Ceramic International, vol. 42, pp. 14222- 4227 (2016)
[4] David Obada, Applied Clay Science, Vols. 132-133, pp. 194-204 (2016)
[5] Meille, Journal of the European Ceramic Society, vol. 32, pp. 3959- 3967 (2012)
[6] Keegan C. Wincewics, Journal of Power Sources, vol. 140, pp. 280-296 (2005)
[7] S. Sulistyo, S. Arifin, S. Mahzan ,International Journal of Science and Engineering, vol. 4, no. 1, pp. 30-33, 2013.

[8] Rowse, Clays and Clay menerals, Pergamon Press, vol. 23, pp. 310-317 (1975)
[9] J.C.F. Segura, et all ,Applied Clay Science, vol. 146, pp. 264-269 (2017)

Polyurethanization temperature effect to the thermal behaviour of cellulose – fiber based hybrid polyurethane

Herald Kurnia Adi [1], *Fauzia* Hanum Ikhwan [2], *Hasyim* Aidilichsan Muliawan [3], and *Mochamad* Chalid*

[1] Department of Metallurgy and Materials Engineering, Universitas Indonesia, UI Depok New Campus, Depok 16424, West Java, Indonesia
[2] Department of Metallurgy and Materials Engineering, Universitas Indonesia, UI Depok New Campus, Depok 16424, West Java, Indonesia
[3] Department of Metallurgy and Materials Engineering, Universitas Indonesia, UI Depok New Campus, Depok 16424, West Java, Indonesia
* Department of Metallurgy and Materials Engineering, Universitas Indonesia, UI Depok New Campus, Depok 16424, West Java, Indonesia

Abstract. Polyurethanes are one of many types of polymers that are currently being developed worldwide. Some studies reported many substances involved in synthesizing polyurethanes (PU), with mostly originated from a biomass substance. Our laboratory has synthesized natural fiber – based PU to obtain hybrid PUs through varying reaction variables such as temperature. This study used cellulose, a biomass substance as well as polyethylene glycol and 4,4' – methylenebis cyclohexyl isocyanate (HMDI) in synthesizing the hybrid PU, and investigated their thermal stability such as decomposition temperature by Differential Scanning Calorimetry (DSC) showed various tendency according to the chain extending temperature for the decomposition temperature. Simultaneous Thermal Analysis (STA) showed glass transition tendency of the product is proportional to the chain extending temperature, where the maximum glass transition obtained from the experimental is 59.8°C at 70°C of polymerization. Evaluation of functional group in the hybrid PUs was conducted with FT – IR spectroscopy measurement indicating urethane presence as part of the hybrid PUs.

1. Introduction

Biomass – based PUs are currently being developed through various studies [1 – 3]. The vast possibilities of many biomass substance to be processed into a polyol for polyurethanization means, and many ways to synthesize PU with a sustainable development. Previous studies about polyurethanes included the use of the product of the self – metathesis of soybean oil (MSBO) based polyol to make polyurethane foams [4], organocatalysis in synthesizing thermoresponsive polyurethanes [5], soy – oil and caster – oil based polyol for novel shape – memory thermoplastic PUs [6], Canola – oil based diol (COBD) for thermoplastic PUs [7], and castor oil based polyol for rigid polyurethanes [8].

Currently, chain extenders in polyurethanization has also being widely developed, using some compounds such as precipitated silica with silanol groups [9], *n* – butanediol (BDO) [10], IDD – diols [11], dithioerythritol [12], and diamine [13] to the biomass – based substance such as cellulose originated from sorghum bicolor plant [14]. Cellulose grafting has also been studied to investigate the effects of graft chain length on thermal behaviour [15]. The availability of cellulose as a grafting substance as well as chain extender in polyurethanization proves the ability of cellulose as a substance available for creating an environmentally friendly PU material [14]. There is also a conducted study focusing on preparation for microfibrilated celluloses based on Arenga Pinnata "ijuk" fiber [16], as well as a study about modified shape memory polyurethane using Fe_3O_4 nanoparticles [17].

Previous study conducting cellulose as a chain extender in synthesizing polyurethane has not yet conduct the temperature optimization in the polyurethanization. The thermal properties of the cellulose – fiber based hybrid PU and expected change in glass transition due to variations in the polyurethanization is the aim of study of the polyurethanization temperature optimization, specifically when reacting the prepolymer and chain extender to synthesize the hybrid PU. The temperature variations in the polyurethanization are 50°C, 70°C, 80°C, and 90°C. The synthesized hybrid PUs with temperature optimization were then investigated with Simultaneous Thermal Analysis (STA) for thermal characterization, and Fourier Transform – Infra Red (FT – IR) to investigate the existence of urethane group in the hybrid PUs.

2. Experimental

2.1 Materials

The main raw materials for polyurethanization are polyethylene glycol with molecular weight of 6000 mole (PEG 6000), 4, 4' – Methylenebis cyclohexyl isocyanate (HMDI) with purity of 90%, dimethyl formamide (DMF) 99.8%, dibutyltin dilaurate (DD) 95%, and cellulose powder. All raw materials but DMF and cellulose powder were obtained from Sigma – Aldrich, with DMF and cellulose powder obtained from Merck. No pre – process conducted before the synthesis of the hybrid PU.

2.2 Synthesis of Hybrid Polyurethane

The total time needed for all process is 5h, and is based on previous study conducted by F. H. Ikhwan, et. al [11]. The first process was to dissolve PEG 6000 with DMF in temperature of 70°C for 2h. The second process was to synthesize the pre – PU by reacting the dissolved PEG 6000 with HMDI in temperature of 70°C for 2h. The last process was to synthesize the hybrid PU by reacting the pre – PU obtained from the previous process with cellulose powder in several temperatures; 50°C, 70°C, 80°C, 90°C. All of the process were conducted in Schlenk tubes, dipped and heated in oil bath. Right after each process were conducted, fluxing was conducted together with nitrogen atmosphere to make vacuum condition inside the Schlenk tubes.

After all of the process were conducted, the sample in the form of white milk solution was obtained. Phase separation was then conducted by terminating the product with 20ml of water and left intact until 24h when the hybrid PU precipitated and separated before dried for 3 days and continued with characterization.

2.3 Characterization

Fourier Transform – Infra Red (FTIR) was used to investigate the existence of urethane group within the hybrid PU. The investigation was conducted in DTMM FTUI using FTIR PerkinElmer UATR Two Spectrum Two which has a limitation of wavelength ranging from 400 – 4000 λ. Simultaneous Thermal Analysis was used to investigate the glass transition (Tg) and thermal stability of the hybrid PU. The investigation was conducted in DTMM FTUI using STA PerkinElmer STA6000 which has a rate limitation of 5 min / °C and max temperature of 1000°C.

3. Result and Discussion

Polyurethanization was carried out by two steps of process. The first being pre – polyurethanization which is the process to synthesize the pre – PU by reacting soft segment and hard segment, and the other being copolyurethanization or chain extending which is the process to synthesize the hybrid PU by reacting pre – PU obtained from the previous step with a chain extender. In this study, PEG 6000 serves as the soft segment, HMDI as the hard segment, and cellulose as the chain extender. The synthesized hybrid PUs were then evaluated by Simultaneous Thermal Analysis (STA) to investigate the glass transition temperature and Fourier Transform – Infra Red (FT – IR) to investigate the existence of urethane group in the hybrid PUs.

3.1. Structure of Hybrid Polyurethanes

Figure 1 shows FTIR spectra for each hybrid PUs, indicating the existence of chemical bonds in the hybrid PUs. Ether (C-O) are found at 1241 cm^{-1} as a typical indicator of successful reaction for pre – PU and cellulose. Amine (NH_3) bonds at 1279 – 1343 cm^{-1}, and secondary amine at 2884 cm^{-1} proved the existence of isocyanate (N=C=O) parts, as urethane group in the hybrid PU. Based on the transmission percentage shown by the figures, HPU 70-50 has the least infrared transmission meaning that the bonds present most in HPU 70-50.

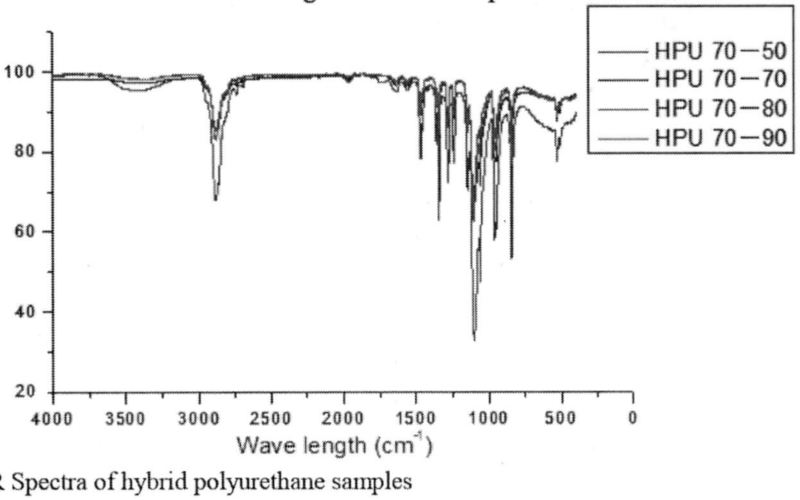

Fig. 1. FTIR Spectra of hybrid polyurethane samples

3.2 Thermal Behaviour of Hybrid Polyurethane

Effects of synthesis temperature in cellulose adding to glass transition thermal characteristics of the hybrid PUs were investigated by Differential Scanning Chromatography (DSC), as shown in Fig. 2. At the beginning the increase in synthesis temperature enhances the glass transition temperature (Tg) which is shown the glass transition temperature of HPU 70-50 and HPU 70-70. However, this tendency undergoes a change into decreasing with the temperature enhancement (see 80°C, 90°C shown in HPU 70-80 and HPU 70-90). The tendency was caused by the thermal degradation that occurred in the cellulose above 70°C.

Table 1. DSC analysis results on hybrid polyurethane

Sample	Tg (°C)
HPU 70-50	52.13
HPU 70-70	59.8
HPU 70-80	59.29
HPU 70-90	58.84

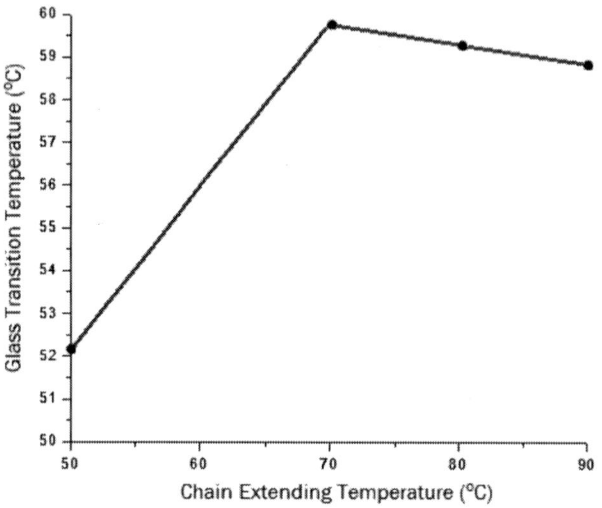

Fig. 2. Glass Transition Temperature Graph

3.3. Thermal Stability Analysis

Thermogravimetric analysis was utilized to analyse furthermore thermal stability of the hybrid PU products. The result of the test was shown as a graph (fig. 3) that shows the reducing of polyurethane mass against temperature. According to the graph, it can be seen that HPU 70-90 has the best thermal stability, followed by HPU 70-80, HPU 70-70, and HPU 70-50. However, the weight reduction shows otherwise. It is predicted that due to the low temperature in the synthesis of HPU 70-50, the bonds present in the HPU did not weaken prior to thermogravimetric analysis. Thermal stability however, increase proportionally to the increase of chain extending temperature for the first decomposition phase, but decrease for chain extending at 80°C and 90°C for the second decomposition phase, and at the third decomposition phase 80°C has the lowest decomposition temperature shown by fig. 4.

Table 2. Decomposition temperature (Td) on hybrid polyurethane

Chain Extending Temperature (°C)	Td_1 (°C) / Weight Reduction (%)	Td_2 (°C) / Weight Reduction (%)	Td_3 (°C) / Weight Reduction (%)
HPU 70-50	302.19 / 8	378.77 / 25	429.69 / 89
HPU70-70	304.27 / 3.2	381.48 / 18.75	432.8 / 92.5
HPU 70-80	320.07 / 16.6	373.73 / 37.37	424.06 / 82
HPU 70-90	325.15 / 5.7	370.35 / 16.67	426.33 / 90

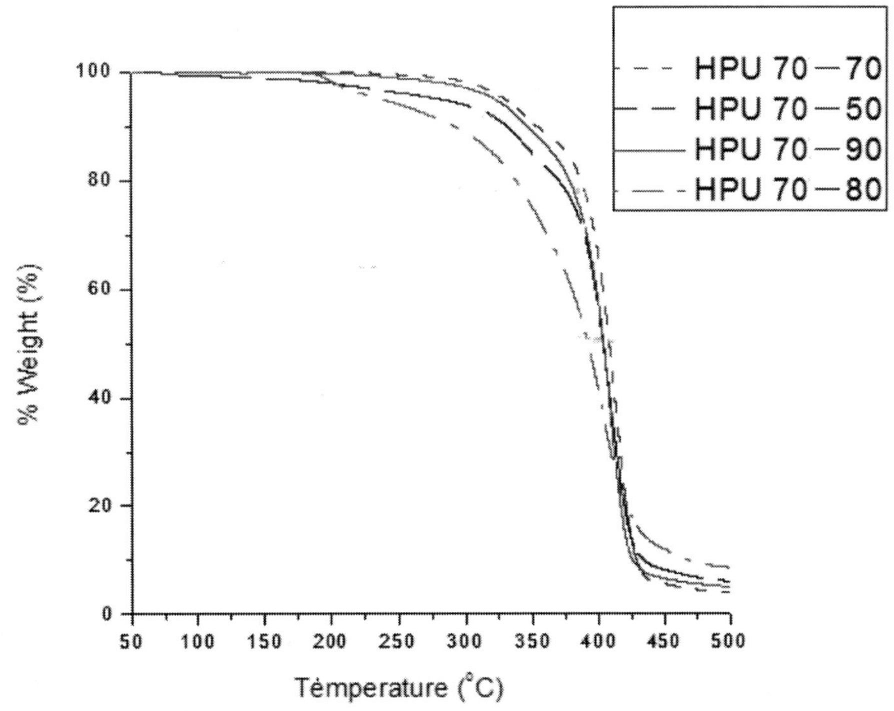

Fig. 3. Variations of Chain Extending Temperature TGA Curve

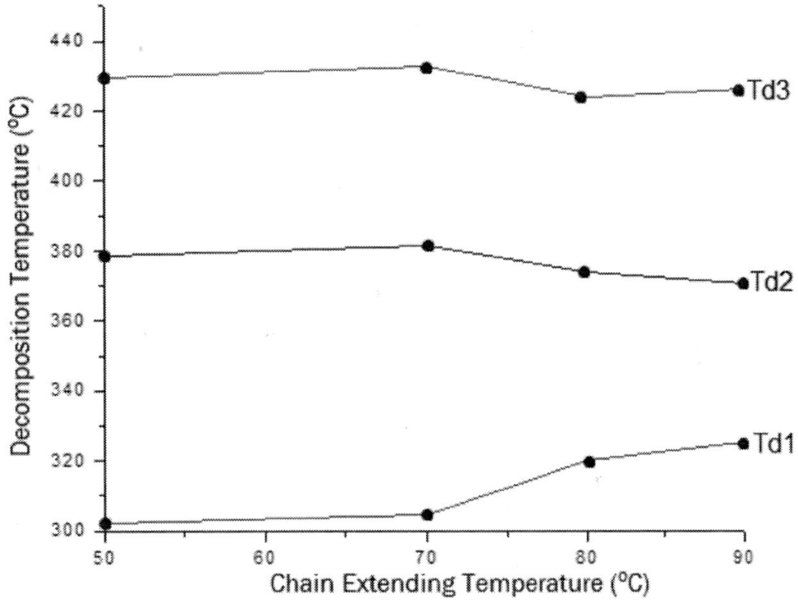

Fig. 4. Decomposition Temperature Graph

The first decomposition phase shows that higher chain extending temperature results in more reaction to take place and increase bonds, this is shown in fig. 4 for 90°C had the highest decomposition temperature in the first decomposition phase. The cause for the phenomenon in second decomposition phase is caused by the less significant effect of pyrolytic resistance.

PEG 6000 has a good thermal stability because it has a relatively high molecular weight, but it also comes with a consequence; pyrolytic resistance becomes less significant. Pyrolytic resistance is a property that gives improvement to thermal stability [17]. The decrease in pyrolytic resistance is the cause of the decrease in decomposition temperature for chain extending at 80°C, and 90°C. The phenomenon in the third decomposition phase was caused by thermal degradation in the cellulose. While higher chain extending temperature triggered more reaction to take place, thermal degradation may occur in cellulose above 70°C.

4. Conclusion

Cellulose – fiber hybrid polyurethanes have been synthesized in different temperatures and evaluated by DSC and TGA. FTIR testing was also conducted to find out the bonds inside the hybrid polyurethanes. All bonds required to acknowledge the urethane group are present in all of the samples. Variation in chain extending temperature give influence in the glass transition temperature and thermal stability.

5. References

[1] M. Chalid, H. J. Heeres, A. A., Broekhuis, Polymer – Plastic Technology and Engineering **54** (2015)
[2] M. Chalid, H. J. Heeres, A. A., Broekhuis, Applied Mechanics and Materials **229 – 231** 297 – 302 (2012)
[3] S. Ilmiati, J. H. Mustafa, A. Yaumal, F. Hanum, M. Chalid, IOP Conference Series: Materials Science and Engineering 223 (2017) 012033
[4] S. Li, L. Bouzidi, S. S. Narine, Europian Polymer Journal **93** 232 – 245 (2017)
[5] L. I. Ronco, A. Basterretxea, D. Mantione, R. H. Aguirresarobe, R. J. Minari, L. M. Gugliotta, D. Mecerreyes, H. Sardon, Polymer **122** 117 – 124 (2017)
[6] Z. S. Petrović, J. Milic, F. Zhang, J. Ilavsky, Polymer **121** 26 – 37 (2017)
[7] M. B. Karimi, G. Khanbabaei, G. M. M. Sadeghi, Journal of Membrane Science **527** 198 – 206 (2017)
[8] M. Ionescu, D. Radojčić, X. Wan, M. L. Shrestha, Z. S. Petrović, T. A. Upshaw, Europian Polymer Journal **84** 736 – 749 (2016)
[9] L. Chen, X. Wang, Z. Jia, Y. Luo, D. Jia, Materials & Design **87** 324 – 330 (2015)
[10] N. Lempesis, P. J. in 't Veld, G. C. Rutledge, Polymer **107** 233 – 239 (2016)
[11] Chien – Hsin Wu, Yu – Ching Chen, S. A. Dai, Su – Chen Chen, Shih – Huah Tung, Rong – Ho Lee, Wen – Chiung Su, Ru – Jong Jeng, Polymer **119** 1 – 12 (2017)
[12] L. H. Chan – Chan, R – Sollis Correa, R. F. Vargas – Corronado, J. M. Cervantes – Uc, J. V, Cauich – Rodriguez, P. Quintana, P. Bartolo – Perez, Acta Biomateriala **6** 2035 – 2044
[13] L. Lei, Y. Zhang, C. Ou, Z. Xia, L. Zhong, Progress in Organic Coatings **92** 85 – 94 (2016)
[14] F. H. Ikhwan, S. Ilmiati, H. K. Adi, R. Arumsari, M. Chalid, IOP Conference Series: Materials Science and Engineering **223** (2017) 012019
[15] M. H. Ryu, J. Park, D. X. Oh, S. Y. Hwang, H. Jeon, S. S. Im, J. Jegal, Polymer Degradation and Stability **142** 226 – 233 (2017)
[16] E. Yuanita, J. N. Pratama, J. H. Mustafa, M. Chalid, Procedia Chemisty **16** 608 – 615 (2015)
[17] D. F. Firdaus, Masrudin, D. A. Lestari, M. R. Arbi, M. Chalid, Indo. J. Chem. **15** 130 – 140 (2017)
[18] C. Yuan, M. Z. Rong, M. Q. Zhang, Polymer **55** 1782 – 1791 (2014)

Fault Diagnostic System Bearing Centrifugal Pump Using K-Means Method For Thermography Image And Signal Analysis Vibrations

Agung Yudhistira Hidayat [1], *Achmad* Widodo [2], *Gunawan* Dwi Haryadi [1]

[1]Departement of Mechanical Engineering Faculty of Engineering, UNDIP, 50275 Semarang, Indonesia

Abstract. Numerous studies reported that infrared thermography and vibration are condition monitoring technology that is important and effective for doing a condition diagnostic of bearing centrifugal pump health without destructing or disturbing machine operational. This paper focuses on thermography image processing based on K-Mean color segmentation which will produce normal and abnormal condition features. Health diagnostic of bearing by processing of digital image, image clustering, segmentation and extraction. Extraction of image pattern is done by calculating the area of heat point and color feature bearing condition of RGB colour space and active contour segmentation in order to process and differentiate between normal and abnormal bearing image by statistical technique. The parameters that can be used as reference to classifying conditions are standard deviation, Mean, Variance, Skewness, Kurtosis, Vibration (RMS) and Shape features (area). Final step is determining the boundary condition between normal and abnormal using statistical logic method.

1 Introduction

Machine condition monitoring and fault diagnostics become a critical issue in maintenance activity to ensure availability, minimize operator hazard, and reduce economic losses [1]. One of method that is being developed is condition monitoring based on temperature measurement using infrared thermography (IRT). IRT is non-destructive test, free from electromagnetic interference, safe, reliable and can provide large inspection coverage. IRT has taken a very important role for predictive and preventive maintenance programs [2]. Health diagnosis of Machine can be done by evaluating the image pattern generated due to the transformation of input energy into waste heats. IRT can use as a valuable tool for fault diagnosis in Machine which relying on infrared data combined with the heat transfer theory.

Image analysis has many different kinds of methods, one of which is a method of digital-image texture evaluation. The main problem of texture analysis is feature extraction, that extraction allows to explain the characteristics of a digital image numerically.

*Corresponding author: yh_agung@yahoo.co.id

Statistical methods of the thermal image analysis can be used for monitoring condition and fault diagnosis during machine operation [3]. Segmentation is an important part in image processing before executing the next stage such as feature extraction and classification [4]. Image segmentation methods can be classified into histogram threshold, clustering, edge based, region growing, Physical model-based, Fuzzy approaches and Neural network methods [5]. Clustering techniques identified homogeneous clusters of pixels in the feature space such as RGB, HSV, YIQ, XYZ, LAB, and LUV color space [6][7]. RGB is an additive color system based on the theory of tri-colored (red, green and blue) and nonlinear visual perception.

This paper proposed condition monitoring and health diagnostics of bearing centrifugal pump based on image of thermal infrared and vibration signals. The aim at this research is to find a new alternative source features and a machine diagnostics technique by utilizing thermal images. Thermal images are processed by the threshold method of image segmentation Using K-Means Clustering which will produce a characteristic of normal and abnormal pattern image. Salient features are then extracted from binary images after images processing. This result based on the size (area, perimeter and eccentricity) and color patterns generated on normal and abnormal condition. The effectiveness of thermographic techniques for fault identification of bearing a centrifugal pump is highly dependent on its accuracy in identifying the hot areas and predicting its severity level. Vibration signals analysis used as validation and strengthen the IRT methods Condition monitoring using vibration signal has information on dynamic behavior that is often used for fault detect in this bearing centrifugal pump. Health diagnosis of bearing centrifugal pump is done based on the characteristics thermal image and graph vibration signals measurement results.

2 Basic Theory

2.1. Image Thresholding.

Image threshold is used for image segmentation [7]. A threshold is a process to separate objects of its background in a digital image. Histogram is the main tool in this separation process. Suppose that the grey level corresponds to an image $f(x, y)$ that is composed of lights objects of a dark background, in such a way that object and background pixels have grey levels grouped into two dominant modes. Both colors are black and white. The colors used for the image are objecting in the foreground. While the rest of the image is the background color. Extracting the object of the background is performed by selecting a threshold T that separates these modes. A threshold image $g(x, y)$ is defined as:

$$g(x,y) = \begin{cases} 0 \ \ jika\, f(x,y) < T, \\ 1 \ \ jika\, f(x,y) \geq T. \end{cases} \quad or \quad g(x,y) = \begin{cases} 0 \ \ jika\, f(x,y) \geq T, \\ 1 \ \ jika\, f(x,y) < T. \end{cases} \tag{1}$$

2.2. Feature Extraction of image

Feature extraction is the process to regain the most important data from the raw data to find parameter that defines shape of a character exactly and uniquely [8]. Feature extraction image can be classified into three groups are color feature extraction, shape and texture features [9]. Extraction feature of RGB is example color feature extraction. The calculation of the perimeter and area are meaningful for binary image. Consider a discrete binary image containing one more object, where **p** (x, y) = 1 if a pixel is part of the object or foreground and **p** (x, y) = 0 for background pixels [10].

3 Methodology

3.1 Data acquisition process

The first task supposedly begins with the image acquisition process. The experimental setups that were carried out by using thermographic camera FLIR E-40 at a distance of 2 metres, measure the temperature on a range scale of -20 ^{0}C to 650 ^{0}C, The emissivity value was set at 0.95 as recommended material of cast iron [11], ambient temperature is 30-41 ^{0}C. The experiment was carried out with the condition of centrifugal pump operation full load in the power generation unit and the condition of bearing is in normal state.Vibration signals data taken with vibscanner. Measurements are made on bearing of centrifugal pump with specifications: 1,000 HP/Type Horizontal Centrifugal, bearing SKF Explorer 6310-2Z and bearing SKF Explorer 6207.

The result of vibration measurement is Root-Mean Square (RMS in mm/s) data record, while Flir E-40 produces thermal images 320 x 240 pixels. After image acquisition from the object, thermal image is cropped to select the observed region of interest (ROI) in MATLAB. The image acquisition processed for normal and abnormal conditions was then conducted.

3.2 Image Segmentation

Segmentation image thermography is done to decompose a thermal image of specific parts of further analysis, resulting in a higher level representation of the image pixels like objects and background image [12]. One of the image segmentation methods is clustering, clustering techniques identified homogeneous clusters of pixels in the feature space such as RGB color space [5]. The pixel features of images are clustered using the K-Means clustering algorithm to group them into five regions of similar color and selected one image of ROI representing temperature increases that is redness image region. This ROI calculates to do RGB characteristic (Mean, standard deviation, kurtosis, skewness and entropy.) and shape feature (area, perimeter and eccentricity).

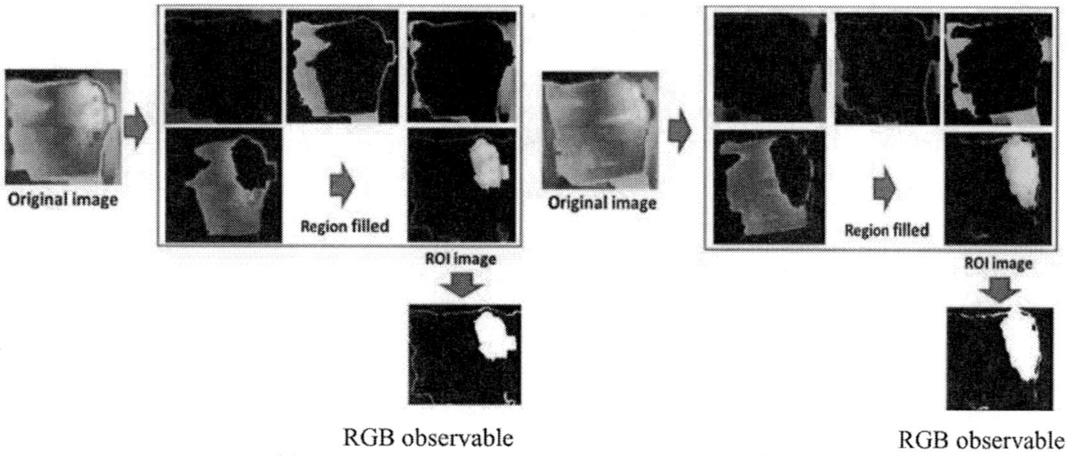

(a) (b)

Fig.1. Color segmentation process, decide to ROI: (a) Normal condition, (b) Abnormal condition

3.3. Image feature extraction

Clustering process will produce one color image area that will be implemented image analysis, calculation of RGB characteristic (Mean, standard deviation, kurtosis, skewness, variance and entropy.), shape feature (area, perimeter and eccentricity) and vibration. Image statistics are investigated and studied based on first and higher order statistics that capture certain statistical regularities of ROI thermal image [7], following equation:

$$Mean(\mu)=\frac{1}{MxN}\sum_{x=1}^{M}\sum_{y=1}^{N}p(x,y),\ Std.\text{dev.}(\sigma)=\sqrt{\frac{1}{MxN}\sum_{x=1}^{M}\sum_{y=1}^{N}(p(x,y)-\mu)^2} \quad (2)$$

$$Entropy(E_n)=\frac{1}{MxN}\sum_{x=1}^{M}\sum_{y=1}^{N}p(x,y)\log_2(p(x,y)) \quad (3)$$

$$Skewness(s)=-\frac{1}{MxN}\sum_{x=1}^{M}\sum_{y=1}^{N}\left|\frac{p(x,y)-\mu}{\sigma}\right|^3 \quad (4)$$

$$Kurtosis(k)=-\frac{1}{MxN}\sum_{x=1}^{M}\sum_{y=1}^{N}\left|\frac{p(x,y)-\mu}{\sigma}\right|^4-3 \quad (5)$$

Next step is plotting the features of statistical RGB and shape feature. Afterwards selected features that can distinguish between normal and abnormal conditions. Final step is determining the boundary condition between normal and abnormal using statistical logic method. Boundary values obtained by calculation value of mean, min, max, standard deviation, delta (D) of selected feature data. Value of the delta (D) is average difference between minimum and maximum value in each class.

$$final\max=\max+D,\ \text{final}\min=min-D \quad (6)$$

4 Result and Discussion

A sentrifugal pump being subjected to this observation according to ISO 10816-3 can be categorized into the group one rigid. Classification based on motor power is more than 300 KW. Normal condition if the root mean square (RMS) values has up 7.1 mm/s. If then RMS $\geq$ 7.1 mm/s been categorized as abnormal condition. From number 50 data can be classified into number 40 data onto normal condition and 10 data of abnormal condition. The acquire images stored in RGB format with size of 256 × 256 pixels. The ROI was determined by red or white colors that represent the hottest temperature area generated by bearings and will be transformed into binary image by threshold method, then extracted to calculate area, perimeter and eccentricity.

The ROI image determination process image clustering are presented figure 1. ROI is an RGB image, extracted to calculate the statistical of each component of RGB. Referring to determination and the resulting feature, not all features can be used as reference to differentiate of bearing centrifugal pump conditions. Figure 2 shows random patterns for the features that can differentiate its condition are standard deviation, mean, variance, skewness, kurtosis of RGB.

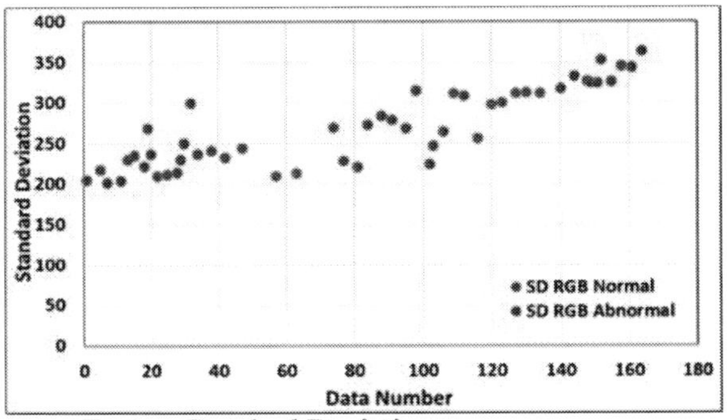

a. Feature Standard Deviation

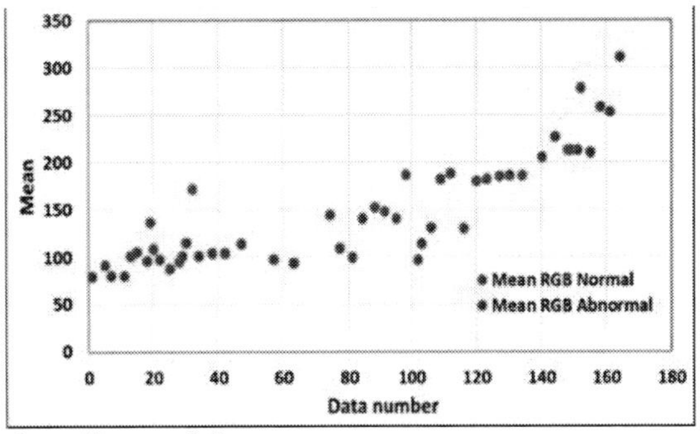

b. Feature Mean

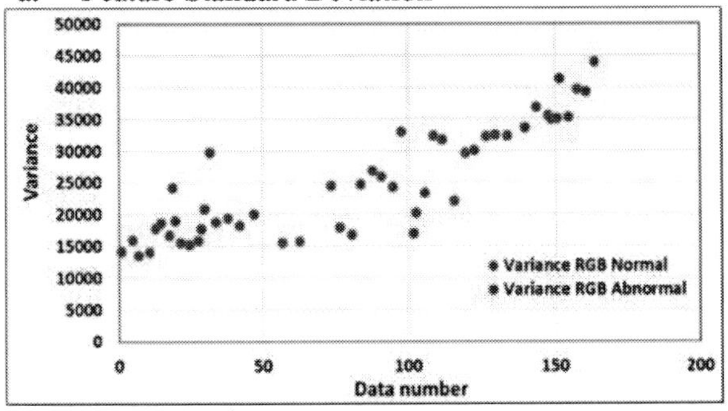

c. Feature variance

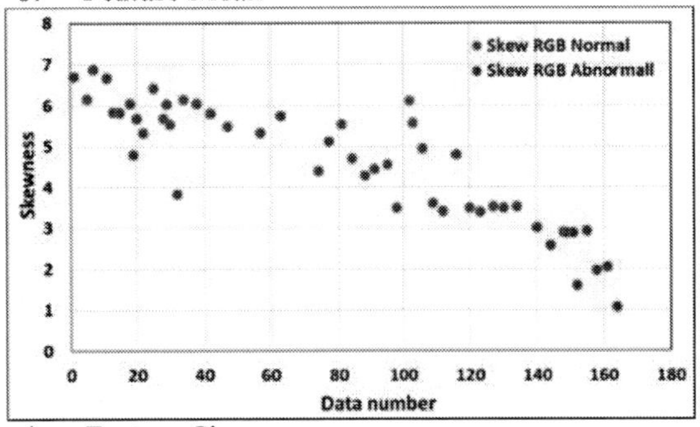

d. Feature Skewness

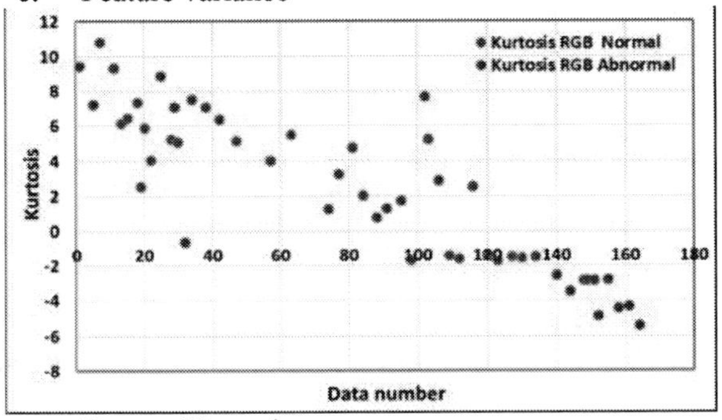

e. Feature Kurtosis

Fig. 2. Feature differentiate between normal and abnormal condition

Based on the result of calculation of equation 6 in Table 1, determination of boundary conditions value (Final maximum/F. Max and minimum/F. Min) does not have same value of boundary condition between normal and abnormal condition.

Table 1. Determination result of boundary condition value of each parameter RGB.

Condition of Bearing Centrifugal Pump	RMS		SD		Mean		Variance		Skewness		Kurtosis	
	Min	Max	Min	Max	Min	Max	Min	Max	Min	Max	Min	Max
Normal	1,83	6,89	200,82	314,77	79,56	187,81	13578,76	33088,85	3,39	6,87	-1,70	10,80
Abnormal	7,17	9,63	317,48	363,16	204,73	311,03	33734,32	43987,12	1,07	3,00	-5,43	-2,53

Conclusion

Based on this study, color segmentation on the thermal image can be used to diagnose normal and abnormal conditions on the bearing of centrifugal pump. Parameters that can be used a good reference to distinguish between normal and abnormal conditions are standard deviation, Mean, Var, Skewness, Kurtosis. Bearing on centrifugal pump can be categorized abnormal if RMS > 7.17 mm/s, standard deviation $\geq$ 317.48, Mean $\leq$ 204.73, Var $\geq$ 33,734.32, Skewness $\leq$ 3 and Kurtosis $\leq$ -2.53.

The author gratefully acknowledges the support of measuring instruments and research objects of PT. Indonesia Power Semarang on this research.

References

1. Lei, Y., He, Z., & Zi, Y. Application of an intelligent classification method to mechanical fault diagnosis. *Expert Systems with Applications*, 36(6), 9941-9948. (2009).
2. Cao, Y., Gu, X., & Jin, Q. Infrared technology in the fault diagnosis of substation equipment. *In Electricity Distribution*, 2008. CICED 2008. China International Conference on (pp. 1-6). IEEE. (2008, December).
3. Fidali, M., An idea of continuous thermographic monitoring of machinery, QIRT 2008 Proceedings, *In 9th International Conference on Qualitative Infrared Thermography*, pp. 61-68. (2008).
4. Solomon, C., & Breckon, T.,, *Fundamentals of Digital Image Processing: A practical approach with examples in Matlab*, John Wiley & Sons. (2011).
5. Chauhan, Poonam, and R. V. Shahabade, Mean shift segmentation on Enhanced Color image, *IOSR Journal of Computer Engineering* (IOSR-JCE) e-ISSN: 2319-8753, Vol.3, (2014).
6. Wang, H., & Suter, D., Color image segmentation using global information and local homogeneity, *In Proceeding of 7th Conf. of Digital Image Computing: Techniques and Applications*. (2003).
7. Ravichandran, D., R. Nimmatoori, and M. R. Ashwin Dhivakar. "A study on Image Statistics and Image Features on Coding Performance of Medical Images." Int. J. of *Advanced Computer Engineering and Communication Technology* (IJACECT) 5,1, p.1-6 (2016).
8. R. C. Gonzalez and R. E. Woods, Digital Image Processing, Prentice-Hall, Upper Saddle River, NJ, USA, 2nd edition, (2002).
9. Kumar, Gaurav, and Pradeep Kumar Bhatia. "A detailed review of feature extraction in image processing systems." Advanced Computing & Communication Technologies (ACCT), 2014 Fourth International Conference on. IEEE, pp. 5-12, (2014).
10. Medjahed, Seyyid Ahmed. "A Comparative Study of Feature Extraction Methods in Images Classification." International Journal of Image, Graphics and Signal Processing 7.3: 16, (2015)
11. Bagavathiappan, S., Lahiri, B. B., Saravanan, T., Philip, J., & Jayakumar, T., Infrared thermography for condition monitoring–a review, *Infrared Physics & Technology*, **Volume 60**, pp. 35-55, (2013).
12. Sural, Shamik, Gang Qian, and Sakti Pramanik. "Segmentation and histogram generation using the HSV color space for image retrieval." *Image Processing, Proceedings. 2002 International Conference* on. Vol. 2. IEEE (2002).

Adomian decomposition method for solving initial value problems in vibration models

Sudi Mungkasi[1,*], and *I Made Wicaksana* Ekaputra[2]

[1]Department of Mathematics, Faculty of Science and Technology, Sanata Dharma University, Mrican, Tromol Pos 29, Yogyakarta 55002, Indonesia
[2]Department of Mechanical Engineering, Faculty of Science and Technology, Sanata Dharma University, Mrican, Tromol Pos 29, Yogyakarta 55002, Indonesia

Abstract. A number of engineering problems have second-order ordinary differential equations as their mathematical models. In practice, we may have a large scale problem with a large number of degrees of freedom, which must be solved accurately. Therefore, treating the mathematical model governing the problems correctly is required in order to get an accurate solution. In this work, we use Adomian decomposition method to solve vibration models in the forms of initial value problems of second-order ordinary differential equations. However, for problems involving an external source, the Adomian decomposition method may not lead to an accurate solution if the external source is not correctly treated. In this paper, we propose a strategy to treat the external source when we implement the Adomian decomposition method to solve initial value problems of second-order ordinary differential equations. Computational results show that our strategy is indeed effective. We obtain accurate solutions to the considered problems. Note that exact solutions are often not available, so they need to be approximated using some methods, such as the Adomian decomposition method.

1 Introduction

Vibration occurs in daily life, such as sounds, acoustics, machines, etc. A mathematical model for vibrations is the second-order ordinary differential equations. The model can be either with or without source terms. A source term is assumed to be an external force involved in the vibration.

The vibration model can have a high degree of freedom, so solving the model can be tedious. A number of researchers have attempted to solve vibration model, such as Nad [1], Ouyang and Zhang [2], and Supriyadi [3]. Nevertheless, it is still an open problem about how to solve the model in an inexpensive computations.

In this paper, we consider vibration models, especially in the scalar form. We use the Adomian decomposition method due to Adomian [4]. The Adomian decomposition method is chosen, as it has some advantages, such as that it is meshless, so solutions can be computed at any time [5-6]. We propose a computational treatment of the source term,

* Corresponding author: sudi@usd.ac.id

when it appears in the vibration model and when we implement the Adomian decomposition method. If the source term is not appropriately treated, the method may lead to a misleading solution.

The rest of the paper is organised as follows. Section 2 provides the mathematical models we consider. Section 3 proposes the computational treatment of the source term of the vibration model. Results and discussion are given in Section 4. We conclude the paper in Section 5.

2 Mathematical models

We consider the vibration problem of a spring-mass system having the vector-matrix form

$$\mathbf{M}\ddot{\mathbf{x}} + \mathbf{C}\dot{\mathbf{x}} + \mathbf{K}\mathbf{x} = \mathbf{s}(t)\,, \tag{1}$$

where $\mathbf{x}$ is the state space vector dependent on the free variable time t, $\mathbf{M}$ is the mass matrix, $\mathbf{C}$ is the friction matrix, $\mathbf{K}$ is the stiffness matrix, and $\mathbf{s}(t)$ is the source-term vector. The source term $\mathbf{s}(t)$ is linearly independent with the $\mathbf{C}\dot{\mathbf{x}}$ and $\mathbf{K}\mathbf{x}$ terms.

When there is no damping, the initial value problem of the vibration model in the scalar form can be written as

$$x''(t) = kx + s(t)\,, \quad x(0) = \alpha\,, \quad x'(0) = \beta \tag{2}$$

defined in a closed domain $[0, b]$. Here, t is the time variable which is free, and x is the position variable dependent on t. In addition, α and β are constants, and b is a positive constant. The short notation $x'(t)$ means dx/dt, which is the first derivative of x with respect to t. The short notation $x''(t)$ is for d^2x/dt^2, which is the second derivative of x with respect to t. We assume that all functions involved in the model are smooth, and the source term $s(t)$ is linearly independent with the kx term, and can be either linear or nonlinear. Here, k is constant.

In this paper, we focus on solving the initial value problem (2) using the Adomian decomposition method. Care should be taken when we have the nonzero source term $s(t)$ in the equation. Otherwise, the Adomian decomposition method may lead to inaccurate results.

3 Adomian decomposition method

In this section, we provide the Adomian decomposition procedure following the work of Al-Khaled and Anwar [7]. This is the complement of the work of Biazar, Babolian, and Islam [8].

We consider the differential operator L defined as $L = d^2/dt^2$. Then the inverse operator L^{-1} is defined as

$$L^{-1} = \int_0^t \int_0^t (\cdot)\,dt\,dt\,. \tag{3}$$

The initial value problem (2) can be written in an operator form as

$$Lx = kx + s(t)\,. \tag{4}$$

Knowing the initial values $x(0) = \alpha$ and $x'(0) = \beta$, we find that equation (4) becomes

$$x(t) = \alpha + \beta t + L^{-1}[s(t)] + L^{-1}[kx]\,. \tag{5}$$

The Adomian decomposition method works by assumption that the function $x(t)$ can be decomposed into a series of functions $x_n(t)$, that is,

$$x(t) = \sum_{n=0}^{\infty} x_n(t) . \tag{6}$$

Substitution of equation (6) to equation (5) results in

$$\sum_{n=0}^{\infty} x_n(t) = \alpha + \beta t + L^{-1}[s(t)] + L^{-1}\left[\sum_{n=0}^{\infty} k x_n(t)\right] . \tag{7}$$

Each term of the series (7) is determined as follows

$$x_0(t) = \alpha + \beta t + L^{-1}[s(t)], \tag{8a}$$

$$x_{n+1}(t) = L^{-1}[k x_n], \quad n \geq 0 . \tag{8b}$$

The nonzero source term $s(t)$ is treated in $x_0(t)$ of equation (8a). This treatment is important to note in order that our approximation is accurate. If the source term is treated in equation (8b) instead, then we call the treatment as a naive treatment. The N-term approximation of $x(t)$ is given by

$$\phi_N(t) = \sum_{n=0}^{N-1} x_n(t) . \tag{9}$$

The Adomian decomposition method is an analytical technique of approximation to a function. We do not need to discretise the given domain. The method converges to the exact solution rapidly for a certain radius of domain.

4 Results and discussion

In this section, we discuss our research results. As has been mentioned, we investigate the performance of the Adomian decomposition method in solving initial value problems of the second-order ordinary differential equations involving source terms. The source term in the model must be handled appropriately. If not, the Adomian decomposition method may not converge. We provide two examples in this section. These examples are extracted from the paper of Al-Khaled and Anwar [7].

4.1 Linear source term

As the first example, we consider the initial value problem [7] having a linear source term

$$x''(t) = x + t , \quad x(0) = 1 , \quad x'(0) = 0 \tag{10}$$

where $t \in [0,1]$. The exact solution to this problem is $x(t) = e^t - t$.

The equation $x''(t) = x + t$ in problem (10) can be written in an operator form as

$$Lx = x + t , \tag{11}$$

for $0 < t < 1$. Knowing the initial conditions $x(0) = 1$ and $x'(0) = 0$, we obtain

$$x(t) = 1 + L^{-1}[t] + L^{-1}[x], \tag{12}$$

where L^{-1} is given by equation (3). Using the Adomian decomposition, equation (12) becomes

$$\sum_{n=0}^{\infty} x_n(t) = 1 + \frac{t^3}{6} + L^{-1}\left[\sum_{n=0}^{\infty} x_n(t)\right]. \tag{13}$$

The recursive formula for the Adomian decomposition method is

$$x_0(t) = 1 + \frac{t^3}{6}, \tag{14a}$$

$$x_{n+1}(t) = L^{-1}x_n, \quad n \geq 0. \tag{14b}$$

Computing the first three components of $\{x_n(t)\}$, we obtain:

$$x_0(t) = 1 + \frac{t^3}{6}, \tag{15a}$$

$$x_1(t) = \frac{t^2}{2} + \frac{t^5}{120}, \tag{15b}$$

$$x_2(t) = \frac{t^4}{24} + \frac{t^7}{5040}. \tag{15c}$$

Note that the N-term approximation of $x(t)$ is given by equation (9).

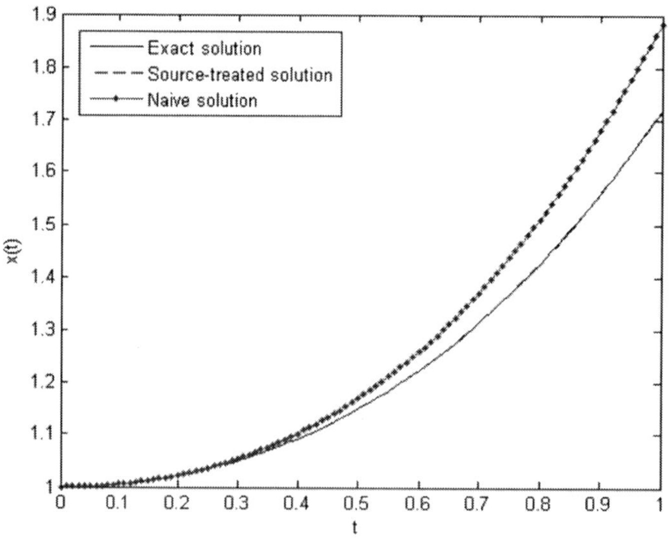

Fig. 1. Plots of the exact solution to the first example, the source-treated solution, and the naive solution. Here the source-treated and naive solutions are computed up to x_2. The exact and the source-treated solutions are almost overlapping, so we cannot see the difference between.

Figure 1 shows the curves of the exact solution to the first example, the source-treated solution, and the naive solution. We have computed the source-treated and naive solutions up to x_2. We observe that the source-treated solution, that is the solution obtained from the proposed computational treatment of the source term, is very accurate. In contrast, the naive solution is not accurate.

4.2 Nonlinear source term

As the second example, we consider the initial value problem [7] having a nonlinear source term

$$x''(t) = -\omega^2 x + (\omega^2 - 1)\sin t, \quad x(0) = 1, \quad x'(0) = 1 + \omega. \tag{16}$$

where $t \in [0, 2\pi]$ and $\omega >> 1$. The exact solution to this problem is

$$x(t) = \cos\omega t + \sin\omega t + \sin t. \tag{17}$$

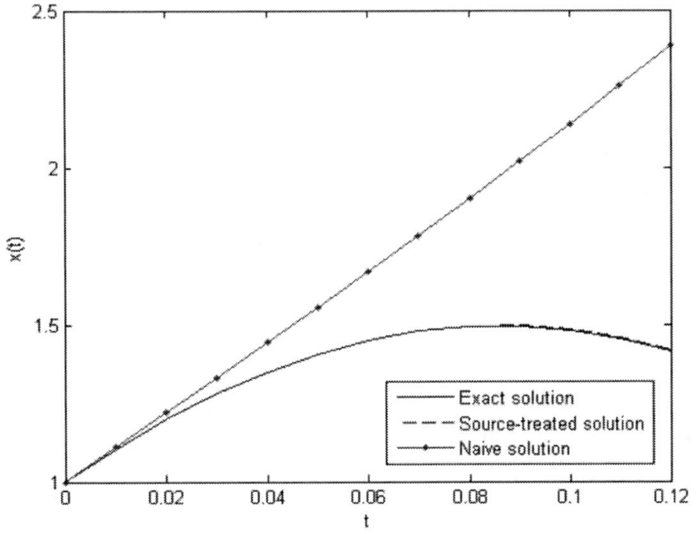

Fig. 2. Plots of the exact solution to the second example, the source-treated solution, and the naive solution. Here the source-treated and naive solutions are computed up to x_2. The exact and the source-treated solutions are almost overlapping, so we cannot see the difference.

The equation $x''(t) = -\omega^2 x + (\omega^2 - 1)\sin t$ in problem (16) can be written in an operator form as

$$Lx = -\omega^2 x + (\omega^2 - 1)\sin t, \tag{18}$$

for $0 < t < 2\pi$. Using the initial conditions $x(0) = 1$ and $x'(0) = 1 + \omega$, we find

$$x(t) = 1 + (1+\omega)t + L^{-1}[(\omega^2 - 1)\sin t] - L^{-1}[\omega^2 x], \tag{19}$$

where, again, L^{-1} is given by equation (3). Using the Adomian decomposition, equation (19) becomes

$$\sum_{n=0}^{\infty} x_n(t) = 1 + (1+\omega)t + L^{-1}[(\omega^2 - 1)\sin t] - L^{-1}\left[\sum_{n=0}^{\infty} \omega^2 x_n(t)\right]. \tag{20}$$

The recursive formula for the Adomian decomposition method is

$$x_0(t) = 1 + (1+\omega)t + L^{-1}[(\omega^2 - 1)\sin t], \tag{21a}$$

$$x_{n+1}(t) = -L^{-1}[\omega^2 x_n], \quad n \ge 0. \tag{21b}$$

The first three components of $\{ x_n(t) \}$ are as follows, for $\omega = 10$:

$$x_0(t) = 1 + 11t + 99\left(\frac{t^3}{6} - \frac{t^5}{120} + \frac{t^7}{5040}\right), \tag{22a}$$

$$x_1(t) = -100\left(\frac{1}{2}t^2 + \frac{11}{6}t^3 + \frac{33}{40}t^5 - \frac{11}{560}t^7 + \frac{11}{40320}t^9\right), \tag{22b}$$

$$x_2(t) = 10000\left(\frac{1}{24}t^4 + \frac{11}{120}t^5 + \frac{33}{560}t^7 - \frac{11}{40320}t^9 + \frac{1}{403200}t^{11}\right), \tag{22c}$$

The 3-term approximation of $x(t)$ is

$$\phi_3(t) = x_0(t) + x_1(t) + x_2(t). \tag{23}$$

Figure 2 shows the curves of the exact solution to the second example, the source-treated solution, and the naive solution. Again, we have computed the source-treated and naive solutions up to x_2. This second example confirms, once again, the effectiveness of the proposed treatment of the source term in the model, when we implement the Adomian decomposition method.

5 Conclusion

In this paper, we have presented how the Adomian decomposition method is applied when it is used to solve second order-ordinary differential equations involving source terms. Our computational treatment of the source term is indeed effective to obtain accurate solutions to the problems. The results are limited to scalar problems. It is projected that the computational treatment will work for problems with higher degrees of freedom.

This research was financially supported by Sanata Dharma University. Both authors thank Sanata Dharma University for the financial support.

References

1. M. Nad, Structural dynamic modification of vibrating systems, *Appl. Comput. Mech.,* **1,** 203-214 (2007)
2. H. Ouyang, J. Zhang, Passive modifications for partial assignment of natural frequencies of mass-spring systems, *Mech. Syst. Signal Process.,* **50,** 214-226 (2015)
3. B. Supriyadi, *Recalage des modeles de batiment* (These de Doctorat, Ecole Centrale de Lyon, France, 1993)
4. G. Adomian, Solutions of nonlinear P.D.E., *Appl. Math. Lett.,* **11,** 121-123 (1998)
5. S. Mungkasi, M.F.S. Dheno, Adomian decomposition method used to solve the gravity wave equations, *AIP Conf. Proc.* **1788,** 030103 (2017)
6. A.M. Wazwaz, *Partial Differential Equations and Solitary Waves Theory* (Springer, Berlin, 2009)
7. K. Al-Khaled, M.N. Anwar, Numerical comparison of methods for solving second-order ordinary initial value problems, *Appl. Math. Model.,* **31,** 292-301 (2007)
8. J. Biazar, E. Babolian, R. Islam, Solution of the system of ordinary differential equations by Adomian decomposition method, *Appl. Math. Comput.,* **147,** 713-719 (2004)

Effect of post curing method on flexural strength of composite friction brake

Herru Santoso Budiono[1], *Eko Surojo*[2], *Nurul Muhayat*[2], *and Wijang Wisnu Raharjo*[2]

[1]Graduate school of Mechanical Engineering, Universitas Sebelas Maret, Surakarta, Indonesia
[2]Department of Mechanical Engineering, Faculty of Engineering, Universitas Sebelas Maret, Surakarta, Indonesia

Abstract: Friction brake material is important component in the braking system of the vehicles. These materials are usually made of asbestos and non-asbestos materials. Some non-asbestos materials use natural fibers as reinforcement and phenolic resin as a binder. The post-curing process is carried out by heating a material with specific temperature and time. During post curing, ammonia gas as the reaction of the phenolic resin and HMTA diffuses out from the friction brake material. The heating rate in post curing was supposed affecting process of removal ammonia gas from composite and it will affect mechanical properties of the composite. Therefore, this paper is concerned with effect of post curing method on the composite flexural strength. Composite specimen were manufactured by mixing the ingredients, preforming, hot molding and followed by post curing. Bending tests performed to determine the effect of post curing the flexural strength. The results showed that heating rate in post curing affected flexural strength of composite. Flexural strength of composite decreased with increasing heating rate. Internal defect that were formed during post curing decreased composite flexural strength.

1 Introduction

The friction brake is important components in the braking system for all kinds of vehicles [1]. Friction brake on the braking system is used to control the speed by converting kinetic energy into frictional energy [2]. The friction brake should have the coefficient of friction as required, high wear resistance, and high stability in coefficient of friction [3]. To fulfill of these characteristics can not be obtained using one type of material and is only possible by combining multiple materials [4]. Composite materials for friction brake is composed by many materials. The ingredients that used in the composite brake affect its characteristics. Therefore, the selection of materials in the manufacture of composite friction brake needs to be done [5]. Phenolic resin is usually used as a binder in the manufacture of the friction brake.

Phenolic resin is fire resistant resin that is superior compared to other types of fire-resistant resin [6]. The material is often used in the industry because of its resistance at high temperatures and its good performance [7]. Those characteristics were produce by cross-linking network structure in the form of three-dimensional arising from the post-curing process of the resin novolac with hexamethylenetetramine (HMTA). The reaction between the phenolic resin and HMTA also generates ammonia gas [8]. The ammonia gas can be removed with a post-curing process [9].

In addition to eliminate the ammonia gas, post curing also increases the density of cross-links in the network phenolic resin [10]. Increasing of cross-link will increase its temperature glass (Tg) [11]. By varying the time, this kind of catalyst, and the ratio of formaldehyde, some characteristics can be produced [12]. The release of gas from composite materials through the diffusion mechanism. Several previous studies have studied effects of post curing on physical and tribological properties. Post curing has an effect on storage modulus [11]. The best wear resistant and friction stability obtain at 6 hours with a temperature of 200°C [13]. Almost of all of those researches were just concerned about physical properties and tribological behaviours. Therefore, this work focused to investigate effect of post curing method on composite flexural strength. Particular attention was directed to changes in flexural strength with respect to variation of heating rate and duration

2 Materials and Methodology

In this study, the composition of the materials used was: fly ash (4.24%), CaCO3 (40.48%), NBR (2.77%), Graphite (30.68%), cashew (3.47%), cantula fiber (2.56%) and phenolic resin (15.79%). Before mixing, cantula fibers was alkali treated to remove the lignin content. Alkali treatment process was carried out by soaking the fiber using 5% NaOH solution and distilled water for 4 hours and then heated in the oven with a temperature of 70°C for 10 hours. After the alkali treated, cantula fibers cut to a length of 3 mm.

The bending test specimen was manufactured by a sequence of process as follows : dry mixing of the ingredients, cold molding, hot molding, and then followed by post curing. The temperatures used in the hot molding process was 160°C for 10 minutes at a pressure of 20 MPa. Bending test was performed by using ASTM standard D790 (three-point bending test) of the size as shown in **Fig. 1.**

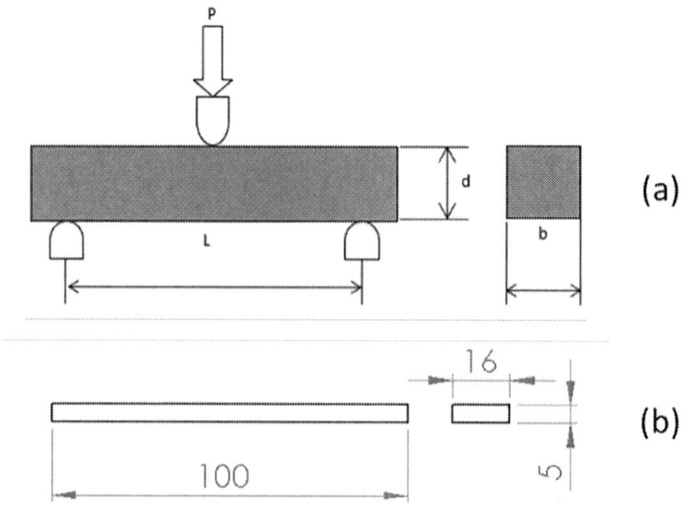

Fig. 1. (a) three-point bending test ASTM D790, (b) Size bending test specimen (mm)

Post curing was carried out by three step of heat treatment. The first step, specimens were continuous heated from room temperature to 140°C in 1 hour. The second step, specimens were heated using method as shown in **Table 1.** The third step, specimens were slowly cooled from 180°C to room temperature in the oven.

Table 1. Variations of the second step of post curing

Heating method	Variation
Continuous heating	Specimens were slowly heated from 140°C to 180°C in : 2, 4, 6, and 8 hours
Isothermal heating	Specimens were rapidly heated from 140°C to 180°C in 5 minutes and then held at 180°C for : 2, 4, 6, and 8 hours

3 Result and discussion

The result of bending test is shown in **Fig. 2.** The specimen of bending test by the continuous heating method with 2 hours of heating rate had an average flexural strength of 26 MPa. The flexural strength of the specimen increased when the heating time was increased from 2 hours to 4 hours and to 6 hours. Meanwhile, the flexural strength did not increased when the heating time was increased from 6 hours to 8 hours. In the continuous heating method, it was found that there was a correlation between the heating duration and the heating rate. The heating rate from 140°C to 180°C decreased as the heating time increased. The specimen with 2 hours and 4 hours of heating time underwent the post curing in heating rate of (180-140)/2 °C/hour and (180-140)/4 °C/hour, respectively.

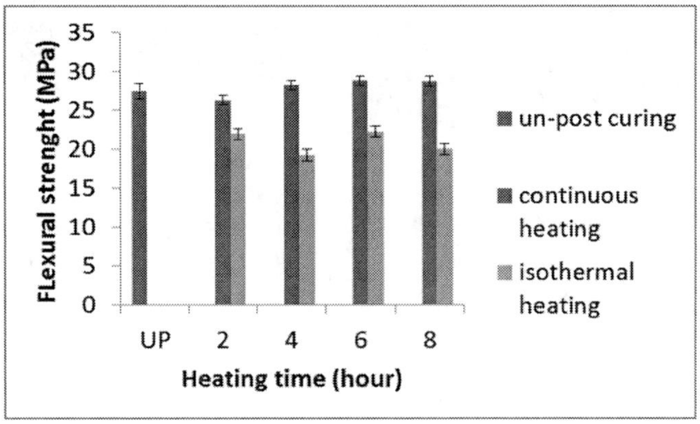

Fig. 2. Effect post curing method on friction brake material (UP = un-post curing)

For each similar heating time, the applied isothermal heating method resulted in a lower flexural strength compared to that of continuous heating. Eventhough the heating time variation was similar, the heating rate in isothermal heating was faster than that found in continuous heating. All specimens that underwent isothermal heating method were heated from 140°C to 180°C with same heating rate i.e (180-140)/(5/60) °C/hour. Therefore, **Fig. 2** indicated that heating rate influence the flexural strength of the composite. Flexural strength tended to decrease as the heating rate increased. Optimal flexural strength was reached in the specimen that underwent continuous heating treatment in 6 hours. Meanwhile, the specimen that underwent continuous heating treatment for 2 hours had lower flexural strength compared to that which was unpost-cured.

The influence of heating rate on the composite flexural strength was related to the formation of internal defects during the post curing process. **Fig. 3** shows the existence of internal defect that supports the hypothesis. The specimen which underwent treatment using continuous heating method in 2 hours of heating time showed internal defect (**Fig. 3a**), while that in 6 hours of heating time showed no internal defect (**Fig. 3b**). The specimen which underwent treatment using isothermal heating method with 2 hours of heating time had a greater internal defect than that which underwent continuous heating with 2 hours of heating time (**Fig. 3c**). There was a possible reason why that phenomenon occurred. The internal defect occurred as the result of ammonia gas which was formed and trapped in the specimen during post curing. It revealed that the heating rate affected the release process of ammonia gas from composite. Increasing of heating rate tended to induce formation of internal defect in composite. The un-post cured specimen did not have any internal defects (**Fig. 3d**) as found in the continuous heating specimen and in the isothermal heating specimen.

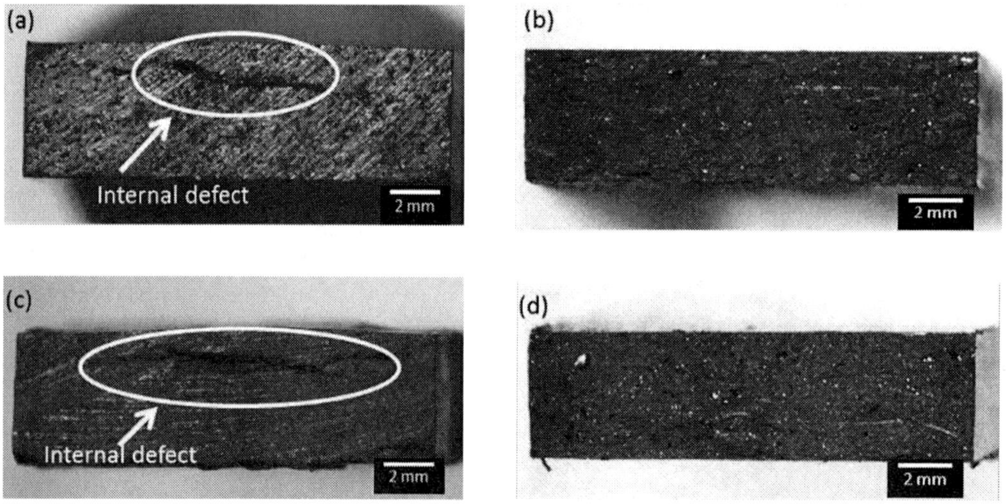

Fig. 3. Fracture surface of specimen that underwent: (a) continuous heating method in 2 hours, (b) continuous heating method in 6 hours, (c) isothermal heating method in 2 hours, (d) un-post curing.

Fig. 4 shows the surface of specimen fracture. The specimen which underwent treatment using the continuous heating method with 2 hours of heating time failed due to pull-out and fiber fracture as well as shown in (**Fig. 4a**). However, in majority it was due to fiber fracture. The same phenomenon also occurred in the specimen which underwent treatment using the isothermal heating method with 2 hours of heating time (**Fig. 4b**). Those phenomena showed the good bonding between fiber and matrix. Fiber pull-out was clearly visible in the un-post curing specimen (**Fig. 4c**). It occurred due to the low bonding strength between the fiber and the matrix. Eventhough the un-post curing specimen had the flexural strength that was nearly as strong as the post-cured with continuous heating one, but the composite was unstable. The un-post curing composite would change in properties when it was exposed to heat.

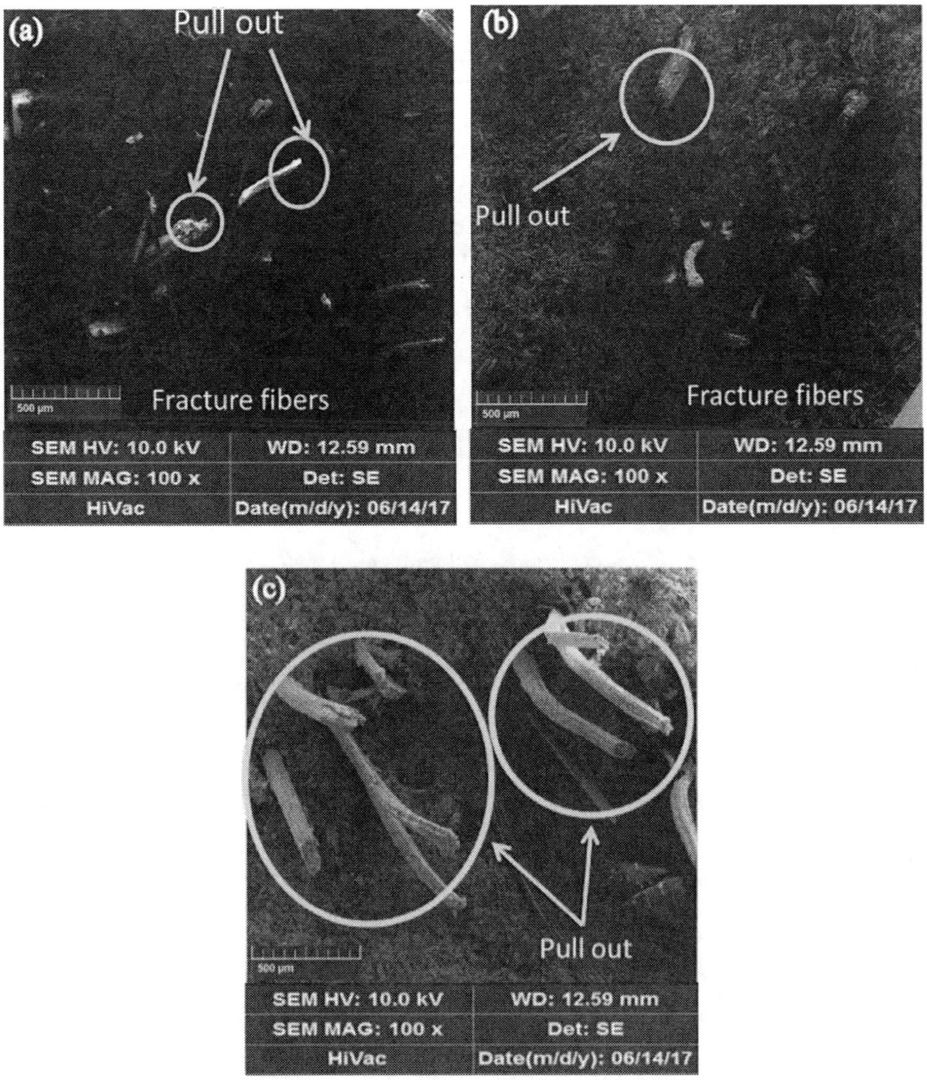

Fig. 4. SEM images of fractures surface of specimen that underwent : (a) continuous heating in 2 hours (b) isothermal heating specimen for 2 hours, (c) un-post curing

4 Conclusion

The post curing tended to lower the flexural strength of the specimen if the heating rate was set too fast. The decrease in the flexural strength was caused by the internal defect in the specimen. Formation of internal defect was influenced by the heating rate in post curing. The increase of heating rate induced the formation of larger internal defect. Internal defect was triggered by the release process of ammonia gas during the post curing in the specimen. The flexural strength in the post-cured specimen using the continuous heating method was higher than that using the isothermal heating method.

The authors gratefully acknowledge the financial support of Ministry of Research, Technology and Higher Education of the Republic of Indonesia through Tim Pascasarjana research grant (contract number: No. 873/UN27.21/PP/2017).

References

1. U. D. Idris, V. S. Aigbodion, I. J. Abubakar, C. I. Nwoye, J. King Saud Univ. - Eng. Sci., **vol. 27**, no. 2, pp. 185–192, (2015).
2. K. K. Ikpambese, D. T. Gundu, L. T. Tuleun, J. King Saud Univ. - Eng. Sci., **vol. 28**, no. 1, pp. 110–118, (2016).
3. J. H. Gweon, B. S. Joo, H. Jang, "*Wear,* **vol. 362–363**, pp. 61–67, (2016).
4. C. Menapace, M. Leonardi, G. Perricone, M. Bortolotti, G. Straffelini, S. Gialanella, Mater. Des., **vol. 115**, pp. 287–298, (2017).
5. E. Surojo, Jamasri, V. Malau, M. N. Ilman, **vol. 37**, no. March 2016, pp. 473–481, (2015).
6. K. Hirano, M. Asami, *React. Funct. Polym.*, **vol. 73**, no. 2, pp. 256–269, (2013).
7. H. Ku, F. Cardona, N. Pattarachaiyakoop, M. Trada, J. *Compos. Mater.*, **vol. 43**, no. 7, pp. 741–754, (2009).
8. A. Izumi, Y. Shudo, T. Nakao, M. Shibayama, *Polym.* (United Kingdom), **vol. 103**, pp. 152–162, (2016).
9. B. Theodore, N. M. Ph, *plast. Eng.* pp. 1–26, (2004).
10. R. M. Rudd, S. R. Ghafarian, A. Taherkhani, **vol. 3**, pp. 36–39, (2013).
11. W. Stark, H. Goering, U. Michel, H. Bayerl, *Polym. Test.*, **vol. 28**, no. 6, pp. 561–566, (2009).
12. J. C. Munoz, H. Ku, F. Cardona, D. Rogers, J. *Mater. Process. Technol.*, **vol. 202**, no. 1–3, pp. 486–492, (2008).
13. S. J. Kim, K. S. Kim, H. Jang, *J. Mater. Process. Technol.*, **vol. 136**, no. 1–3, pp. 202–208, (2003).

The influence of welding speed conditions of GMAW on mechanical properties of 316L austenitic stainless steel

I.M.W. Ekaputra[1,*], *Sudi* Mungkasi[1], *Gunawan* Dwi Haryadi[2], *Rando* Tungga Dewa[3], and *Seon-Jin* Kim[3]

[1]Sanata Dharma University, Mrican, Tromol Pos 29, Yogyakarta 55002, Indonesia
[2]Diponegoro University, Jl. Prof. Sudharto, SH., Tembalang-Semarang 50275, Indonesia
[3]Pukyong National University, 365, Shinsuro, Nam-gu, Busan 608-739, Korea

Abstract. In this study, the influence of welding speed conditions of gas metal arc welding (GMAW) on mechanical properties of 316L austenitic stainless steel (SS) was investigated. The welding speed was applied with three different variations at 175 mm/minute, 190 mm/minute, and 205 mm/minute. The GMAW was equipped by solid electrode wire ER308L with 0.8 mm in diameter. The Vickers micro-hardness and tensile tests were conducted for each GMAW joint, and the results were discussed. It was found that the Vickers micro-hardness showed a dependence on welding speed conditions in weld metal (WM) and heat affected zone (HAZ) areas. A systematic increase in hardness was shown with an increase in welding speed. The ultimate tensile strength and yield strength of GMAW joints were found to decrease systematically with an increase in welding speed. From this investigation, it was found that GMAW joint with the welding speed at 175 mm/minute was suitable to be applied for 316L SS.

1 Introduction

The 316L austenitic stainless steel (SS) is widely used in a lot of applications due to its good mechanical properties at high temperatures, superior corrosion resistance, weldability, and easy of fabrication [1-2]. The major chemical compositions such as *Chromium* (Cr), *Molybdenum* (Mo), and *Nickel* (Ni) make the 316L SS suitable to be applied at elevated temperatures and corrosive environment. The "L" sign following the 316L indicates the low carbon content (<0.03 wt. %) which may responsible to avoid the potential stress corrosion cracking in heat affected zone (HAZ) area [3]. The 316L SS can be found as the components in nuclear reactor, pumps, valves, marine fittings, fasteners, paper and pulp machinery, petro chemical, etc.

Recently, most of the mechanical structures are composed of welded components. The use of welded components cannot be avoided since the construction of designs is becoming more complicated time by time.

* Corresponding author: made@usd.ac.id

In the welded area, the microstructures will have different characteristics with the base area. It has been reported that the weakest links were normally found in welded area and might have an original defect [4]. Hence, the welding work must be one of the important considerations in engineering design since the defects will cause stress concentration.

The general welding problems found for the 316L SS are the risk of hot cracking appearance and the appearance of carbide chrome and oxide chrome during the welding process [4]. A smooth precipitation of carbide chrome will be formed among the grain boundaries in the fusion area, while the oxide chrome may appear on the metal surface. Both of the carbide and oxide chrome will reduce the chrome contents cause of reducing in corrosion resistance. In this study, a gas metal arc welding (GMAW) also called as metal inert gas (MIG) is applied to the 316L SS since the process is versatile, rapid and economical, and good in quality. The GMAW is one of the arc welding processes where a continuous electric arc is generated from the filler (wire electrode) supplied by a roll of wire. The GMAW is able to perform the welding process without concerning about the use of electrode for the limited length [6]. The GMAW can be performed automatically by means of welding robot or semi automatically in various welding speed. Furthermore, the contamination between the oxygen and nitrogen is prevented by Argon gas or active gas, CO_2 as plasma during the welding process [4-6].

It has been reported that the variations of welding parameters in GMAW such as welding speed, torch angle, free wire length, arc welding current, arc voltage, nozzle distance, welding position and direction, and the flow rate gas will have a significant effect to the result of welded material performance [6-9]. A further investigation is still conducted until now to find out the most suitable combination of input process parameters for a desired output in GMAW. Izzatul et al. investigated the effect of variations of input parameters such as arc voltage, welding current and welding speed to the welding penetration, microstructural and hardness measurement in mild steel [6]. Biswajit et al. investigated the characteristics of depth penetration of welded joint due to variations of welding current, arc voltage, and welding speed.

In the present study, the most suitable welding speed of GMAW to the 316L SS was investigated. The welding speed is chosen with three different variations at 175 mm/minute, 190 mm/minute, and 205 mm/minute. The Vickers micro-hardness and uniaxial tensile tests are performed and the results are discussed.

2 Experimental procedures

2.1 Specimen preparation

The complete chemical compositions of the base metal austenitic 316L SS and electrode wire of AWS 5.9/ ASME SFA 5.9 ER 308L are given in tables 1 and 2, respectively. Two plates of base metal were stuck together at the two ends along the width to make a butt joint. The welded plate had a dimension of 300 mm × 125 mm × 3 mm with a V-shaped groove of 60° as shown in figure 1. Prior manufacturing the testing specimens, a non-destructive testing of dye penetrant was applied to the welded plate to identify the possible defects on the specimen surface. The specimens were cut from the welded plate by a power hacksaw and formed into a rectangular shape with the dimensions of 120 mm × 57 mm × 3 mm for tension specimen and 50 mm × 5 mm × 3 mm for the Vickers micro-hardness specimens. A fixed arm of a portable gas cutting machine was applied to the welding torch to control the welding speed.

2.2 Experimental apparatus

MIG/MAG 350 G-KR WELDING MACHINE, manufactured by MULTIPRO is applied to produce the welding specimens. The wire electrode is inserted automatically into the welding gun by a roller drive system. An austenitic stainless steel with the maximum carbon content of 0.02% is used as the electrode wire. An ultra high purity (UHP) of 99.99 % *Argon* (Ar) is used as shielding gas and supplied in regulated manner at a constant flow rate and at constant pressure. The GMAW conditions are given as follows:
- Polarity: DCEP
- Current: 135 A
- Voltage: 20.1 V
- Welding speed: 175 mm/min, 190 mm/min, 205 mm/min
- Pre-heating temperature: minimum of 20 °C

The hardness test was performed using a Vickers micro-hardness machine (model: Mitutoyo HM 122) with 100 gr of load and 15 second of penetration time. The hardness was determined every 500 µm along the base metal (BM), heat affected zone (HAZ), and weld metal (WM) as shown in figure 2. The tension tests were performed by universal testing machine and following the ASTM E8/E8M-13A [10]. A screw-driven load frame at constant crosshead velocity of 100 mm/min was actuated by AC servomotor type.

Table 1. The chemical composition of AISI 316L (wt. %).

Fe	C	Si	Mn	P	S	Cr	Ni	Mo	Fe
bal.	0.022	0.470	1.899	0.046	0.008	17.501	10.018	2.096	67.290

Table 2. The chemical composition of wire ER308L (wt. %).

Fe	C	Si	Mn	P	S	Cr	Ni	Mo	Fe
bal.	0.050	0.400	1.632	0.006	0.006	19.294	9.717	0.481	68.039

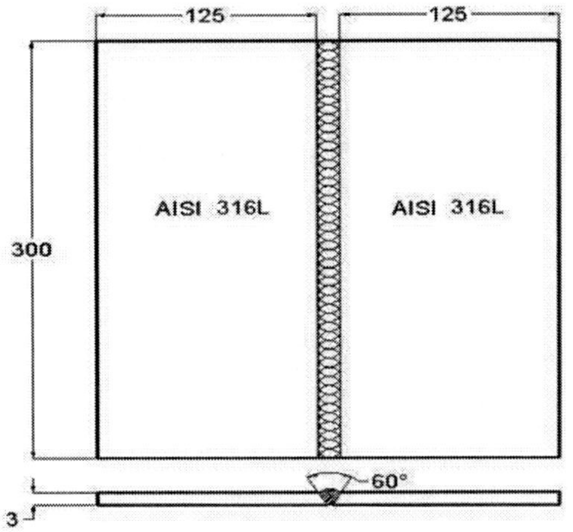

Fig. 1. The welded plate of 316L SS.

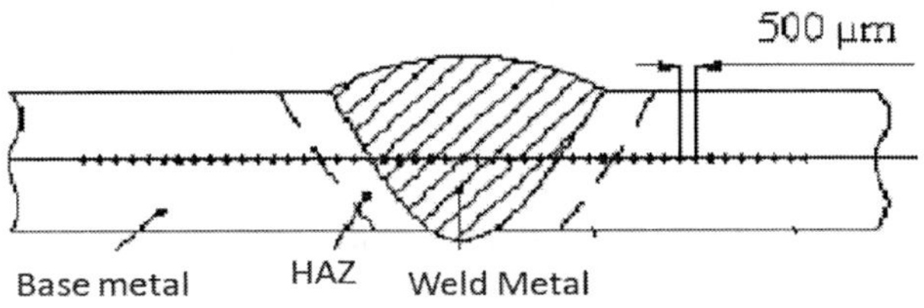

Fig. 2. The Vickers micro-hardness measurement.

3 Results and discussion

3.1 Hardness properties

Figure 3 shows the hardness results of Vickers micro-hardness measurement for GMAW joints of 316L SS under different welding speed conditions. The BM, WM, and HAZ areas are divided by the vertical lines along the sample distance. The BM has the lowest hardness which is approximately from 162 to 168. The hardness starts to increase sharply in HAZ and continues to decrease slightly in the WM. A significant difference in hardness is observed in the HAZ and WM areas among the three different welding speeds. The hardness increases with an increase in welding speed. At the lower welding speed, the grains and grain boundaries restoration may be formed. On the other hand, a precipitation amount will increase following with an increase in welding speed. The precipitation appears from the dissolution or growth of strengthening precipitates during the welding thermal cycle [11]. Chaudari et al reported that an increase in hardness in HAZ and WM is due to over precipitation and increased carbon precipitation amount [12]. As shown in figure 3, it can be seen that the HAZ and WM hardness are nearly close to the BM at 175 mm/min, while the hardness increases significantly at the welding speed above 175 mm/min.

3.2 Tensile properties

The tensile property measurements between BM and GMAW joints are shown in figure 4. Figures 4(a)-4(b) show the ultimate tensile strength (UTS) and yield strength (YS) of BM and GMAW joints for three different welding speeds. It is shown that the UTS and YS in GMAW joints decrease with an increase in welding speed. The YS and UTS of BM are 614 MPa and 414 MPa, respectively. However, the YS and UTS of GMAW joints at 175 mm/min are 498 MPa and 264 MPa, respectively where the strength reduces about 20%. A significant decrease in UTS and YS are observed for the GMAW joints with the welding speed above 175 mm/min. The YS and UTS of GMAW joints at 190 mm/min are 302 MPa and 189 MPa, respectively which are about 50% reduction in strength. With a further increase in welding speed at 205 mm/min, the UTS and YS are about 283 MPa and 196 MPa, respectively which are about 55% reduction in strength.

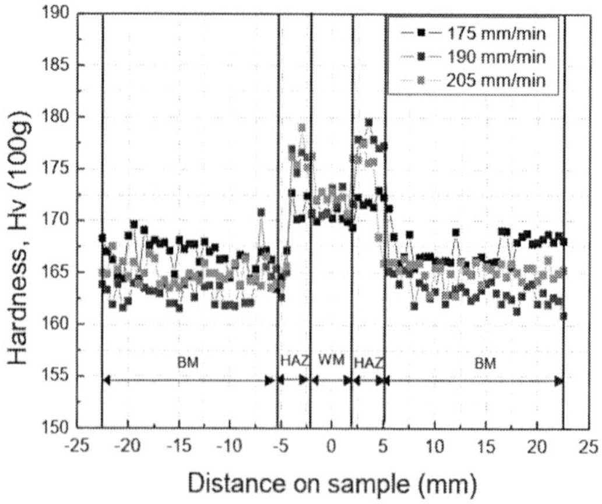

Fig. 3. Vickers micro-hardness result for different welding speed conditions.

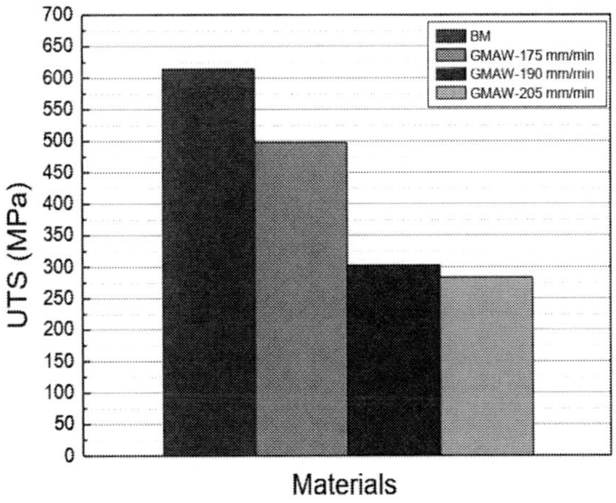

(a)

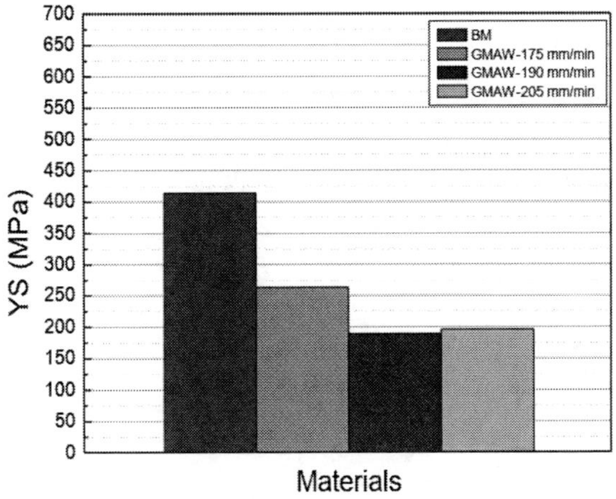

(b)

Fig. 4. Tension test result for different welding speed conditions; (a) ultimate tensile strength, (b) yield strength.

4 Conclusions

In this paper, the influence of welding speed parameter in GMAW for 316L SS has been investigated. The recommended welding speed was proposed based on the following results.

The Vickers micro-hardness results showed that the hardness did not change significantly at the welding speed of 175 mm/min due to the occurrence of restoration process during the welding process. The tensile test results also showed that the UTS and YS did not decrease significantly at 175 mm/min comparing with 190 mm/min and 205 mm/min. According to these results, we therefore have decided to propose the welding speed at 175 mm/min as an optimal value in GMAW operation to be applied to 316L SS. Further investigation is still required to determine the specific parameters of GMAW by varying the parameters such as torch angle, free wire length, arc welding current, arc voltage, nozzle distance, welding position and direction, and the flow rate gas.

References

1. R. K. Desu, H. N. Krishnamurthy, A. Balu, A. K. Gupta, S. K. Singh, J. Mater. Res. Technol. **5(1)**, 13-29 (2016).
2. I. M. W. Ekaputra, W. G. Kim, J. Y. Park, S. J. Kim, E. S. Kim, J. Mech. Sci. Technol. **30 (7)**, 3151-3158 (2016).
3. M. D. Mathew, K. Laha, V. Ganesan, Mater. Sci. Eng A. **535**, 76-83 (2012).
4. A. Nassour, W. W. Bose, D. Spinelli, J. Mater. Eng. Perform. **10 (6)**, 693-698 (2001).
5. S. Kalpakjian, S. R. Schmid, *Pearson pretince hall* (2009).
6. I. A. Ibrahim, S. A. Mohamat, A. Amir, A. Ghalib, Eng. Procedia. **41**, 1502-1506 (2012).
7. B. Das, B. Debbarma, R. N. Rai, S. C. Saha, Int. J. Res. Eng. Technol. **02 (10)**, 220-224 (2013).
8. E. Karadeniz, U. Ozsarac, C. Yildiz, Mater. Des. **28**, 649-656 (2007).
9. E. Zumelzu, J. Sepúlveda, M. Ibarra, J. Mater. Process. Tech. **94**, 36-40 (1999).
10. ASTM E8/E8M-13a, ASM Inter. (2013).
11. A. K. Lakshminarayanan, V. Balasubramanian, K. Elangovan, Int. J. Adv. Manuf. Technol. **40**, 286-296 (2009).
12. P. D. Chaudhari, N. N. More, Int. J. Eng. Res. Technol. **3**, 1937-1942 (2014).

Measurement of injury rate on fish skin and performance comparison based on L*A*B* and HSV color spaces

Minh Thien Tran[1], *Jotje* Rantung[1], *Trong Hai* Nguyen[1], *Hak Kyeong* Kim[1] and *Sang Bong* Kim[1,*]

[1]Department of Mechanical Design Engineering, Pukyong National University, Busan, 48547, Republic of Korea

Abstract. This paper analyses and compares the performance of L*A*B* and HSV color spaces and applies them to calculate the injury rate on fish. To do these issues, the following steps are done. An original image is transformed into L*A*B and HSV color spaces. A channel "a" is separated from L*A*B* color space. In channel a, a formula to adjust "channel a value" is proposed to realize the shapes of injury on fish clearly and a new channel a is obtained by adjusting the channel a. The new channel "a" is converted into injury binary image by manual threshold. Otsu's method is applied converted the original channel a image of fish shape into binary image. Finally, by calculating the number of pixels of areas of shape and total injury of fish, the injury rate is calculated. The steps of image processing of HSV color space is similar to L*A*B* color space. The proposed process are tested on fish.

1 Introduction

"Machine vision", as a crucial part of sorting systems, enables automatic and non-destructive selecting of products that satisfy certain requirements. There are many different methodologies for image processing in recent years. Therefore, choosing a right color space in processing algorithms is crucial. Eyarkai et al. described that mangoes were measured during ripening in 24 hours and evaluated using L*A*B* and RGB color coordinates [1]. Using exactly color space is important for controlling of food quality. Ivana et al. measured color of food products by using L*A*B* and RGB color space[2]. In this paper, fruit quality based on L*A*B* color space is more accurate than RGB color space. On the other hand, Hitesh et al. used image processing technique to detect fish disease [3]. This issue is done by using image processing algorithms based on L*A*B* and HSV color spaces. A method to measure injury rate of fish surface based on K-means clustering image segmentation was represented by Sheng el at.[4]. This method calculated injury rate based on S channel of HSV color space. After that, Otsu's threshold was applied to converted S channel into binary image, and calculate injury rate by counting area pixels of injuries and total shape on fish. The results

* Corresponding author: kimsb@pknu.ac.kr

showed that the injury rate of fish could be measured using image segmentation algorithm. Generally, L*A*B* and HSV color spaces are the two frequently chosen color spaces for pre-processing step in image segmentation.

It is very important to determine which HSV color space or L*A*B* is suitable for calculating injury rate on fish. This paper compares and chooses a right color space for measurment of injury rate on fish and the steps of image processing algorithm is proposed to measure injury rate on fish in L*A*B*color space. Moreover, a value of channel "a" to realize injury on fish clearly is adjusted by the proposed algorithm. The experimental results are verified that the proposed image processing algorithm is more effective in L*A*B* color space than in HSV color space.

2 Proposed process based on L*A*B* color space

L*A*B* is defined by CIE and specified by the International Commission on Illumination [5].

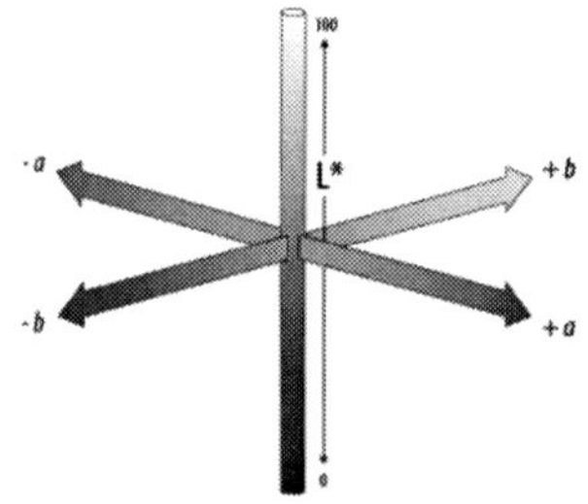

Fig. 1. L*A*B* color space [5]

In the Fig. 1, the central vertical axis is lightness (L*) which have range of value from 0 to 100 (black to white). The coordinate axes follows channel "a". If value of channel" a" is positive value, color of channel "a" is red. Color of channel "a" is green if value is negative. Channel "b" is similar to channel "a". Positive value is yellow and negative value is blue. There are three channels: Luminance (L) and other two color channels are "a" and "b" known as chromaticity layers.

In this section, an image processing algoriyhm based on L*A*B* and HSV color spaces to obtain the injured rate of fish is proposed as the flowchart in Fig. 2. An original color image is transformed into L*A*B* and HSV color spaces sequentially. After separating channel "a" from L*A*B* color space, there are two tasks done in this image:

Firstly, channel "a" is converted into binary image by Otsu's method. After that, area's value of the shape on the binary image is calculated by counting pixels. This step is shown in the flowchart of Fig. 2.

Secondly, a method to adjust value of channel "a" is proposed to realize injury on fish clearly after original channel "a" image applies Gaussian filter to reduce random "noises" – an image applying the adjusted value is called as a new channel "a" image.

The formula for adjusting value of channel "a" is as follows:

$$new\ value(x,y) = h\frac{a(x,y)}{\max(a)} \tag{1}$$

where h is a coefficient which depends on conditional environment, and x *and* y is coordinate of pixels.

The new channel "a" image adjusted by Eq. (1) applies Gaussian filter to filter the rest of noises with Kernel coefficient chosen by manual. For next step, this image is converted into binary image by choosing a manual threshold method. This threshold operation can be expressed as [6]:

$$Binary\ value(x,y) = \begin{cases} 1 & if\ src(x,y) \geq thresh \\ 0 & otherwise \end{cases} \quad (2)$$

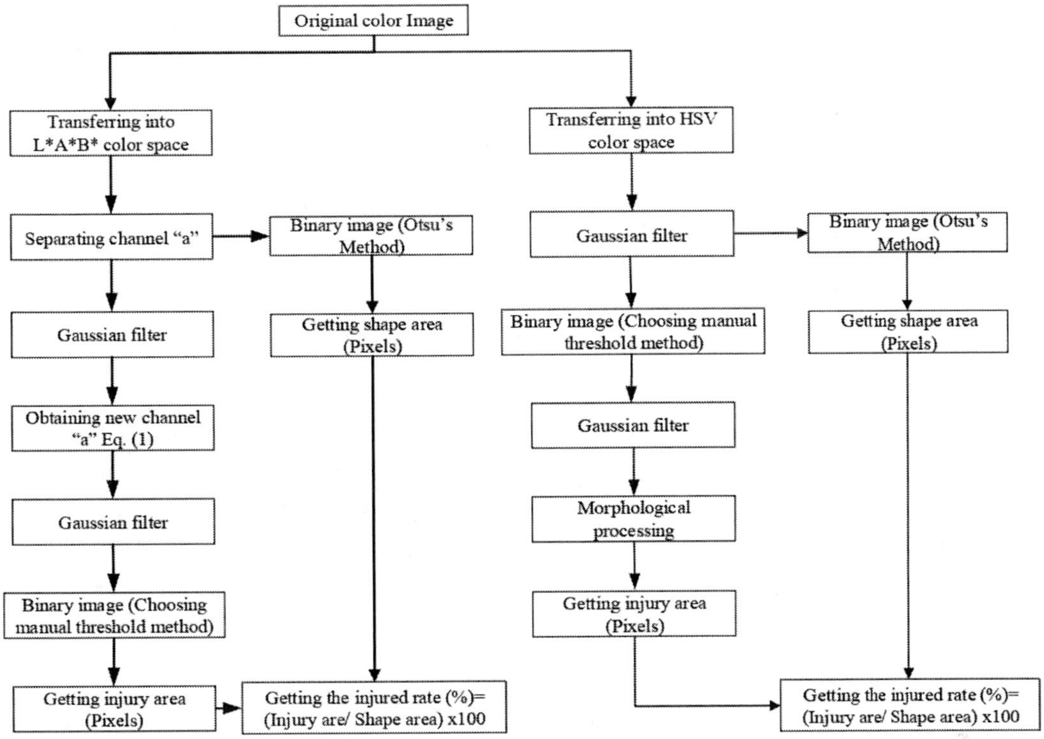

Fig. 2. Flowchart of the proposed image processing algorithm based on L*A*B* color space and HSV color space

Gaussian filter is applied to filter the random of noises in this image converted into L*A*B* color space and HSV color space. That is, the following Gaussian filter is applied to the images with Kernel coefficient chosen by manual [6].

$$G_0(x,y) = Ae^{\frac{-(x-\mu_x)^2}{2\sigma_x^2}+\frac{-(y-\mu_y)^2}{2\sigma_y^2}} \quad (3)$$

where μ_x and μ_y is the mean, and σ_x and σ_y represent variances with respect to the variables x and y, respectively.

Finally, by counting the area pixels of injury and shape of fish, an injury rate of fish is calculated as follows [4]:

$$Injury\ rate = \frac{injury\ area\ (pixels)}{shape\ area\ (pixels)} \times 100\% \quad (4)$$

3 Experiment results and conclusions

An original color image of fish in Fig. 3 is converted into L*A*B* and HSV color spaces in Fig. 4. The original channel "a" image of fish is separated from L*A*B* color space as shown in Fig. 5a. Fig. 5b shows the new channel a image adjusting value of the original channel "a" of fish using the proposed Eq. (1). The original channel "a" image of the shape of fish is converted into binary image from by using Otsu's Method. Likewhile, the new channel "a" image of fish injuries are converted into binary images by choosing a manual threshold method using Eq. (2). Channel "a" of L*A*B* color space is sensitive to the red color, so adjusting value of channel "a" helps clear realization of injury area on fish. All of the results are presented in Fig. 6a and Fig. 6b.

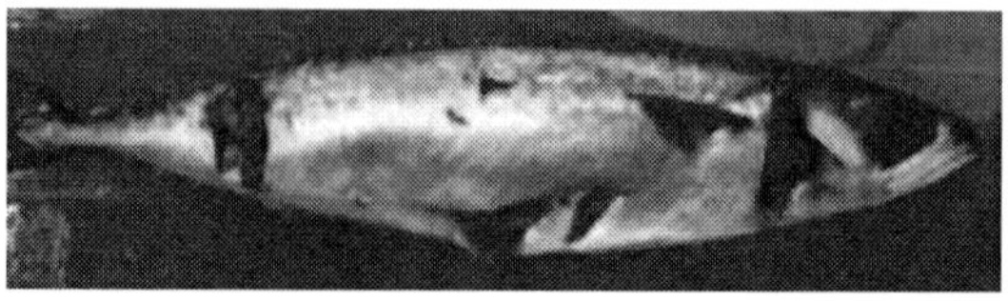

Fig. 3. Original color image.

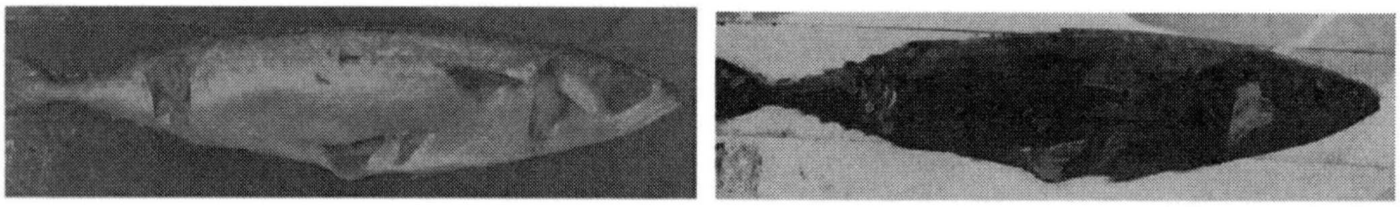

(a) (b)

Fig. 4. (a) L*A*B* color space image; (b) HSV color space image

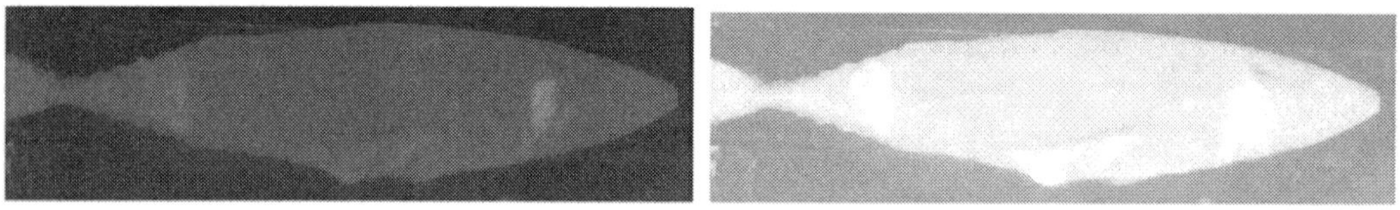

(a) (b)

Fig. 5. (a) Original Channel "a" image based on L*A*B* color space; (b) new channel "a" image based on L*A*B* color space

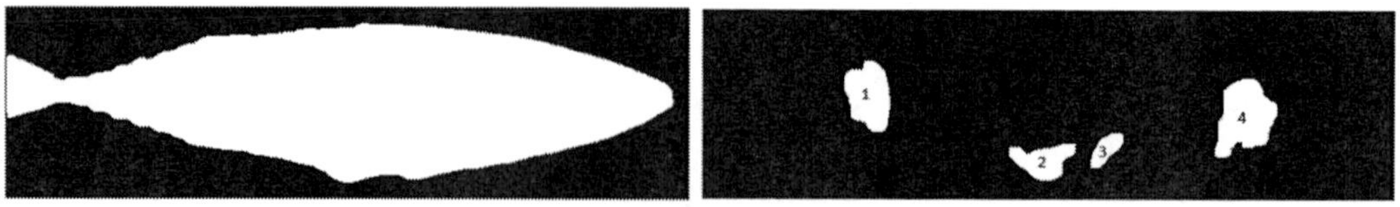

(a) (b)

Fig. 6. (a) Binary image of shape of the fish based on L*A*B* color space; (b) Binary images of injuries of the fish based on L*A*B* color space

The image processing of HSV color space is similar to that of L*A*B* color space. The results are presented in Fig. 7, Fig. 8a and Fig. 8b. The results of counting pixels on areas of binary images of shape and injuries, and their injury rates based on L*A*B* color space and HSV color space are show in Table 1.

Fig. 7. Binary image of injury rate of the fish before morphological processing based on HSV color space image

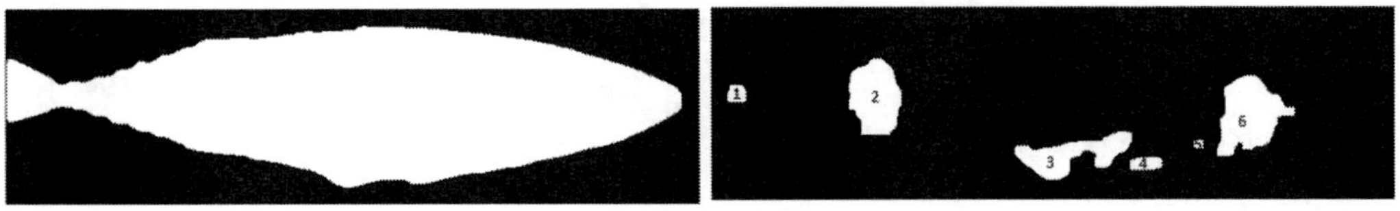

(a) (b)

Fig. 8. (a) Binary image of shape of the fish based on HSV color space image; (b) Binary image of injury rate of the fish after morphological processing based on HSV color space image

Table 1. Table of data images (Pixels).

Object	L*A*B* Color Space	HSV Color Space	Real data
Shape	55492	55580	57323
Injury Object 1	2111	231	1947
Injury Object 2	1319	2264	1209
Injury Object 3	552	1974	573
Injury Object 4	2772	351	1742
Injury Object 5	-	23	-
Injury Object 6	-	3562	-
Total Injury Object	6754	8779	5471
Injury rate	12.17%	15.79%	9.54%

Finally, the injury rate based on areas of shape and total injuries of fish is calculated by Eq. (4). The proposed algorithms are tested by practical experiment on fish. There are four injury objects on the fish in Fig. 3. The proposed process of image processing applied L*A*B* color space also has four injury object as shown in Fig. 6b whereas there are six injury objects on HSV color space as shown in Fig. 6b. It means that process of image processing is effected "noise" on HSV color space. The injury rate of fish using L*A*B* and HSV color spaces is 12.17% and 15.79% on Table 1, respectively. The injury rate using the proposed image processing algorithm based on L*A*B* color space is 12.17% and is close to the real injury rate of 9.54% than that of based on HSV color space.

4. Conclusions

In this paper, comparison in performance using L*A*B* and HSV color spaces and measurement of injury rate on fish skin were presented. A new image processing algorithm based on L*A*B* color space is proposed to measure injury rate on fish. A new formula is proposed to adjust value of channel "a" in detecting injured areas on fish clearly. By calculating the number of pixels of areas of shape and total injury, the injury rate were calculated. By using the proposed image processing algorithm based on L*A*B* color space, the injury rate is 12.17% and close to the real injury rate of 9.54%. The experimental results showed that the proposed image processing method using L*A*B* color space is better in measuring the injury rate than that of HSV color space in this application.

This research was a part of the project titled "Localization of unloading automation system related to Korean type of fish pump", funded by the Ministry of Oceans and Fisheries, Korea.

References

1. V. E. Nambi, K. Thangavel, S. Shahir, V. Geetha, JOAH, **17**, 3 (2015)
2. I. Markovic, J. Ilic, D. Markovic, V. Simonovic, N. Kosanic, JOHEAD, **4**, 8 (2013)
3. H. Chakravorty, R. Paul, IJCSS, **9**, 2 (2015)
4. D. B. Sheng, S. B. Kim, T. H. Nguyen, T. S. Gao, H. K. Kim, KSPSE, **19**, 5 (2015)
5. R. S. Hunter, JTOSA, **48**, 12 (1958)
6. *The OpenCV Tutorials*, Release 2.4.13.2

Flame stability and behavior inside meso-scale combustor with different flame holder

Lilis Yuliati[1,*]

[1]Mechanical Engineering Department, Brawijaya University, Indonesia

Abstract. Flame stability and behavior inside meso-scale combustor with different flame holder was investigated experimentally. Three types of flame holder i.e. wire mesh, flat plate with circular holes and flat plate with narrow slits, were used to improve the flame stability inside the meso-scale combustor. Combustor with flat plate - narrow slits flame holder has the best flame stability, i.e. stable flame established inside the combustor at the highest reactant velocity compared to the other combustor with the different flame holder. Furthermore, combustor with wire mesh and flat plate with narrow slits have a relatively uniform flame colour at the combustor cross section, compared to flame visualization inside meso-scale combustor with flame holder of flat plate with circular holes. This phenomena is related to non-uniform reactant distribution on the combustor cross section.

1 Introduction

The development of micro-power generator (MPG) promotes many researchers about micro- and meso-scale combustor as a main component of the micro-power generator [1-2]. Micro- and meso-scale combustor with a stable combustion is an important component required in MPG systems. However, it is well understood that maintained stable flame inside meso- or micro-scale combustor are difficult, related to short fuel residence time as well as high heat loss to heat generation ratio due to high surface to volume ratio. Especially, it is difficult to obtain stable flame at a high reactant velocity due to shorter fuel residence time, resulting the flame instability such as blow-off. It is important to stabilize the flame at a high reactant velocity to obtain high power density MPG.

Many researches were conducted to improve flame stability inside micro- and meso-scale combustor, such as using backward facing step [3,4], wire mesh [5,6], as well as using plate flame holder [7]. Backward facing step has an important role to stabilize the flame inside micro combustor by prolong fuel residence time in the combustor and improving the fuel mixing in the presence of vortex near the step. Wire mesh has a purpose as flame holder and improves heat recirculation from the flame to the unburned reactant through heat conduction on the combustor wall. As the results, flame stability increased. For the same combustor and reactant, combustor with wire mesh flame holder has more stable combustion than combustor with backward facing step. This is inferred due to wider

* Corresponding author: lilis_y@ub.ac.id

contact area between flame holder – reactant which results better heat recirculation from the flame to the unburned reactant. Therefore, this is important to understand the role of the flame holder type on the heat recirculation, flame stability and behavior in meso-scale combustor.

This research was intended to understand flame stabilization mechanism and flame behavior, such as flame temperature and visualization, related to heat recirculation from the flame to the unburned reactant. In this research, heat recirculation will be optimized by geometry and thermal conductivity of the flame holder. The aim of the research is to obtain meso-scale combustor which has stable combustion at high reactant velocity.

2 Experimental setup

Figure 1 shows a schematic of meso-scale combustor used in this research. The combustor was made from quartz glass tube with an inner diameter of 3.5 mm and an outer diameter of 4.9 mm. It is consisting of two pieces of glass tube with length of 20 and 10 mm, connected with a heat resistant ceramic adhesive (Ceramabond 569, Aremco Products Inc.). Three type flame holders i.e. wire mesh, flat plate with circular holes and flat plate with narrow slits, were used in this research. The wire mesh was made from stainless steel, while the flat plate flame holders were made from copper. For further discussion the wire mesh, flat plate with circular holes and flat plate with narrow slits were named as type-I flame holder, type-II flame holder and type-III flame holder respectively. Detail geometry of each flame holder is indicated in Fig.2. Each piece of these flame holders was inserted between the quartz glass tubes to build the combustors.

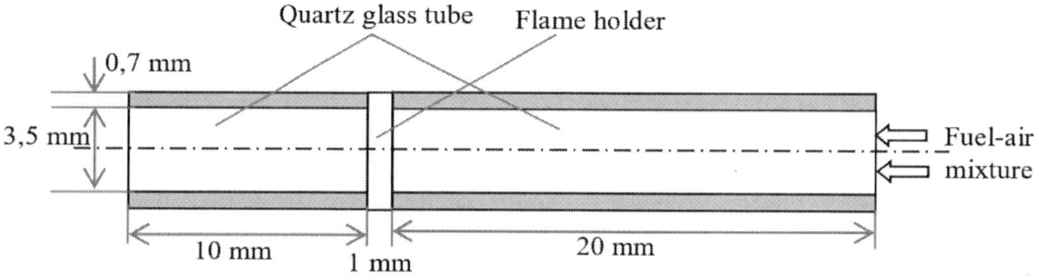

Fig. 1. Schematif of meso-scale combustor.

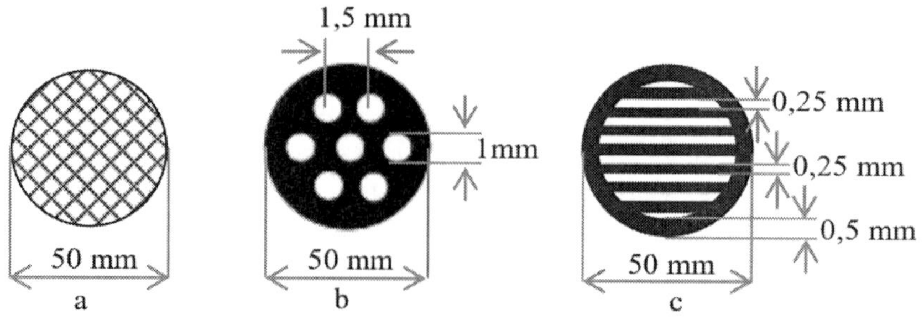

Fig. 2. Detail of a. type-I flame holder, b. type-II flame holder, c. type-III flame holder.

Butane was used as fuel and ambient air was used as the oxidizer, supplied to the combustor from the compressor tank. Fuel and air flow rate were adjusted using flow meter (Koflock, R1250) with measurement range of 2 – 20 mL/minute and 50 – 500 mL/minute respectively. Fuel and air were mixed in mixing chamber before introducing into the combustion chamber. Fuel-air mixture was introduced into the combustor at different flow rate and equivalence ratio to check the flame stability. The mixture was ignited at the end of the combustor using a torch. Flame propagates into the combustor and stabilized near the flame holder at the suitable fuel and air flow rate.

Flame stability is investigated by operating the meso-scale combustor at a constant fuel flow rate, then air flow rate is increased or decreased gradually until flame quenching or blow-off occurred. Furthermore, the same procedure is carried out by varying the fuel flow rate, until it reaches its minimum and maximum value where the flame is stable inside the combustor near the flame holder. Flame visualization is taken from the front and the side of the combustor using a Canon 600D cameras. Flame temperature measurements are conducted on the combustor axis at a certain axial position where the flame temperature achieves the maximum value. Measurement of exhaust gas temperature was conducted at the combustor axis at the end of combustor, 10 mm downstream of the flame holder. The temperature is measured using an R-type thermocouple with a diameter of 0.2 mm connected with data logger (Advantec).

3 Results and discussions

The flame could be stabilized in three combustors with different flame holder. The flame has blue luminosity. Stable flame in this study is combustion with a fixed flame located inside the combustor near the flame holder. Figure 3 indicates an area of flame stabilization limit inside combustors with different flame holder. The region is delineated in the plane of V - ϕ (velocity of reactant - equivalence ratio).

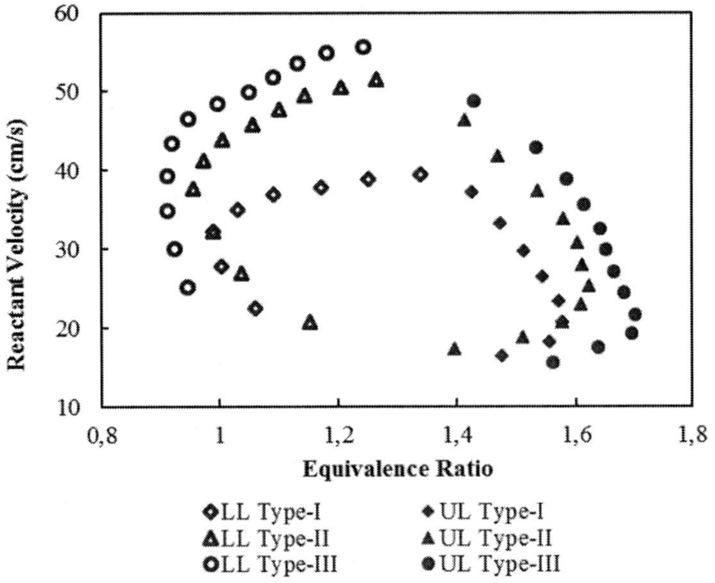

Fig. 3. Flame stabilization limit area in meso-scale combustor with different flame holder.

The figure denotes that meso-scale combustor using type-III flame holder has the most stable combustion. This is indicated by the widest area of flame stability limit on the V-ϕ plane. Flame could be stabilized in this combustor type for leaner or richer mixture, as well as for higher reactant velocity. Vice versa, the meso-scale combustor using type-I flame holder has the narrowest flame stability area and lower reactant velocity in which flame could be stabilized in the combustor. Flame holder has an important role to realize a stable flame within the meso-scale combustor.

Flame stability can be achieved if the reactant velocity could be balanced by flame propagation speed. From this fact could be concluded that the meso-scale combustor using type-III flame holder has the highest flame propagation speed if compared to the other combustor types. The flame propagation speed is affected significantly by flame and reactant temperature. The higher reactant and flame temperature conduce the increasing of flame propagation speed. From these explanations could be expected that the highest flame temperature occurred in meso-scale combustor using type-III flame holder. Moreover, it could be supposed that the reactant temperature entering the combustor with type-III flame holder is higher. This fact will be shown at the next section about flame temperature measurement. Flame visualization inside meso-scale combustor with different flame holder at various reactant velocity and constant equivalence ratio is indicated in Table 1.

Table 1. Flame visualization in meso-scale combustor with different flame holder at constant equivalence ratio.

Condition	Type-I flame holder	Type-II flame holder	Type-III flame holder
$\phi = 1.3$ V = 20 cm/s			
$\phi = 1.3$ V = 37.5 cm/s	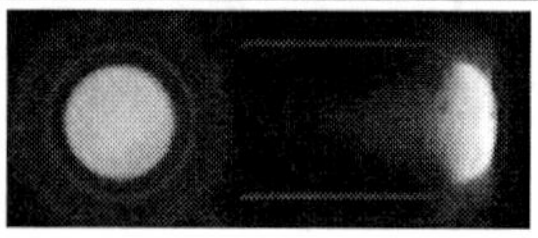		
$\phi = 1.3$ V = 50 cm/s			

The table above shows that the flame does not fulfil the combustor cross section. There is a gap between the flame and the combustor wall. Flame visualizations at the Table 1 show that flame inside combustor with type-I flame holder has the wider flame cross section compared to the other flame holder type. Furthermore, flame visualization shows that combustor with type-I and type-III flame holder have relatively uniform flame colour. Vice versa, non-uniform flame colour occurs in combustor with type-II flame holder, due to uneven reactants distribution on the combustor cross section. This phenomena relates to the shape of holes or gaps in the flame holder that determines the distribution of the reactants when entering the combustion zone, which further determines the flame shape. Brighter flame colour denotes more intense combustion reaction zone due to high concentration reactant.

Table 1 indicates flame visualization in meso-scale combustors with three types flame holder, at constant equivalence ratio and various reactant velocities. Flames luminosity becomes brighter with the increasing of reactant velocity. Likewise, its cross section becomes wider by reactant velocity. As the amount of reactants increases, the volumetric heat release increase, resulting in larger combustion temperatures. It could suppress the flame quenching near the combustor wall due to heat loss as well as radical quenching. Figure 4 indicates flame and exhaust gas temperature in meso-scale combustor with various flame holder types.

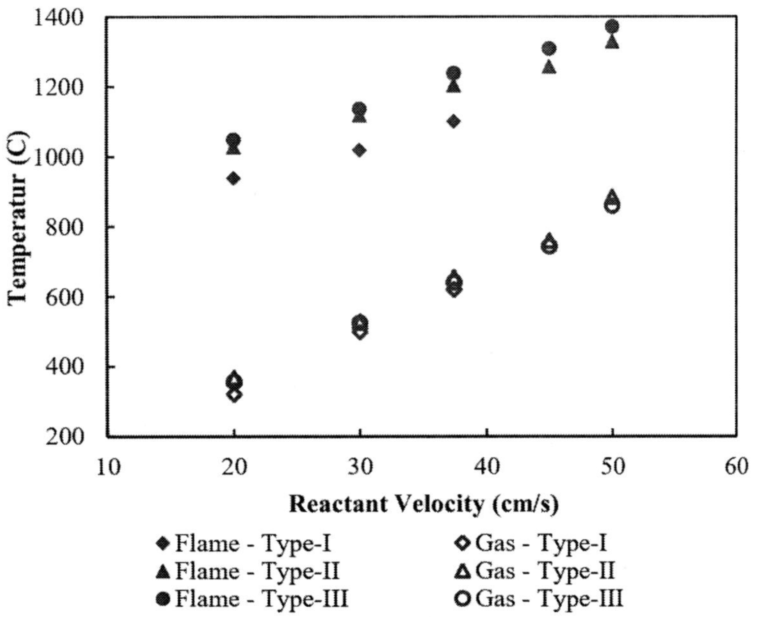

Fig. 4. Flame and exhaust gas temperature in meso-scale combustor with different flame holder.

The figure shows that as the reactant velocity increases, the flame and exhaust gas temperature in each combustor with different flame holder will increase. However, we can see that exhaust gas temperature of all combustors with different flame holder type are relatively same, while the flame temperature on the combustors are significantly different. Especially combustor with type-I flame holder has the lowest flame temperature. This phenomena can be explained as follows. Except to holding the flame, flame holder is also assist heat recirculation from the flame to the reactant. There are two methods of heat transfer from the flame to unburned cold reactant, i.e. direct heat transfer from the flame to unburned reactant by radiation and indirect heat transfer from the flame to unburned reactant through heat conduction on the combustor wall. However, the direct heat transfer is very small and could be neglected due to low radiation of the blue flame. Hence the important heat recirculation mechanism is conduction heat transfer through combustor wall. Heat from the combustor wall is convected to the unburned reactant. Heat convection from the combustor wall to the unburned reactant is affected significantly by the flame holder, its material as well as geometry of flame holder.

Heat flow from the flame to unburned reactant is as follows. A part of heat generated from the combustion process is convected to the combustor wall (H). A fraction of the convected heat will lose to the environment (H_L) and the remainder will be conducted to the upstream direction (Q_R). It can be denoted as $H = H_L + Q_R$. Q_R is conducted to the upstream direction through the flame holder, hence a part of the Q_R is conducted to the flame holder (Q_{RF}) and subsequently convected to the reactant which flows through the small gaps in the flame holder. The remainder is conducted to the upstream direction (Q_{RW}) and eventually convected to unburned reactant. The amount of heat conduction is $Q_R = Q_{RF} + Q_{RW}$. Q_R is an important factor which represent the quantity of heat recirculation from the flame to the unburned reactant. The fraction of Q_{RF} is larger compared to Q_{RW} due to the high thermal conductivity of flame holder which made from copper or stainless steel than the thermal conductivity of the combustor wall. Therefore, heat recirculation to unburned reactant through the flame holder becomes much important than the other one through the upstream combustor wall.

Heat recirculation to the reactant through the flame holder by heat convection is greatly influenced by thermal conductivity of flame holder and contact area between reactant – flame holder surface. So that the geometry and material of flame holder become an important factor in the heat recirculation process and flame stabilization mechanism. Type-III flame holder was made from copper which has high thermal conductivity and it has wider contact area between reactant – flame holder surface. This condition generates the most effective heat transfer to the unburned reactant. This results more heat loss by convection from the flame to the combustor wall, so that the flame cross section becomes narrower. However, Type-III flame holder results more heat recirculation from the flame to the reactant and more reactant pre-heating to produce higher reactant temperature. These conditions promote higher burning velocity and more stable combustion. The higher heat loss from the flame to the combustor wall is the reason why the exhaust gas has almost same temperature, even though the flame temperature in each combustor were significantly different.

4 Conclusions

This study experimentally investigated flame stability and its behavior in meso-scale combustor affected by flame holder geometry. Conclusions of the experiment results are as follow, flame holder geometry affects heat recirculation and preheating of reactants in meso-scale combustor. It results the order of combustion stability from the most stable to less stable are combustion in combustor with type-III flame holder, type-II flame holder and type-I flame holder respectively. Flame visualization indicates a non-uniform flame color due to uneven reactant distribution at the combustor cross section in combustor with type-II flame holder. The last one, combustor with type-III flame holder generates the most effective heat recirculation from the flame to the unburned reactant. This is indicated by higher flame temperature while the exhaust gases temperature is almost same compared to the other combustor.

References

1. Fernandez-Pello, Proc. Combustion Ins. J, **29**, 883–899 (2002).
2. Y. Ju, K. Maruta, Progress in Energy and Combustion Science J, **37**, 669-715 (2011).
3. W.M. Yang, S.K. Chou, C. Shu, Z.W. Li, H. Xue, Applied Thermal Eng. J, **22** (2002) 1777-1787 (2002).
4. M. Baigmohammadi, S. Tabejamaat, Y. Farsiani, Chem. Eng. and Processing J, **95**, 105-123 (2015).
5. L. Yuliati, T. Seo, M. Mikami, Comb. and Flame J, **159**, 462-464 (2012).
6. M. Mikami, Y. Maeda, K. Matsui, T. Seo, L. Yuliati, Proc. Combustion Ins. J, **34**, 3387 – 3394 (2013).
7. J. Wan, A. Fan, Energy Conv. & Management J, **101**, 552-560 (2015).

MATEC Web of Conferences **159**, 02012 (2018) https://doi.org/10.1051/matecconf/201815902012
IJCAET & ISAMPE 2017

Numerical simulation of heat transfer augmentation in fin-and-tube heat exchanger with various number of rows of concave rectangular winglet vortex generator

Syaiful[1,*], *Imam*Syarifudin[2], *Maria F.* Soetanto[3]and *Myung-whan*Bae[4]

[1,2]Mechanical Engineering Department, Diponegoro University, Semarang, Indonesia
[3]Aerospace Department, Bandung Polytechnic, Bandung, Indonesia
[4]Department of Mechanical Engineering for Production, Gyeongsang National University, Jinju, South Korea

Abstract. The passive method by using a vortex generator (VG) is an effective method for the improvement of convective heat transfer. This study is focused on usage of concave rectangular winglet vortex generator (CRW VG) for improving convective heat transfer in a fin-and-tube heat exchanger using numerical simulation. Concave rectangular winglet pairs (CRWP) and rectangular winglet pairs (RWP) VGs were mounted inside the gap between fins (gas side) with variations of the number of VG pairs of rows. Inlet air velocity variations expressed by the Reynolds numbers were ranged from 364 to 689. Augmentation of heat transfer is indicated by the ratio value of heat transfer convection coefficient between cases using VG and that without using VG (baseline). The results show that the convection heat transfer coefficient for cases using CRWP VG is higher than that using RWP VG. Convection heat transfer coefficient increases up to 102% by mounting CRWP VG at Re = 364. However, the increase in convection coefficient is accompanied by a rise in pressure drop to 216.8%.

1Introduction

Fin and tube heat exchanger is widely used in industrial, residential and commercial HVACR (heating, ventilation, air conditioning and refrigeration) systems. Various works have been performed to improve the effectiveness of fin and tube heat exchangers because of their impact on the economics of a product. The passive method using VG is one of the most effective methods to improve the effectiveness of the fin and tube heat exchanger by descending the thermal resistance on the gas side [1]. Besides, VG can be used to improve aerodynamic performance of an airfoil [2]. He et al. studied numerically the thermal-hydraulic performance of fin-and-tube heat exchanger mounted by rectangular winglet pairs (RWP) VG [3].

*Correspondingauthor: syaiful.undip2011@gmail.com

Joardar and Jacobi experimentally tested a delta winglet VG mounted on a compact plain fin heat exchanger in a wind tunnel [4]. Investigation on the effect of delta winglet VGs on heat transfer and pressure drop of novel heat exchanger has been studied numerically by Lei et al.[5]. Numerical investigation of the potential of punched winglet type VG arrays for enhancing heat transfer performance was carried out by He et al.[6]. Gholami et al's work applied the wavy-up and wavy-down rectangular winglet VGs for investigating numerically their effect on the heat transfer enhancement and pressure loss penalty in the fin-and-tube compact heat exchangers [7]. Wang et al. examined experimentally the air side performance of fin-and-tube heat exchangers mounting with semi dimple VG [8]. Li et al. evaluated numerically the use of rectangular and delta winglet VGs punched/mounted on fin surfaces for heat transfer augmentation and pressure drop in the fin-and-tube heat exchanger [9]. Wang et al. analyzed the novel longitudinal vortex generator which combined a modified rectangular wing and trapezoidal wing for predicting the performance of finned-tube heat exchanger [10].

Syaiful et al. studied numerically and experimentally the effect of concave delta winglet on the heat transfer improvement in airflow inside a rectangular channel [11]. Numerical investigation of attack angle effect of concave delta winglet vortex generator on the heat transfer augmentation in the fin-and-tube EGR cooler has been carried out by Syaiful et al. [12]. Experimental study of concave rectangular winglet vortex generator effect on the convection heat transfer coefficient of airflow in the rectangular channel has been performed by Syaiful et al. [13]. Therefore, the present work is focused on the effect of various numbers of rows of concave rectangular winglet on heat transfer enhancement in a fin-and-tube heat exchanger using numerical simulation.

2Physical and numerical models

Longitudinal vortex generator (LVG) is modelled by multiple surface protrusions mounted on the fin surface of fin-and-tube evaporator (see Ref. [4]) in the form of RWP and CRWP as denoted in Fig. 1. RWP and CRWP VGs with 1, 3, 4 and 7 rows were mounted in-line with the first, third, fifth and seventh tubes in attack angle of 15°. For CRWP, the radius of the concave surface curvature is 12 mm. VGs in which the high is 60% of channel high were mounted on the bottom fin surface in the right and left of the tube with the common flow-up orientation.

The numerical model was begun with mesh generation of the model followed by the determination of the boundary condition regions as shown in Fig. 2. Tetrahedral and hexahedral meshes were used for the mesh generation of the model. Fluid flow was assumed as laminar and incompressible with the constant properties. Calculation was performed in the steady state condition. Continuity, momentum and energy equations were applied for solving the physical problems. Conjugate heat transfer was enclosed in the calculation. In order to solve the mathematical equations, boundary conditions were applied to some predetermined regions. Equations of continuity, momentum and energy with some boundary conditions were solved by using computational fluid dynamics code (ANSYS Fluent 14.5). Momentum and energy equations were discretized by using the second order differencing scheme. Pressure and velocity coupling were solved with SIMPLE algorithm. Convergence criteria are revealed by residual in which 10^{-5} was for momentum and velocity and 10^{-8} was for energy.

Some of the investigated parameters in the present study can be referred to [4] and [14]. Validation is required to obtain accurate prediction results from numerical calculations against the real physical phenomena. In order to reduce the effect of the number of mesh on the change of parameter value, the independent grid was determined by comparing the value of the convection coefficient obtained from the experiment with the value of the nu-

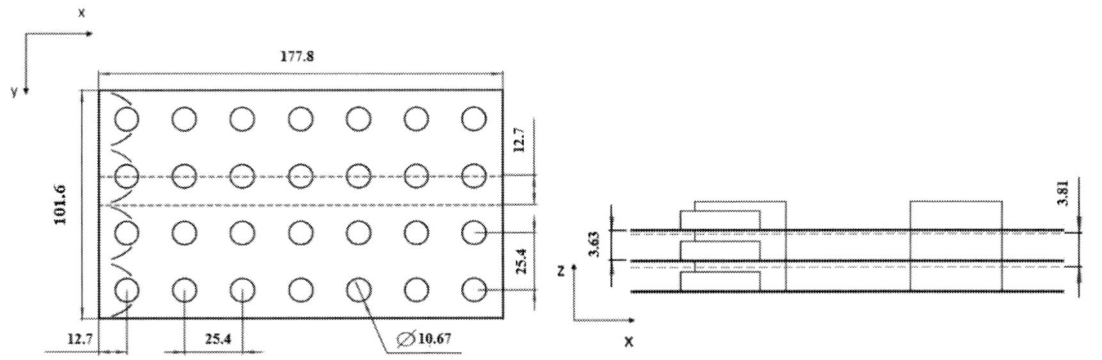

(a) (b)

Fig. 1. Domain computation: a) Top view and b) Side View

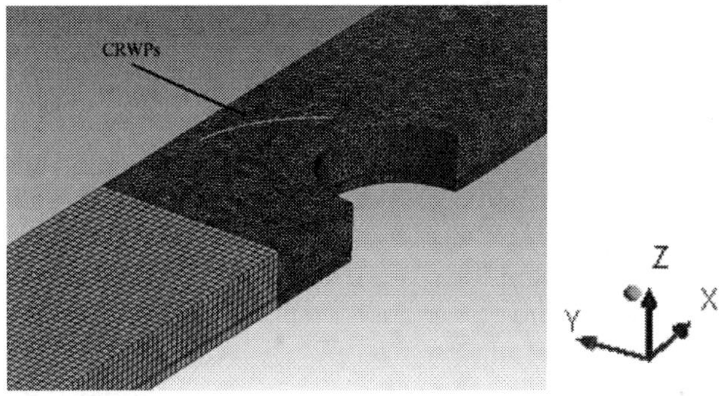

Fig. 2. Mesh generation

merical simulation. Based on the independent grid test, the number of grids was about 1,400,000 with the smallest deviation of 12.6%. Validation was also done by comparing the results of experiments performed by Joardar and Jacobi [4] with results from the present numerical calculations as expressed in Fig. 3.

As shown in Figure 3, there are similar tendencies for convection coefficient and pressure drop values between the results of these simulations with the results of experiments performed by Joardar and Jacobi [4]. The coefficient of convection and pressure drop increase as the Reynolds number increases for both experiment and

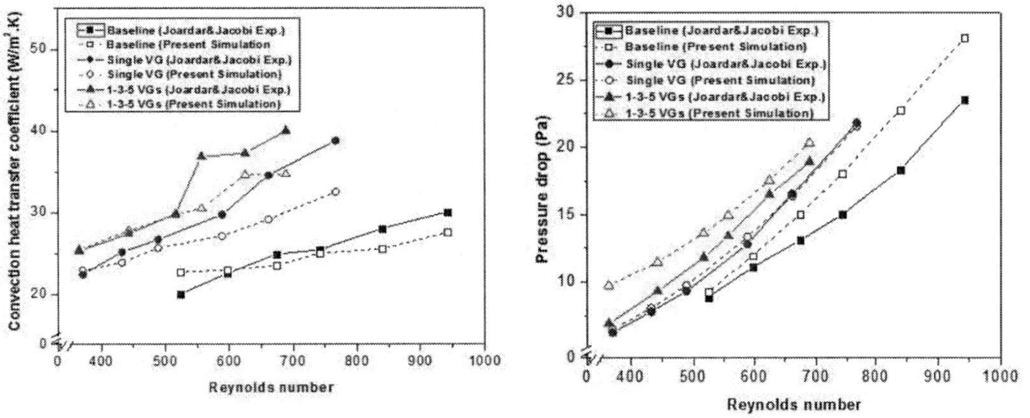

(a) Convection coefficient (b) Pressure loss

Fig. 3. Comparison of convection coefficient and pressure drop values between experimental results and present numerical calculations.

simulation results. By comparing the simulation and Joardar and Jacobi's experimental results, the value of convection coefficients deviate from 8% to 13% in the Reynolds number ranges from 523 to 942. While the pressure drop value deviates from 4% to 16% in the same range of Reynolds numbers.

3Results and discussion

Fig. 4 expresses the value of the convection coefficient to the variation of the Reynolds numbers and the number of rows of VG. Baseline in this case means the fins without VG. In general, the coefficient of convection increases with increasing Reynolds numbers for cases with and without VG. Considering the results stated in Fig. 4, the use of RWP and CRWP VG is capable of increasing the convection coefficient compared to the baseline case of the same Reynolds number. This is caused by the ability of VG to produce longitudinal vortex which promotes swirl flow enhancing the flow mixing [11, 12, 14]. This mixing enhances local heat and momentum transfer from the wall to the main flow. These results indicate that the use of CRWP VG improves the convection coefficient better than RWP VG for the same Reynolds number because of stronger longitudinal vortex generated from the CRWP VG than that of RWP VG. Strong longitudinal vortex is caused by instability of flow due to centrifugal force when fluid flows over the concave surface [15, 16]. At Re = 364, the use of 1 row of RWP and CRWP VGs increases the value of convection coefficients by 11% and 22% respectively to the baseline. The convection coefficient value for case using CRWP VG is doubled than using RWP VG. The highest increase of convection coefficient values are obtained by configuring 7 rows of RWP and CRWP VGs of 38.1% and 102.5%, respectively. From the results of this simulation it can be concluded that the use of CRWP VG on the fin-and-tube evaporator improves thermal performance better than the use of RWP VG at the same flow rate.

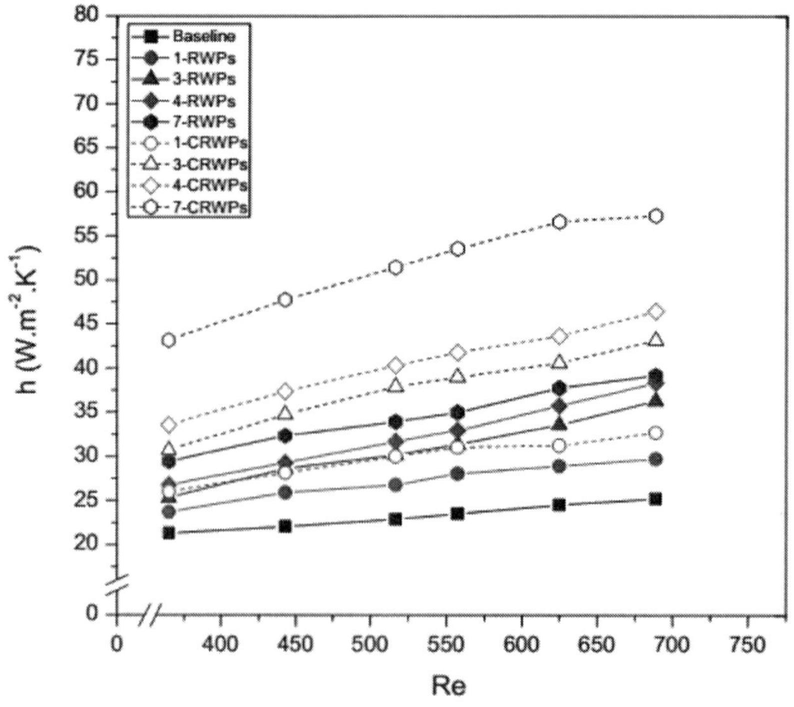

Fig. 4. Comparison of convection coefficient values among baseline, RWP and CRWP VG for variations of Reynolds numbers in difference of number of rows.

Fig. 5 shows the values of pressure drop with the increase of Reynolds numbers for cases with and without VGs on various number of VG rows. Generally, pressure drop increases with increasing Reynolds number for all cases. This is because the pressure drop is directly proportional to the flow velocity. The lowest pressure drop is observed for the baseline case on all Reynolds numbers due to the absence of a surface protrusion obstructing the flow. The values of pressure drop for the case using CRWP VG are greater than the case using RWP VG for all Reynolds numbers. This is due to the higher size of the projected flow contact field for CRWP VG rather than RWP VG [17]. By comparing to the baseline at Re = 364, the values of pressure drop in the configuration of 1 row of RWP and

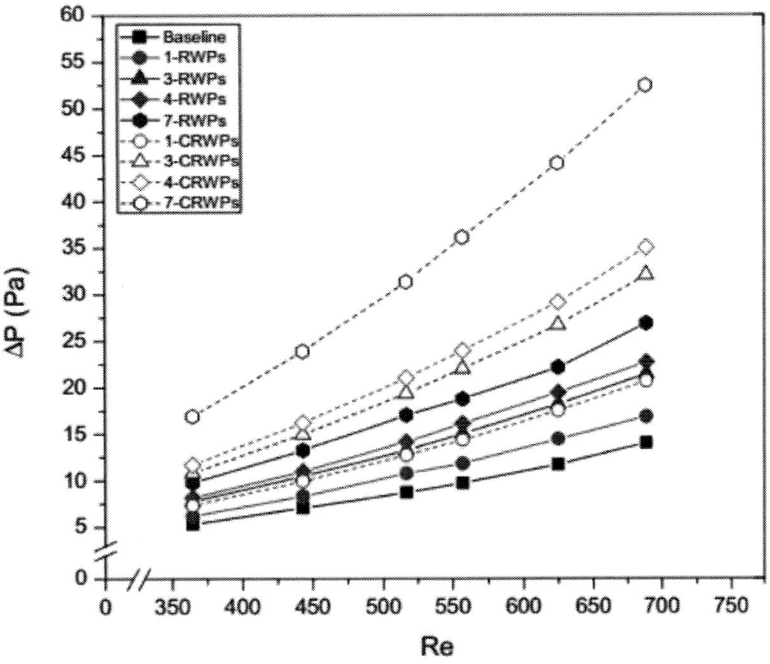

Fig. 5. Comparison of pressure drop values among baseline, RWP and CRWP VG for variations of Reynolds numbers in difference of number of rows

CRWP increase by 16.4% and 37.1%, respectively. Meanwhile, by mounting 7 rows of RWP and CRWP, the pressure drop values increase to 82.6% and 216.8% respectively to the baseline. From the results of this study it can be revealed that the use of CRWP VG in the fin-and-tube evaporator increases flow resistance or decreases of hydrodynamic performance.

4 Conclusions

Numerical simulation to determine the effect of concave rectangular vortex generators on the improvement of heat transfer in fin-and-tube heat exchanger has been completed. The calculation validation has been performed by comparing it with the results obtained by the work of Joardar and Jacobi. Agreement between the simulation and the Joardar and Jacobi's experiments results was obtained. In general, the use of concave rectangular winglet pair (CRWP) was more advantageous in terms of thermal performance compared to the use of rectangular winglet pair (RWP). However, the use of CRWP contributed higher increase in pressure drop than that of RWP.

Acknowledgement

This work was supported by the Fundamental Research of Indonesian Government (KEMENRISTEK DIKTI, contract number: 699-21/UN7.P4.3/PP/2017), Indonesia. The authors are grateful to all research members, especially Lab. Thermofluid of Mechanical Engineering of Diponegoro University Indonesia, Aerospace Department of Polytechnic of Bandung Indonesia and Advanced Combustion Lab. of Mechanical and Aerospace Engineering Faculty of Gyeongsang National University Korea.

References

1. B. Delač, AnicaTrp, K. Lenić, "International Journal of Heat and Mass Transfer", **78** (2014)
2. L. Gao, H. Zhang, Y. Liu, S. Han, "Renewable Energy", **76** (2015)
3. Y. L. He, P. Chu, W. Q. Tao, Y. W. Zhang, T. Xie, "Applied Thermal Engineering", **61** (2013)
4. A. Joardar, A. M. Jacobi, "International Journal of Refrigeration", **31** (2008)
5. Y. G. Lei, Y. L. He, L. T. Tian, P. Chu, W. Q. Tao, "Chemical Engineering Science", **65** (2010)
6. Y.L. He, H. Han, W.Q. Tao, Y.W. Zhang, "International Journal of Heat and Mass Transfer", **55** (2012)
7. A.A. Gholami, M. A. Wahid, H.A. Mohammed, "International Communications in Heat and Mass Transfer", **54** (2014)
8. C. C. Wang, Kuan-Yu Chen, Yur-Tsai Lin, "Applied Thermal Engineering", **88** (2015)
9. Li Li, X. Du, Y. Zhang, L. Yang, Y. Yang, "International Journal of Thermal Sciences", **92** (2015)
10. W. Wang, Y. Bao, Y. Wang, "Applied Thermal Engineering", **86** (2015)
11. Syaiful, R. Jalil, I. Hambali, M. W. Bae, "Proceedings of the Seventh International Symposium on Mechanics, Aerospace and Informatics Engineering", (2015)
12. Syaiful, G. Reviansyah, M. W. Bae, "FISITA world automotive congress", (2016)
13. Syaiful, G. Sugiri, M. F. Soetanto, M. W. Bae, "American Institute of Physics", **1788** (2017)
14. Y. L. He, P. Chu, W. Q. Tao, Y. W. Zhang, T. Xie, "Applied Thermal Engineering", **61** (2013)
15. P. D. McCormack, H. Welker, M. Kelleher, "Journal of Heat Transfer", ASME (1970)
16. V. Malatesta, L. F. Souza, J.T. C. Liu, M. J. Kloker, "Procedia IUTAM", **14** (2015)
17. G. Zhou, Z. Feng, "International Journal of Thermal Sciences", **78**(2014)

Performance Analysis of Sea Water Solar Still To Produce Fresh Water and Salt Using Flat Plate Collectors

Mulyanef *, *Duskiardi* , and *Rahmad* Hidayat

Department of Mechanical Engineering, University of Bung Hatta, 25243 Padang, Indonesia

Abstract. Fresh Water and salt are the most important necessities in human life. The scarcity and the difficulty to get fresh water and salt became main problems that emerge in many places such as for people who live in coastal areas and small islands where their main source of water and salt is the sea water. In order to obtain fresh water, people needs to process seawater into fresh water and salt. The term of fresh water used here means water which is free from dirt, bacteria, and other substances that are harmful for human health. To process the sea water into fresh water and salt has been done by distillation technic. The distillation technic requires heat to evaporate seawater before condensed and produce clean water. Heat for distillation process can be derived from a variety of sources, one of them is solar energy. The purpose of this study is to compare between a flat plate with one slope cover and two slope cover to produce fresh water and salt. The distilled water was tested in laboratory to determine whether the quality of water and salt after distillation meets the consumable requirements. The working principle of the distillator is by filled the basin with sea water and then heated with solar power. The water will be evaporated and due to the different of temperature between inside and out side of cover so that the condensation accurs at inside of cover. The condensed water will flow through surface of glass cover and then the distilled freshwater gathered in storage. The process will continue until water runs out and creates salt crystals in the basin. Solar distillation testing by using one slope cover with a size of 140 cm x 220 cm produce 3,482 ml / day of clean water and 322 grams of salt / 4 days at average intensity of 697 W/m^2. While solar distillation by using two-slope cover with size of 80 cm x 50 cm produce 1,551 ml clean water/day and 199.9 grams of salt / 4 days.

1 Introduction

Solar energy is safe, free, and pollution-free renewable energy (without CO_2 emissions). Opportunities to utilize solar energy in Indonesia, as renewable energy is increasingly to be considered in Indonesia, where geographically is a tropical country and solar energy is emitted during the year. Indonesia also has the most extensive coastal area that supports to

* Corresponding author: mulyanef@bunghatta.ac.id

process of transforming sea water into fresh water and salt-making business, both in small scale and industrial scale. The process of seawater into freshwater is known as desalination process [3]. Desalination is a separation process used to reduce dissolved salt content of brine to certain level so that water can be used. The whole process of desalination requires energy to separate salt from seawater. If desalination is conducted by conventional technology, it would require the burning of large quantities of fossil fuels (clean water production of 1,000 m3 / day requires 10,000 tons of oil per year), while the availability of this fuel is decreasing. So, other sources of energy are needed such as the purification of sea water using solar energy [7]. Solar energy is a promising solution to reduce cost. Indonesia as an archipelagic country with coastline reaching 81,000 km and has high potential to produce large quantities of salt. Although Indonesia is a maritime country, efforts to increase salt production have not been in demand. The necessity of salt in Indonesia increase from year to year, along with population growth and industrial development. One way to process sea water into fresh water and salt is distillation. Distillation requires heat to evaporate seawater before it is condensed so that it can produce clean water.

There are several studies that have been done to treat sea water into freshwater and salt using solar distillation test among others, conducted by; the authors [2] conducted "Experimental study of the enhancement parameters on a single slope of solar still productivity". The productivity of solar still increased by 51% when combined with enhancers such as asphalt basin liner and sprinkler in which has been applied to the still. The condition and temperature of solar still environment also has direct effect on the productivity of distilled water. The productivity of evening distilled water accounts for 16% of daily output. This study also showed that solar daily productivity can still be increased by reducing the water depth in the basin; the authors [5] conducted a study on "the Effect of water depth on productivity of solar still with thermal energy storage". The experiment shows that by decreasing the water depth in the basin, it will increase the productivity of diesel still; the authors [1] conducted a research on "Desalination of water using conventional basin type of solar still ". Field experiment was conducted from June 7, 2011 to June 09, 2012. He found that average distilled water production is 1.80 lit / m2 per day and average production is 0.20 lit / m2 per hour. While the maximum-daily distilled water production obtained 3.76 lit / m2 and the production of the maximum distilled water is 0.46 lit / m2 per hour; the authors [4] examined "the Development and performance analysis of a domestic". The tests were conducted from June, July, August, September, October, November, and December 2011. The highest solar still productivity occurs at June (5738 ml) and the lowest productivity occurs at December (1,510 ml), due to the solar intensity in June is the highest compared with the other months. The average productivity of diesel still from June to December is 3,000 ml and the overall efficiency is 22%; the authors [6] has conducted experiments on "Performance study of sea water distillation by solar energy to generate salt and fresh water"; The results showed that solar collectors using reflectors generate higher productivity compared to not using reflectors. The volume of seawater in each basin is 6,000 ml and the collector area of 80 cm x 50 cm can produce condensate or fresh water of 1,551 ml per day and salt of 199.9 grams for 3 (three) days of testing.

2 Experimental Setup

Experimental equipment used in the research of solar still namely; Basin liner, glass cover, insulating material, distillation channel, and measuring instrument.

2.1 Basin Liner

Basin liner is a major part of solar still, which serves to absorb solar radiation transmitted through a transparent glass cover. Basin Liner must be resistant to hot salt water, has high absorption of solar radiation, strong and easy maintenance. In order to increase the absorptivity of solar still, Basin liner layer is painted black, not to mention that in this research we used aluminum plate.

2.2 Glass Cover

In this study, we use glass cover with a thickness of 4 mm and glass average transmissivity value of 0.89, angle of 35 degrees as shown in figure 1. To reduce heat loss, silicone rubber is installed on the edge of the glass frame cover.

2.3 Insulating Material

Insulating material serves to reduce the heat lost from the bottom and side of the solar still wall. Insulating material used in the solar still is glass wool with a thickness of 5 cm and thermal conductivity of 0.045 W/m^2 °C.

2.4 The Distillate Chanel

The distillate line used to collect condensate from the bottom edge of glass cover and carry it to storage compartment. It is made from aluminum sheets, and the side wall is made from wood as thick as 16 mm.

2.5 Measuring Device

2.5.1 Temperature

Temperature variations in various solar still locations are measured by thermocouple (k-type) merged into data logger. The accuracy of this device is in the range of 0.3 ^{0}C for temperature measurements 1 and 99 ^{0}C. There are five thermocouple points used to measure temperatures in solar still, namely; ambient temperature, absorbent plate temperature, water temperature in basin, evaporation temperature, and glass temperature.

2.5.2 Solar Radiation

In the present study a solarimeter is used to measure the solar radiation. This device measures the instantaneous intensity of radiation [its range from 0 to 1,2 kW/m^2].

3 Work Principle of Solar Still

Solar radiation enters the collector through transparent glass cover to absorbent plate. On the absorber plate, solar radiation is converted into heat. Seawater in the basin will become hot, and the water evaporates because its density is lower than salt and water vapor rises upward at the inside of glass cover. Because there is a temperature difference between inside the basin and the solar still environment, condensation occurs at the inside of glass cover. The condensation result attached to the inside of glass cover flows downward following the slope of the cover glass. Condensation results are held and this is called fresh water productivity. To produce salt, sea water is warmed continuously from morning to evening every day until it evaporates completely and will form salt crystals. Fig. 1 show schematic diagram of the solar still producing fresh water and salt.

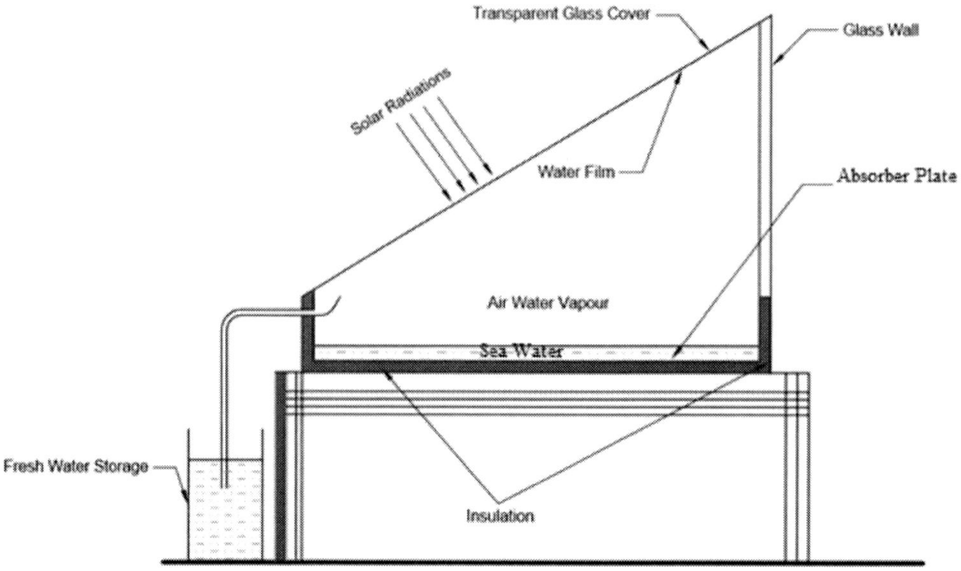

Fig. 1. Schematic diagram of the solar still to produce fresh water and salt

4 Result and Discussions

This section presents the effect of different conditions on solar still productivity, such as ambient temperature, different water depths, different basin coating materials (eg black paint, aphalt), sprinklers (cooling film) on glass and wind speed. Different variables are measured per hour such as glass temperature (Tg), ambient temperature (Ta), sea water temperature (Tw), absorbent plate temperature (Tp), vapor temperature (Tsv), solar Radiation (I), wind velocity (Vw) , Condensate productivity (Ma) and salt productivity (Mg). Fig. 2 shows variations in various solar still temperatures on 15/3/2017.

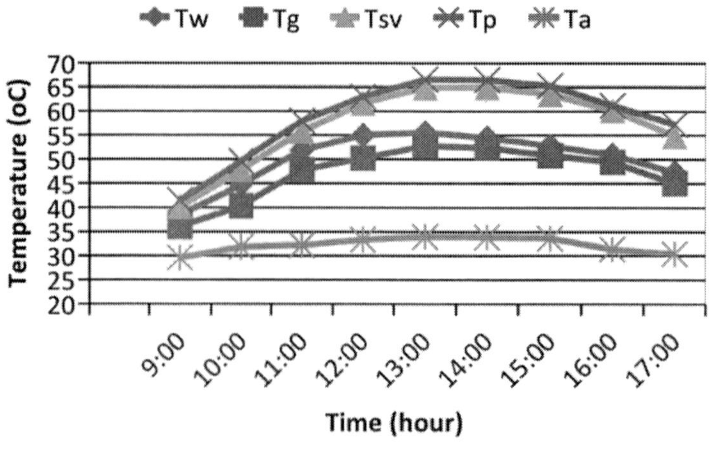

Fig. 2. Different temperature variations on solar still

It can be seen that the initial temperature (09:00) is still low and rises steadily until 12:00. After 12:00 the temperature at the collector begins to fall again, this is caused by the solar radiation that goes into the collector. The maximum temperature on basin obtained on 13.00. At the time, the evaporation temperature is 65 ^{O}C, water temperature 55 ^{O}C and the ambient temperature obtained 36 ^{O}C. The higher solar intensity would increase the temperature in the environmental temperature collector significantly. The highest temperature is generated by an absorbent plate and followed by evaporation temperature, water temperature in basin, glass temperature and ambient temperature.

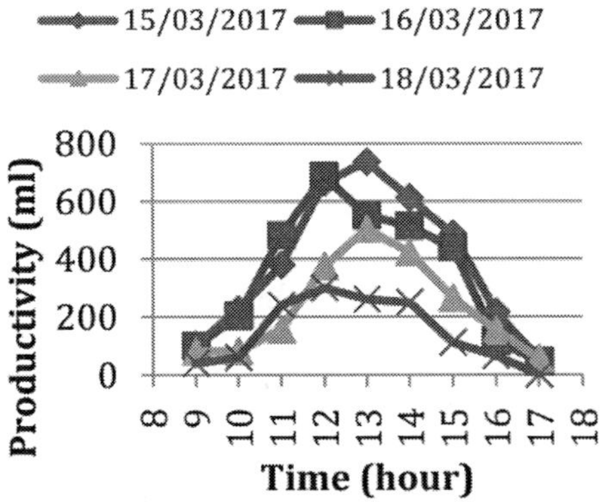

Fig. 3. The relationship between condensate productivity for 4 days of testing and 12,000 ml of water in basin.

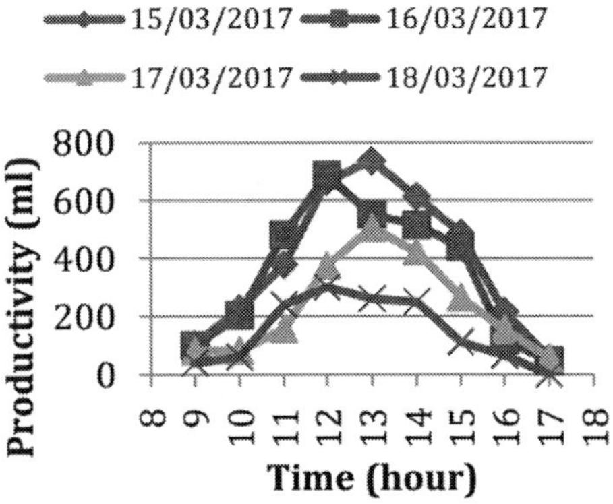

Fig.4. The relationship between condensate productivity for 4 days of testing and 6,000 ml of water in basin

Fig. 3 and Fig. 4 shows the productivity of fresh water condensation for sea water volume in 12,000 ml basin and 6,000 ml in basin. Fresh water productivity in the first day of testing was higher than its productivity on the second, third and fourth test day. This is because during the first test day, the sea water in the basin was still high and the salt content was low. While on the second day, condensate productivity began to decline and the salt content in the basin was higher.

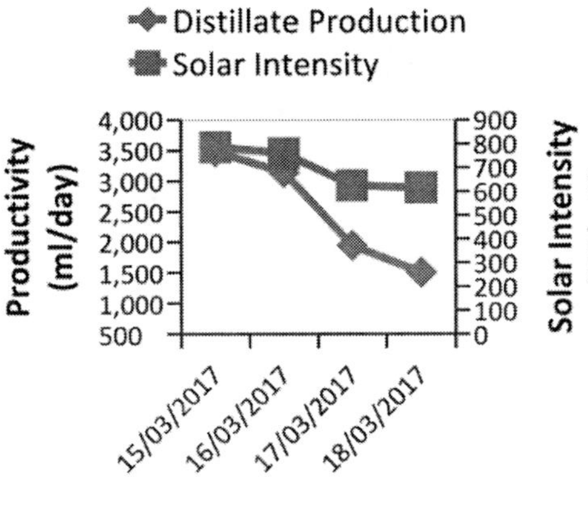

Fig. 5. The relationship between fresh water productivity and solar intensity in the first test with 12,000 ml of water in basin.

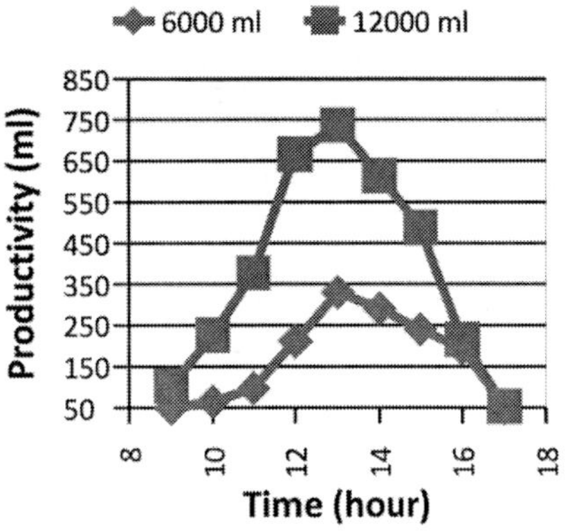

Fig. 6. Relationship of fresh water productivity with time and different size of solar still

Fig. 5 displays the relationship between fresh water productivity and solar intensity on the first test with 12,000 ml of water in basin. Fresh water productivity is closely related to solar intensity. The higher the solar intensity, the higher the productivity of fresh water. The highest productivity was obtained on the first day of testing (March 15, 2017) with 3,482 ml of fresh water and the lowest on the fourth day of the test with 1,505 ml and the average solar intensity during the test of 787 W/m^2.
Fig. 6 Displays the relationship between freshwater productivity and time. High productivity for both sizes of solar still is produced at 13.00, compared to 09.00-12.00 and 14.00-17.00. This is because solar intensity in the morning is still low. As the time goes by, the intensity is getting higher until 12:00 to 13:00, solar intensity drop back. The diagram also showed us that for solar still with 12,000 ml of water in basin, it produces more fresh water compared to 6,000 ml of water.

Table 1. Fresh water and salt productivity from solar still

Productivity	**Volume of sea water in basin**	
	6,000 ml	**12,000 ml**
Fresh Water (ml)	3,815	10,076
Salt (gr)	199.9	322.0

Table 1 displays fresh water and salt productivity for 6,000 ml of sea water in basin. It produced fresh water of 3,815 ml and salt of 199.9 grams. As for 12,000 ml of sea water in basin, it produced 10,076 ml of fresh water and 322 grams of salt. The more of seawater heated in the basin, the more freshwater and salt will be produced. On the contradictory, the more seawater in the basin, the longer time needed to produce salt and fresh water. In this test, it takes four days to obtain salt with average solar intensity of 697 W/m^2.

5 Conclusions

- Total productivity for flat plate collector with area of 140 cm x 220 cm and 12,000 ml of sea water in basin is 10,076 ml of fresh water and 322 gram of salt. As for collectors with area of 50 x 80 cm and 6,000 ml of sea water yields 3,815 ml of fresh water and 199.9 grams of salt. The average solar intensity during the test is 697 W/m^2 and the time required to produce salt is four days.
- The productivity of solar still is influenced by solar intensity, environmental temperature, wind speed, and water depth in the basin.

References

[1] Al Faruq, S.M Abdullah. Akter, Md Shaheen. Alamin Md, Islam, Md Zohrul. Global J. Vol.**13** Issue 3 Ver. 1.0 (2013)
[2] Badran O.O. Desalination. J. **209** (2007)
[3] Deng, Runya. Xie, Lixin. Lin, Hu. Liu, Jie. Han, Wei.. Energy. J. **36** (2010)
[4] Iqbal, I. Ahmad, A. ARPN J. Vol. **9**, No. **7** (2014)
[5] Kumar T R ,Sathish. Bharathi B, Raja. IJSR. J, Volume **2** Issue 3 (2013)
[6] Mulyanef. Duskiardi. Kaidir, Ikhsan, M. *Engineering Science and Technology International Conference 3rd.*.Proc., **153** (2016).
[7] Yilmaz, Ibrahim, Halim, Soylemez, Mehmet, Sait. Desalination J. **291** (2012).

The Influence of Guide Vane to the Performance of Cross-Flow Wind Turbine on Waste Energy Harvesting System

Budi Santoso [1,*], and *Dominicus* Danardono Dwi Prija Tjahjana[1]

[1]Mechanical Engineering Dept., Sebelas Maret University, Surakarta 57126, Indonesia

Abstract. The purpose of this experiment is to know the influence of a single guide vane position and angle to the performance of a cross-flow wind turbine. The cross-flow wind turbine was positioned at the discharge outlet of a cooling tower model to harness the discharged wind for electricity generation. A guide vane was used to enhance the rotational speed of the turbines for power augmentation. Various position and angle of attack of the guide vane were tested in this experiment. To avoid negative impact on the performance of the cooling tower fan and to optimize the wind turbine performance, the turbine position on the discharge wind stream was also studied. The result showed that cross-flow wind turbine with a guide vane attached at the right position had a higher coefficient of power than cross flow turbine without guide vane. A cross-flow wind turbine with the guide vane at the position of 150 mm from the center and 30° angles had the highest coefficient of power of 0.49. Comparing to the wind turbine without guide vane, the coefficient of power of the cross-flow wind turbine was increased about 84.3%.

1 Introduction

Nowadays, the wind energy becomes an energy resource that has the fastest growth in the world. According to Global Wind Energy Council by the end of 2015, the global cumulative installed wind capacity by the end of 2015 was 432,883 MW [1]. Unfortunately, the countries that located near the doldrums cannot harness the wind energy as much as the sub-tropic countries. In the Southeast Asian countries such as Indonesia, the wind speed is low (< 4 m/s) and inconsistent throughout the year [2]. Only in a certain location in the Eastern Islands and along the Coast Line, that may have average wind speed more than 5 m/s.

In the city and urban area, the energy consumption is very high. In Indonesia extracting energy from natural wind in the urban area is not feasible because besides the average wind speed is low, there is strong turbulence that caused by the high rise buildings. On the other hand, most of the high-rise buildings in the big city, such as hotels, malls, and office buildings use air conditioning system. The cooling tower of the air conditioning system usually produces exhaust air that has consistent and strong wind speed. A big cooling tower

* Corresponding author: msbudis@yahoo.co.id

can produce wind speed up to 18 m/s at a distance of 0.3 m above the cooling tower outlet [3,4,5]. A small size of wind turbine generator is possible to be installed on the cooling tower exhaust system.

Although Horizontal Axis Wind Turbine (HAWT) is highly developed and widely used in the world, recent research has indicated that Vertical Axis Wind Turbine (VAWT) has higher power density [6]. The disadvantage of most classical wind turbines is not able to start at low wind speed (2-3 m/s). A Savonius wind turbine can operate at low wind speed, but it has a low maximum efficiency [7,8]. A cross-flow wind turbine that is similar to the Banki Water Turbine is relatively a new design of VAWT that is very fast starting in low wind speed [9,10]. It also has high maximum power coefficient [10]. In this study, a cross-flow wind turbine was installed on the cooling tower model.

Unfortunately, the output wind speed of the cooling tower fan is not uniform. At the center of the fan produce the lowest wind speed, and then increase to the maximum at near to the edge of the fan corresponds to the rotor radius [11]. To overcome this non-uniform wind velocity, in this research the influence of a single guide vane, in various horizontal position and angle, to the wind turbine performance was studied. The power consumption of the cooling tower fan was also studied.

2 Experimental Details

The purpose of this experiment was to determine the optimum angle of a single guide vane and its horizontal position to increase the performance of the wind turbine at the outlet of an exhaust air system. A scaled cooling tower model was fabricated to represent an exhaust air system. The model has an axial flow fan with 900 mm diameter and powered by 630 Watt rated motor. The body of the model is similar to the common counter-flow cooling tower. Figure 1 depicts the cooling tower model with a cross-flow wind turbine installed on the outlet. The design of the model was based on the previous research [2,3,4,5,11], but it used a different type of wind turbine.

The detail configuration of the guide vane is shown in Fig. 2 (a). The guide vane dimensions were 80 mm width and 380 mm length. There were seven horizontal positions of the guide vane, in the range of -150 mm to +150 mm, with a distance of every position of 50 mm. The 0 mm was inline vertically with the axis center of the wind turbine. In every position, the angle of the guide vane was varied to 5 different angles; there were 30°, 45°, 60°, 75° and 90°. The turbine position on the discharge wind stream was also studied. In this experiment, various horizontal positions of the wind turbine were also studied. The variations were 0 mm to 400 mm from the center of the cooling tower fan (Fig. 2 (b)).

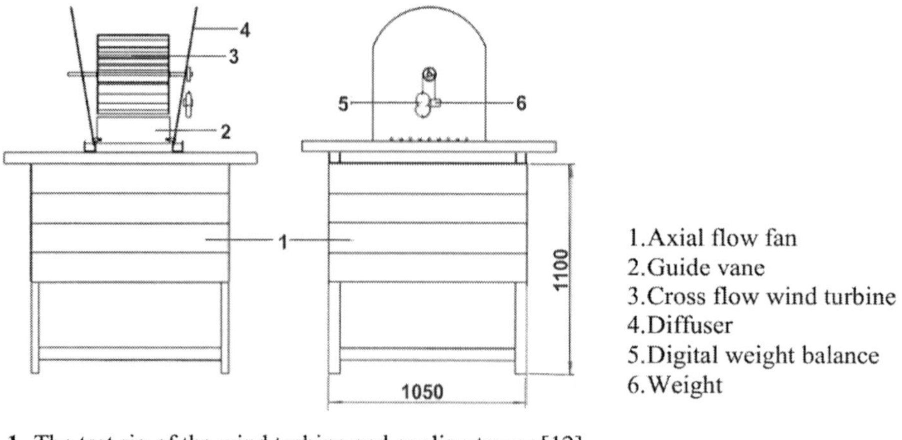

Fig. 1. The test rig of the wind turbine and cooling tower [12].

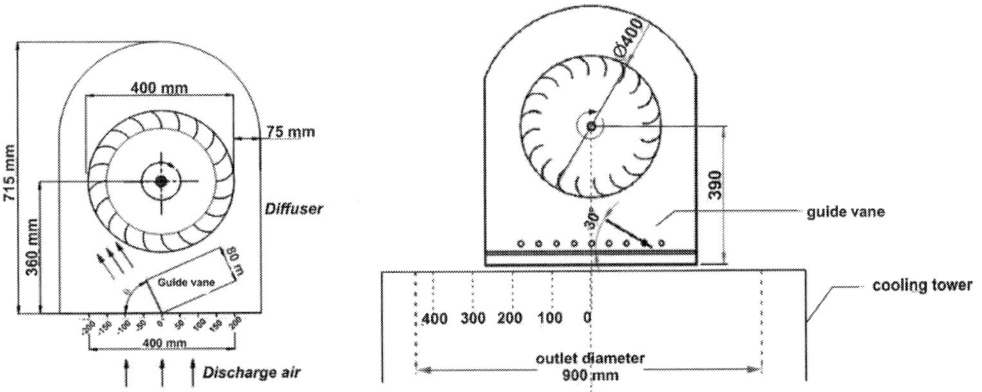

(a) Guide vane position (b) Wind turbine position

Fig. 2. Guide vane position and turbine position about the center of cooling tower fan.

The wind turbine used in the experiment was a cross-flow wind turbine. This type of the wind turbine was selected because it has a high coefficient of power (C_p), simple and compact construction and can work in any direction of wind speed [10]. The outside diameter of the wind turbine was 400 mm, and the inside diameter was 293 mm. The cross-flow wind turbine has 16 blades. The length of the rotor blade was 380 mm and had a radius of 40 mm.

The data taken from the experiment were the power consumption of the fan motor, the average velocity of the fan, the rotation speed and the torque of the wind turbine shaft. The torque of the shaft was measured by using the prony brake.

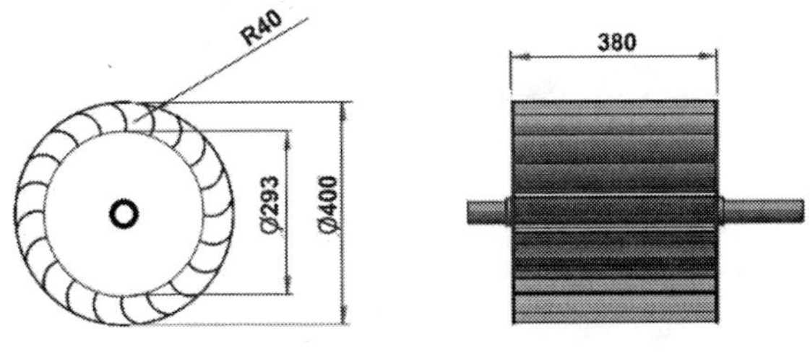

Fig. 3. The cross-flow wind turbine

3 Result and Discussion

The profile of the wind speed output of the cooling tower is shown in Fig. 4. The wind speed profile shows that the minimum speed occurred in the center. The maximum speed achieved at about 400 mm from the center. Due to the no-slip condition, at a position closed to the edge of the fan housing, the wind speed was decreased. The result shows similarity with the previous research [11].

Figure 5 shows the power of the turbine on the various horizontal position of the wind turbine related with the center of the fan. The result shows that without the guide vane the wind turbine produced maximum power of 1.83 W, at 300 mm turbine position. The turbine best position is primarily related with the position of blades that produce positive torque to the turbine, which are about 400 mm from the center. The maximum power of the wind turbine related to the highest wind speed of the fan that occurred near the edge of the fan. The velocity of the wind at this position was 5.97 m/s. Based on the result, the wind turbine was fixed at this position for following guide vane variations.

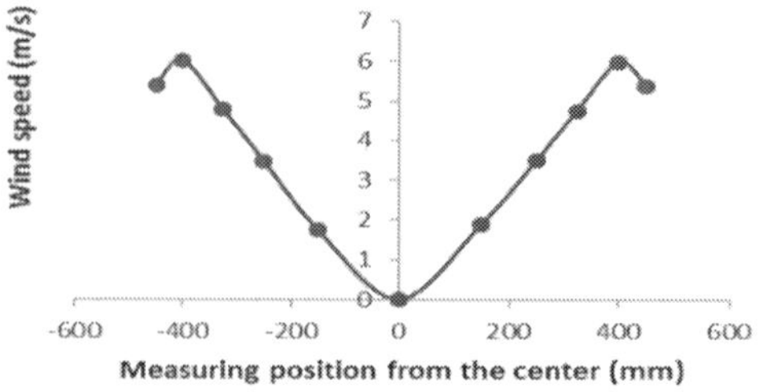

Fig. 4. Wind speed profile at the cooling tower outlet.

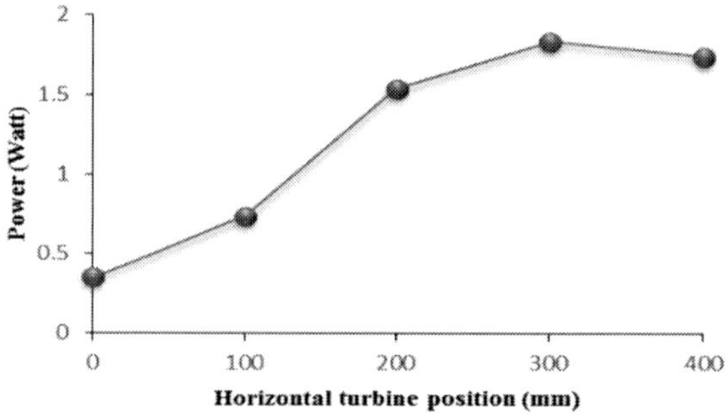

Fig. 5. The power of the wind turbine without guide vane on various horizontal positions.

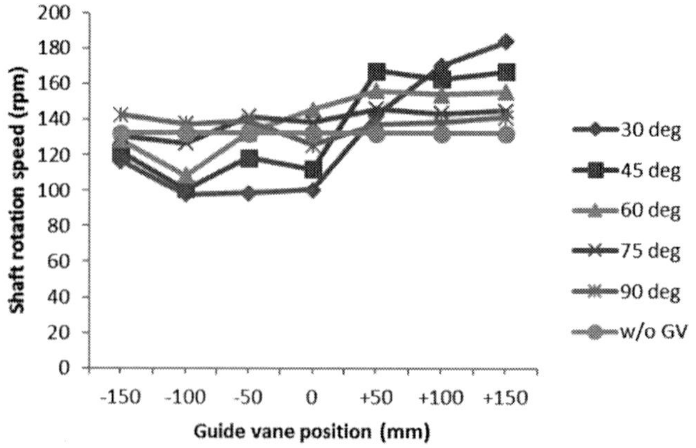

(a) The rotation speed

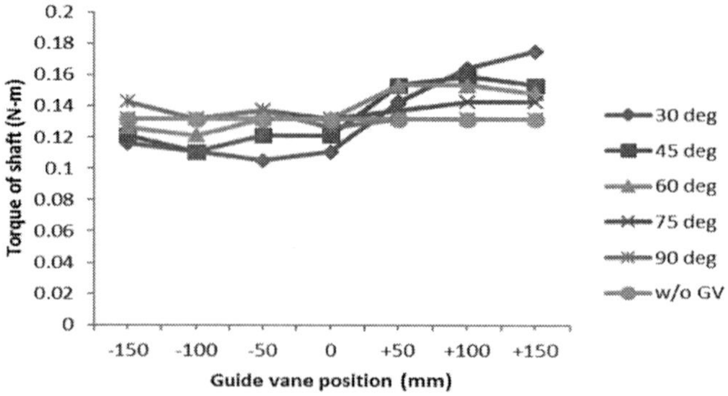

(b) Torque

Fig. 6. The rotation speed and torque of the wind turbine shaft on various guide angle and position.

By setting the position of the wind turbine at 300 mm position, the influence of the guide vane angle and position to the shaft rotation speed and torque is shown in Fig. 6. The coefficient of power (cp) chart depicted in Fig. 7, also shows the same trend. There is no guide vane angle that could be superior in every position. The guide vane with 30° angle shows that it had the lowest rotation speed (Fig. 6.a) and torque (Fig. 6.b) for the position from -150 mm to below +100 mm. However starting from +100 mm position the 30° guide vane angle had the highest rotation speed, torque, and cp. At this position, the 30° guide vane angle was able to deliver maximum wind speed to the blades of the wind turbine. The maximum cp achieved by the cross-flow wind turbine was 0.49.

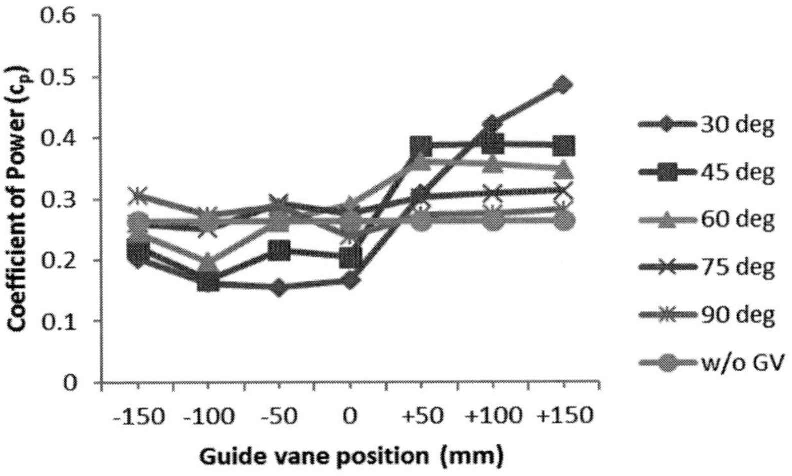

Fig. 7. The power coefficient of the wind turbine on various guide angle and position.

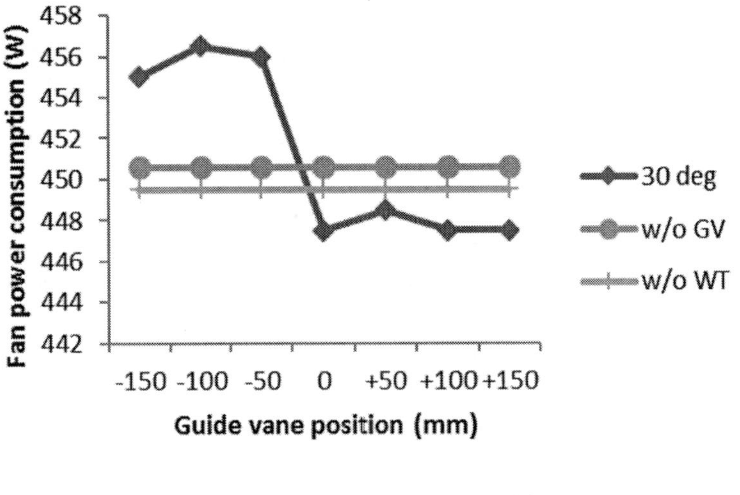

Fig. 8. The cooling tower fan power consumption in various condition.

The influence of the wind turbine installation to the electrical power consumption of the cooling tower fan is depicted in Fig. 8. The installment of the wind turbine increased the power consumption of the fan motor by about 0.24%. By adding the 30° guide vane to the installation, the power consumption of the fan could be reduced at guide vane positions of 0 to +150 mm. At these positions, the average reduction of the power consumption was about 0.39%. At 150 mm, the position for the highest cp, the reduction of the power consumption is 0.45%. It occurred because the speed of the turbine was higher than the incoming wind that created a suction effect. This condition was followed by the increased in wind speed of the intake and power consumption reduction of the fan motor [2]. In the condition, there is not any effect on the cooling tower performance, because the fan power of cooling tower remains constan.

4 Conclusion

As the conclusion, the experiment result showed that some energy from the exhaust of a cooling tower model could be recovered by attaching a cross-flow wind turbine on the cooling tower outlet. The cross-flow wind turbine with a guide vane attached at the right position had a higher coefficient of power than cross flow turbine without guide vane. A cross-flow wind turbine with the guide vane at the horizontal position of 150 mm from the center and 30° angles had the highest coefficient of power of 0.49. Comparing to the wind turbine without guide vane, the coefficient of power of the cross-flow wind turbine was increased about 84.3%.

This work was supported by DP2M DIKTI (Derectorate of Research and Public Service of Derectorate General of Higher Education) Ministry of Research, Technology and Higher Education Indonesia through the grant of Strategi Nasional (Stranas) 2017. The authors appreciate Danang Kurniawan for their experimental cooperation.

References

1. Global Wind Energy Council (GWEC). Global Wind Report – annual market update 2015. http://www.gwec.net/publications/global-wind-report-2/ [accessed August 2016].
2. W. T. Chong, S.Y. Yip, A. Fazlizan, S. C. Poh, W.P. Hew, E.P. Tan and T.S. Lim. Renew. Energ. **67**, 252 (2014)
3. W. T. Chong, S.C. Poh, A. Fazlizan, C.S. Oon and C.C. Tiah, *2011 International Conference on Environment and Industrial Innovation*, IPCBEE vol. **12**, (IACSIT Press, Singapore, 2011)
4. W. T. Chong, S.C. Poh, A. Fazlizan, S.Y. Yip, C.K. Chang and W.P. Hew, Appl. Energ. **112**, 568 (2013)
5. W. T. Chong, W.P. Hew, S.Y. Yip, A. Fazlizan, S.C. Poh, C.J. Tan, H.C. Ong, Energ. Convers. Manage **87**, 145 (2014)
6. M. R. Islam, S. Mekhilef and R. Saidur, Renew. Sustainable Energy Rev. **21**, 456 (2013)
7. M. H. Mohamed, G. Janiga, E. Pap and D. Thevenin, Energ. Convers. Manage. **52**, 236 (2011)
8. J. V. Akwa, H.A. Vielmo and A.P. Petry, Renew. Sustainable Energy Rev. **16**, 3054 (2012)
9. T. Klemm, M. Gabi and J.N. Heraud, JCAM **8**, No. 2, 123 (2007)
10. A. Dragomirescu, Renew. Energ. **36**, 957 (2011)
11. A. Fazlizan, W.T. Chong, S.Y. Yip, W.P. Hew and S.C. Poh, Energies **8**, 6566 (2015)
12. B. Santoso, D. D. D. P. Tjahjana dan P. J. Widodo, *Seminar Nasional Tahunan Teknik Mesin* **XVI**, (to be published)

Development of Tension-Torsion Multiaxial Creep Testing Apparatus for Heat Resisting Steel

Noritake Hiyoshi[1,*], and *Yoshihisa* Iriyama[2]

[1]Division of Engineering, Faculty of Engineering, University of Fukui, 3-9-1 Bunkyo, Fukui-shi, Fukui 910-8507, Japan
[2]Graduate School of Engineering, University of Fukui

Abstract. This paper describes development of a combined tension-torsion multiaxial creep testing apparatus for heat resisting steel. It is essential for high temperature component designing to investigate creep rupture life and creep properties of heat resisting steel. Although high temperature structural components undergo multiaxial stress damage due to complex loading situation or shape discontinuity of the actual structure, there is no commercial testing apparatus which can conduct a creep testing under multiaxial stress conditions. In this study, we developed a combined tension-torsion multiaxial creep testing apparatus which can apply multiaxial stress to a hollow cylinder type testing specimen with 6 kN axial load and 12 Nm torsional load at high temperatures. Since the testing apparatus also has measuring devices for axial and shear displacements of the specimen, relationship curve between testing time and equivalent strain under multiaxial stress conditions of type 304 stainless steel is also discussed.

1 Introduction

High temperature structural components undergo not only cyclic loading damage but also multiaxial creep stress damage due to complex loading of actual service condition. Although it is essential for safety high temperature structure designing to make clear the creep rupture time and the creep properties under multiaxial stress conditions, there is little experimental research paper on multiaxial creep [1-3]. The reason for the small number of experimental study on multiaxial creep may be there is no commercial creep testing apparatus which can conduct experiments under multiaxial stress condition.

There are several methods for multiaxial creep testing. One is using notched bar type specimen with conventional uniaxial tensile creep testing apparatus. Other one is loading bi-axial stress to a cruciform type specimen. In this study, we tried to develop a new type multiaxial creep testing apparatus which can load combined tension-torsion stress to a hollow cylinder type specimen. By comparing the experimental data obtained by using

* Corresponding author: hiyoshin@u-fukui.ac.jp

tension-torsion method with the data obtained by using bi-axial tension method, safety designing for the high temperature structure can be improved.

2 Combined Tension-Torsion Creep Testing Apparatus

Table 1 summarizes a specification of a combined tension-torsion multiaxial creep testing apparatus developed in this study. Figures 1 and 2 show schematic figure and general view photo of the apparatus, respectively. Axial tensile stress and shear stress are generated by deadweight separately. The apparatus has a capacity of 6 kN axial load and 12 Nm torsional load. Load of the deadweight for axial loading is increased tenfold by lever. Torque for torsional loading is generated by deadweight for torsional loading with a pulley which has 184 mm in diameter.

A hollow cylinder type specimen for multiaxial loading was also designed. Figure 3 shows shape and dimensions and general view photo of the specimen. The specimen has 8 mm outer dimeter, 5 mm inner diameter and 8 mm gage length. The shape and dimensions are designed by using finite elements analysis calculation. The specimen is heated with electric resistance furnace of 900 W and the maximum temperature of the furnace is 973 K.

The multiaxial creep testing apparatus is also designed to have measuring sensors for displacement of the specimen in order to investigate multiaxial creep properties. A pair of linear variable differential transformer sensor is used for the axial loading shaft displacement measuring. A rotary encoder and a pair of eddy current sensor are used for the loading shaft twist angle measuring and the specimen shear displacement measuring, respectively.

Table 1. Specification of the developed multiaxial creep testing apparatus.

Maximum axial load	6 kN
Maximum torsional load	12 Nm
Maximum temperature	973 K (700 C)
Axial displacement measuring	Linear variable differential transformer
Shear displacement measuring	Encoder and Eddy current sensor

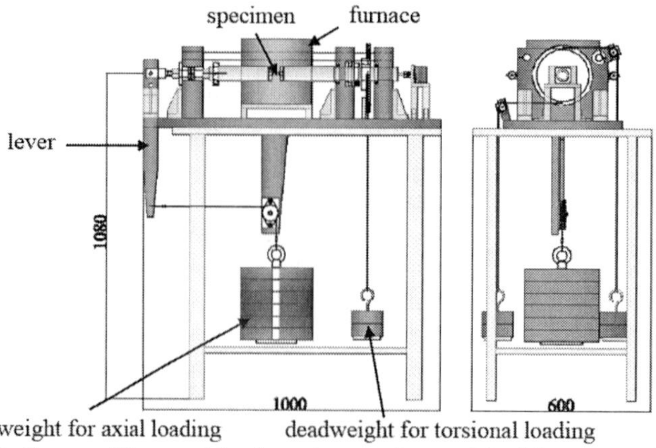

Fig. 1. Schematic figure of the combined tension-torsion multiaxial creep testing apparatus.

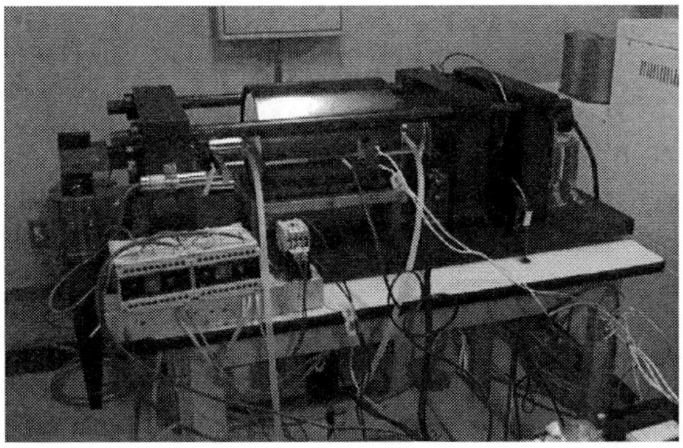

Fig. 2. General view photo of the combined tension-torsion multiaxial creep testing apparatus.

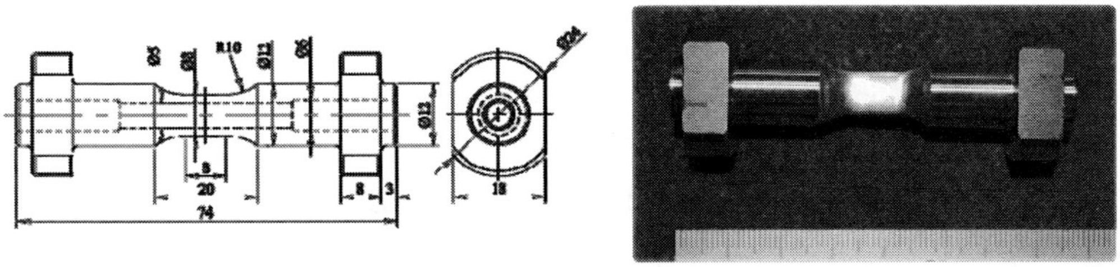

Fig. 3. Shape and dimensions and view photo of the specimen.

3 Experimental Procedure

A combined tension-torsion loading test was conducted at 923 K in order to inspect an ability of the creep testing apparatus and investigate the multiaxial creep properties of heat resisting steel. The material tested in this study was type 304 stainless steel. The chemical composition and heat treatment of the material are listed in Table 2. Mechanical properties of the material at room temperature is listed in Table 3.

Table 2. Chemical composition and heat treatment condition of the tested material (mass%).

C	Si	Mn	P	S	Ni	Cr
0.05	0.32	1.76	0.37	0.27	8.27	18.7

Heat treatment at 1353 K for 11min.

Table 3. Mechanical properties of the tested material at room temperature.

Tensile strength σ_B, MPa	607
Proof stress $\sigma_{0.2\%}$, MPa	280
Elongation δ, %	60
Reduction of area ϕ, %	75
Hardness, HBW	170

In this study, a principal stress ratio λ is used for expressing the stress multiaxiality and is calculated as follows,

$$\lambda = \frac{\sigma_2}{\sigma_1} \tag{1}$$

where σ_1 and σ_2 are maximum principal stress and minimum principal stress, respectively. According to this definition, pure axial tensile loading test is expressed as $\lambda=0$ and pure torsional loading test is expressed as $\lambda=-1$. λ of combined tension-torsion loading tests are in the range of $-1<\lambda<0$. A combined tension-torsion multiaxial creep test with $\lambda=-0.5$ and $\sigma_{eq}=170$ MPa was conducted at 923 K, where σ_{eq} is von Mises equivalent stress and is calculated as follows for combined tension-torsion stress,

$$\sigma_{eq} = \sqrt{\sigma^2 + 3\tau^2} \quad (\sigma\text{: axial stress, } \tau\text{: shear stress}) \tag{2}$$

4 Results and Discussion

Figure 4 shows time versus creep strain curves of the multiaxial creep testing for type 304 stainless steel at 923 K. Symbol in red color and blue color are axial and shear strain, respectively. As axial strain and shear strain are plotted clearly in Fig. 4, these results indicate that strain measuring methods adopted in this study are suitable for the tension-torsion multiaxial creep testing.

Green coloured line in Fig. 4 shows von Mises equivalent strain and is calculated as follows,

$$\varepsilon_{eq} = \sqrt{\varepsilon^2 + \frac{1}{3}\gamma^2} \quad (\varepsilon\text{: axial strain, } \gamma\text{: shear strain}) \tag{3}$$

Primary transient stage, secondary steady stage and tertiary acceleration stage of the equivalent strain are observed clearly in Fig. 4. Strain rate curve calculated with the equivalent strain data is shown in Fig. 5. Since minimum creep strain rate which is essential material data for safety products designing can be obtained in Fig. 5, use of the equivalent strain may be available for the tension-torsion multiaxial creep testing.

Figure 6 correlates the multiaxial creep rupture time with the principal stress ratio in order to discuss about influence of stress multiaxiality on creep fracture properties. Average data of uniaxial creep testing with solid bar type specimen, which were provided by National Institute for Materials Science, is also plotted in the same figure as reference data. The creep rupture time with $\sigma_{eq}=170$ MPa are 2552 h and 194 h for $\lambda=-0.5$ and $\lambda=0$, respectively and the creep rupture time decrease with the principal stress ratio. We need more sufficient experimental data collection and data analysis in order to make clear the influence of the stress multiaxiality on creep rupture time.

Figure 7 is general view photo of the ruptured specimen. Fracture surface direction was perpendicular direction to the specimen axis for $\lambda=-0.5$. Figure 8 shows optical microscope observation results of the fracture surface. Many rubbed marks were observed clearly but there is little axial elongated deformation in Fig. 8. These results imply that torsional loading may give more damage to the specimen than axial loading. We also need more sufficient observation in order to make clear the fracture mechanism under multiaxial loading.

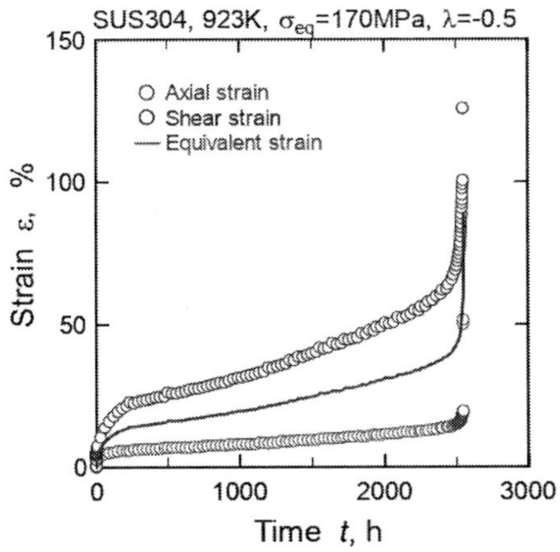

Fig. 4. Creep curves of type 304 stainless steel at 923 K under λ=−0.5.

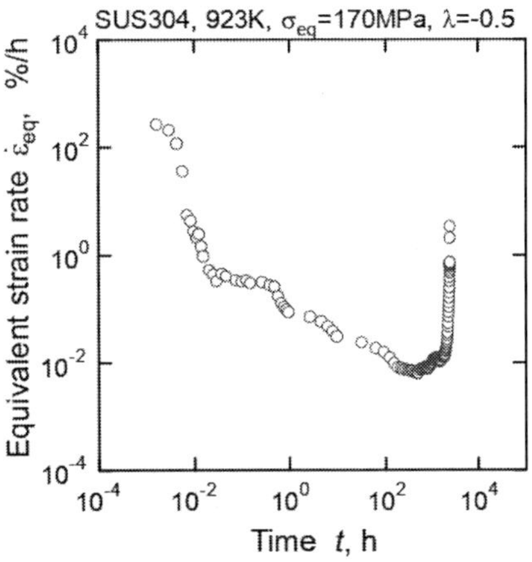

Fig. 5. Equivalent strain rate curve of type 304 stainless steel at 923 K under λ=−0.5.

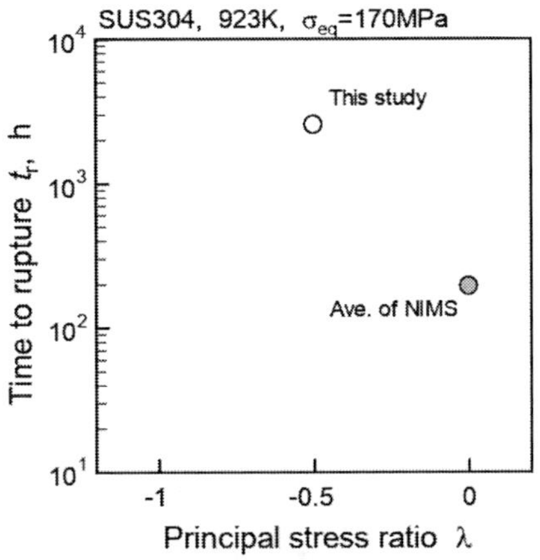

Fig. 6. Correlation of creep rupture time with principal stress ratio.

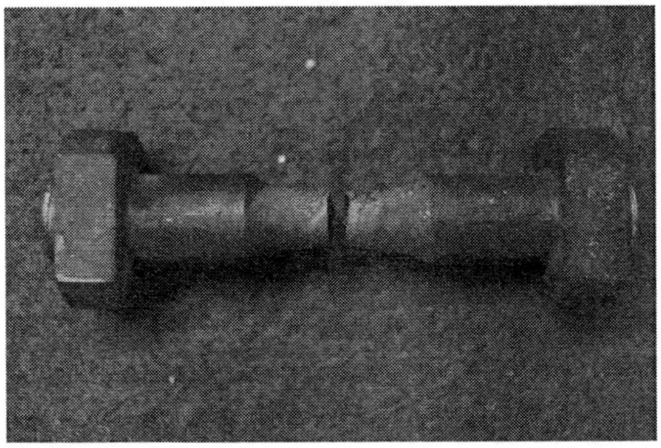

Fig. 7. Specimen surface observation after the test.

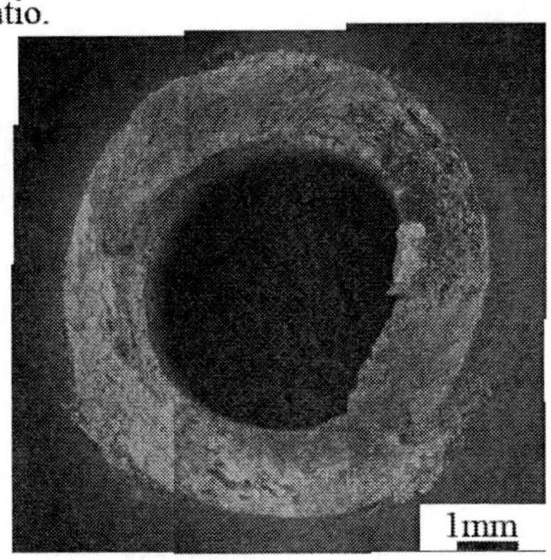

Fig. 8. Fracture surface observation results with optical microscope.

5 Conclusions

Experimental research work on multiaxial creep for heat resisting steel was conducted in this study.

(1) Combined tension-torsion multiaxial creep testing apparatus which has 6 kN axial loading and 12 Nm torsional loading capacity was developed.

(2) Hollow cylinder type specimen for multiaxial creep testing was also designed.

(3) Creep rupture time under $\lambda=-0.5$ was longer than $\lambda=0$ uniaxial stress conditions. There are influence of stress multiaxiality on creep rupture time.

References

1. N. Hamada, M. Maruyama and H. Umeda, Creep strength of Mod.9Cr-1Mo steel under biaxial stress states, Transactions of the JSME (in Japanese), **80**, 1-12 (2014)
2. S. Zhang, T. Wakai, M. Sakane, Creep rupture life and damage evaluation under multiaxial stress state fot type 304 stainless steel, Materials Science and Engineerig A, **510-511**, 110-114 (2009)
3. L. Niu, M. Kobayashi and H. Takaku, Creep Rupture Properties of an Austenitic Steel with High Ductility under Multi-axial Stresses, ISIJ International, **42**, 1156-1161 (2002)

Performance prediction of magnetorheological valves under various type of fluid and flux path

Muhammad Hafiz Idris[1], *Saiful Amri* Mazlan[1*], *Ubaidillah* Ubaidillah[2], and *Fitrian* Imaduddin[3]

[1]Malaysia-Japan International Institute of Technology, Universiti Teknologi Malaysia, 54100 Jalan Sultan Yahya Petra, Kuala Lumpur, Malaysia
[2]Mechanical Engineering Department, Faculty of Engineering, Universitas Sebelas Maret. Jl Ir. Sutami 36A, Kentingan, Surakarta, 57126, Central Java, Indonesia
[3]Faculty of Engineering and Technololgy, Multimedia University, Jalan Ayer Keroh Lama, Bukit Beruang, 75450 Melaka, Malaysia

Abstract. Magnetorheological (MR) valve is one of the key components in regulating the control flow of MR fluid flow in MR devices. Due to MR rheological properties enabling magnetic modifications, many MR valve designs have been introduced and are widely utilized in MR devices. The main objective of this article is to compare the pressure drop generated by each types of MR valves. Five different configurations of MR valves were analyzed in this study. To provide a fair comparison, several parameters were fixed; which is the size of the MR valves, gap size, number of coil turns and power consumption. In order to demonstrate the MR valve performance, the pressure drops were calculated based on the mathematical equations derived from MR valve models and the magnetic fields derived from finite element method based software. The results of the simulations were compared to each type of MR valve and, then, discussed.

1 Introduction

Recently, applications and devices utilizing magnetorheological fluids (MRFs) are experiencing a rise due to the property uniqueness of magnetic controllable and fast-response rheological change with the magnetic field. MRF is a type of smart material where the viscosity of the fluid increases when the magnetic field increases[1]. Currently, MRF devices use several modes of operations, and the most commonly used are the valve mode also known as MR valve [2–4]. A conventional MR valve includes an electromagnetic coil to generate the magnetic field intensity. The pressure drop of the MR valve rises with the raising of the magnetic field strength.

The valve dimensions and configurations has been observed to affect the performances of the MR valve. Researchers have developed several studies on the geometrical arrangements, which utilizes the gaps for the fluid flow path known as annular and radial gaps [5]. An annular gap is a flow channel that allows the MRF flows axially in the MR

* Corresponding author: amri.kl@utm.my

valve and the placement of it is parallel to the inlet and outlet of the valve. Meanwhile, for the radial gaps, the fluid flows transversely in a radial manner, forcing the fluid to flow in perpendicular to the inlet and outlet of the valve. These two gap types, annular and radial, are the most commonly and widely used in studies. The applications have been established in the MR devices [6–8]. The progress of the MR valve studies also continues with new types of configurations which are serpentine configuration and meandering configuration [3,9]. The magnetic flux in the design of serpentine MR valve is weaved into the annular gap by alternating the non-magnetic and magnetic materials. While, in meandering configuration, the multiple annular and radial configurations are combined with orifice flow channel, which also extends the flow path of the valve.

One study reported a methodology in configuring and comparing different MR valve configurations using analytical models and Finite Element Analysis [FEA] [10]. This type of analysis is remarkably time consuming with various processes involved in the evaluation to determine the most suitable design for a particular application. In order to shorten the process, this study aims to compare the pressure drop generated by each types of MR valves. The operating range and the pressure magnitude are also discussed. Thus, the analysis serves as a basis in selecting or designing tool for MR valves.

2 MR valve modelling

The performance of an MR valve is analyzed here based on its capability in producing and flexibility in pressure drop between upstream and downstream sides. Valve ratings are mainly determined by the pressure drop hence the expression of mathematical modelling of pressure drop for each configuration in each section. The basic expression of the pressure drop in the MR valve is observed as [11]:

$$\Delta P = \Delta P_{\text{Viscous}} + \Delta P_{\text{Yield}} \tag{1}$$

where $\Delta P_{\text{Viscous}}$ and ΔP_{Yield} are the viscous and field dependent yield stress of pressure drop. The mathematical expression of viscous and field dependent pressure drop for annular and radial flow path is observed as in Table 1 [12] :

Table 1. Mathematical equations.

	Viscous ($\Delta P_{\text{Viscous}}$)	**Yield** (ΔP_{Yield})
Annular	$\frac{6\eta QL}{\pi d^3 R}$. (2)	$\frac{c\tau\,(B)L}{d}$, (3)
Radial	$\frac{6\eta\, Q}{\pi d^3}\ln\left(\frac{R_0}{R_i}\right)$, (4)	$\frac{c\tau\,(B)}{d}(R_0 - R_i)$, (5)

where d is the valve gap, η is the fluid base viscosity, Q is the flow rate, L is the annular channel length of the valve, R is the channel radius, $\tau(B)$ is the field dependent yield stress value and c is the flow velocity profile coefficient. Also the valve gap size d in equations refers to radial gaps while R_0 and R_i referring to the inner radial gaps and outer radius of radial gaps respectively.

The coefficient of c can be obtained through the calculation of ratio between viscous pressure drop and the total pressure drop as seen following:

$$c = 2.07 + \frac{12Q\eta}{12Q\eta + 0.8\pi R d^2 \tau\,(B)}. \tag{6}$$

Meanwhile, for the equation on the orifice gaps, the expression is slightly different since there is no field dependent yield stress on the MR fluid. The pressure drop equation of orifice gaps solely expressed in viscous resistance as shown [13]:

$$\Delta P = 2\frac{8\eta QL}{\pi R^4} \tag{7}$$

3 MR valve configurations

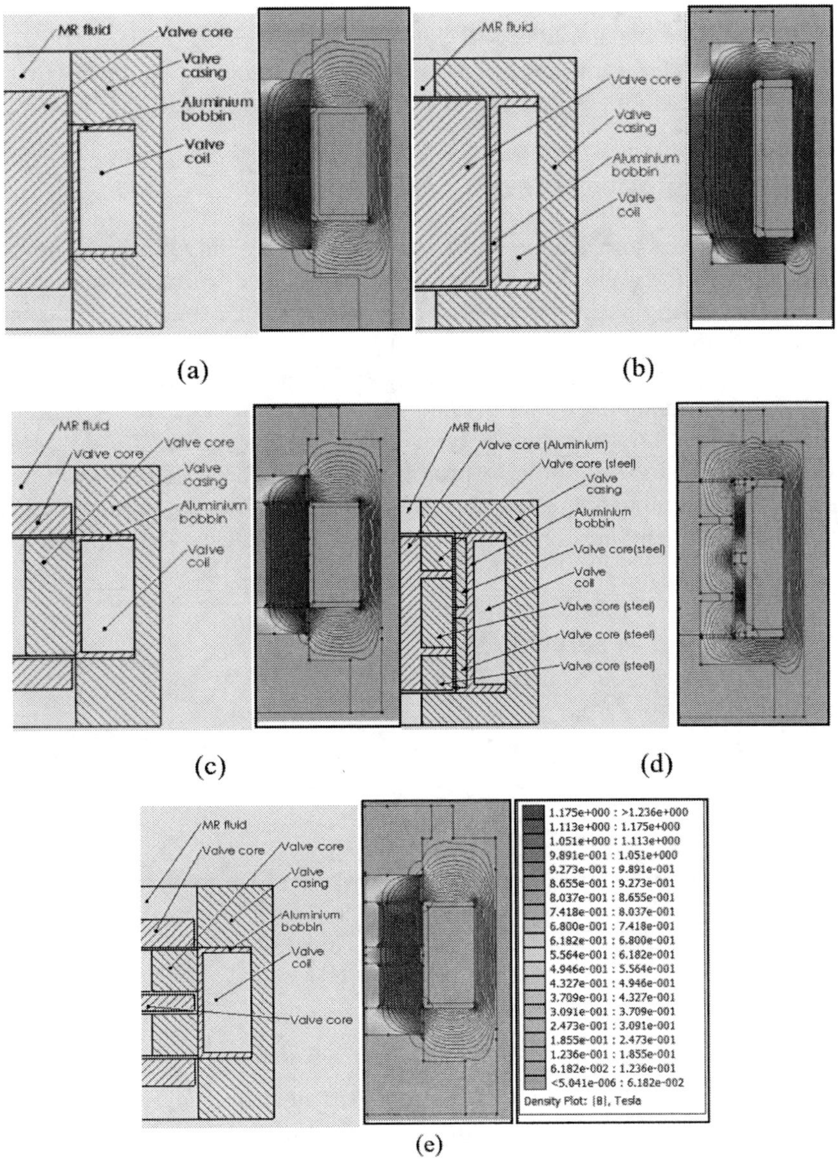

Fig. 1. Configuration of MR valve in FEMM simulation (a) annular, (b) radial, (c) annular-radial, (d) serpentine (e) meandering

In this research paper, the strength of magnetic flux densities was evaluated to predict the performance of MR valve as shown in Figure 1. The magnetic field strength of MR valves are difficult to measure through experimental method thus why the numerical approach was chosen to aid the prediction of MR effect that will be generated by the MR fluid. The author chose the FEMM (Finite Element Method Magnetics) software to analyze magnetic simulations in this study.

Parameters are required when conducting the simulation to obtain the magnetic flux distributions. The coils of each MR valve were designed and fixed with 450 turns of 24 AWG copper wires and a total resistance of 5.5 Ω. Due to the maximum current applied to the coil being 1 A, the power consumption is only 5.5 watt. The simulation was conducted in 2D axis-symmetric with different total elements and node according to the type of MR valves. The permeability of MR fluid was obtained from the manufacturer. The permeability of magnetic material of MR valve is assumed to be similar to the B-H curve of AISI 1020 while the permeability of non-magnetic is assumed to be the B-H curve of AISI 6061-T6. The results will be discussed in each section for each type of MR valve.

4 Pressure drop generated by MR valve

Figure 2 depicts the pressure drop versus current for the MR valve that was proposed in this study. The figure shows the MR valve could experience various pressure drop under different types of configurations because of the effect of the magnetic field intensity that was generated by the coils. During the off-state, the MR valve would work based on the viscosity of the MR fluid like the existing damper. It was identified that the serpentine MR valve can produce a damping force of 0.06 MPa in the absence of the input current and could increase up to 4.4 MPa by applying the input current of 1.0 A. Meanwhile the meandering configuration was able to produce damping force of 0.07 MPa and with the presence of 1.0 A, the range increases up to 3.6 MPa. Specifically, these two configurations portrayed the highest damping force produced by the five MR valve configurations.

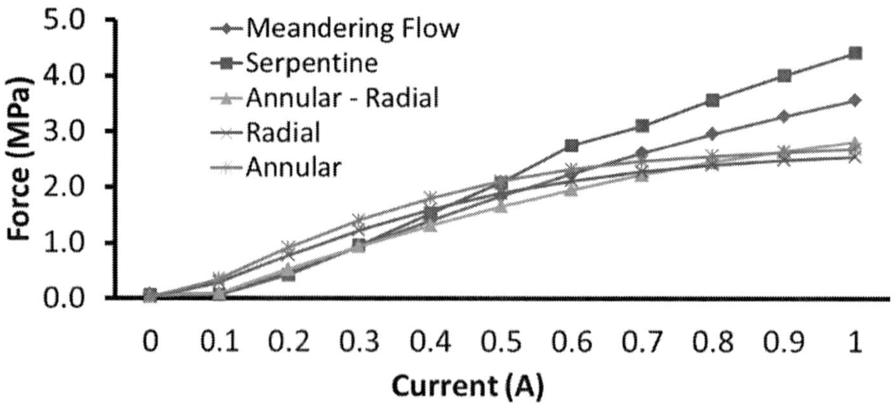

Fig. 2. Pressure drop comparison

Figure 3 depicts the off-state pressure drop generated by each MR valves. The figure shows meandering having the highest force and annular the lowest. Meanwhile, Figure 4 shows the on-state pressure drop generated. Serpentine configuration produced the highest while meandering the second highest in pressure drop generation.

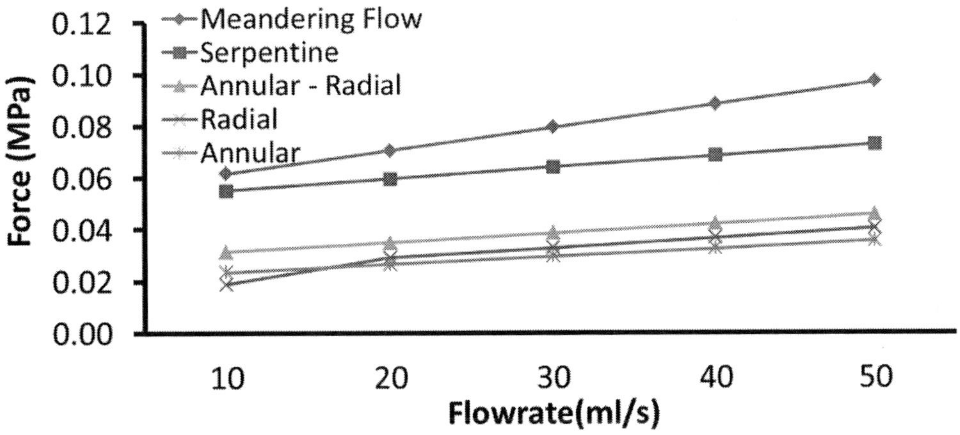

Fig. 3. Off-state pressure drop of MR valve

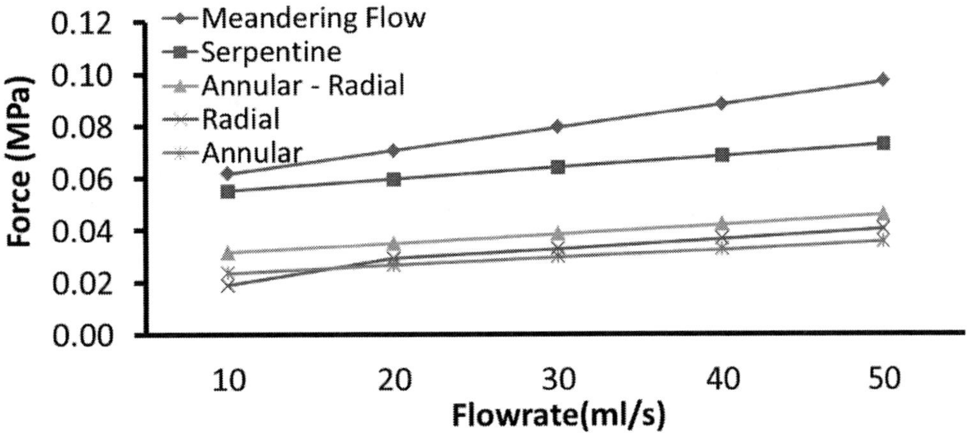

Fig. 4. On-state pressure drop of MR valve

Figure 5 illustrates the widest range of the pressure drop generated from each type of MR valves. Serpentine produced the biggest value compared to the other MR valves. The meandering valve ranked the second highest in pressure drop generated followed by annular-radial. The lowest falls to radial and is followed by the annular configurations.

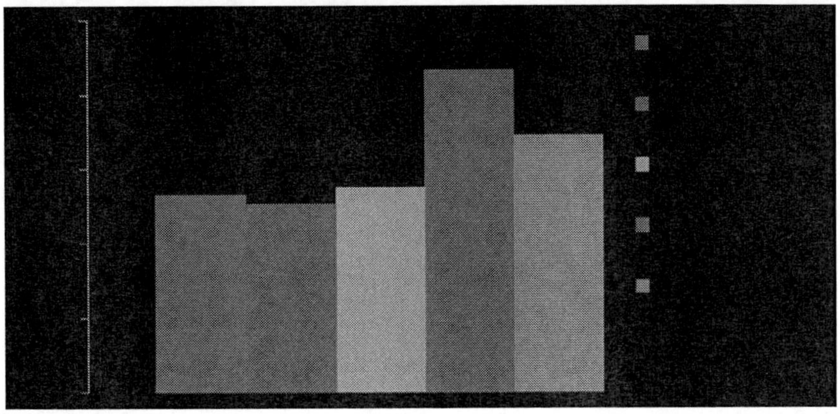

Fig. 5. Dynamic range of MR valve

5 Conclusion

The presented study discussed the comparison of five different types of MR valve: annular, radial, annular-radial, serpentine and meandering valves. For each type of MR valves, individual mathematical modelling of pressure drop for each type of MR valves were developed and the results were compared. The proposed approach aims to study the different configuration of MR valves' performances. The analysis also considers the same size of MR valves, gap size, number of turns and power consumption.

This research was financially supported by the Universitas Sebelas Maret (UNS) through Hibah Kolaborasi Internasional 2017 and Universiti Teknologi Malaysia under PRGS Grant (Vote No: 4L667) and Tier 1 Grant (Vote No: 13H55).

References

1. P.P. Phulé, Smart Mater. Bull. **2001**, 7, (2001).
2. F. Imaduddin, S.A. Mazlan, H. Zamzuri, M.A.A. Rahman, Appl. Mech. Mater. **663**, 685, (2014).
3. A.Y. Abd Fatah, S.A. Mazlan, T. Koga, H. Zamzuri, F. Imaduddin, Int. J. Appl. Electromagn. Mech. **50,** 29, (2016).
4. X. Bai, N.M. Wereley, Y. Choi, J. Intell. Mater. Syst. Struct. **27,** 944, (2016).
5. H. Sahin, F. Gordaninejad, X. Wang, Y. Liu, J. Intell. Mater. Syst. Struct. **23**, 949, (2012).
6. H. Hu, X. Jiang, J. Wang, Y. Li, J. Intell. Mater. Syst. Struct. **23**, 635, (2012).
7. A.Z. Bin Pokaad, K. Hudha, M.Z.B.M. Nasir, N.A. Ubaidillah, Int. J. Struct. Eng. **2,** 164, (2011).
8. F. Gordaninejad, X. Wang, G. Hitchcock, K. Bangrakulur, S. Ruan, M. Siino, J. Struct. Eng. **136,** 135, (2010).
9. F. Imaduddin, S.A. Mazlan, H. Zamzuri, M.A.A. Rahman, Ubaidillah, B. Ichwan, Smart Mater. Struct. **23,** 065017, (2014).
10. T. Xu, M. Liang, C. Li, S. Yang, J. Sound Vib. **355,** 66, (2015).
11. H.X. Ai, J. Intell. Mater. Syst. Struct. **17,** 327, (2006).
12. Q.-H. Nguyen, S.-B. Choi, N.M. Wereley, Smart Mater. Struct. **17,** 025024, (2008).
13. A. Grunwald, A.G. Olabi, Sensors Actuators, A Phys. **148**, 211, (2008)

 https://doi.org/10.1051/matecconf/201815902017

Surface Texture Influences on the Running-in Behavior and Friction Reduction

Zhouyong Hou [1], and *Tomomi* Honda[1*]
[1]Department of Mechanical Engineering, University of Fukui, Fukui, 910-8507, Japan

Abstract. For improving automobile fuel efficiency, the internal combustion engines must be required to reduce the friction and wear. Changing viscosity of lubricant and surface pressure could succeed, but the seizure is easy to happen in engines. However, the surface texture can solve those problems. The running-in behavior affects friction and wear on whole combustion engines. If the running-in is not carefully designed, catastrophic accident can happen. This experiment investigates that the running-in behavior is influenced by textured surfaces and the tested materials are the cast iron and the different area ratio of dimple of aluminum alloy combination. The friction coefficient and the number and size of wear particles are measured by the friction sensor and particle counter. After the tests, the worn surfaces are measured through using surface profile measurement systems, and some significant phenomena are observed and analyzed. The textured surface verifies good consequence and tribological advantages.

1 Introduction

At present, global warming has become increasingly serious. Reducing the exhaust emissions and improving the efficiency of the internal combustion engine have become primary task [1]. Via decreasing the friction and wear can improve these, usually we could succeed to change the viscosity of lubricant and surface pressure, but the seizure is easy to appear on the contact surface, such as a piston and cylinder in the engine, while surface texture can solve those problems [2-3]. In tribology, the advantages of surface texture have been reported by many researchers since 1960s and they also get same conclusion that the utilization of textured surface can improve the tribological performances of rubbing surfaces [4]. In this study, the aluminum alloy and the cast iron combination are tested in the experiment, micro concaving areas called dimples have been machined on the material.

*Corresponding Author : honda@u-fukui.ac.jp

surface, which is the surface texturing technology. In addition, the surfaces of the aluminum alloy have been machined with a different area ratio of dimples. Meanwhile, the friction coefficient and the number and size of wear particles are measured by the friction sensor and particle counter. After the experiment, the worn surfaces are also observed through using surface profile measurement systems, and some significant phenomena are analyzed.

2 Experiment

2.1 Experimental procedure

The schematic diagram of the tribometer is shown in Fig.1, in this experiment. The annular surface of a ring (FC230) was rubbed against the plate (AC8A) under a contact load of 800N as well as sliding velocity of 2.0 m/s. Micro debris was produced during the wear tests, because specimen surfaces were scratched during running-in tests. The sliding distance was 10000 m. The testing surface was immersed in lubricating oil, which is no additive and its viscosity at 60 ℃ is about 14.98 mm^2/s. The lubricating oil was circulated. The particle counter was used to measure the size and distribution of wear particles. During wear tests, the plate was fixed on the free revolution axis in order to transmit friction force to load cell and collected the data of friction force, and then the friction coefficient was calculated.

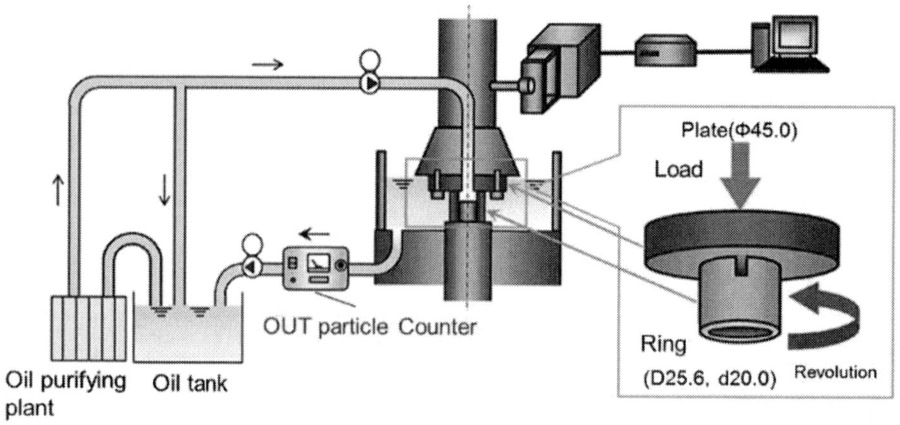

Fig.1 Schematic of wear test apparatus

2.2 Experiment material

Aluminum alloy has been widely used in the automobile industry because of its light weight and mechanical properties [5]. In the experiment, the specimen of the ring (FC230) is cast iron and the aluminum alloy is the plate of AC8A which highly contains Si as shown in Table 1, so that the AC8A has high wear resistance and low thermal expansion in high temperature [6]. As shown in Table 2, the surface texture is applied on the surface of AC8A through machining a great amount of micro concaving areas. For each specimen of AC8A, area ratio of dimples is different from 10 % to 40 %, but the surface roughness of the flat area is same. The Vickers hardness of AC8A and FC230 is 147 HV and 415 HV. Meanwhile, optical microscope images of the original surface are shown in Fig.2.

Table 1 Chemical composition of AC8A (mass%).

	Si	Cu	Mg	Zn	Fe	Mn	Ni	Ti
AC8A	11.0-3.0	0.8-1.3	0.7-1.3	<0.15	<0.8	<0.15	0.2-0.8	<0.2

Table 2 Properties of test materials

Material		AC8A(Plate) A	B	C	D	E	FC230 (Ring)
Arithmetic mean roughness *Ra*, μm		0.050					0.008
Dimple	Diameter, μm	150	175	204	247	302	-
	Depth, μm	2.8	3.9	4.6	6.3	10.0	
	Area ratio, %	10	13	18	28	40	
Vickers hardness HV		147					415

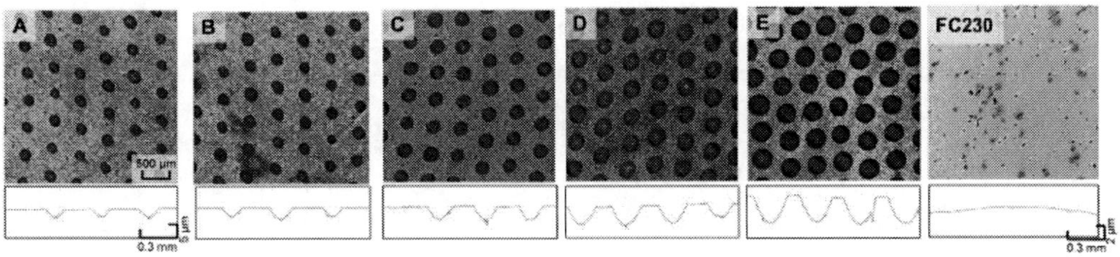

Fig.2 Optical microscope images of the specimens.

3 Results and discussion

3.1 Variations of friction coefficient with sliding distance

During the wear test (see Table 3), the friction coefficient as a function of sliding distance is also shown in the Fig.3 and the AC8A-A, AC8A-C, AC8A-D and AC8A-E show almost similar with friction coefficient, as well as all of them are higher, but AC8A-B shows the lowest friction coefficient in all of the specimens. Running-in distance of AC8A-B is lower than most specimens except for AC8A-E. Compared with the process of running-in, the AC8A-D and the AC8A-E whose dimple area ratios are large early occur the solid contact because of high contact pressure, but the AC8A-A and AC8A-B showed that the running-in distance is different. In addition, AC8A-B is superior to AC8A-A and AC8A-C reducing the distance of running-in [7]. (see Fig.4)

Table 3 Friction coefficient and running-in distance

	A	B	C	D	E
Steady-state friction coefficient μ	0.010	0.006	0.011	0.010	0.011
Running-in distance, m	736	707	794	718	693

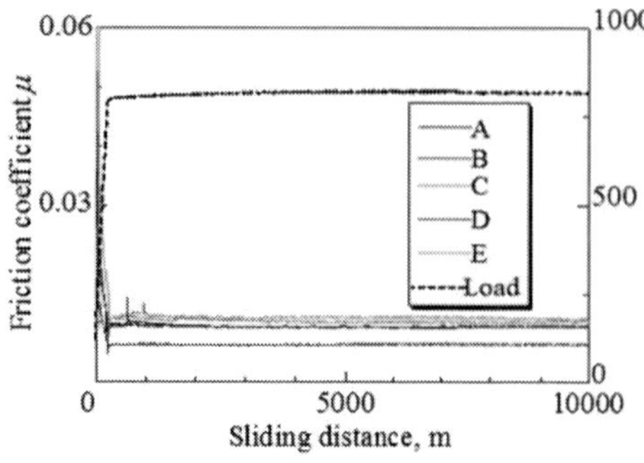

Fig.3 Variation of friction coefficient as a function of sliding distance

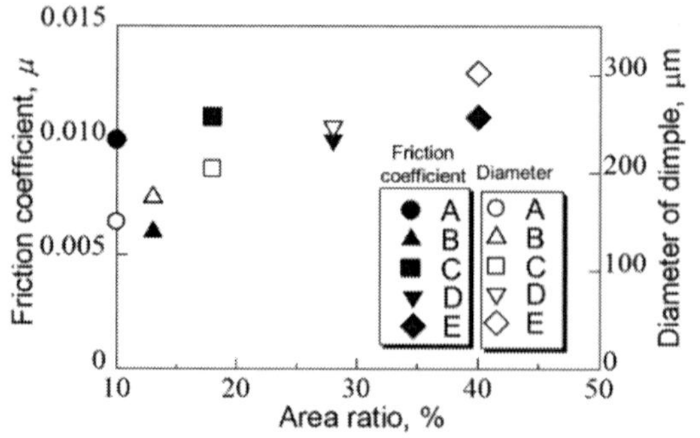

Fig.4 Friction coefficient vs. area ratio

3.2 Distribution of wear particles

In wear tests, the wear particles could include some significant information to realizing more about scuffing and wear. Figure 5 show the variation of the number and size of wear particles as a function of sliding distance. The size of wear particles from 5 μm to 55 μm were measured during the wear test and the number of wear particles at the size of 5-15μm increases during the initial stage of the test, but those of 15-55 μm are far inferior to those of 5-15 μm in the number after the running-in test. From Fig.5, through the variation of wear particles, we can know that plastic deformation takes place at the beginning of the wear test because the real contact area is very small. As we know, the contact pressure at this area is large beyond the yield strength of the AC8A [8-9]. Surface asperities detach and scratch each other. At the running-in process, two-body abrasive wear mainly occurred. However, after the running-in, two-body abrasive wear and three-body abrasive wear occurred concurrently.

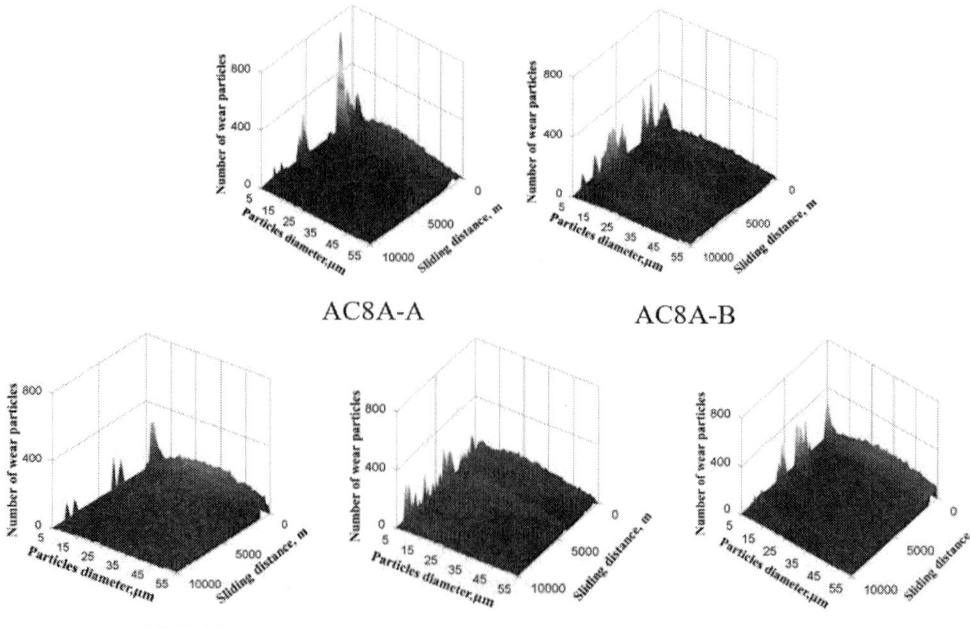

Fig.5 Variation of the number and size of wear particles

3.3 Observation of the specimen surface

Figure 6 presents the variation surface roughness in the plate of AC8A and the ring of FC230 through optical microscope observation. The combination of AC8A plate and FC230 ring, the AC8A-C, AC8A-D and AC8A-E show slighter wear scars than the AC8A-A and AC8A-B. Roughness of the AC8A-C, AC8A-D and AC8A-E are smaller than that of the AC8A-A and AC8A-B. The dimple area ratio of AC8A-A and AC8A-B is small as well as surface contact is large, because of wear particles becoming large, which result to three-body abrasive wear easily appearing and lead to surface becoming rough and surface roughness getting bigger. However, the dimple area ratio of AC8A-C, AC8A-D and AC8A-E is large and the wear particles are easily collected in the dimples so three-body abrasive wear hardly happens. Therefore, surface roughness becomes small.

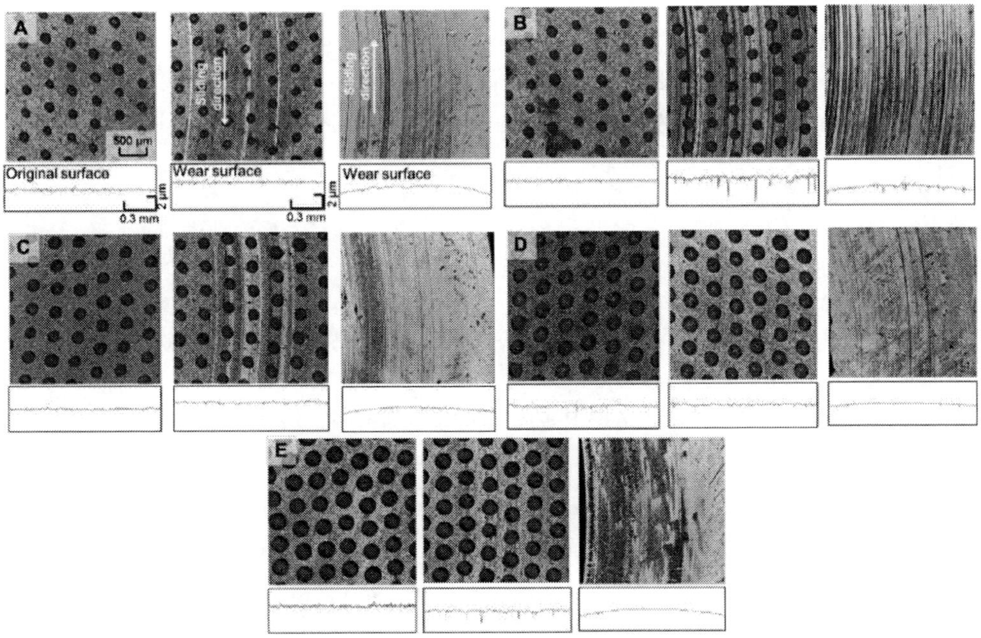

Fig.6 Optical microscope images of AC8A plate and FC230 ring

4 Conclusions

The influence of surface texture on the running-in behavior and friction reduction was investigated. In this study, the wear resistance of AC8A-B specimens performance is better than that of the other specimen. The AC8A-B has also achieved low friction coefficient under the same tribological conditions. Regarding the combination of AC8A and FC230, when the area ratio of dimple becomes large and the contact pressure becomes high, the friction coefficient becomes large because severe plastic deformation and abrasion happen in the interface. So, friction coefficient becomes higher. Meanwhile, regarding the combination of AC8A and FC230, when the area ratio of dimple of specimen is small and contact area increases, the opportunities of surface asperities contact increase as well as the plastic deformation and abrasion easily happen leading to the high friction coefficient. The area ratio of dimple of specimen is large and wear particles are so easy to be collected in the dimple that wear particles are small. So, the friction coefficient becomes low. Besides, at the beginning of running-in test, the number of wear particles whose size is from 5 μm to 15 μm is always large; After the completion of running-in, the numbers of wear particles whose size is from 15 μm to 55 μm steadily decrease. In the process of wear test, two-body abrasion and three-body abrasion occur in the interface and severe plastic deformation also appears. Therefore, regarding the combination of AC8A and FC230, the small area ratio of dimple of specimen leads to large surface roughness. While, the large area ratio of dimple of specimens leads to small surface roughness because the solid contact easily occur, and adhesion as well as abrasion easily happen in the interface. In other words, the study has proved tribological advantage obtained from textured surface. The experiment also proved textured surface is more superior property from the view of the optimal value of the dimple area ratio, it turned out that the dimples lead to the wear reduction and low friction coefficient in the contact surface.

References

1. K Holmberg, P Andersson and B Erdemir. *Global energy consumption due to friction in passenger cars*, Finland, **1-14,** (2011).
2. J.A.Williams, *Engineering tribology*, Department of Engineering, University of Cambridge, (2000).
3. M.priest and C.M Taylor, *Automobile Engine Tribology*, Wear,**193-203** (2000).
4. D.B.Hamilton, J.A.Walowit and CMA.Allen. *Theory of lubrication by microirregularities*. ASME J Basic Eng, 88: **177–185** (1966)
5. G.S.Cole and A.M.Sherman. *Light weight materials for automotive applications*, Wear, **3-9** (1995).
6. M. Okayasu, Y. Miyamoto and K. Morinaka. Department of Materials Science and Engineering, Ehime University, (2015)
7. T. Nagai and T. Honda. *Running-in behavior of diamond-like carbon film using an online particle counter*, University of Fukui, Japan, (2010).
8. P.A.Dearnleya, J.Gummersbachb and H.WeissbA.*The sliding wear resistance and frictional characteristics of surface modified aluminium alloys under extreme pressure*.Wear, **127-134**(1999).
9. N.Bay and T.Wanheim, *Real area of contact and friction stress at high pressure sliding contact*, Wear, **201-209** (1976).

Analysis of chip geometry and surface roughness at low speed turning in various commercial steels

Mudjijanto*, Sulistyo, and Rusnaldy

Department of Mechanical Engineering, Diponegoro University, Semarang, Indonesia 50275

Abstract. Rapid development of technology in manufacturing facilitates the industry to produce mass, high quality and affordable products. The challenge for small industries in manufacturing is to compete for product quality. Small manufacturing industries in controlling the quality of product deal with financial problems. Chip and surface roughness of low speed machining were tested to control the quality of material mechanical properties. The purpose of this study is to investigate the relation between chip characteristic and surface roughness with material hardness. This study used a conventional lathe with low rotation, cutting depth of 1 mm, dry turning process with variations of feed rate of 0.1 mm/rev, 0.2 mm/rev and 0.3 mm/rev, test material of AISI steel (1020, 1045, 1090, D2 and 4340) and tool material from carbides. Stereoscopic Optical Microscopy (SOM) was used to measure temporary shear angle while surface roughness tester was used to measure surface roughness. The results show that the shear angle of the chip and the surface roughness correlated with the hardness of the material.

1 Introduction

The rapid development of technology in manufacturing facilitates the industry to produce mass, high quality and affordable products. The challenge for small industries is to compete particularly for product quality. Small manufacturing industries in controlling product quality deal with financial problems. Chip and surface roughness of machining with low speed were tested to control the quality of material mechanical properties. The result of the machining process is always related to the final surface and chip. The surface finish is an indicator of quality and precision in the machining process and is considered as an important parameter in the industry [1]. Chips are leftover cutting materials in the machining process and are often regarded as waste.

Recent studies of chip dimension measurements aimed to determine optimal cutting conditions and generally used light alloys (magnesium, aluminum, titanium alloys) as the development of the aircraft industry. Research on dimension measurement of titanium chip was similar to shear angle measurement, height of peak, height valley and chip thickness [2]. Measuring the dimensions of chip in aluminum alloys where measurements were performed

* Corresponding author: mudjijanto09@yahoo.com

similarly to Batista et al and added distance measurements between chips (S) and the width of chips (bv) [3]. Analyzing the dimensions of aluminum alloy chips (UNS A97075-T6 (Al-Zn) and UNS A92024-T3 (Al-Cu)) by comparing the dimensions of chips [4]. Whereas in the steel the measurement of chips from machining lathe has not been done, the study still refers to formation of chip by analyzing the shear angle.

The purpose of this study is to investigate the hardness level of steel viewed from shear angle and surface roughness resulting from the process of machining lathe. It is expected to contribute positively to small industries in controlling the quality of material hardness.

2 Experimental

This study used various types of commercial steel. The chemical composition is shown in Table 1.

Table 1. The chemical composition of commercial steel (% wt).

No	Material	Chemical composition (%)							
		C	Si	Mn	P	S	Cr	Mo	Ni
1.	1020	0.10	0.16	0.43	0.016	0.007	0.041	0.006	0.023
2.	1045	0.44	0.21	0.59	0.015	0.008	0.03	-	0.008
3.	1090	0.68	0.29	0.77	0.019	0.15	1.13	0.22	0.17
4.	D2	1.54	0.27	0.32	0.023	0.014	11.46	0.66	0.26
5.	4340	0.34	0.29	0.63	0.012	0.023	1.57	0.214	1.41

The experiment used the following cutting parameters: low rotation of n 65 rpm, cutting depth of p 1 mm and the variations of feed rate f of 0.1, 0.2 and 0.3 mm/rev. Chips were generated on a dry machining process on a cylindrical bars (with the length of 120 mm and diameter of 19 mm) of commercial steel, Figure.1 (a). Testing using the C6127A model KIANGSI lathe, with WC-Co coated with Ti.N and Ti.CN from the Iscar manufacturer TMNG 332-GN ICR3028, Figure. 1 (b). A new cutting insert was used in each test in order to avoid the wear influence on the process and, by that, in the geometry of the chip [2].

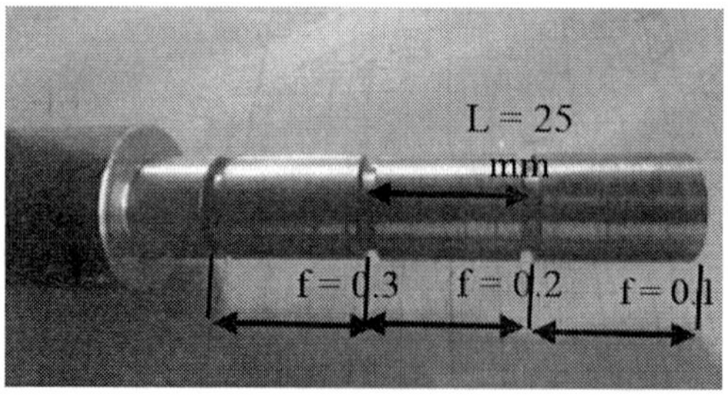

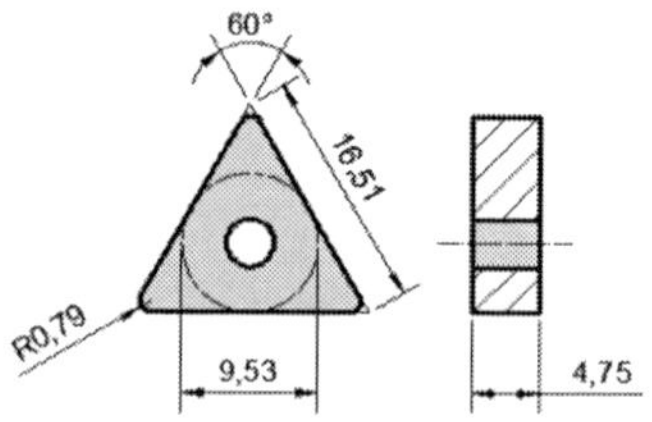

(a) (b)

Fig. 1. (a) Cylinder tests bars; (b) Geometry and dimension that define the tool used.

Chip samples were prepared for observation with Stereoscopic Optical Microscopy (SOM). In the making of metallographic processes, chip samples were mounted on the resin, sanded and lastly polished with aerosol until they were as smooth as glass. Chips were placed in longitudinal positions to show the longitudinal side. A digital microscope with cooling-tech software measured the shear angle. Figure 2 (a) shows the longitudinal chip of commercial steel with a microscope. In this figure the shear angle can be measured and distinguishable, Where φ: angle sliding on grating. The measurement of surface roughness as shown in Fig. 2 (b) used Mitutoyo Portable Surface Roughness Tester SJ_210.

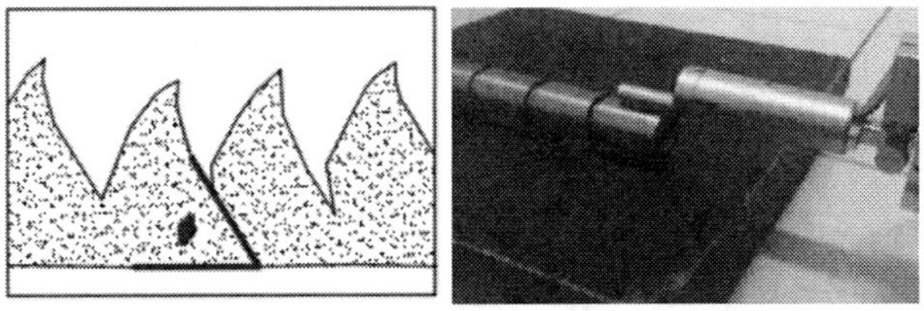

(a) (b)

Fig. 2. (a) Measured shear angle on the chip (b) Measured surface roughness.

3 Results

The chips formation of commercial steel the dry cutting proses are as shown in Table 2.

Table 2. Chips formation of various commercial steel.

	Feed rate 0.1 mm/rev	Feed rate 0.2 mm/rev	Feed rate 0.3 mm/rev	HB
AISI 1020	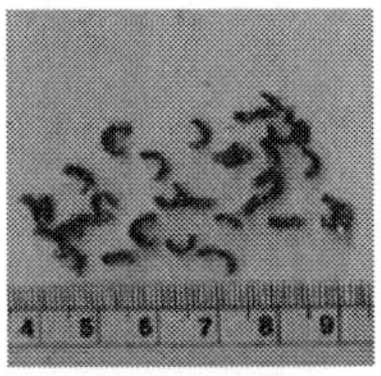	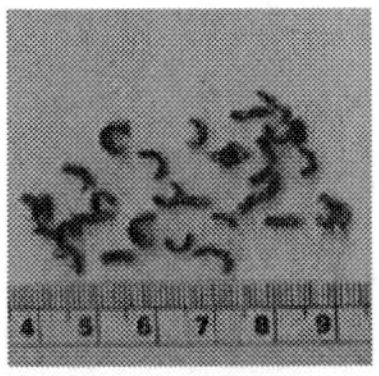	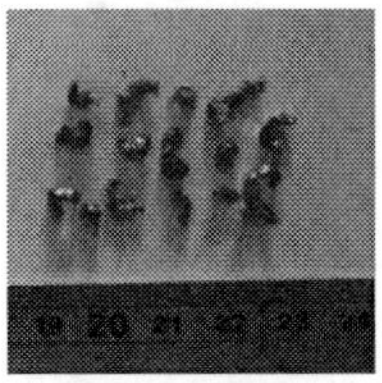	131
AISI 1045	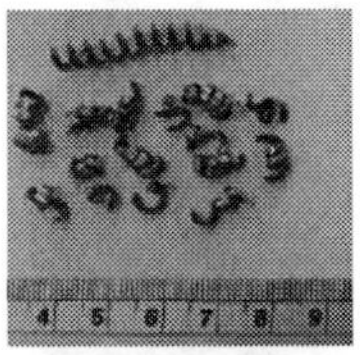	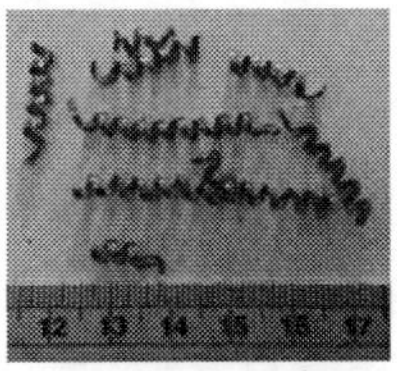	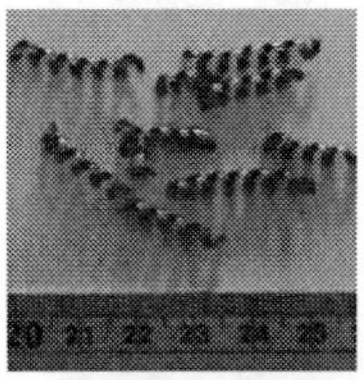	195
AISI 1090	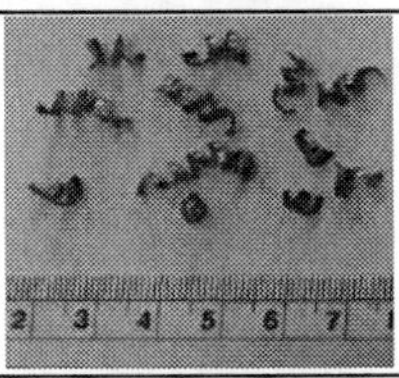	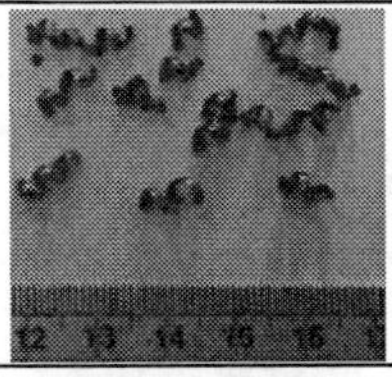		220
AISI D2	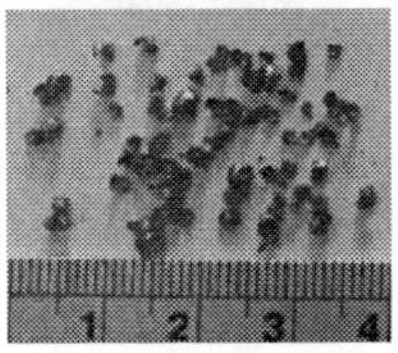			258
AISI 4340	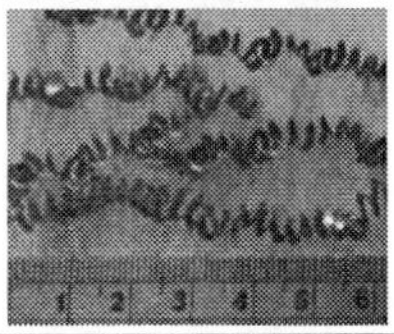	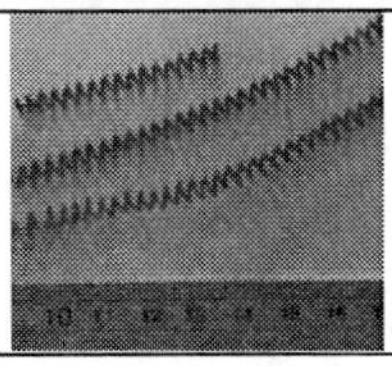		346

The results of the research on the chip formation as shown in Table 2 in general were that ribbon-shaped chips along with increasing spiral feed rates tended to be spiral in accordance with research conducted on aluminum alloy chips. The greater the rate of feed, the shorter the chip, or the smaller and more easily broken the diameter of the spiral [2]. The chip formation of commercial steel as set out in Table 2 is in accordance with ISO standards [5], thus for the AISI 1020 steel all feed rates were categorized in the form of arc type chip, AISI 1045 steel for all feed rates into short conical helical type, AISI 1090 steel for all feed rates into short tabular type, AISI D2 steel for all feed rates into elemental form and AISI 4340 steel for feed rates of 0.1 mm/rev washer type chip, the feed rate of 0.2 mm/rev for long washer type chip and 0.3 mm/rev for chip formation of short type washer. The chip shape for feed rates of 0.1 mm/rev and 0.2 mm/rev of commercial steel, the hardness value of ≤258 HB formed favorable chips while the hardness value of ≥258 HB steel formed unfavorable chip that could harm the operator.

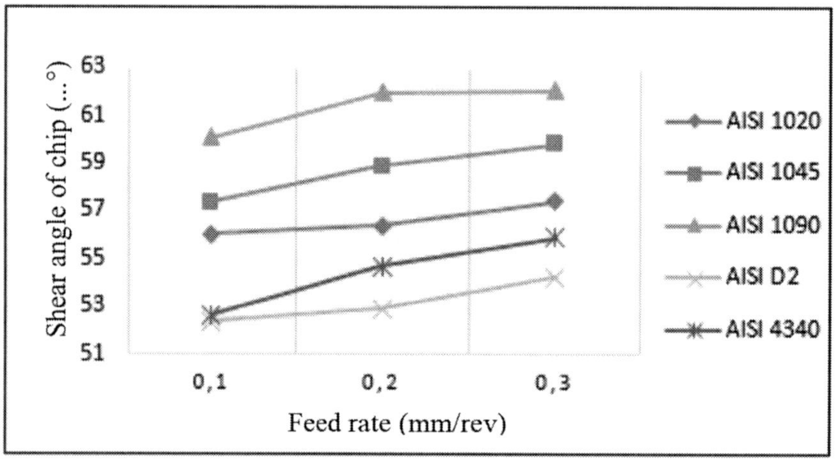

Fig. 3. Graph of the relationship between feeding with the shear angle of chip various commercial steels.

As shown in Figure 3, it can be concluded that higher feed rate will influence the increase of shear angle. In addition, the material hardness also affects the shear angle, the harder the carbon steel and the alloy steel, the increase of shear angle will be more significant.

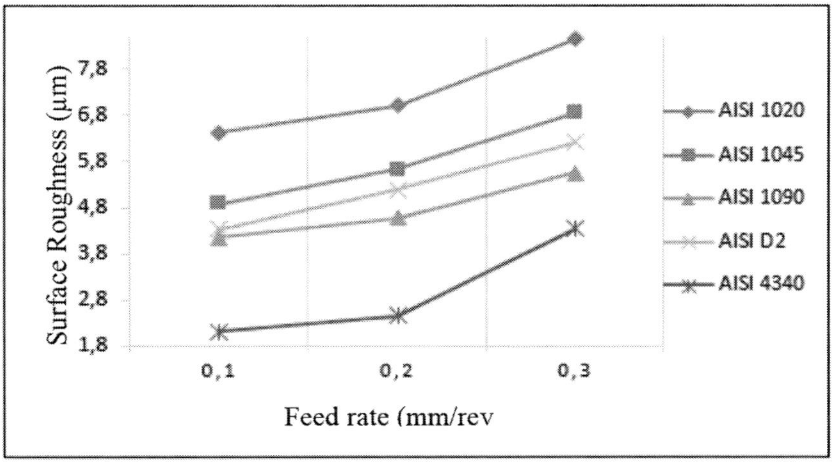

Fig. 4. The surface roughness value of commercial steel.

The study was conducted on the prediction of surface roughness of AISI 4340 steel (69 HRC). One of the testing parameters was feed rate. The process of machining lathe was without cooling and it was found that the increase in feed rate value correlated with the increase in the surface roughness value [6].
Analyzing the surface roughness of dry turning proses. The machining parameters used were cutting depth (ap), feed rate (F), cutting speed (vc) and tool radius (R). The result showed that the cutting depth and feed rate were the most influential surface roughness factors [1]. Figure 4 shows that the increase in feed rate values has implications for surface roughness values. This means that the higher the feed rate, the higher or more coarse the surface roughness.
The study that was conducted on the hardness of stainless steel revealed that the steel hardness would increase the quality of surface roughness, influeced by several factors, such as feed rate and cutting speed [7]. Based on the results shown in the Figure 4, it can be analyzed that the level of hardness of the material takes effect in the surface roughness value. The harder the carbon steel and the alloy steel, the better the quality of the surface roughness in the conventional lathe machining process.

4 Conclusion

Based on the results of data analysis in this study, it can be concluded that:
• The formation of chip generated on the commercial steel from conventional lathe machining processes with low rotation can be characterized as follows: that the increase in feed rates of the chip formation will tend to be shorter or the spiral diameter will be smaller and more easily broken. The chip form for feed rate of 0.1 mm/rev and 0.2 mm/rev of the hardness level of commercial steel ≥258 HB was favourable while the steel hardness value of ≤256 HB the chip was unfavorable.
• The increase in feed rates will affect the increase of shear angle. In addition, the material hardness also affects the shear angle, where the harder carbon steel and alloy steels, the more significant of the shear angle increases.
• The increase in feed rate values will have implications in surface roughness values. It means that the higher the feed rate, the higher or more coarse the surface roughness. The hardness level of the material affects the surface roughness value. The harder the carbon steel and the alloy steel, the better the quality of surface roughness.

• In general, the material hardness level affects the shear angle of the chip and surface roughness. Favourable chip will produce fine machining surface roughness when compared to unfavorable chip.

Reference

1. A. Torres, I. Puertas, and C. J. Luis. Surface Roughness Analysis on the Dry Turning of an Al-Cu Alloy. Procedia Eng, vol. **132**, 537–544 (2015)
2. M. Batista, J. Salguero, A. Gomez-Parra, S. Fernández-Vidal, and M. Marcos. SOM based methodology for evaluating shrinkage parameter of the chip developed in titanium dry turning process. Procedia CIRP, vol. **8**, 534–539 (2013)
3. Y. Sanchez, F. J. Trujillo, L. Sevilla, and M. Marcos. Study of the Influence of the Cutting Parameters on the Chip Geometry during Machining Alloy UNS A97075. Procedia Eng, vol. **132**, 513–520 (2015)
4. F. Vilches, L. Hurtado, F. Fernández, and C. Gamboa. Analysis of the Chip Geometry in Dry Machining of Aeronautical Aluminum Alloys. Appl. Sci, vol. **7**, no. 2, 132 (2017)
5. T. Segreto, A. Simeone, and R. Teti. Chip Form Classification in Carbon Steel Turning through Cutting Force Measurement and Principal Component Analysis. Procedia CIRP, vol. **2**, 49–54 (2012)
6. A. Agrawal, S. Goel, W. Bin, and M. Price. Prediction of surface roughness during hard turning of AISI 4340 steel. Appl. Soft Comput. J, vol. **30**, 279–286 (2015)
7. D. Palanisamy, P. Senthil, and V. Senthilkumar. Science Direct The effect of aging on machinability of 15Cr – 5Ni precipitation hardened stainless steel. Arch. Civ. Mech. Eng, vol. **16**, no. 1, 53–63 (2015)

Comparison of ratio loaded and unloaded foot area of flat foot and healthy foot in younger adults

Gunawan Dwi Haryadi[1,*], *Dwi Basuki* Wibowo[1], *Achmad* Widodo[1], and *Agus Suprihanto*[2]

[1]Center for Biomechanics, Central Laboratory of Research and Services, Diponegoro University, Indonesia
[2]Center for Biomaterials, Central Laboratory of Research and Services, Diponegoro University, Indonesia

Abstract. This study is aimed to investigate loaded and unloaded foot area ratio (RFA, ratio of foot area) as special tests for the basis of clinical examination of flat foot and healthy foot. Type of foot is determined by Cavanagh's arch indexes (AI) which is the ratio between mid foot area to entire footprint area excluding the toes. Type of foot is called high arch when AI<0.21, normal/healthy foot when 0.26>AI ≥ 0.21 and flat foot when AI>0.26. The entire loaded foot and footprint area for evaluating AI derived from a digital footprint is modified from document scanner, while the entire unloaded foot area derived from a 3D scanner. One hundred and two healthy students (87 males and 15 females, average aged 20 years and average BMI 22.51 kg/m^2) is asked voluntarily for doing footprint and scan. From 102 subjects found 63 participants identified as flat foot and 31 subjects are healthy feet. This study proves that the higher the value of AI the higher the value of RFA and foot type can be predicted by the value of RFA. For type of foot is high arch RFA<0.49, for healthy foot 0.55>RFA≥0.49 and for flat foot RFA>0.55.

1 Introduction

The foot supports the body weight and are subjected to many ground reaction forces during daily activities. Foot acts as a lever to resist thrust during walking, running and jumping. Arch is a segmental elevation of the foot which can be classified into three arches: medial longitudinal arch (MLA), lateral longitudinal arch and transverse arch [1]. Variation in the height of MLA leads to two main common foot deformities, pes planus (flat foot) and pescavus (high arch). Pes planus is a flat foot condition in which MLA diminishes, otherwise for high MLA is called pes cavus [2]. A normal foot must be plantigrade, have normal anatomical disposition and physiomechanics, be resilient with proper springiness to provide a rhythmic normal gait [3].

[1] Corresponding author: gunawan_dh@engineer.com

Some parameters are considered as the predisposing factors of flat foot, such as age, sex, body composition, family history, and types of footwear [4]. Males were twice more reliable to have flat foot than females. Overweight and obese children were more likely to have flat foot with proper weight [5].

Cavanagh's and Staheli's arch index were most commonly used for clinical diagnosis by using footprint. Cavanagh's arch index (AI) determine arch type from division of mid foot area to entire footprint area excluding the toes (Fig. 1a). If AI less than 0.21, arch type of foot is called high arch. If AI is between 0.21 and 0.26, arch type of foot is called normal arch, and If AI is greater than 0.26, arch type of foot is called flat foot [6]. Staheli has characterized the width of the foot in the area of the arch and heel and the ratio between these widths was called the Staheli's plantar arch index (Fig. 1b) [7]. A normal plantar arch index (PI), according to the Pediatric Orthopaedic Society is the one comprised within 2 standard deviations (SD) of the population average. If PI values are equal or above the sum of 2 is considered as flat foot.

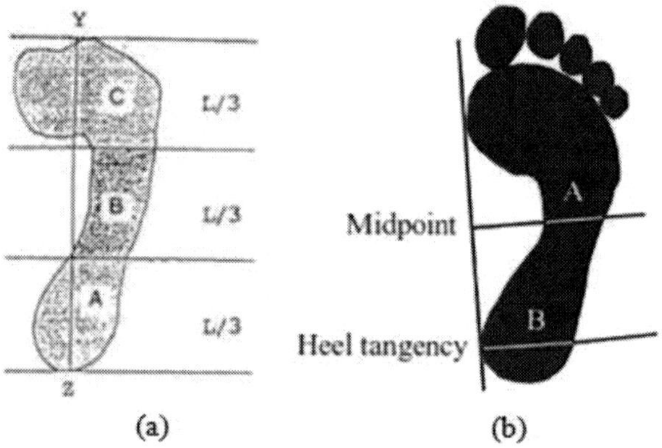

Fig. 1. Cavanagh's and Staheli's arch index

This study is aimed to investigate ratio loaded and unloaded foot area (RFA, ratio of foot area) as special tests for the basis of clinical examination of flat foot and healthy foot. Type of foot determine by Cavanagh's arch index (AI) where healthy foot is equal to normal foot. If this test showed the significant relationship between RFA to AI, than RFA can be used as an alternative tool for the diagnosis of types of foot which is simpler to apply.

2 Methods

This research is a cross sectional study. A total of 102 subjects comprising 87 males and 15 females, average aged 20 years (range 17-26 years) and average BMI 22.51 kg/m2 (range 15.21–41.60 kg/m2) were used for the study. The subjects are students of Mechanical Engineering Department of Diponegoro University, Semarang Indonesia. All subjects had no deformities of the lower limb or history of fractures of the foot based on the check up at Diponegoro National Hospital Semarang. The measurements of AI and foot area were carried out at the Center for Biomechanics, Biomechatronics, Biomaterials, and Biosignals Processing (CBIOM3S) of the Central Laboratory of Research and Services (*UPT Lab. Terpadu*) Diponegoro University.

For each subject, footprints were obtained using digital footprint modified from document scanner (Fig. 2a) [8]. To produce clear image, the amount of light from the environment must be minimized and the foot is cleaned by dipping into warm water (± 30

°C) for about 2-3 minutes and dried with tissue paper. The scanning process occurs when the research subject is really standing in an upright posture above the platform which can be assisted by the operator.

To calculate contact area with and without the toes, uncontacted foot must be removed using MATLAB Software. The procedure is as follows: 1) read the footprint image (Fig.2b) using *imread* function, 2) change the RGB image to gray scale image using *rgb2gray* function, 3) input the index level of image for filtering, 4) change the grayscale image to black and white image with *im2bw* function [9]. The results are obtained only images foot in contact as apparent in Fig. 2c. Calculating the entire loaded foot area (including the toes) is done by using *bwarea* function in MATLAB, but it is still in pixel unit. To change the foot area in pixels to mm^2, scanner resolution data is needed. The scanner is set to scan in 200 ppi (default setting of scanner). The entire loaded foot area can be measured by using equation 1 as shown:

$$\left(\frac{\text{area in pixels}}{\text{scanner resolution}}\right)\times 645.16\ \text{mm}^2 \qquad (1)$$

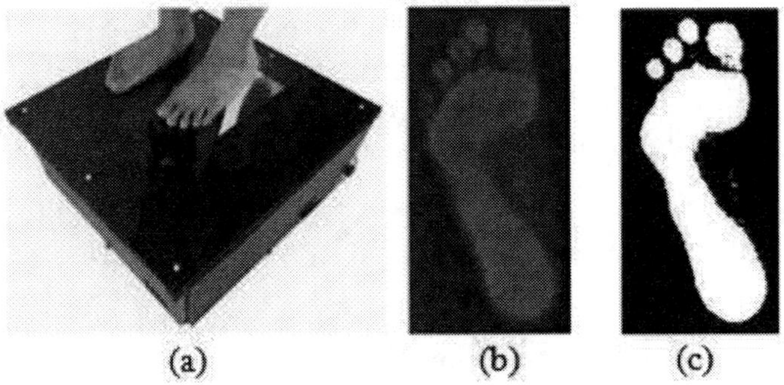

Fig. 2. Scanning of the foot using flatbed document scanner

The Cavanagh's arch index is calculated using MATLAB through the following procedures: 1) read the footprint image Fig. 2c and remove the toes using *bwareaopen* function, 2) use *imcrop* function to cut the image in the upper and lower limits, 3) divide image into 3 regions at the same length (Fig. 1a), 4) calculate the area of each region using *bwarea* and divide the middle third area with all area of footprint without the toes. Equation 2 shows the calculation of AI as follow:

$$\text{AI} = \frac{\text{B}}{(\text{A}+\text{B}+\text{C})} \qquad (2)$$

Where A, B and C are rear foot area, mid foot area and fore foot area respectively.

3 Results

Subjects were divided into 3 groups according to their foot types (Table 1). High archs comprising 7 males and 1 female with average AI 0.182 ± 0.027 (range 0.125 to 0.204) and average RFA 0.46 ± 0.04 (range 0.388 to 0.539), healthy feet comprising 28 males and 3 females with average AI 0.236 ± 0.015 (range 0.210 to 0.260) and average RFA 0.52 ± 0.05

(range 0.437 to 0.607) and flat feet comprising 52 males and 11 females with average AI 0.319 ± 0.040 (range 0.261 to 0.385) and average RFA 0.61 ± 0.09 (range 0.400 to 0.860).

Table 1. Average AI and RFA of the subjects according to foot types.

Foot type and Gender	Amounts	Age (years)	BMI (kg/m^2)	AI	LFA (mm^2)	ULFA (mm^2)	RFA
High arch							
Male	7	20.29 (1.60)	21.00 (4.26)	0.190 (0.015)	9393.29 (1092.45)	20657.71 (1157.99)	0.45 (0.05)
Female	1	23.00	16.85	0.125	7455.00	15535.00	0.48
Healthy foot							
Male	28	20.14 (2.01)	22.01 (3.52)	0.238 (0.015)	10906.46 (1113.55)	20855.93 (1686.97)	0.52 (0.05)
Female	3	21.33 (4.04)	19.69 (3.93)	0.218 (0.011)	7763.33 (573.93)	15799.33 (952.52)	0.49 (0.05)
Flat foot							
Male	52	19.98 (1.90)	23.16 (4.49)	0.318 (0.039)	13134.18 (2152.75)	21354.19 (1966.82)	0.62 (0.09)
Female	11	19.18 (1.17)	22.93 (3.75)	0.319 (0.046)	10574.81 (2068.15)	18272.00 (1269.30)	0.58 (0.10)

Values are expressed as mean ± (standard deviation). AI, Cavanagh's arch index; LFA, loaded foot area; ULFA, unloaded foot area; and RFA, ratio of foot area = LFA/ULFA

Fig. 3a shows the comparison between average AI and average RFA according to foot types. There was a significant relationship between average AI to average RFA which can be presented by linear regression (equation 3) with coefficient of correlation r = 0.99 (Fig. 3b).

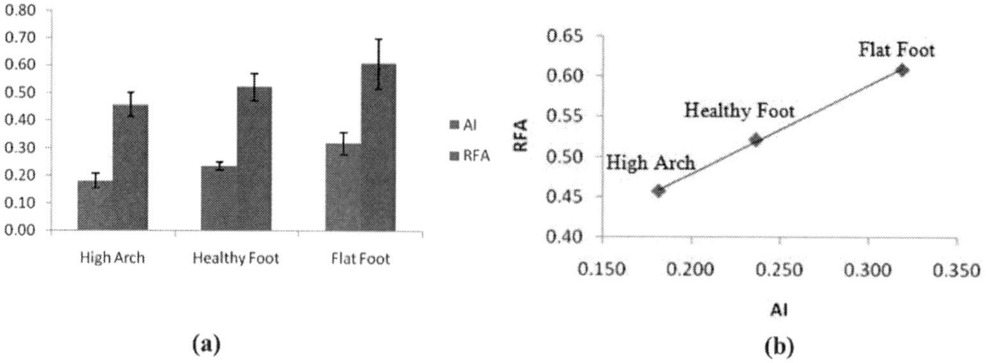

Fig. 3. Comparison and the relationship between AI and RFA according to foot types

The data showed that the overall prevalence of healthy foot and flat foot were 30.39% and 61.76% respectively. The prevalence of flat foot in male and female students are 50.98% and 10.78% respectively (Table 2)

Table 2. The comparison of prevalence of healthy and flat foot (%)

Subjects	High arch	Healthy foot	Flat foot
Male	6.86	27.45	50.98
Female	0.98	2.94	10.78
Total	7.84	30.39	61.76

4 Discussion

The current study shows that the prevalence percentage of flat foot among the total population is 61.76%, two times greater than healthy foot [14-17]. Flat feet were high in males than females, the prevalence of flat feet was 50.98% in males and 10.78% in females. The prevalence percentage of flat feet were also higher 23.50% in males and 7.85% in females compared to the healthy feet [18-20].

The result of foot area found that male had significantly greater foot area than female. This result is corresponding to the research by Lee et al. [19], Periyasamy et al. [20] and Wunderlich et al [21]. The average LFA for normal feet are 10906.46 mm^2 in males and 7763.33 mm^2 in females and the average ULFA are 20855.93 mm^2 in males and 15799.33 mm^2 in females. Offcourse the average foot area of flat foot is greater than normal foot and male had significantly greater foot area than female too. The average LFA for flat feet are 13134.18 mm^2 in males and 10574.81 mm^2 in females and the average ULFA are 21354.19 mm^2 in males and 18272.00 mm^2 in females.

Because the result of measurement of LFA for flat foot are greater than healthy foot and increasing of ULFA for flat foot inversely linear with the increasing ULFA for healthy foot, offcourse RFA of flat foot is greater than healthy foot. The average RFA for healthy foot and flat foot are 0.52 ± 0.05 and 0.61 ± 0.09 respectively.

In this study the technique employed for obtaining footprints is simple, non-invasive, easy to apply and does not use radiation as well. This scanner can tackle shortage of wet foot test which is not accurate because of the difficulty of knowing that the subjects are standing in upright position when printing their foot at a piece of paper [22]. But the method for classifying foot types by Cavanagh's arch index is rather difficult compared to the Staheli's arch index [7][15]. From the result of this study which showed that significant correlation between AI and RFA, classifying foot types can be done simpler only by measuring LFA using flatbed document scanner and ULFA using 3D scanner compared to the Cavanagh's arch index.

5 Conclusion

This study proves that the higher the value of AI the higher the value of RFA and foot type can be predicted by the value of RFA. For type of foot is high arch RFA < 0.49, for normal/healthy foot $0.55 > \text{RFA} \geq 0.49$ and for flat foot RFA > 0.55.

The method for classifying foot types by calculation of the ratio entire loaded and unloaded foot area is simpler and easy to apply compared to the Cavanagh's AI. It is not necessary to divide the footprint image (without the toes) into 3 regions and calculating the area of each region which rather difficult to build in application software. Even the footprint image can be done by wet foot test with a little bit carefully on a piece of paper chart and the entire loaded foot area is calculated by the amount of squares which printed ink on paper.

This work was supported by the Application and Development Research Grant UNDIP Contract No: SP DIPA - 042.01.2.400898/2016

References

1. L.T. Staheli, D.E. Chew, M. Corbett, The longitudinal arch. J Bone Joint Surg Am, **69**, 426-8 (1987)
2. D.H. Van Boerum, B.J. Sangeorzan, Biomechanics and pathophysiology of flat foot. Foot Ankle Clin, **8**, 419 (2003)
3. A.E. Hunt, R.M. Smith, Mechanics and control of the flat versus normal foot during the stance phase of walking. Clin Biomechanics, **19**, 391-7 (Bristol, Avon 2004)
4. T. Abolarin, A. Aiyegbusi, A. Tella, Predictive factors for flatfoot: The role of age and footwear in children in urban and rural communities in South West Nigeria. The Foot 21, 188-192 (2011)
5. A. Evans, K. Rome, Cochrane review of the evidence for non-surgical interventions for flexible pediatric flat feet. Eur. J. Phys. Rehab. Med, **47**, 69-89 (2011)
6. R. Peter, Cavanagh, M.M. Rodgers, The Arch index: a useful measure from footprints, **20(5)**, 547-551 (J. Biomechanics, 1987)
7. A.J. Hernandez, L.K Kimura, M.H.F Laraya, Calculation of staheli's plantar arch index and prevalence of flat feet: a study with 100 children aged 5-9 years, **15(2)**, 68-71 (Acta Ortop Bras, 2007)
8. D.B. Wibowo, G.D. Haryadi, A. Widodo, et al, Estimation of calcaneal loading during standing from human footprint depths using 3D scanner. AIP Conf. Proc. 1788, 030063-1-030063-9, doi 10.1063/1.4968316 (Acta Ortop Bras, 2007)
9. R.C Gonzalez, R.E Woods, S.L Eddins, Digital Image Procesing Using MATLAB. **2**, (Gatesmark Publishing, A Division of Gatesmark, LLC 2009)
10. R.I. Harris, T. Beath, Army foot survey: an investigation of foot ailments in Canadian soldiers, 116 (1947)
11. PS Igbigbi, BC Msamati, MB Shariff, Arch index as a predictor of pes planus: a comparative study of indigenous Kenyans and Tanzanians. J Am Podiatr Med Assoc, **95(3)**, 273-6 (2005)
12. S. Yucesan, H. Dindar, I. Olcay, H. Okur, S. Kilicaslan, Y. Ergoren, et al, Prevalence of congenital abnormalities in Turkish school children, **9(4)**, 373-80 (Eur J Epidemiol, 1993)
13. B. Alamy, An epidemiologic study of flat foot in Iran, **55(3-4)**, 78-83 (Tehran Univ Med J, 1997)
14. A.K Reihaneh, F. Aliabadi, *Prevalence of flat feet: Comparison between male and female primary school students. Iranian Rehabilitation Journal*, **11,** 18 (2013)
15. H.H. Hazzaa, G.H. El-Meniaway, S.E. Ahmed, M.B. Bedier, *Correlation between gender and age and flat foot in obese children*, **10(4),** 207-215 (2015)
16. J.H. Chang, C.L. Wang, Y.W. Shen, L.C Hong, Prevalence of flexible flatfoot in Taiwanese school-aged children in relation to obesity, gender and age, **169**, 447-452 (2010)
17. Lee, Yu-Chi, Lin, Gloria, J Wang, Mao-Jiun, Comparing 3D foot scanning with conventional measurement method. Journal of Foot and Ankle Research, **7**, 44 (2014)
18. R. Periyasamy, et al, Preliminary investigation of foot pressure distribution variation in men and women adults while standing, 142 – 148 (The Foot 21, 2011)
19. R.E. Wunderlich, P.R. Cavnagh, Gender differences in adult foot shape: Implications for shoe design, **33,** 605-611 (Med. Sci. Sports Exer, 2001)
20. R. Stephen, Urry, S.C. Wearing, Comparison of footprint indexes calculated from ink and electronic footprints, **91(4),** 203-209 (J Am Podiatr Med Assoc, 2001)

Optimization of *Ridge Waveguide* Structure for Temperature Sensor Application Using Finite Difference Method

Ian Yulianti[1,*], *Ngurah Made* D. P[1], *Yuni* Lestiyanti[1], *Ojo* Kurdi[2]

[1]Physics Department, Faculty of Mathematics and Natural Sciences, Universitas Negeri Semarang, Central Java, Indonesia
[2]Department of Mechanical Engineering, Faculty of Engineering, Universitas Diponegoro, Central Java, Indonesia

Abstract. This work presents optimization of the dimension of ridge waveguide structure to be used as temperature sensor. The objective of the work is to obtain an optimum dimension of ridge waveguide so that it provides high sensitivity to temperature. Ridge waveguide structure was chosen since it has strong light confinement and low bending loss. The optimization was done by simulating the temperature distribution and electric field distribution inside the waveguide using Finite Difference Method (FDM). The simulations were done for various waveguide width and waveguide thickness. The results showed that as the width increased, the sensitivity decreased. The optimum ranges of waveguide dimensions are between 10μm to 20μm and 15μm to 25μm for width and thickness, respectively.

1 Introduction

Temperature sensors have attracted much attention due to its important role in various applications such as environmental monitoring, chemical industry and automotive industry. Conventional methods have been used to develop temperature sensor such as mechanical and electrical method. However, conventional methods are not suitable for conductive and corrosive environment [1]. Therefore, optical temperature sensors have been proposed to replace the conventional sensors. The most common optical temperature sensor is optical fiber-based sensor with various methods such as interference method [1], wavelength modulation method [2] and intensity modulation method [3]. Optical fiber sensors have advantages of free from electromagnetic interference, suitable for hazardous environment and can be arranged in multiplexed array. However, optical fiber sensor suffers from bending loss that degrades the accuracy, especially for intensity modulation based sensor.

* Corresponding author: ianyulianti@gmail.com

Besides optical fiber-based sensor, optical temperature sensor can be realized by using optical waveguide. Compared to optical fiber, optical waveguide has some advantages such as robust, has high sensitivity, easy light intensity measurement and can be realized in on-chip sensor[4]. Various optical waveguide sensors have been proposed for various applications such as for refractive index measurement [5-7], and biosensor [8, 9]. The published waveguide-based optical sensor were proposed by implementing various method such as surface plasmon resonance [4, 10, 11], Bragg grating [7] and mode coupling [6]. For temperature measurement, waveguide-based optical sensor has been proposed using intensity modulation method by applying curved waveguide [12]. It was shown that by decreasing the curvature radius, the sensitivity increased. However, decreasing curvature radius limits the sensor's working range. Meanwhile, the most recent research proposed a temperature sensor by using prism-waveguide with silicon acrylate as sensing layer [13]. The working principle is based on detection of lateral shift of the reflected light. Despite its high accuracy, the measurement procedure is not straightforward. Simple measurement can be done by adopting intensity modulation method using simple waveguide structure such as ridge waveguide. Hence, in this work, we investigate the sensitivity of ridge waveguide and optimize its dimension for temperature sensor application

2 Measurement Principles and Waveguide Structure

Light propagation in waveguide highly depends on refractive index of waveguide material which are cladding and core material. High refractive index difference between core and cladding will result in high light confinement which is indicated by high propagation constant [14]. Due to thermo-optic nature of material, refractive index is thus changed as the waveguide's temperature change. Therefore, the propagation constant and, consequently, the output light intensity transmitted by the waveguide will also change as temperature changes. Hence, by measuring the light intensity, the waveguide's surrounding temperature can be determined.

Among various waveguide types, ridge waveguide structure is interesting for temperature sensor application since the core is directly exposed to environment i.e. the core is not surrounded by cladding material as shown in Figure 1. Moreover, ridge waveguide has strong light confinement and low bending loss. As material, polymer was chosen since it has high thermo-optic coefficient ($-1{,}7 \times 10^{-4}$/K) that allows high refractive index change due to small change of temperature.

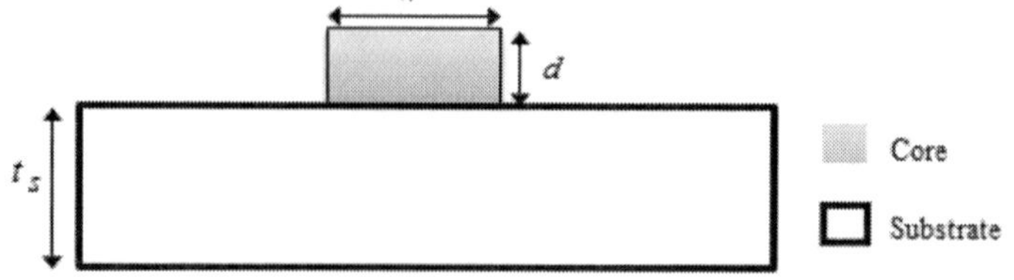

Fig. 1. Cross-sectional view of ridge waveguide

3 Optimization Method

Optimization of waveguide dimension was done by simulating the temperature effect to propagation constant of the waveguide. The first stage of the optimization was performing simulation of temperature distribution inside the waveguide by solving heat transfer equation using Finite Difference Method (FDM). After temperature distribution simulation, the next stage was calculating the waveguide refractive index change due to the temperature change by using thermo-optic coefficient formula. Then, the electric field distribution inside the waveguide and the propagation constant were obtained by conducting modal analysis. The simulations were done for various temperature values and various waveguide widths and thicknesses. The sensitivity of each waveguide dimension was obtained by plotting the propagation constant vs. temperature.

3.1 Temperature Distribution Simulation

Temperature distribution inside waveguide was obtained by solving heat transfer equation in constant thermal conductivity as defined by

$$k\nabla^2 T + Q(x,y,z,t) = \rho c_p \frac{\partial T}{\partial t} \tag{1}$$

where ρ is the material density, c_p is the specific heat, k is the thermal conductivity which is 0.17 W/mK for polymer and $Q(x,y,z,t)$ is the heat generation rate per unit volume. Since the length of the waveguide is much higher than the width, then the heat transfer simulation was done in two dimensional analyses. The simulation was also done for steady state heat conduction, thus equation (1) can be simplified as

$$k\left(\frac{\partial^2 T(x,y)}{\partial x^2} + \frac{\partial^2 T(x,y)}{\partial^2 y^2}\right) + Q(x,y,z,t) = 0 \tag{2}$$

Equation (2) was solved numerically using FDM by applying centered-difference approximation. In the FDM, the domain of analysis is discretized with a rectangular grid of points, which may be of constant or variable spacing. By using constant spacing, Equation (2) becomes

$$4T_{i,j} = T_{i+1,j} + T_{i-1,j} + T_{i,j+1} + T_{i,j-1} + \frac{\mathrm{Q}\,\Delta x\,\Delta y}{k} \tag{3}$$

where Δx and Δy are the grid size. As boundary condition, all sides of the computation window were assumed to be subjected to convection while the bottom surface is considered as heat sink.

The temperature change was then used to calculate the refractive index change of waveguide by using thermo-optic coefficient (*Toc*) formula which is defined as

$$Toc = \frac{dn}{dT} = -\left(\frac{\sigma \partial n}{\partial \sigma}\right)_T \eta + \left(\frac{\partial n}{\partial T}\right)_\sigma \tag{4}$$

where η is the coefficient of volume expansion of material, n is the refractive index, and σ is the material density [15].

3.2 Modal Analysis

In waveguide design process, modal analyses is an important stage since it determines the value of propagation constant (β) and the corresponding modal pattern for each desired mode. The equation to be solved in modal analyses is the Maxwell equation which is defined by

$$\frac{\partial^2}{\partial x^2}E(x,y) + \frac{\partial^2}{\partial y^2}E(x,y) + \left[k_o^2 n(x,y)^2 - \beta^2\right]E(x,y) = 0 \tag{5}$$

for plane wave that propagates along z direction with propagation constant β. $E(x,y)$ is the electric field, k_o is a free space wave number and n is the index of refraction. By double integrating Equation (5) and multiplying with $E(x,y)$, then β can be written as

$$\beta^2 = \frac{\iint E\left[\frac{\partial^2}{\partial x^2}E + \frac{\partial^2}{\partial y^2}E + k_o^2 n^2 E\right]dxdy}{\iint E^2 dxdy} \qquad (6)$$

Like temperature distribution simulation, Equation (5) and (6) were also discretized by using FDM with centered-difference approximation to obtain the propagation constant and electric field distribution inside waveguide.

4 Results and Discussions

The simulations were done for various temperature values which are in the range of 30°C-100°C with increment of 10°C. All stages were then repeated for various waveguide dimensions. In this work, the width of the waveguide (w) was incremented from 10μm to 50μm with increment of 5μm. The thickness (d) was also varied from 10μm to 50μm with increment of 5μm. Meanwhile, the substrate was maintained at thickness of 30 μm. The refractive index of core (n_c) and substrate (n_s) are 2.65 and 1. 489, respectively.

The results showed that as temperature increased, the electric field inside waveguide was less confined in the waveguide as indicated by more scattered electric field distribution as shown in Figure 2. Therefore, it results in low propagation constant. The less confined waveguide was occurred due to the decrease of refractive index which is resulted from negative thermo-optic coefficient of polymer.

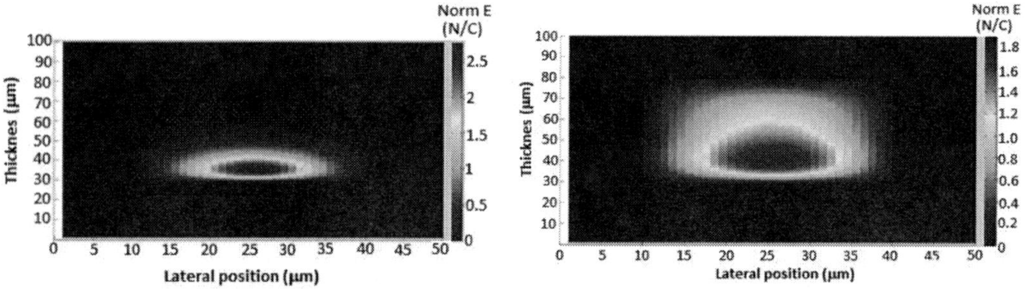

(a) (b)

Fig. 2. Electric field distribution inside waveguide with d=50 μm and w=30 μm for environment temperature of 10°c (a) and 30°C (b).

As shown in Figure 3(a), for waveguide width of 10μm, 30μm and 50μm and thickness of 50μm, the propagation constant decreased as temperature increased which results in negative curve slope. The trend is also valid for all waveguide dimensions. Therefore, the sensors have negative sensitivity. The sensitivity for various waveguide widths and fixed thickness is shown in Figure 3(b). It is shown that by increasing waveguide width up to 25μm, the sensor sensitivity almost maintain at value of around -680 rad/m°C. However, for width of above 25μm, the sensitivity degraded significantly. The highest sensitivity was observed at width of 20μm, which is -685.71 rad/m°C.

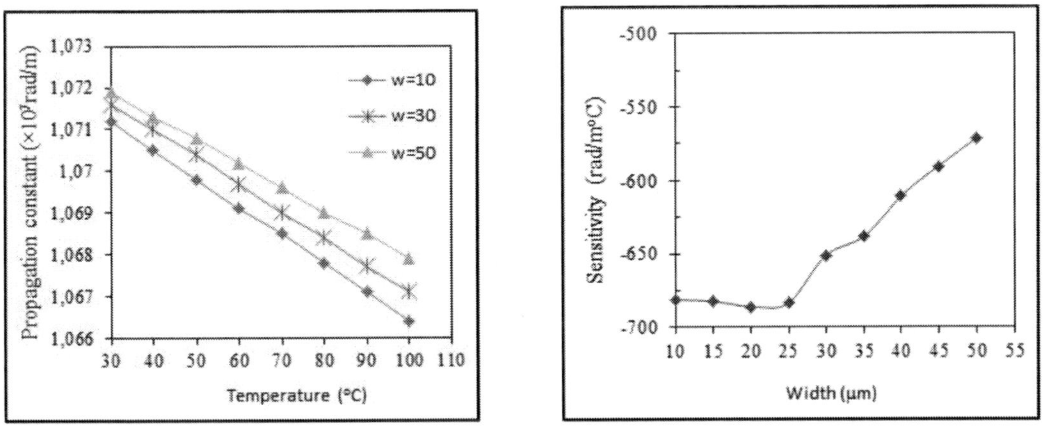

(a) (b)

Fig. 3. Temperature effect to propagation constant for various widths (d=50µm) (a) and sensitivity vs. width for d= 50 µm (b).

Sensitivity's for various widths and various thicknesses are depicted in Figure 4. It is shown that for all thickness values, waveguide widths of below 20µm provide high sensitivity as shown by the black contour on the graph. However, the highest sensitivity was achieved at waveguide width of 10µm and thickness of 15µm, 20µm and 25µm which is -691.67 rad/m°C. Therefore, the optimum ranges of the ridge waveguide dimensions are between 10µm to 20µm and 15µm to 25µm for width and thickness, respectively. Hence, it can be concluded that small waveguide dimensions provide high temperature sensitivity. The results are reasonable since the effect of environment temperature will be higher in small waveguide than that in large waveguide. As consequence, high change of propagation constant will occur in small waveguide.

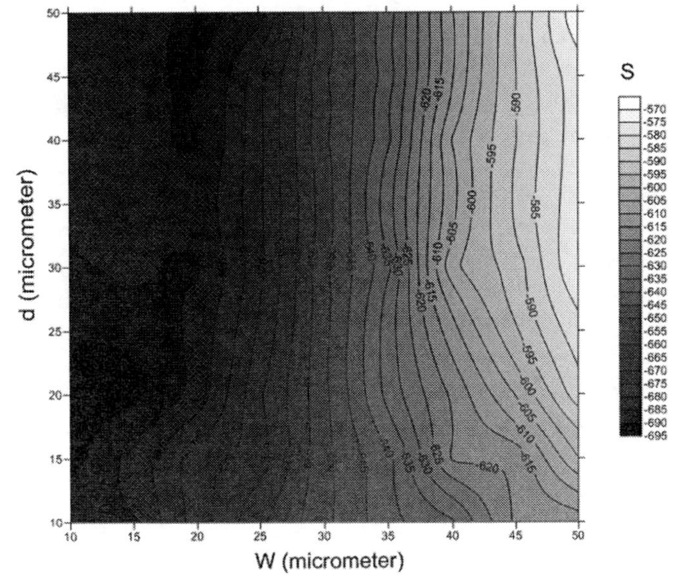

Fig. 4. Temperature sensitivity of ridge waveguide for various widths and various thicknesses.

5 Conclusions

Simulation and optimization of ridge waveguide have been conducted. It was shown that ridge waveguide provide high sensitivity to temperature. The optimum dimension for temperature sensor application is in the range of 10μm to 20μm and 15μm to 25μm for width and thickness, respectively. Therefore, ridge waveguide is a good candidate to be developed as temperature sensor. However, response time of the sensor need to be further investigated which will be carried out in our future work.

References

[1] L. Ma, Z. Kang, Y. Qi, and S. Jian, Optik. **126** (2015).
[2] J. Li, L. Gai, H. Li, and H. Hu, Sensors and Actuators A: Physical. **263** (2017).
[3] A. T. Moraleda, C. V. García, J. Z. Zaballa, and J. Arrue, Sensors. **13** (2013).
[4] W. Du and F. Zhao, Materials Letters. **115** (2014).
[5] B. Hooda and V. Rastogi, Optik - International Journal for Light and Electron Optics. **143** (2017).
[6] G. C. Yadav, G. Sharma, S. Kumar, Deepak, R. Kumar, S. Prasad, *et al.*, Optik - International Journal for Light and Electron Optics. **138** (2017).
[7] N. Saha, A. Kumar, and A. Mukherjee, Optics Communications. **396** (2017).
[8] R. Horvath, H. C. Pedersen, N. Skivesen, and N. B. Larsen, "Optical waveguide sensor for monitoring living cell morphology," in *CLEO/Europe. 2005 Conference on Lasers and Electro-Optics Europe, 2005.*, 2005, pp. 630-630.
[9] Y. Fan, Y. Ding, H. Ma, N. Teramae, S. Sun, and Y. He, Biosensors and Bioelectronics. **67** (2015).
[10] R. Levy and S. Ruschin, Sensors and Actuators B: Chemical. **124**, 2, (2007).
[11] P. K. Teotia and R. S. Kaler, Optics Communications. **395** (2017).
[12] M. Remouche, R. Mokdad, A. Chakari, and P. Meyrueis, Optics & Laser Technology **39** (2007).
[13] Y. Chen, HongRen, and TingtingTang, Optik. **124** (2013).
[14] I. Yulianti, A. S. M. Supa'at, S. M. Idrus, and A. B. Mohammad, "Optimization of buried type waveguide for single-mode operation," presented at the Proceedings of the Fifth IASTED International Conference on Communication Systems and Networks, Langkawi, Malaysia, 2008.
[15] M.B.J Diemeer, Optical Materials. **9** (1998).

Experimental Study Performance R-22 AC Split Retrofitted With Propane

Imam Ridwan Tanaka[1]*, *Berkah* Fajar[1], *Tony Suryo* Utomo[1], *Eflita* Yohana[1], *Mustaqim*[1]

[1]Mechanical Engineering Department, Engineering Faculty, Diponegoro University Jl. Prof. Soedarto, SH, Tembalang, Semarang 50275, Indonesia

Abstract. Refrigerant 22 or called R-22 is still the most widely used for air conditioners. However, since it has a relatively high global warming properties, the use of R-22 should be stopped. Refrigerant 290 briefed as R-290 (propane) is the type of refrigerant recommended by many experts for R-22 replacements because the global warming properties of R290 are much lower than R-22. Therefore, in this study the cooling capacity, flow rate mass, power consumed and COP will be evaluated. The air conditioner engine used in this research has 1 PK capacity which is redesigned by adding separator / ducting on various main components of AC. From the result of the study, there are several conclusions revealed: on the same compressor capacity, R-22 cooling capacity is still greater than R290. Meanwhile, the flow rate of R290 is lower than R-22. For power consumption from R-290, it is smaller than R-22. For the value of COP R-290, the value shows greater ie at R-290 on the variation of Incoming Temperature In Evaporator 20°C = 6.23, 23°C = 6.36, 25°C = 6.86, 29°C = 6.20. While at R-22, it shows 20°C = 5.36, 23°C = 6.02, 25°C = 5.76, 29°C = 6.335

1 Introduction

The vapor compression refrigeration cycle machine is the most widely used between other types of cooling machine today. Generally, the medium used as the working fluid that removes heat from a cooling product to its environment is called a synthetic refrigerant. When it was first discovered around 1930 to mid-1970, the impact of using synthetic refrigerants such as chlorofluorocarbon refrigerants (CFCs) and hydrochlorofluorocarbons (HCFCs) had not made environmental problems. However, this does not mean that the use of those refrigerants had no negative impact to the environment, but rather it was because the limited knowledge and awareness of the environment at that time. With the increase of knowledge and environmental awareness, people now realize that the use of synthetic refrigerant causes problems to the environment; the refrigerant that was originally considered ideal and perfect, is now considered dangerous, and it needs to be removed from its use [1,2].

Later on, that fact encouraged the countries around the world including Indonesia to then jointly carry out prevention and reparation by making an agreement which became known as

*Corresponding author : tanakaunlam@yahoo.co.uk

the Wina Convention in 1985 followed by the Montreal Agreement in 1987. In order to reduce the impact of ODS (Ozone Depleting Substances) program to the ozone layer, in 1992 Indonesia has signed the protocol of Montreal. At the same time, Indonesia has launched the "Indonesian Country for the Phase Out of Ozone Depleting Substances (ODS) under the Montreal Protocol" program so that from 2020 to 2030 the dangerous type of refrigerant such as HCFC would be removed and discontinued [2,7]. Further development in 2007, the countries incorporated in this convention decided to expedite the process of removing all materials that damaged the environment.

There are some alternatives found by the researchers to overcome this phenomenon; one of them is the use of hydrocarbon refrigerant mixed with propane and isobutene as substitutes for CFC-12, while the substitute of HCFC-22 is propane or R-290. Although actually the use of hydrocarbons as refrigerant has been known to the public since 1920 at the beginning of refrigeration technology development alongside with other natural working fluids such as ammonia, and carbon dioxide [2]. Since it was found in 1930, CFC (Chlorofluorocarbon) refrigerant and HCFC played an important role in refrigeration and air conditioning system. This is because the refrigerant has good physical and thermal properties, stable, non-flammable, non-toxic and compatible against most of the material components in the refrigeration system [3,4,7].

As it has been explained from the beginning, the refrigerant vapor compression refrigeration system is a medium for transferring heat. It plays a very important role and it is still widely used for refrigerant materials on various refrigeration machines. One of them is HCFC-22 or generally called as synthetic refrigerant. This refrigerant has high ODP value (ozone depletion potential) and GWP (global warming potential). ODP is a relative measure of the ozone destruction rate from a material against the level of ozone destruction by R-11.

Table 1. ODP and GWP Refrigerant [3]

Refrigerant	ODP	GWP	NBP ((°C)
CFC 11	1.00	4600	23.8
CFC 12	0.86	4000	-29.8
R134a	0.00	1300	-26.2
HCFC 22	0.06	1700	-40.7
HC 600	0.00	2.7	-0.5
HC 290	0.00	1.1	-42.1

GWP is a relative measure of greenhouse gazette potential to greenhouse effect due to CO2, while ODP and GWP values of some refrigerants can be seen in table 1.1. One of the alternative refrigerants replacement of synthetic refrigerant HCFC-22 is propane hydrocarbon refrigerant (HC-290). It is because HC-290 has 0 ODP 1.1 GWP, so it can be ignored. Research on hydrocarbon refrigerant as an environmental friendly material has been started long enough, and it still continues to be developed to obtain the most optimum results [3,7].

The usage of R-290 as R-22 replacement has also been performed by some experts: Lorentzen, Devotta, Xiao et al, and Zhou and Zhang. Based on the studies on thermodynamic properties of R-22 and R290, Lorentzen recommends replacement of R-22 with R-290.

Devotta [4] performed an experimental study in AC windows that originally used R-22 and then it was retrofitted with R-290. The experimental study results reported that there was a 6.6% reduction in cooling capacity. However, COP of R-290 was higher than R-22 which is 7.9%. The COP is a ratio of cooling capacity to the power of the compressor input, that is why the increase in COP is greater than the cooling capacity reduction in the AC window. That concludes retrofitting R-22 with R290 will reduce the compressor input power.

Xiao et al. [5] conducted a retrofit experiment from R-22 to R290 on a small-capacity AC with varieties of R-290 charging. The results of their research reported that the maximum increase in AC performance is achieved when the mass of R-290 is 50% of R-22. In this optimum charging, cooling capacity decreased 2.8% and enhanced EER (energy efficiency ratio) reaching to 12.6%.

Zhou and Zhang [6] conducted a retrofit experimental from R-22 to R290 on a split type air conditioner by varying the diameter of a capillary tube roll. In the experiments, R-290 refrigerant mass filling was 44% and 47% of the R-22 filling mass, and the diameter of the capillary tube rolls improved from 40mm to 120mm. They stated that an increase in the diameter of the capillary tube rolls reflected the changes in the mass flow rate in the system, yet it did not show significant changes in cooling capacity, compressor input power and COP in air conditioners. The results of their study reported that there was a decrease in cooling capacity of 4.7% to 6.7%, and compressor input power by 21.1% to 12.3%, while there was an increase in EER which reached to 8.5%.

2 Experiment Setup and Procedures

2.1 Setup Equipment

The set-up equipment consists of a compressor, condenser, evaporator and capillary pipe. The compressor, condenser, evaporator and expansion valve used are taken from a 1 PK split-AC unit. These main components are separated from its original configuration (split-AC) to obtain data performance on compressor, condenser and evaporator. Then the component is connected with the refrigerant piping. Each component measured the temperature and pressure to determine the level of refrigerant. The mass flow rate of the refrigerant is measured by the refrigerant out of the compressor. In addition to pressure and temperature, the electrical current required by the compressor is also measured. Condenser and evaporator is placed in the ducting. Also, the temperature and relative humidity of the inlet and outflow of the condenser is measured. This measurement is designed to determine the transfer of actual heat that occurs in the condenser and evaporator.

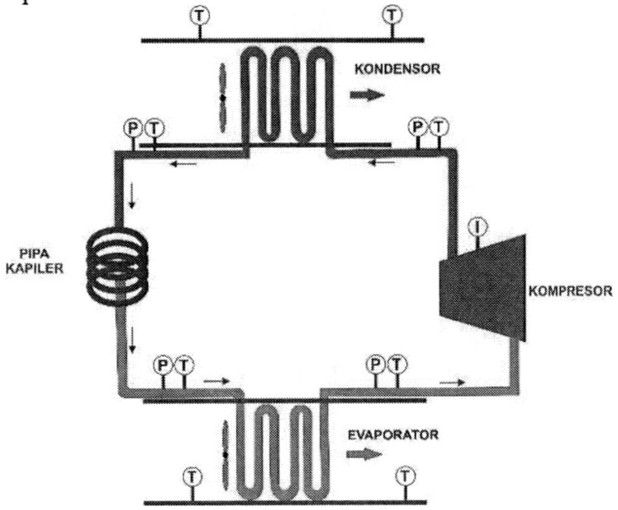

Fig. 1. The Schematic Diagram.

2.2 Recharging Refrigerant

After completion of the vacuum process, the recharging of R-22 as much as 490 grams from the data contains in the nameplate outdoor unit. After completion of data retrieval system using R-22 refrigerant, then it was re-vacuumed as it did before in the initial vacuum process. Furthermore, the system loaded R-290 refrigerant as much as 40% that is fixed as much as 196 grams.

2.3 Procedure

The procedures performed in this research are as follows:
- To test the cooling machine in the laboratory of energy conversion (Training Center) Diponegoro University of Semarang.
- To carry out the test system by using a variation of evaporator entrance temperature as the cooling load. Temperatures used were 20°C, 23°C, 25°C, 29°C, respectively.
- To charge a heavy (mass) charging refrigerant R-22 for 490 grams and R290 around 196 grams.
- To perform R-22 and R-290 test with the same cooling load.

3 Equation

A refrigerating effect is a difference of enthalpy on the outlet and inlet side of evaporator. The refrigerating effect can be calculated as below:

$$\frac{\dot{Q}_{in}}{\dot{m}} = h_1 - h_4 \quad (1)$$

While the work of the compressor can be calculated as follows:

$$\frac{\dot{W}_c}{\dot{m}} = h_2 - h_1 \quad (2)$$

Heat transfer rate per unit mass in condenser is the total heat that is rejected by the condenser onto the surrounding per unit mass of refrigerant flowing.

$$Qc = \dot{m}(h2 - h3) \quad (3)$$

Efficiency of an air conditioner can be stated by Coefficient of Performance (COP or β). It can be inferred as the ratio between refrigerating capacity and work of the compressor.

$$\beta = \frac{\dot{Q}_{in}/\dot{m}}{\dot{W}_c/\dot{m}} = \frac{h_1-h_4}{h_2-h_1} \quad (4)$$

4 Result and Discussion

This refrigerant replacement process from R-22 to R290 does not need replacement of the AC components system. It is because the character of pressure and temperature from HC refrigerant is similar to synthetic refrigerant. Thus, almost all kinds of compressor can accept hydrocarbon refrigerant. The cooling capacity is the output of the expected AC. The larger cooling capacity is, the faster the room temperature will reach to the desired temperature. In Figure 2, it is seen that a higher temperature of the evaporator enters the resulting cooling capacity. It also shows that the cooling capacity of R-22 is higher than R290 for all variations of the inlet temperature of the evaporator. This is one of the reasons why R-22 was still used for decades, given that R-22, in terms of thermodynamic properties is superior than R290. However though, since R-22 still has relatively high ODP and GWP properties, the use of R-22 as a refrigerant must be stopped immediately.

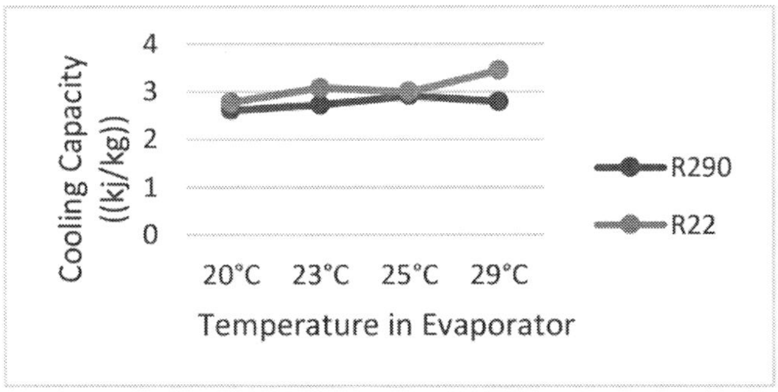

Fig. 2. Graphic Cooling Capacity and Temperature in Evaporator.

In figure 3, the mass flow rate of R290 is smaller than R-22. This shows that the required refrigerant is less than R-22. With a mild lifetime and a lower mass transfer rate than R-22, the compressor's work is not heavy.

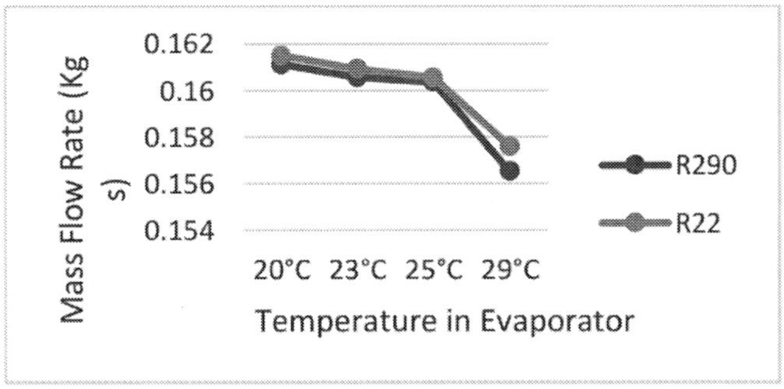

Fig. 3. Graphic Mass Flow Rate and Temperature in Evaporator.

As shown in Figure 4, it shows that the power consumed by the compressor at R290 is lower than R-22. This is due to the lighter compressor work, and because the mass flow rate of R290 is smaller.

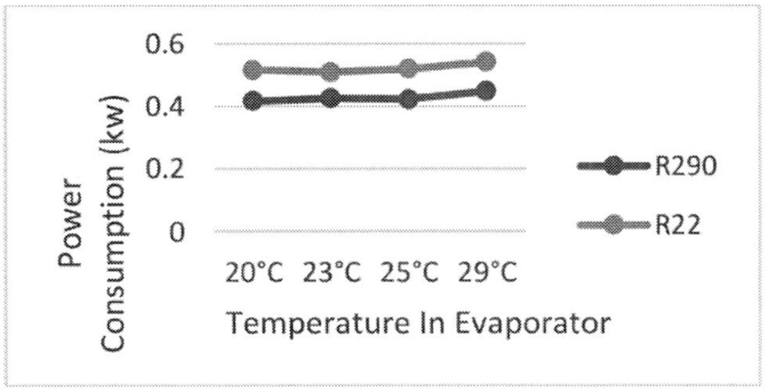

Fig. 4. Graphic Power Consumption and Temperature in Evaporator.

From Figure 5, it can be seen that COP R290 is larger than R-22. In the previous chart we know that the cooling capacity of R290 is smaller than that of R-22. In contrast, here is larger at 5 COP graph, this is because the R290 power consumption is also smaller than R-22, so it is increasing its COP value.

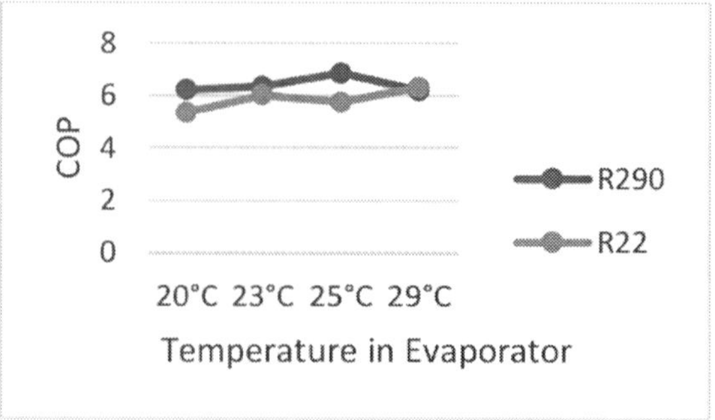

Fig. 5. Graphic Power Consumption and Temperature in Evaporator.

In the previous section, it has explained the advantages and disadvantages of R290 and R-22 from the side of the cooling capacity, thermodynamic properties and power consumption. Moreover, we can draw the conclusion that the performance of R290 is better than R-22, and its use needs to be developed for the better environmental impact. However, due to the flammable and explosive nature of the R290, it is necessary to set standards on how to utilize and socialize them to AC technicians.

5 Conclusions

Based on the above discussion, there are several conclusions made from this study:
• For the same compressor capacity, R-22 cooling capacity is still greater than R290.
• The flow rate of R290 is lower than R-22.
• Power consumption from R290 is less than R-22.
• For the value of COP, R290 value is higher than COP of R-22, R290 on the variation of Incoming Temperature In Evaporator 20 °C = 6.23, 23 °C = 6.36, 25 °C = 6.86, 29 °C = 6.20. While at R-22 the temperature is 20 °C = 5.36, 23 °C = 6.02, 25 °C = 5.76, 29 °C = 6.335.

References

1. Powell, R.L. (2002). CFC phase out; have we met the challenge. *Journal of Fluorine* 114, 237-250.
2. Palm, B. (2008). Hydrocarbons as refrigerants in small heat pump and refrigeration systems – A review. *International Journal of Refrigeration* **Vol.** 31, 552-563.
3. An ECOFRIG, (1997). Refrigeration Appliances using Hydrocarbon Refrigerant, Indian Inststiyute of Technologi Department Of Mechanical Engineering Haus khas, 110016 New Delhi.and INFRAS, CH-8002 Zurich.
4. Devotta, S., Padalkarb, A.S., Sane, N.K. (2005). *Performance assessment of HC290 as a drop in substitute to HCFC22 in a window air conditioner. International Journal of Refrigeration* **Vol.** 28, 594-604.
5. H.H. Xiao, T. Zhang, Y. Hu, (2006), *Experimental research on performance of small room air conditioner with R290, International Journal of Refrigeration,* **Vol.** 49, 26-30.
6. Zhou, G., Zhang Y. (2010). *Performance of a split-type air conditioner matched with coiled adiabatic capillary tube using HCFC22 and HC290*. Applied Energy 87, 1522-1528.
7. Dossat, Roy J., (1990), *Principles of Refrigeration*, Second Edition, SI Version, John Wiley & Son Inc., New York, USA.

The Modeling Of A Conceptual Engineering Design System Using The Decision-Matrix Logic

Safarudin Ramdhani[1,2,*], and *J* Jamari[2]

[1]Industrial Engineering Faculty of Engineering, Dian Nuswantoro University, Semarang, Central Java, 50275, Indonesia
[2]Mechanical Engineering Department Engineering Faculty, Diponegoro University, Semarang, Central java, 50275, Indonesia

Abstract. Decision making a concept the optimal inside a concept engineering design (CED) is a job repeated. Decision matrix based method is the most popular concept selection approach used in engineering design. A matrix is an array that presents an axis of the alternate list being evaluated. The list of weighted criteria depends upon the importance of each of the final decisions to be taken. The decision matrix in this determined some weights and the rank as attributes to evaluate a total scores. Upon the weighting sometimes make the designer confusedly in determine variety. This paper described a decision making computer based, where a logic matrix decisions on a basis in the selection and evaluation. This model presents a logical procedure for concept evaluation considering the specified attribute. In decision-making this model is integrated and intended to improve the ability of beginners in designing.

1 Introduction

Engineering design process is a design planning in formulating solutions to assist a designer in building products with specific performance goals. In an Engineering Design Process (EDP) there are two main phases: initial design or concept and detail design. Techniques in drawing up Concept Engineering Design (CED) require information processing from a variety of sources that serve to meet the functional needs, operational constraints and evaluation criteria associated with the purpose specified. The ultimate goal of CED is to select the most desirable concept. The selected concept is then further developed in the detailed design phase.The conceptual phase broken down into four steps involving concept clarification, generation, selection and development are shown Fig. 1. Concept selection is one of the most critical decision-making exercises in a product-development process.

Concept engineering design has the main goal of determining the most desired concept. The designs that have been selected will be further developed beyond the design detail stage. The selection of design concept is one of the factors of decision making in product development. In this case, it is a recurring evaluation conducted to test the completeness

* Corresponding author: dhannyzee@gmail.com

and comprehension of requirements, which quickly identifies the strongest concepts. A common decision matrix approaches two major weaknesses; (i) some potentially optimal concepts may seem undesirable, since they never receive the highest total value, and (ii) general construction requires that decision makers determine weights and rankings that are not meaningful physically. More specifically, conceptual appraisal is obtained by using a random mapping process, in which the physical criteria value in the concept is mapped to the scoring space using a non-physically significant ranking structure, Messac[1].

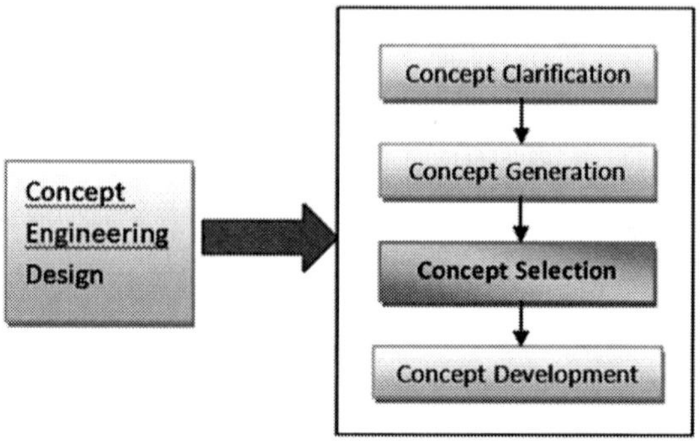

Fig.1. Phases of Conceptual Engineering Design

In the industrial world, there are several methods to do so. These methods include; Matrix decisions, feasibility assessments, intuition, cultivating, numerical and non-numerical charts, pair-wise comparisons, and prototype testing, Ullman [2], Otto [3] and Ulrich [4]. Screening and concept assessment is a popular variant of the matrix-based decision method, Ulrich [4]. The decision matrix method is the most commonly used to the concept of selection in engineering design practice, Mullur et al. [5]. This is an iterative evaluation that tests the completeness and comprehension requirements, which quickly identifies the strongest concepts. Typical versions are Pugh's method, Johnson technique and L-shaped matrix. Several versions of the decision matrix method have been refined for many different case studies. In the selection of suitable raw materials from several proposed feedstock alternatives, the matrix method for this case helps the selection of suitable and relevant material matrix decision ratings, Rao [6]. In addition, we present a theoretical framework for integration of methodology (LCC), Life Cycle Costing (LCC), and Quality Function Deployment (QFD) for selecting alternative product improvement strategies with consideration for Data Uncertainty The focus to the research is how manufacturing companies can be assisted in product design. Where the quality, environmental and cost (QEC) requirements of stakeholders in the life cycle stages within the product system are handled at an early stage where general data inaccuracy is prevalent. Considerations Of these three design requirements lead to multi-attribute decision situations, with respect to the selection of the concept of optimum product system improvement. Ratings of sustainable alternative options related to environmental, cost, and quality are also reported, Halog [7].

This method evaluates and ranks the material and hence chooses the most appropriate material. The Multi Criteria Decision Making (MCDM) approach to review mobile phone choices with respect to a user personal sequence where the most prefer to feature affecting cell phone selections is identified through surveys conducted. Their article then application the Quantitative Strategic Planning Matrix, (QSPM), which is an analytical tool for formulating the strategy used, Gulfem and Gulcin, [8]. The QSPM framework for retail computer stores also highlights the advantage and limitedness of this important strategic planning analysis tool, Meredith, et al., [9]. The importance to the decision evaluation process as it plays a significant role in the flow of scenarios, because variants of the enterprise are interested in introducing new products within a short span of time, Green and Mamtami, [10]. This paper examines the disadvantages of its application over strategic planning techniques in manual form and explores in a typical decision-making approach that potentially effective and easy to use by presenting computer-based models for use in Conceptual Engineering Design (CED).

2 Methodology

2.1 Decision matrix based concept evaluation

A matrix is an array that presents an axis of the alternate list being evaluated. The list of weighted criteria depends upon the importance of each of the final decisions to be taken, Oladejo,et al., [11]. Table 1 shows a typically constructed decision matrix for a beam design problem. It shows beam concepts P^1,P^2, and P^3 rated against two design objectives (criteria); minimal volume and minimal deflection. A higher numerical rating indicates better concept performance. Each criterion is assigned a weight that is intended tocapture its relative importance with respect to the other criteria. For every concept, the weight of each criterion is multiplied by its rating. The summation of all such products is the total score for each concept. The concept that receives the highest score is typically preferred over other concepts, Mullur, et al., [5].

Table 1. Typical construction of a decision matrix, Mullur, et al., [5].

Criterion	Weight	P^1	P^2	P^3
Minimal Volume	0.6	2	5	3
Minimal Deflection	0.4	1	3	5
Total Score	-	1.6	4.2	3.8
Concept Rank	-	3	1	2

The following equation shows the mathematical simple formulation of the evaluation procedure, Mullur, et al., [5].

$$J^i = \sum_{j=1}^{n} w_i \, u_j^i \tag{1}$$

Where, j^i is the total score for concept i,
n is the number of design criteria,
w_j is the weight of the jth criterion,
u_j^i is the rating of concept for the jth criterion.

2.2 Development of computer application package

Computer modeling of dynamic systems is a very important tool for engineering analysis and design. This allows for an active experiment, design modification, and flexibility The model is developed on the Visual BASIC platform running on Microsoft Windows with a graphical user interface (GUI). The GUI determines how various elements look and work. Major activities involved in the development of the model detailed comprises of the design of the programmable algorithm are shown Fig. 2, selection of appropriate programming language, project development, project debugging, validation and implementation of the model.

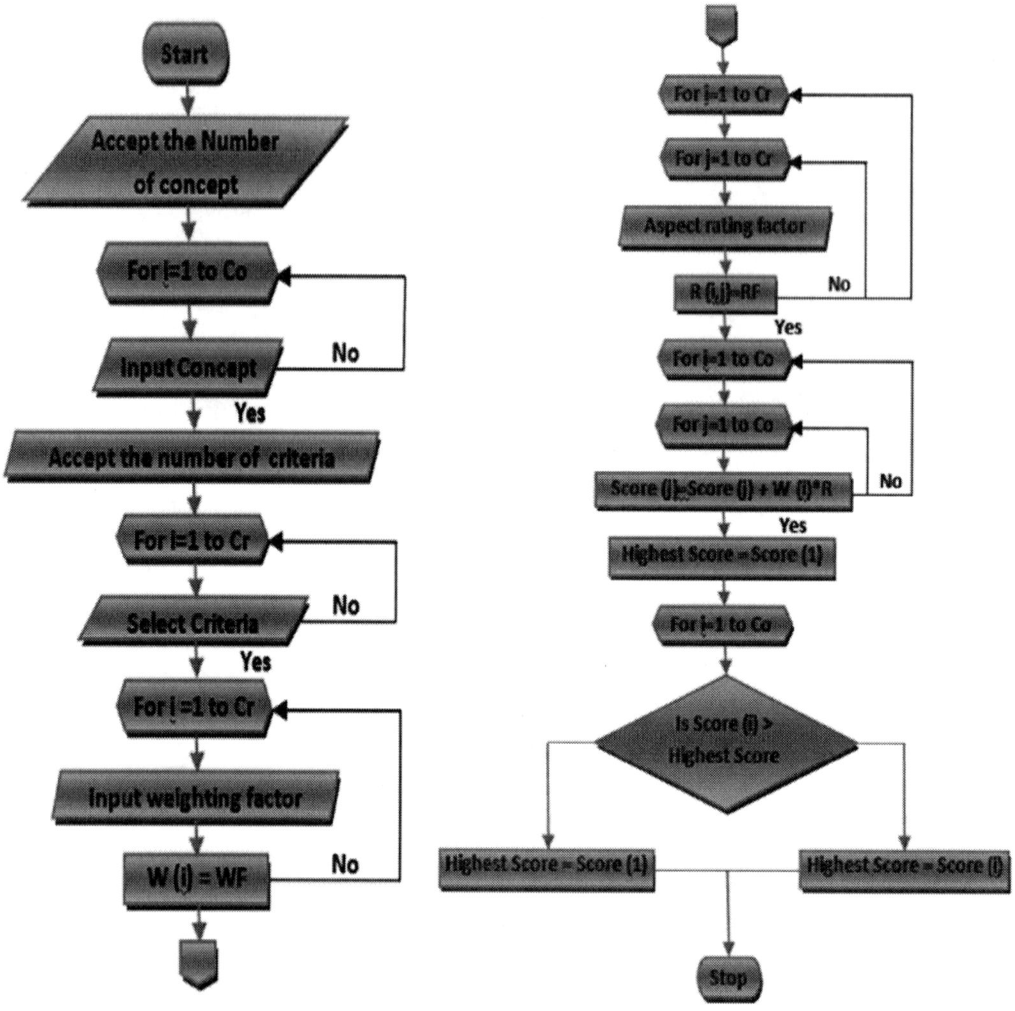

Fig.2. Flowchart of the package computation

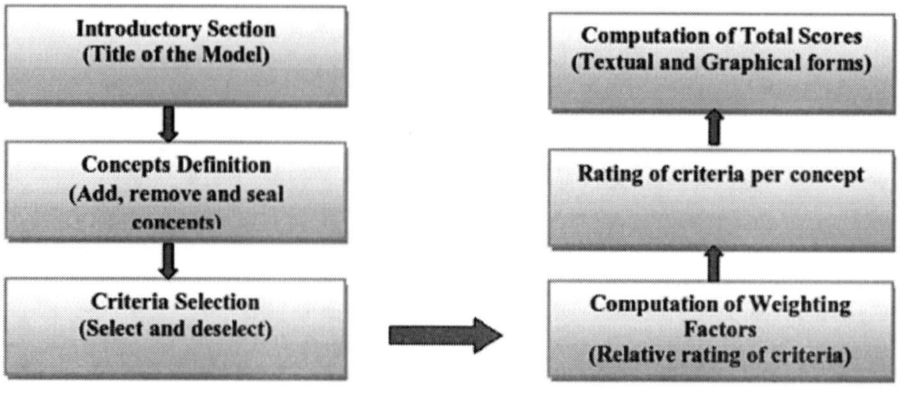

Fig.3. Major Structure of the Package, Oladejo,et al., [11]

The GUI itself consists of six main interfaces are shown Fig. 3, Oladejo,et al., [11]. Introduction Section; This section provides a brief introduction to the model, author and usability. Definition of concept; This section allows precise definition, by title, of all concepts that the model will consider. Selection of criteria; At this stage the designer is allowed to select related criteria from the set of attributes that the model will provide. Computation of weighting factor; The model calculates the weighting factor for each attribute. Rating of criteria per concept; The influence of each performance criterion of each individual concept, determined by the assessment of each criterion and concept. Computation of Total score per concept; The rating factor (RF), for each criterion and for each concept multiplied by the weighting factor (WF) to obtain the final weighting value for each concept. Each GUI is created using a three step process to plan the project and the process is repeated to create the interface. The three-step process involves setting up the GUI, defining properties, and then generating code.

3 Result and Discussion

The model can be used for any type of decision-making situation and has a manual method reference limit. Fig. 4 shows the interface for the designer to define the concept's alternative and criteria to be considered. The criteria is made available on the interface for the user to select. The list of criterion is developed from the engineering specification.

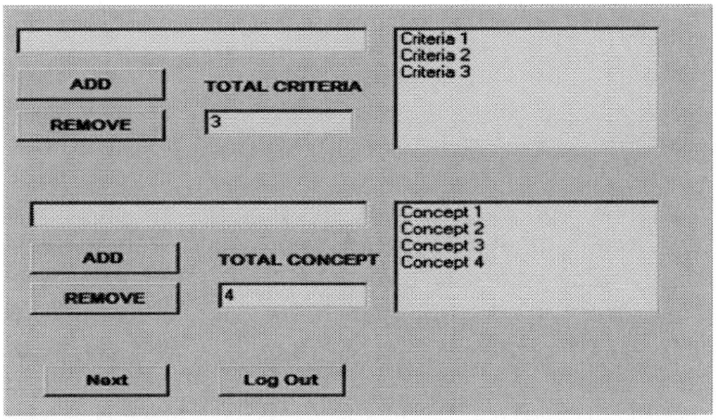

Fig.4. Interface to input the alternative concepts and criteria

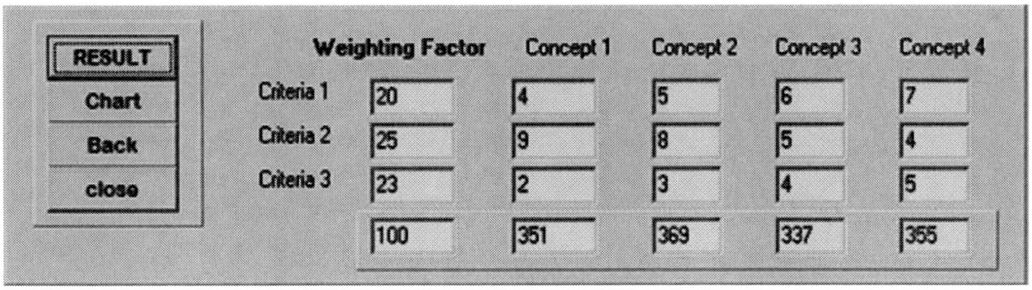

Fig.5. Interface to input the alternative concepts and criteria

Fig. 5 shows the interface for calculating the weight factor of each criterion. This involves the use of a comparison matrix in which all selected criteria are paired and then evaluated against each other in the order of relative importance listed earlier. The weighting factor at the interface is calculated by adding the number of times a certain criterion is selected according to the partner's criteria. The values obtained for each attribute are 20, 25 and 23, as shown in Fig. 5.

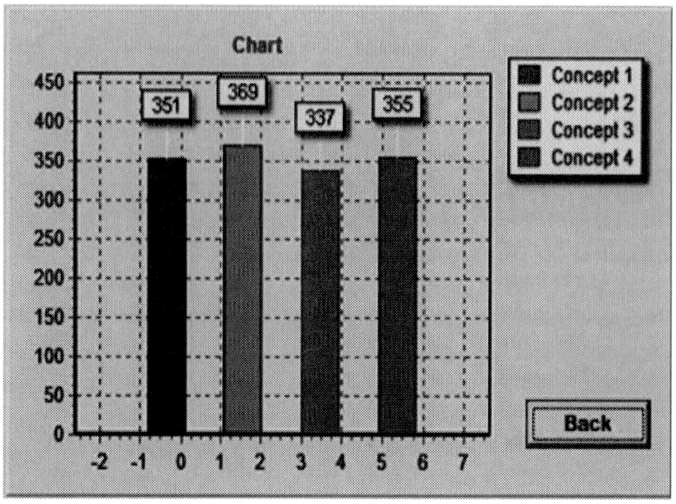

Fig.6. Chart of the concept score

This value is automatically moved to the next slide Fig. 6, where the ranking factor is selected to calculate the total score for each concept and displayed with a chart. In the graph is listed the highest score that is concept 2 with a total score of 369. The highest score or first rank on the graph is the chosen concept to be used.

4 Conclusion

This study presents an optimal selection of concepts developed through computer applications. The selection of the optimal concept comes from a number of alternative concepts available by using certain techniques in CED. The model was developed using matrix decision logic over the Visual Basic platform. The computer app device accepts the concept from the user, evaluates, ranks the ranking of concepts and presents the results in graphical form. This model can be used for all kinds of problem selection design concepts involving a number of attributes and can also serve as teaching aids in engineering design.

References

1. A. Messac, S. Gupta, and B. Akbulut, *Transactions on Operational Research*, **8** (1996).
2. D. G.Ullman, *The Mechanical Design Process*, (McGraw-Hill, New york, 1992).
3. K.N. Otto, Res. Eng. Des, **7** (1995).
4. K.T. Ulrich, and S.D. Eppinger, *Product Design and Development*, (McGraw-Hill, Boston, 2000).
5. A.A. Mullur, C.A. Mattson, and A. Messac, AIAA, **44** (2003).
6. R.V. Rao, *Material and Design*, **29** (2008).
7. A. Halog, J. Sustain. Prod. Des, **4** (2004).
8. I. Gülfem, and B. Gülçin, J.Comp. Stand. Int, **29** (2007).
9. E.D. Meredith, R.D. Forest, and R.D. Fred, CBJ, **8** (2009).
10. G. Green, and G. Mamtani, ACTA Polytechnica, **44** (2004).
11. K.A. Oladejo, D.A. Adetan, M.D. Adewale and B. O. Malomo, IJMSE, **7** (2016).

Characterization chip formation of commercial steel materials at low speed cylindrical grinding processes

Wahyu Isti Nugroho*, *Sri* Nugroho, and *Rusnaldy*

Diponegoro University, Department of Mechanical Engineering, 50275, Jl. Prof Soedharto, Tembalang, Semarang, Indonesia

Abstract. Cylindrical grinding is the process of metal removal using multiple cutting point. The objective of the study is to investigate the chip characteristics (shape and width) and surface roughness materials. The materials are AISI 1020, AISI 1045, AISI 1090, AISI D2, and AISI 4340 with a grinding wheels WA46K8V. The research is using cylindrical grinding processes at low-speed workpiece 100 rpm, depth of cut 0.5 mm, in dry condition. Result of research is comparing characteristic chip with variable feet rate 120 mm/min, 420 mm/min, 1300 mm/min. The result is to show the increasing feedrate lead, the growth of chip width and high surface roughness value. On materials AISI 1020, AISI 1045, AISI 1090, and AISI 4340 show that chips width is related to hardness value.

1 Introduction

Cylindrical grinding is important processes to make part cylindrical with precision and good surface finish. Cylindrical grinding oftens to make the part of cam and shaft from steel materials. Generally, to classify the steels based on hardness and chemical material, it can use hardness tester and chemical material tester. Material properties give influence in machine processes [1]. The mechanical properties of the material include strength, hardness, and ductility correlate to the chip form [2]. By determine the form chip from a cylindrical grinding machine, it can be sure that mechanical properties is formed. The research of chip characteristics attracts many researchers. Because most of them use high spindle speed parameter. It is proven that the characteristic of chip with various grinding wheels and feedrates for different steels can affect the type of chip [3]. Various feedrate and spindle speed for hard steel will give affect to the chip size [4].

Character of chip formation can do to see the quality of the material use. The research shows that the correlation between character of chip formation and mechanical properties materials. To analyze the chip, the variations of feedrates regarding low speed workpiece and the depth of cut should be investigated. Chip from each type of steels with a various feedrates is observed by the perceptions of its formation and dimension.

* Corresponding author: w.istinugroho@yahoo.co.id

By extending the publish result from the main frame of the chip formation, this research aim to investigate the feedrate effect on the chip formation of different types of steel materials experimentally. The research shows the correlation between character of chip formation and mechanical properties of materials.

2 Experimental Method

The experiment performed by a cylindrical grinding machine is shown in figure 1. Various steels that used for this study are AISI 1020, AISI 1045, AISI 1090, AISI D2, and AISI 4340 which have different hardness value. The dimension of the workpiece is Ø30 x 200 mm. The grinding wheel uses WA46K8V Ø405 x 56 x 127 mm. The following is the details of the experimental setup:

(1) Wheel speed : 2900 rpm
(2) Workpiece spindle speed : 100 rpm
(3) Depth of cut : 0,5 mm
(4) Feedrate : a (120 mm/min)
b (420 mm/min)
c (1300 mm/min)
(5) Cooling : dry process

Fig. 1. Set up cylindrical grinding machine.

Cylindrical grinding process low speed without cooling sytem (dry process) has purpose to get an easy chip characterization. As seen in Figure 1 a metal box added in area of chip direction to collect the chip. The chip characterization and measurement using digital microscopes zooms 100x with image processing software.The chip selected based on the criteria good form and dominant.

3 Result and Discussion

The chip from the machine process is collected, captured, and measured. Table 1 shows the chip formation of various steels and feedrates. It is can be classified that types of chips are flowing, knife, slice, and spherical shape show in Table 2. This study characteristics of chip's types are flowing chip Figure 3. The formation of chip in the grinding process is affected by the grit on the grinding stone that incisions, the surface of the material. If further observed that the mechanism of grinding occurs into three processes namely rubbing, sliding, plowing, and cutting/chipping shown as Figure 2.

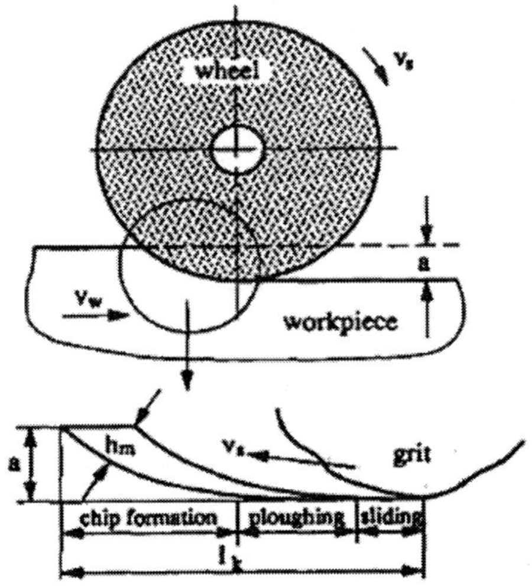

Fig. 2. Mechanism of formation of grinding process grinding [5].

The grinding process is a cutting method using multiple cutting points so that the results from one process of the same parameters produce types forms of chips. The shape and dimensions are influence by the parameters of types of grinding wheel, feedrate, spindle speed, wheel speed, and coolant. Based on the explanation of various types of chip's incisions of the machine process, it can be concluded in Table 2.

Table 1. Chips formation of various steel.

Feedrate / Materials	AISI 1020	AISI 1045	AISI 1090	AISI D2	AISI 4340
120 mm/min	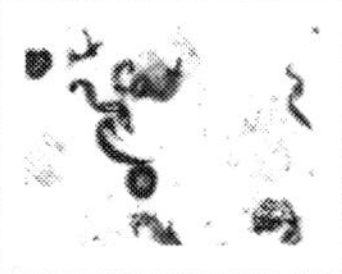	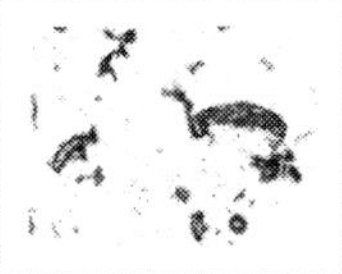			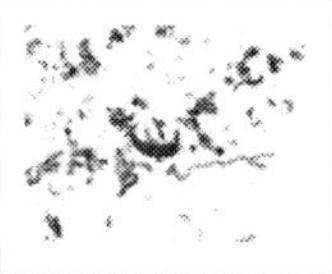
420 mm/min	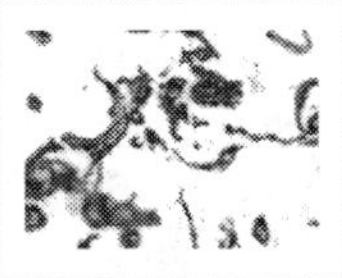		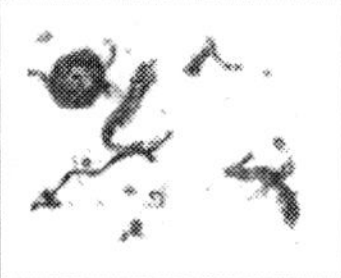		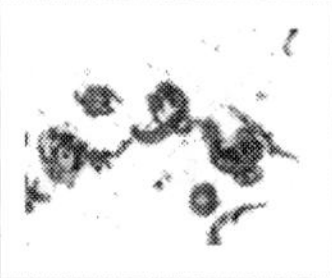
1300 mm/min		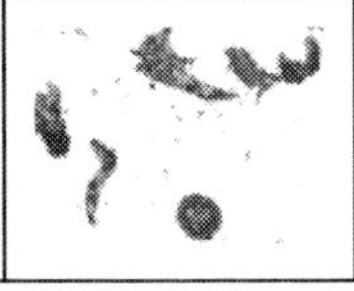	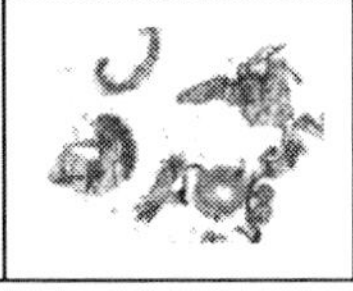	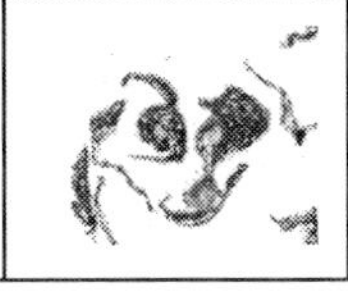	

Table 1 shows the higher feedrate resulting large dimension of chip formation. The higher hardness value of steel then produces a smaller chip formation. In AISI D2, it produces the smallest of all steels chip formations because it has high chromium alloy. Chrome alloys cause steel to have abrasive resistance properties and lowest surface roughness value. There is a correlation between characteristics of machining process of the mechanical properties of the materials.

Table 2. Chip types.

CHIPS TYPES		DESCRIPTIONS
Slice		• Available on WA grinding wheel. • Machining process without coolant. • Indicates high cutting loads. • The cut depth parameter is too large. • Marked with the worst surface roughness among other types of furious.
Knife		• Available on WA grinding wheel. • Machining process without coolant. • Indicates high cutting loads. • The cut depth parameter is too large. • The rash is formed before the formation of slice growl. • Characterized by a bad category surface roughness.
Flowing	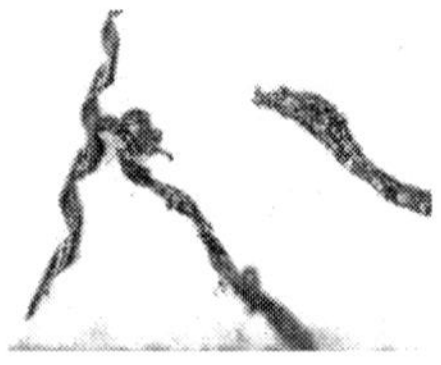	• Available on grinding wheels WA grinding wheel. • Indicates sharp grinding stones. • The shape of the chip is characterized by surface roughness which is expected from the grinding process.
Spherical	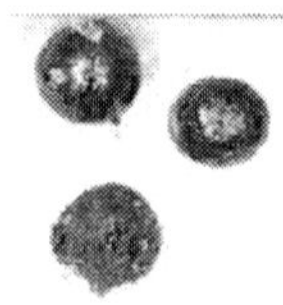	• Available on the grinding process of all types of grinding wheels. • In the process of grinding without cooling liquid potentially more generated melting type. • The uses of high temperatures above 1400o C, will make spark grinding.

Chip types can indicate machine ability of materials. Slice types available on WA grinding wheel during dry process. That indicates high cutting loads and depth of cut are too large. This chip type are showing worst surface roughness quality. Knife chip types are available on grinding process using WA grinding wheel without cooling system (dry process). It is indicating high cutting loads and high depth of cut. This chip's types formed before the formation of slice types. On flowing types indicates sharp grinding stones and best surface roughness quality. Spherical types available on the of all type grinding wheels.

Table 3. Relationship of machining characteristics to material properties.

Materials	Flowing chips types	Hardness Brinell	Chips width (Flowing) µm	Surface roughness µm
AISI 1020	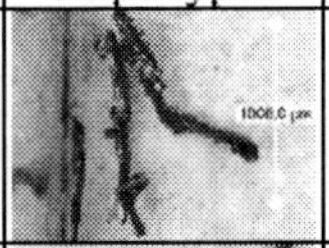	131	108,9	5,949
AISI 1045	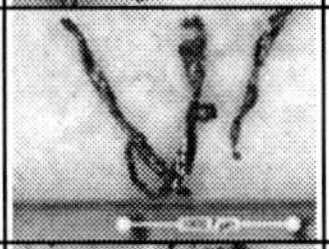	195	101,6	5,736
AISI 1090		220	98	5,043
AISI D2	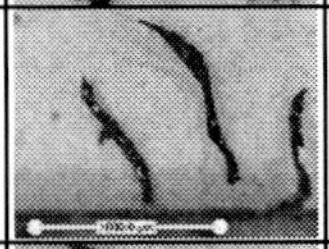	258	80,3	3,978
AISI 4340	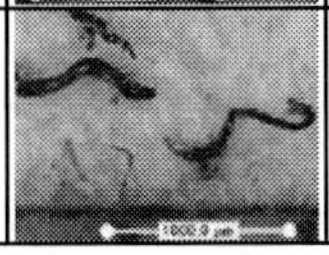	346	91,3	4,145

4 Conclusion

The effect of feedrates on a chip characteristic varying the steel materials (AISI 1020, AISI 1045, AISI 1090, AISI D2, and AISI 4340) has been examined. Based on the discussion, the following conclusion could be drawn. Increasing the feedrate will increase the chip width of all types materials used. The significant result of this study is to make the chip formation which could be considered as the guide to determine the hardness value of the material AISI 1020, AISI 1045, and AISI 1090.

Reference

1. William D. Callister, *Materials science and engineering an introduction*, 8th edition. New Jersey: John Wiley & Sons, Inc, (2009).
2. B. Denkena, J. Kohler, J. Kastner, *Chip formation in grinding : an experimental study*, **6**, 107–115, (2012).
3. P. Tso, *An investigation of chip types in grinding*, *Material process technology*, **53**, 521–532, (1995).
4. P. Tso and S. Wu, *Analysis of grinding quantities through chip sizes*, **95**, 2–8, (1999).
5. S. Malkin, *Grinding technology theory and applications of machining with abrasives*, 2nd Edition. New York, (2008).

The Performance of Pump as Turbine with Machined Impellers

D.L Zariatin[1,*], *Shanti Kumbarasari*[2], and *Dwi Rahmalina*[1]

[1]Mechanical Engineering, Universitas Pancasila, Jakarta, Indonesia
[2]Mechanical Engineering, ATMI Cikarang, Jakarta, Indonesia

Abstract. Pump as turbine (PAT) is a type of micro hydro power plant, which uses a commercial pump as a turbine in order to generate electricity. Most of PAT facilities used centrifugal pump. The main component of the pump is impeller, which commonly produced by using casting process. However, this paper describes a manufacturing process to produce PAT impeller by using machining process. Four impellers were produced from two variations of materials (brass and AA-7075), each with 5 and 6 blades. The impellers were tested in a laboratory scale PAT power-plant facility, and the performance of each impeller was compared to its original impeller. The experiment results show that the impeller with 6 blades made of brass material improves 48.9% of power generated.

1. Introduction

Pump as turbine (PAT) is a type of hydro power plant, which uses a commercial pump as a turbine in order to generate electricity. PAT has several advantages such as low-cost, simple design and easy to assemble, widely application and viability, longer life time, etc. [1]. Therefore, it is suitable for rural and remote area [2, 3]. Numerous research papers about PAT technology have published [4 - 7]. Generally, commercial pump impellers are made by casting process with surface roughness around 50μm- 0.4 μm [8]. However, machining process could be an alternative of manufacturing process to produce an impeller. The main reason of using machining processes to manufacture an impeller is because the surface could be determined.

Young and Chuang [9] developed an integrated 5-axis machining module for a centrifugal impeller. Cutter Location (CL) point was generated based on the geometric model of the blade and the hub by using CAD/CAM software. Lazoglu et al [10] also used CAD/CAM software to generate the tool path by a developed boundary representation that considered as an approach for prediction of cutting forces. Yau et al [11] performed the mathematical calculation in CAD/CAM software simulation of three and five axis milling machine in order to produce complex shape such as an impeller. Meanwhile Tuysuz [12] performed the simulation of the milling process for ball and end-mill cutting tools that includes the cutting forces. Fan and Xi [4] proposed an algorithm to optimize the tool-path generation for three-axis machining of a sculptured impeller blade surface. The algorithm calculated the four tool sub-paths on the sub-surface.

It is known that surface roughness is an important parameter in fluid flow. Taylor et al [13] stated that surface roughness was shown to have a very little effect in the laminar region. However, it played a major role in the turbulent region. The laminar and turbulent region are identified by the

* Corresponding author: dedeliazariatin@univpancasila.ac.id

Reynolds number (Re). The Reynolds number is increased on rougher surfaces [14, 15]. Kaewnai [16] stated that losses in the impeller increase with increasing the surface roughness. Five impeller models with different surface roughness, which is 0 mm, 0.0001 mm, 0.001 mm, 0.01 mm and 0.1 mm, were investigated by using CFD. It is found that the surface roughness has a high effect on loss.

In this research, the machining process was performed to produce impellers for the pump as turbine facilities. The effect of impeller material (brass and AA-7075), numbers of the blade (5 and 6-blades) and impeller surface roughness were studied. The impeller's performances to generate power were tested and analyzed.

2. Manufacture of Impeller

Fig. 1 shows the manufacturing steps to produce the impeller. The impeller dimension was based on the original impeller that measured by using the Coordinate Measuring Machine (CMM). Due to the surface roughness value of the original impeller was exceeded the lower limit of the Mitutoyo Digital Surface Tester SJ-201 range of measurement, therefore the surface roughness of the original impeller was measured by comparing it to standard roughness tester. The original impeller is made of brass and has six blades. In this research two types of material and two variation number of blades were produced. The materials are brass and AA-7075. The number of blades is 5 and 6 blades. So, there is a total of four impellers produced by machining process.

After identifying the geometric properties of the original impeller, CAD software was used to make the impeller model. Because the machining process was performed in 3-axis milling machine, the impeller model was divided into two part section. The machining strategy was generated by using CAM software. The tool path simulation and NC/G code or CL File were generated in this step. The CL file, then transferred and executed on YCM MV66A 3-axis milling machine.

The dimension of machined impellers was inspected by CMM and the surface roughness was measured by Mitutoyo Digital Surface Tester SJ-201. The measurement parameters of Mitutoyo SJ-201 were: length of cut = 0.8x5 mm, drive speed = 0.2 mm/s and sample length = 2.5 mm.

Two section part of the impeller was joint by using pin injector joint and then by the welding process. The next step after the impellers were produced, was to install the impeller in the pump that used as a turbine. Then, install the pump to the PAT test facility.

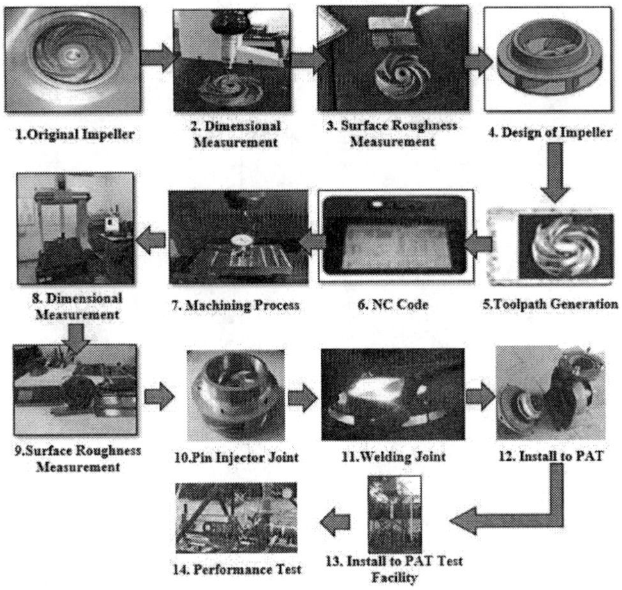

Fig.1. Manufacturing process of machined centrifugal impeller

The machining sequence (simplified) and the cutting parameters for both 5 and 6 blades are shown in Table 1 and Table 2. In order to monitor the cutting tools condition; before, during and after the machining process, the cutting tools were observed by using Dino Lite AM4815 digital microscope camera. The cutting tool monitoring process is performed because improper cutting tool condition such as tool wear, tool crack, etc. would degrade the impeller quality.

Table 1. Machining sequence and cutting parameter for 5 blades impeller

Operation		Roughing			Finishing		

		Rough 1 →	Rest 1 →	Rest 2 →	1 →	2 →	3
Allowance (mm)		0.3	0.3	0.35	0	0	0
Tolerance (mm)		0.05	0.05	0.05	0.005	0.005	0.005
Depth of Cut (mm)		0.5	0.5	0.3	0.1	0	1
Spindle Speed (Rpm)		2500	4500	4500	4500	4500	4500
Cutting Feed Rate (Rpm)		2000	1500	1500	1500	1500	1500
Step Over		16	6	2	0	0	0
Cutting Tool		ZCC 20	End-mill 10	Ball nose 6	Ball nose 6	Ball nose 6	End-mill 10
Machining time (min)	Dural	23	43	179	77	7	34
	Brass	92	24	31	179	78	7

Table 2. Machining sequence and cutting parameter for 6 blades impeller

Operation		Roughing			Finishing					
						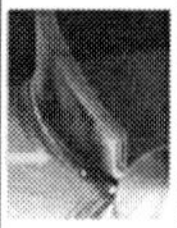	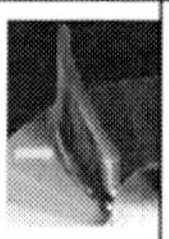			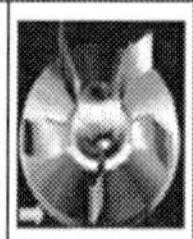
		Rest →	1 →	2 →	1 →	2 →	3 →	4 →	5 →	6
Allowance (mm)		0.35	0.4	0.4	0	0	0	0.02	0.02	0
Tolerance (mm)		0.05	0.05	0.03	0.05	0.005	0.005	0.005	0.005	0.005
Depth of Cut (mm)		0.5	0.3	0.2	0	0.1	0	0	0	0
Spindle Speed (Rpm)		1500	1500	1500	1500	1500	1500	1500	1500	1500
Cutting Feed Rate (Rpm)		1000	1000	1000	1000	1500	1000	1000	1000	1000
Step Over		16	2	13	0.08	0.1	0.1	0	0	0
Cutting Tool		ZCC 20	Ball 6	Ball 4	Ball 4	Ball 6	Ball 6	Ball 4	Ball 2	Flat 10
Machining time (min)	Dural	98	14	7	21	66	67	10	20	6
	Brass	79	14	6	21	74	128	11	28	6

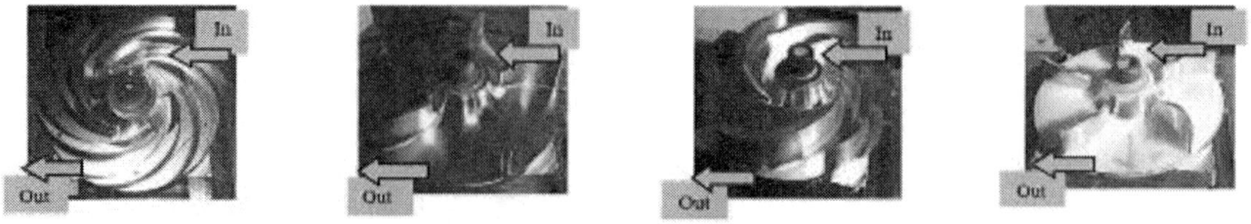

(a) (b) (c) (d)

Fig.2. Cross-section view of the impellers; (a) Brass, 6-blades impeller, (b) Brass, 5-blades impeller, (c) AA-7075, 6-blades impellers, (d) AA-7075, 5-blades impellers.

Fig.2 shows the cross-section view of the impellers. This section part was assembled to the other section by using pin injector and welding. The result of the manufacturing process is shown in Table 3 pictures. Meanwhile the value of surface roughness (Ra) for each impeller is also listed in Table 3.

Table 3. Impeller surface roughness

Picture	Impeller	Outer Ra (µm)	Inner Ra (µm)	Average Ra (µm)
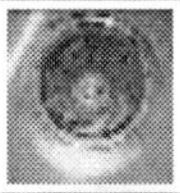	Original Brass, 6-blades	12.5	12.5	12.5
	Brass, 6-blades	0.24	0.33	0.28
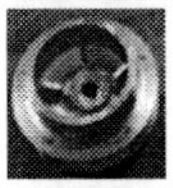	Brass, 5-blades	0.38	0.34	0.36
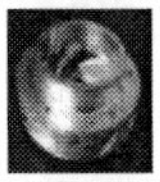	AA-7075, 6-blades	0.46	0.38	0.42
	AA-7075, 5-blades	0.31	0.46	0.31

3. Experimental Setup

The Experimental test was performed by using a Laboratory Scale PAT test facility as shown in Fig.3. In this facility, the top reservoir tank made of PVC, stores the water supply. It is placed 3.7 meters height above the ground. Clean water (assumed as H_2O Constant) flows from the tank into the PAT through PVC pipe. The water flows into the pump and rotates the impeller and its shaft. The shaft rotates the generator and produces electricity. The rotational speed of generator shaft was measured by using a digital tachometer. Meanwhile, the voltage and current produced by the generator was measured by using a Fluke 73 series II digital multi tester with accuracy $\pm$0.4%+1.

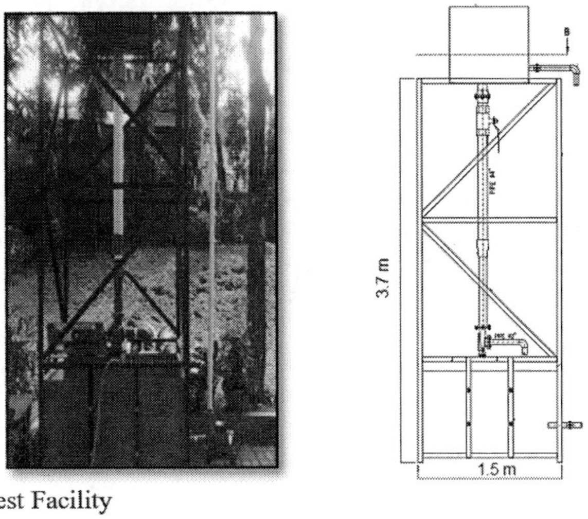

Fig.3. PAT Laboratory Scale Test Facility

4. Result and Discussion

Table 4 shows the experimental results of the shaft generator rotational speed and the power generated for each impeller. The table also listed the rotational speed and power improvement. Fig. 4 (a) shows the chart of the shaft generator rotational speed; meanwhile Fig. 4 (b) shows the electric power generated by the generator. It indicated that impeller made of brass with 6 numbers of blades generates the highest rotational speed and electric power. This impeller has the lower surface roughness among others. It could be concluded that the better quality of surface roughness of PAT impeller would produce more electric power.

Table 4. Experiment Result

No	Type of Impeller	Avg. Surface Roughness (μm)	Shaft Generator Rotational Speed (RPM)	Rotational Speed Improvement (%)	Power (Watt)	Power Improvement (%)
1	Original	12.5	1487	-	10.9	-
2	Brass, 6-blades	0.28	3077	106.9	27.14	148.99
3	Brass, 5-blades	0.36	782.7	-47.36	10.6	-2.75
4	AA-7075, 6- blades	0.42	2190	47.27	13.46	23.48
5	AA-7075, 5 -blades	0.31	640	-56.96	17.28	58.53

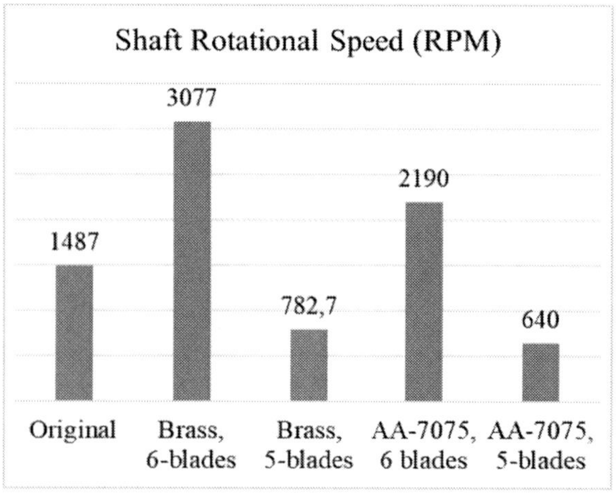

(a)

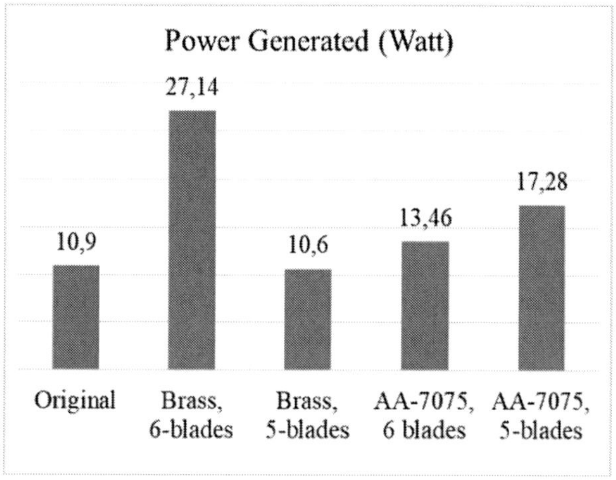

(b)

Fig. 4. (a) The rotational speed of shaft generator, and (b) the power generated

5. Conclusion

In this research, centrifugal impellers manufactured by machining process were used as impellers in a laboratory scale PAT micro-hydro power plant. Four impellers made of brass and AA-7075 with 5 and 6 numbers of blades were manufactured and tested for its performance. The performance is compared to the original impeller performance. Based on the experimental results, it is found that, impeller with 6-blades generates more power than 5-blade impeller. However, the type of material did not have any influence to the power generated. Meanwhile, the smoother impeller surface will increase the impeller rotational speed, up to two times faster than the original impeller.

This research is fund by The Ministry of Research Technology and Higher Education of Indonesia in a grant scheme of "Tim Hibah Pasca Sarjana".

References

[1] A. Elbratran, O. Yaakob, Y. M. Ahmed and H. Shabara, "Operation, performance and economic analysis of low head micro-hydropower turbines for rural and remote areas : A review," *Renewable and Sustainable Energy Reviews,* **43**, no. 1, 40-50, (2015).

[2] A. Mariano, "Pump as turbine - A pico-hydro alternative in Lao People's Democratic

Republic," *Renewable Energy,* **35**, (2010).

[3] A. M. a. S. A. Rusdi, "Research Efficiency Pump as Turbine (PAT) to Support the Planning and Implementation in PLTMH in the District Sumbawa," *Asean Journal of System Engineering,* **2**, no. 1, (2014).

[4] H.-Z. Fan and G. Xi, "Optimizing tool-path generation for three-axis machining of a sculptured impeller blade surface," *Proceeding of the Institution of Mechanical Engineers, Part B : Journal of Engineering Manufacture,* **226**, no. 1, 43-51, (2012).

[5] M. Suarda, N. Suarnadwipa and W. B. Adnyana, "Experimental Work on Modification of Impeller Tips of a Centrifugal Pump as Turbine," in *The 2nd Joint International Conference on "Sustainable Energy and Environment (SEE 2006)"*, Bangkok, Thailand, (2006).

[6] A. S. K. H. M. a. R. N. P. S. V. Jain, "Effect of impeller diameter and rotational speed on performance of pump running in turbine mode," *Energy Conversion and Management,* **89**, no. 1, 808-824, (2015).

[7] H. P. W. J. S. a. P. G. A. D. Giosio, "Design and perfomance evaluation of a pump-as-turbine micryo-hydro test facility with incorporated inlet flow control," *Renewable Energy,* **78**, 1-6, (2015).

[8] S. Kalpakjian and S. R. Schmid, Manufacturing Engineering and Technology, Singapore: Pearson, (2009).

[9] H.-T. Young and L.-C. Chuang, "An integrated machining approach for a centrifugal impeller," *The International Journal of Advanced Manufacturing Technology,* **21**, 556-563, (2003).

[10] I. Lazoglu, Y. Boz and H. Erdim, "Five-axis milling mechanics for complex free form surface," *CIRP Annals - Manufacturing Technology,* **60**, 117-120, (2011).

[11] H. T. Yau, L. S. Tsou and Y. C. Tong, "Adaptive NC simulation for multi-axis solid Machining," *Computer-Aided Design & Application ,* **2**, 95-104, (2005).

[12] O. Tuysuz, "Prediction of cutting forces in three and five-acis ball-end milling tool indentation effect," Thesis - Mechanical Engineering, The University of British Columbia, Vancouver, (2009).

[13] J. Taylor, A. L. Carrano and S. G. Kandlikar, "Characterization of the effect of surface roughness and texture on fluid flow - past, present, and future," *International Journal of Thermal Sciences,* **45**, 962-968, (2006).

[14] J. Qian, Z. Chen, H. Zhan and H. Guan, "Experimental study of the effect of roughness and Reynolds number on fluid flow in rough-wal single fractures : a check of local cubical law," *Hydrological Process,* **25**, 614-622, (2011).

[15] L. Moody, "Friction factores for pipe flow," *ASME Tran* ,**66**, 671-683, (1944).

[16] S. Kaewnai, M. Chamaoot and S. Wongwises, "Predicting performance of radial flow type impeller of centrifugal pump using CFD," *Journal of Mechanical Science and Technology,* **23**, 1620 - 1627, (2009).

A study of tea leaves drying using dehumidification process and regeneration of liquid desiccant in a closed-cycle *dehumidification-humidification*

Eflita Yohana[1,*], *Mohammad* Endy Yulianto[2], *Shofwan* Bahar[1], *Azza* Alifa Muhammad[1], and *Novi* Laura Indrayani[3]

[1]Department of Mechanical Engineering, Faculty of Engineering, Diponegoro University, 50275 Semarang, Indonesia
[2]Department of Industrial Technology, Faculty of Engineering, Diponegoro University, 50275 Semarang, Indonesia
[3]Department of Mechanical Engineering, Faculty of Engineering, Islamic 45 University, 17113 Bekasi, Indonesia

Abstract. Tea plants in Indonesia are derived from *Carmelia sinensis* var. *assamica* which contain catechin in quite high amount compared with other countries tea plant. Green tea is made by inactivating the oxidase / phenolase enzyme that presents in the fresh tea leaf buds from tea garden, by using hot steam to prevent the oxidation of the catechins. Drying process to reduce the moisture of tea, one of the method is by utilizing the dry air from dehumidification process. Liquid desiccant made from 50% concentration of $CaCl_2$, the temperature is lowered to 10 °C and sprayed into the air stream which contains water vapor by using a 0.2 mm spraying nozzle so that mass transfer and latent heat occur in the dehumidifier. The result of air dehumidification process used for drying tea leaves. The air is able to dry the tea leaves from the weight of 58 grams to 47 grams. Then the liquid desiccant dehumidification process will be streamed into the humidifier, where the liquid desiccant regeneration process will have change into the initial concentration. The result of air humidification process has an average absolute humidity rise of 0.07 g/kg. The liquid desiccant regeneration process that happened continuously reaching the saturation point at 280 minutes. It can be concluded that the process of dehumidification-humidification is a fairly effective method for drying the tea leaves.

1 Introduction

Indonesia is a country that has a great wealth of biodiversity, there are approximately 30,000 species of plants that grow in Indonesia [1], which includes the tea plant. Tea plants with Camellia sinensis varieties assamica in Indonesia have a high enough catechin content

* Corresponding author: efnan2003@gmail.com

rather than other tea plants. Green tea is made by inactivates the enzyme oxidase / phenolase present in the fresh tea leaf buds of the tea garden with using hot steam so that oxidation of the catechins can be prevented. Hot steam which used for tea leaves should have a temperature between 80 °C - 100 °C [2]. Furthermore, the tea will undergo a drying process to reduce the water content in the tea leaves.

In the process of drying the tea leaves, temperature and humidity are two things that are important to note. Moist air can inhibit the drying process of the tea leaves and will trigger the growth of bacteria that can be harmful to human life. But if the air is too dry, it will cause some problems, such as case hardening and a condition which the outside of the tea leaves is dry but the inside is still moist. This will cause the growth of mushrooms on the tea leaves [3].

Today research on the air conditioning system has been growing rapidly. There is an alternative to air conditioning systems that are air conditioning system with desiccant as a working fluid. Desiccant has hygroscopic properties, which can absorb water vapor contained in the air. The selection of desiccant as the working fluid in the air conditioning system has some advantages, that is able to achieve low humidity, energy saving and environmentally friendly [4].

The process of reduction of water vapor in the air is called dehumidification process. This process uses low temperature between 22 °C - 25 °C, so there is no oxidation in the tea leaves and the color of the tea leaves still green. After that, a desiccant is sprayed into the air stream containing water vapor which results from mass transfer and latent heat. During the process of absorption of water vapor, sorbent, known as the desiccant liquid reaches a point where the ability of moisture absorption of liquid desiccant will be decreased. To be able to be reused, the liquid desiccant must be regenerated [5]. The tools to regenerate the desiccant liquid called a regenerator or commonly known Humidifier. The working principle of the tools is to eliminate the moisture content in the liquid desiccant, which can be done by heating the liquid desiccant to revert it to original concentration [6].

2 Materials and methods

Liquid desiccant is made by dissolving $CaCl_2$ using distilled water with a concentration ratio of 50% : 50%. This data is collected until the condition of the inlet sample side of the house shows that there are no changes in temperature and humidity or in another word, the sample condition is already as steady as it is. The drying process of tea leaves itself is done by conducting the steaming and rolling process.

The desiccant liquid sampling performs every 20 minutes during the process of dehumidification and regeneration/humidification takes place, namely when the desiccant liquid has a before and after contact with both the dehumidifier and air humidifier.

Scheme research tool in Figure 1 consists of a dehumidifier, humidifier, pumps, chillers, nozzle 0.2 mm, the sensor DHT 11, and $CaCl_2$ with a concentration of 50% as its liquid desiccant.

Using mass and energy balance can calculate dry air for dehumidifying or humidifying

$$\dot{m}_a = \frac{\dot{m}_1 \left(h_{f1} - h_{f2} \right)}{\left(h_{g4} + \omega_4 h_{g4} \right) - \left(h_{a3} + \omega_3 h_{g3} \right)} \tag{1}$$

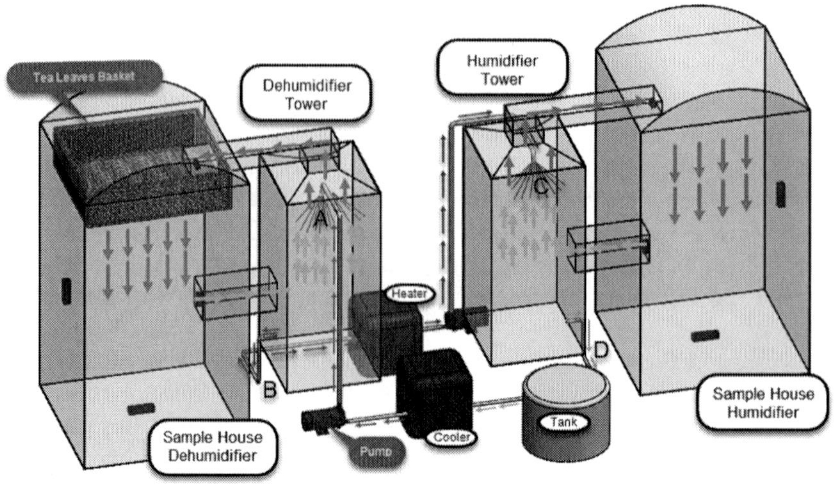

Fig. 1. Analysis tool scheme.

The study was conducted by cooling the desiccant liquid until it can reach a temperature of 10 °C from cooler, and air flow rate of 2.35 m^3/min (without any eliminator) from sample house dehumidifier. In the dehumidifier, the liquid desiccant will be sprayed using a spraying nozzle that has an outlet hole of 0.2 mm and at the same time, the air from sample house dehumidifier will be dry, then the dry air has through pass tea leave basket and remove the moisture content of tea leaves. After the liquid desiccant is sprayed in dehumidifier will be accommodated and fed to the humidifier. Furthermore, liquid desiccant will be going through a regeneration process by raising the temperature from a heating process to heat temperature of 70 °C. In the humidifier, liquid desiccant will be sprayed using a spraying nozzle and then there will be liquid desiccant regeneration process to bring back its original concentration. The occurring changes in humidity and air temperature within the sample house will be read by the sensor DHT 11 that is mounted on four sides, which are at the inlet, the walls, floors, and outlets in the sample house. The use of closed system method considered because less energy consumption throughout the process than if we use open system method.

3 Result and discussion

3.1 The result drying the tea leaves in dehumidification/humidification process

The results of the study will be compared to the international journal which the result of this research was the dry air is the element that is used to dry the leaves. The referenced journal that is used for this comparison is the research conducted by Jun Liu, et al. Beijing Tsinghua University of China in 2015. In June, this research analyzed the results of the research where dehumidification/humidification regeneration is made of a thermally conductive plastic using a liquid desiccant.

In Jun Liu's study, it used lithium bromide (LiBr) as liquid desiccant with a concentration rate of 38.8 to 42.2%, and with a temperature rate of 28.6 to 31.5 °C. Jun Liu, et al. also used additional components such as pipes that drained the water used for the cooling medium air passing through the dehumidifier with a water temperature of 11.1 to 17.0 °C, as well as the heating medium air that passed through the regenerator with a water temperature of 31.1 to 41, 8 °C.

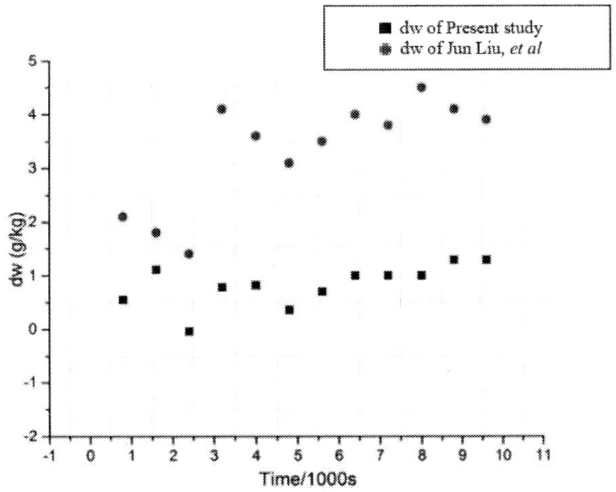

Fig. 2. Comparison of absolute moisture reduction in dehumidification process every 1000 seconds.

When compared to the results of research conducted by Jun Liu, et al. in Figure 2 which has the ability to decrease the average absolute humidity of 3.3 g/kg, the dehumidification tool used in this research still have not achieved the capability yet because it has the ability to decrease the average absolute humidity of 0.78 g/kg. The main cause of the difference between the results of the Jun Liu, et al. study and the results of this current study is due to the different liquid desiccant that was used are LiBr and $CaCl_2$.

The result of the research conducted by Koronaki, et al (2012) on the analysis of thermodynamics in a dehumidifier adiabatic with the type of flow counter flow using different materials, there were namely LiCl, LiBr, and $CaCl_2$ posed as liquid desiccant her, stating that LiBr has the ability to reduce the humidity more nicely compared to $CaCl_2$. This study decided to use $CaCl_2$ which safer to consume than LiBr. Although $CaCl_2$ decrease possibility of humidity will not as good as LiBr.

In the study of Jun Liu, et al. (2015), it used lithium bromide (LiBr) as its liquid desiccant with a concentration of 38.8 to 42.2% a temperature of 28.6 to 31.5 °C. Jun Liu, et al. also used additional components such as pipes that drained the water used for the cooling medium air passing through the humidifier with a water temperature of 11.1 to 17.0 °C, as well as the heating medium air that passed through the regenerator with a water temperature of 31.1 to 41, 8 °C.

When compared to the results of experiments conducted by Jun Liu, et al. (2015), which has an average ability to increase the absolute humidity value of 3.2 g/kg, the regenerator that we use still cannot achieve the right capability since it only has the ability to increase the average absolute humidity values of 0.07 g/kg.

3.2 The weight decrease of tea leaves during the drying process by dehumidification process air

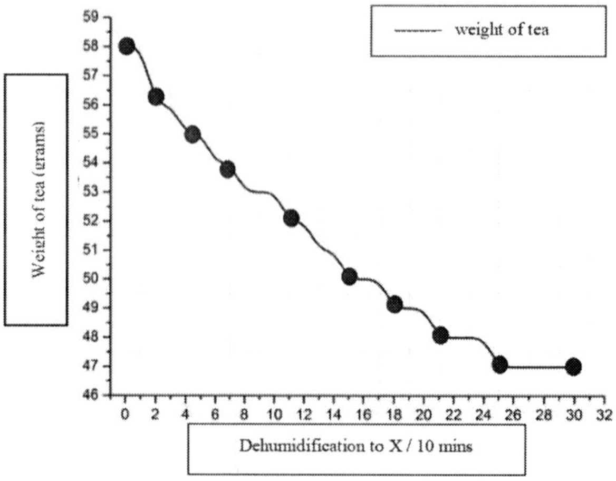

Fig. 3. Graph of tea leaf weight loss during drying process by dehumidification process air.

In Figure 3, it is known that the early tea weight was 58 grams. The weighing is done every 10 minutes during the drying process takes place. It can be seen that the weight loss tea leaves in early weighing less constant over 2 times the weighing (20 minutes) on average by 1 gram and ends on the state of constant weight by 47 grams for 6 times the weighing (1 hour) with the time needed to achieve weight ie constant for 25 times the weighing (4 hours, 10 minutes). This shows that the dry air dehumidification process results capable of drying the tea leaves from the weight of 58 grams to 47 grams, was reduced by 11 grams within 5 hours.

3.3 Changes in liquid desiccant concentration during humidification process

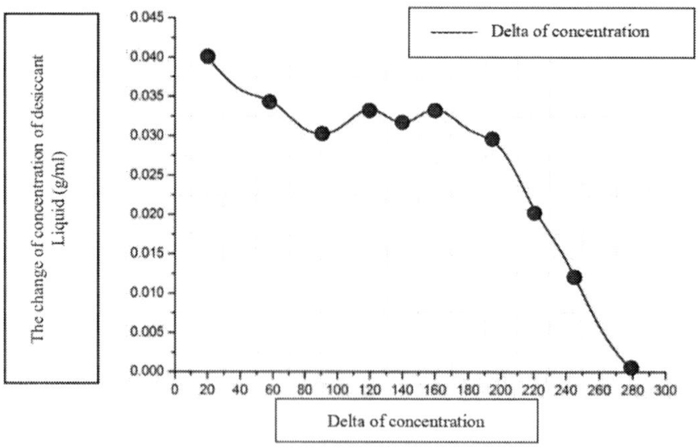

Fig. 4. Graph of change of liquid desiccant concentration humidification process.

In Figure 4, it can be seen that there is a decrease in the concentration of liquid desiccant which is significant at the 200th minute or 1 hour 20 minutes. And at 280th minutes, there is no decline in the concentration of liquid desiccant. This proves that the liquid desiccant has already experienced a saturation point where the rate of moisture absorption comes from the liquid desiccant down. If absorption rate of the air remains stable needed, using another type of liquid desiccant is recommend.

4 Conclusions

The result of air dehumidification process for tea leaves drying process has relative humidity degradation value of 11% with absolute humidity steady degradation value of 1 g/kg and has the capability to reduces 11 grams tea leaves from 58 grams to 47 grams of weight. The result of air humidification process shows absolute humidity increment value of 0.07 g/kg. As can be seen, dry air in the humidifier absorbs water vapor in the liquid desiccant which is evidenced by a significant decrease in the concentration of liquid desiccant in the 200th minute or 1 hour 20 minutes and at 280th minutes there is no more decline in the concentration of liquid desiccant.

Nomenclature

Symbol		Description	Unit
$\dot{m}_a$	:	Flow rate of the dry air	Kg/h
$\dot{m}_1$	:	Flow rate of $CaCl_2$	Kg/h
h_{f1}	:	Specific enthalpy of $CaCl_2$	Kj/Kg
h_{f2}	:	Specific enthalpy of air	Kj/Kg
h_{a3}	:	Enthalpy of atmospheric air	Kj/Kg
h_{a4}	:	Enthalpy of CaCl	Kj/Kg
ω_3	:	Humidity ratio of atmospheric air	-
ω_4	:	Humidity ratio of CaCl	-

References

1. D. Handayani, A. Mun'im, A.S. Ranti, Optimasi ekstraksi ampas teh hijau (camellia sinensis) menggunakan metode microwave assisted extraction untuk menghasilkan ekstrak teh hijau. Tradisional medicine Journal, **19(1)**, 29-35 (2014)
2. E. Yohana, M.S.K.T.S. Utomo, T. Ivantoro, Z. Arifin, M. Tauviqirrahman, Effect of Temperature and Relative Humidity on the Performance of Steamer using Computational Fluid Dynamic (CFD). International Journal of Engineering and Technology (IJET), **8 (4)**, 1737-1749 (2016)
3. E. Yohana, M.E. Yulianto, M.A. Dwisukma, Pengembangan teknik dehumidifikasi absorbsi pada fluidized bed dryer untuk produksi teh hijau kaya polifenol. Semarang: s.n (2015)
4. J. Liu, T. Zhang, X. Liu, J. Jiang, Experimental analysis of an internally-cooled/heated liquid desiccant dehumidifier/regenerator made of thermally conductive plastic. Energy and Buildings, **99**, 75-86 (2015)
5. E. Yohana, C.K. Hwan, Prediction on heat and mass transfer coefficients in a packed layer of a regenerator with a solar desiccant cooling system. Journal of the Korean Solar Energy Society, **30**, 4 (2010)
6. Y. Yonggao, X. Zhang, D. Peng, X. Li, Model validation and case study on internally cooled/heated dehumidifier/regenerator of liquid desiccant systems. International Journal of Thermal Sciences, **48**, 1664-1671 (2009)

Modeling, Identification, and Simulation of Positional Displacement Control for Ribbon Bridges

Van Trong Nguyen[1], *Yong-Woon* Choi[2], *Jung-In* Yoon[3], *Kwang-Hwan* Choi[3], *Chang-Hyo* Son[3], and *Young-Bok* Kim[1,*]

[1] Department of Mechanical System Engineering, Pukyong National University, Busan, Republic of Korea
[2] Department of Information System Science, Soka University, Tokyo, Japan
[3]Department of Refrigeration and Air-conditioning Engineering, Pukyong National University, Busan, Republic of Korea

Abstract. The demand of river-crossing is steadily increasing, the speedy and safety are two most important factors that decide the success of the applied solutions. This paper partially proposes a method for ribbon bridge installation and self-correction by describing the planar motion for a ribbon bridge, including several experimentally identified over a certain range of speeds. The mathematical modeling and system identification aim to provide a model that is sufficient for allowance of model-based control designs. In order to design the stable yaw control system of auto-correction of positional displacement, the Linear-Quadratic Regulator (LQR) was employed for yaw controller. Accordingly, the numerical study was carried out under a variety of disturbances to verify the system modeling and the efficiency of the designed controller. The simulation results demonstrate that the proposed controller has promising feasibility of the yaw placement automatic correct during operation and further extended results of the whole bridge system.

1 Introduction

The river-crossing operation is one of the most necessary operations for emergency restoration in military and other viewpoints. The major objective of river-crossing is transporting the armed forces between two shores of the river, lake, or a canal. In term of bridging, the improved ribbon bridge is one of the most effective solutions because of its transportation ability and the portability. It is constructed on a river by jointing multiple floating bays [1] to connect two river sides. Conventionally, the linearity of Ribbon Bridge is maintained by employing the assistances of a number of erection boats. Therefore, the river-crossing contains a certain risky factors in military standpoint, especially during the combat situation. The safety and speedy are two crucial factors that decide the success of river-crossing process. In reality, a number of operators are required to install and to retain

* Corresponding author: kpjiwoo@pknu.ac.kr

the linearity of the ribbon bridge by adjusting each floating bay. Additionally, it is strongly difficult to maintain the placement of a single floating unit by the erection boats due to many other affections including water flow current, moving load, especially under harsh conditions. Thus, the final target is to achieve the automatic installation and positional correction of the ribbon bridge during the operation time. A recent study of Yasuhiro and co-authors [2] focused on measurement system for measuring the positional displacement of floating units using a multi-copter that enables the feasibility of automatic installation and self-correction of the floating ribbon bridge by applying control theory. Besides, Mohammad and Yoshiyuki [3] and Shixiao and Weicheng [4] analysed the dynamics of floating bridges. However, the needs of a sufficient mathematical study in order to establish the control-based modelling were not satisfied. In this study, we propose an efficient mathematical model based on analytical modelling of floating units, and identify the system parameters. To verify the feasibility of the abovementioned mathematical modelling and etc., a numerical investigation is studied by applying LQR control technique employing computational software MATLAB and Simulink. The numerical results demonstrate that our proposed model and control strategy have potential ability to enable the automated installation and operation of a new ribbon bridge.

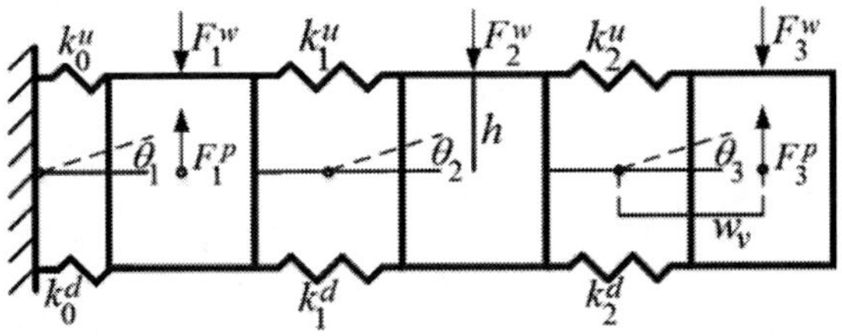

Fig. 1. The structure diagram of Ribbon Bridge.

2 Problem statement

2.1 Mathematical modelling

The modelling process to obtain a model of the floating units is introduced by authors [5]. Then a basic model consists of the kinematic model and kinetic model [5].

$$J\dot{\upsilon}+C_v\upsilon=\tau \tag{1}$$

Where $\upsilon=[u,v,r]^T \in R^3$ describes the surge, sway, and yaw rate of the floating unit motion in body fixed frame; $J \in R^{3X3}$ presents a mass/inertia matrix; C_v is the hydrodynamic damping matrix:

$$J=\begin{bmatrix} m-X_{\dot{u}} & 0 & 0 \\ 0 & m-Y_{\dot{v}} & 0 \\ 0 & -N_{\dot{v}} & I_z-N_{\dot{r}} \end{bmatrix} \quad C_v=\begin{bmatrix} -X_u & 0 & 0 \\ 0 & -Y_v & 0 \\ 0 & 0 & -N_r \end{bmatrix}$$

Then, the mathematic model for the three-bay ribbon bridge is written as:

$$\begin{aligned} &J_1\ddot{\theta}_1+C_v\dot{\theta}_1+C_0\dot{\theta}_1+C_1(\dot{\theta}_1-\dot{\theta}_2)=(F_0^{lu}+F_0^{ld}+F_1^{ru}+F_1^{rd})h+(F_1^p-F_1^w)w_v \\ &J_2\ddot{\theta}_2+C_v\dot{\theta}_2+C_1(\dot{\theta}_2-\dot{\theta}_1)+C_2(\dot{\theta}_2-\dot{\theta}_3)=(F_1^{lu}+F_1^{ld}+F_2^{ru}+F_2^{rd})h+(-F_2^w)w_v \\ &J_3\ddot{\theta}_3+C_v\dot{\theta}_3+C_2(\dot{\theta}_3-\dot{\theta}_2)+C_3\dot{\theta}_3=(F_2^{lu}+F_2^{ld})h+(F_3^p-F_3^w)w_v \end{aligned} \tag{2}$$

Where, C_1, C_2, C_3 are damping coefficient of the corresponding joints (in this case, we consider $C_1 = C_2 = C_3 = C$); θ is the yaw rate; $F_i^{lu}, F_i^{ld}, F_i^{ru}, F_i^{rd}$ are spring forces of upside and downside in left hand side and right hand side, h is the arm of spring forces; F_i^{p}, F_i^{w} are forces generated by actuators and the current flow force, and w_v is the arm of force generated by driving DC motor and current flow.

Table 1. Specifications of experiment apparatus.

No.	Item	Specification
1	Load cell	10N (max) – 5 VDC
2	Incremental Encoder	1024 pulse/rev
3	Motor Driver	KDC248H – 12 ~ 24 V
4	DC Motor	12V/17.200 rpm, Graupner co.
5	Propeller	Ø41 mm, Graupner co.

2.2 Parameter identification

In this study, the 3-bay ribbon bridge pilot model is used for experiment. The modelling of 3-bay ribbon bridge is carried out with regarding individual yaw motion. The identification for the ribbon bridge includes two different steps by conducting several experiments [6].

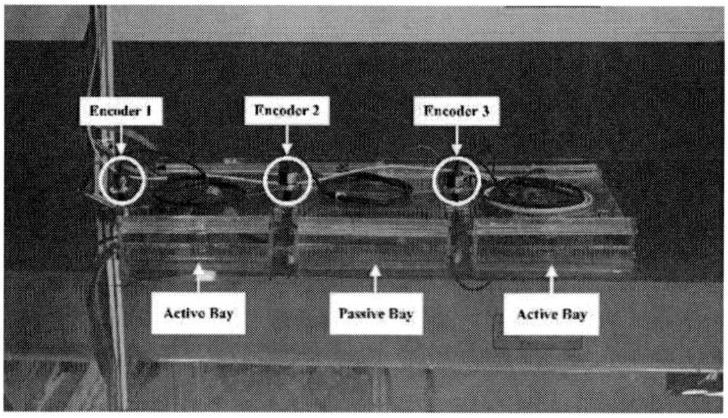

Fig. 2. The ribbon bridge model using for experiment.

The experimental apparatus of ribbon bridge system is constructed as shown in Fig. 2 with 3-bays which are driven by propulsion system with DC motor. Two of three bays have active propulsion system. In Fig. 2, the active bay has a motor driven propulsion system, in other word, there is no propulsion system in the passive bay. The dimensions of each single bay is described by 260mm[L]x440mm[W]x80mm[H]. The mass of each single bay is 3.79kg. As mentioned previously, two active bays are driven by propellers which are controlled by the DC motor respectively. In addition, a camera is employed to measure the position and a load cell is used to obtain the driven force generated by the propulsion system. At first, the relation between the force generated by driving DC motor, propeller, and the input voltage is identified by experiments and employing Matlab Identification Toolbox [7]. The obtained transfer function which approximately describes the input-output relation between the input voltage (V) and the generated force (N) is expressed as:

$$\frac{F_i^p(s)}{U(s)} = \frac{18.5}{s^2 + 6.915s + 55.74} \tag{3}$$

Second, the ribbon bay mass matrix and the damping matrix were determined from simulation and collected data by experiment, as shown in Fig. 3. In the results, the parameters are identified as follows:

$$\begin{aligned} J &= I_z - N_{\dot{r}} = 0.0055, \\ C_v &= -N_r = 0.004, \\ C &= 0.001 \end{aligned} \tag{4}$$

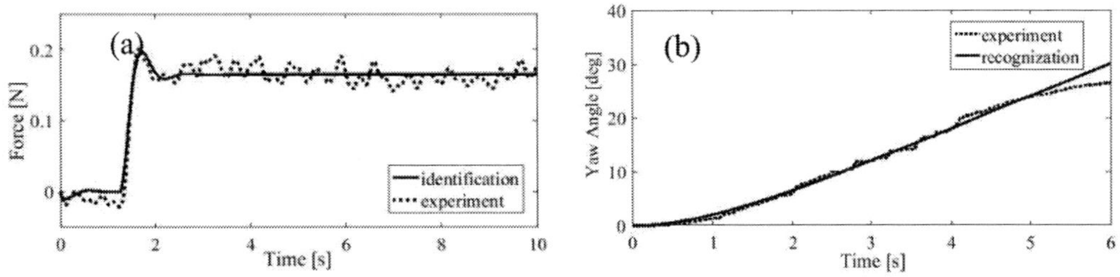

Fig. 3. Comparision studis for system identification by simulation and experiment results. (a)forces produced by propulsion system, (b)free yaw motion responses.

3 Controller design

The 3-bay ribbon bridge with two active bays and one passive driven bay can be modelled in the state-space model with respect to following state variables.

$$[x_1\, x_2\, x_3\, x_4\, x_5\, x_6\, x_7\, x_8\, x_9\, x_{10}]^T = [\theta_1\, \dot{\theta}_1\, x_{f1} \dot{x}_{f1}\, \theta_2\, \dot{\theta}_2\, \theta_3\, \dot{\theta}_3\, x_{f3} \dot{x}_{f3}]^T \tag{5}$$

It can be seen that the yaw angles depend on the input driven forces made from the propulsion system. The linear quadratic regulator (LQR) control [8] is a well known design technique which provides marvellous control performance and reduce the effects of disturbance as well as maintaining the robustness of system stability. In this study, the control objective is minimizing the positional displacement. For this, the LQR controller with integral action is designed based on the mathematical modelling of three-bay ribbon bridge system. Let consider the time invariant system:

$$\begin{aligned} \dot{x}(t) &= Ax(t) + Bu(t) \\ y(t) &= Cx(t) \end{aligned} \tag{6}$$

where $x \in R^n$ is the state vector of the system, $u \in R^m$ control inputs, $y \in R^m$ is output to be regulated, and *A*, *B*, and *C* represent the real constant matrices of system. Consider the problem to regulate the output $y(t)$ to the desired pre-set tracking $r(t)$ under the action of the input disturbance $w_d(t)$. To obtain good tracking performance, the authors design a servosystem shown in Fig. 4 where an integrator is introduced to reject steady state error.

Then, the control system shown in Fig. 4 can be expressed as:

$$\begin{aligned} \begin{bmatrix} \dot{x} \\ \dot{w}_e \end{bmatrix} &= \begin{bmatrix} A - BK_f & BK_w \\ -C & 0 \end{bmatrix} \begin{bmatrix} x \\ w_e \end{bmatrix} + \begin{bmatrix} 0 \\ 1 \end{bmatrix} r \\ y &= \begin{bmatrix} C & 0 \end{bmatrix} \begin{bmatrix} x \\ w_e \end{bmatrix} \end{aligned} \tag{7}$$

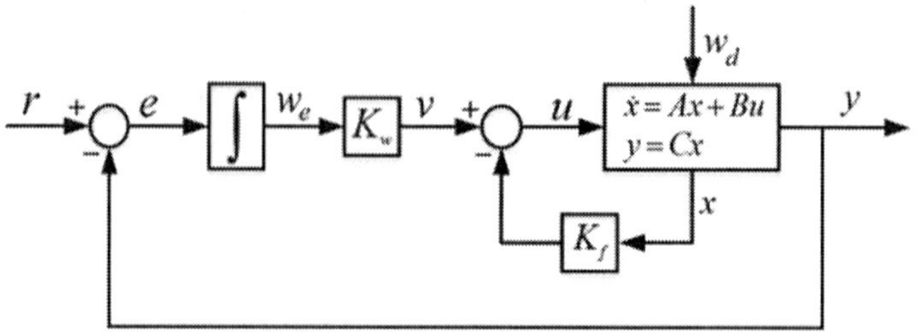

Fig. 4. 3-bay ribbon bridge control system using LQR controller with integral action.

Our goal is to design an optimal tracking controller that tracks the desired angles of each ribbon bay. In this case, the target is maintaining the linearity of the ribbon bridge by minimizing the positional displacement of each bay. That means the error $e = r(t) - Cx(t)$ must approach zero. Following the general optimal control theory, the weighting matrices were chosen as:

$$R = diag\{1.5\ 0\ 0\ 0\ 1.5\ 0\ 1\ 0\ 0\ 0\ 1\ 1\ 1\}; \quad Q = diag\{4501.5\ 4551.5\} \tag{8}$$

In the results, an obtained linear control law is given as follows:

$$u = -K_{fw}\tilde{x} \tag{9}$$

where,

$$K_{fw} = -R^{-1}B_0^{\ T}P = \begin{bmatrix} K_f & K_w \end{bmatrix} \tag{10}$$

And, P is a positive definite symmetric matrix that satisfies the following Riccati equation:

$$PA_0 + A_0^T P - PB_0R^{-1}B_0^T P + Q = 0 \tag{11}$$

Therefore, K_f and K_w are:

$$K_f = \begin{bmatrix} 0.0318 & 0.0127 & 0.0432 & 0.0061 & 0.0191 & 0.0135 & 0.0033 & 0.0102 & 0.0349 & 0.0049 \\ 0.0054 & 0.0101 & 0.0345 & 0.0048 & 0.0257 & 0.0223 & 0.0433 & 0.0311 & 0.1063 & 0.0150 \end{bmatrix}$$

$$K_w = \begin{bmatrix} 0.0136 & 0.0058 & -0.0020 \\ -0.0007 & 0.0063 & 0.0134 \end{bmatrix}.$$

4 Simulation results

At first, in this section, the authors present and discuss the numerical simulation results.

The simulation results are shown in Fig. 5. It is the initial response. In Fig. 5, (a) is yaw motion control responses in which the dotted line is the reference signal and others are yaw motions of three bay, respectively. And (b) represents the control inputs which are supplied to the two propulsion systems. Where, wave disturbance attacks the floating units at 5[s] and 15[s] with 3[cm] height.

The simulation results show that the yaw deviations are quickly approaching to the target point in the presence of disturbance. It confirms that the ability of the proposed controller to maintain the linearity of the bridge system.
From these results, the proposed ribbon bridge control system works well such that the effectiveness of this strategy is verified. This result will be extended to the future study including experiment.

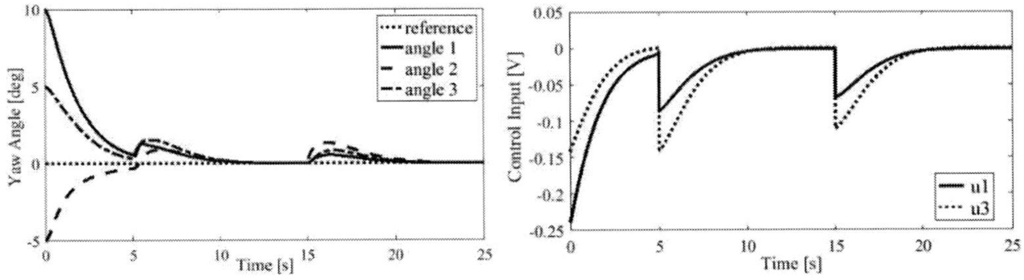

(a) yaw angle responses with disturbance (b) control inputs to the propulsion systems

Fig. 5. Yaw motion control responses and control inputs.

5 Conclusions

In this study, the authors have proposed a control strategy for installating the floating bridge system constructed by connecting multi floating units. For the control system design, a servosystem with an integrator is introduced to maintain the linearity of the bridge system and reject the steady state error. The mathematical model for a ribbon bridge system was established by considering the dynamic characteristics and identifying process. In addition, the LQR control approach was developed for the ribbon bridge motion control.

The numerical simulation results demonstrate the effectiveness of the proposed approaching. And it shows that the promising feasibility of positional linearity control, automated installation and operation of floating bridge system without any human power are possible.

This work was supported by the National Research Foundation of Korea (NRF) grant funded by the Korea Government (Ministry of Education) (No.NRF-2015R1D1A1A09056885).

References

1. Improve Ribbon Bridge, *GENERAL DYNAMICS* **2**-3 (2012)
2. Y. Hirono, Y. Mizuchi, Y-B. Kim, Y-W. Choi, LNEE **135**, 18-27 (2016)
3. M. S. Seif, Y. Inoue, Marine Structures **11** (1998)
4. S. Fu, W. Cui, Marine Structures **29**, 246-256 (2012)
5. T. I. Fossen, *Marine Control System* (Trondheim, Norway, 2002)
6. R. Skjetne, O. N. Smogeli, T. I. Fossen, *Mod. Ident. Con.* **23**-1, 3-27 (2004)
7. Matlab System Identification Toolbox User's Guide (The MathWorks, 2016)
8. C. Hajiyev, H. E. Soken, S. Y. Vural, Linear Quadratic Regulator Controller Design, (Springer, 2015)

Failure Analysis of Shaft Circulating Water Pump (CWP) used in Power Plant

Abdul Hamid[1,*], *Sri* Nugroho[1], *Gunawan Dwi* Haryadi[1], *and Khaeroman*[2]

[1]Department of Mechanical Engineering, Faculty of Engineering, Diponegoro University, Jl. Prof. Soedharto, Tembalang, Semarang 50275
[2]Department of Mechanical, Politeknik Maritime Negeri Indonesia, Semarang Jl.Pawiyatan Luhur 1/1 Bendan Duwur semarang 50233

Abstract. Pump shafts are generally exposed to the liquid being pumped either on a continual basis or at certain locations along the length of the shaft. The shaft material is austenitic stainless steel, description ASTM AU 79 TY 316. The purpose of this study is to determine the failure of the water pump shaft used in the power plant. Metallography is the study of structure metal shaft can used as a means for CWP metal pelleting (Circulating Water Pump), for the purpose of damaged or deeply degraded areas. SEM test is used to know the beginning of the crack (crack initiation). EDS test is used to chemical composition and Vikers hardness test is also used to know the hardness material. These three tests to support in analyzing the failure of the pump shaft. The conclusion of this failure analysis is the shaft material has porosity. Fatigue cracking comes from the outer surface area.

1. Introduction

Pump shafts are generally exposed to the liquid being pumped either on a continual basis or at certain locations along the length of the shaft [1]. During operation, the pump shaft usually occurs degradation due to corrosion and or mechanical degradation, usually in the form of fatigue failure. In many cases corrosion precedes fatigue failure and can actually accelerate the rate of failure [1]. Probably the most common cause of failure on pump shafts is fatigue [1].

Fatigue failures on centrifugal shafts are most commonly of the rotating-bending type. Schematic diagrams of the various fractures, which can be expected from rotating bending conditions, are presented in **Fig. 1**[2]. **Fig. 1**(a) and (c) illustrate the expected appearance of the fracture surface from a single origin under moderate and high degree of stress concentration while **Fig. 1**(b) and (d), describes the appearance of fracture surfaces and experiences moderate and high stress concentrations.

Stress concentration sites on shafts where fatigue cracks may initiate are illustrated in **Fig. 2** [22]. The most common areas of crack initiation are at the stress concentrations occurring at the keyway root radius and sharp changes in cross-sectional area of the shaft and is referred to as a stress concentration factor [3]. Stress concentrations serve to reduce, the fatigue strength of the component [4].

* Corresponding author: hamidsmg80@gmail.com

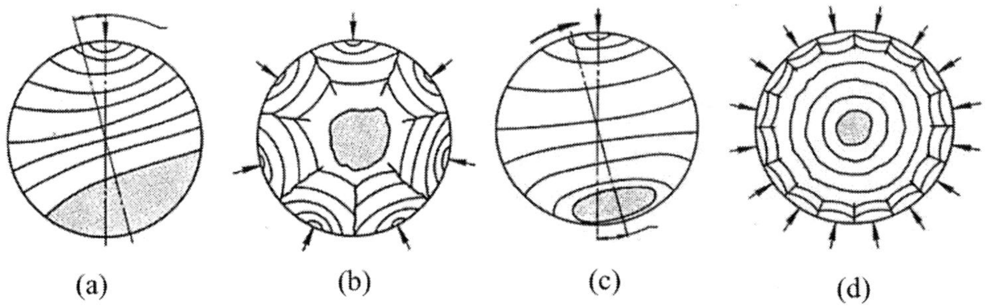

Fig 1. Schematics of fatigue failure fracture surfaces in rotating bending [2].

Thus, the presence of stress concentrators decreases the allowable stress on the shaft at that point, making it more susceptible to fatigue. Stainless steels on the other hand have an endurance limit ie. The material is design to be use for 8 years but is actually after 12 months usage failure.

By means of visual inspection, it was easy to find that most of the fractures occurred on the weld joints rather than the base materials, material property, manufacturing technology, equipment operation, service environment, routine maintenance or other factors, which were the main causes for inducing these abnormal fractures, were urgently investigated.

Consequently, a comprehensive failure analysis including a variety of characterization methods was conducted to identify the root cause based on our previous failure analysis experiences [5–13].

Actually, many researchers have focused on the properties of decision suport system, such as its fatigue behavior, welding property, corrosion resistance [14–20]. This study investigated the failure analysis of a rotating shaft of a circulating water pump injection, located in an power plan.

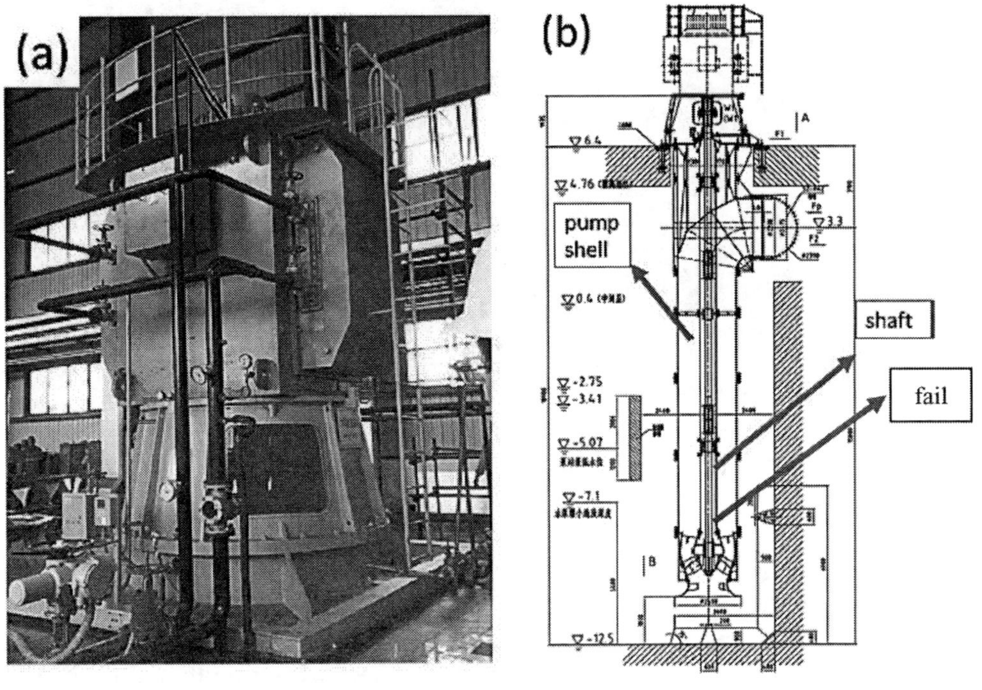

Fig 2. Appearance and structure CWP : (a) Appearance of CWP and (b) Structure of the CWP [22].

2. Materials and methodes

This rotating shaft failure in service under normal operating conditions of the injection pump during the pumping of brackish water wells for power plan. It is known that water in the operatingn component does not contain a significant amount of hydrogen sulfide (H_2S) generated during the decomposition of seaweed.

The shaft material is an austenitic stainless steel, classification ASTM AU 79 TY 316, supplied in forged condition. In the region of fracture of the shaft. Vickers hardness measurements on the HV were taken in twenty different positions in the cross-section of the shaft in a parallel plane close to the fracture of the part, according to standard.

Metallographic polished surfaces, corresponding to the upper shaft and shaft fracture, were observed with chemical attack in an optical microscope. The microstructure of the material was analyzed via scanning electron microscopy (SEM) to determine its phases.

The microstructure of the shaft material was analysed by optical microscopes. The composition of the shaft material was determined by using a standard spectrometer analyser.

3. Result and discussion

3.1 Visual examination

Visual examination of the failed end gave the appearance that the shafts of CWP failed by fatigue. A macroscopic view of the failed region of shaft is shown in **Fig.3**. Signs of smearing and distortion at the key edge were observed.

Areas of damage to the CWP shaft as fatigue failure. Crack propagation begins from an intergranular perlite gap in the yellow arrow. Dynamic layout is displayed on the blue arrow and the red arrow. Static cracking is a very quick propagation in green arrows. The area of damage to the shaft is shown Figure 3.

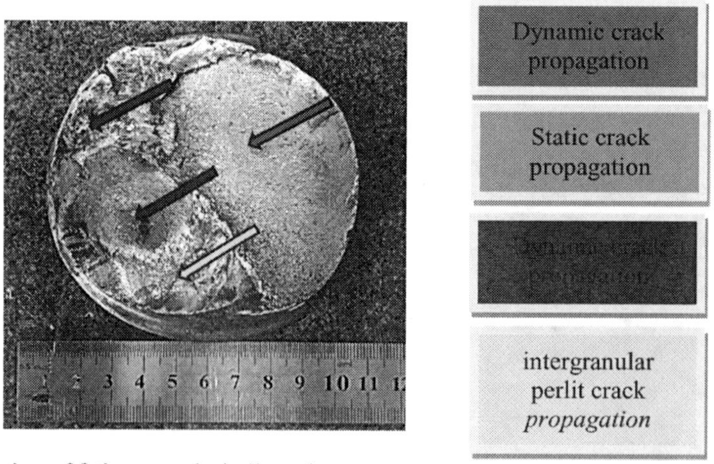

Fig 3. Sites of initiation of fatigue crecks indicated.

3.2 Chemical composition

The shaft material is an austenitic stainless steel, classification ASTM AU 79 TY 316 stainless steel enhanced with an addition of 2.5 % Molybdenum, to provide superior corrosion resistance to type 304 stainless steel. 79 TY 316 has improved pitting corrosion resistance and has excellent resistance to sulphates, phosphates and other salts. 79 TY 316 has better resistance than standard 18/8 types to sea water, reducing acids and solution of chlorides,bromides and iodies. The material's chemical composition upper shaft Fe 67,1%, Cr 17.56%; Ni 9.97%; Mo 2.01%; Mn 1.82%; Si 0.53%; and Cu 0.49% and failure shaft are Fe 67%, Cr 17.45%; Ni 9.93%; Mo 2.08%; Mn 1.82%; Si 0.55%; and Cu 0.5%. From the result of chemical composition test, the value of chemical composition is almost the same on the upper and lower shaft.

3.3 Hardness test

Regarding the hardness test, the top axle has a constant HV value and the failure shaft has a very high HV value. Hardness value of the upper shaft and failure shaft are provided in **Fig 4** and **Fig 5**. From the results of hardness testing (HV) obtained a very high abrasion hardness value this proves that the broken shaft fails because of the residual stress on the material. The residual stress causes the axis to be broken suddenly (unexpectedly).

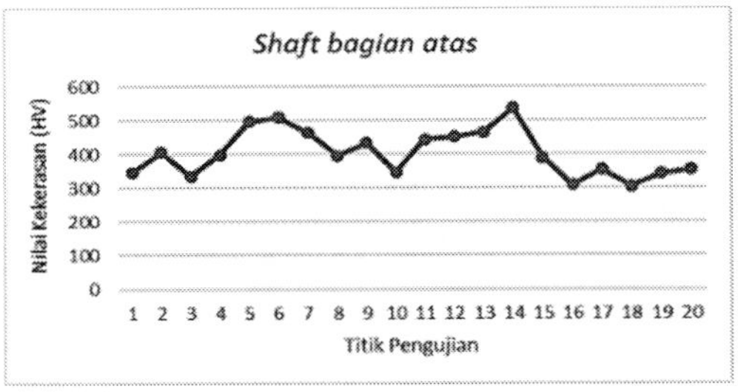

Fig 4. Hardness value upper shaft.

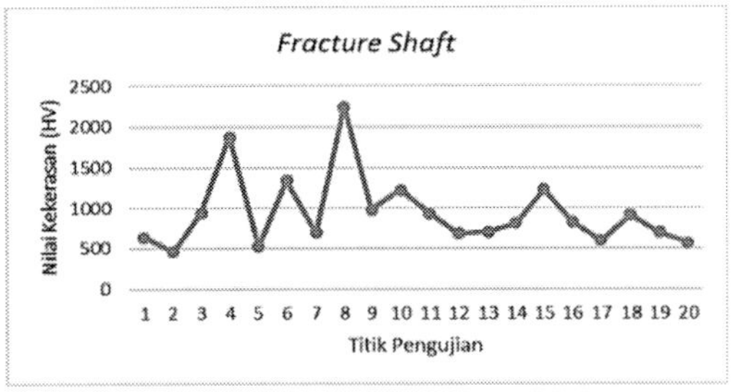

Fig 5. Hardness value failure shaft.

3.4 Metallography analysis

Metallographic test can be explained about microstructure type 316 with description description that type 316 is austenitic stainless steel chromiumnickel containing molybdenum. Type 79 TY 316 is an extra-low type 79 TY 316 carbon version that minimizes the precipitation of harmful carbides. With the influence of heat and rapid cooling rate, the pearlite structure is formed.

Figure 6 shows the microstructure of upper shaft and failure shaft. The failure shaft formed has new pearlite, its means that the failure shaft more hardness than the upper shaft.

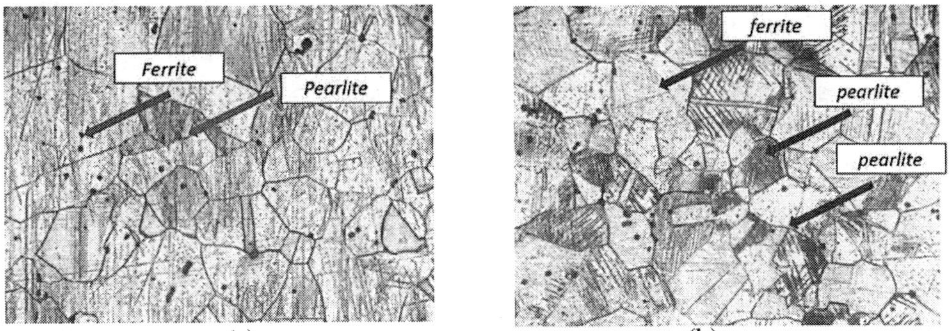

(a). (b).

Fig 6. Microstructure of upper shaft and failure shaft: (a) microstructure of upper shaft and (b) microstructure of failure shaft.

3.5 Fractography

Scanning Electron Microscopy (SEM) is used to fractographic examination. The result of the SEM test we can see the area that became the initial reference of the beginning of the crack. The brittle fracture experienced by the failure shaft can be explained by the contour of a SEM photo marked by a circle, looking very rough because the area has a brittle and high hardness. Figure 7 shows the result SEM in initial crack.

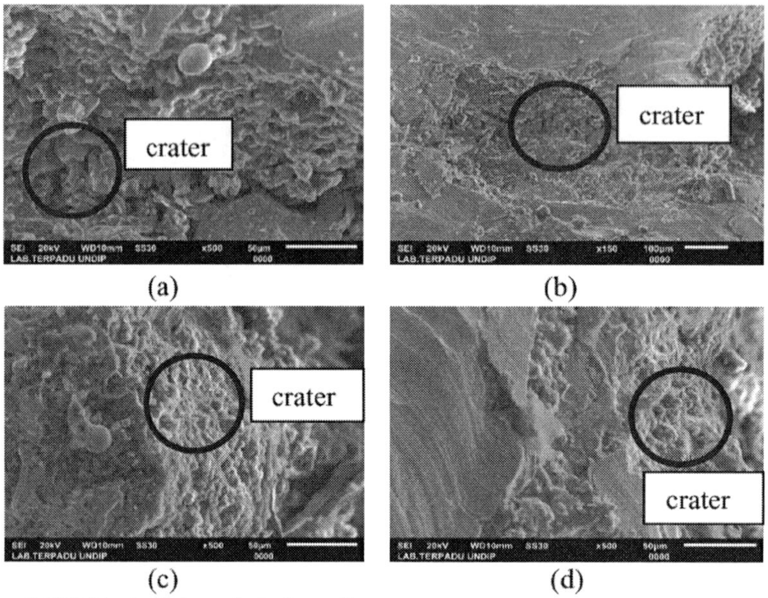

Fig 7. The result SEM in initial crack (a, b, c, d).

4. Conclusion

Chemical composition is almost no change between the shaft failure with a broken shaft. From the composition test, it can be concluded that both shafts include austenite stainless steel type with type AU 79 TY 316. The results of hardness testing on failure shafts tend to be higher than the upper shaft, this may explain that with high hardness levels resulted in rapid crack propagation rates compared with low levels of hardness. On metalography test the influence of heat and the cooling rate are the pearlite structure was formed. With the emergence of the pearlite structure, the properties of the material become hard and brittle. In the SEM test result is the observation reference area is the area experiencing brittle fracture. This can be explained by the contours on the SEM photo looks very rough because the area has a brittle nature and has a high hardness.

The authors gratefully wish to acknowledge the great support provided by Indonesia Power, Reactor Unit in Semarang. The authors extend their sincere thanks to Electrical analysis and engineering measurement reability analysis. Laboratory Management at Center of Research and Services, (UPT. Lab. Terpadu) Diponegoro University, Republic of Indonesia.

References

1. F. Berndta, A..*van Bennekom, Pump shaft failures - a compendium of case studies*, Engineering Failure Analysis 8,135 –144. (2001)

2. ASM Metals Handbook. Failure analysis and prevention, **vol**. 11. 9th ed. pp. 463. (1985)
3. Peterson RE. S.ress concentration factors. New York: Wiley. (1974)
4. ASM Metals Handbook. Mechanical testing, **vol**. 8. 9th ed. 364. (1985)
5. Yang ZG, Gong Y, Yuan JZ. *Failure analysis of leakage on titanium tubes within heat exchangers in a nuclear power plant. Part I: electrochemical corrosion.* Mater Corros;63(1):7–17. (2012)
6. Gong ZG, Yang JZ. *Failure analysis of leakage on titanium tubes within heat exchangers in a nuclear power plant. Part II: mechanical degradation.* Mater Corros;63(1):18–28. (2012)
7. Gong Y, Yang C, Yao C, Yang ZG. *Acidic/caustic alternating corrosion on carbon steel pipes in heat exchanger of ethylene plant*. Mater Corros;62(10):967–78. (2010)
8. Gong Y, Yang ZG. *Corrosion evaluation of one dry desulfurization equipment-circulating fluidized bed boiler*. Mater Des;32(1):671–81. (2011)
9. Gong Y, Cao J, Ji LN, Yang C, Yao C, Yang ZG. *Assessment of creep rupture properties for dissimilar steels welded joints between T92 and HR3C*. Fatigue Fract Eng Mater Struct;34(2):83–96. (2011)
10. Gong Y, Yang ZG, Yang FY. *Heat strength evaluation and microstructures observation of the welded joints of one China-made T91 steel.* J Mater Eng Perform;21(7):1313–9. (2012)
11. Chen Fei-Jun, Yao Chen, Yang Zhen-Guo. *Failure analysis on abnormal wall thinning of heat-transfer titanium tubes of condensers in nuclear power plant Part I: corrosion and wear*. Eng Fail Anal;37:29–41. (2014)
12. Chen Fei-Jun, Yao Chen, Yang Zhen-Guo. *Failure analysis on abnormal wall thinning of heat-transfer titanium tubes of condensers in nuclear power plan*t Part II: erosion and cavitation corrosion. Eng Fail Anal;37:42–52. (2014)
13. Gong Yi, Yang Zhen-Guo, Meng Xin-Hao.(2013) *Failure analysis of one peculiar 'Yin-Yang' corrosion morphology on heat exchanger tubes in purified*. terephthalic *acid (PTA) dryer.* Eng Fail Anal;31:203–10.
14. Iacoviello F, Boniardi M, La Vecchia GM. *Fatigue crack propagation in austeno–ferritic duplex stainless steel 22 Cr 5 Ni*. Int J Fatigue;21:957–63. (1999)
15. Sieurin H, Sandstrom R. *Fracture toughness of a welded duplex stainless steel*. Eng Fract Mech;73:377–90. (2006)
16. Sieurin H, Sandstrom R. *Sigma phase precipitation in duplex stainless steel 2205*. Mater Sci Eng A;444(1–2):271–6. (2007)
17. Lo I-Hsuang, Tsai Wen-Ta. Effect of selective dissolution on fatigue crack initiation in 2205 duplex stainless steel. Corros Sci;49(4):1847–61. (2007)
18. Balbi M, Avalos M, El Bartali A, Alvarez-Armas I. Microcrack growth and fatigue behavior of a duplex stainless steel. Int J Fatigue;31(11–12):2006–13. (2009)
19. Sahu JK, Krupp U, Christ HJ. Fatigue crack initiation behavior in embrittled austenitic–ferritic stainless steel. Int J Fatigue 2012;45:8–14.
20. Chiu PK, Weng KL, Wang SH, Yang JR, Huang YS, Fang Jason. Low-cycle fatigue-induced martensitic transformation in SAF 2205 duplex stainless steel. Mater Sci Eng A;398(1–2):349–59. (2005)

Detection of gear defects using method of wavelet decomposition and cross-correlation - Experiment to develop a feasible detection method

R. Lullus Lambang G Hidayat [1, *], and *Budi* Santoso [1]

[1] Mechanical Engineering Dep. Faculty of Engineering Universitas Sebelas Maret Indonesia

Abstract. Detection of machine component failure is very important to be properly applied in a maintenance program in industries. The objective of this research is to detect gear fault using wavelet transforms. The vibration signal is acquired with accelerometer mounted at bearing houses of 2 parallel shafts with 2 spur gears (28 tooth). The gears are rotated at 1200 RPM and the spectrum is displayed. The spectrum cannot indicate gear mesh frequencies (GMF), because they are covered with frequencies such as natural and harmonic frequencies of rotating shaft. This research have developed a method to obtain GMF using wavelet decomposition and cross correlation. The results showed that with FFT applied to cross-correlation of wavelet detil components,, spectrum of visible and distinctable GMF has been obtained.

1. Background

Gear units are parts of machines serve to transmit speed and torque between two shafts. The units are used in wind turbine [1], automotive and helicopter engines [2]. During using, the gear defects like wear, broken, cracked usually occure. The damage should be detected in a maintenance program. Gear condition monitoring is therefore required in order to improve safety and reduce operating costs.

In this research, a rig consists of 2 shafts with 2 gear is constructed . One gear is normal gear while the other has a partially/full toothcut. Electric motor is used to rotate one shaft to 1200 RPM. Vibration signal is then obtained using accelerometer mounted on the bearing house. The signal processing method uses only FFT has disadvantages, i.e. the frequency associated with fault condition must be known first and if the signal has a low signal to noise ratio or there are many other vibration components then peaks correspond to failure frequency cannot be clearly indicated [3, 4].

This research uses wavelet transformation to indicate GMF features so that gear failure i.e. full / partial toothcut gear can be detected. Wavelet-based signal processing for gear fault detection had been investigated by [4] and [5]. Further studies of wavelet transform applications on vibration signals for gear fault detection have been also performed by [6] and [7] that discuss of wavelet decomposition and feature extraction using PCA.

*lulus_l@staff.uns.ac.id; lulus_l@yahoo.com

2. Theoretical Basis

Gears are used in almost transmission systems, power systems and automotive. The gears have to rotate at a constant speed [4]. If the gear rotates at an unstable speed then one of the gears will spin fast and vibrations as an indicative of damage is produced [2]. Figure 1(a) shows the structure of a spur gear and the terms that are often used in discussions.

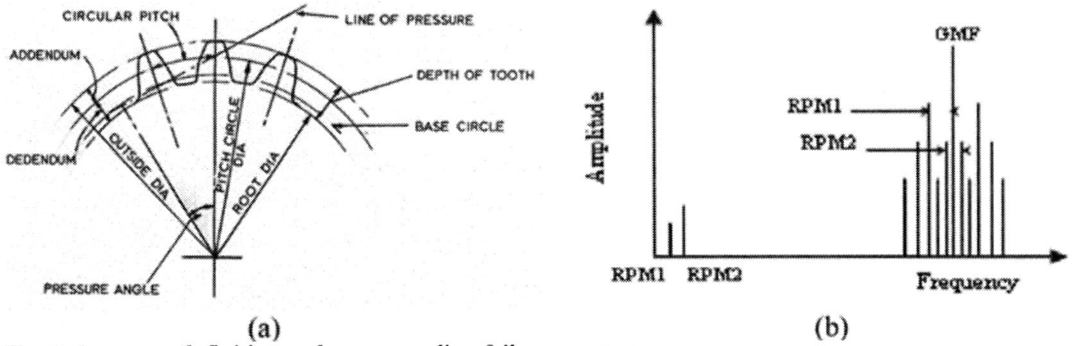

Fig. 1. Spur gear definition and corresponding failure spectrum

2.1 Gear mesh frequencies

The main source of vibration and noise gear is the dynamic excitation of the varying stiffness of the meshed teethgears. The mesh stiffness or elastic bending tooth changes as much as the changes in the number of contacts. As the gear rotates, two teeth meet at a pitch point. GMF is therefore defined as the number of gear teeth multiplied by its rotational speed. GMF and sideband frequencies are shown in Fig. 1(b).

2.2 Wavelet Transformation- Multiresolution Analysis

The wavelet transform of a signal is an interesting and relatively new tool applied for vibration analysis. The basic idea of wavelets is to analyze signals based on scale and dilatation parameters. Multiresolution analysis involves the analysis, processing and synthesis that breaks down a signal into coefficients of the scaling and wavelet functions in different resolution levels. [8].

Wavelet function : Suppose a signal $x = x(t)$ is expressed as a 2-dimensional function, $\psi_{j,k}(t)$, as follows:

$$x(t) = \sum_j \sum_k c_{j,k}\, \psi_{j,k}(t) \tag{1}$$

Wavelet function is defined as:

$$\psi_{j,k}(t) = C \cdot \psi(2^j t - k) \tag{2}$$

where C is a constant, j is dilation parameter and k is shift parameter. Equation (1) can be rewritten as a summation of functions, that is :

$$x(t) = \sum_{i=-\infty}^{j} \sum_k c_{i,k}\, \psi_{i,k}(t) = \sum_{i=-\infty}^{j-1} \sum_k c_{i,k}\, \psi_{i,k}(t) + \sum_k c_{j,k} \psi_{j,k}(t) \tag{3}$$

Scaling function : Given a scaling function $\phi_{j,m}(t)$, then the analysis of signal $x(t)$ is $cA_0(m) = \int_{-\infty}^{\infty} x(t)\phi_{j,m}(t)\, dt$. The synthesis of signal $x(t)$ uses the coefficient $cA_0(m)$ yields an approximation of $x(t)$, that is written as follows:

$$x(t) \approx \sum_m cA_0(m)\, \phi_{j,m}(t) \tag{4}$$

Besides of the approximation of *x(t)* in eq. (4), the signal *x(t)* can be expressed as:

$$x(t) = \sum_k cA_0(k)\phi_{j,k}(t) = \sum_k cA_1(k)\phi_{j-1,k}(t) + \sum_k cD_1(k)\psi_{j-1,k}(t) \quad (5)$$

The approximation coefficient, $cA_1(k)$ and the detail coefficients, $cD_1(k)$, are determined as $cA_{1(k)} = \langle x(t), \phi_{j-1,k}(t)\rangle$ dan $cD_1(k) = \langle x(t), \psi_{j-1,k}(t)\rangle$, respectively. Eq. (5) is rewritten as:

$$x(t) = \sum_k cA_1(k)\phi_{j-1,k}(t) + \sum_k cD_1(k)\psi_{j-1,k}(t) = A_1(t) + D_1(t) \quad (6)$$

Component $A_1(t)$ can be decomposed to the next levels to obtain:

$$\begin{aligned} x(t) &= A_2(t) + D_2(t) + D_1(t) \\ &= A_3(t) + D_3(t) + D_2(t) + D_1(t) \\ &= A_4(t) + D_4(t) + D_3(t) + D_2(t) + D_1(t) \\ &\dots \end{aligned} \quad (7)$$

Eq. (7) implies that the signal *x (t)* can be decomposed into components of lower level, i.e. as a scaling function, $\phi_{j-1,k}(t)$ and wavelet function, $\psi_{j-1,k}(t)$. Signal decomposition is then proceed up to a desired level.

Signal decomposition can be performed using Mallat algorithm. A a low-pass filter (LPF) *h (k)* and a high-pass filter (HPF), $g(k)$ are employed. The filters are defined as $g(k) = (-1)^k h(1-k)$, called Quadrature Mirror Filters (QMF). The wavelet function and scaling function use LPF and HPF are $\phi(t) = \sqrt{2}\sum_k h(k)\phi(2t-k)$ and $\psi(t) = \sqrt{2}\sum_k g(k)\psi(2t-k)$, respectively. Furthermore, the approximation coefficient cA_1 and the detail coefficient cD_1 are determined as $cA_1 = \sum_n h(n-2k)cA_0(n)$ and $cD_1 = \sum_n g(n-2k)cA_0(n)$, respectively. The approximate and detailed signals therefore are $A_1(t) = \sum_k cA_1(k)\phi_{j-1,k}(t)$ and $D_1(t) = \sum_k cD_1(k)\psi_{j-1,k}(t)$, respectively.

The signal analysis using wavelet decomposition includes selection of wavelet function, scaling function, lowpass and highpass filters. These determine the ability of wavelet decomposition to capture transient properties in a signal. This research uses *db4 wavelet* that is suitable for vibration signals analysis [7, 12].

2.3 The signal energy of the wavelet detail component

Parseval theorem states that the signal energy is expressed as the sum of the energy of components of the signal. This also applies to wavelet decomposition as:

$$\sum_{m=0}^{M-1}|x(t)|^2 = \sum_{m=0}^{M-1}|cA_L(m)|^2 + \sum_{l=1}^{L}\sum_{m=0}^{(M-1)/l}|cD_L(m)|^2 \quad (8)$$

where t is the discrete time, l is the l-th decomposition level, m = m-th sample data, L is the number of decomposition level and M is the number of samples. The signal energy of the L-level detail component, cD_L is the dot product of the wavelet detail coefficients, $\sum_{m=0}^{(M-1)/l}|cD_L(m)|^2$. Then De_{max} is determined as the wavelet detail components with the highest signal energy among *1, 2, .. L* level and De_{max-1} is determined as wavelet detail components with signal energy right to lower than detail component energy, De_{max}.

2.4 Cross correlation of wavelet coefficient

The cross correlation of the wavelet decomposition coefficients is calculated as :

$$\langle D_{e_{max}} * D_{e_{max}-1}\rangle[n] = \frac{1}{M}\sum_{m=1}^{M} D_{e\_max}(m)\, D_{e\_max-1}(n+m) \quad (9)$$

where n is the n-th component of the cross correlation. Cross correlation provides information of similarity between 2 signals. This operation is also used to increase signal periodicity. Cross correlation of wavelet coefficients for vibration signals analysis had been performed in the research of [12, 13, 14].

3. Research method

The research was conducted at Laboratory of Vibration and Maintenance of Mechanical Engineering Dept, Faculty of Engineering, Sebelas Maret University. The schematic of rig used in this research is shown in Fig. 2. The figure shows one shaft is connected to motor using a flexible coupling. The driver gear and driven gear are spur gears with 28 tooth. Driver gear is of normal gear whereas driven gear with 25%, 50% 75% and 100% toothcut are used.

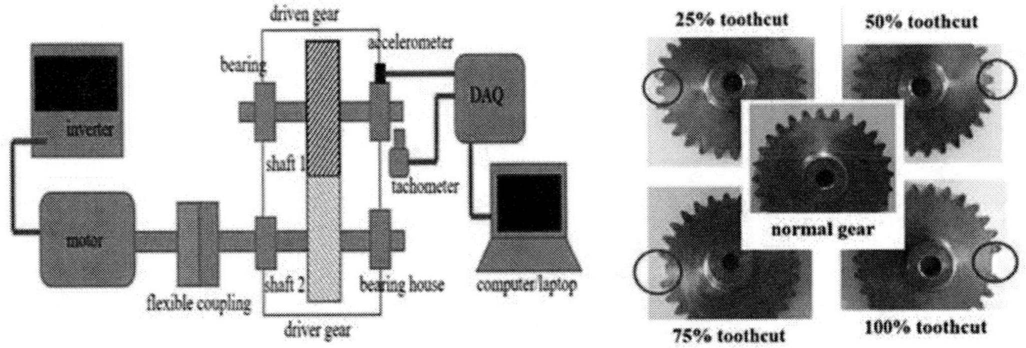

Fig. 2. Schematic of gears rig and driven gears

3.1 Research Implementation

The implementation include the set up of the gear rig and the installation of sensors. The data acquiring steps consists of (1) turning on an electric motor and set it to a constant speed of 1200 rpm ; (2) record vibration signal from accelerometer using DAQ; (3) measure shaft rotation using infrared tachometer, and (4) use computer to store vibration data. The signal data from Dewesoft® DAQ is converted to (.mat) files to be read by MATLAB. The signal is decomposed and the energy of wavelet detil coefficients sets are calculated. The set with largest energy is the most represents the gear defect. This decomposition method has been developed previously for unbalance and misallignment detections [11, 12].

4. Research method

The fault detection is generally performed by observing a particular pattern on the spectrum of recorded signal. The failure pattern is indicated as peaks at frequency of an integer multiplied by 1 X RPM. The proposed method uses wavelet decomposition to obtain transient properties of a signal. Wavelet decomposition have a drawback of low time resolution at high frequency because the frequency resolution multiplied by the time resolution is constant. Because of this, cross correlation is used to improve the performance of wavelet decomposition at high frequencies. Futhermore a spectrum to indicate GMF is constructed.

4.1 Vibration Signals and Wavelet Decomposition Components

Signals of normal, 50% and 75% toothcut gears are shown in Figs. 3(a) and (c). Signals of toothcut gears contains more peaks than the one of normal gear, especially for 75% toothcut gear. This is because impact between defect gear teeth and normal gear teeth is occured more rather than friction contact between the defect teeth with the normal teeth. The wavelet detail coefficients of the signal of toothcut gear also has a magnitude larger than the normal gear, compared using the same scale as shown in Figs. 3(b) and (d).

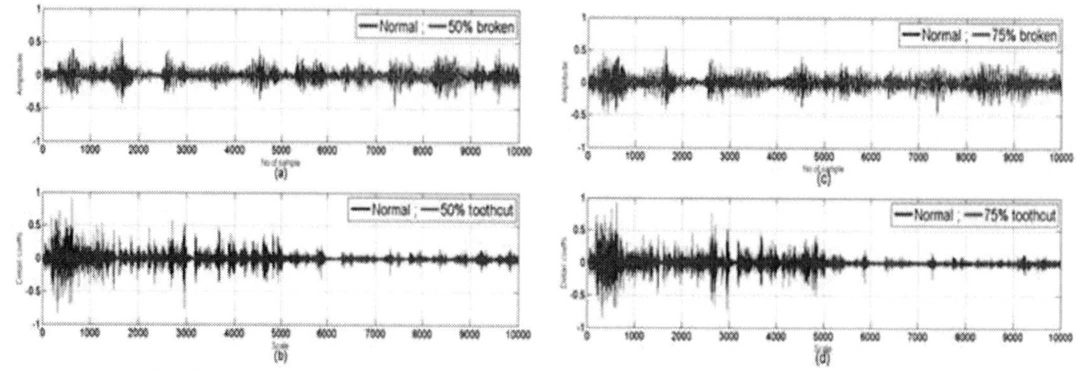

Fig. 3. Gear signal : 50% toothcut (a) time domain; (b) detail coefficients; 75% toothcut (c) time domain; (d) detail coefficients.

4.2 Spectrum of 25% and 100% tooth-cut gears

The spectrum of normal gear, 25% and 100% toothcut gears are shown in Figs 4(a) and (c), respectively. The spectrum cannot indicate amplitudes of the signals of 25% toothcut gear larger than the normal gear signals. The normal and defect gears also cannot be distinguished in time domain signals. The proposed signal processing uses wavelet decomposition-cross correlation-power spectral show fairly different spectrum of normal and 25% toothcut gears as shown in Fig. 4(b). GMF is indicated at 551 Hz (GMF is equal 28 x (1200/60) = 560 Hz) as peak of 0.0081.

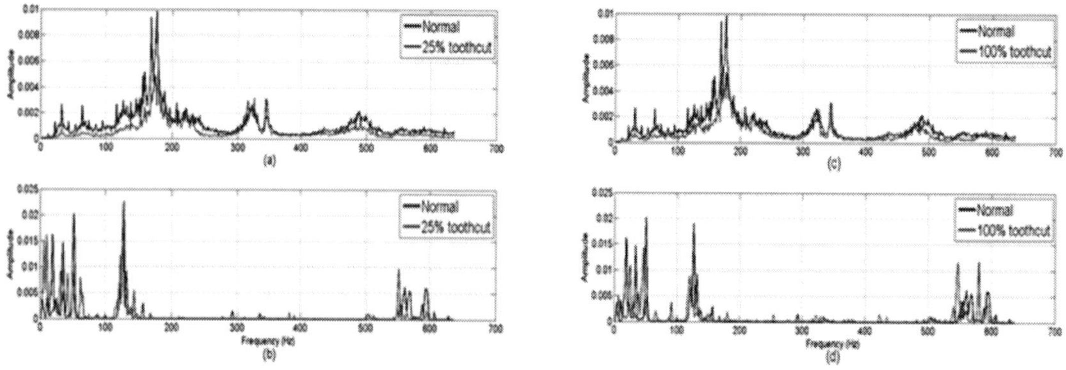

Fig. 4. Gear spectrum: 25% toothcut (a) FFT; (b) proposed method; 100% toothcut (c) FFT; (d) proposed method

Figure 4(c) shows the spectrum of 100% toothcut gear signal with almost all peaks lower than the peaks of the normal gear signals. The missing part of 100% toothcut gear reduces the number of impacts between teeth gear mating. This is indicated by small amplitudes for medium-high scale, compared with normal gear signals in time domain signal and wavelet detail components (not showed). Figure 4(c) does not show GMF while Figure 4(d) shows clearly GMF with peak of 0.0111 at 551 Hz and sidebands with peak of 0.010 and 0.005 at 546 and 568 Hz, respectively.

4.3 Performance of the proposed method

The performance of proposed method is evaluated by comparing to the FFT used to display the gear failure features as shown in Table 1.

Table 1. The features in the spectrum obtained by FFT and proposed method (PM)

25% toothcut gear			**50% toothcut gear**		
Feature	FFT	PM	Feature	FFT	PM
GMF	NE	E	GMF	NE	NCE
Side band frequencies	NE	NCE	Side band frequencies	NE	NE
1X RPM; 2X RPM	E; NCE	E; E	1X RPM; 2X RPM	E;NCE	E; NE
Harmonics freqs of RPM	NE	E (6XRPM)	Harmonics freqs. of RPM	NE	E (6 XRPM)
75% toothcut gear			**100% toothcut gear**		
Feature	FFT	PM	Feature	FFT	PM
GMF	NCE	E	GMF	NE	E
Side band frequencies	NCE	E	Side band frequencies	NE	E
1X RPM; 2X RPM	E; NCE	E; E	1X RPM; 2X RPM	NCE;E	E; E
Harmonics freqs of RPM	E(9X)	E(6XRPM)	Harmonics freqs. of RPM	E(9X)	E(6 XRPM)
Harmonics near GMF	NCE	E	Harmonics near GMF	NCE	E

E = excited, NE = not excited, NCE = Not clearly excited

5. Conclusion

This research has developed a new method for gear failure detection using method of wavelet decomposition-cross correlation-power spectra. The spectrum that shows distinct peaks on GMF and sideband frequencies as an indication of the gear damage is successfuly obtained. The db4 wavelet function for the vibration signal was used. It is highly recommended to use another wavelet function that has the same characteristics as gear fault for the next researches.

References

1. W. Wei, M. Xiaoping, W. Qianjin, Int. J. Control Autom. **8**. No. 11, pp. 127-134 http:/dx.doi.org/10.14257/ijca.2015.8.11.13 (2015)
2. Y. Qu, D. He, J. Yoon, B. Van Hecke, E. Bechhoefer, and J. Zhu, Sensors, **14**, 1372-1393; doi:10.3390/s140101372 (2014)
3. M. H. Lim, and MS. Leong, AIME, Hindawi Publishing Corporation, **2**, Article ID 625863 (2013).
4. P. Gupta, S. B. Wadkar, AVE, **7(2)**, pp 181-195, (2008)
5. J. Chebil a, G. Noelb , M. Mesbah, M. Deriche, JJMIE **3**, No. 4, ISSN 1995-6665 Pages 260 – 267, December (2009)
6. M. Ayad, Dj. Chikouche, N. Boukezzoula, M. Rezki, *Int. Conference on Control, Engineering & Information Technology (CEIT'14) Proceedings* - Copyright IPCO-2014 ISSN 2356-5608
7. H. Bendjama, S. Boucherit, S., JTAM, **54.2** pp 659-670, Warsaw, (2016)
8. Z. Hong-tu, Y. Jing, *Proceeding of the International Symposium on Electronic Commerce* (2010)
9. H. Bendjama, S. Bouhouche, and S. Boucherit, IJMLC, Vol.**2** No. 1 February 2012
10. D. Kanneg, W. Wang, JSIP, 2, 322-329. doi:10.4236/jsip.2011.24046 Published Online November (2011)

Path planning for automatic guided vehicle with multiple target points in dynamic environment

Chang Kyu Kim[1], *Huy Hung* Nguyen[1], *Dae Hwan* Kim[1], *Hak Kyeong* Kim[1], *Sang Bong* Kim[1*]

[1]Department of Mechanical Design Engineering, Pukyong National University, Busan, 48549, Republic of Korea

Abstract. In path planning field, Automatic guided vehicle (AGV) has to move from an initial point towards a target point with capability to avoid obstacles. There are A*, D* and D* lite path planning algorithms in the path planning algorithm. This paper proposes a modified D* lite path planning algorithm using the most efficient D* lite among these algorithms. The modified D* lite path planning algorithm is to improve these D* lite path planning algorithm's weaknesses such as traversing across obstacles sharp corners, or traversing between two obstacles. To do this task, the followings are done. First, a work space is divided into square cells. Second, cost of each edge connecting current node to neighbor nodes is calculated. Third, the shortest paths from the initial point to all multiple target points are computed and the shortest paths from any target point to remaining target points including the goal point are computed by using Hamilton path. Fourth, a cost-minimal path is re-calculated as soon as the laser sensor detects an obstacle and make an updated list of target points. Finally, the validity of the proposed modified D* lite path planning algorithm is verified through simulation and experimental results.

1 Introduction

Automatic guided vehicles (AGVs) are used in many industrial fields for goods delivery, and robot cleaning, health care, etc. In particular, when looking for features in the entire map space, set multiple target points. The path planning of the AGV focuses on how much it costs to reach from the starting point to the target point in the workspace, and how to move to the shortest distance with avoiding obstacles on the route.

M. Likhachev and S. Koenig proposed Lifelong Planning A*(LPA*) algorithm [1]. LPA* is to finds shortest paths from a given start point to a given goal point repeatedly. But, while the edge costs of a graph change, data of a point are added or deleted. S.M. Lavalle proposed the basics of path planning [2]. M.S. Alam and M.U. Rafique et al. proposed optimal path planning algorithm using particle swarm optimization technique [3]. The optimal waypoints computed by the planner lead the robot to the goal position effectively. D* lite algorithm among A*, D*, and D* lite path planning algorithms [1-3] is the most efficient in its path planning. Y.D. Setiawan et al. also proposed D* lite path

* Corresponding author : kimsb@pknu.ac.kr

planning algorithm that was widely used for AGV in static and dynamic environments, and it could replan from the current point when there was an obstacle blocking path [4]. The error of the path replanning algorithm is very small. However, there are weaknesses such as passing over the sharp edges of obstacles or passing between two obstacles.

In this paper, a modified D* lite path planning algorithm is proposed to improve the weakness of the D* lite path planning algorithm. The proposed modified D* lite algorithm is widely used for AGV in static and dynamic environments and it can replan from the current point when there is an obstacle blocking path. The validity of the proposed modified D* lite algorithm is verified through simulation and experimental results.

2 Path planning algorithm

There are A*, D*, and D* lite algorithms in the path planning algorithm. The D* lite path planning algorithm among them is the most efficient in path planning. However, the D* lite path planning algorithm has some weaknesses such as traversing across obstacles sharp corners, or traversing between two obstacles.

In this paper, a modified on D* lite path planning algorithm is proposed to improve these D* lite path planning algorithm's weaknesses and generates an optimal path from a starting point to a goal point through multiple target points with avoiding obstacles. The proposed path planning algorithm also controls an AGV with the laser sensor NAV-200 for detecting obstacles to move from the initial point to a goal point through multiple target points optimally. To do the abovementioned tasks, the proposed path planning algorithm is performed as follows:

Firstly, divide a work space into square cells. Each cell center is assigned as a node. The cell size is greater a bit than AGV size

Secondly, calculate cost of each edge connecting current node to neighbor nodes. It is a combination of the traversal cost of the two cells which it connects and length of the edge is defined as Fig. 1. The costs in horizontal and vertical directions are 1, the cost in diagonal direction is $\sqrt{2} \approx 1.414$ and if AGV moving in a direction detects obstacles, the cost is set ∞ as shown in Fig. 1.b.

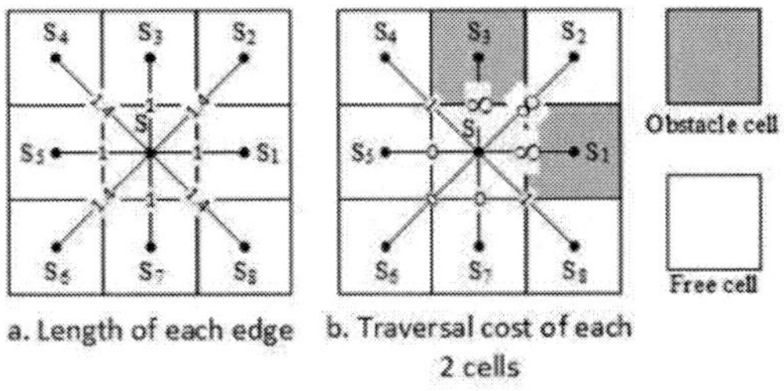

Fig. 1. Cost of each edge connecting current node to neighbor nodes.

Thirdly, compute the shortest paths from the initial point to all target points and the shortest paths from any target point to remaining target points including a goal point and make a Hamilton path with arranged target points based on a nearest neighbor principle. The Hamilton path is defined as follows. In graph theory, the Hamiltonian path is a single path in undirected or directed graph that visits each vertex exactly once. A graph that contains a Hamilton path is called a traceable graph. A graph is Hamilton-connected if for every pair of vertices there is a Hamilton path between the two vertices.

The following Fig. 2 shows the Hamiltonian cycle passing through all the vertices of a graph. The Hamiltonian graph must be a regular polygon, a complete graph of size 3 or greater, and a circulating graph of size 3 or greater. If the graph is discontinuous or not circulating, it is not a Hamiltonian graph. Also, having a vertex with a graph size of 2 or less is not a Hamiltonian graph either.

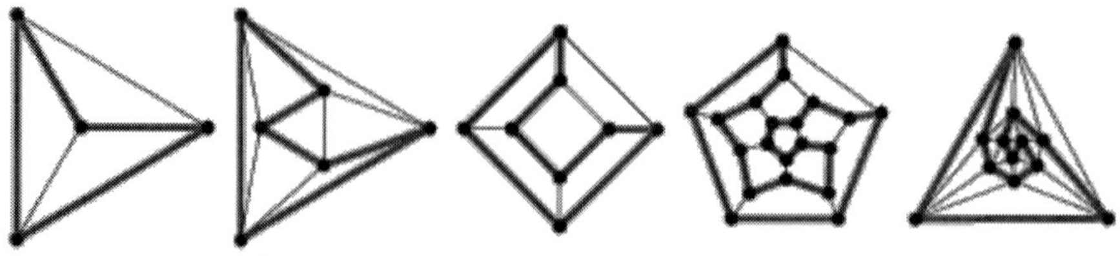

Fig. 2. Hamiltonian cycle.

Fourthly, re-calculate a cost-minimal path as soon as the laser sensor detects an obstacle and make an updated list of target points.

Finally, show simulation results to verify the effectiveness of the proposed path planning algorithm.

The followings show the proposed D* lite algorithms.

2.1 Comparison of D* lite algorithm and proposed D* lite path planning algorithm

In Fig. 3a, the D* lite path planning algorithm can perform incremental retrieval of information in a dynamic environment. Also, the number of lines of code is small and easy to understand. However, Sometimes, the D* lite path planning algorithm has some weaknesses such as traversing across obstacles sharp corners, or traversing between two obstacles. So, a suitable and optimal path planning algorithms should be chosen depending on the requirement of the system. To overcome these disadvantages, the obstacles could be avoided by setting multiple target points.

Therefore, Fig. 3b shows the proposed D* lite path planning algorithm code with four target points. By adding four target points to the existing D* lite path planning algorithm, it does not pass through sharp edges or between obstacles.

Fig. 4a shows the path planning of the D* lite algorithm as a node. As shown in the figure, the D* lite algorithm moves from a start point to a goal point in the shortest distance avoiding obstacles. However, if there exists a path in cell (4, G), the path hits the corner of the obstacle. This is very fatal to the movement of the AGV and can cause damage to the AGV body. Therefore, some measures are needed to prevent this path.

Fig. 4b shows the proposed modified D* lite path planning algorithm in each node. Set the target points which are cell (6,F), cell (3,E) in the middle of moving AGV from a start point to a goal point so that the position of the obstacle can be confirmed clearly. It no longer passes over the edge of the obstacle.

```
key(s)
return[min(g(s),rhs(s))+h(s_start,s); min(g(s),rhs(s))];

UpdateState(s)
if s was not visited before
   g(s)=∞;
if(s ≠ s_goal) rhs(s)=min_{s'∈Succ(s)}(c(s,s')+g(s'));
if(s ∈ OPEN) remove s from OPEN;
if(g(s) ≠ rhs(s)) insert s into OPEN with key(s);

ComputeShortestPath()
while(min_{s∈OPEN}(key(s)) < key(s_start) OR rhs(s_start) ≠ g(s_start))
   remove state s with the minimum key from OPEN;
   if (g(s) > rhs(s))
     g(s)=rhs(s);
     for all s' ∈ Pred(s) UpdateState(s');
   else
     g(s)=∞;
     for all s' ∈ Pred(s) ∪ {s} UpdateState(s');
Main()
g(s_start)=rhs(s_start)=∞; g(s_goal)=∞;
rhs(s_goal)=0; OPEN= ∅;
insert s_goal into OPEN with key(s_goal);
forever
  ComputeShortestPath();
  Wait for changes in edge costs;
  for all directed edges (u, v) with changed edge costs
    Update the edge costs c(u, v);
    UpdateState(u);
```

(a)

```
12. obstacle = [];
if Node_start == 13 % Assuming that an obstacle is detected at the node 13 (target point 1)
        obstacle = [17, 24, 27, 28]; % Node position of actual obstacle
        k_m = k_m + heuristic(Node,Node_last,Node_start);
        Node_last = Node_start;
if Node_start == 32 % Assuming that an obstacle is detected at the node 32 (target point 2)
        obstacle = [34, 43, 44, 45, 46, 47]; % Node position of actual obstacle
        k_m = k_m + heuristic(Node,Node_last,Node_start);
        Node_last = Node_start;
if Node_start == 62 % Assuming that an obstacle is detected at the node 62 (target point 3)
         obstacle = [76, 77]; % Node position of actual obstacle
         k_m = k_m + heuristic(Node,Node_last,Node_start);
         Node_last = Node_start;
if Node_start == 80 % Assuming that an obstacle is detected at the node 80 (target point 4)
         obstacle = [79]; % Node position of actual obstacle
         k_m = k_m + heuristic(Node,Node_last,Node_start);
         Node_last = Node_start;
```

(b)

Fig. 3. Comparison of D* lite and proposed D* lite path planning algorithm; (a) D* lite algorithm; (b) Proposed D* lite path planning algorithm code.

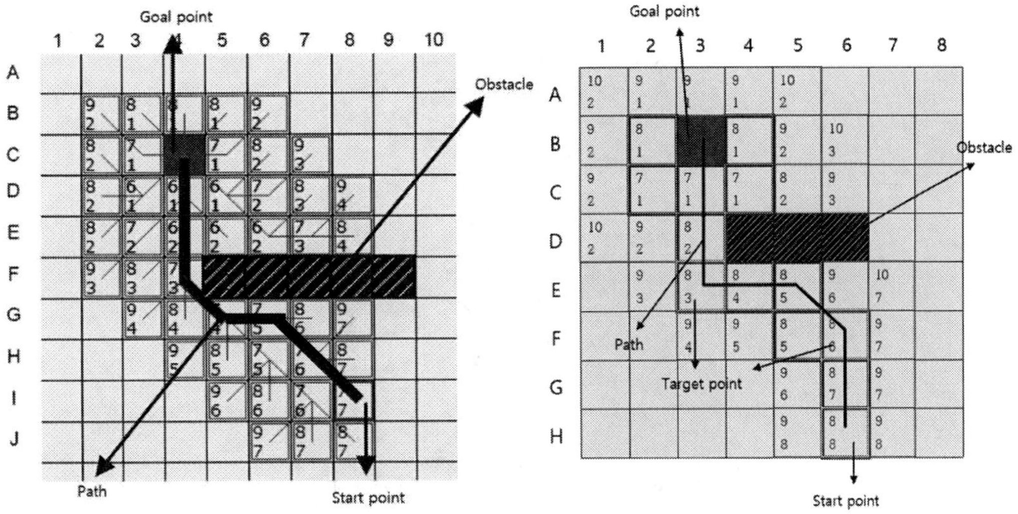

(a) (b)

Fig. 4. Proposed D* lite path planning algorithm in each node; (a) D* lite algorithm path panning algorithm; (b) Proposed D* lite path panning algorithm.

3 Simulation and experimental results

Using the proposed D* lite path planning algorithm, a scenario of Fig. 5 below is set up to confirm the effect of setting four target points, avoiding obstacles, and choosing the shortest path. The test conditions of the proposed D* lite path planning algorithm are as follows : Space size is 3.5m x 7m, cell size is 0.5m x 0.5m, as shown in Fig. 5, a starting point is (7,1) and a goal point is (1,14).

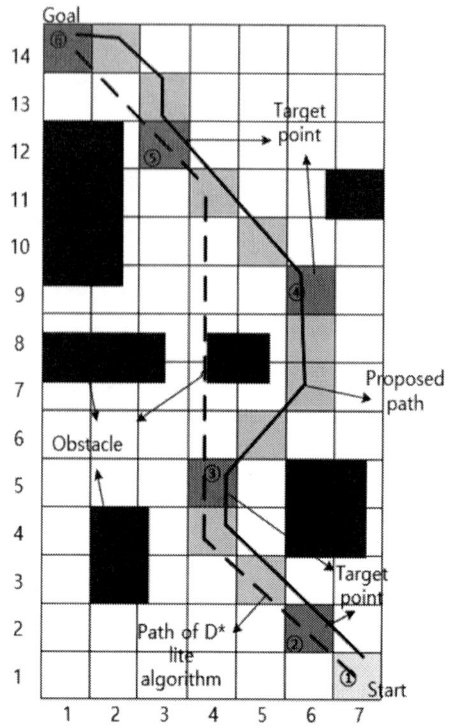

Fig. 5. Path planning in the scenario.

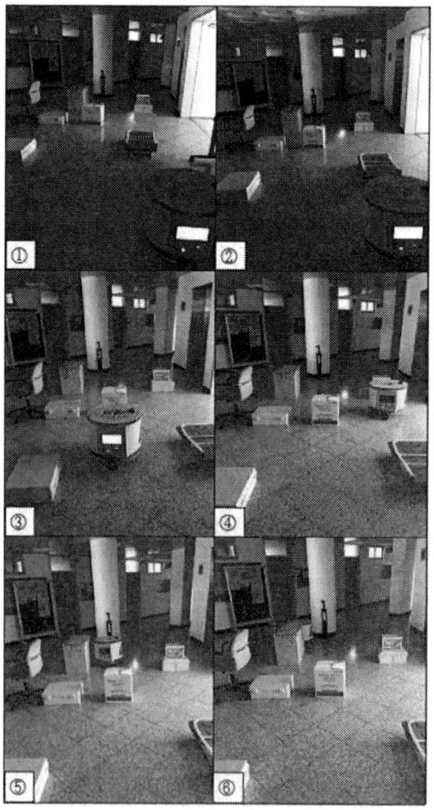

Fig. 6. Experimental results of the proposed modified path planning algorithm.

In the scenario of Fig. 5, the AGV moves from cell (7,1) with start point ① to cell (6,2) with a first target point ②. It detects a obstacle; cell (6,4) and (7,4). AGV keeps going a second target point ③ with avoiding obstacle. In this case, the D* lite path planning algorithm selects the path indicated by the dashed line in Fig. 5. This algorithm is unsafe. However, the proposed D* lite path planning algorithm moves safely to cell (6,9) of a third target point ④ with avoiding obstacles. When the AGV arrives at the cell (6,9), it detects the obstacles of the cells (7,11) and (6,11) and moves to cell (3,12) of a fourth target point ⑤. After the AGV moves to the position of the cell (3,12), it detects the cell (2,12) of the obstacle and reaches the goal point ⑥ with a safe route. As the results, the proposed D* lite path planning algorithm generates a continuous line in Fig. 5. Fig. 6 shows pictures showing the experiment results of the proposed modified path planning algorithm.

4 Conclusion

In this paper, a modified path planning algorithm based on D* lite algorithm was proposed to find the safest path for AGV to move from a starting point to a goal point through predefined multiple target points in dynamic environment. The proposed path planning algorithm was verified by simulation and experiment. The conclusion was described as follows: (i) a modified D* lite path planning algorithm was proposed to improve these D* lite path planning algorithm's weaknesses such as traversing across obstacle's sharp corners (a) and traversing between two obstacles (b). An optimal path from a starting point to a goal point through multiple targets points with avoiding obstacles was proposed. (ii) Simulation and experimental results could be used to identify the surrounding obstacles more clearly by placing multiple targets. So, the AGV was prevented from traversing between two obstacles, and target points was arranged optimally so that the AGV movement from the start point to the final target point was the minimum distance path. Although the D* lite path planning algorithm could give the shortest path, the AGV could collide with obstacles in real environment. Therefore, the proposed path might not be shortest but it is guaranteed to be safe in the proposed algorithm.

This work (Grants No. C0456375) was supported by Business for Cooperative R&D between Industry, Academy, and Research Institute funded Korea Small and Medium Business Administration in 2017.

References

1. M. Likhachev, S. Koenig, *Lifelong planning A* and dynamic A* lite: The proofs. Technical report* (Atlanta, 2001)
2. S.M. Lavalle, *Planning algorithm* (Cambridge university press, 2006)
3. M.S. Alam, M.U. Rafique, *In Control, Automation and Robotics (ICCAR)*, 5 (2015)
4. Y.D. Setiawan, P.S. Pratama, J.W. Kim, D.H. Kim, Y.S. Jeong, S.B. Kim, *International Conference on Advances in Computing, Communications and Informatics (ICACCI)*, (2014)

Design and fabrication of a twist fixture to measure torsional stiffness of a pick up chassis

Ojo Kurdi [1,2*], *Mohd Shukri* Yob [3], *Awisqarni* Haji Ishamuddin [4], *Agus* Suprihanto [1], *Susilo Adi* Widyanto [1], *Dwi Basuki* Wibowo[1], *Ian* Yulianti [5]

[1]Department of Mechanical Engineering, Faculty of Engineering, Universitas Diponegoro, Central Java, Indonesia
[2]National Center for Sustainable Transportation Technology, Indonesia
[3]Faculty of Mechanical Engineering, Universiti Teknikal Malaysia Melaka, Malaysia
[4]Faculty of Engineering and Life Sciences, Universiti Selangor, Bestari Jaya, Selangor Darul Ehsan, Malaysia
[5]Physics Department, Faculty of Mathematics and Natural Sciences, Universitas Negeri Semarang, Central Java, Indonesia

Abstract. Torsional stiffness is important parameter of chassis that affect the handling performance of chassis. Torsional stiffness can be determined using Finite Element Method (FEM) in early stage design of its. In order to validate the FEM result, experimental work needs to be done. The fixture has been design in simpler stucture, flexible for any kind of chassis and using a simple measurement's equipment such as dial indicator and load cell. Twist fixture has been designed for measuring of torsional stiffness of TATA cab chassis indirectly. The fixture measured the deflection caused by torsion subjected to the chassis. The torsional stiffness was calculated based on measured displacement of chassis. The result of comparison shows that the experimental results in agreement with the simulation results. Therefore, the simulation results of TATA cab chassis model are valid.

1 Introduction

Increased torsional stiffness of a truck chassis improves vehicle handling by going away the suspension components to control a larger percentage of a vehicle's kinematics. In summation, a truck chassis must have adequate torsional stiffness so that chassis structural dynamic modes do not adversely couple with the suspension dynamic modes. The torsional stiffness of a vehicle's chassis significantly affects its handling characteristics and is thus an important parameter to evaluate.

Pick up chassis used in vehicles has almost the same appearance since the models were built up more than 30 years ago. Many researches have been carried out to analyze the torsional stiffness of chassis designs using finite element analysis (FEA) [1, 2, 3, 4, 5]. In order to validate these finite element models, an experimental method is needed to directly measure torsional stiffness. The purpose of this work is to design and fabricate a twist fixture for measuring the torsional stiffness of TATA Cab chassis.

[1] Corresponding author: ojokurdi@ft.undip.ac.id

2 Twist Fixture Design

Many factors should be taken into consideration in designing of the test rig or twist fixture in order to get the best design. Some alternatives design also should be considered to fulfill the needs of designer. Three designs were prepared for this project and will be marked based on some parameters such as cost and machineability of each design in order to choose the best design. The design with the highest mark will be chosen as the best design. Table 1 shows the mark of three alternative designs of fixtures.

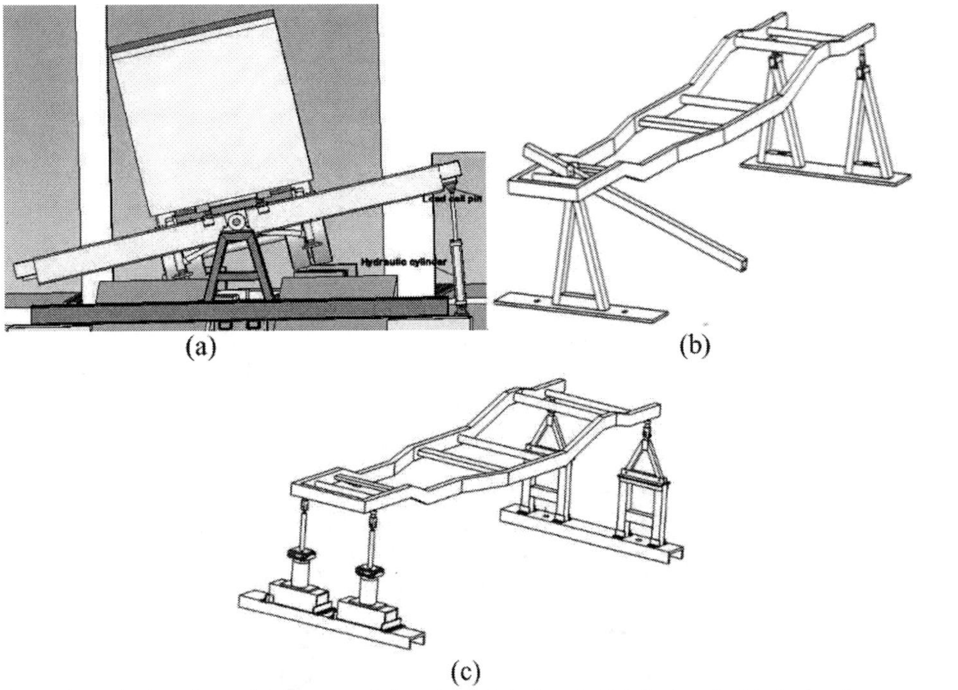

Fig. 1. Alternatives design for twist fixture of chassis; (a) design 1; (b) design 2; (c) design 3

Table 1. Marks for alternatives design of fixtures

Design	Cost	Fabrication	Strength	Weight	Set-up	Total Marks
1	2	2	5	2	2	13
2	3	4	3	4	4	18
3	2	3	3	3	3	17

*Rate scale: 1 (Dismal), 2 (Low), 3 (Moderate), 4 (High), 5 (Exceptionally good)

From the Table 1, design number 1 are low in cost, fabrication, set up and weight, but the advantages is very good in strength. For design 2, cost of the rig is moderate, means cost material not expensive and weight the rig is also high, it means the weight of the fixture is not too heavy. Design 2 also has a high mark for set up, it means it is quite easy to assembly the fixture. Design 3 has a low mark for cost, it means the cost for this design is high. The mark for fabrication, strength, weight, and set up are moderate. Design 2 has the highest mark, therefore design 2 was chosen as the best design and will be proceed for fabrication.

3 Experimental Set up and Torsional Stiffness Calculation

In this section, the procedures used to calculate torsional stiffness using the twist fixture for a typical TATA Cab chassis are discussed. Figure 2 showed the schematic diagram of the experimental set up for measuring the torsional stiffness of TATA Cab Chassis. Figure 3 represented the real experimental set up. Twisting moment has been given to the chassis due to two same single force in opposite direction at front of chassis.

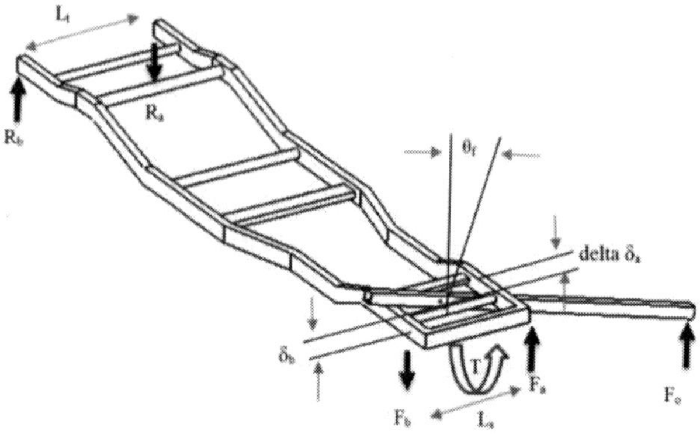

Fig. 2. Schematic diagram of experimental set up

Fig. 3. Real experimental set up

Dial indicators are used to measure the equal and opposite applied vertical deflection δ, at the left and the right chassis. A dial indicator monitored angle chassis displacement, insuring that the mounting fixture was a rigid, zero displacement constrain relative to applied load. The outer face of the dial indicator can be moved so that the zero can be positioned over the pointer. In this experiment, resolution for the indicator is 0.01 mm. The twist angle can be calculated by the formula:

$$\theta_f = \frac{2\delta}{L_f} \quad (1)$$

Where L_f is the lateral distance between the front dial indicators. Force at the left front and right front, denoted *Fa* and *Fb*, respectively, are measured by scales. The torque is calculated from

$$T = \left[\frac{(F_a)}{2} + \frac{(Fb)}{2}\right] x\, L_s \tag{2}$$

The twist angle is adjusted by subtracting the deflection at the rear. The twist angle at the rear θ_t is calculated from the measured displacements in the extremities of the longitudinal rails,

$$\theta_t = \frac{[(\delta_a) + (\delta_b)]}{Lt} \tag{3}$$

where δ_a and δ_b are the measured vertical displacements in the dial indicators, separated laterally by a distance L_t. The torsional stiffness at each increment is calculated by:

$$K = T/\theta\ (Nm/rad)\ and,\ \theta = \theta_f - \theta_t \tag{4}$$

Where $\theta = \theta_f - \theta_t$. The increment is chosen such that at least 10 data points are obtained.

4 Results and Discussions

The FEM simulation results for 60 kg and was presented in Figure 4. It is shown that the maximum deflection occurred at the left side of front chassis. It is also shown that the deflection at the right side has a very close value to the maximum value. The deflection distribution for other loads is similar to that of 60 kg. The calculated twist angles under various loads and the torsional stiffness is then obtained from Equation (4) as tabulated in Table 2.

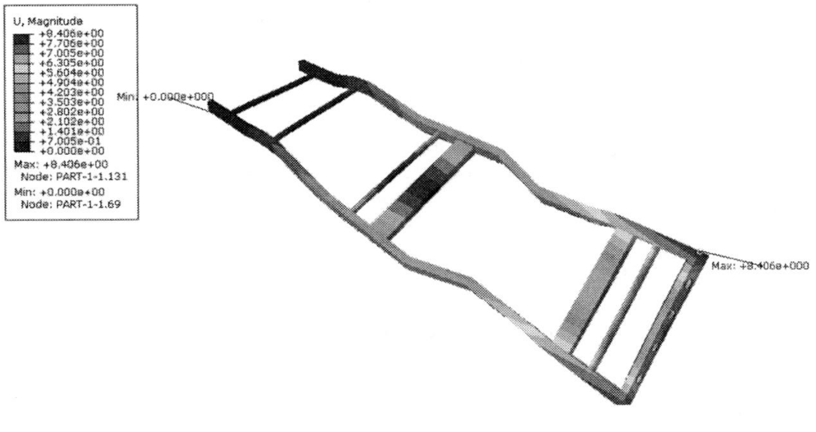

Fig. 4. Deflection of model caused by twisting moment of 60 kg of force

Table 2. Twist angle and torsional stiffness under various load magnitudes obtained from simulation.

Load (kg)	Twist angle (°)	Torsional Stiffness (Nm/deg)
30	0.010252	94071.84
60	0.020485	94162.38
90	0.030801	93937.35
120	0.041287	93438.57
150	0.05163	93401.34
180	0.061032	94814.58

Table 2 shows the average of twist angle of chassis at the right side and left side. The magnitude of load has significant effect to the twist angle. The higher the load magnitude, the larger the twist angle of the chassis. However, the magnitudes of torsional stiffness are mostly similar for all load values.

The calculated twist angle and torsional stiffness from the experiment are tabulated in Table 3.

Table 3. Twist angle and torsional stiffness of various loads obtained from experiment.

Load (kg)	Twist angle (°)	Torsional Stiffness (Nm/deg)
30	0.009770	98717.43
60	0.020010	96396.34
90	0.030569	94648.97
120	0.040514	95222.94
150	0.050379	95719.57
180	0.060051	96364.11

In order to validate the FE simulation result with the experiment, comparison between experiment results and simulations were done. The comparison result is shown in Figure 5.

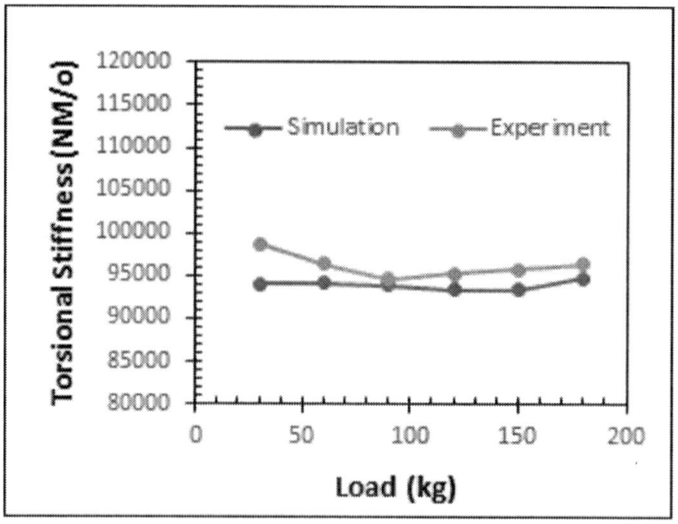

Fig. 5. Comparison of torsional stiffness between experiment and simulation.

From Figure 5, it is shown that the measurement results agree well with the simulation results. The maximum difference between simulation and experimental results is 4.71%. The average difference is 2.28%.

5 Conclusions

Rig test fixture for torsional stiffness measurement has been developed successfully. The measured value of torsional stiffness of TATA Cab chassis was 96178.23 Nm/°. Based on the comparison result, it is shown that for each load value, the torsional stiffness's obtained from the experiment agree well with the simulation results with maximum percentage difference of 4.71 % for load of 30 kg. The average percentage difference between experiment results and simulation results is 2.28%.

This research was partially funded by USAID under SHERA Program, Award no AID-497-A-16-00004.

References

1. Lampert, J.K, *Design and Analysis of a Twist Fixture to Measure the Torsional Stiffness of a Winston Cup Chassis*, Master's Thesis, Department of Mechanical Engineering, Clemson University, August 1998 (1998)
2. Romulo, R.P.F., Jean, C.C.R., Marcus de Freitas Leal, Jose, A.F.B. Automotive Frame Optimization, Sao Paulo, Brazil, SAE Technical Paper Series No. 2003-01-3702, 2-8 (2003)
3. Milliken, W.F. and Milliken, D.L. Journal Race Car Vehicle Dynamics, SAE International, Warrendale, PA (1995)
4. Riley, W.B. and George, A.R., *Design, Analysis and Testing of a Formula SAE Car Chassis,* SAE Technical Paper 2002-01-3300 (2002)
5. Raju, S, *Design and Analysis of a Winston Cup Race Chassis for Torsional Stiffness Using the Finite Element Method.* Master's Thesis, Department of Mechanical

Morphology Evolution Of ZnO Thin Films Deposited By Nitrogen Mediated Crystallization Method

Iping Suhariadi[1,*], *Masaharu* Shiratani[2] , and *Naho* Itagaki[2]

[1]Computer Science Department, School of Computer Science, Bina Nusantara University, Jl. K.H. Syahdan No. 09, Palmerah, Jakarta Barat 11480, Indonesia
[2]Graduate School of Information Science and Electrical Engineering, Kyushu University, Fukuoka 819-0395, Japan

Abstract. We study the surface morphology of ZnO thin films deposited by nitrogen mediated crystallization method utilizing atomic force microscopy as a function of nitrogen flow rates. Initially, the surface morphology of ZnO thin film deposited without nitrogen exhibits a bumpy surface with spiky grains where the skewness and kurtosis values were found to be 0.48 and 4.80, respectively. By addition of small amount of nitrogen, the skewness and kurtosis values of the films significantly decrease associated with a flatter topography. Further increase in nitrogen flow rate to 16 sccm has roughened the surface shown mainly by the increase in kurtosis value to be 3.30. These results indicate that the addition of small amount of nitrogen during deposition process has enhanced the adatoms migration on the surface resulting in a superior film with a larger grain size. Two-dimensional power spectral density analysis reveals that all the films have self-affine fractal geometry with total fractal values in the range of 2.14 to above 3.00.

1 Introduction

ZnO thin film has drawn attention due to has many promising features such as a wide direct band gap energy (~ 3.3 eV) and high room temperature excitonic binding energy (60 meV) [1]. Moreover, the environmentally friendliness and natural abundance of ZnO make them a strong candidate to replace the use of high cost and poisonous indium doped tin oxide (ITO) [2]. However, ZnO exhibits disadvantages because its resistivity is higher than that of ITO and strongly depends on film thickness [3]. We have recently overcome the problems by inserting a very thin ZnO buffer layer deposited by nitrogen mediated crystallization (NMC) between ZnO:Al (AZO) and quartz glass substrate [4]. For example, by inserting 10-nm-thick NMC-ZnO buffer layer, AZO film with resistivity of 5×10-4 Ωcm with much lower thickness dependence can be obtained [5].
Nitrogen flow rate control during the deposition has been reported to be one of keys in determining the physical and chemical properties of the films. The introduction small amount

* Corresponding author: isuhariadi@binus.edu

of nitrogen during the fabrication of the buffer layer have remarkably improved the surface roughness and increased the grain size of the film and hence leading to the improvement of crytallinity, morphology, and electrical properties of AZO films [6]. Nonetheless, the detail morphological evolution of the ZnO buffer layers due to the nitrogen addition has not yet been fully understood. Here we study the morphological evolution utilizing power spectral density and scaled height distribution analysis.

2 Experimental

ZnO thin films were deposited on the cleaned quartz glass substrate using RF-magnetron sputtering. The film thickness of each sample was 10 nm. The Ar/N2 flow rate ratios were 24.5/0, 19.5/5, 18.5/6, and 8.5/16 sccm. The detailed experiment was discussed elsewhere [6]. 2D Power Spectral Density Analysis and Height Distribution Analysis were derived from AFM measurement using Veeco Nanoscope II. The scan size was 1×1 μm^2.

2.1 Height Distribution Analysis

In the surface analysis, there are two very important parameters that are Root Mean Square (RMS) roughness (Rq) and average roughness (Ra) and. Ra is the mean height as calculated over the entire measured area which is described as [7]:

$$R_q = \sqrt{\frac{1}{L}\int\int_0^L z^2(x)dx} \quad (1)$$

While Rq is the square root of the distribution of surface height and is tought to be more sensitive than the average roughness for large deviations from the mean line or plane which is defined as:

$$R_a = \frac{1}{L}\int_0^L |z(x)|dx \quad (2)$$

In real practice, the use of Ra and Rq is interchangable by ignoring the difference between them. However, this is not necessarily correct for surfaces with more peaks than valley (or vice versa). When the Rq become more sensitive to the outliers in this case, however, this completely different profiles of the surface cannot be distinguished by Ra.
To sort this problem out, the height distribution is employed which provides a complete specification of random variable, h(r), at position r. Even though many surfaces have more or less symmetrical height distribution, nonetheless it has been agreed that the well-known Gaussian distribution is a convenient method to represent such surfaces. The hight asymmetry of the Gaussian distribution was described by statistical parameters such as skewness and kurtosis. As the name signifies, skewness (SK) is a measure of the degree of asymmetry of the distribution, which is normalised as [8]

$$S_K = \frac{1}{w^3}\frac{1}{N}\sum_{i=1}^{N}\left(h_i - \bar{h}\right)^4 \quad (3)$$

This SK is of importance to characterize the morphological structure of the surfaces that have the same Ra or Rq. The positive and negative SK represent the asymmetric distribution whose tail extending toward more positive or negative height with refer to the mean surface level. This implies that the positive SK indicates that the surface has more bumps while the negative SK indicates that the surface has more holes. Kurtosis (K) describes the entire shape of the distribution associated with the level of the randomness of the surface profile relative to that of a Gaussian distribution which is defined as:

$$K = \frac{1}{w^4}\frac{1}{N}\sum_{i=1}^{N}(h_i - \bar{h})^4 \quad (4)$$

For Gaussian associated with perfectly random surface, the distribution the K is 3. For K < 3, the surface is classified as platykurtic associated with a bumpy surface. For K > 3, the surface is classified as leptokurtic associated with a spiky surface.

2.2 Power Spectral Density

The scattering theories explain that the amount of optical scatter depends not only on the roughness height of a structure but also on its lateral distribution. Any randomly rough surface can be considered as a Fourier series of sinusoidal waves with different amplitudes, periods, and phases. At a randomly rough surface many different spatial spatial frequency are present. Power spectral density is a technique based on fast Fourier transform algorithm, which is very useful to quantitatively analyze microstructure as a function of spatial frequency. In general, there are two common PSD analyses that are one-dimensional (1D) and two-dimensional (2D) PSD. The 1D PSD is calculated from the scan line profile measurement of AFM result that can be written as [9]

$$PSD(k) = \frac{1}{L}\left[\int_0^L h(x)e^{ikx)}dx\right]^2 \quad (5)$$

Where k and L and are the spatial frequency and the total length of the scanned area. By extracting the scaling behavior, the PSD is then can be written as:

$$PSD(k) \propto k^{-\gamma} \quad (6)$$

Here γ is related to height correlation function exponent through $\alpha = (\gamma\text{-}d)/2$. The unit of 1D-PSD is in $(\text{length})^3$. In the case 2D-PSD the mathematical equation can be written as [10]:

$$PSD_{2D}(f_x, f_y) = \frac{1}{L^2}\left[\sum_{m=1}^{N}\sum_{n=1}^{N} Z_{mn}\, e^{-2\pi i \Delta L(f_x m + f_n n)}(\Delta L)^2\right]^2 \quad (7)$$

where N is the number of data points per line and row, Z_{mn} is the profile height at position (m,n), fx and fy correspond to the spatial frequency in the x- and y- direction, respectively, and $\Delta L = N/L$ is the distance of sampling. For fractals and superstructures characterization, especially for the thin film coating, the 2D-PSD can be written by considering only the pure film as [11]:

$$PSD_{ABC} = \frac{J}{(1+K^2f^2)^{(\gamma+1)/2}} \quad (8)$$

With J defines as the magnitude at low spatial frequency which is related to the amplitude of the rough surface, K is the "knee" which is defined as the slope of a line connecting two point on the surface, being associated with the mean grain size, and γ is the slope at high spatial frequency that gives the nature of the local roughness, respectively. The profile's fractal dimension D_f can be expressed as

$$D_f = \frac{(7-\gamma)}{2} \quad (9)$$

The total fractal dimension then given by [12]

$$D_S = D_f + 1 \quad (10)$$

3 Results and Discussion

Figure 1(a) shows the scaled height distribution of ZnO buffer layers as a parameter of nitrogen flow rate. The kurtosis and skewness values deduced from scaled height distribution are shown in the Figure 1(b) and 1(c), respectively.

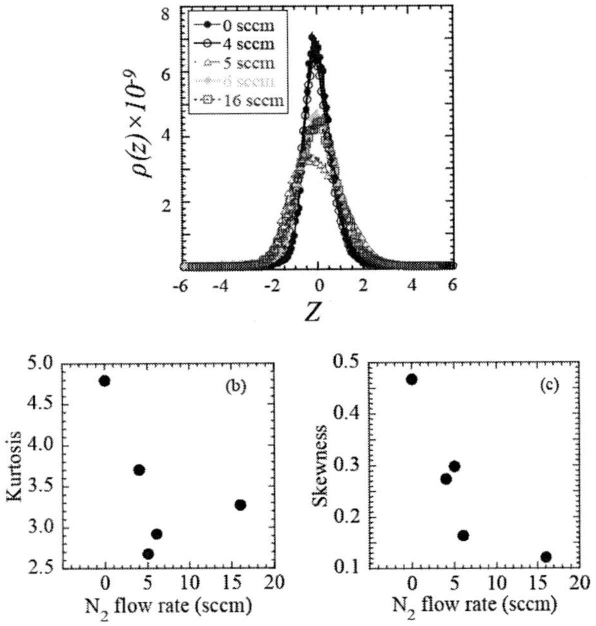

Fig. 1. (a) Scaled height distribution; (b) skewness; and (c) kurtosis of ZnO deposited as a function of nitrogen flow rate.

Initially, the surface morphology of ZnO thin film deposited without nitrogen exhibits more bumps with spiky surface where the skewness and kurtosis values were found to be 0.48 and 4.80, respectively. At this stage, the film is classified as platykurtic. By addition small amount of nitrogen, skewness and kurtosis values of the films decrease to near Gaussian distribution indicating that the surface morphology has transformed from platykurtic to leptokurtic where the bumps decrease and the spiky surfaces become mild surfaces associated with a flatter topography. Nonetheless, further increase in nitrogen flow rate to 16 sccm has roughened the surface shown mainly by the increase in kurtosis value to be 3.30. It is thought that the growth of ZnO follows ballistic deposition mode where the high energetic ablated species from the target sticks whereupon it lands. As a result, the atoms have a low diffusion rate and nucleate instantaneously leading to a rougher surface morphology. The addition of small amount of nitrogen during deposition has inhibited the nucleation by nitrogen adsorption-desorption mechanism enhancing the adatoms migration on the surface resulting in a superior film with a larger grain size.

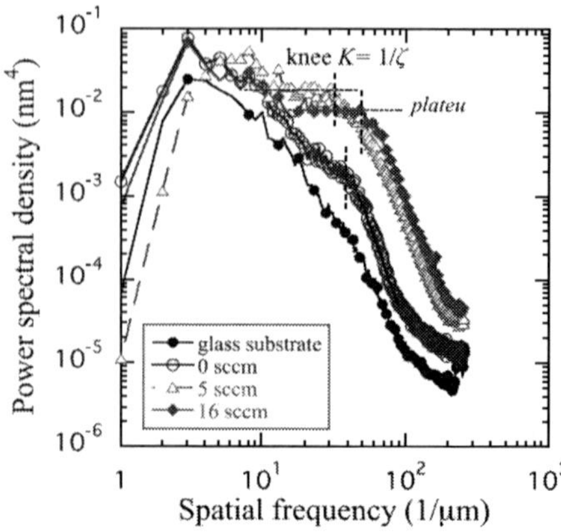

Fig. 2 2D-PSD of ZnO films deposited at various nitrogen flow rate ratio

Figure 2 shows the two-dimensional (2D) power spectral density of the ZnO thin films deposited at nitrogen flow rate of 0, 5, and 16 sccm. The 2D-PSD plot of glass substrate is also presented as comparison. From Fig.2 it can be observed that there is no peak in the high spatial frequency region indicating that all films have self-affine surface structure associated with randomly rough surface. This self-affine structures was expected because amorphous substrate was used in our experiments. 2D-PSDs of all films show power law inverse behavior particularly at high spatial frequency region indicating the existence of fractal component in the surface morphology [11]. The values of fractal dimension D_f and D_S, roughness exponent α, and lateral correlation length ζ deduced from the slope of the spectrum are displayed in Table 1. In the 2D-PSD spectra of Fig. 2, there is no cutoff to zero slope (plateu) for glass substrate indicating that the film topography consists of two surface geometry with D_S values were 2.61 and > 3.00 associated with near extreme and extreme fractals, respectively. The similar pattern to that of glass substrate is observed for ZnO film deposited at nitrogen flow rate of 0 sccm with D_s values of 2.14 and >3.00 associated with marginal and extreme fractals.

Table 1. Fractal, roughness exponent, and lateral correlation length parameters of ZnO deposited at various nitrogen flow rate ratio.

N_2 flow rate (sccm)	γ		D_f		D_S		α		ζ (nm)
	γ_1	γ_2	D_f1	D_f2	D_S1	D_S2	α_1	α_2	
0	4.72	1.61	1.14	2.70	2.14	>3.00	1.86	0.30	26.10
5	4.15	-	1.42	-	2.42	-	1.58	-	31.02
16	4.16	-	1.42	-	2.42	-	1.58	-	20.33

The roughness exponent values $\alpha = 1.86$ and $\alpha = 0.3$ suggest that the anisotropic and the NN ballistic growths were the two main competing mechanisms in the film [13]. In contrast, the ZnO films deposited with nitrogen flow rates of 5 and 16 sccm exhibit only brownian fractal with D_s of 2.42 and roughness exponent α of 1.58 regardless of the amount of nitrogen indicating that local and non-local continuum effects have dominantly contributed to the roughness. Due to the sample holder was rotated during the deposition, it is very unlikely that shadowing effects to exist in the film. Hence, the roughening effect in the ZnO film deposited with nitrogen is inferred mainly due to surface diffussion and grain growth. From the position of knee in Fig.2 it can be seen that ZnO film deposited with nitrogen at 5 sccm has a larger lateral correlation length value ζ associated with a higher surface diffusion and thus a larger mean grain size.

4 Conclusions

We have reported the surface characterization of ZnO thin film deposited at various nitrogen flow rate ratio utilizing scaled height distribution analysis and two-dimensional power spectral density. By this approach, quantitative characterization of the surface morphology can be obtained. We found that the surface morphology of the ZnO film evolved by the addition of nitrogen during the growth. The high quality surface structure was obtained for the film deposited with small amount of nitrogen flow rate.

References

1. A. Janotti and C. G. Vand de Walle. Rep. Prog. Phys. **72**, 126501 (2009)
2. S. Li, J. Wen, X. Mo, H. Long, H. Wang, J. Wang, G. Fang. J. Power Source **256**, 206-211 (2014)
3. T. Yamada, A. Miyake, S. Kishimoto, H. Makino, N. Yamamoto, T. Yamamoto, Appl. Phys. Lett. **91**, 051915 (2007)
4. D. C. Look, K. D. Leedy, A. Kiefer, B. Claflin, N. Itagaki, K. Matsushima, I. Suhariadi. Opt. Eng. **52(3)**, 033801 (2013)
5. N. Itagaki, K. Kuwahara, K. Nakahara, D. Yamashita, G. Uchida, K. Koga, M. Shiratani. Appl. Phys. Exp. **4**, 011101 (2011)
6. I. Suhariadi, K. Oshikawa, K. Kuwahara, K. Matsushima, D. Yamashita, G. Uchida, K. Koga, M. Shiratani, N. Itagaki. Jpn. J. Appl. Phys. **52**, 11NB03 (2013)
7. S. S. Kim and B. T. Lee. Thin Solid Films **446**, 307 (2004)
8. O. Lupan, T. Pauporte, L. Chow, B. Viana, F. Pelle, L. K. Ono, B. R. Cuenya, and H. Heinrich. Appl. Surf. Sci. **256**, 1895 (2010)
9. K. Zhu, V. Kuryatkov, B. Borisov, J. Yun, G. Kipshidze, S. A. Nikishin, H. Temkin, D. Aurongzeb, M. Holtz. J. Appl. Phys. **95**, 4635 (2004)
10. T. Itoh and N. Yamauchi. Appl. Surf. Sci. **253**, 6196 (2007)
11. N. K. Sahoo, S. Thakur, R. B. Tokas. Thin Solid Films **503**, 85 (2006)
12. C. J. Buchko, K. M. Kozloff, D. C. Martin. Biomaterials **22**, 1289-1300 (2001)
13. Nie, Man. 2014. Growth and morphology evolution of semiconducting oxides and sulfides prepared by magnetron sputtering. Dissertation. Technical University Berlin, Berlin, Germany.

Digital integrated instrument for measuring BMI and comparing to the ideal BMI for age as a guidance for Indonesian military recruitment

Dwi Basuki Wibowo[1,*], *Ismoyo*[1], *Agus* Suprihanto[1], *Wahyu* Caesarendra[2], *Mochammad* Ariyanto[2]

[1]Center for Biomechanics, Central Laboratory of Research and Services, Diponegoro University
[2]Center for Biomechatronics, Central Laboratory of Research and Services, Diponegoro University

Abstract. This study aimed to investigate body mass index (BMI) of 102 subjects to be Indonesian military officer candidates. To test the ideal body weight (IBW) used 2 methods for comparison, first from the calculation of IBW by Brodsky's and modified Broca's formula and ideal BMI from WPRO criteria and second from IBW of the Indian Navy and TNI/POLRI recruitment. A digital integrated instrument for measuring BMI was manufactured where BW and body height (BH) are measured using loadcell and ultrasonic sensor respectively and the result is processed by Arduino Mega 2560 microcontroller. From the separate measurement results there are BW error 0.96%, BH error 3.22%, and BMI error 2.88% compared to using digitally integrated measurement.

1 Introduction

Body Mass Index (BMI, kg/m^2) has long been used in US soldier selection criteria [1]. Individuals entering the US army must meet age and sex-specific weight for height screening criteria defined in Army Regulation 40-501: Standard of Medical Fitness [2]. A candidate to become a brigade of Gurkhas of the British army must meet minimum height and weight 158 cm and 50 kg respectively [3]. The Australian Army which used the PULHEEMS recommended BMI guidelines for entry into service i.e minimum and maximum BMI of male and female for age 18+ years are 18 and 28 kg/m^2 respectively, while for age 16 to <18 years are 17 and 27 kg/m^2 respectively [4]. The Indian Navy has published a correlation table between age, height and average weight for the guidance of candidates as shown in Fig. 1 [5]. It should be within permissible limits ± 10% of the ideal weight given in the figure.

To pass the medical test in the Indonesian National Army (TNI) and the Police of the Republic of Indonesia (POLRI) a male candidate must meet the minimum height 165 cm and weight 50 kg [6]. TNI and POLRI have also published the ideal weight guidelines for being soldiers and officers based on the Western Pacific Region of WHO (WPRO criteria, 2000) [7], as shown in Fig. 2 [6].

Corresponding author: rmt.bowo@gmail.com

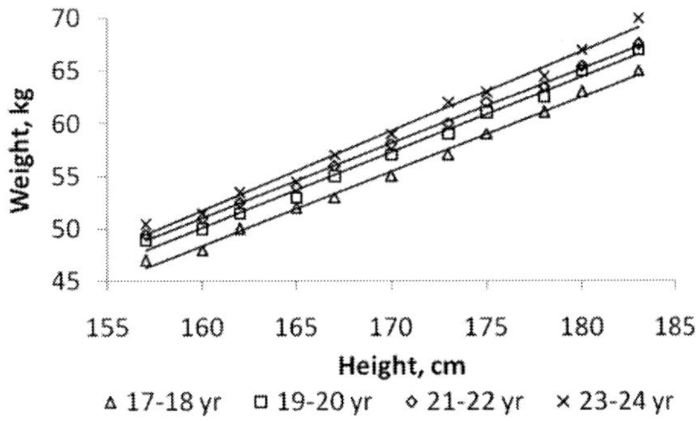

Fig. 1. The ideal weight guidelines for Indian Navy recruitment

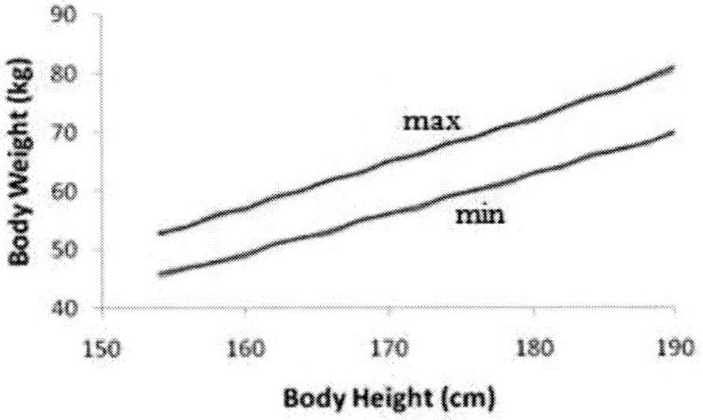

Fig. 2. The ideal weight guidelines for TNI and POLRI recruitment

There are many methods for calculating ideal body weight (BW) or BMI. The ideal BW (IBW) according to Robinson's formula is a 5 feet tall man should be 52 kilograms while that of a woman 5 feet tall should be 49 kilograms. For men who are taller than 5 feet, 1.9 kilograms should be added to get their IBW. In the case of women, 1.6 kilograms should be added for every inch above 5 feet [8]. The Miller's formula is based on the belief that the IBW of a man should be 52 kilograms at 5 feet. For every additional foot on top of the 5 feet, an additional 1.9 kilograms should be added. However, the IBW for women should be 49 kilograms and 1.7 kilograms added for every five inches more [9]. Another method was published by Brodsky et al. [10], which propose a simple approach to estimate IBW based on the BMI as shown in eq. 1. The IBW used by the TNI in candidate recruitment is the modified Broca's formula [11] (multiplied by 0.9, adjusted to Indonesian population BMI) which is calculated in Eq. 2 [6]. WPRO have also categorized BMI into 5 criteria i.e less than 18.4 kg/m^2 as underweight, 18.5 to 22.9 kg/m^2 as normal, 23.0 to 24.9 kg/m^2 as overweight, and over 25.0 kg/m^2 as obese [7].

$$\text{BW (kg)} = 22 \text{ x } \text{BH}^2 \text{ (m}^2\text{)} \quad (1)$$

$$\text{BW (kg)} = [\text{BH (cm)} - 100] \text{ x } 0.9 \quad (2)$$

This study aimed to investigate BMI of 102 subjects to meet the ideal BMI criteria to be Indonesian military officer candidates. In order to obtain the comprehensive results it is necessary to compare the results of IBW calculations from Eq. 1 and Eq. 2, ideal BMI from WPRO 2000 criteria, and IBW of the Indian Navy and TNI/POLRI.

2 Methods

Corresponding author: rmt.bowo@gmail.com

Generally, to obtain BMI of the subjects, BW and BH measurements are taken separately and then calculated the BMI value by dividing BW (in kg) to BH^2 (in m^2). In this study have designed and manufactured digital integrated instrument for measuring BMI (Fig. 3). BW is measured using loadcell, BH is measured using ultrasonic sensor and age entered on touchscreen Liquid Crystal Display monitor (LCD, made by Orbital Matrix) 5 inch and the result is processed by using Arduino Mega 2560 microcontroller made in Italy.

The ultrasonic sensor used is HC-SR04 Micropik brand with the range of 2 to 400 cm and the accuracy can reach to 3 mm. The modules include ultrasonic transmitters, receivers and control circuits. The sensor placed in the upper frame above the subject head and ultrasonic wave reflections with a frequency of 40 kHz will be received at a speed of about 340 m/s. The microcontroller will calculate the time it takes to receive the wave and determine the distance between the sensor with the head as follows: Distance (S, m) = [time (s) x velocity (340 m/s)]/2. Thus the height of H (m) = 2 m - S (m).

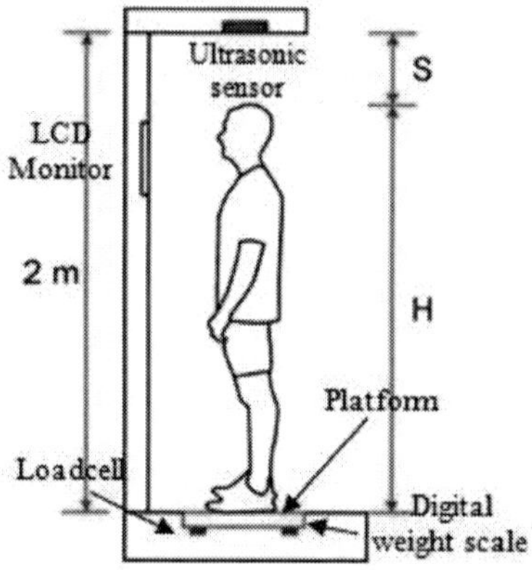

Fig. 3. Digital integrated instrument for measuring BMI

Digital weight scale with an accuracy of ± 0.2 kg (Xiaomi MI Smart made in China) with 4 loadcells (range of 20 to 80 kg) placed on the platform as shown in Fig. 3 and Fig. 4. Output loadcells in the form of analog signals forwarded to Arduino Mega 2560 microcontroller and converted into digital signals by analog to digital converter in microcontroller device. To see the display of weight value in LCD is by using TTL serial communication by entering a certain code.

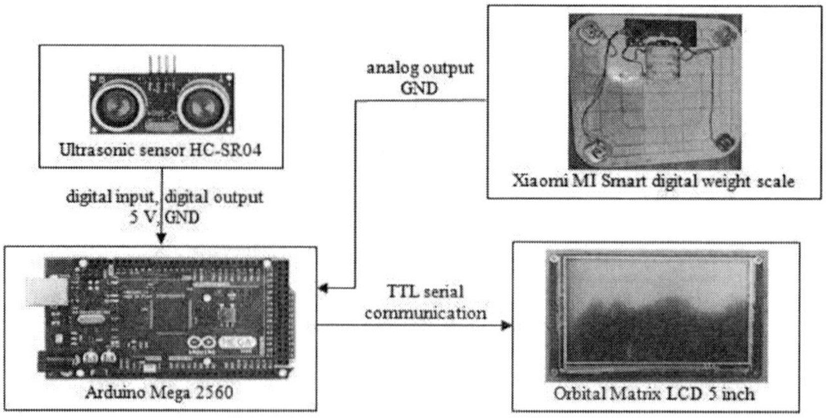

Fig. 4. The connection between hardware in the instrument to measure BMI

Corresponding author: rmt.bowo@gmail.com

One hundred and two healthy students participated in this study (Table 1). Each subject was instructed to measure BW using Xiaomi MI Smart weight scale and BH using height measuring tape and measure BMI using a digital integrated instrument. The BW and BH measurement results separately will be compared to the digital integrated instrument to determine the percentage of error.

Table 1. Subject characteristics

	Age (year)	Gender (male:female)	Weight (kg)	Height (cm)	BMI (kg/m^2)
Range of Value	17 – 24		39 – 130	147 – 185	15.2 – 41.6
Mean and SD	20 ± 1.9	87 : 15	63.3 ± 13.8	167.3 ± 6.8	22.5 ± 4.2

In measurements using digital integrated instrument subjects were asked to use a swimsuit head cover (hijab for women), dressed in very light clothing, and bare feet standing on the platform and enter the age on touchscreen LCD monitor. Automatically displayed on the monitor BH (cm), BW (kg), BMI (kg/m^2), comparison BW subject to Brodsky's and modified Broca's formula (eq. 1 and 2), and comparison BMI subject to WPRO 2000 criteria. Both for Brodsky's and modified Broca's formula the subject BW is an "ideal" category when the weight is still within the 10% bandwidth of the IBW.

To find out BW subject including the ideal category or not based on the Indian Navy and TNI and POLRI can be done by pressing the selection button available on the monitor. The IBW for age charts based on The Indian Navy recruitment (Fig. 1) obtained from correlation table between ideal height and weight for age range 17 to 24 years, which presented as linear regression equation (Table 2) [5]. The subjects who pressed this option on the monitor will show a linear graph of BH vs BW according to the age entered. The subject BW is an "ideal" when the weight is still within the 10% of bandwidth.

Table 2. The linear regression equation of IBW as the basis for recruitment of The Indian Navy

Age range (year)	Linear regression equation	Average ideal BMI (kg/m^2)
17-18	BW (kg) = 0.707 BH (cm) -64.75	19.1±0.2
19-20	BW (kg) = 0.713 BH (cm) -64.09	19.8±0.2
21-22	BW (kg) = 0.707 BH (cm) -62.06	20.1±0.1
23-24	BW (kg) = 0.754 BH (cm) -68.99	20.5±0.3

Conversely, if the subjects selected the IBW based on TNI and POLRI recruitment on the monitor will be shown two linear graphs BH vs BW showing the recommended limit of the maximum and minimum IBW (Fig. 2). These charts obtained from correlation table between ideal height and weight, which presented as a linear regression equation (Table 3) [6]. The subject BW is an "ideal" when the weight is within the recommended limit.

Table 3. The linear regression equation of IBW as the basis for recruitment of TNI and POLRI

Recommended limit	Linear regression equation	Average ideal BMI (kg/m^2)
Max	BW (kg) = 0.769 BH (cm)-65.95	19.1±0.2
Min	BW (kg) = 0.671 BH (cm)-58.02	19.8±0.2

3 Results and Discussion

From the separate measurement results there are BW error 0.96%, BH error 3.22%, and BMI error 2.88% compared to using the digital integrated instrument. This small

Corresponding author: rmt.bowo@gmail.com

percentage error indicates that the digital integrated instrument can be used as an instrument to measure BW, BH, and BMI.

Based on Brodsky's formula (eq. 1) [10] there are 54% of subjects meet IBW criteria (93 male and 9 female), while 24% and 22% are below and above the 10% bandwidth respectively. Otherwise, from modified Broca's formula (eq. 2) [6] there are 57% of subjects meet IBW (96 male and 6 female), while 17% and 26% are below and above the 10% bandwidth respectively. From this result, the modified Broca's formula is more suitable for IBW identification shown by the number of subjects entering the ideal criteria are more than the Brodsky's formula. But for the IBW identification of women the Brodsky's formula is better than the modified Broca's formula, this is consistent with Brodsky et al's claimed that IBW is on eq. 1 applies also to female sex [10]. The use of Broca's formula (without multiplying 0.9) [11] is not suitable for IBW identification because only 37% of the study subjects entered the ideal criteria.

Based on WPRO 2000 [7] there are 56% of subjects in the normal BMI category, 8% underweight, 19% overweight, and 19% obese. These results are close to IBW from the modified Broca's formula which differ only 1% compared to normal weight.

Table 4 shows the average BW and BMI of all subjects for all age ranges. Only 48.1% of subjects who met the ideal criteria based on the Indian Navy recruitment [5]. Most of BW subjects who did not belong to the ideal category are above the 10% bandwidth, that are at the age range of 17-18 years 12 subjects, 19-20 years 19 subjects, 21-22 years 10 subjects, and 23-24 years 5 subjects.

Table 4. The average BW and BMI of all subjects for all age range

Age range (year)	Number of subjects	Average BW (kg)	Average BMI (kg/m^2)	Number of subjects that belong to the ideal category
17-18	21	65.22±16.83	23.09±4.89	9 (8.8%)
19-20	45	61.64±13.21	22.25±3.92	22 (21.6%)
21-22	23	66.64±12.80	22.94±4.17	12 (11.8%)
23-24	13	59.66±11.31	21.71±3.90	6 (5.9%)

Based on IBW of TNI and POLRI recruitment [6] there are only 41.2% of subjects who met the ideal criteria and 42.2% of BW subjects are above the maximum allowable IBW (Fig. 5). This result is close to the IBW on the Indian Navy recruitment for all age range where 45% of BW subjects are above the 10% bandwidth.

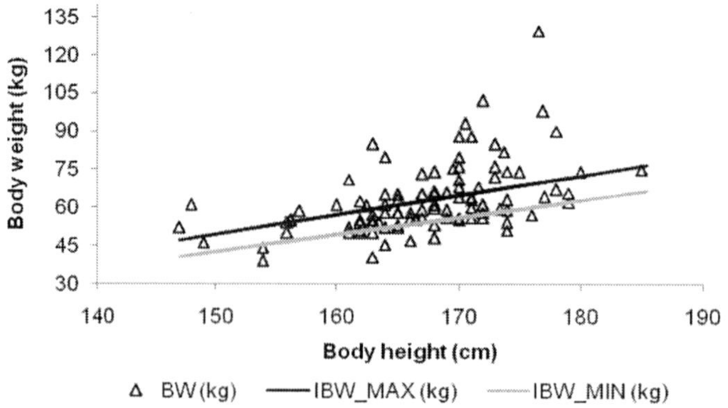

Fig. 5. The BW subjects compared to IBW recommended by TNI and POLRI

Corresponding author: rmt.bowo@gmail.com

4 Conclusion

The digital integrated instrument designed and manufactured in this study can be used as an instrument to measure BW, BH, and BMI. This can be seen from the very small percentage of errors compared to the measurement of BH and BW separately.

The modified Broca's formula is more suitable for IBW identification of Indonesian populations than the Brodsky's formula, but for the IBW identification of women, the Brodsky's formula is better than the modified Broca's formula. The IBW found by modified Broca's formula approaches the normal BMI category of WPRO 2000 criteria.

Both IBW from the Indian Navy recruitment and TNI and POLRI recruitment can be used as medical criteria to determine IBW of the officer candidate.

References

1. K. E. Friedl, *Body composition and military performance: Origin of the army standards. In: Body Composition and Physical Performance; Applications for the Military Services. B.M. Marriott and J. Grumstrup Scot,* (National Academy Press, Washington, DC, 1992)
2. U.S. Department of the Army. *Army Regulation 40-501: Standards of Medical Fitness* Available from: http://armypubs.army.mil/epubs/40_Series_Collection_1.html. (cited 2017 Apr 1).
3. Anonim. http://www.army.mod.uk/gurkhas/27784.aspx. (downloaded on 2017 May)
4. J.M. Sanderson. Defence instructions (Army), *PULHEEMS employment standards*. The Australian Army, Canberra, 1997
5. Indian Navy. Information Brochure. Services Selection Board. 2012
6. Harmen Batubara. *Panduan tes masuk prajurit TNI*. 2015. Published: www.wilayahperbatasan.com
7. WHOIASO/IOTF. *The Asia-Pacific Perspective: Redefining obesity and its treatment*. Health Communications Australia Pty Ltd, 2000
8. Robinson JD, Lupkiewicz SM, Palenik L, Lopez LM, Ariet M. *Determination of ideal body weight for drug dosage calculations*. Am J Hosp Pharm 1983;40:1016–9
9. Miller DR, Carlson JD, Lloyd BJ, Day BJ. *Determining ideal body weight (and mass)*. Am J Hosp Pharm 1983;40:1622–5.
10. Jay Brodsky, Donald P Bernstein, et al. *Estimating ideal body weight – a new formula*. Obesity Surgery, **15**, 1082-1083
11. Pai MP, Paloucek FP. *The origin of the "ideal" body weight equations*. Ann Pharmacother 2000;34:1066–9.

Corresponding author: rmt.bowo@gmail.com

Effects of Different Heel Heights on Heel Pressure Distribution for Calcaneal Spur Patients During Standing: Finite Element Analysis

Dwi Basuki Wibowo[1,*], *Achmad* Widodo[1], *Gunawan* Dwi Haryadi[2], *Agus* Suprihanto[3]

[1]Center for Biomechanics, Central Laboratory of Research and Services, Diponegoro University
[2]Center for Biomechatronics, Central Laboratory of Research and Services, Diponegoro University
[3]Center for Biomaterials, Central Laboratory of Research and Services, Diponegoro University

Abstract. This study was performed to investigate heel height shoe for every patient which peak pressure in the heel area is lower than pain minimum compressive pressure. Heel heights of the shoes are standardized as 0-4 cm. Sixteen patients with symptomatic heel spur participated in this study. Peak pressure due to its own weight is estimated using FEM and compared with the results of measurement using FSR (force sensing resistor). As heel height increased, peak pressure in heel region in the heel height 2 cm, 3 cm and 4 cm are larger 3.86% and smaller 5.04% and 22.11% respectively compared to the smallest pain compressive pressure. These peak pressures are significantly higher than the average pressure in CH (center of heel) measurement results due to the difference in the location of peak pressure to CH at average of 8.64 ± 1.44 mm.

1 Introduction

The pain in patients with calcaneus spur when standing or walking due to the pressure distribution on the calcaneus region large enough. During standing the calcaneal region support about 60% of body weight (BW) [1]. During walking there are peak loads for the Achilles tendon of 3.9 BW, the plantar fascia and the plantar ligaments of 1.8 BW, and the total contact force across the talocalcaneal joint and calcaneocuboid joint of 5.4 BW and 4.2 BW respectively which all contribute to the loads on the calcaneus [2].

To reduce pressure using orthotic shoe can be done by raising heel height shoe [3-5]. Theresa Albon used 10 FSR (Force Sensing Resistor) sensors placed on various parts of the sole of the foot proved that as heel height increases more of the force is placed on the mid-foot or arch of the foot while standing [6].

Many researchers have been used FEM (finite element method) as an experimental approach to predicting the pressure distribution between the foot and different shoe insole, which offer additional information such as the maximum pressure and its location in heel region [7-8]. To developed 3D model of foot, they used MR (Magnetic Resonance) image or CT (Computed Tomography) and MIMICS software to obtain skin surface and the boundaries. The SolidWorks software was commonly used to process the boundary

Corresponding author: rmt.bowo@gmail.com

surfaces and the ABAQUS or Ansys software was used to meshing grid of FE model and assembly the interaction between foot and insole.

This research is a new thing that is comparing the heel pressure between measurement result and FE model when using different heel heights shoe and comparing both results with the measurement of pain pressure around the spur growth using algometer. The method to model 3D solid of a plantar foot is also a novelty that is using a 3D scanner.

2 Methods

Sixteen patients of RSUD (local public hospital) Tugurejo Semarang who suffer from calcanea spur participated in this study (Table 1). Calcanea spur location and dimension of each patient determined by lateral x-ray and assumed the base of the spur is located in the heel center line (Fig. 1a) [9]. Pain compressive points are specified in 7 locations, where the 7th point is set at the base of spur and 6th point at the end of spur, another 5 points are made to be around the spur with the radius equal to spur length added with 1 cm (Fig. 1b). Pain minimum compressive pressure measured using algometer (FDIX 25, Wagner Instruments, Greenwich CT, USA) (Fig. 1c) [10].

Table 1. Subject characteristics

Gender (male : female)	Age (years)	Body Weight (BW, kg)	Body Height (BH, cm)	Foot Length (FL, mm)	Foot Width (FW, mm)
3 : 13	55 ± 12	61.9 ± 9.5	155 ± 7	251 ± 14	102 ± 8

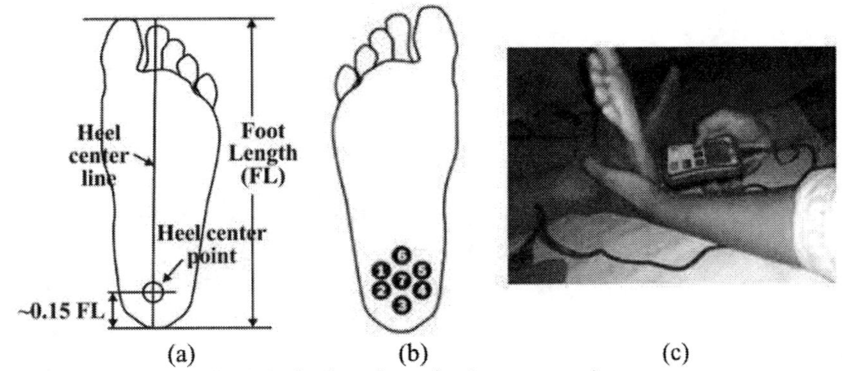

Fig 1. Point location of pain compressive test

The procedure of pain test are as follows: 1) press the skin at 7 points which have been set, 2) increasing pressure gradually and stop until patient's facial expression looks like level 6 or 7 [11], 3) record value of the pressure that causes pain. From the record of quantitative data of pain, test results can be known : pain minimum compressive pressure and point location on each patient.

To obtain the load in heel area, a force measuring device had to be made using 3 FSR sensors (402, made by Interlink Electronics) [12] which attached at the heel area (of the total 8 sensors) these are at medial (MH), lateral (LH) and center of heel (CH). Laying these sensors are determined based on the area of the foot that received the greatest load when the patient stands on the flat surface which is equivalent to the average of the biggest deformation obtained from the 3D scanner (Mini and Scansoft for Foot Orthotic made by Vismach Technology Ltd. China, red color image Fig. 2a) [13]. CH (sensor 7) is a point which is located in the heel center line at a distance 0.15 FL from the heel cup [14]. While LH (sensor 6) and MH (sensor 8) are measurement points in the lateral and medial heel areas that are determined arbitrarily but are still in the red image area (Fig. 2b) [15]. In this study, only the average pressure measurement results in CH compared with peak pressure obtained from FEM analysis because of its exact location. Every patient was asked to stand with barefoot above the shoe insole which has been installed with sensors, starting from heel height 0 cm (Fig. 2c). After the load in CH is recorded, the same procedure is performed for heel height 2 to 4 cm.

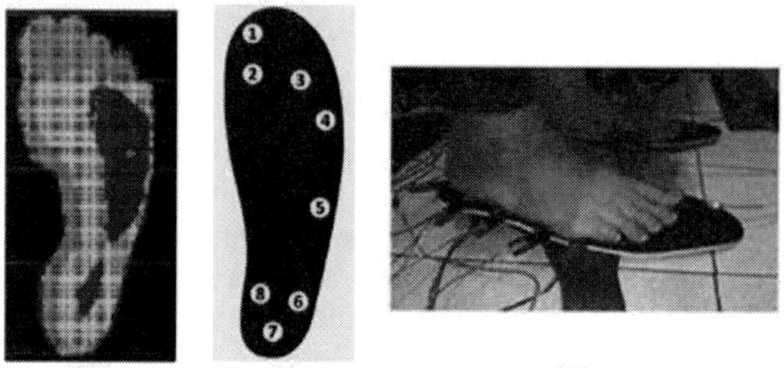

(a) (b) (c)

Fig. 2. The laying pattern of the FSR 402 sensor to measure the load on the foot due to its own weight

In this study the geometry of the patient plantar foot for building the FE model was obtained from a 3D scanner (Fig. 3a) [13, 16]. The output can be in the form of the 3D positive model (Fig. 3b) and 3D foam positive impression (Fig. 3c). To obtain 3D solid model of foot, (.dxf) file from 3D scanner must be imported into Rhinoceros software in order to modify (or reduce) mesh and convert into ABAQUS file format (.jgs).

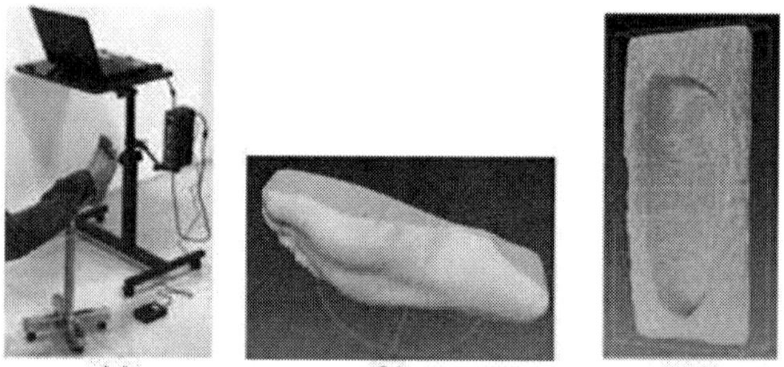

(a) (b) (c)

Fig. 3. The result of foot scanning from 3D scanner for foot orthotic

The measurement of sample some vertical distance along the craftsman-made shoes are used to estimate slope geometry function shoe insole 2 to 4 cm heel heights. This generated function used for FE modeling of the interaction between foot and insole (Fig. 4a) and make simple footwear of wood as a rigid body (Fig. 4b) for pressure measurement in the CH area of each patient.

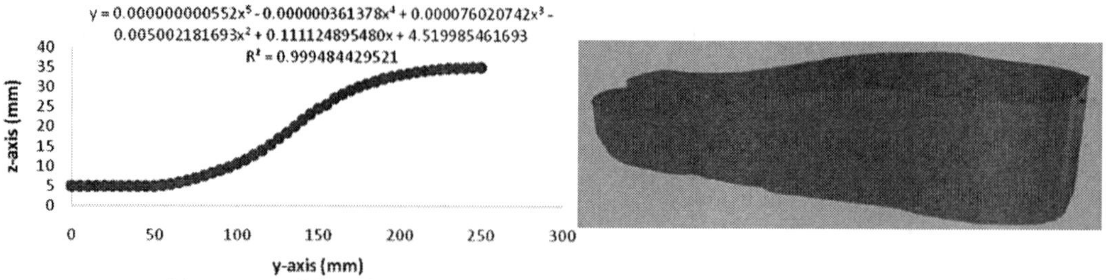

(a) generated function (b) modeling of footwear

Fig. 4. Polynomial function slope of shoe insole for heel height 3 cm

To simulate the surface interaction between the foot (Fig. 3b) and the insole (Fig. 4b), ABAQUS automated surface-to-surface contact algorithm was used where the sole of the foot is defined as slave surface while shoe soles as the master surface. Because of the lubricating nature of the articulating surfaces, the contact behavior between the articulating surfaces can be considered frictionless. The next step is partly given material properties. The material used is human skin with Young's modulus $4.2x10^5 N/m^2$ and Poisson's ratio v = 0.1 [17] while shoe insole is assumed as a rigid body.

In order to simulate the movement of the foot following the contour of the shoe insole in the vertical direction while standing, the bottom of the shoe is restrained and the side of the sole of the foot given the boundary condition which can only move in the direction of the z-axis. Loading in the form of pressure applied on the upper surface of the foot of half the weight (in, N) divided by the area of the upper foot (in, m^2).

The procedure to find peak pressure in heel area (equivalent to the rear foot defined by Cavanagh [9]) for each heel height are as follows: 1) copy all coordinates data and pressures after FEM simulation into MS Excel software, 2) trace the peak pressure value in the heel area (z-value) and record its x-y coordinate. From the result can be known the difference of peak pressure value with pressure in CH and the peak pressure location distance to CH in heel area for every heel height of each patient (Fig.5).

$$L = \sqrt{(\Delta x)^2 + (\Delta y)^2}$$

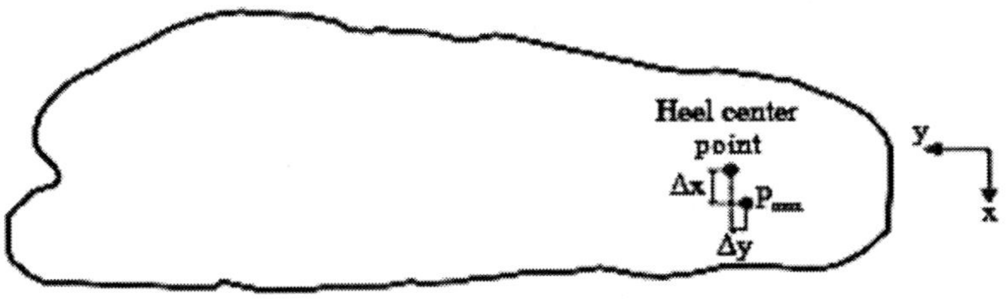

Fig. 5. The differentiation of peak pressure location to CH

3 Results

The result of pain test of 16 patients found the average pain minimum compressive pressure is 183,057.29±78,980.88 N/m^2. The most location point of pain minimum compressive pressure are 11 at 1st point (48%) and 7 at 6th point (30%).

The average pressure in CH obtained by measurements is 142490.53 N/m^2, 33010.04 N/m^2, 21764.39 N/m^2 and 21434.76 N/m^2 for heel height 0 cm, 2 cm, 3 cm, and 4 cm respectively. Meanwhile average peak pressure obtained from FEM analysis are 170323.64 N/m^2, 108259.03 N/m^2, 99079.29 N/m^2, and 85232.55 N/m^2 at the location 9 mm, 8 mm, 8 mm, and 9 mm from CH for heel height 0 cm, 2 cm, 3 cm, and 4 cm respectively.

As heel height increased, average peak pressure in heel region in the heel height 2 cm, 3 cm and 4 cm are larger 3.86% and smaller 5.04% and 22.11% respectively compared to the smallest pain compressive pressure. The average peak pressure values are significantly higher than the average pressure in the CH measurement results amounted to 16.34%, 69.51%, 78.03%, and 74.85% for heel height 0 cm, 2 cm, 3 cm and 4 cm respectively. Except in heel height 0 cm, all the average pressure from measurement results are lower than the smallest pain compressive pressure (Fig. 6a).

The difference in the average peak pressure and average pressure values is due to the location of average peak pressure by the FEM analysis result at about 8.5 mm to the CH. The average pressure in CH obtained by FEM analysis are 162296.87 N/m^2, 39211.50 N/m^2, 34636.37 N/m^2 and 28378.05 N/m^2 or differ at about 12.20%, 15.82%, 37.16% and

24.47% for heel height 0 cm, 2 cm, 3 cm and 4 cm respectively compared to the average pressure in CH obtained by measurements (Fig. 6b).

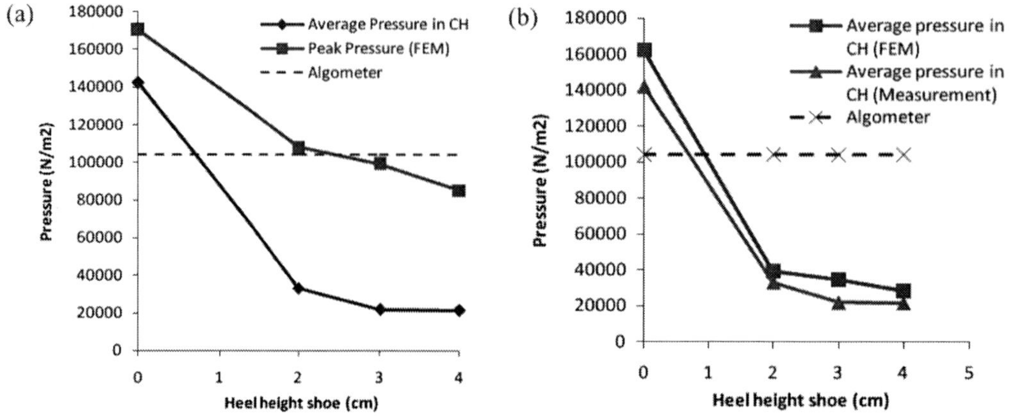

Fig. 6. The comparison of measurements and FEM analysis with pain minimum compressive pressure

4 Discussion

The results of pain compressive tests showed that the smallest pain compressive pressure occurs mostly at the point of the end of spur. This is consistent with the reality that suppression at the end of spur will result in the biggest bending moment at the base of the spur.

During standing as heel height increased, reduce pressure in heel region [3-6]. The significant difference average peak pressure obtained by the FEM analysis and average pressure obtained by the measurement result for all heel height shoe due to the location are not coincided.

Most of the location of patients peak pressure are in the medial heel area (Δx value is positive, Fig. 5) which shows the position of the human foot tend to pronation while standing in an upright posture, but when shoe heel height is raising rearfoot pronation will reduce [3]. It is also seen from the measurement results that the average pressure in MH is greater than in LH. These results are in contrast to the research conducted by Pinto et al. which showed some signs of a supinated foot when standing in a balanced position [8]. Pinto researched only one obese patient feet whose body mass is 121 kg. The differences in human foot position while standing may be caused by the position of the patient's pinto not completely straight when scanned.

The difference average pressure between the FEM analysis to measurement results caused by inaccuracy in load measurement. This is because the FSR 402 sensor is too thin (nominal thickness of 0.46 mm) which does not really support the load perfectly in the CH area. Thus naturally pressure in CH from the measurement is lower than FEM analysis.

5 Conclusion

The differences of average pressure in CH between the FEM analysis to measurement results amounted 22.41% shows that laying of FSR sensors in CH is quite accurate. From this study for reasons of propriety and comfort of use, heel height 2 cm is suitable for male patients and 3 cm is for female patients. For heel height shoe, 2 cm is still a small pain because the value of the average pressure in CH is higher than 3.86% compared to the smallest pain compressive pressure.

This work was supported by the Application and Development Research Grant UNDIP Contract No: SP DIPA - 042.01.2.400898/2016.

References

1) Jason KK Chia, Sanjay S., Angeline K., Jean LJ O., Jessie MT P. and Ai L. S. *Comparative Trial of The Foot Pressure Patterns Between Corrective Orthotics, Formthotics, Bone Spur and Flat Insoles in Patients with Chronic Plantar Fasciitis*. Ann Acad Med Singapore, 2009; 38: 869 – 75
2) Virginia L.G., Gary S.B., Robert T.W. and Dennis R. C. *Calcaneal Loading During Walking and Running*. Med. Sci. Sports Exerc., Vol. 32, No. 3, pp. 627-634, 2000
3) Mika S. and Paul D. A.. *Effect of heel height on the foot in unilateral standing*. J. Phys. Ther. Sci. 11: 95-100, 1999
4) Wei-Hsien Hong, Yung-Hui Lee, Hsieh-Ching Chen et al. *Influence of heel height and shoe insert on comfort perception and biomechanical performance of young female adults during walking*. Foot & Ankle Int. Dec 2005 Vol. 26 no. 12 1042-1048
5) Yan Cong, Jason Tak-Man Cheung, Aaron KL Leung and Ming Zhang. *Effect of heel height on in-shoe localized triaxial stresses*. J. Biom 44 (2011) 2267-2272
6) Theresa Albon. *Plantar force distribution for increasing heel height within women's shoes*. Physics Department, The College of Wooster, Wooster, Ohio, December 2011
7) Jason Tak-Man Cheung and Ming Zhang. *Finite element modeling of the human foot and footwear*. 2006 ABAQUS Users' Conference
8) V. C.Pinto, N. V.Ramos, M. A. P.Vaz and M. A. Marques. *3D Modelling for FEM simulation of an obese foot*. ResearchGate. Conf. Paper, January (2010)
9) Peter R. Cavanagh and Mary M. Rodgers. *The Arch index: a useful measure from footprints*. J. Biomechanics Vol. 20, No. 5, pp. 547-551, 1987
10) Wagner Instruments. *Wagner FPTX Series Economy Manual Pain Threshold Testers*. PAIN TEST™ ALGOMETER, USA
11) Holy Redeemer. *Pain Level Chart*. (https://lane.stanford.edu/portals/cvicu/HCP_Neuro_Tab_4/0-10_Pain_Scale.pdf)
12) Interlink Electronics. *FSR Force Sensing Resistor – Integration Guide and Evaluation Parts Catalog*. 400 Series Evaluation Parts with Suggested Electrical Interfaces, 546 Flynn Road, Camarillo, CA 93012
13) Dwi B.W., Gunawan D.H., and Agus P. *Estimation of Foot Pressure from Human Footprint Depths Using 3D Scanner*. AIP Conf. Proc., ISBN: 978-0-7354-1365-8, Vol. 1717, (2016). doi: 10.1063/1.4943451
14) Rodrigo, Asanka S., Goonetilleke, Ravindra S. and Xiong, Shuping, 2014. *Load distribution to minimise pressure-related pain on foot: a model*. Ergonomics, Vol. 56, no. 7, 1180-1193
15) Dwi B.W., Gunawan D.H., Achmad W., et al. *Estimation of calcaneal loading during standing from human footprint depths using 3D scanner*. AIP Conf. Proc. 1788, 030063-1-030063-9 (2016); doi 10.1063/1.4968316
16) ScanPod3D, 2013. *3D Scanner Mini and Scansoft for Foot Orthotic*. Vismach Technology Ltd., www.scanpod3d.com
17) P. G. Agache., C. Monneur., J. L. Leveque., J. De Rigal. *Mechanical properties and young's modulus of human skin*. Arch Dermatol Res 269, 221-232, 1980

A Review on Sensors for Real-time Monitoring and Control Systems on Machining and Surface Finishing Processes

Tomi Wijaya[1], *Wahyu* Caesarendra[1], *Tegoeh* Tjahjowidodo[2,*], *Bobby K* Pappachan[1], *Arthur* Wee[3], *Muhammad Izzat* Roslan[3]

[1]Rolls-Royce @ NTU Corporate Laboratory, Singapore 639798
[2]School of Mechanical and Aerospace Engineering, Nanyang Technological University. Singapore 639798
[2]Rolls-Royce Singapore, Applied Technology Group, Singapore 639798

Abstract. One of the key components in real-time monitoring and control on machining and surface finishing processes are sensors. The advances of such system have triggered interesting questions on sensor selection that act as the fundamental before starting a project. This paper is made to review and answer the questions surrounding sensor selection. The paper first explains on the type of sensors commonly used in practice for real-time monitoring and control systems. After which, the paper discusses on how often the sensors are used on several machining and surface finishing processes and what are the reasons for the sensor selection. Thereafter, a review on the type features commonly analysed through these sensors is discussed. The paper expects reader would decide better upon selecting sensors and has a better direction in their project. Thus the paper works to guide reader to improve based on what has been completed before.

1 Introduction

The development of manufacturing industry nowadays is pushed to its limit to fulfil the demand of market. The market requires better quality products, more variability, shorter product life cycle, reduced product cost, and globally competitive products [1].

Machining and surface finishing processes develops in the direction of immediate integration with real-time monitoring and control systems to allow users to monitor and control the events happening during machining and surface finishing.

The selection of sensors is one of the fundamental building block for real-time monitoring. This paper will discuss on the sensors used in other real-time monitoring and control projects to better inform reader on suitable sensor application for their project.

* Corresponding author: ttegoeh@ntu.edu.sg

2 Methodology

The development of real-time monitoring and control systems on machining and surface finishing processes consists of several procedural steps as presented in the figure below

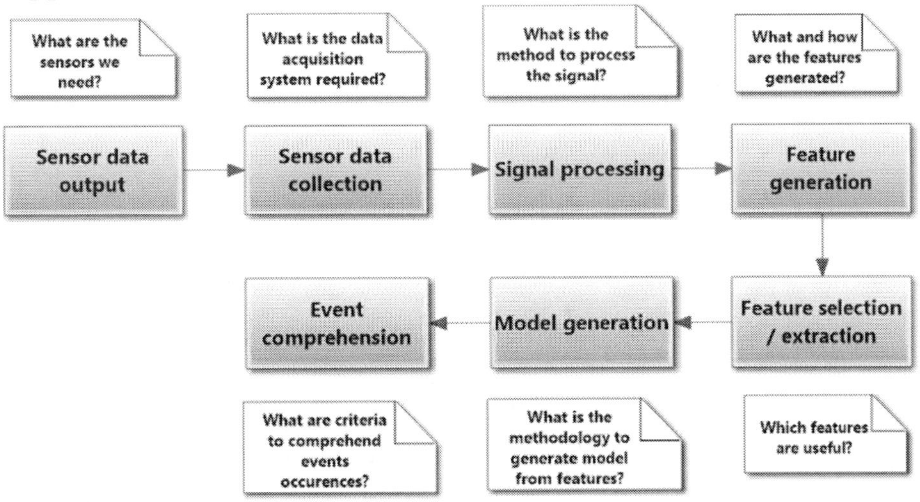

Fig 1. Development methodology

The methodology mentioned above is the generic methodology often used in real-time monitoring and control systems which uses sensors for events interpretation. The scope of this paper covers on "Sensors data output", "Feature generation/calculation" and "Feature selection/extraction". Sensors selection determine the sensors data output. Meanwhile, "Feature generation/calculation" and "Feature selection/extraction" develop by understanding the available features and how these features fit into the application. These three steps is covered considering that the result varies if the three steps were to change.

3 Selection of sensors type system

Commonly, the type of sensors used for real-time monitoring and control for machining and surface finishing processes are acoustic emission sensor, accelerometer, force sensor, and dynamometer/power meter. These 4 sensors are common because they measure the variables involved directly in the processes. Acoustic emission sensor measures the elastic wave emitted from the processes. Accelerometer measures the acceleration of tools or work coupon during the process and detects anomalies or signatures for each event. Force sensor measures the force applied on the interaction between work coupon and tool. Dynamometer / power meter measures the difference in the power behaviour upon different events during the machining or finishing processes. The summary of frequently used sensors for real-time monitoring and control systems in manufacturing processes is presented Table 1.

3.1. Acoustic emission sensor

According to Table 1, the frequently used sensors for manufacturing processes is AE sensor. Caesarendra et.al. [2] mentioned that AE-based method can be applied in wide range frequencies generated from two processes i.e. (i) impacting, and (ii) friction and rubbing. In manufacturing, the impact between work coupon and tools generates instantaneous AE signal. Due to the abrasive process, a transient elastic wave generated by the contact interaction of the work coupon and abrasive tool. Moreover, raw AE signal can be analysed with two general techniques namely: (1) AE-hit parameters and (2) AE signal processing. The AE-hit parameters are calculated from the built-in AE software. The example of AE-hit parameters are amplitude, RMS, energy, counts, events, ASL and AE burst. Aside from AE-hit parameters, some signal processing techniques can be applied for AE signal such as short-time Fourier transform (STFT), wavelet package decomposition-Hilbert Huang transform (WPD-HHT), and Wigner-Ville distribution.

Inasaki [3] used AE sensor on turning, cutting and grinding processes and suggested that AE sensor to be complimented with other sensor such as force sensor due to AE sensor's high sensitivity to the process state.

Y.B Guo et.al. [4] concluded that AE RMS, frequency and count rate can be used to distinguish surface damage, tool wear and surface roughness. In addition, AE frequency strongly correlate to surface finish quality. Surface damage increases the peak of AE amplitude and its corresponding frequency.

Table 1. Frequently used sensors for real-time monitoring and control systems

Surface Finishing & Machining Process	**Sensor System**			
	Acoustic Emission	Accelerometer	Force sensor	Power meter

1.Drilling	[5-7]	[7]	[6-9]	[7, 8]
2.Lapping	[7, 10]	[7, 10]	[7, 9, 10]	[7, 10]
3.Grinding	[3, 7, 10]	[7, 10]	[7, 9, 10]	[7, 10]
4.Milling	[3, 7, 11, 12]	[7, 13]	[7-9, 11, 13]	[7, 8, 13]
5.Broaching	[7, 14]	[7, 14]	[7, 9, 14]	[7]
6.Turning	[4, 7, 10, 15]	[7, 10, 15, 16]	[7-10, 15]	[7, 8, 10, 15]
Total	10	6	9	5

3.2. Accelerometer

Accelerometer is more frequently used than AE in vibration analysis of rolling element bearing [2].This is because of its well-developed data acquisition and analysis tools [17]

Accelerometer can be placed on the work coupon or on the tool and measures the vibration on which it is attached to. The frame of reference for accelerometer is always on the attached surface. If accelerometer is attached to work coupon, it measures vibration when tool is touching the work coupon. However when it is attached to tool, the accelerometer measures vibration from the tool when the tool is in operation. Frame of reference will help to differentiate events occurrence from the accelerometer signal.

Salgado, D.R. et.al [16] used SSA decomposition and LS-SVM to predict the quality of surface roughness. In addition, Salgado, D.R. et.al [16] recommended to use three axis accelerometer for better prediction result

3.3 Force sensor

External force sensors can be applied in two different place. First is on the tool. Once the tool applies force on work coupon, the reacting force to the force sensor will give out the sensor reading. The second one is on work coupon. The sensor detects the reacting force only if the force application area is on the sensor placement.

Force sensor can be applied by integrating it into a direct driven motor spindle using piezo electric force sensor rings. Byrne et.al [6] mentioned that by integrating the force sensor into the motor spindle, it enables the reading of non-process related forces. The higher frequency content received from the sensor reveals spindle signature for monitoring.

M.B. Jun et.al. [9] used Kistler force ring sensor, an 8 uniaxial piezoelectric force sensor. They concluded that the force sensor affected the spindle dynamics slightly through the time-varying temperature differentials, which to be solved using temperature control on the spindle. The signal signatures is similar to Byrne et.al [6] conclusion. There were two signatures, one belong to the spindle internal dynamics and the other to machining. M.B. Jun et.al. compare the machining signature produced by the force sensor and the dynamometer and found similarities in the calibration model resulted

3.4 Power meter / dynamometer

Dynamometer measures the power output of an engine by measuring torque and the rotational speed. When events occur out of the routine, dynamometer indicates power difference. Two placements are possible for dynamometer, embedded and non-embedded.

Axinte, D et. al. [8] uses Kistler dynamometer to measure the spindle power signal in turning (9257B4), milling (9272) and drilling (9272) processes. The aim is for tool condition monitoring. In milling, the force sensor was sensitive enough to detect milling cutter tooth chipped. In turning and drilling, dynamometer helps in detecting such event.

4 Frequency specifications of sensors

In order to obtain more obvious signatures related to the surface finishing stage and quality, the sensor signal processing can be performed in either time-domain or frequency domain. In machining processes, often a rotating tool is employed for material removal and hence frequency domain analysis can help to extract meaningful information from the incoming sensor signal. It is imperative in such applications to identify the frequency range of interest and select suitable sensors accordingly. To extract the frequency range of interest, the typical method used is spectral analysis. Figure below shows commonly used sensors for process monitoring in machining classified based on operating frequency range and from this figure it can be deduced that AE and accelerometer are more commonly used

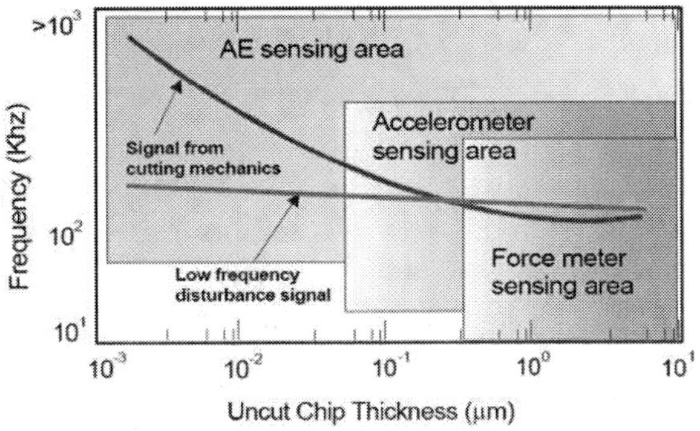

Fig 2. Sensor classification based on operating frequency range and material removal rate [18]

This information aids in identifying and subsequently isolating potential noise that contains no direct information regarding the process.

5 Sensor signal features generation and extraction

In frequency analysis, Fast Fourier Transform (FFT) is the most commonly used method for its ability to identify the dominant frequency event. However, FFT performs effectively in stationary periodic signals and it is less effective for non-stationary signals that arise from time-dependent events. There are a number of features generation in frequency analysis of vibration signal as has been reviewed and discussed in [17]. It is worth noting that the review features are applied in vibration signal of rolling element bearings that may not suitable for other application. The FFT has been applied in manufacturing field as a feature or descriptor that known as signal harmonic (single H) [7]. The following is the summary studies that presented the frequency feature analysis in machining processes.

Table 2. Frequently domain features applied for machining processes

Signal	Frequency-domain Feature		
	Single H	**PSD**	**H ratios**
Acoustic emission	[10, 49]	[9, 49, 32, 36]	[50]
Vibration	[10, 27, 16, 30, 5, 49, 50]	[27, 49]	[23]
Force	[10, 14, 9, 49, 50]	[14, 5, 49]	[NA]
Power	[NA]	[NA]	[NA]

The signal features can be categorised into three domains, time domain, frequency domain, and wavelet domain. The features for each domain is summed up in table 3 below.

Table 3. Signal domains and features for sensors data analysis

Time	RMS, Peak, Mean Value, Standard Deviation, Skewness, Kurtosis, Variance, Autoregressive (AR) models, Time-domain averaging (TDA)
Frequency	Harmonic Ratios, Single harmonic, Power spectral density (PSD)
Wavelet	RMS, Peak

6 Conclusion

The current trend on real-time monitoring and control systems on machining and surface finishing covers different approaches in methodology and type of sensors. This paper has completed the initial steps to align the methodology and given the resources on sensors that have been used before in other researches. The key points that the paper has achieved are:

- Laying down the generic methodology, concluded from the methodologies that have been used before in other projects.
- Gathered the resources of sensors used in other project according to the machining process it is used on. In addition, how the sensors can be used in the project. This includes the possible placement, advantages and disadvantages of the sensor.
- Hinted on frequency analysis on different sensors that leads to different frequency specification per sensor
- Elaborated on the features commonly used in real-time monitoring and control systems, including the equation definition of this feature. In addition, the signal extraction and example of FFT in the signal processing.

References

1. S. Y. Liang, R. L. Hecker, R. G. Landers, J. Manuf. Sci. Eng. **126** (2004).
2. W. Caesarendra, B. Kosasih, A. K. Tieu, H. Zhu, C. A. Moodie, Q. Zhu, Mech. Syst. Signal Process. **72** (2016).
3. I. Inasaki, Ultrasonics, **36** (1998).
4. Y. B. Guo, S. C. Ammula, Int. J. Mach. Tools Manuf. **45** (2005).
5. S. H. Lee, D. A. Dornfeld, J. Manuf. Sci. Eng.**123** (1999).
6. G. Byrne, G. E. O'Donnell, CIRP Ann. - Manuf. Technol. **56** (2007).
7. J. V. Abellan-Nebot, F. Romero Subirón, Int. J. Adv. Manuf. Technol. **47** (2010).
8. D. Axinte, N. Gindy, Int. J. Prod. Res. **42** (2004).
9. M. B. Jun, O. Burak Ozdoganlar, R. E. DeVor, S. G. Kapoor, A. Kirchheim, G. Schaffner, Int. J. Mach. Tools Manuf. **42** (2002).
10. D. A. Dornfeld, Y. Lee, A. Chang, Int. J. Adv. Manuf. Technol. **21** (2003).
11. I. Marinescu, D. Axinte, Int. J. Mach. Tools Manuf. **49** (2009).
12. I. Marinescu, D. A. Axinte, Int. J. Mach. Tools Manuf. **48** (2008).
13. J. C. Chen, B. Huang, Int. J. Adv. Manuf. Technol. **21** (2003).
14. D. A. Axinte, N. Gindy, K. Fox, I. Unanue, Int. J. Mach. Tools Manuf. **44** (2004).
15. R. Teti, K. Jemielniak, G. O'Donnell, D. Dornfeld, CIRP Ann. - Manuf. Technol. **59** (2010).
16. D. R. Salgado, F. J. Alonso, I. Cambero, A. Marcelo, Int. J. Adv. Manuf. Technol. **43** (2009).
17. W. Caesarendra, PhD Thesis, University of Wollongong (2015).
18. D. Dornfeld, Y. Lee, A. Chang, Int. J. Adv. Manuf. Technol. **21** (2003).

Modal Analysis of Replica Boss Hole During the Deburring Process in Aerospace Manufacturing Industry

Claudy Andriani[1] , *Wahyu* Caesarendra[2] , *Tegoeh* Tjahjowidodo[1,*], *Tomi* Wijaya[2] , and *Bobby K* Pappachan[2]

[1]School of Mechanical and Aerospace Engineering, Nanyang Technological University, Singapore 639798
[2]Rolls-Royce @ NTU Corporate Lab, 65 Nanyang Avenue, Singapore 637460

Abstract. The monitoring of surface finishing processes in aerospace manufacturing industry becomes one of key issues to maintain the overall quality of product or part. To date, the surface quality monitoring post machining processes such as deburring, use visual inspection, surface roughness test or laser gap gun. The whole manufacturing process then requires a considerable amount of time as the production line must be halted due to these measurements taking place. This study presents an online monitoring system to measure the chamfer quality of replica boss hole post-deburring process. Vibration signal was measured during the deburring process and the features that correlate to the deburring stages (passes) were extracted. This paper focuses on the validation of actual vibration signal with the modal analysis of work coupon (replica boss hole) to obtain the correlation between the vibration amplitude level on particular region and the mode shape of work coupon during the deburring process.

1 Introduction

Deburring and finishing are the treatment processes for burrs, sharp edges, and rough surfaces removal. They can often be treated as the step-child of a manufacturing process, but its importance is growing as tolerances get tighter and precision devices become the norm [1]. Deburring process was mostly used for safety in the past, however its significance in manufacturing process today has expanded. Other than removing burrs and reducing sharp edges, deburring process also increases robustness by eliminating stress concentrations.

As the automated detection of the machining process malfunctions is able to reduce the production costs and increase the quality of the machine parts, it has become a great interest among engineers and scientists. This use of process monitoring is aimed at eliminating surface anomalies by halting the production when manufacturing conditions indicate they are about to or have arisen. This approach might be of a real benefit for the reduction of the

* Corresponding author: ttegoeh@ntu.edu.sg

costs associated with part inspection after machining as well for the safety of the critical components such as turbine disks for the aero engines.

Acoustic emission refers to the transient elastic stress waves generated due to the rapid release of strain energy from a localized sources within a material. From the analyzed results of the experiment, it is shown that AE sensor feedback can be useful both for controlling metal removal and detecting/tracking the workpiece burr edge. Connected to the whole laser deburring system with a proper control scheme, this AE sensing technology could be a viable means for automating precision deburring operations [2].

Axinte, D. A. et al worked on research which attempts to correlate the quality of the machined surface after broaching and the output signals obtained from multiple sensors, namely acoustic emission, vibration, and cutting forces [3]. The results show that the cutting force signals are sensitive enough to detect the geometrical deviation of the machined profile, burr formation and to a lesser extent the chatter marks.

In another study, Shah, H. et all explored the use of embedded waveguides for monitoring delamination type defects in composites [4]. These waveguides confine the wave transmission in one dimension and waves leak only through the opening provided, which enhances the capability to inspect large composite structures with very low attenuation rate. Inaccessible areas can be inspected and inter-laminar delamination detection can be achieved. Live monitoring and assessment of discontinuities can be accomplished effectively by using this mechanism.

In this paper, an online monitoring system to measure the chamfer quality of replica boss hole post-deburring process will be studied. Vibration signal was measured during the deburring process and the features that correlate to the deburring stages (passes) were extracted. This paper focuses on the validation of actual vibration signal with the modal analysis of work coupon (replica boss hole) to obtain the correlation between the vibration amplitude level on particular region and the mode shape of work coupon during the deburring process.

2 Methods

2.1 Finite Element Method

The Finite Element Method (FEM) has developed into a key indispensable technology in the modelling and simulation of advanced engineering systems in various fields. Engineers and designers go through a sophisticated process of modelling, simulation, visualization, analysis, design, prototyping, testing, and lastly, fabrication. More often than not, much work is involved before the fabrication of the final product or system. This is to ensure the workability of the finished product, as well as for cost effectiveness in the manufacturing process. FEM is one of methods used for solving the natural frequencies and the mode shapes of oscillations.

2.2 Modal Analysis

Natural frequency of the boss hole will be analysed using ANSYS Workbench 17.1. The boss hole model was used to demostrate the vibrations using the parameters set. Natural frequency is found when the model oscillate with the highest amplitude vibration in the absence of damping force.

3 Material and experimental setup

The lab experiment were conducted on a work coupon which was designed to replicate the boss hole of Trent 700 combustor casing as presented in Figure 1. The diameter of the boss hole is 58.5mm, with the depth of 25mm. The deburring tool was able to cut the hole easily and also minimise the vibration caused by the machining torque as it was attached firmly to the T-slot table. The work coupon was attached to the base plate using M8 threaded screws and the base plate is firmly secured to the T-slot table below. The material used for fabrication is mild steel. Tri-axial accelerometer and three work coupons were used in the trials result in six vibration datasets. The machining setup comprised of an ABB IRB 6660 robot mounted with an electrically powered spindle (PDS Colombo RS 90). A standard aluminium oxide brush with a grid size of P80 is used as the deburring brush for removing sharp edges from the chamfer. The RPM of the spindle is controlled by a variable frequency drive set to 166.67 Hz which translates to 10000 RPM and the feed rate programmed to be 30 mm/sec.

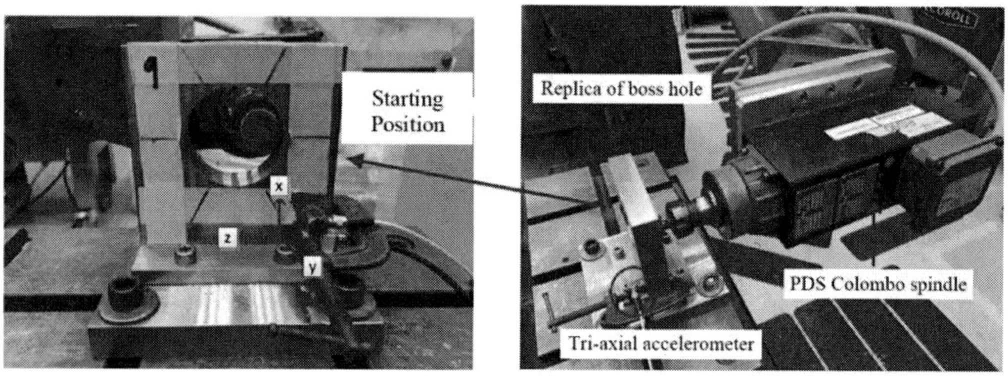

Fig. 1. Replica boss hole and machining setup.

4 Result and discussion

4.1 Vibration signal

The vibration signals were acquired from the following procedure: The DAQ and ABB robot controller were synced to acquire the accelerometer and AE sensor data. The ABB robot controller triggered (digital signal) the DAQ to start acquiring the accelerometer data when the brush first touched the boss hole surface at the start point (left side from the tool view). After the brush made a full circle around the boss hole and about to leave the surface, the ABB robot controller sent a stop signal to the DAQ to stop the sensor data acquisition. Therefore, the data from the accelerometer and AE sensor was acquired throughout actual deburring process. The vibration signals from y and z axis of accelerometer are presented in Figure 2.

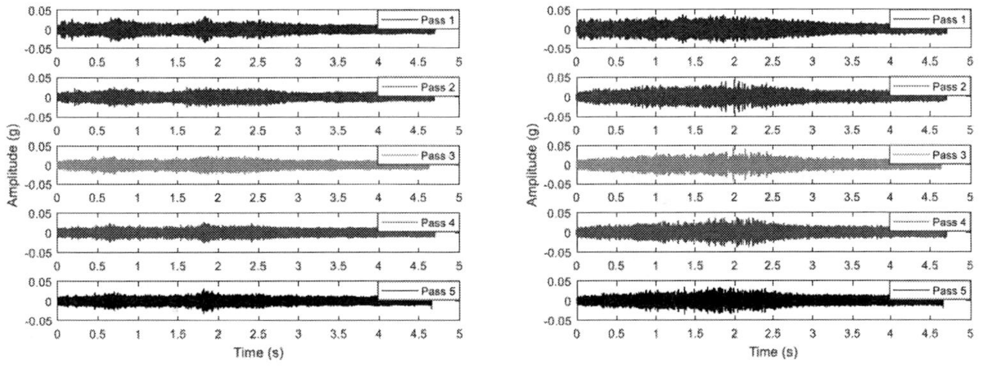

(a) (b)

Fig. 2. The vibration signal in y-axis (a) and the vibration signal in z-axis (b).

The vibration signal in y-axis and z-axis is the highest from 1.5 seconds to 2.5 seconds. This corresponds to when the brush reached the top of the boss hole. Thus, highest amplitude was observed at the free-end of the boss hole.

4.2 Modal Analysis

For solutions of the modal analysis of the replica boss hole, ANSYS Workbench 17.1 was used with toolbox Modal. The replica boss hole in Figure 3 was created using SolidWorks based on the actual measurement of the work coupon used in the lab experiment. The model was then imported to ANSYS Workbench.

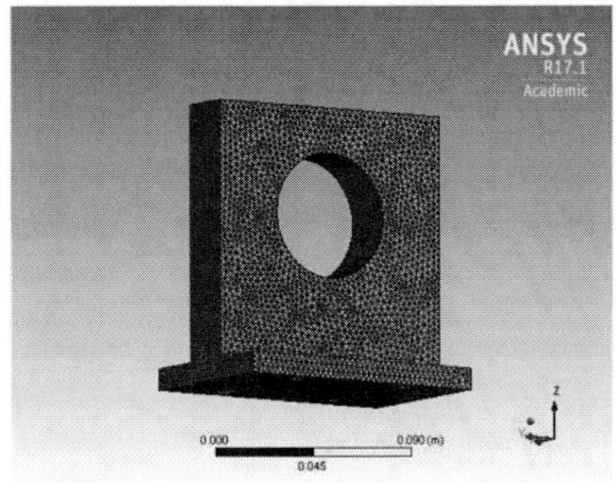

Fig. 3. Replica Boss Hole with Meshing

As a preliminary study, structural steel was chosen as the material used in this modal analysis, although mild steel was used in the lab experiment. Materials properties of the structural steel in the replica boss hole are shown in Table 1.

Table 1. Material Properties of Structural Steel.

Name	*Unit*	*Value*
Young's Modulus	Pa	2×10^{11}
Poisson Ratio	-	0.3
Density	Kg/m^3	7850

Meshing sizing for relevance center was set to "fine", as shown in Figure 4. The bottom of the boss hole work coupon was set as fixed, while the top was a free-end, to simulate the lab experiment performed. Figure 4 shows the boundary conditions, where four supports were fixed. Maximum modes to find was set to 10.

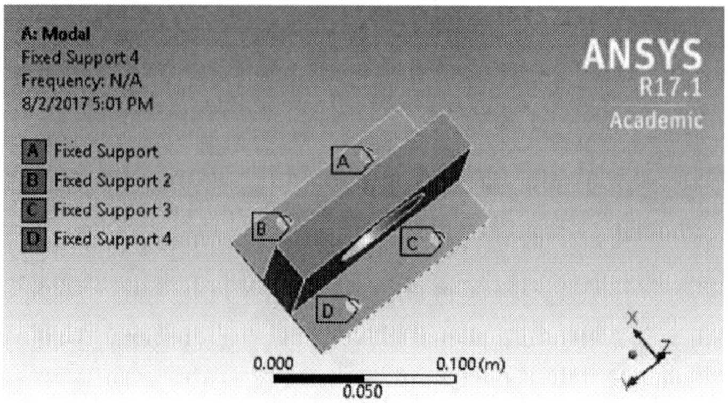

Fig. 4. Boundary Conditions

The first ten natural frequencies are given in Table 2.

Table 2. Natural Frequencies of Replica Boss Hole.

Mode	*Frequency [Hz]*	*Mode*	*Frequency [Hz]*
1.	950.48	6.	7334.
2.	1373.9	7.	7702.5
3.	2243.3	8.	8019.2
4.	3497.3	9.	8789.5
5.	5072.8	10.	8821.3

Next, two frequencies were chosen as they were deemed to suit the actual deformation during the lab experiment. The two frequencies are 950.48 Hz and 5072.8 Hz, correspond to Mode 1 and 5 respectively. The deformations are shown on Figure 5 below.

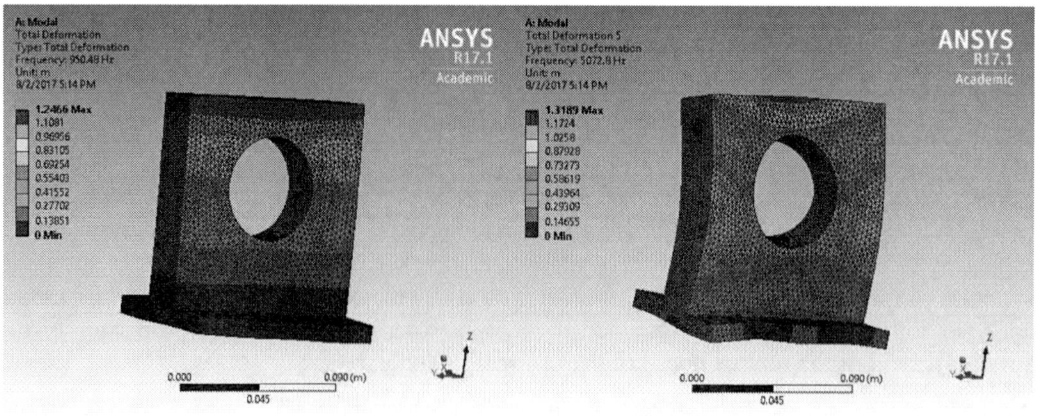

Fig. 5. Deformation of the Boss Hole Model under Natural Frequency 950.48 Hz (left) and 5072.8 Hz (right).

The most suitable model found had the largest deformation at the free-end, which was on the top of the boss hole work coupon. The top of the boss hole work coupon experienced the largest vibration when the boss hole was excited with the natural frequency.

It was found that the pattern of modal analysis of work coupon validated the actual vibration signal resulted from the lab experiment. It is implied that during the deburring process, the vibration amplitude level is related to the modal frequency. The work coupon (replica boss hole) will resonate with higher amplitude, especially at its free-end, when it is excited with its natural frequency. Hence, it is important to note the natural frequencies of the work coupon before deburring process so that it will not interfere with the sensor during data acquisition. However, this comparison holds based on the assumption that the frequency set for the brush in the lab experiment, 166.67 Hz (10,000 RPM), is also another natural frequency of the replica boss hole.

5 Conclusion

In conclusion, the actual vibration signal resulted from the lab experiment was validated by the modal analysis of the replica boss hole, as the vibration amplitude level corresponds to its mode shape. At the free-end of the replica boss hole, it resonated with higher amplitude at natural frequencies. Natural frequency is one important factor to be taken into account before deburring process to prevent it from interfering with the sensor in order to obtain a more reliable and accurate result. Further studies such as impact test and harmonic analysis with certain applied force will be conducted in the future to strengthen this preliminary study.

References

1. Morey, B., The Wide World of Deburring and Refinishing, **157**(4), 69-77 (2016)
2. Lee, S.H. and D.A. Dornfeld, J. Manuf. Sci. Eng. **123**(2), 356-364. (1999)
3. Axinte, D.A., et al., Int. J. Mach. Tool. Manufac. **44**, 1091-1108 (2004)
4. Shah, H., P. Rajagopal, and K. Balasubramaniam, *AIP Conference Proceedings*. **1806**(1): 030013 (2017)
5. R. Teti, K. Jemielniak, G. O'Donnell, and D. Dornfeld, CIRP Ann. Manuf. Tech. **59**, 717-739 (2010)
6. D. A. Dornfeld, Y. Lee, and A. Chang, I. J. Adv. Manuf. Tech, **21**, 571-578 (2003)

Investigation of Mechanical Properties of Al7Si/SiC and Al7SiMg/SiC Composites Produced by Semi Solid Stir Casting Technique

Sulardjaka [1*], Sri *Nugroho* [2], Suyanto [3], Deni Fajar *Fitriana*

[1] Mechanical Engineering Department, Diponegoro University, Semarang, Indonesia
Jl. Prof. Soedharto, Tembalang, Semarang, Indonesia, Phone: 024-7460059.

Abstract. Mechanical characteristic of silicon carbide particle reinforced aluminum matrix composites produced by semi solid stir casting technique was investigated. Al7Si and Al7SiMg were used as metal matrix. High purity silicon carbida with average particle size mesh 400 was used as reinforcement particle. Aluminum matrix composites with variation of SiC: 5 %, 7.5 % and 10 % wt were manufactured by the semi solid stir casting technique. Stiring process was performed by 45 ° degree carbide impeller at rotation of 600 rpm and temperature of 570 °C for 15 minutes. Characteritation of composites speciment were: microscopic examination, density, hardness, tensile and impact test. Hardness and density were tested randomly at top, midlle and bottom of composites product. Based on distribution of density, distribution of hardness and SEM photomicrograph, it can be concluded that semisolid stir casting produces the uniform distribution of particles in the matrix alloy. The results also indicate that introducing SiC reinforcement in aluminum matrix increases the hardness of Al7Si composite and Al7SiMg composite. Calculated porosities increases with increasing wt % of SiC reinforcements in composite. The addition of 1 % Mg also increases the hardness of composites, reduces porosities of composite and enhances the mechanical properties of composites.

1. Introduction

The aim in designing metal matrix composite materials is to combine the desirable attributes of metal and reinfocement materials. The addition of high strength, high modulus refractory particles to a ductile metal matrix produces a material whose mechanical properties are intermediate between the matrix alloy and particle reinforcement [1]. Aluminum metal matrix composites (AMCs) have wide application in aerospace, defence and selected automotive applications such as high performance racing applications. These properties include improved strength, high elastic modulus, hardness, wear resistance, attractive thermal and electrical characteristics [2-5].
Many techniques have been developed for producing particulate reinforced aluminum metal matrix composite, such as powder metallurgy, spray decomposition, liquid metal

*Corresponding author: sulardjaka@undip.ac.id

infiltration, squeeze casting and stir casting [6, 7]. Among these processing techniques, stir casting is one of the methods which is accepted for the production of large quantity commercially practised. The most common problem regarding to production of AMCs in casting routes, including stir casting are porosity and how to achieve the uniform distribution of reinforcement within the matrix [8].
Preparation of low cost SiC particle reinforced MMCs, is very difficult to be prepared due to difficult dispersion SiC particle [9]. In order to produce homogeneously reinforced MMC components, the method chosen for particle incorporation within the matrix becomes important. Some techniques for introducing and mixing the particles have inherent disadvantages. To fabricate AMCs with good mechanical properties, other important factors that need to be considered include the effect of poor reinforcement wettability particulate in the matrix and the propensity on the introduction of porosity during the particle incorporation phase [10]. The wettability of ceramics particle with liquid aluminium alloys is generally poor. The addition of alloying elements can modify the matrix metal alloy by producing a transient layer between the particles and the liquid matrix. Ca and especially Mg were very efficient in increasing the incorporation of ceramic particles into the molten aluminum [11]. Another method to improve the wettability and the distribution of SiC particles are by stirring process in semisolid state rather than in fully liquid state [12, 13]. The aim of the present study is to investigate mechanical characteritics of aluminum Al7Si/SiC and Al7Si1Mg/SiC composites produced by semi solid stir casting methods.

2. Materials and Methods

Aluminum die casting (ADC 11) ingot was used as the matrix. **Table 1** shows the chemical composition of aluminum used in the present study. Micron-sized SiC particles with average particle size mesh of 400 and purity of 99.9% was supplied as the reinforcement of metal matrix composite. Master alloy magnesium was used as alloying element.

Table.1. Chemical composition of ADC 11 ingot

	Al	Si		Fe	Mn	Mg	others
%wt	92.39	7.26		0.147	0.008	0.07	--

Aluminium die casting (ADC 11) was melted in controllable temperature electric furnace. Argon gas was flowed into the furnace to prevent oxidation of aluminum. In this process, approximately, ADC 11 ingot of 850 g was charged into the crucible heated up to 750 °C for melting. To obtain Al7SiMg matrix alloy, master alloy magnesium and aluminium were melted together in crucible furnace.
The SiC addition is carried out between the liquidus and solidus temperatures. SiC particles were preheated until temperature of 250 °C. The preheated SiC particles are added to semi solid molten aluminum at temperature of 570 °C. The semi solid alloy is mechanically stirred using a ceramics impeller driven by an electric motor at 600 rpm for 15 minute. Then the temperature was raised up to 750 °C and the molten product was poured into the low carbon steel mold. This process produces 90 mm x 125 mm x 10 mm square aluminum matrix composite block billets.
Density of the composites was tested by the Archimedes principle. Theoretical density of composites was calculated using the rule of mixture according to the mass fraction of the SiC particles. Hardness of AMCs product was tested by Rockwell B hardness test based on ASTM E18-11. The density and hardness tests were done randomly at the top, middle and bottom of the composite casted products. Microstructural analysis was performed using SEM (JEOL-JSM 5600LV) on etched HF solution polished specimens. The tensile tests were conducted according to ASTM E 8M-04 standard. Impact tests of composites products were tested using Charpy Impact Test Machine.

3. Results and Discusions

The distribution of density of composites products are shown in **Fig. 1.a** and **Fig.1.b.** It can be pointed that there are no significant differenceses of density at bottom, midlle and top of composite plate. **Fig. 2.a.** and **2.b.** shows hardness distribution of composites products. Based on hardness and density observations, it can be concluded that semi solid stir cast produces the composite with homogen distribution of SiC particle. Stiring at semi solid state causes SiC particles are mechanically entrapped. This condition can help to trap the SiC particles and stop them from settling, thus help to achieve the good wettability [14].

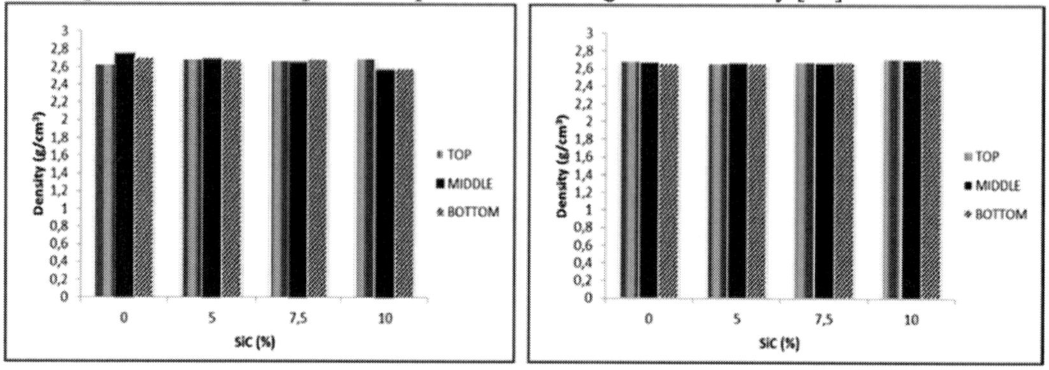

(a) (b)

Fig.1. Density distribution of AMCs reinforced SiC particle. (a) Al7SiSiC (b) Al7Si1MgSiC

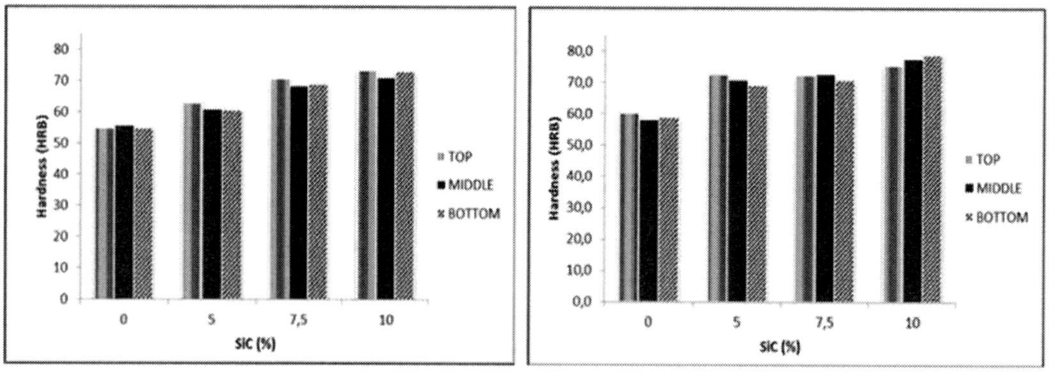

(a) (b)

Fig.2. Hardness distribution of AMCs reinforced SiC particle (a) Al7SiSiC (b) Al7Si1MgSiC

The effect of % wt of SiC on hardness of Al7Si composite and Al7SiMg composites are shown in **Fig. 3**. The hardness of Al7Si composites with 5 %, 7.5 % and 10 % wt SiC are HRB 61, HRB 69 and HRB 72 respectively. The hardness of Al7SiMg composites with 5 %, 7.5 % and 10 % wt SiC are HRB 71, HRB 75 and HRB 80, respectively.The rise of % wt SiC increases the hardness of both Al7Si composite and Al7SiMg composite.

Tensile strength and elongation of composites are shown in **Fig. 4**. It is shown that tensile strength with 5 % wt SiC is higher than aluminum without reinforced particle. The increase in tensile strength of composites is caused by presence of the hard and higher modulus SiC particles embedded in the aluminum matrix. SiC particle acts as a barrier to resist plastic flow when the composite is subjected to strain from an applied load [15]. The rise of % wt from 5 % wt to 10 % wt decreases the tensile strength and elongation of composites. The weakening factors of tensile strength of composite might be responsible for the included particles clusters and porosity [16]. In **Fig. 5**. shows that compare to aluminium monolitic, there isn't significant increasing of porosity of 5% wt SiC composite.

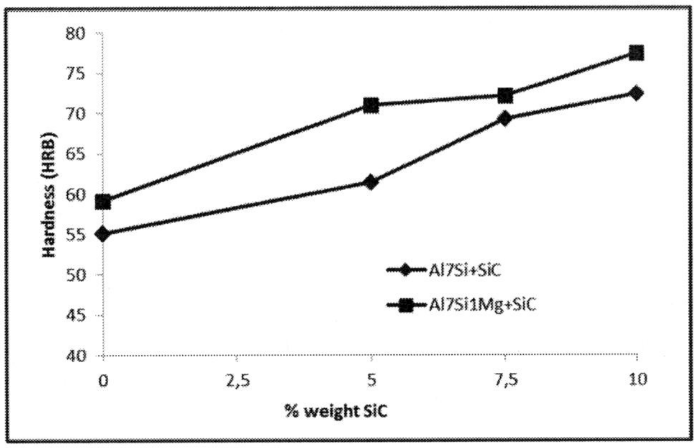

Fig. 3. The hardness of Al7Si and Al7SiMg composites

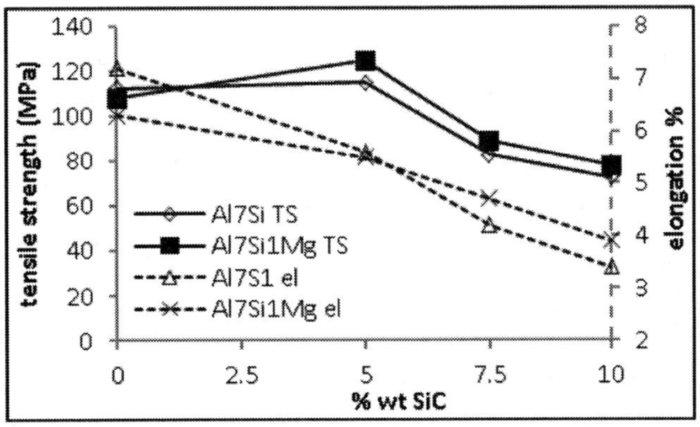

Fig. 4. The tensile strength and elongation of Al7Si and Al7SiMg composites.

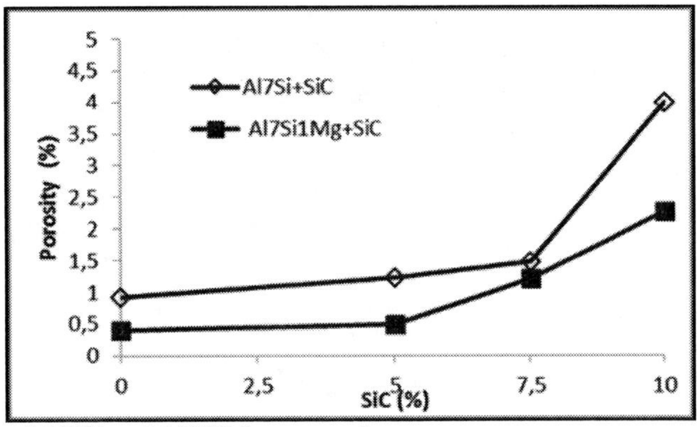

Fig. 5. The porosity of Al7Si and Al7SiMg composites

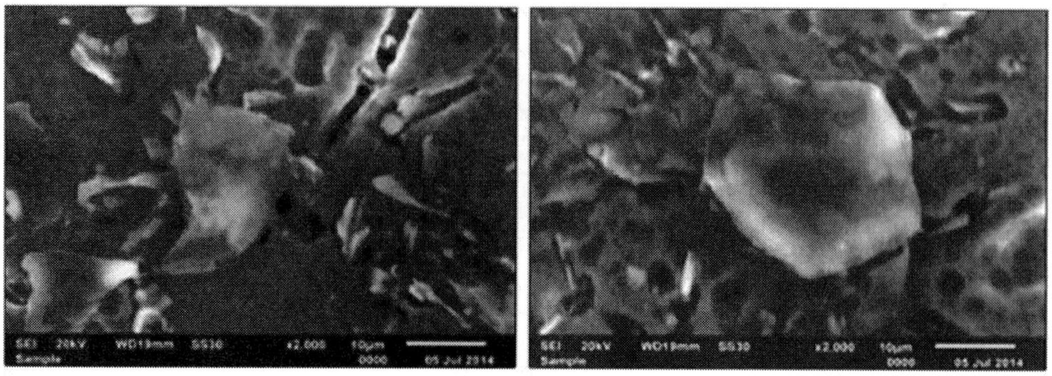

(a) (b)

Fig. 6. (a) Scanning Electron Micrograph Al7Si and SiC particle (b) Scanning Electron Micrograph Al7SiMg and SiC particle

The addition of 1 % Mg increases the hardness and tensile strength of composites. It is observed that the tensile strength of Al7SiMg/SiC is higher than Al7Si/SiC. The increase in the tensile strength of Al7SiMg/SiC can be attributed that Mg increases the bonded SiC reinforcements in Al matrix. Previous authors reported that Mg is known to improve wettablity by reducing the surface tension of the liquid melts [14, 17,18]. **Fig. 5** shows that Mg also decreases the porosity of composites. Mg is also promoting the chemical reaction at the solid liquid interface as shown in **Fig. 6 (a)** and **Fig. 6(b)**. Mg causes less void formation between matrix and particle reinforcements interface which indicates a better matrix–reinforcement interfacial bonding and increasing the impact strength of Al7SiMg/SiC. **Fig. 6.a** shows that some void interfaces formed betwen matrix and SiC particle are found. Porosity causes the stress concentration areas in the composite that leads to the reduce in the tensile and impact strength [18]. The impact strength of composite is shown in **Fig. 7**, and it can be observed that the impact stregth of Al7SiMg/SiC composite is higher than Al7SiMg.

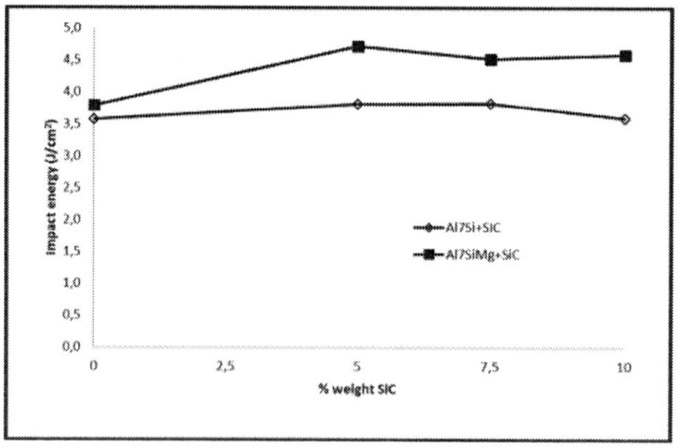

Fig. 7. The impact strength of Al7Si and Al7SiMg composites

4. Conclusions

This study demonstrates the semi solid stir casting technique to enhance the mechanical properties of Al7Si/SiC and Al7SiMg/SiC composites. The following results could be drawn from this study:

1. Semi solid stir casting succesfully produces the homogeneous distribution of SiC particle on aluminum matrix.

2. The addition 1 % Mg decreases the porosity and improves: hardness, tensile strength and impact toughness of the composites.
3. Increasing % wt of SiC from 5 % to 10 % wt will decrease the tensile strength and elongation of composites.

This work was supported by PUPT 2017 research grant (advanced research of higher education 2017), Directorate of research and community service, Ministry of Research, Technology and Higher Education of the Republic of Indonesia.

References

1. N. Chawla, Y.L. Shen, Adv Eng Mater, **3(6)**, (2001)
2. Sulardjaka, Jamasri, M.W. Wildan, Kusnanto, International Journal of Mechanical and Mechanics Engineering, **10(6)**, (2010)
3. Liu Zhang, Hanqing Xu, Zhi Wang, Qinggang Li, Junyan Wu, Journal of Alloys and Compounds, **678** (2016)
4. Lei Wang, Feng Qiu, Qian Zou, De-Long Yang, Jian Tang, Yu-Yang Gao, Qiang Li, Xue Han, Shi-Li Shu, Fang Chang, Qi-Chuan Jiang, Materials Characterization, **131**, (2017)
5. H.G. Prashantha Kumar, M. Anthony Xavior, Procedia Engineering, **174** (2017)
6. S.M.L. Nai, M. Gupta, Compos. Struct., **57**, (2002)
7. S. Kumar, Sai Kiran Reddy, S.V. Joshi, Surface and Coatings Technology, **318**, (2017)
8. J. Hashim, L. Looney, M.S.J. Hashmi, J. Mater. Process. Technol., **123**, (2002)
9. Jufu Jiang, Ying Wang, Materials Science & Engineering **A639**, (2015)
10. Oh S.Y., Cornie J.A., Russell K.C., Metall Trans A, **20**, (1989)
11. M. Mohammadpour, R.Azari Khosroshahi, R.Taherzadeh Mousavian, D. Brabazon, Ceramics International, **40**, (2014)
12. S. Amirkhanlou and B. Niroumand, Trans. Nonferrous Met. Soc. China, **20** (2010)
13. Jufu Jiang, Gang Chen, Ying Wang, Journal of Materials Science & Technology, **32 (11)**, (2016)
14. J. Hashim, L. Looney, M.S.J. Hashmi, Journal of Materials Processing Technology, **119**, (2001)
15. K.K. Alaneme, A.O. Aluko, Scientia Iranica A, **19 (4)**, (2012)
16. Ali Mazahery, Mohsen Ostad Shabani, Trans. Nonferrous Met. Soc. China, **22**, (2012)
17. Shubin Ren, Xinbo He, Xuanhui Qu, Islam S. Humail, Yan Li, Composites Science and Technology, **67 (10)**, (2007)
18. Bin Li, Binghui Luo, Kejian He, Lizhou Zeng, Wenli Fan, Zhenhai Bai, Journal of Alloys and Compounds, **649**, (2015)
19. Sedat Ozden, Recep Ekici, Fehmi Nair, Composites: Part A, **38**, (2007)

Mechanical and microstructural characterization of ST 37 butt joints by friction stir

Kasir[1] and *Sulardjaka*[1*]

[1]Department of Mechanical Engineering, Diponegoro University. Jl. Prof. Sudharto, SH 50275, Semarang, Indonesia

Abstract. The aim of this research is to investigate mechanical and microstructure characteristic of ST 37 Friction Stir Welded (FSW) butt joints. The rolled plate of ST 37 steel was used as the base material. Tungsten carbide (20 mm on diameter) was used as the tool. The FSW process was conducted on a CNC milling machine. FSW joints were produced at 1500 and 2000 rpm of rotational speeds variation and the transverse speed is constan at 40 mm/min. Furthermore, the influence of rotational speed on the microstructure and mechanical properties of the joints were investigated based on parameters mentioned above. The tensile test was done based on JISZ 2241 standard. Microhardness was investigated based on ASTM E 384. Joint microstructure was also analyzed by optic microscopy. As a result, FSW joint at 2000 rpm tool rotation shows higher tensile strength, yield strength and percentage of elongation in compare to 1500 rpm rotaton speed. In addition, its tensile strength is about 5 % lower than Base Material (BM) which its ductility decreases up to 30 %.

1. Introduction

Friction Stir Welding (FSW), one of newly welding processes is unveiled by The Welding Institute (TWI), England at 1991 [1]. It utilizes friction produced by contact between tool and specimens in solid-state. It is might be promising welding process which helds precision joints better than the one encountered in fusion welding [2]. In its development, researchers tend to conduct research in stainless steel [3-5], aluminium [6-7], and steel [8-14]. Its principle is joining two plates which is similar or dissimilar in shape using tool that rotates and passes through along the plate. It performs under conditions which are shoulder is fully contacted and the pin is completely inserted and moves in transverse direction towards the weld. Consequently, the transverse and rotated effects of the pin and shoulder breed heat which softens the specimen in solid-state and welds the plates [15]. Some findings explore microstructure and mechanical properties of dissimilar materials such as joining ST 37 and stainless steel 304 [16]. Due to this, we have intention on delving into further information of mechanical properties and microstructure study of weldments performed in high rotational speed on carbon steel ST 37. The reason why we chose ST 37 because of some considerable advantages such as light in weight and good in formability.

*Corresponding author: sulardjaka@undip.ac.id

2. Experimental procedure

The material used in the research was rolled plate of ST 37 (100x70x30 mm). The shouldered pin tool is cone-pin shaped tool made form Tungsten Carbide as shown in **Fig.1.** FSW processes were done by contacted the rotating tool into the material surfaces with feeding depth 0,02 mm. Then, the rotating tool was moved in transverse direction with velocity 40 mm/min. The FSW products are shown in **Fig.2.** Tensile test of FSW joint was conducted on Shimadzu UH1000 kNI tensile machine based on JIZ 2241 standard test. Microhardness of FSW joining was tested based on ASTM E 384 by Mitutoyo Micro Vickers Hardness Testing Machines HM-210. The microstructure was analyzed by optic microscopy on HNO_3 and CH_3OH etched specimens.

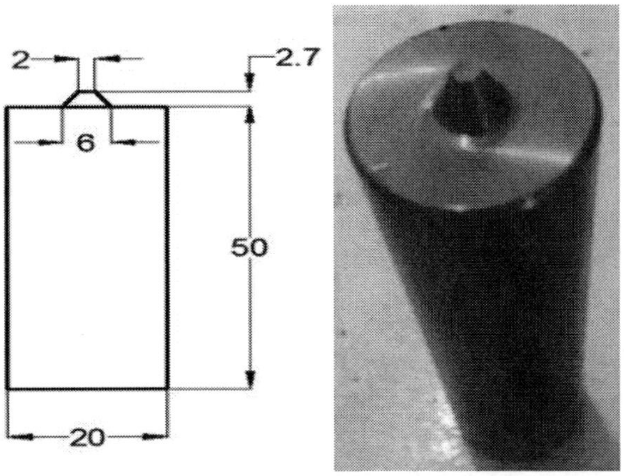

Fig. 1 Dimension of the tool

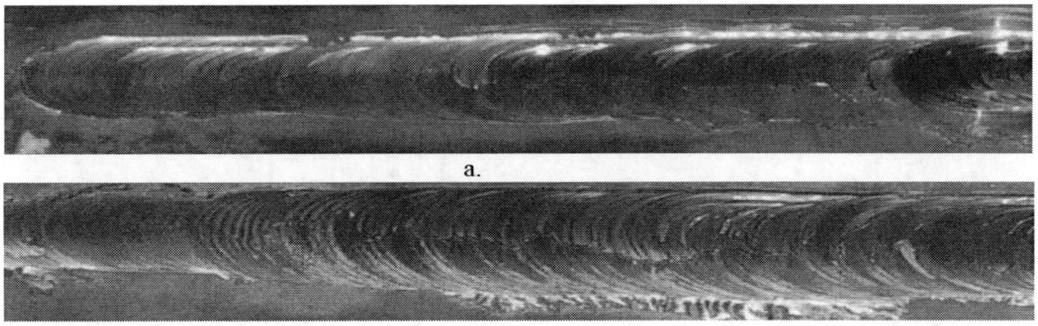

b.

Fig 2. FSW joinning products (a) rotational speed 2000 rpm (b) rotational speed 1500 rpm

3. Result and discussion

3.1. Tensile test

St 37 with a 30 mm thickness were successfully welded at all given welding conditions. The tensile strentgh of raw material, FSW joinning at 2000 rpm and 1500 rpm rotational speed is shown in **Table 1**. Yield strength (YS) and ultimate tensile strength (UTS) of the stirred zone tend to increase with an increasing tool rotating speed. The yield strength of specimens carried out under 2000 rpm reached a peak at 329 MPa which was higher than other specimens were about 315 MPa and 280 MPa respectively. On the other hand, the Ultimate Tensile Strength (UTS) of raw material peaked up at 372 MPa which higher than other performed specimens of different rotational speed which are 350 MPa for 2000 rpm and 331

MPa for 1500 rpm. To sum up, specimen which welded by 2000 rpm has better elongation at value 25% which compares to 1500 rpm at value 10 %.

Table 1. Tensile properties of raw material, FSW 1500 rpm and FSW 2000 rpm

Specimen	YS (MPa)	UTS (MPa)	Elongation (%)
Raw material ST 37	315	372	36
FSW 2000 rpm	329	350	25
FSW 1500 rpm	280	331	10

3.2. Hardness test

Microhardness result was processed in 15 points which were started at a point in the middle of Stir Zone (SZ) and 7 points every 1 mm in range on each side where it was located on Thermo-Mechanical Affected Zone (TMAZ), Heat Affected Zone (HAZ) and Base Metal (BM). It was illustrated in graph that was shown in **Fig. 3.** A SZ with a tool rotating speed of 1500 - 2000 rpm showed a high hardness are around 250 Hv, while the hardness of the BM is around 150 Hv. However, it is found that SZ of 1500 rpm was higher than that 2000 rpm speeds. The SZ points of specimens in variance of rotational speed had highest hardness, they reached peaks at 240.7 HV and 239.0 HV for 1500 rpm and 2000 rpm respectively. Otherwise, the lowest hardness was found in BM.

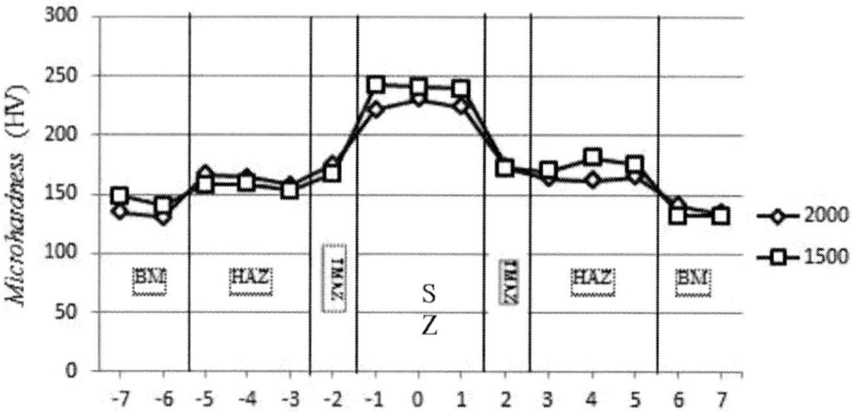

Fig. 3. Hardness distributions of friction stir welded St 37 steel at various tool rotating speeds

3.3. Microstructure study

Optical micrographs of friction stir welded St 37 steels in the transverse cross section are shown in **Fig. 4**. and **Fig. 5**. The microstructure of SZ and WZ defines a change in the shape of ferrite and pearlite grains where the ferrite and pearlite grains are getting smaller than BM.

https://doi.org/10.1051/matecconf/201815902037

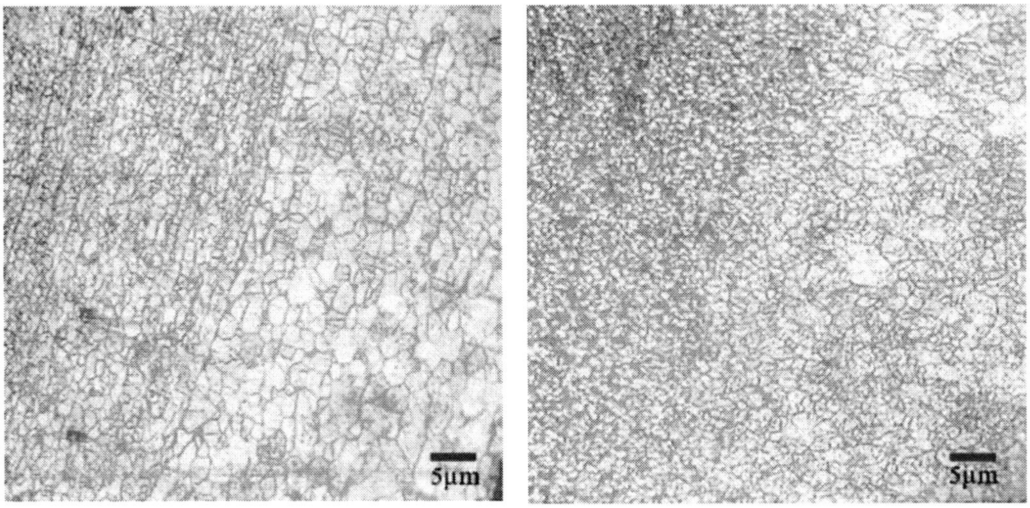

a. b.

Fig. 4. Microstructure images of SZ and HAZ (a) 1500 rpm (b) 2000 rpm

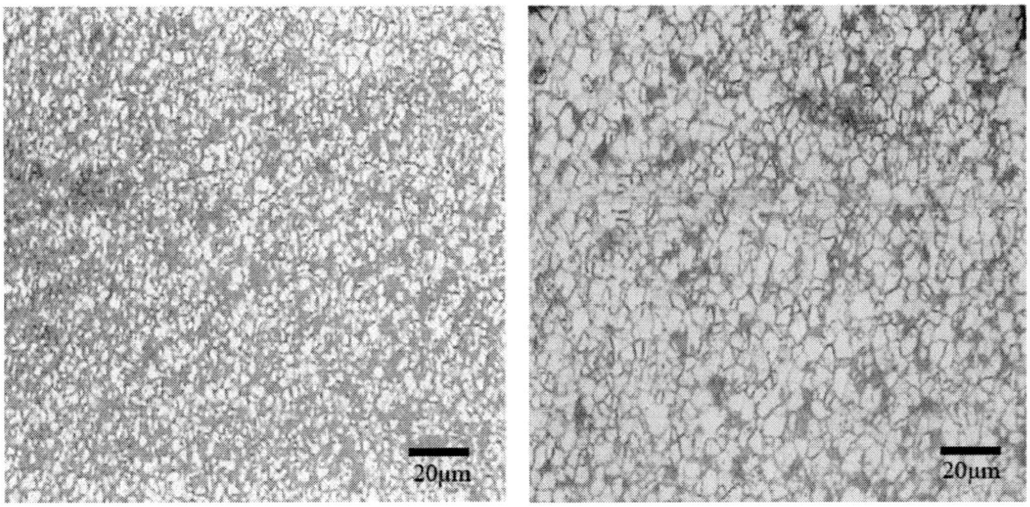

a. b.

Fig. 5. Microstructure images of Weld Zone (a) 1500 rpm (b) 2000 rpm

4. Conclusion

It can be proven that Friction Stir Welding can be performed in Carbon Steel St 37 under parameters: tranverse speed at 40 mm/min and variances of high rotational speed at 2000 rpm and 1500 rpm. FSW with 2000 rpm tool rotation give higher tensile strength than FSW at 1500 rpm rotation. Stir Zones of FSW products have greater value of hardness than base metal.

References

1. T. M. Wayne & Nicholas, E. D., Friction stir welding for the transportation industries, **18**, 269-273 (1997)
2. R. Seerangan, V. Balasubramanian, S. Malarvizhi, A.G. Rao, Def.. Tech., **11**, 308-317 (2015)
3. A.P. Reynolds, W. Tang, T.G. Herold, H. Prask, Scripta Mater, **48**, 1289-1294 (2003)

4. J.C. Lippold, D.J. Kotecki, Welding metallurgy and weldability of stainless steels, (2005)
5. S.H.C. Park, Y.S. Sato, H. Kokawa, K. Okamoto, S. Hirano, M. Inagaki, Sci Tech. Weld. Joint., **10** (2005)
6. S.Y. Tarasov, V.E. Rubtsov, E.A. Kolubaev, Wear, **318**, 130-134 (2014)
7. N. Z. Khan, Z. A. Khan, A. N. Siddiquee, Mater. To. Proceedings **2**, 1450-1457 (2015)
8. V. L. Manugula, K. V. Rajulapati, G. Madhusudhan Reddy, R. Mythili, K. Bhanu Sankara Rao, Met. & Mater. Transc. A., **48**, 3702-3720 (2017)
9. R. Ramesh, I. Dinaharan , Ravi Kumar , E.T. Akinlabi, Mater. Sci. & Eng, **A687**, (2017)
10. A.Pradeep, S. Muthukumaran, Mater & Des., **5**, 25-35 (2013)
11. Y. Liu, D. Dong, L. Wang, X. Chu, P. Wang, M. Jin, Mater & Des., **A627**, 296-305 (2015)
12. L.N. Brewer , M.S. Bennett , B.W. Baker , E.A. Payzant , L.M. Sochalski-Kolbus. Mater. Sci. & Eng., **A647**, 313-321 (2015)
13. A. M. El-Batahgy, T. Miura, R. Ueji, H. Fujii, Mater. Sci. & Eng.,**A651**, 904–913 (2016)
14. J. Rodriguez , A.J. Ramirez, Mater Character, **110**, 126-135 (2015)
15. B. R. Singh, A Hand Book on Friction Stir Welding, (2012)
16. M. Jafarzadegan , A.H. Feng, A. Abdollah-zadeh, T. Saeid , J. Shen, H. Assadi, Mater. Charact., **74**, 23-41 (2012)

CFD analysis on the aerodynamics characteristics of Jakarta-Bandung high speed train

Tony Utomo[1,*], *Berkah* Fajar[1] , and *Hendry* Arpriyanto[2]

[1]Mechanical Engineering Department, Faculty of Engineering, Diponegoro University, Indonesia
[2]College of Mechanical Engineering Department, Diponegoro University, Indonesia

Abstract. In this study, the aerodynamics characteristics of HST type CRH380A which is planned to be operated on the new railway of Jakarta-Bandung are analysed using Computational Fluid Dynamics. The speed of the train in this simulation was varied from 100 km/h to a maximum designed speed of 350 km/h with the increment of 30 km/h. The train was modelled in 3D computational domain with more than 1.7 million cells. The turbulence model employed in this study was standard k-ω. The simulation results show that the drag coefficient (C_D) is slightly decrease by the increase of speed. At the speed of 100 km/h the C_D is 0.216 and decrease to 0.188 at the speed of 350 km/h. The high pressure area is located at the nose of the train. The pressure acting on this location is increase with the increase of the train speed.

1 Introduction

The development of high speed train (HST) in the world is very rapid recently. The aerodynamics characteristic such as drag and lift forces become important in a high speed train, since the flows around train are more disturbed due to turbulence of the increased speed of the train [1, 2, 3]. These aerodynamic performances of a high speed train are strongly influence the energy consumption of the train.

The Indonesian government in cooperation with PT Kereta Api Cepat Indonesia China (PT KCIC) planned to build Jakarta-Bandung railroads special for a high speed train. The total distance is about 142.3 km, starting from Jakarta Halim station up to the station Tegal Luar in Bandung regency. The distance between stations is about 45 km. For this purpose, the consortium has decided to use a CRH380A high speed train. The CRH380A high speed train is able to reach a maximum operating speed of 350 km/h after a distance of 39.01 km [4]. Such speed can possibly be reached on Jakarta-Bandung railroads. The aerodynamic phenomena with regard to high speed train are strongly dependent on the train speed. Thus, the aerodynamic problems become more important as the train speed increase.

In general, the aerodynamic problems are related to the aerodynamic characteristic of a high speed train, and the two of them are aerodynamic drag and lift. The aerodynamic drag and lift are dependent on the geometric parameter of the train, such as cross-sectional area

* Corresponding author: msktonysu@yahoo.co.id

of train body, train length, shape of train fore- and after-bodies, etc [5]. Hence, the aerodynamics characteristics and flow phenomena around high speed train are necessary to understand. In this study, the influence of train speed on aerodynamics characteristics and flow phenomena around a high speed train are analyzed using computational fluid dynamics method.

2 Computational Method

In this study, the computational method is used to solve the three governing equations, which are mass conservation, momentum equation and k-ω standard turbulent model. The method consists of modeling high speed train geometry, generating mesh for computational domain, applying boundary conditions and initial conditions for simulation, and numerical iteration to solve the governing equations.

The train model was created using CAD software as shown in Figure 1. This model is a result of the simplification of the actual CRH380A geometry.

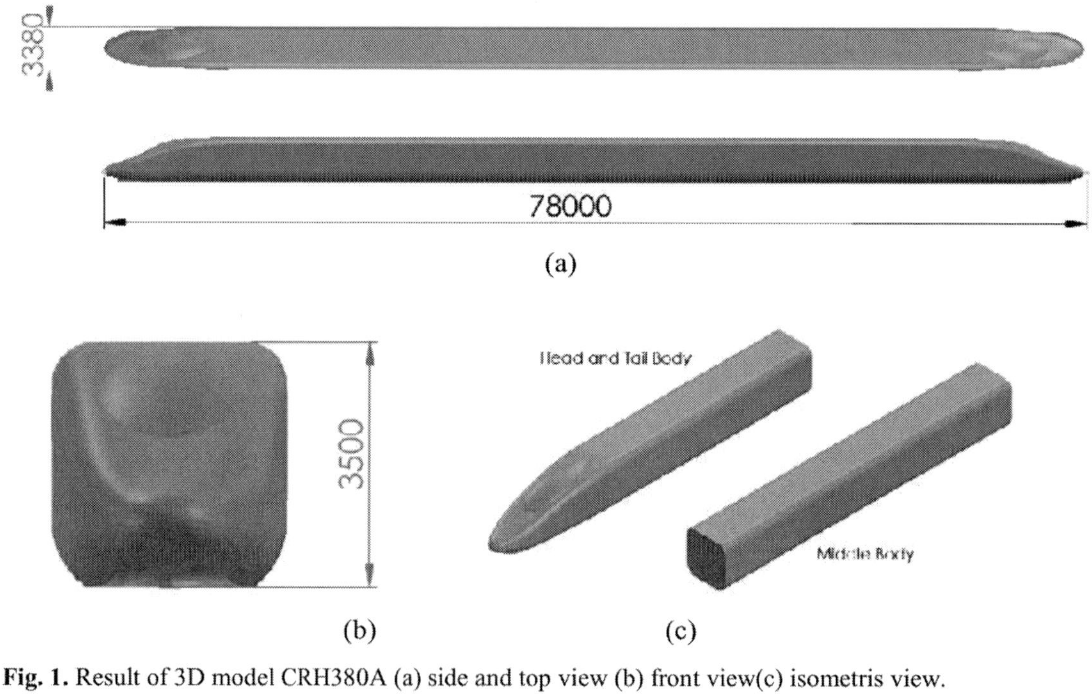

Fig. 1. Result of 3D model CRH380A (a) side and top view (b) front view(c) isometris view.

The computational domain of high speed train in this simulation is shown in Figure 2. In this figure, L is stand for the total length of the train. Inflow has a distance L upstream of the train nose, while the outflow has a distance 2L from the back-end of the train. The width of the computational domain is set to L, while it's height is set as 0,5L. The space between the base of the train to the ground is 0,00235L.

Mesh was divided into three main parts that had different dimension for each mesh. The first part was the surface of CRH380A with an element size of 100 mm , the second part was car box with an element size of 375 mm , while the last part was the outer covering of car box with default meshing size. The result of meshing can be seen in Figure 3. The number element is 1.730.605 cells with the maximum skewness of 0.74 (good quality mesh

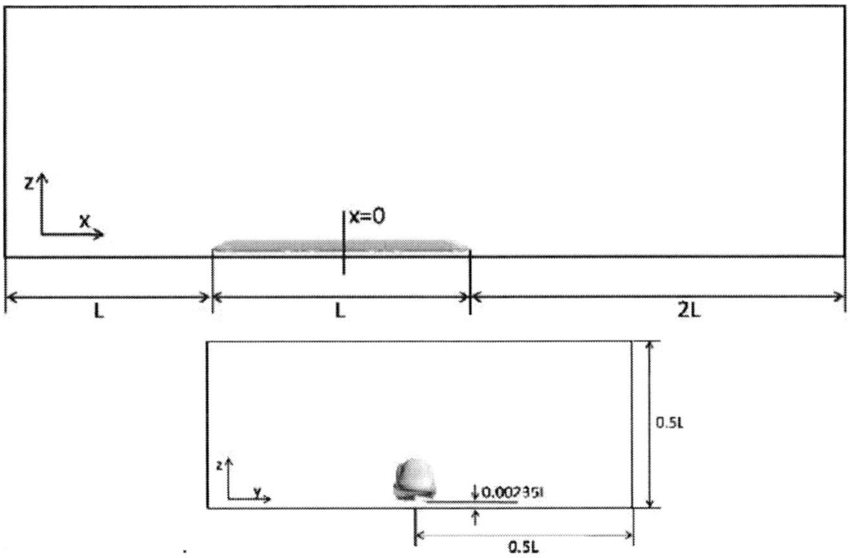

Fig.2. Computational domain

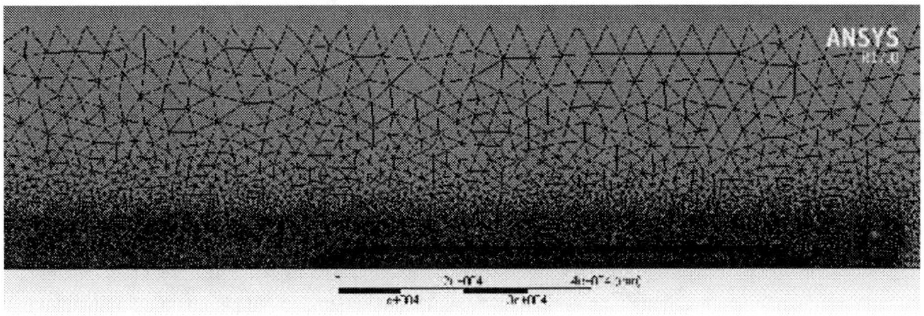

Fig. 3. Meshing of computational domain.

The boundary condition of computational domain is illustrated in Figure 4. The type of boundary condition of surface "a" is set as velocity inlet, surface "b" is set as symmetry, surface "c" is no-slip stationary wall condition, surface "d" is set as pressure outlet, surface "e" is stationary wall, and surface "f" is set as top wall and side wall with symmetry boundary condition type. The simulation was conducted by varying the speed from 100 km/h up to 350 km/h with increments of 30 km/h.

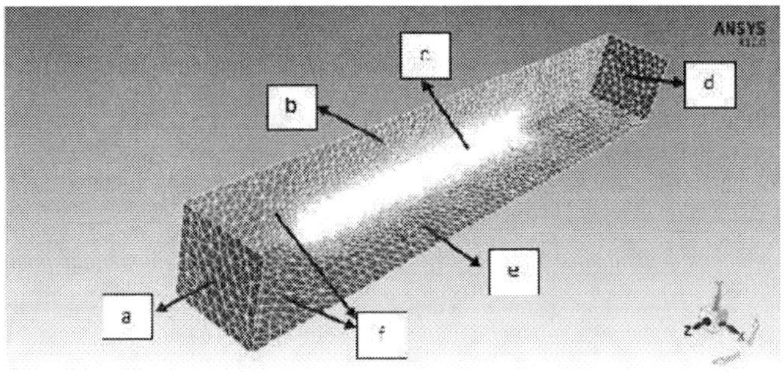

Figure 4. Boundary condition of computational domain.

3 Results and Discussion

Table 1 shows the aerodynamic forces obtained from this simulation. The drag coefficient (C_D) is slightly decrease by the increase of speed. At the speed of 100 km/h the C_D is 0.216 and decrease to 0.188 at the speed of 350 km/h.

Table 1. Drag and Lift of high speed train type CRH380A at various speed.

No	Velocity km/h V	Reynold Number Re	Drag Coefficient C_D	Pressure Drag Coefficient $C_{D\ pressure}$	Viscous Drag Coefficient $C_{D\ Viscous}$	DragForce kN F_D	LiftForce kN F_L
1	100	6.66×10^6	0.216	0.077	0.139	0.584	-0.084
2	130	8.65×10^6	0.210	0.075	0.135	0.958	-0.162
3	160	1.06×10^7	0.204	0.073	0.131	1.415	-0.274
4	190	1.26×10^7	0.201	0.072	0.129	1.956	-0.418
5	220	1.46×10^7	0.197	0.071	0.126	2.579	-0.597
6	250	1.66×10^7	0.194	0.070	0.124	3.283	-0.809
7	280	1.86×10^7	0.192	0.069	0.123	4.067	-1.057
8	310	2.06×10^7	0.190	0.069	0.121	4.930	-1.341
9	350	2.33×10^7	0.188	0.068	0.119	6.202	-1.779

It can be seen that the drag force is increasing proportionally to the increase of train speed. The maximum drag force (6.202 kN) is obtained when the train traveling at it's maximum speed of 350 km/h. The lift force acting on the train is negative. This negative value means a down force. The maximum lift force acting at maximum train speed is -1.779 kN.

Figure 5 shows the contour of static pressure around the HST body at the train speed of 100 km/h, 190 km/h, 220 km/h and 350 km/h. It can be seen that the high pressure area is located at the nose of the train. The pressure acting on this location is increase with the increase of the train speed. At low speeds, the intensity of high pressure located around the nose train is not visible. At the higher train speed, the area of high pressure intensity becomes visible. When the train moves at it's maximum operating speed, the high pressure area is observable at front-end of the train nose and at the obstacle deflector located under the train nose. The maximum pressure on the nose and the obstacle deflector is 5.7 kPa. At maximum speed, the pressure drag is high and so is the drag force. An aerodynamically good design of a high speed train can reduce the pressure drag at the front of the train to reduce total drag force for efficient fuel consumption.

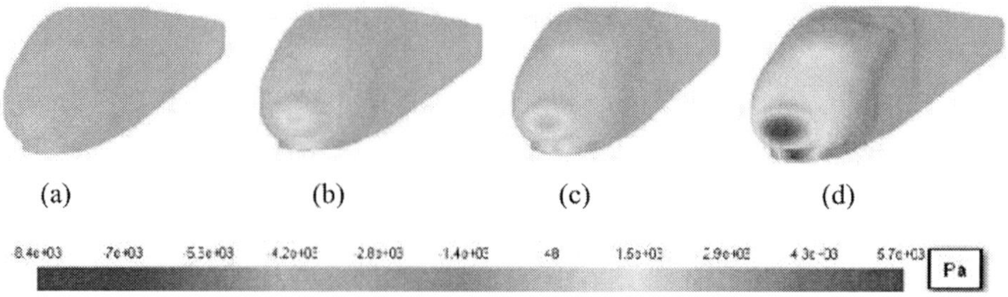

Fig. 5. Static pressure contour around CRH380A body (a)V=100km/h (b)V=190 km/h (c)V=220 km/h (d)V350 km/h.

Figure 6 shows the velocity vector around the back-end of the train at the speed of 100 km/h, 190 km/h, 220 km/h and 350 km/h. Vortex occurs at the speed of 100 km/h and 190 km/h, while it disappear at the speed of 220 km/h and 350 km/h. The disappearance of vortex at higher train speed is affected by the domination of the inertial forces following the high speed of flow around the train body. Large inertial forces drive the fluids flow sequentially on the geometry of the body profile of the train.

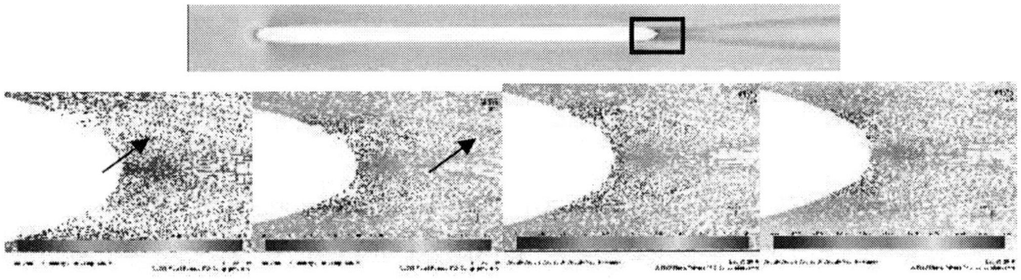

(a) (b) (c) (d)

Fig. 6. Velocity vector around the end of train body on a plane of 0.8 m high from the ground at various speed (a) V=100 km/h (b) V=190 km/h (c) V=220 km/h (d) V=350 km/h.

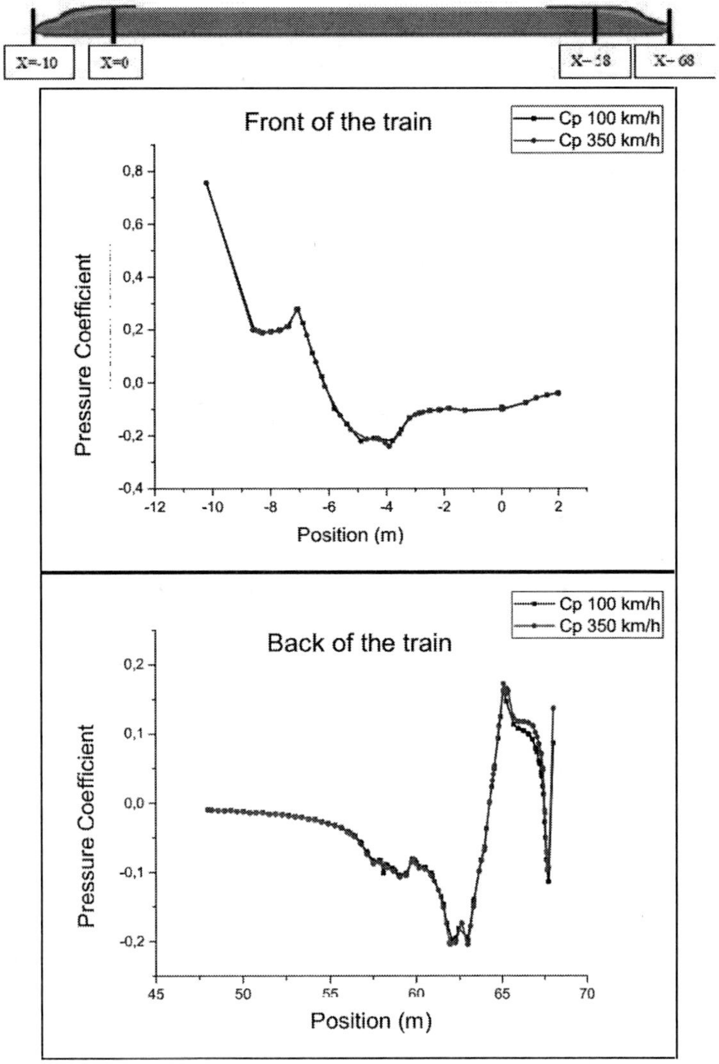

Fig. 7. Pressure coefficient on train geometries profile at traveling speed of 100 km/h and 350 km/h.

Figure 7 shows pressure coefficient on the front and back of the train at speed of 100 km/h and 350 km/h. It can be observed that the pressure coefficient on the front of the train at the speed of 100 km/h is almost the same with that of 350km/h. The highest pressure coefficient is at stagnation point on the front of the train. The pressure coefficient then decrease and at some point increase again caused by the blunt geometries between train and glass body wheelhouse. After passing through the glass rudder, the pressure coefficient tends to have a lower than that at the nose of the train. While at the back of the train, there is different pressure coefficient profile between 100 km/h and 350 km/h of train speed. At the speed of 100 km/h, the pressure coefficient on the back of the train body is lower than the pressure coefficient at the train speed of 350 km/h. The low pressure in the rear of the train makes a large pressure difference between the front and rear of the train, hence the pressure drag coefficient is higher. Therefore, at higher speeds, the pressure drag coefficient is lower than the that at lower speeds.

4 Conclusion

The aerodynamic characteristics of Jakarta-Bandung high speed train has been analyzed using CFD. The results show that the drag force is increase proportionally with the train speed. On the other side, the drag coefficient (C_D) is slightly decrease by the increase of train speed. A negative lift force occurs at all speed, means the high speed train is experiencing a down force in all travelling speed from 100 km/h to 350 km/h. This negative lift force or the down force can be referred that the high speed train travel in safety and the risk of lifted out accident caused by the aerodynamic forces can be reduced or eliminated.

The total drag coefficient consists of pressure drag coefficient and viscous drag coefficient. In the case of a high speed train, the influence of viscous drag coefficient is more dominant than pressure drag coefficient due to the configuration of a very long body of the train compare to its height or width. Both coefficients are decrease versa the increase of speed.

The high pressure area is located at the nose of the train. The pressure acting on this location is increase proportionally to the train speed. At its maximum speed, the pressure drag is high and so is the drag force acting on the train.

At a lower traveling speed, the pressure coefficient on the back of the train body is lower that at a higher traveling speed. The low pressure in the rear of the train makes a large pressure difference between the front and rear of the train, hence the pressure drag coefficient is higher. Therefore, at higher speeds, the pressure drag coefficient is lower than the that at lower speeds.

References

1. C. Baker, J. Wind. Eng. Aerodyn., **98**, 277-298 (2010)
2. R.S. Raghunatan, H. D. Kim, T. Setoguchi, Prog. in Aerospace Sciences, **38**, 469-514 (2002)
3. M. Suzuki, K. Tanemoto, T. Maeda, J. Wind. Eng. Aerodyn., **91**, 209-218 (2003)
4. Y. Shuanbao, G. Dilong, S. Zhenxu, Y. Guowei, C. Dawei, Comp. and Fluids, **95**, 56-73 (2014)
5. J. Paniagua, M. Garcia, A. Crespo, J. Wind. Eng. Aerodyn., **130,** 48-61(2014)

Design and Simulation of Fuzzy Logic Based Temperature Control for a Mixing Process in Therapeutic Pool

Munadi[1,*], *Erwin* F. Ramadhan[1], *Mochammad* Ariyanto[1], *Norman* Iskandar[1], M. *Amirullah* A.[2], *Hari* P. Julianti[3], *Rudy* Setiawan[4]

[1]Dept. of Mechanical Engineering, Diponegoro University, Tembalang, Semarang, 50275, Indonesia
[2]Garuda Maintenance Facility AeroAsia, Soekarno Hatta International Airport, Cengkareng, Indonesia
[3]Dept. of Public Health and Preventive Medicine, Diponegoro University, Semarang, Indonesia
[4]Academy of Marine Engineering-Veteran, Gayamsari-Semarang, Indonesia

Abstract. Stroke is one cause of the most deaths, in which one way of handling stroke can be done with hydrotherapy. Hydrotherapy is a therapy that is done in the therapeutic pool which is filled with warm water in temperature 31°-34° C. When in the warm water, body burden becomes lighter, so the stroke patients with impaired movement of the body can train the nerves and muscles in order to restore the function of these organs. However, the warm water that is used for the therapy is produced by the heater that requires much electrical energy. This research discussed how to use solar energy to produce hot water using Indirect Evacuated Tube Solar Water Heater systems and make the control system for controlling the mixing process between the hot and cold water by using the Fuzzy Logic Control of the Mamdani method. Therapeutic pool system will be modeled using Simulink. Modeling the mixing process is made on the basis of mass balance and energy balance equations. Then the system will be simulated with interrupted condition and gets set value changes in a particular time. The FLC is successfully reached and maintained steady to a set value. The results will be compared to the performance of the one with Proportional Integral Derivative controller.

1 Introduction

Stroke is the leading cause of death in Indonesia. In 2014, according to the Sample Registration Survey (SRS) conducted by Health Research and Development Agency, Ministry of Health shows, that stroke and heart attack becomes the first and second causes of death in Indonesia, with a percentage of 21.1% and 12.9% of 41,590 deaths nationwide throughout 2014. However, at this time the stroke treatment can be done in several ways, one of them is hydrotherapy. Hydrotherapy is an exercise therapy using warm water in the hydrotherapy pool (therapeutic pool). This therapy is one method of healing therapy against a stiff joint disease caused by stroke. The water used has a temperature between 31°-34° C, the temperature is quiet safe and provide a relaxing effect, can expedite the blood circulation, decrease pain and improve the ability of organ motion. The hot water used in

* Corresponding author: munady096@gmail.com

therapuetic pool is produced by Indirect Evacuated Tube Solar Water Heater (IETSWH) system. This system utilizes solar energy, thus it can save the electrical energy. But there is a problem, the existing therapeutic pools do not yet have the ability to control the temperature to be stable at temperatures therapy. Therefore, this study will design the control system that can control the temperature of the pool when the mixing process between hot and cold water. The control system is made using the Fuzzy Logic Control (FLC) with the help of software MATLAB/ Simulink, then the control is simulated and compared with the Proportional Integral Derivative (PID) controller. The result in [1] shows that FLC has better performance than the PID controller. We have also applied the DC motor because it is the most widely applied actuator in automation systems such as in the control position of an arm robot manipulator [2-4].

2 Research Method

2.1 Indirect Evacuated Tube Solar Water Heater System (IETSWH)

The IETSWH is a heater system which produces hot water using thermosyphon method. For the work principle of IETSWH, first the working fluid is heated by solar energy in evacuated tube, while the water will flow in the coiled tube. After the working fluid heated, then heat exchange will happen between the working fluid and the water through coiled tube as heat exchanger. The coiled tube in this research uses the stainless steel 304. By using this system, the output of hot water is more stable because it is operated by water pressure [5]. The reason for using stainless steel 304 in the coiled tube is manufacturing process. Because in the first experiment, we used copper material, but when the rolling process the coiled tube became torn. There are three processes that occur in the system of IETSWH is the collecting, circulating and mixing process. The scheme of these processes is shown in Figure 1 (a), and Figure 1 (b) for 3D.

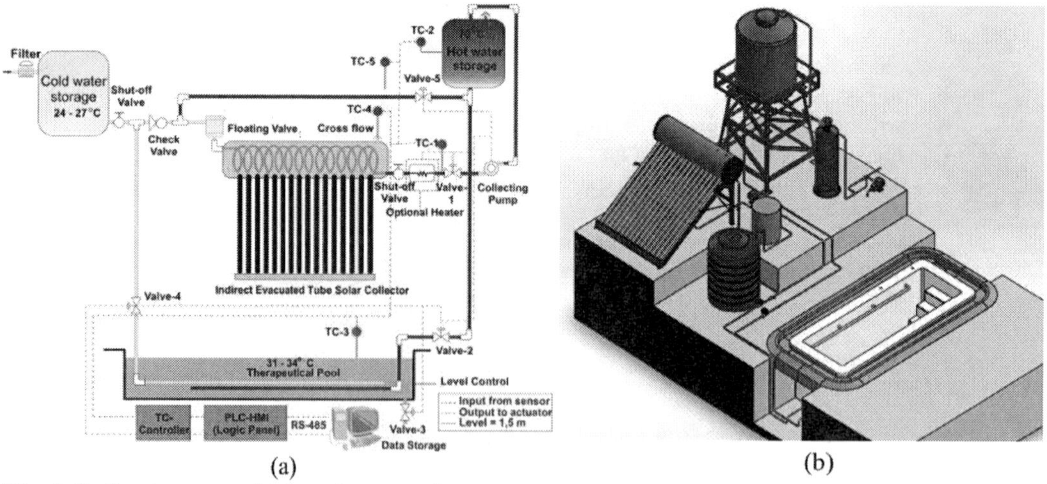

(a) (b)

Fig. 1. Indirect evacuated tube solar water heater system.

Collecting is the process to collect the hot water and store it into the hot water storage. The collecting process begins with the cold water flow from Cold Water Storage (CWS) toward IETSWH and therapeutic pool. In the IETSWH, the cold water heated by working fluids (50% ethylene glycol + 50% water, this refers to [6, 7]) until its temperature reach certain value. After the thermocouple (TC) sensor detects the hot water reaching the setting value, then the solenoid valve opens and the water pump works to make the hot water flow toward Hot Water Storage (HWS). The HWS will save the temperature of hot water.

Circulating is the process to heat the water in hot water storage when the temperature of hot water drops under the setting value. It makes solenoid valve opened, and the hot water flows from HWS to IETSWH for reheating. In the heating process, the system is equipped with a heater, which works when the sun is dim or cloudy weather, so the heater is optional equipment that it can be turned on manually. The mixing process occurs in the therapeutic pool, it begins with cold water from CWS and hot water from HWS flow toward therapeutic pool through motorized valves and then these fluid mix to get temperature 31-34° C. There are several advantages of hydrotherapy systems that we use compared to existing ones, such as cheaper and more efficient electric power because utilizing solar energy to produce hot water. This study focused on the mixing process in the therapeutic pool. The temperature of therapeutic pool will set on 32° C. The position valve of hot water flow will be controlled, but the position valve of cold water flow is kept constant at fully opened.

2.2 Mathematical Model

To build mathematical model, the first thing to do is to make schematic diagram of therapeutic pool as shown in Figure 2. The differential equations is derived from Thermodynamics principle and mass balance equation. The empirical equation of convective heat transfer coefficient can be found in [8]. The therapeutic pool walls are assumed well insulated. The other assumptions to be made are constant properties of water such as specific heat and mass density.

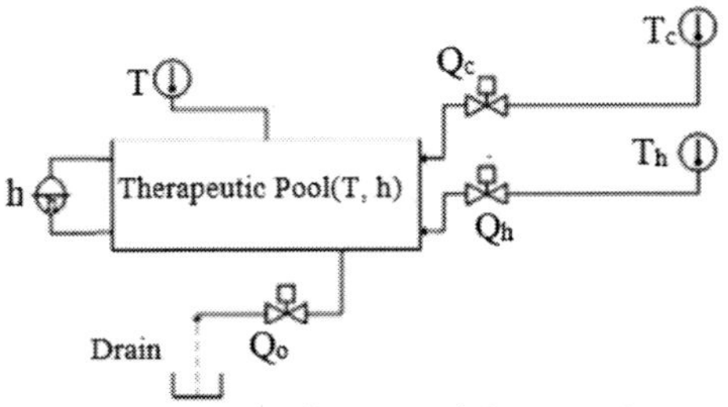

Fig. 2. Schematic diagram of therapeutic pool.

To build mathematical model of therapeutic pool, Thermodynamics principle and and mass balance equation, Equation (1), derived. The equations are written as follows.

$$\frac{dV}{dt} = Q_h + Q_c - Q_o \, ; \; \frac{dE}{dt} = \rho c_p Q_c T_c + \rho c_p Q_h T_h - \rho c_p Q_o T_o - hA(T - T_\infty) \tag{1}$$

To obtain the equation of water that changes over time, then the Equation (1) of thermodynamics principle is derived as Equation (2),

$$\frac{d(AH)}{dt} = Q_h + Q_c - Q_o \text{, then } \frac{dH}{dt} = \frac{1}{A}(Q_h + Q_c - Q_o) \tag{2}$$

Equation (2) shows the water level changes the inlet and outlet water flow. We can derive the temperature rate of change from Equation (1) of mass balance equation, as shown,

$$\frac{d(\rho AHc_p T)}{dt} = \rho c_p Q_c T_c + \rho c_p Q_h T_h - \rho c_p Q_o T - hA(T - T_\infty) \tag{3}$$

Assume that ρ and c_p have constant value, so we can get Equation (4), as shown,

$$\frac{d(AHT)}{dt} = Q_c T_c + Q_h T_h - Q_o T - \frac{hA}{\rho c_p}(T - T_\infty) \tag{4}$$

T and H are the time-varying-variables, thus

$$AT\frac{dH}{dt} + AH\frac{dT}{dt} = Q_c T_c + Q_h T_h - Q_o T - \frac{hA}{\rho c_p}(T - T_\infty) \tag{5}$$

Substitute Equation (2) to above equation, then we get Equation (7)

$$AH\frac{dT}{dt} = Q_c(T_c - T) + Q_h(T_h - T) - \frac{hA}{\rho c_p}(T - T_\infty) \tag{6}$$

$$\frac{dT}{dt}=\frac{1}{AH}\left[Q_c(T_c-T)+Q_h(T_h-T)-\frac{hA}{\rho c_p}(T-T_\infty)\right] \qquad (7)$$

Equation (7) shows the relationship between the temperatures changes with the energy carried with the inlet water. In order to get the convective heat transfer coefficient, we assume the heat transfer over flat plate, written as in the following equations [6].

$$h=\frac{k}{A/P}N_u \;\; ; \;\; N_U=Ra_L{}^{1/4} \; ; \; Ra_L=\frac{g\beta(T-T_\infty)(A/P)^3}{\nu^2}P_r \qquad (8)$$

In this simulation, the values of parameters for the therapeutic pool are defined in Table 1.

Table 1. Therapeutic pool parameters and value.

Parameter	Value	Parameter	Value
Hot water temperature	$T_h = 70^o$ C	Nusselt number	Nu
Cold water temperature	$T_c = 27^o$ C	Rayleigh number	Ra_L
Environment temperature	$T_\infty = 31^o$ C	Prandtl number	P_r
Cold water flow rate	$Q_c = 0.0008$ m^3/s	Gravitational acc.	g = 9.81 m/s^2
Hot water flow rate	$Q_h = 0.0008$ m^3/s	Volumetric expansion coeff	$\beta = 1/^0$C
Drainage flow rate	$Q_o = 0.0008$ m^3/s		
Bottom area	A = 3 m^2	Kinematic viscosity	v, m^2/s
Pool perimeter	A = 3 m^2	Pool Temperature	T, ^{0}C
Air thermal conductivity	k, W/m^0 C	Pool Height	H, m

3 Design of Fuzzy Logic Control

First thing in the design of Fuzzy Logic Control (FLC), it is necessary to determine how the number of inputs and outputs to be used [9]. Next, the system will be designed to use one input of a temperature error and the output of the valve opening value. The FLC has decision-making mechanisms, among others.

3.1 Fuzzification

Fuzzification is a process of change crisp values into fuzzy values, here in the form of fuzzy value is linguistic form. Fuzzification process depends on the design of membership functions created. In this study of input membership function is designed with five members, namely NL (Negative Large), NM (Negative Medium), Z (Zero), PM (Positive Medium) and PL (Positive Large). The fifth member is designed to use members of the triangle. Table 2 shows the input parameters of membership function. So that the membership function input can be seen in Figure 3.

Table 2. Membership function input

Error (Input)	Parameter
NL (Negative Large)	(-inf) - (-0.1)
NM (Negative Medium)	(-0.2) – 0
Z (Zero)	(-0.02) – 0.02
PM Positive Medium)	0 – 0.2
PL (Positive Large)	0.1 – inf

On the output membership function is determined limits the desired output value is 0 to 1, because the output generated is used to adjust the valve opening. Membership output function is designed to use the type of bell member and a triangle. In the ouput membership design, there is MOpen which has the meaning the value of valve position between open and half open. While the MShut means between closed and half open. They are shown in Table 3 and the results are shown in Figure 4.

Table 3. Membership function output

Error (Input)	Parameter
Shut	(-inf) - (-0.9)
MShut	[0.25 2.5 -0.5]
Modulating	[0.25 2.5 0]
MOpen	[0.25 2.5 0.5]
Open	0.9 - inf

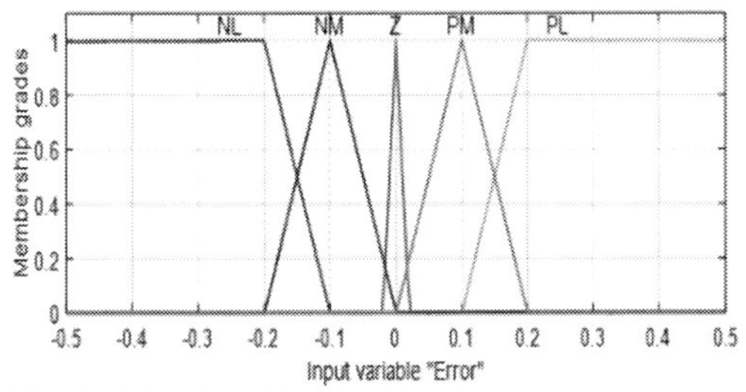

Fig. 3. Membership function input.

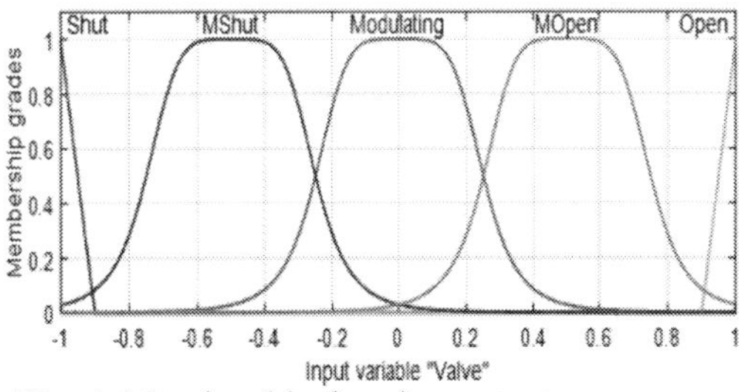

Fig. 4. Membership function output.

3.2 Inference Fuzzy

After the input value fuzzified, then FLC proceeds to the next phase of "decision-making" to bring the temperature according to the temperature set value. In the process of decision making, fuzzy rules are very important because it is the decision-making based on the logic set forth in fuzzy rules. Logic is in fuzzy logic rule is "IF ... THEN ...". IF rule describes the circumstances that have been designed. While THEN elicits responses that will occur in that state [10]. When designing fuzzy rules are very flexible, this meant the rule-making depends on the ability and understanding of the designer for the system [11]. In this study, the fuzzy rules designed consists of five rules. The fuzzy rules in this study are shown as follow: Rule No 1, IF Error is NL, THEN Valve is Shut; Rule No 2, IF Error is NM, THEN Valve is MShut; Rule No 3, IF Error is Z, THEN Valve is Modulating; Rule No 4, IF Error is PM, THEN Valve is MOpen; Rule No 5, IF Error is PL, THEN Valve is Open.

3.3 Defuzzification

Defuzzification is the process of translating the results of the decision-making output signal in the form of linguistic forms into crisp values. Translating it back into a crisp shape is because the hardware can only read signals in the form of crisp signals and can not read the form linguistic input signal. The defuzzification method used is the centroid method.

4 Simulation Results

Simulation results show the FLC performance when controlling the system. To make sure the ability of the FLC, the output response of both the FLC and PID were compared for main parameters and changed parameters as disturbance. The changed parameters would be done to the inputs, cold and hot water temperature. The changed parameters as disturbance cold and hot water temperature to the systems as shown in Figure 5 and Figure 6 respectively. The response systems of the FLC and PID with changed parameters is shown in Figure 7.

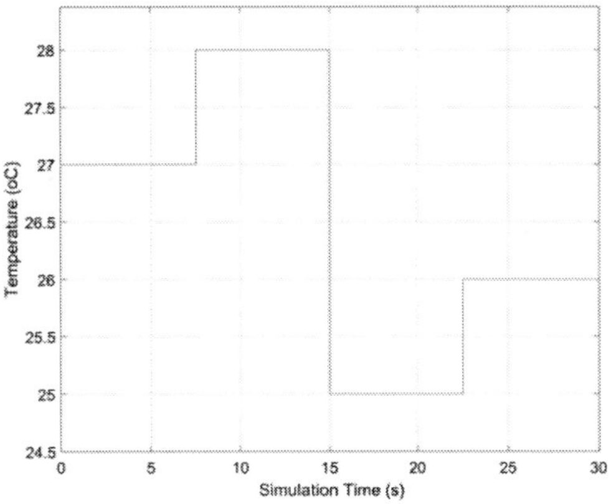

Fig. 5. Cold water temperature as disturbance

The comparison results show that the rise time is 0.23 s for PID and 0.12 s for FLC; overshoot is 3.82 % for PID and there is no overshoot for FLC; settling time is 4.51 for PID and 1.15 for FLC; steady state error is 0.004 s for PID and 0.034 for FLC.

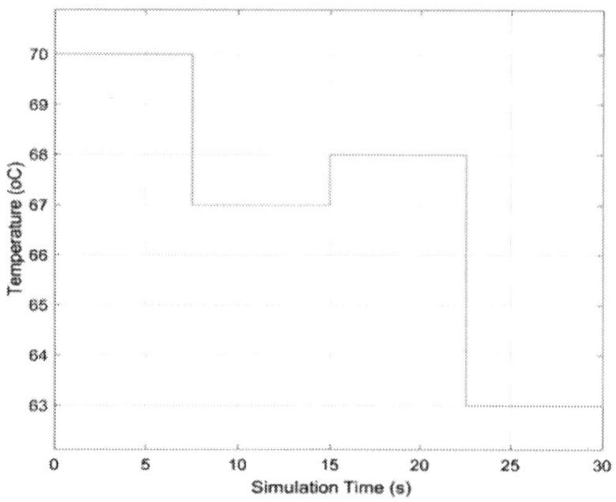

Fig. 6. Hot water temperature as disturbance.

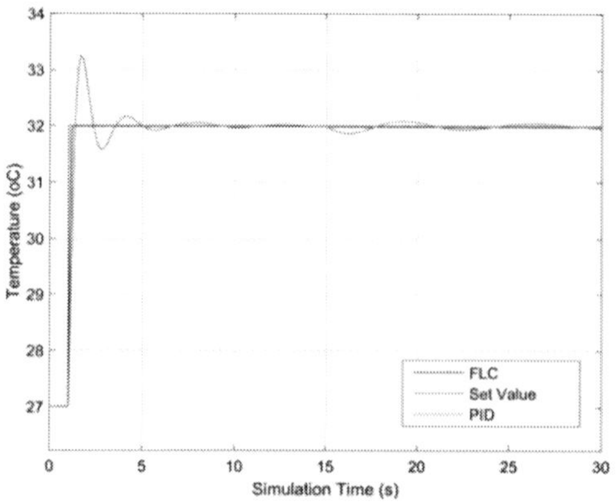

Fig. 7. The response systems of FLC and PID with changed parameters.

5 Conclusion

Simulation results show that Fuzzy Logic Control method has better performance than the conventional PID. The FLC has better adaptation than PID for varying disturbance due to cold and hot water temperature variation. For future work, it will be valuable to implement the FLC to the system, associated with a microprocessor capable of supporting and installation of the actuator.

References

1. A. Zeghoudi, A. Chermitti, A Comparison between a Fuzzy and PID Controller for Universal Motor, International Journal of Computer Applications, **104** (6), 32-36 (2014).
2. Munadi, T. Naniwa, Experimental Verification of Adaptive Dominant Type Hybrid Adaptive and Learning Controller for Trajectory Tracking of Robot Manipulators, Journal of Robotics and Mechatronics, **25** (4), 737-747 (2013).
3. Munadi, M. Amirullah A., Simulation of Fuzzy Logic Control for DC Servo Motor using Arduino Based on MATLAB/Simulink, Intelligent Autonomous Agents, Networks and Systems (INAGENTSYS), 2014 IEEE International Conference, 42-46 (2014).
4. S. Emilian O., M. Dulau, Design and Simulation of Fuzzy Logic Based Temprature Control for a Plasma Nitriding Process, Procedia Technology, **19**, 569-575 (2015).
5. Solartubes, Compact Solar Water Heater, Accessed on Mei 3rd 2016, www.solartubs.com (2016).
6. R. Siegfried, D. Mayer, Ethylene Glycol, Ullmann's Encyclopedia of Industrial Chemistry, **12**, 532 (2005).
7. Engineering Toolbox, Ethylene Glycol Heat-Transfer Fluid, Accessed on May 4th, www.engineeringtoolbox.com (2016).
8. Y. A. Cengel, Heat Transfer: A Practical Approach, 2nd ed, Mcgraw-Hill, 463-469 (2002).
9. S. F. Rezeka, A. H. Attia, A. M. Saleh, Management of Air-Conditioning Systems in Residential Buildings by using Fuzzy Logic, Alexandria Engineering Journal, **54**, 91–98 (2015).
10. J. Mendel, Fuzzy Logic Systems for Engineering: A Tutorial, Proceedings of the IEEE, **83**, 345–377 (1995).
11. B. LaMeres J., N. Nehrir H., Controlling the Average Residential Electric Water Heater Power Demand using Fuzzy Logic, Electric Power Systems Research, **52**, 267–271 (1999).

A Great Achievement of Calculated and Experimental Results of Char Kinetic Rate in Woody Mahogany Pyrolysis

Widya Wijayanti*, *Mega* Nur Sasongko

Department of Mechanical Engineering, Brawijaya University, 65145, Indonesia

Abstract. In the pyrolysis process, it is needed to know the value of k (rate costant) to know the rate of reaction that occurs. It is needed to calculate k caused by T effect due to the reason that it is extremely difficult experimentally to find rate constants. The pyrolysis process was carried out for three hours with the minimum condition of Oxygen using mahogany wood. It was used the pyrolysis temperature variations of 250°C to 800°C. The results showed the increasing temperature and decreasing mass of char as long as the process to obtain, the kinetic rate of char k. We used two calculation approaches method to find the calc ulated k. Fristly, it used the total kinetic rate model equation with the equation $k_T = 60.3\ e^{-2028.4/T}$ and the second one, it was used the local kinetic rate model. In the local kinetic rate approached, it was divided the 3 temperature ranges; k_1 in the range of 28°C - 250°C was $k_1 = 539.1\ e^{-2907.3/T}$, k_2 in the range of 250°C to 500°C was $k_2 = 27.11\ e^{-1424/T}$; and k_3 in the range of at range 500°C to 800°C was $k_3 = 572.4\ e^{-4549.2/T}$. Then, by using the found k values, it was found the great achievement results that the decreasing mass of char in the calculation was coincide with the experimental results. It was used both kinetic rate methods depending on the required energy was needed in the pyrolysis process; at low pyrolysis temperatures, the appropriate kinetic rate model was a local kinetic rate model, and for pyrolysis at high temperatures of 600°C to 800°C, the accurate kinetic rate model was the total kinetic rate model.

1 Introduction

Due the pyrolysis is a chemical reaction, the rate of the reaction depends on the reaction mechanism. It simply breaks apart, in which woody biomass produce char+tar+gas, eventhough in the study; it was only focus on the char production measurement. It was required large enough energy and it takes a lot longer than combustion. One of the crucial parameters to generate the energy is heating to break down the biomass components such as cellulose, hemicellulose and lignin. It needs the differ temperatures to decompose its components to be their products [1]. If the pyrolysis process is conducted in the different temperatures, it will affect not only in the products, but also it will influence in its reaction rate.

If the previous study [2] had examined the kinetic studies of pinewood pyrolysis by using thermogravimetric analysis (TGA), in the study; we use pyrolysis pilot plant by measuring the

* Corresponding author: widya_dinata@ub.ac.id

change of biomass mass during the process. We used the real pilot plant because the actual reaction scheme of wood pyrolysis is extremely complex because of the formation of over a hundred intermediate products. The pyrolysis of wood is, therefore, generally modeled on the basis of apparent kinetics [2]. Furthermore, it is needed easier method to predict the pyrolysis mechanism. It will be understood how optimal the appropriate pyrolysis temperature. Accordingly, there is no excess energy to produce char. It is implementated to find out a detail process occuring in the pyrolysis, such as the biomass changes into the products in a certain time. The study aims at generating kinetic data for biomass pyrolysis and exploring its simplicity to kinetic models and make meaningful kinetic predictions.

Modeling the kinetic rate, it can be estimated the reduction of biomass to be char occurring in elapsed time of pyrolysis. Afterwards, the kinetic rate will be modeled and be validated by using experimental data. Then, the appropriate model and the kinetic rate equations will be found. Numerous works on kinetic studies of wood have been reported in literature [2-5]. Pyrolysis will produce some complex decompositions, so it is necessary to understand how large the kinetic rate is to predict the mass reduction that occurs and how large the energy is needed to break down the biomass elements.

During the pyrolysis process, the change of biomass into char in elapsed time is determined by :

$$\frac{dm}{dt} = -k\, f(x) \tag{1}$$

where $\frac{dm}{dt}$ is the mass change in elapsed time (g/min), k is rate constant (min^{-1}), k for the experimental results is defined by k actual while k obtained by calculation is defined as k total/k local.

Then, it was confirmed by experimental data of the change of mass

$$\frac{dm}{dt} = -k\ \frac{m - m_\sim}{mo - m_\sim} \tag{2}$$

consisted of the initial mass (m_o), the final mass ($m_\sim$), and the calculated mass (m), as well as the rate constant (k). Because the rate constant is influenced by the pyrolysis temperature, k will be replaced by the equation that similar Arhennius equation. It will be obtained some k and found the kinetic rate char mathematical model. It will use the local kinetic rate and the total kinetic rate. The local kinetic rate have the rate constant (k local) obtained from the plotting of each temperature range, at a temperature range of 250°C to 800°C. Here, we found the three constant rates. Meanwhile, the total kinetic rate is the constant rate (k total) gained from the plotting of the k values of all points from a temperature of 28°C to 800°C.

Then, the mathematical model of kinetic rate char will be validated with experimental data. This validation is used to know the truth of the agreement of k, so that it can be used as a suggestion for optimizing parameters in pyrolysis.

2 Methodology

The experimental device was used a vertical pyrolizer [6-9], however we used a container hang up the weight balance [4] to measure the change of solid yield (char) along the pyrolysis process.

The pyrolysis temperature were varied at 250°C, 350°C, 450°C, 500°C, 600°C, 700°C and 800°C. It was started from the ambient temperature. The raw material used was mahogany needle particles with 80 gram mass and 20 μm size. Before putting in the pyrolyzer, the sample was dried using an oven for 1-2 hours so that the water content wood around 2%. Afterwards, the sample was put into the beaker then inserted into the pyrolyzer. Before starting the pyrolisis process, Nitrogen was flowed into pyrolizer to suppress Oxygen until the Oxygen content around 2% furnace volume. Next, the thermocontroller was installed at the pyrolysis temperature. The pyrolysis process was conducted during 3 hours at every the pyrolysis temperature. Every 3 minutes, it would be recorded the change of biomass mass or the formed char. After the pyrolysis process was accomplished, the char was weighed and the final volume was measured.

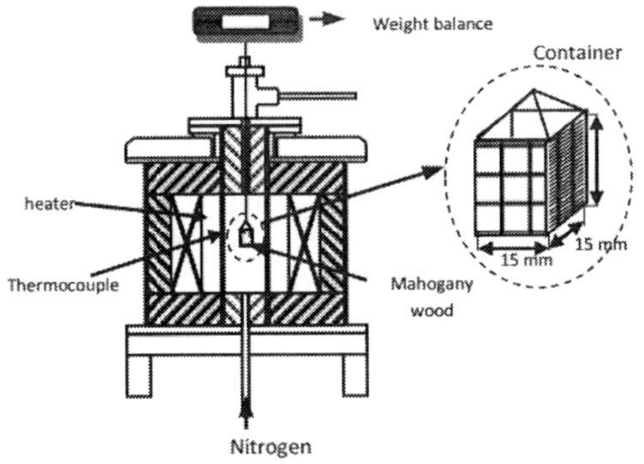

Fig. 1. Pyrolysis experimental run to measure the mass change of biomass

3 Results and Disscussion

Fig. 2 shows the elapsed time against the temperature change with some variations in the setting pyrolysis temperature and the relationship of pyrolysis temperature with char mass changes.. The setting pyrolysis temperature variations are 250°C, 350°C, 450°C, 500°C, 600°C, 700°C, 800°C in which to reach the temperature each experiment takes durimg 3 hours. Then, the graph can be used to determine the actual kinetic rate of each pyrolysis time. Comprehensively, the figure presents that the greater the pyrolysis temperature, the greater the reduction of the biomass mass to be char [9]. It indicates a very rapid pyrolysis reaction because the higher the temperature, the the higher the kinetic energy to decay the mahogany wood components resulting the huge decomposition of biomass components.

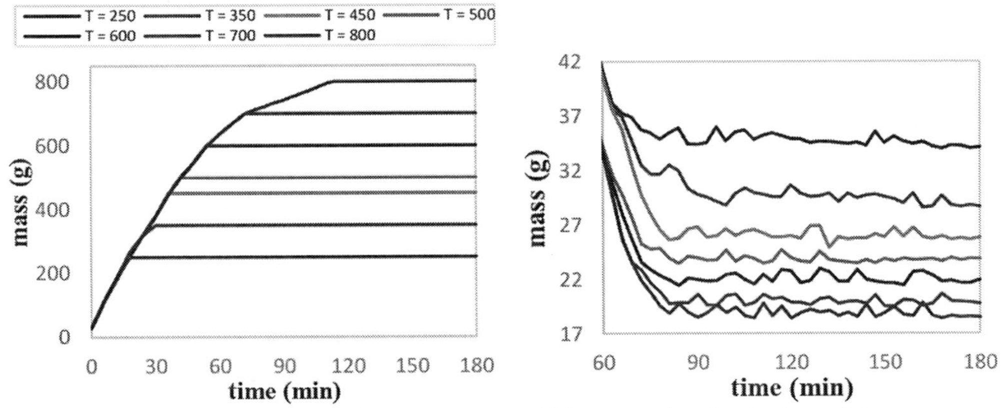

Fig. 2. Increasing temperature and decreasing mass of biomass to be char in elapsed time

The main components of mahogany wood consist of 40%-45% cellulose, 24%-40% hemicellulose, and 18%-25% lignin. These components will decompose at the different temperatures. According to the previous research [10], the hemicellulose will decompose at the pyrolysis temperature of 200°C-260°C, whereas the hemicellulose will decay at 260°C to 350°C, and lignin will decompose at 350°C to 800°C, so it will be divided as some range of temperature as follows:

I. At 250°C, evaporation occurs in the biomass and a little part of the hemicellulose causing a reduction of the biomass mass. It appears that the biomass reduction at 250°C is the least biomass reduction if it is compared to other pyrolysis temperatures because the biomass components except water have not been decomposed yet.

II. In addition to the water evaporation, the hemicellulose is completely decomposed and most of the cellulose will decompose completely at 350°C. As a result, at this temperature, the mass loss will be greater than the lost mass at 250°C.
III. Although there is still a few water evaporation, at temperatures 450°C to 800°C, all biomass components including lignin undergo extreme decomposition. As a result, char formation due to the reduction of biomass occurs with increasing the pyrolysis temperature. The higher the pyrolysis temperature, the lignin undergoes increasingly significant decomposition. Therefore, the highest biomass reduction (char formation) occurs at the highest temperature of 800°C which is the highest temperature of pyrolysis.

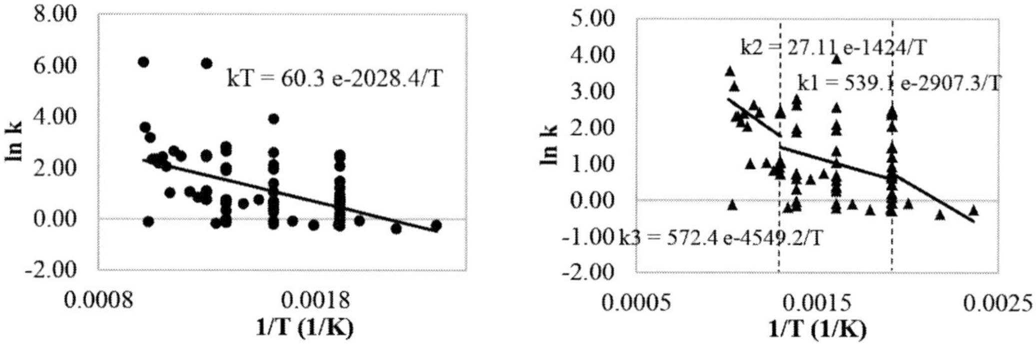

Fig. 3. Total and local kinetic rate in Arrhenius plot

Furthermore, from the obtained experimental results (Fig. 3), the pyrolysis kinetic rate equation can be known. The calculation of the kinetic rate in pyrolysis process in this study use two methods such as by using total kinetic rate and by using local kinetic rate. The total kinetic rate is the kinetic rate obtained from plotting the *k* value obtained from the experimental results during pyrolysis process from temperature 28°C to 800°C. For the total kinetic rate, the obtained equation is $k_T = 60.3\ e^{-2082.4/T}$.

Meanwhile, the local kinetic rate is the kinetic rate that the temperature range during pyrolysis process is divided into 3 ranges based on the decomposition temperature of each of its biomass components, as described in Fig. 3 and in the previous study [4]. In range 1, ie at temperature 28°C to 250°C in which only evaporation process occurs. Then at temperature 250°C to 500°C, it takes place evaporation process + cellulose + hemicellulose decomposition. In the last range, it occurs the decompotition process both of evaporation process and all biomass components. Finally, from these three ranges, it is got the 3 equations of *k* from each temperature ranges. Range I obtained the equation $k_1 = 539.1\ e^{-2907.3/T}$. At temperature range II 250°C to 500°C, it is obtained the equation $k_2 = 27.11\ e^{-1424/T}$. In the range III 500°C to 800°C, it is found the equation $k_3 = 572.4\ e^{-4549.2/T}$. Afterwards, *k* will be used to calculate the mass of char during the pyrolysis process as determined by equation 2.

Before calculating the mass of char occurring at the pyrolysis process, firstly, we will look at the differences in *k* values of the two previously mentioned methods. In general, the higher the pyrolysis temperature, the value of *k* (rate constant) will also be greater. This applies at the equations both using the total and local kinetic rate. It is because the higher the temperature, the greater the kinetic energy to break up the biomass. Consequently, the biomass can decompose quickly. This is in accordance that expressed at the previous research [1, 6, 8, 9]. From the calculation *k*, *k* with using the model of total kinetic rate has a very large range of *k* values and the value of *k* gradient is near constant, ie around 1.3 minutes^{-1}. Meanwhile, the *k* value for the local kinetic rate model has fluctuated values. That is, in the local kinetic rate model, the amount of energy to break the biomass varies in which it depends on the temperature of its biomass decomposition. In contrast, in the total kinetic rate model it is required the same amount of energy to break the biomass component. It means whatever the conditions; the energy used to break the biomass component is the same energy.

From the *k* value obtained, then the *k* will be used to predict the mass char formed during the pyrolysis process at each pyrolysis temperature setting. Then to find out how accurate the value of *k*, it will be compared the calculated results of the mass char both using the *k* local and *k* total) to the experimental results in laboratory. Then, the agreements of the calculated results and the experimental results will be shown in Fig. 4 and Fig. 5.

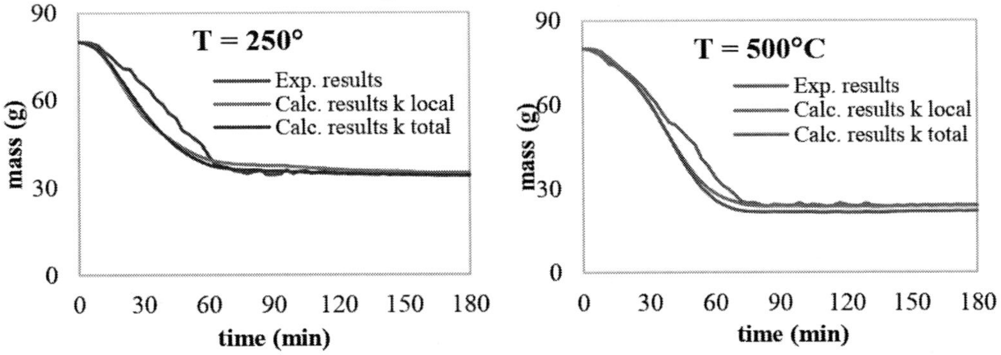

Fig. 4. The decreasing of char at T = 250°C to T = 500°C

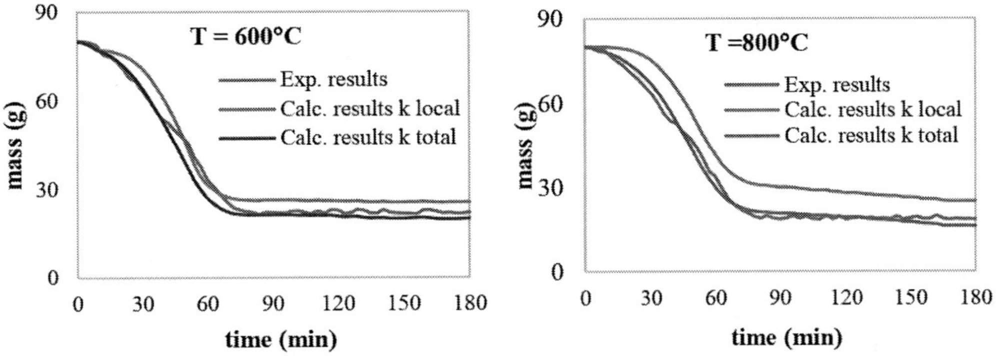

Fig. 5. The decreasing of char at T = 600°C to T = 800°C

From Fig. 4, it can be seen that at temperatures 250°C to 500°C, the calculated results that almost agree with the experimental results are the calculated results using the local *k* model, eventhough around 20-80 minutes pirolysis time, the experimental results have values above the calculated results. It is due to the large evaporation process occurs, so it will affect the mass of measured biomass. As a result, the mass of char in the experimental results tends to be larger. The local *k* model is more appropriate than the total k model, because it is needed a greater energy in the initial pyrolysis process. Therefore, the initial acceleration energy of pyrolysis process is greater than when pyrolysis is already running even for high temperatures. The initial acceleration of energy for pyrolysis is greater than the energy required when pyrolysis is already running even for pyrolysis at the higher temperatures. In addition, when the pyrolysis is at a low temperature, the reaction pyrolysis process is endothermic, in which the biomass only requires heat to break down its biomass compounds [1].

In opposite, at temperature 600°C to 800°C, the calculated results of the total kinetic rate model agree with the experimental results. In Fig. 5, the calculated results of the local kinetic rate model are above the experimental results significantly. It is because the values of k local on the calculation results are smaller than the actual values in the experiment results, so the calculated results (mass reductions) during the pyrolysis process becomes smaller if it uses the local kinetic rate equation model. Afterwards, if it is connected to Fig. 5 and Fig. 6, the rate constant k of the local kinetic rate is greater than the total kinetic rate. It means that the pyrolysis process conducted at above T = 600°C required the same amount of energy to break the biomass component. It means whatever the conditions; the energy used to break the biomass component is the same energy. It is because when the pyrolysis is running, the required energy goes constant. Although theoretically, a great energy is required to decompose the biomass, but the exothermic reaction to the pyrolysis process due to biomass decompositions will give the great energy in pyrolysis at high temperature temperature [1].

4 Summary

The research on predicting the kinetic rate of the pyrolysis reaction has been done. It was presented by determination of mathematical models resulting the kinetic reaction rate constants using two models. The first model used the total kinetic rate model and another applied the local kinetic rate model. From both models, it can be concluded that the validation between the calculated results and the experimental results accomplished with a great achievement. It was obtained that the agreements were very precisely, but with different conditions approach;

1. At low pyrolysis temperatures ranging from 250°C to 500°C, the appropriate kinetic rate model was a local kinetic rate model since the greatest energy was needed at the beginning of pyrolysis.
2. For pyrolysis at high temperatures of 600°C to 800°C, the accurate kinetic rate model was the total kinetic rate model. It was because when pyrolysis was running, the required energy remained constant and coupled with the energy produced by the exothermic reaction due to the biomass decomposition taking place at high temperatures.

References

1. Tanoue, K.-I., Wijayanti, W., Yamasaki, K., Taniguchi, M., Sasauchi, K.-I., Nihon Energi Gakkaishi/Journal of the Japan Institute of Energy **89**, 10 (2010)
2. Di Blasi, Progress in Energy and Combustion Science, **34**, 1 (2008)
3. C.A. Koufopanos, N. papayannakos, G. Maschio, A. Lucchesi, The Canadian J. of Chemical Eng., **69**, 4 (1991)
4. Wijayanti, W., Tanoue, K.-I., Suetomi, T., Taniguchi, M., Sasauchi, K.-I., AIChE Annual Meeting, 11 (2011)
5. Wijayanti, W., Sasongko, M.N., AIP Conference Proceedings **1717**, (2016)
6. Wijayanti, W., Sasongko, M.N., Purnami, ARPN J. of Eng. and App. Sciences **11**, 22016)
7. Wijayanti, W., Applied Mechanics and Materials **493**, 179-185, (2014)
8. Wijayanti, W., Applied Mechanics and Materials, **664**, 215-220, (2014)
9. Wijayanti, W., Tanoue, K.-I., Energy Procedia **32**, 145-152, (2013)
10. D Mohan, CU Pittman, PH Steele, Pyrolysis of wood/biomass for bio-oil: a critical review. Energy & fuels (2006)

Remaining Life Assessment of Superheater Tube in Boiler of Coal Fired Power Plant

Habib Muhajir[1,*], *Gunawan* Dwi Haryadi[1], and *Achmad* Widodo[1]

[1]Departement of Mechanical Engineering Faculty of Engineering, UNDIP, 50275 Semarang, Indonesia

Abstract. Risk Based Inspection (RBI) would be able to establish an effective structural integrity management program, which reduces plant downtime, industry and regulatory burdens, and continue to maintain plant safety. The object of this research is final superheater of boiler coal fired power plant. Both straight tube and elbow tube of final superheater is in a medium risk category. Elbow tube has higher risk ranking because of the higher corrosion rate due to fly ash corrosion. The remaining life of straight tube is about 26 years and elbow tube is only about 7 years using corrosion rate calculations. Base on risk ranking and remaining life assessment, inspection planning can be developed for the next 3 years and prioritize at an elbow tube of the final superheater section. It can be one of the considerations for the company to pending yearly inspection in order to improve cost efficiency and reduce downtime.

1 Introduction

Boiler tube failures are one of the important reasons for unexpected shut down of power plants. The cost penalty is estimated to be in excess of $5 billion a year in replacing power charges and maintenance [1].

Last decade shown a trend where the life management program is globally moving from prescribed/time based towards risk based decision making. RBI would be able to establish an effective structural integrity management program, which reduces plant downtime, industry and regulatory burdens, and continue to maintain plant safety [2].

In Europe and the United States, major oil companies have already implemented RBI technique several years ago. In Japan, the RBI has lately attracted considerable attention as a new technique for maintenance planning of fossils fired power and petrochemical plants that will be deregulated [3].

This research uses data thickness, final superheater section as representative data all of the boiler because of criticality that operate in highest pressure and temperature. It consist of 84 panels which each panel is given in Figure 1 [4]. Data thickness of the tube is taken in front of the flue gas side that the most suffering fly ash erosion at straight tube and elbow tube.

*Corresponding author: habib_pjbs@yahoo.co.id

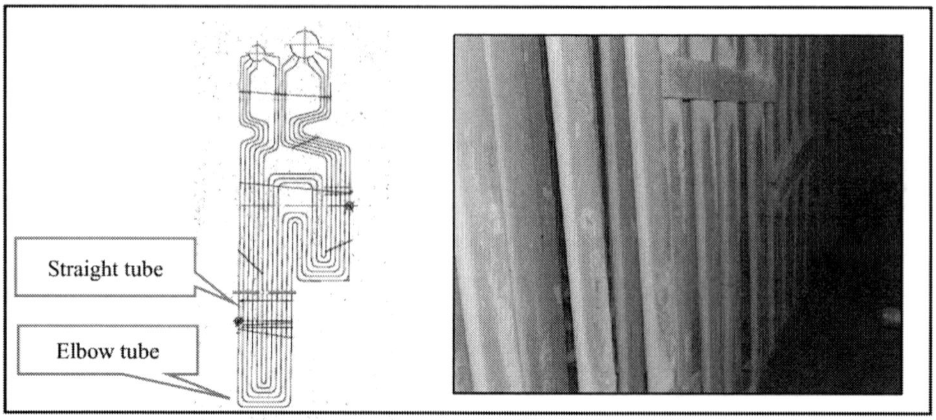

Figure 1. Drawing & picture panel of final superheater [5]

The information about boiler final superheater is shown in Table 1.

Table 1. Operation data of final superheater tube

1[st]Year Operation	Tube Dimension	Material	Steam Pressure	Steam Temperature
April 2011	ϴ51 x 8 mm	SA-213T91	17.26 Mpa	541 °C

2 Methodology

2.1 Risk Based Inspection

The calculation of risk in the RBI methodology involves the determination of a probability of failure combined with the consequence of failure [5]. With the use of the RBI method, high-risk tools get special handling and equipment with lower risk levels can be handled as necessary so as to reduce excessive inspection [6].

There are three types of RBI method that are qualitative, semi quantitative, and quantitative. Qualitative methods are methods that analyze using the general categorization. While the quantitative method is a method that analyzes based on the calculation of complex data. The semi-quantitative method is a mid-method between qualitative and quantitative, in that the result is more accurate than qualitative but the count is not as complex as quantitative. Calculating both elements using Appendix B-Workbook for Semi-Quantitative Risk-Based Inspection Analysis on API 581 Publication. The possibility of failure of the RBI API is calculated using the equation:

$$PF = GFF \times FMS \times DF \quad (1)$$

In the equation, the possibility of failure (PF), general frequency failure (GFF), damage factors (DF), and factor management systems (FMS). There are several steps to get the consequence value are determined the representative fluid and its properties, determine the leak hole, determine the total leakage, determine the rate of leakage, determine the type of leakage, determine the effects of post-leakage. The main result of an RBI method is the risk matrix that places the equipment at a certain level of risk as shown in Figure 2 [7].

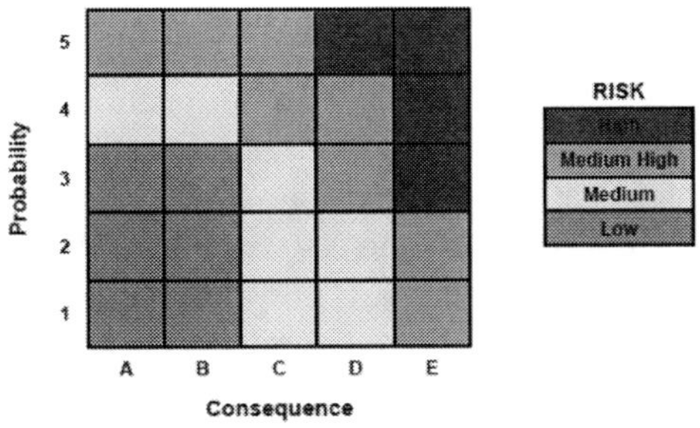

Figure 2. Risk Matrix

2.2 Remaining Life

Determination of equipment remaining life relates to inspection priorities. The method of determining the age of the equipment in the present study was determined by means of metal thinning analysis by T min determination in which the value was the minimum thickness allowed by an operational condition. Data is taken with ultrasonic thickness gauge of GE Type DMS 2 product with a precision of 0.01 mm. According to API BRD 581 the determination of the remaining life of the equipment is calculated according to the equation formula [8]:

$$remaining\ life\ (R_L) = \frac{damage\ tolerance}{damage\ rate} = \frac{T_{act} - Tmin}{corrosion\ rate} \tag{2}$$

2.3 Inspection Planning

From the risk level can be made scheduling for the equipment being analyzed. Scheduling inspection look at the level of risk and also the mechanism of damage that occurs in the equipment The schedule of inspection of a device according to its risk is indicated by the following equation:

$$NID = Ief \times C \times R_L \tag{3}$$

In the equation of the next inspection schedule, NID, Determined as the product of the effectiveness of the inspection, Confidence rating C, and residual life CR_L. The value of the inspection effectiveness is 0 - 1, the value is based on the category of effectiveness of the inspection. While the value of the confidence rating is taken based on the risk level is less than 0.5 for high risk level and more than 0.5 for low risk level. So as to reduce the method of excessive inspection and the results obtained more optimal [9].

3 Result and discussion

3.1 Analisis Risk Based Inspection (RBI) Semi-Kuantitatif

The RBI analysis based on semi-quantitative method can be calculated based on Appendix B-Workbook for Semi-Quantitative Risk Based Inspection Analysis API RBI 581.

Part A: Leak Rate Calculation

The representative fluid in this analysis is classified as Steam, the group has the characteristics as mentioned in table 7.2 API RBI 581.

According to Table B-1 on API BRD 2000, the inventory category identification for superheater with the mentioned capacity is a category C with range between 10,000 – 100,000 lbs. Table 7.6 on API BRD 2000 provides guidance to the user for assigning a qualitative letter rating (A, B, or C) to the unit's detection and isolation systems. The detection and isolation classification system defined as C and B. This is because the boiler is completed with leak detector with marginal coverage and also isolation and shutdown system is operated by an operator in the control room. Therefore, for the evaluated final superheater tube, the leak duration, liquid discharge rate, type and representative fluid phase after release are shown in Table 2.

Table 2. Leak Duration, Liquid Discharge Rate and Type

Hole Size Diameter	Leak Duration	Lb/Sec	Lb/3Sec	Release Type
0.25 Inch	60 Minutes	0.31	56.67	Continuous
1 Inch	40 Minutes	5.03	906.74	Continuous
4 Inch	20 Minutes	80.59	14,507.92	Instantaneous
12 Inch	Rupture	725.39	130,571.24	Instantaneous

Part B: Likehood Analysis
The effectiveness of the inspection of tools is classified in a fairly effective manner because the thickness measurement has been done with spot ultrasonic test (UT) method and visual inspection. The boiler has been conducted twice in inspection, in 2015 and 2016.

Damage factor thinning calculation including minimum wall thickness allowance and corrosion rate is summarized on the Table 3.

Table 3. Thickness and corrosion rate on each part

Section	Actual Thickness 2016 (mm)	Tmin (mm)	Diameter (mm)	Corrosion Rate (mm/year)	Art	Likehood Category
Straight tube	7.75	6.80	51.00	0.04	0.02	1
Elbow tube	7.47	6.80	51.00	0.09	0.06	2

So from the value of the reduction constant and the effectiveness of the inspection obtained the value of TMSF. Adjust obtained by multiplying the TMSF value by the factor value over design and it is found that the over design factor 1 for straight tube and 2 for elbow tube.

Part C: Flammable Consequence Calculation

In this analysis, there is no HF or H2S compound in the analyzed equipment so that the C2 worksheet is negligible and has a value 0. Specifies the general failure frequency value (GFF) based on the type of equipment being analyzed. In this research, the equipment type is pipe with diameter 2 inches.

Table 4. Courses flammable consequence and area

Hole Size	Fraction Contribution	Unit	Flammable Consequence	Flammable Area
Small (1/4)	0.09	ft^2	0.15	0.13
Medium (1)	0	ft^2	2.57	0
Large (4)	0	ft^2	887.04	0
Rupture (12)	0.08	ft^2	3,606.87	306.47

The extent of damage to each hole is added, then get a value of 306,598 ft^2. Based on table B-2 API 581 then the equipment enters the category of the consequences of failure C that is between 100-1,000 ft^2.

The level of risk on a semi-quantitative method API 581 are a combination of categories the possibility of failure and the consequences of failure categories. The final risk ranking is summarized in 5 x 5 matrix that shows Likelihood v's Consequence, in which the values are presented as categories.

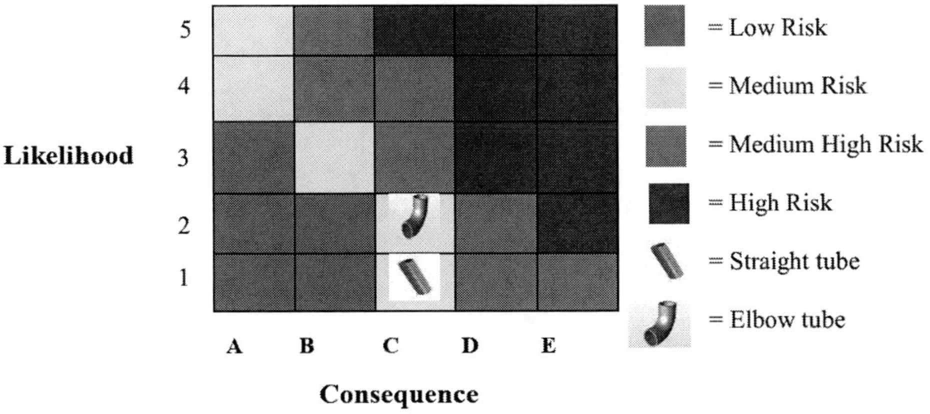

Figure 3. Risk ranking distribution

3.2 Remaining Life

Based on the formula of remaining life calculation using corrosion rate, the following remaining life calculation is shown in Table 5 and Figure 4.

Table 5. Remaining Life Calculation

Section	T act (mm)	T min (mm)	Corrosion Rate (CR) (mm/years)	Remaining Life (RL) (Years)
Straight tube	7.75	6.80	0.04	26.64
Elbow tube	7.47	6.80	0.09	7.31

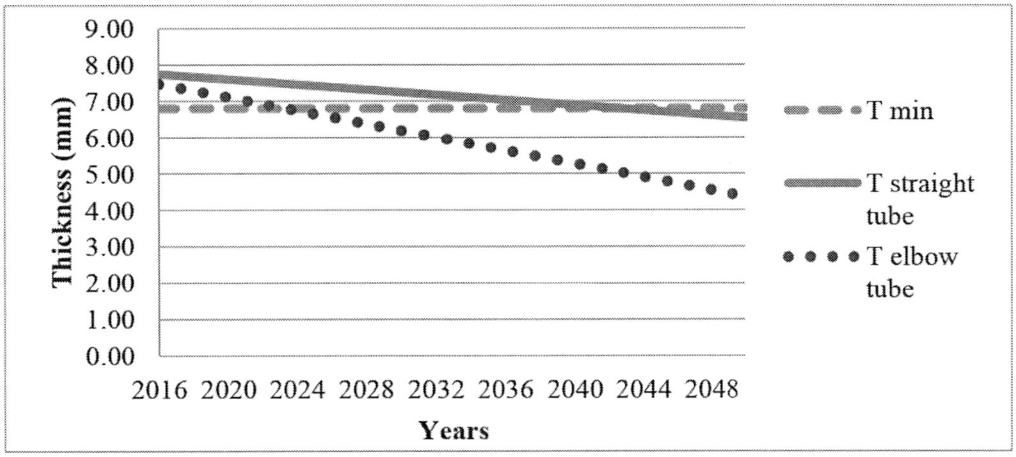

Figure 4. Remaining life tube straight tube and elbow tube

3.3 Inspection Planning

Selected corrosion rates as a reference due to the most influential in residual life use. From the available data, the highest corrosion rate is 0.09 mm / year, then the schedule for the next inspection is as follows:

$$NID = I_{ef} CR_L$$

NID = 0.6 x 0.75 x 7.31 years

NID = 3.28 years

Inspection should be done at least 3 years from the last inspection 2016. In this analysis, the equipment is damaged in the form of thinning wall thickness due to localized corrosion. Some methods that can be used are visual examination, ultrasonic tube thickness, eddy current, EMAT (Electromagnetic Acoustic Transducer), magnetic particle testing. These methods are the most effective method used for this type of thinning damage [10].

4 Conclusions

Both straight tube and elbow tube of final superheater is in a medium risk category. Elbow tube has higher risk ranking than a straight tube because of the higher corrosion rate due to fly ash corrosion. The remaining life of straight tube is about 26 years and elbow tube is only about 7 years using corrosion rate calculations. Base on risk ranking and remaining life assessment, inspection planning can be developed for the next 3 years and prioritize at an elbow tube of the final superheater section. It can be one of the consideration for the company to pending yearly inspection in order to improve cost efficiency and reduce downtime.

References

[1] French DN, *Metallurgical failures in fossil fired boilers*, **22** (New York, Wiley/Wiley–Interscience (1983).
[2] Galang Sandy Prayogo, Gunawan Dwi Haryadi1, Rifky Ismail, Seon Jin Kim et al, *Risk analysis of heat recovery steam generator with semi quantitative risk based inspection API 581* (2016)
[3] Akio Fuji, *Aplication of Risk Based Maintenance with life and Financial Assesment on Fossil Fired Power Plant*, Oral Conference, Research Laboratory, Ishikawajima-Harima Heavy Industries Co., Ltd (2000)
[4] Dongfang, *Operation & Maintenance of Boiler Manual Book for Rembang 2x300 MW Coal FIred Power Plant* (2008)
[5] API RP 581 Risk-Based Inspection Technology (2008)
[6] Assesment Boiler Report 2016, PT PLN (Persero) Research Institute (2016)
[7] Moamar Al Qathafi, Sulistijono, *The Application of Semi-Quantitative Risk Based Inspection Method (RBI) API 581 in Production Separator*, Technical Journal ITS **Vol. 4, No. 1**, (2015)
[8] API Publication 581 Risk-based inspection base resource document, American Petroleum Institute (2000)
[9] Ming-Kuen Chang, et al, *Application of Risk Based Inspection in Refinery and Processing Piping* (2005)
[10] EPRI, Technical Report Boiler Condition Assessment Guideline (2006)

3D Numerical study on the effect of shaft eccentricity on the tribological performance of lubricated sliding contact

Mohammad Tauviqirrahman*, *Bayu* Kurniawan, Jamari

Laboratory for Engineering Design and Tribology, Mechanical Engineering Department, Engineering Faculty, Diponegoro University, Indonesia

Abstract. Recently, a growing interest is given to the wall slip and the artificial texturing for improving the performance of lubricated sliding contact. The use of wall slip, artificial texturing, and the combination of slip and texturing can be the effective approach to enhance the performance of the bearing. The present study examines the effect of shaft eccentricity ratio on the hydrodynamic journal bearing performance. 3D numerical modelling based on modified Reynolds equation is used to analyse the effect of texturing and the wall slip on the characteristics of a hydrodynamically lubricated sliding contact. The analysis results point out that with respect to the load support and the power loss of the bearing, the use of wall slip on smooth surface is the most excellent configuration compared to other patterns (i.e. slip-texturing, pure texturing and conventional patterns). It is also confirmed that the wedge effect due to the shaft eccentricity has a significant role in altering the lubricant behaviour. Thus, a particular care must be taken in choosing the pattern of lubricated sliding contact as well as the shaft eccentricity.

1 Introduction

As a mechanical element, journal bearing is designed to support shaft or journal which rotates freely in a supporting metals or shell. In order to avoid direct contact between the surfaces, the insertion of a lubricant into the region around the interacting devices is carried out, so wear can be reduced. Due to the divergent-convergent gap the pressure gradients that exist in the clearance between the shaft and the housing produce the load support. Recently, along with the development of advanced technology and the measurement, the artificial texturing concept as well as the wall slip concept in various configurations is introduced and explored by numerous researchers in the world focusing on how to improve the hydrodynamic lubrication performance.

The superiority of artificially textured surface over the smooth surface in terms of load support and friction force has been examined by several researchers. Cupillard et al. [1] confirmed that textured journal bearing is superior to smooth journal bearing with thinner

* Corresponding author : mohammad.tauviqirrahman@ft.undip.ac.id

lubricant with respect to the friction force. A comparative study between textured surface and smooth surface of non-recessed hybrid journal bearing configurations was also conducted by Khatri and Sharma [2]. They found that the textured journal bearing provides an improved stability parameter than that of smooth journal bearing. With respect to the texture location, based on the computational fluid dynamic (CFD), Liang et al. [3] investigated the effect of partial texture location and texture depth on load support, friction coefficient and circumferential flow of journal bearing. They revealed that the texture located at leading edge of the contact could enhance the bearing performance, and the effect of shallow dimples is superior to the deep ones. Regarding with the effect of texture shape, Sinanoğlu et al. [4] studied the effect of trapezoidal texture on the performance of journal bearing. They concluded that journal with trapezoidal texture pattern can support more loads than the sawtooth texture.

Recently, in addition to artificial texturing, the wall slip is also considered as a method to improve the lubrication performance. Numerically, Wu [5] proposed the potential of wall slip application to maximize the load support. Wang et al. [6] conducted the optimization of the wall slip region of the combined surface sleeve bearing. Later, the idea of combining the artificial texturing with wall slip has also attracted many researchers. Tauviqirrahman et al. [7] explored the characteristics of the textured/wall slip surface with respect to the load support. It is found that in comparison with a solely textured surface, the load support of the combined textured/wall slip pattern can be increased by around 300%. Later, Aurelian et al. [8] revealed that a particular care should be taken in choosing the texture/slip pattern because an inappropriate choice can lead to a significant decrease in the performance.

In the present work, in order to extend the insight of previously published findings, the artificial texturing and wall slip pattern are considered as a base of design of the journal bearing. Several surface modes of the lubricated sliding contact containing different patterns of the artificial texturing and wall slip (i.e. slip, textured, and textured-wall slip conditions) are investigated. The main aim of the present study is to investigate the effect of shaft eccentricity (i.e. the wedge effect) on the tribological performance for different surface configurations. Here, modified Reynolds equation is developed considering the wall slip surface and solved by means of finite volume method.

2 Analysis

As is known in classical literature, the derivation of traditional Reynolds theory is based on assumption of no-slip between the lubricant and the contacting surfaces. The lubrication model presented here is based on the fact that slip in the lubricant will exist in the interface. Schematic of lubricated sliding contacts with artificial texturing and slip/no slip condition is shown in Figure 1.

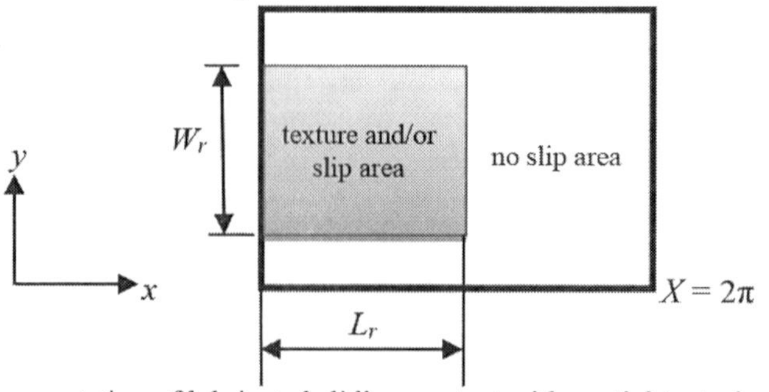

Fig. 1. Schematic representation of lubricated sliding contact with partial texturing and/or wall slip (Note: $W_r = 0.7$, $L_r = 170°$)

In the present work, the artificial texturing zone as well as the wall slip is introduced at the leading edge of the contact. It is defined by W_r x L_r zone where W_r is the width of the modified zone and L_r is the length of modified zone. The numerical analysis is carried out by analysing the behaviour of fluid induced by several patterns studied here. In the developed model, the slip has been treated to allow in certain areas where the shear stress exceeds a critical value. When both criteria are met, the resulting slip velocity is proportional to the difference between the shear stress and the critical value, with constant factors referred as α. Such a lubrication system can be described with solving the modified Reynolds equation as follows [7]:

$$\frac{\partial}{\partial x}\left(h^3\frac{\partial p}{\partial x}\left(1+\frac{3\alpha\eta}{h+\alpha\eta}\right)\right)+\frac{\partial}{\partial y}\left(h^3\frac{\partial p}{\partial y}\left(1+\frac{3\alpha\eta}{h+\alpha\eta}\right)\right)$$
$$=\frac{\partial}{\partial x}\left(-6\eta\frac{\alpha\tau_c}{h+\alpha\eta}h^2+6u_w\left(1+\frac{\alpha\eta}{h+\alpha\eta}\right)\right) \quad (1)$$

The physical meanings of the symbols in Equation (1) are as follows: h the lubrication film thickness (gap) at location, p the lubrication film pressure, u_w the sliding velocity, η the lubricant viscosity, α the slip coefficient and τ_c the critical shear stress of the interface. It can be seen in that if the slip length α and τ_c are set to zero (no-slip condition), Equation (1) reduces to the classical Reynolds equation.

In the present work, the finite volume method combined with tridiagonal matrix algorithm (TDMA) and alternating direction implicit (ADI) method is used to address the nonlinear governing Reynolds equation for lubricated sliding contact based on the critical shear stress model. The Reynolds boundary condition is used to modelling the cavitation. In order to obtain the highest positive effect of the wall slip, the critical shears stress τ_c is set to zero [9, 10].

The primary parameters of the lubricated sliding contact are given as follows: journal rotation is 5100 rpm, total length of bearing is 535.6x10^{-3} m, journal radius is 209.7x10^{-3} m, bearing radius is 210.25x10^{-3} m, clearance is 0.55x10^{-3} m, dynamic viscosity is 0.02785 Pa.s and it is assumed that fluid density is 855.6 kg/m^3. In this study, for modeling the optimal slip characteristic, the dimensionless slip length A of 10 is considered.

3 Results and Discussion

For analysis, 3D lubrication model is developed. In this way, the lubricant behaviour in the y-direction is taken into account to get more realistic result based on the work of [9]. In the present study, the rectangular shape is chosen as a layout for texturing as well as the wall slip. The load support and the power loss are of main particular interest as an indicator of the lubrication performance.

Figure 2 shows the effect of shaft eccentricity ratio on the power loss varying different surface characteristics. Specific features can be identified based on Figure 2. Firstly, for the case of no-slip situation, introducing the artificial texturing on lubricated sliding contact has advantage with respect to the power loss. However, when the eccentricity ratio is increased, the power loss increases significantly and the performances of bearing with texturing is worse than the smooth conventional bearing. It can be seen that for the case of eccentricity ratio ε which is larger than 0.7, the power loss increases significantly with the increase of the eccentricity ratio. Secondly, for the case of slip condition, the lubricated contact with or without texture produces lower power loss compared to the contact with no-slip condition either with texture or without texture. Figure 2 also shows that the best pattern in producing the lowest power loss in lubricated sliding contact is smooth slip surface.

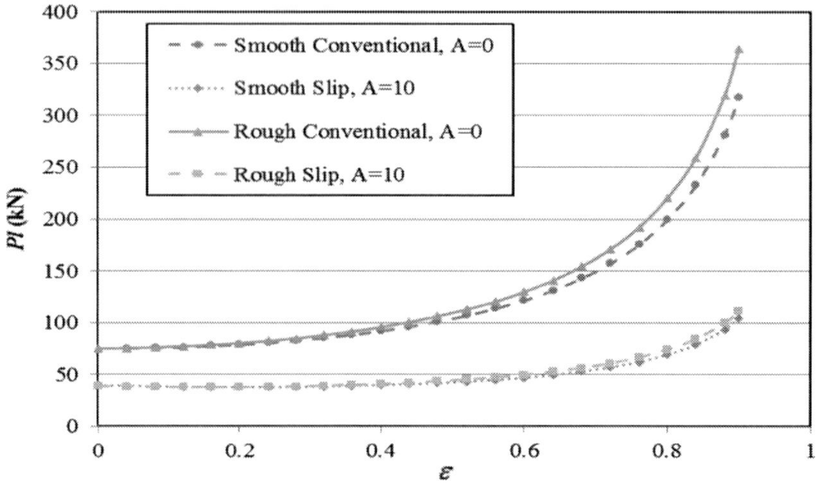

Fig. 2. Effect of the shaft eccentricity ratio ε on the power loss *Pl*

Figure 3 reflects the effect of eccentricity ratio on the load support for different surface configurations. It can be found that the eccentricity ratio has a significant influence on the load support for all configurations. Increasing eccentricity ratio improves the load support significantly especially for high eccentricity ratio ($\varepsilon > 0.8$ in this case). It can also be observed that introducing the modification of contacting surface either with artificial texturing or the wall slip leads to better performance compared to smooth surface. It is obvious in Figure 3 that for $\varepsilon < 0.5$ the rough slip pattern (which refers to the combined texturing and wall slip) generates higher load support than others. However, for higher eccentricity ratio, the smooth slip is the best pattern in generating the high support.

Figure 3 also shows that the introduction of wall slip as well as the artificial texturing in journal bearing will be useless if high eccentricity ratio is used. It means that for high eccentricity ratio ($\varepsilon > 0.8$ in this case), the predicted load support of smooth traditional surface is higher than that of other surfaces. This finding is in a good agreement with the work of [8] for case of slider bearing. It seems that the wedge ratio has more dominant role in altering the lubricant characteristics compared to the texturing and/or slip technique. In detail, the limit of effect of eccentricity ratio on the load support *W* for several patterns considered here is shown in Table 1. Based on this table, it is recommended that for the case of high eccentricity ratio, the smooth surface (without slip) is preferable for real application.

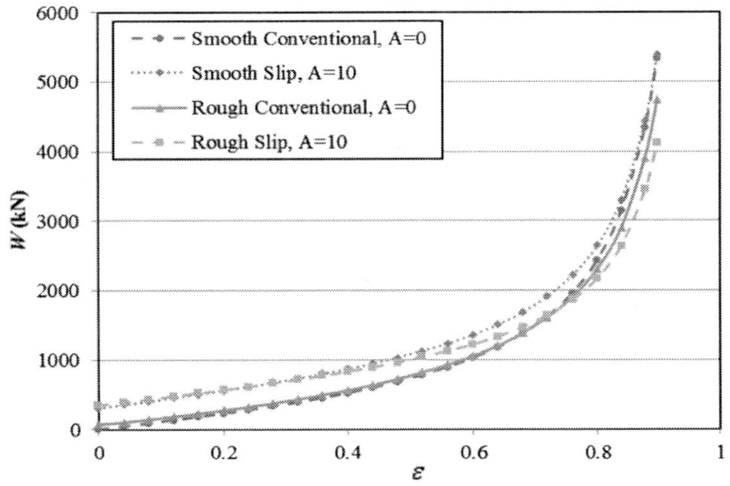

Fig. 3. Effect of the shaft eccentricity ratio ε on the load support W

Table 1. Limit of eccentricity ratio

Model	ε_{lim}
Smooth Slip	0.90
Rough Conventional	0.72
Rough Slip	0.72

Note: ε_{lim} is the eccentricity ratio of the modified journal bearing which generates the same load support with that of traditional journal bearing

4 Conclusion

In the present work, the lubricant behaviour of journal bearing with several configurations (i.e. slip-texturing, pure texturing and conventional patterns) was explored. The effect of the eccentricity ratio on the load support and the power loss of journal bearing is of particular interest. The following conclusions summarize the results of the present study:

1. The eccentricity ratio has a significant effect in altering the lubricant characteristic for all patterns studied here.
2. Introduction of wall slip as well as artificial texturing on the journal bearing is quite effective in improving the lubrication performance especially for case of low eccentricity ratio.
3. The superiority of smooth slip surface is highlighted in terms of the power loss and load support compared to textured surfaces for all range of eccentricity ratios.

References

1. S. Cupillard, S. Glavatskih, M.J. Cervantes, Proc. Inst. Mech. Eng. Part J: J. Eng. Tribol. **222**, 96 (2008).
2. C.B. Khatri, S.C Sharma, Tribol. Int. **95**, 221 (2016).

3. X. Liang, Z. Liu, H. Wang, X. Zhou, X. Zhou, Ind. Lubr. Tribol. **68**, 106 (2016).
4. C. Sinanoğlu, F. Nair, M.B. Karamiş, J. Mater. Process. Technol. **168**, 344 (2005).
5. C. Wu, Ind. Lubr. Tribol. **60**, 293 (2008).
6. L.L. Wang, C.H. Lu, M. Wang, W.X Fu, Tribol. Int. **47**, 100 (2012).
7. M. Tauviqirrahman, Muchammad, Jamari, D.J. Schipper, Tribol. Trans. **57**, 134 (2014).
8. F. Aurelian, M. Patrick, H. Mohamed, Tribol. Int. **44**, 868 (2011).
9. M. Tauviqirrahman, R. Ismail, Jamari, D.J. Schipper, Comput. Fluids **79**, 27 (2013).
10. F. Aurelian, M. Patrick, H. Mohamed, Tribol. Int. **44**, 868 (2011).

Optimal passivity design of a virtual coupling including FIR-type fractional derivatives for a haptic interface

Masayuki Kawai[1,*], *Takuya* Yamagishi[1], and *Shouta* Minami[1]

[1] University of Fukui, Department of Mechanical Engineering, 3-9-1 Fukui, 9108507, Japan

Abstract. This paper describes a haptic interface with a virtual coupling, including fractional derivatives. A haptic interface is a force feedback technology in virtual reality that takes advantage of the human sense of touch. In a haptic interface, virtual impedance, that is called virtual coupling (VC), is commonly used between the virtual and real objects to calculate reaction force. VC generally consists of a virtual stiffness and damper, but the stiffness has to be set low in a system with long sampling periods. In order to increase the virtual stiffness, this paper considers a VC including fractional derivatives approximated by an FIR approximation. First, we theoretically analyse effects of a single fractional derivative term by using the passivity analysis, but the result shows that its effects depend on the occurred frequency. This paper, therefore, proposes a method to combine multiple fractional derivative terms in a VC and a method to optimize parameters in each fractional derivative term. Finally, experiments are performed to measure the maximum value of the stiffness to illustrate the effects of the proposed method.

1 Introduction

A haptic interface is a force feedback technology in virtual reality that takes advantage of the human sense of touch. Operators can intuitively feel and manipulate virtual objects in a virtual environment. In the haptic interface, virtual impedance called "virtual coupling (VC)" is designed between the virtual and real objects, and the force that the operator senses is calculated from the impedance. A common VC generally consists of a virtual stiffness and virtual damper. In a system with long sampling periods, however, the stiffness has to be set low for the stability, and the system only presents virtual objects with soft surfaces. The limitations of virtual stiffness have been examined using passivity analysis [1-4]. Various devices and methods have also been studied for increasing the stiffness [5-8]. On the other hand, applications of fractional derivatives [9] to control methods or model physical phenomena have also been recently studied [10-17]. A fractional derivative is a

* Corresponding author: m_kawai@u-fukui.ac.jp

generalization of an ordinary derivative to a non-integer order. When a fractional derivative is used in control methods, it can easily describe the intermediate effects between ordinal integer-order derivatives.

This research studies the effects of a VC including a fractional derivative, approximated by a finite impulse response (FIR approximation) [13, 14]. An FIR approximation is a less precise approximation than the other types of approximation [15-17], but is the easiest type of approximation to be designed, in which the order of the derivative and filter order should only be fixed. First, this paper considers the effect of a VC including a single term of a fractional derivative theoretically, and derives that its effects depend on the occurred frequency. Then, this paper proposes a method to use multiple terms of a fractional derivative and optimize parameters on each term. Finally, experiments are performed to measure the maximum value of the stiffness to illustrate the effects of the proposed method.

2 Background

2.1 Haptic interface and a common PDVC

In an impedance display in a haptic interface, the force sensed by an operator is calculated using a VC that typically consists of a virtual stiffness and damper. Consider a Pen-type haptic display system and a one-degree-of-freedom modelling, as shown in Fig. 1,

(a) Example of a haptic interface (b) 1-DOF Modeling of a haptic interface

Fig. 1. Haptic display system

where x (m) is the fingertip position and f_a (N) is the force exerted by the device. The ordinary virtual coupling impedance for a virtual wall at $x = 0$ is given by

$$f_a = Kx + B\frac{dx}{dt} \tag{1}$$

where K and B are the coefficients of the virtual stiffness and damper, respectively. As (1) is the same as a common PD feedback control; thus, we call it 'PDVC' herein. In a haptic interface, it is difficult to use the usual stability analysis because it involves human dynamics that are unknown. Therefore, passivity analysis is commonly used instead. Colgate et al. [3][4] have derived the passivity condition for this system. A necessary and sufficient condition for the system except the operator's dynamics to become passive is represented by

$$b > \frac{T}{2}\frac{1}{1-cos\omega T}Re\{(1-e^{-j\omega T})H(e^{j\omega T})\}, \tag{2}$$

where b is the viscous resistance of the device, T is the sampling period, $H(z)$ is a transfer function of the virtual environment and ω is the frequency that satisfies $0 \leq \omega \leq \omega_N$, where ω_N is the Nyquist frequency. For the PDVC case, inequality (2) becomes

$$b > \frac{KT}{2} - Bcos\omega T. \tag{3}$$

Equation (3) shows that the maximum value of K depends on ω. In a system with long sampling periods, K has to be set low and the system can display only virtual objects with soft surfaces.

2.2 Fractional derivatives

A fractional derivative is a calculus used to generalize the ordinary derivative into a non-integer order, which allows us to utilize the 0.5th- or 1.5th- order derivatives. This research uses an FIR approximation that transforms a fractional derivative in the Grünwald-Letnikov form directly to the discrete time domain. The form is described by

$$D^{\alpha}x(t) = \left(\frac{d}{dt}\right)^{\alpha} x(t) = \lim_{h\to 0}\frac{1}{h^{\alpha}}\sum_{m=0}^{\infty}\frac{(-1)^m\Gamma(\alpha+1)}{\Gamma(m+1)\Gamma(\alpha-m+1)}x(t-mh), \tag{4}$$

where α is the derivative order, which is a non-integer number. D^{α} represents the αth-order derivative and Γ is the gamma function. In the FIR approximation, the equation is directly transferred into the discrete time domain with a finite filter order n as

$$Z\{D^{\alpha}x(t)\} = \frac{1}{T^{\alpha}}\sum_{m=0}^{n}\frac{(-1)^m\Gamma(\alpha+1)}{\Gamma(m+1)\Gamma(\alpha-m+1)}z^{-m}. \tag{5}$$

3 Virtual coupling with a single term of a fractional derivative

3.1 VC with a single term of a fractional derivative

This paper considers the replacement of the virtual damper in a PDVC with a fractional derivative. There is a possibility that a VC with fractional derivative feedback makes the virtual stiffness higher because it can use derivatives greater than those of the first order. First, this research uses a new VC which replaces the virtual damper in (1) as follows:

$$f_a = Kx + B_{\alpha}\frac{d^{\alpha}x}{dt^{\alpha}} \tag{6}$$

The first term on the right side of this equation is the same virtual stiffness as in the PDVC. The second term represents the fractional derivative feedback. This method requires the determination of two parameters, i.e., the derivative order α and filter order n of the approximation. This definition is also a generalization of PDVC; obviously, the case of α=1 corresponds to a PDVC. To distinguish this definition from a PDVC, we call it a ‘PD$^{\alpha}$VC’ herein. With the definition (5), the transfer function of a PD$^{\alpha}$VC is described by

$$H(z) = K + \frac{B_{\alpha}}{T^{\alpha}}\sum_{m=0}^{n}\frac{(-1)^m\Gamma(\alpha+1)}{\Gamma(m+1)\Gamma(\alpha-m+1)}z^{-m}. \tag{7}$$

By substituting (7) into (2), the passivity condition becomes

$$b > \frac{KT}{2} - \frac{B_{\alpha}}{2T^{\alpha-1}(1-\cos\omega T)}\sum_{m=0}^{n} Re\left\{\frac{(e^{-j\omega T}-1)(-1)^m\Gamma(\alpha+1)}{\Gamma(m+1)\Gamma(\alpha-m+1)}e^{-jm\omega T}\right\}. \tag{8}$$

To evaluate the influence of the fractional derivative term, we define a new function $f(\omega)$ from the second term on the right side of inequality (8) as

$$f(\omega) = \frac{-B_{\alpha}}{2T^{\alpha-1}(1-\cos\omega T)}\sum_{m=0}^{n} Re\left\{\frac{(e^{-j\omega T}-1)(-1)^m\Gamma(\alpha+1)}{\Gamma(m+1)\Gamma(\alpha-m+1)}e^{-jm\omega T}\right\}. \tag{9}$$

The function can also be used to determine the maximum stiffness K that can be set in the system. When $f(\omega)$ is smaller, we can set the stiffness K higher. For the case of a PDVC, $f(\omega)$ is given as $f(\omega) = -B\cos(\omega T)$.

3.2 Passivity of a single term of a fractional derivative

We calculate the value of $f(\omega)$ for various α. We consider the cases of $0 < \alpha < 1$ and $1 < \alpha < 2$ separately because the nature of $f(\omega)$ is quite different when $0 < \alpha < 1$ than when $1 < \alpha < 2$. Fig. 2 (a) shows the value of $f(\omega)$ for α = 0.3, 0.5, 0.7 and 1. Fig. 2 (b) is the value of $f(\omega)$ α = 1, 1.3, 1.5, and 1.7. B_α is selected as $f(\omega_N) = 1$. In the calculation, we use a sampling period $T = 2$ (msec), and the frequency ω is between 0 and 1570 (rad/s).

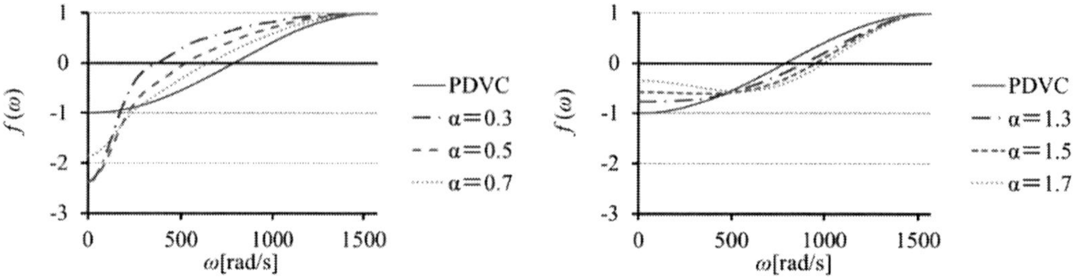

(a) $0 < \alpha < 1$ (b) $1 < \alpha < 2$

Fig. 2. $f(\omega)$ for a single term of a fractional derivative

In Fig. 2, the vertical axis is $f(\omega)$ and the horizontal axis is the frequency ω. When $0 < \alpha < 1$, $f(\omega)$ increases monotonically with ω, and the intersection with the ω axis moves to the right as α increases. As the lower value is more passive, Fig. 2 (a) shows that a PD$^\alpha$VC with $0<\alpha<1$ is more passive in the lowest frequency range but less passive in the middle-frequency range than a PDVC. On the other hand, Fig. 2 (b) shows that when $1 < \alpha < 2$, $f(\omega)$ has a minimum, and the value of $f(0)$ increases monotonically to zero with an increasing α. Fig. 2 (b) also shows that for $1 < \alpha < 2$, a PD$^\alpha$VC is less passive in the low-frequency range but more passive in the middle-frequency range than a PDVC. By these results, it is difficult to keep better passivity in a wide range of frequency using a single term of a fractional derivative.

4 Virtual coupling with multi terms of a fractional derivative

4.1 VC with a linear combination of fractional derivative terms

Next, in order to improve the passivity in a wide range of frequency, we study a VC including multi terms of a fractional deriative, descrived by

$$f_a = Kx + B_\alpha\left(A_1\frac{d^{\alpha_1}x}{dt^{\alpha_1}} + A_2\frac{d^{\alpha_2}x}{dt^{\alpha_2}} + \cdots + A_c\frac{d^{\alpha_c}x}{dt^{\alpha_c}}\right) \tag{10}$$

Where, $A_d(d=1,2,\ldots c)$ is the weight coefficient, c is the number of terms of a fractional derivative. The transfer funtion and the evaluation function $f(\omega)$ are

$$H(z) = K + B_\alpha\sum_{d=1}^{c} A_d\frac{1}{T^{\alpha_d}}\sum_{m=0}^{n}\frac{(-1)^m\Gamma(\alpha_d+1)}{\Gamma(m+1)\Gamma(\alpha_d-m+1)}z^{-m}, \tag{11}$$

$$f(\omega) = B_\alpha\sum_{d=1}^{c}\left[A_d\frac{1}{2T^{\alpha_d-1}(1-\cos(\omega T))}\sum_{m=0}^{n}Re\left\{(1-e^{-j\omega T})\frac{(-1)^m\Gamma(\alpha_d+1)}{\Gamma(m+1)\Gamma(\alpha_d-m+1)}e^{-jm\omega T}\right\}\right]. \tag{12}$$

In (12), $f(\omega)$ depends on the the weight coefficients A_d and the order of derivative α_i.

4.2 Optimization of the weight coefficients and the orders of fractional derivative

In order to obtain a desired system , in which the passivity keeps lower value in a wide range, we optimize the weight coefficients and the order of derivative by using a steepest descent method. In the method, we use a evaluation function as

$$J(\boldsymbol{x}) = \sum_{d=1}^{c}(f(\omega_d) + 1)^2 + (f(\omega_N) - 1)^2 \tag{13}$$

The function means that the desired value $f(\omega)$ at fixed frequencies ω_d is -1, and the desired value $f(\omega)$ at the frequency ω_N is 1. The last term for the ω_N is added for decreasing the passivity in a high-frequency range. The sequence of the steepest descent method is descrived by the following steps using $\boldsymbol{x} = [A_1\ A_2\ \cdots\ A_c\ \alpha_1\ \alpha_2\ \cdots\ \alpha_C]^T$,

Step 1. The initial values of $\boldsymbol{x}$ are given as $\boldsymbol{x}^{(0)}$ and k=0.

Step 2. When Eqn. (14) is satisfied, the sequence finishes.

$$\frac{\partial J(\boldsymbol{x}^{(k)})}{\partial x_d^{(k)}} < \delta \quad (d = 1, \cdots, c) \tag{14}$$

Step 3. The values of $\boldsymbol{x}$ is replaced by the Eqn. (15)

$$\boldsymbol{x}^{(k+1)} = \boldsymbol{x}^{(k)} - \beta \left[\frac{\partial J(\boldsymbol{x}^{(k)})}{\partial x_1^{(k)}} \cdots \frac{\partial J(\boldsymbol{x}^{(k)})}{\partial x_c^{(k)}}\right]^T \tag{15}$$

Step 4. Return to the step 2 with k=k+1.

whrere, δ is a threshold value and β is a positive parameter, that are decided by try and error. The result of the optimization for a VC with two terms of a fractional derivative was A=[1.69, 1.32], α=[1.50, 3.47]. Figure 3 shows $f(\omega)$ of the optimized VC.

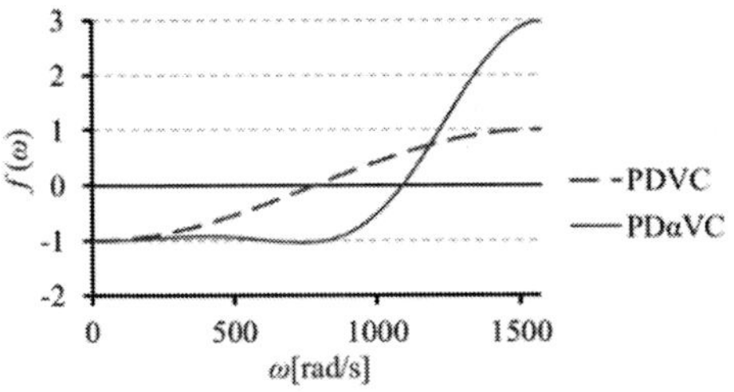

Fig. 3. $f(\omega)$ of the Optimized VC

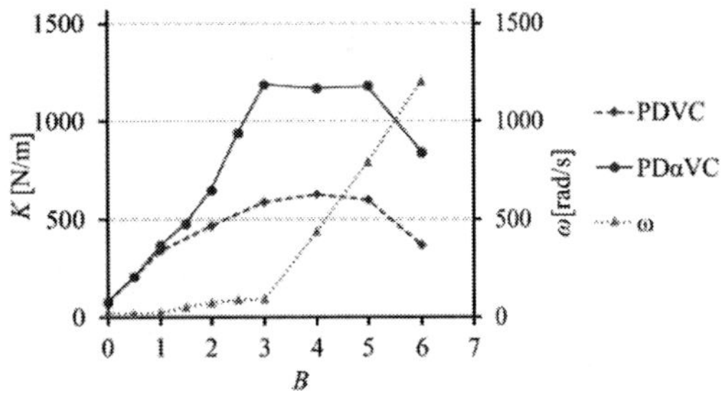

Fig. 4. Experimental results for the PDVC and the optimized VC

The figure shows that the optimized VC keeps lower passivity in the low- and middle- ranges of frequency. The passibity at high-frequency around the Nyquist frequency ω_N, however. becomes worse.

5 Experiment

In order to demonstrate the effect of the optimized VC and the influence of the worse effect at high-frequency range, we conduct experiments. In the experiment, we use a commercialized haptic device, Geomagic Touch X. The virtual environment is a virtual floor, and the task is to push the floor in the vertical direction. In the experiment, we attached a small mass to the fingertip of the device, and the fingertip was dropped onto the virtual floor without any human involvement because it is difficult to repeatedly apply a force of a given magnitude to a virtual wall by human action. We measured the maximum value of K for various B_α. The sampling rate is $T = 2$ (msec).

Figure 4 shows the experimental results for the common PDVC and the optimized VC. In Fig. 4, the solid red line corresponds to the results for the optimized VC, and the dashed blue line is the PDVC. The green line shows the occurred frequency at the experiment on the optimized VC.

6 Conclusion

In this research, we studied a virtual coupling including a fractional derivative approximated by a FIR approximation and assessed its characteristics. We then proposed a method to use multi terms of a fractional derivative and adjust the value of the parameters, based on a steepest descent method. Finally, we performed experiments to demonstrate the effects of the proposed virtual coupling. The analysis and experimental results show that the proposed method is effective and can be used to display virtual objects with harder surfaces.

References

1. R. J. Adams, B. Hannaford, IEEE Trans. on Robotics and Automation, **15(3)**, 465-474 (1999)
2. K. F. Laurin-Kovits, J. E.Colgate, S. D. R Carnes, Proc. of IEEE International Conference on Robotics and Automation, 1476-1481 (1991)
3. J. E. Colgate, J. M. Brown, Proc. of Int. Conf. on Robotics and Automation, 3205-3210 (1994)
4. J. E. Colgate, J. E. Stanley, M. C. Brown, IEEE/RSJ Proc. of the International Conference on Intelligent Robotics and Systems, 140-145 (1995)
5. M. Kawai, T. Yoshikawa, IEEE/ASME Trans. on Mechatronics, **9(1)**, 58-64 (2004)
6. D. W. Weir, J. E. Colgate, M. Peshkin, Symp. on Haptic Interface for Virtual Environment and Teleoperation Systems, 169-175 (2008)
7. B. S. Manohar, H. Vasudevan and M. Manivannan, Proc. of int. conf. on Haptics, 53-62 (2008)
8. J. H. Ryu, Y. S. Kim, B. Hannaford, IEEE Trans. on Robotics, **20 (4)**, 772-776 (2004)
9. I. Podlubny, *Fractional Differential Equations*, Academic Press (1999)
10. I.Petras , L.Dorcak, I. Kostial, Acta Montanistica Slovaca , **3**, 143-148 (1998)
11. S. Sakakibara, JSME International Journal C, **40(3)**, 393-399 (1997)
12. W. Zhang, N. Shimizu, JSME International Journal C, **42(4)**, 1-9 (1999)
13. B. M. Vinagre, Y. Q. Chen, I. Pretras, Journal of the Franklin Institute **340**, 349 -362 (2003)
14. C. Ma, Y. Hori, Trans. of the Institute of Electrical Engineers of Japan D, **124(8)**, 837-842 (2004)
15. Y. Q. Chen, B. M. Vinagre., "A new IIR-type digital fractional order differentiator," Signal Processing, **83**, 2359-2365 (2003)
16. J.A.T. Machado, Anal.-Model.-Simulation **27**, 107-122 (1997)
17. A. Oustaloup, F. Levron, B. Mathieu, F. M. Nanot, IEEE Trans. on Circuits and Systems I, **47(1)**, 25-39 (2000)

MATEC Web of Conferences **159**, 02044 (2018) https://doi.org/10.1051/matecconf/201815902044
IJCAET & ISAMPE 2017

Development of 6 DOF Supernumerary Robotic Fingers Integrated with 3D Animation

Mochammad Ariyanto*, *Joga* D. Setiawan, Munadi, *Rifky* Ismail, and *Zainal* Arifin

Department of Mechanical Engineering, Diponegoro University, Semarang, Indonesia

Abstract. The paper presents the development of wearable robot called supernumerary robotic fingers (SRF) that can be integrated with 3D Animation. The robot comprises of two parts, robotic finger (RF) and robotic thumb (RT). RF is located next to little finger, and RT is located next to the thumb. The robot is driven by modified glove sensor. 3D CAD model is exported into SimMechanics for 3D Animation purpose under MATLAB/Simulink environment. Forward kinematics of the robot is studied using SimMechanics and the trajectory of each robotic fingertip will be calculated. Based on the experimental result, the robot and its 3D animation can be operated by a healthy person in real time. The robot can assist in object manipulation with ten-inch tablet using single human hand.

1 Introduction

In wearable robot, a supernumerary robot is an additional robotic limb that is used to help users in holding, lifting, manipulating objects, and making it easier to do a job. Supernumerary robotic limbs (SRLs) can be fingers, arms, or legs. Supernumerary robotic fingers (SRF) has been extensively researched by the team from MIT [1-4]. The glove worn by user comprises of three stretch sensor. Based on the research works, the SRF and the human finger can work together to perform object manipulation tasks with single hand. SRF has been developed as an alternative device for prostheses and exoskeletons.

In this research, we develop a six degree of freedom (DOF) SRF as can be shown in figure 1. The robot comprises of two parts, robotic finger (RF) and robotic thumb (RT). RF is located next to little finger and RT is located next to the thumb as shown in figure 1 (b). The robot is designed by using SolidWorks CAD software. Both of the RF and the RT use six servos as the actuator and have 3 DOF. RT moves like a thumb in human hand having motion for circumduction, abduction, and flexion, while the RF moves like one of the human hand fingers having motion for abduction, and two flexion. The SRF will be attached on the wearer wrist/user. The human hand will wear modified glove sensor comprising three bend/flex sensor. The flex sensor has been utilized in previous research and can be used as input signal for robotic hand [5]. The sensor will read the bending motion of thumb, index, and middle finger of the human hand. The signal from sensor will be processed by microcontroller. It will drive the SRF and 3D animation simultaneously.

* Corresponding author: ari_janto5@yahoo.co.id

2 Design and Prototype

The posture of SR fingers is driven by modified glove sensor. It uses three flex sensors that are attached on thumb, index, and middle on the finger gloves. The flex sensors read the bending of the user's fingers and send it to the microcontroller. The virtual robot of SR fingers is developed in SimMechanics exported from CAD model in SolidWorks as shown in figure 1 (a). The weight of the final SRF is about 650 grams excluding battery. The battery is placed on the body of user to reduce the weight of SRF. Metal servo bracket is utilized to join between two medium servos.

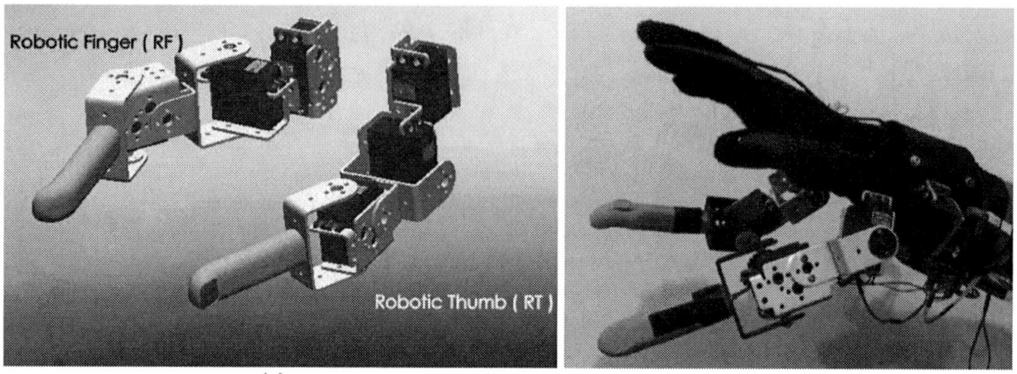

(a) (b)

Fig. 1. SRF design (a) CAD model, (b) Prototype.

3 Kinematics

For forward kinematics of RT and RF, Denavit-Hartenberg (DH) parameter is utilized to study the forward kinematics from each joint input to the fingertip position of RT and RF. Each of RT and RF on SRF has three link and three revolute joints. The D-H parameter of RT and RF can be summarized in Table 1 and Table 2 respectively.

Table 1. D-H parameter of robotic thumb

Link	a_i(mm)	α_i (degree)	d_i(mm)	θ_i(degree)
1	25	90°	0	θ_1
2	75	90°	0	θ_2
3	116	0°	0	θ_3

Table 2. D-H parameter of robotic finger

Link	a_i(mm)	α_i (degree)	d_i(mm)	θ_i(degree)
4	25	90°	0	θ_1
5	75	0°	0	θ_2
6	116	0°	0	θ_3

Based on the D-H parameter summarized in Table 1 and Table 2, the position of end effector in RT can be expressed from equation (1) to (3), and the end effector position of RF can be written from equation (4) to (6) as follows

$$X = a_1 \cos\theta_1 + a_2 \cos\theta_1 \cos\theta_2 + a_3 \sin\theta_1 \sin\theta_3 + a_3 \cos\theta_1 \cos\theta_2 \cos\theta_3 \tag{1}$$

$$Y = a_1 \sin\theta_1 + a_2 \sin\theta_1 \cos\theta_2 - a_3 \cos\theta_1 \sin\theta_3 + a_3 \sin\theta_1 \cos\theta_2 \cos\theta_3 \tag{2}$$

$$Z = a_2 \sin\theta_2 + a_3 \sin\theta_2 \cos\theta_3 \tag{3}$$

$$X = a_4 \cos\theta_4 + a_5 \cos\theta_4 \cos\theta_5 + a_6 \cos\theta_4 \cos\theta_5 \cos\theta_6 - a_6 \cos\theta_4 \sin\theta_5 \sin\theta_6 \tag{4}$$

$$Y = a_4 \sin\theta_4 + a_5 \sin\theta_4 \cos\theta_5 + a_6 \sin\theta_4 \cos\theta_5 \cos\theta_6 - a_6 \sin\theta_4 \sin\theta_5 \sin\theta_6 \tag{5}$$

$$Z = a_5 \sin\theta_6 + a_6 \cos\theta_5 \sin\theta_6 + a_6 \sin\theta_5 \cos\theta_6 \tag{6}$$

In this section, each joint angles in both RT and RF will be given with predefined angles as shown in figure 2 and figure 3. The trajectory of the end effector in both RT and RF can be calculated using equation from (1) to equation (6), or it can be computed by using SimMechanics Software. The resulted trajectory of the end effector can be seen in Fig. 4 (a) for robotic thumb tip and figure 4 (b) for robotic fingertip. Based on the simulation result, RT tip and RF tip do not contact to each other.

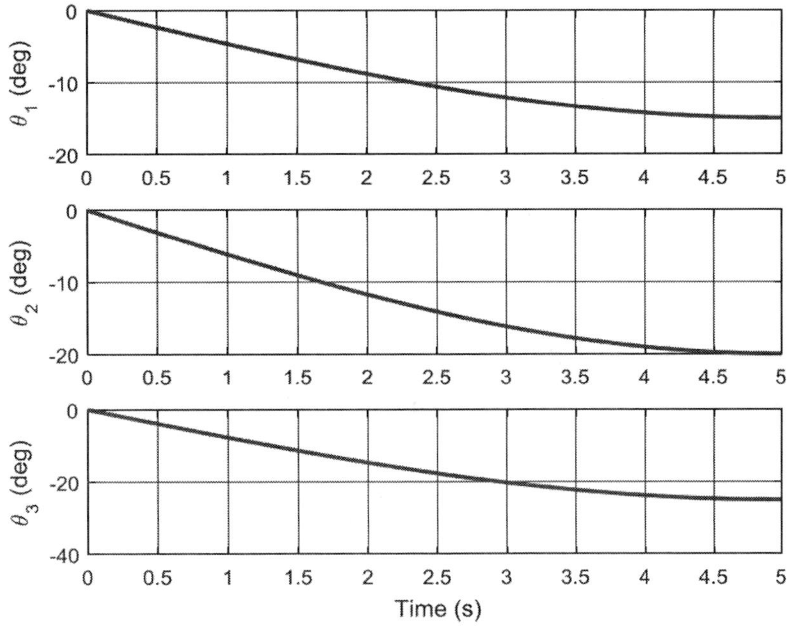

Fig. 2. Robotic thumb angle inputs.

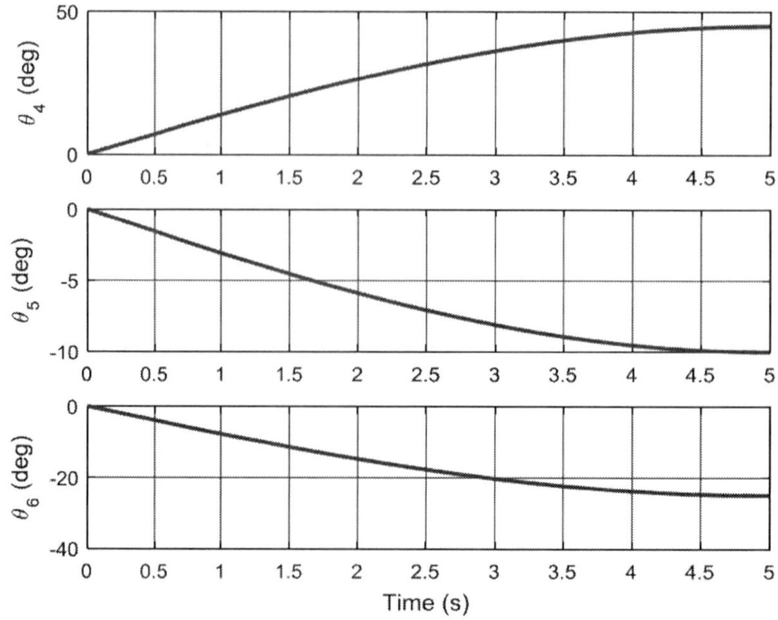

Fig. 3. Robotic finger angle inputs.

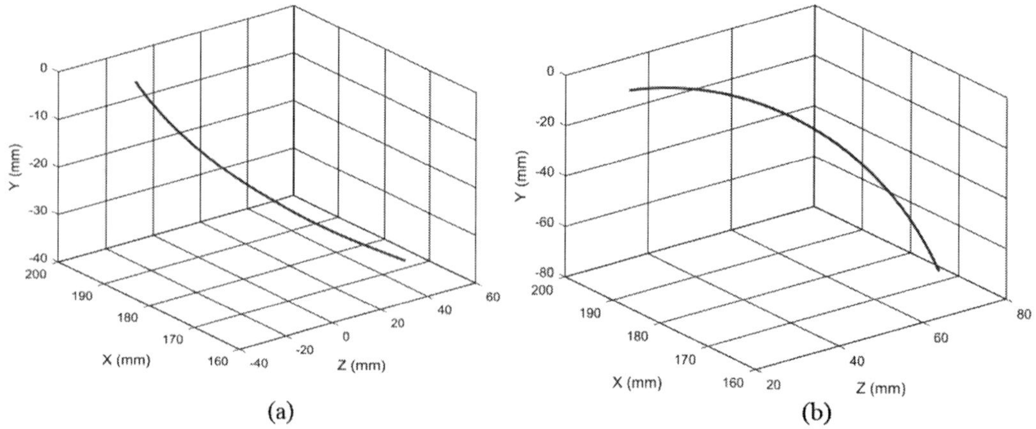

Fig. 4. Trajectory of robotic fingertips, (a) robotic thumb tip, (b) robotic fingertip

4 SRF with 3D Animation

For visualizing the motion both RF and RT, we develop 3D Animation of SRF that is built in SimMechanics environment. For fast transmission data between SRF and host computer, the selected baud rate is 115.200 bit/s. Based on the previous research, the virtual hand or 3D animation of robotic hand can run simultaneously with robotic hand [6,7]. The 3D CAD model of SRF as shown in figure 1 (a) is exported from SolidWorks software to SimMechanics First Generation. The resulted of SRF block diagram is presented in figure 5. This model is run in a host computer. The test of SRF integrated with 3D animation is presented in figure 6. Based on the test result, SRF and 3D animation can work simultaneously in real time.

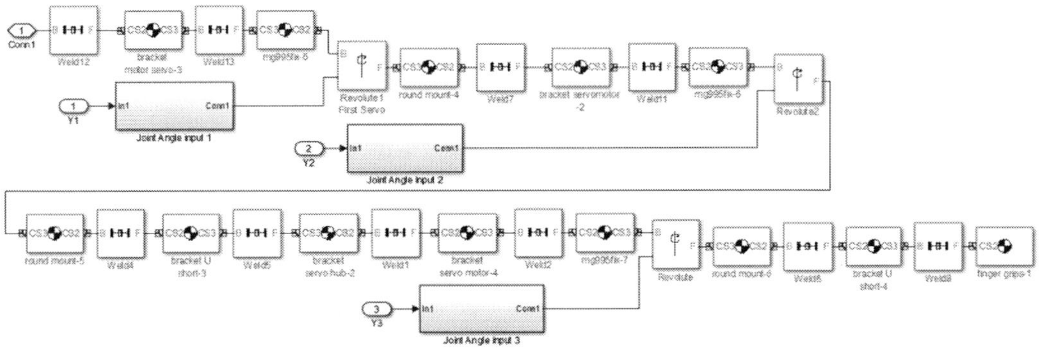

Fig. 5. SRF block diagram in SimMechanics.

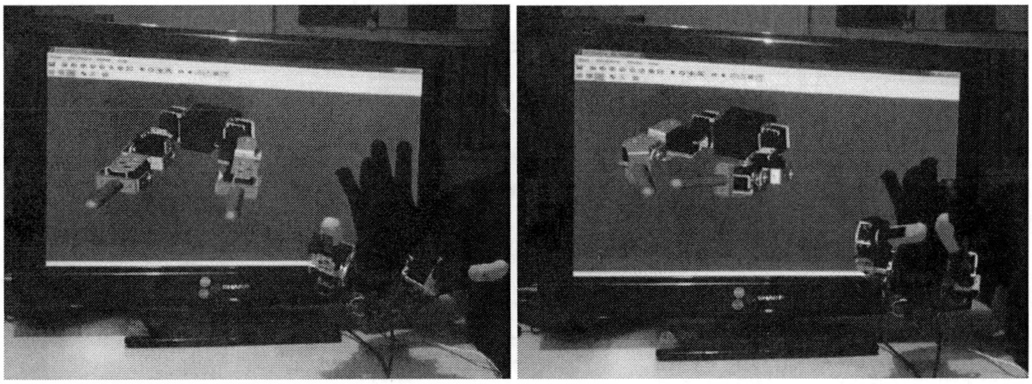

Fig. 6. SRF integrated with 3D Animation SimMechanics.

In this research, the SRF will be tested to perform object manipulation with ten-inch tablet. The test shows a promizing result in bimanual task. The SRF can hold a tablet while the human fingers operate the tablet.

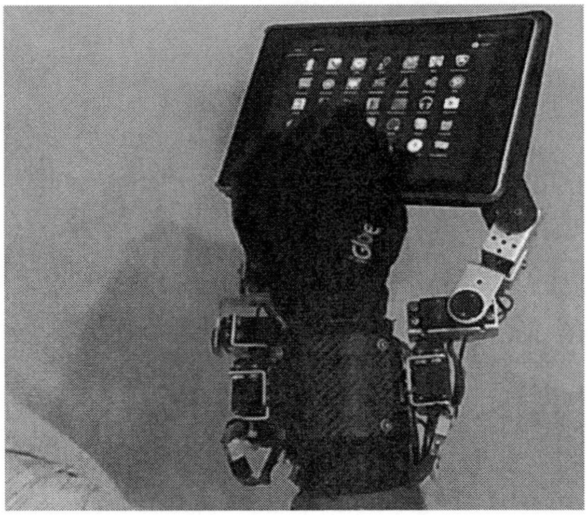

Fig. 7. Possible application of SRF for object manipulation.

5 Conclusion

Forward kinematics can be used to study the trajectory of the end effector in both RT and RF. Based on the experimental result, the robot and its 3D animation can be operated by a healthy person in real time. The robot also can do object manipulation with ten-inch tablet using single healthy hand. For future study, the robot will be explored for bimanual object manipulation using single human hand with the assistance of the SR fingers.

References

1. F. Y. Wu, H Asada, ASME Dynamic Systems and Control Conference, 1-8, (2014)
2. F. Y. Wu and H. H. Asada, 2015 IEEE International Conference on Robotics and Automation (ICRA), 6205-6212, (2015)
3. F. Y. Wu and H. H. Asada. IEEE Trans. Robot., **32**,176-186 (2016)
4. F. Y. Wu, and H. Asada. in Robotics: Science and Systems X, University of California, Berkeley, USA (2014)
5. M. Ariyanto, R. Ismail, A. Nurmiranto, W. Caesarendra, Paryanto and J. Franke, 2016 IEEE EMBS Conference on Biomedical Engineering and Sciences (IECBES), Kuala Lumpur, 341-346,(2016)
6. R. Ismail, M. Ariyanto, W. Caesarendra, I. Haryanto, H. K. Dewoto and Paryanto, 2016 IEEE EMBS Conference on Biomedical Engineering and Sciences (IECBES), Kuala Lumpur, 335-340, (2016)
7. R. Ismail, M. Ariyanto, W. Caesarendra and A. Nurmiranto, 2016 International Seminar on Intelligent Technology and Its Applications (ISITIA), Lombok, 629-634, (2016)

Design of a Low-Cost Fixed Wing UAV □

Mochammad Ariyanto*, *Joga* D. Setiawan, *Teguh* Prabowo, *Ismoyo* Haryanto, *Munadi*

Department of Mechanical Engineering, Diponegoro University, Semarang, Indonesia

Abstract. This research will try to design a low cost of fixed-wing unmanned aerial vehicle (UAV) using low-cost material that able to fly autonomously. Six parameters of UAV's structure will be optimized based on basic airframe configuration, wing configuration, straight wing, tail configuration, fuselage material, and propeller location. The resulted and manufactured prototype of fixed-wing UAV will be tested in autonomous fight tests. Based on the flight test, the developed UAV can successfully fly autonomously following the trajectory command. The result shows that low-cost material can be used as a body part of fixed-wing UAV.

1 Introduction

Recently utilization of Unmanned aerial vehicle (UAV) has grown significantly especially for fixed wing and rotary wing type. The UAV is controlled by the pilot (operator) in the ground control station. UAV can fly autonomously based on programmed flight plans into the flight controller board. Many users/hobbyist build fixed-wing Unmanned Aerial Vehicle (UAV) by themselves, they have to design, calculate aerodynamic performance, and manufacture the UAV by themselves. Furthermore, they have to decide which mechanical and electrical components needed in order to build the UAV that can perform as intended. Building UAV is cheaper than buying factory-built UAV. Design and manufacture of UAV require design process, manufacturing, and numerous flight tests until the UAV has good tracking performance both in manual and autonomous flight.

This research will focus on the design optimization of six parameters to achieve low cost and easy to manufacture fixed-wing UAV. The parameters are basic airframe configuration, wing configuration, straight wing, tail configuration, fuselage material, and propeller location. The selection of structural configuration and material of the UAV is presented in section 2. The fixed-wing UAV is equipped with available commercial flight controller board. After the prototype is built, it will be tested to fly autonomously following the predefined trajectory.

2 UAV Fixed Wing Design

In this section, six design configuration structures will be considered i.e., basic airframe configuration, wing configuration, straight wing, tail configuration, fuselage material, and propeller location. Three basic airframe configurations are considered. There are four

* Corresponding author: mochammad_ariyanto@ft.undip.ac.id

things to consider, namely the ease of manufacturing process, stability, ease of hand launch, and mass. Table 1 shows the comparison between the three basic airframe configurations, where configuration five is the best, and configuration one is the worst. In the conventional configuration, horizontal tail is at the back of the fuselage. Horizontal tail and vertical tail can be attached to each other. The back of the fuselage can be simplified in the form of an aluminum rod to reduce the weight of the system. Because of the aluminum rod, horizontal tail and vertical tail can be fitted strongly enough without the need for a special cradle, just by using a clamp. □

Table 1. Comparison of basic airframe configuration.

Design Parameters	**Weight**	**Conventional**	**Canard**	**Flying Wing**
		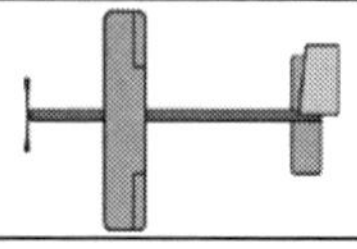	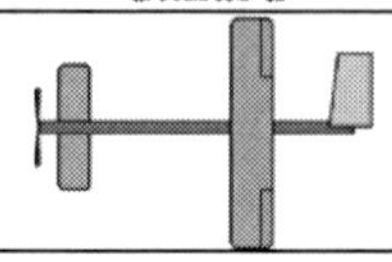	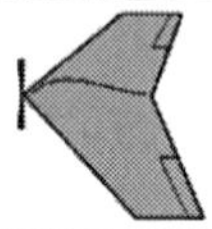
Ease of manufacturing process	0,3	4	3	5
Stability	0,3	5	4	2
Ease of hand launch	0,2	5	5	4
Mass	0,2	4	4	5
Total Σ (weight x value)	1	4,5	3,9	3,9

Three vertical wing positions against the fuselage will be considered. There are four things to consider, namely the ease of manufacturing process, stability, ease of hand launch, and ground clearance. Table 2 shows the comparison of the three vertical positions of the wings to the fuselage. In this research, the selected configuration is high wing configuration.

Table 2. Comparison of wing configuration.

Design Parameters	**Weight**	**High Wing**	**Mid Wing**	**Low Wing**
		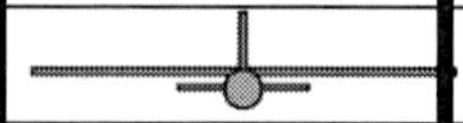	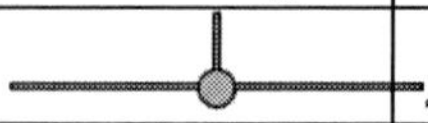	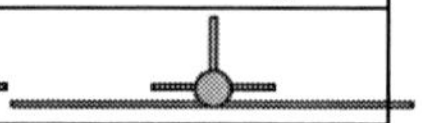
Ease of manufacturing process	0,3	5	3	4
Stability	0,3	5	4	3
Ease of hand launch	0,2	5	4	3
Ground Clearance	0,2	5	4	3
Total Σ (weight x value)	1	5	3,7	3,3

Planform is a wing shape when the wing is viewed from above. This study does not require a UAV that prioritizes speed. The wing planform suitable for low-subsonic planes is straight wing [1]. Three planform of straight wing are considered. There are three things to consider, namely the ease of manufacturing process, structural strength, and aerodynamic performance. Table 3 shows the comparison between the three planform of straight wing. In terms of aerodynamics and structure, rectangular wing tends to be inferior to two other wing planform. However, in terms of manufacturing, this wing is the easiest and cheapest to make because it has only one size air foil [1].

Table 3. Comparison of straight wing.

Design Parameters	Weight	Rectangular	Tapered	Prismatic Mid-Section
		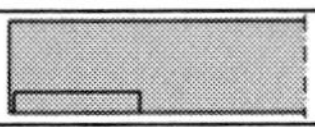	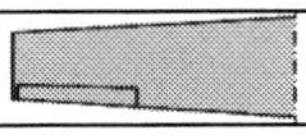	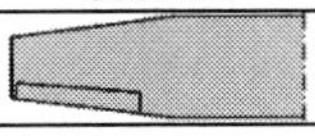
Ease of manufacturing process	0,6	5	3	4
Strength of structure	0,2	4	5	5
Aerodynamic performance	0,2	3	5	4
Total Σ (weight x value)	1	4,4	3,8	4,2

Three tail configurations are considered. The tails are conventional, T-Tail, and V-Tail. There are three things to consider, namely the ease of manufacturing process, weight structure, and aerodynamic performance. Table 4 shows the comparison between the three tails. Conventional tail configurations are the most common configuration used in UAVs. This configuration is easy to build because it has a simple, lightweight, and powerful structure. In this setting, the selected tail configuration is a conventional configuration.

Table 4. Comparison of tail configuration

Design Parameters	Weight	conventional	T-Tail	V-Tail
		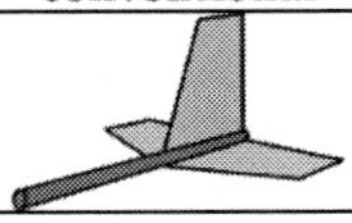	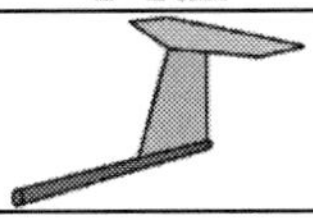	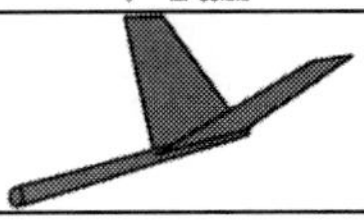
Ease of manufacturing process	0,4	5	4	4
Weight of structure	0,4	5	3	5
Aerodynamic performance	0,2	4	5	3
Total Σ (weight x value)	1	4,8	3,8	4,2

Three types of fuselage materials are presented in this study. There are five things to consider, namely the ease of manufacturing process, structural strength, price, availability, and weight of the structure. Table 5 shows the comparison between the five types of fuselage materials. Based on Table 5, the selected fuselage material is plastic tube and aluminum cylinder.

Table 5. Comparison of fuselage material.

Design Parameters	Weight	Plastic tube and aluminum cylinder	Wood Balsa	Foam
		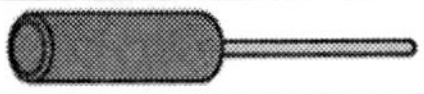	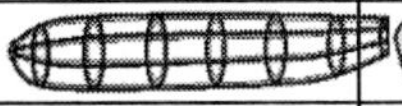	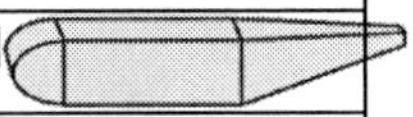
Ease of manufacturing process	0,2	5	3	4
Strength of structure	0,2	5	2	3
Price	0,2	5	2	5
Availability	0,2	5	3	5
Weight of structure	0,2	4	5	4
Total Σ (weight x value)	1	4,8	3	4,2

Four propeller locations are considered in the selection of propeller locations. Table 6 shows the comparison between the four propeller locations. The pusher configuration has a motor located behind the fuselage. Similar to the tractor configuration, the position of the propeller axis is parallel to the fuselage symmetry axis (longitudinally) so that the distance from the edge of the propeller to the ground (ground clearance) becomes minimal. This increases the risk of propeller breaks during landing. Since the propeller is at the back of the UAV, the risk of throwing hands is exposed to the edge of the propeller when the hand launch is enormous. In addition, since the UAV in this study will use aluminum cylinder as the rear fuselage, the installation of the motor on the aluminum cylinder requires a special cradle, thus increasing the difficulty of manufacturing the UAV.

Table 6. Comparison of propeller location.

Design Parameters	Weight	Tractor	Twin Engine	Mid Engine	Pusher
		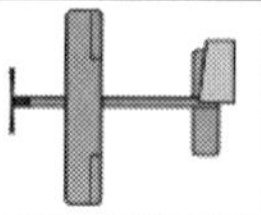	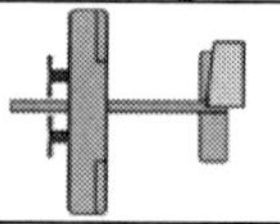	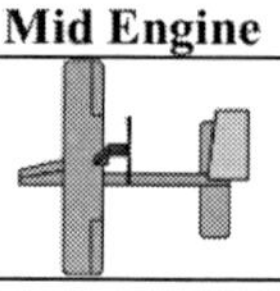	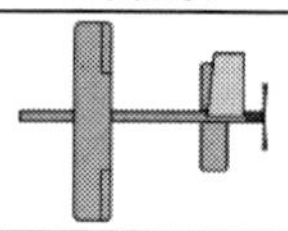
Ease of manufacturing process	0,3	5	3	3	4
Safety when hand launch	0,3	5	5	5	3
Mass	0,2	5	3	4	5
ground clearance	0,2	3	4	5	3
Total Σ (weight x value)	1	4,6	3,8	4,2	3,7

In the fuselage design, the selected fuselage material is a plastic drawing tube and an aluminum cylinder. The plastic drawing tube is for the front fuselage and aluminum cylinder for the rear fuselage. The plastic drawing tube used in this study has a diameter of 8 cm. The aluminum cylinder has a diameter of 1.2 cm. Figure 1 shows the fuselage design. From its original size, the plastic drawing tube is modified so that its length becomes 62 cm. Aluminum cylinder has a length of 59 cm but put 10 cm into the drawing tube. The final 3D design is presented in Figure 2.

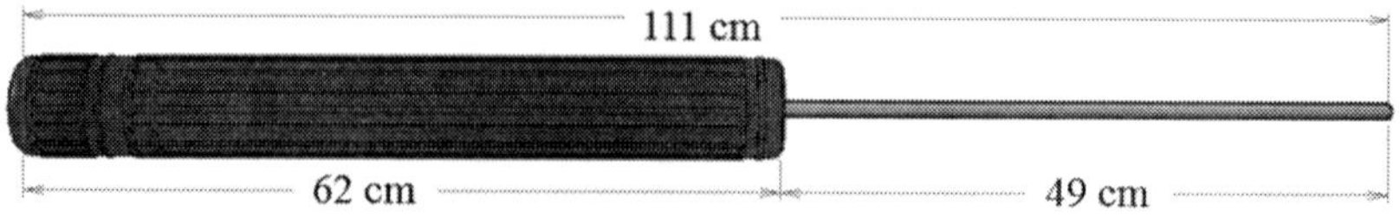

Fig. 1. Design of fuselage

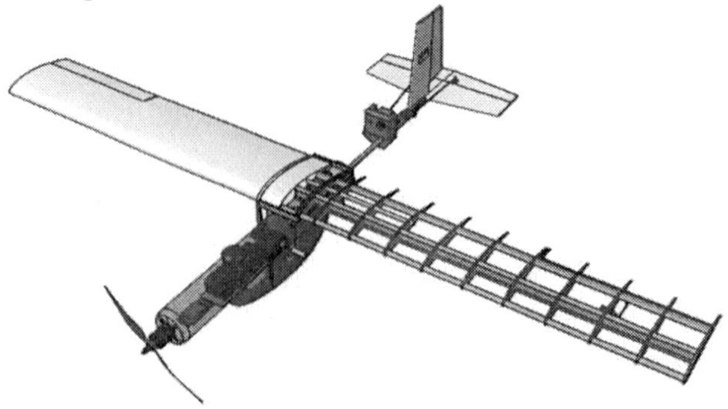

Fig. 2. Isometric view of the final design of fixed-wing UAV

3 Prototype

After the design process is complete, next step process is to manufacture the fixed-wing UAV. The prototype produced in this study can be seen in Figure 3. The camera is mounted on an aluminum cylinder using a stand made of EVA (ethylene-vinyl acetate) foam. The stand is also useful for damping vibrations from the UAV structure to the camera for better video recording quality. The motor used in this research is brushless DC motor. The brushless DC motor has better efficiency and power to weight ratio than the brushed DC motor. In this study, the brushless DC motor used is SunnySky X2820-800KV. Based on the data specifications SunnySky X2820-800KV, the motor can produce a maximum thrust of 2.24 kg when paired with a 12 x 6 propeller and 14.8 V battery. So that, the propeller used is propeller 12 x 6 made of nylon. Propeller made from nylon is cheaper than wood or carbon fiber. To obtain a thrust of 2.24 kg, the specifications of the SunnySky X2820-800KV motor require the use of 14.8 V batteries. The batteries used in this study are the most commonly used batteries on RC plane, the lithium polymer batteries. Four servos are used to drive the ailerons elevator, and rudder.

4 Flight Test

In this research, improving tracking performance in UAV is conducted gradually. Before improving tracking performance, fixed-wing UAV performance in the entering waypoints is improved first. In the first Flight test, the UAV is unable to enter the waypoint accurately. On the second flight test, K_p, K_i, and K_d on roll and pitch controller still use default settings, same as the first flight. However, to improve the performance of the UAV in entering the waypoint, the Turn Control Period and Waypoint Radius parameters must be tuned on the second flight.

PID is the widely used compensator for controlling the stability of quadrotor [2], and also PID is commonly employed in fixed-wing UAV [3]. In this research, PID control is utilized to control roll and pitch angles of fixed-wing UAV. In the first flight test, the Turn Control Period parameter value is 20. The UAV must turn sharper in order to enter waypoints with better accuracy. Therefore, the value of the Turn Control Period parameter must be decreased. In the second flight test, the value of the Turn Control Period parameter is decreased by 25% so that the value becomes 15. In the first flight test, the value of Waypoint Radius parameter is 30 m. In the second flight test, the value of the Waypoint Radius parameter is reduced to 20 m. This is done so that when entering the waypoint, UAV is not too far from the center of Waypoint Radius and can turn sharper. Table 7 shows the setting values on the roll, pitch, and navigation controller.

Table 7. Control parameters for first and second flight test.

Controller	PID parameters	Values	
		First flight test	Second flight test
Roll	[K_p K_i K_d]	[0.415 0.05 0.02]	[0.415 0.05 0.02]
Pitch	[K_p K_i K_d]	[0.405 0.04 0.02]	[0.405 0.04 0.02]
Navigation	Turn Control Period	20	15
Navigation	Waypoint Radius (m)	30	20

In the second flight test, the UAV is capable of autonomous flight and video recording as shown in Figure 3. Figure 4 shows a two-dimensional view of the UAV flight path in Auto mode on the second flight test. On the second flight test, it appears that the UAV can enter the waypoint accurately. However, it appears that the flight path does not coincide with the track. That is, the tracking performance is still not accurate enough.□

(a) (b)

Fig. 3. Fixed-wing UAV in autonomous flight, (a) Flight above 100 m from terrain, (b) First person view display

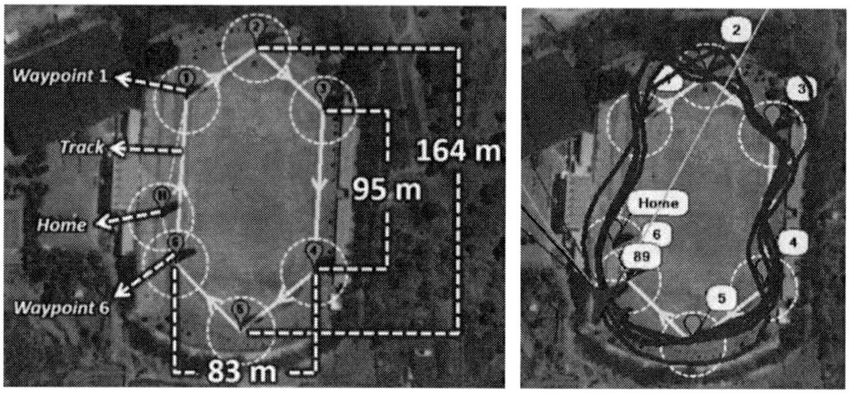

Fig. 4. UAV fixed wing trajectory, (a) Trajectory command, (b) Trajectory response

5 Conclusion

In this research, a low-coconlst fixed-wing UAV has been successfully designed and built. The material that was used for creating a fixed-wing UAV should be lighter, stronger, cheaper, and easier to manufacture. For future flight test, the control parameters will be tuned to get the better result. The final selected structure configurations for low cost fixed-wing UAV are conventional airframe, high wing, rectangular wing, conventional tail configuration, tractor propeller, plastic tube and aluminum cylinder for fuselage material. Based on the flight test, the developed UAV can successfully fly autonomously following trajectory command. The result shows that low-cost material can be used as a body part of fixed-wing UAV.

References

1. E. Torenbeek. *Synthesis of Subsonic Airplane Design*. Springer Netherlands, (1982)
2. J. D. Setiawan, Y. D. Setiawan, M. Ariyanto, A. Mukhtar and A. Budiyono, *Development of real-time flight simulator for quadrotor*, 2012 International Conference on Advanced Computer Science and Information Systems (ICACSIS), Depok, 59-64, (2012)
3. F. A. Warsi et al., *Yaw, Pitch and Roll controller design for fixed-wing UAV under uncertainty and perturbed condition*, 2014 IEEE 10th International Colloquium on Signal Processing and its Applications, Kuala Lumpur, 151-156, (2014)

The effect of acid leaching time in modifying natural zeolite as catalyst for toluene steam reforming

Joko Waluyo[1,2], *I.G.B.N.* Makertihartha[1], *Herri* Susanto[1,*]

[1]Department of Chemical Engineering, Institut Teknologi Bandung, Jalan Ganesha 10 Bandung, Jawa Barat, Indonesia
[2]Department of Chemical Engineering, Universitas Sebelas Maret, Jalan Ir. Sutami 36A Surakarta, Jawa Tengah, Indonesia

Abstract. Tar accumulation can result in blockage, corrosion, and poisoning of catalyst. One of the methods applicable for tar removal is by using a catalyst that enhances steam reforming reaction. In this study, natural zeolites were modified for the use as tar steam reforming catalyst. This study also covers characterizing the modified natural zeolite by ion exchange and acid leaching, and measuring its activity as toluene steam reforming catalyst. The XRD pattern showed that natural zeolite contains three crystalline phases, i.e. mordenite, clinoptilolite, and heulandite. The main structure of the mordenite phase was also detected by using FTIR analysis on T-O stretching band with the range of 1045-1075 cm^{-1}. The zeolite modification by ion exchange and acid leaching resulted in decreasing content of impurities, such as Na and K, and the decrease in crystallinity from 68.8% to 61.8%. The surface area increased from 16 m^2/g to 31.4 m^2/g and 161.8 m^2/g for ZMH (after ion exchange), ZM3 (after acid leaching for 3 hour), and ZM6 (after acid leaching for 6 hour) respectively. Theses catalyst gave toluene conversion of 30.5% to 42.6% and 54% respectively.

1 Introduction

Gasification is a thermal conversion process that can converts wastes or biomass into syngas. The syngas obtained can be utilized for electricity production, synthesis of valuable chemicals such as methanol, DME, H_2, other Fischer-Tropsch fuel, etc. However, syngas still contains impurities such as particulates, ammonia, H_2S and tar [1] that must be removed prior to its utilization.

Tar can be defined as complex compounds of polyaromatic hydrocarbons with higher molecular weight as compared to benzene. The tar content in syngas ranges from 10-150,000 mg/Nm^3 depending on the type and operating condition of the gasifier [2]. Tar accumulation may cause blockage, corrosion in pipeline, and also deactivation of catalyst

* Corresponding author: herri@che.itb.ac.id

for chemical synthesis. The presence of tar leads to both operational and maintenance problems in gasification processes.

The advantages in using a catalytic process for tar removal includes the avoidance of producing hard to handle liquid waste, high thermal efficiency as it can be carried out at the temperature in the range that the gasifier operates, and the increase in H_2 and CO concentration [3]. Steam reforming reactions for tar removal have additional advantages such as increasing the H_2 content in syngas and reduce the amount of carbon deposits on the catalyst, which extends the catalyst lifetime [4]. Nickel is widely used as a catalyst for tar steam reforming with various supports. Natural zeolites are relatively inexpensive and so the use of natural zeolite catalyst is economically feasible if the produced syngas is to be utilized for small-scale power generation. The natural zeolite has been widely developed for industrial use in various applications, such as cracking, hydrocracking and isomerization[5]. Natural zeolite has diverse crystal phase, mordenite, clinoptilolite and mesolite are commonly used as a catalyst.

Mordenite can be used as a steam reforming catalyst [6, 7]. Mordenite can be obtained naturally or through synthesis. Natural zeolite has several disadvantages for use as a catalyst such as containing impurities and having low surface area. The existing impurities such as Na, K have a low melting point and when the modified zeolite is used as a catalyst at high temperature, it may decrease the activity of the catalyst. It is, therefore, necessary to modify or improve these properties.

Zeolite modification can be performed by conducting ion exchange, acid leaching, alkaline leaching or hydrothermal treatment. In this study, the natural zeolite is modified by ion exchange and acid leaching. Acid leaching are performed with the intention of removing impurities and also possibly increasing the surface area through dealumination. The dealumination process causes increased thermal resistance and decreases the acidity of the zeolite. The purpose of this study is to determine the effect of acid leaching duration, the characteristics of zeolite and its activity.

2 Material and method

2.1 Material

Natural zeolite collected from Wonosari, Yogyakarta was used as raw material. The zeolite used has a size of 5-10 mesh. Solution of NH_4NO_3 and HNO_3 were used as ion exchange and acid leaching agent to remove impurities and improve catalyst characteristics. NH_4NO_3, HNO_3, and toluene were purchased from Merk. Natural zeolite pulverized into a powder then characterized using XRD to determine the crystal phase. XRF and EDX are used for characterizing elemental composition.

2.2 Modification of zeolite

Natural zeolite was washed using demineralized water, followed by drying. The first modification was an ion exchange to make H-zeolite. One gram of zeolite stirred and refluxed with 10 mL 2M NH_4NO3 at 90°C for 2 hours. Ion exchange produces NH_4-zeolite and then washed with distillates water followed by calcination at 550°C for 12 hours to remove NH_3 and obtain H-zeolite. Acid leaching was carried out by reflux using 6M HNO_3 at 90°C with the variation of 3 hours (ZM-3) and 6 hours (ZM-6) leaching time and 20 wt. / wt. At the end of a modification of sample washed with demine water and calcined at 550°C for 12 hours. Table 1 shows the condition for modification process.

Table 1. Modification condition of natural zeolite

No	Zeolite	ion exchange	acid leaching
1	ZMH	NH_4NO_3 2M, 2hour	-
2	ZM3	NH_4NO_3 2M, 2hour	HNO_3 6M, 3hour
3	ZM6	NH_4NO_3 2M, 2hour	HNO_3 6M, 6hour

2.3 Characterization of catalyst

The sample pulverized into powder with particle size of less than 100 microns. The crystalline morphology structure was characterized using X-Ray Diffraction with Bruker D8 Advance type. The diffractometer uses CuKα radiation by recording 2θ at room temperature of 5°-50°, scan rate 76.8°/minute and step size 0.019. For surface and pore structure, liquid nitrogen absorption was measured at 77.35 K using Nova 3200e. The Brunauer-Emmet-Teller (BET) method used to measure surface area, pore volume and pore size using the Barrett-Joyner-Halenda (BJH) method. The Composition of the zeolite element was characterized by field-emission electron microscope (FE-SEM) and energy disperse X-ray (EDX) and also XRF. Infrared absorption was performed using FTIR spectrophotometer with a wavelength in the range 400-4500 cm^{-1}.

2.4 Activity test of zeolite

Two grams of zeolite incorporated into tar cracking fixed bed reactor with a diameter of 2 cm and a length of 50 cm. The reactor conditions were set at 1 atm and cracking temperature 750°C. Nitrogen as a carrier gas is passed through the tar model (toluene) at ambient condition and regulated by using a needle valve and measured using a flow meter. Water flowed using a syringe pump. The flow rate was adjusted to tar flowed having a concentration of 50 g/Nm^3 and has a ratio of steam/ tar 10. The tar rate used was 0.21 grams/hour and the rate for N_2 8400 mL/hr. Tar cracking was done for 8 hours. The toluene concentration in and out of the reactor was analyzed using GC-FID Shimadzu GC-14B. Spent cracking catalysts were tested using TGA to determine the possibility of carbon deposits.

3 Result and discussion

3.1 Characteristic of zeolite

The XRD results show that natural zeolite of Wonosari containing Mordenite, Clinoptilolite and Heulandite crystal phase. Unique diffraction for mordenite at 2θ 6.5°; 9.76°; 13.43°; 19.59°; 22.18°; 25.6°; 26.22°; 27.65°; 28.25°, while 2θ at 25.7°; 26.02°; 28.57° and 30.02° are unique diffraction for clinoptilolite and heulandite. Ion exchange results in their missing of diffraction peaks of zeolite (Figure 1). This indicates the loss of some zeolite impurities especially Na and K which have been replaced with H^+ cations, such as the results of elemental composition analysis presented in Table 2. In modification with acid leaching, there is a slight decrease from the relative height of the peak on the diffraction. This is due to the destruction of heulandite structures. Kim, Hwang [8], states that the heulandite is easily destroyed by acid leaching while mordenite and clinoptilolite are stronger against acid leaching. The decrease in intensity indicates a decrease in the number of Al atoms, resulting in a decrease in the crystallinity of the zeolite (Table 2).

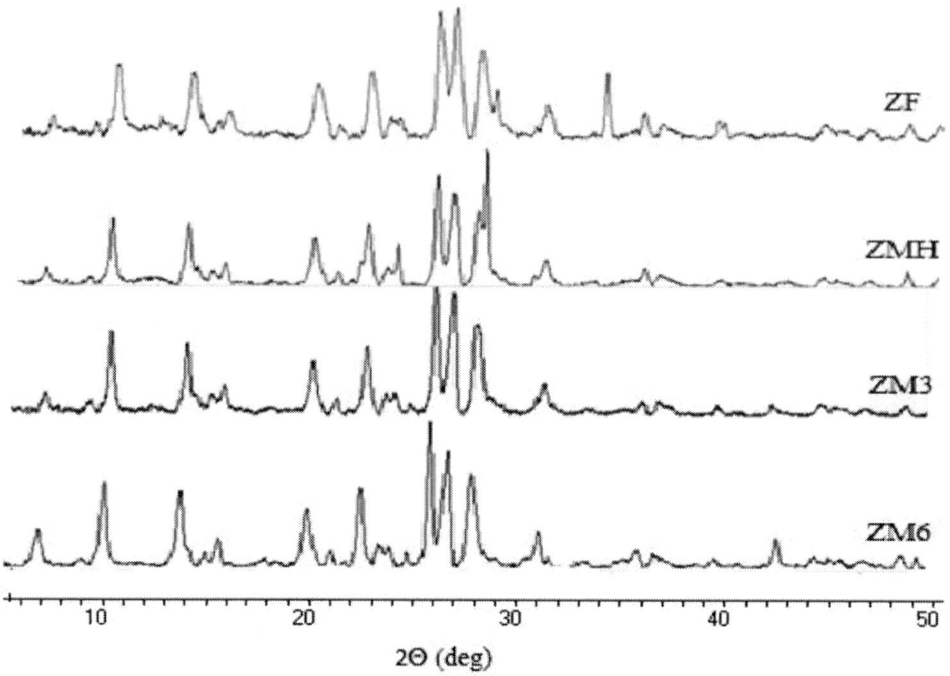

Fig. 1. XRD pattern of parent zeolite and after modification

Table 2. Elemental composition as determine by EDX analysis

No	Samples	ZF	ZH	ZM3	ZM6
1	Elemental composition (EDX analysis)				
	Si	29.75	32.47	31.48	31.68
	Al	7.09	7.85	7.23	6.71
	Na	0.72	0.42	0.22	0.22
	K	0.65	0.17	0	0
	Mg	0.87	0.53	0.53	0.58
	Ca	3.06	2.15	1.72	1.75
	Fe	2.57	2.21	1.66	2.01
	Ti	0.27	0.34	0.27	0.34
	O	55.02	51.31	51.21	51.19
	C	0	2.55	5.68	5.52
2	Psychochemical properties				
	Si/Al ratio*	-	5.09	5.02	5.61
	A_{BET} (m^2/g)	-	16	31.4	161.8
	V_t (cm^3/g)	-	0.094	0.086	0.159
	Crystallinity	-	66.3%	63.7%	61.8%

*Analyzed by XRF

The zeolite structure for the mordenite phase can also be detected by using FTIR analysis (Figure 2) on a T-O stretching band with a range of 1045-1075 cm^{-1}. In that range, the higher the wavenumber indicates that there is a dealumination or decreasing number of Al atoms in the tetrahedral framework [9]. FTIR spectra also shows the presence of amorphous or quartz SiO_2 silicate on a waveband of about 795 cm^{-1}. While at wavenumber about 450 cm^{-1} is another Si-O-Si bonding.

Vibration at the wavelength of 1600-3700 cm^{-1} is the wavelength for zeolitic water that indicates an interaction between hydroxyl and cation. FTIR analysis results in ZF and ZMH show the appearance of a wavelength at 2854 cm^{-1} and 2928 cm^{-1}. After acid leaching (ZM3 and ZM6) the wavelength disappears indicating that the hydroxyl bond has been removed. In the dealumination process, the loss of Al from the tetrahedral framework can be an extra framework octahedral Al, this is apparent at the wavelengths formed in the 3700-3650 cm^{-1} wavelength range.

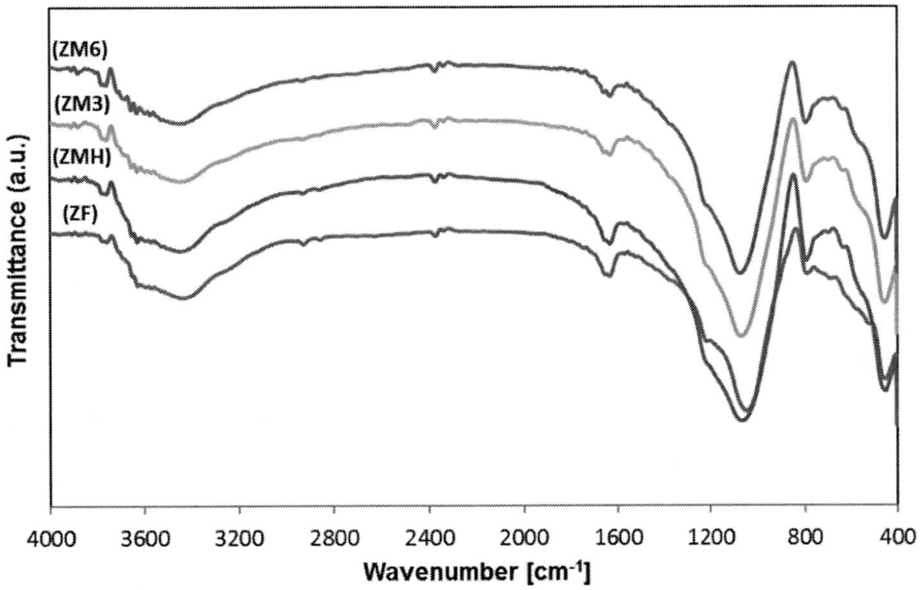

Fig. 2. FTIR spectrum of parent zeolite and modified zeolite

Acid leaching in natural zeolite can remove impurities and some aluminum atoms from the zeolite framework and lead to changes in pore structure such as an increase in pore volume at 38 Å pore diameter. In acid leaching with HNO_3, the XRF results show that the Si/Al ratio does not increase significantly, this is due to the formation of extra framework aluminum (EFAL) as described in FTIR characterization above. The low Si/Al ratio makes the acidity of zeolite is also relatively not too high. This will be advantageous for reactions related to cracking or breaking of bonds.

3.2 Catalytic activity test

Modifying the natural zeolite with mordenite as the main phase, through ion exchange and acid leaching can increase the conversion of toluene cracking by reaction of steam reforming. Conversion of toluene with the modified natural zeolite can reach up to 54% (Figure 3), this is due to modification by acid leaching can increase the surface area of zeolite mordenite, and the toluene will easily react with steam. Acid leaching time affects the surface area of the zeolite, the longer the leaching makes the surface area become larger as presented in Table 2.

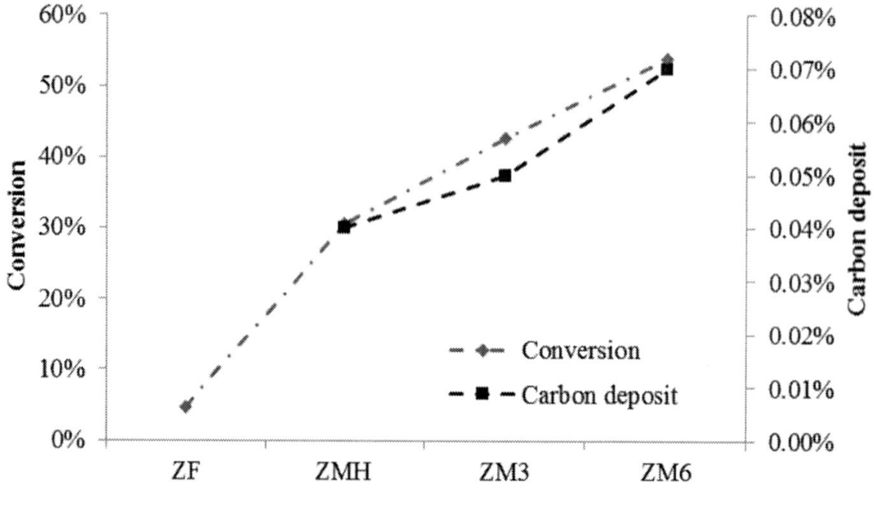

Fig. 3. Conversion of toluene and carbon deposits on various zeolites modification

4 Conclusions

The mordenite crystal phase of the natural zeolite is unchanged after modification by acid leaching using the HNO_3 solution, but the heulandite crystal phase undergoes a reduction. The longer the acid leaching can increase the surface area of the natural zeolite because the loose aluminum from the zeolite framework forms EFAL so that the catalyst activity increases.

This topic is a part of the doctoral research in Chemical Engineering Dept. ITB. The authors would like to acknowledge the financial support from *Hibah P3MI, Institut Teknologi Bandung 2017.* Thanks to Dr Jenny Rizkiana for valuable discussion.

References

1. N. Abdoulmoumine, S. Adhikari, A. Kulkarni, and S. Chattanathan, *Applied Energy*, **155**, 294-307 (2015)
2. M. Asadullah, *Renewable and Sustainable Energy Reviews*, **40**, 118-132 (2014)
3. S. Schmidt, S. Giesa, A. Drochner, and H. Vogel, *Catalysis Today*, **175**(1), 442-449 (2011)
4. M. Kaewpanha, G. Guan, X. Hao, Z. Wang, Y. Kasai, S. Kakuta, K. Kusakabe, and A. Abudula, *Journal of the Taiwan Institute of Chemical Engineers*, **44**(6), 1022-1026 (2013)
5. J. Waluyo, T. Richards, I. Makertihartha, and H. Susanto, *Asean Journal of Chemical Engineering*, **17**(1), 37-45 (2017)
6. J.F. Da Costa-Serra, M.T. Navarro, F. Rey, and A. Chica, *International Journal of Hydrogen Energy*, **37**(8), 7101-7108 (2012)
7. T. Matsumoto, T. Nishiguchi, H. Kanai, K. Utani, Y. Matsumura, and S. Imamura, *Applied Catalysis A: General*, **276**(1–2), 267-273 (2004)
8. M.H. Kim, U.-C. Hwang, I.-S. Nam, and Y. Gul Kim, *Catalysis Today*, **44**(1–4), 57-65 (1998)
9. K. Elaiopoulos, T. Perraki, and E. Grigoropoulou, *Microporous and Mesoporous Materials*, **134**(1–3), 29-43 (2010)

Effect of quenching agent on dimension stability of Al 6061-Al_2O_3 composite

Hammar Ilham Akbar[1,*], *Eko* Surojo[2], and *Dody* Ariawan[2]

[1]Graduate School of Mechanical Engineering, Universitas Sebelas Maret, Indonesia
[2]Department of Mechanical Engineering, Faculty of Engineering, Universitas Sebelas Maret, Indonesia

Abstract. Al 6061-Al_2O_3 composite is a lightweight materials that can be used as an alternative material in automotive parts. T6 heat treatment is generally used on aluminium alloy series 6xxx . On T6 heat treatment process, material heated to temperature for obtaining a single phase of α, quenched and then continued by artificial ageing. Quenching is critical step because distortion can be occur in material. The aim of this research is to investigated effect of quenching agent and electroless coating treatment of Al_2O_3 in distortion at Al 6061-Al_2O_3 composite . The distortion was measured by dimensional change in the specimen. Coordinate Measuring Machine (CMM) was used to measured distortion based on coordinate change. Lower distortion are occur on 10E and 10N specimens. Electroless coating process on the reinforcement is more effective to obtain lower distortion on Al 6061-Al_2O_3 composite. The severity (H) of quenchant influenced distortion on specimen. Higher H value produce higher distortion on Al 6061-Al_2O_3 composite.

1 Introduction

Environmental problem encourage the development of lightweight materials. Lightweight materials can be used to reduce fuel consumption and overcome environmental issue. Composite is one of solution for that issue.

Aluminium metal matrix composite (AAMc) are widely used in aerospace, ground and water public transportation. AAMc demand are increased due to their strength to weight ratio, corrosion resistance, high thermal capacity [1]. One of AAMc is Al 6061-Al_2O_3 composite. This AAMc with particle reinforcement materials is widely used as structure and automotive parts. Boeing have been applied Aluminium 6xxx in their structural parts [2].

The mechanical properties of composite materials with aluminium matrix can be increased by using heat treatment in general. Heat treatment is the treatment consist of two different heat treatment that are solution heat treatment and precipitation hardening [3]. Solution heat treatment is carried out by heating Al alloy to a temperature that produce single phase of α and then quenching to room temperature. Heat treatment with various of quenching agent affected mechanical properties of AA6060 material [4]. Quenching rate must be fast to obtain maximum strength of material. On the other hand, quenching rate must be slow to avoid the distortion [5].

Tianbin et al [6] reported that heat treatment in addition changed mechanical properties, can caused distortion in material. Higher heating temperature can cause distortion effect on material [7]. The distortion in quenching is caused by inhomogeneity of cooling rate [8] and material thickness [9].

At the present, many of quenching agent are used in heat treatment, such as water, oil and salt solution. Water was widely employed as quenching agent with beneficially cheap and high cooling rate until 300°C. With addition of 5-10% salt, the solution can be increased cooling rate until 500°C on steel [9]. Compared with water, oil caused dangerous smoke and expensive, but oil has lower cooling rate than water [10]. Higher cooling rate tends to distortion occurred. High cooling rate resulted high residual in stress that lead the crack, instability in machining and dimensioning [11].

In heat treatment, different quenching agent can be produce different mechanical properties and distortion of materials. The distortion could change the dimension of the material, thus eliminate the used value of materials. Therefore, it is important to know for investigated effect of quenching agent and electroless coating treatment on distortion of Al 6061-Al_2O_3 composite.

2 Materials and Methodology

Aluminium alloy 6061 and Al_2O_3 powder with size <1μm was used in this research. The properties of Al_2O_3 powder is shown in Table 1. Al 6061-Al_2O_3 composites were manufactured by stir casting method as shown in Figure 1. and then followed by T6 heat treatment process. Stir casting is general method on manufacturing metal matrix composite because it has low cost, flexible, and large quantity production [12].

Table 1. Properties of Al_2O_3 particle

Chemical composition	
Al_2O_3 (%)	99.6
Na_2O (%)	0.3
SiO_2 (%)	0.02
Fe_2O_3 (%)	0.02
Physical composition	
Ave. Grain Dia μ	50
Specific Gravity	3.92

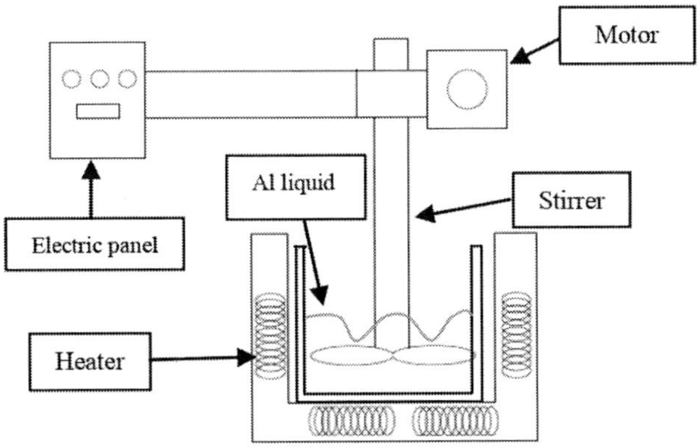

Fig. 1. Stir casting equipment

Al 6061 ingot was melted at a temperature of 700-750° C, then Al_2O_3 was added with a weight fraction as shown in Table 2. This experiment using Al_2O_3 particles with electroless coating and without electroless coating. Particles of Al_2O_3 were electroless coated in order to increase their wetting ability in liquid aluminium [13]. The electroless coating process of Al_2O_3 was conducted with 40 ml of 65% HNO_3 solution, 0.1 gram of Mg fine powder, and 0.5 gram of fine Al powder.

Table 2. Specimen tabulation

Specimen	Weight fraction of Al_2O_3 (%)	Treatment of Al_2O_3
0X	0	—
10N	10	Without electroless coating
10E	10	Electroless coating

Stirring casting of Al 6061-Al_2O_3 composite was carried out at 500 rpm and 700-720°C and then poured into the mould. The heat treatment of T6 in Al 6061-Al_2O_3 composite was conducted with solution heat treatment at 540° for 6 hours, then quenched on various quenching agent (water, oil, and salt solution 10%) and followed by artificial ageing treatment at 200° C for 4 hours.

Dimensional stability of Al 6061-Al_2O_3 composite specimen was measured by CMM Mitutoyo Bright-STRATO 707 with 5 point measurement method as shown in Figure 2. The Al 6061 matrix and Al_2O_3 interface bonding was examined using SEM JEOL JSM6510LA.

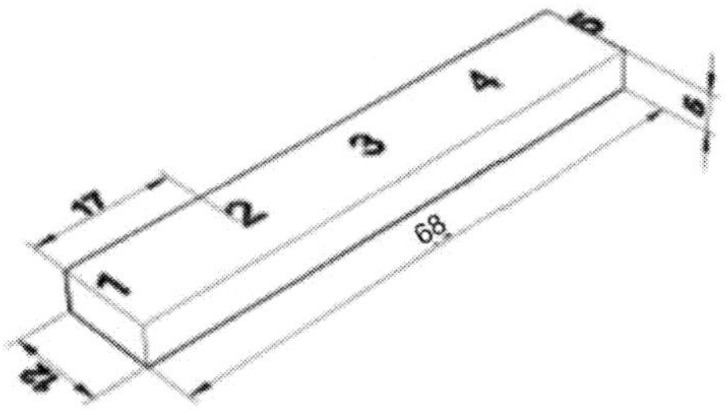

Fig.2. Specimen test

3 Result and discussion

In this research the difference between the highest and the lowest deflection was defined as distortion, an example deflection on 10N with brine quenching as shown in Figure 3. Distortion on various quenching agent was shown in Figure 4.

The specimen without reinforcement had the biggest distortion. It decreased by addition of 10% Al_2O_3 ceramics. The electroess coating process on Al_2O_3 ceramics also affected distortion. It produce the lowest distortion.

The distortion also affected by quenching agent. Higher cooling rate of quenchant produce higher distortion. Brine with highest cooling rate produce highest distortion. The lowest distortion was obtained by using oil quenching.

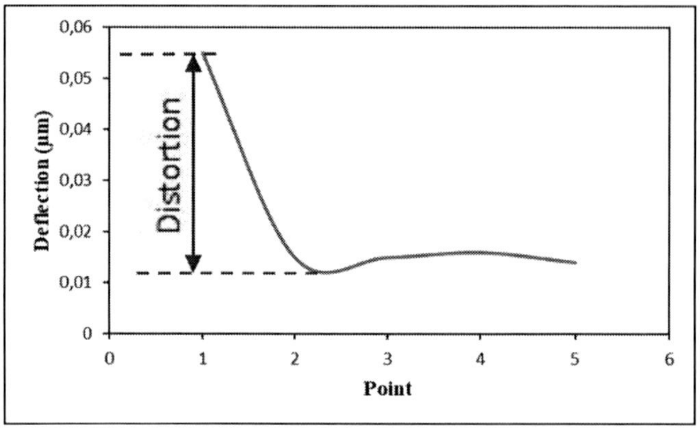

Fig. 3. Deflection on 10N brine quenching agent.

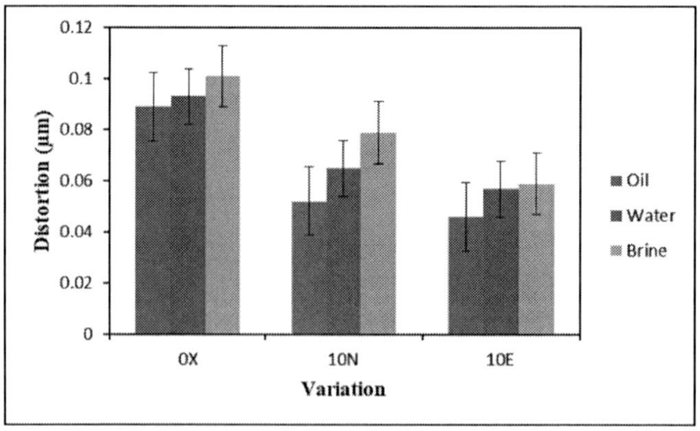

Fig. 4. Distortion of Al 6061-Al_2O_3 composite.

Figure 4 shows that the effect the quenching agent on distortion have similar trend. The quenching process with brine medium in all variation has the largest distortion, followed by water and oil quenching medium. That phenomenon was related to severity quenching coefficient (H) of quenching agent. It because brine has the highest severity quenching coefficient (H) with 2.0-2.2. Water has 1.0-1.1 of H value, and oil has 0.30-0.35 of H value [9,14]. The characteristic of water quenching produce non-uniform geometry and high temperature gradient on specimen, which generated the residual stress and the distortion [5].

Al_2O_3 particle affected distortion of the specimen. Specimen of Al 6061-10% Al_2O_3 have lower distortion than specimen without Al_2O_3. The Addition of ceramics particles (e.g Al_2O_3) on the metal matrix composite can decrease thermal conductivity [15]. The electroless coating treatment on Al_2O_3 can affected the distortion of specimen. Specimen with electroless coating process had lower distortion result than specimen without electroless coating process. Al_2O_3 with electroless coating has higher wettability in Al matrix than Al_2O_3 without electroless coating. Therefore, electroless coating that was applied on Al_2O_3 able to increase interface bonding caused the oxide film is formed to despotition the metal film so the wettability increased.

The specimen without electroless coating process formed agglomeration of Al_2O_3 reinforcement. The agglomeration of Al_2O_3 reinforcement on specimen was shown in Figure 5a is caused by weak interface bonding between Al matrix and Al_2O_3. The weak interface bond is an obstacle to achieving the desired properties of the composite [16] . The agglommeration did not occur on specimen with electroless coating process as shown in Figure 5b

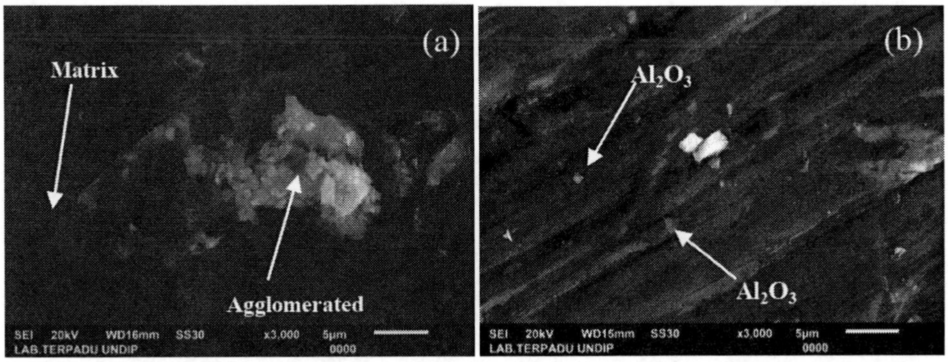

Fig. 5. SEM image (a) non-electroless coating specimen; (b) electroless coating specimen

The agglomeration of reinforcement can generate bigger distortions. That phenomenon was due to the fact that the agglomeration of reinforcement have lower interface bonding strength between matrix and reinforcement. It show that main problem in Al 6061-Al_2O_3 composite is wettability of Al to Al_2O_3 ceramics. Moreover, the agglomeration reinforcement increase opportunities to obtain bigger porosity on the composite material.

The addition of reactive Mg to HNO_3 at electroless coating process will produce Al_2O_3 with $MgAl_2O_4$ spinel phase that have higher wettability [17]. Moreover, the addition of Mg at electroless coating process increase the capillarity of penetration to the passive layer of oxide and also facilitate the interface bonding formation. It produces Al 6061-Al_2O_3 composite with good interface bonding. The good bonding interface between Al 6061 and ceramics Al_2O_3 could increase the heat capacity of composite, so it reduce distortion in quenching process.

4 Conclusion

Increasing of H value of quenching agent produce higher distortion on the specimen. Addition of Al_2O_3 reinforcement decrease the distortion on the specimen. Application of electroless coating on the Al_2O_3 reinforcement more effectively decrease distortion on Al 6061-Al_2O_3 composite.

References

1. Yasphal, Sumankant, C.S. Jawalkar, A.S. Verma, N.M. Suri. Mat. Today : P **4**, 2927-2936 (2017)
2. A. Abundez, I. Pereyra, B. Campillo, S. Serna, E. Alcudia, A. Molina, A. Blanco, and J. Mayén. Mat. Sci. & Eng. A.(to be published)

3. S.K. Tiwari, S. Soni, R.S. Rana, A. Singh. Mat. Today **4,** 3458-346 (2017)
4. K. Strobel, M.D.H. Lay, M.A Easton, L. Sweet, S. Zhu, N.C. Parson, A.J. Hill. Mat. C **111** (2016) 43-52.
5. B. Milkereit, H. Fröck, C. Schick, O. Kessler. Trans. N. Met. Soc. China, **24**, 2025-2033 (2014)
6. H. Tianbin, L. Huiqu, T. Pengun, L. Peiyong.. Mat. C **118** (2016) 554-552.
7. K.F. Rahman, M.M. Benal.. IOSR J. of Mech. and Civ. Eng. **Volume 1**. 2278-1684 (2012)
8. K-E. Thelning. *Steel and Its Heat Treatment* (Butterworths, London, 1984)
9. J.S. Robinson, D.A. Tanner, C.E. Truman, A.M. Paradowska, C.R. Wimpory, Mat. C **65,** 73-85. (2012)
10. X. Chen, L. Zhang, X. Jie, Y. Li, X. Huang, Int. J. of Heat and Mass Trans. **109**, 209-214 (2017)
11. G.P. Dolan, J.S. Robinson. J. of Mat. P. Tech, **153-154**, 346-351 (2004)
12. M.T. Sijo, K.R. Jayadevan, Pro. Tech, **24**, 379-385 (2016)
13. A. Zulfia, M.K. Ajiriyanto, Ad.. Mat. Rsch Vols **328-330**, pp 1552-1555 (2011)
14. C.R. Brooks. *Principles of the Heat Treatment of Plain Carbon and Low Alloy Steels*. (ASM International, 1996)
15. P. Ma, Z.J. Wei, Y.D. Jia, S.Z. Yu, K.G. Prashanth, S.L. Yang, C.G. Li, L.X. Huang, J. Eckert, J. of A. and Comp. **709** 329-336 (2017)
16. B. Guo. M. Song, J. Yi, S. Ni, Y. Du, Mat. and Dsgn **120** 56-65 (2017)
17. G. Sahari, A. Zulfia, E. Siradj, M. J. of Sci, **Vol 13 No.1**, 39-44 (2009)

Design of Fractal Features-Based Partial Discharge Pattern Recognition using Multi Support Vector Machine Method

Rahayu, *G.T.* Anuraga, *H.* Prasetia, *Umar* Khayam

Bandung Institute of Technology, School of Electrical Engineering and Informatics, Bandung, Indonesia

Abstract. Partial Discharge (PD) is one of the causes of insulation deteriorisation mode and impacts on the reliability of high voltage equipment. Therefore, PD measurement is used for diagnostic technique of high voltage equipment. Diagnostic output of high voltage equipment contain information about PD type, PD cause, PD location and PD severity. after identification, a proper preventive maintenance pattern can be performed. Therefore PD pattern recognition system is very important on PD diagnostic system to recognize the PD pattern and determine the level of hazard that occurs in specimen object or high voltage equipment.. In this paper, PD pattern recognition system is designed with fractal geometry approach and support vector machine (SVM) algorithm. The coding and programming of graphical user interface of the application is done. Each PD type and hazard level on various insulating materials (solid, liquid and gas) have the dimensions of the fractal and the lacunarity. The type of PD (void, corona) and its danger level (bad, fair and good) can be identified with the support vector machine (SVM)

1 Introduction

Diagnosis of power apparatus based on partial discharge (PD) measurement requires PD pattern recognition and judgement system. PD pattern recognition and judgement are actually analysis and evaluation stages which produce information of insulating qualities or insulation deterioration level or health conditions of power apparatus. At first, the stages of analysis and evaluation is done entirely by human (expert). The PD pattern interpretation relies mainly on the knowledge and the experience of a human expert. However, the development of computer technology and the need for speed and accuracy of analysis require the application of pattern recognition and judgement system without human assistance (expert), which can easily be done by anyone, anytime [1-4].

Meanwhile, SVM has been popular as a method of pattern recognition in the world of bioinformatics, such as fingerprint image recognition, voice recognition in security systems, as well as disease classification automatically based on the diagnosis of the patient medical record [5]. On the other hand, mathematicians often use fractal geometry to describe natural phenomena and complex shapes. PD signal is natural and random occurrence, so that it can be extracted in the form of a unique fractal. The patterns are unique. PD can be further

* Corresponding author: unar@hv.ee.itb.ac.id

identified by the SVM algorithm so that in the end PD source, PD type and the level of danger can be predicted.

The purpose of this paper is to design PD recognition and judgement system. The system recognizes patterns of partial discharge (pattern recognition) and determine the level of hazard that occurs on the object specimen / high voltage equipment under test (judgement system), with fractal geometry approach and SVM algorithm.

2. Partial Discharge, Fractal Geometry and Support Vector Machine (SVM)

2.1 Partial Discharge

Partial Discharge (PD) occurs when the electric field in the void beyond the field strength at the gas permeability of the void. According to Standard IEC 60270: 2000, partial discharge is defined as discharge which only partially connects the insulation between the electrodes [6]

2.2 Fractral Geometry

Fractals are geometric structures with dimensions of fractions [7-10]. To explain the pattern of a natural phenomenon and complex shapes, as well as stochastic. Examples of natural phenomena and other complex forms between clouds, snow crystal, batik and leaves. Fractal is a pattern within a pattern, in a pattern. Some fractal objects can be broken down into several parts that are similar to the original fractal.

An object that has a fractal dimension is something that has the same variation with self-similarity in a variety of scales, so maximum detail will never be achieved by increasing the scale. Lacunarity, first introduced by Mandelbrot (1982) to describe the characteristics of fractals with the same dimensions with different textures. In general, a low value indicates Lacunarity texture similarity, whereas a high value indicates a different texture of lacunarity.

2 .3 Support Vector Machine (SVM)

SVM concepts can be explained simply as an attempt to find the best hyperplane which serves as a separator are two classes in the input space. Hyperplane in the vector space dimension d is affine subspace dimension d-1 which divides the vector space into two parts, each of which corresponds to the different classes.The best separator hyperplane between the two classes can be found by measuring the hyperplane margin.

3. Design of PD Recognition and Judgement System

3 .1 System Block Diagram

Data input in the form of *csv is sample of PD signal and AC voltage signal in 10 cycles. Which is derived from the output of signal processing stage, so that it has been filtered from the noise.

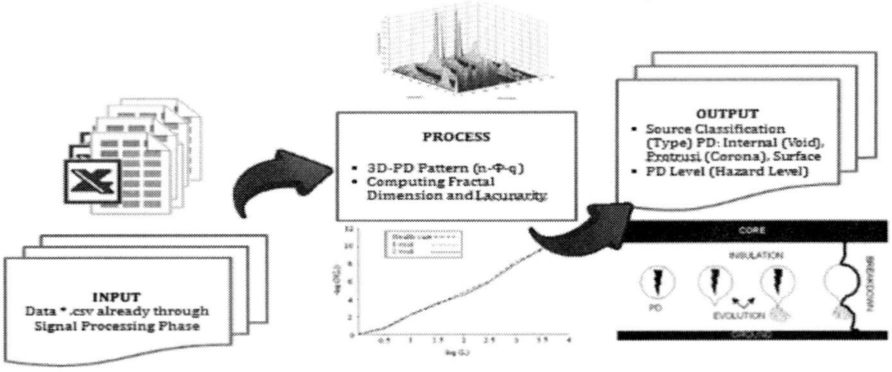

Fig. 1. Block Diagram of PD Pattern Recognition and Judgement System

3.2 Flowchart of the Application

In performing its function as partial discharge pattern recognition application, this application requires a database with known data, to be used as a reference or benchmark for identifying unknown data. Thus, the workings of this application, consists of 2 pieces flowchart, which consists of building database and pattern recognition.

3.2.1 Flowcharts Database Creation

To build a database of pattern recognition, it takes the known test data. The data form * csv file, which consists of two pieces of data, namely data of applied voltage, hereinafter referred to as an AC signal and partial discharge measurement data from sensors, hereinafter referred to as the signal PD.

Both of these data will have normalized, so that the value in conformity with the actual conditions. In addition, the normalization phase, is also used to retrieve the required data. Fig. 2 shows the flowchart of the database creation process. At signal processing stage, each data will experience different processing. In the AC signal data, the data has a lot of noise, visible from the non-smooth waveform. The AC signal data will pass through low pass filter with frequency passed by 50 Hz. So the output of this stage, the fundamental sine wave is smooth and less noise amount.

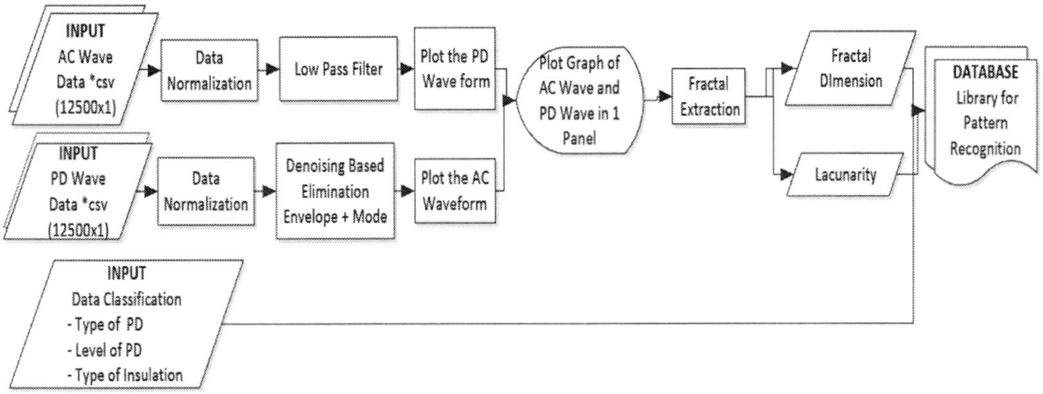

Fig. 2. Flowchart Database Creation

After going through the stages of signal processing, the data is then plotted into a graph, and then combined in a single panel. The graph is then converted into image data. Data conversion chart into image data, aiming to facilitate the extraction process to form a fractal.

The image data, and then will be fractal extraction resulting data in the form fractal dimension and lacunarity. This data was to be used as database applications partial discharge pattern recognition.

3.2.2 Flowchart Pattern Recognition

Basically, the pattern recognition process is similar to the process of making the database. To perform pattern recognition, must be prepared the test data that want classified its type and its PD level. The data should be a * csv file, which consists of two pieces of data, namely : Data source voltage injection, hereinafter referred to as an AC signal and PD measurement data from sensors, hereinafter referred to as the signal PD.

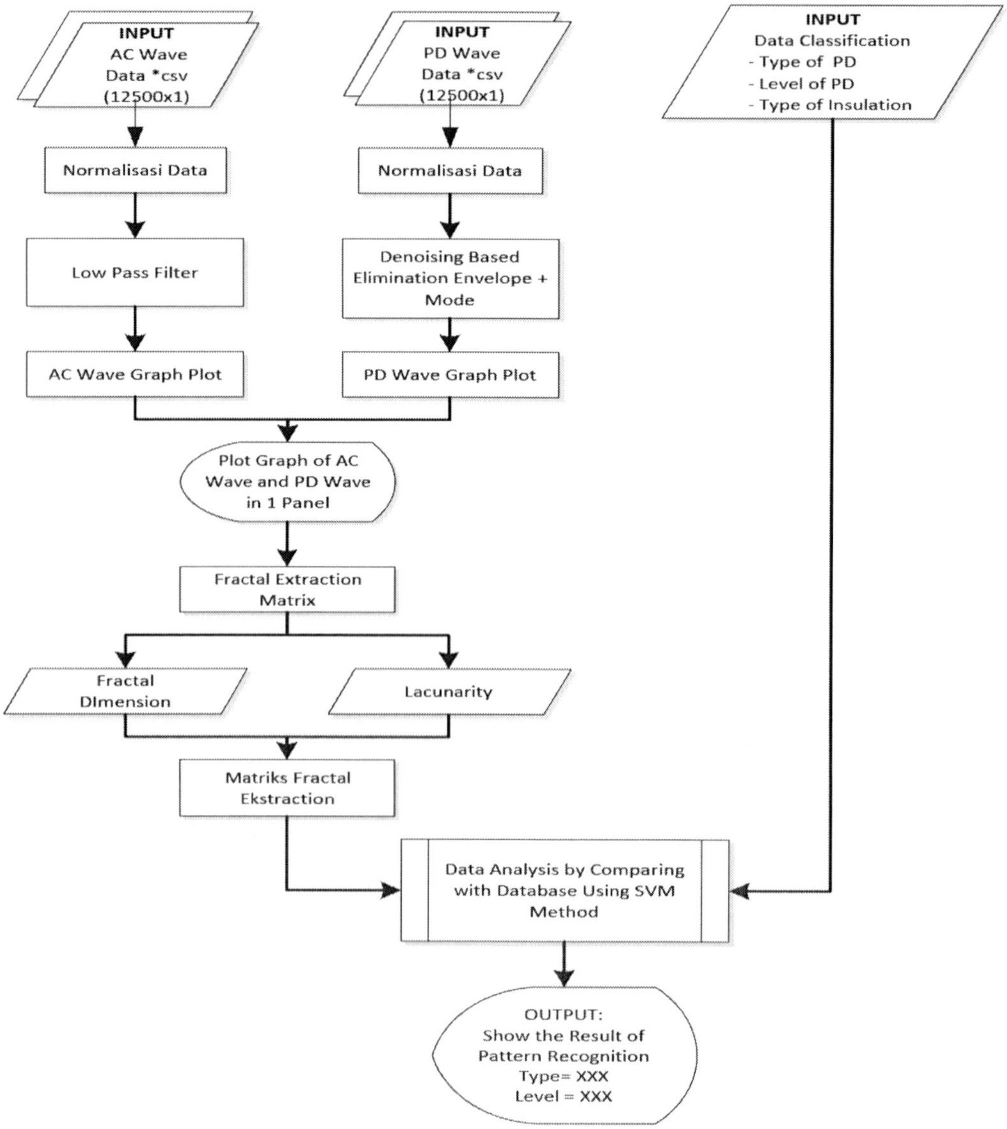

Fig 3. Flowchart Pattern Recognition Applications

In the signal processing stages, each data will undergo a different treatment. In the AC signal data, the data of crude has a lot of noise, seen from the waveform is not smooth. AC signal data going through the Low Pass Filter with a frequency of 50 Hz are bypassed. So that the output of this stage, the fundamental sine wave that is smooth and less noise amount.

While the PD signal data, the data of crude has a lot of noise complicates the introduction of actual PD signals. In this research, PD signal will pass through the stages of denoising. Stages denoising is done by our group, slightly different from the denoising in general (wavelet or high pass filter with a reference Background Noise). Stages of our denoising, performed by eliminating the reference value of the mode value (value with the highest frequency of occurrence) and the maximum value of the data.

After going through the stages of signal processing, the data is then plotted into a graph, and then combined in a single panel. The graph is then converted into image data. Data conversion chart into image data, aiming to facilitate the extraction process to form a fractal.

The image data, and then will be fractal extraction resulting data in the form fractal dimension and lacunarity. This data was to be compared with the existing database, using Support Vector Multi Macchine, which will be the output of this process is the classification of the types and levels of PD.

3.3 GUI Design

In making the application of pattern recognition, we use MATLAB software with GUIDE (Graphical User Interface Design Environment) and assisted with the syntax of programming. This application is to get input in the form of 2 pieces * csv file.

GUI application designed into 4 main modules, namely: Display module for displaying graphs and PD parameters, database creation module to add a database of patterns of PD, Processing module to ensure the process is running normally as well as to know its not a database, pattern recognition module, to perform a partial discharge pattern recognition, as well as featuring level and its type.

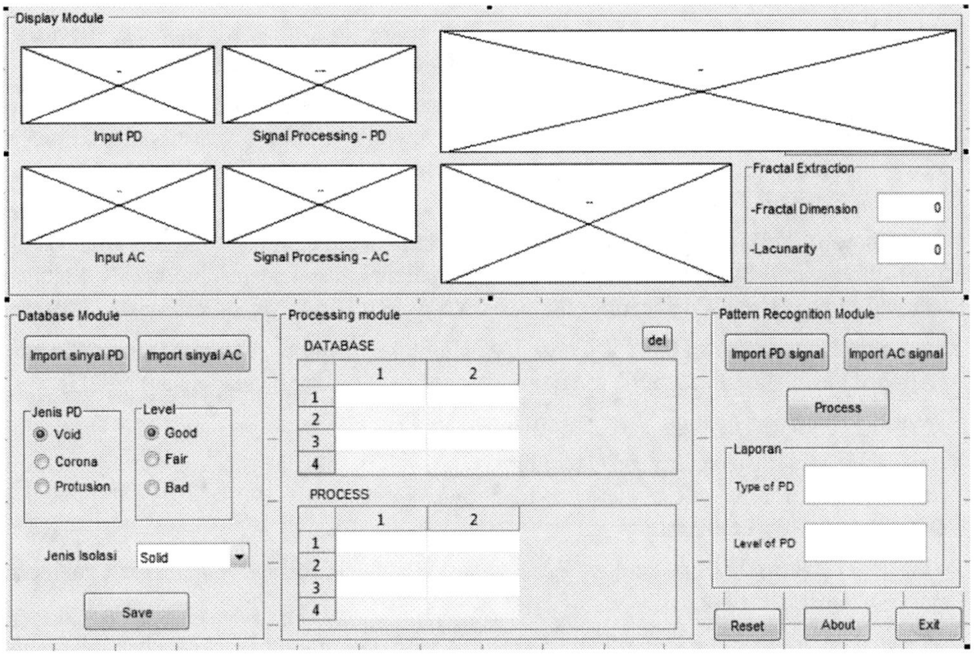

Fig. 4. The Concept of the GUI

4. Conclusion

Applications with the pattern and judgement of PD systems with fractal extraction and SVM methods have been designed. In performing its function, then this application requires database with known data. The workings of this application, consists of 2 pieces of flowchart, which consists of database creation and pattern recognition. To build a pattern recognition database, it takes the data of a known test result. The data is a csv * file, consisting of 2 pieces of data, ie data source injection voltage (AC signal) and PD measurement data from the sensor (PD signal). Both of these data will be normalized, so the inside is in accordance with the actual conditions. In addition, the normalization stage, also used to retrieve the required data.

Next in the signal processing stage, each data will experience different processing. In the AC signal data, the raw data has a lot of noise, visible from the non-smooth waveform. The AC signal data will pass through low pass filter with frequency passed by 50 Hz. From this result, the fundamental sine wave is smooth and the amount of noise is less. While in the PD signal data, data has a lot of noise that complicates the introduction of actual PD signals. The PD signal will pass through the denoising stage. After going through the Signal Processing step, the data is then plotted into the graph, then merged into one panel. The graph then becomes the image data.

Reference

1. Roro Roudhotul Jannah, Umar Khayam," Design, Implementation, and Testing of Partial Discharge Signal Pattern Recognition and Judgment System Application Using Statistical Method", Joint International Conference on Electric Vehicular Technology and Industrial, Mechanical, Electrical and Chemical Engineering (ICEVT & IMECE), Surakarta, Indonesia, (2015)
2. Umar Khayam, Tria Kasna Lestari,"Application of Wavelet Transform on Partial Discharge Signal Denoising", The Second International Conference Industrial, Mechanical, Electrical and Chemical Engineering (IMECE), Yogyakarta, Indonesia, October 6-7, (2016)
3. Berkah Suprayogi, Umar Khayam,"Design of Partial Discharge Location Identifier Software for High Voltage Generator Using Artificial Neural Network ", Joint International Conference on Electric Vehicular Technology and Industrial, Mechanical, Electrical and Chemical Engineering (ICEVT & IMECE), Surakarta, Indonesia, (2015)
4. Umar Khayam, Yoga Aji Surandaka,"Design, Implementation, and Testing of Partial Discharge Signal Processing System", The Second International Conference Industrial, Mechanical, Electrical and Chemical Engineering (IMECE), Yogyakarta, Indonesia, October 6-7, (2016)
5. Anto Satrio Nugraha, Arief Budi Widarto, Dwi Handoko, "Support Vector Machine, Theory and its Application in Bioinformatics", ilmukomputer.com, (in Indonesian language). (2003)
6. Suwarno, "Diagnosis of High Voltage Equipments", ITB, (2010)
7. Sherif Ghoneim, "Investigation on Using Fractal Geometry for Classification of Partial Discharge Patterns", IOSR Journal of Electrical and Electronics Engineering (IOSR-JEEE) Volume 6, Issue 5 (Jul. - Aug. 2013), PP 50-57.
8. Li Jian, Sun Caixin, Li Xin, Du Lin, Zhou Quan, "Partial Discharge Pattern Recognition using Fractal Dimension"
9. A. Krivda, E. Gulski, "The Use of Fractal Features for Recognition of 3-D Discharge Patterns"
10. Hung-Cheng Chen. "Fractal Features-Based Partial Discharge Pattern Recognition Using Extension Method". International Conference on Intelligent Systems Design and Engineering Application (2012)

A review of low-cycle fatigue of Alloy 617 for use in VHTR components: Experimental outlook

Rando Tungga Dewa[1], *Jeong-Hun* Park[1], *Seon-Jin* Kim[1,*], *Woo-Gon* Kim[2], *Eung-Seon* Kim[2], *Gunawan D.* Haryadi[3], and *I. M. W.* Ekaputra[4]

[1]Pukyong National University, Department of Mechanical Design Engineering, Busan 48547, Korea
[2]Korea Atomic Energy Research Institute (KAERI), Daejeon 34057, Korea
[3]Diponegoro University, Department of Mechanical Engineering, 50275 Semarang, Indonesia
[4]Sanata Dharma University, Department of Mechanical Engineering, 55002 Jogjakarta, Indonesia

Abstract. The effect of strain range and temperature on the low-cycle fatigue behaviour and microstructure change during cyclic deformation of Alloy 617 for use in very high temperature gas-cooled reactor components were studied at elevated temperature starting from ambient condition. Increasing the strain range and the temperature was noticed to reduce the fatigue resistance of nickel-based Alloy 617 due to facilitating the transformation behavior of the carbides in the grain interior, precipitates along the grain boundary, and oxidation behavior inducing surface connected precipitates cracking. Initial hardening behavior was observed at room temperature condition during cyclic due to the pile-up dislocation of micro-precipitates. The grain size was also taking a role due to the formation of an obstacle in the matrix. In the high temperature regime, the alloy 617 was found to soften for its entire life due to the fast recovery deformation, proved by its higher plasticity compared with lower temperature. The deformation behavior also showing high environmentally assisted damage. Oxidation behavior was found to become the primary crack initiation, resulting in early intergranular surface cracking.

1 Introduction

Low-cycle fatigue (LCF) is an important design consideration in structural components operating at high temperatures. Repetition (cyclic) of thermal stresses are generated as a result of temperature gradients which release on heating and cooling during start-up, shut-down and thermal transient conditions. LCF resulting from startups and shut-downs occurs under essentially strain-controlled conditions, since the surface region is constrained by the bulk of the component. For example, a large steam turbine component may be undergoing power under peak-load conditions during its operation. Under these circumstances, the component must quickly respond from its stand-by state, and severe thermal stresses may result from the thermal transients induced by the start-up. In the other case of aircraft gas turbines, the normal operation of an engine on airlines is one implicating start and stop operation.

* Corresponding author: sjkim@pknu.ac.kr

In this study, the authors are focusing in the nuclear field area, such nuclear components are subjected to cyclic thermal transient and mechanical stress due to the high pressure and temperature of the coolant or of the fuel [1-2]. Therefore, nuclear plant life extension therefore remains a challenge for the material scientists and low emission as well as better performance remain as a worldwide issue. The Next Generation Nuclear Plant (NGNP) being developed in the US and Republic of Korea is a Very High Temperature Nuclear Reactor (VHTR) with helium as the primary coolant. In the VHTR, some of the major components such as the reactor internals, reactor pressure vessel, piping, hot gas ducts (HGD), and intermediate heat exchangers (IHX) are key components, with helium as a primary and secondary coolant. The VHTR may merge diversities of this baseline design to allow eventual operation at gas outlet temperatures up to 950°C. Alloy 617, a Nickel-base superalloy, is a primary candidate material for a VHTR because of its excellent thermal stability for components of power generating plants with high temperature environment [3]. The American Society of Mechanical Engineers (ASME) Boiler and Pressure Vessel Code (BPVC) allows use of Alloy 617 for construction of non-nuclear pressure vessels, however, Alloy 617 is now being developed for use in VHTR plant. From these understanding, it is clear that the problem of high-temperature LCF involves a broad cross section of our advanced technology and that a deeper study of proper application is essential. These considerations can be used to determine material resistance against the cyclic loading to ensure the reliability of material structures.

This work deals with a series of LCF tests on Alloy 617, investigating for a fully reversed strain-controlled (strain ratio, $R\varepsilon = -1$) regarding to different applied total strain ranges, i.e., 0.6, 0.9, 1.2, and 1.5% at elevated temperatures, room temperature, 900, 950°C in an open air environment. The main aim is to study the basic characteristic of low-cycle fatigue properties during temperature change. Cyclic deformation phenomena such as hardening/softening behavior, cyclic stress-strain curves, and fatigue life dependent in a representation of a plastic deformation and ductility resistance to failure were obtained.

2 Experimental sections

2.1 As-received materials

A commercial grade Alloy 617 is approved for non-nuclear construction in the ASME Code. Thus the composition (wt%) of the Alloy 617 used for material chosen in this study is 53.11Ni, 22.2Cr, 12.3Co, 9.5Mo, 1.06Al, 0.08C, 0.949Fe, 0.4Ti, 0.084Si, 0.029Mn, 0.027Cu, 0.003P, <0.002S, and <0.002B. Fig. 1 shows the microstructure of as-received Alloy 617. The average grain size of the alloy is approximately 100 μm for coarser grain and 30 μm for finer grain. Intergranular carbides are present in banded regions aligned with the rolling direction.

The cylindrical LCF specimens with 6.0 mm in diameter in the reduced section with a parallel length of 18 mm and gauge length of 12 mm were made from alloy plate. The long axis of the specimen aligned with the rolling direction. Low stress grinding was performed during the final machining to avoid notches for the LCF test.

2.2 Low-cycle fatigue test

A closed loop 100kN servo hydraulic testing machine was used and equipped with a tube furnace for heating the specimens of Alloy 617 in dry air, therefore the temperature was remained within ± 2°C of the nominal temperature. We performed a fully reversed (strain ratio, $R\varepsilon = -1$) axial strain controlled LCF tests for Alloy 617 regarding to four different total strain ranges, i.e., 0.6, 0.9, 1.2, and 1.5%. A triangular waveform and constant strain rate of 10^{-3}/s was applied. The failure criteria was defined as the number of cycles which means a 20% reduction in the stress ratio (peak tensile-compressive stress ratio). Microstructures were observed by optical microscopy (OM) and scanning electron microscopy (SEM) equipped with an energy-dispersive X-ray spectroscopy (EDS).

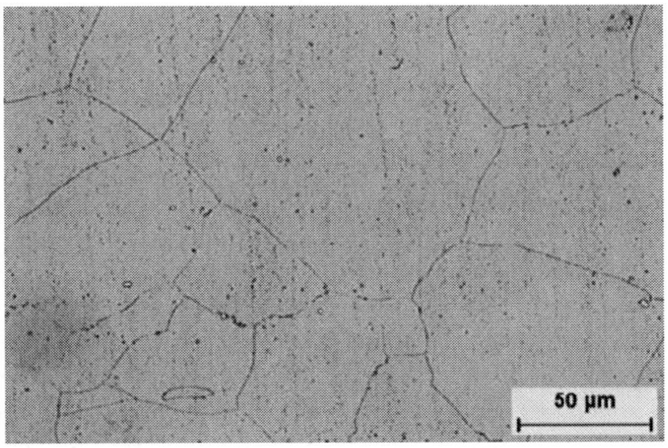

Fig. 1. Metallography of carbides and grain structure in the as-received Alloy 617.

3 Results and Discussions

3.1. Low-cycle fatigue results

LCF testing was completed in air at elevated temperatures to provide a baseline of understanding in cyclic deformation behavior. Table 1 shows the test results in term of strain range and temperature. Four different applied strain ranges were shown from 0.6% - 1.5% total strain. The number of cycles to failure, elastic and plastic strain at a half-life cycle are also listed in Table 1.

Table 1. LCF test results of Alloy 617 at elevated temperatures.

Temperature	Total strain range (%)	Cycles to failure (cycles)	Plastic strain range at half-life (%)	Stress range at half-life (MPa)
Room temperature	0.6	42360	0.078	474.2
	0.9	13042	0.205	505.5
	1.2	3410	0.321	552.4
	1.5	1409	0.435	583.0
900°C	0.6	1179	0.176	193.5
	0.9	913	0.323	188.0
	1.2	487	0.470	186.2
	1.5	405	0.625	181.2
950°C	0.9	823	0.343	148.9
	1.2	481	0.480	139.9
	1.5	392	0.641	136.6

The stress amplitude response curves at four total strain ranges as a function of number of cycles to failure are illustrated in Fig. 2 under continuous cyclic loading. The cyclic stress response at room temperature exhibited a cyclic softening regime for the major portion of the life after a brief period of initial hardening. The short period of cyclic initial hardening was observed about 2–200 cycles and remained softening phase until failure. Almost all of the stress amplitude paths of Alloy 617 at high temperature ranges exhibit a cyclic softening region for the major portion of the life. At the end of the test, the stress amplitude was decreased rapidly as a formation of macro-crack initiation or just prior to failure. Peak tensile and compressive stresses as a function of cycle reached a stable value within less than 20 cycles.

3.2. Fracture behaviour

Fig. 3 shows microstructure evolution of LCF fracture behaviour under wide range of temperatures. At room temperature crack initiation and propagation occurred in transgranular nature (Fig. 3a). With the help of the cross section of OM photograph, the fatigue crack propagation site was essentially characterized by typical fatigue striations which were perpendicular to the fatigue loading direction. Fig. 3b and 3c showing crack initiation and propagation behavior was similar to that exhibited at 900°C and 950°C that is the transition from intergranular oxide cracking to a mixed mode transgranular propagation occurred at a slightly greater depth from the surface. This is confirmed in [7] when over exposed to high temperature applications, metal alloys are oxidized, which formed precipitates and cavities. Surface brittle oxides were found in outside layer of specimen and became potential sites for crack initiation where cyclic strain was localized and generated a critical zone of fatigue failure.

Fatigue damage process of the propagation mode is identified by the result of the slip band. Slip character is described as a measure of the degree to which dislocations tend to disperse during plastic deformation [8]. Therefore, it can be drawn that at high temperature process, fatigue damage is more obvious due to the presence of more parallel slip band.

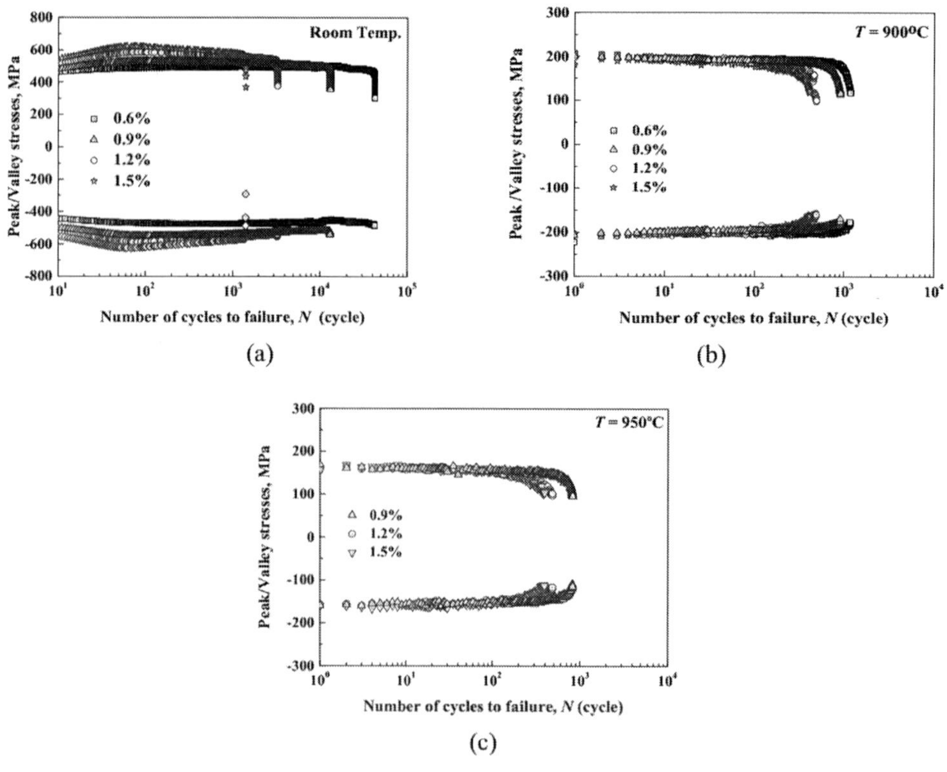

Fig. 2. Peak and tensile compressive stresses response as a function of the number of cycles of Alloy 617 at elevated temperatures: (a) Room temperature; (b) 900°C; and (c) 950°C.

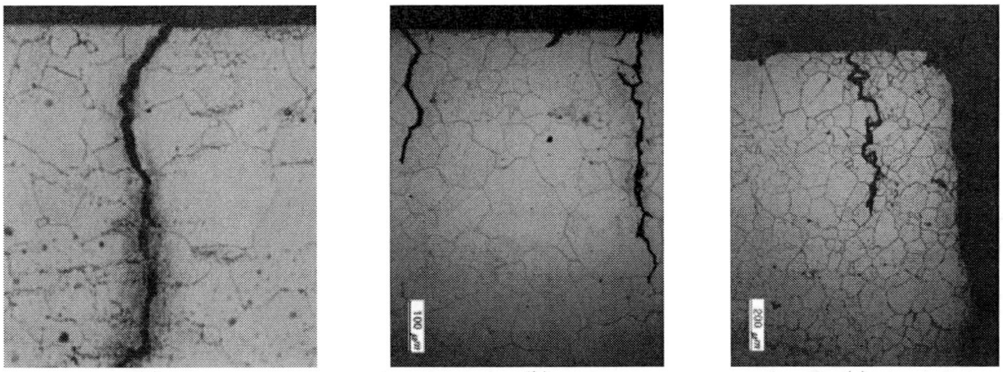

(a) (b) (c)

Fig. 3. Micrographs illustrating failure mechanism of Alloy 617 at elevated temperatures: (a) Room temperature; (b) 900°C; and (c) 950°C. The stress axis is from left to right in the plane of the page.

3.3. Discussion

Failure mechanism at room temperature occurred in nature way of cyclic slip. The slip band from fatigue mechanism that contain extrusions and intrusions leads to the production of point defects that will develop into micro-cracks within the grain in gauge section, finally, at the critical stage, coalescence of the micro-cracks again develops into the macro-crack that the stress rapidly drops until failure.

The high temperature condition has been well known to induce the heavily damaged by environment. Fig. 4 shows EDX spectroscopy results, the primary carbides of Ti-rich M_6C, with large precipitations, were formed in the body of the grain structure and the carbides of Mo, Cr-rich $M_{23}C_6$ were developed mainly on the grain boundary. These precipitates (especially Mo, Cr-rich $M_{23}C_6$) were diffused with oxygen ions simultaneously and adapted with metal ions to create a protective oxide layer, namely Cr_2O_3. These brittle cracks on the surface-connected grain boundary were found to generate crack nucleation and accelerate the propagation rate, which the dislocation movement become easier around the grain boundary. This phenomenon at high temperature was also known to reduce the applied stress during cyclic loading, otherwise the plastic strain accumulation was continuously increasing.

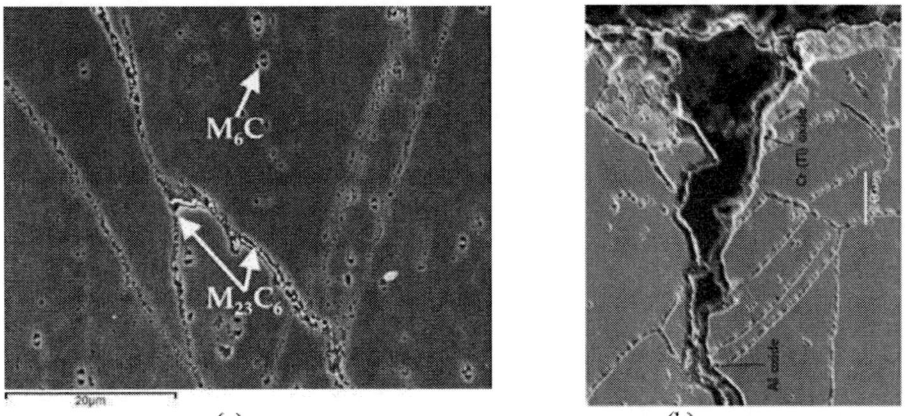

(a) (b)

Fig. 4. Typical SEM images of Alloy 617 at high temperature environment illustrating: (a) Precipitates formed in the grain interior and its boundary; and (b) Surface brittle Cr-oxides cracks.

4 Conclusion remarks

Low-cycle fatigue behaviours were investigated by LCF tests with different temperature at room temperature, 900, and 950°C and strain ranges at 0.6-1.5%. The cyclic stress response at room temperature exhibited a cyclic softening regime for the major portion of the fatigue life after a brief period of initial hardening. The short period of cyclic initial hardening was observed about 2–200 cycles (pile-up dislocation across the slip barrier) and remained softening phase until failure. Meanwhile, at high temperature the softening phase governed the major mechanism due to the recovery of precipitates during fatigue. It was observed that the reduction in LCF resistance with increasing temperature yields partly from homogenization of slip mechanism during fatigue and simultaneous increase in plastic deformation generated and partly from the intergranular crack initiation obtaining from oxidation of surface connected grain boundaries and environmentally assisted mixed mode propagation.

This work was granted from Nuclear Research and Development Program, NRF, and Ministry of Science, ICT, & Future Planning through Korea Atomic Energy Research Institute, Republic of Korea (NRF-2016M2A8A2902895).

References

1. L. F. Coffin, Jr., Experimental Mechanics **8**, 5 (1968)
2. H. Mughrabi, Phil. Trans. R. Soc. A, **373** (2015)
3. S. J. Kim, Rando T. Dewa, W. G. Kim, and M. H. Kim, Adv. In Mater. Scie. Eng., **2015**, 20749 (2015)
4. J. K. Wright, L. J. Carroll, T. –L. Sham, N. J. Lybeck, R. N. Wright, *Proc. of the ASME 2016 PVP Conference*, **PVP2016-63704** (2016)
5. T. S. Jo, J. H. Lim, Y. D. Kim, J. of Nuclear Mater., **406** (2010)
6. V. Shankar, A. Kumar, K. Mariappan, R. Sandhya, K. Laha, A. K. Bhaduri, N. Narasaiah, Int. J. of Fatigue, **100** (2017)
7. Rando T. Dewa, S. J. Kim, W. G. Kim, and E. S. Kim, Metals **7**, 54 (2017)
8. K. B. S. Rao, H. Schiffers, H. Schuster, and H. Nickel, Metallurgical Trans. A ,**19A** (1988)

Development of force feedback in steering systems for virtual driving simulator

Joga Dharma Setiawan[1,*], *Masri B.* Baharom[2], and *M. Ammar* Bin Abdul Wali[2]

[1]Department of Mechanical Engineering, Faculty of Engineering, Diponegoro University, Indonesia
[2]Department of Mechanical Engineering, Universiti Teknologi Petronas, Malaysia

Abstract. In steer-by-wire (SBW) vehicles, the elimination of the rigid mechanical column shaft would require the system to generate an artificial feedback torque which should produce similar driving feeling and behavior as to the conventional steering system. The objective of this study is to evaluate the characteristics of force feedback inside SBW vehicle for driving simulator utilizing Logitech G27 steering wheel. The model of the system is developed in Matlab/Simulink/3D Animation. A J-turn test is performed to see the resulting handwheel torque and its effect on the vehicle dynamic. The evaluation shows that the results are reasonable such that the driver of the simulator can feel the similar forces coming from the real road.

1 Introduction

A steering wheel is one of the contributing factors that plays an important role in enhancing the safety of the passenger beside maintaining the vehicle stability. Inside a typical steering system, the driver's steering input is transmitted by a steering shaft through a type of gear mechanism to generate steering motion at the front wheels. Nowadays, power steering assist has become a standard built-in feature inside most of road vehicles. The power steering amplifies the driver's-applied torque at the steering wheel such that steering effort is reduced.

Steer-by-wire (SBW) system utilizes the by-wire technology to perform and also augment existing functions of mechanical steering systems. The term "by-wire" can also be referred to as the replacement of mechanical controls to the electronic system. SBW systems have no mechanical linkage (steering column) between the steering wheel and the rack and pinion gear system [1]. SBW system omits the intermediate shaft available in conventional vehicles, which allows the possibility of independently commanding the roadwheels through both driver and lane-keeping assistance system [2]. This study produces an algorithm for producing steering feedback torque using Simulink for virtual experimentation.

2 Research Methodology

2.1 Bicycle Model with Artificial Force Feedback Model

A bicycle model of a car is used as a reference model for three degree-of-freedom (DOF) in the driving simulator as shown in Fig. 1(a) and represented by Eqs. (1-3). An artificial force

* Corresponding author: joga.setiawan@ft.undip.ac.id

 https://doi.org/10.1051/matecconf/201815902050

feedback model for a simple conventional steering system shown in Fig. 1(b) will be integrated with the bicycle model. The last step is to develop the simulation on the Simulink for different maneuver specified later on, inputted by the Logitech G27 steering wheel with the 3D car animation.

$$\dot{V}_x = \frac{F_x}{m} + rv_y \tag{1}$$

$$\dot{V}_y = \frac{1}{mv_x} - (-aC_{af} + bC_{ar})r - \frac{1}{mv_x}(C_{af} + C_{ar})v_y + \frac{1}{m}C_{af} + C_{ar}\delta_F - rv_x \tag{2}$$

$$\dot{r} = \frac{1}{I_{zz}v_x} - (-a^2C_{af} - b^2C_{ar})r - \frac{1}{I_{zz}v_x}(aC_{af} - bC_{ar})v_y + \frac{1}{I_{zz}}aC_{af}\delta_F \tag{3}$$

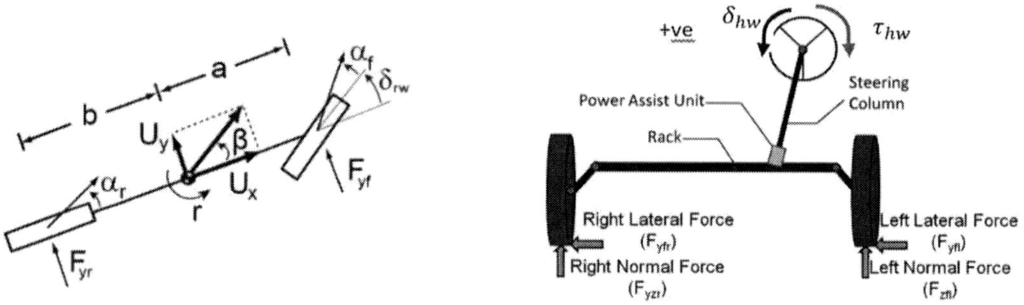

(a) (b)

Fig. 1. (a) Simple bicycle model extracted from [3], (b) Simple conventional steering system taken from [4] with some modifications to signify the steering wheel angle and handwheel torque.

Table 1. Parameter values for simulation

Parameter	**Value**	**Units**	**Parameter**	**Value**	**Units**
α	1.02	m	k_{db}	0.01	-
α_f	0.10	rad	k_{jack}	250	-
b	1.48	m	k_{tire_mom}	0.035	-
b_{system}	0.015	N.m/rads^{-1}	m	1200	kg
C_{af}	110000	N.rad^{-1}	μ	0.90	-
$\Delta_{damping}$	0.10	N.m/rads^{-1}	σ_{ps}	0.20	-
$\Delta_{inertia}$	1.00	kg.m^2	t_m	1.00	cm
δ_{db}	0.05	rad	t_p	0.58	cm
δ_{rw}	1.00	rad	t_{po}	1.00	cm
F_{yf}	-7030	N	τ_{allign}	11100	N.m
F_{zf}	9680	N	$\tau_{asstd\_tire\_mom}$	-9820	N.m
F_{yf}	9.81	N	τ_{damp}	0.10	N.m
G	17	-	τ_{hw}	-343	N.m
γ	0.20	-	$\tau_{inertia}$	1.00	N.m
I_{add}	0.01	-	τ_{jack}	-250	N.m/rad
I_{zz}	1552	kg.m^2	τ_{motor}	-343	N.m
J_{system}	0.0014	kg.m^2	τ_{tire_mom}	10800	N.m
k_{damp}	0.10	-	Vx_{mph}	31.07	mph

The steering sense can be developed in an SBW system utilizing force feedback (FFB) interface, depicted in Fig. 1(b) where it utilizes Eqs. (4-5) generate the handwheel torque, τ_{hw}. The handwheel torque is felt by a driver once the driver turns the steering wheel. Three dominant torques which play a role in producing the steering feel [5]: torques due to damping and inertia, and tire moment torque due to power assist.

$$\tau_{hw} = \tau_{motor} + J_{system}\ddot{\delta}_{hw} + b_{system}\dot{\delta}_{hw} \quad (4)$$

$$\tau_{motor} = \tau_{damp} + \tau_{inertia} + K\tau_{assisted\ tire\ moment} \quad (5)$$

The details of Eqs. (4-5) are available in [4] and [5]. The parameter values for simulation are given in Table 1. The calibration of the Logitech G27 steering resulted in the torque T in Newton shown in Eq. (6),

$$T = 3.0156\ U_s + 0.1161 \quad (6)$$

where U_s is the value specified in Simulink for the force feedback port of Joystick block.

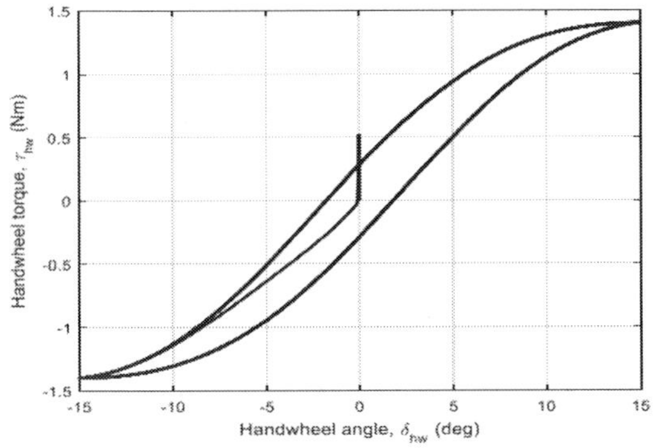

(a)

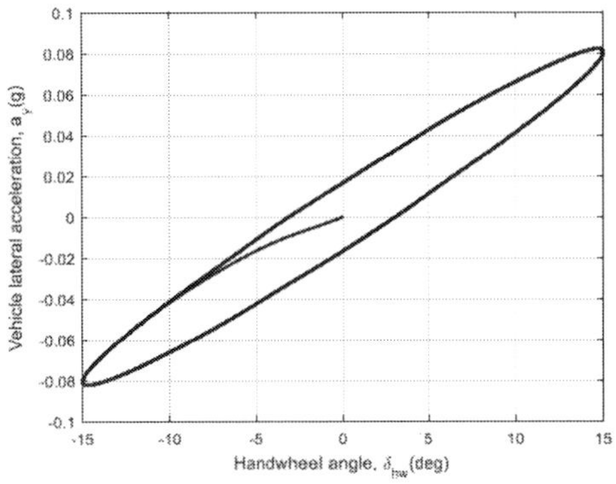

(b)

Fig. 2. (a) Hand wheel torque as function of handwheel angle, (b) Vehicle lateral acceleration as function of handwheel angle.

2.2 Validation test

For validation purposes, weave test is performed to obtain several measures which includes the handwheel angle (δ_{hw}), vehicle lateral acceleration (a_y) and handwheel torque (τ_{hw}) as well as to evaluate on-center handling. On-center handling can be defined as the steering feel generated of a vehicle during straight-line driving and in negotiating large radius bends at some high speeds. These conditions were obtained from the input parameters performed by Balachandran and Gerdes [6], where weave manouver tests were conducted.

Weave test generates the characteristics to evaluate the steering feel and steering precision around the central position. During this test, steering angle oscillates sinusoidally with constant amplitude and frequency around set vehicle speed of 50 km/h as shown in Fig. 2 which has a similar pattern as in [7]. Validation of the result is performed by a weave test showing the characteristics of the steering feel/force feedback produced. It is shown in Fig. 3 in which the graphs consist of handwheel angle (δ_{hw}), vehicle lateral acceleration (a_y) and handwheel torque (τ_{hw}) as function of time at a constant speed of 50km/h.

3 Analysis of J-Turn Test

This J-turn was selected to study the performance of the force feedback inside a SBW steering system. The virtual experiment is conducted with the help of 3DView of the vehicle as shown in Fig 4. Four different kinds of graph, starting with the steering wheel angle, throttle, brake as the function of time and lastly the position of the car on the x and y position as shown in Fig. 5. This test illustrates the dynamic vehicle behavior when entering a curve. The car has an initial velocity, v_x. In this test, this initial velocity is set to 60 km/hr. Five seconds after the simulation is started, input to the steering angle is set to 85° to the right, and 10 seconds later the steering angle is set back to zero. The behavior of the vehicle during the virtual experiment is shown in Fig. 6 which includes the handwheel torque, yaw velocity, longitudinal acceleration and lateral acceleration.

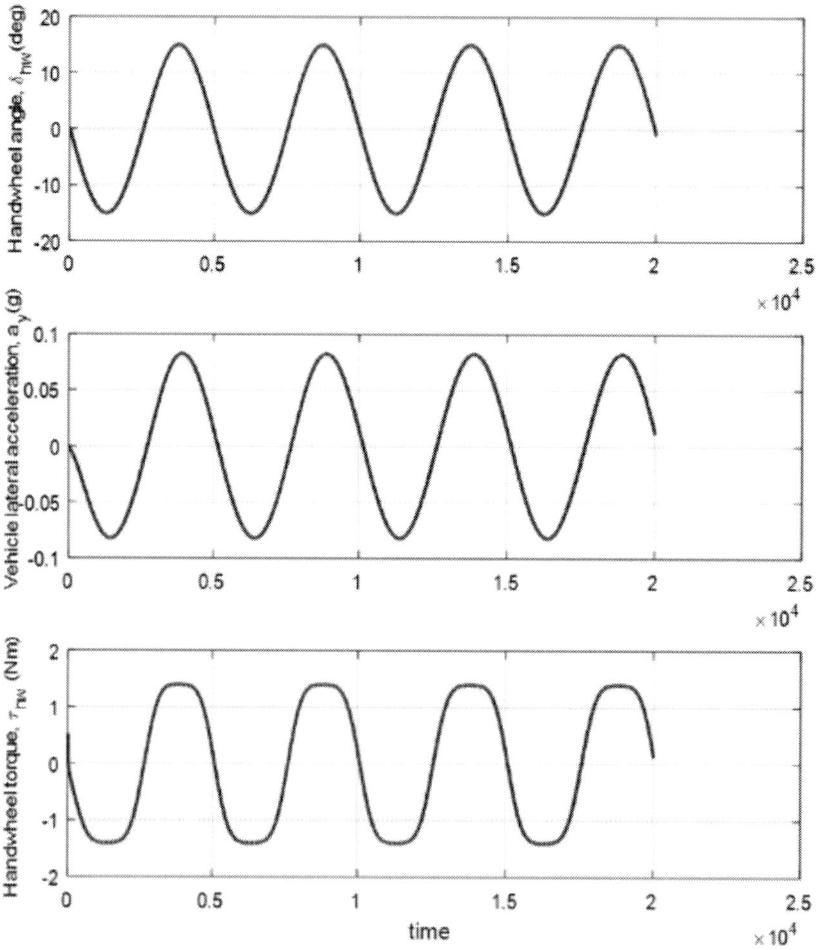

Fig. 3. Handwheel angle, vehicle lateral acceleration and handwheel torque as function of time.

Fig. 4. View of Car in 3D animation of Matlab

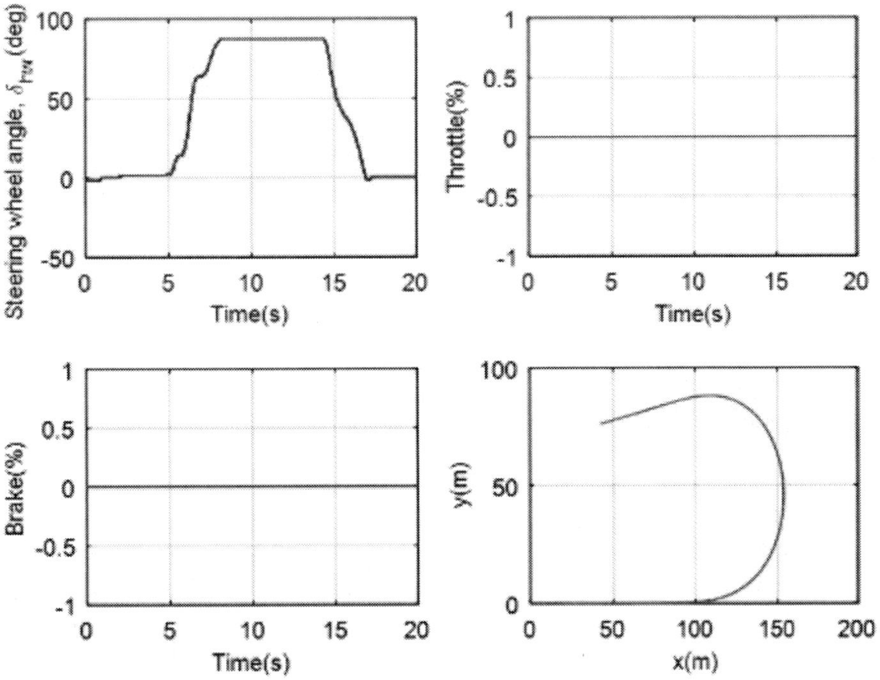

Fig. 5. Steering wheel angle, throttle, and brake as function of time, and x-y position of vehicle during a J-turn test.

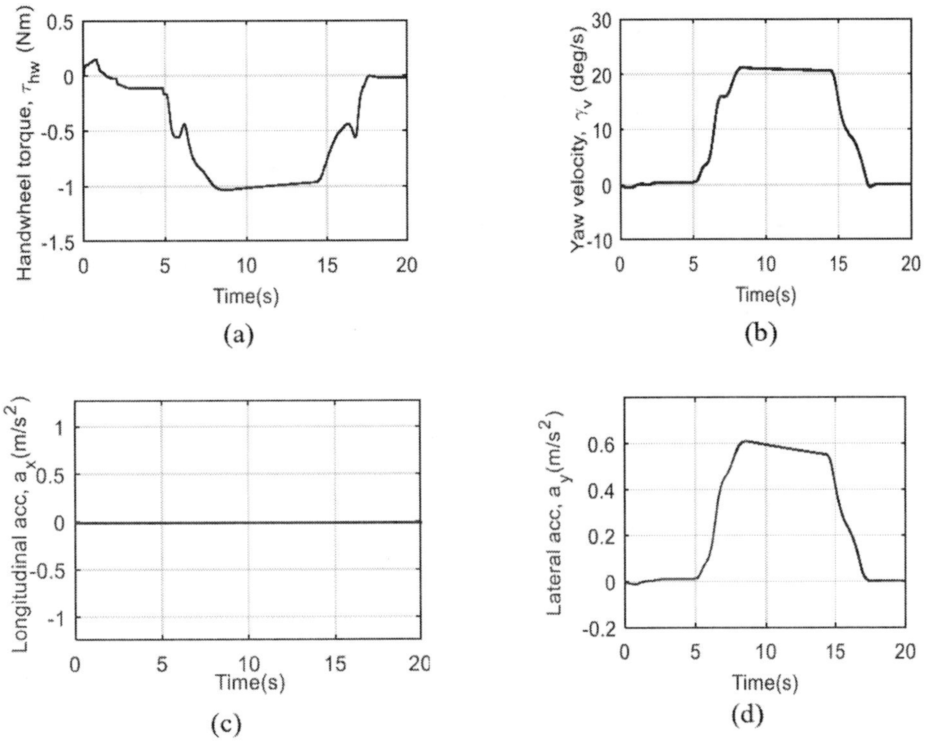

Fig. 6. (a) Handwheel torque, (b) Yaw velocity, (c) Longitudinal acceleration, (d) Lateral acceleration.

4 Conclusions

In this study, handwheel torque that contributed the biggest value to the force-feedback had been modeled. An artificial force-feedback behavior in a steer-by-wire vehicle had been successfully produced using Simulink for a virtual vehicle simulator. An experiment had been performed to show vehicle dynamic behavior during a J-turn test. Calibration test played a crucial role to identify the appropriate amount of constant gain in generating force-feedback signal. The driving simulator being developed has reasonably made the driver of the simulator feel the similar forces coming from a real road.

References

1. S.M.H. Fahami, H. Zamzuri and S.A. Mazlan, Development of Estimation Force Feedback Torque Control Algorithm for Driver Steering Feel in Vehicle Steer by Wire System: Hardware in the Loop, *International Journal of Vehicular Technology*, (2015)
2. U.B. Mandhata, M.J. Jensen, J.R. Wagner and F.S. Switzer , Evaluation of a customizable haptic feedback system for ground vehicle steer-by-wire interfaces, *Proc. American Control Conference (ACC)*, 2781-2787, (2012)
3. R.A. Rashid, J.D. Setiawan, M. Maharun and M. B. Baharom, Investigation Into the Effects of Maximum Deflection in Low Stiffness Resilient Shaft of Semi Active Steering Systems, *ARPN Journal of Engineering and Applied Sciences,* **11** (22), (2016)
4. M. B Baharom, K. Hussaina and A. J. Day, Design concepts and analysis of a semi-active steering system for a passenger car, *Journal of Multi-body Dynamics* **223**(4), 283-291, (2009)
5. E.S. Mohamed and S.A. Albatlan, Modeling and Experimental Design Approach for Integration of Conventional Power Steering and a Steer-By-Wire System Based on Active Steering Angle Control, *American Journal of Vehicle Design*, **2**(1), 32-42, (2014)
6. A. Balachandran and J.C. Gerdes, "Designing Steering Feel for Steer-by-Wire Vehicles Using Objective Measures, *IEEE/ASME Transactions on Mechatronics* **20**(1), 373-383 (2015)
7. B. Avinash and C. G. J, Artificial Steering Feel Design for Steer-by-wire Vehicles, in *7th IFAC Symposium on Advances in Automotive Control*, Tokyo (2013)

Enhancement Surface Coating Stainless Steel And Copper Using Ultrasonic Batch

Sendie Yuliarto Margen*, *Sulistyo* S, *Sri* Nugroho, *Yoga* Setiawan Adi Nugroho

Mechanical Engineering Department, Engineering Faculty, Diponegoro University, Semarang, Central Java, 50275, Indonesia.

Abstract. Electrodeposition is considered an easy and cost effective technique for the manufacture of alloy layers. Various properties for coatings can be achieved by selecting various electrodeposition parameters. These materials have the advantage of corrosion resistance but relatively expensive. They will need alternative materials that can easily obtained and replace these materials. This paper discusses the process of coating stainless steel AISI 304 and copper with Ni. Electric current and coating time is used in the electroplating coating process of Ni to Ni thickness supported by ultrasonic batch. AISI 304 stainless steel and copper coated using nickel (Ni) with variable time and fixed electric current parameters. The electric current used 1.5 ampere, the time of plating process of 60 seconds, 120 seconds, and 180 seconds. The composition of nickel electrolyte solutions using nickel sulfate 300 gr/L, nickel chloride 30 gr/L, and sour borax 30 gr/L with brighteners added one (I) 15 mL and brighteners two (II) 1 mL. Measurement of thickness is done by using optical microscope assisted by imageraster software. The best process used 1.5 amperes of electric current and time of 180 seconds. The value of copper thickness is 26.50 (μm) using ultrasonic batch.

1 Introduction

Coating is a technology that is well known in the field of industry. The advantage of coating technique is to improve the quality of the material. Its quality that can be improved include mechanical properties, physical properties, chemical properties and decorations. Increased mechanical properties are resistance to wear and increased hardness, chemical properties of coated material can be corrosion resistant, while the physical properties is to smooth the surface of the sighting side. Coating technology can be carried out by various process methods such as Physical Vapor Deposition (PVD), Chemical Vapor Deposition (CVD), spraying, deep coating and electroplating. [1,2].

The method of coating that has been developed has its own unique advantages. PVD was developed to produce a thin layer with a range of nanometer thickness up to a micrometer in a vacuum. The formed layer is obtained from the conversion of materials in the form of atoms or molecules from the evaporation process of a precipitated material

* Corresponding author : sendiemargen@gmail.com

under vacuum conditions [3]. Electroplating is the process of solving the coating in a solution assisted by the flow of electricity. The quality of the electroplating coating is influenced by the type of solution and the current flowed through the electrode [4].

Principally of the electroplating process, the electricity flows between the anode and the cathode, in aqueous electrolyte solution and the chemical reaction on the metal surface occurs. When given the potential difference or voltage, ions move toward the electrode. The cation moves toward the cathode and anion toward the anode. This process occurs by catalyzing a chemical reaction to reduce the metal salt (anode). The thickness range that can be obtained by electroless process is 1-120 μm. The coating process is always through several stages : surface cleaning to be coated, surface activation, coating process, and after coating treatment [5,6].

The electroplating process is based on Faraday's law that (1) the mass is released to an electrolyte solution which is proportional to the current through a large electrolyte solution and (2) the released mass is proportional to the electrochemical equivalent (atomic valence ratio). The workpiece was coated by electroplating included iron, steel, bronze, copper, plastic, and stainless steel. The transfer of metal ions was helped by an electric current through the electrolyte solution. The metal ions precipitates on solid objects. The metal ion is obtained from the electrolyte solution as well as derived from the dissolving metal anode in the electrolyte. Deposition occurs on the workpiece acting as a cathode [7,8]. AISI 304 stainless steel and copper have been selected as substrate. Nickel (Ni) is used as a coating as a material because it can improve corrosion resistance at high temperature of AISI 304 stainless steel and copper as an alternative material. Electroplating process can be improved coating quality by using ultrasonic batch [9].

Ultrasonic batch is process that use ultrasound waves (typically from 20-400 kHz). The ultrasonic process can improve the coating due to fluid agitation. Agitation produces high strengths on contaminants attached to substrates such as metals, copper, plastics, glass, rubber, and ceramics. This action also penetrates the holes, cracks, and niches of the blind. The ultrasound transducer generator is installed indoors, or lowered into the liquid. Ultrasonic waves are produced in liquids by changing the particle size inside the tank by using an oscillating electrical signal at ultrasonic frequencies. This creates compression waves in the fluid tank that helps the coating process [10,11].

The method using ultrasonic can improve the agitating process during electroplating process. Transfer of electric current metal ions can be through an electrolyte solution with the help of ultrasonic batch. The particles from the electrolyte solution can be divided into smaller and dissolved electrolyte ions in the electrolyte will settle evenly on the solid to be coated. The quality of coating substrate can be improved. Conventional coating electroplating method has been done but its quality is relatively worsh. This paper discusses improved electroplating processes supported by ultrasonic batch.

1.1 Nickel As Metal Coating

The coating process there are several elements found in the periodic table can be used as a coating material. Nickel is a chemical element whose symbol is Ni and atomic number 28. Nickel is a metal element with solid phase. At this time, nickel plating in iron is widely performed both for the purpose of preventing rust and beauty. The properties possessed by nickel are contained in table 1 [12].

Table 1. Properties of the element nickel

Criteria	Specification
crystal structure	FCC
atomic mass	58.7 gr/mol
Density	8.907 gr/cm^3
melting point	1453 °C
boiling point	2730 °C
Tensile strength	317 MPa
Yield strength	59 MPa
Hardness	64 *HV (annealed)*

2 Materials and Methods

2.1 Stainless Steel AISI 304 As a metal coated (Substrate)

Stainless steel AISI 304 is an austenitic stainless containing chromium and nickel elements of 18-20% wt and 8-10.5% wt respectively. These elements are the alloying elements most commonly produced and used amongst all stainless steel [13].

2.2 Copper (CU)

Copper is a transition metal (group I B) is reddish, easy to filter and easy to work metal. Brass or copper (Cu) is generally crystalline and reddish color. One of the metal from the IB with atomic number 29; the atomic weight is 63.546; melting point 1083 ° C; Boiling point 2310 ° C [14].

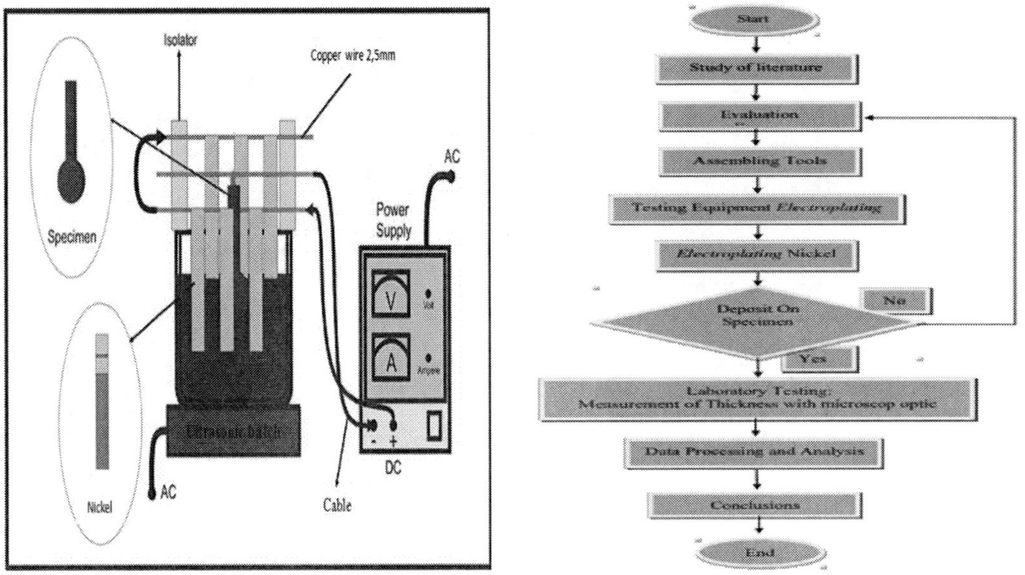

a)   **b)**

Fig 1. a). The illustration of experiment tools and b). Flow chart of process.

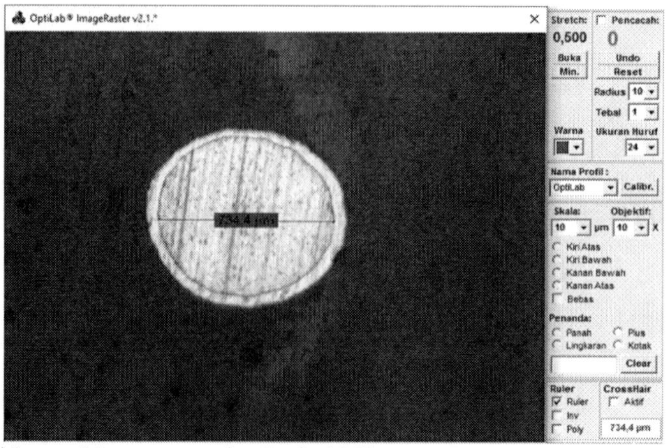

Fig 2. Dimensional Image Measurement by Optical Microscopes.

3 Result and Discussion

The weight and thickness of the specimen (deposit) is obtained by a nickel electroplating process with 1.5 amperes of electric current and time layer (60, 120, 180 seconds). This section also displays tabular data and images on the nickel micro layer thickness. Tables and figures can shows the effect of electric current and the thickness of the coating layer of time on nickel.

3.1 Measurement of weight and thickness

The measurements of the weight specimens and the thickness layer of nickel on the research done by destructive methods. Weight measurement with digital scales and thickness measurements made using optical microscope assisted by software imageraster.

Table 2. 304 stainless steel weight data on the process of electroplating nickel.

No	ELECTRO PLATING	WITHOUT ULTRASONIC BATCH			USING ULTRASONIC BATCH		
		60 s	120 s	180 s	60 s	120 s	180 s
1	Before	10.7573 g	10.7498 g	10.7624 g	10.7778 g	11.7893 g	11.7629 g
2	After	10.7811 g	10.7988 g	10.8395 g	10.8281 g	11.8631 g	11.8705 g

Table 3. Copper weight data on the process of electroplating nickel.

No	ELECTRO PLATING	WITHOUT ULTRASONIC BATCH			USING ULTRASONIC BATCH		
		60 s	120 s	180 s	60 s	120 s	180 s
1	Before	11.7578 g	11.7593 g	11.7629 g	11.7598 g	11.7693 g	11.7711 g
2	After	11.7811 g	11.8031 g	11.8305 g	11.8011 g	11.8431 g	11.8875 g

Table 2 and 3, the weight measurement using digital scales shows that the copper has a heavier specimens. That's because copper has heat conductivity and electrical conduction better than others. The copper also has a low chemical reactivity. During the process, the nickel particles are dissolved in a solution which will more easily stick to copper specimens. The nickel layer on copper specimens is thicker and increases heavy specimens. The nickel layer on copper specimens is thicker and increases weight specimens.

Table 4. The thickness stainless steel 304 using electroplating 1.5 Ampere.

No	Electric Current(A)	Temperature (°C)	Time (s)	Thickness coating without us (μm)	Thickness coating using us (μm)
1	1.5	45	60	4.36	7.57
2	1.5	45	120	6.66	10.82
3	1.5	45	180	8.27	14.90

Table 5. The thickness copper using electroplating 1.5 Ampere.

No	Electric Current (A)	Temperature (°C)	Time (s)	Thickness coating without us (μm)	Thickness coating using us (μm)
1	1.5	45	60	5.76	11.34
2	1.5	45	120	8.58	18.75
3	1.5	45	180	11.60	26.50

Table 4 and 5 show that the effect of time on coating thickness. The longer time will produce the thicker coating. It was caused by the dissolved metal having time to finish evenly and accumulate on the coated object. Electroplating process is supported by using ultrasonic, can improve the agitating electrolyte process during the process.

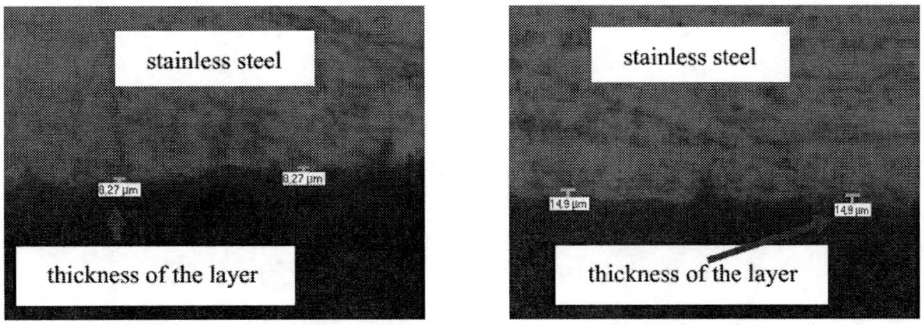

a). Without ultrasonic b). Using ultrasonic

Fig 3. a). Thickness coating of 6.27 μm at SS 304 and b). The thickness of 14.9 μm using ultrasonic at SS 304.

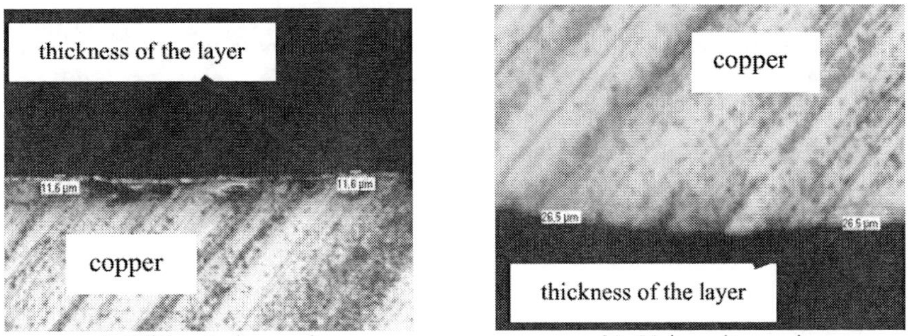

a). Without ultrasonic b). Using ultrasonic

Fig 4. The coating of copper substrate a). The thickness of 11.6 μm without ultrasonic and b). The thickness 26.5 μm using ultrasonic.

Fig. 3 and 4 show the electroplating thickness using ultrasonic and without ultrasonic. Fig. 3 shows the electroplating thickness on the stainless steel substrate. The thickness is 11.6 μm without using ultrasonic batch, when using ultrasonic thickness 14.9 μm. Fig. 4 shows the electroplating thickness of the copper substrate. The thickness is 11.6 μm without ultrasonic and 16.5 μm using ultrasonic. Electroplating processes supported by ultrasonic produce a better thickness coating than without ultrasonic. Ultrasonic equipment during the process can improve the process especially agitating electrolytes which can increase the addering of the nickel particles in the substrate.

4. Conclusions

Several conclusions can be presented as follows :
The composition of nickel electrolyte solutions using nickel sulfate 300 gr/L, nickel chloride 30 gr/L, and sour borak 30 gr/L with brighteners added one (I) 15 mL and brighteners two (II) 1 mL. The electroplating process can be enhanced by using ultrasonic for stainless stell and copper substrate. The process of electric current and electroplating time is directly proportional to the thickness of the layer (deposit). The best process electrode is obtained on the copper substrate. The best process was obtained using 1.5 A and 180 seconds and a thickness higher than 26.50 μm.

Reference

1. M. Kumar, S. Mishra, R. Mitra, *Surf. Coat. Technol.* **228** (2013).
2. A.M. Pagon, E.D. Doyle, D.G. McCulloch, *Surf. Coat. Technol.* **235** (2013).
3. S.S. Djokic, *Theory and Practice, Springer Science & Business Media*, (2010).
4. Yli-Pentti, A. *Electroplating and Electroless Plating*, (Finlandia 2014).
5. K.H. Kim, T.Lim, K.J. Park, H.C. Koo, M.J. Kim, J.J. Kim, **151** (2015).
6. Y. Jie, H. Fan, J.R. Niskala, W. You, *Surf. A-Physicochem. Eng. Asp,.* **434** (2013).
7. Mordechay Schlesinger, M.P. *Modern Electroplating*, (John Wiley & Sons Metallography test 2010).
8. B. Díaza, J. S´wiatowska, V. Maurice, A. Seyeux, E. Härkönen, M. Ritala, S. Tervakangas, J. Kolehmainen, P. Marcus, **90** (2013).
9. K. Asa Deepthi, R. Balachandran, B.H. Ong, K.B. Tan, H.Y. Wong, H.K. Yow, S. Srimala, *Appl. Surf. Sci.*, **360** (2016).
10. Fuchs, F. John. *Ultrasonic Cleaning: Fundamental Theory and Application.* (Blackstoney Ultrasonic 2010).
11. AO L, HAO H, JI S H, FANG C F, ZHANG X G. *Transactions of Nonferrous Metals Society of China,* **6** (2011).
12. Bardet, W.P. *ASM HANBOOK Surface Engineering*, (ASM International 1994).
13. DiBari, G., *Nickel Plating Handbook*, (Canada 2014).
14. Marco Boniardi, A.C., *Stainless steels.* (Gruppo Lucefin Research & Development 2014).
15. I.V. Blinkov, S.G. Tsareva, A.V. Zentseva, A.O. Volkhonskii, V.I. Buzanov, N.N. Stepareva, Russ. J. *Non-Ferrous Met.*, **51** (2010).

Fabrication Dye Sensitized Solar Cells (DSSCs) Using *β-Carotene* Pigment Based Natural Dye.

Wa Ode Nirwana Sari Halidun[1], *Eka* Cahya Prima[1,3], *Brian* Yuliarto[1,2,*], and *Suyatman*[1]

[1]Advanced Functional Materials Laboratory, Departement of Engineering Physics, Faculty of Industrial Technology, Institut Teknologi Bandung, Bandung 40132, Indonesia
[2]Research Center for Nanoscience and Nanotechnology (NRCN), Institut Teknologi Bandung, Bandung 40132, Indonesia
[3]Science Laboratory, Departement of Science Education, Fakultas Pendidikan dan Ilmu Pengetahuan Alam, Universitas Pendidikan Indonesia, Bandung 40154, Indonesia

Abstract. The adsorption characteristic of natural dye has been studied on harvesting light using purification process. The purification process used for separation pigment photosynthesis of syngonium leafs was called fractionation. The first fractionation of the dye contains pigment of *β-carotene* . The absorpton spectrum of *β-carotene* has been investigated by using Uv-Vis spectroscopy. The Fourier Transform Infrared revealed that the binding of the dye was on the surface of TiO_2. Analysis absorbance and energy gap of the dye has been investigated by using Uv-Vis spectroscopy and Tauc relation. The cells were illuminated under 100 mW/cm^2 AM 1.5 condition. The performance of the cells with sensitized the fractionation dye has a short circut current density (J_{sc}) of 0.27 mA/cm^2, an open circuit photovoltage (V_{oc}) of 0.4V, and Fill Factor (FF) of 68.89%. Furthermore, the overall efficiency of DSSC is 0.074%. The low efficiency for the fractionation cell based dye sensitized solar cells due to the weakness of interaction between dye molecule and the surface of TiO_2.

1 Introduction

Dye sensitized solar cells (DSSCs) is the third generation of photovoltaic device that attracted considerable attention due to the low cost fabrication, enviromental friendly and simple fabrication [1]. The first concept of DSSCs was introduced by Grätzell and his co-worker in the year 1991 [2]. The device can convert the light energy into electrical energy by using sensitizer. The sensitizer can absorb visible light and then inject an electron to the semiconductor to generate photocurrent.

The process of photosensitization of the dye on the surface of semiconductor has an important role to increase the performace of DSSCs. The photosensitization is the process of harvesting light and electron injection of the dye from low level energy to the conduction band of semiconductor [3]. Thus, the efficiency of DSSC is mainly determained by the using ofsensitizer.

* Corresponding author: brian@tf.itb.ac.id

Generally, transition metal coordination compound such as ruthenium transition metal polypyridyl complex is used as an effective sensitizer. The highest efficiency of DSSC using ruthenium complex is 11 – 12% [4]. Although, the ruthenium has given the best performance as sensitizer, ruthenium polypyridyl contains heavy metals, which makes this kind of sensitizer unfriendly for the enviroment, and limited in amount. Moreover, the synthesis of this complex is very complicated and expensive [5]. In order to replace the rare and expensive ruthenium material, the natural dye from pigment of leaf, seed, and fruit can be used as a low cost, easy extraction and friendly enviroment sensitizer.

Photosyntetic pigments that commonly consists of chlorophyll, and its derivative, and carotenoid, have an important role in harvesting light and electron transport. Those pigment contain carboxylate group that influence photosensitization process on the surface TiO_2.

In this work, pigment *β-carotene* from *syngonium sp* has been separated by using column chromatography technique. The *β-carotene* pigment extract from *Syngonium sp* was used as natural dye in DSSC. The absorption spectrum, the band gap analysis, and the interaction of the dye onto surface of TiO_2 have been investigated by using Uv-vis spectroscopy, Tauc relation and FTIR spectroscopy, respectively.

2 Experimental

2.1 Preparation of Natural dye solution

Syngonium sp was washed, crushed in a small pieces and dried in room temperature over night. The leaf was blended to obtain powder of *Syngonium sp*. Then, the pigment of *Syngonium sp* was extracted by using maserasi method. The leaf was immersed in acetone solutian with ratio at 1:5 (v/v).

Separation of *β-carotene* pigment in the *Syngonium sp* is performed using column chromatography tecnique. The technique follows the procedure of Shioi [6]. The *β-carotene* pigment in the purification process has been obtained from the first fractionation. The **Fig.1** shows the result of the first fractionations (F1) from *Syngonium sp*. The F1 dye will be used as natural photosensitizer for this work as natural dye.

Fig.1 The first fractionation pigment from *Syngonium sp*

2.2 Preparation of dye sensitized solar cells

Thin film semiconductor TiO_2 and the counter electrode, Pt, were coated onto FTO glass substrate that obtained from dyesol. These electrode were washed by using isopropil alcohol solution. Semiconductor TiO_2 was immersed into 0,1 M dye solution for 3 days. Then, TiO_2 that was sensitized dye, was rinsed by using ethanol solution.

TiO_2 sensitized dye photoelectrode and Pt counter electrode were assembled by using suryline polymer that followed the procedure of Prima [7]. The electrolyte solution, EL HSE, that obtained from Dyesol was dropped into the hole of counter electrode and then the hole was sealed by another suryline to prevent electrolyte leakage.

2.3 Characterization and Measurement of dye sensitized solar cells.

To investigate physical properties of natural dyes, the dye of *Syngonium sp* absorption spectrum was investigated by using Uvi-visible spectroscopy of Hewlett Packard 8453 Agilent Technologies. The binding vibration of dye solution and TiO_2 photoelectroda, after and before immersed in dye solution, were investigated by using Fourier Tranform Infrared spectroscopy of Prestige 21 Shimadzu Japan. To calculate energy gap of dye solution was investigated by using Tauc realation, as given **Eq.**(1) [8].

$$\alpha h\upsilon \approx (h\upsilon - E_g) \quad (1)$$

$$FF = V_{max} \times I_{max} / V_{oc} \times I_{sc} \quad (2)$$

$$\eta = V_{oc} \times I_{sc} \times FF / P_{in} \quad (3)$$

To investigate performance of dye sensitized solar cells, the device was iluminated under standard solar simulator (AM 1.5; 1000 mW/cm^2). The photocurrent- voltage (J-V) curve was used to determine the short-circuit photocurent (J_{sc}) and open-circuit voltage (V_{oc}). The **Eq.**(2) was used to calculate the Fill Factor (FF). The overall efifciency of DSSC has been observed by using **Eq.**(3).

3. Results and Discussion

3.1 Uv-vis Spectroscopy

The uvi-vis absorption spectra of F1 dye was shown in **Fig.2**. According to the **Fig.2**, the absorption peaks of F1 dye was observed between 414 nm and 502 nm that confirmed as carotene pigment that indicated transition energy from S_0 to S_1 [9-13]. The *β-carotene* pigment is the part of the carotene pigment that mainly obtained in the leaf that have absorption characterization of 440 – 480 nm [14]. It corresponds to the results that was showed in **Fig.2**.

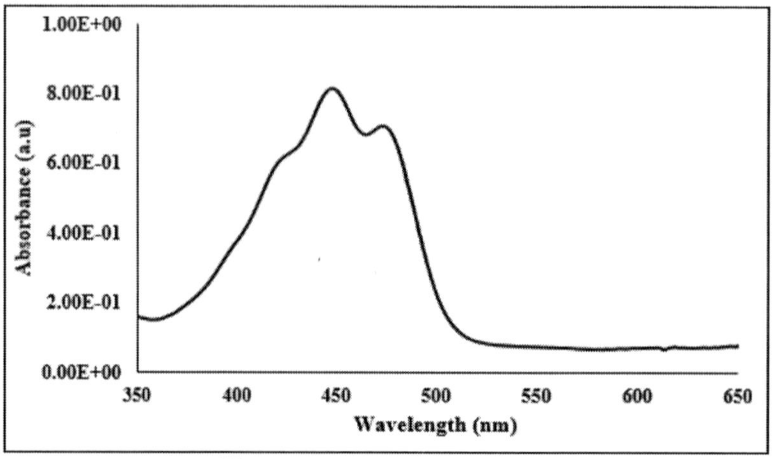

Fig 2. The absorption spectrum of The *β-carotene* pigment.

3.2 Determination Gap Energy of Natural Dye from pigment *Syngonium sp.*

The gap energy of each pigment can be evaluated by using Tauc relation that is described on **Eq.1. Fig. 3** shown the plot of hυ versus $(\alpha h\upsilon)^2$ for *β-carotene* dye. The gap energy of *β-carotene* based natural dye is 2.47 eV.

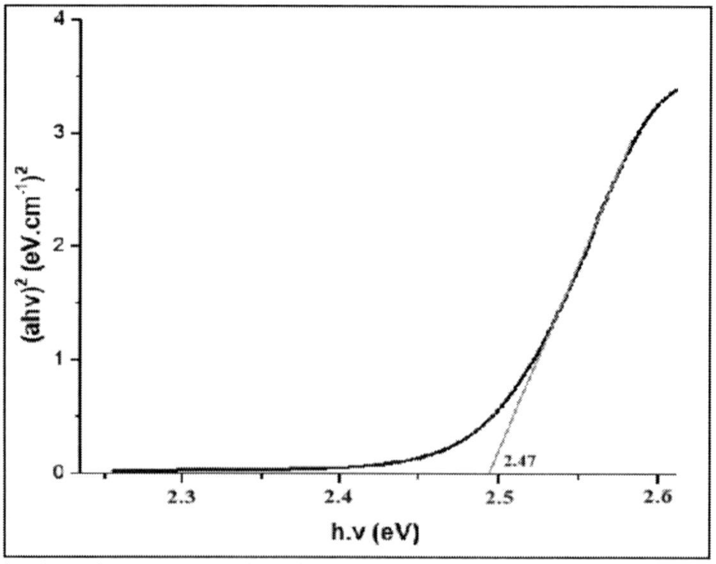

Fig.3 Determmination of gap energy using Plot Tauc relation

3.3 The Fourier Transform Infrared (FTIR) Characterization

Fig.4 shows the fourier transform infrared of TiO_2 before and after soaking in dye solution. According to the **Fig.4**, the peaks of TiO_2 presented at 3429.53 cm^{-1} is affected by O-H stretching vibration bonding of hydrogen water molecule. The peak shown at 1625.21 cm^{-1} is assigned to O-H bending, and the peak at 454.3 cm^{-1} is due to the vibration bonding of Ti-O [15]. After TiO_2 was immersed in dye solution, it appeared two new peaks at 2917.3 cm^{-1} and 2846.6 cm^{-1}, both of peaks were confirmed as peak of natural dye that bond onto surface TiO_2.

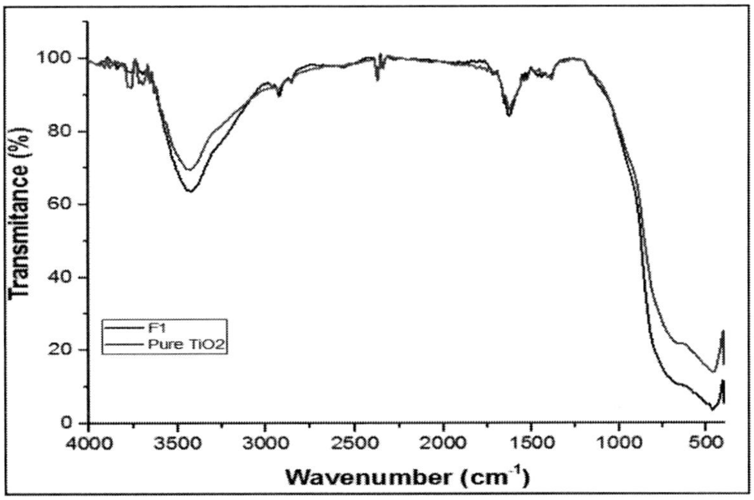

Fig.4 The Fourier Transform Infrared of TiO_2 before and after immersed into dye Solution

3.4 The Performance of Dye Sensitized Solar Cells

The performance natural dye from pigment photosynthesis of *β-carotene* was performed by measuring the *current density-voltage* (J-V) under illumination 100 mw/m^2 AM, 1.5 standard. The parameter of short circuit current density (J_{sc}), open circuit voltage (V_{oc}), fill factor (FF), and efficiency (η) were listed in **Table 1**.

Table 1 Parameter performance natural dye using natural pigment of *β-carotene* from *syngonium sp.*

Natural Dye	J_{sc} (mA/cm^2)	V_{oc} (V)	FF (%)	η (%)
F1	0.27	0.4	68.89	0.074

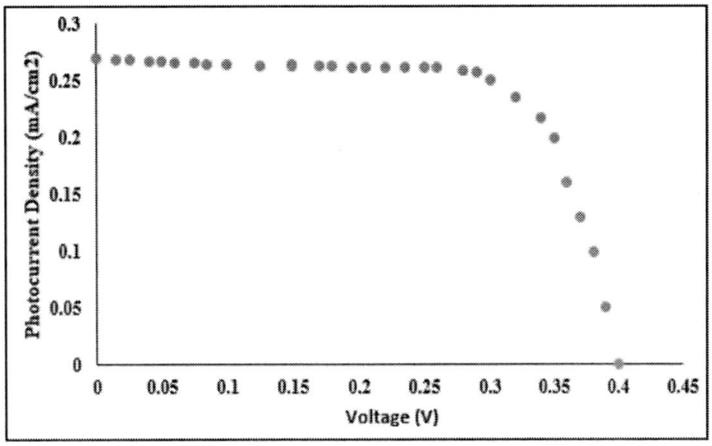

Fig.5 The current – voltage curve for F1 dye based DSSC

The **Fig.5** shown the J-V curve of *β-carotene* pigment based natural dye of DSSC. The low efficiency (η) was probably influenced by some factors such as the concentration of dye was low, the gap energy of dye was wide and the absorption of dye in visible light spectrum was limited. Therefore, they cause the weakness of interaction between dye molecule and the surface of TiO_2.

4. Conclusion

Natural dye from natural pigment photosynthesis such as *β-carotene* from the leaf of *syngonium sp* through the purfication process can be used as natural photosensitizer in DSSC because it is ability to harvest light in visible spectrum. Moreover, the performance *β-carotene* based natural dye DSSC is still low. It is caused by the concentration and the limited harvesting visible light of the dye is low due to the low of photosensitization prosess between dye and the surface of semiconductor TiO_2.

This research was supported by Advance Functional Material Laboratory (AFM), Bandung Institute of Technology (ITB) and funded by Indonesia Endowment Fund for Education (Lembaga Pengelola Dana Pendidikan (LPDP), Indonesia).

References

1. S. Shalini, R.B. Prabhu, S. Prasanna, S.P. Mallick, S. Senthilarasu. Spectrochimica Acta A: Molecular and Biomolecular Spectroscopy **148** (2015).
2. B. O'Reagen, M. Grätzel. Nature **353** (1991)
3. J. Spikes. Photosensitization. In: Andery Lim.,et al. Spectrochimica Acta Part A; Molekular and Biomolecular Spectroscopy **138** (2015)
4. H. Zhou, L. Wu, Y. Gao, T. Ma. Chemistry **219** (2012)
5. H. Chang, H.M. Wu, T.L. Chen, K.D. Huang, C.S. Jwo, Y.J. Lo. Alloys and Compounds **495** (2010)
6. Y. Shioi. Biochemistry, Biophysics, Function and Aplication, pp. **123-133** (2006)
7. E.C. Prima, N. N. Hidayat, B. Yuliarto, Suyatman, H.K. Dipojono. Spectrochimica Acta A: Molecular and Biomolecular Spectroscopy **171** (2017).
8. J. Tauc. Mat. Ress. Bull, **Vol. 3**, pp. 37 – 48 (1968)
9. G. Calogero, A. Bartolotta, G.D. Marco. **DOI 10.1039/c4cs00309h** .
10. E. Yamazaki, M. Murayama, N. Nishikawa, N. Hashimoto, M. Shoyama. Solar Energy **81** (20017)
11. N.M. Gomez-ortiz, I.A. Vazquez-Maldonado. A.R. Perez-Ezpadas, G.J. Mena-rejon, J.A. Azamar-Barrios, Oskam. Solar Energy Materiar and Soar Cells **94** (2010).
12. S. Hao, J. Wu. Y. Huang, J. Lin. Solar energy **80** (2006)
13. K.V. Helmalata, S.N. Karthick, C. Justin Raj, N.Y. Hong, S.K. Kim, H.J. Kim,. Spectrochimica Acta A: Molecular and Biomolecular Spectroscopy **96** (2012)
14. X.F. Wang, A. Matsuda, Y. Koyama, H. Nagae, S. Sasaki, H. Tamiaki, Y. Wada. Chemical Physics Letter **423** (2006).
15. S. Ananth, P.Vivek, T. Arumanayagam, P. Murugakoothan. Spectrochimica Acta A: Molecular and Biomolecular Spectroscopy **128** (2015)

Aluminum corner joint using friction stir welding

Djarot B. Darmadi[1,*], *Widia* Setiawan[2] , *Anindito* Purnowidodo[1] and *Eko* Siswanto[1]

[1]School of Mechanical Engineering, Brawijaya University, 65145 - MALANG
[2]Sekolah Vokasi, Gadjahmada University, 55281 - YOGYAKARTA

Abstract. This article studies the feasibility of applying Friction Stir Welding (FSW) to an aluminum 6061 corner joint. The friction tool rotated at constant speed (2000 RPM) whilst feed rate varied in 10, 15 and 30 mm/minutes. Generally the lower feed rate produces higher tensile strength due to higher heat is embedded in the base metal, but the 15 mm/minute feed rate exhibited the least tensile strength. Lower feed rate reduces Mg alloy in base metal which in turn decreases the tensile strength. The final tensile strength is the compromise between higher heat input and lower Mg content that is why the 15 mm/minute has lowest tensile strength.

1 Introduction

Metals with high strength to weight ratio will be more broadly used due to their economical benefits [1]. The metal with higher strength to weight ratio such as aluminium will replace steel as the most applied metals [2]. Unfortunately, when fusion welding is applied, aluminum encounters problems such as: porosity due to the big difference of hydrogen solubility at liquid and solid forms [3], less compatibility owing to Al_2O_3 film [4], hot cracking because of the alloy element which is added intentionally to pure aluminium [5] and the strength loss while melting takes place [6]. The solid state welding is hoped able to overcome those problems. Most of solid state welding researches observed lap and butt joints that is why this article discusses corner joint to underline the wide opportunities of FSW application.

There are three major streams of welding study: analytical, numerical and experimental methods [7, 8]. This study used a true experimental method with independent variable is feed rate. The quality of the joint was represented quantitatively by the tensile strength as a dependent variable. SEM, hardness profile, XRD and ongoing temperature tests were performed to support the analysis.

2 Experiment setup

The base metal was a 10mm plate of aluminum 6061 series with chemical content shown in Table 1. The plate was cut in 50mm x 150mm square and the edge where the FSW will be

* Corresponding author: b_darmadi_djarot@ub.ac.id

applied is trimmed to 45°. Two plates then attached at a jig as shown schematically in Figure 1. The probe was attached at 2000rpm rotating chuck. The pressure around 500 kg and idle rotation around 10 seconds then applied; produce enough heat to soften the material in the nugget zone. After the idle rotation the feeding movement was applied with three different feeding speeds: 10, 15 and 30 mm/minutes. Certain specimen was equipped with a thermocouple to monitor the temperature histories at certain position as shown by Figure 2.

Table 1. Chemical composition of 6061 aluminum alloy (wt %).

Composition	Mg	Si	Cu	Mn	Fe	Cr	Ti	Zn	Al
Content	0.9	0.6	0.25	0.086	0.18	0.1	0.192	0.01	Bal.

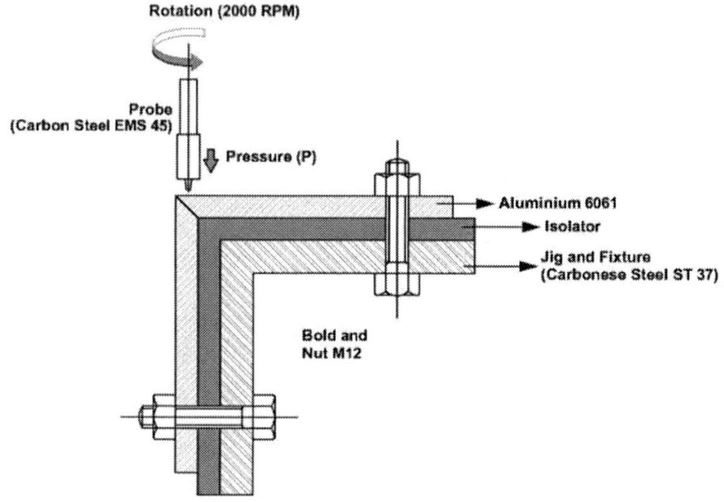

Fig. 1. Friction Stir Welding on The Corner Joint

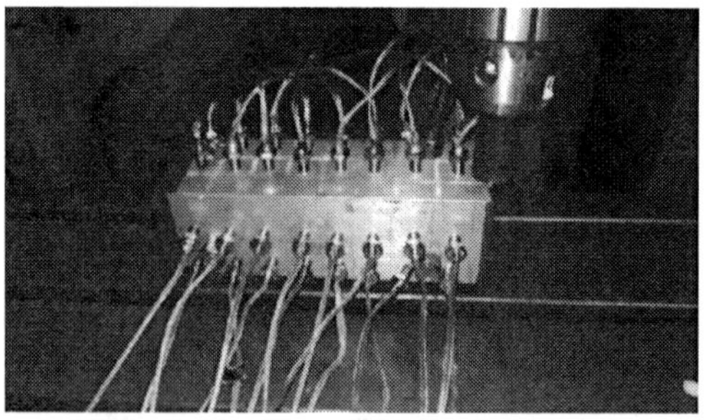

Fig. 2. Embedded thermocouple to record temperature histories.

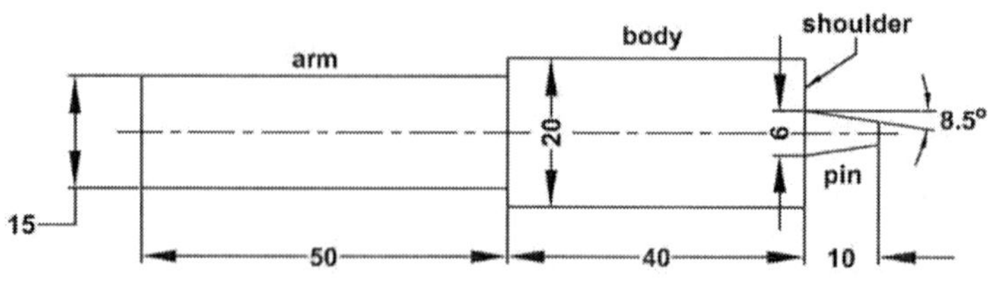

Fig. 3. Tool/probe design

The probe (friction tool) was made from hardenable EMS 45 steels. First an EMS 45 cylinder bar is formed using turning machine with geometry and measurement shown in Figure 3. All measurements in Figure 3 are in millimetres. After the probe was shaped then heat hardened by heating it in a furnace up to 1100°C and held in two hours. Suddenly, from the furnace the probe was quenched in oil and cooled down to room temperature. The tempering was applied following the hardening by heating the friction tool up to 300°C and held for two hours. From the furnace the tool was exposed to the ambient air to be naturally cooled down to the room temperature. With these two heat treatments the probe has enough hardness and ductility to carry out FSW on the aluminum plate.

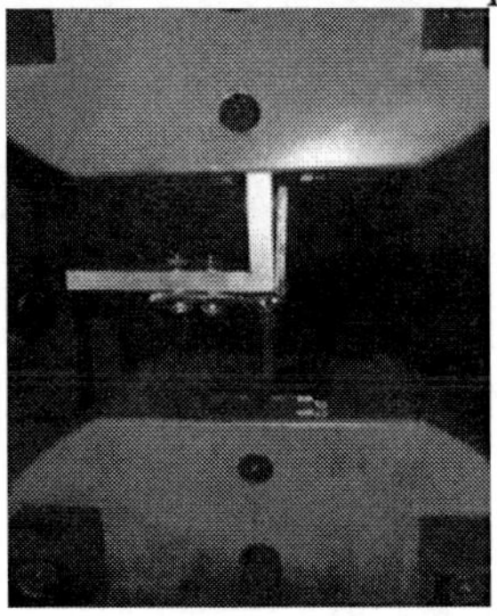

Fig. 4. Tensile test for the corner joint using special jig

In this article the rotating speed of the probe was kept constant at 2000 RPM whilst the feed rates (welding speed) were varied at 10 mm/min, 15 mm/min and 30 mm/min. After the two plates have been joined the tensile test specimens were formed. A specifically jig was used as auxiliary equipment as shown in Figure 4. The tensile strength test is the main result. Temperature histories, hardness distribution, XRD and chemical composition data are used to explain and enhance the result obtained from tensile test.

3 Results and discussion

Visual inspection shows that the 10 mm/min feed rate provides neat upper surface while for the 15 mm/min the upper surface is a little bit rough. For the 30 mm/min the rough surface is found and some holes are found in certain places. The FSW surface for varied feed rates are shown visually in Figure 5.

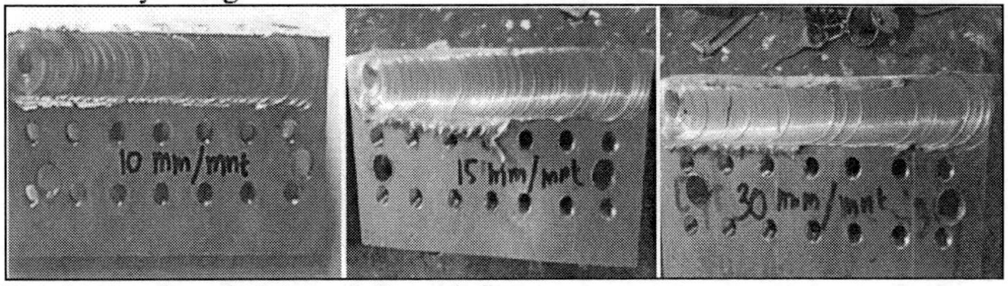

Fig. 5. Visual inspection of 10, 15 and 30 mm/mins

Tensile test results are shown graphically in Figure 6 that it can be said the lower feed rate produce higher tensile stress. With the less feed rate the heat input (heat embedded for each length) will be higher. The higher heat input provides enough soft material (nugget zone) which is needed to perform mechanical bonding when the mixing force from pin is applied. The higher heat input is also confirmed by recorded temperature by thermocouple as shown in Figure 7. Figure 7 shown temperature histories of thermocouple 11 for varied feed rate, those are 10, 15 and 30 mm/sec. Higher temperature indicates the higher the heat is inputted to the work piece. For lower heat input longer time is needed to complete the FSW process and since the heat embedded to work piece for certain time is equal, as a result of equal RPM, the longer FSW process means higher heat input and in turn increase the peak temperature recorded by thermocouple 11 as it is demonstrated by Figure 7. This longer time also describes more intensive mixing since the RPM is kept constant.

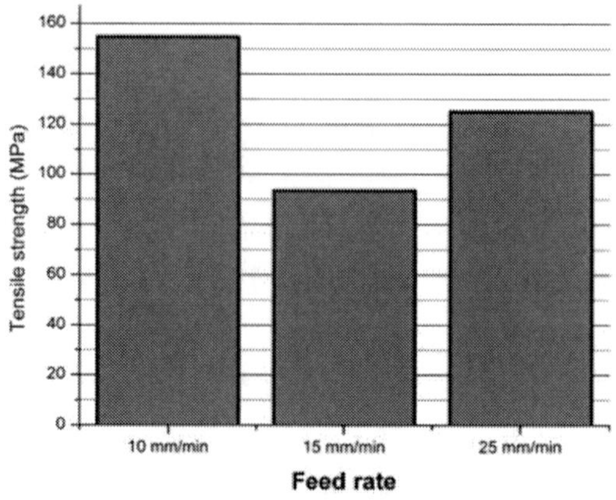

Fig. 6. Tensile test results.

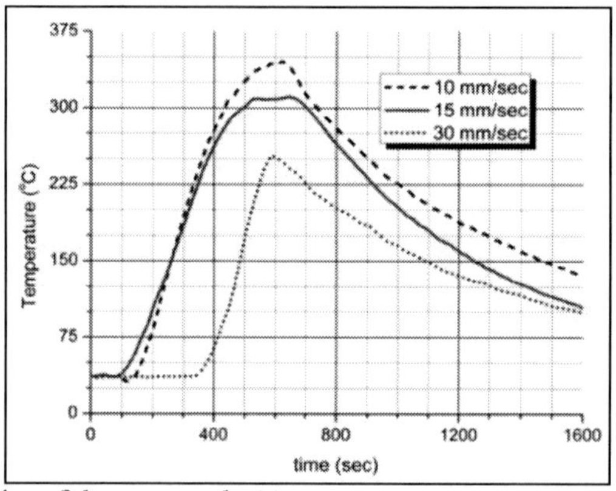

Fig. 7. Temperature histories of thermocouple 11.

It was hoped with a lower feed rate, which also means higher heat input and intensive mixing forces, the higher tensile strength will be obtained, but Figure 6 showed that the lowest tensile strength is shown by feed rate equal to 15 mm/sec. The hardness profile as shown in Figure 8, also confirmed the tensile test result. The distance between nodes in Figure 8 is 500 μm.

From X-Ray diffraction in Figure 9, it can be shown that MgO is formed while FSW was applied. The content of the MgO is shown in Table 2. From the table it can be said that with the lower feed rate, more MgO is produced. MgO is formed when the Mg is exposed to temperature above 265°C [9, 10]. From Figure 7 it can be concluded that the slower the feed rate, the material will be exposed in temperature above 265°C in the longer period.
It should be noted that the thermocouple is embedded at a certain distance from the higher temperature in a FSW process (under probe or in nugget zone). However, the Figure describes that with the lower feed rate, a certain point will experience higher peak temperature and will be exposed to temperature above 265°C in the longer period. While they are exposed to temperature above 265°C the MgO will be formed and since the lower feed rate exposed a certain point above 265°C in longer period, the MgO content will be higher which is confirmed by table 2. The method used to obtain the Table 2 is atomic absorption spectroscopy.

The Mg alloy in aluminum alloy increases the strength, including tensile strength [11, 12]. Since the total Mg in a certain aluminum plate is constant, increasing MgO means decreasing Mg content in the aluminum alloy. This depleting Mg in the alloy in turn decreases the tensile strength which reverse to the expectation that with the slower feed rate it is expected higher tensile strength will be obtained.

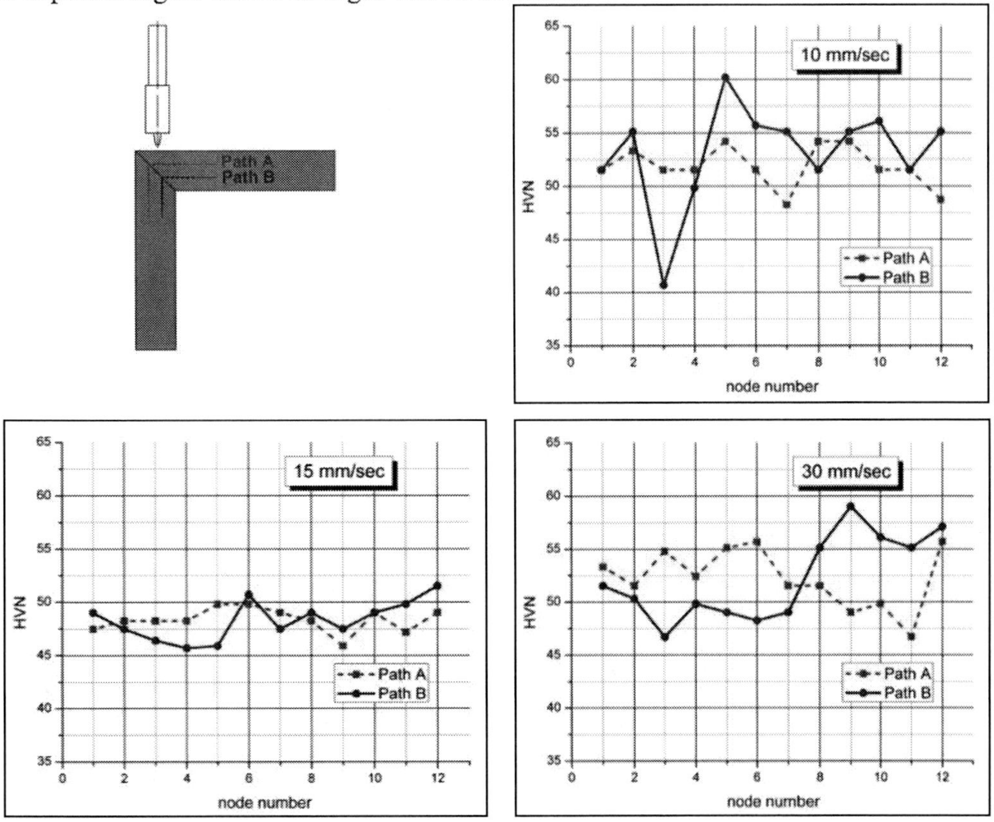

Fig. 8. Hardness profile for varied feed rate.

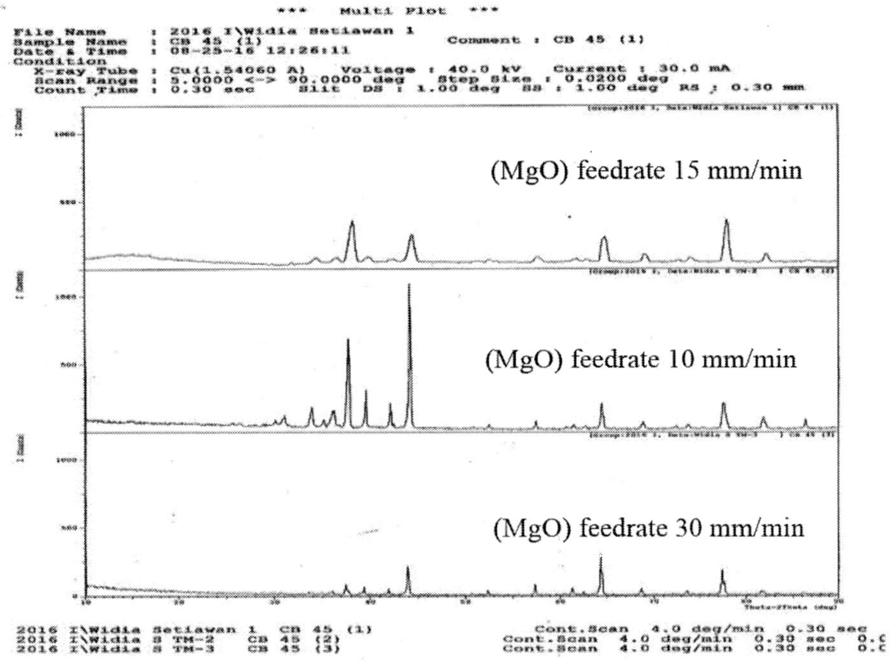

Fig. 9. X-ray defraction.

Table 2. The Mg and MgO contents

No.	Sample Code	Content	Metrology result (ppm)		
			I	II	III
1.	CL-45 (1)	Mg	1868.412	1868.412	1868.825
2.	30mm/min	MgO	3097.828	3097.828	3128.355
3.	CL-45 (2)	Mg	1997.299	1978.886	3097.828
4.	15mm/min	MgO	3311.521	3280.993	3280.993
5.	CL-45 (3)	Mg	1960.474	1960.474	1997.299
6.	10mm/min	MgO	3250.466	3250.466	3311.521

The final result of slower feed rate is the compromise between the increasing strength due to higher temperature and intensive mixing force and the decreasing strength due to the lower Mg content in the aluminum alloy. The final results are as it is shown in Figure 6, the 15 mm/sec feed rate provides the lowest tensile strength which also supported by micro hardness profiles as shown in Figure 8.

4 Conclusion

Although the principle of mechanical bonding in the FSW process has been established, the contribution of each parameter still needs further study to describe the process comprehensively. The effect of feed rate on the resulted tensile strength was demonstrated here. It was expected that with the lower feed rate the tensile strength will be higher as the result of high temperature and intensive mixing forces, but the high input also oxidizes the Mg alloys which in turn decrease the tensile strength.

References

1. Djarot B. Darmadi, International Journal of Engineering Systems Modelling and Simulation, **8**, 3 (2016).
2. Francois J. Dessart and Laurent Bontoux, European Union (2017).
3. Andrei C. Popescu, Christophe Delval and Marc Leparoux, Metals, **7**, 11 (2017).
4. S. Brumm and G. Burkner, Materials Today : Proceeding 2S : S179 – S187 (2015).
5. J. Enz , C. Carrarin, S. Riekehr, V. Ventzke and N. Kashaev, The International Journal of Advanced Manufacturing Technology (*in print*).
6. Lan-Qiang Niu, Xiao-Yan Li, Liang Zhang, Xiao-Bo Liang and Mian Li, Acta Metallurgica Sinica, **30**, 5 (2017).
7. Djarot B. Darmadi, Anh Kiet Tieu & John Norrish, International Journal of Materials and Product Technology, **48**, 1-4 (2014).
8. Djarot B. Darmadi, International Review of Mechanical Engineering, **11**, 8 (2017).
9. Teng-Shih Shih and Zin-Bou Liu, Materials Transactions, **47**, 5 (2006).
10. Nur Syazeila Samadi, Mohd. Khairul Asyraf Amat Mustajab and Abdul Rahim Yacob, International Journal of Basic & Applied Sciences, **10**, 2 (2010).
11. Annalisa Fortini, Mattia Merlin, Elettra Fabbri, Stefano Pirletti and Gian Luca Garagnani, Procedia Structural Integrity, **2** (2016).
12. S. Morin, E.M. Elgallad, H.W. Doty, S. Valtierra and F.H. Samuel, Advances in Materials Science and Engineering, doi.org/10.1155/2016/7841380 (2016).

MATEC Web of Conferences **159**, 02054 (2018) https://doi.org/10.1051/matecconf/201815902054
IJCAET & ISAMPE 2017

The Structural Response Investigation of Modular Pontoon Collar Floating Cage due to Current Load to Support Fish Farming Activities

Aulia Windyandari[1], and *Ahmad Fauzan* Zakki[2,*]

[1]Industrial Engineering Department, Vocational School, Diponegoro University, Indonesia
[2]Naval Architecture Department, Engineering Faculty, Diponegoro University, Indonesia

Abstract. The size of net cages increases rapidly, reaching volumes above current experience. More exposed locations are used for fish production, introducing higher loads on the net cage and fish farm from strong water currents and large waves. Strength analysis can be a useful tool for development of net-cage designs to avoid escape of fish and ensure sufficient volume for good fish welfare and water quality. The main objectives of this research are to investigate the structural response of modular pontoon collar floating cage due to the current load to support the fish farming activities. Strength analysis was performed using commercial explicit finite element software to calculate the distribution of loads in the net cage due to current, weights and gravity. The net cage was modeled using truss elements that represented several parallel twines. Sub-elements allowed the trusses to buckle in compression, and only negligible compressive forces were seen in the numerical results. Resulting drag loads and cage volume were shown to be dependent on the net cage size and collar shape.

1 Introduction

Good practical cage designs should resolve engineering and cost considerations with the requirements of the species being farmed and those of the farmer who must operate the system. Cage structures must withstand the forces of winds and waves while holding stock securely. The design is important to the fish in that it both circumscribes living space and, within the site chosen, influences environmental quality inside the cage. From the farmer's point of view, the cage must be safe, secure and easy to manage. All of this must be achieved cost-effectively.

Net cages for aquaculture have traditionally been dimensioned and produced based on empirical data, [1]. However, the requirements for documentation of net-cage strength and volume increase, and the need for the development of methods for structural analysis arises. In addition, fish farming is developing outside the borders of experience. The size of net cages increases rapidly, reaching volumes above current experience. More exposed

* Corresponding author: ahmadfzakki@undip.ac.id

locations are used for fish production, introducing higher loads on the net cage and fish farm from strong water currents and large waves. Strength analysis can also be a useful tool for development of net-cage designs, [2], to avoid escape of fish and ensure sufficient volume for good fish welfare and water quality.

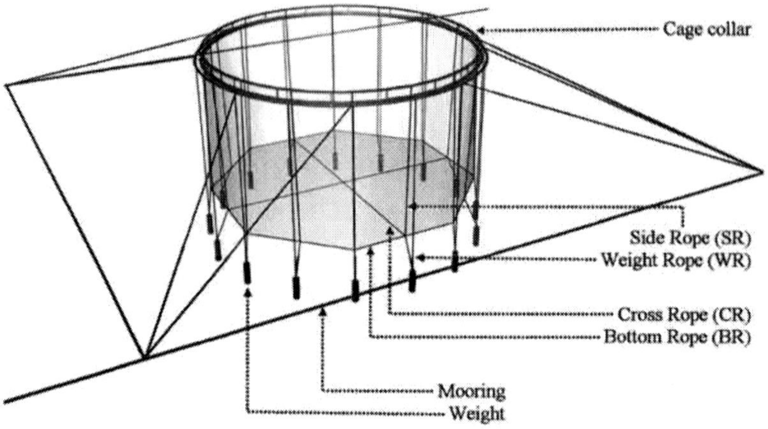

Fig. 1. Illustration of circular collar floating cage.

It is not straightforward to analyze a fish farm or a single net cage (**Fig. 1**). The loads acting on the structure will be the result of a fluid-structure interaction between moving sea-water and the deformed net. Analyses of net cages involve a high degree of non-linearity, both in loads, deformation and sometimes also material properties. In addition, loads from waves and current, damping and inertia loads are complex to model for netting materials in a general Finite Element Analysis (FEA) program, while programs with appropriate load modules often have shortcomings in the structural model. There is ongoing work to verify and develop CFD methods for flow around net structures. This work reveals the complexity of such flows and the need of new knowledge and methods, thus, the only current option is to use methods to calculate loads from moving fluids on netting structures, such as Morison's equation or equations based on tank test results, [3]. However, both these methods currently have limitations concerning amongst other the solidity of the net panel (ratio between the area of the net and total area), deformations and current velocity.

Based on the condition the structural response investigation of Modular Pontoon collar floating cage was proposed to estimate the structural behavior of the floating cage due to the current load. Our approach to structural analyses was to apply commercial FEA software, which could include various material models and ensure effective modelling, processing and post-processing. In this work, truss elements were applied to model the net cage. The fact that netting and ropes do not take compression was modelled by introducing sub-elements, which combined with an explicit finite element code allowed the elements to buckle in compression.

2 Research Method

In the study of the development of modern cage technology, the articles reviewed were related to the mechanics of floating cages. Tiao-Jian Xu, [4] studied the hydrodynamic characteristics of submerged cages (submersible net cage) and the mooring system on currents and wave. Taeho Kim, [5] conduct engineering studies on the system of cages embedded in the cultivation of abalone shells. The flow characteristics of submerged cage structure were analyzed using the approach of computational fluid analysis (CFD analysis). Hydrodynamic response on the net binding structure investigated using Morison equation and the finite element model (FE model) is used for analysis of fluid-structure interaction (FSI analysis). Trygve Kristiansen, [6] performed experiments and numerical studies on the cage buoyed by considering the influence of waves and currents. Formerly, Trygve also did modeling studies of the current load on the cage, [6]. Li Li, [7] conducted a study hydroelastic nonlinear analysis of the floating cages on a wave of irregular (irregular waves). Heidi Moe, [8] conducted a study on the structural analysis of the influence of the current net cages. Previously, Heidi Moe has conducted studies of the tensile strength of the material used nets to cages, [9].

2.1 Structure Modeling of Floating Cage

Two different models, described in **Fig. 2**, were analyzed.The circular shaped collar model was assembled from modular pontoon unit using a rod hook in forming a circular pattern. For all models, the netting dimensions were equal to typical nets for Nila fish (Oreochromis niloticus). The floating cage net was modeled with truss element. This model was also analyzed using a detailed mesh with one bar for each twine to verify the model simplification. Two full-scale models represented the floating cage using rectangular and circular collar had 131992 and 77563 elements around the circumference, which should be sufficient to capture the global deformations in the net cage.

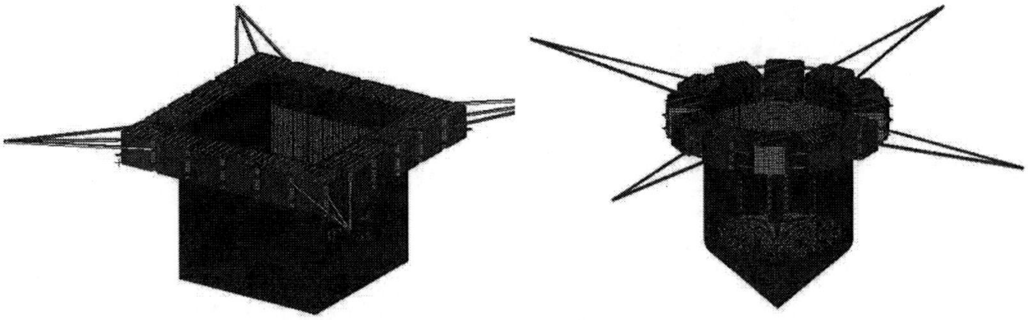

(a) (b)

Fig. 2. FE Models of the floating cages using modular pontoon collar: (a) Rectangular shaped collar; (b) the circular collar.

2.2 Boundary conditions and loads

The numerical models were subjected to loads representing a uniformly distributed current with constant velocity, U, acting in the x-direction. The cross-flow principle was applied assuming that the current could be separated into flow tangential and normal to the element axis. The first step in the current load calculation for an element was to decompose U into tangential and normal velocity components, U_T and U_N, as shown for an element in the XZ-plane. However, most elements were oriented in the XYZ-space and subjected to three-dimensional velocity vectors. Each truss element was considered as individual, friction-free cylinders, and tangential force was ignored. The resulting load acting on each truss element was the normal load, F_N, calculated using the drag term of Morison's equation:

$$F_N = 0.5\, \rho\, C_D\, A_p\, U_N^{\,2} \tag{1}$$

where $\rho = 1025\ kg/m^3$ (density of seawater) and C_D is the drag coefficient. A_p is the projected area of the truss element, which is the sum of the projected area for all twines represented by the truss element. $A_p = ndL$, where n is a number of twines represented by the truss, d the twine diameter and L the truss length. F_N was decomposed in the X-, Y- and Z-directions and applied as concentrated loads in adjacent nodes. Drag load (F_D) and lift load (F_L) were found as the x and z components of F_N, respectively.

Rectangular collar shaped was analyzed using four different bottom weight configurations: *4* weights of *400*, *600* and *800 grams* (*3*, *4.5* and *6N* submerged weight) and the circular shaped using *1* weights of *6N*. The weights were attached directly to the netting and equally distributed along the circumference. The floating cage net was subjected to a concentrated load of $1.2\times106\ N$ at the end of each truss element.

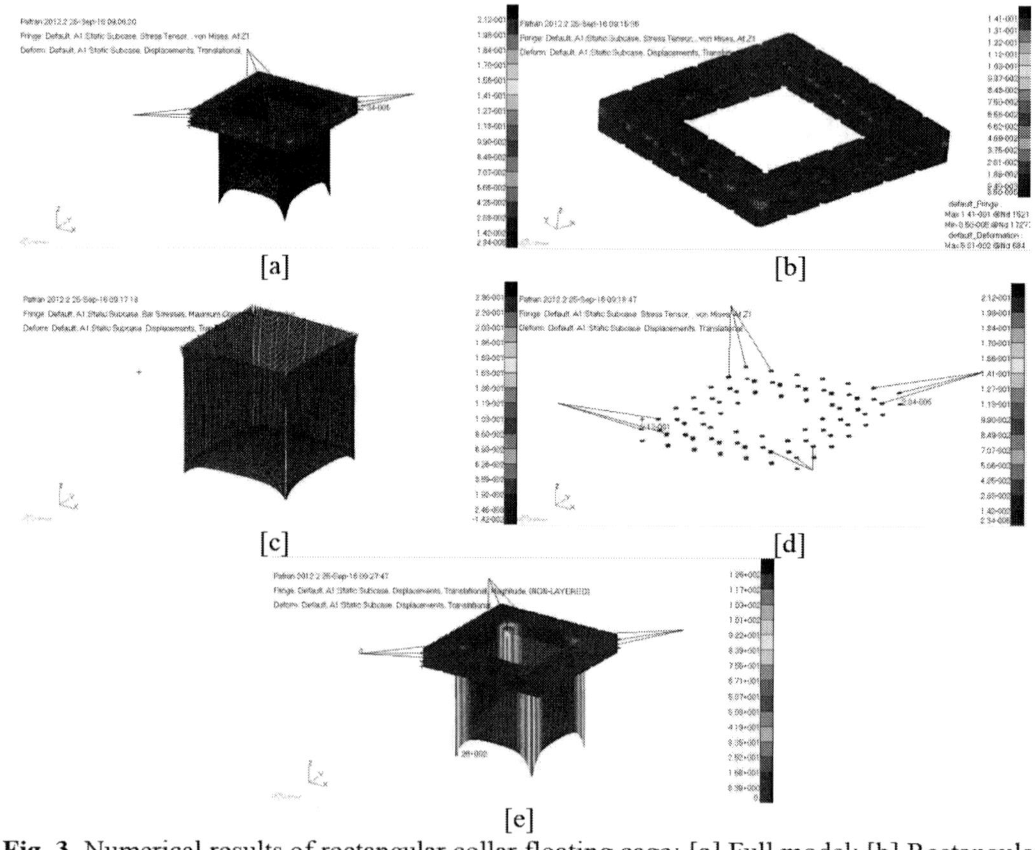

Fig. 3. Numerical results of rectangular collar floating cage: [a] Full model; [b] Rectangular collar; [c] Cage net; [d] Collar connector; [e] Displacement of full model

3 Results and Discussion

3.1 Strength analysis of full-scale net cages

The FE model that has been defined with the loading and boundary condition was analyzed using finite element method. The linear static analysis was chosen to solve the problem case. The structure of the rectangular shaped floating cage was modeled using *310,021* shell elements and *100,971* bar elements. The shell elements were mixed using quadrilateral and triangular elements. The illustration of the results of numerical analysis of rectangular shaped floating cage can be seen in **Fig. 3**. Additionally, the structure of the circular shaped floating cage was modeled using *18,612* shell elements and *58,951* bar elements. The illustration of the results of numerical analysis of circular shaped floating cage can be seen in **Fig. 4**.

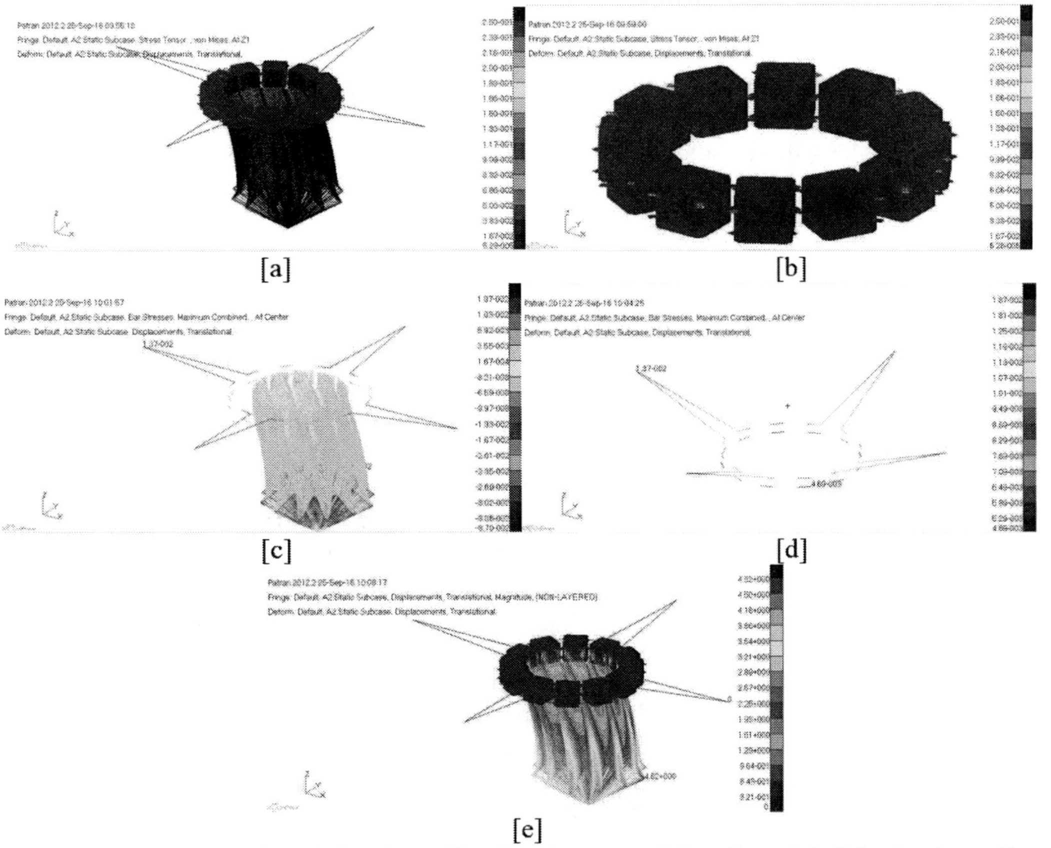

Fig. 4. Numerical results of circular collar floating cage: [a] Full model; [b] Circular collar; [c] Cage net; [d] Collar connector; [e] Displacement of full model

In **Fig. 3** the structural response of the rectangular floating was shown. The maximum stress appeared on the connection between the mooring cable and the floating collar. It can be explained that the stress was induced by the current loads has been responded by the mooring system. These stresses are about 0.212 MPa, significantly smaller than the permissible stress (68.9 MPa). The maximum stress of modular pontoon collar in **Fig. 3[b]** is about 0.141 MPa which was occurred in the same region of the connector of the pontoons (bolted clevis). The maximum stress of the cage net can be seen in **Fig. 3[c]**. The results show that the net has 0.236 MPa that was induced by the anchoring weight of the cage. The exact location of the maximum stress which is located at the connector can be seen in the **Fig. 3[d]**. Finally, the displacement of the floating cage, **Fig. 3[e]** shows that the maximum displacement appeared at the cage net which the anchoring weight is applied.

In **Fig. 4** the structural response of the circular floating was shown. The maximum stress appeared on the connection between the mooring cable and the floating collar. It can be explained that the stress was induced by the current loads has been responded by the mooring system. These stresses are about 0.250 MPa, significantly smaller than the permissible stress (68.9 MPa). The maximum stress of modular pontoon collar in **Fig. 4[b]** is about 0.250 MPa which was occurred in the same region of the connector of the pontoons (bolted clevis). The maximum stress of the cage net can be seen in **Fig. 4[c].** The results show that the net has 0.0137 MPa that was induced by the anchoring weight of the cage.

The exact location of the maximum stress can be seen in the **Fig. 4[d]**. Finally, the displacement of the floating cage, **Fig. 4[e]** shows that the maximum displacement appeared at the cage net which the anchoring weight is applied.
As expected, the results of the strength analysis show that both structure of modular pontoon collar floating cage is reliable to support the current load and anchoring load. The maximum stress occurred at the connector of the pontoons, the stresses are about 0.212-0.250 MPa essentially smaller than the permissible stress 68.9 MPa. It is indicated that the structure of the modular pontoon collar floating cage is reliable for supporting the current load in the aquaculture activities.

4 Conclusion

The evaluation of the floating cage strength was calculated using finite element method comprising, to build FE model (meshing), to define the load and boundary conditions, to define the material properties and structure scantlings. Full load condition was selected for the strength analysis since the condition was the most severe condition for the pontoon structure.

In the case of rectangular shaped collar floating cage, the maximum stress was 0.212 MPa which was occurred at the connector of the pontoons, significantly smaller than the permissible stress (68.9 MPa). The maximum stress of the cage net was 0.236 MPa which is induced by the anchoring weight. Additionally, the maximum displacement of the floating cage was 126 mm

In the case of the circular shaped collar floating cage, the maximum stress was 0.250 MPa which is occurred in the similar region as the rectangular shaped. The maximum stress of the cage net was 0.0137 MPa which is smaller than the stress of the rectangular shaped floating cage net. Finally, the maximum displacement of the floating cage was 4.82 mm. Accordingly, it can be concluded that the structure of rectangular and circular shaped collar of the floating cage is reliable for fish farming activities.

References

1. Standards Norway, *NS 9415 Marine Fish Farms—Requirements for Design, Dimensioning, Production, Installation, and Operation*, (2003)
2. Moe, H., Fredheim, A., Heide, M., *IMAM 2005*, Lisbon, Portugal, (2005)
3. Løland, G., *Doctoral Thesis*, (The Norwegian Institute of Technology, Norway, 1991)
4. Taeho Kim, Jihoon Lee, David W. Fredriksson, Judson Decew, Andrew Drach, Kiyoung Moon, Aquaculture Eng., **63** (2014)
5. Kristiansen, T., Faltinsen, O.M., J. Fluids and Struct. (2014)
6. Kristiansen, T., Faltinsen, O.M., J. Fluids and Struct., **34** (2014)
7. Li Li, Fu, Shixiao, Xu, Yuwang, Marine Structures, **34** (2013)
8. Heidi Moe, Fredheim, A., Hopperstad, O.S., J. Fluids and Struct., **26** (2010)
9. Heidi Moe, Olsen, A., Hopperstad, O.S., Jensen, O., Fredheim, A., Aquaculture Eng., **37** (2007)

Impact phenomena assessment: Part II – Buffer container as a measure to reduce cargo leakage in collision

Aditya Rio Prabowo[1], *Seung Jun* Baek[2], *Seung Geon* Lee[2], *Dong Myung* Bae[2], *Jung Min* Sohn[1,2,*], *Aldias* Bahatmaka[1] and *Samuel* Samuel[2]

[1]Pukyong National University, Marine Convergence Design, 48513 Busan, Korea
[2]Pukyong National University, Naval Architecture and Marine Systems Eng., 48513 Busan, Korea

Abstract. The safety of ship under collision has been developed in terms of navigational and structural aspects. These efforts are aimed to provide more safety from the internal part of a ship. However, in a collision, the external dynamics also needs to be considered, for instance velocity and location. In this study, these parameters are applied and observed regarding on a series of ship-container collision scenario. A *freight container* is used in the first stage and behaviours such as internal energy and effective stress are summarized. An alternative design so called *buffer container* is presented to offer better protection to a liquid cargo. The second stage is started by applying similar collision scenario as the first stage to the buffer container and comparing structural behaviour of this analysis. It is obtained that leakage possibility can be reduced on the buffer design as the strength has become approximately twice better than the freight container.

1 Introduction

Rapid development of human race affects every branch of science and technology to fulfil various societies' demands, for example, cargo distribution from production center to both national and international markets. One of the most important thing is to keep cargos safe during the process. Shipping container is developed to carry out this mission which the container can be multi-purpose carrier including for fast-degradable commodities and dangerous fluids, such as oil and gas. During the distribution process, the container is mostly transported by ship since it is acceptable in terms of cost and time efficiencies. However, safety for the container could be threatened by a collision. On the sea, this phenomenon is possibly triggered by arrangement failure of container column on the container ship, which leads to containers fall down to sea and collides with the ship hull.

In this study, objective is determined to propose a new container concept which is called *buffer container* as a measure to reduce cargo leakage in collision. A freight container is considered as representative of the existed design which to be analysed with parameters of the external dynamic of ship collision, namely velocity and location. The buffer container is developed based on the analysis result of the freight container.

* Corresponding author: jminz@pknu.ac.kr

2 Literature review

The operational system of every engineering design has a risk for unexpected phenomena such as an accidental impact. This situation is taken seriously in field of marine structures which structural failure due to an impact may trigger further disastrous events. Among of all impact forms, a ship collision is ranked as top-three dangerous evens for maritime accident. Certain parties across the globe have paid their intention to this impact form. Zhang [1] proposed a mechanical expression to assess a structural response after a collision. These methodologies were re-conducted in recent researches by comparing the numerical method using finite element (FE) approach, such as presented by researchers from Indonesia [2], and South Korea [3]. In terms of freight container incidents since late 1990s until 2017, APL China is considered as the worst one when records up to 100 million US Dollars financial loss. Other casualties in forms of a damaged container and cargo lost are main concerns for this incident type. Most of container incidents, for instance container collapse of the CSAV Shenzen and a recent accident of San Rafael in Vigo, Spain, produced damaged container and cargo lost. Terrible chain reaction in form of marine pollution was also recorded during ship grounding of the MSC Napoli in South-West England. Condition of container cargo especially in a fluid-hazardous state has to be taken as priority.

Arrangement failure of a container ship may lead to collision with other marine structure, e.g. ship and rupture of container walls and internal cargo (such as drum which carries dangerous cargo) cannot be avoided. Worse situation may occur if the cargo gets spilled inside the container which may initiate explosion or other dangerous chain reaction. Considering demands to ensure cargo safety against impact load, an alternative design of a freight container to ensure hazardous materials are still well contained during spontaneous impacts on ship, is considered as a research opportunity and determined as the state-of-the-art for this study. Impact analysis considers ship-container collision model as the ship is denoted as representative of marine structure. In modelling stage, a freight container is used as conventional model which will be analysed and developed. Several structural responses are used parameter to evaluate the containers against impact.

3 Numerical models

The double hull vessel is idealised with the Length L = 85.92 m, Breadth B = 15 m and Depth H = 10.4 m. In the finite element model, this ship is modelled as four node-shell elements with a deformable material and the hull part is applied with the high-strength low-alloy steel with Density ρ = 7850 kg/m^3, Young's modulus E = 210,000 MPa, Poisson's ratio n = 0.3 and Yield strength σ_Y = 315 MPa.

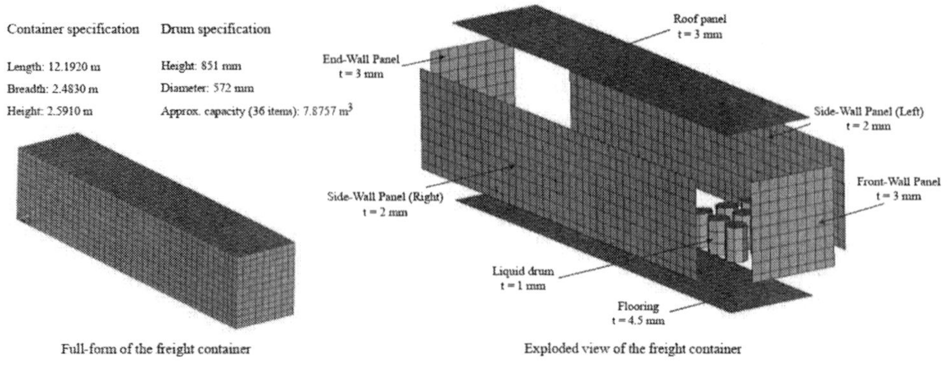

Fig. 1. Numerical model for the freight container.

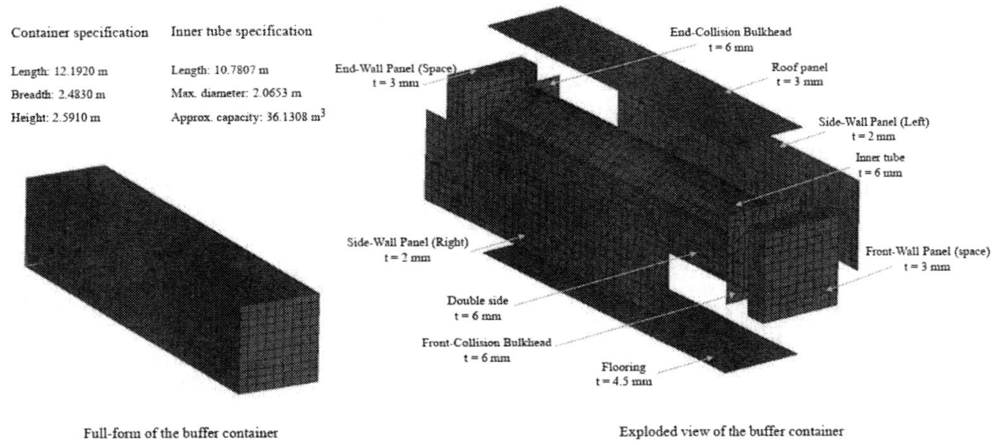

Fig. 2. An alternative design: the buffer container.

In other hand, the container model is generated into two geometries. The first container is denoted as the *freight container* (Fig. 1). The model is built after the regulation of the Det Norske Veritas (DNV) for a 40-foot-container [4]. The applied material on the container is similar with JIS SPA-H Corten steel according to ISO regulation for 1AAA type cargo container [5]. The properties is given as follow: Density $\rho = 7850$ kg/m^3, Young's modulus $E = 200{,}000$ MPa, Poisson's ratio $n = 0.3$ and Yield strength $\sigma_Y = 345$ MPa.

The second container is denoted as the *buffer container* (Fig. 2) and described as the modified version of the freight design. An inner tube is added inside the container as a large tank to contain fluid-hazardous cargo. On the both sides of the buffer container, double side is provided with size 200 mm in width and on front and end walls, 700 mm safety space is applied. Such containment system is considered as a simplified geometry based on construction of the tanker vessel [6]. The fundamental material properties for this model are similar to the freight container. However, structure of the inner tube was applied by a thicker plate than the outer walls, which a steel with thickness 6 mm will be applied. Failure criterion of the Peschmann [7] will be appended on both container and ship structure. During collision, the ship is determined to stay on the centerline. The end of the ship model is fixed on the transverse frames and rotational displacement is restrained on the shell plating. Detail illustration of the collision scenario is presented in Fig. 3 and boundary conditions on the ship are applied based on pioneer work in ship collision [8].

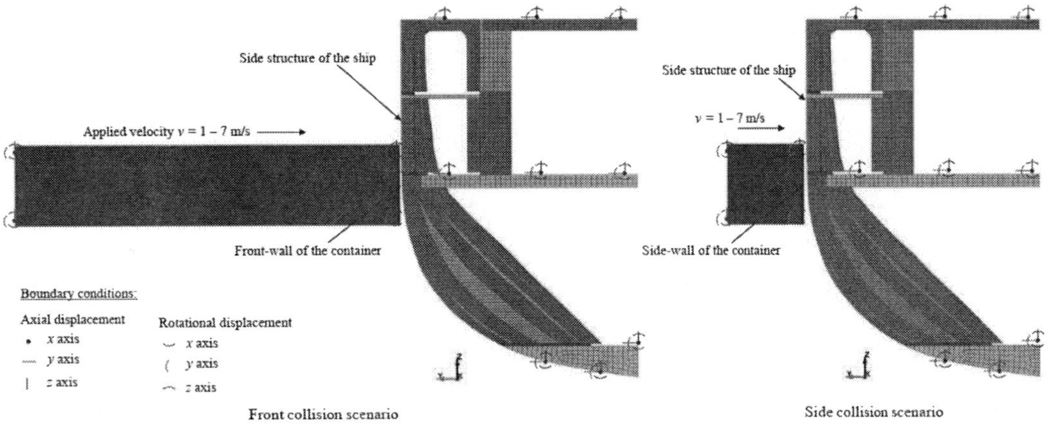

Fig. 3. Scenario for the container collision. Notations highlight boundary conditions on the models.

4 Results and Discussion

The calculation results are presented in terms of the internal energy and other structural responses, such as von Mises stress (effective stress) will be discussed in further parts.

Firstly, the internal energy represents amount of energy which is needed to plastically deform the containers. During collision with the freight container, the energy (Fig. 4) reached peak point and went down after surpassing this point. This result indicates that the container rebounded to opposite of the penetration direction (y axis) after it collided with the side hull. As side part of the container has wider area than the front-wall, the side collision (Fig. 4b) was concluded produce higher energy level than the front collision (Fig. 4a). Besides in terms of energy level, behaviour of two scenarios presented notable distinction. During the front collision, after reaching peak point, the energy reduced and stayed in stable condition. This phenomenon indicated that impulse of the collision (affected by kinetic energy during collision) did not ignite further deformation process when the container experienced rebounding. In other hand, fluctuations were spotted on the side collision especially in V_4 = 7 m/s. It is predicted that due to wider impact area than the front collision, collision impulse expanded to wide range and made minor deformation was still experienced by the container in rebounding. Minor was denoted as the fluctuation was not as high as peak point in the first contact. Effect of the collision impulse can be seen in more convenient way during discussion is addressed to the effective stress.

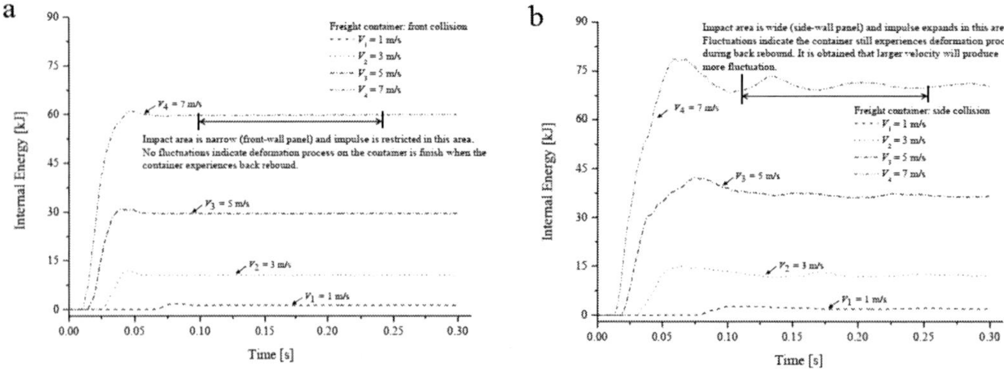

Fig. 4. The internal energy on the *freight container*: (a) front collision and (b) side collision.

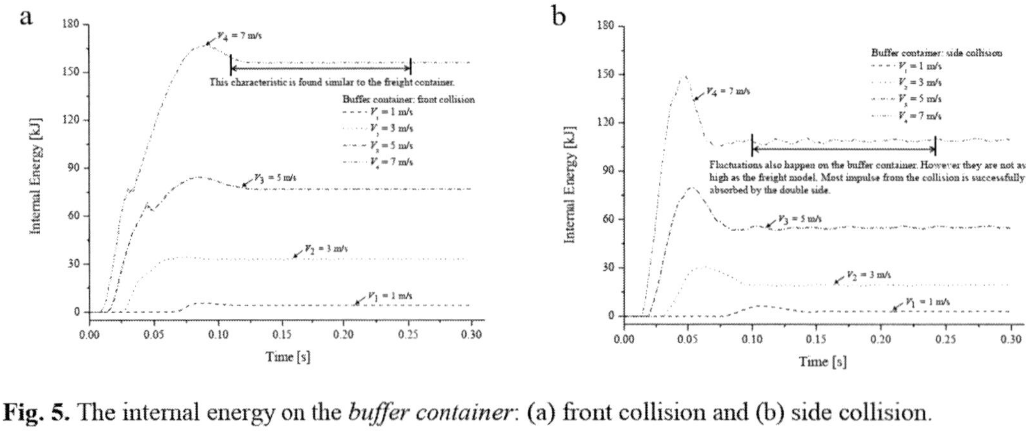

Fig. 5. The internal energy on the *buffer container*: (a) front collision and (b) side collision.

Next discussion in terms of the internal energy is pointed to the buffer container. As shown in Fig. 5, the buffer design successfully produced higher energy level than the freight container during impact. This result provided a solid indication that resistance capability of a conventional (freight) container had been increased by implementation of the double space on side and front panels. It was obtained that the energy increased approximately 160% in the front collision and more than 85% in the side collision during the buffer design was implemented on the freight container. Contributions of the double spaces' size was predicted as a major factor in difference of the strength increment percentage. The front panel was applied by larger space than the side space which provided better resistance in collision. The spaces on the buffer design are concluded satisfyingly present adequate safety distance which allowed the outer structure of the container to absorb the impulse as generated by the kinetic energy during collision. Contribution of two designs to the inner cargo condition to be explained in the next part.

In this study, other structural response besides the internal energy is also considered, namely the effective stress (Figs. 6 and 7). Energy-stress relationship (the hardest material to be plastically deformed is expected experience higher stress) is achieved based on these results which the freight container produced lower peak stress than the buffer one. In comparison between the front and side collisions, it was obtained that stress expansion on the side collision was significantly wider than the front scenario. Stress contours have also successfully verified fluctuation behaviour in Figs. 4 and 5. The front collision produced stable condition after peak as the impact area was narrow, deformation during rebounding was unlikely and impulse was limited in this area. It was match with the stress contour that focussed on the front wall (Figs. 6a and 7a). However, the side part of container was wide and stress expanded to wider area (Figs. 6b and 7b) than the front collision. Fluctuation happened as during this expansion, impulse ignited minor deformation process on the container and appeared as stress contour in these figures.

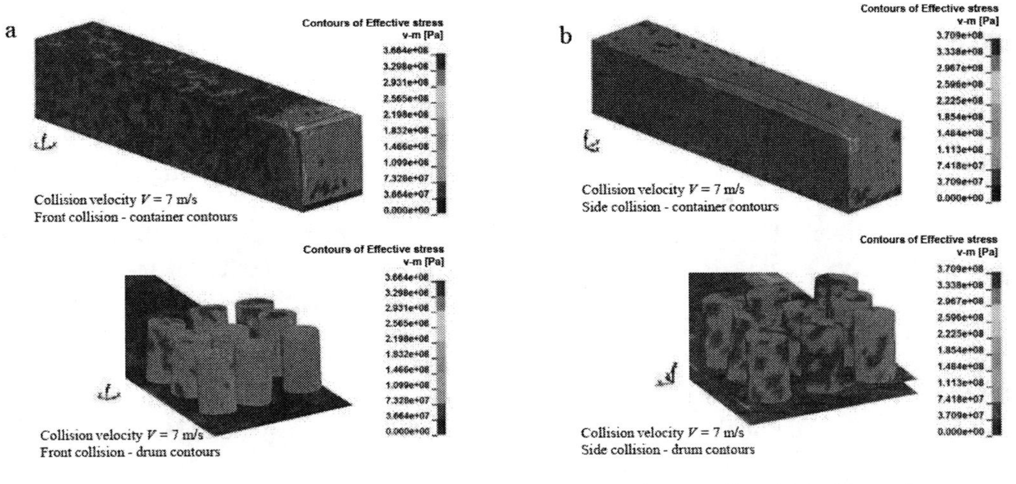

Fig. 6. Structural response of the *freight container* collision: (a) front-wall and (b) side-wall.

After that, discussion is directed to the inner container. As shown in Fig. 1, the freight container was assumed to carry fluid cargo using drums, and the buffer container (Fig. 2) was embedded by an inner tube to transport similar cargo. Comparison of the liquid container capacity for both freight and buffer designs was presented in same figures (Figs. 1 and 2). The freight (36 drums - unstacked) could carry liquid cargo approximately 7.8757 m^3 or 2080 US liquid gallons and capacity of the buffer design was estimated 144.5232 m^3 or equal with more than 38,100 US liquid gallons. Maximum capacity of the freight container was increased 1700% if the buffer design was applied. Containment system of each container model is found influence structural response under collision. As presented in Fig. 6 that the drums on the freight container which contained fluid cargo were observed collide each other under front and side collisions. Furthermore, contact between the drum and structure was also unavoidable. The drums experienced stress in this situation and displaced from its original position. In other situation where belt and plastic are used to ensure fixation on the drums, possibility of displacement is predicted still high and may lead to liquid leakage due to accidental load. Besides its advantage in terms of capacity, effectiveness of the buffer design was also explicitly presented by the stress contour (Fig. 7). As adopted of the double hull concept, stress expansion was reduced under front and side collisions.

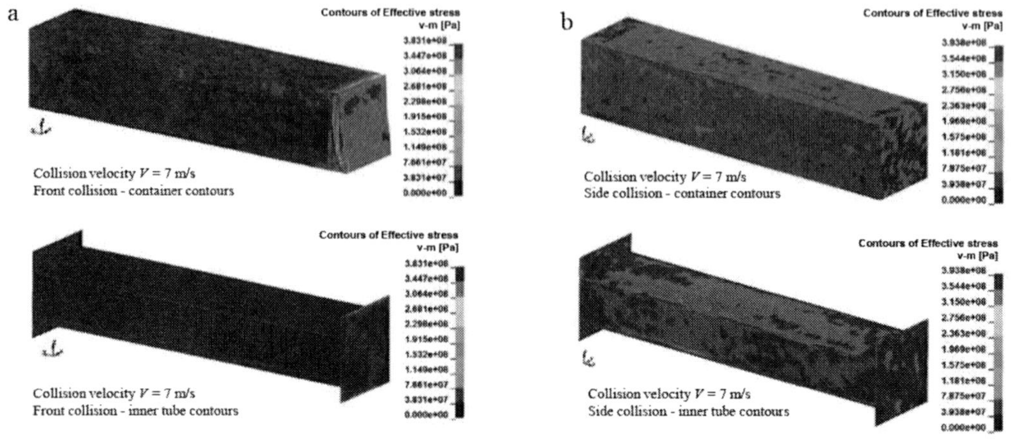

Fig. 7. Structural response after collision to the *buffer container*: (a) front-wall and (b) side-wall.

5 Conclusion

This study aimed to analyse structural responses of containers under collision with other marine structure. It was obtained that the buffer container provided better resistance subjected to impact loads than the freight container. Possibility of cargo leakage was considered unlikely as the safety space provided adequate space to absorb collision impulse. Condition of the inner tube was stable and did not experience displacement as the drums inside the freight container. In damage progression, contribution of the safety space was considered powerful enough to reduce deformation process in rebounding and stress expansion as generated by the kinetic energy after ship-container interaction. This statement was also validated by fluctuation in terms of the internal energy which was almost vanished after the peak point was achieved on the buffer design. Besides offering better structural strength, the buffer design also satisfyingly increases cargo capacity.

Reference

1. S. Zhang, R. Villavicencio, L. Zhu, P.T. Pedersen, Mar. Struct. **52**, 69 (2017)
2. A.R. Prabowo, D.M. Bae, J.M. Sohn, A.F. Zakki, B. Cao, J.H. Cho, Thin-Walled Struct. **115**, 225 (2017)
3. A.R. Prabowo, D.M. Bae, J.M. Sohn, A.F. Zakki, B. Cao, Q. Wang, Theo. Appl. Mech. Lett. **7**, 6 (2017)
4. DNV, *Rules for certification of freight container* (Det Norske Veritas, Høvik, 1981)
5. ISO 668, *Series 1 freight containers - Classification, external dimensions and ratings* (International Organization for Standardization, Geneva, 2013)
6. D.J. Eyres, *Ship construction, fifth edition* (Butterworth Heinemann, Oxford, 2001)
7. E. Lehmann, J. Peschmann, Mar. Struct. **15**, 429 (2002)
8. A.R. Prabowo, S.J. Baek, H.J. Cho, J.H. Byeon, D.M. Bae, J.M. Sohn, Lat. Am. J. Sol. Struct. **14**, 1345 (2017)

Investigation on the performance of the traditional Indonesian fishing vessel

Aldias Bahatmaka[1,*], *Dong-Joon* Kim[2], *Samuel*[2], *Aditya Rio* Prabowo[1], *May-Thu* Zaw[2]

[1]Marine Design Convergence Department, Pukyong National University, Republic of Korea
[2]Naval Architecture and Marine Systems Engineering Department, Pukyong National University, Republic of Korea

Abstract. In this study, numerical investigation of ship resistance and ship motions at the traditional Indonesian fishing vessel is presented. The Computational Fluid Dynamic (CFD) code is used to calculate three dimensional, incompressible, and RANS equations. Different types of the fishing vessel hull form were performed. In this research, the data were collected chosen in north and south coast of Java island. The models were drawing in three dimensional, required for performing the analysis, were developed using Rhinoceros. The present study, the open-source computational fluid dynamics library, OpenFOAM was used to predict the resistance with the interFOAM solver. For the motion analysis, using strip theory in the Maxsurf motions. The probability of deck wetness analysis and ship motion were performed for comparing the models. Both analyses of ship Response Amplitude Operators (RAOs) are performed at three types of sea state (slight, moderate, and rough water). The comparisons of the hull form design will be evaluated to get the best performance based on the sea state condition of the java island.

1 Introduction

The traditional fishing vessel is one of transportation and livelihood support in Indonesia. The vessel was made based on the knowledge gained from generation to generation. Due to the experience in the field and instinct adapt to the environment. Thus, the traditional fishing vessel of region or island is one of the developed means product based on the ability to adapt to the natural environment in the region. The process of adaptation and culture of each island will determine the various forms of the traditional fishing vessel in terms of various size and style of the hull form.

Based on the territory, Indonesia has 17.508 islands, which is 70% of the sea with an area 5.8 million km^2. Indonesian sea is consists of 2.3 million km^2 marine territory and million km^2 Exclusive Economic Zone Area (EEZ)[1].

The common of numerical models for predicting the ship resistance performance in preliminary step is using CFD method. CFD is the computational fluid dynamics which based on potential flow theory, the Navier-Stokes equation to solve. The recent developments in computing technology have big problems to solve in shipbuilding industry

* Corresponding author: aldias@pukyong.ac.kr

because of the license of the software. The open-source of CFD is one of the solutions to engineering problems, especially in CFD problems. OpenFOAM provides solver that can be used to predict the CFD approach and can be compared with others. Many studies on the hydrodynamics ship performance have been investigated by numerical models [2–4]. The analysis of flow around mono-hull form has been conducted with the traditional snake boats of Kerala [5] by CFD, while the prediction of ship resistance also performed based on empirical approach has been investigated [6]. The proof numerical analysis in comparing the experiments result has been discussed for the planning hull [7]. In addition, the numerical approach with strip theory in ship motion to determine optimal short-range routing of vessels in a seaway has been analyzed [8].

2 Model description and simulation conditions

The traditional fishing vessel of the present study was a mono-hull designed by Indonesian Shipyard in North and South Coastal of Java island. The South coastal design has the character of hull shape with a larger deadrise angle than the North coastal. Table 1 shows the principal dimensions of the models.

Table 1. Principal dimensions of the traditional Indonesian fishing vessel.

Principal Dimensions	Unit	Designs	
		North Coastal	South Coastal
Length overall (LOA)	*m*	17.65	20.00
Length waterline (LWL)	*m*	15.75	18.09
Breadth (B)	*m*	4.00	4.80
Depth (H)	*m*	2.50	1.80
Draft (T)	*m*	1.75	1.20
Gross Tonnage (GT)	***Ton***	30.00	30.00
Speed (Vs)	***Knot***	10.00	10.00

(a) North coastal design (b) South coastal design

Fig. 1. Model in Rhinoceros: (a) North coastal design; (b) South coastal design

The models in Fig.1 should be scaled in the same dimension and can be compared to the resistance simulations. The North coastal is scaled 1:5 and South coastal design scaled 1:5.74 then, both the designs will be at same Wetted Surface Area (WSA) and the Froude Number (FN).

The numerical simulation for ship motion in Maxsurf motions was carried out in the 4 types of speed (2.50; 5.00; 7.50; 10.00 knots), with the heading parameters are stern on 0deg, stern quartering on 45deg, beam on 90deg, bow quartering on 135deg, and head on 180deg. The spectra parameters are JONSWAP spectra in three sea conditions that are significant wave height (H1/3) are 1.25 m; 2.5 m and 4 m, peak period 7.5, 8.8, 9.7 s. for slight water, moderate water, and rough water.

In the present study, the open-source CFD libraries were used to predict the ship resistance. To predict free-surface flow around the hull form, interFOAM is one of the simplest ways to solve the problem. This solver is a multiphase flow that uses the VOF method to solve the free surface for incompressible flow [9].

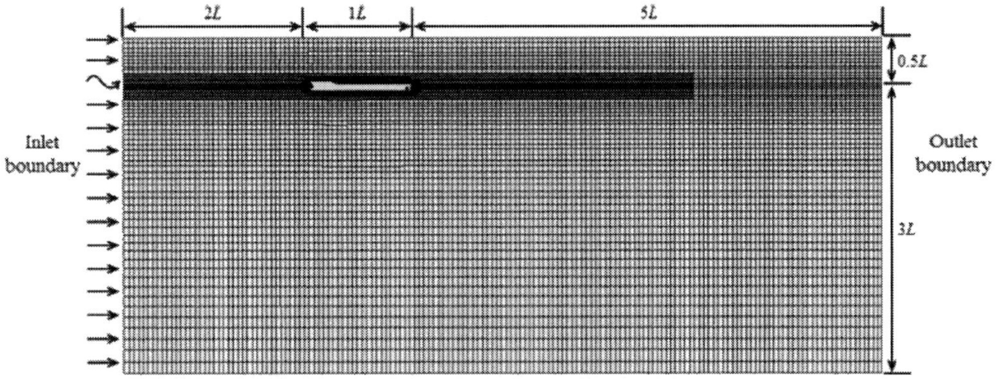

Fig. 2. Domain grid and boundary condition in OpenFOAM [6]

In the Fig.2 is the domain grid and boundary conditions that should be created in OpenFOAM [6]. The boundary conditions for this research are:

Inlet $$U = U, \frac{\partial p}{\partial n} = 0, \alpha = \int_0^1, \text{ if z} < 0 \text{ otherwise} \quad (1)$$

Outlet $$\frac{\partial p}{\partial n} = 0, p = 0, \frac{\partial \alpha}{\partial n} = 0 \quad (2)$$

Wall $$U = 0, \frac{\partial p}{\partial n} = 0, \frac{\partial \alpha}{\partial n} = 0 \quad (3)$$

Top $$\frac{\partial U}{\partial n} = 0, p = 0, \alpha = 0 \quad (4)$$

Where n is the outer normal vector [2].

In resistance test, the main non-dimensional parameters in the flow around the hull are Froude number (Fn) which can be expressed as:

$$Fn = \frac{V}{\sqrt{gL}} \quad (5)$$

The probability of deck wetness is often important to be able to predict in a particular cycle of motion, and it was explained [10]. The probability is defined as:

$$P = \exp\left(\frac{-F^2}{2m_{0z}}\right) \quad (6)$$

Where F is the freeboard and m_{0z} is the mean square relative vertical motion. The roll motion of the vessel was analyzed using the software Maxsurf motions to obtain the RMS roll value. This program also uses the linear ship-motion strip theory to calculate the motion in the frequency domain.

3 Numerical results

The traditional Indonesian fishing vessel for the north and south coastal design was performed in this research.

3.1 Ship resistance results

The results of the ship resistance for scale model can be compared as shown in Table 2.

Table 2. Resistance results for scale model

	Froude number (*Fn*)	Velocity of models (*m/s*)	North coastal resistance (*N*)	South coastal resistance (*N*)
V1	0.20	1.11	5.23	5.23
V2	0.40	2.22	48.90	36.10
V3	0.60	3.33	203.00	102.00
V4	0.80	4.45	262.00	151.00

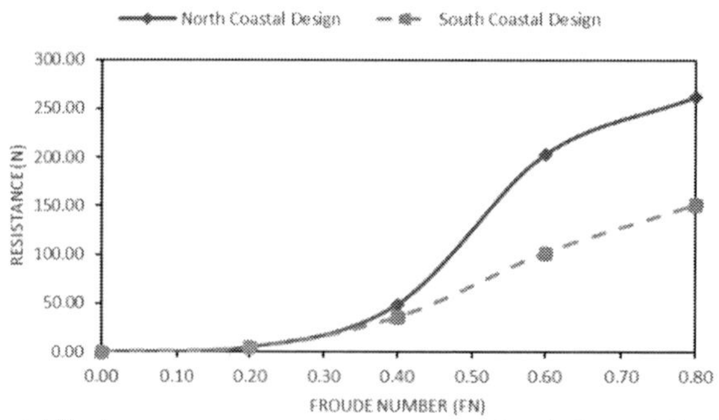

North Coastal Design South Coastal Design

(a) Resistance result

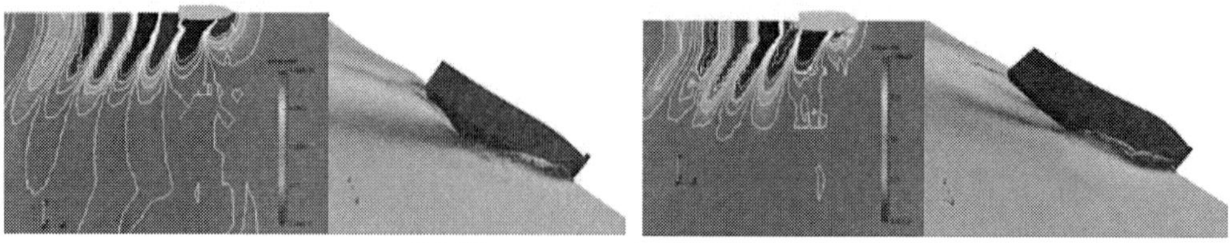

(b) Wave pattern in OpenFOAM

Fig. 3. Post-processing of resistance simulation in OpenFOAM

In the Fig.3 the computed results from OpenFOAM produce the difference wave pattern both the designs. From the wave pattern and results, the north coastal design has a stronger pattern and greater resistance than the south coastal design.

3.2 Ship motion results

In this study, both the designs will be compared for the added resistance and the probability of the deck wetness. From the simulation, the results of ship motions for full scale can be seen in Table 3.

Table 3. Ship motion results for full scale

VS (*Knots*)	Added Resistance (*N*)		Deck Wetness (%)	
	North	South	North	South
2.50	13358.18	14525.509	0.0141	0.0117
5.00	14613.09	15742.903	0.0143	0.0118
7.50	15158.42	16264.678	0.0146	0.0120
10.00	15436.10	16434.678	0.0149	0.0121

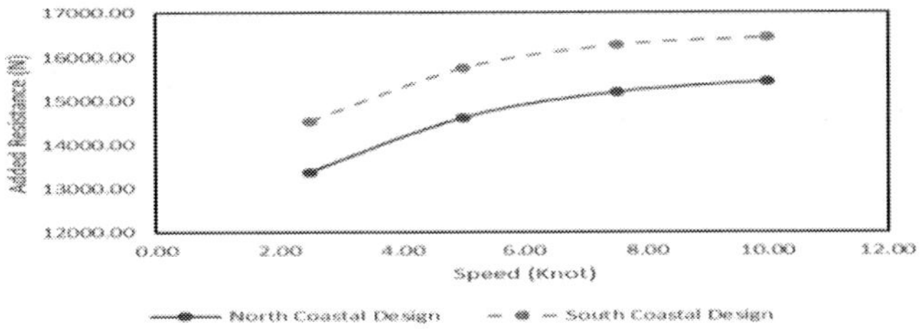

Fig. 4. Added resistance in ship motions

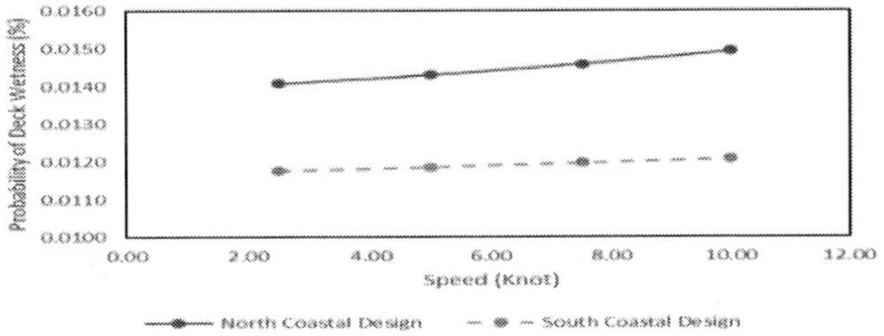

Fig. 5. Probability of deck wetness

The input of the simulation condition in this study was the same. From the Fig.4 and Fig.5 show the result the north coastal design has the added resistance smaller and the probability of deck wetness is greater than south coastal design. Thus, the north coastal design is better than the south coastal design.

The summary of ship motion analysis for both the designs can be seen in Table 4.

Table 4. Summary of ship motion performance

Vessel motion: 10kts; head seas	**North Coastal Design**	**South Coastal Design**
MSI at bridge	approx. 20% after 2hrs	approx. 20% after 2hrs
Propeller emergence	41.76% (0.156 per hr)	34.11% (0.008 per hr)
Deck Crew	36.75% (0.01 per hr)	31.61% (0.000 per hr)
Working Deck	40.19% (0.05 per hr)	35.03% (0.003 per hr)
Deck wetness	0.0149%	0.0121%
Calm water resistance	14.40 *kN*	13.71 *kN*
Added resistance (Slight, Moderate, Rough Sea)	(14.61; 15.18; 15.44 *kN*)	(15.35; 16.22; 16.43 *kN*)

4 Conclusions

In this work, hydrodynamic analysis of the traditional Indonesian fishing vessel was performed using the open-source CFD software OpenFOAM. The analysis was performed by interFOAM solver from the hull form design has an effect on resistance performance. Due to the results, the north coastal design has greater resistance than the south coastal design for the calm water conditions. In other work, the ship motion was analyzed using strip theory in Maxsurf motions. The result was performed that the north coastal design has smaller added resistance than the south coastal design. It means for the same condition is better to use the north coastal design. The probability of deck wetness for the north coastal design is greater than the south coastal design. It is caused by the south coast is more fierce than the north coast of Java, thus the south coastal design has characteristic to reduce the deck wetness by the hull form. The future research is the development of bow height calculation. The designs should be modified using several scenarios and point of deck wetness. The ship length and wave heights are parameters to get an optimum result and best performances of the ships.

This paper is the result of project work titled 'Next Generation Korea-model Fishing Vessel Development & Test-bed Application' funded by the Ministry of Oceans and Fisheries, Korea and the BK21 plus MADEC human resources development group, Pukyong National University for the support and providing the facilities.

References

1. Dishidros. Potential of Indonesia. *Indonesian Navy* (1987).
2. Axner, L., Gong, J., Chiarini, A. & Mascellaro, L. Partnership for Advanced Computing in Europe SHAPE pilot Monotricat SRL : Hull resistance simulations for an innovative hull using OpenFOAM, 1–8 (2014).
3. Hakan, Y. & Barlas, B. ScienceDirect Numerical study of ship motions and added resistance in regular incident waves of KVLCC2 model. *Int. J. Nav. Archit. Ocean Eng.* **9,** 149–159 (2017).
4. Jeong, S., Choi, K., Kang, K. & Ha, J. ScienceDirect Prediction of ship resistance in level ice based on empirical approach. *Int. J. Nav. Archit. Ocean Eng.* (2017).
5. Subbaiah, B. V., Thampi, S. G. & Mustafa, V. Modelling and CFD Analysis of Traditional Snake Boats of Kerala. *Aquat. Procedia* **4,** 481–491 (2015).
6. Seo, S., Park, S. & Koo, B. Effect of wave periods on added resistance and motions of a ship in head sea simulations. *Ocean Eng.* **137,** 309–327 (2017).
7. Marco, A. De, Mancini, S., Miranda, S., Scognamiglio, R. & Vitiello, L. Experimental and numerical hydrodynamic analysis of a stepped planing hull. *Phys. Procedia* **64,** 135–154 (2017).
8. Dolinskaya, I. S., Kotinis, M., Parsons, M. G. & Smith, R. L. Optimal Short-Range Routing of Vessels in a Seaway. **53,** 121–129 (2009).
9. Emmanuel, C., Mesina, L., Emmanuel, C. & Mesina, L. The Conceptual Design of a Ballast-Free Ship by The Conceptual Design of a Ballast-Free Ship Ballast Free (2017).
10. Combatant, S. & Damage, A. in *14th International Ship Stability* 1 (UTM, Malaysia, 2014).

Bulbous bow applications on a catamaran fishing vessel for improving performance

Samuel[1,2,*], *Dong-Joon* Kim[1], *Muhammad* Iqbal[2], *Aldias* Bahatmaka[3] and *Aditya* Rio Prabowo[3]

[1]Department of Naval Architecture and Marine Systems Engineering, Pukyong National University, Republic of Korea

[2]Department of Naval Architecture, Diponegoro University, Republic of Indonesia

[3]Department of Interdisciplinary Program of Marine Design Convergence, Pukyong National University, Republic of Korea

Abstract. This study aims to prove the usefulness of a bulbous bow traditional catamaran fishing vessel. Bulbous bow applications are widely used to reduce wave resistance. The fishing vessel resistance was calculated using CFD (Computational Fluid Dynamic) approach and combined with classical slender body theory. In this research, the application of bulbous bow was being adopted on traditional catamaran fishing vessel to reduce the total resistance. The bulbous bow has adopted based on Kracht method. Modeling procedure was developed by using linear form coefficients. The ship resistance was simulated using Tdyn software which basically Finite Element Method. The results show that adding bulbous bow can decrease or increase the ship's total resistance.

1. Introduction

The conversion of fishing vessels from monohull to catamaran form was carried out in 2015. The ship was launched using crane (fig. 1a). This conversion aims to increase the payload capacity up to 2 times since the ship has twin hull (fig. 1b). However, this conversion became bad news for the fishermen because the resistance of catamaran form increases up to almost four times compared to monohull form. Several studies on catamaran resistance have been conducted in the past, including earlier experiments by Everest, Oving, Pien [1]–[3] as well as a theoretical study Doctors [4]. Molland has generated the form factor of catamaran form which includes the interference factor as part of viscous resistance component [5]. Jamaluddin et al. has modified the Molland's form factor by adding S/L variable [6]. The calculation of ship resistance using Tdyn has been carried out by Samuel et al. [7] and Iqbal et al. [8] to estimate multihull ship resistance. Julio Garcia et al. using finite element method based on the finite calculus for analysis of free surface flow problems [9]. A bulbous bow that has been adopted widely in a monohull should only be fitted if it will reduce the ship resistance which its effect will increase the ship speed or reduce the fuel consumption [10]. Therefore, the main objective of this research is to

* Corresponding author: undip_samuel@yahoo.com

develop catamaran ship resistance using a bulbous bow. This proposed method can generate the new shape of a bulbous bow. The ship resistance will be simulated using Tdyn software which is basically CFD approach.

a b

Fig. 1. Catamaran fishing vessel MV. Laganbar.

2. Bulbous bow design

The main function of the bulbous bow is to reduce resistance when the ship is operating. The bulbous bow shape affects the hydrodynamics of the velocity field around the bow, that is, on the rising wavelength of the ship, which is usually accompanied by a reduction of the wave resistance. By smoothing the flow around the body, the resistance will decrease.

Therefore, the addition of bulbous bow will be advantageous depending on the size, position, and shape of the bulbous bow. Generally, at low speeds, the effect of the bulb is negative. When the Froude Number increases, its effect becomes positive and increases up to a maximum value. In this study, Kracht's method was used to get an initial bulbous bow. Kracht [11] classified the bulbous bow types :

a. ∇ - Type: The nabla-type also has a drop-shaped sectional area A_{BT} (fig. 2c), but its center of the area is situated in the upper-half part, indicating a volume concentration near the free surface.
b. O – Type: This type (fig. 2b), with an oval sectional area A_{BT} and a center of the area in the middle, has a central volumetric concentration.
c. Δ – Type: (fig. 2a) Shows the drop-shaped sectional area A_{BT} of the delta type with the center area in the lower half part.

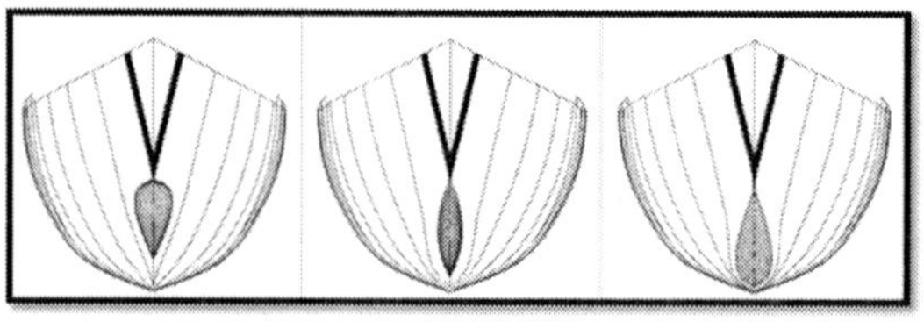

a b c

Fig. 2. Modeling bulbous bow section.

Kracht [11] has three linear bulb parameter :

a. The breadth parameter: the maximum breadth B_B of the bulbous bow sectional area at the FP is divided by the beam B_{MS} of the ship, which is denoted by C_{BB} ($C_{BB} = B_B/B_{MS}$)
b. The length parameter: the bulbous bow length L_{PR} is divided by the L_{PP} of the ship; it is denoted by C_{LPR} ($C_{LPR} = L_{PR}/L_{WL}$)
c. The depth parameter: the height Z_B of the bulbous bow tip is divided by the draft T_{FP} at the FP; it is denoted by C_{ZB} ($C_{ZB} = Z_B/T_{FP}$)

3D modeling of the bulbous bow is assisted using Rhinoceros software. The solid model generated from multiple curves and surface. The linear bulb ratio of the Kracht parameter is used to obtain an initial bulbous form. The definition of the parameters of the linear coefficients can be seen in fig. 3. The purpose of this research to make a bulbous bow shape with type Δ-Type, O-Type, and ∇-Type. In this study, using variable Zb to produce the desired bulbous bow shape, while at L_{PR} and B_B become fixed variable. Comparison of $L_{PR}/LPP = 0.018$; $B_B/B = 0.170$ and $Zb/draft = 0.274$; 0.572; 0.838 as shown in fig. 2. This model that uses the parameter scenario will generate the shape Δ-Type, O-Type, and ∇-Type.

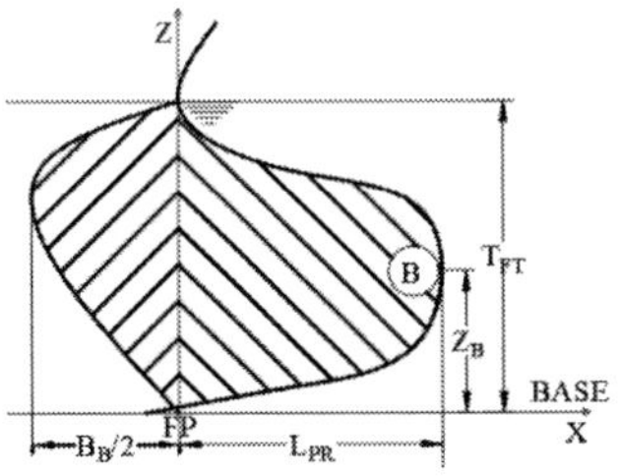

Fig. 3. Linear bulb quantities.

3. Methodology

The subject of this study is a traditional catamaran fishing vessel, the main dimensions measure obtained from the results of the direct measurement, which was modified by fishermen. The main dimensions result are presented in table 1. The slender body method was applied for the calculation of monohull and multihull hull resistance, assuming a thin hull where the ratio between the length and width of the large vessel (L/V1/3> 4.0). The slender body method is based on Michell's theory approach [12] which expresses the wave resistance of the thin wall of the hull and was developed by Couser [13]. The effects of resistance due to spray and wave-breaking cannot be calculated properly [14]. Therefore, The viscous resistance was calculated based on the skin friction formula ITTC'57, and then the calculation of form factor (1+βk) for catamaran using the empirical formula given by Molland and Jamaluddin [6],[15]. This is because Michell's theory has not been able to calculate breaking-waves and wave interference. Insel and Molland [16] proposed that the total resistance of a catamaran expressed by :

$$CT_{CAT} = (1+ \beta K)\, CF + \tau\, CW \tag{1}$$

Where CT_{CAT} is the total resistance coefficient; (1+ βK) is the hull form coefficient; CF is the frictional resistance coefficient; τ is the wave interference factor and CW is the wave resistance coefficient.

Table 1. Main dimensions of traditional catamaran fishing vessel MV. Laganbar.

Length of waterline (LWL)	8.72 m
Draft (T)	0.50 m
Beam demi hull (B)	1.01 m
Beam overall (BOA)	2.88 m

Then, Jamaluddin et al. modified hull form coefficient by considering S/L effect, can be calculated from equation 2.

$$(1+\beta k) = 3.03\ (L/V^{1/3})^{-0.40} + 0.016\ (S/L)^{-0.65} \quad (2)$$

The value of τ is the wave interference factor caused by the fluid flow around the catamaran at a certain distance (S) of the ship length (L). The resistance component of catamarans has a more complex phenomenon than a monohull, caused by the influence of the interaction between the two-ship's hull and the interference wave. Which gives effect to the formation of the boundary layer and the longitudinal vortices. While the wave interference due to the interaction of waves generated by each catamaran hull. That is difficult to predict the total resistance of the catamaran vessel, this study is assisted by Tdyn.

Tdyn software is simulation program, which capable to compute the resistance of ship by potential flow method, boundary layer method and Reynold Average Navier-Stokes (RANS). The total resistance was calculated in every element based on finite element method. The analysis was carried out with computer code SHYNE, based on the algorithm and pre/post processor GID developed at CIMNE. CFD formulation presented can be found in Tdyn [17].

The NURBS definition is shown in fig. 4b, nonuniform rational B-splines (NURBS) are mathematical representations of 2D or 3D objects, NURBS can generate curves and surfaces to make 3D modeling.

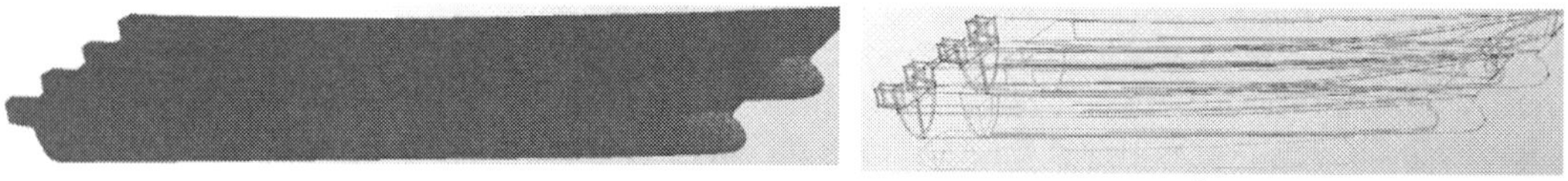

a b

Fig. 4. Geometrical definitions.

The computation domain is discretized using an unstructured surface mesh grid. The mesh around the catamaran consists of about 140.130 tetrahedra and 39.610 triangle elements distributed. The surface mesh catamaran used in the analysis is presented in fig. 4a.

4. Results and discussions

The total resistance of a catamaran can be calculated using equation 1-2. It was validated using a combination of calculations, the empirical formula, and slender body. The result of numerical computation and Jamaluddin's formula shows that the total catamaran resistance value does not exceed the 5% error, as shown in fig. 5. Interference of wave resistance between two catamaran hulls can be contribute negative or positive as presented from the theoretical and experimental results.

The fisherman modified a monohull form to a catamaran with the S/L ratio being 0.2, that modification can increase the payload capacity up to two times with the same draft. In fig. 6 shows the total catamaran resistance and two times the monohull.

The comparison of resistance between original catamaran and catamaran with a bulbous bow was presented in fig. 7. Total resistance catamaran based on Froude number used in the Jamaluddin's formula, i.e. the Froude number 0.28-0.65. The reason to use the same Froude number is to simplify and to reduce errors in the calculations performed.

The nabla-type form has the lowest resistance at fr 0.47, but at Fr 0.65 is relatively larger than the other. At low speed, the bulbous bow will give bad impact, but at high speed, will have a good effect. The applications of the bulbous bow are complicated to apply at several speeds, this is because each bulb has different characteristics. However, the naval architect may choose the bulb which used at service speed.

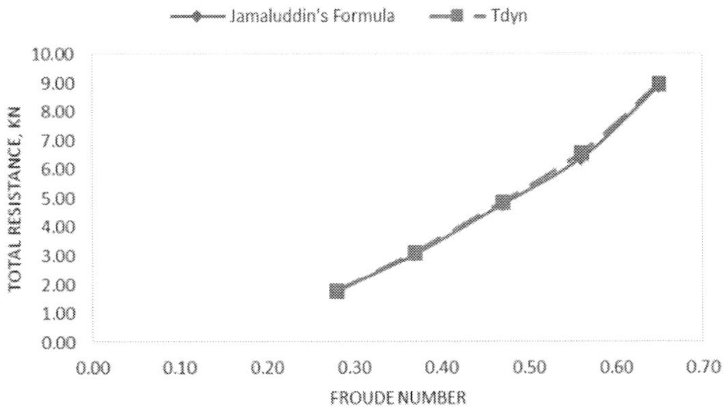

Fig. 5. Comparison of resistance numerical results with Jamaluddin's formula.

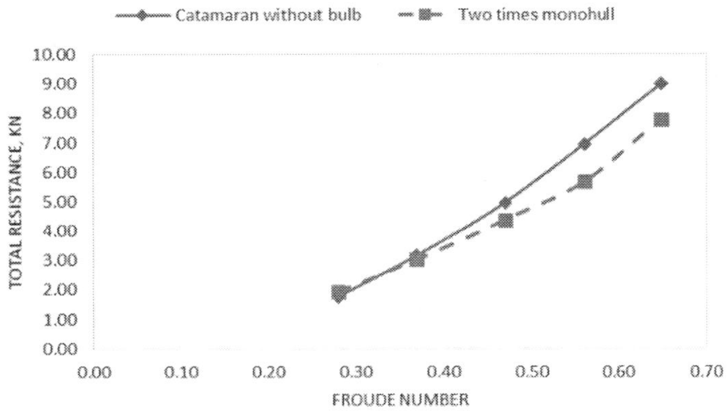

Fig. 6. Comparison of catamaran resistance.

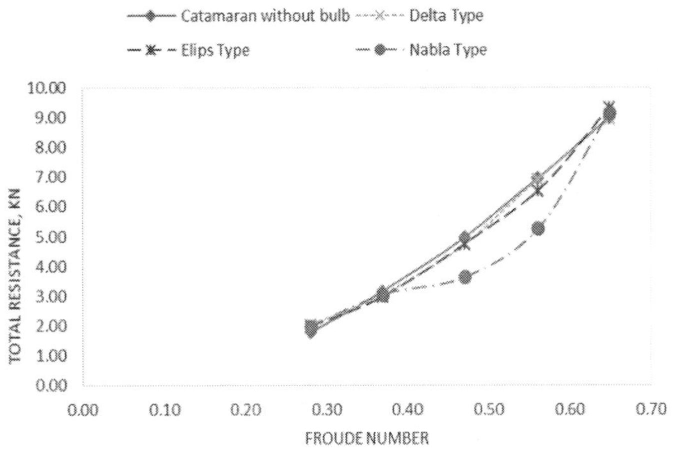

Fig. 7. Result of resistance

5. Conclusion

Efforts to improve ship performance vary widely, one of which is to add a bulbous bow with several parameters. This effort was successfully done by using three forms of a bulbous bow. Nabla type was selected as the most influential bulbous bow to be applied on a traditional catamaran boat. Nabla type can reduce the resistance by an average of 10%. At low speeds, the bulbous bow has no significant effect. However, at high speed, its effect becomes positive and increases up to a maximum value.

The authors wished to thank the Ministry of Oceans and Fisheries Korea for funding this research under the scheme 'Next Generation Korea-model Fishing Vessel Development & Test-bed Application' and Indonesia Ministry of Finance Lembaga Pengelolaan Dana Pendidikan (LPDP).

References

1. J. T. Everest, *Some research on the hydrodynamics of catamarans and multi-hulled vessels. North East Coast Inst. Eng. Shipbuild.* (1968).
2. P. C Pien, *Int. Semin. wave Resist. Soc. Nav. Archit. Japan* (1976).
3. A. J. Oving, *Resistance prediction method for semi planing catamarans with symmetrical demihulls.* (1985).
4. L. J. Doctors, J. Eng. Australia **4,** 295–302 (1991).
5. A. F. Molland, P. A. Wilson, D. J. Taunton, S. Chandraprabha & P. A. Ghani, The Int. J. of Maritime Eng **146,** 19–38 (2004).
6. A. Jamaludin, I. K. A. P. Utama, B. Widodo & A. F. Molland, J. of Eng. for the Maritime Env. **227,** 51–60 (2012).
7. Samuel, M. Iqbal & I. K. A. P. Utama, Int. J. of Tech **6,** 432–441 (2015).
8. M. Iqbal, Samuel, Int. J. of Tech **8,** 85–91 (2017).
9. J. Garcia, A. Valls & E. Oñate, Int. J. for Num. Methods in Eng **76,** 1297–1327 (2008).
10. D. Chrismianto, & D. J. Kim, J. of Marine Sci. and Tech **19,** 479–492 (2014).
11. A. M. Kracht, SNAME Trans **86,** 197–217 (1978).
12. J. H. Michell, Phil. Magazine **5,** 106–123 (1989).
13. P. R. Couser, J. F. Wellicome & A. F. Molland, Inter. Shipbuilding Prog **45,** 444 (1998).
14. Murdijanto, I. K. A. P. Utama & A. Jamaluddin, Makara Seri Teknologi **15,** 25–30 (2011).
15. A. F. Molland, J. F. Wellicome & P. R. Couser, Trans. RINA **138,** 59–71 (1996).
16. M. Insel & A. F. Molland, The RINA **134,** 1–11 (1991).
17. Tdyn. in Tdyn Theory Manual 1–33 (2014).

Analysis of Effect of Bulbous Bow Shape to Ship Resistance in Catamaran Boat

Deddy Chrismianto[1,*], *Kiryanto*[1] , and *Berlian* Arswendo Adietya[1]

[1] Department of Naval Architecture, Faculty of Engineering, Diponegoro Univerity, Indonesia

Abstract. Ship resistance is one of the main factors affecting the design of a ship. Catamaran boat is a ship with small wet surface area that able to reduce drag and improve ship power. Generally, a bulbous bow is implemented to reduce wave resistance because the bulbous shape is believed to attenuate the bow wave system. Additionally, the bulbous bow also tends to reduce viscous resistance. When the flow around the body is smooth, the total ship resistance can be reduced significantly if the optimum bulbous bow is obtained. In this study, the main purpose is to get the bulbous bow shape in catamaran boat which produces the smallest ship resistance by using computational fluid dynamic (CFD). Generating the variation of the bulbous bow shapes apply the one-to-one correspondence of the cross section parameter (A_{BT}) and lateral parameter (A_{BL}). The result of investigation shows that application of bulbous bow on catamaran boat can reduce about 11-13% of total resistance of ship.

1 Introduction

A bulbous bow is a protruding bulb at the bow of a ship just below the waterline. The bulb modifies the way the water flows around the hull, reducing drag and thus increasing speed, range, fuel efficiency, and stability. Ships with bulbous bows generally have a 12 to 15 percent better fuel efficiency than similar vessels without them.

During the last decade catamaran boats have rapidly evolved into a dominant mode of sea transportation in Indonesian sea. Their particular area of proliferation is the short sea shipping where they show considerable superiority over competitive designs in attributes such as space availability and speed quality. However, resistance or drag characteristics are principal aspects of the catamaran boat design spiral as they are strongly coupled with speed and ship fuel and, consequently, the ship operating cost is one of problem on the catamaran boat that has to be solved.

Development of bulbous bow design for catamaran boat is a proposed model to solve the fuel problem. CAD-integrated CFD method is applied to obtain the better bulbous bow form of catamaran boat with minimum ship resistance. Recently, some of the studies CAD-integrated CFD and optimization method can be used simultaneously to solve any problems, including the optimization of hull shape. The use of the control points of the cubic B spline to generate the parametric of ship hull design as well as the parametric of

* Corresponding author: deddy.chrismianto@ft.undip.ac.id

bulbous bow design has been introduced with successful results [1, 2, 3, 4]. Deddy, C [5, 6] has developed the parametric bulbous bow design and submarine design are constructed using the cubic Bezier curve method and the curve-plane intersection method. The methods can be used to generate a parametric bulbous bow shape automatically using solid modeling for the optimization calculation in the effort to minimize the ship's resistance.

The non-linear bulbous bow parameter is investigated and varied to modify the bulbous bow form of catamaran boat by the cross section parameter (A_{BT}) and lateral parameter (A_{BL}). Generating the variation of the bulbous bow shapes apply the one-to-one correspondence of both parameters. The best of bulbous bow of catamaran boat is obtained with good performance that is able to save energy or fuel.

2 Bulbous Bow Design

Generally, a bulbous bow is implemented to reduce wave resistance because the bulbous shape is believed to attenuate the bow wave system. Additionally, the bulbous bow also tends to reduce viscous resistance. When the flow around the body is smooth, the total resistance can be reduced significantly if the optimum bulbous bow is obtained. There are some parameters that affect the dimension and form of the bulbous bow (Fig.1).

Kracht [7] classified the bulbous bow parameters into six quantitative parameters, including three linear parameters and three nonlinear parameters as follows:

- The breadth coefficient C_{BB}: the maximum breadth B_B of the bulbous bow sectional area at the FP is divided by the beam B_{MS} of the ship, which is denoted by

$$C_{BB} = B_B/B_{MS} \tag{1}$$

- The length coefficient C_{LPR}: the bulbous bow length L_{PR} is divided by the L_{PP} of the ship; it is denoted by

$$C_{LPR} = L_{PR}/L_{WL} \tag{2}$$

- The depth coefficient C_{ZB}: the height Z_B of the bulbous bow tip is divided by the draft T_{FP} at the FP; it is denoted by

$$C_{ZB} = Z_B/T_{FP} \tag{3}$$

- The cross section coefficient C_{ABT}: the cross sectional area A_{BT} of the bulbous bow at the FP is divided by the midship section area A_{MS} of the ship, and it is denoted by

$$C_{ABT} = A_{BT}/A_{MS} \tag{4}$$

- The lateral coefficient C_{ABL}: the area of the bulbous bow in the longitudinal plane at center line A_{BL} is divided by the midship section area A_{MS} of the ship , and it is denoted by

$$C_{ABL} = A_{BL}/A_{MS} \tag{5}$$

- The volumetric coefficient $C_{\forall PR}$: the volume of the bulbous bow $\forall_{PR}$ is divided by the volume of displacement $\forall_{WL}$, and it is denoted by

$$C_{\forall PR} = \forall_{PR} / \forall_{WL} \tag{6}$$

Some previous research have analyzed the effect of linear parameters of bulbous bow. The nabla type of bulbous bow can reduce ship resistance significantly. This type has bigger value of dept coefficient C_{ZB} than other types. So, in this study, the non-linear parameters of bulbous bow is chosen to be investigated its effect on the nabla type of bulbous bow. The cross section coefficient C_{ABT} and lateral coefficient C_{ABL} is applied to obtain the minimum ship resistance on the nabla type. C_{ABT} has value around 0.064 – 0.122

and C_{ABL} is 0.068 – 0.146. These values are combined in several bulbous bow models. (Table 1)

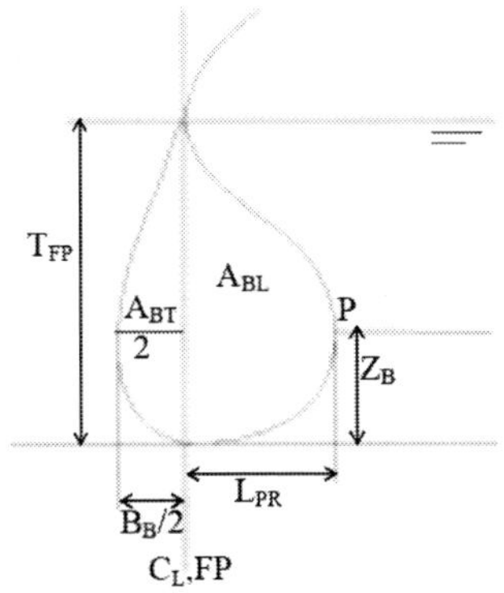

Fig. 1. Bulbous bow dimension

Table 1. Bulbous bow models based on C_{ABT} and C_{ABL} Values

No	C_{ABT}	A_{BT} (m^2)	C_{ABL}	A_{BL} (m^2)
Model 0	-	0.065	-	0.123
Model 1	0.064	0.126	0.068	0.134
Model 2	0.064	0.126	0.107	0.211
Model 3	0.064	0.126	0.146	0.288
Model 4	0.093	0.183	0.068	0.134
Model 5	0.093	0.183	0.107	0.211
Model 6	0.093	0.183	0.146	0.288
Model 7	0.122	0.240	0.068	0.134
Model 8	0.122	0.240	0.107	0.211
Model 9	0.122	0.240	0.146	0.288

3 Resistance calculation and results

Catamaran boat without a bulbous bow owned by PT. Sanur Marindo Shipyard is modified and fitted a bulbous bow. This boat has main dimension, such as length perpendicular (Lpp) about 23.45 m, breadth 9.00 m, and draft 1.00 m. One to one correspondency method is used to create several models with diiference of cross section coefficient C_{ABT} and lateral coefficient C_{ABL}. There are 9 (nine) of models that be generated with the different linesplans of bulbous form (Fig.2).

Total resistance of ship is calculated using CFD with 3 (three) various of speed, that are: 16 knot, 20 knot, and 24 knot. Table 2 shows that applied bulbous bow on the catamaran boat is able to reduce drag significantly in all of speed. Model no. 8 is the best of bulbous bow form in which it has total resistance about 32.31 kN in 16 knot speed, 44.94 kN in 20 knot spped, and about 59.53 kN in 24 knot speed. It shows that the total ship resistance value can be reduced up to 11 - 13%.

In Fig. 3 described the wave pattern of catamaran boat between original model (model no. 0) and the best model (model no. 8). Comparison of both show several change of color in wave pattern, especially around the bow areas. The yellow color of wave pattern on model no. 8 has reduced compared with original model. This model is able to reduce the wave resistance significantly around bow areas. This fact explained that the wave resistance also shows reducing drag significantly about 21 – 22,5%.

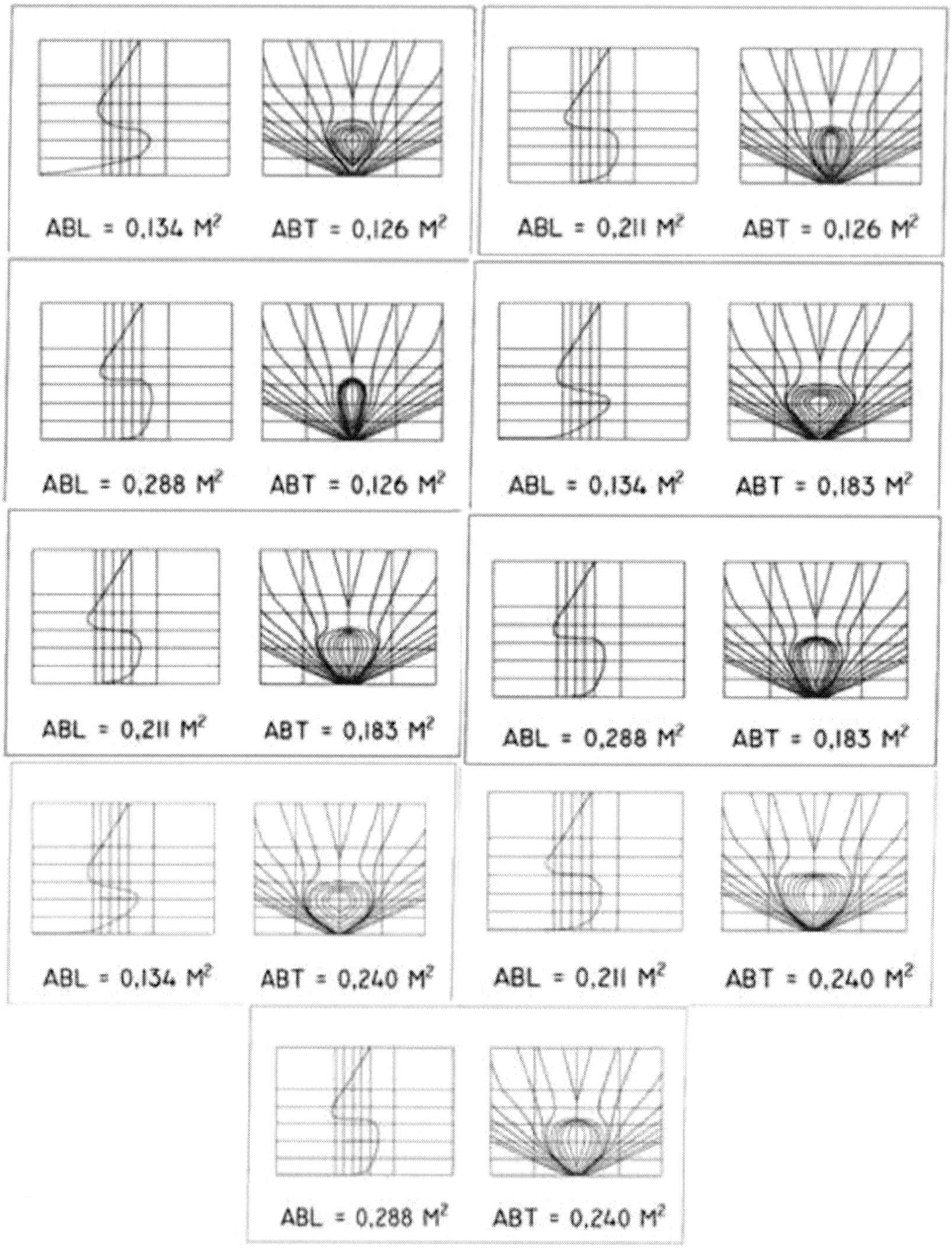

Fig. 2. Linesplan of bulbous bow models based on A_{BT} and A_{BL}

Table 2. Calculation of total resistance in each model

Model	Total Resistance Value (kN)		
	16 Knot	**20 Knot**	**24 Knot**
Model 0	36.45	51.49	67.36
Model 1	35.17	50.27	64.83
Model 2	36.94	50.68	65.35
Model 3	35.90	50.22	63.50
Model 4	33.65	48.12	62.37
Model 5	34.29	48.06	62.67
Model 6	33.79	46.90	60.75
Model 7	33.98	46.89	60.60
Model 8	32.31	44.94	59.53
Model 9	35.72	49.67	64.42

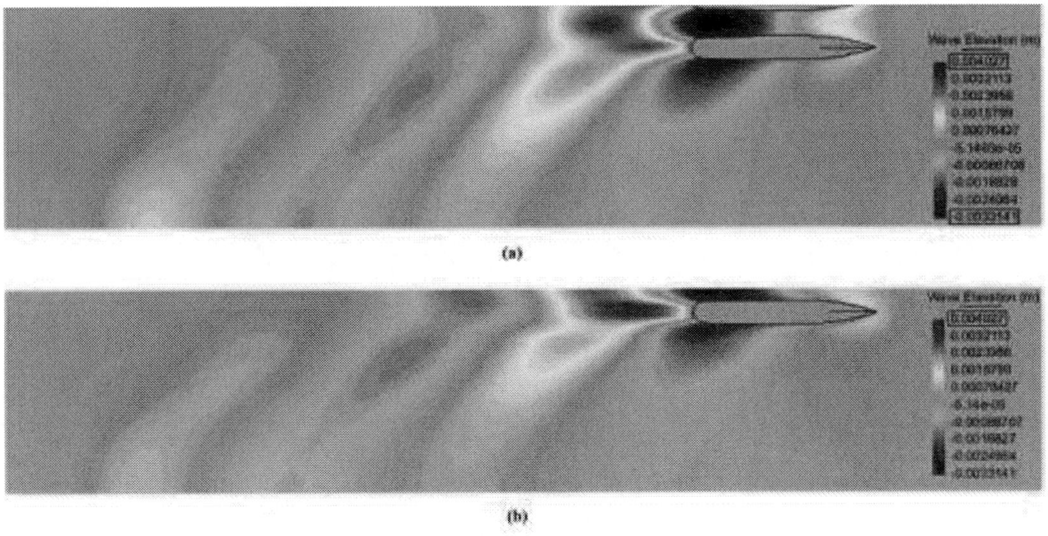

Fig. 3. Comparison of wave resistance at original model (a) and model no. 8 (b)

I would to thank to Faculty of Engineering Diponegoro University, Research Grant of DIPA FT - UNDIP, Penelitian Strategis Tahun 2017.

Conclusion

1. Based on the CFD analysis result, it was found that ship resistance value for various variations of bulbous bow shape in catamaran boat. Model no. 8 has the smallest value of ship total resistance.
2. The total ship resistance value can be reduced up to 11 - 13% on model no. 8 of catamaran boat, this value occurs at a speed of 16 - 24 knots. The bulbous bow shape has A_{BT} = 0.211 m2 and A_{BL} = 0.240 m2.

3. Model no. 8 (best model) has fewer yellow color at the bow area than model no. 0 (original model). This fact explained that the wave resistance also shows reducing drag significantly about 21 – 22,5%.

References

1. F. Perez, J.A. Suarez, J.A. Clemente, A.Souto, J Mar Sci Technol **12**, 83-94 (2007)
2. Z. Ping, Z.D. Xiang, L.W. Hao, J. Hydrodynamics **20**, 804-810 (2008)
3. F. Perez, J.A. Clemente, Computer-Aided Design **43,** 1829-1840 (2011)
4. E.F. Campana, D. Peri, Y. Tahara, F. Stern, Comput. Methods Appl. Mech. Engrg, **196**, 634-651 (2006)
5. D. Chrismianto, D.J. Kim, J. Mar. Sci. Technol. **19**, 479-492 (2014)
6. D. Chrismianto, A. F. Zakki, B. Arswendo, D.J. Kim, J. Marine Sci. Appl. **14**, 399-405 (2015)
7. Kracht, M. Alfred, SNAME Transactions **86**, 197-217 (1978)
8. Lewis, V Edward, *Principles of Naval Architecture Vol. II: Resistance and Propulsion* (SNAME, New Jersey, 1988)

Numerical approaches in idealizing mass of fluid in tank for ship vibration analysis

Jung Hoon Byeon[1], *Hyun Jin* Cho[1], *Seung Jun* Baek[2], *Aditya Rio* Prabowo[1], *Dong Myung* Bae[2], and *Jung Min* Sohn[1,2*]

[1]Pukyong National University, Marine Convergence Design, 48513 Busan, Korea
[2]Pukyong National University, Naval Architecture and Marine Systems Eng., 48513 Busan, Korea

Abstract. Vibration analysis is a crucial process in ship design to guarantee crew's comfortability and to save operating costs by predicting problematical resonance in advance. This study focuses on idealizing fluid mass in tanks by positioning point mass element and quad elements inside the tanks. Also, the added mass effect induced sea water surrounding a ship was included in the present study. The natural frequency obtained from free and forced ship vibration analysis were compared and assessed in the manner of ISO regulation. Finally, it was concluded that the quad element was more adequate to idealize fluid entities in tanks since it was difficult in arranging point mass elements at each grid points properly.

1 Introduction

Ship design is a sustainable process which each stage correlates each other. A design assessment should be conducted to achieve the optimum level and to satisfy demands of the ship owner for the purpose of operational costs. Ship vibration analysis is one of assessment categories to reduce additional expenses before sea trial by solving resonance problems which is related to fatigue damage which has potential to cause a remarkable financial loss.

A ship is accelerated in the ocean, sea water surrounding the ship simultaneously has to be accelerated. This is called an inertial effect known as added mass effect which has a tendency to make natural frequencies decrease [1]. In the early 20th century, added mass used to be approximated through using a submerged circular cylinder [2]. After computational performances had been developed, more accurate methodologies were calculated by using finite element method.

The virtual mass method in NASTRAN using fluid-structure interaction has been used as an adequate technique to consider an added mass effect caused by sea water [3]. However, it is not appropriate to analyse the situation which hydrodynamic forces are applied on inner and outer sides simultaneously because this method is proper to solve a simply connected domain [4]. However, ship vibration analysis should reflect the contact with a liquid at both sides to take into account fluid entities in tanks, e.g. fresh water, heavy fuel oil and lubricating oil. Hence, additional methodologies are needed: point mass elements and/or quad elements.

* Corresponding author: jminz@pknu.ac.kr

This study investigates the global ship vibration of a 1500-ton fishery research vessel using finite element method and summarizes the dynamic structural response depending on methodologies. Added mass induced from sea water was calculated by the virtual mass method which utilizes MFLUID card in NASTRAN. Fluid in tanks was dealt with additional two methods to remedy a mentioned disadvantage of the virtual mass method. Also, free and forced vibration analysis were conducted to compare two methodologies and its results were evaluated based on the Guidance Note and ISO standards [5-6].

2 Methods for considering added mass

An added mass matrix is obtained by using MFLUID card in NASTRAN [4] which solves Laplace's equation by distributing a set of sources over the outer boundary. In this paper, MFLUID card was used to consider the added mass effect induced by the sea water contacted with the outer shell plates. All flow is assumed inviscid and irrotational. The velocity $\dot{\boldsymbol{u}}_i$ at any point is defined as follows:

$$\dot{\boldsymbol{u}}_i = \sum_j \int_{A_j} \frac{\sigma_j \boldsymbol{e}_{ij}}{|\boldsymbol{r}_i - \boldsymbol{r}_j|^2} dA_j \tag{1}$$

where $\boldsymbol{\sigma}_i$ is the value of a point source of fluid (units are volume flow rate per area) located at location $\mathbf{r}_i$ and the point source is assumed acting over an area $\mathbf{A}_j$ while $\mathbf{e}_{ij}$ is the unit vector in the direction from point j to point i. The pressure, $\mathbf{p}_i$, at any point is following:

$$\dot{\boldsymbol{p}}_i = \sum_j \int_{A_j} \frac{\rho \dot{\sigma}_j \boldsymbol{e}_{ij}}{|\boldsymbol{r}_i - \boldsymbol{r}_j|^2} dA_j \tag{2}$$

where ρ is density.

After that two matrices, $[\chi]$ and $[\Lambda]$, can be obtained by integrating (1) and (2) equations over the surfaces adjacent to fluid,

$$\{\dot{u}\} = [\chi][\sigma] \tag{3}$$

$$\{F\} = [\Lambda]\{\dot{\sigma}\} \tag{4}$$

where F are the forces converted from the pressures at the grid points. Then, if $[\chi]$ is not singular, an added mass matrix, $[M_f]$, is derived by combing (3) and (4) as follows:

$$\{F\} = [\Lambda]\{\chi\}^{-1}\{\ddot{u}\} = [M_f]\{\ddot{u}\} \tag{5}$$

Then, an advanced added mass matrix when $[\chi]$ is singular or nearly singular is obained by introducing a set of Lagrange multipliers [4] as follows:

$$\{F\} = [\Lambda\chi^T \quad \Lambda_2] \begin{bmatrix} \chi\chi^T & \chi_2 \\ \chi_2^T & 0 \end{bmatrix}^{-1} \begin{Bmatrix} \ddot{u} \\ 0 \end{Bmatrix} = [M_f]\{\ddot{u}\} \tag{6}$$

The added mass matrix is combined with the ordinary mass matrix, $[M_s]$, in the characteristic equation to solve an eigenvalue problem as follows:

$$\det\left([K] - \omega^2([M_s] + [M_f])\right) = 0 \tag{7}$$

where ω is natural freequency which is calculated by Lanczos method [7].

3 A simple cargo tank

The fluid entities in the tanks can be modeled by two kinds of methods: point mass elements and quad elements. The former one is that weight of the fluid entities is equally distributed on every node point. The other is to model additional quad elements at the same location as the the fluent entities. It is important to generate quad elements which has neligible stiffness. Fig. 1 shows the effect of fluid stiffness on the results. In the results, it was verified that stiffness of quad elements for fluid should be a million times smaller than the stiffness of structural elemenets.

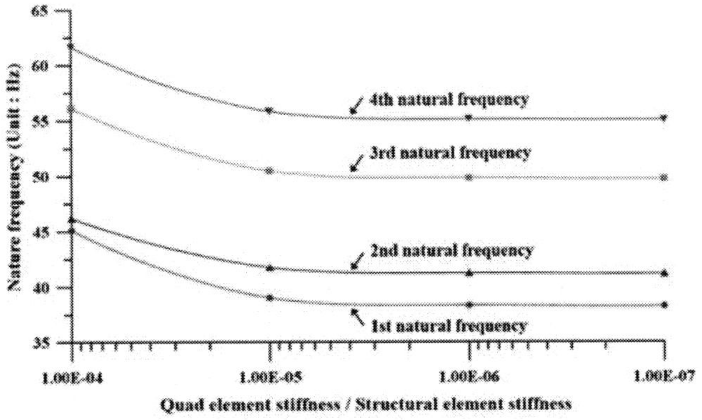

Fig. 1 Influence of stiffness.

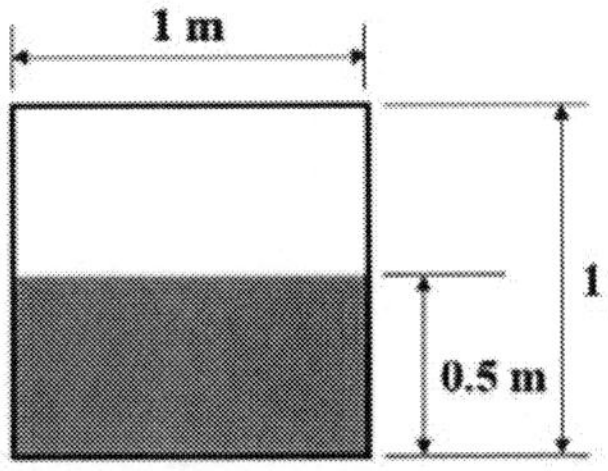

Fig. 2. Principal dimension of half-filled tanks.

It is easily expected that they show the identical results. However, the results varied depending on mass arrangement. In the case of quad elements, mass is automatically calculated by considering the area, and the identical mass is placed on each node which composes an element. If one element is adjacent to the other elements, the mass of adjacent node is overlapped. Therefore, the mass of node in quad elements is determined depending on the area of the element, the number of node in one element and the number of contiguous elements. However, the mass of nodes using point mass does not consider those items, every node are the same. Mass on one node is calculated that the total mass is divided by the total number of node.

In regard to this problem, the tank with half-filled water were used as a numerical example, as shown in Fig. 2. Density of tank and water were given as ρ_t=7850 kg/m^2 and ρ_w=1000 kg/m^2. Point mass method is divided into two parts : one is equally distributed, the other is well arranged by considering the mentioned items. Table 1 shows comparisons of methods. A little gap is shown between uniformly posed point mass and quad element. Otherwise, the same result is obtained between well-arraged point and quad element. Therefore, it is verified that mass arrangement should be considered the area of structural elements. It is not easy to consider various shaped element as the structure becomes more complicated.

Table 1. Influence of mass arrangement (Unit: Hz).

Method		1st	2nd	3rd	4th	5th
Point element	Uniformly posed	38.684	39.928	48.585	53.893	69.440
	Well-arranged	38.250	41.190	49.762	55.049	69.397
Quad element		38.250	41.191	49.762	55.049	69.397

4 Ship vibration analysis

A 1500-ton fishery research vessel was used for the target for ship vibration analysis. Principal dimensions and the main information are summarized in Table 2. A half of finite element model is shown in Fig. 3, and weight of the fluide entities is modelled like as Fig. 4. In the case of point mass method, mass of fluid entities was uniformly distributed in each node to avoid expending time.

Loading conditions for vibration analysis were normal ballast arrival condition and scantling departure condition. Details of conditions are specified in Table 3.

Table 2. Principal dimensions of the target vessel.

Item	Description
Type of ship	Fishery research vessel
Displacement	1500.0 Ton
Length over all	75.7 m
Length between perpendicular	64.5 m
Breadth (Moulded)	13.8 m
Depth (Moulded)	7.4 m
Draft (Scantling)	4.8 m
Number of blades	4/5 EA (front/aft)
1st harmonic frequency	13.2/16.4 Hz (front/aft)

Fig. 3. Finite element model.

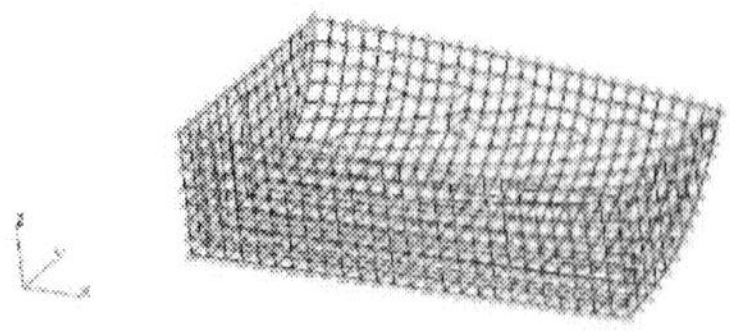

Fig. 4a. Point mass elements.

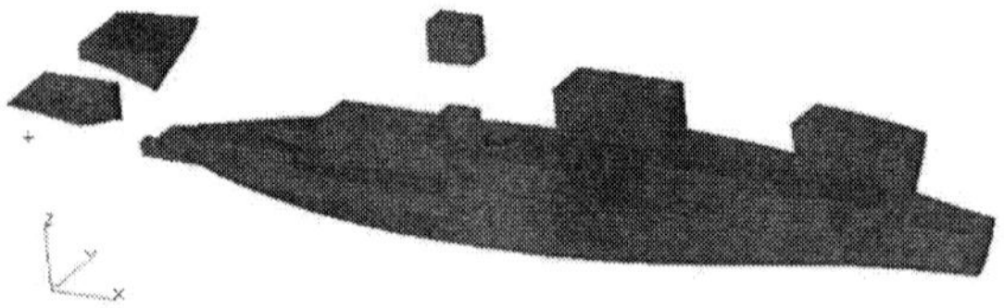

Fig. 4b. Quad elements.

Table 3. Details of conditions.

	Normal ballast arrival	Scantling departure
Deadweight	306.4 Ton	717.2 Ton
Lightweight	1830.0 Ton	1830.0 Ton
Displacement	2136.4 Ton	2547.2 Ton
Draft F.P.	3.8 m	4.9 m
Draft A.P.	4.8 m	4.9 m
Draft equivalent	4.4 m	4.9 m

Free vibration analysis is conducted to avoid problomatical resonance with major excitational sources such as propellers and the main engine. In this study, vibration analysises were conducted by NASTRAN [4]. Table 4 represents the results of free vibration analysis. The possibility of resonance induced by the propellers was significantly low since the natural frequencies of global modes was notably lower than 1st harmonic frequencies of prepellers. On the other hand, the difference of natural frequencies between two methods was under 0.06 percent. It is marginal because tank weight is a small portion of the whole ship.

Table 4. Natural frequency obtained from free vibration analysis (Unit: Hz).

Order	Normal ballast arrival condition			Scantling departure condition		
	Point	**Quad**	**Err. (%)**	**Point**	**Quad**	**Err. (%)**
Ver. 2-Node	3.1639	3.1638	0.0032	2.9919	2.9932	0.0434
Ver. 3-Node	5.6129	5.6142	0.0232	5.3474	5.3501	0.0505
Ver. 4-Node	8.0029	7.9986	0.0538	7.5196	7.5227	0.0412
Hor. 2-Node	5.4149	5.4137	0.0222	5.1881	5.1888	0.0135

The ship used in this study is propelled by the electric propulsions known as Azimuth thruster. Since main power is created from generator engines, an axial force conveyed by shaft does not exist. Thus, major excitational sources are both guide forces induced the generator and fluctuation pressure induced by the proppellers. The engine equipped in the target ship does not have any external force or moment. Hence, the engine only induces guide forces such as 4th and 8th order moments known as X and H-moments. 4th order moment, given as 20.8 kN·m, was only considered since 8th order moment, given as 3.0 kN·m, was negligible. Fluctuation pressure induced by the propeller was considered as 4 kPa which figure is commonly used as an upper limit for special ships. Damping coefficients normally expressed as a percentage of critical damping are applied according to Guidance notes published by Lloyd's Register [5]. Its value linearly increases from 1 percent of critical damping to 3 percent as the frequency rises from 5 Hz to 20 Hz.

Forced vibration analysis was conducted by the modal superposition method utilizes the results of free vibration analysis of NASTRAN [3]. The values, specified in Table 5 is maximum values on each observation point. The observation points were selected to verify comfortability in the manner of ISO regulation [6]. Root Mean Square(RMS) velocity is principal criteria for assessment as stated by the ISO 6954:2000 which was enhanced in the manner of considering average vibration velocities. All dynamic structural responses by the global modes were satisfied the lower limit for working comfortability of crews in ISO regulation [6], given as 4 mm/s. The maximum response of funnel top was 6.3655 mm/s. Its value is lower than the recommended limit, given as 30mm/s, to avoid fatigue fracture [5].

Moreover, the difference of dynamic responses between point elements and quad elements were within 5 percent. The error is higher than the results of free vibration analysis because these values are reflected in the whole modes of vibration.

Table 5. RMS velocities in both conditions (Unit: mm/s).

Method	Dir.	Propulsion rm		AFT. end		Funnel top		Eng. control rm		Wheel house	
		Nor.	Scant.	Nor.	Scant.	Nor.	Scant.	Nor.	Scant.	Nor.	Scant.
Point element	V.	0.899	2.844	3.306	0.946	2.206	0.867	1.876	0.494	0.683	0.217
	L.	0.054	0.153	0.518	0.185	2.377	0.608	0.414	0.135	0.235	0.127
	T.	0.409	1.550	0.789	0.228	6.367	2.658	1.019	0.273	0.627	0.153
Quad element	V.	0.907	2.860	3.327	0.947	2.205	0.865	1.926	0.490	0.685	0.218
	L.	0.055	0.155	0.521	0.186	2.385	0.607	0.419	0.140	0.236	0.121
	T.	0.414	1.576	0.789	0.230	6.366	2.680	1.036	0.277	0.638	0.154
Error (%)	V.	0.845	0.549	0.634	0.116	0.045	0.173	2.591	0.789	0.263	0.368
	L.	0.552	1.033	0.538	0.542	0.319	0.066	1.169	3.402	0.424	4.341
	T.	1.346	1.637	0.063	0.877	0.024	0.801	1.631	1.427	1.740	0.786

5 Conclusion

The vibration analysis of the fishery research vessel has been conducted to compare point mass elements with quad elements. The difference of natural frequencies was under 0.06 percent. It verifies that domination from uniform distribution of point mass was negligible. However, the maximum error of dynamic structural responses was about 80 times higher than the error of free vibration analysis because errors triggered by uniformly distributed point mass elements were accumulated. Hence, using quad elements is more suitable methodology for ship vibration analysis. Also, not only there was not possibility of resonance problems but also all dynamic responses were acceptable to the IMO regulation.

This study was successfully presented and published with the support from BK21 plus MADEC Human Research Development Group, Republic of Korea. Authors gratefully acknowledged to Korea Maritime Services Company for research fund. This paper has been presented at the 2nd International Joint Conference on Advanced Engineering and Technology, 24-25 August 2017, Bali, Indonesia.

Reference

1. T.S. Kim, H.N. Rhee, D.B. Yoon, J.H. Park, Analysis of vibration characteristics of submerged tubes using computational fluid structure interaction analysis, The Proc. of Korean Soc. for Noise and Vib. Eng., 684-685 (2013)
2. Lesser, M. B., J. A. Lewis, Scattering of surface waves by a submerged circular cylinder, J. Math. Phys., **6**(8), 1231-1243 (1972)
3. S.H. Choi, K.S. Kim, S.W. Son, Analytic and experimental study on vibration characteristics for rectangular tank structure filled with fluid, The J. of Korean Soc. for Noise and Vib. Eng., **12**(3), 195-203 (2002)
4. SIEMENS, *Advanced dynamic analysis user's guide* (2014)
5. Lloyd's Register, *Guidance notes general overview of ship structural vibration problems* (2015)
6. International Organization for Standardization, *ISO 6954: Mechanical vibration – guidelines for the measurement, reporting and evaluation of vibration with regard to habitability on passenger and merchant ships* (2000)
7. B.W. Kim, J.H. Kyoung, S.Y. Hong, S.K. Cho, I.W. Lee, Investigation of efficiency of starting iteration vectors for calculating natural modes, The J. of Korean Soc. for Noise and Vib. Eng., **15**(1), 112-117 (2005)

Design and Construction of Spud Housing for Floating Dock

Hasanudin Hasanudin[1,*], and *Ardi Nugroho* Yulianto[1]

[1]Department of Naval Architecture, Institut Teknologi Sepuluh Nopember, 60111, Surabaya, Indonesia

Abstract. A floating dock is a type of pontoon for dry docking ships, possessing floodable buoyancy chambers. The Floating docks have more advantages than the other type of ship docking for maintenance. The purpose of the research study is to design and calculate of spud housing construction for floating dock. The spud housing is used to keep of floating dock in the safe condition. Mild steel material is used to construction of spud housing with yield point 235 x 10^6 N/m^2. Design calculations were done by finite element analysis (FEA). The minimum stress is 1042.36 N/m^2 and minimum deflection is 0.447 mm. The maximum stress is 0.142 x 10^9 N/m^2 and maximum deflection is 4.019 mm. The maximum stress is not more than yield point of material.

1 Introduction

1.1 Floating dock

The floating dock is also modern in concept when compared with the history connected with graving docks and marine railways. The earliest known use of the floating dock principle was about 1700 when a British Royal Navy Captain, lacking other means of docking facilities, bought an old hulk, gutted it and fitted it with a watertight stern gate, docked his ship therein and pumped the water from the hulk [1].

A floating dock is a type of pontoon for dry docking ships, possessing floodable buoyancy chambers and a "U"-shaped cross-section that show on figure 1. The Floating docks have more valuable than the other type of ship docking. The walls are used to give the dry dock stability when the floor or deck is below the surface of the water. When valves are opened, the chambers fill with water, causing the dry dock to float lower in the water. The deck becomes submerged and this allows a ship to be moved into position inside. When the water is pumped out of the chambers, the dry dock rises and the ship is lifted out of the water on the rising deck, allowing work to proceed on the ship's hull. A typical floating dry dock involves multiple rectangular sections. These sections can be combined to handle ships of various lengths, and the sections themselves can come in different dimensions. Each section contains its own equipment for emptying the ballast and to

* Corresponding author: hasanudin@na.its.ac.id, ardi.nugroho@na.its.ac.id

provide the required services, and the addition of a bow section can facilitate the towing of the dry dock once assembled. For smaller boats, one-piece floating dry docks can be constructed, potentially coming with their own bow and steering mechanism [1].

The Floating docks have more advantages than the other type for ship maintenance. They offer large construction advantages and thereby contribute greatly to faster mobilization. They can be constructed in areas where materials and labor are available and then moved to the working site which may be considerably removed from the site of construction. Furthermore, the time required for construction of floating dock is less than that for a comparable size graving dock. In addition, our present pattern for all large docks is to construct them in sections. This method of building lends itself to the use of building contractors at different sites working upon the same dock simultaneously and thereby further reducing considerably the time of construction [1].

Fig. 1. Floating Dock.

1.2 Spud housing

Floating docks are moored in any one of several ways. In general, docks used in a yard are moored to a pier by means of spuds. This method of mooring consists essentially of at least 3 jaws connected to a pier or to dolphins. Attached to the dock in a vertical position and riding up and down in the jaws as the dock is flooded and pumped, are three long mooring spuds in the general form of H-beams. These spuds extend far enough down the side of the dock so that they will not come out of the jaws when the dock is at light draft and they extend far enough up the side of the deck so that the jaws remain engaged at maximum submerged. The centre spud resists forces parallel to the longitudinal centreline of the dock while the two end spuds resist only forces in a transverse direction. Use of the spud mooring is limited to docks where the required depth of water can be obtained next to a pier or dolphin. Figure 2 shows spud housing of floating dock [1].

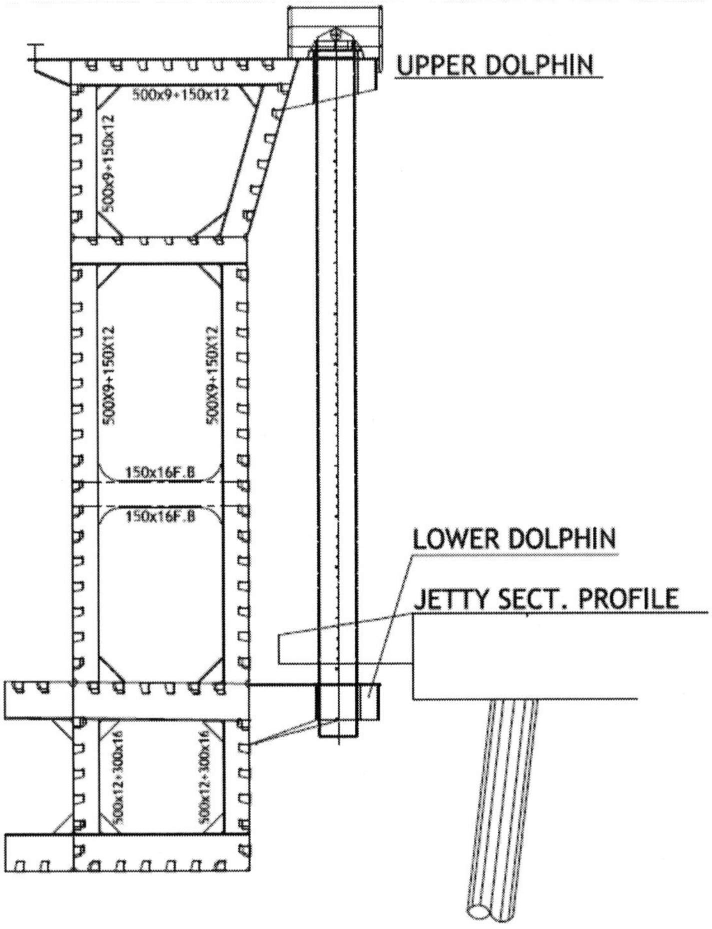

Fig. 2. Spud housing of floating dock.

1.3 Finite Element Analysis

Finite element analysis (FEA) is a computational technique used to obtain approximate solutions of boundary value problems in engineering. A boundary value problem is important for this analysis. It is a mathematical problem in which one or more dependent variables must satisfy a differential equation everywhere within a known domain of independent variables and satisfy specific conditions on the boundary of the domain. It is also sometimes called field problems. Figure 3 shows the general techniques and terminology of finite element analysis. Figure 3 depicts a volume of some material or materials having known physical properties. The volume represents the domain of a boundary value problem to be solved. For example, at this point, it assumes a two-dimensional case with a single field variable P(x, y) such that a known governing equation is satisfied exactly at every such point [2].

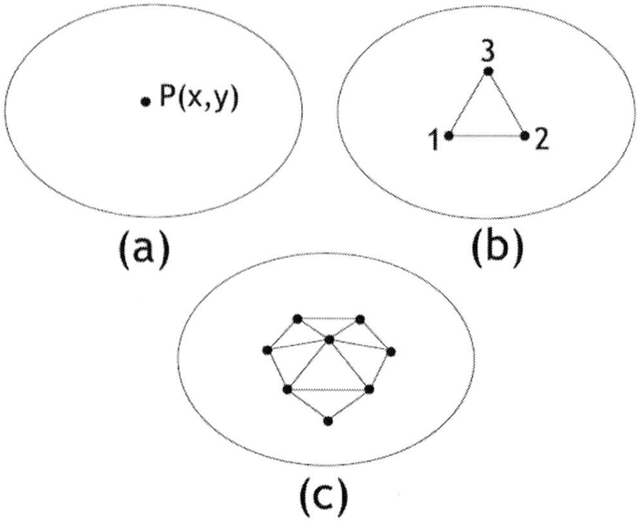

Fig. 3. (a) A general two-dimensional domain of field Variable P(x, y). (b) A three-node finite element defined in the domain. (c) Additional element.

2 Method

2.1 Model preparation

Preparation of the model is done with the help of software modeling with AutoCAD. Construction of spud housing will be placed in the side of floating dock as seen in figure 4. There are some modification from original design of floating dock to support spud housing structure that will be installed on the deck of floating dock. The structure model is shown in Figure 5.

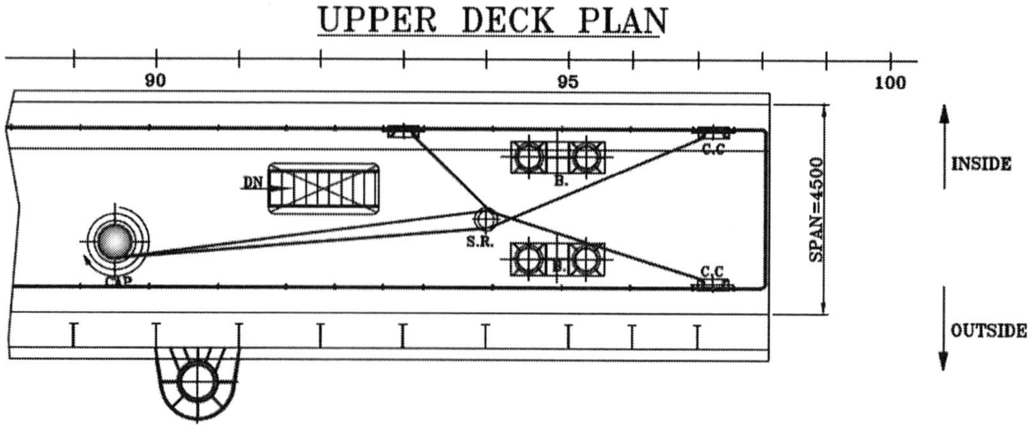

Fig. 4. Deck arrangement of floating dock for spud housing.

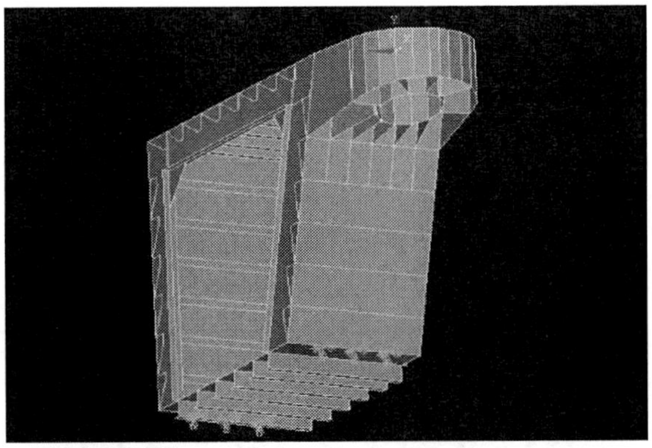

Fig. 5. Construction of spud housing.

2.2 Simulation

Finite element methods are numerical methods for approximating the solutions of mathematical problems that are usually formulated so as to precisely state an idea of some aspect of physical reality [3,4]. A finite element method is characterized by a variational formulation, a discretization strategy, one or more solution algorithms and post-processing procedures. Examples of variational formulation are the Galerkin method, the discontinuous Galerkin method, mixed methods, etc. There are various numerical solution algorithms that can be classified into two broad categories; direct and iterative solvers. These algorithms are designed to exploit the sparsity of matrices that depend on the choices of variational formulation and discretization strategy. There are some very efficient post-processors that provide for the realization of super-convergence. There are some step of this method. It has been divided by Pre-processor, Processor/Solver, Post-processor. Pre-processor phase is the process that prepare for dimension of model for 3D model that use to calculation in the solver phase [5]. Processor/Phase is the process that calculate based on model preparation, boundary, load and the other factor. Post-processing procedures are designed for the extraction of the data of interest from a finite element solution. In order to meet the requirements of solution verification, postprocessors need to provide for a posteriori error estimation in terms of the quantities of interest. When the errors of approximation are larger than what is considered acceptable then the discretization has to be changed either by an automated adaptive process or by action of the analyst. Post-processor shown graph, curve, contour [6]. For Finite Element Analysis, element model should follow as the following requirements:

a. The global Finite element model should be made of shell elements for the hull plating and frame in the hull.
b. Stiffeners may be modelled as a beam element provided the effective plate flange width comply with rules.

2.3 Material properties

Below the material properties and load of the hull structure which applied in the strength analysis:

a.	Modulus of Elasticity	= 210 x 109	N/m^2
b.	Poisson's Ratio	= 0.32	
c.	Steel Density	= 7850	kg/m^3
d.	Material Grade	= A (Mild Steel)	

e.	Yield Point	= 235 x 10^6	N/m^2
f.	Elongation Min	= 22 %	
g.	Pressure load (20T)	= 3001192	N/m^2

3 Analysis of spud housing structure

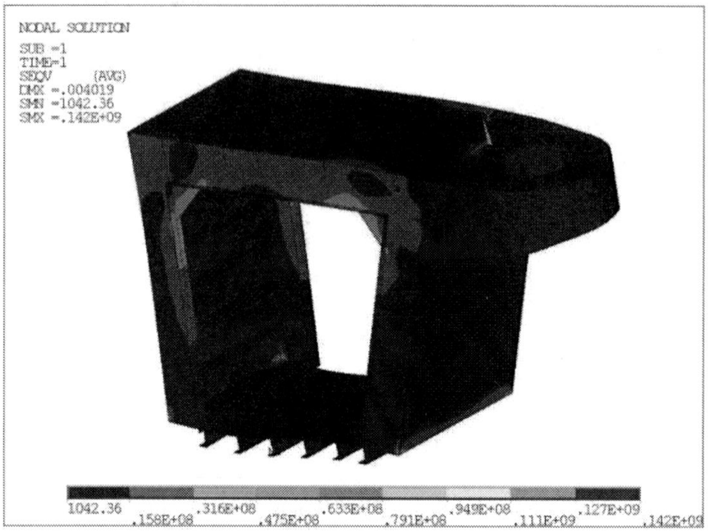

Fig. 6. Stress of spud housing structure

Figure 6 shows stress of spud housing structure from finite element calculation. It can be seen the result of finite element analysis in the post-processing for contour of von mises stress and deflection. In this simulation, it gives for load 20 ton based on self-weight of spud housing structure and additional load from environment factor. For this model, it uses material grade A with yield point 235 x 10^6 N/m^2. It can be seen that the value of minimum stress 1042.36 N/m^2, the value of maximum stress 0.142 x 10^9 N/m^2, the value of minimum deflection 0.447 mm, the value of maximum deflection 4.019 mm.

4 Conclusion

In this paper, finite element method is used to get the maximum stress of spud housing construction. The maximum stress of spud housing construction is not more than yield point of mild steel (grade A). It has minimum stress 1042.36 N/m^2, maximum stress 0.142 x 10^9 N/m^2, the minimum deflection 0.447 mm and the maximum deflection 4.019 mm.

References

1. V. E. Cook, *General discussion of floating drydocks* **1**, 289-306(1957)
2. D. V. Hutton, *Fundamentals of finite element analysis*, **2** (2004)
3. K. J. Bathe, *Numerical methods in finite element analysis*, **11** (1976)
4. J. Chaskalovic, *Finite elements methods for engineering sciences*, **2** (2000)
5. O. C. Zienkiewicz, R. L. Taylor, *The finite element method*, **1** (2000)
6. A. N. Yulianto, D. Atmono, M. S. Arif, *Design and construction of crawler crane seating for floating dock*, **3** (2015)

Impact phenomena assessment: Part I – Structural performance of a tanker subjected to ship grounding at the Arctic

Aditya Rio Prabowo[1], *Jung Hoon* Byeon[1], *Hyun Jin* Cho[1], *Jung Min* Sohn[1,2,*], *Dong Myung* Bae[2] and *Joung Hyung* Cho[1]

[1]Pukyong National University, Marine Convergence Design, 48513 Busan, Korea
[2]Pukyong National University, Naval Architecture and Marine Systems Eng., 48513 Busan, Korea

Abstract. The remarkable influence of the global warming to Arctic environment opens a possibility to conduct a voyage from Asia to Europe through the Northern Sea Route (NSR). This option is considered as a decent solution to reduce fuel consumption and increase time efficiency in delivering cargo to the designated destination. However, this alternative comes with a challenge to keep safety of ship structures against an impact with obstacles which are possibly encountered at the Arctic. Solid ice can be considered as a serious threat to the double bottom of ship structures in impact phenomena, especially ship grounding. In this work, a series of grounding calculations are conducted to produce estimation of structural crashworthiness during interaction between double bottom and conical type ice. Material characteristics based on tensile testing of polar class material are applied to calculation and compared with non-polar steel.

1 Introduction

In recent decades, a global warming has become an attractive issue for many countries. High increment of carbon dioxide (CO_2) concentration in the air is followed by the remarkable changes in temperature. In southern region, this phenomenon has a strong influence to human activity. During summer, the greenhouse effect influences water crisis in Asia and Africa. On the other hand, a serious challenge is faced to the society in northern region as high temperature during summer reduces Arctic sea-ice extent [1]. International shipping societies and parties show their interest to observe this opportunity as an alternative route to reach Europe from Asia. A comparative study indicates that the operational cost and sailing time can be considerably reduced using Arctic Sea route [2]. Despite of this positive tendency, a negative influence of ice and other obstacles cannot be neglected. Maritime accidents in forms of collision and grounding impacts are considered as high risks since casualties (e.g. human life loss, oil spill, cargo lost) are massive. Several studies [3-4] for maritime accidents and impact phenomena have been conducted to observe structural crashworthiness of ship under various conditions. However, the analysis conditions have not been addressed to reflect the condition of the Arctic.

* Corresponding author: jminz@pknu.ac.kr

The aim of this study is to assess crashworthiness of a tanker ship against the grounding. A temperature level of the Arctic is considered based on the survey and reports of related parties. Series of tensile tests in the low temperatures [5] are used as a reference data to define mechanical properties of ship materials. Numerical analysis is conducted under determined scenarios to provide structural behaviour during and after impacts.

2 Water temperature and polar classification

The effect of global warming on sea temperature at the Arctic has caused changes in almost every physical climate system. As reported by Hassol [1], there are at least three climate trends in this region. The first is the rising temperature. In winter, the temperature is increased in Alaska and western Canada approximately 3-4°C over the past half-century. The second is declining snow cover. It has declined about 10% over the past 30 years. The third is diminishing lake and river ice. The tendency is concluded that later freeze-up and earlier break-up of river and lake ice have combined to reduce ice season one to three weeks in several areas. Emissions of daily activity also cause heat up and more than 93% of this temperature increment goes to the ocean. In the investigation by Northern Europe countries, it was concluded that sea temperature is -40°C in January and can drop until -50°C over large parts of the Arctic in winter. Emerson and Lahn [6] presented that winter temperature reaches the most extreme temperature -50°C at the Laptev Sea.

In the initial requirement, International Association of Classification Society (IACS) divides the Polar Class into seven classes [7] depending on the operating season and ice conditions. A simpler classification is presented in Fig. 1 as recommended by International Maritime Organization (IMO) which divides ships based on the water condition and structural risks [8]. Based on this regulation, both IACS's polar class designation and IMO's C-class which is not applied by ice-strengthening can operate at the Arctic.

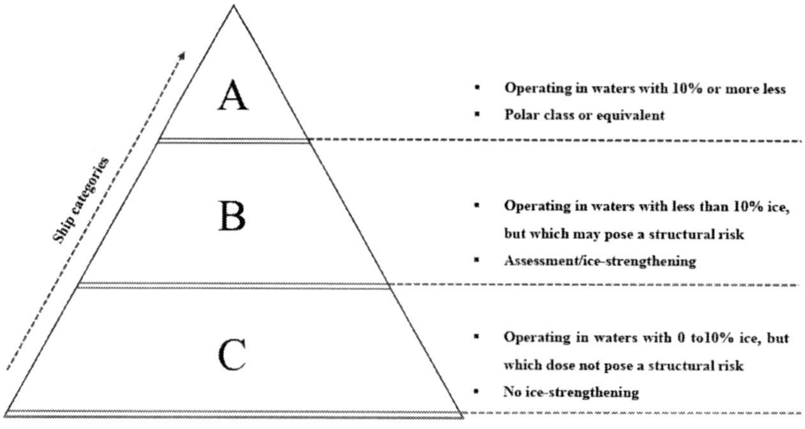

Fig. 1. Ship classification on the Polar Regions. Illustration is presented based on reference [8].

3 Numerical model and analysis configuration

3.1 Ship dimension and finite element setting

A C-class tanker vessel based on IMO is deployed in this study as the target ship. The tanker has principal dimensions as follows: Length L = 144 m, Width B = 22.6 m and Depth H = 13.95 m. Double bottom structure is designed for a numerical simulation with approximately 15 m or equal with two cargo tanks in order to avoid interference of boundary condition. A grounding scenario is simulated by finite element codes ANSYS LS-

DYNA [9]. Considering complex behaviour of the grounding process, the element formulation with fully integrated type of the Belytschko-Tsay is implemented on the structure. The material characteristics for the ship are idealized as plastic-kinematic model with kinematic hardening. Properties of a typical HSLA steel are given as follows: Elastic modulus E = 210000 MPa, Yield strength σ_Y = 315 MPa, Poisson's ratio n = 0.3, and Density ρ = 7850 kg/m^3. Nonlinear parameters are described using the Cowper-Symonds strain rate c = 3200 /s and P = 5 [10]. These properties are denoted as the *non-polar class*.

As introduced in previous sections, the study is addressed to observe structural crashworthiness at the Arctic. Therefore, results from tensile tests in extreme temperatures [5] are used to investigate the structural behaviour under grounding impacts. The material is tested under three different temperatures, including room temperature (RT), (-30°C) and (-50°C). The results are adequate to represent the Arctic conditions as the most extreme temperature during winter is included in this test. The material of the tensile tests will be denoted as the *polar class* and for the temperatures RT, (-30°C) and (-50°C), ordinal numberings I, II and III are given consecutively. Variety of material strength in each temperature is summarized as: the polar class I ($\sigma_{Y-\mathrm{I}}$) = 385.05 MPa, polar class II ($\sigma_{Y-\mathrm{II}}$) = 425.27 MPa and polar class III ($\sigma_{Y-\mathrm{III}}$) = 448.01 MPa. During numerical simulation, an indenter with applied ice material is deployed. The ice is treated as a solid-rigid material. This approach is a conventional but it is considered good enough to estimate the most extreme damage extents in the worst-case scenario. It is because that the indenter does not undergo deformation and can produce steady-damage geometrically during penetrations. Properties of the ice are given as follows: Poisson's ratio n = 0.27, Young's modulus E_x = 1000 MPa and density ρ = 960 kg/m^3. As indicated in pioneer works [3-4], the failure criterion is necessary to be involved in calculation. The Peschmann criterion [11] is applied in this work which concludes the element length of the engineering models is 150 mm and failure strain for the double-bottom structure is 0.21. The friction between ice and steel is considered very low with the static friction coefficients η_s = 0.06 and kinematic friction coefficient η_k = 0.05 are applied.

3.2 Collision scenario

Grounding is defined as the interaction process between the structure and obstacle during the movement process. In this process, the involved models are classified as the *structure* for the double bottom and *obstacle* for the conical ice. A specific target on the structure is denoted as the *impact area*. The structure is given a constant velocity 7 m/s to move on the positive longitudinal direction (x *axis*) according to the Cartesian coordinate system and approaching the obstacle.

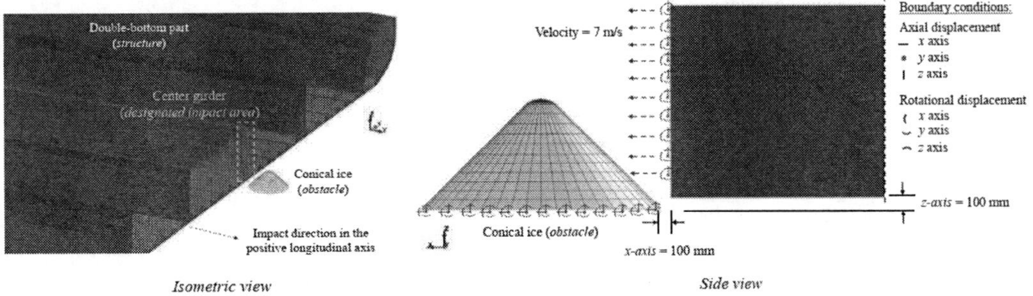

Fig. 2. Model of the double bottom tanker. The yellow highlights the target part in grounding process.

Initial distance (Fig. 2) between the structure and obstacle is determined 100 mm in both longitudinal and vertical directions (*x* and z *axes*). Axial constraint is applied on the end of the structure model. Rotational movement of the bottom and inner bottom plates in this location is also restrained to keep the models stand-still on the centerline. Evaluation is focused on the occurred energy and strain behaviour on the structure due to the grounding.

4 Results and Discussion

During grounding, it is fundamentally assumed that the most remarkable damage will be observed in the bottom plate. Its position on the lowest part of the double bottom makes the bottom plate experiences various damages, such as tearing damage in head-on grounding. Even though the center girder was included as the designated impact area, Fig. 3a indicates that the highest energy was achieved by the bottom plate. In this discussion, the internal energy is determined as a main parameter which reflects the amount of energy that is necessary to plastically deform the involved structures in impact [3,4,10]. More detail observation is conducted on the polar class materials while failure took place. At the first contact between the obstacle and structure (Fig. 3a), both center girder and bottom shell showed similarity in pattern and magnitude of the internal energy. In terms of the center girder as the designated impact area, a difference was observed after 0.09 s which indicated that the transition from a deformation state to the failure condition. On the other hand, the bottom shell experienced higher damage than the center girder, the transition was began earlier than the center girder. This phenomenon occurred due to the increasing size of lower part of the conical ice. The larger diameter produced earlier contact and made the structure experienced early failure. In extreme temperatures, the Polar classes II and III showed relatively distinction approximately after 0.12 s for the center girder and 0.075 s for the bottom shell. It is concluded that the structural resistance against the impact for the ship using Polar class material is more effective during operation in low temperature region.

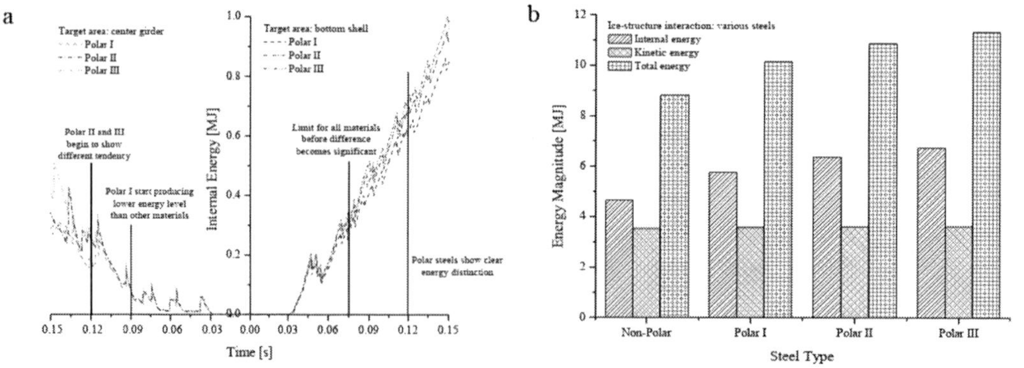

Fig. 3. Crashworthiness criteria: (a) internal energy of the girder and shell under several temperatures and (b) statistical data of the calculated energies after impact with different material properties.

The statistical comparison in terms of energy is presented in next discussion for the non-polar and polar materials. There are three main energies which are shown in Fig. 3b, namely internal, kinetic and total energies. As mentioned in fundamental concept of physic, the kinetic or kinematic energy is an amount of the energy to move an object from one location to others. Since constant velocity was applied in the methodology, the kinetic energy is observed similar for all materials. This result is a satisfactory in confirming grounding process since the kinetic energy should not have significant difference if the constant velocity is applied. The total energy is roughly translated as the summary of the kinetic and internal energies in impact. As the constant velocity was implemented, the total energy highly depends on by the internal energy. According to tensile test in extreme temperatures, polar class materials have a characteristic to increase their strength as a temperature goes down [5]. This characteristic was indicated a match with the energy magnitude that the total energy for the polar class III in the lowest temperature (-50°C) was the highest and it was harder to be plastically deformed in grounding than other polar material. The overall results in terms of total energy also conclude that the polar class material is harder than the marine steel (non-polar class) which is intended for ship in sea with normal temperature.

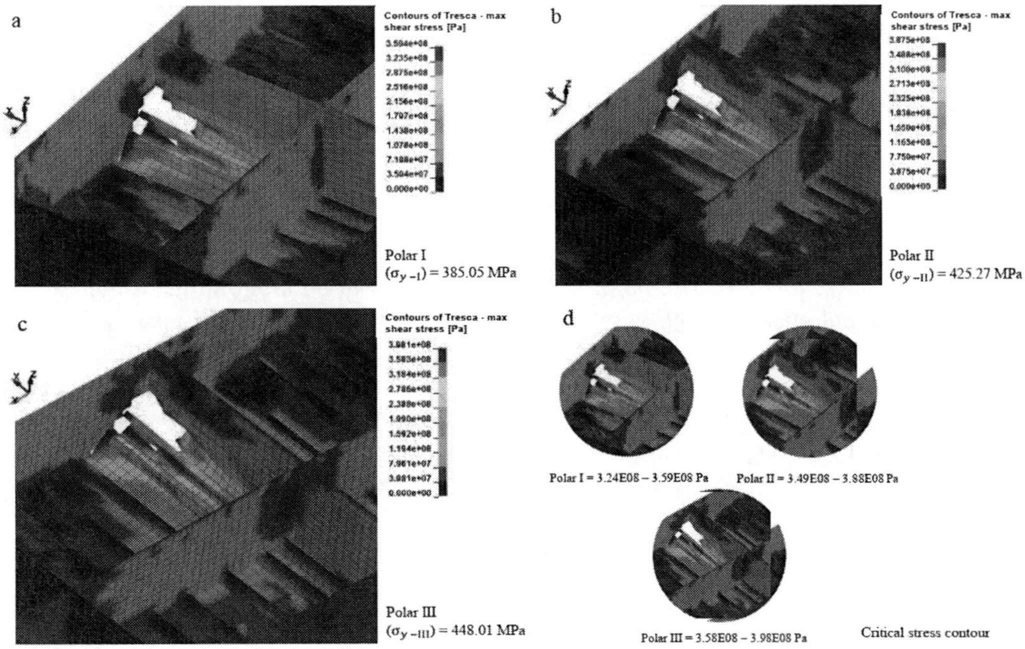

Fig. 4. Shear stress (Tresca) accounting for the selected materials: (a) Polar I, (b) Polar II, (c) Polar III and (d) summary of the maximum stress values.

After comparison in terms of energy, the discussion is directed to the extent of damage after grounding. In order to describe this behavior, Tresca stress contour is chosen to be discussed and presented in Fig. 4. Fundamental expression of the Tresca is also used as failure criterion for structures under various loads together with the von Mises formulation. Historically, the Tresca was considered more fundamental, but the von Mises was seen more appealing for convenient approximation [12]. Similar works have used the von Mises as found in [4,13] which lead this work to observe Tresca contour. In comparing the material strength (Fig. 4a-c), the Polar III is superior to the Polar I and II with difference approximately 9.45% and 5.06% consecutively. The highest stress on the structure indicates that Tresca stress reached its peak on the bottom shell and bottom stiffener. This behavior made these components would experience early failure if the grounding was continued. Distinction on the peak value as shown in Fig. 4d is found 7.47% and 2.51% for Polar I-III and Polar II-III. The highest stress on the bottom area has successfully confirmed pioneer discussion on the internal energy that the shell produced higher energy than the center girder in grounding. Correlation between these two behaviors can be concluded that harder an object to be plastically deform in impact will experience higher stress level. Accounting for the failure prediction, if grounding time is expanded and similar indenter with this work is applied, early failure of the bottom shell is highly expected. Since conical indenter was applied in analysis, the lower part of the structure (in this case is bottom shell and bottom stiffener) will contact with the largest diameter of the indenter. Thus, the highest stress occurred on the bottom shell and among of all applied materials, the experienced stress was produced during a ship voyage was conducted on the most extreme temperature at the Arctic. Based on this study, it can be stated that contribution of the polar class material is relatively high to the double bottom strength depends on the environment temperature. In order to increase structural strength against impact, polar class material can be applied on non-polar class ships (C-class) which operate through the North Sea passage, as adequate replacement of ice-strengthening requirement as explicitly stated for A and B-class on the polar operating ship categories [8].

5 Conclusion

This work was addressed to observe crashworthiness criteria and failure progress of the applied polar class material on the double bottom structure. Comparison with non-polar steel was presented in the initial discussion which was followed by a series of discussion in terms of temperature influence to the energy and stress. It was concluded that the material for operation in extreme regions such as Northern Sea Route has stronger characteristic than a ship material for sailing in general sea territory as can be found in the Southern Sea Route. The internal energy which is related to structural strength against impact provides a conclusion that resistance by the polar class material reaches its peak during ship operation in winter since sea temperature drops to 50°C below zero in this season. Stress behaviour of three materials on the double bottom structure satisfyingly confirmed strength-stress relationship in the grounding. Consideration to apply polar class material on the c-class ships which operate at the Arctic is encouraged to increase its safety against impact.

Reference

1. S.J. Hassol, *Impact of a warming arctic* (Cambridge University Press, Cambridge, 2004)
2. W. Østreng, K.M. Eger, B. Fløistad, A.J. Dahl, L. Lothe, M.M. Larsen, T. Wergeland, *Shipping in arctic waters, A comparison of the northeast, northwest and trans polar passages* (Springer-Verlag, Berlin, 2013)
3. A.R. Prabowo, D.M. Bae, J.M. Sohn, A.F. Zakki, B. Cao, J.H. Cho, Thin-Walled Struct. **115**, 225 (2017)
4. A.R. Prabowo, B. Cao, D.M. Bae, S.Y. Bae, A.F. Zakki, J.M. Sohn, Lat. Am. J. Sol. Struct. **14**, 1106 (2017)
5. D.K. Min, C.K. Shim, D.W. Shin, S.R. Cho, J. Soc. Nav. Arc. Korea **48**, 171 (2011)
6. C. Emmerson, G. Lahn, *Arctic opening: opportunity and risk in the high north* (Lloyd's, London, 2012)
7. IACS, *Requirements concerning polar class, Rev.2* (International Association of Classification Societies, London, 2016)
8. D.K. Park, D.K. Kim, J.K. Seo, B.J. Kim, Y.C. Ha, J.K. Paik, Ocean Eng. **102**, 197 (2015)
9. ANSYS, *ANSYS LS-DYNA user's guide* (ANSYS, Inc., Pennsylvania, 2017)
10. O. Ozguc, P.K. Das, N. Barltrop, Mar. Struct. **18**, 511 (2005)
11. E. Lehmann, J. Peschmann, Mar. Struct. **15**, 429 (2002)
12. A.R. Prabowo, S.J. Baek, H.J. Cho, J.H. Byeon, D.M. Bae, J.M. Sohn, Lat. Am. J. Sol. Struct. **14**, 1345 (2017)
13. A. AbuBakar, R.S. Dow, Int. J. Sol. Struct. **50**, 623 (2013)

Calcium soap from palm fatty acid distillate for ruminant feed: $Ca(OH)_2$ as calcium source

Lienda A. Handojo*, *Antonius* Indarto, *Dian* Shofinita, *Anggina* Meitha, *Rakhmawati* Nabila, *Harry* Triharyogi, *Maulana* G.A. Hakim, and *Muhammad* R. Saadi

Department of Chemical Engineering, Faculty of Industrial Technology, Institut Teknologi Bandung, 40132 Bandung, Indonesia

Abstract. PFAD can be utilized as raw material for producing calcium soap, a kind of ruminant feed supplement. It gives benefits to the ruminants' health and increases the quantity and quality of cattle's milk. This research evaluated the effect of stoichiometric mole ratio of $Ca(OH)_2$ to PFAD and water temperature on calcium soap production using PFAD and calcium hydroxide as raw materials by applying modified fusion method. The result showed that the acid value of products decreased as the mole ratio of $Ca(OH)_2$ to PFAD increased and the low water temperature lowered the acid value of products. The highest fatty acid conversion (98%) was achieved at the stoichiometric mole ratio of 3. However, the use of such high mole ratio is not preferable because the product will have excessive calcium, which may disturb the absorption of other minerals in ruminant's digestion. Thus, further research related to operating condition and the use of other calcium sources in the calcium soap production is still required.

1 Introduction

Indonesia is the largest world palm oil producer, with 32.5 million tonnes of palm oil annual production (Indonesian Palm Oil Producers Association, 2016). This is equal to approximately 54% of world's total palm oil production (USDA, 2016). Refining process of crude palm oil produces palm fatty acid distillate (PFAD), a by-product which contains free fatty acid. Countries such as USA and UK used PFAD as animal feed, laundry soap, and raw materials for oleochemistry and biofuel industries [1]. Utilization of PFAD in Indonesia is not optimum and it is predominantly exported without further processing.

PFAD can be utilized as raw material of animal feed supplement after reacted with calcium in the form of protected fat [2]. It functions as an energy source for cattle and has positive effects on milk productions [3-5] as well as the fertility of cattle [5, 6]. Giving calcium soap 0.45 kg/day to the cows can increase milk production by 3-8% [4]. Fatty acids need to be converted into soap before being consumed, as the direct consumption of fatty acids by cattle can disrupt the activity of digestive bacteria in cattle rumen [2].

* Corresponding author: lienda@che.itb.ac.id

Based on above information, this study was conducted to evaluate the production of calcium soap from PFAD. Modified fusion reaction allows fatty acid to react with calcium source in the form of its oxide and hydroxide [7]. This research evaluated the effect of calcium source of $Ca(OH)_2$ and the operating conditions of saponification reaction on the acid value of resulting product. Acid value was used as the basis of reaction evaluation as it indicates the amount of fatty acid converted into product.

2 Working procedures

2.1 Materials

The main materials used in this research were technical grade $Ca(OH)_2$ as the source of calcium and PFAD from PT Tunas Baru Lampung Tbk. as the source of fatty acid.

2.2 Methodology

In general, research conducted by reacting PFAD with $Ca(OH)_2$ using modified fusion method [8, 9]. $Ca(OH)_2$ was added to 50 grams of molten PFAD and stirred until homogeneous. Then followed by the addition of 10 g of water, or equivalent to 20% of the mass of PFAD.

Stoichiometric mole ratio of $Ca(OH)_2$ to PFAD were varied from 1 to 3, while water temperature was between 60 and 80°C. The products were evaluated through the acid value measured using ISO 660: 1990, and moisture content of the product based on ASTM D3173. Reaction conversion was estimated by calculating the amount of free fatty acid as palmitic acid of PFAD that was consumed at the saponification reaction, or simply by calculating the percentage difference of the acid value of the product and PFAD.

3 Analysis

3.1 The effect of the variation of stoichiometric mole ratio and water temperature on the acid value of product

The acid value is one of the parameters for determining the extent of the saponification reaction and representing the percentage of conversion reaction. The smaller the acid value of calcium soap products, means higher fatty acids have been converted to soap. Results of the study showed that PFAD used in this experiment had an average acid value of 180-205 mg KOH/g sample [10]. Figure 1 shows the effect of stoichiometric reactant mole ratio and water temperature to acid value.

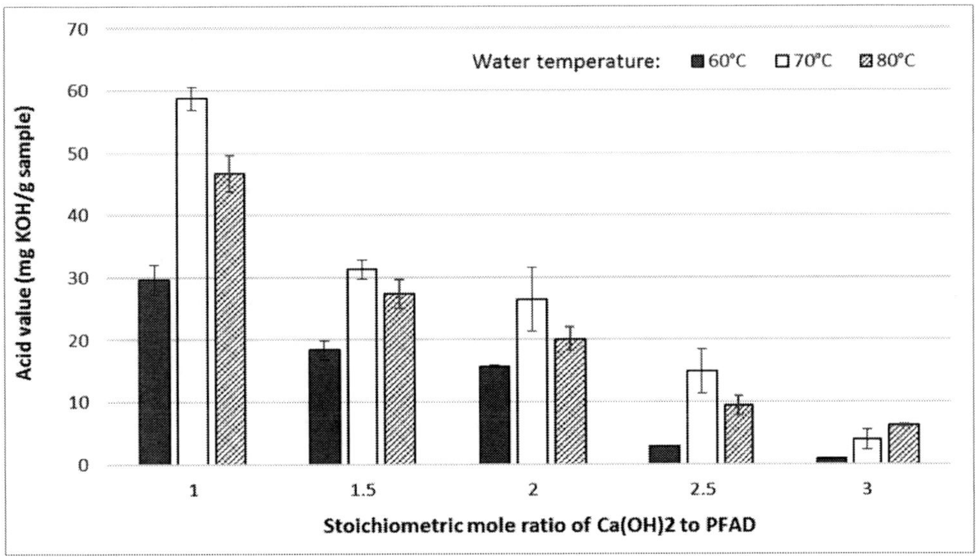

Fig. 1. Acid value of the calcium soap in variations of reactant mole ratio and water temperature.

In the mole $Ca(OH)_2$ to PFAD ratio of 1.5 with water temperature of 60˚C, the acid value of the soap was high (29.6 mg KOH/g sample), which means only about 84% fatty acid in PFAD was converted into calcium soap. The highest conversion occurred at the highest mole ratio of calcium to PFAD. With a mole ratio of 3, the conversion of fatty acids reached an average of 98%. These showed a consistence with the equilibrium reaction theory that the equilibrium shifts to the right as the reactant concentration increases.

However, feeding the excessive calcium content can inhibit the absorption of other minerals such as zinc and phosphorus in cattle digestion [11]. One of the factors affecting phosphorus absorption efficiency is the ratio of calcium to phosphorus. Based on the study, the recommended supplemental calcium for cattle is not more than 50.7 g/day [12]. With an optimum calcium-to-phosphorus ratio of 1.75, feeding excessive calcium without offsetting the addition of phosphorus can inhibit the phosphorus uptake [13]. Calcium soap with mole ratio of 1 and 1.5 that would meet the recommended amount of calcium consumption for cattle [4].

Water temperatures did not show a regular trend to the acid values of product. The lowest acid value was obtained at 60˚C, then followed by 80˚C and 70˚C respectively. This means that the highest conversion occurred at a lower temperature. This was related to the saponification reaction which is an exothermic reaction. Lower water temperatures will lead the reaction going towards the product, resulting a higher conversion of reaction.

3.2 The effect of the variation of stoichiometric mole ratio and water temperature on the moisture content of product

The added water and resulted water from the saponification reaction affected the moisture of the product. High water content will cause the soap textured wet and soft, while the small moisture content made the soap textured dry, hard, and be easily grinded into powder. The profile of moisture content of the calcium soap product is shown in Figure 2.

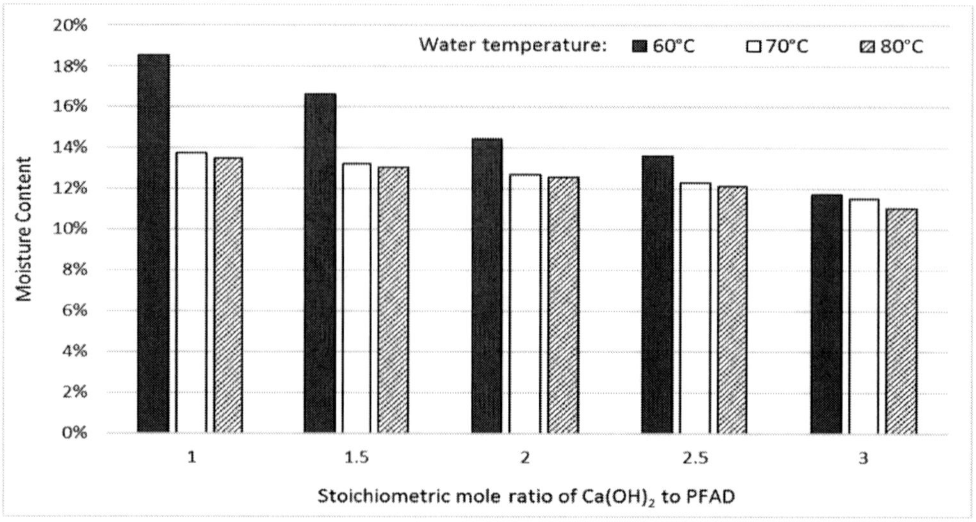

Fig. 2. Moisture content of the calcium soap in variations of reactant mole ratio and water temperature.

Based on Figure 2, the more amount of added calcium, the lower the moisture content of the soap. It was related to the saponification reaction that is exothermic. Larger amount of calcium would convert more fatty acids and produced more heat, which in turn gave rise to the temperature of the system. As a result, the water in the system evaporated with a larger amount along with the rise in temperature. Based on Figure 3, it can be seen that high peak temperature causes the lower moisture content.

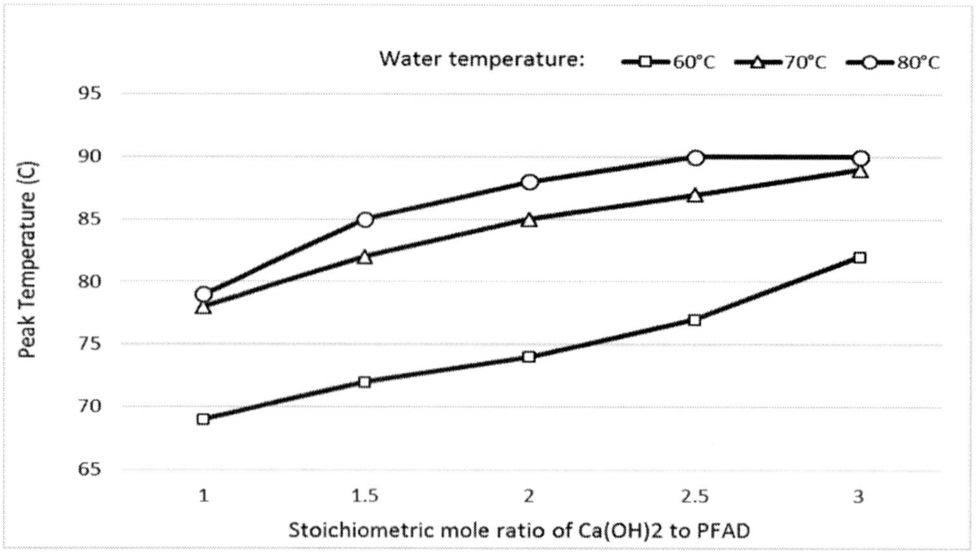

Fig. 3. Peak temperature in variations of reactant mole ratio and water temperature.

In variation in water temperature, the higher water temperature would produce soap with lower moisture content. This relates to the amount of heat available to evaporate water. Higher water temperatures provided more heat, characterized by the peak temperature values. As a result, water evaporated in a larger amount along with the presence of more heat.

The moisture content of calcium soap was still higher than commercial calcium soap which is in the range of 3-5% because of the absence of drying process. This indicated that the mole ratio of $Ca(OH)_2$ and water temperature could not produce soaps with moisture content close to the commercial product. Another study showed that when the mole ratio of calcium to PFAD was 1, soap with CaO as the calcium source has an average moisture content of 8.7%, relatively lower than the soap using $Ca(OH)_2$ that was 18.5% [10]. The use of calcium oxide allowed water to react with CaO forming $Ca(OH)_2$, which then react with the fatty acids. So, when water is used in the same amount, the fatty acid reaction with calcium oxide will produce water in the final product less than using calcium hydroxide.

4 Conclusion

This experiment evaluated the process of producing calcium soap from PFAD using $Ca(OH)_2$ as calcium source. The reactants mole ratio and water temperature could influence the characters of calcium soap, noticed from the value of acid number and moisture content. The increasing amount of calcium lowered the acid value, indicating the high conversion of the saponification reaction. In this experiment, the highest fatty acid conversion was obtained at the stoichiometric mole ratio of $Ca(OH)_2$ to PFAD mole of 3, with an average conversion value of 98%. The stoichiometric mole ratio of 1 and 1.5 meets the recommended supplemental calcium. The characteristic analyzed, physical appearance, and the texture of all experimental products showed a low similarity to the commercial soap. Therefore, further research is necessary to be conducted especially regarding the effect of other sources of calcium on the character of soap.

Acknowledgements

The authors would like to extend their sincere gratitude to BPDPKS Indonesia 2016 for supporting and enable the authors to conduct this research and PT Tunas Baru Lampung Tbk for providing PFAD for this research.

References

1. A.G.M. Top, Lipid Tech. J. **22**, 13 (2010)
2. J. Waters, *Fat supplements for dairy cows* (2010)
3. E. Block, W. Chalupa, E. Evans, T. Jenkins, P. Moate, D. Palmquist, C. Sniffen, *Calcium salts are highly digestible* (2005)
4. Volac International Ltd. *Megalac product sheets* (2016)
5. E. Wina, Susana IWR, WARTAZOA **23,** 176 (2013).
6. U. Moallem, M. Katz, H. Lehrer, L. Livshitz, S. Yakoby, J. Dairy Sci. **90**, 1243 (2007)
7. R. H. Rogers, W.R. Blew, US Patent 2 890 232 (1984)
8. H.L. Blachford, US Patent 4 316 852 (1982)
9. W. McAskie, US Patent 4 826 694 (1989)
10. L. Handojo, A. Indarto, D. Shofinita, A. Meitha, R. Nabila, H. Triharyogi, Medwell J. (to be published)
11. D.L. Palmquist, T.C. Jenkins, J. Dairy Sci. **67**, 978 (1984)
12. NRC, *Nutrient requirements of dairy cattle* (2001)
13. A.C. Field, J. Kamphues, and J.A. Woolliams, J. Agr. Sci. **101**, 237 (1983)

Effect of MgO to Fatty Acid Molar Ratio on the Production of Magnesium Salt of Fatty Acid from Palm Fatty Acid Distillates (PFAD) for Food Additives

Listianingrum[1], *Reni* Yuniarti[1], *Rd.H.R.M.T.* Al-Aziz[1], *Defri* Rizaldy[2], *Muhamad* Insanu[2], *Ardiyan* Harimawan[1], and *Dianika* Lestari[1*]

[1]Institut Teknologi Bandung, Department of Chemical Engineering, 40132 Labtek X Building, Jalan Ganesha 10, Bandung, Indonesia
[2]Institut Teknologi Bandung, School of Pharmacy, 40132 Labtek VII Building, Jalan Ganesha 10, Bandung, Indonesia

Abstract. Palm Fatty Acid Distillate (PFAD), a by-product of CPO industry, contains approx. 82%-wt of free fatty acids, which can be utilized as raw material for Magnesium Salts of Fatty Acids (Mg-PFAD). The objectives of the experiment were to produce Mg-PFAD salts through saponification-fusion reaction of PFAD at low temperature and ambient pressure and investigate the effect of MgO to fatty acid molar ratio on reaction conversion and yield. Next, washing Mg-PFAD salts with ethanol was able to facilitate the recovery of vitamin E (tocopherol and tocotrienol) from PFAD. Composition of Mg-PFAD were determined by AAS and GC analysis. Based on the data, yield of Mg-PFAD was increased by the increased of MgO to PFAD molar ratio. Mg-content of the product was within the standard according to FAO reference (4-5%-wt). Reaction was completed within 5-7 min after the addition of H2O (as catalyst).

1 Introduction

Indonesia is one of the largest Crude Palm Oil (CPO) producers in the world, where the production was reported to reach 33.4 million tons in 2015 [1]. Palm Fatty Acid Distillate (PFAD) is by-product of the CPO refining process from the deodorization unit, with a high fatty acid content of 82% and is estimated to be about 4% -w CPO [2]. This fatty acid can be utilized to produce metal salts, such as magnesium stearate, which is abundantly used in food and pharmaceutical industries as emulsifier, lubricants, anti-caking agent. In addition, thermal treatment in deodorization unit will transfer most of vitamin E content from CPO to PFAD. Reduced mass flow and massive stacking of vitamin E will increase the vitamin E content in PFAD up to 5 times higher than in CPO, which is about 4000-5000 ppm or about 0.5%-w [3]. This will increase the feasibility of vitamin E extraction from PFAD.

Metal salts are defined as products of reaction from saturated or unsaturated fatty acid with alkali, alkaline or transition metals. There are various methods usually applied for the commercial manufacture these metallic salts, namely decomposition process, fusion process,

*Corresponding Author : dianika@che.itb.ac.id

and modified reaction process. The fusion reaction was known to have advantages such as the low temperature reaction and atmospheric pressure, fast reaction (3-5 minutes), and the possibility to utilize its exothermic reaction heat [4]. This research aims to investigate the effect of molar ratio between free fatty acid (FFA) in PFAD and MgO on reaction conversion and its magnesium salts' yield, which are substantial for process optimization at later stages.

2 Materials and Methods

This research adapted the method of Rogers patent (1959), US Patent number 2890232 for PFAD [5]. 100 grams of PFAD and 10.6 grams of MgO was mixed at 80 °C, then 3.2 grams of water was added while maintaining the heating process. The saponification reaction will occur at 98 °C and completed when the temperature reaches around 105 °C [6]. The experimental scheme of research is shown in Fig.1. In this experiment, we varied the molar ratio of fatty acids in PFAD to MgO, as shown in Table 1. Amount of water added as a catalyst was equally set of 3.2 grams in all variations.

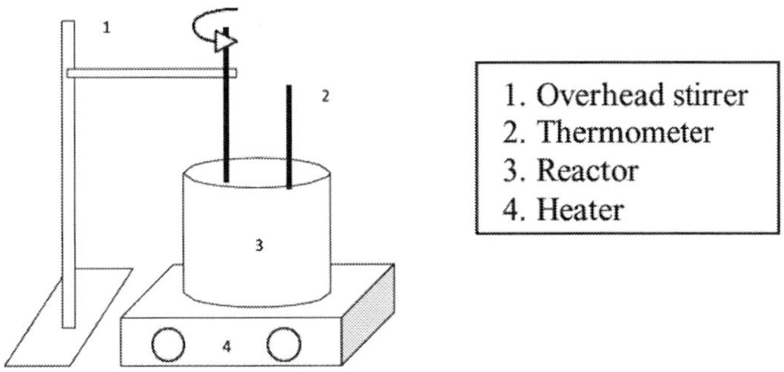

Fig. 1. Schematic experiment diagram

Table 1. Variation of molar ratio of fatty acids contained in PFAD and MgO

No.	Molar ratio of fatty acids : MgO	Temperature feeding of MgO
1	1.4 : 1.00*	
2	2 : 1**	
3	1 : 1.00	80 °C [5]
5	1 : 1.25	
6	1 : 1.50	

The flow diagram of experiments is shown in **Fig 2**. Reaction yield was calculated based on the reaction's mass balance. Atomic Absorption Spectroscopy (AAS) was used to measure Mg content of Mg-PFAD salts and Gas Chromatography – Mass Spectrometry (GC-MS) was used to observe Mg-PFAD salt formed as well as its quantity. The data from GC-MS was showed as percent area detected related to the percent composition of the sample.

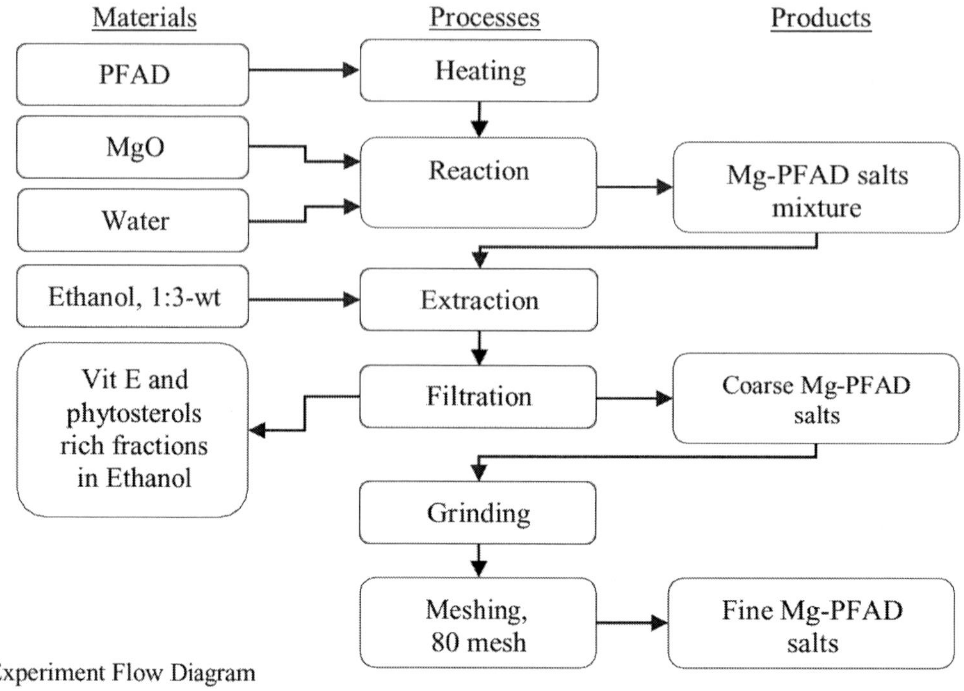

Fig. 2. Experiment Flow Diagram

3 Results and Discussions

3.1 Palm Fatty Acid Distillate Analysis

Table 2 presents that PFAD used in this experiment has PFAD content of 94.35%, which was obtained from acid value analysis. Later, it was required to determine the conversion of reaction, through Equation 1. Most types of FFA content of PFAD are palmitic and oleic acid by 51.38% and 39.85% consecutively.

Table 2. Fatty Acid Composition of PFAD

	Fatty acid content (%)
PFAD	94.35 ± 0.35 (AV = 198.2 mg KOH/ g sample)
Palmitic acid	51.38
Linoleic acid	4.44
Oleic acid	39.85

The number of FFA in PFAD obtained from different by-products of CPO processing may vary. FFA composition in PFAD may range from 65.50% to 95.55% depends on the quality of purified CPO and the process condition of rafination.

3.2 Effect of FFA to MgO molar ratio on fatty acid conversion and Mg-PFAD salts yield

Figure 3 shows the effect of FFA to MgO molar ratio on reaction yield. From this experiment, the reaction yield ranges from 88-92% for FFA to MgO molar ratio of 1:1 to 1:1.5.

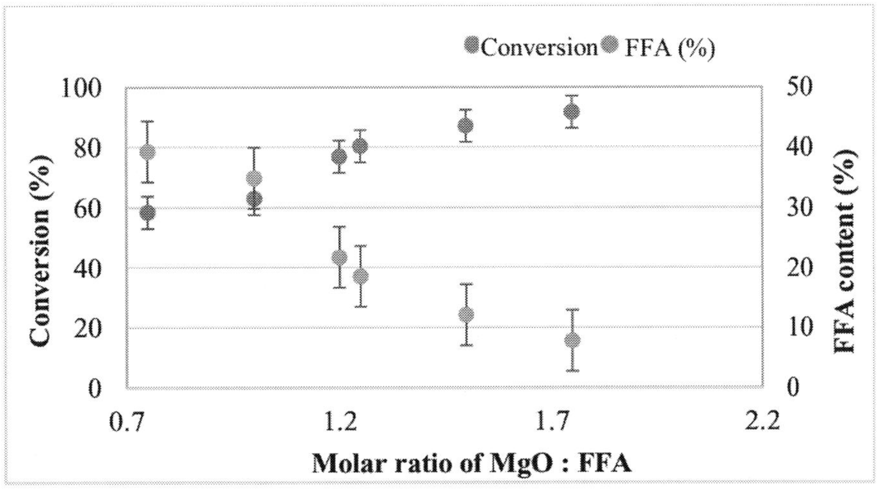

Fig. 3 Profile of reaction conversion and FFA percentage on different MgO to FFA molar ratio

Based on the results of the experiment (**Fig.3**), it was conducted an analysis of factorial design of percent reaction conversion against the ratio of moles of MgO to PFAD. Based on the analysis of the data, the molar ratio of feed (A) greatly affects the conversion of reaction (with a P-value of 0.000), whereas the initial temperature of the reaction (B) and the interaction between the A&B has no effect against a conversion reaction (with P-value 0.931 and 0.961). The largest conversion obtained by conducting molar ratio of MgO : FFA as 1:1.75, when MgO molar ratio excess of stoichiometric requirement. Excess MgO suspected to drive the reaction to the right (product side) so that the magnesium salts products increased and thus reduced the levels of FFA content in the product. It can be seen that the more the MgO fed, then the lower the conversion reaction. However, the conversion of 91.75% did not meet the standard commercial magnesium stearate product because it still has FFA of 7.79%. Based on Ph Eur 8th Ed 2014, magnesium stearate must meet the limits of maximum FFA content by 4% or less [7]. Therefore, further processing is required to reduce the content of FFA in the final product.

$$MgO_{(s)} + 2\ RCOOH_{(l)} \rightarrow Mg(OOCR)_{2(s)} + H_2O_{(l)} \quad (1)$$

3.3 Effect of FFA to MgO molar ratio on Mg-PFAD salts composition

Mg-PFAD salt composition represents characteristic of the salt and determines its properties and functionality. Fatty acid composition in Mg-PFAD salts showed similar composition to fatty acid composition on PFAD (Fig. 4). Briefly, PFAD is composed by low content of stearic acid (C18:0) (4-5%) and higher content of palmitic acid (C16:0), oleic acid (C18:1), linoleic acid (C18:2) and linolenic acid (C18:3) [8]. Furthermore, the result showed that FFA to MgO molar ratio had no significant effect on Mg-PFAD salts composition.

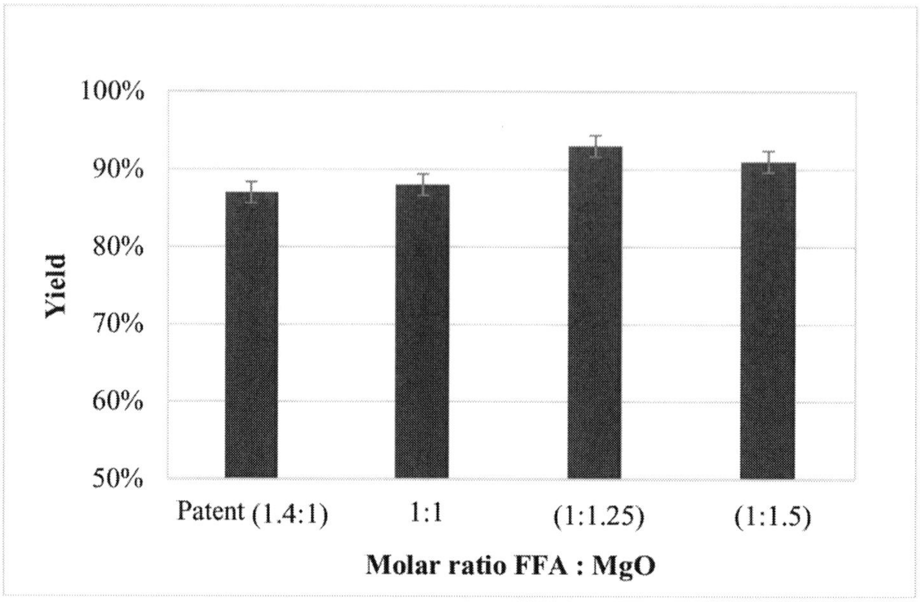

Fig. 4. Reaction yield on different FFA to MgO molar ratio

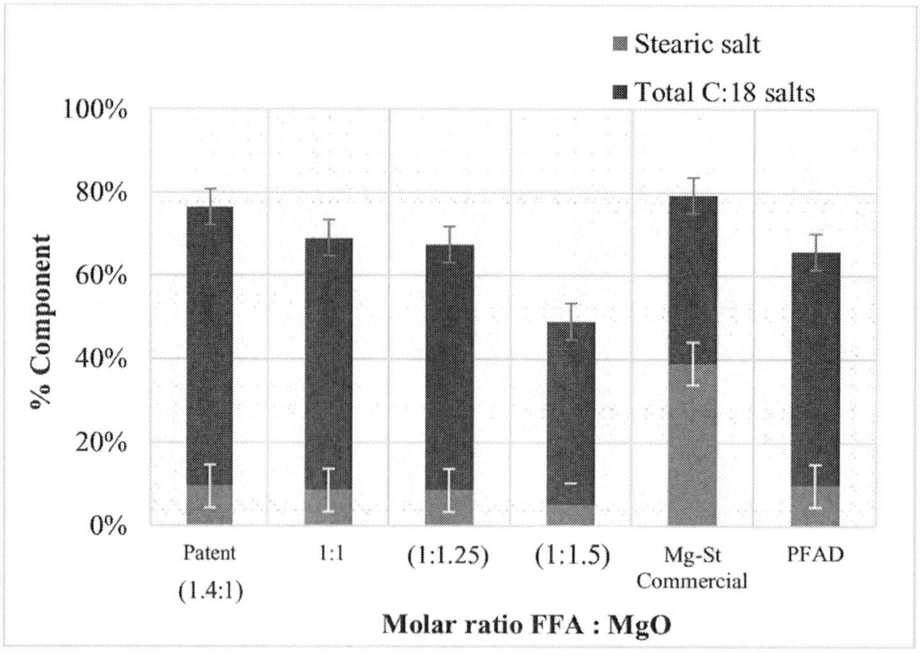

Fig. 5. Mg-PFAD salts composition on different FFA to MgO molar ratio

3.4 Effect of FFA to MgO molar ratio on Mg content of Mg-PFAD salts

Fig. 6 shows magnesium content in Mg-PFAD salts on different FFA to MgO molar ratio. Based on this result, FFA to MgO molar ratio showed no significant effect on Mg content of Mg-PFAD salts.

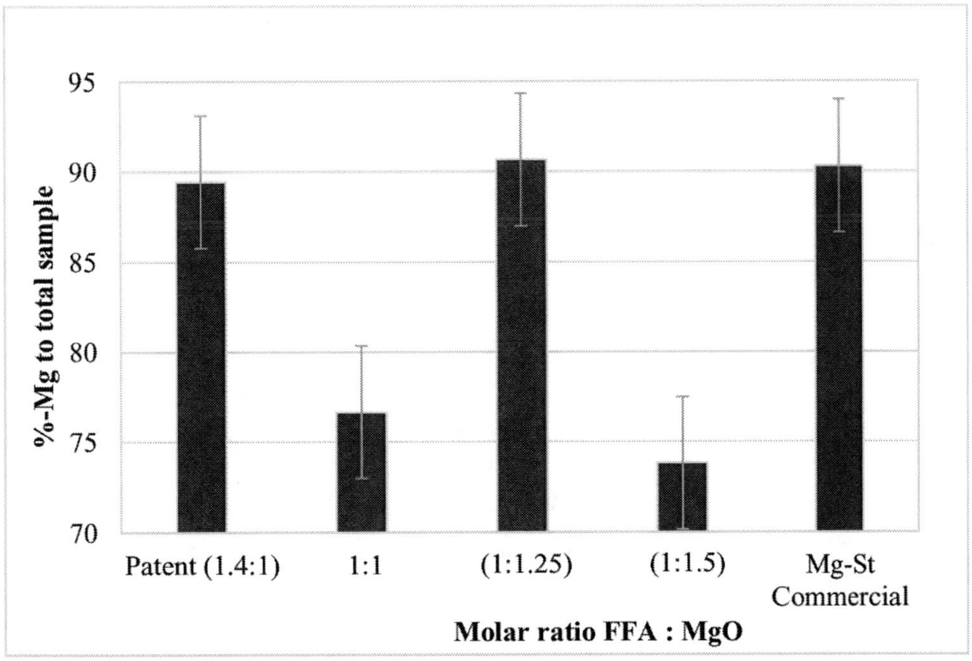

Fig. 6. Mg content in Mg-PFAD salts on different FFA to MgO molar ratio

4 Conclusions

Based on the experimental data, we conclude that fatty acid to MgO molar ratio affect reaction yield and composition of Mg-PFAD salts. Mg-content of the product was within the standard according to FAO reference (4-5%-wt). Reaction was completed within 5-7 min after the addition of H2O (as catalyst).

This research was funded by Grant Research Sawit (GRS, K-16) Programme from BPDPKS, Indonesian Ministry of Finance, 2016

References

1. Indonesian Palm Oil Producers Association and Indonesian Ministry of Agriculture, (2016).
2. A.G.M. Top, Lipid Tech, **22**, 11-13 (2010)
3. Y. Basiron, B.S. Jalani, K.W. Chan, Malaysian Palm Oil Board, **2** (2000)
4. Manufacture of Metal Soaps, US2890232 (1959)
5. J.D. Opem, R.H. Rogers, US2890232 (1959)
6. R.H. Rogers, P. Park, J.D. Opem, US3047496 (1962)
7. *Ph. Eur*, **8** (2014)
8. Maarasyid, Cici, I.I. Muhamad, E. Supriyanto, Teknologi **69**, 4 (2014)

An innovative method to produce drop-in fuel by alkaline earth-transition metals basic soap decarboxylation

Godlief Fredrik Neonufa[1,2,*], *Meiti* Pratiwi[1], *Astri* Nur Istyami[1], *Lidya* Elizabeth[1], *Sri* Suminar Dewi[1], *Ronny* Purwadi[1], *Tirto* Prakoso[1], and *Tatang* H. Soerawidjaja[1]

[1]Department of Chemical Engineering, Institut Teknologi Bandung, Bandung 40132, Indonesia
[2]Department of Agriculture Product Technology, Universitas Kristen Artha Wacana, Kupang 85000, Indonesia

Abstract. Decarboxylation of metal soap is an emerging method to produce drop-in fuels as an alternative to the expensive hydrotreatment process. In this study, the Mg-Fe basic soap produced from palm kernel fatty acid had been successfully decarboxylated into jet-fuel type biohydrocarbons. The Mg-Fe basic soap with ratio of 8:2 mol was decarboxylated for 5 hours at atmospheric pressure and temperature varied up to 370°C; it produced a liquid product whose yield was around 60 %-weight. The resulting hydrocarbon product was a complex mixture consisted of normal paraffins in the range of carbon chain length C_8 – C_{17}, iso- and cyclo-paraffins, and various olefin products.

1 Introduction

Interest in renewable fuels has been encouraged in recent years due to environmental, economic and geopolitical considerations, including global warming, dwindling petroleum deposits, rising crude oil prices and a desire for energy independence. Biofuels have been proposed as a renewable alternative to fossil fuels. Renewable hydrocarbon biofuels (also called drop-in biofuels) are fuels produced from biomass resources through a variety of biological and thermochemical processes. Biomass such as vegetable oils and fats are the feedstock candidates that are very potential and promises to produce renewable fuels [1-2]. Approximately 201.1 million tons of oils and fats were produced worldwide in 2014, from animal and vegetable feedstocks [3].

The production of renewable fuel become important nowadays due to the achievement of the goal set by the Indonesian government, i.e. replacing 20% of the nation's petroleum consumption with renewable alternatives by 2020 [4-5]. Drop-in biofuels are liquid transportation fuels made from oils and fats which are expected to replace FAME biodiesel. These fuels can be used as whole fuel as well as additive to the current petroleum gasoline, diesel, or jet fuel, thus they are considered as fully infrastructure-compatible fuels i.e. they can be used in vehicles without engine modifications and can utilize existing petroleum distribution systems [6]. Jet biofuel meets or exceeds the most rigorous jet fuel standards for

* Corresponding author: godliefneonufa@gmail.com

performance. For example, blending the jet biofuel in a 50/50 ratio with petroleum-based jet fuel offers significant advantages over petroleum jet fuel, such as reducing greenhouse gas emission by 65-85% compared to petroleum-based fuels. Moreover, it has shown higher energy density in flight, which allows aircraft to fly farther on less fuel. Furthermore, it meets or exceeds critical jet fuel specifications, such as flash point, freeze point, stability and heat of combustion [7]. Fatty acids and their derivatives such as fatty acid methyl esters of palm kernel oil can be converted to renewable and carbon-neutral fuel-like hydrocarbons (e.g. jet biofuels) that are entirely fungible with fossil fuels by deoxygenation methods. This method can be conducted via decarboxylation, decarbonylation and hydrodeoxygenation pathways, as illustrated below for lauric acid.

Decarboxylation yields CO_2 and n-undecane:

$$C_{11}H_{23}COOH \rightarrow n\text{-}C_{11}H_{24} + CO_2 \quad (1)$$

Decarbonylation yields CO, water and undecenes:

$$C_{11}H_{23}COOH \rightarrow C_{11}H_{22} + CO + H_2O \quad (2)$$

Hydrodeoxygenation yields H_2O and n-dodecane:

$$C_{11}H_{23}COOH + 3H_2 \rightarrow n\text{-}C_{12}H_{26} + 2\ H_2O \quad (3)$$

In the present commercial method, hydrodeoxygenation is directly applied to vegetable oils through liquid phase hydrotreatment. This method is expensive because it requires the use of problematic sulfide catalysts and high pressure of hydrogen; it also requires severe operating conditions. In view of this, an alternative has been sought in the deoxygenations of triglycerides and fatty acids or their derivatives via basic soap decarboxylation. The literatures report that soaps made from vegetable oils or fats can be processed into products that are rich in hydrocarbons by pyrolysis reaction with high yields and temperatures not so high as expected [8]. Decomposition of the alkali and alkaline earth metals soaps of the higher fatty acids generally gives better yields of hydrocarbons than that of obtained by decomposition of the corresponding fatty acids or fats [9]. Chang and Wan have been studied the formation of hydrocarbons which correspond to its oil fraction. The products are obtained from catalytic and thermal cracking of vegetable oils and soaps. Pyrolysis calcium soaps derived from rapeseed oil, peanut oil and Tung oil [10] was conducted and appears to be effective to produce hydrocarbons whose physical-chemical properties similar to the specifications of fossil oil fractions. Basic soap decarboxylation as illustrated below, has not been studied intensively to be developed on an industrial scale.

$$M(OH)(C_{11}H_{23}COO) \rightarrow n\text{-}C_{11}H_{24} + MCO_3 \quad (4)$$

When compared with hydrodeoxygenation via hydrotreating and hydro-decarboxylation of the fatty acids or their derivatives, basic soap decarboxylation permits the use of simpler catalysts and requires less hydrogen, which makes deoxygenation via basic soap decarboxylation appealing from an economic standpoint [11]. Therefore, basic soaps decarboxylation has emerged as a route that is very promising to be developed as the basis of a commercial decarboxylation process.

The objective of this study was to produce alkane that can be considered to be drop-in replacements for petroleum-based jet fuels, by catalytic thermal decarboxylation of Mg-Fe basic soap. In particular, this study investigated the decarboxylation of the basic soap at mild operating conditions without external supply of H_2.

2 Experiments

$Mg(OOC_2H_5)_2.4H_2O$ and $Fe(OOC_2H_5)_3.2H_2O$ supplied by Merck (analytical grade), were used to make the soap. The palm kernel oil was obtained from PT. Ecogreen, Batam. Mg-Fe basic soap was obtained by metathesis process and it was used as a reactant model compound. Decarboxylation reaction was performed in a glass batch reactor at temperatures up to 370°C and atmospheric pressure for 5 hours. The liquid reaction products were analyzed by gas chromatograph (Shimadzu 2010) equipped with a capillary column (rtx-1) with dimensions of 30 m x 0.25 mm x 0.25 μm and flame ionization detector. One microliter of sample was injected into the GC with a split ratio 1:50 and the carrier gas (helium) flow rate was 42.9 ml/min. The injector and detector temperature were 340°C and 340°C, respectively. The following chromatographic temperature program was used for analysis: 40°C (at first) – 300°C (5°C/min) – 340°C (1°C/min, constant 45 min). A number of chemical standards were purchased to enable product identification and calibration. Determination of the freezing point had been tested with ASTM D-2500 procedures.

3 Results and Discussion

This section presents information about the product observed from the decarboxylation reactions of Mg-Fe basic soap from palm kernel oils.

3.1 Product Yield from Decarboxylation

Table 1 shows various products that arise from Mg-Fe basic soap decarboxylation at 370°C for 5 hours and at atmospheric pressure.

Table 1. The results of Mg-Fe basic soap decarboxylation.

Types of product	Yields [%-wt]
liquid bio-hydrocarbons	59.98
water	2.31
solid residues	25.12
others (including the gas)	15.59

These result shows that liquid bio-hydrocarbons yield to 59.98 %-weight were produced via decarboxylation of Mg-Fe basic soap from palm kernel oils. It proved that liquid bio-hydrocarbons could be synthesized from basic soap as ~~a~~ reactants. It also shows that the combination of Mg-Fe metals had good catalytic activity for converting the reactants into hydrocarbons via decarboxylation reaction at 370°C for 5 hours. This result shows that the renewable hydrocarbon synthesis via basic soap decarboxylation did not depend on the presence of Pt and Pd catalysts and high pressure operation, as reported by previous researchers [11-13].

3.2 Physicochemical Properties of Liquid Bio-hydrocarbons Product

Table 2 shows some selected physicochemical properties of liquid bio-hydrocarbons product of Mg-Fe basic soap from palm kernel oils at 370°C for 5 hours. Acid value in liquid bio-hydrocarbons product was very small (0.4 mg KOH / g sample). It shows that all free fatty acid (from palm kernel oils) had been converted into basic soap which then was subsequently converted by decarboxylation reaction into bio-hydrocarbons. For the freezing point, Table 2 shows that liquid bio-hydrocarbons product obtained from Mg-Fe basic soap were not just dominated by n-undecane (whose freezing point is 21.0 °C) as expected from palm kernel oils (C_{12} predominantly fatty acids).

Table 2. Physicochemical properties of liquid biohydrocarbons.

Components	Value
Acid value (mg KOH/gr sample)	0.5
Freezing point (°C)	< 11
Ketones/aldehydes group (−/+)	−

It also consisted of a mixture of bio-hydrocarbons with various range of carbon chain length, particularly short chain molecule bio-hydrocarbons with lower freezing point.

3.3 Distribution of Liquid Bio-hydrocarbons Product

GC-FID chromatogram patterns associated to distribution of liquid bio-hydrocarbons fraction were obtained from decarboxylation of Mg-Fe basic soap from palm kernel oils at 370°C for 5 hours; they are shown in Figure 1.

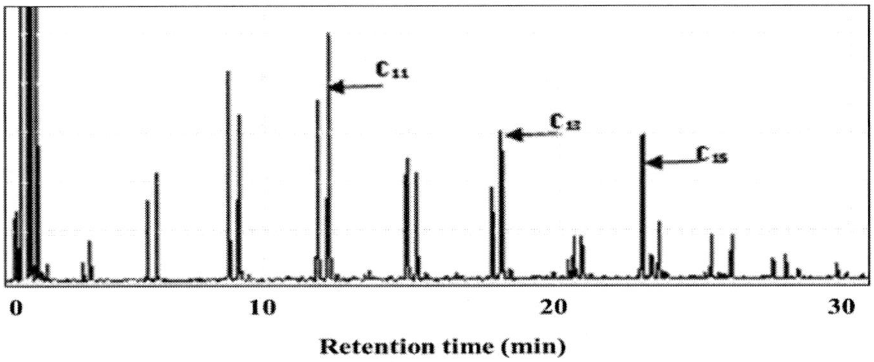

Fig 1. GC-FID chromatograms patterns for liquid bio-hydrocarbons fraction.

Figure 1 shows that decarboxylation of Mg-Fe basic soap produced liquid bio-hydrocarbons with the boiling point ranging from C_8 to C_{17}. The normal undecane (n-C_{11}) was suspected as the dominant component in the liquid bio-hydrocarbons product with retention time around 13 minutes, followed by normal decane (n-C_{10}) with retention time around 9 minutes. It indicated that the basic soap decarboxylation reaction had perfectly occurred, because it had produces higher component of normal undecane (n-C_{11}). The palm kernel oil has more fatty acids with C_{12} (lauric acids) that is around 66.57%-wt and 23.56%-wt of myristic acid/C_{14} and 7.34%-wt of palmitic acid/C_{16} as well. Therefore, when the basic soap decarboxylation is normal, it would produce a high amount of n-undecane.

The result of GC-FID chromatogram analysis related to the distribution of the liquid bio-hydrocarbons fraction obtained from Mg-Fe basic soap decarboxylation is summarized in Table 3. Table 3 shows that approximately 12.37 %-mole of n-undecane were obtained from decarboxylation of Mg-Fe basic soap of palm kernel oil. The rest of the fraction was dominated by n-decane and lower n-paraffin molecules, including olefin and iso-paraffin molecules.

The composition group of n-paraffin, olefins and iso-paraffin in the liquid bio-hydrocarbons obtained from Mg-Fe basic soap decarboxylation of Mg-Fe basic soap of palm kernel oil is shown in Figure 2. Figure 2 shows that, approximately 73%-mole of liquid bio-hydrocarbons obtained from Mg-Fe basic soap decarboxylation were paraffin molecules which are expected in jet fuel. The rest (around 27%-mole) were various 1-alkenes molecules.

Table 3. The results of GC-FID chromatogram analysis of liquid bio-hydrocarbons.

Bio-hydrocarbon compounds	formulas	Product Yields (%-mole)
n-octane	C_8H_{18}	0.94
1-octenes	C_8H_{16}	0.00
n-nonane	C_9H_{20}	2.67
1-nonenes	C_9H_{18}	1.81
n-decane	$C_{10}H_{22}$	6.58
1-decenes	$C_{10}H_{20}$	2.82
n-undecane	$C_{11}H_{24}$	12.37
1-undecenes	$C_{11}H_{22}$	7.58
n-dodecane	$C_{12}H_{26}$	3.30
1-dodecenes	$C_{12}H_{24}$	3.29
n-tridecane	$C_{13}H_{28}$	6.54
1-tridecenes	$C_{13}H_{26}$	4.06
n-tetradecane	$C_{14}H_{30}$	1,46
1-tetradecenes	$C_{14}H_{28}$	1.46
n-pentadecane	$C_{15}H_{32}$	3.50
1-pentadecenes	$C_{15}H_{30}$	3.49
n-heptadecane	$C_{17}H_{36}$	0.22
Total i-paraffins		21.78

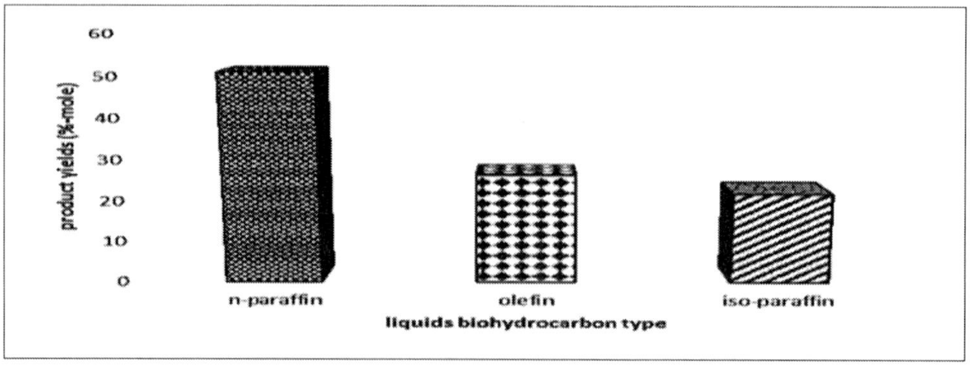

Fig. 2. Liquid bio-hydrocarbon types Mg-Fe basic soap decarboxylation.

The presence of 1-alkene in significant amount shows that there had been a dehydrogenation reaction during the basic soap decarboxylation in the batch reactor. In particular, Figure 3 shows the carbon chain length distribution of n-paraffin type in liquid bio-hydrocarbons was produced from Mg-Fe basic soap of palm kernel oil at 370°C for 5 hours. Figure 3 shows that bio-hydrocarbon with carbon chain length C_{11} (n-undecane) was most dominant (around 12 %-mole) in the liquid product of Mg-Fe basic soap decarboxylation up to 370°C for 5 hours. The bio-hydrocarbons with further carbon chain length were C_{10} (n-decane, 6.5%-mole) and C_{13} (n-tridecane, 6.1%-mole).

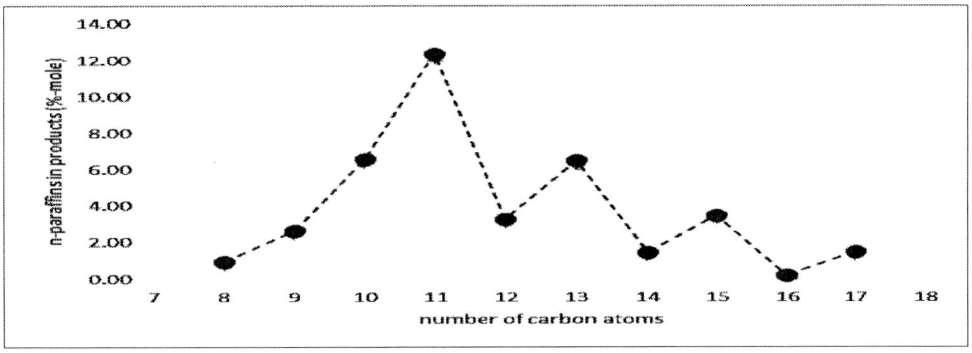

Fig. 3. Distribution of liquid n-paraffin plot according to carbon chain length.

4 Conclusion

The drop-in fuels in jet fuel type can be synthesized via basic soap decarboxylation, without the aid of hydrogen from the outside and without the using of commercial catalyst. The Mg-Fe basic soap was derived from palm kernel oil and decarboxylated for 5 hours at atmospheric pressure and temperatures up to 370°C; it effectively produced a jet fuel-type biohydrocarbons with a liquid product yield around 60 %-weight. The resulting hydrocarbon product was a complex mixture consisting of normal paraffins in the range of carbon chain length $C_8 - C_{17}$, iso-paraffins and various olefin products.

This work was supported by Indonesian Oil Palm Estate Fund Agency (BPDPKS) under GRS-16 Grant. The authors gratefully acknowledge the contributions of Prof. Subagjo to the GC-FID analysis work in his laboratory.

References

1. I. Kubickova, Catal. *Today*, **106,** 197-200 (2005)
2. P. Ma'ki-Arvela, *Energy & Fuels*, 30-41 (2007)
3. FAO and Oil World, *Food Outlook*, New York (2014)
4. J. Fu, *The Journal of Chem Sus Chem*, **4**, 481-486 (2011)
5. T. H. Soerawidjaja, *Posisi strategis minyak-lemak nabati di dalam panorama teknologi BBN cair dan ketahanan energy* nasiona, Jakarta, Balitbang ESDM (2013)
6. U.S.Departement of Energy, *Energy Efficiency and Renewable Energy Alternatives fuels data center,* http:/www.afdc.energy.gov, 10/11/2016 3:07 PM (2016)
7. Honeywell UOP, *Honeywell Green Jet Fuel,* https://www.uop.com/processing-solutions/renewables/green -jet-fuel/, 10/11/2016 2:59 PM (2016)
8. A. Demirbas, H. eKara, *Energy Sources,* **28,** 619-626 (2006)
9. A. Ralston, *Fatty acids and derivatives,* London: Chapman Hall Limi (1948)
10. C. C. Chang, S. W. Wan, *Ind. and Eng. Chem,* **39,** 1543-1548 (1947)
11. E. Santilan-Jimenes, M. Crocker, *J Chem Technol Biotechnol*, **87,** 1041-1050 (2012)
12. J. Fu, *The Journal of Energy and Env. Sci.*, **3,** 311-317 (2010)
13. M. Snare, *Fuel,* **87**, 933 – 945 (2008)

Crystallization of Barium Sulfate ($BaSO_4$) in a Flowing System: Influence of malic acid on Induction Time and Crystal Phase Transformation

F. Sariman[1,*]; *A.P.* Bayuseno[1]; *S.* Muryanto[1,2]

[1] Department of Mechanical Engineering, Faculty of Engineering, Diponegoro University, Tembalang Campus, Semarang 50275, Indonesia
[2] Department of Chemical Engineering, UNTAG University in Semarang

Abstract. Barium sulfate scaling usually occurs in pipelines of offshore oil drilling wells, which has become a persistent problem. This scale can create technical problems, including the increased energy in consumption and unscheduled equipment shutdown. This paper presents the crystallization of barium sulfate and this experiment was focused on evaluating the effect of solution concentration and malic acid to the induction time and barium sulfate crystallization. In this research, the experiments on the growth of barium sulfate scale in the pipe test was conducted by reacting $BaCl_2$ and Na_2SO_4 at concentrations of 3500, 4000, 4500 and 5000 ppm Ba2+, flow rate (40 ml / min), temperature (50 ^{o}C), while malic acid (0, 5, 10 ppm) was added to the solution as a scale inhibitor. $BaSO_4$ scale formation was observed with the declining value of the conductivity of the solution. $BaSO_4$ crystal shape was observed from the results of morphological studies performed using SEM, further analysis of the micro-crystal composition using EDX and XRD analysis was used to prove the crystal produced. The SEM imaging shows that the morphology crystals have a star-like particle, while XRD analysis confirmed that the pure barite crystal (barite) was produced during the experiments. Obviously, the malic acid inhibited the crystal growth of barite.

1 Introduction

Barium sulfate (barite) has been known as a scale deposit that is particularly prevalent in the pipeline of oil and gas industry [1, 2]. This happening mineral scale deposition in the injection wells is imposed during the production process operation [3]. This mineral can occur when seawater (brine that does not fit with plenty of water containing cations such as Ca^{2+}, Ba^{2+}, Sr^{2+}) is mixed with SO_4^{2-} yielding precipitation of $CaSO_4$, $BaSO_4$ and/or $SrSO_4$ [4]. However, the deposition of barium sulfate (barite) commonly occurred in the oil fields [5].

* Corresponding author: wap_mesin@yahoo.co.id

Furthermore, barium sulfate solubility increases with rising temperatures, of which ionic strength of the salt water increases with pressure. Thus precipitation of barium sulfate is influenced by temperature [6].

Moreover, factors that affect the size of the resulting crystals are the speed of nucleation and crystal growth [7]. While the nucleation and growth are greatly influenced by the condition of supersaturation, as well as by the acidity, temperature and the presence of seeds or impurities and/or surfactant in the crystallizer [8]. Therefore, understanding of these factors is needed to develop inhibition strategy of scaling in the industrial pipeline of offshore drilling well.

In the present study, the experimental scaling of $BaSO_4$ in pipes was conducted using the rig to build in-house by researchers previously [9]. In this work, an effort has been made to inhibit crystal growth with varying the concentration of the stock solution (3500, 4000, 4500, 5000 ppm Ba^{2+}), flow rate (40 ml/min) and temperature (50°C). This research also studied the effect of adding malic acids (0, 5, 10 ppm) with the different concentration on the induction time and the crystallization of barium sulfate. XRD and SEM/EDX were performed for material characterization of the scale obtained.

2 Materials and Methods

2.1 Material Research

The raw materials used for this study were $BaCl_2{\cdot}2H_2O$ powder, crystal (barium chloride dihydrate) with grade analytic, and Na_2SO_4 powder, crystal with grade analytic, distilled water (aquades) and malic acid ($C_4H_6O_5$) as the additive. The stock solution was prepared by diluting $BaCl_2{\cdot}2H_2O$ with concentrations of 3500, 4000, 4500, 5000 ppm Ba^{2+} in the water. Consequently, the stock solution of Na_2SO_4 was prepared according to the equimolar of Ba^{2+} as follows:

$$BaCl_2\,(s) + Na_2SO_4\,(s) \rightarrow BaSO_4\,(s) + 2NaCl\,(aq) \qquad (1)$$

Each stock solution of 500 ml volume was then prepared according to the calculation Eq.(1) and stored in the vessel prior to carrying on the experiments.

2.2 Characterization of barium sulfate scale

In this study, an experimental tool had been designed for the formation of barium sulfate scale. The schematic tool can be seen in the Figure 1 below.

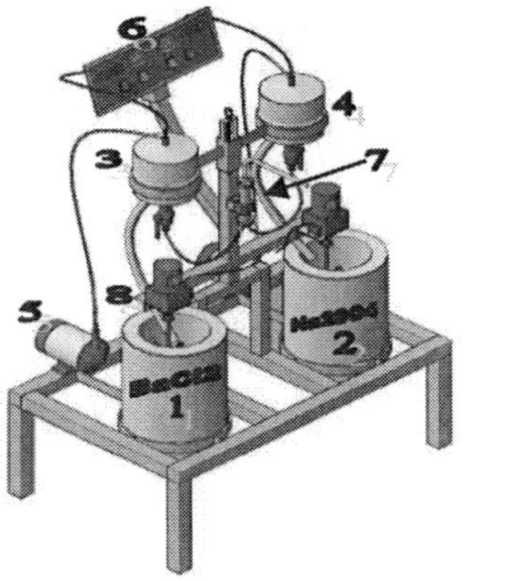

Description of the tool names in the picture above:

1. Vessel for collecting $BaCl_2$ solution
2. Vessel for collecting Na_2SO_4 solution
3. Vessel for intake of $BaCl_2$ solution
4. Vessel for intake of Na_2SO_4 solution
5. Water pump
6. Temperature control
7. Coupons and the house of coupon
8. Outlet of solution

Fig. 1. Research tools for the formation barium sulfate scale.

Barium chloride and sodium sulfate solutions respectively in vessels 1 and 2 were mixed with water in as much as 2.5 liters. During the experiments, each stock solution was stirred at 30 rpm. The barium chloride and sodium sulfate solution were pumped into the vessel of 3 and 4, respectively. To maintain the temperature and flow rate was regulated by temperature and flow rate controls (6). Subsequently, the barium chloride and sodium sulfate solutions were met in the house of coupon (7). To measure the induction time in each experiment employed the conductivity meter. The conductivity of the solution was measured every two minutes, so that the time for experimental trials were finished in two hours and thirty minutes. After finishing the experiments, the crystal attached to the coupon wall was dried at room temperature and then characterized by SEM/EDX and XRD method for morphology and phase composition evaluation.

3 Results and Discussion

3.1 Induction time

The induction time is the time taken by the ions in solution to react to form the first crystal nucleus. The induction time occurring in each experiment varies according to the parameter variable. Figure 2 presents the result of induction time with parameters of barium sulfate solution concentration of 3500, 4000, 4500 and 5000 ppm. Ba2+ using 10 ppm malic acid additives each occurring in 28, 32, 34, 37 minutes.

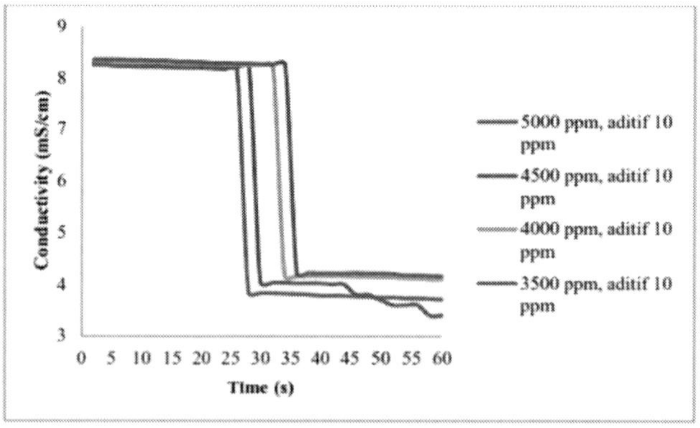

Fig. 2. Plot of conductivity versus time of scaling at flow rate 40 ml/min.

3.2 Mass scale barium sulfate

This study used variation parameters of barium sulfate solution concentration of 3500, 4000, 4500, and 5000 ppm Ba^{2+} and variation of concentration of malic acid 0, 5, 10 ppm as the additive. Table 1 shows the mass of scale barium sulfate formed by the variation of barium sulfate solution concentration and the variation of malic acid as the additive. From the present results, the more malic acid was added, then the reduced mass scale of barium sulfate was formed. Apparently, the presence of malic acid inhibited the crystal growth of barium sulfate.

Table 1. The mass of scale produced from the solution with malic acid additives.

No	Concentration Ba^{2+} (ppm)	Flow Rates (ml/second)	Temperature (⁰C)	Malic Acid (ppm)	Mass of $BaSO_4$ (gram)
1	3500	40	50	0	0.145
2	3500	40	50	5	0.140
3	3500	40	50	10	0.132
4	4000	40	50	0	0.156
5	4000	40	50	5	0.148
6	4000	40	50	10	0.142
7	4500	40	50	0	0.163
8	4500	40	50	5	0.157
9	4500	40	50	10	0.153
10	5000	40	50	0	0.176
11	5000	40	50	5	0.171
12	5000	40	50	10	0.164

3.3 Characterization of the $BaSO_4$ scale

The barium sulfate crystals formed were characterized by SEM/EDX. Figure 3 shows the morphology of barium sulfate crystals with and without adding the malic acid. It can be seen that in the presence of malic acid, barium sulfate crystals were formed smaller than that formed without malic acid. This observation was in close agreement with data of the reduced mass of barium sulfate.

Fig. 3. Morphology of barite crystals precipitated in the absence of additive (left), and in the presence of 10 ppm citric acid (right).

The chemical composition of the $BaSO_4$ crystal can be examined using the EDX data (Figure 4). EDX spectrum also shows that the major elements of Ba, S, and O belongs to the BaSO4 crystal in the presence 10.00 ppm malic acid. The weight percent (wt. %) of chemical elements can be calculated as follows: Ba = 35.05 wt. %, S = 8.11 wt. %, O = 16.23 wt. %. From EDX analysis results, it can be concluded that the crystals formed was BaSO4 (barite) scale, because of each weight percent of Ba, S, and O.

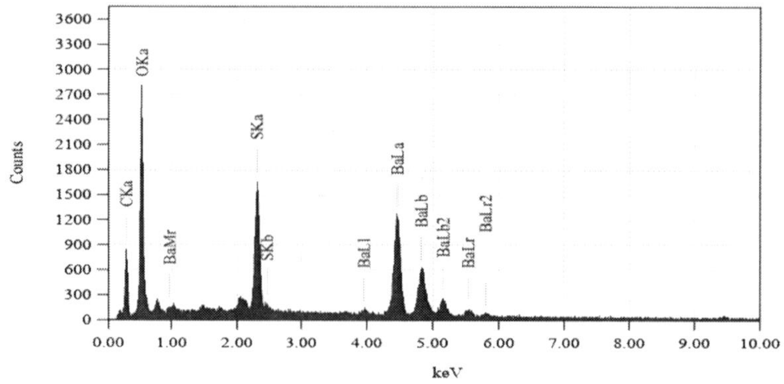

Fig. 4. EDX analysis results of $BaSO_4$ crystals in the presence of 10.00 ppm malic acid.

Further, XRD test results were further processed by the XRD Rietveld method using the fullprof software. Figure 5 shows barium sulfate crystals formed present 10 ppm malic acid as the additive and in the absence malic acid. The Powder Diffraction File (PDF) numbers of 04-012-5411 matched with the measured all XRD data obtained in the present study and proved that the pure barite was precipitated from the solution.

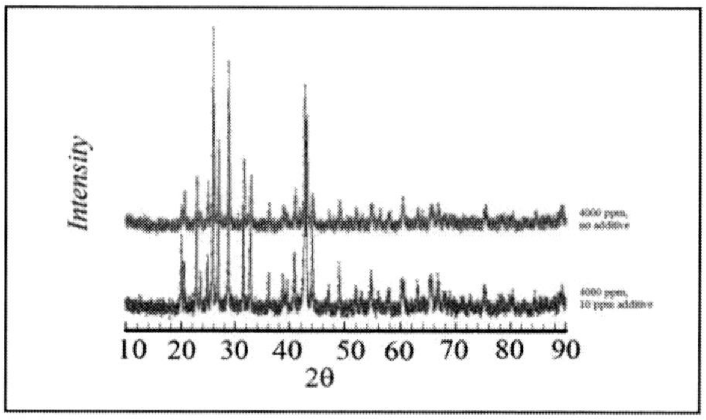

Fig. 5. XRD patterns of the samples in the presence of malic acid.

4 Conclusion and Remarks

Scale formation of barium sulfate ($BaSO_4$) with variations in concentration of Ba^{2+} (3500 ppm, 4000 ppm, 4500 ppm and 5000 ppm), temperature 50°C, a flow rate of 40 ml/min and the different concentrations of malic acid (0 ppm, 5 ppm, 10 ppm) was performed. From the result of induction time, the greater the concentration of barium sulfate solution used, the faster the induction time occurred, while the greater the concentration of malic acid additives used, the longer induction time that means by using malic acid additives then the formation of barium sulfate crystals can be inhibited. SEM testing concluded that the morphology of barium sulfate crystals by using malic acid additives smaller means crystal was difficult to stick to the walls of the coupon. From the results of EDX, that the crystal formed was $BaSO_4$ scale, because each weight percent of Ba, S and O. The XRD results also confirmed that the crystal formed was barite crystals.

Authors grateful to the Laboratory of Physical Metallurgy Diponegoro University that has given the place during the experiment. The research grant provided by the DIKTI through the post-graduate scholarship is also much appreciated. The authors would like to appreciate the comments useful and the kind suggestions from the anonymous reviewers.

References

1. Sorbie K S and Mackay E J. Petroleum Sci Eng. **27** 85.
2. Benton W J, Collins I R, Grimsey I M, Parkinson G M, Rodger S A, and Chem J. Soc Faraday Disc. **95** 281. (1993)
3. Bin Merdhah, A.B., Yassin, A.A.M. and Muherei, M.A., Laboratory and prediction of barium sulfate scaling at high-barium formation water. Journal of Petroleum Science and Engineering, **70**(1), pp.79-88. (2010)
4. Amer Badr Bin Merdhah, Abu Azam Mohd Yassin. Barium Sulfate Scale Formation in Oil Reservoir during Water Injection at High-Barium Formation Water. Journal of Apllied Sciences **7** (17), 2393-2403. (2007)
5. Isopescu, R., Mateescu, C., Mihai, M., & Dabija, G. The effects of organic additives on induction time and characteristics of precipitated calcium carbonate. Chemical Engineering Research and Design, **88**(11), 1450-1454. (2010)
6. Moghadasi, J., Jamialahmadi, M., Muller-Steinhagen, H., Sharif, A., Scale Formation in Oil Reservoir and Production Equipment during Water Injection (Kinetics of $CaSO_4$ and $CaCO_3$ Crystal Growth and Effect on Formation Damage). The SPE European Formation Damage Conference. Hague, Netherlands, SPE 82233, pp. 1–12. (2003)
7. Zahid A. Inhibition of Barium Sulfate Preception: Effects of Additives, Solution pH, and Supersaturation. China Ocean Press. (1994)
8. Fachry, A.R. and Tumanggor, J., Effect of Crystallization Time With The Cooling Process Of Ammonium Sulphate Crystallization From The Solution. Journal Chemical Engineering, **15**(2). (2008)
9. Muryanto, S., Bayuseno, A.P., Sediono, W. and Mangestiyono, W. Development of a versatile laboratory project for scale formation and control. Education for Chemical Engineers, **7**(3), 78-84. (2012)
10. Muryanto S, Bayuseno A. P, Ma'mun H, Usamah, and M Jotho. Procedia Chemistry. **9** 69–76. (2014)

Improvement of magnetorheological greases with superparamagnetic nanoparticles

Norzilawati Mohamad[1], Ubaidillah[2,3], *Saiful Amri* Mazlan[1,*], *Seung-Bok* Choi[4] and *Naim Abdul* Halim[1]

[1]Malaysia Japan International Institute of Technology, Universiti Teknologi Malaysia (UTM), Jalan Sultan Yahya Petra, 54100 Kuala Lumpur, Malaysia
[2]Mechanical Engineering Department, Faculty of Engineering, Universitas Sebelas Maret (UNS), Jl. Ir. Sutami 36A, Kentingan Surakarta, 57126, Central Java, Indonesia
[3]National Center for Sustainable Transportation Technology (NCSTT), Indonesia
[4]Mechanical Engineering Department, Inha University, Incheon, Republic of Korea

Abstract. Magnetorheological greases (MRGs) is a viscoelastic suspension comprised of carbonyl iron (CI) particles in a thixotropic medium; grease, which able to solve the instability problem that occurs in oil medium like MRFs. The primary purpose introduction of superparamagnetic nanoparticles γ-Fe_2O_3 in MRGs suspension is to investigate the effect of additives toward MRGs characteristics. Two types of MRGs suspension with and without nanoparticles γ-Fe_2O_3 were prepared in the same CI particles concentration for comparison purpose. It was observed that MRGs with incorporated of 1 wt% nanoparticles γ-Fe_2O_3 showed an increment regarding magnetic properties. In the meantime, the off-state viscosity of MRGs containing nanoparticles γ-Fe_2O_3 was reduced, however, increased during on-state condition. This fact indicates that the interspaces between CI particles are filled by the nanoparticles γ-Fe_2O_3. Therefore, the introduction of the nanoparticles indirectly improved the MRGs properties.

1 Introduction

Magnetorheological (MR) materials have been attracting researcher intention in last six decades. This material comprises of non-magnetic viscoelastic materials such as fluids, greases, plasticine, and elastomer, filled with micron-sized soft magnetic materials [1-3]. The intrinsic properties of this smart material are responsive to the occurrence of magnetic fields and reversible to the initial state rapidly in the absence of magnetic fields [4]. Among several types of MR materials, MR fluids (MRFs) have been most attractive due to their broad application. However, according to Chand et al. [5], MRFs has been facing some problems such as settling rate and fluid leaking. These problems can be overcome by introducing thixotropic materials such as greases as the matrix carrier [6].

There is a very limited number of studies related to MR greases (MRGs) have been reported. Park et al. [7] stated that there are no sedimentation drawbacks in MRGs due to the greases property which able to suspend the particles against gravity. As has been reported by Mohamad et al. [8] MRGs have lower the off-state viscosity and yield stress

* Corresponding author: ubaidillah_ft@staff.uns.ac.id ; amri.kl@utm.my

compared to the MRFs. Meanwhile, Kim et al. [9] stated that by adding an additive like kerosene oil in the grease medium have reduced the viscosity of MRGs and at the same time lower the yield stress.

Therefore, in this study, superparamagnetic nanoparticles γ-Fe_2O_3 was introduced to the MRGs as an additive. The primary goal of this study is to investigate the effect of superparamagnetic nanoparticles γ-Fe_2O_3 towards MRGs characteristics. The preparation of MRGs with nanoparticles (MRGN) was purely mechanical mixing. The amount of nanoparticles γ-Fe_2O_3 incorporated in MRGs was about 1 wt%. Some tests were undertaken to evaluate their morphology, magnetization saturation, and rheological properties.

2 Materials and testing

Soft magnetic carbonyl iron (CI) particles (standard OM grade, BASF, Germany) with an average diameter of 6 µm and density 7.86 g ml^{-1} were used as received without any further chemical treatment. Commercial grease (NPC Highrex HD-3 Grease, Nippon Koyu Ltd, Japan) with NLGI 3 was selected as suspending medium. The nanoparticles of γ-Fe_2O_3 were synthesized thru co-precipitation method described by Leong et al. [10]. Two types of MRGs were prepared which were CI particles based MRGs with and without nanoparticles γ-Fe_2O_3 as an additive. Firstly, CI particles (70 wt%) and nanoparticles γ-Fe_2O_3 (1 wt%) were dispersed in ethanol. The suspension was stirred using mechanical stirrer and mixed for two hours to produce a more homogenized suspension. The particles size distribution of MRGN were determined using environmental scanning electron microscope (ESEM). Meanwhile, the magnetic properties and magneto-induced viscosity of MRGs and MRGN were examined using vibrating sample magnetometer (VSM) and rheometer, respectively. All the samples characterizations were conducted at room temperature (25 °C).

3 Result and discussion

Figure 1 depicts a morphological photograph of MRGN in which the nanoparticles can be seen dispersed within the micron-sized particles. However, some micron-sized particles are still aggregated due to the carrier phase.

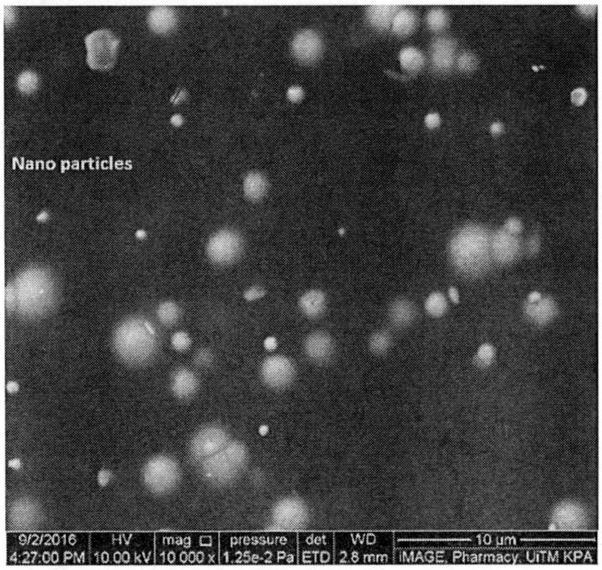

Fig. 1. Distribution of CI particles and nanoparticles γ-Fe_2O_3 in suspending medium of grease.

The magnetization saturation (Figure 2) was measured through vibrating sample magnetometer (VSM) showing that there is a significant improvement of magnetization saturation. The MRGs has magnetization saturation of about 130 emu/g while the MRGN has a much higher value of about 201 emu/g. Also, the hysteresis loop of MRGN is very narrow with a small coercivity compared to MRGs which indicates that nanoparticles γ-Fe_2O_3 also behave as soft materials that important for reversibility of MR performance [11]. It was proven that the superparamagnetic nanoparticles had improved the magnetic properties.

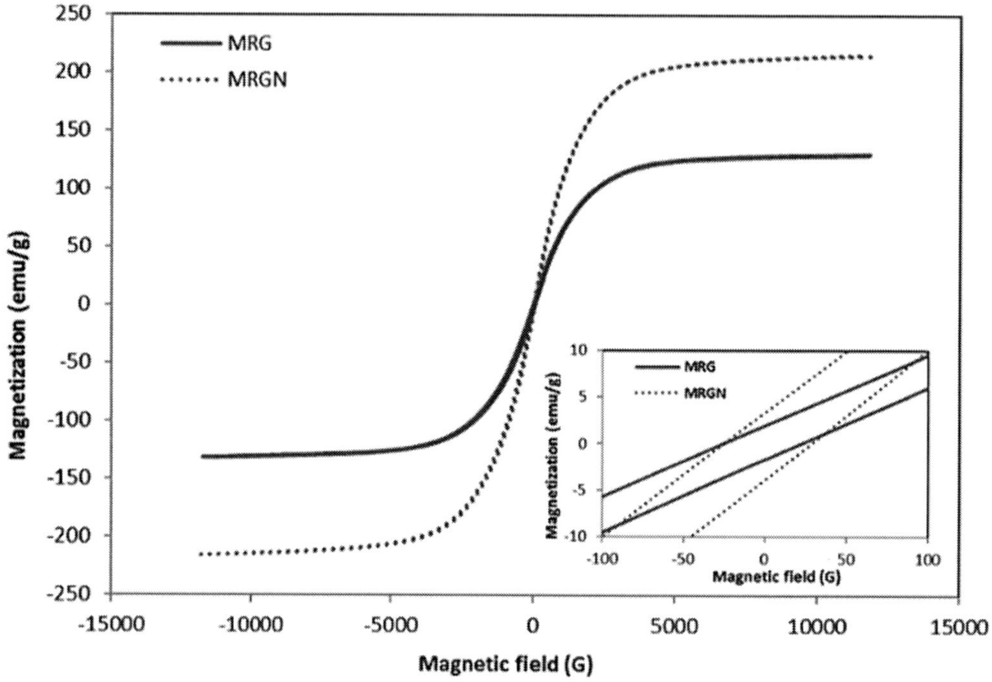

Fig. 2. Comparison of magnetic properties between MRGs and MRGN.

Moreover, rotational rheometer was used to evaluate the magneto-induced viscosity from 0 to 2000 1/s shear rate. As seen in Figure 3, the magneto-induced viscosity of MRGN and MRGs demonstrate a typical shear-thinning behavior. The introduction of nanoparticles γ-Fe_2O_3 has lowered the off-state viscosity of MRGs. However, during on-state which by applying current 4 A or equivalent to 658 mT, the viscosity of MRGN is higher than that of MRGs. As expected, this phenomenon is due to the nanoparticles γ-Fe_2O_3 had attached and filled the interspaces between the micron-sized particles when the magnetic field is applied [12,13]. Thus, the formation of regular and compact chains structure in the MRGN is attained, indirectly, enhance the MR performance.

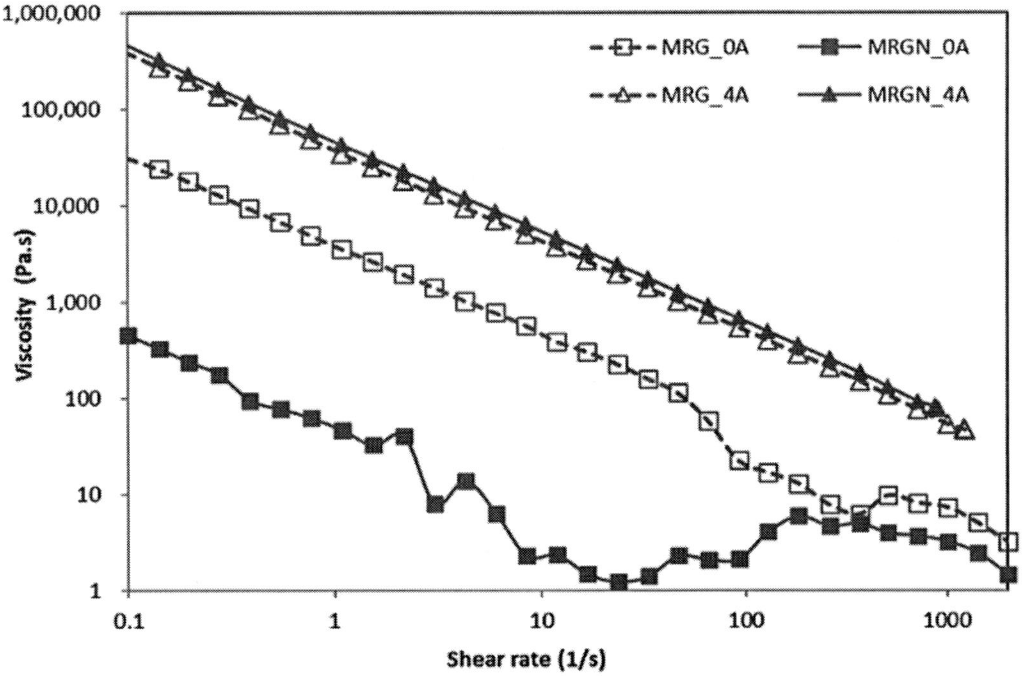

Fig. 3. Comparison of magneto-induced viscosity between MRGs and MRGN.

4 Conclusion

In this work, the influence of superparamagnetic nanoparticles in the MRGs have been investigated. Two types of MRGs were prepared with and without incorporation of nanoparticles γ-Fe_2O_3 for comparison purposes. It has been observed that MRGN showed a significant increment in magnetization saturation compared to the MRGs. It has also been observed that the MRGs incorporated of superparamagnetic nanoparticles γ-Fe_2O_3 has reduced the off-state viscosity but slightly increased the on-state viscosity. This fact is due to the nanoparticles γ-Fe_2O_3 has filled in the voids between micron-size particles in grease medium. As a conclusion, the introduction of superparamagnetic nanoparticles γ-Fe_2O_3 has been proven for improving the intrinsic properties of the MRGs.

This research was financially supported by: SHERA Project through Prime Award: AID-497-A-16-00004, USAID as well as the Universiti Teknologi Malaysia under Tier 1 Grant (Vote No: 13H55)

References

1. K. Hairuddin, Ubaidillah, S.A. Mazlan, H. Zamzuri, and N.M. Nor, Appl. Mech. Mater. **660**, 763 (2014).
2. N. Rosenfeld, N.M. Wereley, R. Radakrishnan, and T.S. Sudarshan, Int. J. Mod. Phys. B **16**, 2392 (2002).
3. A.Z. Pokaad, K. Hudha, M.Z.B.M. Nasir, and N.A. Ubaidillah, Int. J. Struct. Eng. **2**, 164 (2011).
4. B. Ichwan, S.A. Mazlan, F. Imaduddin, Ubaidillah, T. Koga, and M.H. Idris, Smart Mater. Struct. **25**, 37001 (2016).

5. M. Chand, A. Shankar, K. Jain, and R.P. Pant, RSC **4**, 53960 (2014).
6. Ubaidillah, J. Sutrisno, A. Purwanto, and S.A. Mazlan, Adv. Eng. Mater. **17**, 563 (2015).
7. B.O. Park, B.J. Park, M.J. Hato, and H.J. Choi, Colloid Polym. Sci. **289**, 381 (2011).
8. N. Mohamad, S.A. Mazlan, Ubaidillah, S. Choi, and M.F.M. Nordin, Smart Mater. Struct. **25**, 95043 (2016).
9. J.E. Kim, J.-D. Ko, Y.D. Liu, I.G. Kim, and H.J. Choi, IEEE Trans. Magn. **48**, 3442 (2012).
10. S.A.N. Leong, P. Mohd Samin, A. Idris, S.A. Mazlan, and A.H.A. Rahman, Smart Mater. Struct. **25**, 25025 (2016).
11. S.A.N. Leong, S.A. Mazlan, P.M. Samin, A. Idris, and Ubaidillah, AIP Conf. Proc. **1710**, (2016).
12. N.M. Wereley, A. Chaudhuri, J.-H. Yoo, S. John, S. Kotha, A. Suggs, R. Radhakrishnan, B.J. Love, and T.S. Sudarshan, J. Intell. Mater. Syst. Struct. **17**, 393 (2006).
13. E.F. Burguera, B.J. Love, R. Sahul, G. Ngatu, and N.M. Wereley, J. Intell. Mater. Syst. Struct. **19**, 1361 (2008).

MATEC Web of Conferences **159**, 02067 (2018) https://doi.org/10.1051/matecconf/201815902067
IJCAET & ISAMPE 2017

A Quantitative Analysis of System Readiness Level Plus (SRL⁺): Development of Readiness Level Measurement

Novi Marlyana[1,2,*], *Alva Edy* Tontowi[1], *and Hari Agung* Yuniarto[1]

[1]Department of Mechanical and Industrial Engineering, Faculty of Engineering, Universitas Gadjah Mada, Jl. Grafika No.2 Kampus UGM, Yogyakarta, Indonesia, 55281
[2]Department of Industrial Engineering, Faculty of Industrial Technology, Universitas Islam Sultan Agung, Jl. Raya Kaligawe KM. 4, Semarang, Indonesia, 50221

Abstract. The research explains the development of Readiness Level measurement. The initial concept of Readiness Level is Technology Readiness Level (TRL). TRL is an important metric used by U.S. government agencies such as NASA and the Department of Defense and is designed to quantify the maturity of a new technology and to enable comparisons with alternatives. System technology assessment evolves with the presence of Integration Readiness Level (IRL) and System Readiness Level (SRL). A quantitative combination of levels of readiness can be made and open the potential for expanding the other sizes of readiness levels, such as the Manufacturing Readiness Level (MRL). A measurement tool to measure the development of readiness level by involving MRL is called System Readiness Level Plus (SRL^+). This research focuses on quantitative analysis of SRL^+ model. It consists of the mathematical properties method and readiness reversal method. Several steps can be conducted to design the SRL^+ model. This model was developed from the System Readiness Level metric by Ross, combined with Incidence Matrix Approach by London. The first step is developing the function of the SRL^+ model. The second step is conducting computation using a development model that is SRL^+. The third step is carrying out validity of SRL^+ model. The result indicates that SRL^+ model can be mathematically proven.

1 Introduction

A product starts from a technology before it is introduced to the market. New technologies have more prospects to be accepted in the market because technology is the key to growth [1]. But often in many companies, products with new technology are not necessarily ready to be commercialized. This is because there is no certainty about the readiness of technology when heading to the market. The possibility that happens is the technology ready to be accepted by the market or emerged problems that will lead to waste of time and cost, the

* Corresponding author: novi@unissula.ac.id

emergence of consumer dissatisfaction and loss of consumers into the hands of competitors. Technology Readiness Assessment (TRA) is required as part of product development [1].

TRA is needed when an organization tries to determine the maturity of new technologies and / or capabilities (including technical or economic performance levels required) [2,3]. This method is applied in many domains such as military system [3], defense acquisition, aeronautics and astronautics and energy resources [4], and automotive industry [5]. The Technology Readiness Level (TRL) method is widely used for TRA [4]. TRL is used to measure the readiness of equipment (hardware) [1,2,4,6].

TRL is an important measure used by US Government agencies such as NASA and the Department of Defense and is designed to measure maturity of new technologies and enable comparison of alternatives. TRL has proven useful in identifying whether new components developed in technology programs are ready to operate as mature products. The TRL scale starts from 1 (basic principles observed) and ends at 9 (mission / proven activities) [1,2,4,6,7]. Furthermore, to evaluate the interconnection of components within a complex system, it is used Integration Readiness Level (IRL) [8].

The System Readiness Level (SRL) was developed to evaluate whole-system development risk and to support program acquisition decisions. SRLs mathematically combined component TRL values with system integration IRL and created a separate measure of system technical progress [8–10]. The underlying motivation of SRL was that complex systems and System-of-Systems (SoSs) require a more robust readiness assessment than TRL component evaluations alone could provide. SRL values are measured for individual components (both technology components/TRL and integration elements/IRL) and these subsystem component values produced a comprehensive system level SRL value.

A quantitative combination of readiness levels can be made and opens the potential for expanding the SRL by integrating other measurement of readiness levels, such as the Manufacturing Readiness Level (MRL) [8]. A measurement tool to measure the development of readiness level by involving MRL is called System Readiness Level Plus (SRL^+). This tool was developed from the System Readiness Level metric by [11], combined with Incidence Matrix Approach by [9,10]. The three TRA components described at the product level illustrate the relationship between individual components [9,10], then on measurement of technology at the company level will involve the readiness of the technology of its manufacturing technologies known as Manufacturing Readiness Level (MRL) [5]. A combination of TRL, IRL, and MRL was improved by [11]. The combination was developed from the previous concepts of the System and Integration Readiness Level by Sauser [3,8–11], combined with the concepts of Technology, Integration, and Manufacturing Readiness Levels, into a single metric—System Readiness Level. In the Sauser formula, TRL value and IRL value are multiplied to form an SRL value [8,11].

This research will emphasize on proving mathematical model SRL^+ through quantitative analysis of the mathematical model of SRL^+ by using mathematical properties method [9,12] and readiness reversal method [9,10]. The purpose of performing quantitative analysis is to increase the validity of the calculation of SRL^+.

2 Methods

This study will involve the concept of technology readiness on the product (TRL, IRL and SRL), will also involve technological readiness at manufacturing technologies (MRL). The Fig.1 below illustrates the schematic relationship between technological readiness assessments. The overall relationship between TRL, IRL, SRL and MRL will produce SRL^+.

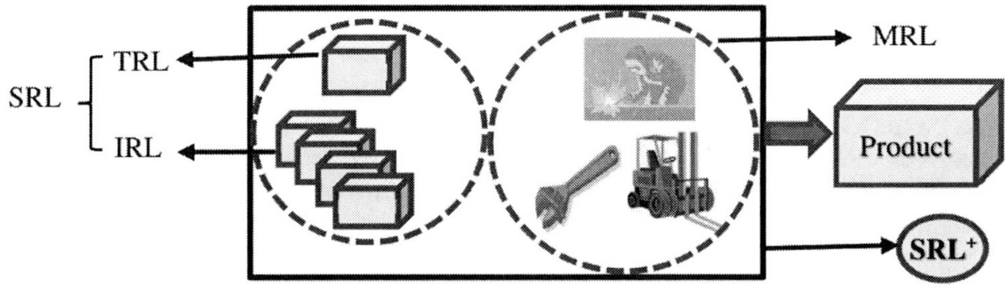

Fig. 1. Schematic relationship between TRL, IRL, MRL and SRL$^+$

Table 1 demonstrates the comparison of the existing system technology assessments.

Table 1. The comparison of the existing system technology assessments

System Technology Assessment	TRL	IRL	SRL	MRL
The focus of measure-ment	To measure technology components or critical systems in a period of time [5, 6]	To measure the ability of technology elements/ interrelated discrete components [5, 6]	• To evaluate whole-system development risk and to support program acquisition decisions [6] • SRLs mathematically combined component TRL values with system integration IRL and created a separate measure of system technical progress [7]	To measure technology readiness level of production facilities of an industry [8]
The level of measurement	9 levels	9 levels	involves levels of TRL and of IRL.	10 levels

3 Result and discussion

3.1 Development of SRL (SRL$^+$)

Several steps can be conducted to design the SRL$^+$ model.

- The first step is developing the function of the SRL$^+$ model. Early SRL involves two functions: TRL and IRL, while in SRL$^+$ the function of MRL is added. The SRL is then termed IMSRL (Incidence Matrix System Readiness Level). The IMSRL is defined as a novel method for calculating sub-system and whole-system readiness levels using an incidence matrix minimum-value approach [10].
- The second step is conducting computation using a development model that is SRL$^+$.
- The third step is carrying out mathematical properties and SRL$^+$ readiness reversal. Both are adopted from SRL model testing [9,10,12]. The mathematical properties is used to help build mathematical frameworks for the further development of readiness metrics for systems and SoSs (System of Systems). Readiness reversal generates inaccurate system development assessments and encourages a system to a developmental phase before the system has demonstrated the required characteristics.

3.1.1 The first step:

The basic concepts of SRL involves 2 functions, namely TRL and IRL [8].

$$SRL = f\,(TRL,\ IRL) = TRL \times IRL \qquad (1)$$

While SRL⁺ adds MRL function becomes:

$$SRL^{+} = f\{(TRL, IRL), MRL\} \tag{2}$$

Based on the SRL metric developed by [11], SRL is formed by multiplying the normalized MRL, the TRL, and the mean of the normalized IRL in a scalar contraction (dot product)[11]. And to make the SRL metric by [11] is different from the basic concept from [8], so this article using the term of SRL plus or SRL^{+}.

$$SRL^{+} = (TRL \; x \; IRL) \; x \; MRL = SRL \; x \; MRL \tag{3}$$

where, TRL is a technological readiness of the measured component; IRL is a technological readiness of interconnected components; SRL is a technological readiness of the measured system; and MRL = technological readiness of an industrial manufacturing technologies.

SRL hereinafter called IMSRL (Incidence Matrix System Readiness Level), is described as follows:

$$IMSRL = [min(ITRL_{min})] \tag{4}$$

$$ITRL_{min} = \left[min\left(min\left(\{IRL_{i,j}^{T} > 0\}_{i=1}^{m} x\{TRL^{T}\}\right)\right)\right] \tag{5}$$

Where, IMSRL is a novel method for calculating sub-system and whole-system readiness levels using an incidence matrix minimum-value approach; and ITRL = Integration-Technology Readiness Level, which determines the relationship between TRL and IRL [10]. So:

$$SRL^{+} = f\{(TRL, IRL), MRL\} = f\{IMSRL, MRL\} \tag{6}$$

If the level scales on IRL and SRL are 1-9, then the MRL is 1-10, so it will be normalized to make it feasible.

3.1.2 The second step:

The computation is carried out using example in [10], as indicated in Fig. 2. The figure below gives the example of IMSRL assessment. Consider the general graph of Fig. 2 with three TRL component values {TRL1=6, TR2=4, TRL3=8} and five IRL edges {IRL1->2=5, IRL2-3=9, IRL1<->3=7, IRL3->2=9, IRL2->2=5}. The IRL edge set contains both undirected {IRL2-3=9} and directed edges {IRL1->2=5, IRL1<->3=7, IRL3->2=9}, as well as a vertex loop at TRL2 {IRL2->2=5}.

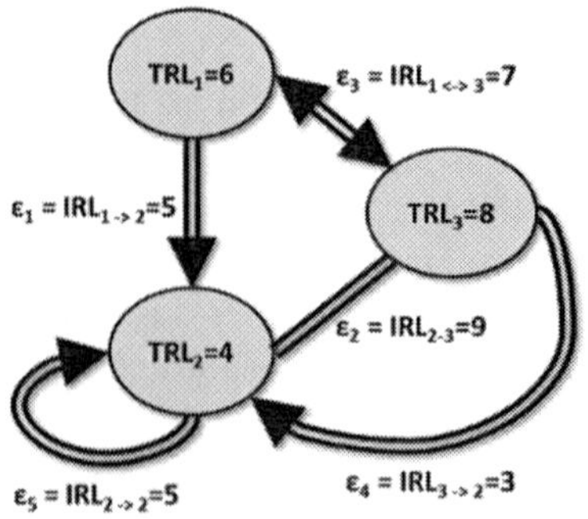

Fig. 2. An example of IMSRL assessment; adopted from [10]

The following examples of IMSRL calculations taken from [9,10]:
There are 3 TRL component values, namely TRL-1 = 6, TRL-2 = 4; TRL-3 = 8.
There are 3 IRL connections, namely: IRL 1-2 = 5, IRL 2-3 = 9, IRL 1-3 = 7

$$TRL = \frac{1}{9}\begin{bmatrix}6\\4\\8\end{bmatrix} = \begin{bmatrix}0.67\\0.44\\0.89\end{bmatrix} ; IRL = \frac{1}{9}\begin{bmatrix}5 & 0 & 7\\5 & 9 & 0\\0 & 9 & 7\end{bmatrix} = \begin{bmatrix}0.56 & 0 & 0.78\\0.56 & 0.44 & 0\\0 & 0.44 & 0.78\end{bmatrix}$$

$$ITRL_{min} = \begin{bmatrix}0.44\\0.44\\0.44\end{bmatrix} ; IMSRL = min\begin{bmatrix}0.44\\0.44\\0.44\end{bmatrix} = 0.44$$

3.1.3 The third step:

The mathematical properties and SRL^+ readiness reversal are adopted from SRL model testing [10,12]. The mathematical properties is adopted through a testing on four model properties [12].

- Property # 1: The $ITRL_i$ and SRL^+ cannot exceed the maximum available value for TRL or IRL.

$$SRL^+ = min\ (min\ (IRL\ x\ TRL) \leq 1 \qquad (7)$$

- Property # 2: raising TRL without changing other components will not decrease ITRL or SRL^+ values
- Property # 3: The existence of new technology with TRL and IRL is equal to or greater than the initial SRL, then the new SRL will be greater or equal to the initial SRL.
- Property # 4: ITRL component or subsystem does not exceed the maximum TRL or ITRL component or subsystem.

We use the example above and using Equation (6):
If : SRL^+ = f {(TRL, IRL), MRL} = f {IMSRL, MRL}. Then:

$$SRL^+ = min\ (IMSRL\ )\ x\ (\tfrac{1}{10})MRL \qquad (8)$$

- If MRL = 3, so: $SRL^+ = 0.44 \text{ x } \frac{3}{10} = 0.132$ (0.132 ≤ 1); property 1 is fulfilled.
- If MRL is raised to 4, then: SRL^+ = 0.44 x 4/10 = 0.44 x 0.4 = 0.176 (0.176 > 0.132); property 2 and 3 are fulfilled.
- ITRL = 0.44 < max TRL = 0.89; property 4 is fulfilled.

Table 2 shows the acquisition stage for the value of a particular SRL which established according to the scale of the mapping SRL. This process is referred to as SRL^+ readiness reversal.

Table 2. IMSRL values; adopted from [10]

	TRL/IRL Scale Value				
	1 2 3 4	5 6	7	8	9
Acquisition Phases	Material Solution Analysis (MSA)	Technology Demonstration (TD)	Engineering & Manufacturing Development (EMD)	Production & Deployment (P&D)	Operations & Support (O&S)
IMSRL Value Range	[0.00,0.44]	[0.44,0.55]	[0.56.0.79]	[0.80,0.89]	[0.90,1.00]

This study defined the SRL^+ readiness reversal as a calculation of certain higher SRL values than certain TRL and IRL combinations that can be supported. By referring to the Table 2, then in the example : $SRL^+ = 0.132$. This means the cluster is in the MSA phase. With a minimum value of TRL = 4, then the system is still in the MSA phase.

4 Conclusion

The value of the four properties are fulfilled and confirmed that SRL^+ is a stable and mathematically robust model capable of assessing complex systems. Readiness reversal method is utilized as a method of SRL^+ computation evaluation [10]. It is clear from the method that the value of SRL^+ will be higher than the values of TRL, IRL, and MRL which support it. In reference to table of SRL scale, the value of TRL= 4, 6, 8; IRL = 5,7,9; IMSRL = 0.44; and MRL = 3, so the value of SRL^+ is 0.132. It is meaning that the clusters are in phase of MSA. With TRL minimum value of 4, the system remains in the phase of MSA. This indicates that SRL^+ model can be mathematically proven.

References

[1] D. Clausing, M. Holmes, Res. Manag. **53**, 52–59 (2010)

[2] J.C. Mankins, Acta Astronaut, **65**, 1216–1223 (2009)

[3] N. Azizian, T. Mazzuchi, S. Sarkani, D.F. Rico, Syst. Eng., **14**, 410–425 (2011)

[4] C. Wei-gang, L. Wo-ye, G. Yan, H. Fei, Int. Conf. Manag. Sci. Eng. - Annu. Conf. Proc., 1993–1998 (2013)

[5] S. Peters, Prod. Eng., **9**, 647–654 (2015)

[6] M.S. Seablom, L.A. Lemmerman, Int. Geosci. Remote Sens. Symp. 5646–9 (2012)

[7] A.E. Gudanowska, Bus. Theory Pract., **17**, 243–250 (2016)

[8] W. Tan W, J. Ramirez-marquez, B. Sauser, Syst. Eng., **14**, 279–293 (2011)

[9] M.A. London, T.H. Holzer, T.J. Eveleigh, S. Sarkani, J. Syst. Sci. Syst. Eng. **23**, 377–403 (2014)

[10] M.A. London, Evaluating System Readiness Level Reversal Characteristics Using Incidence Matrices, Dissertation (2015)

[11] S. Ross, Def. ARJ, **23,** 248–273 (2016)

[12] E. Mcconkie, T.A. Mazzuchi, S. Sarkani, D. Marchette, Syst. Eng., **16**, 391–400 (2013)

Design of Key Performance Indicators (KPI) for Sustainable Supply Chain Management (SSCM) Palm Oil Industry in Indonesia

Elisa Kusrini*[1], *Rangga* Primadasa[1]

[1]Department of Industrial Engineering Indonesia Islamic University, Jl Kaliurang Km 14,5Yogyakarta 55584 , Indonesia

Abstract. This paper aims to design key performance indicators (KPIs) for evaluating sustainable supply chain management (SSCM) for palm oil processing industry in Indonesia. Supply chain sustainability is the management of environmental, social and economic impacts, and the encouragement of good governance practices, throughout the lifecycle of goods and services. The method in designing KPIs using a triangulation method, ie combining of theory (literature study) with field surveys and validate with the opinion of the expert (expert judgment) as well as compatibility with existing regulations (Indonesian sustainable palm oil / roundtable on sustainable palm oil). There are 29 proposed KPIs for measuring SCM's sustainable palm oil in Indonesia based on literature and expert judgment. In order to obtain a more precise performance, then the importance level of KPI will be measured using Analytic Hierarchy Analysis (AHP) method. According to the preferences of some managers of palm oil using AHP method showed that the economic factor is the most dominant indicator (62,92%), followed by environmental (18,93%) and social factors (18,15%). In subsequent studies, the results of the KPIs will be used to measure the index of sustainability in the palm oil's supply chain industry in Indonesia.

1 Introduction

Palm oil industry is an important industrial sector for Indonesia since it contribute significant revenue for Indonesia. It can be seen from the total exports of CPO (Crude Palm Oil) for the last three year that remain high. In year 2016, total export reached 25.1 million tons, while in 2015 total export reached 26.40 million tons and in 2014 total export reached 21.76 million tons [1]. The nominal value of CPO exports in 2016 reached 18.1 billion dollars and 2015 reached 18.67 billion dollars [1]. Palm oil industry face many problems and getting increasingly higher. At late 90's palm oil sector started facing criticism due to environmental and social issue such as forest fire and land ownership [2]. Meanwhile, Indonesia's CPO exports to the EU in 2015 reached 4.4 million tons, an

* Corresponding author: elisakusrini@uii.ac.id

increase of 3 percent compared to 2014 which reached 4.34 million tons [1]. The challenges facing the palm oil industry in Indonesia in 2020 is the European Union plans to apply the rules of obligation 100 percent sustainable supply chain for palm oil-based products. Also, this time there is a tendency of people will spend more money on a product that is a good impact on the environment. To face the challenges mention above, The palm oil industry must improve their sustainability performance throughout their supply chain.

The principle of sustainability can be implemented well, if it is integrated in all the actors in the industry. Hence the sustainable supply chain management (SSCM) is required. SSCM can be used to address issues of socio-environmental and improve the performance of community [3]. Performance measurement systems make an important contribution towards decision making in supply chains [4]. It also helps in organizational change and improvement [5]. Moreover, SSCM performance measurements is required to identify critical indicators that can affect the performance of SSCM in achieving its objectives. Therefore in this research, key performance indicators of SSCM palm oil industry will be identified.

A literature study conducted to determine the indicators that are typically used in the measurement of sustainability. The indicators used are base on study conducted by Veleva et al [6], Faulkneer et al [7], Sparks.D.T [8], Bai&Sarkis [9], Hall et al [10]. These indicators are then selected based on conformity with the principles, criteria and guidelines of the RSPO (Roundtable on Sustainable Palm Oil) and or ISPO (Sustainable Indonesian Palm Oil). RSPO is an international multi-stakeholder organization established in 2004, adopted the Millennium Development Goals (MDGs) related to 3P (People, Profit, Planet) on the principles and criteria. RSPO is a business initiative whose members voluntarily committed themselves to the RSPO mechanism with the aim to produce and use sustainable palm oil. Principle plantations adhering to the principles of legality prioritize sustainability, environmental and socio-economic feasibility of the long term [11]. In March 2011, the Government of Indonesia through the Ministry of Agriculture launched guidelines for sustainable oil palm plantations Indonesia is often called ISPO. ISPO aims to ensure the implementation of laws and regulations relating to palm oil plantations so that it can be produced sustainable palm oil and support the Indonesian president's commitment to reduce greenhouse gases [12]. While RSPO is voluntarily, ISPO itself is compulsory (mandatory) for plantation companies except small plantation business. This regulation in March 2015 became Certification Systems of Sustainable Palm Oil [13]. This paper proposed a set of Key Performance Indicators (KPIs) to measure the performance of sustainable supply chain management (SSCM) corresponding to the palm oil industry in the three factors of sustainability, namely economic, social and environmental.

2 Methodology

There are five stages in identifying KPI. The first step is to identify sustainability KPI based on criteria of ISPO and RSPO. The second step is identifying KPI from literature, and then combine KPI from RSPO/ISPO and literature. The third step is intended to validate KPI from previous step through expert survey. The last step is to weighting KPI using Analytical Hierarchy Process (AHP) based on the preferences of the experts. The results of this research is a set of weighted KPI that can be used to measure sustainability index for the palm oil industry.

2.1 Identify KPIs from Principles and Criteria of ISPO/RSPO

This study begins by identifying sustainability KPI's through reviewing the ISPO and RSPO criteria. Then, KPI of ISPO and RSPO are grouped into three sustainability factors include economic, environmental, and social.

2.2 Identify KPIs from Literature and Expert Survey

Reviews of the literature conducted to obtain sustainability KPIs for evaluating sustainable supply chain management (SSCM) palm oil mills, as seen in Table 1. A total of 10 palm oil industry actors/expert (an expert is someone who has knowledge or extensive skills in palm oil more than 7 years) are selected to give opinion regarding the importance level of the KPIs criteria to evaluate the results of KPI from the literature study. The expert is from some positions including : mill manager, head mill assistant, assistant process, quality assurance assistant, SHE assistant, transportation assistant, plantation manager, head plantation assistant, plantation assistant, accounting assistant. The questionnaire using Likert scale with a range of 5, from 1 (very important) to 5 (very important) is employed as a measuring tool. The average value of the level of importance is presented in Table 1. The lowest KPI will be removed from the list.

Table 1.KPI from Literature Review and Expert Survey

No	Indicator	Literature	Mean
1	Work In Process	Faulkneer et.al.[7]	3.7
2	Cost Associated with EHS	Veleva et.al. [6], Bai&Darkis .[9],	3.7
3	Rate of Customer Complaint and Return	Veleva et.al. [6]	2.8
4	Material Usage	Sparks. D.T [8] Hall et al [10]	3.6
5	Acidification Potential	Veleva et.al. [6]	3.6
6	Percent Product with Take-back Policies	Veleva et.al. [6]	3
7	Physical Load Index	Faulkneer et.al.[7]	4.2
8	Electrical System Hazard	Faulkneer et.al.[7], Bai&Darkis .[9],	3.7
9	Noise Hazard	Faulkneer et.al.[7], Bai&Darkis .[9],	3.7
10	Average Length of Service of Employees	Veleva et.al [6], Hall et al [10]	4
11	High-Speed Component Hazard	Faulkneer et.al.[7], Bai&Darkis .[9],	3

2.3 Proposed KPIs base on ISPO/RSPO And Expert Judgment

The KPI selected from ISPO / RSPO and assessment from industry players and experts are then combined as shown in Table 2.

Table 2. The Proposed KPI base on ISPO/RSPO and Expert Judgment

No	Category	KPI
1	Economic	OER (Extraction), Oil Losses, FFA, Moisture, % Breakdown, Lead Time/Plant Time, Value Added Time, Work In Process, Cost Associated with EHS
2	Environmental	Water Usage, Energy Usage, Waste Generated Before Recycled, Global Warming Potential, BOD, COD, % Waste Reused, % CPO Certified , Material Usage, Acidification Potential, Percent Product with Take-back Policies
3	Social	Lost Workday and Injury Illness, Hour of Employee Training, Worker Job satisfaction, Local Community Hiring, Physical Load Index, Electrical System Hazard, Noise Hazard, Average Length of Service of Employees, High-Speed Component Hazard

3 SSCM Model Evaluation Using Analytic Hierarchy Analysis Model (AHP) for Palm Oil Industry

In this study, SSCM performance evaluation using KPI selected from RSPO/ISPO and actors/expert judgment is proposed. In order to obtain a more precise performance, then the importance level of KPI will be measured using Analytic Hierarchy Analysis (AHP) method. Analytic Hierarchy Process (AHP) was first introduced by Thomas L. Saaty in 1971 and became the most widely method for multiple criteria decision making (MCDM). It is a decision-making approach that is designed to find solutions to complex problem with many criteria into multiple domains. It is also known as an essential tool for practitioners and academics to conduct research in decision making and measure management theory. AHP as a flexible problem-solving methods and systematic tools where can represents elements of a complex problem [14,15].

Cheng.et.al [16] mentioned that AHP has several benefits. First, it helps to breakdown unstructured problems into a hierarchy of rational decision making. Second, it can display a lot of information from experts and decision makers using pair wise comparisons of different groups of individuals. Third, it has calculation procedure to weight the importance level of each element. Fourth, it has consistency measurements to validate the rating given the consistency of experts and decision makers. The following are the stages of development of AHP model for the measurement of sustainable supply chain management (SSCM) of palm oil industry.

Step 1: Creating Hierarchy of Indicators

The hierarchy of indicators is created from the KPI in the previous stage. The KPI is grouped in to three category, ie, economic, environmental, and social.

Step 2: Weighting the KPIs

After the hierarchy has been created, the importance of KPIs can be calculated. A questionnaires of pair wise comparisons among indicators is prepared and a total of ten senior managers from palm oil mills were asked to give their preferences on the questionnaires. All managers chosen carefully based on their experience in the palm oil industry. Pair wise comparisons conducted on factor between KPIs. The scale of 1 to 9 (1 = equally, 3 = moderate, 5 = strong, 7 = very strong, 9 = extreme) is used to rate the preferences of expert. Consistency ratio (CR) is used to check the consistency of the comparison. CR less than 0.1 means that comparison is accepted.

Based on the calculation of CR, It has value between 0.017696 and 0.027408, which means all pair wise comparisons are consistent because they are on the acceptable range recommended. This indicates that the experts provide consistently in weighting the importance of KPI for sustainable supply chain management palm oil mills. According to the preferences of some managers of palm oil using AHP method showed that the economic factor is the most dominant indicator (62,92%), followed by environmental (18,93%) and social factors (18,15%).

4 Discussions

Palm oil mill processing of fresh fruit bunches (FFB) into Crude Palm Oil (CPO) and kernel. This plant also produces waste in the form of empty bunch, palm oil mill effluent (POME), liquid waste and greenhouse gas emissions. The oil industry in general is also facing major challenges in the social such as land ownership conflicts and demands of the welfare of the surrounding community. The application of sustainability have tended to be fragmented and incomplete along the supply chain. Therefore, this study determined indicators to measure the success of the supply chain of palm oil from sustainability's view. There are 29 indicators from KPI selection process based on criteria ISPO and RSPO, literature and judgment from actors/experts. According to the preferences of the manager of palm oil using AHP method showed that the economic factor is the most dominant indicator, followed by environmental and social factors. The most important economic indicators are OER and Oil Losses. The most important indicator in the environmental factor is energy usage, water usage, as well as BOD. The most important indicator of the social factor is the loss of social workday injury and illness and hiring local community. This indicator can be used as guidance for prioritizing action in performance improvement in SSCM palm oil industry.

5 Conclusions

There are 29 valid for measurement of SCM sustainable palm oil in Indonesia. KPI's sustainability consists of three categories, namely 9 KPI's economics (OER/extraction), oil losses, FFA, moisture, % breakdown, lead time/plant time, value added time, work in process, cost associated with EHS), 11 kpi's environment (water usage, energy usage, waste generated before recycled, global warming potential, BOD, COD, % waste reused, % CPO certified, material usage, acidification potential, percent product with take-back policies), 9 kpi's social (lost workday and injury illness, hour of employee training, worker job satisfaction, local community hiring, physical load index, electrical system hazard, noise hazard, average length of service of employees, high-speed component hazard). According to the preferences of the manager of palm oil using AHP method showed that the economic factor is the most dominant indicator, followed by social and environmental factors. Further research is to measure performance based on KPI's and determine strategies for improvement.

References

1. Gapki.id, *Press release: reflection of palm oil industry 2016 and 2017 outlook*, Accessed 2 February (2017)
2. Moyano-Fuentes. J., Martinez-Jurado. P.J, JCP, **85**, 134-150,(2014)
3. Kusi-Sarpong. S., Sarkis. J., Wang. X., Int. J. PE, (2015)
4. Kurien G and Qureshi M N, SRE, Vol. **7**, pp. 3149-3161 (2012)
5. Tangen S, *Analysing the Requirements of Performance Measurement Systems, Measuring Business Excellence*, Vol. **9**, No. 4, pp. 46-54.,(2005)
6. Veleva. V, Ellenbecker. M., JCP, Vol **9,** pp 519-549,(2001)
7. Faulkner. W., Badurdeen. F., Templeton. W., Gullet. David.,*Proceedings of International Conference on Industrial Engineering and Operation Management Istanbul, Turkey, July 2-6, (2016*)
8. Sparks. D.T., *Thesis and Dissertations University of Kentucky UKnowledge*, (2014)
9. Bai .C and Sarkis.J, An International Journal 19/3 (2014) 275–291, Emerald Group Publishing Limited [ISSN 1359-8546], [DOI 10.1108/SCM-12-2013-0441],(2014).
10. Hall,J. , Matos,S. and Silvestre,B., IJPR, Vol. **50**, No. 5, 1332–1348,(2012)
11. Nedyasari.et.al. *920150 Joint Study of The Same and The Differences ISPO and RSPO Certification*, (2015)
12. Ministry of Agriculcure Policy of Republik Indonesia No.**11** Year (2015)
13. Principles and Criteria of RSPO, (2006)
14. Saaty TL., Eur JPAM,Vol **1,** 122-96, (2008)
15. Chan FTS, Chan HK, Lau HCW, Ip RWL, An International Journal **13** (6): 636-61,(2006)
16. Cheng EWL,Li H, Ho DCK..,*Analytic Hierarchy Process : A defective tool when used improperly. Measuring Business Excellence Vol* **6,** 33-37, (2006)

Usability Evaluation of Anthropomorphic Prosthetic Hand Based on ISO 9241-11

Novie Susanto[1,*], *Wiwik* Budiawan[1], *Rifky* Ismail[2], *Mochammad.*Ariyanto[2] and *Ahmad* Sahal[1]
[1]Industrial Engineering Department, Faculty of Engineering, Diponegoro University, Semarang
[2]Centre for Biomechanics, Biomaterials, Biomechatronics and Biosignal Processing (CBIOM3S), Central Laboratory Building, 5th Floor, Diponegoro University, Semarang

Abstract. This paper aims at evaluating an anthropomorphic prosthetic hand based on usability concept. It can be controlled for pre-defined five motions. The assessment is conducted by design approach. Users are given the task according to the rules of hand benchmarking in Southampton Hand Assessment Procedure (SHAP) by using Anthropomorphic Prosthetic Hand to perform daily activities. The quantitative data is obtained by a test that based on the criteria of learnability, memorability, satisfaction, efficiency and errors. The qualitative data is obtained by giving USE (Usefulness, Satisfaction and Ease of Use) questionnaires. The recommendation design is prepared by using Quality Function Deployment (QFD) method to combine suggestions from respondents and product developers. Assessment by USE questionnaires shows that the usability level of Anthropomorphic Prosthetic Hand on the aspect of usefulness is 68.3% and 63.8%; satisfaction is 59.5% and 55.6%. The improvement recommendations based on the result of (QFD) are Variant material to connect to the arm, System to connect to arm, Dimension of connecting to arm, Connection information to the arm, and Instructions for the exercise stage. So that improvement is done in accordance with ISO 9241-11: 1998.

1 Introduction

The success of industry is determined by two things. First, success in designing and developing products that suit the consumer's wishes. Second, the speed of the industry is in adapting/responding to changes in consumer desires [1]. Various product innovations are developed as a manufacturer's step to get customers from various sectors. One sector that can not be separated from the creation of various innovative products is the health field. This field is considered quite potential as one of the media product development in accordance with one's personal needs.

Aids in the health field can be said to be an instrument, apparatus, machine, and implant containing no medication. The tools are used to prevent, diagnose, cure and alleviate illness, treat the sick, restore health to humans, and structure and improve body function [2].

* Corresponding author: novie.susanto@ft.undip.ac.id

One type of tool that is often used is a tool of human motion. This tool is commonly used for patients who have physical deficiencies due to various things that have occurred such as accidents, diseases, and heredity [3]. In general, the main function of the tool is to help and ease the patient's complaints [4]. Manufacturers are required to create a tool that suits the needs of the patient. The development of research on hand tool products for disabled persons in Indonesia is not much. This is due to the high cost of purchasing a variety of sensors used to make tools to have the same function with human hands [5]. Therefore, this tool is designed to help develop research on tools for hand disability patients although in the manufacture of the product it is only equipped with some features of hand movement.

Anthropomorphic Prosthetic Hand (Fig. 1) is one of the motion aids developed by Diponegoro University Mechanical Engineering which is specially designed for disabled persons [5]. It is designed based on a custom dimension of Indonesian. It controls some pre-defined movements as designed by the product developer. Anthropomorphic Prosthetic Hand uses a drive motor with a linear type Actuonic PQ-12 acuator mounted inside an artificial hand component. The materials used are PLA (Polylactid Acid) with brown color and printed using 3D printer. Microcontroller or brain control on Anthropormophic Prosthetic Hand is Arduino nano which is one type of microcontroller developer board with a small size and complete function. As an indicator of the current movement, Anthropormophic Prosthetic Hand using RGB LED lights. Each color emitted to inform what movement is active [5].

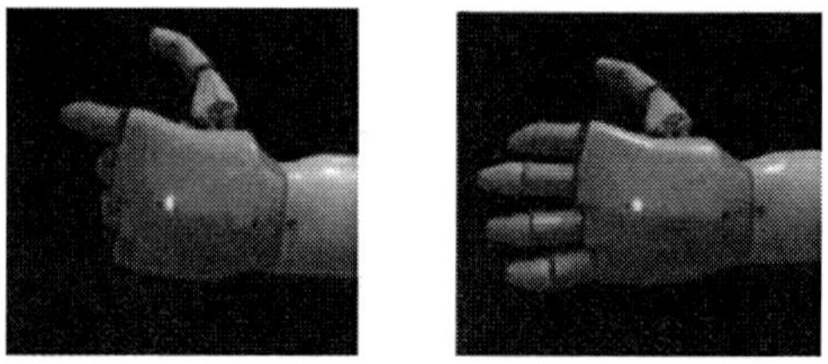

Fig. 1 Anthropomorphic Prosthetic Hand

This tool is designed based on the human dimension of Indonesia. The designer can control this tool for some predetermined movement. Anthropormophic Prosthetic Hand uses a motor with a linear type Actuonic PQ-12 actuator mounted inside an artificial hand component. The material used is PLA (Polylactid Acid) with brown color and printed with 3D printer. The microcontroller used in Anthropormophic Prosthetic Hand is Arduino nano. Arduino nano is one type of microcontroller developer board that has a small size and complete. As an indicator of movement, this tool use RGB LED light. The color of the emitted lamp informs what movement is active [5]. The majority of people with disabilities in Indonesia use artificial hands and feet that are passive [6]. Passive aids cannot perform the main function of patients with disabilities. America, China, and Britain have developed an active artificial hand. The Aids can move automatically according to what the patient wants. The active tool can move with electromyograph (EMG) sensor or electroencephalograph (EEG) sensor. EMG and EEG are sensors used to capture the input signal and muscle contraction of the patient's arm.

Currently, the increasing demand for medical equipment has not been followed by the development of domestic industry. This causes 90% of health equipment circulating in Indonesia is imported products [7]. The price provided by the manufacturer is also high because of the active functionality of the disability tool. Manufacturers must install several sensors used to capture the input signal from the patient's muscle contraction. Therefore, the people of Indonesia prefer to use passive motion tools (cheaper).

Development of Anthropomorphic Prosthetic Hand is expected to suit the needs of users (dimensions and prices). In support of the creation of products that suit the needs of users, testing and evaluation needs of the product. One of the evaluations that can be done is the usability of the product [8]. This evaluation is needed to assess how far the product is made to meet the needs of consumers so that later consumers will feel satisfied when using the product. Based on these conditions, it can be formulated problem about evaluation Usability use of Anthropomorphic Prosthetic Hand. Evaluation is needed to measure how easily the product can be understood and used by the patient. The measured Usability Criteria are Learnability, Memorability, Errors Efficiency and Satisfaction. The data was collected by experiment according to the Southampton Hand Assessment Procedure (SHAP) handling procedure and completed with USE (Usefulness, Satisfaction and Ease of Use) questionnaire.

2 Methods

The purposes of this research are to identify the factors that influence the level of ease of use of Anthropomorphic Prosthetic Hand product, to measure the usability level of Anthropomorphic Prosthetic Hand product on handicapped patient with USE (Usefulness, Satisfaction and Ease of Use) questionnaire, and to propose improvement on Anthropomorphic Prosthetic Hand products use QFD (Quality Function Deployment) method. To answer the purpose of research, there are several stages of research done. Stages of research done are Research Design, Identify Variables, Experiment, Result Analysis, and Product Improvement [10].

2.1 Research Design

In this study, testing the tool to the handicapped persons respondents does the assessment of product development performance. Respondents will be given information about the research step. Taking video as a tool for measuring the time when respondents do the task does research. The data of the research were collected for analysis of usability level. The analysis results are then used as the basis for the input of the QFD method. Research data used for this research is quantitative and qualitative data. For quantitative measurements this study refers to Jacob Nielsen's theory of measuring the usability of a product or system. Usability parameters are used because there are appropriate criteria for new products or systems [9]. There are 5 usability criteria: learnability, efficiency, memorability, errors and satisfaction. Qualitative data collection was collected by questionnaire. There are several types of questionnaires that can be used to measure the level of ease of use of products or systems, namely USE (Usefulness, Satisfaction and Ease of use), CSUQ (Computer System Usability Questionnaires) and PUTQ (Purdue Usability Testing Questionnaire). Of the several types of questionnaires are mostly used to measure the product or digital system with consideration of human computer interaction system. Thus, the questionnaire used in this study is USE. USE can be used for evaluation of usability of non-digital products.

2.2 Identify Variables

There are 2 variables in this research that is Independent Variable and Dependent Variable. The independent variable in this study is the size and features of Anthropomorphic Prosthetic Hand products. The dependent variable in this study is the level of Learnability, Efficiency, Memorability, Errors and Satisfactions. It also complies with the standard Ergonomic requirements for office work with visual display terminals (ISO 9241-11:1998) [14]: Learnability (The level of ease of learning of Anthropomorphic Prosthetic Hand products is measured by interviewing the respondents); Efficiency (The level of efficiency is measured by measuring the respondent's time doing the task); Memorability (The level of Memorability is measured by repeating the tests performed with the same task without instruction); Errors

(The level of errors is measured by how the respondent performs the assigned task); and, Satisfaction (The level of user satisfaction Anthropomorphic Prosthetic Hand is measured by interviews shortly after the responder experiments.)

2.3 Experiment

Anthropomorphic Prosthetic Hand products were then tested on handicapped persons to measure these factors. Tests performed are the standard daily activities defined by the SHAP (Southampton Hand Assessment Procedure). Then the experiment is measured based on the criteria of Learnability, Efficiency, Memorability, Errors and Satisfactions. Based on the five functions (Tripod, Active Index, Precision Closed, Power grip, Hook) that exist in the Anthropomorphic Prosthetic Hand, the activities in the SHAP are capable of: Taking coin (Active index); Moving objects Cylindrical light (Power grip); Moving the tray / tray (Precision closed); Opening the door with the handle (Hook); Pour water from a carafe into a glass (Tripod); Cut with a knife (Power grip); and Opening / closing the zipper (Active index).

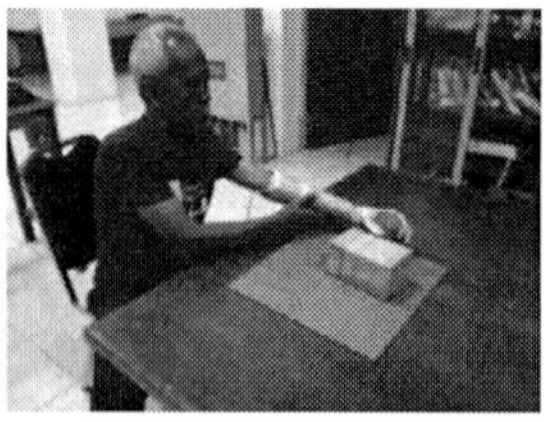

Fig. 2 Activity Moving Objects

Fig. 3 Activity Take a Coin

2.4 Data Analysis

In this section the data collected from the respondents is analysed more deeply. The results of data recapitulation were analysed to the measured criteria for later design of proposed improvement on Anthropomorphic Prosthetic Hand products.

2.5 Product Improvement

There are several methods used as the basis for the preparation of product evaluation. Among them are: QFD (Quality Function Deployment), QUAD (Quality and Usability Assured Design), and UCD (User Cantered Design). For design improvement, QFD (Quality Function Deployment) method is used in this research. The QFD method is chosen because it considers the respondent's voice as a consumer of the Anthropomorphic Prosthetic Hand product [10]. The stages of QFD are:

1. Voice of Customer: In the early stages is the interview after the product is tested to the respondent. Respondents were given a USE questionnaire to obtain qualitative data. Questionnaire results are seen aspects that have a low scale (scale below the middle value) of the assessment respondents. This low-scale aspect is an unmet consumer need when using Anthropomorphic Prosthetic Hand
2. House of Quality: House of Quality is a useful matrix for transforming consumer needs into a technical system for a product. The process is to set priorities for each of the characteristics and target development of the product [11][12][13].
3. Analysis and interpretation: The HoQ matrix is compiled using template matrix (QFD Method). The HoQ matrix is used as the design benchmark. Then, the HoQ Matrix

Results are analyzed and interpreted to obtain product improvement recommendations. Product improvement recommendations are generated according to customer needs.

3 Result and Discussion

In this study the respondents who successfully collected a number of two people. Two respondents have similar characteristics. Characteristics include: Disability on the left hand with elbow borders/ Disability limit. The number of persons with disabilities with similar characteristics is not much. This should be answered with the flexibility of Anthropomorphic Prosthetic Hand products. Furthermore, Anthropomorphic Prosthetic Hand is used on both respondents. Respondents then run daily activities determined by SHAP. When running SHAP activities, respondents assessed the achievement of the time. Time measurements other than using stopwatch, also re-verified with video recording. Some of the daily activities defined by SHAP include: Open the Door; Take a Coin; Moving Objects; and Opening Zippers. During the five experiments, the average time required to perform the activity decreased (both respondents). Table time decrease performance can be seen in Table 1.

Table 1. Time Decrease Performance (Seconds)

No.	Open The Door	Moving Objects	Take a Coin	Opening The Ziper
1	12.06	3.89	39.05	26.98
2	9.35	3.67	50.47	19.59
3	2.02	3.19	47.11	10.59
4	2.29	3.07	33.75	10.01
5	3.16	3.05	27.5	2.35

After the experiment, then the respondents fill out the questionnaire. The questionnaire is used to answer some aspects of the product. Aspects of these products include: usefulness, ease of use, learnability, and satisfaction. There are 30 questions in the questionnaire. The assessment results of both respondents, Anthropomorphic Prosthetic Hand products overall status of the four aspects is "GOOD" (Total usability level Respondent 1: 71.8% and Respondent 2: 73.2%). Based on the recapitulation of four aspects (usefulness, Ease of use, Learnability, and Satisfaction), there are two aspects that have an average of less than 70%. Both aspects are Usefulness and Satisfaction. Both aspects are then verified to both respondents. The interview results found some obstacles in the aspects of Usefulness and Satisfaction.

1. Early use of Anthropomorphic Prosthetic Hand product cannot be directly used. Users should familiarize themselves when using the product.
2. At the time of connection need a flexible hook mechanism because the dimensions of the human body variant.

Furthermore the results of such data are used as the basis for product improvement (QFD). To determine priorities for improvement of technical characteristics, each technical characteristic is ranked based on the value of the contribution of technical characteristics. Considering the fulfilment of attribute needs and achievement targets also does the priority determination of the development of technical characteristics. Based on the results of the ranking and the fulfilment of the target technical characteristics, then selected 5 attributes that have not met the target and can represent all the attributes of customer needs: (1) Variant material to connect to the arm; (2) System to connect to the arm; (3) Dimension of connecting to arm; (4) Connection information to the arm; and, (5) Instructions for the exercise stage. Of the 5 attributes selected, then the discussion of concepts. The concept aims to focus on improvements that will be done by the developer. By conceptualizing, the improvements made will have more variation in the improvement. The design of the concept will select all the attributes found in the previous experiment. However, alternative concepts will be discussed in subsequent research.

4 Conclusion

Excellent status on Anthropomorphic Prosthetic Hand products must be achieved so that customers feel satisfied. The downside of the Anthropomorphic Prosthetic Hand product is found in the aspects of usefulness and satisfaction. Then the discovery of these two aspects is used in product improvement. Future product improvements will focus on five attributes. The five attributes are: Variant material to connect to the arm, System to connect to arm, Dimension of connecting to arm, Connection information to the arm, and Instructions for the exercise stage. These improvements are made to obtain the standard Ergonomic requirements for office work with visual display terminals (ISO 9241-11: 1998).

References

1. Flores, J., Gonçalves-Coelho, A., Mourão, A., Cavique, M. Theory of practices as a means to uncover the customer needs. *9th International Conference on Axiomatic Design – ICAD 2015*. Elsevier Publisher (2015)
2. Clarckson, Douglas M, *Medical Device Guidebook: A browser information resource for medical device users*. Elsevier Publisher (2017)
3. Lazar, J.,Feng, HJ., Hochheiser, H. *Working with research participants with disabilities : Research Methods in Human Computer Interaction* **(Second Edition)**. Elsevier Publisher (2017)
4. Wang, N., Lao, K., Zhang, X. Design and Myoelectric Control of an Anthropomorphic Prosthetic Hand. *Journal of Bionic Engineering*. **14**, 47. Elsevier Publisher (2017)
5. Primarezky, E. Development of 5 DOF Myoelectric Prosthetic Hand in Hand Patients With Disability Semarang. Department of Mechanical Engineering Diponegoro University (2017)
6. Riskesdas. *Data riskesdas*. Access: March 2017, Ministry of Health: www.depkes.go.id/resources/download/general/Hasil%20Riskesdas%202013. Pdf (2013)
7. Kemenkes. *Health Profile*. Access March 2017, Ministry of Health: www.depkes.go.id/resources/download/.../profil-kesehatan-indonesia-2013.pdf (2013)
8. Chisnel, D. *Hand Book of Usability Testing*. Indianapolish. Wiley publisher (2008)
9. Xiangwei, L.,Yun, B.,Yuhui, Q. *Measuring Usability: Use HMM Emotion Method and Parameter Optimize*. Springer Publisher (2006)
10. Revelle, Jack B. *Quality Essentials: A Reference Guide from A to Z*. ASQ Publisher. (2004)
11. Gazpersz, V. *Analysis Methods to Improve Quality*. Gramedia Indonesia. (2001)
12. Jonnasen, David H. *Problem-Based Learning: Encyclopedia of the Sciences of Learning*. Springer Publisher (2012)
13. Cohen, L. *Quality Function Deployment: How to Make QFD Work for You*. Addison-Wesley, Inc. (1995)

Integration between Green Quality Function Deployment, Modularity Concept and Life Cycle Assessment Toward Sustainable Product Design

Heru Prastawa, *Sri* Hartini, *Mohamat* Anshori, *Siechara* Hans, *Christoper* Wimba

Industrial Engineering Program, Diponegoro University, Prof. H. Soedarto, SH. Semarang, Indonesia

Abstract. The design phase is recognized as a key phase in the application of sustainable manufacturing concepts. Green Quality Function Deployment (GQFD) and modularity play an important role in product design. Green Quality Function Deployment produces technical parameters that represent the needs of consumers while taking into account environmental aspects. Modularity benefits manufacturing and flexibility in facing adjustments and changes. Integration of GQFD and modularity is expected to generate synergistic gains from both. The results are measured by life cycle assessment (LCA) to determine the impact of the product on the environment. This study shows that GQFD, modularity and LCA integration in realizing sustainable product design is worthy of consideration. The case study was conducted with the fan because the product is very needed in the tropics, such as Indonesia.

1 Introduction

Increased public awareness and government regulation on environmental protection challenge product designers to consider environmental aspects of product design goals [1]. The design phase has been widely recognized as a key phase in the product life cycle in the application of sustainable manufacturing [2]. The design of a product will affect the next stage, such as material requirements, manufacturing process, energy consumption, and activity at the product disposal stage [1]. 80% of the cost of development, production, and use of the product is determined by the initial design stage [3]. The earlier the design team considers the environmental factors in the product life cycle, the greater its potential for environmental benefit and cost reduction.

Quality Function Deployment (QFD) is a method that transforms consumer needs into engineering parameters in design and production. From 1993 to 2009, there were 17 eco-design methods developed based on QFD [4]). The first method was developed to improve the quality and cost of eco-design with integrating environmental aspects into the product specification matrix [5]. Then QFD integrated with other methods, such as 3D-QFDE, QFDE/LCA, QFDE/LCA/TRIZ [6-7] and adding environmental criteria as consumer needs [8]. But this integration has not involved the concept of modularity. Modular design can increase the flexibility and availability of adjustments and changes [9], reducing the number of components [10]), reducing cost and development time [11-12], reducing environmental impact [7,12].

This research intends to develop the design concept by integrating Green QFD and modularity design. The results of the integration will be compared using the LCA method. The LCA has the ability to discover important environmental issues [14-15] and is a highly representative tool for analyzing and quantifying resource consumption and environmental impacts on throughout the product life cycle [1]. LCA is often used to compare related product alternatives based on environmental impacts making it easier to know which alternatives are the most environmentally friendly [14]. Case studies were conducted on a fan, electronic products among the lower and middle class in hot climates. Electronic products produce many the environment impact [16]. About 50 million tons of electronic waste is produced annually and less than 10% are recycled [17]. If this is allowed to happen without a precautionary measure, electronic waste can be a threat to the preservation of the earth's environment

2 Research Method

The purpose of product design development by integrating Green QFD, modularity and LCA is to gain synergistic advantages from the three methods. The product design that meets the needs of the consumers is profitable for the manufacturer and also environmentally friendly. Fig 1 is a framework of development model to achieve goals.

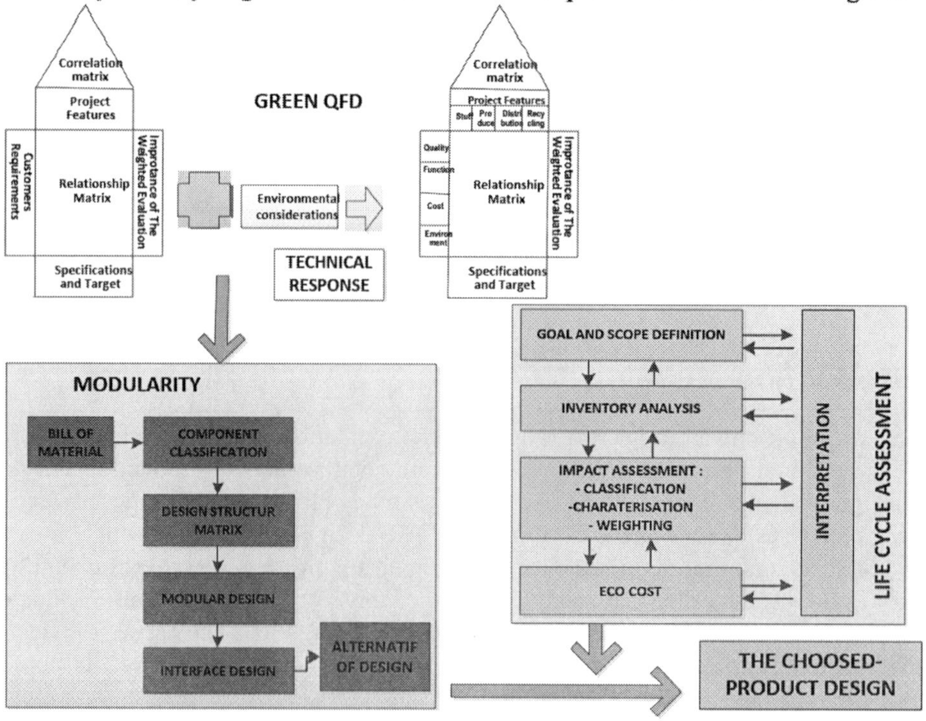

Fig 1. Framework integration GQFD-modularity-LCA

The Green QFD consists of voices of customers matrix, engineering matrix, the house of quality, benchmark, and roof. Voice of customers contains the questionnaire data. The engineering matrix contains technical response factors. The quality hypothesis is a matrix that shows the relationship each indicator of the voice of customers and each factor of the technical responses. Benchmark shows the comparative data of respondents' assessment of the products on the market. The Roof is a matrix that shows the influence of a factor in engineering matrix on other factors. Green QFD was developed by adding environmental criteria in exploring consumer desires. Table 1 presents the dimensions and variables used. This data will then become the voice of customers. Questionnaire method is done to the consumer to know the needs of the consumer. Consumers are selected randomly in the city of Semarang and surrounding areas. Respondents are confirmed as users of fans who have experienced damage. The Proposed design is made to develop products based on the most important criteria according to consumers. House of quality would be developed based on the results of this questionnaire. Benchmark did on stand fan as the best-selling fan of three companies with demand greatest. Consumers are required to provide an assessment of the 3 products based on the assessment variable. The weight calculation for each model is taken based on the most choice of respondents. This weight is then multiplied by the weighted value of the relationship between the voices of customers with technical responses (weight of the house of quality).

Modular design concept can simplify the change of product design as a whole. As an independent module, in case of design change, it is not necessary to change the design core in each module. There are three main features of the module, namely: separable, isolable and re-combinable [18].

The LCA method calculation is performed using SimaPro v 7.1.8 software for evaluating the environmental impact value of the fan product. The LCA has 4 stages, the goal and scope definition, inventory analysis, impact assessment and calculate eco-cost. Eco-Cost is a cost that must be spent to reduce environmental pollution according to the ability of the earth. The Eco-Cost calculation uses 2012 eco-cost method using Eco-Invent V3 2014.

Table 1. Dimensions and Variables used in Green QFD

No	Dimensions	Variable
1	Appearance	Size [19];[20], Color [20]
2	Service	Price [19];[21], Easy to assemble/disassemble [20]
3	Functionality	Multifunction [22], Features replacement speed of rotation [20]
4	Responsiveness	Weight [23], Ergonomic [23]
5	Reliability	Not noisy [23], Stable Usage [23], Easy to clean [8], Easy to maintain [20]
6	Environmental	Recyclable [20], Save energy [19], Not using harmful materials (toxic) [20], The life of the material [23], Easy to reuse [23], Number of components (less material use) [23]

3 Result and Discussion

Based on the questionnaire, the top five criteria are durable, energy efficient, not noisy, easy to repair, and do not use hazardous materials. The life time of the fan does not match the expectations of consumers during this. Besides, to repair a broken fan is not an easy thing. 63% of respondents state that it is difficult to replace the broken component of the fan. This condition triggers behavior for buying a new fan (24%). Oftenly, the damage of one vital component causes the fan becomes disturbed for example motor. If the motor is damaged, although the other components are still good condition, the fan can't use. The life time of the fan becomes more shorter than standard. Furthermore, the increase in the volume of waste will also cause economic loss and a negative effect on the environment [24].

The factor of engineering matrix that needs to be considered is using the non-permanent connection. This factor makes the fan easy to be disassembled and re-installed. It certainly facilitate the user in performing maintenance or minor repairs. This design also has a positive impact on the environment. The use of screws and bolts can reduce chemicals in the connection technique and more easily dismantled without damaging the connection.

The next important factor is the selection of materials. The correspondent wants a fan that is free from hazardous materials. Some ingredients that are proven toxic and should not be used are listed in the Restriction of Hazardous Substances directive. Aluminum may be considered as having high recyclability, low cost, light weight, corrosion resistance, and are not harmful to humans and the environment. Polypropylene is a mild and sufficient plastic strong; not dangerous, easy to recycle, and not easy to absorb dirt so easy to clean. It is in accordance with the wishes of the consumer, which is easy to clean.

Table 2. The comparison between the current fan with the proposed.

	Existing: fixed and single function	**Modular, single function**	**Fixed, multifunction*)**	**Modular Multifunction*)**
Maintenance	Difficult, especially the dynamo as a framework that integrates with the main electrical installation	Easier, because the modular system makes it easy to uninstall.	Difficult, the dynamo system is still the same as the current fan.	Easy, because of the modular concept and system connector
Disposal	All products have the potential to become waste, even if only the damaged dynamo.	The coil, if the dynamo is damaged, just by replacing the coil.	All products have the potential to become waste, even if only the damaged dynamo.	Damaged module only

*) Adding function: Emergency lights, electric mosquito repellent, USB slot, air freshener and backup electrical energy storage.

To provide an eco-fan based on consumer criteria, modularity concept is proposed to facilitate disassembly and improve product lifetime. To reduce energy consumption, the multifunction concept will also be an alternative. Finally, there are three alternatives to be studied, namely the modular concept, multifunctional, and a combination of modular-multifunctional. Based on the survey, respondents were very interested in the double function concept (62.5%) with the desired additional features are room deodorizer (37%), electric mosquito (30%), emergency lamp (17%), and others (16%). The comparison between the current fan with the proposed in Table 2, Table 3 and Table 5.

Table 3. Comparison between the environmental impact value of existing and proposed fan

Impact category	**Unit**	**Current Fan**	**Modular Fan**	**Multifunction Fan**	**Fan+Feature Separated Additional *)**
Climate change	kg CO2 eq	23,235.50	5,958.51	34,437.45	110,614.98
Acidification	kg SO2 eq	406.19	282.81	834.45	1,111.16
Eutrophication	kg P eq	1.05	0.25	1.47	7.88
Photochemical	kg NMVOC	32.68	7.58	36.17	189.41
Fine DUST	kg PM2.5 eq	10.43	1.53	12.18	46.34
Human toxic	CTUh	0.00	0.00	0.00	0.01
Ecotoxicity	CTUe	29,641.99	6,244.77	32,702.71	197,742.07
Metals Depletion	euro	1,154.19	272.01	1,181.39	5,486.89
Waste	MJ	412,182.38	402,081.56	1,218,527.40	412,585.39

Based on Table 3, the modular fan has much less environmental impact than the original fan, especially in ecotoxicity and climate change categories. So is the multifunction fan when compared with the fan along with other multifunction when designed separately. However, the multifunction fan has a very high environmental impact. The additional features on the fan will increase the use of materials and reduce the consumption of electrical energy.

Table 4. Comparison of the cost (1 batch = 1000 unit) (Rupiah)

Impact category	Initial Fan	Modular Fan	Multifunction Fan	Fan+Feature Separated Additional
Total	**219,285,682**	**135,097,294**	**451,316,542**	**767,070,279**
Climate change	48,915,570	12,543,788	72,497,365	232,865,548
Acidification	52,256,200	36,384,063	107,352,477	142,950,774
Eutrophication	194,129	46,120	270,478	1,452,292
Photochemical OF	2,924,659	678,076	3,237,034	16,952,242
Fine dust	4,821,548	708,709	5,633,280	21,425,608
Human toxicity	15,543,154	6,243,954	18,576,658	185,420,497
Ecotoxicity	658,457	138,719	726,447	4,392,574
Metals Depletion	17,998,483	4,241,756	18,422,524	85,562,563
Waste	75,973,901	74,112,107	224,600,001	76,048,185

The decrease in the initial environmental cost of the fan to the modular fan is approximately eighty million rupiah. However, the multi-function fan increases about two hundred thirty millions rupiah with the notes that multi-functional fan capacity reduces the eco cost by about three hundred million rupiahs compared to when the additional functionality is designed separately (Table 4). With the declining value of eco-cost, this will provide social impacts of prevention and maintenance of ecosystem quality, human health quality, and other natural resource aspects. Decreasing the environmental impact is one of sustainable production design criteria.

Products with the modular concept are easy to maintenance and repair, increase product life time, reduce waste and cost of environmental impact. The disadvantage is to increase the production process stage because it requires the interconnection component. The multifunctional product will increase the intensity of product usage and decrease the amount of product to be purchased. The disadvantage is to increase the price of the fan that affects consumer buying decisions, especially if consumers do not need the added features. The proposed alternative design is asked again to the respondent. The result is the majority chose stand fan with modularity concept without dual function (modularity 41%, multi-function 24 % and multifunction with modularity 35%). From the environmental aspect, the first alternative has the least risk. The comparison between the existing fan and proposed fan is shown in Fig 2.

4 Conclusion

Integration GQFD with modular concept results in the product design that meets consumer needs and reduces the environmental impact. Although the concept of modularity must increase production process for the manufacturer, it enables the manufacturer to sell the product as the components. This will make it easier for consumers to do their own replacement of broken parts and add product innovation to the manufacturer. The study also reveals the fact that the level of consumer interest in fans will no longer be the price but is increasing to the need for quality and environmental protection (product life, energy saving, non-hazardous materials). And the concept of modularity can increase the life time and reduce the environmental impact. To improve the sustainability of the product, further study about Product Service-System (PSS) of the fan is needed.

The existing design :
- Design of motor with fixed design using welding
- The body has not added function

Motor design with modularity concept with nuts and bolts

Adding function: emergency lights, electric mosquito repellent, USB slot, air freshener and backup electrical energy storage

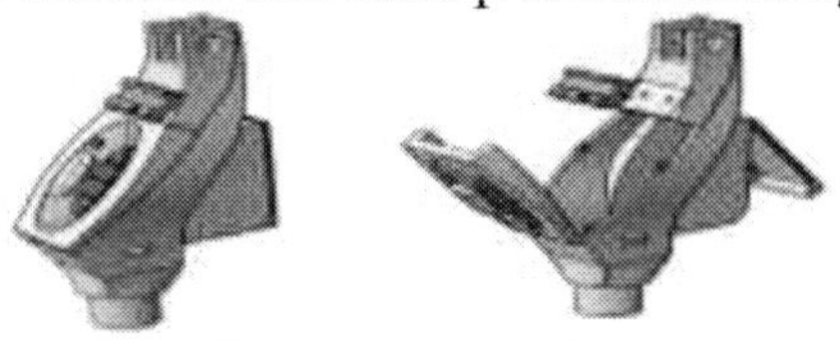

The body of proposed fan have space for the adding component multifunction

Fig 2. The comparison between the existing fan and proposed fan

References

1. G. Rebitzer, T. Ekvall, R. Frischknecht, D. Hunkeler, G. Norris, T. Rydberg, W.-P. Schmidt, S. Suh, B. P. Weidema, and D. W. Pennington, Environ. Int., **30** (2004).
2. K.T. Ulrich and D. Eppinger, Product Design and Development, **2nd** (1995)
3. C. Mascle and H. P. Zhao, Int. J. Prod. Econ., **112**, 5–17 (2008).
4. F. N. Puglieri, A. Ometto, and P. a C. Miguel, Prod. Manag. Dev., **9,** 23–29 (2011).
5. S.D. Hochman and P.A. O'Connell, IEEE Int. Symp. On Elect. and Env (1993)
6. K. Masui, T. Sakao and A. Inaba, IEEE (2001)
7. S. Vinodh and G. Rathod, J. Clean. Prod., **18**, 8, (2010).
8. C.S. Wang, P.Y. Lin, and T.R. Chang, Proc. 2010 14th Int. Conf. CSCWD **1** (2010)
9. J. K. Gershenson and G. J. Prasad, Int. J. Agil. Manuf., **1**, no. 1, 1997.
10. T. AlGeddawy and H. ElMaraghy, CIRP Ann. - Manuf. Technol., **62**, 1 (2013).
11. D. Collado-ruiz and S. F. Capuz-rizo, J. Mech. Des., **132** (2013).
12. Cacherat, N. Lecocq, and G. Kremer, Congr. Int. surl'Analyse du Cycle Vie (2011).
13. F. Salvador and V. H. Villena, J. Supply Chain Manag., **49**, 1 (2013).
14. K. Kuan, D. C. Y. Foo, R. R. Tan, S. Kumaresan, and R. A. Aziz, CTEP, **9**, 3(2007).
15. S. G. Andrae and O. Andersen, Int. J. Life Cycle Assess., 15, 8 (2010).
16. Q. Song, Z. Wang, J. Li, and X. Zeng, Waste Manag., **32**,10 (2012).
17. Olowu, Law Environ. Dev. J., **8**, 1 (2012).
18. Cabigiosu, F. Zirpoli, and A. Camuffo, Res. Policy, **42**, 3 (2013).
19. M. K. Daws, Z. A. Ahmed, and A. A. Moosa, Jordan J. Mech. Ind. Eng., **3,** 1 (2009).
20. H. Hsu, A. Y. Chang, and H. M. Kuo, WSEAS Trans. Comput., **11,** 1 (2012).
21. K.J. Kim, H. Moskowitz, A. Dhingra, and G. Evans, Eur. J. Oper. Res., **121**, 3(2000).
22. R. R. Hischier Inge, LCA Case Stud., **8**, 4 (2003).
23. Bereketli and M. Erol Genevois, J. Clean. Prod., **54**, (2013).
24. A. Susanty, S. Hartini, D. Puspitasari, P. Arsiwi. Mediter. J. of Soc. Sci., **6**, (2015)

Improving performance measurement efficiency in public service using lean six sigma methodology

Purnawan Adi Wicaksono* , *Heru* Prastawa , *Ghabyanza* Pitaloka, and *Nurana* Putri

Industrial Engineering Department, Diponegoro University, Indonesia

Abstract. This research aims to conduct performance measurement efficiency in government and recommends an improvement to reduce the delay of submitting Functional Worksheet (FW) through Lean Six Sigma methodology. At recent system, the causes of delay in submitting FW are human factors i.e. the employee's negligence and technical factor i.e. duplication work in inputing data. The employee's performance measurement takes many times so that the leader can evaluate the employee's FW after 35 days and 195 minutes. In this research, Web-Based Functional Worksheet Information System was designed as the result of the improvement stage of Lean Six Sigma. The Web-Based Functional Worksheet Information System improved the recent system in submitting FW and reduced the time of finishing the submitting FW. Moreover, the leader could evaluate the employee's FW in a day.

1 Introduction

From year to year, the growth of service sectors became higher. Therefore, many organizations started to pay attention to the efficiency of operations in term of improvement [1]. This improvement covered several matters such as product improvement (goods and/or services), defects or service errors reducing, more compact operation system, good employees' skills, and others [2].

Government as the public service provider was under the pressure to improve the efficiency and the quality of public service [3]. It still attempts to improve the quality. Nowadays, most people rarely give positive assessment to the public service performance [4]. Government should initiate the improvement to remove the negative image of community by the effective movement. Therefore, it is important for public service organization to measure its performance systematically and consistently to improve its organizational management and its community satisfaction of the given services.

*Corresponding author: purnawan@undip.ac.id

Lean Six Sigma concept can be used by an organization to increase the performance as well as the efficiency as it still considers the quality, rate, customer satisfaction, and costs [5]. Lean Six Sigma, a combination between lean and six-sigma, is defined as the business philosophy to identify and eliminate the waste or non-value added activities through continuous improvement [6]. This research tries to apply Lean Six Sigma methodology to improve the process quality in governmental sector.

2 Literature Review

On previous research, Lean Six Sigma was used as a method to shorten the process, to improve the service quality, and to decrease the dissatisfying services in the case of IT help-desk service [7]. The result was that the average time of service process in IT help-desk service was reduced to 79.5 minutes or 47.5% after the previous performance. Total Quality Management (TQM) and Lean Six Sigma were used to improve the quality and the punctuality of local government services[8]. Local Financial Department might significantly reduce the times to process wages, purchases and accounts' payable, accounts' receivable and monthly reconciliation. The times to process wages, purchases and accounts' payable, accounts' receivable, and monthly reconciliation process were reduced to 60%, 40%, 90%, and 87% respectively. Lean Six Sigma aimed to illustrate its application in call center of a services company[5]. The results were the increase of first call resolution ratio, decrease of operator earnings, and streamline process. Lean Six Sigma methodology was used well to improve the surgery room efficiency in an academic children hospital [9]. Lean Six Sigma methodology could be applied appropriately in UK public sector to get the benefits and better return of investment[10]. Lean Six Sigma methodology is a good tool to be applied in all public organizations to make the efficient and effective processes [11].
After the previous researches, this paper aims to conduct an efficient measurement process of government position and recommends improvements in order to reduce the delay of submitting Functional Worksheet (FW) through Lean Six Sigma method.

3 Methodology

The case study focused on the goverment performance measurement in Boyolali, Central Java, Indonesia. In recent system, each employee must measure his/her performance by writing his/her efficiency in the Functional Worksheet paper. The Functional Worksheet paper should be submitted to the Financial and Asset Management Sub Division at least on the 5th of the following months. However, more than 50% of employees did not submit the Functional Worksheet papers on time. It was caused by the human factor i.e. the employees' negligence. After that, it was verified. The Financial and Asset Management Sub Division should manually re-input the results to Microsoft Excel which would take more time. The input of the same data showed that the recent system was not efficient. Lean Six Sigma containing DMAIC (Define, Measure, Analyze, Improve, Control) was used to improve the process efficiency of government employees' performance. Lean Six Sigma methodology help us to analyze the performance measurement process and to improve the performance measurement efficiency.

4 Result and Discussion

4.1 Definition

The recent performance measurement process was analyzed which would be accurately improved and determine anything needed for the better process. The performance measurement efficiency was analyzed by using SIPOC chart (Suppliers, Inputs, Process, Outputs, and Customers) in Figure 1.

Three types of waste were identified in the process as follows:

- Waiting, the delay of submitting Functional Worksheet caused the Financial and Asset Management Sub Division waiting when it must input the data.
- Transportation, there was an unnecessary stage of transporting the Functional Worksheet since it is a non-value added activity.
- Defect, the Financial and Asset Management Sub Division inputted the Functional Worksheet manually, so its error possibility was high.

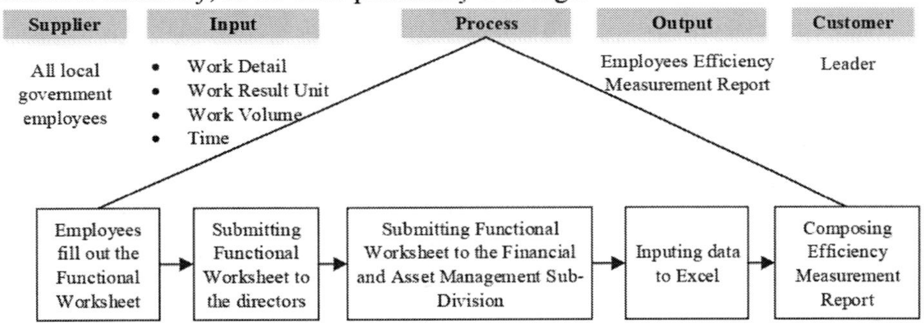

Fig. 1. SIPOC Chart

4.2 Measurement

The recent performance was measured to be compared with the target. The writers used Value Stream Mapping to describe the value added and non value added activities for the customers, so the significance and appropriate improvement could be implemented (Figure 2). At this time, the leader might evaluate the employees' efficiency value after 35 days and 195 minutes. The activity details of recent performance measurement can be seen in table 1. Table 1 describes the mapping process that consists of Value Added (VA) and Non Value Added (NVA), Input (I) and Output (0) item, Controllable (C) and Uncontrollable (U).

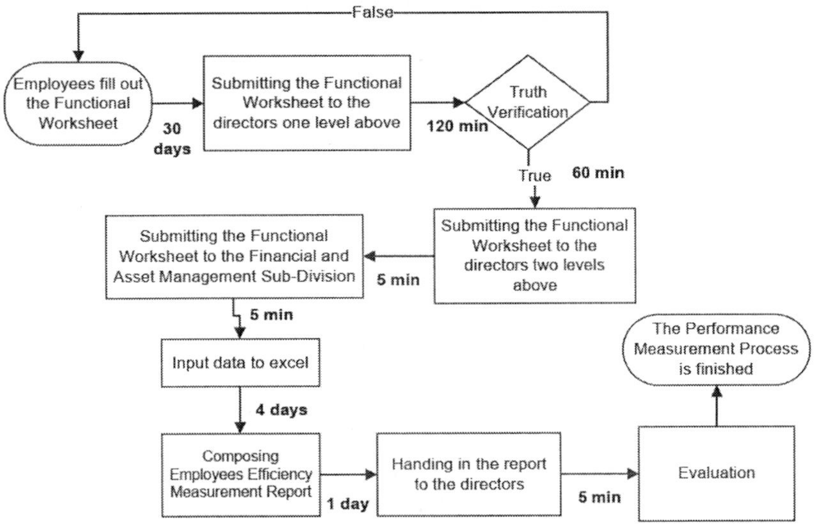

Fig. 2. Previous Value Stream Mapping System

4.3 Analysis

The data was analyzed and determined the possible improvement. The cause and effect chart was used to analyze the problem (Figure 3). The purpose of this chart is to identify the various potential causes of the problem.

Table 1. Process Mapping

Process Mapping	VA/ NVA	Input/output items	I/O	C/U
Employees fill out the Functional Worksheet	VA	Functional Worksheet Form	I	C
	VA	The employees write down the activity details, work result unit, work volume, time	I	C
	VA	Measuring the Functional efficiency	I	C
	VA	Employees Functional Worksheet which has been filled out	O	C
Submitting the Functional Worksheet to the leader one level above	VA	Functional Worksheet of each employee	I	C
	VA	Functional Worksheet collected based on the department	O	C
Functional Worksheet verification	VA	Functional Worksheet of each employee	I	C
	VA	Functional Worksheet verification	O	U
Revision	VA	Functional Worksheet Form	I	C
	VA	Functional Worksheet Revision	O	C
Submitting the Functional Worksheet to the leader two levels above	NVA	Employees Functional Worksheet on a particular department	I	C
	NVA	Functional Worksheet collectedbased on the department	O	C
Submitting the Functional Worksheet to the Financial and Asset Management Sub Division	VA	Functional Worksheet of each employee	I	C
	VA	The Functional Worksheet is submitted	O	C
Inputing data to Excel	NVA	Functional Worksheet of each employee	I	C
	NVA	Functional Worksheet is input to Excel	O	C
Composing efficiency measurement report	VA	The employees efficiency value	I	U
	VA	Efficiency Measurement Report	O	C
Submitting report to the leader	VA	Efficiency Measurement Report	I	C
	VA	Efficiency Measurement Report is submitted to the leader	O	C
Evaluation	VA	Efficiency Measurement Report	I	C
	VA	The leader and evaluation of the employees efficiency	O	U

4.4 Improvement

The improvement aims to make changes in the process by eliminating defects, waste, and cost. From the previously mentioned problems, it could be found that the performance measurement system was not well integrated yet. There were duplication works in inputing data. A web-based information system was designed which might improve the performance measurement process (Figure 4). It is dynamic application–a combination between database and world wide web – to compact the process so the data processing and distribution become faster and easier [12]. By the web-based information system, the leader may evaluate the employees' performance on the day (Figure 5). In table 2, comparison between the recent system and the web-based information system were presented.

4.5 Control

The control is conducted to improve the process capabilities. One way in developing the control planning is making the standard operating procedures (SOP) as the guidance of the performance measurement process implementation [7].

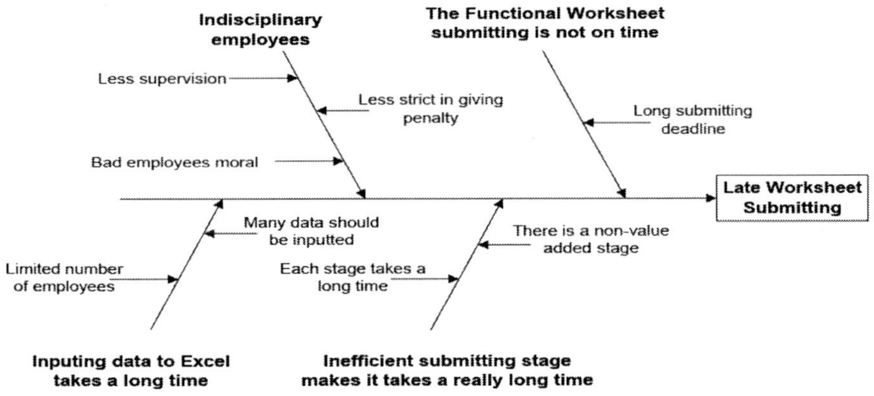

Fig. 3. Fishbone Chart

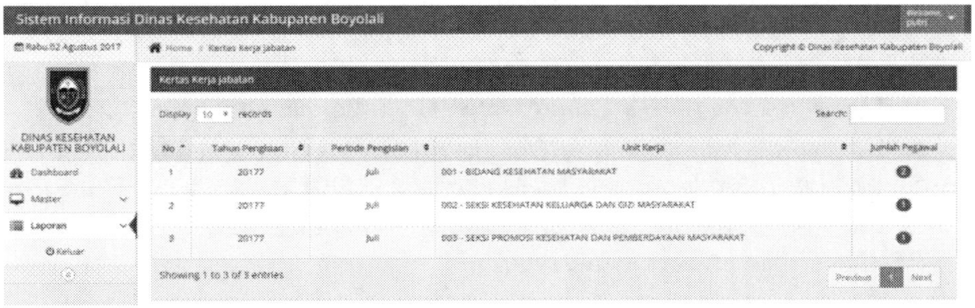

Fig. 4. Image of Information System

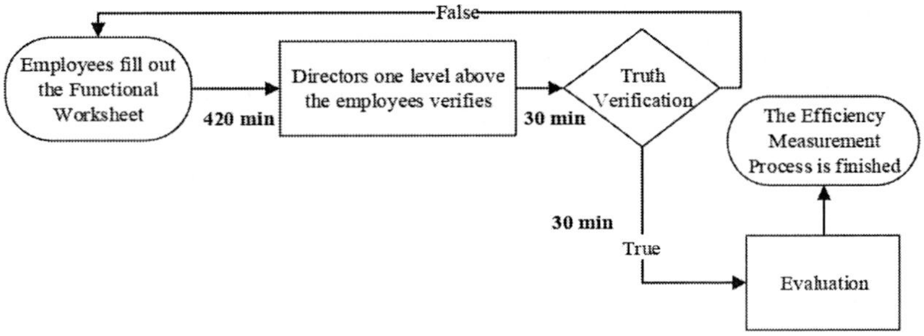

Fig. 5. Value Stream Mapping of Recent System

Table 2. Time Comparison between the Recent System and Web-Based Information System

Comparative Factor	Recent System	Web-Based Information System
Recording and storage media	Microsoft Excel	Database is integrated to the system
Estimated time until the evaluation	35 days 195 minutes	1 day

5 Conclusion

The problem of employees' performance measurement was the delay of submitting the Functional Worksheet. It was caused by the waste and non value added activities. Lean Six Sigma was used to identify and eliminate the waste and non value added activities. Lean Six Sigma used simple method called DMAIC (Define, Measure, Analyze, Improve, Control). The recent performance measurement process takes 35 days and 195 minutes in order to be able to be evaluated. The time of performance measurement can be decreased by implementation of Web-Based Functional Worksheet Information System. Moreover, the leader can evaluate the employees' performance on the day. Further research is to make the appropriate standard operating procedure as the guidance of the performance measurement implementation process.

References

1. G. H. Cavaness, J.P., & Mannochehri, "Building Quality into Services.pdf," *SAM Adv. Manag. J.*, **3**, 6 (1993).
2. J. R. Evans and W. M. Lindsay, *An Introduction to Six Sigma and Process Improvement*, Second Edi. Jakarta: Penerbit Salemba Empat, (2014).
3. A. Agus, S. Barker, and J. Kandampully, "An Exploratory Study of Service Quality in the Malaysian Public Service Sector," *Int. J. Qual. Reliab. Manag.*, **24**, 2(2007).
4. B. Teeuwen, *Lean for the Public Sector*. USA: Productivity Press, (2010)
5. A. Laureani, J. Antony, and A. Douglas, "Lean Six Sigma in a Call Centre: a Case Study," *Int. J. Product. Perform. Manag.*, **59**, 8 (2010)
6. V. Gaspersz, *Lean Six Sigma for Manufacturing and Service Industries*. Jakarta: PT. Gramedia Pustaka Utama, (2007).
7. C. Su, "Improving Service Quality by Capitalising on an Integrated Lean Six Sigma Methodology," **2**, 1 (2006).
8. S. Furterer and A. K. Elshennawy, "Implementation of TQM and Lean Six Sigma Tools in Local Government: a Framework and a Case Study," *Total Qual. Manag. Bus. Excell.*, **16**, 10 (2005).
9. E. P. Tagge, A. S. Thirumoorthi, J. Lenart, C. Garberoglio, and K. W. Mitchell, "Improving operating room efficiency in academic children' s hospital using Lean Six Sigma methodology," *J. Pediatr. Surg.*, **52**, 6 (2017)
10. J. Antony and B. Rodgers, "Can Lean Six Sigma make UK public sector organisations more efficient and effective," *Int. J. Product. Perform. Manag.*, **65**, 7 (2016)
11. J. Antony, B. Rodgers, and Cudney, "Lean Six Sigma for public sector organizations: is it a myth or reality?," *Int. J. Qual. Reliab. Manag.*, **34**, 9 (2017)
12. S. O. Cheung, H. C. H. Suen, and K. K. W. Cheung, "PPMS: A Web-based construction Project Performance Monitoring System," *Autom. Constr.*, **13**, 3 (2004)

Effect of blanching-brine-calcium pretreatment on red chili pepper drying

Uma Fadzilia Arifin[1,*], *Mohamad* Djaeni[1] and *Febiani Dwi* Utari[1]

[1]Department of Chemical Engineering, Faculty of Engineering, Diponegoro University Jl. Prof. Soedarto, SH, Tembalang, Semarang, Indonesia 50275

Abstract. The red chili pepper drying aims to remove moisture content up to 11% (wet basis). Dried chilies can be easy to distribute and have longer shelf life because it can be protected from decaying by enzymatic and microbial activity. Pretreatment in drying is one of alternative to decrease drying time. This present study was carried out on effect of blanching-brine-calcium pretreatment toward drying kinetics in chili. In this case, chili provided from the local market was pretreated with blanching, dehydrated osmotic with brine, calcium dipping pretreatment and combination of them. The chili products were dried in tray dryer at 40°C, 50°C, 60°C and 70°C for 8 hours afterward. Result showed that Lewis model is suitable drying model. The blanching-brine-calcium pretreated chili can be well and faster dried with activation energy was 36.4290 kJ/mol and k_0 was 7.8712. At 70°C, red chili pepper needed 6.467 hours to dry. Thus, the blanching-brine-calcium pretreatment can be potential to reduce the energy cost for drying as well as retain the quality.

1 Introduction

Red chili pepper (*Capsicum frutescens*) is one of the most tropical agricultural crop widely used food flavouring or food seasoning. It is the most abundantly consumed spice because of its hot sensation [1]. It is important commodity and highly commercial demanded by the food industry, not only based on its economic value, but also for the combination of aromatic, colour, flavour, hotness properties and nutritional values [2, 3]. The post-harvest red chili pepper has high moisture content ranging 70-80% (wet basis) [4]. So, it is one of perishable vegetable due to decay easily and very susceptible by organisms, enzymes, fungal and insect attack during storage. Drying may be an alternative method in order to prevent deterioration of food and agricultural products [5].

Drying is one of the agricultural preservation to prevent decay and spoilage [6]. Drying aims to reduce microbiological activity and to improve the stability of moist materials with removal of moisture content until desired level by introduction of heat [7, 8, 9]. So, it can provide longer shelf-life, easy to storage and distribution [5]. However, physicochemical quality degradation of products is major challenges during drying process. The longer drying time result the higher quality degradation. As a result, pretreatments have been

* Corresponding author: umafadzilia@gmail.com

applied prior to drying fruits and vegetables to control the physicochemical quality changes [10].

Pretreatment is an essential step in the processing of food materials such as drying. Pretreatment can speed up drying rate, improve the quality of dried product, prevent browning, keep texture and help retain volatile compounds [11]. There are various pretreatments in drying which includes chemical pretreatment, blanching, and osmotic dehydration [12]. Blanching and osmotic dehydration can increase the drying rate by removal moisture content before drying. Chemical pretreatments such as calcium chloride dipping can increase the drying rate by removing the surface resistance of fruits or vegetables and retain firmness [12]. Thus, it is important to predict and simulate the drying behaviour. Drying kinetics model is one of ways that can be used to estimate drying time to reach desired level moisture content of materials at various temperatures for retaining quality of product [11]

The objective of this research was to investigate the effect of blanching-brine-calcium pretreatment on red chili pepper drying kinetics at various temperatures. It is also used to estimate drying time for red chili pepper drying. The information obtained could be usefull to finding proper drying condition in order to minimize the quality degradation.

2 Materials and Methods

2.1 Materials

Fresh red chili peppers (ghost chili) with moisture content 78.08 % (wet basis) or 3.56 gram water gram^{-1} dry basis (db), were purchased from a local market in Semarang, Central Java, Indonesia. They were sorted and cleaned. Calcium chloride (food grade with purity 94.0%, Merck Germany) and sodium chloride (purity 97.99%, UniChemCandi Indonesia) were used for pretreatment chili drying.

2.2 Methods

2.2.1 Pretreatment drying of red chili pepper

Three pretreatments before drying for red chili pepper were blanching, osmotic dehydration with brine dipping and calcium solution dipping. Blanching was done by immersing red chili peppers in hot water (distilled water) at 90°C for 3 minutes [13]. They were cooled in cold water and removed the excess water with tissue paper before drying [10]. The chilies were osmotically pretreated by brine solutions dipping in 10% (w/v) NaCl solution at room temperature for 30 minutes. Sodium chloride was chosen as osmotic dehydration agent because it has high osmotic pressure [10, 14]. The chilies were chemical dipped in 1% $CaCl_2$ solution process to remove surface resistance and retain firmness in chili, afterwards [15]. The last sample was pretreated by combination of blanching-brine-calcium pretreatment. So, the red chili peppers with various pretreatments have been ready to dry.

2.2.2 Red chili pepper drying

The red chili pepper drying was conducted in a laboratory tray dryer. The non pretreated and pretreated red chili pepper was dried under 40°C, 50°C, 60°C and 70°C for 8 hours. The constant air flow rate was operated equal to 0.22 ms^{-1}. As the responses, the total moisture contents were observed every 30 minutes by gravimetric method.

2.2.3 Drying kinetics model development

Drying kinetics model is one of alternative to predict and simulate the drying behaviour. It can be proposed to evaluate the rate of moisture with time. So, drying time needed to reach desired level moisture content of materials can be estimated. There have been several drying models to describe moisture removal in food materials presented in literatures [5]. In this research, to select a suitable drying model for describing the drying process of chili, drying curves were fitted in drying models as in Table 1. The correlation between temperature and constant of drying rate is described by Arrhenius equation. In these models, the values of the equilibrium moisture content (Me) to calculate moisture ratio were obtained from sorption isotherm of dried chili under various temperature that is affected by relative humidity. The equilibrium moisture content in chili was assumed refer to Modified Oswin equation as derived by [16]

$$HR = 1/[\{(a + bT)]/Me\}c + 1] \tag{1}$$

Where, HR is the equilibrium relative humidity in decimal; Me is the equilibrium moisture content in % (dry basis); T is the temperature in Celsius. a, b and c are parameter constants. The value of parameter constants are a = 19.299; b = -0.19499 and c = 1.517 [16]. The parameter values used to determine other indefinite variables such as equilibrium moisture content at various temperatures and relative humidity.

3 Results and Discussion

3.1 Drying kinetics at various pretreatments

Pretreatment is one of alternatives to increase drying rate of products. In this research, the various pretreatments in red chili pepper drying were fitted on drying curves. Moisture ratio and drying time are fitted as drying curve to determine drying kinetics of red chili pepper. Effect of various pretreatments on drying kinetics can be seen in Fig. 1.

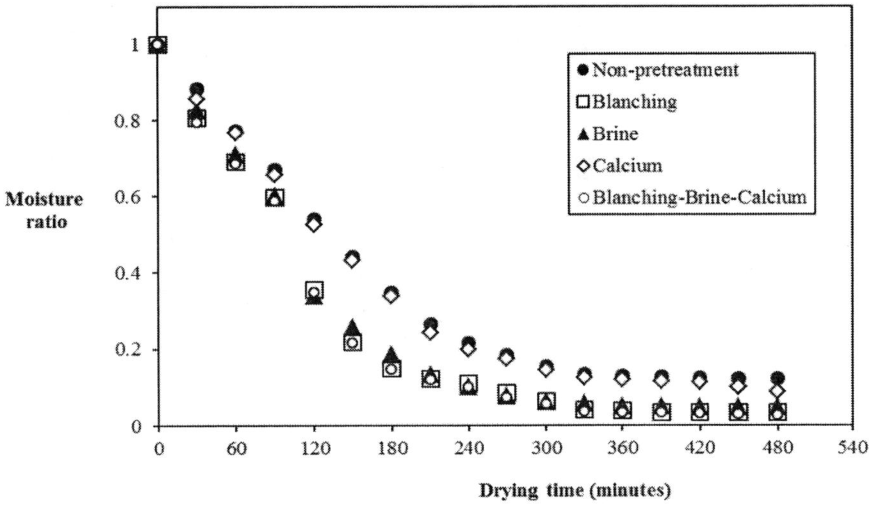

Fig. 1. Effect of various pretreatments on drying kinetics red chili pepper at 70°C

Based on Fig.1, the blanching-brine-calcium pretreated samples showed the faster drying rate. It is apparent that moisture ratio decreases significantly with drying time. On the other hand, unpretreated chili has slower drying rate. Blanching-brine-calcium pretreatment for red chili pepper drying can be well and faster dried. Tunde-Akintunde [12] also reported that blanching and osmotic dehydration can increase the drying rate by removal moisture content before drying. Blanching can made soften tissues and enhance mass transfer during drying. Osmotic pretreatment is achieved by dipping the product in a brine solution prior to drying [10]. In addition, chemical pretreatments such as calcium chloride dipping can increase the drying rate by removing the surface resistance [15].

The drying model can be used to predict drying behaviour and the drying time to reach desired level moisture content of material. The moisture ratio and drying time were fitted in drying curve. Several models in exponential regression can be used to obtain parameter value and statistical result of the drying model such as constant of drying rate (k) and coefficient of determination (R^2) values. Table 1 shows values of the drying constants and drying coefficients of the several models.

Table 1. The values of the drying constants and drying coefficients of the several models.

Kinetics Model	Pretreated	Drying Constants	R^2
Lewis MR = exp(-kt)	Non	k = 0.0055	0.9836
	Blanching	k = 0.0084	0.9750
	Brine	k = 0.0081	0.9759
	Calcium	k = 0.0058	0.9866
	BBC	k = 0.0087	0.9763
Page MR = exp(-kt^n)	Non	k = 0.0032; n = 1.1050	0.9866
	Blanching	k = 0.0019; n = 1.2987	0.9874
	Brine	k = 0.0020; n = 1.2800	0.9874
	Calcium	k = 0.0031; n = 1.1209	0.9903
	BBC	k = 0.0020; n = 1.3014	0.9888
Modified Page MR = exp[-$(kt)^n$]	Non	k = 0.0055; n = 1.1057	0.9866
	Blanching	k = 0.0082; n = 1.3016	0.9874
	Brine	k = 0.0079; n = 1.2834	0.9874
	Calcium	k = 0.0057; n = 1.1216	0.9903
	BBC	k = 0.0083; n = 1.3039	0.9888
Handerson and Pabis MR = a exp(-kt)	Non	k = 0.0059; a = 1.0414	0.9861
	Blanching	k = 0.0089; a = 1.0551	0.9780
	Brine	k = 0.0086; a = 1.0582	0.9794
	Calcium	k = 0.0060; a = 1.0382	0.9885
	BBC	k = 0.0091; a = 1.0521	0.9790
Logarithmic MR = a exp(-kt) + c	Non	k = 0.0062; a = 1.0226; c = 0.0288	0.9868
	Blanching	k = 0.0085; a = 1.0650; c = -0.0159	0.9786
	Brine	k = 0.0084; a = 1.0613; c = -0.0052	0.9794
	Calcium	k = 0.0062; a = 1.0328; c = 0.0083	0.9886
	BBC	k = 0.0086; a = 1.0644; c = -0.0196	0.9800
Two term model MR = a exp(-k_0t) + a exp(-k_1t)	Non	k_0 = 0.0056; k_1 = -0.0178; a = 1.0522; b = 1.351 x 10^{-5}	0.9907
	Blanching	k_0 = 0.0089; k_1 = 0.0089; a = 0.9970; b = 0.0580	0.9780
	Brine	k_0 = 0.0086; k_1 = 0.0086; a = 1.0129; b = 0.0452	0.9794
	Calcium	k_0 = 0.0060; k_1 = 0.0060; a = 0.8290; b = 0.2092	0.9885
	BBC	k_0 = 0.0066; k_1 = -0.0057; a = 3.1122; b = -2.0723	0.9815
Approximation of diffusion MR = a exp(-kt) + (1-a) exp(-kbt)	Non	k = 0.0419; a = -0.1239; b = 0.1488	0.9890
	Blanching	k = 0.0220; a = -0.9810; b = 0.5659	0.9881
	Brine	k = 0.0187; a = -1.6418; b = 0.6858	0.9884
	Calcium	k = 0.0309; a = -0.1560; b = 0.2153	0.9919
	BBC	k = 0.0062; a = 3.2412; b = -0.8644	0.9800
Wang and Singh MR = 1 + at + bt^2	Non	a = -0.0046; b = 5.820 x 10^{-6}	0.9956
	Blanching	a = -0.0057; b = 8.100 x 10^{-6}	0.9707
	Brine	a = -0.0057; b = 7.986 x 10^{-6}	0.9762
	Calcium	a = -0.0046; b = 5.882 x 10^{-6}	0.9942
	BBC	a = -0.0058; b = 8.218 x 10^{-6}	0.9707
Midilli MR = a exp[-k$(t)^n$] + bt	Non	k = 0.0007; n = 1.4297; a = 0.9840; b = 0.0002	0.0002
	Blanching	k = 0.0009; n = 1.4674; a = 0.9771; b = 8.192 x 10^{-5}	0.9909
	Brine	k = 0.0008; n = 1.4815; a = 0.9818; b = 0.0001	0.9936
	Calcium	k = 0.0009; n = 1.3855; a = 0.9801; b = 0.0002	0.9976
	BBC	k = 0.0009; n = 1.4551; a = 0.9739; b = 6.843 x 10^{-5}	0.9914

These values show that the tested drying models predict drying behaviour of red chili pepper adequately. In all cases, the R^2 value of the model was maximum, indicating a good fit [17]. Generally, Midilli model gave a higher R^2. Thus, in the empirical case the Midilli model may be assumed to represent the drying behaviour of red chili pepper.

The model was validated for estimating drying time in laboratory experiment. It is aimed to find out the model that suitable with the real condition of red chili drying process. The result of experimental validation that is compared the Lewis and Midilli model to know suitable model can be seen in Table 2.

Table 2. Drying Time to Dry Red Chili Pepper until 11% Dry Basis at 70°C and Various pretreatments using Lewis and Midilli Models

Pretreatment	Drying Time (hours)		
	Lewis Model	Midilli Model	Experimental Validation
Non	10.39	7.06	10.67
Blanching	6.81	6.47	6.83
Brine	7.10	6.27	7.63
Calcium	9.91	7.51	9.45
BBC	6.66	5.99	6.47

Based on Table 2, we can know that the analysis results using the lewis model shows closer to the experimental validation result. This suggests that Lewis model is more accurate than the Midilli model although the R^2 value of the Midilli kinetics model is higher. In addition, Lewis model is a basic and simple kinetics model so it is easy to calculate and manual analysis. Whereas, the Midilli model is a complex model and requires relatively complex calculations.

3.2 Drying time estimation at blanching-brine-calcium pretreatment under various temperatures

The moisture ratio was influenced by temperature and drying time. The higher temperature and the longer drying time, resulted the lower of moisture ratio in chili. Based on experiment under various temperatures, the relative humidity can be determined by psychometric chart. Thus, equilibrium moisture content of red chili pepper in experiment can be determined based on sorption isotherm. Additionally, the drying time when moisture content decreased from 78.08 % to 11% (wet basis) also can be calculated. The value of constant of drying rate (k), relative humidity, equilibrium moisture content and drying time estimation of red chili pepper under various temperatures using Lewis Model analyses were summarized in Table 3.

Table 3. Parameter Value and Drying Time Estimation of Drying Red Chili Pepper

Temperature (°C)	H_R	Me	k (minutes^{-1})	Drying Time (hours)
30	0.80	0.0907	0.0013756	56.44
40	0.45	0.0714	0.0024916	28.11
50	0.28	0.0563	0.0029622	22.25
60	0.17	0.0236	0.0043249	13.74
70	0.12	0.0117	0.0086579	6.65
80	0.07	0.0042	0.0106660	5.30
90	0.048	0.0013	0.0150140	3.74

The result showed that the higher temperature caused the higher constant of drying rate and shorter drying time. Based on Table 3 we can know that drying was greatly affected by the drying temperature. The drying rate constant (k) increased with the drying air temperature. The dependence of the rate constant on drying air temperature was correlated by the Arrhenius. The plot depicting the relationship between k and 1/T was found to be exponential regression in the range of temperatures investigated. Here, the activation energy (Ea) was about 36.4290 kJ/mol and value of k_0 was 7.8712. The activation energy value of red chili pepper drying is in the range to the values reported (15–40 kJ/mol) by Rizvi [18] for various foods. Higher activation energy values signify greater temperature sensitiveness [19]. Thus, the drying rate constant on blanching-brine-calcium pretreatment chili can be estimation at various temperatures.

4 Conclusions

The red chili pepper was dried at various pretreatments and temperatures to reach 11% (wet basis) moisture content. The Lewis model is suitable drying model to estimate drying time in chili drying. The blanching-brine-calcium pretreated chili can be well and faster dried with activation energy was about 36.4290 kJ/ mol and value of k_0 was 7.8712. The temperature dependence of drying was correlated by Arrhenius. Thus, blanching-brine-calcium pretreatment under high temperatures can be potential to reduce the time for drying as well as energy cost.

The first author gratefully acknowledges the Indonesian Endowment Fund for Education (LPDP), the Indonesian Ministry of Finance for providing the financial support during study and research in Diponegoro University.

References

1. A. Vega-Galvez, R. Lemus-Mondaca, C. Bilbao-Sainz, P. Fito, A. Andres, J. Food Eng **85**, 1, 42–50 (2008)
2. C. K. Kouassi, R. Koffi-Nevry, L. Y. Guillaume, Z. N. Yesse, M. Koussemon, T. Tablan, K. K. Athanase, Innov. Rom. Food Biotechnol **11**, 23–31 (2012)
3. A. Fadhel, S. Kooli, A. Farhat, and A. Belghith, IJREB **2014** (2014)
4. N. C. Wade, S. S. Wane, S. M. Kshirsagar, Ind. J. Sci. Res. Tech **2**, 3, 105–110 (2014)
5. E. O. M. Akoy, Int. Food Res. J **21**, 5, 1911–1917 (2014)
6. T. Y. Tunde-Akintunde, T. J. Afolabi, J. Food Process Eng **33**, 649–660 (2010)
7. M. Djaeni, U. F. Arifin, S. B. Sasongko, AIP Conference Proceedings **1823**, 1 (2017)
8. I. Alibas, Food Sci. Technol **32**, 2, 394–401 (2014)
9. M. Beigi, Food Sci. Technol **36**, 1, 145–150 (2016)
10. P. T. Akonor and C. Tortoe, Br. J. Appl. Sci. Technol **8**, 4, 1215–1229 (2014)
11. E. E. Abano, H. Ma, W. Qu, E. Teye, African J. Food Sci **5**, 7, 425–435 (2011)
12. T. Y. Tunde-Akintunde, J. Food Process. Preserv **34**, 595–608 (2010)
13. N. Toontom, W. Posri, S. Lertsiri, M. Meenune, Int. Food Res. J **23**, 1, 289–299(2016)
14. W. Phomkong, S. Soponronnarit, P. Thammarutwasik, Dry. Technol **28**, 1466–1476 (2010)
15. P. P. Lewicki, E. Michaluk, Dry. Technol **22**, 8, 1813–1827 (2004)
16. S. Kaleemullah, R. Kailappan, Biosyst. Eng **88**, 1, 95–104 (2004)
17. S. Kaleemullah, R. Kailappan, J. Food Eng **76**, 531–537 (2006)
18. S. S. H. Rizvi, *Thermodynamic Properties of Foods in Dehydration* (2005)
19. S. Arora, S. Bharti, V. K. Sehgal, Dry. Technol **24**, 2, 189–193 (2006)

Optimization foam mat drying of roselle *(Hibiscus sabdariffa L.)* extract

Mohammad Djaeni[1,*], *Febiani Dwi* Utari[1] and *Uma Fadzilia* Arifin[1]

[1]Department of Chemical Engineering, Faculty of Engineering, Diponegoro University Jl. Prof. Soedarto, SH, Tembalang, Semarang, Indonesia 50275

Abstract. In this study, foaming condition of roselle was optimized using response surface methodology (RSM) and the effect of drying characteristic was investigated. Roselle extract was foamed by addition of 1-5% w/w foaming agents (ovalbumin). The foaming stabilizer, glycerol mono stearate (0-1% w/w) was used to remain mechanic and thermodynamic stability of foam. As the response foam density and drainage volume was determined. The optimum foam variable was then dried at various drying temperatures (50-70°C). The moisture content was observed by gravimetry every 10 minutes for 90 minutes. Result showed that optimum formulation was 3.31% egg albumin and 1% GMS. The constant rate of the foam mat drying (temperature 50°C) was 3 times higher than non foam mat drying. Higher drying temperature can speed up the driving force but lead to color degradation.

1 Introduction

The roselle plant is cultivated in China, India, Sudan, Uganda, Indonesia, Malaysia, and Mexico. The total roselle plant in Indonesia around 6.23% of worldwide total production. The main bioactive compounds in Roselle calyces are phenols and anthocyanins that its concentration according to the roselle cultivation and extraction process [1]. Previous research observed that roselle extraction with roselle:water ratio of 1:10 resulted in 502.33±0.52 mg of anthocyanins. Roselle extract has the potential aspect as natural colorant and antioxidant [2].

Drying roselle extract can result stable powder and enhance the storage life. Foam mat drying can shorten the drying time and retain the quality of the product by transforming liquid and semisolid materials into stable foam, with addition of foaming agent and incorporation of air [3]. A foaming agent is a surfactant materials that can reduce the surface tension and improve the foam formation. Egg albumen (EA), whey protein, soy protein, guar foaming albumin are the most common foaming agent [4]. The egg albumin can produce lower foam diameter, enhanced the surface area, reduced drying time and increased the water diffusivity [3,5]. GMS as the foaming stabilizer are polysaccharides that were added to thicken the aqueous solution. Addition of GMS in foam mat drying can form the stable foam [4,6].

* Corresponding author: moh.djaeni@live.undip.ac.id

The previous observation studied about the optimum foaming condition of shrimp using xanthan gum [7]. Result showed that 0.19% (w/w) xanthan gum and whipping time 5.89 minutes were optimum foaming condition. In foam drying of muskmelon, 11.59% w/w EA and 0.59% w/w CMC were optimum foaming condition [8]. There was no study about foam optimization using egg albumin and GMS. The aim of this study was to investigate the optimum parameter to foam mat drying using egg albumin and GMS, determine the constant rate of drying in various temperature, and determine the change of Total Color Ratio (TCR) in foam mat drying of roselle extract.

2 Materials and Methods

2.1 Roselle extract preparation

The red roselle *(Hibiscus sabdariffa Linn.)* calyces (dry) were obtained from a herbal medicine market in Solo, Indonesia. The roselle calyces were ground and sieved until the size 0.25 mm. The roselle extraction used water as the solvent and roselle calyces mass to water volume ratio was 1:10 [2]. The extraction was operated at temperature 50°C under continuous agitation. The extract was then filtered through a Buchner filter.

2.2 Foam formulation

Eggs were purchased from the local market in Semarang, Indonesia. The egg albumin was separated from the yolk. One hundred ml of roselle extract was mixed with foam agent (egg albumin 1-5% w/w) and foam stabilizer (GMS 0-1% w/w). Then the solution was whipped with a hand domestic mixer (360 W power) at maximum speed for 3 min. Before the drying process, the foam was analyzed for the foam characteristic. There was two characteristic of the foam: foam density and foam drainage volume.

Foam density was determined using previous method [9]. One hundred ml of the foam was transferred into 100 ml measuring cylinder and weighed. The foam density was calculated using the following equation:

$$foam\ density = weight\ of\ foam\ (g)/volume\ of\ foam\ (cm^3) \quad (1)$$

Drainage volume was determined using previous method with modification [7]. Fifteen grams of foam was poured into a Buchner filter (80 mm diameter) and was placed in a 100-mL graduated cylinder. The drainage volume was liquid volume of the collapsed foam during 30 min. Each experiment was carried out in triplicate and the average was reported.

2.3 Optimization

Design Expert 7.0.0 software was used to find optimized levels of independent factors, analysis of variance (ANOVA), and regression models. The goals of the optimization process were minimum value of foam density and foam drainage volume. The independent variables were placed in the experimental range and the levels can be seen in Table 1.

Table 1. Levels of the independent variables.

Independent variables	units	Coded levels		
		-1	0	1
(A):Egg albumin	% w/w	1	3	5
(B):GMS	% w/w	0	0.5	1

2.4 Drying process

The foams were then dried in tray dryer. The foams were placed in aluminium tray (length 20 cm, width 15 cm and thickness of 0.2 cm) under drying temperature (50°C-70°C). The water loss was determined by weighing the samples every 15 minutes. The drying process took place for 90 minutes with superficial velocity of 0.18 m s^{-1}. As a comparison, the control sample (without addition of foam agent and foam stabilizer) was dried.

Thin layer model was used to determine the constant rate of the drying. Based on the literature, the newton model (Eq. 2) has been proven to predict the constant rate of foam drying [10,11]. The moisture ratio in Eq. 3 was simplified [7,12,13].
Refer to thin layer drying model, the correlation between moisture ratios with time can be represented by Eqn (1) [9].

$$kt = -\ln(MR) \qquad (2)$$

$$MR = M_t/M_0 \qquad (3)$$

Where k was the drying rate constant ($minute^{-1}$), t was drying time (min), M_t was the moisture content at time t, M_0 was the initial moisture content. Arrhenius equation (Eq. 4) was used to correlate the dependency of k on temperature (T)

$$k \;= k_0 exp^{\frac{-Ea}{R(T+273)}} \qquad (4)$$

with k_0 was the pre-exponential factor ($minute^{-1}$), Ea was the activation energy (J mol^{-1}), and R was the gas constant (8.32 J mol^{-1} K^{-1}).

2.5 Color measurement

Color measurement was done by using Chroma Meter (CR-300, KONICA MINOLTA SENSING INC, Japan) in terms of observing L (lightness), a (redness and greenness) and b (yellow and blueness) as previously used by Kumar *et al.* [10]. The changes of total color ratio (TCR) was used to evaluate the total color degradation during drying process (Eq. 5) [16]:

$$TCR=(L_s a_s\, b_s)/(L_o a_o b_o) \qquad (5)$$

Where, L_s, a_s, and b_s were measured on a sample at a given time, while L_o, a_o, and b_o were measured on raw roselle extract (before the drying process).

3 Results and Discussion

3.1 Foam density and foam drainage volume

Foam density (FD) and foam drainage volume (FDV) are commonly used to evaluate the foam characteristic. The density of the foam varied from 0.118 g cm^{-3} to 0.419 g cm^{-3}. Based on ANOVA analysis, the P value less than 5% indicated that the addition of egg white and GMS give the significant effect on FD. The correlation was related in quadratic model (Eq. 6). The foam density decreased with increased concentration of egg albumin and GMS (Fig 1.a). A similar result was reported by previous observation, increasing egg albumin from 1 to 3% make the FD decreasing[17]. The decreased of FD indicated that addition of foam agent reduces the surface tension and interfacial tension to form the sufficiently interfacial film and reducing the droplet size [18,4]. The decreased FD lead to increased the surface area for drying and speed up the drying rate [15].

$$FD = 0.16 - 0.10A + 0.018B - 0.043AB + 0.11A^2 - 0.022B^2 \quad , R^2=0.943 \tag{6}$$

$$FDV = 9.06 + 1.42A - 1.08B - 0.13AB - 0.58A^2 - 0.083B^2 \quad , R^2=0.997 \tag{7}$$

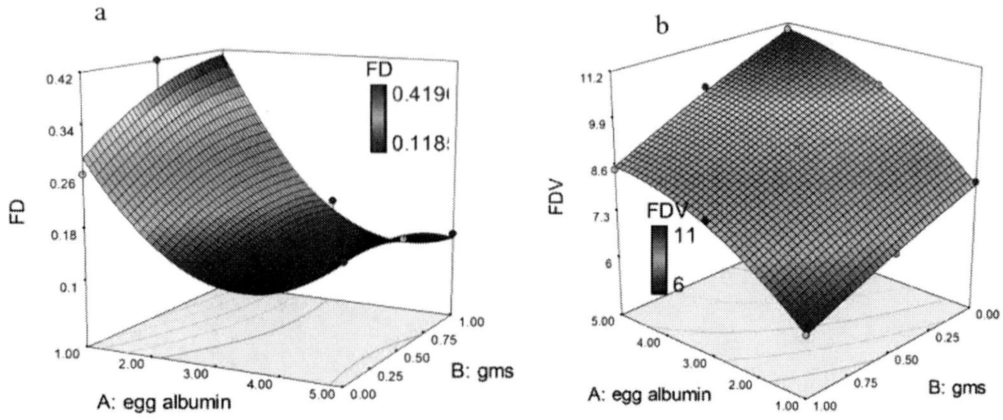

Fig. 1. a. Effect of addition egg albumin and GMS on Foam Density (FD) b. Effect of addition egg albumin and GMS on Foam Drainage Volume (FDV)

The FDV indicated the foam stability. The FDV varied from 6 to 11 ml. Based on ANOVA analysis, the P value less than 5% indicated that the addition of egg albumin and GMS give the significant effect on FDV. The correlation was related in quadratic model (Eq. 7). FDV decreased with increased concentration of and GMS (Fig 1.b). The GMS stabilization mechanism by form the lamellae and trapping the droplet in the microgel matrix [4].

3.2 Optimization

The goals of the optimization were to form the lower FD and lower FDV. Based on the optimization using software Design Expert, the formulation using 3.31% egg albumin and 1% GMS resulted in the lowest FD (0.135 g cm^{-3}) and the lowest FDV (8.07 ml) with desirability 86.2%. The optimum formulation was then validated and resulted FD 0.162 g cm^{-3} and FDV 9 ml. The results were more significant, compare with another foam formulation. The formulation resulted in the lower FD compare to foam formulation in muskmelon pulp [8]. In muskmelon foam formulation, using 11.59% egg albumin and 0.59% CMC resulted in FD 0.3390 g cm^{-3}. Bag et al. observed foam optimization in bael fruit using 3.1% w/w GMS and 0.32% w/w MC resulted higher FD (0.658) and lower FDV (1.75 ml) [18].

3.3 Drying Characteristic

The roselle extract was dried with foam agent and foam stabilizer at 50-70°C. As the comparison, the extract dried without foam agent and foam stabilizer. The drying curves in the form of moisture content versus time were depicted in Figure 2. The moisture content of roselle extract reduced drastically with foam mat drying than non foam mat drying. The moisture content in foam mat drying until 0.426 (dry basis) while the moisture content of non foam mat drying 6.085 (dry basis). The constant rate of the foam mat drying (temperature 50°C) was 3 times higher than non foam mat drying (Table 2). The foam formulation forms the decreased droplet size and increased the total surface area. The larger surface area leads to increased significantly moisture evaporation rate [3,4].

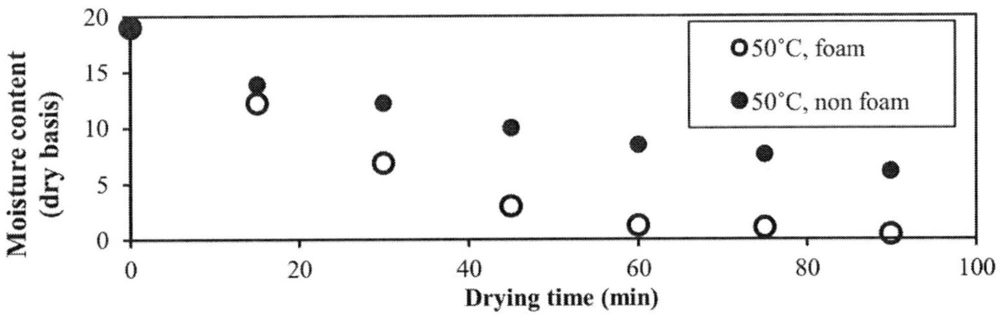

Fig. 2. Moisture ratio of foam mat roselle extract drying

The constant rate of the drying calculated using Eq. 2. The Newton model resulted the higher value of R^2 . The R^2 value varied from 0.853 to 0.981. The constant rate of the drying increased with increased drying temperature. The constant rate of foam mat drying in temperature 70ºC was 1.7 times higher than constant rate of foam mat drying in temperature 50ºC (Table 2). In higher drying temperature, the relative humidity decrease. The driving force for the drying higher and produced lower moisture content [19]. Arrhenius model was used to correlate the dependency of k on temperature (T) (Eq. 4). The value of Arrhenius constants were listed in Table 2. Using the value of Ea and K_0 , the value of constant rate of the drying in various temperature can be predicted.

Table 2. Constant rate of foam mat roselle extract drying

T (°C)	Foam mat				Non foam mat			
	k (min^{-1})	R^2	$k_o \times 10^2$	Ea (kJ mol^{-1})	k (min^{-1})	R^2	$k_o \times 10^7$	Ea (kJ mol^{-1})
50	0.041	0.984	5.22	25.35	0.013	0.971	7.3	60.72
60	0.057	0.853			0.016	0.889		
70	0.071	0.941			0.049	0.947		

3.4 Color Quality

The total color ratio (TCR) between fresh roselle extract and dried roselle extract in different drying temperatures was depicted in Figure 3. After 90 minutes drying, the TCR decreased. The TCR reduction was higher at higher drying temperature. In drying temperature 70ºC, the TCR was the lowest (0.49). The color in roselle extract was from anthocyanins pigments. The constant rate of anthocyanin degradation depends on pH, temperature, copigmentation, ascorbic acid, oxygen, etc [2]. In this dying process, the degradation of anthocyanin because of the thermal degradation. The color change from red to brown indicated the thermal degradation in the drying process. [20].

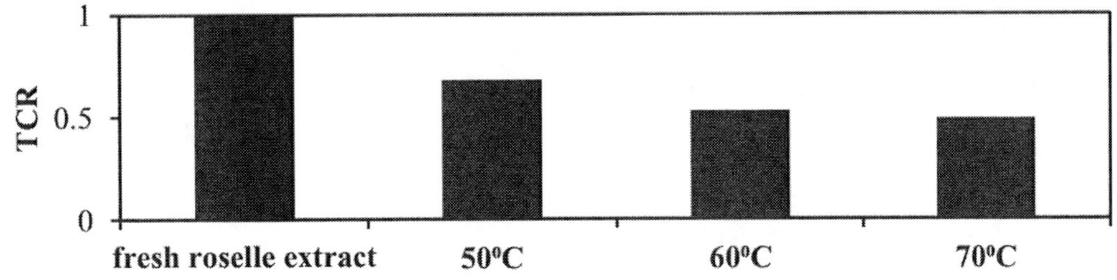

Fig. 3. Effect of drying temperatures on Total Color Ratio (TCR)

4 Conclusions

The roselle formulation using 3.31% egg albumin and 1% GMS resulted in the lowest FD (0.135 g cm^{-3}) and the lowest FDV (8.07 ml). Egg albumin contributed to form the sufficiently interfacial film and reducing the droplet size. GMS contributed to stabilizing the foam by during the drying process. The constant rate of the foam mat drying (temperature 50°C) was 3 times higher than non foam mat drying. Higher drying temperature can speed up the driving force but lead to thermal degradation.

This study was supported by Directorate General of Higher Education (DIKTI) under Competency Grant (Hibah Kompetensi) for 2017-2018.

References

1. S. Cid-Ortega, J.A. Guerrero-Beltran, J. Food Res **3**, 5, 83 (2014)
2. P. Chumsri, A. Sirichote, A. Itharat, Songklanakarin J. Sci. Technol **30**, Suppl.1, 133–139 (2008)
3. M. Djaeni, A. Prasetyaningrum, S.B. Sasongko, W. Widayat, C. L. Hii, J. Food Sci. Technol **52**, 2, 1170–1175 (2013)
4. A. Sangamithra, S. Venkatachalam, S.G. John, International Journal of Food Engineering **39**, 3165–3174 (2015)
5. A. Muthukumaran, C. Ratti, V.G.S. Raghavan, Dry. Technol **26**, 4, 508–512 (2008)
6. K.O. Falade and J.O. Okocha, Food Bioprocess Technol **5**, 1173–1180 (2012)
7. M. Azizpour, M. Mohebbi, M.H.H. Khodaparast, M. Varidi, Agric. Eng. Int. CIGR J **15**, 3, 159–165 (2013)
8. A. Sangamithra, V. Sivakumar, K. Kannan, S.G. John, Int. J. Food Eng **11**, 1, 127–137 (2015)
9. A.A. Karim, C.C. Wai, Food Hydrocolloids **13**, 203–210 (1999)
10. P. Rajkumar, R. Kailappan, R. Viswanathan, G.S.V Raghavan, J. Food Eng **79**, 1452–1459 (2007)
11. R.A. Wilson, D.M. Kadam, S. Chadha, M. Sharma, Int. J. Food Sci. Nutr. Eng **2**, 4, 63–69 (2012)
12. I. Doymaz, J. Food Eng **61**, 3, 359–364 (2004)
13. E. Meisami-asl, S. Rafiee, CIGR Ejournal **XI** (2009)
14. T.S. Franco, C.A. Perussello, L.D.S.N. Ellendersen, M.L. Masson, J. Food Eng **158**, 48–57 (2015)
15. S.S. Kumar, P. Manoj, N.P. Shetty, P. Giridhar, Journal of the Science of Food and Agriculture **95**, 9, 1812–1820 (2014)
16. J. Ahmed, U. S. Shivhare, G. S. V Raghavan, Eur. Food Res. Technol **218**, 525–528 (2004)
17. M. R. Salahi, M. Mohebbi, M. Taghizadeh, J. Food Process. Preserv **39**, 6, 1798–1808 (2015)
18. S.K. Bag, P.P. Srivastav, H.N. Mishra, Food Bioprocess Technol **4**, 8, 1450–1458 (2011)
19. M. Djaeni, D. Anggoro, G.W. Santoso, D. Agustina, N. Asiah, C.L. Hii, Adv. J. Food Sci. Technol **6**, 7, 833–838 (2014)
20. M. Cisse, F. Vaillant, A. Kane, O. Ndiaye, M. Dornier, Journal of the Science of Food and Agriculture **92**, 6, 1214–1221, (2011)

The Effect of Shear Wall Configuration on Seismic Performance in the Hotel Building

Daud Rahmat Wiyono[1,a], Roi Milyardi[1] and Cindrawaty Lesmana[1]

[1]*Department of Civil Engineering, Universitas Kristen Maranatha, 40164 Bandung, Indonesia*

Abstract. Earthquake effects on the buildings must be evaluated within the current standard provision. The shape of building gives a unique seismic performance on the structure. In typical hotel building, the lobby area in the first floor usually has some slender columns due to the needs of higher clearance to give a widely space area. The slender columns in the big hall tends to create asymmetric building and torsional behaviour on seismic performance. This behaviour is one of the most frequent source of structural damage and failure. One of the solution is to add shear wall in elevator area. The purpose of this paper is to seek the effect of shear wall configuration in elevator area on the seismic performance through numerical analysis. There are some requirements for structural analysis under seismic load, such as: time period, modal analysis, story drift, and other details. In building with dual system, story shear in frame at each level must carry over 25 % of total story shear at that level. In this study, an eleven-storey hotel building located in Tanjung Pinang City, Indonesia was evaluated due to gravity and seismic load. For the building, the requirements of the time period from the standard are 1.21 sec (minimum) and 1.70 sec (maximum). As results, two-sided shear wall in X direction and two-sided shear wall in Y direction is recommended because it has the best seismic performance, time period below the minimum, story drift below allowable, the dynamic lateral load has meet minimum requirement (85% Static Load), and frame structure has carry more than 25% lateral load in dual system building.

1 Introduction

The earthquake resistant buildings design codes help the designer to improve the behaviour of structures to withstand the earthquake effects without significant loss. Seismic load on the building must be evaluated within the unique standard provision that particular to the country. The shape of building gives a unique seismic performance on the structure. In public buildings, such as: hotel, mall, apartment, the needs of higher clearance in the entrance are common to be found to give a widely space area sensation for the user. This widely space usually includes several slender columns in the lobby area at the first floor. The slender columns in the big hall tends to create asymmetric building and torsional behaviour on seismic performance. This behaviour is one of the most frequent source of structural damage and failure [1]. One of the solution is to add shear wall in elevator area. The configuration of shear wall must be analysed in accordance some requirements for structural analysis under seismic load in earthquake resistant buildings design code, such as: time period, modal analysis,

[a] Corresponding author: daud.rw@eng.maranatha.edu; hwiesiong1@gmail.com

story drift and dynamic lateral force must be larger than 85% static lateral force. In dual system, story shear in frame at each level must carry over 25 % of total story shear at that level. In this paper, the hotel building is analysed based on the requirements for structural analysis under seismic load that referred to Indonesian earthquake resistant buildings design code (SNI 1726:2012).

2 Seismic Calculation

The seismic load was calculated based on Indonesian earthquake resistant buildings design code, SNI 1726:2012, The requirements for structural analysis under seismic load included time period, modal analysis, story drift, dynamic lateral force that must be larger than 85% static lateral force ($V_d \leq 85\% V_s$) in dynamic analysis, and story shear in frame at each level must carry over 25 % of total story shear at that level in dual system [2].

The time period of the building must be met the requirement on SNI 1726:2012 Section 7.8.2. The minimum time period ($T_{a\ min}$) can be calculated as:

$$T_{a\ min} = C_t \cdot h_n^x \tag{1}$$

where h_n is the height of building, C_t and x is period parameter based on building system (for dual system: $C_t = 0.0488$ and $x = 0.75$).
The time period should less than:

$$T_{a\ max} = C_u \cdot T_{a\ min} \tag{2}$$

where $T_{a\ max}$ is the maximum time period, and C_u is maximum period parameter based on SNI 1726:2012.

The modal results of the building must be met the requirement of SNI 1726:2012 Section 7.7.3., which the first mode and the second mode are not allowed to control by rotation, hence, the modes should be in two directions of the orthogonal floor plan. The rotation control is allowed for the third mode and so on.

The story drift of the building should meet the requirement on SNI 1726:2012 Section 7.8.6. The story drift of the building (δ_x) is defined as a difference of deflection in floor centre mass in each level, which should be met as follows:

$$\delta_x = \frac{C_d \times \delta_{xe}}{I_e} \tag{3}$$

where C_d is amplification factor for deflection, δ_{xe} is deflection at center mass, and I_e is seismic importance factor based on building function.

The story drift maximum requirement (δ_a) for dual system and importance factor (I_e) = I or II is defined as follows:

$$\delta_a = 0.020\ h_{sx} \tag{4}$$

where h_{sx} is the story height.

In dynamic analysis, the minimum dynamic lateral force should be more that 85% of the static lateral force ($V_d \geq 0.85\ V_s$), which is specified on SNI 1726:2012 Section 7.9.4.1. Special requirement from SNI 1726:2012 Section 7.2.2 for dual system is the story shear from the frame elements at each level must carry over 25 % of total story shear at that level.

3 Detail of Buildings

Eleven-storey Reinforced Concrete building is considered for the study. The function of the building is for hotel which located in Tanjung Pinang City, Indonesia. The structures are treated as a discrete system having lumped masses at each floor level. The loads considered on each floor are the permanent loads consisting of all the dead loads on each floor, weight of one-half of the columns and walls above and below the floor, and an appropriate portion of the live load that always act on the structure [3]. Geometrical detail of the building is illustrated in Fig. 1. In total, nine proposed configurations for the shear wall in elevator area are evaluated. Furthermore, the numerical simulation using ETABS [4] is created to obtain the seismic performance of the building.

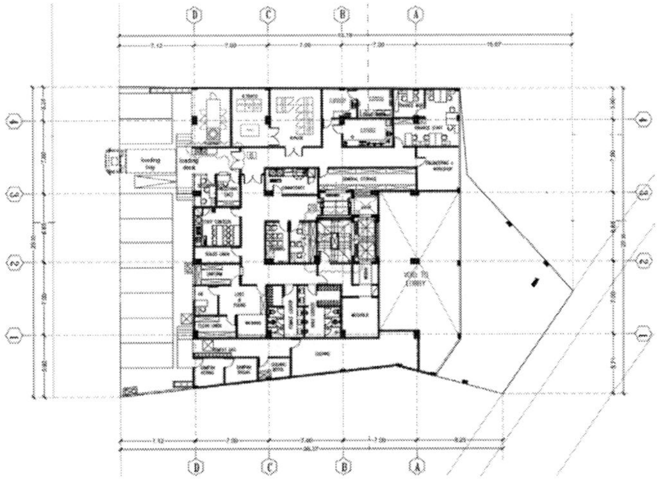

Figure 1. Building plan.

The seismic load is included to the model based on the local provision standard [5] which the seismic coefficients for the Tanjung Pinang City are S_S=0.061 and S_1= 0.088. All the dimension of the structural elements meets the requirement on the ACI building code [6] and Indonesian Building Code for Reinforced Concrete Structure (SNI 2847:2013) [7]. In average, the column dimension were 400 mm x 600 mm to 600 mm x 800 mm and the beam dimension was 300 mm x 500 mm.

4 Analysis and Results

The analysis of the hotel buildings was carried out using ETABS. Fig. 2(a) shows the 3D model of the hotel building. Fig. 2(b) illustrated one of the configuration of the shear wall from the total of nine proposed evaluated configurations for the shear wall in elevator area. Fig. 2(c) showed the plan view of the best configuration of the shear wall.

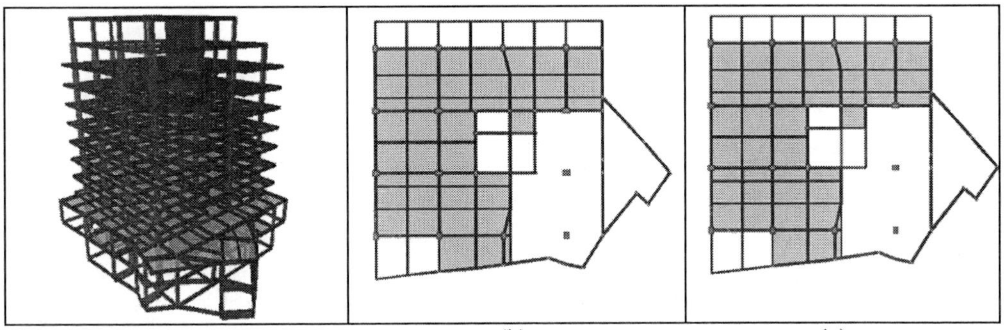

(a) (b) (c)

Figure 2. Hotel building model: (a) 3 d image (b) shearwall one sided X direction (c) shear wall two sided X and two sided Y direction.

There were 9 cases of different configuration of shear wall shown in Table 1. The minimum time period ($T_{a\,min}$) of building requirement was $T_{a\,min} = C_t \cdot h_n^x = 0.0466 \times 37.4^{0.9} = 1.21$ second and the maximum time period ($T_{a\,max}$) of building was $T_{a\,max} = C_u \cdot T_{a\,min} = 1.61 \times 1.21 = 1.95$ second. In the result, two-sided shear wall in X direction and two-sided shear wall in Y direction was satisfied the period requirement and direction control from the early mode which the configuration has translational direction control in first and second mode.

Table 1. Time period and direction of mode in some configuration shear wall placement.

No.	Case		Period*	Direction in Mode 1, 2, and 3 respectively
1	No Shear wall		1.76	Translational Y, Rotational Z, Translational X
2	One Sided Shear wall in X-Direction		1.74	Translational Y, Rotational Z, Translational X
3	Two Sided Shear wall in X-Direction		1.72	Translational Y, Rotational Z, Translational X
4	Two Sided Shear wall in X-Direction and one side in Y-Direction		1.6	Rotational Z, Translational X, Translational Y
5	One Sided Shear wall in Y-Direction		1.13	Rotational Z, Translational X, Translational Y
6	Two Sided Shear wall in Y-Direction		1.66	Rotational Z, Translational X, Translational Y
7	One Sided Shear wall in X-Direction and one sided in Y-Direction		1.64	Rotational Z, Translational X, Translational Y
8	Two Sided Shear wall in Y-Direction and one sided in X-Direction	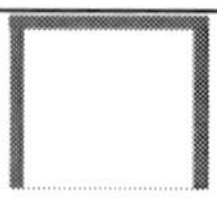	1.65	Rotational Z, Translational X, Translational Y
9	Two Sided Shear wall in X-Direction and Two Sided Shear wall in Y-Direction		1.63	Translational Y, Translational X, Rotational Z

Note: * maximum period

Furthermore, the dynamic lateral force was analyzed only on building model in case 9 which had meet all requirements of the time period and mode direction control. Both lateral force in X-direction (Fig. 2) and lateral force in Y-direction met the dynamic lateral force requirement ($V_d \geq 0.85.V_s$). The maximum story drift for the building model in case 9 was 13.23 mm (X-direction) and 18.11 mm (Y-direction) which satisfied the requirement of the allowable story drift 76 mm in the top story as shown in Table 2.

Table 2. Story Drift Analysis for building model case 9.

Storey	Story	X Direction	Y Direction

	Height (mm)	Story Drift δ_x (mm)	Allowable Story Drift δ_a (mm)	$\delta_x < \delta_a$	Story Drift δ_y (mm)	Allowable Story Drift δ_a (mm)	$\delta_x < \delta_a$
Story 11	3800	13.23	76	OK	18.11	76	OK
Story 10	3200	9.48	64	OK	4.97	64	OK
Story 9	3200	10.41	64	OK	7.39	64	OK
Story 8	3200	10.96	64	OK	9.80	64	OK
Story 7	3200	11.37	64	OK	10.57	64	OK
Story 6	3200	11.49	64	OK	10.94	64	OK
Story 5	3200	11.14	64	OK	10.94	64	OK
Story 4	3200	10.25	64	OK	11.08	64	OK
Story 3	3200	8.63	64	OK	9.55	64	OK
Story 2	4000	6.91	80	OK	7.74	80	OK
Story 1	4000	0	80	OK	0	80	OK

In Table 3, the story shear for frame and shear wall elements were presented. From the standard, frame at each level has to carry over 25 % of total lateral force in each level. The results showed 49.76% story shear in X-direction and 31.57% story shear in Y-direction were carried out by the frame that met the requirement for dual system building.

Table 3. Dual System Base Shear Analysis for building model case 9.

Spectrum Response	X-Direction		Y-Direction	
	Frame	Shear Wall	Frame	Shear Wall
Reaction Force (Kg)	8502.68	8584.18	3352.14	7265.80
Total Force (Kg)	17086.86		10617.94	
Percentage (%)	49.76%	50.24%	31.57%	68.43%

Lateral force in X-direction taken 50,24% by shear wall and distributed in each level that show in Fig. 3. In Y-direction, 68.43% lateral force taken and distributed in each level by shear wall (Fig. 4). Both figures show that frame can resist 25% shear force at each level. Fig. 5 and Fig. 6 present the comparison of lateral force that taken by frame system in hotel building have compared by different lateral load, 25% earthquake lateral load and 100% earthquake lateral load in each direction. The lateral force that taken by frame system in 25% earthquake lateral load is smaller than lateral force that taken in 100% of combination of the dual system earthquake lateral force in each direction. So, the lateral force taken by frame system with 100% earthquake lateral load of the dual system is used for further load design concrete reinforcement.

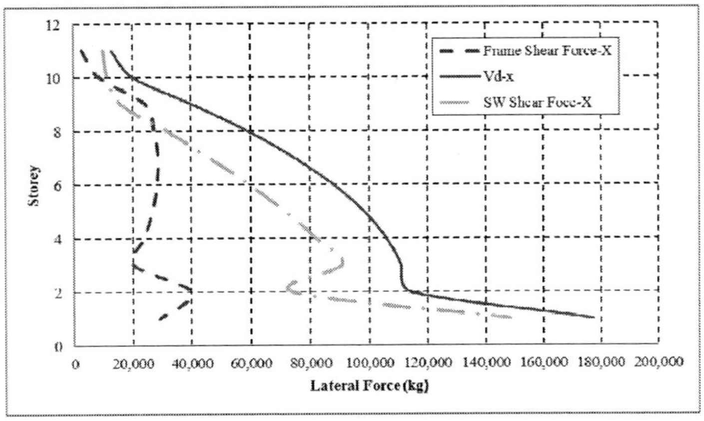

Figure 3. Dual system lateral force analysis in X-direction for building model case 9.

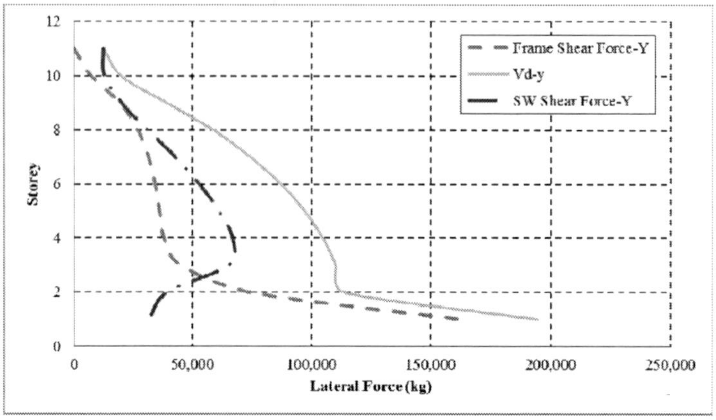

Figure 4. Dual system lateral force analysis in Y-direction for building model case 9.

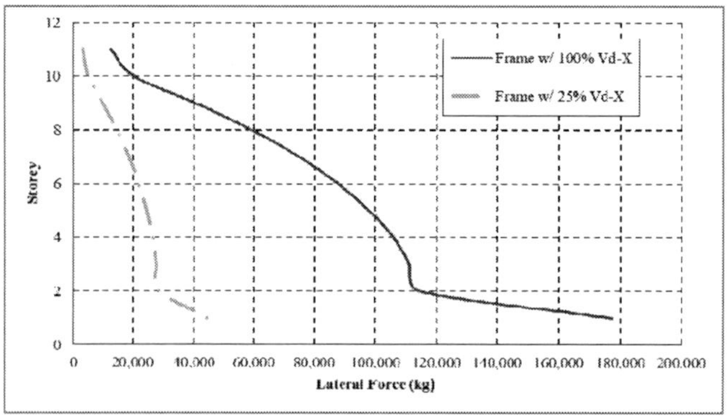

Figure 5. Comparison 25% to 100% Vd that taken by frame system in X-Direction for building model case 9.

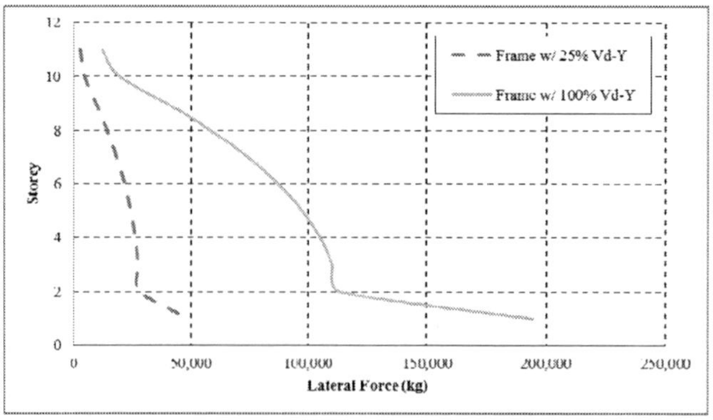

Figure 6. Comparison 25% to 100% Vd that taken by frame system in Y-direction for building model case 9.

5 Conclusion

Two-sided shear wall in X direction and two-sided shear wall in Y direction is recommended because it has the best seismic performance, time period below the minimum, story drift can be accepted, the dynamic lateral load has met minimum requirement (85% Static Load), and frame structure has carry more than 25% lateral load in dual system building.

References

[1] S. G. Maske and D. P. S. Pajgade, *Torsional Behaviour of Asymmetrical Buildings*. Int. J. Mod. Eng. Res., **3**(2), 1146–1149 (2013)
[2] P. W. Somers, Reinforced Concrete, *FEMA P-752 CD, NEHRP*, (2009)
[3] A. S. C. E. Institute, Minimum Design Loads for Buildings and Other Structures. *ASCE 7 – 05*, (2005)
[4] Computers and Structures Inc., Concrete Shearwall Design Manual. *CSI ETABS*, (University Avenue Berkeley, California, 2007)
[5] Badan Standarisasi Nasional, Tata cara perencanaan ketahanan gempa untuk struktur bangunan gedung dan non gedung. *SNI 1726:2012,* (Badan Standarisasi Nasional (BSN), Indonesia, 2012)
[6] PCI, Building Code of Requirements for Structural Concrete (ACI 318-08) and Commentary. *ACI 318-08*, (2008)
[7] Badan Standarisasi Nasional, "Persyaratan beton struktural untuk bangunan gedung. *SNI 2847:2013*, (Badan Standarisasi Nasional (BSN), Indonesia, 2013)

Performance of current transformer operate under harmonic condition and their effects on transformer differential protection

Indra Nisja

Department of Electrical Engineering, Faculty of Industrial Technology, Bung Hatta University, Gajah Mada Street, Padang, West Sumatera, 25142 Indonesia

Abstract. This paper focused to determine the performance of Current Transformer (CT) operates under harmonics condition and their effects on transformer differential protection. A laboratory test has been implemented to determine the error produced by both CT and power transformer when operating under harmonic condition. The test was performing with the actual condition, where the power transformer is connected to an adjusted nonlinear load, so that the test can be conducted with several levels of total harmonic distortion current (THD_i). The results shows, for THD_i ranging from 16.70% to 40.88% the maximum errors occurred on CT at secondary power transformers is 27.21% and CT at primary power transformers is 8.12%. This error resulted in differential current flow 0.17A and relay trip without any fault. In this study it was found that the relays started to operate incorrectly on THD_i 31.5%.

1. Introduction

The advanced technologies in power electronics development over the past decade, the application of power electronics in industrial, commercial, office building, educational institutions and residential areas have increase. Power electronic equipments are harmonic producing equipments due to its switching devices and called nonlinear load. Nonlinear loads are broadly classified as loads, which draw non-sinusoidal current even when the supply voltage is perfectly sinusoidal. Because of their extraordinary gains in efficiency and control, power electronics loads are expected to be significant in the future. Currently, power electronic loads can be found at all level of power system, ranging from low voltage appliances up to a high voltage converter. Non-linear load typically comprise more than 50%, and in some case as much as 90% of total building load [1]. Heavy use of power electronics equipment in the industrial, commercial and office building can lead to considerable harmonic distortions in the distribution feeder. The higher harmonic distortion level in the distribution system has caused a serious problem in power system quality and stability [2].

Increased the levels of harmonic currents in distribution systems, creating concern for electricity distribution network service providers will face malfunction of protection system components. Harmonics current and voltage can distort the operating characteristic of

* Corresponding author: drindra765@gmail.com

protective relays which is depending on the design features and principles operation of the relay [3]. During fault or normal conditions, the harmonic in distribution system might cause to miss operation of differential protection as well as relay calibrations inaccurate [4], [5], [6].

2. CT operates under harmonic condition

When a CT operates under harmonic condition, the distorted excitation current will flow in the excitation circuit, so that the secondary current produced by this CT will distorted as well. These phenomena will cause degradation of CT performance due to its core saturation. While CTs it self are constrained in their performance by the magnetic flux limitations of the core. Ferromagnetic materials cannot support infinite magnetic flux densities. High content of harmonic currents in power distribution systems will lead to the phenomenon of CTs saturation that can cause some errors in the operation of differential relays.

Two current transformers at both sides of protected equipment might not perform equally even when they are from the same brand, same ratio and type. The excitation current of both current transformer are also not identical due to remnant magnetic fluxes in the core may not identical. The different level saturation of both sides current transformers will affect to differential current, whether in normal condition or fault condition. The error signal produced by current transformer at both sides of protected equipment will cause to miss operation of differential relay.

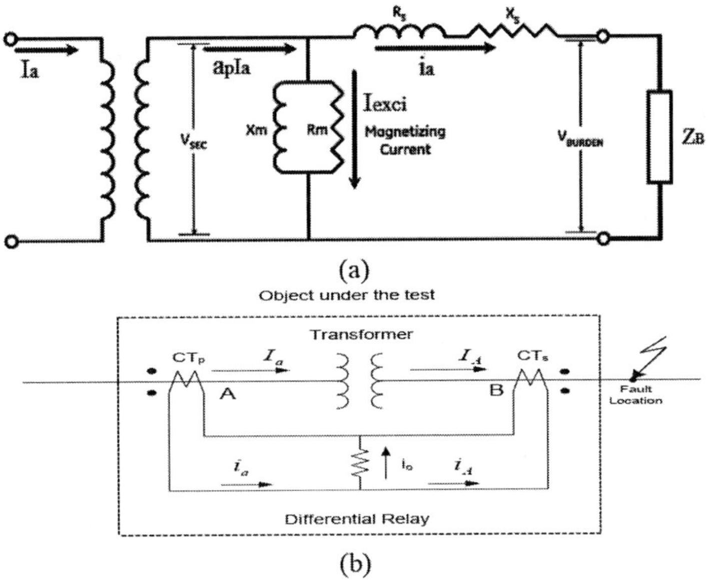

Fig. 1. (a) Current transformer equivalent circuit , (b) Differential relay scheme for power transformer protection

Fig. 1 (a) shows current transformer equivalent circuit at primary side of power transformer (CT_p). The secondary current in the pilot wire of CT_p is equal to:

$$i_a = a_p I_a - I_{exci(p)} \tag{1}$$

where,
a_p is the transformation ratio of CT_p, I_a is line current at primary side of power transformer, $I_{exci(p)}$ is the excitation current of CT_p

For the current transformer at secondary side of power transformer (CT_s), the equation of secondary current in pilot wire of CT_s is equal to:

$$i_A = a_s I_A - I_{exci(s)} \quad (2)$$

where,
a_s is the transformation ratio of CT_s, I_A is line current at secondary side of power transformer, $I_{exci(s)}$ is the excitation current of CT_s

Fig. 1 (b) shows single line diagram of a differential protection system under normal or external fault condition (no fault) [7]. It can be observed that the protection zone is delimited by a couple of CTs. Current entering the protected unit (I_a) would be equal to the current leaving protected unit (I_A). So the current i_a is equal to current i_A and differential current (i_d) is equal to zero (relay not operate). Assuming equal transformation ratios, a_p = a_s = a, the relay operation current i_o is given by:

$$i_o = i_a - i_A \quad (3)$$

During normal system operation, subtitute (1) and (2) into (3):

$$i_o = -I_{exci(p)} + I_{exci(s)} \quad (4)$$

The differential relay should not operate for this differential current.

For internal fault,

$$i_o = i_a + i_A \quad (5)$$

Subtitute (1) and (2) into (5) :

$$i_o = a(I_a + I_A) - I_{exci(p)} - I_{exci(s)}) \quad (6)$$

The differential relay must operate for this differential current.

3. Experimental results and discussion

Current Transformer Data:

Ratio	:15/1A	Acc.Class	:1
Rated Burden	: 5VA	Frequency	:50/60Hz
Voltage	: 0.72/3.0kV	Type	: SES

Fig. 2 shows experimental result of a CT by directly connected to system with 15.2 A line current. The secondary current error is 5.3% for THD_i 20%. As for the THD_i increased to 30% and 40% the secondary current deviations increase to 13.5% and 21.7% respectively. This clearly indicates that, the errors occur at the secondary side of CT as a result from the harmonics at the primary side of CT.

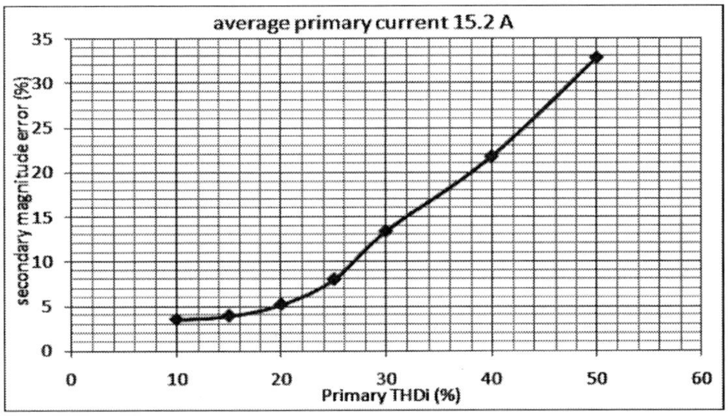

Fig. 2. Primary THD_i vs secondary current error for THD_i 10% to 50%

3.1 Performance of CT at primary side of power transformer

Table 1 (left side) shows the relationship between secondary current error of CT_p and THD_i at primary side of power transformer. The primary current of power transformer was remain constant at 15.2 A. In this experiment, CT_p has been connected to the primary sides of power transformer. The THD_i at primary side of power transformer increased from 16.7% until 32%. When THD_i at primary side increased until 32% the secondary current error occurs is 20.58%.

3.2 Performance of CT at secondary side of power transformer

Table 1 (right side) shows the relationship between secondary current error of CT_s and THD_i at secondary side of power transformer. The secondary current of power transformer was remain constant at 8.8 A. In this experiment, CT has been connected to the secondary sides of power transformer. The THD_i at secondary side of power transformer increased from 31.55% until 69.67%. With increasing of THD_i will cause to reduce the secondary current produced by current transformer. As THD_i at primary side of CT_s increased until 69.67% the secondary current error occurs is 55.77%.

Table 1. Percentage of CT_P and CT_S secondary current error

CT at primary transformer		CT at secondary transformer	
THD_i (%)	**Secondary current error (%)**	**THD_i(%)**	**Secondary current error (%)**
16.70	5.50	31.55	17.13
19.62	8.12	40.88	27.21
23.40	12.13	51.72	37.40
27.50	16.24	59.52	46.14
32.00	20.58	69.67	55.77

3.3 Effect to Differential Protection operation

Fig. 3 shows the characteristics of bias differential relay operate at purely sinusoidal. Without harmonics conditions, the internal fault operating point are in operation region, so the differential relay will trip. As for the external fault, the operating point of differential relay at restraining region, so the differential relay will not trip. At no fault conditions, current flow through operating coil is 0.006A and the operating point of differential relay is inside restrain region. So, it means the differential relay operates correctly.

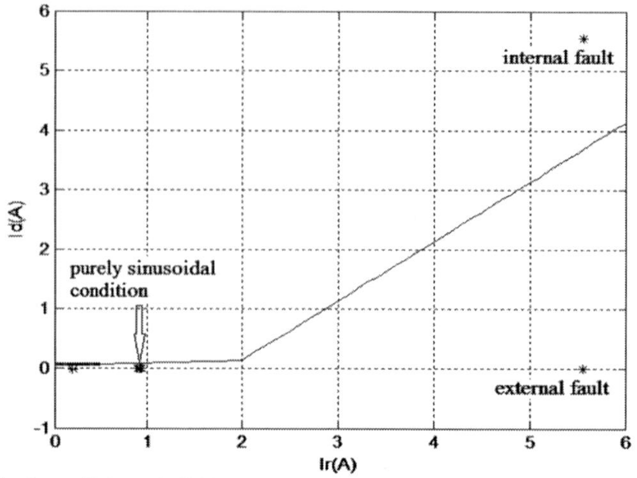

Fig. 3. Characteristics of biased differential relay operate under purely sinusoidal condition

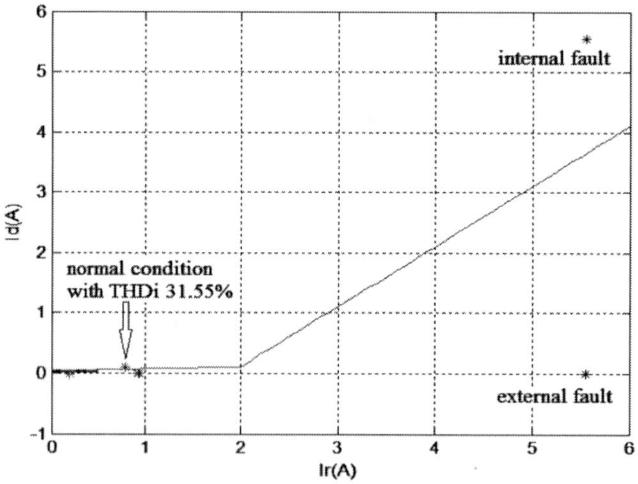

Fig. 4. Characteristics biased differential relay for normal condition with THD_i 31.55%

Fig. 4 shows the operating point of a differential relay at several conditions such as internal fault, external fault and normal condition. For internal fault that occurs at THD_i of 31.55%, the differential relay operating point is at operating region. As for external fault that occurs for THD_i of 31.55%, the differential relay operating point is at restraining region. These shows that the differential relays operate normally. However, when no fault occurs and the differential relay operates in harmonic condition for THD_i of 31.55%, the differential current flow is 0.109A. At this condition the operating point of the differential relay has been moved to the operating region. That means, the differential relay will trip at no fault condition (malfunction). This is something not allowed for the differential relay protection because the relay trip without any internal fault. At normal condition and THD_i 40.88%, differential current flow is 0.176A. The operating point of differential relay has been moved to operation region and then the differential relay will operate as shown in Fig. 5. It is evident that, this relay operate is caused by harmonics.

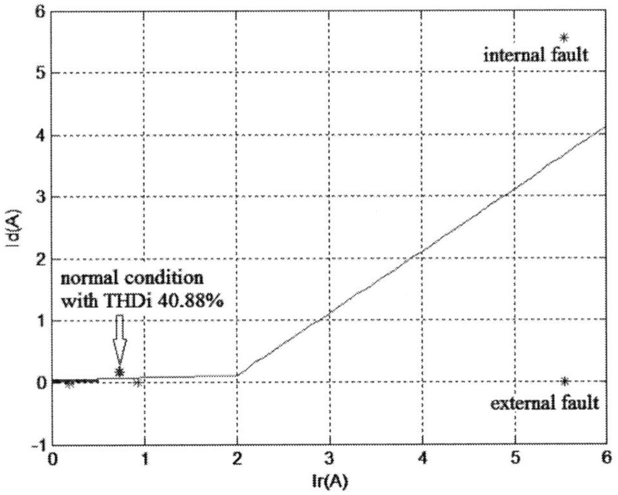

Fig. 5. Characteristics biased differential relay for normal condition with THD_i 40.88%

4. Conclusion

From the research that has been carried out can be concluded that an increasing of percentage THD_i in power system will increase the magnitude of current errors that occur on the secondary side of the CT. When THD_i at primary side power transformer increased until 32% the secondary currents errors on CT_p is 20.58% and if the THD_i on the secondary side power transformer rises to 40.88%, a secondary currents error on CT_s is 27.21%. This will result in reading error by the relay that connected to the CT. For THD_i at load side of power transformer of 31.55% and 40.88% and without an internal fault the differential current flow is 0.109A and 0.17A respectively which causes the relay to operate.

References

1. M. Aiello, A. Cataliotti, V. Cosentino, and S. Nuccio, *IEEE Transaction on Instr. and Measur.*, Vol.**54**, No.1, pp. 15-23, February (2005).
2. A. Medina, and C. M. Martinez, *Power Engineering Society General Meeting, IEEE,* Page(s):741 - 745 Vol. **1**, 12-16 June (2005).
3. Schweitzer and H. Daqing, *Schweitzer Engineering Laboratories, Inc. Pullman, Washington.47th Annual Georgia Tech Conference*, April 28-30 (1993).
4. J. Arrillaga, B. C. Smith, N. R. Watson, and A. R. Wood, *Power System Harmonic Analysis,* John Wiley & Sons (1997).
5. C. Sankaran, *Power Quality,* CRC Press (2002).
6. S. P. Kennedy, and W. Barry, *Mc Graw Hill* (2000).
7. J. C. Tan, *Ph.D Dissertation,* FAMU-FSU of College of Engineering, Florida State University (2000).

A Comparative Analysis of Mental Workload between Train and Bus Drivers

Hartomo Soewardi [1] and *Perdana Suteja* Putra [1*]

[1] Industrial Engineering Department, Islamic University of Indonesia, Yogyakarta, Indonesia

Abstract. Human error is one of the crucial problems in land transport, sea transport, and air transport that inflicting some undesirable occurrence such as delay, cancellation, or accident. This incident causes inconvenience and apprehension for passengers. Hence, drivers should avoid some errors in driving by improving alertness and reducing the excessive workload. This study presents a comparative study about mental workload between train and bus drivers based on subjective and objective criteria. NASA-TLX method is used to measure subjective variables by distributing questionnaires. While the content of a drivers' Salivary α-amylase (SAA) is measured as an objective parameter before and after driving. 30 drivers are participated in this study who have experience for more than 3 years. Statistical analysis is conducted to test the hypothesis for a difference. The result of this study shows that bus drivers' SAA level is significantly higher (Asymp. Sig. 0.49) than train drivers for both before (91.99 kU/l) and after (83.19 kU/l) driving activity. Similarly, weighted workload subjectively of bus drivers' mental workload (1051) is higher than train drivers (1031) insignificantly (Asymp. Sig. 0.561).

1 Introduction

Human error is one of the causes of an accident in modes of transportation [1-5]. Risyapudin in [5], a traffic director of Polda Metro Jaya, stated that the main factor in an accident on land transportation such as bus, car, motorcycle, and a train is human error. This condition is an undesirable incident for all passengers because it can cause delay, cancellation, rescheduling, unclear information even accident particularly in public transportation such as bus and train [6].

The fact, accident level of train transportation is still in the high ratio [7]. Indonesian Ministry of Transportation stated that there are 42 accidents of train in 2010 (i.e. 3 crashing accidents, 25 fallen trains, 4 rolled trains, 6 floods and erosion cause, and 4 other reasons). In 2011, there are 33 accidents as if 1 crashing accidents, 23 fallen trains, 2 rolled trains, 1 floods and erosion cause, and 6 other reasons. While 33 accidents have occurred in 2012 i.e. 2 crashing accidents, 21 fallen trains, 2 rolled trains, 4 floods and erosion cause, and 2 other reasons. Most of the accidents take place in 2013 including 25 fallen trains, 1 rolled trains, 7 floods and erosion cause, 6 other reason. Moreover, [8] and [9] found that as much as 394 people have died that was caused by 23 bus accidents in 2012 and 14 bus accident in

* Corresponding author: perdanasuteja@gmail.com

2014. Refers to traffic corps of Indonesia's data, Bus accidents have happened in 2013 as many as 13 incidents. It brings out passengers into apprehend public transportation usage.

Based on the result of investigation by the transportation department of Indonesia in 2013 and also the result of study in 2002 found that some causes of these accidents are mostly a lack of concentration, a lack of anticipation, an internal condition by drivers, and the mileage of a public transportation [10-11]. Other studies identified that such causes of accidents were produced by the excessive mental and physical workload of drivers in driving [12-15]. These study conducted was based on the subjective considerations only. As for the objective consideration are still a few of study investigated. Thus, it is significant to investigate the mental workload of drivers with considering both subjective and objective criteria. The objective is to investigate the comparison of mental workload between train and bus drivers by using two criteria that are NASA-TLX and Salivary Alpha-Amylase.

2 Research Method

2.1 Subject

Thirty respondents were recruited for this study which consists of fifteen bus drivers and fifteen machinists. Their age is in range between 27 to 47 years old where all of them has no physic or mental health problem. Drivers and machinist worked for 8 hours/day in route of Yogyakarta-Jakarta and they have experience for more than three years respectively.

2.2 Apparatus

Cocorometer and disposable test strips by Nipro Inc. were used to measure salivary alpha-amylase (SAA) level. And a National Aeronautics and Space Administration-Task Load Index (NASA-TLX) questionnaire implemented to assess mental workload level subjectively. While IBM SPSS software version 21 and Microsoft Excel software to count up statistical data for analysis.

2.3 Procedures

2.3.1 Objective measurement
Disposable test strips inserted into the mouth (for 20 seconds under the tongue), and later were attached back into the instrument. It is assessed for before and after driving. After a few minutes, the display would read the value (in kU/l), with moderate and highly stressful conditions were indicated by a value of greater than 45 and 60 KU/l, respectively [16].

2.3.2 Subjective measurement
NASA-TLX assess six indicators influence on mental workload which is Mental Demand (MD), Physical Demand (PD), Temporal Demand (TD), Effort (EF), Own Performance (OP) and Frustration Level (FR). Unlike the SAA measurement, the NASA-TLX was administered after a driver had just finished a trip by fulfilling pair-wise comparisons and rating scale from 0 (Low) to 100 (High), Unless OP is 100 (Low)to 0 (High).

2.4 Statistical Analysis

The statistical analysis used is non-parametric Wilcoxon Signed-Rank and Mann-Whitney U test. Wilcoxon Signed Rank in the significant level of 5% takes account to differentiate the SAA level between before and after driving and 5% significant level of Mann-Whitney U test differentiate the SAA level and weighted workload between train and bus Drivers.

3 Result and Discussion

3.1 Objective Analysis

In prior to driving bus activity, the SAA result is 63.78-139.29 kU/l (mean: 91.99 kU/l; SD ± 30.25) which is classified in high stress. It produced by constrained time to complete preparation and bus check that leads to drivers' anxious on time departure [17]. Moreover, the high level also gained for after driving which is 83.19 kU/l (SD ± 15.96). It identified that 93.3% of respondents are in high stress that influenced by driving effects such as high focus, boredom, traffic jam and fatigue [18].

Moreover, the high SAA average also obtained by machinists (70.33 kU/l; SD±23.46). Mostly 53% of respondents are at high-stress level. A complex of standard operational procedures causes the drivers' apprehensive in missing stages that lead to human error, absenteeism, complaint, and sickness [18]. A large percentage of the drivers had fairly long commuting time (one to two hours) that could have introduced a source of stress since they had to deal with traffic jams and uncertainty [15]. Meanwhile, after driving stress of machinist classified at a high level which is 72.28 kU/l (SD±12.39). A total of 73.3% of respondents is at the high stress that produced by after driving impact such as boredom and high concentration, occurrence report and overview the next task. [15] stated that deal with many level train crossings, a large percentage of which were unguarded and illegal could create additional mental burden since the drivers had to watch the traffic and made difficult decision to cross safely. It was exacerbated by the fact that many houses and local activities were very close (often done on) the railway track.

In statistical analysis demonstrates insignificant difference on stress felt by bus drivers (Asymp. Sig. 0.256) as well as train drivers (Asymp. Sig. 0.733) between prior to driving and after driving activity as shown in Figure 1.

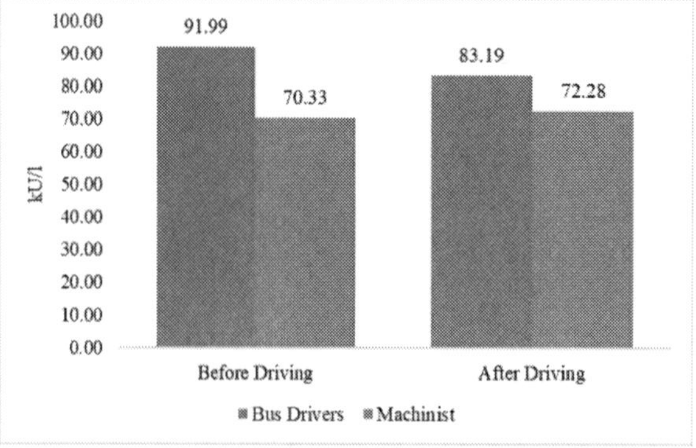

Fig. 1. Comparison of SAA Average between Before and After Driving.

When examined bus drivers individually, changes in SAA level demonstrates that mostly respondent (57%) are decreasing from before to after driving. This condition influenced by their satisfaction to usher passengers safely [19]. Besides, the change of salivary level from before to after driving found that 53% of respondents are rising. Generally, it affected by some after procedures that have to be finished in fatigue condition [20]. Then, drivers' anxious increased for missing procedures.

Thence, Figure 1 describes the SAA average of bus drivers (91.99 kU/l) is higher than machinists (70.33 kU/l) significantly (Asymp. Sig. is 0.049) in prior to driving activity. The higher uncertainty and complexity is main factor influenced on this condition. Moreover, in after driving described SAA level of bus drivers (83.19 kU/l) is higher than machinists (72.28 kU/l) significantly (Asymp. Sig. is 0.049). It mostly affected that bus driving activity needs more focus, boredom, concentration and uncertainty traffic jam.

3.2 Subjective Analysis

In deeper analysis, subjective assessment of bus drivers evinces temporal demand gains the highest level (203.933) whereupon followed by mental demand which is 199.87. Then, physical demand, effort and frustration product is 192.733, 185.133, and 185.1 respectively. Lastly, the lowest product is performance indicator which is 83.13. Hence, total WWL obtained is 1051. Furthermore, the frustration level is the most dominant indicator of subjective criteria which is 211.067. Secondly, temporal demand (203.9) and effort (182.3) indicator dominates the result of stress assessment. It is followed by mental demand, physical demand and performance indicator whereas the product is 182.3, 179, and 77 respectively.

Bus drivers commonly worked overtime since the job rotation is changed based on their own will that leads to the lack of resting time and produce human error [20]. Unpredicted complexity (e.g. traffic jam, trouble machine) also leads the pressure in time management [22-23]. Drivers need to estimate a traffic jam, distance, fuel, and budget. Hence, improvisation, thinking and deciding to solve the problem are needed by drivers [24]. [25] and [18] has supported that drivers need to focus to operate the vehicle and concentrate on the road such as a sign, lamp, announcement, etc. In addition, the brake is not lightly handled and traffic jam rises fatigue and physical demand [14, 18, 23]. Additionally, drivers' frustration originates from feelings of uncertainty which stems from a sense of inability to fulfill the need. [26] has explained when these needs are constantly ignored or unsatisfied, anger, depression, loss of self-confidence, annoyance, and sometimes violence in relation to the partner, passengers, manager, etc. However, Management has offered assurance for drivers and bus to decrease an insecure condition. Lastly, performance is the lowest indicator due to the perspective that tasks as a responsibility that leads to the stable satisfaction on work performance [27-28]. Moreover, [29] stated compensation, reward, and facilities provided would improve drivers' satisfaction.

On the other result, machinist often disguises occurrence such as driving for more than a certain speed and did not take a regular break induce anxious for work communication among workers thereupon increase frustration level indicator [11]. Meanwhile, the time pressure is influenced by the arrival on time schedule but drivers are not allowed to drive for more than a certain speed in uncertainty condition. They also need to carry out 21 slogans, point and mention it. Moreover, the daily routine such as dynamic braking (every 30 m) and locomotive examination (if stops more than 5 m) have to be remembered by drivers cause mental and physical demand. It is relevant to study conducted by [19] and [30] in mental workload influence. Lastly, performance indicator is the lowest level since the management implement bonus, reward and facilitate them satisfactorily [29].

Furthermore, subjective assessment asserts that weighted-workload of bus drivers is higher than machinist insignificantly (Asym. Sig. is 0.561) as described in Figure 2. In indicator analysis, bus drivers dominate on temporal demand (199.87), physical demand (192.73), temporal demand (203.93), own performance (186.3), and effort (83.13) rather than machinists' product. Moreover, the machinist has higher product once on frustration level (211.067) as described in Figure 3.

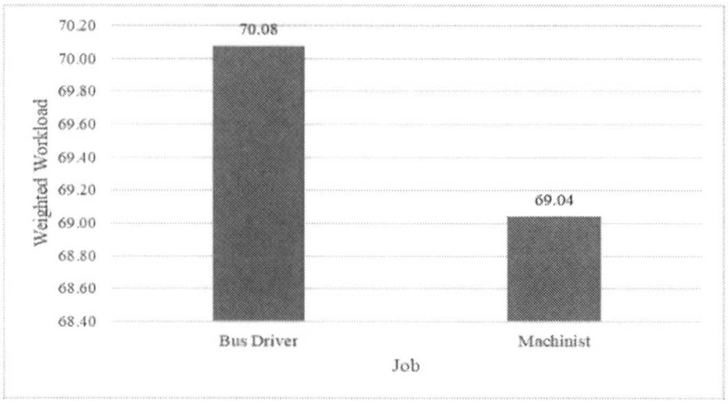

Fig. 2. Comparison of Weighted-workload between Bus Drivers and Machinists.

As explained that bus drivers involve higher uncertainty, complexity, focus, time management skills, mentally and physically effort, and unsatisfactorily work performance. Meanwhile, machinists involve higher frustration to disguise occurrence that produces high level of frustration and inconvenience in communication both drivers and management.

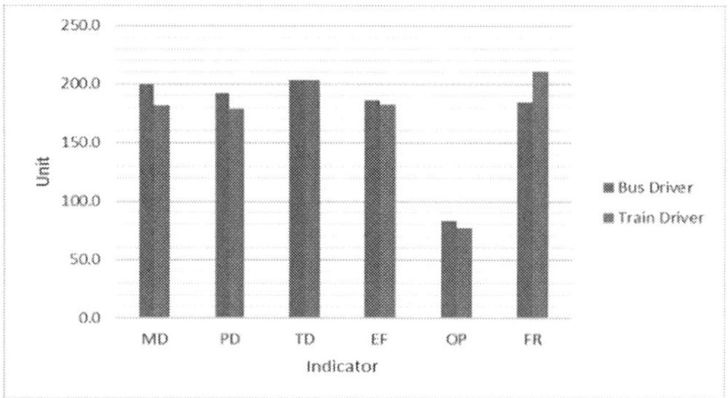

Fig. 3. Comparison of Indicator Product between Bus Drivers and Machinists.

Hence, the result and analysis evince that bus drivers' mental workload is higher than machinist in both objective criteria and subjective criteria. Nonetheless, it significantly differences in salivary-alpha amylase and insignificant difference in weighted-workload result. It happened due to the higher mentally factors felt by bus drivers.

4 Conclusion

It concluded as follow:

1. Based on the objective analysis (Salivary alpha-amylase level), bus drivers and machinist have high stress before and after driving. They are 91.99 kU/l and 83.19 kU/l for before and after driving of bus drivers. Meanwhile, machinists' mental workload is 70.33 kU/l and 72.28 kU/l in prior to and after driving activity.
2. On the subjective analysis, bus drivers have average weighted-workload in 70.08 level while machinists have 69.04 level of weighted-workload mean.
3. The result evinces on both objective and subjective parameters; it is valid that bus drivers have higher mental workload than the machinist at 5% significance level.

References

1. A. Smiley, K.A.Brookhuis, *Road Users and Traffic Safety*. 83–105(1987)
2. P. C Cacciabue, Gui.App.Hum.Fac.Meth, **1**(2004)
3. D. Woods, Dekker,R.I.Cook,L.Johannesen,N.B.Sarter, *Behind human error,* (2010).
4. A. Noroozi,N.Khakzad,F.Khan,S.Mackinnon,R.Abbassi,Rel.Eng.Sys.Saf **119**, (2013)
5. G. Lazuardi,Hum.Err.Fak.Dom.Pen.Kec.Lal.Lin, (2015)
6. J.A. Michon, Hum.Beh.and.Traf.Saf 485–520,(1985)
7. Biro Komunikasi dan Informasi Publik, Dal.Lim.Tah.Ter.Kes.Per.Men,(2014)
8. E.S. Budi, Lag.Kec.Bus, *Kompasiana*, (2012)
9. E. Rusyanto, Cat.Dar.Kec.Bus.Mem, (2014)
10. Dirjen Perhubungan Darat, Pet.Tek.Pem.Aw.Ken.Um.Tel.Ting.Nas.Th.2013, (2013)
11. G. Matthews, *Issues in Ergonomics Science*, 195(2002)
12. K. Brookhuis, D. Waard, *Virtual Reality*, 191–205(2003)
13. Tarwaka, S.H. Bakri, Sudiajeng, *Ergonomi Untuk Keselamatan, Kesehatan Kerja dan Produktivitas,* (2004)
14. E.T. Solovey, Zec, Garcia, Reimer, Mehler, Cla.dri.wo.usg.phy.dri.perf.d, **4057** (2014)
15. C.H. Moerti, Iridiastadi, Sutalaksana, *International Journal of Science and Research (IJSR)*, 2014–2017 (2016)
16. M. Yamaguchi, Kanemori, Kanemaru, Takai, Mizuno, Yoshida, *Biosensors and Bioelectronics*, 491–497 (2004)
17. M. A. J. Kompier, Martino, Rev.bu.dri.occ.st.st.pre, 253–262 (1995)
18. M. A. J. Kompier, *Bus drivers: Occupational stress and stress prevention*, (1996)
19. S.G.Hart,L.E.Staveland, Dev.NAS.Res.Emp.The.Res, 139–183 (1988)
20. R.Schlegel, *Automotive Ergonomics*, 359–382(1989)
21. J.L.M. Tse, R.Flin, K.Mearns, Bu.dr.wel.rev.50.y.res,(2006)
22. H. Soewardi, E.B.Wibowo, An.Men.Wor.Pub.Bus.Dri, 335–339(2014)
23. V. Faure, R. Lobjois, Benguigui, *Traffic Psychology and Behaviour*, **40**, 78–90 (2016)
24. J.Paxion, E.Galy, C.Berthelon, Men.wor.driving. *Frontiers in Psychology*,**5**, 1–11 (2014)
25. T.Heine, Lenis, Reichensperger, Beran, Doessel, Deml, *Applied Ergonomics*, **61**, 31(2017)
26. A. De Botton, The consolations of philosophy, (2000)
27. M.S.P.Hasibuan, *Manajemen sumber daya manusia*, (2005)
28. S.Martoyo, *Manajemen sumber daya manusia (5th ed.)*, (2000)
29. A.Kinicki, Kreitner, *OrganizationalBehavior:KeyconceptsskillsandbestPractice*, (2008)
30. S.G.Hill, Lavecchia, Byers, Bittner, Zaklad, Christ, *Human Factors*, **34**, 429–439. (1992)

AUTHOR INDEX

AUTHOR INDEX

AUTHOR INDEX

AUTHOR INDEX